D0984068

FRENCH-ENGLISH
SCIENCE AND TECHNOLOGY
DICTIONARY

FRENCH-ENGLISH SCIENCE AND TECHNOLOGY DICTIONARY

LOUIS DE**VRIES**

Formerly of Iowa State University

Revised and Enlarged by

STANLEY HOCHMAN

Fourth Edition

McGRAW-HILL BOOK COMPANY

New York   St. Louis   San Francisco   Auckland   Düsseldorf
Johannesburg   Kuala Lumpur   London   Mexico   Montreal
New Delhi   Panama   Paris   São Paulo   Singapore
Sydney   Tokyo   Toronto

Library of Congress Cataloging in Publication Data

De Vries, Louis, date.
  French-English science and technology dictionary.

  First-3d ed. 1940–62 published under title:
French-English science dictionary for students in
agricultural, biological, and physical sciences.
  1. Science—Dictionaries—French. 2. Tech-
nology—Dictionaries—French. 3. French language
—Dictionaries—English. I. Hochman, Stanley.
II. Title.
Q123.D37 1976   503   75-45091
ISBN 0-07-016629-3

CONTENTS

INTRODUCTION TO THE FOURTH EDITION

This new edition has been revised to include some 4,500 terms that have found their way into scientific and technical literature in the decade or more since the last revision. The new entries have been incorporated into the Supplement which follows the main body of the dictionary.

Special emphasis has been placed on new developments in the vastly expanded fields of automotive technology, astronautics, electronics, and electronic data processing. In addition, all the terms published by the French Ministry of National Education in January of 1973 as part of its vocabulary enrichment program have been included. Most of these terms are now obligatory—and others are strongly recommended—in all government publications, correspondence, and contracts.

Many people both here and in France contributed invaluable advice and suggestions, but I would like to extend special thanks to my son, David Hochman, and to Louis-Avit de Puylaroque.

STANLEY HOCHMAN

NEW YORK CITY
October, 1975

INTRODUCTION TO THE THIRD EDITION

This dictionary has again been revised to include more than 5,000 new terms and newly recognized translations of terms that have become important in scientific literature since the Second Edition was published. These new entries have been incorporated in the Supplement which follows the main body of the dictionary. The new terms will, it is hoped, make the book much more useful to students and others who read or translate material published in French since 1951, of which there has been an increasing volume in all fields of science.

In addition to the revised Supplement, there has been added a Grammatical Guide for Translators at the back of the dictionary. The Guide, which covers a number of grammatical principles and matters of syntax frequent in scientific French prose, should be helpful to all users of the dictionary, and particularly to students who are preparing for certification examinations in reading scientific French. The Guide cannot of course be an adequate substitute for a textbook, but it provides a convenient reference and set of reminders about a number of "tricks of the trade" to aid in recognizing certain constructions and idioms that can give trouble to the inexperienced reader of French; it has been prepared with special reference to scientific facts, methods, and ideas.

The author wishes once more to express his gratitude to the many colleagues at Iowa State University, elsewhere in this country, and also abroad, whose advice and suggestions have meant so much to him in the preparation of a useful dictionary.

Louis DeVries

Ames, Iowa
September, 1962

INTRODUCTION TO THE FIRST EDITION

The reception of the German-English Science Dictionary was so encouraging as to warrant the completion of this volume, planned and started years ago, to assist candidates for advanced degrees in acquiring a reading knowledge of the language.

Research within the various departments of science has gradually developed during these years into complicated interdepartmental problems. No entomological vocabulary, for example, can today dissociate itself from the many aspects of biology in general; even the physical sciences are embraced. Terms must be included covering not only entomology and the sciences into which it enters, such as embryology, cytology, physiology, morphology, genetics, ecology, but also chemistry, physics, botany, and medicine, all of which enter into modern treatises on insects.

Since the French irregular verb is always of some concern to the average graduate student in the sciences, the author has given many forms of the present, past, and future tenses as well as the past participles with at least one meaning and the infinitive form in parentheses in case other meanings are desired. In addition to this, some five hundred common idioms have also been added, most of them based on verb forms. These idioms will facilitate the translation problems of a student who is not majoring in French.

This dictionary of 43,000 entries is, like the German, the first of its kind including terms of the agricultural, biological, and physical sciences, as well as many literary terms. Not all names of animals, insects, plants, or chemical compounds have been included, since each subject would make up a dictionary of its own.

Many words have been included even though the French and the English spellings are the same or practically so. If these words were not included, the average science student would easily give up, whereas now he will take the meaning, even though spelled like the French, and refer to Webster's Unabridged or the New Standard Dictionary for further definitions. The author uses both dictionaries constantly.

This dictionary in its compact size cannot claim completeness. The author would be grateful if its users would send him any useful suggestions with regard either to significant words that have been omitted or to incorrect meanings and equivalents.

This volume could not have become a reality without the inspiration from Dean R. E. Buchanan of Iowa State College and the constant aid of his graduate faculty, including the assistance of Dr. Charles H. Brown and Professor Robert W. Orr of the College Library.

Grateful acknowledgment is also made of the valuable support and sympathetic encouragement by President Charles E. Friley and Dr. Harold V. Gaskill, Dean of Science.

LOUIS DE VRIES

AMES, IOWA
August, 1940

FRENCH-ENGLISH
SCIENCE AND TECHNOLOGY
DICTIONARY

FRENCH-ENGLISH SCIENCE DICTIONARY

A

a (avoir), has; **il y a,** there is, there are.

à, to, at, with, in, on, of, by, for, from, after, about; **à la . . . ,** in the style; **un livre à moi,** a book of mine; **c'est à vous,** it is your turn.

abaca, *m.* Manila hemp.

abaction, *f.* abortion.

abaissé, depressed, flattened.

abaissement, *m.* lowering, falling, depression, dropping, dip, reduction, subsidence, abatement, diminution, abasement, debasement.

abaisser, to lower, pull down, drop, let fall, reduce (an equation), abate, diminish, depress, humble, abase; **s'—,** to become lower, fall (away), subside, demean, humble or abase oneself, sink, stoop; **— une perpendiculaire,** to draw a perpendicular.

abaisseur, *m.* depressor (muscle).

abajou, *f.* cheek pouch, gill.

abandon, *m.* surrender, abandonment, relinquishment, unreserve, abandon; **à l'—,** neglected, at random, adrift, derelict.

abandonnement, *m.* abandonment.

abandonner, to forsake, leave, let go of, renounce, abandon, give up, throw over, surrender; **s'— à,** to give way to.

abaque, *m.* chart, graph, table.

abas, *m.* blight (of wheat).

abasourdir, to deafen, daze, astound, dumbfound.

abasourdissant, **-e,** deafening, stunning, astounding.

abat (abattre), (he) slaughters, knocks down.

abat, *m.* downpour, killing, slaughtering, slaughtered animal.

abâtardir, to render degenerate, debase, deteriorate.

abâtardissement, *m.* degeneracy.

abat-foin, *m.* hay trap, hay shoot.

abat-jour, *m.* lamp shade, shade, skylight, slanting shutter, awning.

abats, *m.pl.* offal, giblets.

abattage, abatage, *m.* killing, slaughtering, demolition, felling, cutting, throwing down, lowering, casting; **— à la ferme,** slaughter at the farm; **— à la mine,** blasting; **faire l'—,** to overturn, cant (up), turn up.

abattement, *m.* prostration, dejection, low spirits, perpendicular depth.

abattis, abatis, *m.* fellings, killing, slaughter, offal (of animals), giblets, things pulled down; **— d'arbres,** felled trees.

abattoir, *m.* slaughter-house.

abattre, to knock or beat down, throw down, fell, hew, cast, cut down, pull down, demolish, slaughter, kill, bring down, depress, humble, dishearten, lop off, cut away, round (angle), reduce (surface); **s'—,** to fall, crash down, abate, subside, become discouraged, despond; **— à la scie,** to fell with the saw; **— à la hache,** to fell with the ax.

abatture, *f.* beating down, trace (of a stag).

abat-ven', *m.* penthouse, shed, windboard, (window) shutter.

abbé, *m.* abbé, abbot, priest.

abcédation, *f.* formation of an abscess.

abcéder, to suppurate.

abcès, *m.* abscess.

abcisse, *f.* abscissa.

abdiquer, to resign, give up, renounce, surrender, abdicate.

abdomen, *m.* abdomen, paunch.

1

abducteur, abducent, leading away, abductor; **muscle —,** abductor (abducent) muscle; **tube —,** delivery (exit) tube or pipe.

abécédaire, *m.* primer.

abécédaire, alphabetical.

abecquement, *m.* feeding (of young birds).

abeille, *f.* bee, honey bee; **— mâle,** drone; **— mère,** queen (mother) bee; **— ouvrière,** worker.

aberrant, aberrant.

aberration, *f.* aberration; **— moyenne,** constant of aberration.

aberrer, to wander from the point.

abêtir, to render (become) stupid.

abhorrer, to abhor, loathe, detest.

abiétées, *f.pl.* abieteae (Abietineae).

abiétin, referring to fir trees.

abiétine, *f.* abietene.

abiétinée, *f.* fir, conifer.

abiétique, abietic (acid).

abîme, *m.* abyss, chasm, depth, gulf, tallow vat.

abîmer, to engulf, destroy, injure, spoil, overwhelm; **s'—,** to spoil, be engulfed, swallowed up.

abiogénèse, *f.* abiogenesis.

abiose, *f.* abiosis.

abiotique, abiotic, characterized by the absence of life.

ablaction, *f.* weaning.

ablation, *f.* excision, removal, denudation.

ablativo, promiscuously, in confusion.

ablepsie, *f.* blindness.

ablette, *f.* bleak, ablet.

abluant, abluent, detergent.

abluer, to wash (off), clean.

ablution, *f.* ablution, washing.

abnormité, *f.* abnormality.

aboi, *m.* bark, barking, bay; **aux —s,** at bay, hard-pressed.

abolir, to abolish, repeal, suppress, annul.

abolissement, *m.* abolition.

abolition, *f.* abolishment, repeal, annulment, suspension, abolition.

abominable, detestable.

abomination, *f.* abhorrence.

abondamment, abundantly, plentifully, copiously.

abondance, *f.* abundance, plenty; **d'—,** offhand, extemporaneously.

abondant, abundant, copious, plentiful, high, ample; **peu —,** scanty.

abonder, to abound, be plentiful, be full.

abonné, *m.* subscriber.

abonnement, *m.* subscription, season ticket, contract.

abonner(s'), to subscribe.

abonnir, to improve; **s'—,** to become better.

aboquage, *m.* cramming, forcible feeding.

abord, *m.* access, approach, landing, sight, arrival; **à l'—,** at first sight; **au premier —,** at first glance; **d'—,** (at) first; **d'— que,** as soon as; **dès l'—,** from the first; **tout d'—,** at once, to begin with, first of all.

abordable, approachable, accessible.

abordage, *m.* boarding.

aborder, to accost, approach, deal with (a question) take up, enter on, attack, come near, board, land.

aborigène, aboriginal, native, indigenous.

abornement, *m.* demarcation, boundary settlement.

aborner, to mark out (field), demarcate, limit.

abortien, abortient, becoming abortive.

abortif, -ive, abortive.

abot, *m.* fetter, hobble, clog.

abouchement, *m.* interview, conference, anastomosis, joining, insertion, welding, opening of one vessel into another.

aboucher, to join by insertion, bring together; **s'—,** to join, come together, inosculate, confer.

about, *m.* butt(-end), bedplate, butt-joint graft; **— de jante,** rim joint; **— de traverse,** butt end of sleeper.

abouté, placed end to end.

abouter, to join end to end.

aboutir, to end in, result in, converge on, lead to, terminate in, succeed, abut, come to a head, burst, bud, come out, join.

aboutissant, bordering upon, abutting.

aboutissants, *m.pl.* borders, circumstances.

aboutissement, *m.* effect, outcome, result, suppuration, abutment.

aboyer, to bark, bay.

abrasif, *m.* abrasive.

abrasin, *m.* Chinese wood; **huile d'—,** Chinese wood oil, tung oil.

abre, *m.* see ABRUS.

abréaction, *f.* catharsis.

abréagir, to vent off.

abrégé, *m.* outline, summary, abstract.

abrégé, abridged, short.

abréger, to shorten, abridge, abbreviate, cut short; s'—, to grow or become short, shorten.

abreuvage, abreuvement, *m.* watering, soaking, priming.

abreuver, to water, soak, irrigate, saturate, prime, size, steep lye.

abreuvoir, *m.* rotten hollow, watering place, drinking(-trough) fountain, horsepond.

abréviation, *f.* shortening, abbreviation, contraction.

abri, *m.* shelter, (crown) cover, screen, refuge, protection, leaf canopy; à l'— de, protected from, under cover, secure from, sheltered from, away from; **mettre, placer à l'—,** to shelter, protect.

abricot, *m.* apricot.

abricotier, *m.* apricot tree (*Prunus armeniaca*).

abrine, *f.* abrin.

abrité, sheltered.

abriter, to shelter, give cover, shield, protect, shade, screen; s'—, to take shelter.

abrivent, *m.* mat, matting.

abroger, to abrogate, annul, repeal.

abrouti, browsed.

abroutir, to browse.

abroutissement, *m.* browsing (by game).

abrupt, rugged, abrupt, steep (upward), bluff(y) (downward), precipitous.

abrus, *m.* Indian licorice (*Abrus precatorius*).

abrutir, to brutalize.

abscisse, *f.* abscissa.

abscission, *f.* excision, ablation.

absence, *f.* absence.

absent, absent, missing.

absenter(s'), to absent oneself, stay away.

absinthe, *f.* wormwood (*Artemisia absinthium*), absinthe; — **pontique,** Roman wormwood (*Artemisia pontica*).

absinther, to mix with absinthe.

absolu, absolute, unconditioned, peremptory, positive.

absolument, absolutely, unconditionally, positively.

absolvant, (absoudre), absolving.

absorbant, absorbent, absorbing, absorptive.

absorber, to absorb, take up, suck up; s'—, to become absorbed.

absorbeur, *m.* absorber.

absorption, *f.* absorption.

absorptivité, *f.* absorptivity, absorptiveness.

absoudre, to acquit, absolve, forgive.

absous (absoudre), absolved.

abstenir(s'), to abstain, refrain (from), forego.

absterger, to absterge, clean.

abstersif, -ive, abstergent, abstersive.

abstient (abstenir), (he) abstains.

abstractif, -ive, removed by distillation.

abstraction, *f.* abstraction; faire — de, to take away, exclude, disregard.

abstraire, to abstract, separate, detach.

abstrait, abstracted, absorbed, abstruse, separated.

abstraitement, abstractedly.

abstrich, *m.* the impure oxide (scum on molten lead).

abstrus, abstruse.

absurde, absurd, irrational, preposterous.

absurdement, absurdly, nonsensically.

absurdité, *f.* absurdity, nonsense.

abus, *m.* abuse, misuse, overindulgence, error, mistake.

abuser, to misuse, abuse, make ill use of, deceive, delude; s'—, to be mistaken, mistake.

abusif, -ive, improper, excessive.

abuter, abutter, to abut.

abyssal, -aux, abysmal, abyssal.

abzug, *m.* first scum, cupellation of lead.

ac., *abbr.* (acide), acid.

acabit, *m.* nature, stamp, quality.

acacia, *m.* — **vrai,** acacia; — **aquatique,** swamp locust, water locust; — **vulgaire,** locust tree (*Robinia pseudacacia*).

académie, *f.* academy (of learned men), university.

académique, academic, of the Academy.

acajou, *m.* mahogany tree (*Swietenia mahogani*), cashew (*Anacardium occidentale*); — **à pommes,** cashew, acajou (*Anacardium occidentale*); — **moiré,** mottled mahogany; — **moucheté,** spotted mahogany; — **veiné,** veined mahogany; **noix d'—,** cashew nut.

acaliculé, acalycinous, without a calyx.

acanthe, *f.* Acanthus; — **sauvage,** Scotch thistle.

acanthocarpe, acanthocarpous, having spiny fruit.

acanthoïde, acanthoid, spiny.

acare, *m.* acarien, mite.

acariose, *f.* acariasis, acarinosis, infestation with mites.

acarocécidie, *f.* gall (nut).

acarophile, acarophilous, harboring mites.

acarpe, acarpous, sterile.

acarpotropique, acarpotropic, not throwing off its fruit.

acaude, acaudate, tailless.

acaule, acaulous, stemless, stalkless.

acaulescent, acaulescent, becoming stemless.

accablant, overwhelming, oppressive, crushing.

accablement, *m.* pressure, dejection, prostration.

accabler, to overpower, overwhelm, crush, load, bear down.

accalmie, *f.* lull.

accaparement, *m.* monopolizing, cornering.

accaparer, to corner, hoard, monopolize.

accéder, to accede, agree, reach, arrive, enter.

accélérateur, *m.* accelerator.

accélérateur, -trice, accelerative, accelerating.

accélération, *f.* acceleration, quickening; — dans la chute libre, acceleration due to gravity; — de la vitesse de la charge, load acceleration; — de la vitesse de levage, acceleration of lift.

accéléré, accelerated, quick, high-speed.

accélérer, to accelerate, quicken, hasten.

accentuer, to stress, accentuate, accent, emphasize; s'—, to stand out, be conspicuous, be accentuated, set off or emphasized.

acceptable, acceptable, reasonable.

acceptation, *f.* acceptance; — à découvert, blank acceptance.

accepter, to accept, agree, receive, honor.

accepteur, *m.* acceptor, acceptant.

acception, *f.* favoring, regard, respect, preference, acceptation, meaning, sense.

accès, *m.* access, approach, outburst, entrance, admittance.

accessoire, *m.* accessory, fitting, attachment, additional.

accessoire, secondary; **produit —.** by-product.

accident, *m.* accident, symptom, unevenness, chance, irregularity, mishap; —s de terrain, undulations of the ground; par —, by accident, accidentally; sans —, safely.

accidenté, uneven, rough, hilly, broken.

accidentel, -elle, accidental, casual, adventitious, occasional.

accidenter, to break into ridges, make uneven.

accise, *f.* excise, accise.

acclamation, *f.* acclamation.

acclamer, to acclaim, applaud.

acclimatation, *m.* acclimatization, acclimation, acclimatement.

acclimater, to acclimatize.

accoinçon, *m.* jack rafter.

accolade, *f.* brace, embrace.

accolage, *m.* training, uniting.

accolé, coupled, bracketed, united, conjugated, joined together, stuck.

accolement, *m.* joining, uniting, union, bracketing, brace, embrace.

accoler, to join side by side, place against each other, prop or tie up; s'—, to intertwine, cling to, prop, tie up, join.

accolure, *f.* band (of straw), osier twig.

accombant, accumbent.

accommodant, good-natured, accommodating.

accommodation, *f.* adjustment.

accommodement, *m.* compromise, arrangement.

accommoder, to suit, fit, serve, adapt, accommodate, arrange; s'—, to be adjusted, adapt oneself, agree, suit, put up (with), get on with; s'— de, to accept readily, get along with; il s'accommode facilement, he is easily satisfied.

accompagnement, *m.* accompaniment, appendage.

accompagner, to accompany, go or come with, suit, follow.

accompli, accomplished, thorough, complete, perfect.

accomplices, *f.pl.* bacteria of a secondary infection.

accomplir, to accomplish, achieve, effect, finish, complete, fulfil; s'—, to be accomplished, be fulfilled, be done.

accomplissement, *m.* accomplishment, completion, performance, execution.

accord, *m.* agreement, settlement, chord, pitch, accordance; d'—, agreed, in accordance with, in union; **être d'—**, to agree, be agreed.

accordailles, *f.pl.* espousals, betrothals.

accordé, approved.

accorder, to reconcile, grant, concede, bestow, harmonize, give, admit; **s'—**, to agree, be in accord with, be suited, go with, correspond.

accoter, to lean against.

accotoir, *m.* prop, armrest.

accouchée, *f.* the delivered one.

accouchement, *m.* confinement, childbirth, delivery, dropping (of young).

accoucheur, *m.* obstetrician.

accoucheuse, *f.* midwife.

accoudoir, *m.* window sill, elbowboard.

accouplé, coupled.

accouplement, *m.* pairing, mating, coupling, joining, coitus, linking, connecting, clutch; **— à bague et clavette,** annular wedge coupling; **— à friction,** friction coupling; **— à griffes,** claw or dog coupling; **— d'aiguille,** coupling of the tongues; **— des arbres,** shaft coupling; **— par manchon,** muff coupling.

accoupler, to couple, mate, pair, join, connect.

accourcir, to shorten, abridge.

accourir, to run hither, flock, rush, run to, hasten.

accoutumé, accustomed, used to, usual.

accoutumance, *f.* habituation, habit, wontedness.

accoutumer, to accustom, inure, familiarize, be wont.

accouvage, *m.* artificial incubation.

accouver, to set (hen), brood.

accréditer, to open a credit.

accréditif, *m.* letter of credit, (*pl.*) credentials.

accrescent, accrescent, increasing in size with age.

accreusement du lit de fleuve, *m.* erosion or deepening of river bed.

accrochant, hooked (fruits, etc.).

accrocher, to hook, catch, hang up, delay, hang upon, hook to; **s'—**, to fasten to, cling, catch, stick to, hang on; il s'accroche, it gets caught.

accroissement, *m.* growth, increase, increment, accretion; **à — mince,** fine-zoned; **— annuel,** current annual increment (*abbr.* c.a.i.); **— en diamètre,** diameter

accretion; **— dû à l'état,** girth increment; **— dû à l'état clair,** open stand increment.

accroître, to increase, enlarge, augment, inflate, extend, grow, rise.

accru, *m.* sucker, shoot.

accrue, *f.* accretion (of land), alluvial deposit.

accueil, *m.* reception, welcome.

accueillir, to receive, greet, welcome, meet, honor, entertain, accept.

accul, *m.* blind alley, lair, burrow, corner.

aculer, to bring (animal) to bay, corner.

accumulateur, *m.* accumulator, storage battery, storage cell, hoarder.

accumuler(s'), to accumulate, amass, store up.

accusateur, -trice, *m. f.* accuser.

accusé, *m.* acknowledgment, accused, prisoner, defendant; **— de réception,** acknowledgment, receipt.

accusé, marked, indicated, pronounced, prominent, conspicuous.

accuser, to accuse (of), charge, define, show, indicate, blame, acknowledge.

acénaphtène, *m.* acenaphthene.

acéphale, headless, acephalous.

acéracées, *f.pl.* aceraceae (Aceraceae).

acérage, *m.* steeling.

acérain, steely (iron).

acéras, *m.* green man-orchis.

acerbe, sharp, harsh, acrid, tart, sour (taste).

acerbité, *f.* tartness, sourness, sharpness, harshness.

acerdèse, *f.* manganite.

acère, hornless, without antennas.

acéré, steeled, steely, sharp-edged, keen (blade), needle-shaped.

acérer, to point, steel, give a keen edge to.

acéreux, -euse acerose, acerous, acicular, needle-shaped.

acescence, *f.* acescence.

acétabule, *m.* acetabulum.

acétabuliforme, saucer-shaped.

acétal, *m.* acetal.

acétamide, *f.* acetamide.

acétanilide, *f.* acetanilide.

acétate, *m.* acetate; **— d'alumine,** aluminium acetate; **— d'ammoniaque,** ammonium acetate; **— de chaux,** calcium acetate. **— de cuivre brut,** crude basic copper acetate, verdigris; — **de plomb,** lead acetate, sugar of lead; — (sous-) **de plomb liquide,** solution of lead subacetate.

acéteux, -euse, acetous, acetic, sour.

acétification, *f.* acetification, acetifying.

acétifier, to acetify.

acétimetrie, *f.* acetimetry.

acétique, acetic, acetous; **acide** — **concentré,** concentrated (glacial) acetic acid; **acide** — **cristallisable,** glacial acetic acid; **acide** — **trichlore,** tri-chloroacetic acid.

acétocellulose, *f.* cellulose acetate.

acétol, *m.* acetol.

acétolé, *m.* acetic medication.

acétomel, *m.* oxymel.

acétomellé, mixed with oxymel.

acétométrie, *f.* acetometry.

acétone, *f.* acetone.

acétonémie, acétonhémie, *f.* acetonemia.

acétonique, acetonic, of or containing acetone.

acétonitrile, *m.* acetonitrile.

acétonurie, *f.* acetonuria.

acétophénone, *f.* acetophenone.

acétoselle, *f.* wood sorrel (*Oxalis acetosella*).

acétosité, *f.* acetosity, acetous quality.

acétoxime, *m.* acetoxime.

acét-phénétidine, *f.* phenacetin, acetphenetidin.

acetum fermentatif, *m.* fermentative vinegar or acetum.

acétylacétate, *m.* acetoacetate, acetylacetate.

acétylacétique, acetoacetic, acetylacetic.

acétylacétone, *f.* acetylacetone, 2,4-pentanedione.

acétylant, acetylating.

acétyle, *m.* acetyl.

acétylène, *m.* acetylene.

acétylénique, pertaining to acetylene, acetylenic.

acétyler, to acetylate.

acétylure, *m.* acetylide.

achaine, *m.* see AKÈNE.

acharné, strenuous, desperate, fierce, bitter.

acharnement, *m.* relentlessness, obstinacy, desperation. desperate eagerness, stubbornness.

acharner, to flesh, excite, infuriate; s'—, to be infuriated, persist in, slave at, apply oneself, strive, be set upon, bent upon.

achat, *m.* purchase, buying.

ache, *f.* smallage, water parsley, wild celery; — **d'eau,** water parsley; — **de montagne,** skirret (*Sium sisarum*); — **des chiens,** fool's-parsley (*Aethusa cynapium*); — **des marais,** smallage, wild celery.

acheminé, forwarded, sent off.

acheminement, *m.* step, preliminary measure, advance(ment).

acheminer, to direct, dispatch, send, forward; s'—, to proceed, progress, set out.

achérontia, *f.* death's-head moth.

acheter, to buy, purchase, bribe.

acheteur, -euse, *m.&f.* purchaser, buyer, customer, consumer.

achevage, *m.* comp'etion, finishing.

achevé, accomplished, perfect, finished.

achèvement, *m.* completion, finishing, conclusion.

achever, to end, conclude, finish, complete, dispatch; s'—, to end, reach completion.

acheveur, *m.* finisher.

achevoir, *m.* finishing tool or shop.

achillée, *f.* milfoil, nosebleed, yarrow (*Achillea millefolium*); — **sternutatoire,** sneezewort, goosetongue (*Achillea ptarmica*).

achit, *m.* wild vine.

achlamydé, achlamydeous, destitute of perianth.

achlorophyllose, *f.* decomposition of chlorophyll.

acholie, *f.* acholia, absence of bile.

achoppement, *m.* obstacle, stumbling.

achopper, to stumble, kick, fail.

achorion, *f.* achorion.

achroie, *f.* lesser wax moth.

achromatine, *f.* achromatin.

achromatique, achromatic, free from color.

achromatiser, to achromatize.

achromatopsie, *f.* color blindness.

achrome, colorless, achromous.

achromique, colorless, achromic.

achroodextrine, *f.* achroödextrin, not colored by iodine.

achropsie, *f.* color blindness.

aciculaire, needle-shaped, acicular.

acicule, *f.* acicula, a bristle or prickle.

aciculé, aciculate(d).

acide, *m.* acid; — **acétique concentré,** concentrated (glacial) acetic acid; — **acétique cristallisable,** glacial acetic acid;

— **acétique trichloré,** trichloroacetic ᴛ.id;
— **acétylacétique,** acetoacetic acid; —
aloïque, aloetic acid; — **aminé,** amino
acid; — **azoteux,** nitrous acid; — **azotique,**
nitric acid; — **bromhydrique,** hydrobromic
acid.

acide, — **carbonique,** carbonic acid; — **chlorhydrique,** hydrochloric acid; — **chloro-
carbonique,** carbonyl chloride, phosgene;
— **cinchonique,** cinchonic (quinic) acid;
— **cyanhydrique,** hydrocyanic acid; — **du
citron,** citric acid; — **gras,** fatty acid; —
iodhydrique, hydriodic acid; — **marin,** hydrochloric (marine) acid; — **muriatique,**
hydrochloric (muriatic) acid; — **oxygène,**
oxygen acid, oxyacid; — **phénique,** carbolic
acid; — **sulfhydrique,** hydrogen sulphide,
hydrosulphuric acid.

acide, sharp, tart, acid, sour.

acide-acétone, *m.* ketonic acid, acid ketone.

acide-alcool, *m.* hydroxy acid, alcohol-acid.

acide-cétone, *m.* ketonic acid, acid ketone.

acideur, *m.* acid-liquor man.

acidifiant, *m.* acidifier.

acidifiant, acidifying.

acidificateur, *m.* acidifier.

acidification, *f.* acidification, acidulation,
adding a starter.

acidifier, to acidify, acidulate, sour; s'—,
to become acid, turn sour.

acidimètre, *m.* acidimeter, acid tester,
battery hydrometer.

acidism, *m.* acid dyspepsia.

acidité, *f.* acidity, sourness, tartness.

acidophiles, *f.pl.* acidophilic bacteria.

acido-résistance, *f.* acid-fast character.

acido-résistant, *m.* acid-fast organism.

acido-résistant, acid fast.

acidose, *f.* acidosis.

acidule, subacid, acidulous.

acidulé, acidulated, made acid, somewhat
acid.

aciduler, to acidulate, render acid.

acier, *m.* steel; — **affiné,** refined steel,
shear steel; — **à outils,** tool steel; — **à
ressort,** spring steel; — **au chrome,** chrome
steel; — **au creuset,** crucible steel;
— **boursousslé,** blister steel; — **brut,** crude
(raw) steel.

acier, — **cémenté,** cement steel, blister
steel; — **cémenté à la surface,** casehardened steel; — **corroyé,** shear steel; —
coulé, cast steel; — **de forge,** forge steel;
— **de lingot,** — **en lingot,** ingot steel; —

doux, mild (soft) steel; — **dur,** — **durci,**
hard steel; — **fondu au creuset,** crucible
(cast) steel; — **forgé,** forged steel, cold
steel.

acier, — **indien,** wootz; — **inoxydable,** stainless (rustless) steel; — **laminé,** rolled steel;
— **malléabie,** malleable steel; — **manganésé,** manganese steel; — **moulé,** cast
(molded) steel; — **nickelé,** nickel steel,
nickel-plated steel; — **poule,** blister steel;
— **puddlé,** puddled steel.

acier, — **raffiné à deux marques,** double-
shear steel; — **raffiné à une marque,**
single-shear steel; — **recuit,** tempered steel;
— **sauvage,** wild (fiery) steel; — **soudable,**
welding steel; — **soudé,** weld steel; —
tiré, rolled steel; — **trempé,** hardened
steel; — **une fois corroyé,** single-shear
steel; — **vif,** hard steel.

aciérage, *m.* steeling, steel facing.

aciération, *f.* steeling, case-hardening,
converting of iron into steel.

aciérer, to steel, convert into steel.

aciéreux, -euse, steely.

aciérie, *f.* steelworks.

acier-nickel, *m.* nickel steel.

aciforme, acicular, needle-shaped.

acinacifolié, acinacifolious, having scimitar-
shaped leaves.

acinaciforme, scimitar-shaped.

acinaire, acinarious, with vesicles resembling grape seeds.

acine, *m.* acinus, a drupelet.

acineux, -euse, acinous, like grapes.

aciphylle, aciphyllous, with pointed leaf.

aclaste, aclastic.

acompte, *m.* installment, payment (on
account).

aconit, *m.* aconite, wolfsbane (Aconitum);
— **napel,** monkshood (*Aconitum napellus);*
— **tue-chien,** dogbane, yellow wolfsbane
(*Aconitum lycoctonum).*

aconitine, *f.* aconitine.

aconitique, aconitic.

acore, *m.* sweet flag, sweet rush (*Acorus
calamus);* — **odorant,** — **vrai,** (sweet) calamus, sweet flag (*Acorus calamus).*

acotylédone, *f.* acotyledon.

acotylédone, acotyledonous.

à-coup, *m.* jerk, shock, start.

acoustique, *f.* acoustics, the science of
sounds.

acoustique, acoustic.

acquérable, acquirable.

acquérir, to acquire, obtain, secure, get, reclaim.

acquis, *m.* acquirement, attainment.

acquis, acquired, established.

acquisition, *f.* acquisition, purchase, achievement.

acquit, *m.* discharge, receipt; — **de fermage,** remission of rent; **pour** —, paid, received.

acquit-à-caution, *m.* permit, excise bond.

acquittement, *m.* discharge, payment, acquittal, adjustment.

acquitter, to release, acquit, discharge, fulfill, pay, receipt, pay off, perform; **il s'acquitte,** he fulfills his obligations.

acre, *f.* acre.

âcre, acrid, tart, bitter, pungent, sharp, sour, acrimonious.

âcreté, *f.* acridity, pungency, sourness, sharpness.

acridine, *f.* acridine.

acridinique, acridine, of acridine.

acridique, acridic, acridinic.

acrocarpe, acrocarpous, terminal fruited.

acrocome, *m.* macaw palm or tree (Acrocomia).

acrodrome, acrodromous, running to a point, convergent.

acrofuge, acrofugal, basipetal.

acrogène, *m.* acrogen.

acrogène, acrogenous.

acrogénèse, *f.* acrogenesis, terminal fructification.

acrogyne, acrogynous, stem terminated by female organs.

acroléine, *f.,* **acrol,** *m.,* acrolein.

acroléique, acrylic.

acrone, (ovary) not widening at base.

acronécrose, *f.* (de la pomme de terre) top necrosis (of potatoes), acronecrosis of potatoes, healthy potato virus.

acropete, acropetal.

acroscope, acroscopic, looking towards the summit.

acrospire, *f.* acrospire, first sprout.

acrospore, *f.* acrospore, a spore at summit of sporophore.

acrotone, acrotonous, reaching to the top.

acte, *m.* action, act, deed, certificate, document; **prendre — de,** to record or take a note of.

actée, *f.* baneberry (Actaea); — **à grappes,** black cohosh, bugbane (*Cimicifuga racemosa*).

acteur, -trice, *m.&f.* actor, player.

actif, *m.* credit, outstanding debt, assets, favor; **à l'— de,** to the credit of.

actif, -ive, active, quick, agile.

actinium (-iom), *m.* actinium.

actinomorphe, actinomorphous, radially symmetrical.

actinomycose, *f.* wooden tongue, actinomycosis.

actinotropism, *m.* inflexion toward a source of light.

action, *f.* action, act, share, working, activity, operation, effect; — **à distance,** action of distance, distant effect; — **de dresser,** — **de jointer les douves,** stave jointing; — **d'une force,** stress, strain; — **privilégiée,** — **de priorité,** preference share; **hors d'—,** out of action, out of gear; **sans —,** ineffective.

actionnable, movable, actionable.

actionnaire, *m.* shareholder, holder of stock.

actionné, busy, brisk, occupied; — **par moteur,** motor driven.

actionnement, direct —, *m.* direct driving.

actionner, to set in motion, operate, sue, drive, work, move.

activement, actively, briskly, vigorously.

activer, to accelerate, activate, urge on, hasten, push, rouse, stir up, press.

activité, *f.* activity, action, quickness, dispatch, operation; — **optima,** optimum activity.

actualité, *f.* actuality, reality.

actuel, -elle, actual, real, present (day), for the time being.

actuellement, now, at the present time.

acuité, *f.* acuteness, keenness.

aculé, aculeate, with prickles.

aculées, *f.pl.* Aculeata.

aculéiforme, spine-shaped.

acuminé, acuminate, attenuate, long tapering.

acuminifolié, acuminifolious, with acuminate leaves.

acutangle, acute-angled, sharp-edged.

acutangulé, acutangular.

acutesse, *f.* sharpness, keenness.

acutiflore, acutiflorous, with acute perianth segments.

acutifolié, acutifolious, with acute leaves.

acutipenne, acutipennate.

acyclique, acyclic.

acyler, to acylate.

adamine, *f.* adamite.

adaptation, *f.* adaptation.

adapter, to fit, apply, adjust, adapt, make suitable; **s'—,** to fit, adapt, apply, be suitable, adjust oneself.

additif, -ive, additive.

addition, *f.* accession, addition

additionnel, -elle, additional.

additionner, to add up, sum up, mix or treat with, supplement, add; **— de,** to treat with; **lait additionné d'eau,** watered milk.

adducteur, *m.* adducent (muscle), adduction; **tube —,** delivery tube.

adduction, *f.* conveying (water); **— d'eau,** water supply.

adèle, *f.* adela (butterfly).

adelphe, adelphous, having brotherhoods of stamens.

adelphie, *f.* adelphia, bundle of stamens.

adelphogamie, *f.* adelphogamy (fertilization).

adénine, *f.* adenine.

adénite, *f.* adenitis.

adénites, *f.pl.* glanders.

adénophylle, adenophyllous, glandular-leaved.

adenter, to indent, dovetail.

adepte, *m.&f.* adept.

adhérence, *f.* adherence, adhesion, adsorption, traction, attachment, spreading capacity.

adhérent, *m.* follower.

adhérent, adherent, adhesive, adnate, sticky.

adhérer, to adhere, stick, cling, approve, stick fast.

adhésif, *m.* binding material (agent), fixing agent, agglutinant.

adhésif, -ive, adhesive.

adhésion, *f.* adhesion, adherence.

adiabatique, adiabatic.

adiabatisme, *m.* adiabatic state (of gas).

adiante, *m.* Adiantum; **— capillaire,** maidenhair (fern) (*Adiantum pedatum*).

adiaphorèse, *f.* adiaphoresis.

adieu, farewell.

adipeux, -euse, adipose, fatty, fat.

adipique, adipic.

adipocire, *f.* adipocere.

adipogénie, *f.* production of fat.

adipose, *f.* adiposis.

adiposité, *f.* adiposeness, adiposity, fatty condition.

adiscal, adiscalis, without a disc.

adjacent, adjacent, accumbent.

adjoignant, joining, uniting.

adjoindre, to adjoin, unite, associate, attach, add.

adjoint, *m.* associate, assistant, deputy, adjunct.

adjoint, attached, united, assistant.

adjonction, *f.* adding, adjoining, joining.

adjudicataire, *m.* purchaser, highest bidder.

adjudication, *f.* sale by auction, adjudging, public tender, submission.

adjuger, to adjudge, award, allocate, sell, knock down.

adjuvant, adjuvant.

admettre, to admit, let enter, accept, grant, allow, acknowledge, own. let in.

administrateur, -trice, *m.&f.* administrator, director, manager, receiver, trustee.

administratif, -ive, administrative.

administration, *f.* trusteeship, administration, direction; **— forestière,** forest administration.

administrer, to administer, manage, direct, conduct, govern, work, exploit.

admirablement, admirably, famously, wonderfully.

admirer, to admire, wonder at, be surprised at.

admis, admitted.

admissibilité, *f.* admissibility.

admissible, admissible, allowable.

admission, *f.* admission (passage), intake, inlet, induction, introduction, inclusion; **— radiale,** radial admission; **— tangentielle,** tangential admission.

admit (admettre), admitted.

admittance, *f.* admittance.

admixtion, *f.* admixture.

admonester, to admonish, warn.

admotif, -ive, admotivus, in germination albumen remains with sheath.

adné, adnate, attached the whole length.

adonc, then.

adonné, addicted, given.

adonner(s'), to devote oneself, give oneself up, take up; — **à**, to addict oneself to, apply.

adopter, to adopt, embrace.

adoption, f. adoption.

adorable, charming.

adorer, to adore, worship.

ados, m. slope, talus, banked up bed or border, sloping bed (in gardens), ridge, tillage in ridges, bank, sloping warm border.

adossé, back to back, supported on to something.

adosser, to place back to back, lean against, lean or rest something against.

adoucir, to soften, tone down, sweeten, relieve, calm, moderate, mollify, smooth down, polish, burnish, furbish, anneal, temper; s'—, to become soft, grow sweet or mild, be mitigated, calm down.

adoucissage, m. smoothing, rough polishing, softening, grinding, rubbing, glazing.

adoucissant, m. polishing substance, emollient.

adoucissant, softening, lenitive, emollient, demulcent.

adoucissement, m. softening, alleviation, smoothing off.

adoucisseur, -euse, m.&f. polisher.

adoué, coupled, paired.

adoux, m. woad.

adpresse, adpressed, appressed.

adragant, m. tragacanth; **gomme —e, gomme d'—,** gum tragacanth.

adrénaline, f. adrenaline.

adresse, f. address, speech, direction, destination, skill, dexterity, craftiness, cunning.

adresser, to address, label, send, recommend, forward; s'— (à), to address oneself, apply to, speak to, have recourse; **il adresse la parole à,** he speaks to.

adroit, m. sunny side (mountain).

adroit, dexterous, skillful, ingenious, clever, adroit.

adroitement, shrewdly, skillfully, cleverly.

adsorber, to adsorb.

adsorption, f. adsorption, adherence.

adulte, adult, grown up, full grown, able to bear fruit.

adultérant, m. adulterant.

adultérant, adulterating.

adultération, f. adulteration.

adultéré, adulterated.

adultérer, to adulterate, falsify.

aduste, (sun)burnt, adust, scorched.

adustion, f. cauterization.

advenir, to occur, happen, chance, befall, turn out, come to pass.

adventice, adventitious, casual.

adventif, -ive, adventive, casual.

advenu, occurred, happened.

advers, opposed, adverse, opposite.

adversaire, m. adversary, opponent.

adverse, opposite, contrary, unfavorable, adverse (leaves).

adversité, f. adversity, misfortune.

advient (advenir), (it) occurs, is happening.

adynamandrie, f. adynamandry, self-sterility.

adynamie, f. prostration, adynamy.

aérable, capable of ventilation.

aérage, m. ventilation, aeration, airing, renewal of air.

aérateur, m. fan, ventilator.

aération, f. ventilation.

aérenchyme, f. aerenchyma, a secondary respiratory tissue.

aérer, to ventilate, air, aerate, give air.

aéricole, aerial.

aérien, -enne, aerial, airy, overhead, elevated.

aérifère, air-conducting; **voies —s,** air (wind) passages.

aérifier, to ventilate, aerify.

aériforme, aeriform, gaseous.

aériser, to air, aerify.

aérobie, m. aerobe.

aérobie, aerobic.

aérobiose, f. aerobiosis, life in atmospheric air.

aérobisme, aerobism.

aérocarpé, f. aerocarpy, producing fruit aboveground.

aérocyste, f. air bladder.

aérodynamique, aerodynamic.

aéro-électrique, aeroelectric.

aérogames, f.pl. aerogams, phanerogams.

aérogène, aerogenic.

aérolithe, m. aerolite, aerolith, air stone.

aérolithique, aerolitic, meteoritic.

aéromètre, m. aerometer.

aérométrie, *f.* aerometry.

aéromorphose, *f.* aeromorphosis.

aéromoteur, *m.* windmill.

aéronef, *m.* airship, aircraft.

aérophagie, *f.* aerophagia.

aérophytes, *m.pl.* epiphytes, air plants.

aérostat, *m.* lighter-than-air craft, balloon.

aérostation, *f.* ballooning, aeronautics.

aérotactisme, *m.* aerotaxis, stimulus by oxygen.

aérotropisme, *m.* aerotropism, influence of gases on growth.

aéschynomène, *f.* sola (*Aeschynomene aspera*).

aescule, *m.* horse chestnut (*Aesculus hippocastanum*).

aesculine, *f.* esculin, aesculin, an alkaloid from the horse chestnut.

aethalie, *f.* aethalium, a compound sporiferous body.

aéthéogame, aetheogamic, cryptogamic.

affable, gracious.

affadir, to render insipid, tasteless or dull, flatten; s'—, to become insipid, lose flavor.

affadissant, sickening, insipid, tiresome.

affadissement, *m.* insipidity, dulling, loss of flavor.

affaiblir, to weaken, dilute, lessen, reduce, impair, tender, enfeeble, debase; s'—, to become weak, lose strength, grow faint.

affaiblissement, *m.* weakening, diminution, enfeeblement, (state of) weakness, debility, lowering.

affaiblisseur, -euse, weakening, reducing (agent).

affaire, *f.* business, affair, concern, thing, question, matter, deal, venture, engagement, fight, case, difficulty, transaction; avoir — à, to deal with, have to do with, have business with; tirer d'—, to get out of difficulty; voici son —, that's what he wants.

affairé, busy.

affaires, *f.pl.* business.

affaissement, *m.* collapse, prostration, sinking, depression; — plat, shallow subsidence.

affaisser, to cause to sink or collapse, subside, depress, weigh down, sink, settle, sag, bear down; s'—, to collapse, sink, sag, give way, become downhearted.

affaler, to unwind lower; il s'affale, he slides down.

affamé, hungry, starving, eager.

affamer(s'), to follow a starvation diet.

affectable, easily affected.

affectation, *f.* periodic block; — en aménagement, block with crop of one age class.

affecter, to affect, take on, appropriate, move, touch, pretend, assume, attach, provide, destine.

affection, *f.* affection, infirmity, love, liking; — du foie, hepatitis, inflammation of the liver; — mammaire, garget; — saturnine, lead poisoning.

affectionné, affectionate, loving.

affectionner, to love, like, be fond of.

affectueux, -euse, affectionate.

affenage, *m.* feeding.

affenager, to feed.

afférent, afferent, assignable, belonging, relating to.

affermage, *m.* farming out, leasing.

affermataire, *m.* tenant farmer.

affermer, to lease, rent, farm out, put out to lease.

affermir, to strengthen, establish, make firm, harden, consolidate, fasten; s'—, to become stronger, firmer, be consolidated.

affermissement, *m.* consolidation, strengthening, support.

afféterie, *f.* affectation.

affichage, *m.* billposting, placarding.

affiche, *f.* placard, poster, bill.

afficher, to post (up), display, parade, expose, show.

affidé, trustworthy.

affilage, *m.* sharpening, whetting.

affilée, *f.* cinq heures d'—, five hours at a time.

affiler, to sharpen, set, whet, hone.

affilié, *m.* affiliated member, associate

affilié, affiliated.

affilier, to affiliate.

affiloire, *f.* rubber, slip, whetstone, hone.

affin, related, closely connected.

affinage, affinement, *m.* improvement, refining, ripening, maturing, cropping.

affiner, to improve, refine, make better, ripen, mature, heckle, sharpen, crop, shear, point; s'—, to become pure or refined, ripen.

affinerie, *f.* refinery.

affineur, -euse, *m.&f.* refiner, finisher.

affinité, *f.* affinity, similarity, relationship.

affirmatif, -ive, affirmative, positive.

affirmation, *f.* assurance, assertion, statement.

affirmer, to affirm, assert, state, assure.

affleurage, *m.* mixing (of paper pulp).

affleurement, *m.* leveling, outcrop.

affleurer, to level, make flush, be even, mix, crop out, brush.

affleurie, *f.* fine flour.

affligé, afflicted, grieved.

affligeant, distressing, painful, afflicting.

affliger, to afflict, pain, distress, grieve.

affluence, *f.* flow, flood, influx, affluence, crowd, abundance, flowing together.

affluent, affluent, tributary, confluent.

affluer, to flow (in or into), rush, crowd, abound.

afflux, *m.* rush (of blood), influx.

affodile des forêts, *f.* Lancashire asphodel, bog asphodel (*Narthecium ossifragum*).

affolé, distracted, disturbed, mad, infatuated, demented.

affoler, to infatuate, madden, derange.

afforestage, *m.* right to firewood, estovers.

afforestation, *f.* (re-) afforestation.

affouage, *m.* fuel supply, right to cut firewood.

affouagiste, *m.* firewood rightholder.

affouillement, *m.* undermining, underwashing, erosion.

affouiller, to undermine, erode, wash (away); s'—, to cave in, fall in.

affourager, affourrager, to fodder (cattle).

affranchir, to (set) free, affranchise, liberate, deliver, repay, exempt, pay (for), prepay, trim (off), castrate, release.

affranchissement, *m.* freeing, release gelding (horse), postage.

affréter, to freight, charter (ship).

affreusement, terribly, frightfully, horribly.

affreux, -euse, frightful, horrible, hideous, shocking, detestable, dreadful, terrible.

affriandant, tempting, appetizing.

affricher, to let (land) lie fallow; s'—, to run to seed or to waste.

affront, *m.* affront, insult, disgrace, reproach.

affronter, to face, confront, encounter, join face to face, bring into apposition, join flush, insult, affront.

affruiter, to bear fruit.

affusion, *f.* pouring on, affusion.

affût, *m.* hiding place, watch, lookout, gun carriage, lurking place, ambush, stand, stalking; être à l'— de, to be on the lookout for.

affûtage, *m.* sharpening, setting; — à biseaux, alternating bevel teeth; — droit des dents, sharpening the teeth square across.

affûter, to sharpen, whet, smooth.

affûteur, *m.* sharpener, deer stalker.

affûteuse, *f.* grinding machine, sharpener.

afin de, in order to, so as to; — que, so that, in order that.

africain, African.

Afrique, *f.* Africa; l'— du Sud, South Africa.

Ag, *abbr.* (argent), silver.

agace, *f.* magpie.

agacer, to incite, irritate.

agacerie, *f.* incitement, teasing, allurement.

agalaxie, *f.* agalaxis, lack of milk.

agamandroecie, *f.* agamandroecism, with male and neuter flowers.

agame, agamous, agamic, asexual.

agamogynoecie, *f.* agamogynaecism, with female and neuter flowers.

agamogynomonoecie, *f.* agamogynomonoecism, with neuter, female, and perfect flowers.

agamotrope, *f.* a flower remaining open.

agar-agar, *m.* agar-agar.

agaric, *m.* agaric (Agaricus), toadstool, mushroom; — amadouvier, tinder agaric (*Polyporus igniarius*); — blanc, — purgatif, white agaric, purging agaric (*Polyporus officinalis*); — miellé, honey fungus, tree root rot, palisade fungus, shoestring rot (*Agaricus melleus*).

agaricine, *f.* agaricin.

agate, *f.* agate; — mousseuse, moss agate

agaté, agaty.

agatifier, to agatize.

agave, *m.* agave; fibre d'—, aloe fiber, sisal grass.

age, *m.* plow beam.

âge, *m.* age, generation, period, epoch, century, time; — de fertilité, ability to bear fruit, age of fertility; — d'exploitabilité, final age, exploitable age, age of maturity: moyen —, Middle Ages, mean age, average age.

âgé, aged, old, elderly; **il est — de dix ans,** he is ten years old.

agence, *f.* agency, bureau, business, representation.

agencement, *m.* arrangement, fitting up; *pl.* fixtures.

agencer, to arrange, dispose, fit up, adjust, order.

agenda, *m.* memorandum book, diary.

agenouiller(s'), to kneel.

agent, *m.* agent, medium, causal organism, transactor, mean; — **de change,** stockbroker, foreign-exchange broker; — **de gestion,** executive officer; — **de liaison,** condensing agent, means of union, connecting file; — **fertilisant,** fertilizer; — **forestier,** forest officer; — **réducteur,** reducing agent, reducer; — **révélateur,** developer; — **transmetteur,** carrying agent, carrier.

agglomérant, *m.* binding material.

agglomération, *f.* agglomeration, aggregation; **grand —,** crowded center (of population), community, mass.

aggloméré, *m.* compressed fuel, briquette, conglomerate.

aggloméré, conglomerate.

agglomérer, to agglomerate, bring together, mass, cohere, bind, heap together.

agglutinabilité, *f.* agglutinability.

agglutinage, *m.* agglutination, fritting, sintering.

agglutinant, agglutinant, adhesive, agglutinating.

agglutinatif, *m.* agglutinant, binding material.

agglutination, *f.* agglutination.

agglutiner, to agglutinate, bind; **s'—,** to stick together.

agglutinine, *f.* agglutinin; — **spécifique,** specific agglutinin.

aggravant, aggravating.

aggraver, to aggravate, increase, augment, worsen.

agile, agile, nimble, active, fleet, quick.

agilité, *f.* agility, nimbleness.

agio, *m.* premium, agio.

agiotage, *m.* speculation in stocks.

agioteur, *m.* stock jobber.

agir, to act, be in action, take action, operate, work, do, behave, carry on; — **en,** to act as (the part of), behave as; **faire —,** to set in motion; **il s'agit,** he moves about, is restless; **s'— de,** to be a question of,

matter; **il s'agit de,** it is a question (matter) of, it concerns, the question is.

agissant, active, busy.

agitable, debatable, excitable.

agitant, agitating, disquieting.

agitateur, *m.* agitator, stirrer, stirring rod, mixer, stirring axle; — **à bouton,** stirring rod with a flat end; — **en verre,** glass stirring rod; — **mécanique,** mechanical stirrer or agitator; — **va-et-vient,** reciprocating shaker.

agitation, *f.* agitation, stirring, uneasiness, excitement, shaking, discussing.

agité, restless, troubled.

agiter, to agitate, stir, excite, shake, disturb, vex, disquiet; **s'—,** to be agitated, come into play, be in motion, stir, move, wiggle, wriggle.

aglomérant, agglomerant.

aglutinage, agglutinage.

aglutiner, see AGGLUTINER.

aglykon, *m.* aglucon, aglycon.

agneau, *m.* lamb, male lamb, lamb (meat).

agnelage, *m.* lambing.

agneler, to lamb, yean, ean, ewe.

agneline, laine —, lamb's wool.

agnelle, *f.* ewe lamb.

agoniser, to be dying.

agoraphobie, *f.* agoraphobia.

agrafage, *m.* hooking, attaching.

agrafe, *f.* hook, fastener, clasp, buckle, clip, cramp, clamp, catch, snap, saw cap; — **à chaud,** shrunk dowel; — **de courroie,** belt clip; — **de scie,** blade holder.

agrafer, to fasten with, clamp, hook, clip, clasp, cramp.

agraire, agrarian.

agrairien, *m.* agrarian.

agrandir, to enlarge, make greater, extend, increase, exaggerate, reclaim, magnify, amplify, aggrandize, raise, ennoble, exalt; **s'—,** to become greater, increase, expand, grow larger, extend.

agrandissement, *m.* enlargement, increase.

agréable, agreeable, pleasant, pleasing, gratifying.

agréablement, agreeably, acceptably, pleasantly.

agréation, *f.* approval.

agréer, to accept, recognize, relish. approve, agree to, suit, please.

agrégat, *m.* aggregate.

agrégation, *f.* aggregation, binding, aggregate, admission, admittance, fellowship.

agrégé, *m.* fellow, teacher, professor.

agrégé, aggregate, clustered, gregarious.

agréger, to admit, aggregate, incorporate.

agrément, *m.* pleasure, agreeableness, pleasantness, charm, ornament, consent, approval, attraction, liking.

agrès, *m.pl.* apparatus, equipment, rigging.

agressif, -ive, aggressive.

agressine, *f.* aggressin.

agreste, rustic, uncultivated, rural, wildgrowing.

agricole, agricultural.

agriculteur, *m.* agriculturist, farmer, agricultor, husbandman.

agriculture, *f.* agriculture, farming, husbandry.

agriffer, to claw, grip.

agrile, *m.* — de chêne, oak borer (*Agrilus angustulus*); — du hêtre, pear flat-headed borer (*Agrilus viridis*).

agrion, *m.* dragonfly.

agripaume, *f.* motherwort (*Leonurus cardiaca*).

agripper, to grip, snatch up.

agrologie, *f.* soil science, agrology.

agronome, *m.* agronomist, agriculturist.

agronomie, *f.* agronomy, agricultural science.

agronomique, agronomic(al), agricultural.

agrostème, *f.* corn cockle (*Agrostemma githago*).

agrostide, *f.* bent grass (Agrostis); — eventée, corn grass, silky bent grass.

agrotide, *f.*, **agrotis,** *f.* agrotis; — des moissons, turnip moth, common dart moth.

agrume, *f.* cedrat(e) fruit.

aguerri, warlike.

aguerrir, to harden, accustom, inure, war.

ahan, *m.* deep breath, panting.

ai, j'ai, I have.

aicher, to bait with worms.

aidant, *m.* helper.

aidant, aiding.

aide, *m.&f.* assistant, helper, support.

aide, *f.* help, succor, assistance, aid; à l'— de, with the aid of, by means of.

aider, to help, assist, aid; s'—, to make use of, help each other, avail oneself of.

aïeul, *m.* grandfather; —e, *f.* grandmother.

aïeux, *m.pl.* ancestors, forebears.

aiglantine, *f.* common columbine, capon's-feather (*Aquilegia vulgaris*).

aigle, *m.* eagle, genius; **grand — criard,** larger spotted eagle; **petit — criard,** lesser spotted eagle.

aiglefin, *m.* haddock.

aigre, *m.* acidity, sourness, tartness.

aigre, sour, sharp, acid, tart, shrill, harsh, brittle, acrid, musty; **tourner à l'—,** to turn sour; **sentir l'—,** to smell sour.

aigre-doux, -ce, between sweet and sour, sourish, bitterish.

aigrelet, -ette, sourish, tart, subacid.

aigrelier, *m.* wild service tree (*Sorbus torminalis*).

aigrement, sourly, harshly.

aigremoine, *f.* agrimony (Agrimonia).

aigremore, *m.* charcoal dust.

aigret, -ette, tartish, slightly sour, subacid, acetose.

aigrette, *f.* aigrette, egret, white heron, plume, tuft, thistledown, pappus; — **lumineuse,** brush-light (of generator).

aigretté, bearing a pappus, having an aigrette, tufted, crested.

aigreur, *f.* sourness, tartness, sharpness, acidity, brittleness, cold-shortness, harshness; *pl.* acidity of the stomach, heartburn.

aigri, embittered, soured.

aigrin, *m.* young pear or apple tree.

aigrir, to turn sour, sour, irritate, exasperate; s'—, to become irritated or embittered.

aigrissement, *m.* souring.

aigu, sharp, pointed, acute, acuminate, shrill, intense, keen, piercing.

aiguayer, to rinse, wash, water (horse).

aigue-marine, *f.* aquamarine.

aiguillage, *m.* throwing the points.

aiguillat, *m.* piked dogfish.

aiguille, *f.* needle, index, pointer, hand, spire, switch tongue (blade), point, acicular, hand of a clock, (bridle) pile; — **à coudre,** sewing needle; — **aimantée,** magnetic needle, compass needle; — **à pointiller,** dotting needle; — **à tricoter,** knitting needle; — **de boussole,** compass needle; — **de changement de voie,** switch blade or tongue; — **de fusil,** firing pin; — **du pin,** pine needle.

aiguillé, needle-shaped, acicular.

aiguiller, to switch off, turn off.

aiguillette, *f.* aiguillette, lady's-comb, shepherd's-needle (*Scandix pectenveneus*).

aiguilleur, *m.* pointsman.

aiguillon, *m.* goad, prickle, thorn, spine, sting, stinger, incentive, stimulus, trichome.

aiguillonner, to goad, sting, urge on, inspire.

aiguisage, aiguisement, *m.* sharpening, acidification.

aiguisant, *m.* sharpening.

aiguisé, ground, sharp-set.

aiguiser, to sharpen, whet, grind, acidify, stimulate.

aiguisoir, *m.* whet steel.

ail, *m.* garlic, leek (*Allium porrum*); — **des champs,** field garlic (*Allium oleraceum*).

ailante, *m.* tree of heaven (*Ailanthus altissima*).

aile, *f.* wing, vane, blade, sail, flange, plane, mudguard, ala; *pl.* flies (Diptera); — **du nez,** wing of the nose; **battre de l'**—, (of bird) to beat its wings, flutter.

ailé, winged, feathered, alate.

aileron, *m.* pinion (of bird), fin (of shark), aileron, blade, paddle, balancer, poisers (of diptera).

ailette, *f.* small wing, blade, fin, wheel paddle, fan, vane, lug, flange, tenon, stud.

aille (aller), (may) go.

aillent (aller) (they) (may) go.

ailleurs, elsewhere, somewhere else, anywhere else; **d'**—, moreover, besides, furthermore, though; **nulle part** —, nowhere else; **par** —, by another way, in other respects, besides, in addition.

aimable, amiable, agreeable, kind, attractive.

aimant, *m.* magnet, loadstone; — **en fer à cheval,** horseshoe magnet; **pierre d'**—, loadstone, magnet.

aimantaire, magnetic.

aimantation, *f.* magnetization.

aimanter, to magnetize; **s'**—, to be magnetized.

aimantin, magnetic.

aimer, to like, love, be fond of, care for; — **mieux,** to like better, prefer; **s'**— **à,** to be fond of.

aimez-moi, *f.* forget-me-not (Myosotis).

aine, *f.* groin.

-aine, dizaine, about ten, ten or so; **vingtaine,** score, some twenty; **centaine,** about a hundred.

aîné, eldest, elder, senior.

ainsi, so, thus, in this way, hence, therefore; — **que,** as, as well as, just as, so as; **et** — **de suite,** and so forth, and so on; **pour** — **dire,** so to say, as it were; **s'il en est** —, if (it is) so.

air, *m.* air, wind, draught, blast, manner, way, look, appearance; — **ambiant,** atmosphere; — **carburé,** carbureted air; — **chaud,** hot blast; — **comprimé,** compressed air; — **supérieur,** overgrate blast; — **vicié,** vitiated air foul air, choke damp; **au grand** —, **en plein** —, in the open air; **avoir l'**—, to look, seem.

airain, *m.* bronze, brass.

aire, *f.* area, flat space, surface, floor, threshing floor, platform, bed, face (of hammer), aerie, Aira, tufted hair grass, hassock grass (*Deschampsia caespitosa*), eyrie (of eagle), group, hurst, horst; — **blanchâtre,** gray hair grass; — **de vent,** point of the compass; — **d'habitation,** habitat.

airelle, *f.* whortleberry, bilberry (*Vaccinium myrtillus*), huckleberry, blueberry (Vaccinium); — **canche,** — **rouge,** mountain cranberry, red bilberry (*Vaccinium vitis-idaea*); — **canneberge,** — **coussinette,** cranberry (Oxycoccus); — **des marais,** — **uligineuse,** marsh bilberry (*Vaccinium uliginosum*).

ais, *m.* plank, board, stave; — **de carton,** pasteboard; — **de contre-marche,** riser; — **de marche,** tread or step board; — **d'entrevous,** sound boarding.

aisance, *f.* ease, freedom, *pl.* water closet.

aise, *f.* ease, pleasure, joy, comfort, gladness. (*pl.*) comforts; **à l'**—, with ease; **mieux à l'**—, more comfortable.

aise, glad, well-pleased; **bien** —, very glad, well-pleased.

aisé, easy, free, comfortable, convenient, well off.

aiselle, *f.* axil (of leaf).

aisément, easily, readily, comfortably, convenient.

aissante, *f.* **aisseau,** *m.* shingle.

aisselier, *m.* shoulder belt, strut, brace.

aisselle, *f.* armpit, axilla, axil; — **de la feuille,** axil of leaf, axil angle formed by leaf and stalk.

aitiomorphose, *f.* aitiomorphosis, change in shape.

ajointer, to join up, fit end to end.

ajonc, *m.* furze, gorse, whin (*Ulex europaeus*).

ajouré, perforated, pierced, latticed.

ajourer, to perforate, pierce.

ajournement, *m.* postponement.

ajourner, to postpone, adjourn, defer, delay, summon.

ajouté, *m.* addition (to MS).

ajouter, to add, join; s'—, to be added; — foi à, to put faith in, believe; sans — que, to let alone that.

ajoutoir, *m.* delivery tube.

ajustage, *m.* adjusting, finishing, fitting.

ajustement, *m.* adjustment, arrangement, adjusting.

ajuster, to adjust, set, fit (up), arrange, settle, tune, gage, assemble, regulate, adapt, aim (at), attire, adorn, fit; s'—, to be adjusted, fit, agree, dress, be reconciled.

ajusteur, *m.* adjuster, fitter.

ajustoir, *m.* precision (coin) balance.

ajutage, *m.* delivery or outlet tube, discharge pipe, tip, jet, joint; — à pointeau, needle nozzle; — de caoutchouc, rubber-tube connection.

ajutoir, *m.* (ajutage), delivery tube.

akène, *m.* achene, achenium, dry fruit, nut.

Al, *abbr.* (aluminium), aluminum.

alaire, alar, winged.

alambic, *m.* still, alembic, retort; — d'essai, testing still.

alambiqué, strained, fine spun, far-fetched.

alambiquer, to distill, refine (too much), puzzle.

alandier, *m.* alandier, mouth of a furnace.

alanine, *f.* alanine.

alarmant, alarming, startling.

alarme, *f.* alarm.

alarmer, to alarm, frighten; s'—, to be alarmed.

alaterne, *m.* alatern, evergreen buckthorn (*Rhamnus alaternus*).

albâtre, *m.* alabaster, compact gypsum; — calcaire, calcareous alabaster.

albatros, *m.* albatross.

alberge, *f.* clingstone peach or apricot.

albergier, *m.* clingstone peach or apricot (tree).

albication, *f.* albication, becoming blanched.

albicaule, with whitish stem.

albin, white, unpigmented.

albinervé, with whitish veins (nerves).

albinisme, *m.* albinism.

albinos, albino.

albite, *f.* albite, clevelandite.

albugineux, -euse, albugineous, whitish.

albumen, *m.* albumen, endosperm.

albuminate, albuminate.

albumine, *f.* albumin, white of egg; — glycerinée, egg white and glycerol.

albuminé, albumineux, -euse, albuminous, albuminized (paper).

albuminoïde, albuminoid.

albuminurie, *f.* albuminuria.

albumoses, *f.pl.* albumoses, similar to albuminates.

alc., *abbr.* (alcalin), alkaline.

alcalescence, *f.* alkalescence, alkalescency.

alcali, *m.* alkali; — carbonaté, alkali carbonate; — de terre, alkaline earth; — volatil, volatile alkali, ammonia.

alcalifiant, alkalifying.

alcaligène, alkaligenous.

alcalimètre, *m.* alkalimeter.

alcalimétrique, alkalimetric.

alcalin, alkaline.

alcalinisation, *f.* alkalization.

alcaliniser, to alkalize, render alkaline, alkalinize.

alcalinité, *f.* alkalinity.

alcalino-salé, alkalino-saline.

alcalino-terreux, -euse, alkaline earth.

alcaliotropisme, *m.* alcaliotropism, chemotropism induced by alkalies.

alcalisation, *f.* alkalization.

alcaliser, to alkalize, render alkaline, make alkaline.

alcaloïde, *m.* alkaloid.

alcaloïdique, alkaloid, alkaloidal.

alcarazas, *m.* alcarraza, (porous) water cooler.

alcedo, *m.* kingfisher.

alcée, *f.* hollyhock, rose mallow (*Althaea rosea*).

alchémille, *f.* alchemilla; — des champs, parsley piert (*Alchemilla arvensis*); — vulgaire, lady's-mantle (*Alchemilla vulgaris*).

alchimie, *f.* alchemy.

alchimique, alchemic(al), alchemistic(al).

alchimiste, *m.* alchemist.

alcogel, *m.* alcogel, gel form of an alcosol.

alcool, *m.* alcohol, spirit(s), ethyl alcohol; — à bruler, methylated spirit, denatured alcohol; — allylique, allyl alcohol; — amylique, amyl alcohol; — amylique de fermentation, amyl alcohol of fermentation; — butylique, butyl alcohol; — campholique, borneol, camphol; — camphré, spirit of camphor, camphorated spirit of wine; — carburé, carbureted alcohol.

alcool, — de betteraves, beet alcohol, beetroot spirit; — de fécule, starch spirit, fecula brandy; — de menthe, spirit of peppermint; — de vinaigre, triple vinegar, concentrated vinegar; — enhydre, absolute alcohol; — éthal que, cetyl alcohol, ethal; — éther, mixture of alcoho' and ether.

alcool, — éthylénique, glycol, ethylene glycol; — éthylique, ethyl alcohol; — méthylique, methyl alcohol; — phényléthylique, phenyl ethyl alcohol; — propylique, propyl alcohol; — sulfuré, mercaptan, thiol; — vinique, wine alcohol, ethyl alcohol.

alcoolase, f. alcoholase.

alcoolat, m. spirit, alcoholic preparation (of aromatic herbs).

alcoolate, m. alcoholate.

alcoolature, f. alcoholature, tincture of fresh plants.

alcoolé, m. (alcoholic) tincture.

alcoolique, alcoholic, spirituous.

alcoolisable, alcoholizable.

alcoolisation, f. alcoholization.

alcooliser, to alcoholize, fortify (wine).

alcoolisme, m. alcoholism.

alcoolomètre, m. alcoholometer.

alcoolyse, f. alcoholysis.

alcoolyser, to alcoholyze.

alcôve, f. alcove, recess.

alcoylation, f. alkylation.

alcoyle, m. alkyl.

alcoyle, alkyl.

alcoyler, to alkylate.

aldéhydate, m. — d'ammoniaque, aldehyde ammonia.

aldéhyde, m. aldehyde; — acétique, acetaldehyde; — allylique, acrolein; — anisique, anisaldehyde; — cuminique, cumic aldehyde, cuminol; — formique, formaldehyde; — salicylique, salicylaldehyde, salicylic aldehyde.

aldol, m. aldol.

aldoliser, to aldolize.

aldose, n. aldose, an aldehyde sugar.

aldoxime, m. aldoxime (the oxime o' an aldehyde).

ale, f. ale.

aléa, m. risk, hazard, chance.

aléatoire, hazardous, risky, uncertain.

alembic, m. still, alembic.

alêne, f. awl, acuminate bristle; feuille en —, acuminate leaf.

aléné, awl-shaped, pointed, acuminate.

alénois, m. garden peppergrass, garden cress (Lepidium sativum).

alentour, around, roundabout.

alentours, m.pl. environs, neighborhood, associates.

alerte, f. alarm, warning.

alerte, alert, watchful, brisk, lively, quick.

alésage, m. bore, internal diameter, reaming, boring (out), hole.

aléser, to bore, ream out, drill.

aléseuse, f. drilling machine, boring machine.

alésoir, m. borer, reamer(-bit), broach, boring machine, polisher, finishing or polishing bit.

alésure, f. borings.

alétophytes, f.pl. aletophytes, ruderal or wayside plants.

alétris, m. colicroot, star grass, blazing star, blackroot (Aletris farinosa).

aleuriosporés, m.pl. Aleuriosporeae.

aleurone, f. aleurone, proteid grains.

alevin, m. fry.

alevinage, m. stocking with (young fish).

aleviner, to stock with fry.

alevinier, m. breeding pond, nursery, fry pond.

alexine, f. alexin.

alexinique, of the nature of alexins.

alezan, m. sorrel (horse); — rouge, red fox.

alfa, m. esparto (grass), alfa (Stipa tenacissima).

alfange, f. long-leaved lettuce.

alfénic, m. alphenic, sugar candy.

algarobille, f. algarrobilla.

algaroth, m. (powder of) Algaroth.

algèbre, f. algebra, literal calculus.

algébrique, algebraic.

algébriquement, algebraically.

algébriste, m. algebraist.

Alger, n. Algiers.

algérien, -enne, Algerin(e).

algue, f. alga, seaweed, tang; — des mers, seaweed.

aïidade, f. alidade.

aliénation, _f._ alienation.

aliéné, alienated, insane.

aliéner, to alienate, give (a)way, part with, sell, transfer; **s'—,** to become insane, estranged.

alifère, aliferous, wing-bearing.

aliforme, wing-shaped, alate.

alignement, _m._ alignment, (straight) line, row, building line.

aligner, to align, lay out, line up, arrange, straighten, range, square, put in a straight line, dress, sight out; — **des planches,** to square timber.

aliment, _m._ food, sustenance, nutrition, supply, medium; — **d'épargne,** protein-sparing food; **—s crus,** raw food (diet).

alimentaire, alimentary, nutritious, feed, feeding.

alimentateur, _m._ feed(er).

alimentateur, -trice, alimentary, nourishing.

alimentation, _f._ alimentation, feeding, feed, nutrition, foddering, supply, delivery, fodder; — **irrationelle,** malnutrition.

alimenter, to feed, nourish, supply, sustain.

alimenteux, -euse, alimental, nourishing, nutritive.

alinéa, _m._ paragraph, break; **en —,** indented.

alios, _m._ layers of iron pan, bog ore, moor bed pan, hard pan.

aliphatique, aliphatic.

aliquote, aliquot (part).

alise, _f._ sorb apple.

alisier, _m._ whitebeam (_Sorbus aria_); — **nain,** dwarf medlar (_Sorbus chamaemespilus_); — **torminal,** checker tree, service tree (_Sorbus torminalis_).

alisma, _m._ water plantain (_Alisma plantago_).

alizari, _m._ madder root.

alizarine, _f._ alizarin, madder dye.

alizé, trade (of wind).

alkékenge, _m._ alkekengi, winter cherry (_Physalis alkekengi_).

alkermès, _m._ alkermes.

alkyler, to alkylate.

alla, (he) went.

allaitement, _m._ suckling, nursing.

allaiter, to suckle, nurse.

allantoïde, _f._ allantois.

allantoïne, _f._ allantoin.

allassotonique, allassotonic.

allécher, to allure, decoy.

allée, _f._ going, lane, walk, avenue, drive.

allégation, _f._ assertion, statement.

allégement, _m._ lightening, relief, mitigation, removal (discharge) of load, balancing.

alléger, to lighten, unburden, ease, alleviate, relieve, unload, soothe, disburden; **s'—,** to become lighter.

allégir, to reduce.

allègre, lively, gay, brisk, jolly.

allégrement, briskly, cheerfully, gaily.

allégresse, _f._ briskness, liveliness, cheer, joy, mirth, gladness, joyfulness.

alléguer, to allege, urge, plead, cite, quote; — **une preuve,** to bring forward a proof.

alléluia, _m._ wood sorrel (_Oxalis acetosella_).

Allemagne, _f._ Germany.

allemand, German.

aller, _m._ journey, going; **pis —,** last resort, poor substitute; **au long —,** in the long run.

aller, to go, be going, be about to (with infinitive), move, be in motion, proceed, extend, advance, reach, act, work, run, continue, agree with, fit, be becoming; **s'en —,** to go away, depart, leave, end, disappear, dwindle, evaporate, die.

aller, allons, come! let us (go), make haste; **cela ne va pas,** that will not do; **cela va bien,** that's coming along fine; **cela va de soi,** that's understood; **cela va sans dire,** that is a matter of course.

aller, — au-devant de, to go to meet; — **au feu,** to go into action, endure fire; — **chercher,** to go and fetch, go for; — **de soi,** to go without saying, be self-evident; — **ensemble,** to go together, match; — **en diminuant,** to decrease, diminish, taper; — **en pente,** to fall, incline, slope.

aller, — et retour, round trip, back and forth; — **et venir,** to come and go, go to and fro; — **voir,** to go and see; — **trouver,** to go to, go and see; **faire —,** to make go, drive, work, run; **faire en —,** to send away; **y —,** to lay, stake, go at it, be a matter (question) of.

alliable, miscible.

alliacé, of or pertaining to garlic, alliaceous.

alliage, _m._ alloy, blending, alligation.

alliaire, _f._ alliaria (_Sisymbrium alliaria_).

alliance, _f._ alliance, union, association.

allié, _m._ ally, relation by marriage.

allié, allied, alloyed.

allier, to alloy, mix, unite, harmonize, blend, ally, combine, marry; **s'—,** to alloy, mix, unite, harmonize, blend.

allocarpie, *f.* allocarpy, fruiting from cross-fertilized flowers.

allocation, *f.* allowance, dole, grant.

allocinnamique, allocinnamic.

allogamie, *f.* allogamy, cross-fertilization.

allogène, of another race.

allomorphe, allomorphic.

allonge, *f.* adapter, percolator, extension piece, eking piece, coupling rod, elongation, tube.

allongé, lengthened, long, elongated, extended.

allongeable, extensible.

allongement, *m.* lengthening, extension, elongation, dilatation, expansion, protraction, damping stretch, natural elasticity.

allonger, to lengthen, elongate, eke out, dilute, prolong, stretch out, draw out, extend; s'—, to lengthen, grow longer, be lengthened.

allons, let us go; (nous) —, (we) are going.

allopathe, allopathic.

allotir, to allot.

allotropie, *f.* allotropy.

allotropique, allotropic.

allotropisme, *m.* allotropism.

allouchier, *m.* whitebeam (*Sorbus aria*).

allouer, to allow, grant, allocate.

alloxane, *m.* alloxan, mesoxalyl urea.

alloxantine, *f.* alloxantin.

alluchon, *m.* cog, tooth, cam.

allumage, *m.* lighting, switching on, ignition, illumination, kindling.

allume-feu, *m.* fire lighter, kindling, piece of kindling.

allume-gaz, *m.* gas lighter.

allumer, to light, kindle, ignite, inflame, set fire to, prime or start (pump), stir up, excite: s'—, to light, kindle, take fire, light up, be lighted, be excited.

allumette, *f.* match; — amorphe, safety match; — de sûreté, safety match, fuse; — soufrée, sulphur match, brimstone match; — suédoise, Swedish match, safety match.

allumette-bougie, *f.* wax match, vesta.

allumeur, *m.* lighter, igniter, primer.

allumière, *f.* match factory, match box.

allumoir, *m.* lighter, lighting apparatus.

allure, *f.* course, direction, bearing, working, gait, pace, conduct, behavior, rate (of speed), speed, motion, manner, way, walk, trend; — des fibres, grain; —s forestières,

silvicultural characteristics, relation of a tree to the forest.

allusion, *f.* allusion.

alluvial, alluvien, -enne, alluvial.

alluvion, *f.* alluvium, alluvion, alluvial (deposit); *pl.* accretion, accession, warp.

alluvionnaire, alluvial.

alluvionnement, *m.* alluviation, shifting of alluvial deposits.

allyle, allyl.

allylique, allyl, allylic (alcohol).

almanach, *m.* almanac, calendar.

almandine, *f.* almandite.

aloès, *m.* aloe (*Aloe vulgaris*); — caballin, caballine (horse) aloes; — dépuré, purified aloes; — hépatique des Barbades, Barbados aloes (*Aloe vera*).

aloétique, aloetic.

aloi, *m.* degree of fineness, standard, quality, kind, alloy; de bon —, sterling.

aloïne, *f.* aloin, a purgative glucoside.

aloïque, acide —, aloetic acid.

alopécie, *f.* baldness, alopecia.

alors, then, at that time, so, in such a case; — quoi, and what then? — que, when, whereas, while, even if; — même que, even when, even though; d'—, of that time; depuis —, since then, from that time; jusqu'—, until then.

alose, *f.* shad.

alouchier, *m.* checker tree, service tree (*Sorbus torminalis*).

alouette, *f.* lark; — des champs, skylark.

aloumère, *m.* sweet-smelling agaric.

alourdir, to make heavy, load, weigh down, render dull or stupid.

alourdissement, *m.* rendering heavy, growing heaviness, dullness.

aloyage, *m.* alloying, alloy.

aloyau, *m.* sirloin (of beef).

aloyer, to give legal quantity of alloy to gold or silver, alloy.

alpaca, alpaga, *m.* alpaca.

Alpes, *f.pl.* Alps.

alpestre, Alpine, alpestrine (plant).

alphabet, *m.* alphabet.

alphabétique, alphabetic(al).

alphénic, *m.* alphenic, white barley sugar.

alpin, alpine (plant).

alpique, alpine.

alpiste, *m.* canary grass (*Phalaris canariensis*).

alquifoux, *m.* alquifou, potter's ore.

alsacien, -enne, Alsatian.

alsine, *f.* chickweed (*Stellaria* (*Alsine*) *media*).

altérabilité, *f.* alterability.

altérable, alterable, unstable, perishable.

altérant, alterant, causing thirst, alterative.

alterateur, -trice, adulterating, adulterator.

altération. *f.* alteration, adulteration, change, deterioration, misinterpretation, impairing, excessive thirst; — du coeur, decaying heartwood.

altercation, *f.* dispute.

altéré, faded, drawn, haggard, thirsty.

altérer, to alter, corrupt, change (for the worse), impair, adulterate, spoil, debase, tamper with, disturb, make thirsty, excite thirst, injure, mar, lessen; s'—, to change, alter, deteriorate, weather, spoil.

alternance, *f.* alternation; — des générations, metagenesis.

alternant, alternating, rotating.

alternateur, *m.* alternator.

alternatif, -ive, alternate, alternating, reciprocating.

alternation, *f.* alternation, rotation, reciprocation, permutation.

alternative, *f.* alternation, alternate, interchange.

alternativement, alternately, in succession, one after the other.

alterne, alternate (leaves, angles).

alterner, to alternate, follow in turn, rotate (crops).

alterniflore, bearing alternate flowers.

alternomoteur, *m.* alternating-current motor, alternomotor.

althée, *f.* marsh mallow (*Althaea officinalis*).

althéine, *f.* altheine (asparagine).

altimètre, *m,* altimeter, height gage; — enregistreur, altigraph, recording altimeter.

altise, *f.* flea beetle (Altica).

altitude, *f.* altitude, height (above sea level).

alucite, *f.* alucita, plume moth; — des céréales, Angoumois (grain) moth.

alude, *f.* aluta.

aludel, *m.* aludel.

aluine, *f.* absinthe, wormwood.

alule, *f.* alula, bastard wing.

alumelle, *f.* blade (of knife).

alumen, *m.* alum, alumen.

aluminage, *m.* alumination, aluming.

aluminaire, aluminiferous.

aluminate, *m.* aluminate.

alumine, *f.* alumina.

aluminé, aluminous.

aluminer, to aluminate, aluminize, alum.

alumineux, -euse, aluminous.

aluminium, *m.* aluminum.

aluminothermie, *f.* aluminothermy.

alun, *m.* alum(en); — ammoniacal, — d'ammoniaque, ammonium alum, ammonia alum; — brulé, — calciné, burnt alum; — de fer, iron alum; — de plume, feather alum; — de potasse, potassium alum; — de roche, — de Rome, rock alum, Roman alum; — de soude, soda alum; — desséché, burnt alum; — ordinaire, common alum, potassium alum.

alunage, *m.* aluming.

alunation, *f.* formation of alum.

aluner, to alum.

alunerie, *f.* alum works.

alunière, *f.* alum (pit) works.

alunifère, aluminiferous, alum-bearing.

alute, *f.* aluta, colored sheepskin.

alvéolaire, alveolar, cell-like, cellular.

alvéole, *m.&f.* alveolus, cell, socket, seat(ing) cavity, pit.

alvéolé, pitted, honeycombed, favose.

alvier, *m.* cembra pine, Swiss pine, stone pine (*Pinus cembra*).

alvin, alvine.

alysse, alysson, *m.* Alyssum, German madwort (*Asperugo procumbens*).

alyte, *m.* obstetrical toad, nurse frog.

amabilité, *f.* amiability, kindness.

amadou, *m.* touchwood, amadou, punk, German tinder, pyrotechnical sponge.

amadouerie, *f.* amadou (tinder) works.

amadouvier, *m.* tinder fungus (*Polyporus fomentarius*), tinder agaric (*Polyporus igniarius*).

amaigri, thin, lean, attenuate.

amaigrir, to make thin, emaciate, thin down, reduce, impoverish; s'—, to become thin, lose flesh, shrink.

amaigrissant, reducing, making thin.

amaigrissement, *m.* reducing, growing thin, loss of flesh, emaciation.

amalgamation, *f.* amalgamation.

amalgame, *m.* amalgam, mixture.

amalgamer, to amalgamate, blend.

amalgameur, **-euse,** *m.&f.* amalgamator.

amande, *f.* almond, kernel, nucleus; —
amère, bitter almond; — **de semence,**
sperm; — **de terre,** chufa, earth almond
(*Cyperus esculentus*); — **douce,** sweet almond; — **d'une drupe,** kernel of a drupe;
— **pilé,** ground almond; **en** —, almond-
shaped.

amandé, *m.* almond milk, milk of almonds,
amygdalate.

amandé, containing (flavored with) almonds.

amandier, *m.* almond tree.

amandine, *f.* amandin.

amanite, *f.* amanita.

amant, **-e,** *m.&f.* lover, sweetheart.

amarante, *f.* amaranth (Amaranthus);
— **crête-de-coq,** cockscomb.

amarelle, *f.* pale sour cherry.

amarescent, bitterish, slightly bitter.

amarine, *f.* amarine.

amarrage, *m.* mooring.

amarrer, to make fast.

amas, *m.* heap, pile, accumulation, hoard,
mass, crowd.

amasser, to heap (up), hoard, collect,
gather, amass, accumulate.

amateur, *m.* lover, amateur.

amateur, fond (of).

amati, mat, unpolished.

amatir, to mat, dull, deaden.

ambassade, *f.* embassy.

ambiance, *f.* environment.

ambiant, surrounding, ambient, circum-
ambient, atmospheric; **température** —**e,**
room temperature.

ambigène, ambigenous.

ambigu, *m.* mixture, medley.

ambigu, **-e,** ambiguous.

ambiguïté, *f.* ambiguity, ambiguousness.

ambigument, ambiguously.

ambipare, ambiparous.

ambitieux, **-euse,** ambitious, pretentious.

ambitionner, to aspire to, desire earnestly,
be ambitious.

amble, *m.* amble; — **rompu,** broken amble.

ambocepteur, *m.* amboceptor.

ambre, *m.* amber; — **blanc,** spermaceti; —
gris, ambergris; — **jaune,** yellow amber,
ordinary amber.

ambré, perfumed with amber(gris), amber-
colored, amber(ed).

ambrette, *f.* musk seed, abelmosk.

ambrin, amberlike.

ambroïde, *m.* amberoid.

ambroïne, *f.* ambroin(e).

ambrosie, *f.* ragweed (Ambrosia).

âme, *f.* soul, core, pith, bore, clack, valve,
web, mind, spirit, heart, feeling; — **du
bois,** heart or core of wood; **rendre l'**—, to
give up the ghost.

amélanchier, *m.&f.* shadberry, shadbush
(Amelanchier).

améliorable, improvable.

amélioration, *f.* amelioration, improvement,
betterment.

améliorer, to ameliorate, make better,
improve, appreciate, purify, selectionnate.

aménage, *m.* transport, carriage.

aménagement, *m.* management, preparation
(for service), care (of a forest), equipping.

aménager, to manage, prepare, arrange, fit
up, divide, parcel out, care, spare, organize,
cut up (wood), prepare a working plan.

aménagiste, *m.* (forest) organizer.

amendable, improvable.

amendage, *m.* enriching (of soil).

amende, *f.* fine.

amendement, *m.* improvement, betterment,
amendment, ameliorator (of soil).

amender, to make better, improve, amend,
reclaim, manure, improve (soil); **s'**—, to
grow better, become more fertile.

amenée, *f.* bringing, leading, inlet, intake,
inflow.

amener, to bring (about), lead, conduct,
bring in, induce, introduce, cause, admit,
lower.

amentacé, catkin-bearing, catkinlike.

amentacée, *f.* amentaceous plant.

amentiflore, amentiflorous, wind-pollinated.

amenuiser, to reduce, make thin.

amer, *m.* bitter substance, gall; — **de boeuf,**
ox gall.

amer, **-ère,** bitter, sour, painful, sad.

amèrement, bitterly.

américain, American, of the United States.

Amérique, *f.* America; — **du Nord, du Sud,**
North, South America.

améristique, ameristic, unsegmented, un-
differentiated.

amertume, *f.* bitterness.

améthyste, *m.* amethyst.

ameublement, *m.* furniture.

ameublir, to loosen, stir, mellow (of soil).

ameublissement, *m.* mellowing, breaking up, loosening (of soil).

ameulonner, to stack (hay).

ami, -e, *m.&f.* friend.

ami, friendly.

amiable, friendly, amicable.

amiantacé, amianthoid.

amiante, *m.* asbestos, amianthus; — **platiné,** platinized asbestos.

amianté, covered or treated with asbestos; **toile métallique** —**e,** asbestos wire gauze.

amiantin, asbestine, made of asbestos, amianthine.

amiantine, *f.* asbestos cloth.

amibe, *f.* amoeba.

amibiase, *f.* amoebiasis.

amibien, -enne, amoebic.

amiboïde, amoeboid, amoeba.

amical, -aux, amicable, friendly, kind.

amicrobien, -enne, nonbacterial.

amidasses, *f.pl.* amidases, enzymes occurring in the mycelium of Aspergillus.

amide, *f.* amide; — **acétique,** acetamide, acetic amide.

amidé, amido, amino.

amidification, *f.* conversion into an amide.

amidine, *f.* amidine (an aminoimino base), amidin (from starch).

amidoazoïque, aminoazo.

amidogène, *m.* amidogen (the group NH_2).

amidol, *m.* amidol.

amidon, *m.* starch, amylum, fecula; — **de froment,** wheat starch; — **iodé,** starch iodide; — **nitré,** nitrated starch; **empois d'**—, starch paste; **fécule d'**—, starch powder.

amidonner, to starch, treat with starch.

amidonnerie, *f.* starch works.

amidonnier, *m.* starch maker, larger spelt, starch wheat.

amidonnière, *f.* kneading trough employed in starch manufacture, starch maker.

amidophénol, *m.* aminophenol.

amidure, *m.* (metallic) amide, a compound in which a metal has replaced hydrogen of ammonia.

amihou, *m.* wild fig.

aminci, thin, slender, attenuate.

amincir, to make thinner, thin down, reduce in thickness; **s'**—, to become thin, more slender.

amincissement, *m.* thinning down, reduction.

amine, *f.* amine; — **acide,** amino acid, acid amine.

aminé, amino.

aminoïde, aminoid, scents with amine as foundation.

amique, amidic.

amitié, *f.* friendship, affection, favor.

amitose, *f.* amitosis, a degenerate mitosis.

amméline, *f.* ammeline.

ammodyte, living in sand.

ammonal, *m.* ammonal, a mixture of ammonium nitrate and powdered aluminum.

ammoniac, -aque, ammoniacal, ammoniac.

ammoniacal, aux, ammoniac(al), relating to ammonia.

ammoniacé, ammoniated.

ammoniaco-cobaltique, of or pertaining to ammonium and cobalt.

ammoniaque, *f.* ammonia, volatile alkali; ammonium; — **liquide,** liquid ammonia.

ammoniate, *m.* ammoniate.

ammonicosodique, of or pertaining to ammonium and sodium.

ammonié, ammoniated.

ammonio-, ammonio-, of ammonia or ammonium.

ammonio-iridique, of or pertaining to ammonia and iridium.

ammonium, *m.* ammonium.

ammoniure, *m.* ammoniate.

ammophile, *f.* sea reed, mat grass.

amnésie, *f.* amnesia.

amniotique, amniotic.

amodiataire, *m.* lessee (of farm etc.).

amodier, to farm out, lease.

amoindrir, to lessen, diminish, decrease, lower.

amoindrissement, *m.* reduction, diminution.

amoitir, to dampen, moisten.

amollir, to soften, mellow; **s'**—, to grow tender.

amollissement, *m.* softening.

amonceler, to heap up, accumulate.

amoncellement, *m.* heaping up, accumulation.

amont, up, upstream, up the river; **en—,** upstream.

amorçage, *m.* priming, beginning, starting.

amorce, *f.* beginning, primer, priming, fuse, starting, bait, allurement, tinder; **— cirée, wax** primer; **— de quantité,** quantity fuse, low-tension fuse; **— de tension,** (high-) tension fuse.

amorcement, *m.* priming, beginning.

amorcer, to prime, start (a reaction), begin, prepare, initiate, bait, lure, mark with the center punch.

amorceur -euse, *m.&f.* baiter, starter, primer.

amorçoir, *m.* auger, boring bit, primer box, ground-baiting appliance.

amordancer, to mordant.

amorphe, amorphous, structureless, shapeless.

amorphie, *f.* amorphism, amorphia.

amortir, to deaden, damp, tone down, diminish, subdue, weaken, dull, allay, absorb, redeem, **pay** (off), extinguish, amortize, write off.

amortissement, *m.* paying off, damping, subduing, absorption, writing off, redemption, amortization.

amortisseur, *m.* bumper, buffer, damping device; **— de chocs,** bumper, shock absorber.

amouille, *f.* first milk after calving, colostrum of cows, beestings.

amouler, to sharpen, grind (on a stone).

amour, *m.* love, passion, heat (of animals); **—s,** *f.* amours, flame; **en —,** in heat, in a state of fermentation.

amourette, *f.* quaking grass (*Briza media*), toothwort (*Lathraea squamaria*).

amoureux, -euse, amorous, loving.

amour-propre, *m.* self-respect, (legitimate) pride.

amoussément, *m.* apathy.

amovible, revocable, removable, detachable.

ampélidées, *f.pl.* Ampelideae.

ampélis, *m.* waxwing, chatterer, ampelis.

ampérage, *m.* amperage, current (intensity or strength).

ampère, *m.* ampere.

ampère-heure, *m.* ampere-hour.

ampèremètre, *m.* ammeter; **— thermique,** hot-wire ammeter.

ampère-tour, *m.* ampere-turn.

amphante, *f.* amphanthium, clinanthium, receptacle of an inflorescence.

amphibie, *m.* amphibian.

amphibie, amphibious.

amphibiens, *m.pl.* amphibia.

amphibole, *f.* amphibole, hornblende.

amphicarpe, amphicarpic.

amphidermis, *m.* cuticle.

amphigame, amphigamous.

amphigastre, *f.* amphigastria, stipular organs in Hepaticae.

amphigène, amphigenous.

amphigonie, *f.* amphigony, sexual reproduction.

amphimixie, *f.* amphimixis, sexual reproduction.

amphisarque, *f.* amphisarca, an indehiscent multilocular fruit.

amphitecium, *m.* amphithecium, peripheral layer of cells.

amphithéâtre, *m.* amphitheater, auditorium, lecture room.

amphitrope, campylotropous, amphitropous, incurved.

amphore, *f.* amphora.

amphotère, amphoteric.

ample, ample, large, broad, copious.

amplectant, amplectant, clasping, embracing.

amplement, amply, largely, fully.

ampleur, *f.* breadth, amplitude, spaciousness, fullness.

amplexicaule, amplexicaul, stem-clasping, embracing.

ampliatiflore, ampliatiflorus, the rayflorets enlarged.

ampliation, *f.* copy, exemplification; **pour —,** true copy.

amplifiant, magnifying (lens), amplifying (power).

amplificateur, *m.* enlarger, amplifier, intensifier.

amplificateur, -trice, magnifying, amplifying, enlarging.

amplificatif, -ive, amplifying, magnifying (lens).

amplification, *f.* amplification, enlargement, increase, magnification.

amplifier, to amplify, enlarge, magnify, exaggerate.

amplitude, *f.* amplitude, width, range.

ampoule, f. (incandescent) bulb, tube, lamp, ampoule, blister, ampulla, vessel, vial, bubble; — à **brome,** dropping funnel; — à **décantation,** separatory funnel; — à **robinet,** dropping funnel; — **compte-gouttes,** medicine dropper.

ampoulé, blistered, turgid, inflated.

ampouler, to blister.

ampullacé, ampullaceous.

amputer, to amputate.

amulet, f. amulet.

amusant, amusing, diverting.

amusement, m. amusement, diversion.

amuser, to amuse, entertain, deceive; s'—, to enjoy oneself.

amygdale, f. tonsil, amygdule.

amygdalé, amygdalin, amygdaline, resembling an almond.

amygdalées, f.pl. amygdaloid or rose plants.

amygdalique, amygdalic.

amygdalite, f. tonsillitis.

amygdaloïde, amygdaloid (rock).

amylacé, amylaceous, starchy, amyloid.

amylacétique, éther —, amyl acetate.

amylamine, f. amylamine.

amylase, f. amylase, an enzyme.

amyle, m. amyl.

amylène, m. amylene.

amyline, f. starch cellulose.

amylique, amyl(ic); **alcool —,** potato spirit, fusel oil, amyl alcohol.

amylite, f. amylite, · skeleton of starch granules.

amyloïde, starchy, amyloid.

amyloleucite, f. amyloplast, leucoplast.

amylolyse, f. saccharization of starch.

amylolytique, amylolytic.

amylopectine, f. amylopectin.

amylose, f. amylose, waxy degeneration.

an, m. year.

anabionte, f. anabiont, perennial.

anabiose, f. reanimation.

anabolisme, m. anabolism, assimilation.

anacarde, m. anacard, cashew nut.

anacardier, m. cashew tree (*Anacardium occidentale*).

anacrogyne, anacrogynous, archegonia not arising near apex.

anadrome, anadromous.

anaérobe. m anaerobe, anaërobe.

anaérobie, m. anaerobe; — **facultatif,** facultative anaerobe; — **strict,** strict anaerobe.

anaérobie, anaerobic.

anaérobien, -enne, anaerobic.

anaérobiose, f. anaerobiosis, the state of living without oxygen.

anaérobisme, m. anaerobism.

anagallide, f. anagallis, pimpernel.

anagyre, m. bean trefoil (*Anagyris foetida*).

anagyrine, f. anagyrine, an alkaloid from anagyris.

anal, -aux, anal.

analgésie, analgie, f. analgesia.

analgésine, f. analgesine, antipyrine.

analgésique, analgique, m. analgesic.

analgésique, analgique, analgesic, analgetic, analgic.

analgre, f. analgia.

analogie, f. analogy, similarity, relation.

analogique, analogic(al).

analogiquement, analogically.

analogue, m. analogue, parallel.

analogue, analogous, similar, like.

analysable, analyzable.

analyse, f. analysis, assay(ing), outline; — **chimique,** chemical analysis; — **contra-dictoire,** check analysis; — **des denrées,** food examination; — **élémentaire,** elementary analysis; — **immédiate,** proximate analysis; — **par (la) voie humide,** wet analysis; — **spectrale,** spectrum analysis; **en derniére —,** to sum up.

analyser, to analyze, assay; s'—, to be analyzed.

analyseur, m. analyzer.

analyste, m. analyst.

analytique, f. analytics.

analytiquement, analytically.

ananas, m. ananas, pineapple (*Ananas sativa*).

anandre, anandrous, having no stamens.

ananthe, ananthous, flowerless.

anaphase, m. anaphase, formation of daughter nuclei.

anaphylactique, anaphylactic.

anaphylactisé, anaphylactized.

anaphylaxie, f. anaphylaxis; — **sérique,** anaphylaxis following serum injection.

anas, m. chaff, awn, shore, boon.

anasarque, f. anasarca.

anastigmate, -matique, anastigmatic.

anastomose, *f.* anastomosis, inosculation.

anastomoser(s'), to anastomose, inosculate.

anate, *f.* annatto (a red dye to color cheese and butter).

anatomie, *f.* anatomy.

anatomique, anatomical.

anatomiser, to anatomize, dissect.

anatomiste, *m.* anatomist.

anatoxine, anatoxin.

anatrope, anatropous.

ancêtre, *m.* ancestor, forefather.

anchois, *m.* anchovy.

anchuse, *f.* bugloss, oxtongue, alkanet (Anchusa).

anchusine, *f.* anchusin, coloring matter from Anchusa.

ancien, -enne, ancient, former, late, old, ex-; les —s, the ancients.

anciennement, anciently, formerly, of old, of yore.

ancienneté, *f.* old age, seniority, oldness, antiquity, primitiveness.

ancipité, ancipitous, two-edged (stem).

ancolie, *f.* columbine (Aquilegia).

ancrage, *m.* anchoring, mooring; — de pieux, pile strutting.

ancre, *f.* anchor, grapple (the boiler).

ancrer, to anchor, brace, tie, fix, moor.

andain, *m.* swath, windrow.

andalous, Andalusian.

andésite, *f.* andesite.

andorite, *f.* andorite.

andouiller, *m.* antler.

androcée, *f.* androecium, the stamens collectively.

androdioecie, *f.* androdioecism.

androdyname, androdynamous, stamens highly developed.

androgyne, androgynous, bisexual.

androgyniflore, androgyniflorous, head of composite bearing hermaphroditic flowers.

andromède, *f.* wild rosemary, marsh andromeda, moorwort (*Andromeda polifolia*).

andromonoïque, andromonoecious.

andromorphose, *f.* andromorphosis, alterations from excitation of pollen tubes.

andropétalaire, andropetalous.

androphore, *f.* androphore, support of column of stamens.

andropogon, *m.* beard grass (Andropogon).

androspores, *f.pl.* androspores, swarm spores of Oedogoniae.

âne, *m.* ass, donkey.

anéantir, to annihilate, destroy, crush.

anéantissement, *m.* annihilation, destruction.

anélectrique, anelectric.

anémiant, impoverishing, weakening.

anémie, *f.* anemia.

anémier, to render anemic, weaken.

anémique, anemic.

anemochore, *f.* anemochore, plant distributed by wind.

anemogame, anemogamous, wind pollinated.

anémographe, *m.* anemograph; — à moulinet, rotating wheel or wind vane anemograph.

anémomètre, *m.* anemometer, wind gage; — à ailes hémisphériques, — à tasses tournantes, cup anemometer; — à ailettes, — à pales, windmill anemometer.

anémone, *f.* anemone, windflower (Anemone).

anémophile, anemophilous, wind-pollinated.

anémophobe, anemophobe, shunning wind.

anéroïde, aneroid.

ânesse, *f.* she-ass, jenny ass; lait d'—, ass's milk.

anesthésiant, anesthetic.

anesthésie, *f.* anesthesia.

anesthésier, to anesthetize.

anesthésique, anesthetic.

aneth, anet, *m.* dill (*Anethum graveolens*); — odorant, dill (*Anethum graveolens*).

anéthol, *m.* anethole.

anfractuosité, *f.* anfractuosity, sinuosity, winding and turning.

ange, *m.* angel; — (de mer), angel fish.

angélique, *f.* angelica; petite —, goutweed, ashweed, dogelder, (*Aegopodium podagraria*).

angélique, angelic(al).

angine, *f.* angina, inflammation in the throat, quinsy, tonsillitis; — membraneuse, croup.

angiocarpe, *f.* angiocarp, an angiocarpous plant.

angiogastres, *f.pl.* conidia produced in interior of tissues.

angiosperme, *m.pl.* angiosperms, Angiospermae.

angiosperme, angiospermous.

angiospore, angiosporous.

anglais, *m.* Englishman, English (language).

anglais, English.

angle, *m.* angle, corner; — **aigu,** acute angle; — **critique,** critical (stalling) angle; — **d'ascension,** angle of altitude; — **d'attaque de la dent,** angle of cutting edge; — **de l'aube de la couronne directrice,** guidewheel blade angle; — **d'équerre,** right angle; — **des aubes,** vane angle; — **de biseau,** cutting angle.

angle, — **dièdre,** dihedral angle; — **d'inclinaison,** angle of slope, gradient of slope; — **droit,** right angle; — **oblique,** bevel rule, miter square; — **plan,** plane angle; — **rapporteur,** protractor; — **rectiligne,** plane angle; — **saillant,** salient angle; — **trièdre,** trihedral angle.

anglésite, *f.* anglesite, lead vitriol.

Angleterre, *f.* England.

angleux, **-euse,** imbedded in angular cavities.

angoissant, distressing, alarming, tense.

angoisse, *f.* anguish, distress, agony, spasm; **poire d'—,** choke pear.

angousse, *f.* dodder.

anguiforme, anguine, snakelike.

anguille, *f.* eel.

anguillule, *f.* eelworm; — **de blé,** wheatworm, wheat nematode.

angulaire, angular.

angulairement, angularly.

angulé, angular, angulate(d).

anguleux -euse, angular, bony, hooked, rugged.

angulinervé, angulinerved, veins at angle with midrib.

angustifolié, angustifoliate, narrow-leaved.

angusture, *f.* angostura bark.

anheliotrope, *m.* organ insensitive to direct light.

anhydrate, *m.* anhydride.

anhydre, anhydrous.

anhydride, *m.* anhydride; — **acétique,** acetic anhydride; — **azoteux,** nitrous anhydride; — **carbonique,** carbon dioxide, carbonic acid gas; — **d'acide,** acid anhydride, acidic oxide; — **sulfureux,** sulphur dioxide,

sulphurous anhydride; — **sulfurique,** sulphur trioxide, sulphuric anhydride.

anhydrifier, to become anhydrous.

anhydrisation, *f.* dehydration.

anhydrite, *f.* anhydrite, karstenite.

anil, *m.* anil, indigo plant.

anilidure, *m.* anilide.

aniline, *f.* aniline; — **chlorhydrique,** aniline hydrochloride.

anille, *f.* tendril.

anillé, tendriled.

animal, *m.* animal, beast; — **reproducteur,** breeding animal, breeder; — **sans vertèbre,** invertebrate; — **d'expériences,** experimental animal.

animal, animal, sensual.

animalcule, *m.* animalcule.

animaliser, to animalize, convert (food) into animal matter; **s'—,** to become animalized.

animalité, *f.* animality, animal nature.

animateur, *m.* prime mover, animator.

animateur, -trice, animating, quickening.

animaux, *m.pl.* see ANIMAL.

animé, *f.* animé, resin.

animé, animated, living, animate.

animer, to animate, quicken, give life, actuate, move, propel, drive, incite, enliven, hearten, excite; **s'—,** to become animated, be excited.

anion, *m.* anion.

anis, *m.* anise (*Pimpinella anisum*); — **étoilé,** star anise (Illicium); — **vert,** aniseed; **graine d'—,** aniseed.

anisette, *f.* anisette (cordial).

anisine, *f.* anisine.

anisique, anisic; **aldéhyde —,** anisaldehyde.

anisogone, anisogonous, hybrids not equally combining characters of parents.

anisogyne, anisogynous, with fewer carpels than sepals.

anisoïde, (plant odor) of anise.

anisol, *m.* anisole.

anisomère, anisomerous, anisomeric.

anisopétale, anisopetalous.

anisophylle, anisophyllous, with non similar leaves.

anisophyllie, *f.* anisophylly, inequality of leaves.

anisostémone, anisostemonous, with unequal stamens.

anisotropic, *f.* anisotropy.

anisotropique, anisotropic.

ankylostome, *m.* hookworm.

annales, *f.pl.* annals.

anneau, *m.* ring, body-ring, somite, link, hoop, collar, rim, coil, shackle; — **annuel,** annual ring; — **benzénique,** benzene ring; — **de caoutchouc,** rubber band; — **de joint,** — **de garniture,** joint or packing ring.

année, *f.* year, twelvemonth; — **de rapport,** fruit-bearing year; — **de repos,** barren year; — **d'essaimage,** swarm year (insects); **mauvaise** —, nonseed year, fail year; **pendant de longues** —**s,** for many years.

annélation, *f.* girdling, deadening.

annelé, annulated, having rings, ringed.

anneler, to girdle, deaden.

annelet, *m.* small ring, annulet.

annélides, *m.pl.* Annelida.

annexe, *f.* annex, appendage, supplement.

annexer, to annex, attach, append.

annihiler, to annihilate, destroy, cancel, annul.

anniversaire, anniversary.

annonce, *f.* announcement, publication, notification, advertisement, sign.

annoncer, to announce, declare, foretell, proclaim, publish, advertise, show.

annoter, to annotate, make an inventory.

annuaire, *m.* annual, yearbook, almanac, (trade) directory.

annuel, -elle, annual, yearly.

annuellement, annually, yearly.

annuité, *f.* rent.

annulaire, annular, ring-shaped, ring (finger).

annulation, *f.* annulment, invalidation, cancellation.

annuler, to annul, nullify, repeal, abolish, cancel.

anobie, *f.* anobiid, anobium, esp. death-watch beetle.

anoblir, to ennoble, make void.

anode, *f.* anode.

anodin, anodyne, soothing, paregoric.

anodique, anodic.

anomal, anomalous.

anomalie, *f.* anomaly, irregularity.

anomaloecie, anomaloecious, polygamous.

anon, *m.* ass's foal.

anonacé, annonaceous.

anone, *f.* Annona.

anonym, anonymous; **société** —, joint-stock company.

anophèles, *m.pl.* Anophelinae.

anorganique, inorganic, anorganic.

anormal, abnormal, irregular, exceptional.

anormalement, abnormally.

anotto, *m.* annatto, a red dye for coloring cheese and butter.

anoure, an(o)urous, tailless.

anoxémie, anoxhémie, *f.* anoxemia, anoxaemia, deficient aeration of the blood.

anse, *f.* handle (of jug or basket), loop cove, bay, grip, ears, hilt, tiller; — **à vis** screw eye (screw-bolt); — **de panier,** basket handle.

anser, to furnish with a handle.

ansérine, *f.* goosefoot (Chenopodium); — **blanche,** fat hen, lamb's-quarters, white pigweed (*Chenopodium album*).

ansette, *f.* small handle.

anspect, *m.* shod bar, lever, crowbar.

antagonisme, *m.* antagonism.

antagoniste, antagonist(ic).

antécédemment, antecedently, previously.

antédiluvien, -enne, antediluvian.

antéhistorique, prehistoric.

antenne, *f.* antenna, feeler, aerial, catch; — **en massue,** clavate antenna; — **noueuse,** nodose antenna.

antennule, *f.* short feeler.

antérieur, anterior, front, former, prior, previous.

antérieurement, anteriorly, previously, before.

antériorité, *f.* anteriority, priority.

antesporophylle, *f.* antesporophyll, primitive structure of spore-bearing organ.

antetrophophylle, *f.* antetrophophyll, the ancestral form of leaf.

antetrophosporophylle, *f.* antetrophosporophyll, ancestral leaflike organ.

anthèle, *f.* anthela, a form of cymose inflorescence.

anthelminthique, anthelmintic, vermifuge.

anthère, *f.* anther; **depourvu d'**—**s,** anantherous.

anthéridie, *f.* antheridium.

anthéridiophore, _f._ antheridiophore, a unisexual gametophore.

anthérifère, antheriferous.

anthérizoïde, _m._ antherozoid, spermatazoid.

anthèse, _f._ anthesis, full bloom.

anthochlore, _m._ anthochlorin, xanthein, the yellow coloring of flowers.

anthocyane, _m.,_ **anthocyanine,** _f._ anthocyanin, anthocyan.

anthode, _m._ head of flowers.

anthodion, _m._ anthodium, head, capitulum, calyx.

anthoïdique, anthoid, flowerlike.

anthologie, _f._ anthology, a collection of flowers.

antholyse, _f._ antholysis, retrograde metamorphosis.

anthophore, _m._ anthophore, an elongated internode.

anthoxanthine, _f._ anthoxanthin, xanthophyll, the coloring matter of yellow flowers.

anthracène, _m._ anthracene.

anthracénique, anthracene, of anthracene.

anthracifère, anthraciferous.

anthracine, _f._ anthracene.

anthracite, _m._ anthracite, stone coal, blind coal.

anthraciteux, -euse, anthracitous, anthracitic.

anthracnose, _f._ anthracnose.

anthracomètre, _m._ anthracometer.

anthragallol, _m._ anthragallol.

anthranilique, anthranilic.

anthranol, _m._ anthranol, 9-anthrol.

anthraquinone, _f._ anthraquinone.

anthrarufine, _f._ anthrarufin.

anthrax, _m._ carbuncle, anthrax.

anthrène, _m._ anthrenus, carpet beetle.

anthrol, _m._ anthrol.

anthropochore, _m._ anthropochore, ruderal and adventitious plant.

anthropoïde, anthropoid.

anthropologiste, _m._ anthropologist.

anthropomorphe, anthropomorphous, anthropoid.

anthropomorphon, _m._ mandragora.

anthure, _f._ anthurus, a cluster of flowers.

anthyllide, _f._ kidney vetch (_Anthyllis vulneraria);_ — **vulnéraire,** lady's-finger, kidney vetch (_Anthyllis vulneraria_).

antiabrine, antiabrin.

antiacide, antacid, acidproof, acid-resisting, noncorrosive.

antialcalin, antalkaline.

antianaphylaxie, antianaphylaxis.

antiar, antiaris, _m._ antiar.

antibacterien, -enne, antibacterial.

antibiose, _f._ antibiosis, antipathy.

anticalcaire, _m._ scale-preventer, scale-removing composition.

anticathode, _f._ anticathode.

anticellulaire, anticellular.

anticharbonneux, -euse, antianthrax.

antichlore, _m._ antichlor.

anticholérique, anticholeraic, cholera immune.

anticipation, _f._ anticipation; **payer par —,** to pay in advance.

anticiper, to anticipate, forestall, encroach.

anticlaveleux, -euse, anti-sheeppox.

anticlines, _f.pl._ anticlines, anticlinal walls or planes.

anticombustible, incombustible, fireproof.

anticongélateur, _m._ antifreezing mixture.

anticorps, _m._ antibody.

anticryptogamique, fungicidal, anticryptogamic.

antidéperditeur, -trice, antiwaste (food).

antidérapant, nonskid(ding); **pneu —,** nonskid tire.

antidiphtérique, antidiphtheritic.

antidromie, _f._ antidromy, the course of a spiral reversed.

antiendotoxine, antiendotoxin.

antiendotoxique, antiendotoxic.

antifébrine, _f._ antifebrin (acetanilide).

antiferment, _m._ antiferment.

antifermentescible, _m._ antifermentative, nonfermenting substance.

antifermentescible, antifermentative, nonfermenting.

antifriction, _f._ antifriction metal, babbitt.

antigène, _m._ antigen.

antihémolytique, antihaemolytic.

anti-incrustant, scale-preventing; **graphite —,** boiler graphite.

antiméningococcique, antimeningococcic.

antimicrobien, -enne, antibacterial.

antimoine, _m._ antimony; — **cru,** antimony sulphide, crude antimony; — **sulfuré,** antimony (glance) sulphide, stibnite.

antimonial, antimonial.

antimoniate, *m.* antimon(i)ate.

antimonico-potassique, potassium antimonyl.

antimonié, containing antimony, antimonial, antimoniureted (hydrogen).

antimonieux, -euse, antimonious.

antimonique, antimonic.

antimonite, *n.* antimonite.

antimoniure, *m.* antimonide.

antinéphrétique, antinephritic.

antinévralgique, antineuralgic.

antiparasitaire, antiparasitic.

antipathie, *f.* antipathy, repugnance.

antipathique, averse, repugnant, antipathetic.

antipesteux, -euse, antiplague.

antipied, *m.* forefoot.

antiputride, antiputrefactive.

antipyrétique, antipyretic.

antipyrine, *f.* antipyrine.

antique, ancient, old, antique, antiquated.

antiquité, *f.* antiquity; de toute —, since the earliest times.

antirabique, antirabic.

antirhumatisant, antirheumatic.

antirouille, *m.* antirust composition, rust preventive.

antirrhine, *f.* snapdragon (Antirrhinum).

antisensibilisme, *f.* antisensibilism.

antisepsie, *f.* antisepsis.

antiseptique, antiseptic.

antisérum, *m,* antiserum.

antitétanique, *m.* antitetanic (serum).

antithermique, antithermic, antipyretic, antifebrile.

antithèse, *f.* antithesis.

antitoxine, *f.* antitoxin.

antitoxique, antitoxic.

antitrope, antitropous, antitropal, (radicle) directed away from the hilum.

antitryptique, antitryptic.

antityphique, antityphoid.

antityphoïdique, antityphoidal.

antivénéneux, *m.* antidote, antipoison.

antivénéneux, -euse, alexipharmic, antitoxic.

antivénérien, -ienne, antivenereal, antisyphilitic.

antivermineux, -euse, anthelmintic, vermifuge.

antizymique, antizymotic, antizymic.

antracène, *m.* anthracene.

antre, *m.* cave, cavern, retreat, den, antrum, sinus.

antrors, antrorse, directed forward or upward.

anuclée, anucleate.

anurie, *f.* anuria.

anus, *m.* anus.

Anvers, *n.* Antwerp.

anxiété, *f.* anxiety, concern.

anxieux, -euse, anxious, uneasy.

aorte, *f.* aorta.

aortique, aortic, aortal.

août, *m.* August, harvest (time).

aoûtat, *m.* harvest bug, tick, mite.

aoûtement, *m.* ripening (of fruit).

aoûter, to ripen (wood), lignify.

apagyne, apagynus, monocarpic.

apaiser, to appease, pacify, soothe, calm, allay; il s'en apaise, it grows calm.

apanage, *m.* appanage, lot, property, prerogative.

apandrie, *f.* apandry, loss of function in male organs.

apepsie, *f.* dyspepsia.

apercevable, perceptible.

apercevoir, to perceive, notice, see; s'—, to perceive, be perceived, notice, be aware of.

aperçoit, (he) perceives; il s'en —, he perceives it.

aperçu, *m.* glance, rough estimate, sketch, glimpse, thought, hint.

aperçu, perceived.

apérianthacé, without a perianth.

apériodique, aperiodic.

apérispermé, aperispermic, exalbuminous, without perisperm.

apéritif, *m.* appetizer, stomachic, aperient.

apéritif, -ive, aperient, diuretic, sudorific.

apétale, apetalous.

apétalie, *f.* apetaly, wanting petals.

apex, *m.* apex, anther.

aphanite, *f.* aphanite.

aphidiens, aphidés, *m.pl.* plant lice, aphides.

aphigénose, *f.* disease produced by lice.

aphis, *m.* aphis, plant louse.

aphlebie, *f.* aphlebia, pinnae in certain fossil ferns.

aphlogistique, aphlogistic, flameless.

aphotique, aphotic.

aphotométrique, aphotometric, turning to light.

aphrodisiaque, aphrodisiac.

aphteux, -euse, aphthous; **fièvre aphteuse,** foot-and-mouth disease, aphthous fever.

aphylle, aphyllous, leafless.

aphyllies, *f.pl.* aphyllae, having only rudimentary leaves or none.

apical, apical, anterior.

apicale, dehiscence of certain anthers.

apicifixe, apicifixed, like a suspended anther.

apiciflore, having terminal flowers.

apiciforme, in the form of bundles.

apicule, *f.* apicula, a sharp and short point.

apiculé, apiculate.

apiculteur, *m.* beekeeper, apiarist.

apiculture, *f.* beekeeping, apiculture.

apiéceur, -euse, *m.&f.* pieceworker.

apiol, *m.* apiol(e).

aplaigner, to raise, teasel (nap of cloth).

aplanir, to level, smooth, plane, remove; **s'—,** to become level, grow easier or smoother.

aplanissement, *m.* flattening, smoothness, planing.

aplanogamètes, *f.pl.* aplanogametes, nonciliated gametes.

aplanospore, *f.* aplanospore, a nonmotile asexual spore.

aplastique, aplastic.

aplat, *m.* flat tint.

aplati, flattened, flat, deflated, oblate.

aplatir, to flatten, debase; **s'—,** to become flat, flattened out, fall.

aplatissage, *m.* flattening, crushing down.

aplatissement, *m.* flattening, flatness, oblateness.

aplatisserie, aplatissoire, *f.* flatting mill.

aplatisseur, *m.* smooth roller, bruiser mill.

aplomb, *m.* perpendicularity, plumb, equilibrium, self-possession, bearing, demeanor, personality; **à l'— de,** straight above or below; **d'—,** upright, vertical(ly), plumb, perpendicularly; **prendre l'—,** to plumb.

apoatropine, *f.* apoatropine, atropamine.

apobatique, apobatic, repulsive.

apocarpe, *m.* apocarp.

apochromatique, apochromatic.

apocyn, *m.* dogbane (Apocynum).

apocyte, apocytial, multinucleate and unicellular.

apode, footless, without ventral fins.

apogame, apogamic, apogamous.

apogamie, *f.* apogamy.

apogénie, *f.* apogeny, loss of reproductive function.

apogéotropisme, *m.* apogeotropism, growing away from the earth.

apogyn e, *f.* apogyny, loss of reproductive power in female.

apolaire, apolar (cell).

apomorphine, *f.* apomorphine.

aponévrose, *f.* aponeurosis.

apophyse, *f.* apophysis, process, offshoot.

apophytes, *f.pl.* apophytes, autochthonous plants.

apoplexie, *f.* (stroke of) apoplexy, apoplectic fit, palsy, (paralytic) stroke.

aporogamie, *f.* aporogamy, pollen tube not passing through micropyle.

aposporie, *f.* apospory.

apostasie, *f.* apostasis, monstrous disunion of parts normally united.

apostrophe, *f.* apostrophe.

apothèce, *f.* shield (of lichen).

apothicaire, *m.* apothecary.

apothicairerie, *f.* apothecary's shop, pharmacy, dispensary.

apotoxine, apotoxine.

apôtre, *m.* apostle.

apotrope, apotropic (used of ascending axis).

apozème, *m.* infusion, apozem(a).

apparaître, to appear, become evident, come in sight.

appareil, *m.* apparatus, machine, device, appliance, instrument, (formal) preparation, pomp, set, show, display, dressing (on wound), bond; — **à chargement,** charging apparatus; — **à cliquet,** pawl coupling; — **à équilibrer les fers à raboter,** cutter balancing machine; — **à ranger les tiges,** splint-leveling machine; — **à** (or **de**) **ruissellement,** trickling apparatus.

appareil, — d'alimentation, feed apparatus; — **d'allumage,** lighting apparatus, ignition device; — **d'arrosage,** watering apparatus; — **d'aspiration,** suction apparatus, exhauster; — **de bord,** accessory; — **de chauffage,** heating apparatus, heater, stove; — **d'éclairage**, lighting apparatus; — **de déviation**, deflector, arrangement to catch the jet; — **de fixation**, cramping apparatus.

appareil, — **de lavage**, washing (apparatus) machine; — **de levage**, lifting appliance; — **de lessivage**, leaching apparatus, lixiviation apparatus; — **de mesure de la dérive**, drift indicator; — **de prise de vues cinématographiques**, cinematograph camera, motion-picture camera; — **de soutien**, supportive tissue; — **de succion**, suction apparatus; — **de sûreté**, safety (device) appliance; — **de suspension**, suspension gear, hinge plate.

appareil, — **de vidange**, skidder; — **de voie**, switch; — **d'extinction**, fire extinguisher; — **digestif**, digestive organs, digestive system; — **frigorifique**, refrigerating apparatus, refrigerator; — **moteur**, driver, driving gear, driving mechanism; — **protecteur**, protecting devire, safety guard; — **radiculaire**, root system; — **sexuel**, reproductive organs; — **soufflant**, blast engine, blower; — **tendeur**, stretching or clamping device.

appareillage, *m.* equipment, installation, preparation, setting-up, bonding (of bricks, stones).

appareillement, *m.* selection of uniform breeding animals, pairing.

appareiller, to install, fit, bond, pair, mate, match, dress.

apparemment, apparently, likely.

apparence, *f.* appearance, semblance, trace, likelihood.

apparent, apparent, evident, conspicuous, visible, that shows, obvious, plain.

appariation, *f.*, pairing, mating.

apparié, paired.

appariment, *m.* pairing, mating.

apparition, *f.* appearance, publication, apparition.

appartement, *m.* apartment, flat, suite of rooms.

appartenances, *f.pl.* appurtenance.

appartenant, appertaining.

appartenir, to belong, (ap)pertain, concern, behave, become.

appartient (il), (it) belongs, etc.

apparu (apparaître) appeared.

apparut (il), (apparaître) (it) appeared, etc.

appât, *m.* bait, lure.

appauvri, impoverished.

appauvrir, to impoverish, exhaust; **s'—**, to become poor(er), impoverished.

appauvrissement, *m.* impoverishment; — **du sol**, exhaustion of the soil.

appel, *m.* call, appeal, tension, pull (of air), draft, ventilation, roll call, challenge; — **d'air**, intake of air, suction; — **limité à la concurrence**, limited submission.

appelé, called.

appeler, to call, name, term, call for, appeal, require, draw, drive; **s'—**, to be called, named, call oneself.

appendant, appendent, hilum directed towards upper part of seed.

appendice, *m.* appendix, appendage, process, projection, stud.

appendicule, *m.* small appendage.

appendre, to hang up.

appesantir, to make heavy, weigh down, dull; **s'—**, to become heavy, lay stress, dwell, insist (upon).

appétence, *f.* impulse, trend.

appétissant, tempting, appetizing.

appétit, *m.* appetite, desire.

applaudir, to applaud, commend.

applaudissement, *m.* applause, approval.

applicabilité, *f.* appropriateness.

applicable, applicable, appropriate.

applicateur, *m.* pasteurizer.

application, *f.* application, applying, posting (of bills), use, attention.

applique, *f.* application against wall, wall bracket, mounting, setting.

appliqué, applied, attentive, intent, perseverant.

appliquer, to apply, put, lay, fit, post (bills); **s'—**, to apply, apply oneself, devote, attribute to oneself.

appoint, *m.* added portion, balance, contribution, difference.

appointements, *m.pl.* salary, wages, allowance.

appointer, to point, sharpen (pencil), put on salary basis.

appontement, *m.* landing stage.

apport, *m.* contribution (of capital), deposit, application.

apporter, to bring, contribute, adduce, apply, use, bring to bear, cause, occasion, convey.

apposer, to affix, place, append, add.

apposition, *f.* affixing, apposition, accretion.

appréciable, appreciable.

appréciateur -trice, *m.&f.* appraiser, valuer, appreciator.

appréciation, *f.* valuation, estimating, measurement, appreciation, valuing, judging.

apprécier, to estimate, determine, value, tax, rate, appreciate, appraise, assess.

appréhender, to apprehend, fear, seize.

appréhension, *f.* fear, dread.

apprendre, to learn, teach, tell, inform (of), apprise (of), acquaint (with).

apprennent (ils), (they) are learning.

apprenti, -e, *m.&f.* apprentice.

apprentissage, *m.* apprenticeship.

appressé, appressed, lying flat.

apprêt, *m.* preparation, affectation, dressing, finishing, priming, sizing.

apprêtage, *m.* dressing, finishing, sizing.

apprêter, to prepare, get ready, dress, finish; **s'— à,** to get oneself ready to.

apprêteur, *m.* finisher, dresser.

apprêteuse, *f.* finisher, dresser, calenders.

apprimé, appressed, lying flat.

appris (apprendre), learned.

apprivoiser, to tame, domesticate.

approbation, *f.* approval, approbation.

approchant, approaching, like, somewhat like, akin, similar, approximate, near(ly), about.

approche, *f.* approach, coming near.

approché, approximate.

approcher, to bring or draw near, bring up, approach; **s'— (de),** to approach, come near, draw near.

approfondi, thorough, searching, exhaustive, profound.

approfondir, to deepen, go deeply into, examine (study) thoroughly; **s'—,** to get or grow deeper, deepen.

approfondissement, *m.* deepening, investigation, profound study.

approprié, suitable, appropriate, made for the purpose, fit.

approprier, to appropriate, adapt, suit, fit.

approuver, to approve, approve of, consent, authorize.

approvisionnement, *m.* supplying, store, stock, supplies.

approvisionner, to supply, provision.

approximatif, -ive, approximate, rough (estimate).

approximation, *f.* approximation.

approximativement, approximately, about, roughly.

approximer, to approximate.

appui, *m.* support, prop, stay, rest, sill, fulcrum, rail, emphasis, pressure, bearing.

appuie (appuyer), (he) supports.

appui-bras, *m.* arm rest.

appui-main, *m.* maulstick, hand rest.

appuyé, supported, resting.

appuyer, to support, prop, lean, rest, press: **s'—,** to be supported, recline upon, lean upon, rely (on), support oneself, depend, trust to, bear on, press, prop up, strengthen, insist (on), stress.

âpre, rough, harsh, sour, bitter, acrid, tart, hirsute.

âprement, harshly, acridly, with asperity.

après, after, behind, next to, about, for, at, against, upon, afterward, later; **— que,** after, when; **— quoi,** after which, afterward; **d'—,** according to, after, next, from.

après-demain, day after tomorrow.

après-midi, *m.&f.* afternoon.

âpreté, *f.* acridity, asperity, harshness, tartness.

à priori, a priori, in advance.

à-propos, *m.* seasonableness, pertinency.

apte, qualified, fit, apt, suited.

aptère, apterous, wingless.

aptéryx, *m.* apteryx, kiwi.

aptitude, *f.* aptitude, fitness, disposition, inclination, capability.

apyre, fireproof, incombustible.

aquarelle, *f.* water color (painting).

aquatique, aquatic, marshy.

aqueduc, *m.* aqueduct, culvert.

aqueux, -euse, aqueous, watery, water, waxy (potato).

aquiculture, *f.* water conservancy, fish breeding.

aquifère, water-bearing, aquiferous.

aquifoliacé, resembling holly.

aquilaire, *f.* aloe.

aquilégie, *f.* columbine (Aquilegia).

aquosité, *f.* aquosity, wateriness, aqueousness.

arabe, Arabic, Arabian.

arabine, *f.* arabin (arabic acid).

arabinose, *n.* arabinose.

arabique, arabic (acid).

arable, tillable, arable.

arabonique, arabonic.

arac, *m.* arrack.

arachide, *f.* peanut, peanut plant, earthnut (*Arachis hypogaea*); **huile d'—,** peanut oil.

arachidique, arachic.

arachine, *f.* arachin.

arachnides, *m.pl.* Arachnida.

arachnoïde, arachnoid (membrane), cobwebby.

arack, *m.* arrack.

aragonite, *f.* aragonite.

araignée, *f.* spider — **de mer,** sea spider, weaver; **toile d'—,** cobweb, spider's web.

araire, *m.* drill plow, swing plow.

aralie, *f.* Aralia; — **épineuse,** Hercules'-club (*Aralia spinosa*); — **nue,** wild sarsaparilla (*Aralia nudicaulis*).

aranéeux, -euse, like a cobweb.

araser, to level, make flush with.

aratoire, agricultural, farming, aratory; **instruments, —s,** farm instruments.

arbacine, *f.* arbacin.

arbitrage, *m.* arbitrage, arbitration.

arbitraire, *m.* discretion, arbitrariness.

arbitraire, arbitrary, absolute.

arbitrairement, arbitrarily.

arbitre, *m.* arbitrator.

arbois, *m.* laburnum (Laburnum).

arborer, to raise, erect, hoist.

arborescent, arborescent, branched (chain), arborious, treelike, dendriform.

arboretum, *m.* arboretum, forest garden.

arboriculteur, *m.* nurseryman.

arboriculture, *f.* arboriculture.

arborious, treelike, dendriform, dendroid.

arborisation, *f.* arborization, dendrite.

arborisé, arborized, dendritic.

arbouse, *f.* arbutus (fruit) berry; — *m.* arbute, strawberry tree (*Arbutus unedo*); **huile d'—,** arbutus seed oil.

arbousier, *m.* cane-apple, strawberry tree (*Arbutus unedo*); — **commun,** strawberry tree (*Arbutus unedo*).

arbre, *m.* tree, arbor, spindle, shaft, axis, axle, mast; — **abandonné,** tree marked for felling; — **à cames,** camshaft; — **à cire,** wax tree, wax myrtle (Myrica); — **à coude,** — **à manivelle,** crankshaft; — **à gomme,** eucalyptus, gum tree; — **à la perruque,** fustet, Venetian sumac, wig tree (Cotinus); — **à suif,** Chinese tallow tree (*Sapium sebiferum*), **wax** myrtle (Myrica).

arbre, — **au poivre,** chaste tree (*Vitex agnus-castus*); — **commandé,** driven shaft; — **coudé,** crankshaft, cranked axle; — **creux,** hollow shaft; — **d'arbri,** nurse tree. shelter tree; — **d'avenir,** thriving (promising) tree; — **de commande du couteau,** cutter spindle; — **de couche,** driving (horizontal) shaft; — **de Diane,** arborescent silver, silver tree; — **de distribution,** camshaft.

arbre, — **de haute futaie,** lofty tree, timber tree; — **d'émonde,** lopped tree; — **de franc pied,** seedling; — **de futaie,** crop of formed trees (produced from seed); — **de Noël,** Christmas tree; — **dépérissant,** dying tree; — **de reserve,** standard in high forest, final crop tree; — **de Saturne,** arborescent lead.

arbre, — **de tour,** lathe spindle, mandrel; — **de vie,** arborvitae, white cedar (Thuja); — **droit,** tail erect; — **du castor,** beaver tree (*Magnolia virginiana*); — **du communicateur,** gearing shaft; — **d'une charrue,** plow beam; — **du tiroir,** valve stem, eccentric shaft; — **en buisson,** bush; — **en espalier,** wall tree; — **en grume,** tree with bark, wood in the round; — **feuillu,** broad-leaved tree; — **forestier,** forest tree; — **fruitier,** fruit tree.

arbre, — **généalogique,** family tree, pedigree, genealogical tree; — **intermédiaire,** middle (counter-) shaft; — **moteur,** drive shaft; — **moyen,** mean tree, average stem; — **nain,** dwarf tree; — **provenant de plantation,** — **provenant de semis,** tree grown artificially, seedling tree; — **rabougri,** stunted or dwarfed tree; — **résineux,** coniferous tree; — **sans aubier,** heartwood tree; — **vert,** evergreen tree.

arbre-piège, trap tree (for destroying beetles).

arbrisseau, *m.* shrub, shrubby tree, young tree.

arbue, *f.* clay used as a flux.

arbusculaire, arbuscular, shrubby.

arbuste, *m.* shrub, bush, arborescent shrub.

arbustif, -ive, shrubby, frutescent.

arbutine, *f.* arbutin, a crystalline glucoside in plants.

arc, *m.* arc, arch, (of a bridge) bow; — **embrassé,** arc of contact; — **d'une scie,** saw frame; — **voltaïque,** voltaic (electric) arc; **en —,** arcuate, curved; **—s branchiaux,** visceral arches.

arcade, *f.* arcade, arch, bow, bridge.

arcadien, -enne, Arcadian.

arcane, *m.* arcanum.

arcanne, *f.* ruddle, red ocher.

arcanson, *m.* common rosin, colophane.

arc-boutant, *m.* (flying) buttress, stay, prop; — d'une arbalète, hipstrut.

arc-bouter, to buttress, prop, support.

arceau, *m.* arch(way).

arc-en-ciel, *m.* rainbow.

archaeophytes, *f.pl.* archaeophytes.

archal, *m.* fil d'—, brass wire.

archangélique, *f.* (arch)angelica.

arche, *f.* furnace, arch (vaulting), ark.

archée, *f.* archeus.

archéen, -enne, Archean.

archégone, archegonium.

archéologue, *m.* archaeologist.

archespore, *m.* archespore.

archet, *m.* curved part, arch, violin bow, (fracture-) cradle; — de scie, saw bow.

archibuse pattue, *f.* rough-legged buzzard.

archicarpe, *m.* archicarp.

archidie, *f.* sporangium.

archiduc, *m.* archduke.

archipel, *m.* archipelago.

archiplasma, *m.* kinoplasm, part of cytoplasm involved in spindle formation.

archisoblaste, *f.* embryo germinating without producing radicle.

architecte, *m.* architect.

archives, *f.pl.* archives, record office.

archocléistogamie,, *f.* archocleistogamy, flowers closed when sexual organs ripen.

arçon d'une scie, *m.* saw frame.

arctie, *f.* tiger moth.

arctique, arctic.

ardemment, ardently, zealously.

ardent, hot, burning, blazing, glowing, ardent, eager, fervent, bright (colors), vivid, red (hair), fiery (horses), sandy, spirited.

ardeur, *f.* ardor, vividness, fervor, mettle, heat, burning, passion, intensity; — d'estomac, heartburn.

ardillon, *m.* tongue, catch (of buckle), barb (of hook).

ardoise, *f.* slate, schist, slate color; — faitière, ridge slate.

ardoisé, slate-colored, slated.

ardoiser, to render slate-colored, slate (roof).

ardoiseux, -euse, ardoisier, -ère, slaty.

ardoisière, *f.* slate quarry.

ardoisin, resembling slate.

ardu, arduous, steep, abrupt, difficult.

arduité, *f.* arduousness, difficulty.

are, *m.* are (100 square meters).

area, *f.* cell, space.

arec, aréca, *m.* areca palm (Areca) betal nut, areca nut.

aréfaction, *f.* drying, desiccation.

arénacé, arenaceous, sandy.

arénation, *f.* sanding, arenation.

arène, *f.* sand, gravel, shingle, arena; *pl.* amphitheater.

aréner, to sink.

aréneux, -euse, sandy, arenous.

aréolaire, areolar.

aréole, *f.* areola, incrustation (in the form of a ring or halo).

aréolé, bordered (pit).

aréomètre, *m.* hydrometer, areometer.

aréométric, *f.* hydrometry, areometry.

aréométrique, hydrometric, areometric.

arête, *f.* edge, crest, ridge, arris, groin, hip, awn, beard, arista (of grain), (fish) bone, prickle; — de poisson, herringbone (work); — latéral, lateral border; — médiane, splitter (of a Pelton wheel); — mousse, rounded edge; — taillée, trimmed edge; — vive, sharp edge; à vives — s, sharp-edged, full-edged, square; muni d'une —, awned.

arêtier, *m.* hip rafter; — de noue, valley rafter.

argémone, *f.* prickly poppy (Argemone).

argent, *m.* silver, money; — comptant, ready money, cash; — fulminant, fulminating silver; — raffiné, refined silver; donner de l'—, placer de l'—, to invest money.

argentage, *m.* silvering, silver plating.

argental, -aux, containing silver, argental.

argentan, *m.* argentan, German silver.

argenté, silvered, silvery, silver plated.

argenter, to silver.

argenterie, *f.* silver (ware) plate.

argenteur, *m.* silverer, silver plater.

argentifère, argentiferous, silver-bearing.

argentin, silvery, silver, Argentine.

argentine, *f.* argentine, goosegrass, silverfish, silverweed.

argentique, argentic, silver.

argenture, *f.* silvering, silver plating; — **galvanique,** electrosilver.

argilacé, argillaceous, clayey.

argile, *f.* clay, argil, loam; — **à blocaux,** tillite, boulder clay; — **à façonner,** potter's clay; — **à fougère,** slate clay; — **à porcelaine,** porcelain clay; — **cuite,** terra cotta; — **grasse,** rich (greasy) clay; — **pauvre,** lean clay; — **plastique,** plastic (fatty) clay; — **réfractaire,** fire clay, refractory clay; — **schisteuse,** shale, slate clay, bedded clay; — **smectique,** fuller's earth.

argileux, -euse, clayey, argillaceous, clayish.

argilière, *f.* clay pit.

argilifère, argilliferous, clay-bearing.

argilite, *f.* argilite.

argillacé, argillaceous.

argilo-calcaire, argillocalcareous.

argiloïde, argilloid, like clay.

argilolithe, *m.* clay stone.

arginine, *f.* arginine.

argol, *m.* crude tartar.

argon, *m.* argon.

argousier, *m.* sea buckthorn (*Hippophaë rhamnoides*).

argue, *f.* wire-drawing appliance (for gold and silver).

arguer, to wire-draw (gold and silver), argue, deduce, conclude.

arguérite, *f.* arguerite.

argument, *m.* argument, proof.

argumenter, to argue (with).

argyrophylle, with silvery leaves.

argyrose, *f.* argentite, argyrite.

argyrythrose, *f.* pyrargyrite.

arhize, arrhizal, rootless.

aricine, *f.* aricine.

aride, arid, dry, barren, sterile.

aridité, *f.* aridity, dryness, barrenness, drought.

arillaire, arilloid, like an aril.

arille, *m.* aril, mace, arillus.

arillode, *m.* arillode, arillodium, a false aril.

aristé, aristate, awned.

aristerostyle, aristostylous, an exserted style bent toward the left.

aristoloche, *f.* birthwort (Aristolochia); — **en arbre,** Dutchman's-pipe (*Aristolochia macrophylla*).

arithmétique, *f.* arithmetic.

arithmétique, arithmetical.

arithmétiquement, arithmetically.

armagnac, *m.* Armagnac brandy.

armaturage, *m.* bracing.

armature, *f.* frame, brace, truss, fastening, clamps, sheathing, facing, fitting, coating (Leyden jar), armature, magnet, plate, bracing, hanging truss; *pl.* fittings, mountings.

arme, *f.* arm, weapon, arm (of army), corps; — **de jet,** missile weapon; — **du génie,** engineer corps; — **petite,** side arm; — **rayée,** rifle; — **s à feu,** firearms; — **s blanches,** side arms, swords.

armé, armed, reinforced (of cement), equipped; **verre** —, wired glass.

armée, *f.* army, force(s); — **de mer,** navy, fleet, marine forces; — **de terre,** land forces; — **navale,** naval forces.

armement, *m.* armament, arming, providing equipment.

armer, to arm, provide, equip, fit out, reinforce, strengthen, stiffen, insert (a fuse), wind (a dynamo), cap (a magnet); **s'**—, to arm oneself.

armistice, *m.* armistice.

armoire, *f.* closet, wardrobe, clothes-press, cupboard; — **à dessiccation,** drying closet; — **de séchage,** drying chamber.

armoise, *f.* Artemisia; — **absinthe,** — **amère,** wormwood (*Artemisia absinthium*); — **commune,** mugwort (*Artemisia vulgaris*).

armoisin, *m.* sarsenet (a woven silk of soft texture).

armure, *f.* armor, frame, braces, sheathing, arming, armature.

arnica, *f.* arnica, mountain tobacco.

arnicine, *f.* arnicine.

arnique, *f.* arnica.

arnoséride, *f.* swine succory (Arnica).

arolle, *f.* cembra fir, zirbel, stone pine.

aromate, *m.* aromatic, flavoring compound.

aromatique, aromatic, fragrant.

aromatiser, to aromatize, give aroma to.

arome, *m.* aroma, smell, odor.

aronde, *f.* swallow.

arpent, *m.* acre.

arpentage, *m.* land measurement, land surveying; — **des bois,** forest survey.

arpenter, to survey.

arpenteur, *m.* land surveyor.

arpenteuse du pin, *f.* pine looper moth.

arqué, arched.

arquer, to bend, arch, curve.

arrachage, *m.* pulling up, rooting up.

arrache clou, *m.* nail nippers, nail puller.

arrachement, *m.* lifting, plucking, pulling up.

arrache-pied (d'), without interruption.

arracher (à), to tear, pull (up), dig (up), root up, lift, tear away (from), strip (off), take (away), wrest, snatch (away), extract, force out of, extort.

arracheuse, *f.* lifter, digger.

arracho-racine, *m.* spud.

arrak, *m.* arrack.

arrangement, *m.* arrangement, order, method, planning, disposition, terms.

arranger, to arrange, set in order, settle, accommodate, contrive, trim (up), dress (off); **s'—,** to place oneself, arrange, agree, manage, get along, be arranged; **— pour,** to devise a way to, so that; **il s'en arrange,** he puts up with it.

arrestation, *f.* arrest.

arrêt, *m.* stopping, stop, stoppage, shifter rest, catch, check, arrest, judgment, edict, decree, order, distrain, attachment, danger (of a signal), point (of a dog); **— sur les versements,** stopping of payment; **chien d'—,** setter, pointer; **être en —,** to point (of a dog).

arrêtage, *m.* stop, check, arrest, catch.

arrêt-barrage, *m.* dam, barrier.

arrête-boeuf, *m.* restharrow, cammock (*Ononis repens*).

arrêté de compte, *m.* clearing.

arrête-gaz, *m.* gas check.

arrêter, to stop, halt, clamp, delay, check, detain, fix, fasten, arrest, seize, engage, hire, decide; **— avec de coins,** to quoin; **— un compte,** to balance, settle, square; **s'—,** to stop, stay, dwell (on), pay attention to, insist on (a subject).

arrêtoir, *m.* stop, lug, catch.

arrhize, arrhizal, arrhizous, wanting true roots.

arrière, *m.* rear, back (hind) part, rump, stern, aft, arrears; **en —,** behind, back, backward, behindhand, ago; **en — de,** behind.

arrière, back, backward, behind, in the rear, reverse, astern.

arriéré, backward, behindhand, delayed, left undone, ancient.

arrière-bois, *m.* backwoods.

arrière-dent, *f.* last molar, wisdom tooth.

arrière-dique, *f.* back dike, inner dike.

arrière-essaim, *m.* afterswarm.

arrière-faix, *m.* afterbirth.

arrière-gorge, *f.* pharynx.

arrière-goût, *m.* aftertaste.

arrière-main, *f.* hind quarters.

arrière-molaire, *f.* back molar (tooth).

arrière-panage, *m.* after pannage.

arrière-pensée, *f.* afterthought.

arriérer, to delay, defer, put off; **s'—,** to stay or get behind.

arrière-saison, *f.* close of autumn.

arrière-train, *m.* trailer.

arrimage, *m.* stowage, stowing.

arrimer, to stow (away).

arrivage, *m.* arrival.

arrivée, *f.* feed pipe, inlet, influx, admission, arrival; **— d'air chaud,** blast main; **— d'eau,** water inlet; **— du gaz,** gas supply; **d'—,** at first, at once, on arrival; **tube d'—,** inlet tube, arrival platform, coming.

arriver, to arrive, come, approach, reach, succeed, happen, occur; **où veut-il en —,** what is he getting at?; **il arrive à,** he succeeds in.

arroche, *f.* goosefoot, orach, mountain spinach (*Atriplex hortensis*).

arrondi, round(ed), rotund.

arrondir, to make round, round (off); **s'—,** to become round, round out; **— un nombre,** to approximate.

arrondissage, *m.* rounding (off), making round.

arrondissement, *m.* rounding (off), roundness, ward, district.

arrosage, *m.*, **arrosement,** *m.* watering, wetting, sprinkling, irrigation.

arroser, to water, sprinkle, spray, wet, moisten, shower, irrigate.

arrosoir, *m.* watering pot, sprinkler, watering can.

arsendiméthyle, *m.* arsenic dimethyl (cacodyl).

arséniate, *m.* arsenate; **— de baryte,** barium arsenate; **— de plomb,** lead arsenate, **— de soude,** sodium arsenate.

arsenic, *m.* arsenic; — **blanc,** arsenious oxide, white arsenic; — **sulfuré rouge,** realgar, disulphide of arsenic.

arsenical, *m.* arsenical compound or substance, arsenical.

arsenical, -aux, arsenical.

arsenicophile, habituated to strong doses of arsenic.

arsénié, arsenicated, arsenical.

arsénieux, -euse, arsenious.

arsénifère, arseniferous.

arsénio-sulfure, *m.* sulpharsenide.

arsénique, arsenic, arsenical.

arsénite, *m.* arsenite; — **de cuivre,** copper arsenite.

arséniure, *m.* arsenide; — **d'hydrogène,** arsine.

arsénolithe, *m.* arsenolite.

arsine, *f.* arsine.

art, *m.* art, dexterity, ability; **beaux-arts,** fine arts.

artémisie, artémise, *f.* artemisia, sagebrush, wormwood.

artère, *f.* artery; — **à gaz,** gas main; — **alimentaire,** feeding (cable) wire.

artérialiser, to arterialize.

artériel, elle, arterial; **tronc —,** axis.

arterieux, euse, arterial.

artériosclérose, *f.* arteriosclerosis.

artésien, -enne, artesian, of Artois; **puits —,** artesian well.

arthrite, *f.* joint evil, arthritis.

arthrospore, *f.* arthrospore, a bacterial resting cell.

arthrosterigmates, *m.pl.* arthrosterigmata, jointed sterigmata.

artichaut, *m.* artichoke (*Cynara scolymus*); — **des Indes,** sweet potato (*Ipomoea batatas*); — **des toits,** houseleek (*Sempervivum tectorum*).

article, *m.* article, entry, item, subject, joint, ring, link, articulation; — **mâle,** male cell; —**s en bois,** woodenware; —**s de pulpe de bois,** wood-pulp articles; —**s de vannerie,** wickerwork, basketware.

articulaire, articular, of the joints.

articulation, *f.* articulation, node, link, joint, hinge, junction of the bones, hinge, wrist; — **sphérique,** ball-and-socket joint.

articulé, articulate(d), jointed, vertebrate.

articuler, to articulate, join, pronounce, joint. assert, utter.

articulés, *m.pl.* Articulata.

artifice, *m.* artifice, contrivance, skill, cunning; — **de guerre,** artifice of war, stratagem; — **de rupture,** any explosive device used for demolition; — **incendiare,** incendiary firework, incendiary; **feu d'—,** fireworks, pyrotechnic display.

artificiel, -elle, artificial, imitation, false.

artificiellement, artificially.

artificier, *m.* fireworks maker, pyrotechnist, artificer.

artificieux, -euse, crafty, artful.

artillerie, *f.* artillery, ordnance.

artiodactyle, cloven-footed, artiodactylous.

artisan, *m.* skilled laborer.

artison, *m.* wood (worm) moth, clothes moth.

artisonné, moth-eaten, worm-eaten.

artiste, *m.&f.* artist; — **vétérinaire,** veterinary surgeon.

artistement, artistically, cleverly, skillfully.

artistique, artistic.

artocarpe, *m.* breadfruit tree.

arum, *m.* arum.

arundinacées, *f.pl.* Arundinaceae.

arundinaire, *f.* canebrake.

arvin, arvensis, growing in fields.

as, *m.* ace; **tu as,** (avoir) you have.

As, *abbr.* (arsenic), arsenic.

asa-foetida, *m.* see ASSA FOETIDA.

asaret, *m.* Asarum.

asarone, *f.* asaron, asarum camphor.

asbeste, *m.* asbestos.

asbolane, *m.* asbolite, bog manganese, earthy cobalt.

asboline, *f.* asbolin (contained in wood soot).

ascaride, ascaris, *m.* ascaris, threadworm.

ascendant, *m.* ascendant, ancestor.

ascendant, ascending, assurgent, upward, rising, return, reflux (condenser).

ascenseur, *m.* elevator, lift, hoist.

ascension, *f.* rise, ascent, ascension, rising, upstroke (of piston).

ascensionnel, -elle, ascensional, upward, lifting.

ascidie, *f.* ascidium.

ascites, *f.pl.* ascites, abdominal (peritoneal) dropsy.

asclépiade, *f.* asclepiad.

ascogène, ascogenous, producing asci, asciferous.

ascomycète, *f.* ascomycete; *pl.* large group of fungi, forming ascospores and stylospores.

ascomycète, ascomycetous.

ascophore, *m.* ascophore, ascus-bearing hyphae.

ascospore, *f.* ascospore.

asepsie, *f.* asepsis.

aseptique, aseptic.

aseptiquement, aseptically.

aseptiser, to asepticize.

aseptol, *m.* aseptol.

asexué, sexless.

asexuel, -elle, asexual, vegetative.

Asie, *f.* Asia; — **Mineure,** Asia Minor.

asile, *m.* refuge, place of refuge, shelter, asylum, retreat, asilus, hornet fly.

asomatophyte, *m.* asomatophyte.

asopies, *f.pl.* pyralids.

aspalax, *m.* spalax, mole rate.

asparagine, *f.* asparagine.

asparaginique, asparaginic, aspartic.

aspartique, aspartic, asparagic.

aspect, *m.* sight, aspect, appearance, look, phase, point of view.

asperge, *f.* asparagus (shoot); **—s en branches,** stick asparagus, spear asparagus.

aspergement, *m.* sprinkling, spraying.

asperger, to sprinkle, sparge.

aspergillaire, aspergillary.

aspergille, *f.* Aspergillus.

aspergillomycose, *f.* aspergillomycosis.

aspergillose, *f.* aspergillosis.

aspérité, *f.* unevenness, roughness, asperity, harshness.

asperme, aspermous, seedless.

aspermie, *f.* aspermatism.

aspersion, *f.* sprinkling, spraying, aspersion.

aspersoir, *m.* sprinkler.

aspérule odorante, *f.* (sweet) woodruff (*Asperula odorata*).

asphaline, *f.* asphaline.

asphalte, *m.* asphalt, asphaltum.

asphalteux, -euse, asphaltic, asphalt.

asphaltique, asphaltic.

asphodèle, *m.* asphodel (Asphodelus).

asphyxiant, asphyxiating, suffocating; **obus —,** *m.* gas shell.

asphyxie, *f.* asphyxia, suffocation.

asphyxier, to asphyxiate, suffocate.

asphyxique, asphyxial.

aspic, *m.* great lavender, spike aspic (*Lavandula spical*), asp.

aspidie, *f.* shield-fern, buckler fern (Dryopteris).

aspidospermatypus, *m.* aspidospermotype, a wind-dispersed seed.

aspirail, *m.* air hole, vent hole, vent, flue.

aspirant, *m.* aspirant, candidate.

aspirant, suction, sucking, aspiring.

aspirateur, *m.* aspirator, exhauster, aspirating, exhausting, suction (device); **— de copeaux,** shavings separator.

aspiration, *f.* aspiration, inspiration, intake, inlet, admission, suction.

aspirer, to aspirate, inspire, draw by suction, suck up, suck in, draw off, exhaust, inhale, aspire (after).

aspléniacé, asplenioid, like the fern genus, Asplenium.

asporogène, asporogenous.

asporulé, sporeless.

asque, *m.* ascus.

assa dulcis, *f.* asa dulcis.

assa foetida, *f.* asafetida, a fetid gum resin.

assailir, to assail, assault, attack.

assainir, to make healthful, purify, make wholesome, render fertile, productive, drain.

assainissement, *m.* cleansing, purification, sanitation, wholesomeness, fertility, healthfulness, drainage.

assaisonnement, *m.* seasoning, flavoring, condiment, relish, herb, spices.

assaisonner, to season, flavor.

assamar, *m.* assamar.

assassiner, to assassinate.

assaut, *m.* assault, attack, onset, storm.

asséché, drained, reclaimed.

assèchement, *m.* drying, drainage.

assécher, to drain, dry up, run dry, reclaim.

assemblage, *m.* assemblage, assembling, set, erection, joint, group, collection, coupling, connection, joining up, junction; **— à contre-fiches,** truss, strut frame; **— à entailles,** cogging joint; **— à goujons,** pegging, doweling; **— à joint saillant,** halving, halved joint; **— à languette,** tonguing; **— à manchon,** spigot and socket joint; **— à mi-bois,** halving together; **— à onglet,** miter joint.

assemblage, — à tenon et entaille, notching; **— à vis,** screw joint; **— croisé, — à**

croisettes, cross brace or bond; — de charpente, timber framing; — des bois, wood joints; — du sommet, hip rafters; — en crémaillère, tabled joint; — en fausse coupe, bevel joint; — en quantité, parallel connection; — en tension, — en serie, joining up in series; — par embrèvement, tongue and groove joint.

assemblée, *f.* assembly, meeting.

assembler, to join, assemble, adjust, collect, gather, put together, connect, join up; — à entailles, to cog, join by cogging; — à mi-bois, to scarf; — à tenon et entaille, to join timbers by cogging; — à tenon et mortaise, to mortise; — en queue d'aronde, to dovetail; s'—, to meet together, congregate.

assener, to deal (a blow), strike hard.

assentiment, *m.* assent.

asseoir, to seat, set, put, place, lay, found, base, establish; s'—, to sit down, be seated; — une coupe, to locate a felling.

assertion, *f.* assertion, statement.

asservir, to bring under control, subdue.

asseyant (asseoir), seating, placing.

assez, enough, sufficient(ly), rather, passably, fairly, quite, tolerably; — bien, pretty well.

assidu, assiduous, diligent

assiduité, *f.* assiduity, diligence, industry.

assidûment, assiduously, constantly.

assied (asseoir); il s'—, he sits down.

assiette, *f.* plate, position, state (of mind), assessment, seat, situation, formation level; — de la coupe, location of fellings; — du ballast, subgrade formation; — d'une coupe, felling area.

assignation, *f.* assignation, assignment, check, draft.

assigner, to assign, appoint, allot.

assimilabilité, *f.* capability of assimilating.

assimilable, assimilable, comparable.

assimilateur, -trice, assimilative, assimilating.

assimilatif, -ive, assimilative.

assimilation, *f.* assimilation, comparison, constructive metabolism.

assimiler, to assimilate, make like, compare.

assis (asseoir), seated.

assise, *f.* bed, course, layer, stratum, base; — de (par) boutisse, course of headers; — de panneresse, course of stretchers; — des rails, rail seat.

assistance, *f.* attendance, audience, assistance.

assistant, *m.* person present, onlooker, assistant.

assister, to assist, attend; — à, to be present (at), attend, witness.

association, *f.* association, company, (partnership) society; — de propriétaires forestiers, forestry association.

associé, *m.* associate, partner, honorary member.

associé, associated.

associer, to associate, unite, make someone a party to, take into partnership; s'—, to enter into partnership with someone, be connected with, share, join.

assoit (asseoir), il s'—, he sits down.

assolement, *m.* rotation (of crops).

assoler, to rotate (crops).

assombrir, to darken, obscure, render gloomy; s'—, to become dark, cloud over.

assommer, to brain, knock down, beat, stun, overpower, bore, tire to death; — un boeuf, to fell an ox.

assomptif, -ive, assumptive.

assomption, *f.* assumption.

assorti, assorted, mixed, matched, stocked, furnished.

assortiment, *m.* assortment, matching, sorting, grade, class.

assortir, to assort, match, stock, sort, grade, pair; s'—, to harmonize, match, choose each other, pair off.

assoupir, to make drowsy, allay, deaden, quiet; s'—, to become drowsy, doze, nap, grow sleepy, wear away.

assoupissant, soporific, narcotic, drowsy.

assoupissement, *m.* slumber, coma.

assouplir, to make supple, make flexible; s'—, to become supple.

assouplissage, *m.* breaking, beating, suppling.

assouplissement, *m.* limbering.

assourdir, to deafen, deaden, muffle, tone down, soften.

assouvir, to satiate, sate, appease, satisfy.

assouvissement, *m.* satisfying (of hunger).

assoyant (asseoir), seating, placing.

assujettir, assujétir, to subject, fix, fasten, make fast or firm, bind, subdue, constrain, wedge in; — avec des coins, to drive in the coins, quoin.

assumer, to assume.

assurance, *f.* assurance, confidence, insurance, security.

assuré, firm, sure, assured, confident, certain, secure, insured.

assurément, assuredly, confidently, certainly, securely.

assurer, to assure, make sure, provide, affirm, insure, secure, assert, render safe; **s'—,** to procure, ascertain, rely on, assure or satisfy oneself, secure, make sure (of).

assyrien, -enne, Assyrian.

aster, *m.* aster, Michaelmas daisy (Aster).

astéré, asteriate, asteroid, star-shaped.

astérisque, *m.* asterisk.

asternale, not attached to sternum.

astéroïde, *m.* asteroid, meteorite.

asthénie, *f.* debility, asthenia.

asthénique, pertaining to a diminution (partial or general) of an organic action.

asthme, *m.* asthma, pursiness, broken wind.

asti, *m.* Asti(wine), silk (made in Asti).

astic, *m.* polisher, polishing stick or paste.

asticot, *m.* mite, maggot.

astigmate, astigmatic.

astiquer, to polish, furbish.

astome, astomous, without orifice.

astragale, *m.* astragalus, anklebone, milk vetch (Astragalus).

astre, *m.* heavenly body, star; **— du jour,** sun.

astreignant, exacting (work).

astreindre, to restrict, limit, compel, oblige.

astreint, restricted, (he) restricts.

astrictif, -ive, binding, astrictive, astringent.

astringence, *f.* astringency.

astringent, astringent.

astrologie, *f.* astrology.

astronome, *m.* astronomer.

astronomie, *f.* astronomy.

astronomiquement, astronomically.

astrosphère, *f.* astrosphere, attraction (directive) sphere.

astucieux, -euse, astute, cunning.

asymblastie, *m.* asymblasty, various periods of germination.

asymétrie, *f.* asymmetry, dissymmetry.

asymétrique, asymmetric(al).

asymptote, *f.* asymptote.

asymptotique, asymptotic(al).

asynchrone, asynchronous.

asyngamie, *m.* asyngamy.

atakamite, *f.* atacamite.

atavisme, *m.* atavism, ancestral resemblance.

atavistique, atavistic, reverting to former type.

atelier, *m.* workshop, working party, repair gang, studio, shop, mill; **— de fonderie,** casting room, foundry; **— de fonte,** foundry, smelting works; **— de granulation,** graining room; **— de rabotage,** planing mill.

athanor, *m.* athanor.

athermane, athermanous.

athérome, *m.* atheroma, encysted tumor.

athérosperme, *m.* plume nutmeg (*Atherosperma moschata*).

athlétique, athletic.

Atlantique, Atlantic Ocean.

atlas, *m.* Indian satin.

atmidomètre, *m.* atmometer, evaporimeter.

atmolyse, *f.* atmolysis.

atmomètre, *m.* atmometer.

atmosphère, *f.* atmosphere, weather.

atmosphérique, atmospheric(al).

atome, *m.* atom.

atomicité, *f.* atomicity, atomic value, valency.

atomique, atomic.

atomisme, *m.* atomism.

atomiste, *m.* atomist.

atomistique, *f.* atomistics.

atomistique, atomistic(al).

atoxique, atoxic, nonpoisonous.

atoxyl, *m.* atoxyl.

atrabilaire, splenetic, atrabiliary.

âtre, *m.* hearth, fireplace.

atroce, atrocious, cruel, excruciating.

atromarginé, black(-bordered)-edged.

atropamine, *f.* apoatropine, atropamine.

atrope, atropal, orthotropous, ovule with straight axis.

atrophie, *f.* atrophy, dwarfing, emaciation, abortion, degeneration.

atrophier, to atrophy, waste away.

atrophique, atrophic, aplastic, not exhibiting growth.

atrophytes, *f.pl.* atrophytes, fungi causing atrophy.

atropine, *f.* atropine.

atropique, atropic.

attachant, interesting, attaching, engaging (personality).

attache, *f.* attachment, assent, consent, fastening, string, tie, bond, brace, clip, band, setting on; — **à broche filetée,** vice coupling; — **à cliquet,** pawl coupling; — **à vis,** screw coupling.

attaché, *m.* attaché, devotee, adherent.

attaché, attached, tied up.

attachement, *m.* attachment, affection, fondness.

atiacher, to attach, fasten, connect, endear, bind, tie, stick, interest; **s'—,** to be fastened, stick, adhere, be fixed, encounter, become attached to.

attache-vache, *f.* halter rope.

attaquable, attackable, contestable, assailable; **peu —,** refractory.

attaque, *f.* attack, assault, infestation, corrosion.

attaqué, infested.

attaquer, to attack, assault, assail, drive, corrode, infest; **s'—,** to pitch into.

attarder, to delay, keep late; **il s'attarde,** he delays.

atteignant, (atteindre), attaining.

atteindre, to attain, attack, arrive at, reach, strike, hit, harm, injure, overtake, come up to.

atteint (atteindre), (he) attains.

atteinte, *f.* reach, stroke, attack, blow, tread, harm, shock, touch, contact, striking, ray; **hors d'—,** beyond reach, out of reach.

attelage, *m.* team, coupling, clutch, set, draught; — **à chevaux,** horse draught.

atteler, to hitch, couple, put on (horse).

attenant, adjoining, next, near by; **tout —,** close by.

attendant, waiting; **en —,** meanwhile, in the meantime; **en — que,** till, until.

attendre, to wait (for), await, look for, tarry for, expect, keep (fruit); **s'— à,** to expect, rely.

attendrir, to make tender, soften, move; **s'—,** to become tender, be softened or moved.

attendrissement, *m.* softening, making tender, tenderness, sensibility, emotion.

attendu, waited for, considering, expected, on account of; — **que,** considering that, since, seeing that, inasmuch as, whereas.

attente, *f.* waiting, expectation, hope; **dans l'— de,** in the expectation of, looking forward to.

attenter, to make an attempt.

attentif, -ive, attentive, heedful, mindful, careful.

attention, *f.* attention, care, heedfulness.

attentivement, attentively.

atténuant, attenuating, extenuating, attenuant.

atténuation, *f.* attenuation, diminishing, extenuation.

atténué, attenuate(d), weakened, tapering, acuminate, narrowed, tapered.

atténuer, to attenuate, diminish, extenuate, weaken, lessen; **s'—,** to become attenuated or diminished.

atterrissage, *m.* landing, alighting.

attester, to attest, avouch.

attiédi, made tepid, lukewarm.

attiédir, to make tepid or lukewarm, cool, warm, take chill off; **s'—,** to become tepid or lukewarm.

attiédissement, *m.* making tepid, cooling off, lukewarmness, coolness.

attier, *m.* sweetsop (*Annona squamosa*).

attique, Attic, attic, Athenian.

attirable, attractable.

attirail, *m.* apparatus, requisite, appliances, implements, accoutrement, paraphernalia.

attirant, attractive, engaging, attracting.

attirer, to attract, allure, incite, lead to, draw; **s'—,** to attract each other, draw upon oneself, incur.

attiser, to stir (up), poke or stoke (a fire).

attiseur, -euse, *m.&f.* stoker, poker.

attisoir, attisonnoir, *m.* poker, fire rake.

attitude, *f.* attitude, posture.

attouchement, *m.* touch(ing), contact, feeling.

attoucher, to touch.

attracteur, -trice, attracting, attractile.

attractif, -ive, attractive, attracting.

attraction, *f.* attraction, gravitation, charm.

attractivement, attractively.

attraire, to attract, entice, allure.

attrait, *m.* attraction, charm, allurement, bait, attracted.

attrape, *f.* tongs, catch, trap, snare (birds), life line; — **mouche,** *m.* catchfly.

attrape-poussière, *f.* dust catcher or collector.

attraper, to catch, get, cheat, dupe, take in.

attrayant, attractive, charming, winsome, inviting.

attrempage, *m.* gradual heating (furnace).

attremper, to heat gradually, anneal (glass), temper (steel).

attribuable, attributable.

attribuer, to attribute, assign, ascribe; s'—, to assume, claim, be attributed.

attribut, *m.* attribute, symbol, emblem.

attrister, to sadden.

attrition, *f.* attrition, wearing away.

atypique, atypical.

au (à le), to the, at the, in the.

aubage, *m.* blading.

aube, *f.* dawn, float, paddle, bucket (of wheel); — **directrice,** (guide) vane, guide blade.

aubépine, *f.*, **aubépin,** *m.* hawthorn (Crataegus), hedgethorn (*Crataegus oxyacantha*); — **commune,** hawthorn, whitethorn, Maybush.

auberge, *f.* inn, tavern.

aubergine, *f.* eggplant.

aube-vigne, *f.* virgin's-bower (Clematis).

aubier, *m.* sapwood, alburnum.

aubifoin, *m.* bluebottle, cornflower, bachelor's-button, blue poppy (*Centaurea cyanus*).

aubour, *m.* laburnum (Laburnum).

aucubine, aucubin.

aucun, any (with negation expressed or understood), no, none, no one, not any.

aucunement, in any way, at all (with negation expressed or understood), in no way, not at all, in no wise, not in the least.

aucuns, some (with negation expressed or understood), no one, not one, none, not any; d'—, some.

audace, *f.* audacity, daring, boldness.

audacieux, -euse, audacious, daring, bold.

au deçà, on this side.

au dedans, inside, within.

au dehors, outside.

au-delà, on the other side of, beyond, over.

au-dessous, below, beneath, underneath, under.

au-dessus, above, upward, on top of, over, beyond.

au-devant, in front (of), before.

audience, *f.* audience, hearing, court.

audion, *m.* audion, vacuum tube.

auditeur, -trice, *m.&f.* auditor, hearer.

audition, *f.* hearing, audition, auditing.

auditoire, *m.* auditorium, audience.

auge, *f.* trough, bucket, flume, canal, hod, tray, spout, manger.

augée, *f.* troughful, mangerful.

auget, *m.* small trough, bucket, groove, cartridge hopper; — **basculant,** tipping bucket.

augmentation, *f.* increase, augmentation; — de la qualité, quality increment.

augmenter, to increase, augment, raise, advance, rise, grow, multiply; s'—, to increase, be increased.

augure, *m.* omen, augur.

aujourd'hui, today, nowadays, at present now.

aulne, *m.* see AUNE.

aulx, *m.pl.* see AIL.

aune, *f.* ell (old measure, 1.188 meters).

aune, *m.* alder (tree or wood), (Alnus); — **blanc,** gray alder, Alnus; — **glutineux,** common alder (*Alnus vulgaris*); — **grisâtre,** hoary or speckled alder.

aunée, *f.* elecampane (*Inula helenium*).

auparavant, before, previously, first, ere now, heretofore.

auprès, near (by), hard by, with, to, close to; — **de,** near (to), about, with (persons), in comparison with, to, at the side (of); tout — de, close beside.

auquel (à lequel), to whom, to which.

aura (avoir), (he) will have.

aurait (avoir), (he) would have.

auramine, *f.* auramine.

aurantia, *n.* aurantia.

aurantiacées, aurantiées, *f.pl.* Aurantiaceae.

auréole, *f.* areola, gas cap, testing flame, ring, corona, aureola, halo.

auréosine, *f.* aureosin.

aureux, aurous.

aurez (avoir), (you) will have.

auriculaire, auricular; le **doigt** —, the little finger.

auricule, *f.* auricle (heart), auricula, bear's-ear.

auriculé, auriculate, eared.

aurifère, auriferous, gold-bearing.

aurifique, aurific.

aurique, auric.

aurochs, *m.* aurochs, bison.

aurone, *f.* southernwood (*Artemisia arbrotanum*); — des champs, field wormwood (*Artemisia campestris*).

auront (avoir), (they) will have.

aurore, *f.* dawn, morn, aurora, (color) at dawn, saffron, golden, yellow; — boréale, aurora borealis, northern lights.

aurure, *m.* auride.

aussi, also, too, as, besides, likewise, accordingly, so, therefore, and so; — bien, therefore, as well, moreover, besides, the more so, as, in fact, for; — bien que, as well (as), both . . . and; — . . . que, as . . . as; moi —, so am I, I too, so can I, etc.; pas —, not so, not as.

aussitôt, immediately, (at) once, forthwith, directly, as soon as; — dit — fait, no sooner said than done.

austère, austere, severe, sharp, harsh.

austérité, *f.* harshness, strictness.

austral, austral, southern.

Australasie, *f.* Australasia.

Australie, *f.* Australia.

austro-hongrois, Austro-Hungarian.

autant, as much, as many, as far, so much, so many; — . . . —, as much . . . so much, as . . . as; — de, as (so) much, as (so) many; — de moins, so much the less; — de pour que de contre, as many for as against; — que, as (much) many as, as far as, in so far as; — . . . que, as . . . as.

autant, — vaut, as much as; — vaudrait dire que, one might as well say that . . . ; d'—, to the same extent, accordingly; d'— mieux que, all the better because; d'— moins, the less so, so much the less; d'— plus, the more so, so much the more; d'— plus (moins) que, all the more (less), the more (or less) because, the more so, the less.

autant, d'— que, in as much as, since, as, seeing, as far as, more especially so; encore —, twice as much; en dire —, to say the same thing; en faire —, to do the same thing; tout —, quite as much; une fois —, as many (much) again.

autécologie, *m.* autecology, ecology of the species.

autel, *m.* altar, fire bridge, fire stop, Ara, (Altar); — de foyer, flame bridge, fire box.

auteur, *m.* author, authority, maker, writer.

authenticité, *f.* genuineness.

authentique, authentic, original (text).

auto, *m.* automobile.

auto-allogamie, *m.* autoallogamy, flowers adapted for self-fertilization and cross-fertilization.

auto-ballon, *m.* dirigible.

autobasidie, *f.* autobasidiomycete, basidium not divided.

autoblaste, *m.* autoblast, independent bioblast.

auto-camion, *m.* motortruck.

autocarpie, *m.* autocarpy, product of autogamy.

autocarpien, -enne, autocarpous.

autochenille, *f.* caterpillar tractor.

autoclave, *m.* autoclave.

autocopiste, *m.* (self-) copying (apparatus), hectograph.

autocycle, *f.* all the tissues surrounding the pericycle.

autoécique, autoecious.

autoépuration, *f.* self-purification.

auto-excitateur, -euse, self-exciting.

auto-excitation, *f.* self-excitation.

autofécondation, *f.* self-fertilization.

autofertilité, *f.* autofertility.

autogamie, *f.* autogamy, self-fertilization.

autogène, autogenic, autogenous.

autogénétique, autogenetic, produced independently.

autointoxication, *f.* autointoxication.

autolyse, *f.* autolysis.

autolytique, autolytic.

automate, *m.* automatic, automaton.

automatic, automaticité, *f.* automatic working.

automatique, automatic, self-acting.

automnai, produced or flowering in fall.

automne, *m.&f.* autumn, fall.

automobile, automobile.

automobilisme, *m.* automobile manufacture, motoring, use of automobiles.

automobiliste, *m.* automobilist.

automorphose, *f.* morphosis caused by internal factors.

automoteur, -trice, self-acting, self-propelling.

automotrice, *f.* self-propelling railway carriage.

autonome, autonomous, independent, complete.

autonomie, *f.* autonomy, independence.

autonyctitropique, autonyctitropic, spontaneously assuming night position.

autophyte, *m.* autophyte.

autoplaste, *m.* autoplast, chlorophyll granule.

autopollinisation, *f.* self-pollination.

autopsie, *f.* autopsy; — **cadavérique,** postmortem examination, autopsy.

autorégulateur, -trice, self-regulating.

autorisation, *f.* authorization.

autoriser, to authorize, warrant, empower; **s'—,** to be authorized.

autorité, *f.* authority, weight.

autostérilité, *f.* autosterility.

autotrophique, autotrophic, capable of self-nourishment.

autotropisme, *m.* autotropism.

autour, *m.* goshawk, hawk.

autour, around, round, about, roundabout; — **de,** around, about.

autre, other, another, else, different; **d'—s,** others; (nous) (vous) —**s,** (we) (you) people; **quelque chose d'—,** something else.

autrefois, formerly, of old, of yore.

autrement, otherwise, differently, else; — **dit,** in other words.

Autriche, *f.* Austria.

Autriche-Hongrie, *f.* Austria-Hungary.

autrichien, -enne, Austrian.

autruche, *f.* ostrich.

autrui, others, other people, another.

auvergne, *f.* tan liquor.

auvergner, to dip or soak in tan (liquor).

auvernat, *m.* auvernat (a red wine).

aux (à les), to the, at the.

auxanomètre, *m.* auxanometer.

auxdits (à les dits), to the aforesaid, to the above.

auxèse, *f.* auxesis, new formation of organs.

auxiliaire, auxiliary.

auxiliairement, in an auxiliary way.

auxomètre, *m.* auxometer.

auxosis, *m.* auxosis, variations in growth of organ.

auxospore, *m.* auxospore, a form of reproductive cell.

auxotonique, auxotonic.

auxquelles (à lesquelles), to whom, to which, at which.

auxquels (à lesquels), to whom, etc.

avait (avoir), (he) had.

aval, *m.* guaranty, endorsement.

aval, downstream, down; **eau d'—,** downstream water; **en —,** downstream; **en — de,** below.

avalanche, *f.* avalanche, landslip.

avalé, drooping.

avaler, to guarantee, swallow, lower, cut off (branch) close to trunk; **s'—,** to be swallowed, hang down, go downstream.

avalies, *f.pl.* pelt wool, skinner's wool.

avaloire, *f.* crupper.

avance, *f.* advance, start, projection, lead, feed; — **à découvert,** unsecured loan; — **par cylindres,** roller feed, **à l'—,** in advance; **d'—, par —,** in advance; **en —,** beforehand.

avancé, advanced, far ahead, spoiled, tainted (meat), forward, early, late.

avancement, *m.* advancement, promotion, advance, feed, forward motion, progress, protrusion.

avancer, to advance, hold, hasten, bring forward, put forward, bring nearer, move on, progress, be fast, gain (time); **s'—,** to advance, move, progress, approach.

avant. before, far, front, fore (part), head; — **de,** before; — **projet,** preliminary project; — **que,** — **que de,** before; — **tout,** above all; **en —,** on(ward), forward, ahead, in front, forth; **en — de,** in advance of, before, in front of.

avantage, *m.* advantage, odds, gain, profit.

avantageusement, advantageously.

avantageux, -euse, advantageous, becoming profitable, cheap.

avant-bras, *m.* forearm.

avant-creuset, *m.* forehearth.

avant-dernier, -ère, last but one, next to last.

avant-garde, *f.* advance guard.

avant-goût, *m.* foretaste.

avant-guerre, *m.* prewar period.

avant-hier, day before yesterday.

avant-main, *m.* flat of the hand.

avant-projet, *m.* preliminary project.

avant-propos, *m.* preface, forward, introduction.

avant-veille, *f* two days before.

avare, *m.&f.* miser.

avare, avaricious, miserly, greedy.

avarice, *f.* avarice, covetousness, stinginess.

avarie, *f.* damage, injury; *pl.* average.

avarier, to damage, spoil.

avec, with, among, along with, together, with it, them; **d'—,** from.

avelanède, *f.* acorn cup, valonia.

aveline, *f.* filbert.

avelinier, *m.* red filbert tree (*Corylus maxima*).

avénacé, avenaceous, relating to oats.

avenant, pleasing, comely; **à l'—,** in keeping, correspondingly; **mal —,** unseemly, uncouth.

avènement, *m.* coming, arrival, event, happening, accession.

avénière, *f.* oat field.

avenir, to happen, occur.

avenir, *m.* future; **à l'—,** in the future, hereafter.

aventure, *f.* adventure, chance, venture, luck, hazard; **à l'—,** at random; **d'—,** **par —,** by chance, perchance.

aventurer, to venture, chance, risk; **s'—,** to venture, expose oneself.

avenu, non —, canceled.

avenue, *f.* avenue.

avérer, to prove, aver, establish.

averse, *f.* shower.

aversion, *f.* aversion, antipathy, dislike.

avertir, to inform of, warn, advise, caution, give notice of.

avertissement, *m.* warning, notice, advertisement, publication, admonition, forward (to books).

avertisseur, *m.* warning signal, alarm, annunciator, call bell, (motor) horn.

avertisseur, warning, announcing.

avet, *m.*, **avette,** *f.* silver fir (*Abies picea*).

aveu, *m.* consent, admission, confession, acknowledgement, avowal.

aveuglant, blinding, glaring.

aveugle, *m.&f.* blind person.

aveugle, blind, deluded.

aveuglément, blindly, rashly.

aveugler, to blind, darken, dazzle, stop (up).

avez (avoir), (you) have.

aviation, *f.* aviation.

aviculteur, *m.* poultry keeper.

avide, avid, desirous, greedy, eager, covetous; **— d'eau,** having strong affinity for water, absorbent of water.

avidité, *f.* avidity, greed, eagerness, covetousness.

avient (avenir), (it) happens.

avilir, to depreciate, debase, lower, degrade; **s'—,** to lose value, come down (in price), depreciate.

avilissement, *m.* debasement, depreciation.

aviner, to season or soak with wine; **s'—,** to get drunk.

avion, *m.* airplane; **— sans moteur,** motorless plane, glider.

avionnette, *f.* small airplane.

aviron, *m.* oar.

avis, *m.* opinion, judgment, mind, decision, advice, counsel, notice, information, warning, intelligence, sentiment, preface, vote; **— autorisé,** official notice; **à mon —,** in my opinion, to my mind; **avoir — de,** to be informed of.

aviser, to advise, inform, consider, think upon, perceive, see (about); **— de,** to think of, see to; **s'— à,** to see to; **s'— de,** to think of, bethink oneself, take into one's head.

avitaminose, *f.* vitamin deficiency, avitaminosis.

avivage, *m.* brightening, touching up.

aviver, to brighten, heighten, polish, quicken, poke, stir, stoke, tin (a surface), put keen edge on (tools), touch up, irritate, refresh (wound), enliven; **s'—,** to brighten, quicken, revive, become sharper or more acute.

avocat, *m.* lawyer, advocate, barrister, avocado, alligator pear.

avocatier, *m.* alligator-pear tree.

avocette, *f.* avocet.

avoine, *f.* oat, oats, oat plant (*Avena sativa*); **— d'Orient,** side oats, tartarian oats (*Avena orientalis*); **— folle,** wild oat; **— jaunâtre,** yellow oat grass (*Trisetum pratense*); **— odorante,** sweet grass, holy grass, vanilla grass (*Savastana odorata*); **— rude,** meager oat, bristle-pointed oat; **— sauvage,** wild oat (*Avena fatua*).

avoinerie, *f.* oat field.

avoir, to have. get, have on, be the matter with; **— à,** to have to, be obliged to; **— affaire à,** to have dealings with, have to do with, be busy with; — **beau** (crier), (to shout) in vain; — **besoin de**, to be in need of, need; — **chaud**, to be warm; — **coutume de**, to be accustomed to; — **dix ans**, to be 10 years old; — **dû faire**, must have done.

avoir, — **faim**, to be hungry; — **froid**, to be cold; — **la haute main**, to have the greatest authority; — **lieu**, to take place, occur; — **quelque chose à faire**, to have something to do; — **raison**, to be right; — **soif**, to be thirsty; — **soin**, to be careful; — **tort**, to be wrong; — **vent de**, to get wind of; **y** —, to be; **en** — **à**, to have a grudge against.

avoir, **il y a**, there is, there are, ago; **il y a beau temps**, long ago; **il y a cinq ans**, five years ago; **il y a de quoi**, there is a reason; **il y a eu**, there has been, there have been; **il y a lieu de**, there is reason to; **il y a quelque chose**, something is the matter; **il n'y a pas de quoi**, never mind; **il n'y a qu'à** (écouter), one has only to (listen); **il y avait**, there was, there were.

avoir, *m*. property, possessions, credit.

avoisinant, neighboring, near-by.

avoisiner, to be adjacent to, border (upon); **s'**—, to approach, be adjacent, draw near.

avorté, retarded in its development.

avortement, *m*. abortion.

avorter, to miscarry, fail.

avorton, *m*. miscarriage, abortion.

avoué, *m*. attorney.

avoué, avowed, acknowledged.

avouer, to acknowledge, approve, confess, admit, grant.

avrelon, *m*. rowan tree, mountain ash (*Sorbus aucuparia*).

avril, *m*. April.

avrillet, *m*. spring wheat.

avron, *m*. wild oat grass (Danthonia).

axe, *m*. axis, axle, arbor, spindle, center line, pin, shaft; — **d'une fleur**, axis, rachis; — **floral**, axis; — **principal**, main stem; **grand** —, major axis; **petit** —, minor axis.

axial, axial.

axilé, having, (growing around) an axis.

axillaire, axillary.

axillante, axillant, subtending an angle.

axiomatique, axiomatic.

axiome, *m*. axiom.

axis, *m*. axis (of second vertebra).

axonge, *f*. lard, hog's fat.

axospermé, axospermous, with axile placentation.

axuel, **-elle**, axial.

ayant (avoir), having.

Az, *abbr*. (azote), nitrogen.

azalée, *f*. azalea.

azédarach, *m*. China tree, bead tree (*Melia azedarach*).

azimidé, azimido-, azimino-.

azimut, *m*. azimuth.

azine, *f*. azine.

azoamidé, **azoaminé**, aminoazo.

azobenzène, *m*. azobenzene.

azobenzoique, azobenzoic.

azoïde, *m*. azo compound.

azoïque, azoic.

azotate, *m*. nitrate; — **d'ammoniaque**, ammonium nitrate; — **d'argent**, silver nitrate, — **de fer liquide**, solution of ferric nitrate; — **de potasse**, — **de potassium**, potassium nitrate, niter, saltpeter; — **de soude**, sodium nitrate.

azotation, *f*. nitrogenation.

azote, *m*. nitrogen, azote; — **restant**, residual nitrogen.

azoté, nitrogenous, nitrogenized; **engrais** —**s**, nitrate fertilizers.

azoter, to nitrogenize, azotize.

azoteux, **-euse**, nitrous.

azotine, *f*. azotine, an explosive of the gunpowder type.

azotique, nitric, azotic.

azotite, *m*. nitrite.

azoture, *m*. nitride.

azoturie, *f*. azoturia.

azotyle, *m*. nitro group, nitryl, nitroxyl.

azoxybenzène, *m*. azoxybenzene.

azur, *m*. azure, (sky) blue, sky color; — **de cuivre**, verditer blue, blue carbonate of copper.

azurage, *m*. bluing.

azuré, azure, sky-colored.

azurer, to blue, tinge blue, azure.

azurin, pale blue.

azurite, *f*. blue carbonate of copper, azurite.

azyme, unleavened, azym.

azymique, azymic.

B

babeurre, *m.* buttermilk.

babillarde, *f.* whitethroat; — ordinaire, lesser whitethroat.

babiole, *f.* toy, bauble, trifle, curio.

bâbord, *m.* port side.

babouin, *m.* baboon, pimple (on lip).

bac, *m.* ferryboat, vat, tank, jar, pan; — à carbonatation, — à carbonater, carbonating vat, saturator; — d'attente, crystallizer; — décanteur, decanting tank; — refroidisseur, cooling tank.

baccharide, *f.* baccharis.

baccifère, berry-producing.

bacciforme, berrylike, bacciform, baccate.

baccivore, berry-eating, baccivorous.

bâche, *f.* tank, cistern, forcing frame, cover (of canvas), chamber, tarpaulin, case, casing, awning.

bachelier, *m.* bachelor (of a college).

bacile, *m.* samphire.

bacillaire, bacillary.

bacille, *m.* bacillus; porteur de —s, germ carrier.

bacille-virgule, *m.* comma bacillus.

bacillifère, bacillus-laden.

bacilliforme, bacilliform, rod-shaped.

bacillose, *f.* tuberculosis.

bactériacées, *f.pl.* bacterials, Bacteriaceae.

bactéricide, bactericidal.

bactérie, *f.* bacterium; *pl.* bacteria, microorganisms.

bactérien, -enne, bacterial.

bactériocécidie, *f.* bacterial tumor of plants.

bactériochlorine, *f.* pigment encountered in purple bacteria.

bactériologie, *f.* bacteriology.

bactériologique, bacteriological.

bactériologiquement, bacteriologically.

bactériologiste, *m.* bacteriologist.

bactériolyse, *f.* bacteriolysis.

bactériolysine, *f.* bacteriolysin.

bactériolytique, bacteriolytic.

bactériophage, bacteriophagous.

bactérioprécipitine, *f.* bacterial precipitin.

bacteriopurpurine, *f.* red pigment found in purple bacteria.

bactérioscopique, bacterioscopic.

bactériotropine, *f.* bacteriotropin.

bactérium, *m.* bacterium, microorganism.

bactériurie, *f.* bacteriuria.

bacteroïdes, *f.pl.* bacteroids.

bactrioles, *f.pl.* gold beatings.

badamier, *m.* Terminalia.

badiane, *f.* badianier, *m.* Chinese anise, star anise (*Illicium verum*).

badigeon, *m.* whitewash, wash (of any color), make-up, badigeon.

badigeonnage, *m.* coloring, painting, whitewashing.

badigeonner, to whitewash, paint (with iodine), plaster.

badiner, to jest, play, trifle.

bagace, *f.* bagasse, cane trash, husks.

bagage, *m.* baggage, luggage.

bagarre, *f.* hurly-burly, scuffle.

bagasse, *f.* bagasse, cane trash, fruit of the bagassier.

bagassier, *m.* an ulmaceous tree of Guiana.

bague, *f.* ring, band, collar, sleeve, collector; — d'appui, washer; — d'assemblage, collar, sleeve, thimble coupling; — d'écartement, expanding ring; — de garniture, piston (packing) ring, gasket; — de réglage, governor ring; — de serrage, clamping ring.

baguenaude, *f.* bladdernut, bladder senna pod.

baguenaudier, *m.* bladder senna (*Colutea arborescens*).

baguer, to ring (trees, pigeons).

baguette, *f.* rod, wand, ramrod, small stick, drumstick; — de nettoyage, cleaning rod; — d'étalement, glass-rod spreader for surface plates; —s ouvrées, carved moldings; en forme de —, virgate.

bai, bay (horse); — châtain, chestnut (horse).

baie, *f.* berry, bay, gulf, window; — de fenêtre, aperture of a window; — de genévrier, juniper berry; — de laurier, bayberry; — de porte, door bay; — de sureau, elderberry.

baignage, *m.* bathing, irrigation.

baigner, to bathe, steep, lave, soak, wash, float, swim, dip, be immersed.

baignoire, *f.* bath tub, box (in theater).

bail, *m.* lease; — **de chasse,** *m.* shooting lease; **donner à —,** to give on lease; **prendre à —,** to take on lease.

bâillement, *m.* yawning, yawn, fissure.

bâiller, to yawn, gape, open (in fissures) stand ajar.

bailleur, *m.* lessor.

bâillon, *m.* gag, muzzle.

bain, *m.* bath, bathing (place), (sheep) dip; — **à sensibiliser,** sensitizing bath; —**d'air,** air bath; — **d'argent,** silver bath; — **de Barèges,** sulphur bath; — **de cendres,** ash bath; — **de développement,** developing bath; — **de fixage,** fixing bath; — **de fonte,** bath of melted cast iron.

bain, — **d'électrolyse,** electrolytic bath; — **de mortier,** bed of mortar; — **de plomb,** lead bath; — **de sable,** sand bath; — **de sel,** salt bath; — **de soleil,** sun bath; — **de teinture,** dye bath, liquor; — **de trempe,** tempering (hardening) bath.

bain, — **de vapeur,** vapor (steam) bath; — **de Vénus,** wild teasel (*Dipsacus sylvestris*); — **de virage,** toning bath; — **d'huile,** oil bath, white bath; — **fixateur,** fixing bath; — **réducteur,** reducing (bath) solution; — **révélateur,** developing bath; **en —,** melted, fused; **mettre en —,** to melt, fuse.

bain-marie, *m.* water bath.

baïonnette, *f.* bayonet.

baiser, to kiss; **se —,** to touch, osculate.

baiser, *m.* kiss.

baisse, *f.* drop, decline, fall in value.

baisser, to lower, bend (head), crop, let down, turn down, fall, sink, decline, deteriorate; **se —,** to stoop, bow (bend) down, let down, lower, droop.

baisser, *m.* setting (of the sun), droop.

baissière, *f.* bottom (of a cask), depression, dip (where rain collects).

baissoir, *m.* brine tank or reservoir.

baladeur, *m.* sliding gear.

baladeuse, *f.* trailer.

balai, *m.* broom, brush; — **à laver,** mop; — **de communication,** commutator brush; — **de sorcière,** witches'-broom, hexenbesen; — **électrique,** dynamo brush; — **mécanique,** carpet sweeper.

balaie (balayer), (he, it) sweeps.

balance, *f.* balance, scales, annual statement; — **à court fléau,** short-beam balance; — **à ressort,** spring-balance valve; —

bascule, platform balance; — **de précision,** analytical balance; **être en —,** to be in suspense.

balancé, balanced, weighed, considered, swayed, rocked.

balancement, *m.* balancing, swinging, rocking, tremolo.

balancer, to balance, swing, rock, weigh, examine, clear, settle; **se —,** swing, rock, balance, counterbalance.

balancier, *m.* balance maker, beam, lever, pendulum, flyer, balance (wheel), press, gimbal, walking (cross) beam, balancing pole, equalizer; — **à vis,** fly (screw) press; — **compensateur,** equalizing spring; — **découpoir,** fly (cutting) press.

balancier, *m.pl.* balancers, poisers (of diptera).

balançoire, *f.* seesaw, swing.

balane, *m.* balanus, acorn shell.

balanifère, balaniferous, bearing acorns.

balanin, *m.* balaninus, nut weevil.

balanite, *f.* balanitis.

balanophage, acorn-eating.

balanophore, acorn-bearing.

balantidien, -enne, balantidial, appertaining to Balantidium.

balauste, *f.* balausta, fruit of pomegranate with firm rind.

balaustier, *m.* wild pomegranate (tree), balaustium.

balayage, balayement, *m.* sweeping.

balayer, to sweep, blow.

balayette, *f.* small broom.

balayures, *f.pl.* sweepings.

balbutiement, *m.* stammering, stuttering.

balbutier, to stammer, lisp.

balcon, *m.* balcony.

bale, bâle, balle, *f.* glume, husk, chaff.

Bâle, *f.* Basel, Basle.

Baléares, *pl.* Balearic Islands.

baleine, *f.* whale, whalebone, baleen, rib (of umbrella).

baleineau, *m.* whale calf.

baleinier, *m.* whaler.

baleinier, -ère, whaling.

balèvre, *f.* scale, scab, under lip.

baline, *f.* packing cloth, wrapping.

balise, *f.* canna seed, beacon; — **flottante,** buoy.

balisier, *m.* Indian shot (*Canna indica*).

balistique, *f.* ballistics.

balistique, ballistic.

balistite, *f.* ballistite.

balivage, *m.* marking standards, marking reserve trees.

baliveau, *m.* sapling, staddle; — de l'âge, tree of one rotation.

baliver, to mark standards in reserve, staddle.

balkanique, Balkan.

ballage, *m.* balling up (of iron).

ballant, swinging, dangling, loose, slack.

ballast, *m.* ballast.

ballastage, *m.* ballasting.

ballaster, to ballast.

ballastrière, *f.* gravel pit.

balle, *f.* ball, bale, pack, bullet, shot, glume, husk, chaff; — à éclairer, light ball; — à feu, fireball; — à fumée, smoke ball; — conique, conical bullet; — luisante, light ball; — multiple, fragmentation bullet.

baller, to ball up (iron), hull (corn).

ballon, *m.* spherical or round (-bottomed) flask, balloon, carboy, inflated ball, football; — à col court, short-necked balloon; — à distillation, — à distiller, distilling flask; — à fond plat, flat-bottomed flask; — à fond rond, round-bottomed flask; — à gaz, gas balloon.

ballon, — à long col, long-necked balloon, receiver, bolthead; — d'artifices, fire balloon; — de mesure, graduated flask; — de Pasteur, Pasteur flask; — enregistreur, — explorateur, sounding or registering balloon; — jaugé, graduated flask; — ordinaire, flask (usually of flat-bottomed type).

ballonné, distended, swollen.

ballonnement, *m.* swelling.

ballonner, to distend, inflate, swell, bulge.

ballonnet, *m.* ballonet, cell; — à gaz, gas bag.

ballospore, *f.* ascospore, produced by an ascus.

ballot, *m.* bundle, package, bale.

ballote, *f.* (Ballota) horehound.

ballotte, *f.* small ball, a wine vessel.

ballotter, to toss (about), shake, pack (in parcels and bales), ballot (for), rattle, wobble, tumble.

balnéaire, pertaining to baths.

balnéation, *f.* use of baths.

bâlois, of Basel.

balsamier, *m.* balsam (tree).

balsamifère, balsamiferous.

balsamine, *f.* touch-me-not (*Impatiens balsamina*); — des bois, yellow balsam, touch-me-not; — à petites fleurs, balsam; — sauvage, Impatiens.

balsamique, balsamic.

balustrade, *f.* railing.

bambou, *m.* bamboo.

bambusacé, bambusaceous, resembling bamboo.

banal, commonplace, vulgar; phrase —e, commonplace.

banaliser, to vulgarize, render banal or commonplace.

banane, *f.* banana.

bananier, *m.* banana (tree).

banc, *m.* bed, layer, stratum, bench, frame, band, ledge, seat, shoal, school (of fish); — de tourbe, peat layer, turf bed.

bancoulier, *m.* candlenut (tree) *Aleurites moluccana*); huile de —, candlenut oil.

bandage, *m.* tire, hoop, bandage, bandaging, belt; — caoutchouc, rubber tire; — creux, pneumatic tire; — de roue, tire; — plein, solid tire; — pneumatique, pneumatic tire.

bande, *f.* band, strip, strap, belt, bandage, tire, bar, compression (of spring), side, shore, gang, company, flock (of birds), school (of fish); — de frein, brake band; — sans fin, endless ribbon or band; — transporteuse, belt conveyor; par —s, in strips.

bandé, bandaged, taut.

bandeau, *m.* headband, bandage.

bandelette, *f.* narrow band or strip, small bandage, flat bar iron, coil; — agglutinative, adhesive strip.

bander, to bandage, bind up, compress, tighten, stretch, wind up (a spring), bend, cock (a gun).

banderole, *f.* banderole.

bandine, *f.* buckwheat flour, buckwheat.

bandit, *m.* bandit, highwayman.

bandoir, *m.* spring, mainspring.

bandoulière, *f.* shoulder belt.

bandure, *f.* pitcher plant (Sarracenia).

bang, bangh, *m.* bhang, Indian hemp.

banlieue, *f.* outskirts, suburbs.

banne, *f.* hamper, tarpaulin, awning, corf, bucket, tilt of a wagon.

banneau, m. fruit basket.

bannière, f. banner.

bannir, to banish, expel.

banque, f. bank, banking, exchange.

banqueroute, f. bankruptcy, insolvency.

banquette, f. bench, seat, banquette, ledge, berm, terrace.

banquier, m. banker.

baptiser, to baptize.

baquet, m. tub, bucket, trough.

baquetier, m. barrelmaker.

baraque, f. barrack.

barattage, m. churning (of milk).

baratte, f. churn.

baratte-malaxeur, m. a combined churn and kneader.

baratter, to churn.

Barbade, f. Barbados.

barbare, m. barbarian.

barbare, barbarous, rude, savage.

barbarée, f. winter cress (Barbarea).

barbaresque, of Barbary.

barbarie, f. barbarity, rudeness.

barbe, f. beard, rough edge, bur (on casting), deckle edge (of paper), barb, wattle, gill, mildew, awn, down, whiskers; — de plume, feather of a quill.

barbé, bearded, barbed.

barbeau, m. bluebottle, cornflower (Centaurea cyanus), barbel (a fish); bleu —, cornflower (blue).

barbe-de-bouc, f. goatsbeard (Tragopogon pratense).

barbe-de-capucin, f. chicory (Chicorium intybus).

barbe-de-Jupiter, f. red valerian (Centranthus ruber).

barbe-de-moine, f. dodder (Cuscuta).

barbelé, barbed, bearded.

barbelure, f. beard, awn.

barbet, m. poodle, large rough water dog.

barbiche, f. goatee, bishop's-wort, goatsbeard.

barbigère, barbigerous, bearded, hairy.

barbillon, m. barb, barbel, little barbel, wattle, lobe gill.

barbinervé, with hairy veins.

barbone, f. septic pleuropneumonia of cattle.

barbotage, m. bubbling, splash, dipping, (bran) mash.

barboter, to bubble, paddle, splash, dabble, mumble.

barboteur, m. bubbler, mixer, washer (with paddles), gas washer, stirrer, duck.

barboteur, -euse, bubbling, mixing.

barbotin, m. sprocket wheel, chain pulley.

barbotine, f. slip, slop, santonica, wormseed, (common) mugwort (Artemisia pauciflora).

barbotteur, m. bubbler, mixer.

barbouiller, to daub, smear, soil, mumble, blot, scribble.

barbu, bearded, barbate, moldy, mildewed, barbelate, barbed.

barbule, f. barbule.

bardane, f. bur, burdock (Arctium lappa); huile de —, burdock oil; petite —, burweed.

bardeau, m. shingle, lath.

bardelle, f. arm of support for glass.

baril, m. barrel, cask, keg, flask.

barille, f. barilla (a soda derived from plants).

barillet, m. small barrel or keg, hydraulic main (of gas works), condenser, drum, cylinder, exit tube.

bariolage, m. variegation, medley, motley.

bariolé, variegated, parti-colored, motley, gaudy.

barioler, to speckle, variegate.

baritel, m. winding (hoisting) machine, whim.

barium, m. barium.

barle, f. fault.

barlong, -gue, of unequal sides, lopsided.

baromètre, m. barometer; — à cadran, wheel barometer; — à cuvette, cistern (cup) barometer; — à siphon, siphon barometer; — enregistreur, recording barometer, barograph; — étalan, standard barometer.

barométrique, barometric(al).

barométrographe, m. barograph, recording barometer.

baron, m. baron.

baronne, f. baroness.

baroque, irregularly shaped, baroque, quaint, odd.

barque, f. bark, craft, (small) boat.

barquieux, m. lye tank for soap manufacture.

barrage, m. dam, barrier, embankment, weir, palisade, damming.

barras, m. barras, scrap resin, galipot.

barre, *f.* bar, bank, railing, bolt, dash, stripe, line, stroke; — **à T, T iron;** — **à vannes,** sluice dam or weir; — **d'appui,** handrail; — **de fourneau,** — **de foyer,** fire **bar, grate bar;** — **de jointure,** connecting rod; — **de retour,** negative bus bar; — **de tirage,** draft rod, draft bar; — **d'excentrique,** eccentric rod; — **directrice,** guide bar; — **témoin,** trial bar, tap bar.

barreau, *m.* (small) bar; — **aimanté,** bar magnet; — **de grille,** grate bar.

barrer, to bar (up), dam, obstruct, fence up, stop up, strike out, cross out.

barrette, *f.* (small) bar, rib, stay, crossbar; — **d'essai,** test (piece) bar.

barricader, to barricade, obstruct.

barrière, *f.* barrier, railing.

barrique, *f.* barrel, cask, hogshead.

barymorphose, *m.* barymorphosism, morphosis due to gravitation.

baryte, *f.* baryta (baryte); — **caustique,** barium oxide, caustic baryta; — **hydratée,** hydrated baryta, barium hydroxide.

barytine, *f.* barite.

barytique, barytic.

baryum, *m.* barium.

bas, *m.* bottom, lower part, foot, stocking; **à —,** down, low, off; **à — de,** out of, down from, off; **au — de,** at the bottom of; **de — en haut,** upward; **en —,** down, downward, below, downstairs; **en — de,** at the foot (bottom) of; **par —,** below, in the lower part; **par en —,** downward, at the bottom, from below.

bas, basse, low, shallow, mean, base, cheap, inferior, low(er), low (down), off; — **foyer,** low hearth; **basse pression,** low pressure; **basses terres,** lowlands; **mettre —,** to overthrow some one, bring forth, drop (young); **mise —,** dropping, throwing (young).

basal, basal, basic.

basalte, *m.* basalt.

basaltigène, growing on basaltic rock.

basaltique, basaltic.

basane, *f.* basil, sheepskin, sheep.

basané, sunburnt, tanned, swarthy.

bascule, *f.* apparatus with a seesaw motion, weighing scales, rocker, lever, plier, swing, tilting, balance; — **romaine,** platform (Roman) balance; **à —,** swinging, rocking; **balance à —,** weighing machine; **faire la —,** to seesaw, rock, overbalance; **mouvement de — rocking motion.**

basculer, to rock, rotate, swing, teeter, seesaw, tilt, dip, dump.

basculeur, *m.* rocker, dump, rocking lever.

base, *f.* base, basis, foundation, foot (of an abutment); — **antimoniée,** one of several substituted stibines; — **cadavérique,** cadaveric base, ptomaine; — **colorante,** basic stain; — **minérale,** underlying rock; — **nucléinique,** nuclein base.

baselle, *f.* Malabar nightshade (*Basella alba*), lowland, shallow.

baser, to base, ground, fix, found; **se —,** to be based, be fixed, be founded; **se — sur,** to be based on, take as a basis.

bas-fond, *m.* low ground, deep bottom, shoal, bottom land, depression, valley; — **de fleuve,** — **fluvial,** river valley.

bas-foyer, *m.* low hearth.

basicité, *f.* basicity.

baside, *m.* basidium.

basidie, *m.* sterigma.

basidiomycètes, *f.pl.* Basidiomycetes.

basidiospore, *m.* a spore produced by a basidium.

basification, *f.* basification, basifying.

basifixe, attached by the base.

basifuge, basifugal, from the base upward.

basigamie, basigamy, fertilization by the chalaza.

basigyne, *f.* gynophore, basigynium.

basilaire, basilar.

basile, *m.* basil.

basilé, elevated on a base.

basilic, *m.* basil, sweet basil (*Ocimum basilicum*), basilisk.

basilique, *f.* basil.

basilique, basilic (vein).

basinervé, basal-nerved.

basipète, basipetal.

basiplaste, *f.* basiplast, leaves with permanent tissue first at the apex.

basique, basic.

basisoluté, with nonadhering base.

bas-jointé, low-pasterned (horse).

basophile, basophilic, basophilous.

bas-perchis, *m.* small pole crop.

basse, *f.* shoal, bass (music).

basse, see **BAS.**

basse-cour, *f.* farmyard, poultry yard.

basse-courier, -ière, *m.&f.* farm hand.

basse-étoffe, *f.* base metal (esp. alloy of lead and tin).

bassement, *m.* plumping (hides).

basse-mer, *f.* low tide.

basserie, *f.* plumping (hides).

bassesse, *f.* baseness, lowness, mean action.

basset, *m.* badger dog, dachshund.

basse-tige, *f.* nursery-grown deciduous tree.

bassin, *m.* basin, pond, reservoir, (scale) pan, dock, pelvis, pelvic cavity; — de **filtration,** filtering basin, — de réception, collecting area, catchment basin, watershed, river basin; — **de repos,** settling basin, settling tank; — **d'une source,** head waters, water-collecting area; — **d'un fleuve,** drainage area; — **filtrant,** filtering basin, filter bed; — **houiller,** coal field, coal basin.

bassinage, *m.* bathing (wound), sprinkling (seedlings).

bassine, *f.* an open vessel, pan.

bassiner, to bathe, spray, sprinkle, moisten.

bassinet, *m.* (small) basin, cup, pelvis of the kidney, buttercup; — **d'or,** buttercup, tall crowfoot; petit —, figwort, lesser celandine.

bassorine, *f.* bassorin.

bassotin, *m.* indigo vat.

bastardocarpie, *m.* bastardocarpy, production of fruits by hybrids.

basting, *m.* plank, thick board, balk.

bastringue, *m.* black-ash furnace, spokeshave.

bas-ventre, *m.* lower part of the belly.

bat (battre), (he) beats.

bataille, *f.* battle, fight.

batailler, to fight, struggle, battle.

bataillon, *m.* battalion.

bâtard, bastard, mongrel, hybrid, illegitimate, spurious, crossbred; race —e, degenerate race.

batardeau, *m.* cross dike, spit, weir, caisson.

batate, *f.* sweet potato, batatas.

batavique, larme —, Rupert's drop.

bateau, *m.* boat, bark, barge; — à vapeur, steamer.

bateau-citerne, *m.* tank boat, tanker.

bateau-glisseur, *m.* hydroplane boat.

batée, *f.* pan, wash trough (gold mining).

bat-flanc, *m.* bar, barrier.

bathymétrique, bathymetrical.

bâti, built, basted.

bâti, *m.* frame(work), carcase, support, bed, stand, basting (thread).

bâtiment, *m.* building, structure, vessel, ship, boat; — à **voiles,** sailing vessel; — de **graduation,** graduation house, graduation tower; en —, upon a building.

bâtir, to build, erect, raise, baste; se —, to be built, build (for oneself).

bâtisse, *f.* construction, structure, masonry.

batiste, *f.* batiste, cambric.

bâton, *m.* stick, staff, rod, cane, club; — de **Jacob,** yellow asphodel (Asphodelus); — de **Saint-Jacques,** hollyhock (*Althaea rosea*); — **d'or,** wallflower (*Cheiranthus cheiri*); travailler à —s rompus, to work without method, by fits and starts.

bâtonnage, *m.* titillation of the gums of cattle, casting into stick form.

bâtonner, to beat (with a stick), cudgel, cross.

bâtonnet, *m.* small stick, peg, wooden pin, rod bacterium, rodlike cell; — en **verre,** glass rod.

batraciens, *m.pl.* Batrachia.

battage, *m.* beating, churning, threshing, pile driving.

battant, *m.* leaf (of table or door), clapper (of bell), lift (of latch), batten (of a loom).

battant, beating, swinging or folding (door), fighting, slamming, going, working.

batte, *f.* beater, beetle, rammer; — à **beurre,** dasher, plunger (of churn).

battée, *f.* amount (of wool) beaten at one time.

battement, *m.* beat(ing), stamp(ing), clapping, flapping.

batterie, *f.* battery, fight.

batterie-tampon, *f.* floating (balancing) battery.

batteur, *m.* beater; — **d'or,** goldbeater; — en **grange,** thrasher.

batteuse, *f.* beater, churn, threshing machine.

battitures, *f.pl.* iron (hammer) scale.

battoir, *m.* beater.

battre, to beat, strike, batter, thrash, thresh, drive (piles), coin (money), dash against, churn, mill, throb, fight, be loose (horseshoe), forge.

battu, beaten, heavy (eyes).

battue, *f.* battue, game drive, beat.

batture, *f.* gold lacquer; — de **roches,** reef.

baudet, *m.* (he-) ass, donkey.

baudruche, *f.* goldbeater's skin.

bauge, *f.* clay and straw mortar, lair, cover of a wild boar.

baume, *m.* balsam, balm, mint; — **calédonien,** a solution of kauri gum in an equal weight of 90 per cent alcohol; — **de Carthagène,** — **de Tolu,** balsam of Tolu; (*Myroxylon balsamum*); — **de cheval,** horse balm (*Collinsonia canadensis*); — **de copahu,** copaiba; — **de Giléad,** — **de la Mecque,** balm of Gilead, balsam of Mecca (*Commiphora meccanensis*); — **de Pérou,** — **de Perou** noir, — **des Indes,** — **de Sonsonate,** balsam of Peru (*Myroxylon pereirae*); — **du Canada,** Canada balsam; — **sauvage,** wild mint; — **vert,** spearmint (*Mentha spicata*).

baumier, *m.* balsam tree, balm tree; — **de Giléad,** balm of Gilead (*Commiphora meccanensis*).

baux, *m.pl.* see BAIL.

bauxite, *f.* bauxite.

bavard, talkative, loquacious, gossiping.

bave, *f.* slaver, slobber, slime, foam (of horse).

baver, to slaver, slobber, dribble (at mouth), run (of ink), foam (at mouth).

bavette, *f.* bib, drip flap.

baveux, -euse, slobbering.

bavoché, smeared, blurred, mackled (proof).

bavure, *f.* seam, rough edge (of mold), fin, bur, beard.

bayer, to gape.

bdellaire, suctorial.

bdellaires, *m.pl.* leeches.

bdellium, *m.* bdellium, gum resin.

béance, wide open (mouth), gaping (wound).

beau, *m.* best, beautiful, beau, fop.

beau, bel, belle, beautiful, handsome, fair, fine, lofty, noble, proper, becoming; **avoir** — **faire quelque chose,** to do something in vain, uselessly; **de plus belle,** louder (worse) than ever; **il fait** —, the weather is fine, fair.

beaucoup, much, a great (good) deal, many, far; **à** — **près,** by a great deal, by far, near; — **de,** much, (a great) many; **de** —, by far, much; **un peu** —, rather much, a good bit.

beau-fils, *m.* stepson, son-in-law.

beau-père, *m.* father-in-law.

beauté, *f.* beauty, comeliness.

beaux-arts, *m.pl.* fine arts.

bebirine, *f.* bebeerine, buxine.

bec, *m.* burner, beak, bill, neck (of retort), nose, nozzle spout, lip, stud, catch, arm, vise jaw, mouthpiece, tip; — **à gaz,** gas burner; — **à incandescence,** incandescent gas burner; — **à queue de poisson,** fishtail burner; — **Auer,** incandescent burner; — **croisé,** crossbill.

bec, — **d'âne,** crosscut chisel, cold chisel; — **de cigogne,** stork's-bill (Pelargonium); — **d'étain,** beak of tin; — **figue,** pied flycatcher; — **fin,** warbler; — **papillon,** butterfly burner; — **renversé,** inverted burner; **l'oiseau se fait le** —, the bird is sharpening its beak.

bec-à-cuiller, *m.* spoonbill.

bécasse, *f.* woodcock, silly person.

bécasseau, *m.* small woodcock, sandpiper, stint.

bécassine, *f.* common snipe; **petite** —, jacksnipe.

bec-de-cire, *m.* waxbill.

bec-de-grue, *m.* stork's-bill (Pelargonium); — **tacheté,** spotted crane's-bill (*Geranium maculatum*).

bec-dur, *m.* common grosbeak, hawfinch.

bêchage, *m.* digging.

bêche, *f.* spade; — **plantoir,** planting wedge, notching spade.

bêcher, to dig, spade.

becherglas, *m.* beaker.

béchique, cough-relieving.

becquée, *f.* beakful.

becqueter, to eat, feed, pick up (crumbs).

bedaine, *f.* stomach, belly, paunch.

bédane, *m.* mortise chisel, bolt.

bedegar, *m.* bedeguar, a fibrous gall.

bedonnant, barrel-bellied, pursy.

bédouin, Bedouin.

bégaiement, *m.* stammering.

bégayer, to stutter, stammer, lisp.

Beggiatoées, Beggiatoaceae.

bégonie, *f.* begonia (Begonia).

béguètement, *m.* bleating (goat).

beige, natural, raw (wool); **étoffe en** —, cloth in natural color.

bel, see BEAU.

bêler, to bleat.

belette, *f.* weasel.

belge, Belgian.

Belgique, *f.* Belgium.

bélier, *m.* ram, buck.

belladone, *f.* belladonna, deadly nightshade (*Atropa belladonna*).

belladonine, *f.* belladonnine.

belle, *f.* belle, beauty, fair one.

belle, see BEAU.

belle-de-jour, *f.* (*Convolvulus tricolor* or *C. minor*).

belle-de-nuit, *f.* four-o'clock, sedge warbler

bellide, resembling the daisy (Bellis).

belliqueux, -euse, warlike, bellicose, quarrelsome.

bellite, *f.* bellite.

bellon, *m.* lead colic.

beltête, *f.* wood betony.

belvédère, *m.* belvedere, sightly place.

ben, *m.* ben (nut), ben tree (*Moringa zeylanica*).

bénédiction, *f.* blessing, consecration.

bénéfice, *m.* benefit, profit, gain, advantage; **au — de,** for the benefit of; **part de —,** bonus.

bénéficiel, -elle, beneficial, advantageous.

bénéficier, to make a profit, take advantage (of), gain, profit by, work at a profit.

bénévole, benevolent, kindly; **auditeur —,** nonmatriculated person.

bengale, *m.* Bengal; **feu de —,** Bengal light.

béni, bénit, consecrated, blessed.

bénin, -igne, benign, kind(ly), mild.

bénir, to bless.

benjoin, *m.* benzoin.

benne, *f.* bucket, tub, hutch, basket, cage, car; **— prenante, — preneuse,** grab.

benoîte, *f.* avens, bennet; **— d'eau,** avens, bennet; **— des ruisseaux,** water avens.

benthos, *m.* benthos, sea bottom.

benzaldéhyde, *m.* benzoic aldehyde.

benzaldoxime, *m.* benzaldoxime.

benzène, *m.* benzene; **— diazimide,** triazobenzene.

benzène-sulfoné, benzenesulphonic.

benzénique, pertaining to benzene, aromatic.

benzényle, *m.* benzenyl.

benzhydrol, *m.* benzohydrol, benzhydrol.

benzidine, *f.* benzidine.

benzidinique, benzidine.

benzile, *m.* benzil.

benzine, *f.* benzene, benzol, benzine, turpentine, gasoline, petrol; **— de la houille,**

benzene (or benzol); **— légère,** light benzol; **— lourde,** heavy benzol; **— rectifiée,** oil of turpentine.

benzoate, *m.* benzoate.

benzobleu, *m.* benzo blue.

benzoïne, *f.* benzoin.

benzoïné, benzoinated.

benzoïque, benzoic.

benzol, *m.* benzol, benzene.

benzoloïde, *f.* benzoloid, scent from aromatic bodies.

benzonaphtol, *m.* benzonaphthol, B-naphthyl benzoate.

benzonitrile, *m.* benzonitrile.

benzopurpurine, *f.* benzopurpurin.

benzothiophène, *m.* thionaphthene, benzothiophene, benzothiofuran.

benzoyle, *m.* benzoyl.

benzylamine, *f.* benzylamine.

benzyle, *m.* benzyl.

béquettes, *f.pl.* pliers.

béquille, *f.* crutch, handle, prop, support, bar, rod, lever, tail skid, grubbing hoe, stilt.

berbérine, *f.* berberine.

berbéris, berbéride, *m.* **— commun,** barberry.

berce, *f.* cow parsnip (Heracleum).

berceau, *m.* cradle, vault, bower, arbor; **— de la vierge,** hedge clematis, traveler's-joy (*Clematis vitalba*).

bercement, *m.* rocking.

bercer, to rock, cradle, lull, delude.

béret, *m.* cap.

bergamote, bergamotte, *f.* bergamot (orange).

bergamotier, bergamottier, *m.* bergamot orange tree (*Citrus bergamia*).

berge, *f.* steep river bank, berm, setoff, rampart.

berger, *m.* shepherd.

bergerie, *f.* sheepfold.

bergeronnette, *f.* wagtail.

béribéri, *m.* beriberi.

berlue, *f.* dimness of vision, false vision.

berme, *f.* berm, bench, terrace.

béryl, béril, *m.* beryl; **— noble,** aquamarine.

béryllium, *m.* beryllium.

besaigre, turning sour, sourish, tartish.

besicles, *f.pl.* spectacles.

besogne, *f.* work, task, labor, job.

besogner, to work hard.

besoin, *m.* want, need, necessity, distress, passion, instinct, requirement; **au —**, at need, if necessary, on a pinch, if need be; **avoir — de,** to have occasion for, need, require; **— de force,** want of power, power consumption; **— de lumière,** light requirement, intolerance of shade; **est-il —?**, is it necessary?

bestial, -aux, bestial.

bestiaux, *m.pl.* cattle, livestock.

bestiole, *f.* little beast, beastie.

bétail, *m.* cattle, livestock.

bétaïne, *f.* betaine, a nonpoisonous crystalline base.

bête, *f.* beast, animal, aversion, beast of chase, blockhead; **— à feu,** glowworm; **— de compagnie,** sounder; **— de la Vierge,** ladyfly; **— de proie,** beast of prey; **— fauve,** red deer; **— noire,** wild boar; **— puante,** polecat; **— rousse,** young wild pig; **petite —,** insect.

bête, stupid, foolish.

bétel, *m.* betel (nut).

bêtise, *f.* stupidity, folly, blunder.

bétoine, *f.* betony (Stachys).

bétol, naphthalol.

béton, *m.* concrete, (rarely) colostrum; **— aggloméré,** ordinary cement concrete; **— armé,** reinforced concrete, ferroconcrete; **— gras,** rich concrete; **— maigre,** poor concrete.

bétonie, *f.* wood betony.

bétonnage, *m.* concrete work.

bétonner, to construct, concrete.

bétonneur, *m.* concrete worker.

bétonnière, *f.* concrete mixer.

bette, *f.* beet, Swiss chard (*Beta cicla*), mangold, mangel-wurzel (*Beta vulgaris*); **— poirée,** white beet.

betterave, *f.* beet(root); **— à sucre,** sugar beet; **— blanche,** sugar beet; **— fourragère,** forage beet, mangel-wurzel (*Beta vulgaris*); **— potagère,** beet for cooking; **— rouge,** red beet; **— sucrière,** sugar beet; **sucre de —,** beet sugar.

beurre, *m.* butter; **— de cacao,** cacao (cocoa) butter; **— de coco,** coconut butter, coconut oil; **— de muscade,** nutmeg butter; **— de vache,** cow butter, ordinary butter; **— frais,** fresh butter, unsalted butter.

beurrerie, *f.* creamery, churn room.

beurrier, -ère, butter-producing.

bévau, *m.* miter rule.

bevilacque, *m.* Indian pennywort (*Centella asiatica*).

bévue, *f.* blunder.

bézoard, *m.* bezoar.

biacétyle, *m.* biacetyl.

biacide, biacid, diacid.

biacuminé, biacuminate, two diverging points.

biailé, with two wings.

biaiométamorphose, *m.* biaiometamorphosis, a disadvantageous change.

biaiomorphose, *m.* biaiomorphosis, form produced under normal influences.

biais, *m.* skew, slant, bias, obliquity, expedient.

biais, skew, oblique, slanting; **de —, en —,** sloping obliquely, slantwise, askew, on the bias.

biaiser, to slant, slope (off), lean.

bianthérifère, having two anthers.

biapiculé, split or grooved at top.

biaristé, ended in two silky extensions.

biatomique, diatomic.

biauriculé, biauriculate, with two auricles.

bibasique, bibasic, dibasic.

bibassier, bibacier, *m.* loquat, Japanese medlar (*Eriobotrya japonica*).

bibelot, *m.* small ornament, knickknack.

biberon, *m.* sucking bottle, feeding cup.

bibion, *m.* bibio.

bibliographe, *m.* bibliographer.

bibliographie, *f.* bibliography.

bibliographique, bibliographical.

bibliothécaire, *m.* librarian.

bibliothèque, *f.* library, reading room, bookcase.

bibromé, dibrom.

bibromer, to insert two bromine atoms.

bicalcique, dicalcium, dicalcic.

bicarbonate, *m.* bicarbonate; **— de potasse,** potassium bicarbonate; **— de soude,** sodium bicarbonate.

bicarboné, bicarbureted, carbureted; **hydrogène —,** *m.* marsh gas.

bicarbure, *m.* bicarbide, dicarbide.

bicariné, bicarinate, with two keels.

bicarré, biquadratic, raised to the fourth power.

biche, *f.* hind, doe; **petite —,** little horn bug.

bichlorure, *m.* dichloride, bichloride; — de mercure, mercuric chloride.

bichromate, *m.* dichromate, bichromate; — de potasse, potassium dichromate; — de soude, sodium dichromate.

bicilié, biciliate, with two cilia.

bicipité, bicipital.

bicollatéral, bicollateral.

biconcave, biconcave, double concave.

biconjugué, biconjugate.

biconvexe, biconvex, double convex.

bicorne, two-horned.

bicotylédones, *f.pl.* dicotyledons.

bicuspidé, bicuspid.

bicyclette, *f.* bicycle.

bident, *f.* bur marigold (Bidens).

bident, *m.* pitchfork — **partagé,** double tooth, bidens.

bidenté, bidentate, double-toothed, bi-serrate.

bidet, *m.* pony.

bidigité, bidigitate, with petiole terminating in two folioles.

bidon, *m.* can, drum, tub; — à essence, petrol tin, can; — à huile, oil can.

biductuleux, -euse, border (limb) of leaf with two veins.

bief, *m.* millrace, reach, level; — d'aval, tailrace, tail bay.

bielle, *f.* rod, link, connecting rod, strut; — d'accouplement, track link, coupling rod; — de réglage, governor rod; — de tiroir, (slide) valve rod; — motrice, main rod, driving rod; tête de —, crank head.

biellette, *f.* link.

bien, *m.* good (thing), blessing, property, possession, welfare; *m.pl.* goods, lands; —s immeubles, real estate, real property; —s meubles, personal property, movables.

bien, well, very, (very) much, really, quite, entirely, indeed, right, many, rather, proper; — d'autres, many others; — de, much, many; — entendu, of course, well understood, certainly; — entendu que, provided that; — peu, very little; — plus, more than that, what is more; — que, although, though; — sûr, most certainly.

bien, à —, successful(ly); **aussi** —, in any case, after all; **aussi** — que, as well as, besides; **eh** —, well! **en** —, for the better, favorably; **ou** —, or else, either; **ou** — . . . **ou** —, either . . . or; **si** — que, so that, and so, however, with the result that;

tant — que mal, somehow (or other); un peu —, somewhat.

bien-être, *m.* welfare, comfort.

bienfaisant, beneficent, benevolent, kind(ly), helpful.

bienfait, *m.* benefit, kindness, favor, advantage, utility.

bienfaiteur, -trice, beneficent.

bien-fonds, *m.* estate, landed property.

biennal, -aux, biennial.

bienséance, decency, propriety.

bientôt, soon, shortly; à —, I'll see you soon.

bienveillance, *f.* benevolence.

bienveillant, kind, benevolent.

bienvenant, thrifty, flourishing, fast-growing.

bienvenu, welcome.

biéperonné, provided with two spurs.

biépillé, with two ears or spikes.

bière, *f.* beer, bier, coffin; — de Brunswick, mum; — de gingembre, ginger beer; — de malt, malt liquor.

biérémé, bieremus, a two-celled fruit.

bièvre, *m.* beaver.

bifactorielle, *f.* hybridization of dihybrid nature.

bifarié, bifarious, distichous.

biffage, *m.* cancellation.

biffer, to erase, cross out, cancel.

bifide, bifid, two-parted, bifidous.

bifidité, *m.* bifidity, twice-cleft.

biflore, biflorate, having two flowers.

bifolié, bifoliate.

bifoliolée, bifoliolate, having two leaflets.

bifollicule, *m.* bifolliculus, a double follicle.

biforé, biforate, with two perforations.

bifurcateur, *m.* dividing piece, forked tube.

bifurcation, *f.* bifurcation, forking in two's, branching, junction, crossover, dichotomy.

bifurqué, bifurcate(d), forked, branched, dichotomous.

bifurquer (se) to be bifurcated, be forked, fork.

bigarade, *f.* bitter orange.

bigarré, mottled, parti-colored, variegated.

bigarreau, *m.* bigaroon.

bigarreautier, *m.* sweet cherry (tree), bigaroon.

bigemmé, with two buds or shoots.

bigène, growing twice a year.

bigénère, bigener, a hybrid between two species.

biglanduleux, -euse, biglandular, with two glands.

biglobuleux, -euse, having aspect of two spheres.

bigorne, *f.* beakiron.

biiodure, *m.* diiodide, biiodide, biniodide; — **de mercure,** mercuric iodide.

bijon, *m.* pine resin.

bijou, *m.* jewel, gem.

bijouterie, *f.* jewelry, jeweler's trade or shop.

bijugué, grouped in pairs.

bijumeau, muscles bijumeaux, biceps muscles.

bilabié, two-lipped, bilabiate.

bilamelle, bilamellar, bilamellate(d).

bilan, *m.* balance sheet; — **des eaux,** water distribution; **dresser un —,** to strike a balance.

bilatéral, bifacial, isolateral, arranged on opposite sides.

bile, *f.* bile, gall; — **de boeuf,** oxgall, ox bile; — **répandue,** jaundice.

biliaire, biliary, bile; **calculs —s,** bilestones; **vesicule —,** gall bladder.

bilieux, -euse, bilious, choleric.

bilinéaire, bilinear.

billard, *m.* billiards, billiard table or room.

billarder, to dish.

bille, *f.* small ball, glass ball, billiard ball, marble, sleeper, log, saw block, butt, block, billet, sucker, tie, shoot, cutting, ball bearing; — **de bois,** log of wood; — **de chemin de fer,** railway sleeper.

billet, *m.* note, bill, ticket, permit, certificate; — **de banque,** bank note; — **de santé,** certificate (bill) of health; — **simple,** single (one way) ticket, promissory note.

billion, *m.* one thousand million.

billon, *m.* ridge of earth, balk (of squared timber), land, copper (debased) coin.

billonnage, *m.* plowing in lands.

billonner, to castrate.

billot, *m.* block, short butt; — **de batte,** rammer log; — **de sciage,** saw log.

bilobé, bilobed.

biloculaire, bilocular, two-celled.

bilomentum, *m.* bilomentum, a double lomentum.

biloupe, *f.* a magnifying glass (with two lenses).

bimane, bimane, two-handed.

bimanes, *m.pl.* Bimana.

bimétallique, bimetallic, containing two metals, dimetallic.

binage, *m.* weeding, second tilth, hoeing.

binaire, binary, double.

bine, *f.* hoe.

biner, to weed, hoe.

bineuse, *f.* mechanical hoe.

binitré, *m.* dinitro compound.

binitré, dinitrated, dinitro.

binocle, *m.* eyeglasses, field glass.

binoculaire, binocular, of both eyes.

binoter, to prepare, dress (soil).

bioblaste, *m.* bioblast, a cell.

biochimie, *f.* biochemistry.

biochimique, biochemical.

biochimiste, *m.* biochemist.

biocyte vivante, *f.* parenchyma found between the elements of the protoxylem.

biogène, *f.* biogen (living protoplasm).

biogénèse, *f.* biogenesis, biogeny.

biographe, *m.* biographer, biograph.

biographie, *f.* biography.

biologie, *f.* biology.

biologique, biologic(al).

biologiste, biologue, *m.* biologist.

biomécanique, biokinetic.

biométrie, *f.* biometry.

biomorphose, *f.* morphosis.

bion, *m.* sucker.

bionner, to plant the suckers (of artichoke).

bionomie, *f.* bionomics, physiology.

biophore, *m.* biophore.

bioplasme, *m.* bioplasma, bioplast.

biotique, biotic(al).

biotonus, *m.* ratio of assimilation to disassimilation.

biotype, *m.* biotype, an elementary stable form.

biovulé, with two ovules.

bioxyde, *m.* dioxide; — **d'azote,** nitrogen dioxide; — **d'étain,** tin dioxide, stannic oxide; — **d'hydrogène,** hydrogen peroxide.

bipaléole, bipaleolate, having two paleae.

bipalmée, bipalmate.

bipare, biparous.

biparietal, -aux, biparietal.

bipartit, bipartite.

bipartition, *f.* segmentation.

bipède, biped, two-footed.

bipennatifide, bipinnatifid.

bipenné, bipennate(d).

biperforé, with two openings.

bipétalé, bipetalous.

biphasé, two-phase.

bipinnatiséqué, bipinnatisect(ed).

bipliqué, biplicate, twice-folded.

bipolaire, *f.* bipolar.

bipolarité, *m.* bipolarity, with two polar processes.

biprisme, *m.* biprism.

biréfringent, birefringent, having a double refraction.

bis, grayish brown, brown, swarthy, twice, a second time, encore.

bisaccharide, *m.* disaccharide, bisaccharide·

bisage, *m.* second dyeing.

bisaille, *f.* coarse flour, mixed peas and vetches.

bis-ancien, *m.* first-class standard, veteran.

bisannuel, -elle, biennial, semiperennial.

bisazoique, *m.* diazo compound or dye.

biscuit, *m.* biscuit.

biscuit, twice-baked, twice-burned.

bisé, re-dyed, dyed a second time.

biseau, *m.* bevel, feather edge.

biseautage, *m.* beveling.

beseauter, to bevel.

bisegmentation, *f.* bisection.

bisel, *m.* dibasic salt.

biser, to redye, dye over again, darken, deteriorate (grain).

bisérié, biserial, biseriate.

biset, *m.* carrier pigeon.

bisexué, bisexuel, -elle, see BISSEXUÉ.

bisillonné, with two furrows.

bismuth, *m.* bismuth.

bismuthifère, bismuthiferous.

bismuthisme, *m.* bismuth poisoning.

bismuthine, *f.* bismuthine, bismuthinite.

bismuthique, bismuthic.

bisoc, *m.* two-furrow plow.

bissecter, to bisect.

bissecteur, -trice, bisecting (line).

bissection, *f.* bisection.

bissexué, bissexuel, -elle, bisexual.

bissoc, *m.* double-furrow plow.

bistélique, bistelic, having two steles.

bistorte, *f.* bistort, snakeweed (*Polygonum bistorta*).

bistortier, bistotier, *m.* a wooden pestle.

bistouri, *m.* bistoury, lancet.

bistre, *m.* bister.

bistrer, to color with bister, brown.

bistrié, bistriate.

bisulfate, *m.* bisulphate.

bisulfitage, *m.* operation of adding bisulphite.

bisulfite, *m.* bisulphite, acid sulphite; — de soude, sodium bisulphite.

bisulfitique, bisulphitic, bisulphite.

bisulfure, *m.* disulphide, bisulphide.

bisulque, bisulcate(d), cloven-footed. bisulcous.

bisymmetric, bisymmetric(al).

bitartrate, *m.* bitartrate, acid tartrate; — de potasse, potassium bitartrate.

bitemps, *m.* two-cycle engine.

biterné, biternate.

bitter, *m.* bitters, appetizer.

bitumage, *m.* asphalting, covering with bitumen.

bitume, *m.* bitumen, asphalt; — de Judée Jew's pitch, bitumen of Judea, asphalt; — liquide, naphtha, rock or mineral oil; — solide, asphalt.

bituminer, bitumer, to bituminize.

bitumineux, -euse, bitumeux, -euse, bituminous, asphaltic.

bituminifère, bituminiferous.

bituminisation, *f.* bituminization.

bituminiser, to bituminize.

biuret, *m.* biuret.

bivalence, *f.* bivalence.

bivalve, bivalved, bivalvular.

bizarre, fantastic(al), odd, whimsical, strange, queer.

blafard, dull, pallid, dim, pale, wan.

blaireau, *m.* badger, brock, brush (of badger's hair); — puant, Indian ratel.

blâme, *m.* blame, reproach.

blâmer, to blame, censure, find fault with.

blanc, *m.* white, blank (space), margin, white heat, incandescence, whiteness; — d'argent, white lead; — d'eau, water lily; — de baleine, spermaceti; — de baryte, baryta white; — de bismuth, bismuth white; — de céruse, white lead; — de Champagne, whiting, whitening; — de chapon, breast; — de chaux, whitewash, coat of whitewash.

blanc, — de fard, pearl white (basic bismuth nitrates); — de feu, candent; — de lait, milky white; — de l'aune, powdery mildew of alder; — de Meudon, a form of whiting; — de neige, snow white (zinc white); — de plomb, white lead; — de poulet, breast (of chicken); — des groseilles, gooseberry mildew; — d'Espagne, chalk, whiting, whitening.

blanc, — de Troyes, a form of whiting; — de zinc, zinc white; — d'oeuf, egg white, white of egg; — du pommier, powdery mildew of apple; — fixe, blanc fixe, barium sulphate; — mat, dull white; — salé, dirty white, grayish or yellowish white; — soudant, welding heat, white heat; — vierge, spawn; en —, in white, unpainted, in blank.

blanc, blanche, white, clean, pure, blank.

blanc-cul, m. bullfinch.

blanchâtre, whitish (color).

blanche, see BLANC.

blanchet, m. cloth filter, strainer, white wool cloth, (variety of) agaric.

blancheur, f. whiteness.

blanchi, whitened.

blanchiment, m. whitening, refining, bleaching, scalding; — chimique, chemical bleaching; — naturel, — sur (au) pré, grassing, grass bleaching.

blanchir, to whiten, bleach, blanch, wash, whitewash, limewash, clean up, polish (surface), refine (pig iron), scald, parboil, blaze (tree), turn white, pale, fade; — à la chaux, to whitewash; — au pré, to grass; — un manuscrit, to revise a manuscript.

blanchis, m. blaze; marquer d'un —, to blaze.

blanchissage, m. washing, laundering, whitewashing, refining (of sugar).

blanchissant, whitening, foaming.

blanchisserie, f. laundering, laundry, bleachery, bleaching trade.

blanchisseur, -euse, m. bleacher, launderer.

blanchisseur, bleaching, laundering.

blanchoyer, to begin to whiten, go white-hot.

blanc-soudant, m. white heat, welding heat.

blanquette, f. blanquette, white grape, fig, pear, stew of white meat.

blaste, m. blastus, plumule.

blastématique, blastemal, rudimentary.

blastesis, m. blastesis, microconidium.

blastocarpe, m. blastocarpous, germinating in the pericarp.

blastocèle, f. blastocoele, chalaza, base of the nucellus of an ovule from which the integuments arise.

blastocolle, f. blastocolla, balsamic varnish on buds.

blastomycose, f. blastomycosis.

blastophore, m. blastophore, embryonic origin of plumule.

blattaires, m.pl. Blattidae.

blatte, f. cockroach.

blé, m. wheat, grain, cereal, corn; — barbu, bearded wheat; — cornu, — ergoté, ergot (Claviceps purpurea); — de printemps, spring wheat; — des sables, lyme grass; — de Turquie, — d'Inde, Indian corn, maize (Zea mays); — d'hiver, winter wheat; — dur, Algerian hard-grained wheat.

blé, — égrugé, coarsely ground unbolted corn; — en herbe, corn in the blade; — froment, wheat; — méteil, wheat mixed with rye; — niellé, purples, cockles, false nematode gall of cereals, peppercorn, earcockle; — noir, buckwheat; — renflé, duckbill wheat; — seigle, rye.

bleime, f. corn.

blême, pale, wan, dun, lurid.

blende, f. blende, zinc blende, sulphide of zinc, sphalerite.

blendeux, -euse, blendous, of or containing blende.

blennorrhagie, f. blennorrhea.

blennorragique, blennorrhagique, gonorrheal.

blépharoplaste, f. blepharoplast.

blessé, wounded injured.

blesser, to wound, injure, hurt, offend.

blessure, f. wound, hurt, injury; — à la cuisse, hit in the haunches; — aux entrailles, body hit; — à l'épine dorsale, hit in the spine; — par décortication, injury caused by peeling.

blet, -ette, bletted, overripe, half-decayed, mellow.

blète, blette, f. blite, scale.

blettir, to become overripe, blet.

blétissement, m. bletting, beginning of decay on certain fruits.

bleu, m. blue, blueprint; — azoïque, azo blue; — azur, — céleste, azure, sky blue; — clair, light blue; — Coupier, Coupier's blue, induline; — **d'acier,** steel blue; — **d'azur,** smalt; — **de Berlin,** Prussian (Berlin) blue; — **de ciel,** sky blue; — **de cobalt,** cobalt blue; — **de gallamine,** gallamine blue.

bleu, — **de Kuhne,** carbol methylene blue; — **de Löffler,** Löffler's alkaline methylene blue; — **de Lyon,** ' Bleu de Lyon"; — **d'émail,** smalt; — **de méthylène,** methylene blue; — **de méthylène phénique,** carbol methylene blue; — **de naphtol,** naphthol blue; — **de nuit,** night blue; — **de phénylène,** phenylene blue; — **de Prusse,** Prussian blue; — **de résorcine,** resorcin blue.

bleu, — **de Saxe,** Saxon blue, Saxony blue; — **de toluidine,** toluidine blue; — **de toluidine phénique,** carbol toluidine blue; — **de toluylène,** toluylene blue; — **de Unna,** Unna's alkaline methylene blue; — **d'indigo,** indigo blue; — **foncé,** dark blue; — **minéral,** mineral blue; — **Nil,** Nile blue; — **outremer,** ultramarine blue, ultramarine; — **patenté,** patent blue; — **sombre.** dark blue.

bleu, blue, cerulean.

bleuâtre, bluish, livid.

bleuet, *m.* kingfisher.

bleuîr, to blue, make blue, become blue.

bleuissage, *m.* bluing.

bleuissant, turning blue.

bleuissement, *m.* turning blue, bluing, blue rot, blue stain, blue timber.

bleuté, bluish, tinged with blue.

bleuter, to blue slightly, tinge with blue.

bleu-violet, blue-violet.

blindage, *m.* protection, armor (plating), casing, blindage, lining, timbering.

blinder, to protect, armor, case, blind.

bloc, *m.* block, lump, log, bulk, boulder, ball, unit, butt; — **coulissant,** crosshead or sliding block; — **de bois,** log, plank, timber; — **de plâtre,** gypsum block (for yeast); — **de sciage,** saw log; **en —,** in a lump, in one piece.

blocage, *m.* blocking, stopping.

blocaille, *m.* ballast; — **d'empierrement,** ballasting or boxing material.

bloc-film, *m.* film pack.

blochet, *m.* dragon beam or piece; — **d'arêtier,** — **de recrue,** jack rafter.

blockhaus, *m.* log hut, log cabin.

bloc-moteur, *m.* transmission gear case.

blocus, *m.* blockade, embargo.

blond, *m.* blond(e), pale color, light color; — **ardent,** auburn.

blond, fair, flaxen, blond, light(colored), beige.

blondeur, *f.* fairness, light color, golden hue (of corn).

blondir, to turn light yellow, golden, bleach.

blondissant, yellowing.

blondoyer, to have a yellow reflection, gleam yellow.

bloom, *m.* bloom.

bloquer, to stop, block, lock, clamp, blockade, fill up (of cavities), arrest.

blossissement, *m.* bletting, decay.

blottir se, to squat, cower, hide.

blouse, *f.* blouse, frock, noil, comber waste.

bluet, *m.* cornflower, bluebottle, bachelor's-button (*Centaurea cyanus*).

bluette, *f.* spark, sparkle.

bluetter, to spark, sparkle.

blutage, *m.* bolting.

bluter, to bolt, sift.

bluterie, *f.* bolting mill, bolting house.

blutoir, *m.* bolter, bolting machine, bolting cloth, duster.

boa, *m.* boa.

boarmie, *f.* boarmia (moth); — **du chêne,** great oak beauty.

bobèche, *f.* socket, sconce, steel edge.

bobinage, *m.* winding, spooling.

bobine, *f.* bobbin, spool, coil, reel; — **de dérivation,** shunt coil; — **de réaction,** — **de réactance,** choking coil, reactance coil; — **de résistance,** resistance coil; — **d'induction,** — **inductrice,** induction coil; — **primaire,** primary coil; — **secondaire,** secondary coil.

bobiner, to wind, reel.

bobinet en bois, *m.* wooden spool or bobbin.

bobineur, *m.* winder.

bocage, *m.* scrap, grove, copse, coppice.

bocal, *m.* (wide-mouthed, short-necked) bottle, show bottle, vase, globe, jar, large vial (phial), glass bowl, mouthpiece.

bocard, *m.* stamp (mill), stamping mill, crusher.

bocardage, *m.* crushing, stamping.

bocarder, to crush, stamp.

bocaux, *m.pl.* see BOCAL.

boeuf, *m.* ox, bull; *pl.* cattle, beef; — **salé,** salt (corned) beef.

boghead, *m.* Boghead coal, Boghead.

bogie, boggie, *m.* truck.

bogue, *f.* chestnut bur.

boguette, *f.* buckwheat.

bohé, *m.* bohea (tea).

boire, to drink, absorb, imbibe, drink up, suck in, soak up, blot (paper).

boire, *m.* drink, drinking, channel; **faire —,** to soak (skins, etc.).

bois, *m.* wood, timber, forest, horns, antlers, head of stag; **— à aubier,** alburnum tree; **— abattu,** felled timber; **— à éclisses, — d'éclisses,** wood chipped off in squaring timber, wood for shingles; **— alterne,** protoxylem; **— amer,** bitterwood, quassia, stavewood; **— à ouvrer,** timber, lumber; **— blanc,** soft wood, white deal, spruce wood.

bois, — bombé, compass timber, curved (crooked) wood; **— chablé, — chablis,** rolled timber, windfall; **— colorant,** dyewood, sumac, **— contreplaqué,** plywood; **— coudé,** knee timber; **— courbant,** curved wood, compass timber; *pl.* figured wood, knees; **— courbé,** kneewood, compass timber, knee timber; **— couvert,** unbarked oakwood; **— d'absinthe,** quassia(-wood).

bois, — d'acajou, mahogany; **— d'affouage,** firewood for rightholders; **— d'aigle,** eaglewood, agalloch; **— d'arcole et de bourrellerie, — d'atteiles,** wood for horse collars and bullock yokes; **— d'aubier,** sapwood; **— d'automne, — d'été,** summer wood; **— d'ébénisterie,** cabinet-makers' wood, cabinet wood; **— de boissellerie,** hoop wood, sieve wood.

bois, — de bout, cross-grained timber, crosscut wood; **— de Brésil,** brazilwood; **— de brin,** timber in the round, unhewn timber; **— de broussin,** curled wood, speckled wood; **— de buis,** flowering dogwood; **— de campêche,** logwood, campeachy wood; **— de cerf,** horns, antlers (of deer); **— de charpente,** timber, lumber; **— de charronage,** wheelwrights' wood; **— de chauffage,** firewood.

bois, — de chevron(s), plywood, quarter(s); **— de chien,** flowering dogwood; **— de cintrage,** wood bent by steam; **— de construction,** timber, lumber; **— de constructions civiles,** building timber; **— de constructions navales,** ship timber; **— de corde,** cordwood; **— de daim,** horns, antlers (of deer); **— de déchet,** offal timber, timber refuse.

bois, — déchirage, old timber; **— de délit,** stolen wood; **— de démolition,** old timber; **— de fascinage,** fascine wood, brushwood;

— de fente, cloven wood; **— de fer blanc,** ironwood (*Sideroxylon cinereum*); **— de fil,** wood cut with the grain, wood cloven lengthwise, long log; **— de frêne,** ash wood, ash; **— de gaïac,** guaiacum wood, pock wood, lignum vitae.

bois, — de haute futaie, forest timber, timber forest, tall or lofty wood; **— déjeté,** warped wood; **— de lambrissage,** wainscoting; **— de mai,** hawthorn; **— de maronge,** building timber for rightholders; **— de mine,** pit props, mining timber; **— de moule,** cordwood; **— de paille,** straw board; **— de Panama,** soapbark, quillaja bark, Panama bark.

bois, — d'épicea, spruce wood; **— de placage,** veneer; **— de printemps,** spring wood; **— de quartier,** split firewood; **— de quassie,** quassia (wood); **— de raclerie,** carved wood; **— de rebut,** refuse wood; **— de refend,** sawn or split timber, timber split twice, cloven wood; **— de réglisse,** licorice root; **— de rose,** rosewood (the cabinet wood), rhodium wood.

bois, — déroulé, peeled veneer; **— de sapan,** sapan wood (a red dyewood); **— de sciage,** sawn timber; **— de service,** timber, lumber, building timber; **— desséché,** dry wood, seasoned wood; **— de Ste-Lucie,** Rockcherry wood; **— de teinture,** dyewood; **— de tour, — de tourneur,** turner's wood; **— de travail,** timber, lumber, **— d'oeuvre,** timber, lumber; **— douvain,** cask or stave wood.

bois, — droit, straight-grained wood; **— du centre,** heartwood; **— d'un beau brin,** straight timber; **— dur,** hard wood, hardwood; **— durci,** artificial wood, wood pulp; **— échauffé,** wood beginning to rot, dry-rotten wood; **— écorcé,** barked wood, disbarked wood; **— en grume,** wood in round, unhewn timber; **— en retour,** decaying wood; **— équarri,** hewed wood, scantlings.

bois, — factice, xylolite, xylogen; **— fendable,** cleavable wood; **— feuillard,** hoopwood; **— feuillu,** broad-leaved wood; **— fin,** fine wood, close-grained wood; **— flacheux,** wany wood, dull-edged timber; **— franc,** knee holly; **— gâté,** punky wood, rotten wood; **— gauchi,** warped wood; **— gélif,** frost-cracked wood; **— gentil,** mezereum, spurge olive; **— gras,** soft wood, highly resinous wood, torchwood, fir or pine wood.

bois, — gris, unbarked oak wood; **— imparfait,** sapwood; **— madré,** figured wood, curled wood; **— maillé,** silver-grained wood; **— malandreux,** knaggy wood, knotty wood; **— marbré,** veined wood; **—**

méplat, half-round wood; — **mouliné,** worm-eaten wood; — **nerveux,** durable timber, hardwood; — **neuf,** unfloated wood.

bois, — **noueux,** knotty wood, gnarled wood; — **parfait,** heartwood, duramen, core; — **passé,** brittle or rotten wood; — **pelard,** barked wood; — **piqué,** decayed wood; — **pouilleux,** — **pourri,** dry-rotten wood; — **pour cercles,** bind wood; — **précoce,** spring wood; — **primaire de seconde formation,** metaxylem; — **punais,** dogwood.

bois, — **rabougri,** dwarfed wood, stunted wood, copse wood; — **résineux,** gymnosperm wood, resinous wood; — **rouge,** redwood, red deal, Scotch pine timber; — **roulé,** cup-shapen wood; — **sain,** sound wood; — **santal,** sandalwood; — **satiné,** snakewood; — **sec,** — **sèche,** seasoned wood or timber; — **soufré,** sulphurated wood; — **sûragé,** — **suranné,** overseasoned wood; — **sur le retour,** overseasoned wood.

bois, — **sur mailles,** silver-grained wood; — **tapiré,** curled wood, speckled wood; — **tardif,** summer wood; — **teint,** dyed wood; — **tendre,** soft wood; — **tors,** — **virant,** torse wood, crooked wood; — **tranché,** wood with crooked fibers; — **vermoulu,** worm-eaten wood; — **versé,** rough-edged timber; — **vif,** green timber, green wood; **en —,** wooden.

boisage, *m.* timbering, casing, framing, woodwork.

boisé, wooded, wainscoted, paneled.

boisement, *m.* afforestation.

boiser, to timber, panel, wainscot, afforest.

boiserie, *f.* wainscot(ing), paneling.

boiseux, -euse, woody, ligneous.

boisseau, *m.* bushel, drain tile, faucet pipe; — **robinet,** cock casing, cock shell.

boissellerie, *f.* hollow ware.

boisson, *f.* drink, beverage, drunkenness; — **fermentée,** fermented liquor.

boite, *f.* maturity (of wine), footsoreness (of cattle).

boîte, *f.* box, case, Petri dish, chest; — **à étoupe,** stuffing box; — **à feu,** firebox; — **à fumée,** smokebox; — **à noyau,** core box; — **à réactifs,** reagent box; — **à réfrigération,** ice chest; — **à sécher,** drying chamber; — **à soupape,** valve chest; — **à vapeur,** steam chest; — **à vent,** wind box or chest.

boîte, — **de culture,** culture plate, culture flask; — **de distribution,** connecting box, switchboard; — **de garde,** screening box; — **de Koch,** Koch's dish for plate culture; — **de la vis sans fin,** worm casing; — **de moyeux,** wheel boxing; — **de Pétri,** Petri dish: — **de résistance,** resistance box; — **de Roux,** Roux "plate" culture bottle; — **d'essieu,** axle bearing; — **de vitesses,** transmission gear case; — **du ressort,** spring case.

boitement, *m.* limping, irregular action (of machine).

boiterie, *f.* lameness, claudication.

boiteux, -euse, lame, limping.

boitiller, to hobble.

boivent (boire), (they) are drinking.

bol, *m.* bole, bolus, pellet, bowl.

bolaire, *f.* bolary.

bolaire, bolar, bolary, clayey.

bolet, *m.* boletus.

bolide, *m.* bolis, fireball.

Bolivie, *f.* Bolivia.

bolonais, Bolognese.

bombace, *m.* bombax.

bombage, *m.* bending, rendering convex, blown cans, swells.

bombardement, *m.* bombardment.

bombarder, to bombard, shell.

bombe, *f.* bomb; — **fumigène,** smoke bomb; — **incendiaire,** incendiary bomb.

bombé, convex, curved, bulged.

bombement, *m.* bulging, convexity.

bomber, to make or render convex, cause to bulge (out) or swell out, become convex.

bombeur, *m.* convex-glass maker; — **de verre,** maker of convex glass.

bombonne, *f.* carboy, demijohn.

bombyce, *m.* bombyx; — **disparate,** gipsy moth; — **du saule,** willow moth, atlas moth; — **moine,** nun moth, black arches moth; — **neustrien,** lackey caterpillar; **pinivore,** pine procession moth; — **processionnaire,** procession moth; — **pundibond,** vaporer moth.

bon, *m.* good, best, order, check, draft, bond; **à quoi —,** what is the use; — **de caisse,** cash voucher.

bon, bonne, good, simple, foolish, kind, fine, proper, convenient, well, stoutly, tightly; **à bon compte,** cheaply; **à bon marché,** cheap, cheaply; **à la bonne heure!** well done! that's right! very well! good! **bon marché,** cheap; **bon teint,** fast color, fast dye; **de bonne heure,** soon, early, in good time; **tenir bon,** to hold firmly.

bonbon, *m.* bonbon, sweetmeat.

bonbonne, *f.* carboy, demijohn.

bond, *m.* bound, skip, gambol, leap.

bonde, *f.* opening, vent, outlet, bunghole, bung, sluice.

bonder, to fill very full, cram.

bondir, to (re)bound, skip, bounce.

bondissant, bounding, skipping, frisking.

bondon, *m.* bung, stopper, bunghole, bondon cheese.

bondonner, to bung, tap, broach.

bondrée apivore, *f.* honey buzzard.

bon-henri, *m.* allgood, goosefoot.

bonheur, *m.* good fortune, luck, success, happiness.

bonhomme, *m.* good-natured man, fellow, man, mullen, bolt, catch, pin.

bonhommie, *f.* joviality, placidity.

boni, *m.* surplus, bonus.

bonification, *f.* improvement, allowance, reduction.

bonifier, to improve, ameliorate, make up, make good.

bonjour, *m.* good morning, good day.

bonne, *f.* maid (servant).

bonne, see BON.

Bonne-Espérance, *f.* **cap de —,** Cape of Good Hope.

bonnement, tout —, simply, sincerely, plainly.

bonnet, *m.* cap, prickwood, honeycomb stomach; **— de fou,** foolscap; **— de prêtre,** squash(-melon).

bonneterie, *f.* hosiery, knitted goods.

bonnette, *f.* shade.

bonsoir, *m.* good evening.

bonté, *f.* goodness, kindness, favor.

boqueteau, *m.* small wood, spinney.

boracique, boracic, boric.

boracite, *f.* boracite; **— de chaux,** calcium borate; **— de soude,** sodium borate, borax.

borate, *m.* borate; **— de soude,** borax.

boraté, borated.

borax, *m.* borax.

bord, *m.* edge, border, margin, rim, bank; brink, shore, binding, side, shipboard-skirt; **— à —,** edge to edge, alongside, **— de la mer,** seashore; **— rabattu,** flanging; **— supérieur,** coronet, crest; **à — de,** on board; **au — de,** beside; **de haut —,**

having several decks; **le long du —** alongside.

bordage, *m.* planking, ship's planks, lining with planks.

bordé, bordered.

bordeaux, *m.* Bordeaux (wine).

bordelaise, *f.* Bordeaux cask of 225 liters, Bordeaux bottle of ¾ liter.

border, to border, edge, line, hem, fence in, flange.

bordereau, *m.* memorandum. statement.

bordure, *f.* border, edging, frame, bordering, edge, binding, rim.

bordurer, to border, bind.

bore, *m.* boron.

boré, containing boron.

boréal, north, northern, boreal; **aurore —e,** aurora borealis, northern lights.

borgne, one-eyed, unintelligible, disreputable.

borique, boric.

boriqué, treated with (containing) boric acid.

bornage, *m.* boundary settlement.

borne, *f.* landmark, boundary, limit, binding post, bound, confine, terminal, boundary pillar; **plaque à —s,** terminal plate.

bornéol, *m.* borneol.

borner, to limit, confine, restrict, bound. beacon, mark out, stake out; **se — à,** to keep within bounds, limit oneself.

bornyle, *m.* bornyl.

boro-tungstique, borotungstic.

borraginées, *f.pl.* Boraginaceae.

borure, *m.* boride.

bosquet, *m.* grove, thicket.

bossage, *m.* (em)bossing, boss, swelling.

bosse, *f.* hump, bump, lump, boss, protuberance, knob, prominence, relief; **— rentrante,** inward bulging.

bosselé, bunched, embossed, fasciculate.

bosseler, to emboss, dent, bruise

bossu, hunchbacked.

bossuer, to dent, emboss.

bostriche, *m.* **— à six dents,** pine-tree beetle; **— bidenté,** little pine-tree beetle; **— chalcographe,** six-toothed spruce bark beetle; **— curvidenté,** silver-fir bark beetle; **— disparate,** shot borer, dissimilar bark beetle; **— liséré,** lineate bark beetle; **— sténographe,** pine-tree beetle; **— strié,**

lineate bark beetle; — **typographe**, typographer bark beetle.

bostrichide, *m.* bark beetle.

bostryx, *m.* bostryx, a uniparous, helicoid cyme.

bot, clubfooted.

botanique, *f.* botany.

botanique, botanical.

botaniquement, botanically.

botaniser, to botanize.

botaniste, *m.* botanist.

botrychion, botrychium, *m.* Botrychium; — **lunaire,** moonwort.

botryocyme, *m.* botrycymose, raceme.

botryoïde, botryoidal.

botrytiqúe, botryoid, indefinite.

botte, *f.* bunch, bundle, bale, coil, clod, tuft, hank, boot, butt, cask, horses' feed, feeding time, flyweevil; — **d'écorces,** bundle of bark.

bottelage, *m.* tying up (hay).

botteler, to bundle.

bottillon, *m.* bunch, small bundle (herbs).

bottine, *f.* half boot, high shoe.

botulinique, pertaining to *Bacillus botulinus.*

botulisme, *m.* botulism.

bouc, *m.* he-goat, buck goatskin (bottle), skin.

boucage, *m.* anise, burnet saxifrage.

boucanage, *m.* smoking (meat, fish).

boucanager, to smoke, cure by smoking, drying meat.

boucaner, to smoke, dry, cure, hunt wild animals.

boucanière, *f.* smokehouse.

boucaut, *m.* hogshead, barrel, cask.

bouchage, *m.* stoppering, stopping, obstruction, plugging, corking, stopper; — **à l'émeri,** stopper grinding; — **au coton,** plugging with cotton wool.

bouchain, *m.* bilge.

bouchant à l'émeri, with ground-in glass stopper.

bouche, *f.* mouth, opening, aperture, orifice, lips, muzzle; — **à feu,** gun, cannon; — **d'eau,** hydrant; **de —,** orally.

bouché, plugged up, choked; — **au coton,** plugged with cotton wool; **flacon — à l'émeri,** glass-stoppered flask.

bouchée, *f.* mouthful.

bouchement, *m.* stopping up, filling.

boucher, to stopper, stop (up), obstruct, plug (up), cork, close; — **à i'émeri,** to provide with a ground-in glass stopper.

boucher, *m.* butcher.

boucherie, *f.* butchery, butcher trade. butcher's board.

bouche-trou, *m.* stopgap, substitute.

bouchon, *m.* stopper, plug, bung stopple, cork, tampon, wad, cap, cover, bank (wool), bundle (linen), wisp; — **à deux trous,** two-holed stopper; — **à l'émeri,** ground stopper; — **à vis,** screw stopper; — **d'amorce,** fuse plug, primer plug; — **de liège,** cork.

bouchon, — **de pissette,** wash-bottle fittings; — **de sûreté,** safety plug; — **de verre rodé,** ground-in glass stopper; — **de vidange,** drip cock, emptying plug; — **d'ouate,** cotton-wool plug; — **en caoutchouc,** rubber stopper; — **en liège,** cork stopper, cork; — **en verre,** glass stopper; — **fusible,** fusible plug.

bouchon-écrou, *m.* screw plug.

bouchonnier, *m.* cork cutter, maker of, dealer in corks.

boucle, *f.* loop, ring, nose ring, eye, bend (U tube), elbow, turn, buckle, curl, lock; — **d'évitement,** loop (at the end of a railway line).

boucler, to buckle, ring, nuzzle, curl.

bouclier, *m.* buckler, shield.

bouddhisme, *m.* Buddhism.

bouder, to pout, sulk, be stunted.

bouderie, *f.* pouting, sulking.

boudin, *m.* (black, blood) pudding, sausage, (explosive) fuse, roller, glass bulb, roll, twist (tobacco), coil, flange.

boudine, *f.* bull's-eye (in blown sheet glass).

boudoir, *m.* boudoir, lady's bower.

boue, *f.* mud, slime, mire, dirt, sediment, sludge, ooze, pus.

bouée, *f.* buoy.

boueux, -euse, muddy, miry, dirty, foul.

bouffée, *f.* puff, whiff, outburst, breath.

bouffer, to puff (out), swell, bulge, (of bread) rise, eat greedily.

bouffi, puffy, puffed, swollen, bloated.

bouffir, to swell, blow out, inflate, puff up, puff out, bloat.

bouge, *m.* bulge, swell, convexity, recess (room), closet, tub, vat; — **d'un fût,** bulge of a cask.

bougeoir, *m.* candlestick, candle holder.

bougeonnier, *m.* bullfinch.

bouger, to budge, stir, wiggle.

bougie, *f.* candle, taper, filter candle, bougie, spark plug, candle power; — Berkefeld, Berkefeld filter; — **d'allumage,** spark plug; — décimale, decimal candle, bougie decimale; — de cire, wax candle; — de porcelaine, porcelain "candle"; — de stéarine, — stéarique, stearin candle (made of stearic acid); — d'oribus, resin torch; — électrique, electric-light carbon, electric candle; — normale, standard candle.

bougie-heure, *f.* candle hour.

bougran, *m.* buckram.

bouillage, *m.* boiling (of clothes).

bouillaison, *f.* fermentation (of beer, cider, etc.).

bouillait (bouillir), (it) was boiling.

bouillant (bouillir), boiling, seething, fiery, hot.

bouillerie, *f.* distillery (brandy).

bouilleur, *m.* distiller, still, boiler (pipe) tube, heater; — de colle forte, glue boiler; — de cru, one who distills privately; — d'os, bone boiler.

bouilli, *m.* boiled beef.

bouilli, boiled.

bouillie, *f.* pap, gruel, porridge, paste, pulp; — bordelaise, — cuprique, Bordeaux mixture; —s cupriques, copper sprays, Burgundy mixtures.

bouillir, to boil, seethe, bubble, simmer.

bouillissage du bois, *m.* boiling the wood.

bouilloire, *f.* boiler, kettle.

bouillon, *m.* bubble (given off by boiling liquid), bleb, bubbling, broth, ripple, wave, liquid manure, air bubble, air hole, mullein, high taper; — ascite, ascitic fluid and broth; — blanc, mullein, Aaron's-rod *Verbascum thapsus);* — de boeuf, beef broth; — de culture, culture medium; — gras, clear (meat-) soup, beef tea; — noir, dark mullein.

bouillonnement, *m.* bubbling (up), effervescence, ebullition.

bouillonner, to bubble, boil up, effervesce, gush out, seethe.

bouillonneux, -euse, containing bubbles, blebby.

bouillotte, *f.* boiler, kettle.

bouillotter, to boil gently, simmer.

boulange, *f.* unbolted flour.

boulanger, to bake (bread).

boulanger, *m.* baker.

boulangerie, *f.* baking, bakery, bakehouse.

boule, *f.* ball, bulb, globe, bole, rot (in sheep); — à gaz, gas connection; — à robinets, a ball-shaped connection, stopcock fixture; — à teinture, dye ball; — d'acier, — de Mars, — de Nancy, ball of iron and potassium tartrate; — de bleu, blue ball, ball of bluing; — de verre, glass bulb; — d'or, mountain globe flower.

bouleau, *m.* birch (tree), birchwood (Betula); — pleureur, drooping birch, weeping birch *(Betula pendula);* — verruqueux, common birch.

boule-de-neige, *f.* guelder-rose, snowball-tree *(Viburnun opulus).*

boule-d'or, *f.* globe flower *(Trollius europaeus).*

boulée, *f.* cracklings (sediment) of tallow, greaves.

bouler, to swell, pout, puff out, (of dough) rise, roll, thrash.

boulet, *m.* ball, cannon ball, bullet, fetlock joint.

boulette, *f.* small ball, pellet, bolus.

bouleversement, *m.* overthrow, overturning, commotion, panic, revolution, disruption, convulsion.

bouleverser, to overturn, upset, unsettle, overthrow, bewilder, throw into confusion.

boulimie, *f.* morbid hunger.

boulin d'échafaudage, *m.* pullock, putlog.

boulingrin, *m.* bowling green.

bouloir, *m.* stirrer, beater.

boulon, *m.* bolt, pin, peg, screw; — à ailettes, thumb (winged) screw; — à clavette, key bolt, cotter bolt; — à croc, — à crochet, hook bolt or screw; — à écrou, screw bolt; — à goupille, pinned bolt; — à oeil, eyebolt.

boulon, — à oreilles, wing bolt, thumb bolt; — à vis, screw bolt; — d'articulation, pivot; — de ressort, spring bolt, pin bolt; — de retenue, safety screw; — fendu, slotted-head screw; — taraude, threaded bolt, tap bolt.

boulonner, to bolt (down), screw on.

bouquet, *m.* bunch, cluster, clump, tuft, wisp (straw) bouquet, aroma (of wine); — d'arbres, clump (of trees), group, cluster of trees; par —s, in clumps or groups.

bouquetier, *m.* flower vase.

bouquetin, *m.* wild goat, ibex, Caucasian goat.

bouquin, *m.* buck hare, buck rabbit.

bourache, *f.* borage.

bourbe, *f.* dredgings, mud.

bourbeux, -euse, miry, muddy.

bourbonnaise, *f.* clammy lychnis.

bourdaine, *f.* alder buckthorn (*Rhamnus frangula*).

bourdillon, *m.* oak stave.

bourdon, *m.* bumblebee; — **de St. Jacques,** marsh mallow, white mallow.

bourdonnement, *m.* buzzing, humming, murmur.

bourdonner, to buzz, hum, murmur, sound (bell).

bourg, *m.* small market town.

bourgade, *f.* small town, village, market town.

bourgène, *f.* alder buckthorn (*Rhamnus frangula*).

bourgeois, middle-class, bourgeois, private, simple, plain, common.

bourgeon, *m.* bud, adventitious bud, shoot, sprout; — **à bois,** leaf bud; — **adventif,** adventitious bud; — **à fruits,** flower bud; — **proventif,** dormant bud, resting bud.

bourgeonné, budded.

bourgeonnement, *m.* budding, putting forth shoots; — **cellulaire,** budding, formation of conidia.

bourgeonner, to bud, shoot (of plants), sprout, granulate.

bourgépine, *f.* buckthorn (*Rhamnus*).

Bourgogne, *f.* Burgundy.

bourgogne, *m.* Burgundy (wine).

bourlet, *m.* see BOURRELET.

bourrache, *f.* borage (*Borago officinalis*).

bourrade, *f.* blow, cuff, thrust.

bourrage, *m.* tamping, choking.

bourro, *f.* wad, (a fibrous mass or material), wadding, stuffing, padding, pad, plug, short hair, tag wool, silk, waste, trash, refuse, down, floss (of buds), tamping.

bourré, packed.

bourrée, *f.* fagot.

bourreler, to torment, flange, tamp (powder).

bourrelet, *m.* padding, pad, swelling (on a stem), flange, rim, cushion, ridge on a stem, excrescence, corky layer between scion and stock; — **de moyeu,** bead of the boss; — **soufflé,** enlargement.

bourrer, to stuff, pad, cram, pack, tamp, ram, wad.

bourrier, *m.* chaff.

bourroir, *m.* tamper.

bourru, unfermented (wine), rough, surly, crabbed.

bourse, *f.* purse, bag, pouch, stock exchange, bursa, sac, uterus, scholarship; — **à berger,** — **à pasteur,** shepherd's-purse (*Capsella bursa-pastoris*).

boursette, *f.* lamb's-lettuce, corn salad.

boursouflé, swollen, inflated, bloated.

boursouflement, *m.* swelling, inflation, blistering, bubble (in metals), enlargement, abnormal fermentation (of cheese with production of large cavities).

boursoufler, to swell, puff up, bloat, blister, inflate.

boursouflure, *f.* swell(ing), bloatedness, turgidity.

bousage, *m.* dunging of cloth.

bouse, *f.* cow (ox) dung.

bousier, *m.* dung beetle.

boussole, *f.* galvanometer, (prismatic) compass, dial; — **des sinus,** sine galvanometer; — **des tangentes,** tangent galvanometer; — **d'inclinaison,** dipping needle; — **marine,** mariner's compass, sea compass.

bout (bouillir), it boils.

bout, *m.* end, extremity, tip, nipple, top, bit, fragment, side (of the chain or rope); — **en caoutchouc,** rubber teat (for pipette): **au — de,** at the end of, after; **de — en —,** from end to end; **être à —,** to be exhausted; **le — du monde,** the utmost, very far away; **petit —,** top end.

boute, *f.* hogshead, barrel, tub.

boutefeu, *m.* igniter.

bouteille, *f.* bottle, cylinder, bubble, bottle gourd, rot (in sheep); — **à gaz,** gas cylinder; — **à goulot étroit,** narrow-necked bottle; — **de Leyde,** Leyden jar; — **de savon,** soap bubble.

bouteiller, to fill with bubbles (glass).

bouteloue, *m.* grama (grass), mesquite grass (*Bouteloua oligostachya*).

bouter, to become thick or ropy (of wine).

boutique, *f.* shop, store, stock, tools, implements.

boutoir, *m.* snout, paring knife, parer.

bouton, *m.* bud, button, knob, stud, pin, lug, handle, plug, vesicle, pimple, pustule; —**d'argent,** sneezewort (*Achillea ptarmica*); — **de bielle,** crank (wrist) pin; — **de pommier,** bur (of burdock); — **de sonnerie,** push button, bell push; — **d'or,** tall buttercup (*Ranunculus acris*); — **noir,** deadly nightshade; **en —,** in bud.

boutonner, to button, bud.

boutonnière, *f.* buttonhole, incisior; **en —,** gapingly.

bouturage, *m.* propagation by cuttings, disbudding.

bouture, *f.* slip, cutting, sucker, shoot, set.

bouturer, to set, cut.

bouverie, *f.* ox stall.

bouvet, *m.* — **à appronfondir,** grooving plane.

bouvetage, *m.* grooving and tonguing.

bouveter, to groove and tongue.

bouvette, *f.* groove and tongue.

bouvier, *m.* ox driver, drover.

bouvière amère, *f.* bitterling.

bouvillon, *m.* one-year-cld ox calf.

bouvreuil, *m.* bullfinch; — **carmoisi,** scarlet grosbeak; — **commun,** common bullfinch; — **dur-bec,** pine grosbeak.

bovides, *m.pl.* bovine animals.

bovin, bovine, pertaining to cattle.

boxe, *f.* boxing.

boyau, *m.* intestine, bowel, gut, hose, tubing, inner tube, catgut, narrow place, pipe, long tunnel, branch.

boyauderie, *f.* gut-dressing works.

bracelet, *m.* metal band, ring, bracelet, node, knot (grasses).

brachial, brachial (artery).

brachyblaste, *f.* brachyblast.

brachystyle, short-styled, brachystylic.

braconnage, *m.* poaching.

braconner, to poach.

braconnier, *m.* poacher.

bractéaire, bracteal, bracted.

bractée, *f.* bract.

bractéifère, bracteate.

bracteole, *f.* bractlet, a prophyll.

bractéolé, bracteolate, having bractlets.

bradsot, *m.* bradsot, braxy, sheep disease.

brai, *m.* pitch, tar, rosin; — **gras,** soft (moist) pitch, tar; — **sec,** dry pitch, resin, rosin.

braire, to bray.

braise, *f.* wood cinders, glowing embers, bed of coals, live coal, live charcoal, breeze, braised meat; — **chimique,** tinder-box paste; — **de charbon de bois,** small charcoal.

bran, *m.* bran; — **de scie,** sawdust.

brancard, *m.* handbarrow, stretcher, litter, shaft, thill.

branchage, *m.* branches, boughs, ramification, limb, crown of a tree.

branche, *f.* branch, bough, arm, division, leg (of tripod), limb, part; — **de levier,** arm or crank of lever; — **gourmande,** epicormic branch, water sprout.

branchement, *m.* branching, branch pipe.

brancher, to branch, connect.

branchette, *f.* small branch.

branchia¹, branchial.

branchies, *f.pl.* gills.

branchu, branchy, ramose.

brande, *f.* heather.

branderie, *f.* brandy distillery.

brandevin, *m.* brandy (made from wine).

brandevinier, *m.* brandy distiller or seller.

brandir, to brandish, swing.

brandon, *m.* straw wisp, (fire-)brand.

branlant, shaking, shaky, tottery.

branle, *m.* swinging, motion, oscillation, impulse.

branler, to swing, shake, move, totter, waver, joggle, jog.

branloire, *f.* seesaw, handle, (of bellows), rockstaff.

braque, *m.* harrier, beagle, hound.

braquer, to point, set, level, direct, train, fix.

bras, *m.* arm, hand, handle, lever (arm) rod, brace, bracket, extension arm; **à —,** by hand; **sur les —,** on one's hands.

brasage, *m.* brazing, hard-soldering.

brase, *f.* cinders of wood.

brasement, *m.* brazing.

braser, to braze, hard-solder.

brasier, *m.* fire of live wood coals, brazier, fireplace.

brasiline, *f.* brasilin, a coloring matter from Brazilwood.

brasiller, to grill, broil, sizzle, shine, sparkle.

brasque, *f.* brasque, lining, luting (of crucible).

brasquer, to brasque, lute, line with brasque.

brassage, *m.* mixing, mashing, mintage, brewing, stirring.

brasse, *f.* fathom, two yards.

brassée, *f.* armful.

brasser, to brew, mash, mix, stir (up), rabble, puddle.

brasserie, *f.* brewery, brewing, beer shop.

brasseur, *m.* brewer. mixer, puddler, potching stick.

brassicaire, *f.* cabbage moth.

brassin, *m.* mash tub, quantity brewed, boiling, brewing.

brasure, *f.* brazing, brazed seam or joint, hard solder.

braunite, *f.* braunite.

brave, brave, bold, gallant, fine, good, honest, worthy, intrepid.

braver, to brave, dare, defy, face.

bravoure, *f.* bravery, valor.

braye, *f.* puddle, mud, clay.

brayer, to pitch, tar.

brayer, *m.* truss, belt.

brebis, *f.* ewe, sheep.

brecciolaire, brecciated.

brèche, *f.* gap, breach, flaw, hole, rupture, break, notch (in blade), breccia.

bréchet, brechet, *m.* breast bone, brisket.

bredouillement, *m.* stuttering.

bref, -ève, brief, short, compact, succinct, in short; **en —,** in brief, briefly.

bréhaigne, bréhaine, barren, sterile.

brème, *f.* bream; — **bordelière,** *f.* white bream.

brenthe, brente, *m.* brentid.

Brésil, *m.* Brazil.

brésil, *m.* brazilwood, brazil, logwood.

brésiller, to dye (red) with brazilwood, crumble, shatter.

brésillet, *m.* brasiletto, (Caesalpinia).

bresilleur, *m.* brazilwood dyer.

Bretagne, *f.* Brittany.

bretelle, *f.* strap, brace, sling, suspender, crossover (between two parallel tracks).

breton, -onne, Breton, of Brittany; (fauvette) **bretonne,** *f.* warbler.

bretté, toothed, indented, jagged.

bretter, to indent.

breuvage, *m.* beverage, drink, draft, drench.

brève, see **BREF.**

brevet, *m.* patent, warrant, license, certificate, diploma, commission; — **de perfectionnement,** patent for improvement; — **d'invention,** patent (for an invention); **droit de —,** copyright.

brevetable, patentable.

breveté, *m.* patentee.

breveté, patented; — **S.G.D.G.,** — **sans garantie du Gouvernement,** patented without government guarantee.

breveter, to patent, grant a patent (certificate) to.

brévicaude, brevicaudate, short-stalked or -tailed.

brévicaule, short-stemmed.

brévicolle, short-stalked or -tailed.

brévicorne, short-horned.

bréviflore, with short flowers.

brévifolié, short-leaved.

brévipède, short-footed.

brévirostre, short-billed, brevirostrate.

brévistyle, with short style.

brévité, *f.* shortness.

bréviuscule, a little short.

bride, *f.* bridle, rein(s), clamp, cramp, strap, stay, curbing, flange.

brider, to bridle, check, restrain, flange.

bridon, *m.* bridle bit, bridoon.

brie, *m.* (soft) Brie cheese.

brier, to knead.

brièvement, briefly, shortly, soon, in short.

brièveté, *f.* brevity, conciseness, shortness.

brigade, *f.* brigade, gang, party, body.

brigadier, *m.* foreman, corporal, chief guard, forester.

brigand, *m.* brigand, highwayman.

brigandage, *m.* robbery.

brillamment, brilliantly.

brillant, *m.* brilliancy, silky luster, polish brilliant (diamond).

brillant, brilliant, bright, shining, sparkling, glossy.

brillanter, to glitter, gloss, cut (diamond) into a brilliant.

brillement, *m.* brilliance, luster.

briller, to shine, glisten, gleam, sparkle, glitter.

brin, *m.* blade, shoot, sprig, twig, stalk, spear, branch, young seedling, fragment, strand, fiber (of wool); — **de câble,** side of rope; — **de semence,** seedling; **jeune —,** seedling plant.

brindille, *f.* branchlet, spring, twig, spray wood; *pl.* small dead branches.

bringé, brindled (cat).

brique, *f.* brick, bar (of soap); — **blanche,** firebrick; — **creuse,** hollow brick; — **crue,** air- (sun-) dried brick; — **cuite,** burned brick; — **de savon,** bar of soap; — **de**

scories, slag brick, slag stone; — **émaillée,** enameled brick, glazed brick; — **pilée,** brick dust; — **pleine,** solid or full brick; — **réfractaire,** firebrick; — **vernissée,** glazed brick.

briquet, *m.* steel (for striking light), tinder box.

briquetage, *m.* brickwork, embedding.

briqueté, imitating brick, bricked, brickcolored, lateritious.

briqueter, to brick.

briqueterie, *f.* brick field, brickmaking, brick factory, match factory.

briqueteur, *m.* bricklayer.

briquetier, *m.* brickmaker, matchmaker.

briquettage de tourbe, *m.* peat briqueting.

briquette, *f.* briquette, briquet.

bris, *m.* breaking, shattering; — **de givre,** ice-break, rime break; — **de neige,** snowbreak; — **de verglas,** icebreaker.

brisant, breaking, shattering, iisruptive.

brise, *f.* breeze; — **de mer,** sea breeze; — **de terre,** land breeze.

brisé, broken, folding (door).

brise-circuit, brise-courant, *m.* circuit breaker.

brisées, *f.pl.* broken boughs, deer's track. (forester's) mark.

brise-jet, *m.* antisplash tap nozzle.

brisement, *m.* breaking, shattering.

brise-mottes, *m.* clod crusher.

briser, to break, break to pieces, shatter, smash, shiver, crack, interrupt, wear out; **se —,** to break, be or become refracted, fold up.

brise-vent, *m.* windbreak, screen.

bristol, *m.* Bristol board, Bristol, perforated cardboard.

brisure, *f.* break, crack, fissure, flaw, fragment, folding joint, difference.

britannique, British.

brize, *f.* common quaking grass (Briza).

brocard, *m.* brocket, roebuck.

brocatelle, *f.* brocatel.

broche, *f.* spike, peg, pin, spindle, spigot, plug; — **de fraisage,** cutter or milling spindle.

broché, (of books) sewn, paper-bound, (of stuffs) figured, brocaded.

broches, *f.pl.* tusks; *m.pl.* knobbers.

brochet, *m.* pike.

brochette, *f.* skewer, pin, peg, scrow.

brochure, *f.* stitching, sewing (of books), pamphlet, brochure, inwoven pattern (in fabrics).

broder, to embroider.

broderie, *f.* embroidery.

broie (broyer), (he) crushes, breaks.

broie, *f.* brake (for hemp), brake harrow.

broiement, broiment, *m.* grinding, pounding, breaking, crushing, laceration.

bromacétique, bromacetic, bromoacetic.

bromal, *m.* bromal.

bromate, *m.* bromate.

brome, *m.* bromine, brome grass (Bromus).

bromé, brominated, containing bromine.

bromer, to brominate.

bromhydrate, *m.* hydrobromide.

bromhydrique, hydrobromic.

bromidrose, *f.* bromidrosis, fetid perspiration.

bromique, bromic.

bromocyanuration, *f.* bromocyanidation.

bromocyanure, *m.* bromocyanide.

bromocyanurer, to bromocyanide.

bromoforme, *m.* bromoform.

bromo-ioduré, containing bromide and iodide.

bromure, *m.* bromide; — **d'argent,** bromide of silver, silver bromide; — **de calcium,** — de chaux, calcium bromide; — **de fer,** iron bromide, ferrous bromide; — **de magnésie,** magnesium bromide; — **de méthyle,** methyl bromide, bromomethane; — **de potassium,** potassium bromide; — **d'éthyle,** ethyl bromide, bromoethane; — **ferreux,** ferrous bromide.

bromuré, brominated, containing bromine.

bronche, *f.* bronchus, bronchial tube.

broncher, to stumble, shy, flounder, falter, budge, stir.

bronchial, -aux, bronchique, bronchial.

bronchite, *f.* bronchitis.

bronchocèle, *m.* goiter.

bronzage, *m.* bronzing, browning, blueing (of steel).

bronze, *m.* bronze, hard brass; — **à canon,** gun (cannon) metal; — **à cloches, bell metal;** — **au manganèse,** manganese bronze; — **au zinc,** zinc-bronze; — **d'aluminium,** aluminum (gold) bronze; — **de cloches,** bell metal.

bronze, — de nickel, nickel bronze; — **dur, — durci,** hard bronze, hardened bronze;

— mouiu, bronze powder; — **phosphoreux,** phosphor bronze; — **pour robinetterie,** cock metal, cock brass; — **siliceux,** — silicié, silicon bronze.

bronzé, bronzed, browned, sunburnt.

bronzer, to bronze, brown, blue (gun barrels), tan, sunburn, harden.

broquart, *m.* see BROCARD.

broquette, *f.* tack.

brossage, *m.* brushing.

brosse, *f.* brush, bristle, scopa (on bee's leg); — **à cheval,** currycomb; — **douce,** soft brush; — **en fil de métal,** wire brush: — **métallique,** wire brush; — **passe-partout,** small dusting brush; — **rude,** stiff (hard) brush.

brosser, to brush.

brosserie, *f.* brushmaking, brush trade.

brossier, *m.* brushmaker.

brou, *m.* hull, shell, husk, shuck (of walnut).

brouet, *m.* gruel, broth.

brouette, *f.* wheelbarrow; — **à deux roues,** handcart.

brouettée, *f.* barrowful, barrowload.

brouetter, to wheel, convey, in a barrow, barrow.

brouetteur, brouettier, *m.* barrowman.

broui, *m.* enameler's blowpipe.

broui, blighted.

brouillard, *m.* fog, mist, haze; — **fin sur le sol,** ground fog.

brouillement, *m.* mixing.

brouiller, to mix up, embroil, jumble, tangle up, throw into confusion, confuse, beat, scramble (eggs), blunder.

brouillerie, *f.* disagreement, discord, coolness, trouble.

brouir, to blight, nip (plants).

broussaille(s), *f.(pl.)* brushwood, underbrush, bush, tangled vegetation brush, scrub.

broussin, *m.* gnarl (on tree), bur, curled wood.

brout, *m.* young sprout, browse.

brouter, to browse, graze, bite off, feed on, nibble, (of tools) cut unevenly, chatter.

broutille(s), *f.(pl.)* twigs or small shoots, brushwood, small fagot wood, trifle.

brownien, Brownian, molecular (motion).

broyage, *m.* grinding, crushing, brake, breaking.

broyement, *m.* grinding.

broyer, to pound, grind, crush, pulverize, powder, bruise, break.

broyeur. *m.* grinder, crusher; — **à boulets,** ball mill; — **à cylindres,** — **à rouleaux,** roll crusher; — **pulvérisateur,** crusher and pulverizer.

broyeuse, *f.* grinder, flax-breaking machine.

broyeuse-sécheuse, *f.* grinder and drier.

bruant, *m.* — **des roseaux,** reed bunting; — **jaune,** yellow bunting, yellow hammer; — **ortolan,** ortolan bunting; — **proyer,** corn bunting.

brucelles, *f.pl.* tweezers, (spring) forceps, pincers.

bruche, *f.* beetle.

brucine, *f.* brucine.

brugnon, *m.* nectarine.

bruine, *f.* drizzle, ground fog, mist.

bruire, to make noise, rustle, roar, brawl, hum.

bruissant, noisy, humming, roaring.

bruissement, *m.* rumbling, humming, buzzing.

bruit, *m.* noise, uproar, sound, rumor, report; **sans** —, noiselessly, silently.

brûlable, burnable.

brûlage, *m.* burning (of weeds), sod burning, scorching (clothes).

brûlant, burning, on fire, glowing, hot, eager.

brûlé, *m.* burn, burning, burnt smell or taste.

brûlé, burned, burnt, calcined.

brûlement, *m.* burning.

brûler, to burn, blast, heat, scorch, scald, cauterize, destroy by oxidation, ruin (reputation), distill (wine); — **en dedans,** to burn at the base (of a burner); **se** —, to burn oneself, be burnt.

brûlerie, *f.* (brandy) distillery.

brûleur, *m.* burner, heater, (brandy) distiller, (coffee) roaster; — **à couronne,** gas ring, rose burner; — **à gaz,** gas burner, gas jet; — **ambulant,** traveling distiller; — **à pétrole,** oil burner; — **(de) Bunsen,** Bunsen burner; — **d'os,** bone calciner; — **électrique,** electric heater.

brûlis, *m.* patch of burnt land.

brûlot, *m.* burnt brandy, half-burned charcoal; *pl.* brands.

brûlure, *f.* burn, scald, frost nip, blight, smut (on corn).

brumaille, *f.* slight fog or mist.

brume, *f.* thick fog or mist.

brumeux, -euse, foggy, misty.

brun, brown, brunette, dark; — **châtain,** chestnut (brown); — **clair,** light-brown; — **de montagne,** umber; — **de sapin, fir** leaf blight; — **foncé,** dark-brown; — **marron,** chestnut-brown, chestnut, maroon; — **noire,** dark-brown; — **rouge,** brown ocher; avec une touche de —, tawny; **d'un** — grisâtre, grayish brown.

brunâtre, *m.* brownish color.

brunâtre, brownish.

brune, *f.* dusk, nightfall.

brune, see BRUN.

brunelle, *f.* selfheal (*Prunella vulgaris*).

bruni, *m.* burnish, polish.

bruni, browned.

brunir, to brown, burnish, turn brown, polish, tan, darken.

brunis, *m.* burnish, polish.

brunissage, *m.* burnishing, polishing.

brunissement, *m.* browning, darkening, burnishing.

brunisseur, -euse, *m.&f.* burnisher (person).

brunissoir, *m.* burnisher (tool).

brunissure, *f.* burnish, browning, polish, (potato) blight or rot.

brun-mat, dull-brown, mat-brown.

brun-rouille, rusty-brown, rust-colored.

brusque, sudden, abrupt, gruff, rough, sharp, blunt.

brusquement, suddenly, roughly, unexpectedly, abruptly, brusquely.

brut, *m.* crude oil; — **de coulée,** — **de fonte,** rough-cast, as cast.

brut, brute, brutish, crude, raw, gross, coarse, uncouth, rough, unpolished, uncut, unfinished, uncultivated (land), unorganized, inorganic, uneducated.

brutal, -aux, brutal, surly.

brutalement, brutally, churlishly.

brutaliser, to use brutally.

brute, *f.* brute, beast.

Bruxelles, *n.* Brussels.

bruyamment, noisily.

bruyant, noisy, loud, humming.

bruyère, *f.* heath, heather, ling (Erica), moor; — **à clochettes,** small heath; — **arborescente,** *f.* tree heather, brierwood (*Erica arborea*); — **cendrée,** Scotch heather (*Erica cinerea*); — **des marais,** — **quaternée,** cross-leaved heather (*Erica tetralix*);

callune —, ling, common heather (*Calluna vulgaris*).

bryon, *m.* Bryum (moss).

bryone, *f.* bryony (Bryonia).

bryonine, *f.* bryonin.

bu (boire), drunk.

buandier, -ère, *m.&f.* bleacher, launderer.

bube, *f.* pimple.

bubon, *m.* bubo; —**s** pesteux, plague buboes.

bubonique, bubonic (plague).

bucail, *m.* **bucaille,** *f.* buckwheat.

buccal, buccal, relating to the mouth.

bucco, *m.* buchu.

bûche, *f.* billet (of wood), stick, log, block, (of coal) lump, butt, sheave, **grosses —s noueuses,** knotty firewood.

bûcher, to hew, cut.

bûcher, *n.* woodhouse, woodshed, pile of fagots.

bûcheron, *m.* woodcutter, lumberman.

bûchette, *f.* stick (of dry wood), small log.

bûchille, *f.* small log; *pl.* borings, turnings.

budget des forêts, forest budget.

buée, *f.* damp vapor, moisture, lye, washing.

buer, to steam (of hot bread), reek.

buffle, *m.* buffalo, buff (-leather), buff stick.

buffleterie, *f.* shoulder belts, leather equipment.

bugle, *f.* bugle.

buglosse, *f.* bugloss, alkanet, oxtongue (*Anchusa officinalis*).

bugrane, *f.* restharrow, cammock.

buis, *m.* box, boxwood, box tree (Buxus).

buisson, *m.* bush, thicket.

buisson-ardent, fire thorn, evergreen thorn (Pyracantha).

buissonnant, bushy.

buissoneux, -euse, bushy, woody.

bulbe, *m.&f.* bulb, corm, root, offset, bulbotuber; — **de colchique,** colchicum corm; — **pileux,** root of hair.

bulbeux, -euse, bulbous, bulbose, bulbed (plant).

bulbiculteur, *m.* bulb grower.

bulbifère, having bulbs, bearing bulbils, bulblets.

bulbiforme, *m.* the form of a bulb.

bulbille, *m.* bulblet.

bulbine, *f.* bulbous root.

bulle, *f.* (gas) bubble, blister, airhole, bleb, vesicle, drop(let), bulla, bull (edict).

bullé, bullate, blistered or puckered.

bullescence, *f.* bullescence (being blistered).

bulletin, *m.* bulletin, report, notice, receipt, ticket, statement; — **de livraison,** delivery note; — **du marché,** market report.

bulleux, -euse, vesicular, covered with bubbles.

bullule, *f.* a small blister.

bullulé, blistered, puckered.

buphtalme, *m.* oxeye (Buphthalmum).

bupreste, *m.* buprestis (beetle), metallic wood borer (beetle); — **du tilleul,** linden burncow; — **du tremble,** aspen burncow; — **vert,** pear flat-headed borer.

bureau, *m.* bureau, office, table, desk, committee, board.

burent (boire), (they) drank.

burette, *f.* burette, cruet, oil can, oiler, oil-feeder; — **à huile,** oil can, oiler; — **à pince,** burette with pinchcock; — **à robinet,** burette with (glass) stopcock; — **de Dupré,** Dupré's pipette; — **graduée,** graduated burette.

burgeon, *m.* bud, shoot; — **adventif,** adventitious bud.

burin, *m.* chipping chisel, chisel.

burinage, *m.* (en)graving, chiselling, chipping, cutting (metal).

buriner, to engrave, chisel, chip.

burineur, *m.* chiseler, chipper.

bursaire, purselike, saclike.

bursicule, *f.* bursicule, small sac, pouchlike expansion of stigma.

bursiculé, bursiculate, purselike.

busard, *m.,* **buse,** *f.* buzzard; — **des marais,** — **harpaye,** marsh harrier.

buse, *f.* nozzle, nose, nose-piece, snout, (blast) pipe, channel, common buzzard; — **de soufflet,** blast pipe; — **féroce,** long-legged buzzard; — **rousse,** red buzzard.

busette, *f.* blast pipe, gas nozzle.

bussent, (boire), (they) might drink, (they) drank.

busserole, *f.* red bearberry (*Arctostaphylos uva-ursi*).

but (boire), (he) drank.

but, *m.* object, purpose, aim, end, mark, goal, butt; **dans ce —,** for this purpose; **dans le — de,** with the object of.

butalanine, *f.* butylanin, butalanine.

buté, fixed, set, determined, butted.

butée, *f.* fixed piece, lug, stop, shoulder, support, abutment, striking, tapping.

buter, to butt, strike, aim (at), hit, oppose, prop up, rest against, earth (hill) up.

butin, *m.* booty, prey, spoil.

butiner, to pillage, plunder.

butineuse, *f.* working bee.

butor, *m.* bittern, booby.

buttage, *m.* ridging, tumping, earthing up.

butte, *f.* hillock, knoll, mound.

butter, to heap earth round, ridge, tump, mold.

buttoir, butteur, *m.* catch, stop, projection, tappet, driver, double-breasted plow; ridging plow, buffer, bumper.

butyle, *m.* butyl.

butylène, *m.* butylene.

butylique, butyl (alcohol).

butyracé, butyraceous.

butyrate, *m.* butyrate.

butyreux, -euse, buttery, butyr(ace)ous.

butyrine, *f.* butyrin.

butyrique, butyric.

butyromètre, *m.* butyrometer.

buvable, drinkable, fit to drink.

buvant (boire), drinking.

buvard, *m.* blotter, blotting paper.

buvard, blotting (of paper), absorbent.

buxine, *f.* buxine.

C

C, *abbr.* (carbone), carbon.

c., *abbr.* (centimètre, compte), centimeter, account.

c', *abbr.* (ce), this.

ca, *abbr.* (calcium), calcium.

ça, *abbr.* (cela), this, that, it.

çà, here; — **bas,** down here; — **et là,** here and there.

cabalistique, cabalistic, magical.

caballin, caballine.

cabane, *f.* cottage, cabin, hut.

cabané, living in cabins.

cabaner, to cant (up), overturn.

cabaret, *m.* wild ginger, hazelwort, asarabacca (*Asarum canadense*).

cabère, *f.* cabera, wave.

cabestan, *m.* capstan, windlass, winch.

cabine, *f.* cabin, room, tower, cage; — **de navigation,** navigating room.

cabinet, *m.* cabinet, closet, office, study.

câblage, *m.* twisting.

câble, *m.* cable, rope, cord; — **à noeuds,** knotted rope; — **d'alimentation,** feeder, supply circuit; — **de transmission,** power cable; — **métallique,** wire rope; — **porteur,** suspension cable; — **sous-marin,** submarine cable.

câbler, to twist, lay strands into a cable, cable.

cabochon, *m.* cabochon, polished, uncut gem, stud nail; **en —,** head-shaped.

cabrer (se), to rear.

cabrillon, *m.* goat's-milk cheese.

cabron, *m.* kid(-skin), burnishing tool, buff stick.

cacao, *m.* cacao, cocoa.

cacaotier, *m.* cacao tree (*Theobroma cacao*).

cacaotière, *f.* cacao plantation.

cacaoyer, *m.* cacao tree (*Theobroma cacao*).

cacaoyère, *f.* cacao plantation.

cachalot, *m.* cachalot, sperm whale; **huile de —,** sperm oil.

caché, hidden, unseen, unperceived, secret.

cachemire, *m.* cashmere, Cashmere, Kashmir.

cacher, to hide, conceal, secrete.

cachet, *m.* seal, stamp, ticket, cachet; — **d'aspirine,** dose of aspirin.

cacheter, to seal.

cachexie, *f.* cachexia, general debility.

cachibou, *m.* cachibou (resin).

cachiment, cachiman, *m.* soursop.

cacholong, *m.* cacholong.

cachot, *m.* dungeon.

cachou, *m.* catechu.

cachouter, to dye or treat with catechu.

cacodylate, *m.* cacodylate.

cacodyle, *m.* cacodyle.

cacodylique, cacodylic.

cactacées, cactées, *f.pl.* cacti, Cactaceae.

cactus, cactier, *m.* cactus.

cacumen, *m.* cacumen, the apex of an organ.

c.-à-d., *abbr.* (c'est-à-dire), that is to say, *i.e.,* that is.

cadastre, *m.* land registry, survey.

cadavéreux, -euse, corpselike, cadaverous, deathlike.

cadavérine, *f.* cadaverine.

cadavérique, cadaveric.

cadavre, *m.* corpse, dead body.

cade, *m.* cade, Spanish juniper (*Juniperus oxycedrus*).

cadeau, *m.* present, gift.

cadence, *f.* cadence; **en —,** rhythmically, at the same time.

cadencé, in cadence.

cadet, *m.* younger (son) brother, junior, cadet, caddie.

cadet, younger, junior.

cadette, *f.* paving stone, flagstone, younger daughter (or sister).

cadetter, to pave with hewn stones.

cadmie, *f.* cadmia.

cadmique, cadmic.

cadmium, *m.* cadmium.

cadran, *m.* dial, face, card.

cadrané, heart-shaken.

cadranure, *f.* star(-heart)-shake.

cadranuré, heart-shaken, shaken.

cadre, *m.* frame, border, compass, limits, framework, outline, plan, skeleton, cadre, loop, loop aerial, gallow; — **subérisé,** radial dot (endodermis).

cadrer, to tally, agree, suit, fit in, square with.

caduc, -uque, decaying, broken down, declining, null and void, caducous, deciduous, dropping off early; **à feuilles caduques,** deciduous.

caduciflore, with flowers dropping off early.

caducité, *f.* caducity, decay, senility.

caecal, aux, caecal (appendage).

caecum, *m.* caecum.

caemospores, *f.pl.* spores originating in a caeoma.

caenogénèse, *f.* cenogenesis.

caesium, *m.* cesium.

cafard, *m.* cockroach.

café, *m.* coffee, café, coffee berry; — **au lait,** coffee and milk, light-brown (color); — **en poudre,** ground coffee; — **torréfié,** roasted coffee.

caféier, *m.* coffee tree.

caféière, *f.* coffee plantation.

caféine, *f.* caffeine.

caféique, caffeic (acid).

caféone, *f.* caffeone, caffeol (aromatic coffee oil).

cafétannique, caffetannic.

caffût, *m.* scrap iron.

cafier, *m.* see CAFÉIER.

cafique, see CAFÉIQUE.

cage, *f.* cage, coop, case, casing, frame, housing, staircase, shaft.

cageot, *m.* crate, hamper.

cagette, *f.* bird trap.

cagneux, -euse, toe narrow, base wide, pigeon-toed, knock-kneed.

cagot, *m.* cretin.

cahier, *m.* record book, copybook, notebook; — **d'aménagement,** working-plan report; — **de** (des) **charges,** statement of works, specifications.

cahot, *m.* jolt(ing), bump, obstacle.

cahoter, to jolt, shake.

caïeu, *m.* offset bulb, clove (of garlic), sucker.

cail-cédra, *m.* bastard mahogany.

caillage, *m.* curdling, clotting (of milk).

caillasse, *f.* (gravelly) marl.

caille, *f.* quail.

caillé, *m.* curdled milk, curd, curds.

caillé, coagulated, curdled, clotted.

caillebot, *m.* guelder-rose, snowball tree.

caillebottage, *m.* curdling, clotting.

caillebotte, *f.* curdled milk, curd(s).

caillebotté, curdled, coagulated, clotted.

caillebotter, to curdle, clot, coagulate.

caille-lait, *m.* cheese rennet, yellow bedstraw.

caillement, *m.* curdling, clotting, coagulation.

cailler, to curdle, curd, clot, coagulate, congeal.

caillette, *f.* rennet, abomasum, fourth stomach.

caillot, *m.* coagulum, clot (of blood).

caillotis, *m.* kelp.

caillou, *m.* pebble, flint (stone), small stone; **cailloux roulés,** boulders, pebbles.

cailloutage, *m.* fine stoneware, pebble work.

caillouteux, -euse, pebbly, flinty, stony.

cailloutis, *m.* road metal, broken stone, macadam.

cailloux, *m. pl.* see CAILLOU.

caïnca, *f.* cahinca root.

caïre, *m.* coir, cocoanut fiber.

caisse, *f.* case, box, chest, tank, crate, tub (for shrub), bank, cashier's desk, cash, fund, shell (of pulley), body (of vehicle), drum, frame, coffer, cylinder; — **à claire-voie,** crate; — **à eau,** water tank; — **à feu,** firebox; — **à outils,** tool (chest) box; — **à secousses,** shaking box; — **d'emballage,** packing case; — **en fonte,** cast-iron box.

caissier, *m.* cashier.

caisson, *m.* case, caisson; — **de fonçage,** caisson.

cajeput, cajeputier, *m.* cajuput.

cal, *m.* callosity, callus.

cal., *abbr.* (calorie) calorie.

calage, *m.* wedging, tightening, chocking, adjustment, leveling, propping up, timing.

calamandrie, *f.* germander.

calamarié, sedgelike.

calambart, calambac, calambour, *m.* aloes-wood, agalloch.

calame, *m.* calamus.

calament, *m.* calamint.

calaminaire, calamine bearing.

calamine, *f.* calamine, mineral of zinc carbonate.

calamité, *f.* calamity, distress.

calamiteux, -euse, calamitous, dilapidated

calamus, *m.* calamus, rattan(palm).

calandrage, *m.* calendering.

calandre, *f.* calender, roller, calandra, grain weevil, glazing machine.

calandrer, to calender, roll, press, glaze satine.

calandreur, -euse, *m.&f.* calenderer.

calant, chocking, propping up, leveling.

calathide, *f.* calathidium, calathium, head of a composite.

calathidiflore, calathidiflorus, having a capitulum.

calathidiphore, *m.* the stalk of a capitulum.

calathifère, having a calthide.

calathiforme, cup-shaped.

calathin, with cup-shaped crown.

calcaire, *m.* limestone, chalk; — **carbonifère,** carboniferous limestone; — **coquillier,** shell limestone, coquina; — **jurassique,** Jura limestone.

calcaire, calcareous, chalky.

calcanéum, *m.* heel bone, calcaneum.

calcareus, calcareous, growing on limestone.

calcarifère, calciferous, lime-bearing, calcareous.

calcariforme, spur-shaped.

calcarone, *m.* calcarone (large sulphur kiln).

calcédoine, *f.* chalcedony.

calcédoinieux, -euse, chalcedonic, chalcedonous.

calcéiforme, calceolate, slipper-shaped.

calcicole, calcicolous, dwelling on chalky soil.

calcicoles, *f.pl.* bacteroids adapted to calcareous soils.

calcifère, calciferous, calcareous.

calcifuge, calcifugous.

calcifuges, *f.pl.* bacteroids suitable for acid soils.

calcilithe, *f.* (compact) limestone.

calcin, *m.* boiler scale.

calcinable, calcinable.

calcinage, *m.* calcination, calcining.

calcination, *f.* calcination.

calcine, *f.* calcine, product of calcination, an enameling mixture of lead and tin oxides.

calciner, to calcine, calcinate, burn (as lime).

calcique, calcic, of or containing calcium; **chlorure —,** calcium chloride.

calcite, *f.* calcite, calc-spar.

calcium, *m.* calcium.

calcul, *m.* calculus, calculation, estimate, arithmetic; **— biliaire,** gallstone; **— des aires,** calculation of areas; **— du rendement,** determination of annual yield; **— urinaire,** vesical calculus.

calculable, that may be calculated.

calculateur -trice, *m.&f.* calculator.

calculateur, calculating.

calculatoire, calculatory, calculating.

calculer, to calculate, estimate, compute; **se —,** to be calculated.

calculeux, -euse, calculous, calculary (affection).

cale, *f.* wedge, chock, block, key, prop, hold (of a ship), cove.

calé, wedged, fixed, leveled.

calebasse, *f.* calabash, gourd.

caleçon, *m.* (pair of) drawers, pants.

calédonien, -ienne, Caledonian, Scotch

calendrier, *m.* calendar, almanac.

calendule, *f.* marigold, calendula.

calepin, *m.* notebook, memorandum book.

caler, to set, time, adjust, support, chock, block, fix, jam, stick, prop up, wedge, key, level, draw (of ship); **— le frein,** to brake, lock, put on the brake.

calfat, *m.* calking.

calfater, to calk.

calfeutrer, to stop up (chinks), make tight, close calk.

calibrage, *m.* calibration, grading by size.

calibre, *m.* caliber, caliper, (wooden) gage, template, pattern, mold, model, bore (of gun); **— à vis,** gage, caliber; **— de filetage,** thread gage.

calibrer, to calibrate, gage, measure.

calice, *m.* cup, chalice, calyx, calix, egg calyx; **— dialysépale,** polysepalous calyx; **— gamosépale,** gamosepalous calyx.

calicé, provided with a calyx.

caliche, *m.* caliche, Chile saltpeter.

caliciflore, calyciflorate, stamens and petals adnate to calyx.

calicin, calicinal, calicine.

calicinal, -aux, calycine, pertaining to calyx.

calicot, *m.* calico.

calicule, *m.* involucre, calicle, epicalyx.

caliculé, calyculate, bearing a calycle.

Californie, *f.* California.

californien, -enne, Californian.

califourchon, *m.* hobby(horse); **à —,** astride, astraddle.

calige, *m.* sea louse.

calin, *m.* calin (Chinese alloy of lead, tin, etc., for tea chests).

calleux, -euse, callous, horny; **corps —,** corpus callosum.

callose, *f.* callose.

callosité, *f.* callosity, callus.

callusmetaplasie, *f.* metaplastic alteration of cells through injury.

calmant, calming, soothing, sedative.

calme, *m.* calmness, calm, quiet, still, stillness.

calme, calm, quiet, tranquil.

calmer, to calm, soothe, quiet, still, appease, pacify; **se —,** to become calm.

calmouque, Kalmuck.

calomel, *m.* calomel, mercurous chloride.

calophylle, having beautiful leaves.

calopode, *m.* spathe, a large bract enclosing a flower cluster.

caloricité, *f.* caloricity.

calorie, *f.* calorie, unit of heat; **grande —,** large calorie, kilogram calorie; **petite —,** small calorie, gram calorie.

calorifère, *m.* heater, heating apparatus or installation, radiator; **— à air,** hot-air heater; **— d'eau,** hot-water heater.

calorifère, heat conveying, **tuyaux —s,** heating pipes.

calorifique, calorific, heating.

calorifuge, *m.* heat insulation.

calorifuge, heat-insulating.

calorimètre, *m.* calorimeter.

calorimétrie, *f.* calorimetry.

calorimétrique, calorimetric(al).

calorique, *m.* heat, caloric.

caloritropisme, *m.* caloritropism, curvature produced by heat.

calosome, *m.* ground beetle; **— sycophante,** ground beetle.

calot, *m.* block, wedge.

calotte, *f.* calotte, (skull) cap, dome, knob, head, cup, case (of a watch), brain cap, headpiece, neck, nose (of a Bessemer converter); **— sphérique,** segment of a sphere.

calpa, *f.* calpa, the capsule of Fontinalis.

calquage, *m.* tracing, copying.

calque, *m.* tracing, copy.

calquer, to trace, copy, imitate.

calquoir, *m.* tracing point.

calumet, *m.* calumet, reed.

calus, *m.* callosity, callus.

calvados, *m.* cider brandy.

calvitie, *f.* baldness.

calvus, calvous, naked, bald.

calybion, *m.* calybio, a hard one-celled dry fruit (acorn).

calycanthe, *m.* calycanthus.

calycanthémie, *f.* calycanthemy, abnormal petalody of the calyx.

calyce, *m.* calyx.

calyptratus, calyptrate, bearing a calyptra.

calyptre, *f.* calypter (of mosses).

calyptré, (root) with cap at its lower end.

calyptriforme, calyptriform, shaped like an extinguisher.

calyptrogène, *f.* calyptrogen.

cam, *m.* **bois de —,** camwood.

camaïeu, *m.* cameo.

camarade, *m.&f.* comrade, fellow, mate.

camare, *m.* camara, the cells of a fruit, follicle.

camarine noire, *f.* crowberry (*Empetrum nigrum*).

cambium, *m.* cambium.

cambouis, *m.* dirty oil or grease, coom (grease).

cambre, *f.* camber, curve, convexity.

cambré, cambered, arched, curved.

cambrement, *m.* cambering, bending.

cambrer, to bend, camber, curve, arch, warp.

cambrien, -enne, Cambrian.

cambrure, *f.* camber, curve, cambering, sweeping, curving.

came, *f.* cam, lifter, wiper, cog.

camée, *m.* cameo, camaieu.

caméléon, *m.* chameleon.

caméléonage de lait, alternation of reaction in litmus milk.

camélia, *m.* camellia (*Thea japonica*).

caméline, *f.* wild (false) flax, gold-of-pleasure (*Camelina sativa*).

camelle, *f.* a heap (pyramid) of sea salt.

camérine, *f.* crowberry (*Empetrum nigrum*).

camérule, *f.* small loculus.

camion, *m.* dray, truck, camion; **— citerne,** oil truck, gasoline truck.

camionage, camionnage, *m.* carriage, truckage, cartage.

camionner, to transport by truck or dray.

came, camme, *f.* cam, wiper, nipper, lifter, arm.

camomille, *f.* camomile, (*Anthemis sativa*); **— d'Allemagne,** German (wild) camomile (*Matricaria chamomilla*); **— des champs,** dog's fennel; **— fausse,** field (corn) camomile; **— ordinaire,** camomile; **— puante,** mayweed, stinking camomile (*Anthemis cotula*); **— pyréthre,** pellitory (*Anacyclus pyrethrum*); **— romaine,** Roman (common) camomile (*Anthemis nobilis*).

camouflage, *m.* artificial concealment, camouflage.

camoufler, to camouflage, disguise.

camourlot, *m.* cement (for filling the joints).

campagnard, *m.* countryman.

campagne, *f.* rural district, country, field, campaign, season, voyage, countryseat, region; **en —,** in the field.

campagnol, *m.* field vole, meadow mouse.

campane, *f.* campana, pasqueflower, capped hock.

campaniflore, having bell-shaped flowers.

campaniforme, campanulate, bell-shaped.

campanulacé, campanulate, bell-shaped.

campanule, *f.* harebell, bellflower, Canterbury bell (Campanula); — **gantelée,** throatwort.

campanule, campanulate, bell-shaped.

campanuliné, bell-shaped.

campêche, *m.* logwood, campeachy wood.

campement, *m.* encampment.

camper, to encamp, tent.

camphène, *m.* camphene.

camphine, *f.* camphine.

camphocarbonique, camphocarboxylic.

camphoglycuronique, camphoglucuronic, camphoglycuronic.

camphol, *m.* camphol, borneol.

camphorate, *m.* camphorate.

camphorique, camphoric.

camphre, *m.* camphor; — **anisique,** fenchone; — **de menthe,** mint camphor, menthol.

camphré, camphorated.

camphrer, to camphorate.

camphrier, *m.* camphor tree (*Cinnamomum camphora*).

campimètre, *m.* perimeter.

campong, *m.* kampong, native village.

campos, *m.* campo, Brazilian savannas.

camptodrome, camptodrome, having a bent course.

camptotrope, campylotrope, camptotropous, campylotropous, amphitropous, incurved.

campylidium, *m.* campylidium, an accessory fruit in certain lichens.

çampylospermé, campylospermous, seeds grooved lengthwise.

campylothèque, *f.* plant with contoured fruit.

can, *m.* edge; **sur —,** on edge, edgewise.

canadien, -enne, Canadian.

canaigre, *f.* canaigre, a large dock (*Rumex hymenosepalus*).

canal, *m.* channel, duct, canal, pipe, conduit, tube, culvert, flue, sluice, groove, fluting, passage; — **aérifère,** air passage; — **cholédoque,** biliary duct; — **d'aérage,** air passage, ventilating flue; — **d'amenée,** feeder, inlet pipe; — **de coulée,** runner; — **déférent,** vas deferens; — **de fuite, waste** (pipe) channel, tailrace; — **de**

graissage, lubrication or oil hole; — **résinifère,** resin duct.

canalicule, *f.* canaliculus, a small channel.

canaliculé, pitted, channeled, canaliculate, grooved.

canalifère, canaliferous, having a canal.

canaliforme, channeled.

canalisation, *f.* canalization, piping, system of pipes, pipe line, water main, drain (of house), sewerage, conduits, mains, wiring; — **d'éclairage,** lighting leads.

canaliser, to canalize, pipe (gas), connect, wire.

canamelle, *f.* sugar cane.

canard, *m.* drake, mallard, false report, hoax, nag (horse), feeding cup, air (passage) pipe; — **sauvage,** wild duck; — **siffleur,** widgeon; — **souchet,** shoveler.

canardière, *f.* decoy for wild ducks, duck pond.

canarine, *f.* canarin.

canaux, *m. pl.* see CANAL.

cancellé, cancellate, latticelike.

cancer, *m.* cancer, carcinoma.

cancéreux, -euse, cancerous.

canche, *f.* hair grass.

cancre, *m.* crab, cancer.

candélabre, *m.* candelabrum, chandelier, branched candlestick, espalier fruit tree.

candeur, *f.* candor, frankness.

candi, *m.* (sugar) candy, candied fruit.

candi, candied, candy.

candidat, *m.* candidate.

candidement, openly, frankly.

candir, to candy, crystallize, become candied.

candiserie, *f.* candy manufacture.

cane, *f.* duck.

canéfice, *f.* cassia.

canelle, *f.* cinnamon.

canelure, *f.* flute, furrow, groove.

canepetière, *f.* little bustard.

canescent, hoary, grayish-white.

caneton, *m.* duckling.

canette, *f.* duckling, cop, quill (of loom).

canevas, *m.* canvas, sketch, outline, skeleton, survey, triangulation; — **d'ensemble,** general triangulation.

cani, beginning to rot, unsound (timber).

canidés, *m.pl.* Canidae.

canif, *m.* penknife.

canillée, *f.* duckweed.

canin, canine.

canines, *f.pl.* canine teeth.

caniveau, *m.* gutter, gully, channel, conduit, trough, main, duct.

cannabin, *m.* cannabin, a poisonous resin.

cannacées, *f.pl.* Cannaceae.

cannaie, *f.* cane field, canebrake.

canne, *f.* cane, reed, blowing iron; — **à sucre,** sugar cane.

canneberge, *f.* cranberry (*Vaccinium oxycoccus*).

cannelas, *m.* cinnamon.

cannelé, channeled, grooved, fluted, ribbed, serrated, corrugated.

canneler, to flute, channel, groove.

cannelier, *m.* cinnamon tree.

cannelle, *f.* cinnamon (bark), an aromatic substance, channel, groove, tap, faucet.

cannellé, cinnamon-colored.

cannelure, *f.* groove, channel, fluting, grooving.

cannette, canette, cannelle, *f.* fassett, mug, spool, jug, cap, quill (of loom), common reed.

cannibale, *m.* cannibal, man-eater.

canon, *m.* cannon, gun, barrel (of rifle), gun barrel, body (of syringe), the nozzle, spout, pipe (of bellows), a roll or stick (of sulphur), a reel for yarn, canon, cannon (bone), shank, shin (of horse), measuring glass; — **de fusil,** gun (rifle) barrel; — **rayé,** rifled (barrel) gun, rifle; **soufre en —s,** roll sulphur.

canot, *m.* canoe, launch, boat.

cantaloup, *m.* cantaloup.

cantatrice, *f.* public singer.

cantharide, *f.* cantharis, Spanish fly

cantharide, iridescent like the cantharis.

cantharidé, cantharidated, of cantharides.

cantharider, to cantharidate, sprinkle with cantharides.

cantharidine, *f.* cantharidin.

cantharophiles, *m.pl.* cantharophilae, plants fertilized by beetles.

canthus, *m.* canthus.

cantine, *f.* canteen, chest, case.

canton, *m.* canton, district, forest range; — **de chasse,** shooting district.

cantonné, divide into sections or cantons.

cantonnement. *m.* range, dividing into sections, forest-range isolating (of sick animals); — **de chaleur,** accumulation of heat.

cantonner, to have the habitat, be confined.

canule, *f.* cannula, small tube, nozzle (of syringe).

canyon, *m.* canyon, gulch.

caoutchouc, *m.* (India) rubber, caoutchouc, (rubber) tire; — **durci,** hard rubber; — **en feuilles,** sheet rubber (in rough leaves); — **factice,** artificial rubber, substitute; — **mélangé,** mixed rubber; — **pneumatique,** pneumatic tire; — **régénéré,** reclaimed (regenerated) rubber; — **sulfureux,** — **vulcanisé,** vulcanized rubber; **tube de —,** rubber tubing.

caoutchoutage, *m.* treating with rubber, rubberizing.

caoutchouté, rubber, with a rubber tire.

caoutchouter, to treat with rubber, rubberize.

caoutchoutier, *m.* rubber plant, rubber worker.

caoutchoutifère, rubber-bearing.

cap, *m.* cape, head, promontory, headland, ship's head, foothill.

capable, capable, able, fit, competent.

capablement, skillfully.

capacité, *f.* capacity, ability, extent, volume, labor, efficiency, output, area, cubicle contents; — **cube,** cubic contents.

capcique, *m.* capsicum.

capelet, *m.* capped hock.

capillaire, *m.* capillary (tube), maidenhair (fern), adiantum; *pl.* capillaries.

capillaire, capillary, filiform, piliform, hairlike.

capillarité, *f.* capillarity, capillary attraction.

capillite, capillitium, *m.* capillitium.

capistrale, capistral.

capistrum, *m.* capistrum, mandibular region.

capitaine, *m.* captain, master.

capital, *m.* capital, principal thing; — **d'apport,** opening capital; — **foncier,** — **immeuble par nature,** soil capital; — **social,** registered capital (of firm), social capital.

capital, -aux, capital, chief, leading, great, important.

capital-action(s), *m.* capital stock.

capitale, *f.* capital.

capitalement, chiefly, absolutely.

capitaliser, to capitalize.

capitaliste, *m.* capitalist.

capitan, *m.* swaggerer, bully.

capité, capitate, pin-headed, growing in heads.

capitel, *m.* extract of lye of ashes and lime.

capitellé, capitellate.

capiteux, -euse, heady, strong, alcoholic.

capitule, *m.* capitulum, head, cluster.

capitulé, capitular, capitate.

capituler, to capitulate.

capituliforme, like a capitulum.

capnofuge, smoke-preventing.

capnoide, smoke-colored.

capoquier, *m.* kapok tree, silk-cotton tree (*Ceiba pentandra*).

caporal, *m.* corporal.

capot, *m.* cover, hood, cowling, conning tower; — **de cheminée,** smokestack top; — **de la manivelle,** crank guard.

câpre, *f.* caper.

caprice, *m.* caprice, whim, fit, humor, offshoot (vein).

capricieusement, capriciously, whimsically.

capricieux, -euse, capricious, whimsical.

capricorne, *m.* cerambycid, capricorn beetle.

câprier, *m.* caper tree, caperbush.

caprification, *f.* caprification, artificial fecundation.

caprifolié, resembling honeysuckle.

caprin, caprine, goatlike.

caprinate, *m.* caprate.

caprinique, capric.

caprique, capric (acid).

caproique, caproic (acid).

capron, *m.* large strawberry.

capryle, *m.* capryl.

caprylique, caprylic (acid).

capselle, *f.* shepherd's-purse (*Capsella bursa-pastoris*).

capsulage, *m.* capping (of bottles).

capsulaire, capsular (fruit).

capsule, *f.* capsule, cup, evaporating dish, cap, detonator, pod, round dish; — **à trous,** perforated dish, strainer, colander; — **de (en) porcelaine,** porcelain dish, evaporating dish; — **des mousses,** pyxis, pyridium; — **d'évaporation,** porcelain evaporating dish; — **fulminante,** detonator, blasting cap; — **surrénale,** suprarenal capsule.

captage, *m.* piping of water, recovery of waste products.

captation, *f.* catching, collecting (of water or current), tapping.

capter, to collect, catch, pick up, make available, pipe (water), tap, captivate, win over, recover, save (by-product).

captif, -ive, *m.&f.* captive, prisoner.

captif, -ive, captive.

captiver, to captivate, subdue, enslave.

captivité, *f.* capitivity.

capture, *f.* capture, seizure.

capuchon, *m.* cap, cover, hood, growth of tissue above the funiculus (of Euphorbiaceae), gas mantle; — **de caoutchouc,** rubber cap.

capuchonné, capped, hooded.

capucine, *f.* nasturtium (*Tropaeolum*), band (of rifle).

caquage, *m.* barreling, curing and barreling (of herrings).

caque, *f.* keg, herring barrel.

caquer, to (cure and) barrel.

caquesangue, *f.* dysentery.

caquetage, *m.* cackling (of hens).

caqueter, to cackle, squawk.

car, for, because, as.

saraba, *m.* cashew-nut oil, cardol.

carabe, *m.* carabid, ground beetle; — **dore,** gilt ground beetle.

carabé, *m.* a variety of amber.

carabidés, carabiques, *m.pl.* carabids, ground beetles.

caracoli, *m.* caracoli (an alloy).

caractère, *m.* character, characteristic, spirit, nature, letter, type, stamp, print, mark; —**s sexuels primaires,** primary reproductive organs.

caractérisant, characteristic.

caractériser, to characterize, mark, distinguish.

caractéristique, characteristic.

caracul, *m.* karakul (fur).

carafe, *f.* (wash) bottle, carafe, decanter.

Caramanie, *f.* Kerman.

caramel, *m.* caramel, burnt sugar.

caramélique, of or pertaining to caramel.

caraméliser, to caramelize, convert into caramel, mix with caramel.

caramine, *f.* crowberry (*Empetrum nigrum*).

carapace, *f.* shell, carapace.

carassin, *m.* crucian carp; — **doré,** goldfish.

carat, *m.* carat, small diamond.

carature, *f.* alloying (of gold), gold alloy.

caravane, *f.* caravan.

carbamate, *m.* carbamate.

carbamide, *f.* carbamide, urea.

carbamique, carbamic (acid).

carbanile, *n.* carbanil, phenyl isocyanate.

carbazide, *f.* carbazide.

carbazol, *m.* carbazole.

carbazotique, carbazotic, picric (acid).

carbet, *m.* ship shed.

carbimide, *f.* carbimide, carbonimide isocyanic acid.

carbinol, *m.* carbinol, methyl (alcohol).

carbinolique, carbinol, of or pertaining to a carbinol.

carboazotine, *f.* carboazotine.

carbodynamite, *f.* carbodynamite.

carbol, *m.* carbolic acid.

carbolique, carbolic.

carbonade, *f.* washing soda.

carbonado, *m.* carbonado.

carbonatation, *f.* carbonation, carbonatation.

carbonate, *m.* carbonate; — acide, bicarbonate; — d'ammoniaque, ammonium carbonate; — de chaux, calcium carbonate, carbonate of lime; — de fer, iron carbonate; — de magnésie, magnesium carbonate; — de plomb, white lead, lead carbonate; — de potasse, potassium carbonate, carbonate of potash; — de soude, sodium carbonate, carbonate of soda. — neutre, normal carbonate; — niccolique, nickel carbonate.

carbonaté, carbonated, containing carbonate.

carbonater, to carbonate; se —, to become carbonated.

carbonateur, *m.* carbonator.

carbone, *m.* carbon; — combiné, combined carbon; — de cémentation, — de carbure, cement carbon, carbide carbon, combined carbon; — de recuit, annealing carbon, temper carbon; — de trempe, temper carbon, hardening carbon; — fixe, fixed carbon; — graphitique, graphitic carbon; — volatil, volatile carbon.

carboné, carbonaceous, carbureted, carbonated, of or pertaining to carbon.

carboneux, -euse, carbonous, oxalic acid.

carbonide, *m.* carbide, carbonide.

carbonifère, carboniferous, coal-bearing.

carboniférien, -enne, carboniferous.

carbonimide, *f.* carbimide, carbonimide.

carbonique, carbonic, carboxylic.

carbonisation, *f.* carbonization, charring; — du bois, carbonization (of wood), charcoal making, charring (wood).

carbonisé, carbonized, charred.

carboniser, to carbonize char.

carbonite, *m.* carbonite.

carbonoïde, resembling carbon.

carbonométrie, *f.* carbonometry.

carbonyle, *m.* carbonyl.

carborundum, *m.* carborundum, silicon carbide.

carbosulfure, *m.* carbon disulphide.

carbosulfureux, -euse, containing carbon disulphide.

carboxyle, *m.* carboxyl.

carburant, *m.* carburetant, (motor) fuel.

carburant, carbureting.

carburateur, *m.* carburetor.

carburateur, -trice, carbureting.

carburation, *f.* carburization, carburation.

carbure, *m.* carbide, hydrocarbon; — acétylénique, acetylene hydrocarbon; — à noyau fermé, ring hydrocarbon; — de fer, carbide of iron; — d'hydrogène, hydrocarbon; — éthylénique, ethylene hydrocarbon; — homologue, homologous hydrocarbon; — saturé, saturated hydrocarbon; — térébénique, terpene; — volatil, volatile hydrocarbon.

carburé, carburized, carbureted.

carburer, to carburize, carburet, carbonize.

carburet, *m.* carburet.

carbylamine, *f.* carbylamine, carbamine, isonitrile.

carbyle, *m.* carbyl.

carcaise, *f.* annealing kiln (for plate glass).

carcas, *m.* refined cast iron, salamander.

carcasse, *f.* skeleton, frame, casing, carcass, dead body, shell; — en bois, wooden stand.

carcel, *m.* carcel (lamp).

carcérule, *m.* lime-seed capsule.

carcinomateux, -euse, carcinomatous.

carcinome, *m.* carcinoma, cancer.

carcithe, *f.* soft part of trees decomposed by fungi.

cardage, *m.* carding, combing (wool), teaseling.

cardamine, *f.* cardamine; — des prés, cuckooflower, meadow cress.

cardamome, cardamone, *m.* cardamom (tree), cardamon.

cardan, *m.* Cardan shaft, universal joint.

carde, *f.* card, carding brush, teasel frame, chard; — **poirée,** chard of beets.

carder, to card (wool), teasel (cloth).

cardère, *f.* teasel (Dipsacus).

cardiaire, *f.* motherwort (*Leonurus cardiaca*).

cardiaire, cardiac.

cardialgie, *f.* cardialgia, heartburn.

cardiaque, *f.* motherwort (*Leonurus cardiaca*).

cardiaque, cardiac.

cardinal, -aux, cardinal, dominating.

cardiopétale, with petals in the heart.

cardiophylle, with leaves in the heart.

cardium, *m.* kind of mollusk, cockle.

cardol, *m.* cardol (cashew-nut oil).

cardon, *m.* cardoon (*Cynara cardunculus*), thistle, shrimp.

carduacé, resembling the thistle.

carémage, *m.* March sowing.

carence, *f.* deficiency, lack (in diet), lack of assets.

carène, *f.* keel, carina, streamline body.

caréné, carinate, keeled.

caresse, *f.* caress.

caresser, to caress, fondle.

carex, *m.* carex, sedge (Carex).

cargaison, *f.* cargo, shipping.

cargo, *m.* cargo vessel, freighter

caricé, resembling carex or sedge.

caricine, *f.* caricin, papain.

carie, *f.* caries, decay (of bone), heart-shake, blight smut, stem rot; — **des bois,** heart-shake, heart rot; — **des os,** caries, cariosis; — **sèche,** humid, dry or wet rot (in timber).

carié, rotten at the core, carious.

carillon, *m.* rod (of iron), chimes.

carinal, carinal, resembling a keel.

carinifère, carinate, having carina (keel).

carinulé, with small carina.

carline, *f.* carline thistle (Carlina).

carmin, *m.* carmine.

carminatif, -ive, carminative.

carminé, carminated, carmine-colored.

carminer, to mix or color with carmine.

carnage, *m.* carnage, slaughter, raw meat.

carnallite, *f.* carnallite.

carnassier, -ere, carnivorous, flesh-eating.

carnassière, *f.* game bag.

carné, flesh-colored, of meat; **diète —e,** meat diet.

carneau, carnau, *m.* (boiler) flue or pipe, vent, draught; — **à fumées,** chimney flue.

carnet, *m.* notebook, memorandum book.

carnillet, *m.* bladder campion.

carnine, *f.* carnine.

carnivore, *m.pl.* carnivora.

carnivore, carnivorous.

caronculaire, *f.* aril, arillus formed of several caruncles.

caroncule, *f.* caruncle.

carossable, passable, practicable.

carossier, *m.* harness horse.

carotide, carotid (artery).

carotte, *f.* carrot, trick, core sample.

carottine, carotine, *f.* carotin.

caroube, *f.* carob (bean), St.-John's-bread; — **à miel,** honey locust, sweet locust.

caroubier, *m.* carob (tree), sweet pod tree (*Ceratonia siliqua*).

carouge, *f.* see CAROUBE.

carpadelle, *f.* an indehiscent and multi-locular fruit.

carpe, *f.* carp; *m.* wrist, carpus.

carpellaire, carpellary.

carpelle, *m.* carpel, seed pod.

carpien, carpal (bone).

carpinicole, living in the wood of *Carpinus betulus*.

carpobalsame, *m.* carpobalsamum.

carpogénèse, *f.* production of fruit.

carpogone, *m.* carpogonium.

carpogonie, *f.* female sex organs, female cell of red algae.

carpologie, *f.* classification of fruits.

carpolyse, *f.* dissociation of carpels.

carpomanie, *f.* carpomania, grittiness in fruit.

carpomorphe, carpomorphous.

carpophage, carpophagous, fruit-eating.

carpophore, *m.* fruit body, sporophore, carpophore.

carpophylle, *m.* carpel.

carposome, *f.* carposoma, the fruit-body of fungi.

carpospore, *f.* carposphore.

carpotropisme, *m.* carpotropism, the movements of fruits before or after pollination.

carpozygote, *m.* carpozygote, zygospore.

carquaise, *f.* see CARCAISE.

carquois, *m.* quiver.

carre, *f.* back and shoulders (of man), crown (of hat), corner (of field), square toe (of shoe), groove, resin blaze, streak, face.

carré, *m.* square, any square figure, landing place, printing demy, medium, floor, quadrate muscle; — **des officiers,** officers' messroom, officers' quarters.

carré, squared, square, quadratic, perpendicular, square-built, straightforward, plain-spoken, demy.

carreau, *m.* small square, tile, flagstone, pane (of window), paving brick, platform, floor, check, checker, tailor's goose, square cushion (of church), stretcher (masonry), diamond (cards), thunderbolt, dump (at mine); — **de poêle,** stove tile; — **en ciment,** cement slab; — **glacé,** stove tile.

carrelage, *m.* tiling, laying (of floor), flag pavement.

carreler, to pave, tile, lay (floor).

carrelet, *m.* square ruler, packing needle, plaice.

carrelier, *m.* tilemaker.

carrément, squarely, plainly, bluntly.

carrer, to square.

carrière, *f.* racecourse, arena, quarry, career, grit (in pear), course, play, scope; — **de gravier,** gravel pit; — **forestière,** forest career.

carrossable, for carriages.

carrosserie, *f.* coach building, body.

carrossier, *m.* coach (carriage) builder, body builder.

carrot vulgaire, *m.* goldeneye.

carroyage, *m.* squaring (of map).

carrure, *f.* breadth (of man) across shoulders.

cartable, *m.* instrument (herbarium) for collecting plants.

cartacé, chartaceous, papery, tegument of a dry seed.

carte, *f.* map, card, (piece of) cardboard, chart, bill, account, bill of fare; — **aérienne,** airway map; — **couchée,** glazed cardboard; — **de pain,** bread card; — **de visite,** visiting card; — **postale,** postal card.

carte-fiche, *f.* index card.

cartel, *m.* cartel, trust, challenge.

cartelle, *f.* veneer.

carter, *m.* case, casing, housing, crankcase.

carthame, *m.* saffron, safflower (*Carthamus tinctorius*); **huile de —,** saffron or safflower oil.

carthamine, *f.* carthamin.

carthamique, carthamic.

cartilage, *m.* cartilage, gristle.

cartilagineux, -euse, cartilaginous, gristly, hard, tough.

carton, *m.* pasteboard, (paper) board, card, cardboard, cardboard box, carton, portfolio, cartoon, small sketch, octavo part (print), four page, cancel; — **bitumé,** tarred board, tar (paper) felt, cardboard; — **collé,** pasteboard; — **couché,** glazed board; — **cuir,** leatherboard; — **d'amiante,** asbestos (mill) board; — **de collage,** sized pasteboard; — **de deuxième moulage,** pasteboard from wastepaper.

carton, — **de montagne,** asbestos board; — **de moulage,** millboard; — **de paille,** strawboard — **de paquetage,** strawboards; — **de tourbe,** peat board; — **en pâte de bois,** wood board, cellular pasteboard; — **glacé,** glazed board; — **gondronné,** asphalted board; — **minéral,** asbestos board; — **ondulé,** corrugated cardboard, cellular board; — **porcelaine,** glazed (enameled) board.

carton-cuir, *m.* leatherboard.

cartonnage, *f.* making of paperboard articles, cardboard boxes, cases, (binding in) paperboards.

cartonné, bound in cardboard.

cartonner, to bind (book) in boards.

cartonnerie, *f.* cardboard manufactory, pasteboard trade.

cartonneux, -euse, resembling cardboard.

cartonnière, guêpe —, *f.* paper wasp.

carton-paille, *m.* strawboard.

carton-pâte, *m.* papier-mâché, paper pulp.

carton-pierre, *m.* stone board (for roofing).

cartouche, — *m.* case, frame; *f.* cartridge, shell; — **à blanc,** blank cartridge.

cartoucherie, *f.* cartridge factory.

carus, *m.* coma, profound sleep.

carvacrol, *m.* carvacrol.

carvène, *m.* carvene, a terpene.

carvi, *m.* caraway (*Carum carvi*).

carvol, *m.* carvol.

caryer, *m.* hickory; — **amer,** bitternut hickory, swamp hickory; — **blanc,** shagbark hickory; — **tomenteux,** mockernut hickory, black hickory.

caryocarpe, with fruit similar to a nut.

caryocinèse, *f.* karyokinesis, nuclear division.

caryogamie, *f.* karyogamy, fusion of male and female nuclei.

caryoïde, *m.* albuminous substance in conjugate leaves.

caryomixie, *f.* fusion of two nuclei.

caryophylle, caryophyllaceous, caryophylleous.

caryophylline, *f.* caryophyllin.

caryopse, *f.* caryopsis.

cas, *m.* case, accident, event, circumstance, affair, matter, value; **au — où,** in the event; **dans tous les —,** in any case; **en — que,** in case, if; **en tout —,** in any case, at any rate; **hors le — de,** barring the case of; **le — échéant,** should the occasion arise, in case of need; **selon le —,** as the case may be.

cascade, *f.* cascade, waterfall; **en —,** in a series.

cascarille, *f.* cascarilla.

case, *f.* hut, compartment, box, square, division, pigeonhole.

caséase, *f.* casease.

caséate, *m.* caseate.

caséation, *f.* caseation.

caséeux, -euse, caseous, cheesy, casein like.

caséification, *f.* caseation, coagulation.

caséifié, casefied.

caséifier, to convert into cheese.

caséiforme, cheeselike, cheesy.

caséine, *f.* casein, curd.

caséinerie, *f.* casein factory, cheese making.

caséinogene, *m.* caseinogen.

caséinogene, casein-producing.

caséique, caseic.

caser, to place, put (stow) away, find a place.

caserette, *f.* **caserel,** *m.,* cheese mold (sieve).

caserne, *f.* barracks.

caset, *m.* caseworm, caddis worm.

casette, *f.* little (house) case, sagger.

caséum, *m.* caseum, casein.

casier, *m.* set of pigeonholes, filing case, bin, rack.

casilleux, -euse, brittle (glass).

casoar, *m.* cassowary.

casque, *m.* helmet, headset, hood, casque, helm.

casquette, *f.* cap.

cassable, breakable.

cassage, *m.* breaking, crushing (stone).

cassant, breaking, brittle (china), short (steel), curt, abrupt; — **à chaud,** red-(hot-) short (iron).

cassation, *f.* annulment, cancellation, invalidation.

cassave, *f.* cassava.

casse, *f.* breaking, breakage, break, ladle, pan, crucible, boiler, (letter) case, cassia, senna; — **aromatique,** cassia bark, Chinese (bastard) cinnamon (*Cinnamomum cassia*); — **en bois,** cassia bark; — **ligneuse,** cassia bark; — **noisettes,** nutcracker; — **vraie,** purging cassia, pudding-pipe tree (*Cassia fistula*), cassia pods.

cassé, *m.* casse paper, damaged paper, crack, snap.

cassé, broken, worn out.

casse-coke, *m.* coke breaker.

cassement, *m.* breaking.

casse-motte, casse-mottes, *m.* clod breaker.

casse-noisettes, *m.* nutcracker.

casse-noix, *m.* slender-billed nutcracker.

casse-pierre, casse-pierres, *m.* stone breaker, stone crusher, stone hammer, saxifrage (Saxifraga), samphire.

casse-poitrine, *m.* fiery liquor.

casser, to break, breakdown, snap, crush, crack (nuts), annul, shiver, peel.

casserole, *f.* casserole, saucepan.

casseur, *m.* breaker.

casseur, -euse, breaking, crushing.

casside, *f.* — **nébuleuse,** small green tortoise beetle.

cassier, *m.* cassia tree, case-rack.

cassis, *m.* black currant, black-currant liqueur.

cassissier, *m.* black currant shrub (*Ribes nigrum*).

cassitérite, *f.* cassiterite.

casson, *m.* piece (of broken glass), (rough) lump of sugar, broken cocoa nibs.

cassonade, *f.* muscovado sugar, brown (moist) sugar.

cassure, *f.* fracture, break, crack, breaking, rupture; — **à éclats,** splintering fracture;

— à **fibres,** fibrous fracture; — à **froid,** cold break; — à **grain(s),** granular fracture; — à **nerf,** fibrous fracture; — de **bois,** woody fracture; — **lamelleuse,** lamellar fracture; — **terreuse,** earthy fracture.

castanocarpe, with fruit of chestnut tree.

castine, *f.* limestone flux.

castor, *m.* beaver.

castoréum, *m.* castoreum.

castorine, *f.* castorin.

castration, *f.* castration, removal of anthers.

casuarines, *f.pl.* Casuarinaceae.

casuel, -elle, casual, accidental.

casuellement, casually, by chance.

catabolisme, *m.* catabolism, dissimilation.

cataclysme, *m.* cataclysm, deluge, subversion, overthrow, douche.

cataclysmique, cataclysmic.

catacombes, *f.pl.* catacombs.

catacorolle, *f.* catacorolla.

catagéotropique, positively geotropic.

cataire, *f.* catnip (*Nepeta cataria*).

cataire, purring.

catalase, *f.* catalase (an enzyme).

catalepsie, *f.* catalepsy.

catalogue, *m.* catalogue, list.

cataloguement, *m.* cataloguing.

cataloguer, to catalogue, list.

catalpa, *m.* common catalpa, smoking bean.

catalyse, *f.* catalysis.

catalyser, to catalyze.

catalyseur, *m.* catalyzer, catalytic (agent).

catalyte, *f.* catalyst, catalyzer.

catalytique, catalytic.

cataménial, catamenial, menstrual.

catapétale, catapetalous.

catapetalisme, *m.* adherence of petals to the androphore.

cataphorèse, *f.* cataphoresis.

cataphorétique, cataphoretic.

cataplasme, *m.* cataplasm, poultice.

catapuce, *f.* caper spurge (*Euphorbia lathyris*).

cataracte, *f.* cataract.

catarrhe, *m.* catarrh.

catastrophe, *f.* catastrophe.

catéchine, *f.* catechol, catechin.

catéchique, catechuic.

catéchu, *m.* catechu.

catégorie, *f.* category; — de **bois,** wood assortment; — de **grosseur,** diameter class, girth class.

catégorique, categorical.

caténaire, catenary, chainlike.

caténé, catenate, the coherency in a connected chain.

caténulé, catenulate.

cathartine, *f.* cathartin (cathartic acid).

cathartique, cathartic.

cathédrale, *f.* cathedral.

cathérétique, catheretic.

cathète, *f.* cathetus, perpendicular.

cathéter, *m.* catheter.

cathétomètre, *m.* cathetometer.

cathode, *m.* cathode.

cathodique, cathodic, cathode.

catholicon, *m.* catholicon, panacea.

cati, *m.* gloss, luster.

cati, pressed, lustered.

cation, *m.* cation.

catir, to press, gloss, give luster to (cloth).

catissage, *m.* glossing, pressing, lustering (of cloth).

catizophyte, *f.* plant with stamens inserted on the disc.

catoclésie, *f.* a heterocarpous fruit.

catoptrique, *f.* catoptrics.

catoptrique, catoptric, catoptrical.

catotrope, *f.* free hymenium.

cauchemar, *m.* nightmare, incubus.

caucher, *m.* parchment form (for gold leaf).

caudal, caudal.

caudé, caudate, tailed.

caudex, *m.* stock, stem, caudex.

caudicule, *f.* caudicle, hypocotyl.

caudifère, caudate, tailed.

caulescent, caulescent, with leafy stem aboveground.

caulicole, caulicolous.

caulicule, *f.* small stem.

caulidium, *m.* caulidium, pseudo stem of cellular plants.

cauliflore, cauliflorous, flowers spring up on a stem.

cauliflorie, *f.* cauliflory.

cauliforme, *f.* cauliform, shape of a stalk.

caulinaire, *f.* cauline.

caulinaire, cauline, arising from stem.

caulobulbe, _m._ pseudobulb.

caulocalice, _f._ protective covering of the sporogone.

caulocarpien, caulocarpous, bearing fruit repeatedly.

cauloide, _m._ cauloid, emulating a stem.

caulome, _m._ caulome, stem structure.

causalité, _f._ causation.

causant, causatif, -ive, causing, causative.

causativement, causatively.

cause, _f._ cause, consideration, reason, case, grounds, motive, subject; **à — de,** because of, owing to, on account of; **à — que,** because; **en —** concerned.

causé, caused, for value (received).

causer, to cause, occasion, be the cause of, talk, chat.

causticité, _f._ causticity, caustic humor.

caustificateur, _m._ causticizer.

caustifier, to causticize, render caustic.

caustique, _m._ caustic (curve), stain, mordant; **— au chlorure de zinc,** zinc caustic.

caustique, caustic, biting.

caustiquement, caustically.

cautère, _m._ cautery, cauterization, cauterizing iron, issue, ulcer.

cautérisant, _m._ cauterizing-iron.

cautérisant, cauterizing.

cautérisation, _f._ cauterization, cautery.

cautérisé, cauterized, when bleeding has ceased.

cautériser, to cauterize, sear.

caution, _f._ security, guarantee, bail, surety.

cautionnement, _m._ security, bond, bail.

cautionner, to guarantee, go bail, give security for.

cavalerie, _f._ cavalry.

cavalier, _m._ cavalier, rider (of balance), staple, spoil bank (of road), size of paper 64 by 62 cm.; **— crampon,** staple; **— curseur,** sliding (adjustable) rider.

cave, _f._ cellar, cave.

cave, hollow, concave.

caveau, _m._ small (wine) cellar, vault.

caveçon, _m._ barnacle, cavesson.

caver, to hollow (out), excavate, undermine, stake, bet; **se —,** to wear away.

cavernaire, cavernarius, growing in caves.

caverne, _f._ cave, cavern, den, (lung) cavity.

caverneux, -euse, cavernous, hollow, spongy (tissue).

caviar, _m._ caviar.

cavicorne, hollow-horned.

cavité, _f._ cavity, hollow; **— pour antennes,** antennal scrobe.

cazette, _f._ sagger.

ce, this, that, he, she, it; **à ce que,** from what, from this; **ce dont, ce que, ce qui,** that which, that of what; **c'est-à-dire,** that is, that is to say; **si ce n'est,** except, unless.

céanothe, céanote, _m._ New Jersey tea (_Ceanothus americanus_).

ceci, this.

cécidie, _f._ cecidium, gall, bacterial tumor of plants.

cécidomye, _f._ gall midge; gall gnat; **— destructive,** Hessian fly; **— du blé,** wheat midge; **— du saule,** willow-tree midge.

cécité, _f._ blindness, amaurosis; **— des neiges,** snow blindness.

cédat, _m._ natural steel.

céder, to give up, part with, cede, yield, grant, surrender, transfer, assign, sell, give in, submit, give way, resign; **le — à,** to be inferior to.

cédrat, _m._ citron (_Citrus medica_), lemon (_Citrus limonia_).

cédraterie, _f._ citron plantation.

cédratier, _m._ citron tree, lemon tree.

cèdre, _m._ cedar (wood or tree); **— blanc,** Northern white cedar, arborvitae; **— de Siberie,** Siberian cedar, stone pine (_Pinus cembra_); **— de Virginie,** red cedar (_Juniperus virginiana_); **— du Liban,** cedar of Lebanon (_Cedrus libani_).

cédrel, _m._, **cédrèle,** _f._ Cedrela.

cédréléon, _m._ oil of cedar.

cédrie, _f._ cedar resin, cedar pitch.

cédron, _m._ cedron (_Simaba cedron_).

ceignant (ceindre), girding, surrounding.

ceignit (ceindre), (he) girded, surrounded.

ceindre, to surround, gird on (a sword), enclose, bind, encircle.

ceint, surrounded, inclosed, encircled, (he) encloses.

ceinturage, _f._ girdling, deadening, banding.

ceinture, _f._ belt, band, girdle, zone, circle, enclosure; **— de colle de chenille,** grease band.

ceinturer, to girdle.

cela, that, that (person), he, him, her, they, them; à — près, for that all; c'est — même, the very thing; c'est ça, that's it; n'est-ce que —? is that all? par —, for that reason.

céladon, pale (sea) green.

célastre, m. Celastrus; — grimpant, (climbing) bittersweet, (Celastrus scandens).

célèbre, celebrated, famous, renowned, well-known.

célébrer, to celebrate, solemnize.

célébrité, f. fame, celebrity.

celer, to conceal, hide.

céleri, m. celery; — frisé, curled celery.

céleri-rave, m. celeriac, turnip-rooted celery, German celery.

célérité, f. celerity, swiftness, quickness.

céleste, celestial, heavenly; bleu —, sky blue.

célestine, f. celestite.

célibat, m. celibacy.

cellase, f. cellase.

celle, she, her, this one, that one, the one, that.

celle-ci, this (one), she, her, the latter.

celle-là, that (one), she, her, the former.

celles, they, them, those.

celles-ci, they, them, these, the latter.

celles-là, they, them, those, the former.

cellier, m. storeroom on ground floor.

celloïdine, f. celloidin.

cellose, f. cellose.

cellulaire, cellular.

cellule, f. cell, cellule, small cell; — à compter, — à fond divisé, counting chamber; — adipeuse, fat body; — annexe du stomate, subsidiary cell of stomate; — aréolée, cell with bordered pits; — compagne, companion cell; — composée, composite cell; — de base, stalk cell of hair; — de bordure, guard cell; — de canal du col, neck canal cell; — de l'assise mécanique, mechanical cell.

cellule, — de Thoma, Thoma counting chamber; — disloque, cell of complementary tissue of lenticel; — du bois, wood parenchyma cell; — durable, resting cell; — grillagée, scalariform cell; — laticifère, latex vessel; — libérienne, phloem parenchyma cell; — ligneuse, tracheid; — mère, mother cell; — mère terminale, apical cell; — pierreuse, stone cell; — ponctuée, pitted cell.

cellulé, cellular, cellulated.

celluleux, -euse, cellular.

cellulite, f. cellulite, cellulitis.

celluloïd, celluloïde, m. celluloid.

cellulose, f. cellulose, rude fiber.

cellulosique, cellulosic, cellulose.

celui, he, him, she, her, (the) one, that, this, this one, that one.

celui-ci, this, this one, he, him, the latter.

celui-là, that, that one, he, him, the former; — qui, the one who, he or him who.

cembre, cembro, m. stone pine (Pinus cembra).

cément, m. cement.

cémentation, f. cementation, casehardening.

cémentatoire, cementatory.

cémenter, to caseharden, face-harden, cement.

cémenteux, -euse, cementlike.

cendrage, m. ashing over of molds, blacking.

cendraille, f. ashy debris.

cendre, f. ash, ashes, cinder, dust; — bleue, blue ashes, mountain blue; — de plomb, dust shot; — d'os, bone ash; — mouvante, — volante, flue dust; — verte, green verditer; —s de bois, wood ashes; —s de levure, ash of yeast; —s d'orfèvre, gold(smith's) ash; —s gravelées, crude pearlashes, wine lees ashes; —s provenant des mottes gazonnées, turf ashes; mettre (réduire) en —, to reduce to ashes, incinerate.

cendré, cinerous, ash(-colored), gray, ashen, grayish.

cendrée, f. lead ashes, dross, refuse. litharge, dust shot, small shot; — de Tournay, a hydraulic cement (from coal dust and lime).

cendrer, to mix with ashes, color ash-gray.

cendreux-, euse, ashy, covered with ashes, brittle or flawy (from ash).

cendrier, m. ashpit, ash box, ashpan.

cendrière, f. peat, a woman selling ashes.

cendrure, f. spots, flaws (in steel due to ashes), ash spot, cinder spot.

cenelle, f. hawthorn berry.

cénobie, f. colony of blue-green algae cells.

cénobion, m. dry indehiscent many-celled fruit.

cénocentre, m. coenocentrum.

cénocyte, f. coenocyte.

cénomonécies, f.pl. coenomonoecia, polygamous plants.

cénopode, *m.* coinopodus, terminating downwards in a cone.

cénosphère, *f.* coenosphere, coenocentrum, cytoplasmic structure.

censé, considered, supposed, reputed, regarded, thought.

censeur, *m.* censor, proctor, auditor; **de —,** censorious.

censure, *f.* censorship, censure.

cent, (a, one) hundred; **pour —,** per cent.

cent., *abbr.* (centime) (one 100th part of a franc), (centième), hundredth.

centaine, *f.* about a hundred.

centaurée, *f.* centaury (Centaurea).

centenaire, *m.* centennial, centenarian.

centenaire, centenary, centenarian.

centenille, *f.* false (bastard) pimpernel, chaffweed (*Centunculus minimus*).

centennal, centennial.

centésimal, centesimal, per cent, percentage.

centiare, *m.* centiare (1 square meter).

centième, hundredth, hundredth part; **en —s,** in hundredths, in per cent.

centigrade, centigrade.

centigramme, *m.* centigram.

centilitre, *m.* centiliter (10 ml.).

centime, *m.* centime (one 100th part of a franc).

centimètre, *m.* centimeter (0.3937 inch), tape measure; **— carré,** square centimeter; **— cube,** cubic centimeter.

centinode, *m.* knotweed, knapweed (*Polygonum aviculare*).

cent-millionième, hundred-millionth.

centrage, *m.* centering, adjusting, centration.

central, central, centric, in the center, main.

centrale, *f.* central establishment, plant, or power station.

centralement, centrally.

centralisateur, -trice, centralizing.

centralisation, *f.* centralization.

centraliser, to centralize.

centration, *f.* centering.

centraux, *pl.* see CENTRAL.

centre, *m.* center, middle, centrosome, a minute protoplasmic body; **— de triage,** railroad yard, **placé au —,** centered.

centrer, to center, adjust.

centreur, *m.* centering tool, center punch, finder.

centrifugateur, *m.* see CENTRIFUGEUR.

centrifugation, *f.* centrifugation, centrifuging.

centrifuge, *f.* centrifuge.

centrifuge, centrifugal, (of a turbine) outward-flow, inflorescence.

centrifuger, to centrifugalize, centrifugate, centrifuge.

centrifugeur, -euse, *m.&f.* centrifugal machine, centrifuge, separator.

centripète, centripetal, (of a turbine) inwardflow, inflorescence.

centrochondres, *m.pl.* kinetic centers, attractive spheres.

centrosome, *f.* centrosome, a minute protoplasmic body.

centrospermé, cyclospermous.

centrosphère, *f.* centrosome.

centroxyle, *f.* centroxyly, centrifugal primary woody structure.

centuple, centuple, hundredfold.

centupler, to centuple, increase (augment) a hundredfold.

centurie, *f.* century, hundred.

cénure, *m.* coenurus, coenure, many-headed bladder worm.

cep, *m.* vine-stock, -plant; sole (of plow).

cépage, *m.* vine plant, (variety of) vine.

cèpe, *m.* flap mushroom.

cépée, *f.* clump of stool shoots, group of coppice shoots, cluster.

cependant, meanwhile, in the meantime, however, still, nevertheless, yet; **— que,** while.

céphalalgie, *f.* cephalalgia, headache.

céphalé, cephalate.

céphalée, *f.* violent and persistent headache.

céphalique, cephalic.

céphalode, *m.,* **céphalodie,** *f.,* cephalodium, granular outgrowth on the lichen thallus.

céphalophore, *m.* cephalophorum, receptacle, stipe of some fungi.

céphalo-rachidien, -enne, cerebrospinal.

cèphe, *m.* cephus, stem sawfly.

cephoeline, *f.* cephaëline.

céracé, ceraceous, waxy.

ceraïste, *m.* chickweed (Cerastium).

cerambyx du noisetier, *m.* hazelnut capricorn beetle.

cerame, *m.* an earthenware vessel.

cérame, pertaining to earthenware vessels.

céramide, *f.* ceramidium.

céramique, *f.* ceramics, ceramic ware.

céramique, ceramic.

céramist, *n.* ceramist.

céramographie, *f.* ceramography.

cérasine, *f.* cerasin, metarabic acid.

cérat, *m.* cerate, ointment; — **de Goulard,** Goulard's cerate, cerate of lead sub-acetate; — **de résine anglais,** rosin cerate.

ceratoïde, ceratoid, horny.

cerce, *f.* curved template, circle, ring, hoop of a sieve, sieve frame.

cerceau, *m.* hoop.

cerche, *f.* sieve frame, sieve hoop.

cerclage, *m.* encircling, hooping, binding.

cercle, *m.* circle, hoop, ring, band, sphere, tire, orbit, club, collar; — **de fût,** cask hoop; — **de roue,** tire; — **de visée,** goniometer; **en —s,** in circles, in casks; **vin en —s,** wine in the wood.

cerclé, (yellow leaves) rolled up.

cercler, to encircle, ring, hoop.

cercope, *f.* froghopper.

cercueil, *m.* coffin, shell.

céréale, *f.* cereal, grain, corn grain, corn crops.

céréaline, *f.* cerealin, aleurone.

cérébelleux, -euse, cerebellar (artery).

cérébrine, *f.* cerebrin.

cérémonie, *f.* ceremony, formality.

cérémonieux, -euse, ceremonious, formal.

céréolé, *m.* cerate.

cérésine, cérésite, *f.* ceresin.

céreux, -euse, cerous (oxide).

cerf, *m.* stag, hart, red deer; — **sans bois,** stag without antlers, flat head.

cerfeuil, *m.* wild chervil, cow parsley (*Anthriscus sylvestris*); — **frisé,** curled chervil, double chervil; — **hérissé,** hemlock chervil (*Torilis anthriscus*).

cerf-volant, *m.* stag beetle.

cérides, *m.pl.* cerium metals.

cérifère, ceriferous, wax bearing.

cérification, *f.* a waxy coating.

cérine, *f.* cerin, cerotic acid.

cérique, of or pertaining to cerium, ceric; **oxyde —,** ceric oxide; **terres —s,** ceric earths.

cerisaie, *f.* cherry orchard.

cerise, *f.* cherry; *m.* cherry (color), cerise (dye).

cerisette, *f.* dried cherry, winter cherry.

cerisier, *m.* cherry (tree or wood); — **à fruits acides,** sour cherry (*Prunus cerasus*); — **à fruits doux,** — **merisier,** sweet cherry, gean (*Prunus avium*); — **à grappes,** bird cherry (*Prunus padus*); — **anglais,** duke cherry; — **des bois,** sweet cherry, mazzard cherry; — **de Virginie,** (wild) black cherry (*Prunus serotina*); — **mahaleb,** mahaleb, perfumed cherry (*Prunus mahaleb*); — **tardif,** black cherry.

cérite, *f.* cerite, cererite.

cérium, *m.* cerium.

cerne, *m.* circle, (annual) ring.

cerneau, *m.* green walnut (kernels), half of a kernel.

cernement, *m.* surrounding, ringing (of tree).

cerner, to encircle, surround, cut round, encompass; **se —,** to become encircled.

céroïde, resembling wax.

céroléine, *f.* cerolein.

céromel, *m.* ceromel (mixture of wax and honey).

cérotique, cerotic, cerotyl; **acid —,** cerin; **alcool —,** ceryl alcohol.

céroxyle, *m.* wax palm (*Ceroxylon andicolum*).

céroxyline, *f.* palm wax ceroxyle.

cerques, *m.pl.* cerci.

certain, certain, sure, some, positive; **—s,** *pl.* some, certain.

certainement, certainly, surely, indeed.

certes, certainly, surely, indeed, forsooth, in sooth.

certificat, *m.* certificate, pedigree.

certificateur, *m.* certifier, guarantor.

certifier, to certify, attest, guarantee, testify.

certitude, *f.* certainty, certitude, steadiness.

cérulé, céruléen, -enne, cerulean (blue), azure.

céruléine, *f.* cerulein (a coal-tar dyestuff).

céruléum, *m.* ceruleum (a pigment).

céruline, *f.* soluble indigo blue.

cérumen, *m.* cerumen, ear wax.

cérumineux, -euse, ceruminous (gland).

céruse, *f.* white lead, ceruse, basic lead carbonate.

cérusier, *m.* white-lead worker.

cérusite, *f.* cerussite (mineral lead carbonate).

cervaison, *f.* stag season.

cerveau, *m.* brain, mind, intellect, cerebral ganglion.

cervelet, *m.* cerebellum.

cervelle, *f.* brain (substance), brains.

cervical, -aux, cervical (artery).

cervoise, *f.* cervisia, barley beer.

cervule, *m.* cervulus, muntiacus.

céryle, *m.* ceryl.

cérylique, ceryl (alcohol).

ces, these, those, they.

césalpinie, *f.* brazilwood.

césium, *m.* see CAESIUM.

cespiteux, -euse, cespitose, growing in tufts.

cessant, ceasing, suspended.

cessation, *f.* cessation, ceasing.

cesse, *f.* ceasing, cessation, intermission; **sans —,** without ceasing, incessantly, continually, constantly.

cesser, to cease, stop, discontinue, end, leave off.

cession, *f.* assignment, transfer, cession.

cessionnaire, *m.* transfere, assignee, endorser.

c'est, it is, he is, she is.

c'est-à-dire, that is to say, that is, *i.e.*

cestodes, *m.pl.* Cestoda, tape worms.

cestoïde, cestoid.

cet (see CE), this, that.

cétacé, *m.* cetacean (whale).

cétazine, *f.* ketazine.

cétimine, *f.* ketimine.

cétine, *f.* cetin.

cétohexose, *f.* ketohexose.

cétoine, *f.* cetonia; **— darée,** rose beetle.

cétone, *f.* ketone.

cétonique, ketonic.

cétose, *m.* ketose.

cette (fem. of ce), this (one), that (one).

cétyle, *m.* cetyl.

ceux (pl. of celui), they, them, those.

ceux-ci, they, them, these, the latter.

ceux-là, they, them, those, the former.

cévadille, *f.* cevadilla, sabadilla.

cévadine, *f.* cevadine (from sabadilla seeds).

ch., chap., *abbr.* (chapitre), chapter.

chable, *m.* breakage.

chablis, *m.* Chablis (wine), windfallen wood, wood, breakage, windfall.

chabot, *m.* bullhead, miller's-thumb.

chabotte, *f.* anvil block, anvil (of power hammer).

chacal, *m.* jackal.

chacrille, *f.* cascarilla (bark of *Croton eluteria*).

chacun, each, each one, everyone, apiece.

chaf(f)ée, *f.* wheat bran (starch manufacture).

chagrin, *m.* grief, sorrow, affliction, trouble, chagrin, vexation, shagreen, grain (leather).

chagrin, sad, downcast, peevish.

chagriné, covered with small granulations.

chagriner, to grieve, distress, afflict, vex, chagrin, shagreen (skins).

chagrinier, chagrineur, *m.* shagreen maker.

chai, *m.* storage room, wine and spirits store.

chaille, see CHALOIR; **non qu'il m'en —** not that I care.

chaînage, *m.* chain measuring.

chaîne, *f.* chain, range, ridge, mountain chain, bonds, warp; **— à godets,** chain pump, bucket chain; **— à rouleaux,** block chain, roller chain; **— d'arpentage,** surveyor's chain; **— de collines,** range or chain of hills; **— de transmission,** driving chain, power chain; **— d'idées,** train of thought; **— fermée,** closed chain, ring; **— ganglionnaire ventrale,** ventral nerve cord; **— latérale,** side (lateral) chain; **— sans fin,** endless chain; **— transporteuse,** chain conveyor; **à — ouverte,** open chain; **pas de —,** chain pitch.

chaînée, *f.* length of land chain.

chaîner, to chain, measure with the chain.

chaînette, *f.* small chain, catenary, surveyor's chain.

chaînon, *m.* link (of chain); **— sulfoné,** sulphonic (sulpho) group.

chair, *f.* flesh, flesh side (of hide), meat, fiber (of iron), chair (of rail), pulp; **— fossile,** a kind of asbestos.

chaire, *f.* chair (in a university), professorship, pulpit.

chais, *m.* see CHAI.

chaise, *f.* chair, seat, frame(work), hanger, rail chair; **— console,** wall bracket; **— pendante,** hanger bracket.

chalait (chaloir), (it) mattered.

chaland, *m.* barge, lighter, customer.

chalaze, *f.* chalaza, base of ovule nucellus.

chalazogame, *m.* chalazogam.

chalazogamie, *f.* chalazogamy.

chalcanthite, *f.* mineral sulphate of copper, chalcanthite.

chalcédoine, *f.* chalcedony.

chalcographie, *f.* chalcography.

chalcopyrite, *f.* chalcopyrite, copper pyrites.

chalcosine, *f.* chalcocite.

chale, *f.* wood pile or stack.

châle, *m.* shawl.

chalef, *m.* oleaster (*Elaeagnus angustifolia*).

chalet-hydromètre, *m.* weather box.

chaleur, *f.* heat, warmth, hot weather, heated period, ardor, rut (animals); — blanche, white heat; — de combinaison, heat of (formation) combination; — latente, latent heat; — sombre, dark-red heat; — suante, — soudante, welding heat.

chaleureux, -euse, warm, ardent, passionate, glowing.

chalkolite, *f.* chalcolite (torbernite).

chalkosine, *f.* chalcocite.

chaloir, to matter, be important.

chaloupe, *f.* launch.

chalu, see CHALOIR.

chalumeau, *m.* blowpipe, pipe, reed, straw, flute; — à bouche, mouth blowpipe; — à gaz, gas blowpipe; — à gaz hydrique, oxyhydrogen blowpipe; — articulé, blast lamp; — à soufflerie, blowpipe with bellows; — oxhydrique, oxyhydrogen blowpipe.

chalybé, chalybeate.

chambertin, *m.* Chambertin (a Burgundy wine).

chambourin, *m.* a kind of white (glass) sand, strass (for false jewelery), common green glass.

chambranle, *m.* frame, casing, case; — de la porte, door lining.

chambre, *f.* chamber, room, apartment, bedroom, cavity, space, honeycomb; — à air, air chamber, inner tube; — à air comprimé, compressed-air chamber; — à compter, counting chamber; — aérienne, air chamber of stomate; — à poussière, dust catcher, dust chamber; — à vapeur, steam chamber or chest; — claire, camera lucida.

chambre, —de chauffe, stokehold, fireroom; — de décantation, screen door; — de nymphose, pupal chamber; — de plomb, lead chamber; — étuve, cool incubator; — humide, moist chamber; — noire, — obscure, camera, camera obscura; —

photographique, camera; — sous stomatique, inner air chamber of stomate.

chambré, chambered, honeycombed.

chambrer(se), to become honeycombed, pitted.

chambreur, *m.* holder.

chameau, *m.* camel.

chamécyparis, *m.* white cedar (*Chamaecyparis thyoides*).

chamelle, *f.* female camel.

chamois, *m.* chamois.

chamoiser, to chamois.

chamoiserie, *f.* chamois-leather factory, chamois leather.

chamoiseur, *m.* chamois dresser

chamotte, *f.* chamotte, fine clay, refractory clay.

champ, *m.* field, camp, range, scope, compass, matter, edge, cell, space; *pl.* country, croft; — de force, field of force; — des charges, area of loading; — d'essai(s), experimental plot, demonstration plot; — d'inondation, flood area; — labouré, tilled soil; — magnétique, magnetic field; — tournant, rotating field; — visuel, visual field, field of vision, field of view; —s maudits, fields where anthrax is endemic; de —, edgewise, on edge; sur le —, immediately.

champagne, *m.* champagne (wine); *f.* liqueur brandy.

champagniser, to treat (wine), aerate, give sparkle to (wine).

champelure, *f.* frostbite (of fruit trees).

champêtre, rural, rustic, country.

champignon, *m.* fungus, mushroom, round head (of bolt), countersink, head of rail, toadstool; — des charpentes, dry rot meruliose; — des maisons, house fungus (*Merulius lacrymans*); — du bois, wood fungus, dry rot; — filamenteux, hyphomycete.

champignonnière, *f.* mushroom bed.

champignonniste, *m.* mushroom grower.

champlé, frostbitten, nipped.

champlevé, champlevé, enamel.

chance, *f.* chance, luck, fortune.

chanceler, to stagger, totter, waver.

chanceux, -euse, hazardous, fortunate, lucky.

chanci, moldy, mildewed.

chancir, to mold, become mildewed.

chancissure, *f.* mold, moldiness, mildew.

chancre, *m.* canker, chancre, ulcer, crayfish, crawfish; — **du mélèze,** larch blister, larch; — **du sapin,** silver-fir canker; — **mou,** soft sore, chancroid.

chandelier, *m.* candlestick, chandelier, (branched) fixture, stump, snag; — **de fer,** iron stanchion; — **de jauge,** a gauge for measuring the width of a vessel.

chandelle, *f.* (tallow) candle, candlelight, stay, prop, shore, upright, stud, vertical support, icicle; — **à la baguette,** dipped candle; — **moulée,** molded candle; — **romaine,** Roman candle; **arbre à —,** candleberry tree (*Aleurites moluccana*).

chandellerie, *f.* candle works or shop, (tallow) chandlery.

chanfrein, *m.* chamfering, slopping.

chanfreiner, to chamfer, bevel, deaden, dull, rabbet.

change, *m.* change, exchange.

changeant, changing, changeable, variable, fickle.

changement, *m.* change, alteration, changing, variation, switch; — **d'avance,** feed change; — **de couleur,** discoloring; — **de marche,** reversing gear.

changer, to change, exchange, alter, convert, transform, replace, interchange; **se —,** to be changed, converted, change; — **de,** to change; — **de couleur,** to change color, become discolored; — **le sens de marche,** to reverse, change direction; **il change d'avis,** he changes his (mind).

chanlatte, *f.* chantlate.

chanson, *f.* song.

chant, *m.* singing, song, melody, air, canto, warble, crowing (cock).

chanter, to sing, chirp, warble.

chanterelle, *f.* chanterelle, decoy bird.

chanteur, -euse, *m.&f.* singer, songster.

chantier, *m.* stand (for barrels), gantry, (timber) yard, wood depot, slipway; — **d'ouvriers,** gang of woodcutters.

chantignole, *f.* bracket.

chantonné, (of paper) lumpy, defective.

chantournement, *m.* sweeping, curving.

chantourner, to cut in profile, saw round.

chanvre, *m.* hemp oakum, tow; — **d'eau,** double tooth, bidens; — **de Madras,** sunn, false hemp (*Crotolaria juncea*); — **de Manille,** Manila hemp, wild plantain (*Heliconia bihai*); — **sauvage,** hemp nettle.

chanvreux, -euse, hempen, hemplike.

chanvrière, *f.* hemp field.

chanvrin, *m.* hemp beetle, hemp agrimony, water hemp.

chaos, *m.* chaos.

chaparral, *m.* chaparral, dry shrubby regions.

chape, *f.* cover, lid, cap, capping, head, bearing (of balance), strap, clasp, loop, cope (of mold), pileus; — **de tête,** head.

chapeau, *m.* hat, cap, cover, lid, top, bonnet, head (of still or beer), cowl, hood (of chimney), felt, pileus; — **de cloison,** coping piece; — **d'ecrou fileté,** cap nut, screw; — **de fer,** gossan, "eisenhut."

chapelet, *m.* chaplet, string of beads, rosary, string, rope, beads (on liquors), chain (pump), series.

chapelle, *f.* valve box, valve case, cylinder casing, vault, crown of oven, chapel.

chapellerie, *f.* hat making, hat trade, hat shop.

chaperon, *m.* hood, cover, capstone, coping, chaperon.

chapiteau, *m.* head, cap, cornice, hood, roof, top, head (of still), capital (of column).

chapitre, *m.* chapter, subject, head(ing), item.

chapon, *m.* capon, young vine plant.

chaponner, to capon, castrate.

chaque, each, every.

char, *m.* car, chariot, wagon; — **d'assaut,** tank; — **de Venus,** monkshood, helmet-flower, aconite (*Aconitum napellus*).

characées, *f.pl.* Characeae, stoneworts.

charançon, *m.* weevil; — **de la patience,** willow weevil, willow borer; — **des glands,** acorn weevil; — **grand du pin et du sapin,** pine weevil; **petit — du pin,** white-pine weevil.

charbon, *m.* charcoal, char, coal, carbon, anthrax, carbuncle, charbon, smut, black rust, blight; — **à coke,** coking coal; — **à dessin(er),** drawing charcoal, charcoal crayon; — **à lumière,** electric-light carbon; — **à mèche,** cored carbon; — **bactéridien,** anthrax; — **bitumineux,** bituminous coal; — **blanc,** anasarca; — **de bois,** wood charcoal, charcoal; — **de cornue,** gas-(retort) carbon; — **de lumière,** electric-light carbon; — **de Paris,** briquettes; — **de saule,** drawing (willow) charcoal.

charbon, — **des tiges,** stripe smut of rye, stem smut; — **de terre,** (mineral) coal, pitcoal, bituminous coal; — **de tourbe,** peat coal; — **du lin,** foot rot of flax, dead stalk; — **en poudre,** charcoal dust; — **gras,** fat coal, soft coal; — **luisant,** anthracite, lustrous coal; — **maigre,** lean coal; — **plastique,** plastic coal; — **roux,** hydrogenous (or red) charcoal; — **symptomatique,** symptomatic anthrax; — **végétal,** vegetable charcoal.

charbonnage, *m.* coal mining, coal mine, charcoal burning.

charbonnaille, *f.* small coal, coal cinders, slack.

charbonné, charred, blackened, drawn with charcoal, smutty (of corn).

charbonnée, *f.* layer of charcoal, charcoal sketch, grilled steak.

charbonner, to char, carbonize, blacken with charcoal, polish with charcoal; **se** —, to char, deposit carbon.

charbonnette, *f.* charcoal wood, branch wood.

charbonneuse, *f.* anthracoid erysipelas, symptomatic anthrax.

charbonneux, -euse, carbonaceous, coaly, charry, carbuncular, affected with anthrax.

charbonnier, *m.* charcoal burner, (char)coal dealer or trade, coal cellar, collier.

charbonnière, *f.* charcoal kiln, charring furnace, charring place.

charbucle, *f.* blight.

charcuterie, *f.* pork, pork butchery.

charcutier, *m.* pork butcher.

chardon, *m.* thistle; — **à foulon,** teasel (Dipsacus); — **bénit,** blessed thistle (*Cnicus benedictus*).

chardonner, to teasel, bur.

chardonneret, *m.* goldfinch, thistle finch.

charge, *f.* charge, load, burden, pressure, office, responsibility, charging, accusation, pack, filling, tax; — **admise,** assumed load; — **admissible,** working load, safe load; — **amorce,** — **d'amorçage,** priming charge; — **brute,** total (gross) load; — **de marche,** useful load, working load; — **d'épreuve,** test (charge) load; — **de rupture,** breaking load, tensile strength, breaking stress; — **utile,** useful load, carrying capacity, live weight; **en** —, under load; **mettre en** — **le moteur,** to put motor under load.

chargé, charged, loaded, burdened, trusted with, turbid, saturated, overcast; — **de cours,** university reader (not a professor), deputy lecturer; — **de résine,** resinous, rosinous.

chargement. *m.* charging, load, cargo, charge, registered letter, loading.

charger, to charge with, load, overdo, exaggerate, register (a letter), burden, encumber, entrust, destine, accuse; **se** —, to become turbid (liquids), become overcast (weather); **se** — **de,** to charge (load) oneself with, undertake, take charge; **il s'en charge,** he takes the responsibility for (it).

chargeur, *m.* loader, charger, stoker, feeder.

chariot, *m.* chariot, wagon, carriage, trolley, cart, car, truck, carriage (typewriter), carrier, slide (rest), traveler; — **à couteaux,** cutter slide, knife slide; — **inférieur,** bottom carriage.

charioter, to cut down, round on lathe.

charité, *f.* charity.

charlatan, quack, charlatan.

charlatanisme, *m.* quackery, charlatanism.

charmant, charming, delightful, pleasant.

charme, *m.* charm, enchantment, delight, hornbeam, yoke elm (*Carpinus betulus*); — **commun,** *m.* hornbeam (*Carpinus betulus*); — **houblon,** hop hornbeam (Ostrya).

charmer, to charm, delight, bewitch, please.

charmille, *f.* arbor, hornbeam, yoke elm (*Carpinus betulus*).

charnière, *f.* hinge, joint; — **inférieure,** inner hinge of stomate; — **supérieure,** outer hinge of stomate; **à** —, with hinge(s), hinged.

charnu, fleshy, plump, meaty, pulpy.

charogne, *f.* carrion.

charpente. *f.* frame(work), framing, woodwork, carpentry, timberwork, skeleton.

charpenter, to frame, cut for framing.

charpenterie, *f.* carpentry.

charpentier, *m.* carpenter.

charpie, *f.* lint; — **de bois,** wood fiber, wool.

charquer, to dry, desiccate (meat).

charrée, *f.* (buck) ashes, lye ashes.

charretée, *f.* cartload, cartful.

charretier, practicable for wagons.

charrette, *f.* cart, wheelbarrow.

charriage, *m.* cartage, carriage, carrying.

charrier, to cart, carry, transport.

charroi, *m.* carriage, drayage, wagon traffic.

charron, *m.* wheelwright, cartwright.

charroyer, to cart, transport.

charrue, *f.* plow; — **fouilleuse,** subsoil plow; — **polysoc,** gang plow.

charte, *f.* charter.

chartré, chartered.

chartreuse, *f.* Chartreuse (liqueur).

chas, *m.* starch (paste) size, eye (of needle).

chaseret, *m.* cheese mold, cheese frame.

chasmanthère, *f.* cleistogamic flower.

chasmanthérie, *f.* partial cleistogamy.

chasmochomophyte, *f.* a plant of a rock crevice.

chasmogame, chasmogamic.

chasmogamie, *f.* chasmogamy, the opening of perianth at maturity.

chasmopétalie, *f.* persistent opening of corolla.

chasmophyte, *f.* chasmophyte, a plant growing in rock crevices.

chasse, *f.* chase, hunting, pursuit, flooding flushing, play, driver, set hammer, setter, drift; — à courre, — aux chiens courants, hunting, coursing; — à tir, — au fusil, shooting, stalking; petite —, small-game shooting.

châsse, *f.* knife edge (of balance), mounting, frame (of spectacles).

chasser, to drive out or away, expel, drive, propel, drive in, chase, pursue, hunt, go shooting, shoot; — au furet, to ferret.

chasseur, *m.* hunter, chasseur, sportsman.

chasseur, hunting.

châssis, *m.* frame, framework, chassis, sash, flask, plate holder, truck; — à semis, wire netting for seed beds.

chassoir, *m.* drift punch driver.

chat, *m.* cat, tomcat.

châtaigne, *f.* (sweet) chestnut (Castanea); — amère, — de cheval, — d'Inde, horse chestnut (*Aesculus hippocastanum*).

châtaigneraie, *f.* chestnut crop.

châtaignier, *m.* chestnut tree, sweet chestnut (Castanea).

châtain, chestnut (brown), nut-brown, auburn.

châtaire, *f.* catnip (*Nepeta cataria*).

château, *m.* castle, chateau; — d'eau, water tower.

chat-huant, *m.* screech owl, tawny owl.

châtiment, *m.* punishment.

chatoiement, chatoîment, *m.* chatoyancy, iridescence, sheen, variable luster (of colors).

chaton, *m.* kitten, catkin, ament, setting (of stone), bezel, husk (of nuts).

chatonné, incarcerated.

chatonnement, *m.* incarceration.

chatonner, to set (a stone).

chatouillement, *m.* tickling.

chatouiller, to tickle.

chatoyant, chatoyant, iridescent.

chatoyement, *m.* see CHATOIEMENT.

chatoyer, to be chatoyant, shimmer, sparkle.

châtrer, to castrate, prune, top, geld.

châtrure, *f.* castration.

chatte, *f.* she cat.

chaud, *m.* heat, warmth; à —, in a heated state, hot; tout — de . . . , straight from . . . .

chaud, hot, warm, fresh, recent, ardent, passionate.

chaude, *f.* heat, heating, brisk fire; — blanche, white heat; — grasse, melting heat; — rouge, red heat; — sombre, dark-red heat; — soudante, — suante, welding heat.

chaudière, *f.* boiler, copper, kettle, pan, evaporator, still; — à basse pression, low-pressure boiler; — à brai, pitch pot; — à brasser, (brewer's) copper; — à cuire, boiler, evaporator; — à moût, wort copper; — à retour de flamme, return-flame boiler; — à tubes d'eau, watertube boiler; — à vapeur, steam boiler; — de fusion, smelting crucible; — d'évaporation, evaporating kettle; — fixe, stationary boiler.

chaudra (chaloir), (it) will matter.

chaudron, *m.* kettle, vessel, caldron, copper, canker, daffodil (*Narcissus pseudo-narcissus*).

chaudronnée, *f.* kettleful.

chaudronnerie, *f.* coppersmith's work and wares, tinsmith's trade and wares.

chaudronnier, *m.* coppersmith, brazier.

chauffable, heatable.

chauffage, *m.* firing, fuel, heating, warming, stoking, combustible; — à vapeur, steam heating; —s discontinus, discontinuous sterilization, fractional sterilization.

chauffant, heating; plaque —e, hot plate.

chauffe, *f.* heating, firing, distilling, fire-place, stokehole, firehole, furnace.

chauffer, to heat, warm, fire (up), urge on, get hot, boost, excite, be heated, get up steam; se —, to get hot or warm, warm oneself; — à blanc, to heat to whiteness; — à la vapeur, to heat with steam; — au gaz, to heat with gas.

chaufferette, *f.* electric heating plate, foot warmer, chafing dish.

chaufferie, *f.* chafery, stokehole, boiler rooms, forge.

chauffeur, *m.* fireman, stoker, driver, chauffeur.

chauffure, *f.* scaling of iron or steel due to overheating, burnt iron.

chaufour, *m.* limekiln.

chaufournerie, *f.* lime burning.

chaufournier, *m.* lime burner.

chaulage, *m.* liming (of ground, trees, hides), whitewashing, manuring with lime.

chaulé, limed.

chauler, to sprinkle with lime, lime, whitewash.

chaulier, *m.* lime burner.

chaumage, *m.* digging up of the stubble.

chaume, *m.* stubble, stubble straw, thatch, culm, stem, haulm (of grasses), stubble field; —s, *pl.* pasture grounds.

chaumière, *f.* thatched cottage.

chausse, *f.* bag filter, straining bag.

chaussée, *f.* road, highway, roadway, ground floor, causeway, dike, turnpike.

chausser, to put on (stockings and shoes), fit, suit, hill, earth up (a tree).

chausse-trape, *f.* trap, star thistle (*Centaurea calcitrapa*), caltrop.

chaussette, *f.* sock, half hose.

chaussure, *f.* footwear.

chaut (chaloir), (it) matters.

chauve-souris, *f.* bat, flittermouse.

chaux, *f.* lime, calx, salt, limestone; — **anhydre,** quicklime; — **calcinée,** quicklime; — **caustique,** quicklime, caustic lime; — **éteinte,** slaked lime; — **éteinte à sec,** air(dry)-slaked lime; — **fibreuse,** fibrous gypsum, satin spar; — **fluatée,** fluorite, mineral calcium fluoride; — **fusée,** (air-) slaked lime.

chaux, — **grasse,** fat lime, lime from calcium-rich rocks; — **maigre,** poor lime, quiet lime, lime from magnesium-rich rocks; — **métallique,** metallic calx, china blue, blue pigment; — **morte,** dead-burned or over-burned lime; — **phosphatée,** phosphate of lime, calcium phosphate; — **pour engrais,** manuring lime; — **sodée,** soda lime; — **vive,** quicklime.

chaux-azote, *f.* calcium cyanamide.

chavibétol, *m.* chavibetol.

chavica, *n.* pepper (Piper).

chavirage, chavirement, *m.* turning upside down, upsetting, capsizing.

chavirer, to turn upside down, upset, capsize.

cheddite, *f.* cheddite.

chef, *m.* head, chief, chieftain, leader, foreman; — **d'atelier,** (shop)foreman; — **de bataillon,** major; — **de cantonnement,**range officer; — **de chantier,** — d'escouade, foreman, head workman; — **de file,** leader, first member, leader of file; — **de place,** foreman glass blower; **de ce —,** on that account; **en—,** in charge.

chefferie, *f.* small forest district.

chef-ouvrier, *m.* (shop) foreman.

chéilanthe, with labiate flowers.

cheilodrome, cheilodromous, lateral veins of leaf run from midrib to margin.

chélidoine, *f.* celandine (*Chelidonium majus*), swallowwort.

chélidonine, *f.* chelidonine (a bitter alkaloid).

chélidonique, chelidonic.

chemin, *m.* way, path, trail, track, road, passage; — **de fer,** railway, railroad; — **de fer d'intérêt général,** main line, trunk line; — **de parcours,** drift road; — **de vidange,** wood (forest) road; — **faisant.** on the way; — **ferré,** metaled road, macadamized road.

cheminée, *f.* chimney, flue, stack, shaft, smokestack, funnel, fireplace, nipple (of gun); — **d'appel,** air (ventilating) shaft; — **de meule,** flue (of charcoal kiln).

cheminement, *m.* advancing, progress (of thought), traveling, passage.

cheminer, to travel, advance, walk, go on.

chemise, *f.* shirt, chemise, case, casing, jacket, wrapper, lining, facing, envelope (of bullet); — **d'eau,** water jacket; — **de cylindre,** sleeve (cylinder) jacket; — **de vapeur,** steam jacket.

chemiser, to jacket, case, cover, line, coat

chemokinèse, *f.* chemokinesis.

chemomorphose, *f.* chemomorphosis.

chemotropism, *m.* response to chemical stimulus.

chênaie, *f.* oak-crop.

chenal, -aux, *m.* channel, gutter, (metal) drain.

chêne, *m.* oak (tree or wood) (Quercus); — **à écorce,** tanbark oak (*Lithocarpus densiflora*); — **à fleurs sessiles,** sessile oak; — **à la galle,** gall-bearing oak; — **blanc,** white oak (*Quercus alba*); — **chevelu,** —

de Bourgogne, Turkey oak (*Quercus cerris*); — des Indes, Indian oak, teak (*Tectona grandis*); — écarlate, scarlet oak (*Quercus coccinea*).

chêne, — kermès, kermes oak (*Quercus coccifera*); — liège, cork oak (*Quercus suber*); — occidentale, Western cork oak; — pédonculé, — rouvre, robur, British oak (*Quercus robur* var. *pedunculata*); — saule, willow oak (*Quercus phellos*); — tauzin, Pyrenean oak (*Quercus toza*); — vert, — yeuse, evergreen oak, holly, holm oak, (*Quercus ilex*).

chènevis, *m.* hempseed, hemp; huile de —, hempseed oil.

chènevotte, *f.* woody part of hemp, hemp stalk, boon.

chenille, *f.* caterpillar, worm, larva, chenille.

chenillette, *f.* Scorpiurus.

chénopode, *m.* goosefoot (Chenopodium).

chénopodées, chénopodiacées, *f.pl.* Chenopodiaceae.

chenu, bleached, hoary, gray-headed.

chepel, *m.* (live-)stock.

chèque, *m.* check.

cher, -ere, dear, costly, expensive.

chercher, to seek (for), look for, search (for), try, endeavor, get, attempt; aller —, to go and bring; — à faire, to try, endeavor to, do something.

chercheur, *m.* finder (of telescope), investigator, research man, seeker, searcher.

chère, *f.* cheer, fare, living.

chèrement, dearly, dear.

chérir, to cherish.

chersophytes, *f.pl.* chersophytes, dry waste plants.

cherté, *f.* dearness, expensiveness.

chervis, *m.* skirret, caraway.

chester, *m.* Cheshire cheese.

chète, *m.* seta, bristle.

chétif, -ive, weak, poor, miserable, wretched, mean, paltry, sorry, puny, thin, emaciated.

cheval, -aux, *m.* horse, donkey engine, metric horsepower; — de force, horsepower; — de trait, draught horse; — effectif, actual horsepower (a.h.p.); — électrique, electric horsepower; à — on horseback, astride, on both sides.

cheval-an, *m.* horsepower year.

chevalement, *m.* gallows frame, headgear.

chevalet, *m.* (wooden) horse, frame, support, trestle, rack, stand, jack, beam, pedestal (for bearings), easel, sawhorse, saw jack, felt board.

cheval-heure, *m.* metric horsepower hour.

chevalier, *m.* knight; — gambette, common redshank.

cheval-vapeur, *m.* metric horsepower, nominal horsepower.

chevanne, *f.* chub, chavender.

chevauchant, equitant (leaves).

chevauchement, *m.* lapping, overlapping, crossing.

chevaucher, to ride, overlap, cross.

chevaux, *m.pl.* see CHEVAL.

chevé, hollowed out.

chevêche, *f.* little owl.

chevelu, *m.* fine root fibers, beard, tuft, assemblage of fine roots.

chevelu, long-haired, hairy, bearded, comose, comate (seed).

chevelure, *f.* (head of) hair, coma, foliage, tail (of comet).

chever, to hollow out.

chevet, *m.* bed head, bolster.

chevêtre, *m.* trimmer.

cheveu, *m.* hair; les —, the hair.

cheville, *f.* peg, pin, bolt, spike, dowel, plug, ankle.

cheviller, to peg, pin, bolt.

chevillette, *f.* small peg, pin.

chevillure, *f.* tray antler, pegging, doweling.

chèvre, *f.* goat, she-goat, derrick, gin, crane, jack, trestle.

chevreau, *m.* kid, young goat.

chèvrefeuille, *m.* honeysuckle (Lonicera); — des bois, woodbine (*Lonicera periclymenum*); — des haies, European fly honeysuckle (*Lonicera xylosteum*).

chevrette, *f.* kid, doe, roe, (small) gin, tripod, jack, firedog, andiron.

chevreuil, *m.* roe deer, roebuck, roe.

chevrillard, *m.* fawn of the roe deer.

chevron, *m.* small rafter; bois de —s, quarters.

chevrotain, *m.* musk deer.

chevrotant, quivering, tremulous.

chevrotin, *m.* kid leather, fawn, musk deer, goat's-milk cheese.

chevrotine, *f.* buckshot.

chez, at the house of, at the home of, in the case of, with, among; — (lui), in (his) case; — moi, — lui, at home; — soi, in one's own home; chez . . . , care of . . . (on letters).

chiasse, f. dross, scum, rubbish; — de mouche, flyspeck(s).

chibou, m. chibou, cachibou (resin).

chicane, f. baffle, baffle plate, obstacle, chicanery, quibble, pettifoggery; joints en —, staggered joints.

chicané, baffled, provided with baffles.

chicoracées, f.pl. Chicuraceae, Liguliflorae.

chicorée, f. chicory (Chicorium intybus); — frisée, endive (Chicorium endivia).

chicot, m. snag, stub, stump.

chien, m. dog, sporting dog, hound, a small (trolley) car (for ore transportation), hammer, cock (of gun); — courant, (fox) hound; — d'arrêt, — couchant, setter, pointer; — de quête, bloodhound.

chiendent, m. couch grass, quack grass (Agropyron repens), triticum; — de mer, — marin, wrack, sea wrack; — fossile, amianthus; — officinal, couch grass (Agropyron repens); — queue-de-renard, hunger grass, hungerweed, slender foxtail (Alopecurus agrestis).

chienne, f. bitch, female dog.

chiffon, m. scrap, dress, chiffon; — d'essuyage, wiping (dust) cloth; —s de nettoyage, cotton waste; papier de —, rag paper.

chiffonier, m. ragman, junkman, chiffonier.

chiffrage, m. numbering, figuring, estimate.

chiffre, m. figure, numeral, amount, number, monogram, digit, cipher, signature; — comparatif, comparative value; — d'affaires, volume (amount) of business.

chiffrer, to number, cipher, calculate, write in cipher.

chignon, m. nape.

Chili, m. Chile.

chimère, f. chimera, imagination, illusion.

chimérique, chimerical, fantastical, fanciful.

chimiatre, m. iatrochemist.

chimiatrie, f. chemiatry, iatrochemistry.

chimiatrique, chemiatric, iatrochemical.

chimico-bactérien, chemicobacterial.

chimico-légal, -aux, chemicolegal.

chimico-minéralogique, chemicomineralogical.

chimico-physique, chemicophysical.

chimie, f. chemistry; — alimentaire, food chemistry; — analytique, analytical chemistry; — appliquée, applied, practical chemistry; — de l or, chemistry of gold; — industrielle, industrial, technical chemistry; — métallurgique, metallurgical chemistry; — minérale, inorganic, mineralogical chemistry; — pure, pure, theoretical chemistry; — végétale, plant chemistry.

chimiomorphose, f. chemomorphosis.

chimiosynthèse, f. chemosynthesis.

chimiotactique, chemotactic.

chimiotactisme, m. chemotaxis.

chimiotaxie, f. chemotaxis.

chimiotaxique, chemotactic.

chimiothérapie, f. chemotherapy.

chimiotropisme, m. chemotropism.

chimiquage, m. dipping (matches) into the inflammable compound.

chimique, chemical.

chimiquement, chemically.

chimiqueur, m. phosphorus composition attendant.

chimisme, m. chemism, chemical phenomena or process.

chimiste, m. chemist; — analyste, analytical chemist; — conseil, consulting chemist; — essayeur, analytical assayer, chemist; — indienneur, chemist in a calico printery; — métallurgiste, metallurgical chemist; — organicier, organic chemist.

chimiste-coloriste, m. color, dye chemist.

chimitypie, f. chemitype, chemitypy.

chimoine, m. Indian cement for stones, a kind of imitation marble.

china, m. chinaroot, (rhizome of Smilax china), cinchona, Peruvian bark.

chinage, m. yarn clouding, tinting of paper.

china-grass, m. ramie, China grass (Boehmeria nivea).

chinchilla, m. chinchilla.

Chine, f. China; papier de — rice paper.

chiner, to cloud (fabric), render chiné.

chinois, Chinese.

chinoline, f. quinoline.

chinone, f. quinone.

chio, m. taphole.

chiocogne, m. (West-Indian) snowberry (Chiococca racemosa).

chionante, m. fringe tree, snowflower tree.

chionophobe, f. chionophobe, plant shunning snow.

chipage, m. infiltration, puring, bating (of skins).

chipeau bruyant, *m*. gadwall.

chiper, to pure, bate (skins), infiltrate with tan liquor.

chique, *f*. sand flea, chigoe.

chiquer, to chew.

chirette, *f*. chirata, chiretta.

chiromancie, *f*. chiromancy, palmistry.

chirurgical, -aux, surgical.

chirurgie, *f*. surgery.

chirurgien, *m*. surgeon.

chirurgique, surgical.

chitine, *f*. chitin, entomolin.

chitineux, -euse, chitinous.

chiure, *f*. flyspeck.

chlamydé, *m*. chlamydia, bud scales, floral envelopes.

chlamydospore, *f*. chlamydospore.

chloracétate, *m*. chloracetate.

chloracétique, chloracetic.

chlorage, *m*. chloring (of wool).

chloral, *m*. chloral.

chloralamide, *f*. chloralamide.

chloralimide, *n*. chloralimide.

chloraliser, to chloralize.

chloralose, *n*. chloralose.

chloramidure, *m*. chloramide; — de mercure, mercury chloramide, aminomercuric chloride.

chloranile, *m*. chloranil.

chloranthe, *f*. chloranthy.

chlorate, *m*. chlorate; — de potasse, potassium chlorate, chlorate of potash; — de soude, sodium chlorate.

chloration, *f*. chlorination.

chlore, *m*. chlorine, yellowwort; — liquide, liquid chlorine, chlorine water.

chloré, chlorinated, containing chlorine, chlora; eau —e, chlorine water.

chlorenchyme, *f*. chlorenchyma.

chloréthane, *f*. chloroethane.

chloréthyle, *m*. ethyl chloride, chloroethane.

chloreux, -euse, chlorous.

chlorhydrate, *m*. hydrochloride, hydrochlorate, chlorhydrate; — d'ammoniaque, ammonium chloride.

chlorhydraté, muriated.

chlorhydrine, *f*. chlorohydrin.

chlorhydrique, hydrochloric, muriatic.

chloride, *m*. chloride.

chlorique, chloric.

chlorite, *m*. chlorite.

chlorité, chloritic, chloritous.

chloroamidure, *m*. chloroamide.

chloroamylite, *m*. chloramylite, chlorophyll granules.

chloro-anémie, *f*. chloroanemia.

chloro-aurique, chloroauric.

chlorobromé, containing, or combined with, chlorine and bromine, bromochloro.

chlorobromure, *m*. papier au —, chlorobromide paper.

chloro-carbonique, acide —, carbonyl chloride, phosgene.

chlorocarpe, green printed.

chlorocéphale, green- or yellow-headed.

chlorochromique, chlorochromic.

chloroforme, *m*. chloroform.

chloroformer, to chloroform.

chloroformiate, *m*. chloroformate.

chloroformique, chloroformic.

chloroformisation, *f*. chloroforming.

chloroformiser, to chloroform.

chlorogonimique, couche —, layer of brood cells.

chloroleucite, *f*. chloroplast.

chloromètre, *m*. chlorometer.

chlorométrie, *f*. chlorometry.

chlorophyllane, *m*. chlorophyllan.

chlorophyllase, *f*. chlorophyllase.

chlorophylle, *f*. chlorophyll, leaf green.

chlorophyllien, -enne, of or pertaining to chlorophyll.

chlorophyte, *f*. plant with green assimilating tissues.

chloropicrine, *f*. chloropicrin.

chloroplastide, *f*. chloroplastid.

chloroplatinate, *m*. chloroplatinate.

chloroplatinique, chloroplatinic.

chlororaphine, *f*. chlororaphin.

chlorose, *f*. chlorosis, greensickness, etiolation.

chlorosel, *m*. chloro salt.

chlorostachyé, with greenish spikes.

chlorosudation, *f*. exudation of watery liquid at sunset.

chlorosulfure, *m*. chlorosulphide.

chlorotique, chlorotic.

chlorotranspiration, *f*. chlorovaporization.

chlorurage, *m*. chlorination.

chlorurant, *m*. chlorinating agent.

chlorurant, chlorinating.

chloruration, *f.* chlorination.

chlorure, *m.* chloride; — **acide,** — **d'acide,** acid chloride; — **cuivrique,** cupric chloride; — **d'ammonium,** ammonium chloride, sal ammoniac; — **d'antimoine,** antimony chloride; — **d'argent,** silver chloride; — **d'azote,** nitrogen chloride; — **calcique,** calcium chloride; — **de calcium,** calcium chloride; — **de carbone,** carbon tetrachloride; — **de chaux,** chloride of lime, bleaching powder.

chlorure, — **de cuivre,** copper chloride, cupric chloride; — **de fer,** iron chloride, ferric chloride; — **de manganèse,** manganous chloride; — **de mercure, (bi),** mercuric chloride; — **de mercure, (proto),** mercurous chloride; — **de méthyle,** methyl chloride; — **de phosphore,** phosphorus chloride; — **de platine,** platinum chloride; — **de plomb,** lead chloride; — **de potassium,** potassium chloride; — **de sodium,** sodium chloride.

chlorure, — **de soude,** — **de soude liquide,** solution of chlorinated soda, Labarraque's solution; — **de soufre,** sulphur chloride; — **d'étain,** tin chloride; — **d'éthylène,** ethylene chloride; — **de zinc liquide,** zinc chloride solution; — **d'or,** gold chloride, auric chloride; — **ferreux,** ferrous chloride; — **ferrique,** ferric chloride; — **ferrique liquide,** solution of ferric chloride; — **mercureux,** mercurous chloride; — **mercurique,** mercuric chloride, corrosive sublimate.

chloruré, chlorinated, chloridized.

chlorurer, to chlorinate, chloridize, chlorinize.

choc, *m.* impact, shock, collision, knocking, blow, clash, jolt, encounter.

chocard, *m.* alpine chough.

chocolat, *m.* chocolate; — **au lait,** milk chocolate.

chocolaterie, *f.* chocolate factory, trade.

chocolatier, *m.* chocolate maker or seller.

choeur, *m.* chorus, choir, chancel.

choir, to fall, tumble.

choisi, *m.* choice article or person.

choisi, select, selected, chosen, elected, choice.

choisir, to choose, select, pick.

choix, *m.* choice, selection, option.

cholagogue, *m.* cholagogue (causing evacuation of bile).

cholémie, *f.* cholemia (presence of bile).

choléra, *f.* cholera; — **du porc,** hog cholera; — **des poules,** chicken cholera.

cholérique, *m.* cholera patient.

cholérique, cholera, choleraic.

cholestérine, *f.* cholesterol, cholesterin.

choline, *f.* choline.

cholinhémie, *f.* cholehemia, cholemia (flowing of bile into the blood).

cholurie, *f.* choluria.

chômage, *m.* stoppage, standing idle, inactivity, laying off, unemployment.

chômer, to suspend work, shut down, stand idle, be out of work, be short of (money).

chomophyte, *f.* chomophyte, a plant growing on ledges.

chon, *m.* plank with sloping sides.

chondrine, *f.* chondrin (a protein).

chondrogène, *m.* chondrigen.

chondrogène, chondrigenous.

chondroïde, chondroid, cartilaginous.

chondrome, *m.* chondrome, granular masses in the fluid cell-contents.

choquer, to strike, attack, shock, beat against, jolt, dash against; **se —,** to come in collision with.

chordorrhize, with filiform root.

chorée, *f.* chorea, Saint Vitus's dance.

chorion, *m.* corium (of skin).

choripétale, choripetalous, having petals separate, polypetalous.

chorise, *f.* doubling, separation.

chorisépale, chorisepalous, with separate sepals, polysepalous.

choroïde, *f.* choroid.

chortomie, *f.* procedure to preserve dry plants.

chose, *f.* thing, matter, affair, goods, property, deed; **pas grand'—,** nothing much; **peu de —,** a trifle; **autre —,** anything else; **quelque —,** something, anything; **—s diverses,** miscellaneous articles.

chou, *m.* cabbage; — **au vin,** wine cabbage; — **blanc,** white cabbage; — **frisé,** kale. borecole; — **frisé-pommé,** savoy; — **marin,** sea kale, sea cabbage (*Crambe maritima*); — **vert,** kale, green cabbage; —**x de Bruxelles,** Brussels sprouts.

choucas, *m.* daw, jackdaw.

choucroute, *f.* sauerkraut, pickled cabbage.

chouette, *f.* tawny owl.

chou-fleur, *m.* cauliflower.

chou-navet, *m.* turnip, rutabaga.

chou-palmiste, *m.* palm cabbage.

chou-rave, *m.* kohlrabi, turnip cabbage.

choux, *m.pl.* see CHOU.

chrême, *m.* chrism, holy oil.

chrétien, -enne, Christian.

christe, *f.* samphire (*Salicornia herbacea*).

christe-marine, *f.* (sea) samphire (*Crithmum maritimum*), common glasswort (*Salicornia europaea*).

christianisme, *m.* Christianity.

chromage, *m.* chromium plating.

chromate, *m.* chromate; — **acide,** bichromate; — **(bi) de potasse,** potassium dichromate, potassium bichromate; — **de baryte,** barium chromate; — **de fer,** iron chromate; — **de plomb,** lead chromate; — **de potasse,** potassium chromate.

chromater, to chromate, chrome.

chromatie, *f.* chromatism, chromatic aberration.

chromatine, *f.* chromatin.

chromatique, chromatic(al), of chromatin.

chromatiser, to render chromatic.

chromatisme, *m.* chromatism, coloration.

chromatogène, *m.* chromogen.

chromatogène, chromogenic.

chromatophore, *f.* chromatophore.

chrome, *m.* chromium, chrome.

chromé, containing chromium; **acier** —, chrome steel, chromium steel; **cuir** —, chrome-tanned leather.

chromeux, -euse, chromous.

chromidie, *f.* chromidia, discrete chromatin granules.

chromifère, chromiferous.

chromique, chromic; **acide** —, *m.* chromic acid.

chromisme, *m.* chromatism.

chromite, *m.* chromite, chrome iron; — **de fer,** chrome iron ore.

chromocinématographie, *f.* color cinematography.

chromocre, *m.* chrome ocher.

chromogène, *m.* chromogen.

chromogène, chromogenic, color-producing.

chromogenèse, *f.* chromogenesis.

chromogénique, chromogenic.

chromophile, chromophilous.

chromoplaste, *m.* chromoplast.

chromoplastide, *f.* chromoplastid, chromoplast.

chromosensible, color-sensitive.

chromosomes, *m.pl.* chromosomes.

chromosphère, *f.* chromosphere, color sphere.

chromule, *f.* chromule, coloring matter other than chlorophyll.

chromyle, *m.* chromyl, chromophyll.

chronique, *f.* chronicle.

chronique, chronic.

chronizoospore, *f.* chronispore, resting spore;

chronologique, chronological, of time.

chronomètre, *m.* chronometer.

chrysalide, *f.* chrysalis, pupa.

chrysalider (se), to pupate.

chrysaloïde, resembling a chrysalis.

chrysaniline, *f.* chrysaniline.

chrysanthème, *m.* chrysanthemum.

chrysarobine, *f.* chrysarobin.

chrysène, *m.* chrysene.

chrysobéryl, *m.* chrysoberyl, cymophane.

chrysocale, chrysochalque, *m.* pinchbeck (imitation gold of copper, tin, and zinc).

chrysocarp, chrysocarpous, having gold or yellow colored (fruit).

chrysocéphale, with head or bud of yellow gold.

chrysocolle, *f.* chrysocolla, copper or mountain green.

chrysoïdine, *f.* chrysoidine (producing orange and yellow colors).

chrysolithe, *f.* chrysolite.

chrysomèle de l'aune, *f.* alder leaf beetle.

chrysomélide, *m.* leaf beetle.

chrysophanique, chrysophanic (acid).

chrysoprase, *f.* chrysoprase.

chrysorrhize, with yellow roots.

chrysos, aureus, glowing yellow.

chu, (choir) fallen.

chuchotement, *m.* whispering.

chuchoter, to whisper.

chute, *f.* fall, drop, falling, failure, lapse, downfall, ruin, decline, prolapse; — **d'eau,** waterfall, head of water; — **de montagnes,** landslip; — **de potentiel,** fall (drop) of potential; — **des feuilles,** leaf fall, defoliation.

chylaire, chylous, chylaceous.

chyle, *m.* chyle.

chyleux, -euse, chylous, chylaceous.

chylifère, chyliferous (vessel).

chylifier, to chylify.

chylocaule, *f.* chylocaula, with succulent stems.

chylophylle, chylophyllous, with succulent leaves.

chylurie, *f.* chyluria.

chyme, *m.* chyme.

chymifier, to chymify.

chymose, *f.* chymification.

Chypre, *f.* Cyprus.

chypre, *m.* Cyprian wine.

chytridinées parasites, *f.pl.* parasitic Chytridiaceae.

ci, here, now; **celui-ci,** this, this one, the latter; **celui-là,** that, that one, the former; **de ci, de là,** here and there, in every direction; **par ci, par là,** here and there, now and then; **faire ci et ça,** to do this, that and the other.

ci-après, below, hereafter, later.

ci-bas, (signature affixed) below.

cible, *f.* target.

ciboule, *f.* scallion, green onion, Welsh onion, eschalot, shallot.

ciboulette, *f.* chive(s).

cicatrice, *f.* scar, cicatrix, stigma; — **de la feuille,** leaf scar.

cicatriciel, -elle, cicatricial (tissue).

cicatricule, *f.* cicatricule, tread (of egg).

cicatrisant, *m.* cicatrizant.

cicatrisant, cicatrizant (lotion).

cicatrisation, *f.* cicatrization, occlusion of wounds.

cicatriser, to cicatrize, heal, scar, mark.

cicérole, *f.* chick-pea (*Cicer arietinum*).

cicindèle, *f.* Cicindela, tiger beetle.

cicinnus, *m.* cincinnus, a uniparous scorpioid cyme.

ci-contre, opposite, at the side, in the margin.

cicutaire, *f.* water hemlock (*Cicuta virosa*).

cicuté, containing hemlock.

cicutine, *f.* cicutine (a liquid alkaloid).

cidariforme, having the form of a cap.

ci-dessous, below, undermentioned.

ci-dessus, abovementioned.

ci-devant, previously, formerly, former, late.

cidre, *m.* cider; — **doux,** unfermented fruit juice, sweet must.

cidrerie, *f.* cider making.

Cie, *abbr* (Compagnie), Company, Co.

ciel, *m.* sky, heaven, roof, crown, canopy, firmament; — **couvert,** clouded sky, cloudy weather; **à — couvert,** under cover; **à — ouvert,** open, in the open air, out of doors.

cierge, *m.* candle, wax candle, taper, cereus, mullein (Verbascum); — **à grandes fleurs,** night-blooming cereus (*Cereus grandiflorus*); — **géant,** giant cactus, saguaro (*Carnegiea* (Cereus) *giganteus*).

cierger, to wax.

ciergier, *m.* maker of or dealer in wax candles.

cieux, *m.pl.* sky, heaven.

cigale, *f.* grasshopper.

cigare, *m.* cigar, weed.

cigarette, *f.* cigarette.

cigogne, *f.* stork, crank, crank lever; **bec-de —,** *m.* crane's-bill.

ciguë, *f.* poison hemlock, conium (*Conium maculatum*); — **d'eau,** water hemlock (Cicuta); — **officinale,** — **ordinaire,** poison hemlock, cowbane (*Conium maculatum*); — **vireuse,** water hemlock, cowbane (*Cicuta virosa*); **petite —,** fool's-parsley.

ci-inclus, enclosed (copy).

ci-joint, herewith, annexed, enclosed.

cil, *m.* eyelash, cilium, lash, hair, flagellum, filament; — **vibratile,** flagellum.

ciliaire, ciliary.

ciliatifolié, with ciliate leaves.

ciliatopétale, with ciliate petals.

cilice, *m.* haircloth.

cilié, ciliate(d), flagellated, fringed.

cilié, *m.* flagellate.

ciliés, *m.pl.* Ciliata.

cilifère, ciligère, ciliferous, ciliated.

ciliolé, ciliolate, minutely ciliate, fringed with hairs.

ciller, to wink, blink.

cimaise, *f.* cyme, ogee.

cime, *f.* summit, top, cyme, peak, crown, crown structure.

cimeaux et branchages, *m.pl.* top and lop, tops and branches.

cimeen, cymose.

ciment, *m.* cement, concrete; — **à prise lente,** slow-setting cement; — **à prise (prompte) rapide,** quick-setting cement; — **armé,** reinforced cement or concrete; — **de laitier,** — **de scories,** slag cement; — **hydraulique,** hydraulic cement; — **réfractaire,** fireproof cement; — **romain,** Roman cement.

cimentage, *m.* cementing.

cimentaire, pertaining to cement, cement.

cimentation, *f.* cementation.

cimenter, to cement, strengthen.

cimentier, *m.* cementmaker.

cimetière, *m.* cemetery.

cimicaire, *f.* cimicifuga, bugbane.

cimicide, cimicicide, bug-destroying (powder).

cimicifuge, *f.* bugbane.

cimicifuge, bug-expelling (process).

cimier, *m.* sirloin, rumpsteak.

cimolée, *f.* cimolite, pipe clay, cutler's dust.

cinabarin, vermilion, cinnabarine, cinnabaric.

cinabre, *m.* cinnabar, red mercuric sulphide, vermilion.

cinchonamine, *f.* cinchonamine.

cinchone, *m.* cinchona.

cinchonicine, *f.* cinchonicine.

cinchonidine, *f.* cinchonidine.

cinchonine, *f.* cinchonine.

cinchonique, cinchonic; acide —, cinchonic (quinic) acid.

cinéfier, to incinerate.

cinelle, *f.* oak gall, oak apple.

cinéma, *m.* moving pictures.

cinématique, *f.* kinematics.

cinématique, kinematic(al).

cinématiquement, kinematically.

cinématographe, *m.* cinematograph.

cinématographie, *f.* cinematography.

cinène, *m.* cinene (a liquid terpene).

cinéol, *m.* cineol.

cinération, *f.* cineration, incineration.

cinériforme, resembling ashes.

cinèse, *f.* karyokinesis, changes in cell division.

cinésique, kinetic.

cinétique, *f.* kinetics.

cinétique, kinetic.

cinétospore, *f.* spore endowed with movement.

cinglage, *m.* (metal) shingling, sailing, run (of ship).

cinglard, *m.* shingling hammer.

cingler, to shingle, forge (bloom), lash, switch.

cingleresse, *f.* shingling tongs.

cingleur, *m.* shingler, squeezer; — rotatif, shingling rollers, squeezer.

cingleur, shingling.

cingleuse, *f.* shingler, squeezer.

cinnabre, *m.* cinnabar.

cinnamate, *m.* cinnamate.

cinname, *m.* cinnamon (tree) (Cinnamomum).

cinnaméine, *f.* cinnamein (benzyl cinnamate).

cinnamène, *m.* cinnamene, styrene, styrol.

cinnamique, cinnamic.

cinnamome, *m.* see CINNAME.

cinnamyle, *m.* cinnamyl.

cinq, five.

cinquantaine, *f.* about fifty, fiftieth anniversary.

cinquante, fifty, fiftieth.

cinquantenaire, *m.* fiftieth anniversary.

cinquantième, fiftieth, the fiftieth part.

cinquième, fifth, the fifth part.

cintrage, *m.* bending, deflection.

cintre, *m.* bend, curve, bow, arch, semicircle, centering; en plein —, full-centered, semicircular.

cintré, bent, curved, arched.

cintrement, *m.* bending, cambering (of timber), sagging (of beam).

cintrer, to bend, curve, sag, arch, camber.

cioche de verre, bell glass, glass cover.

cipolin, *m.* cipolin (marble).

cirage, *m.* waxing, polishing, blacking, varnish.

circée, *f.* enchanter's nightshade (Circaea).

circiné, circinate.

circoncis, circumscissile (dehiscence).

circoncision, *f.* circumcision.

circonférence, *f.* circumference, girth (of tree), perimeter.

circonflexe, circumflex, crooked.

circonlocution, *f.* circumlocution.

circonscription, *f.* circumscription, division, district, area.

circonscrire, to circumscribe, encircle.

circonscrit, circumscribed, limited.

circonspect, circumspect, cautious, wary.

circonspection, *f.* circumspectness, caution, wariness.

circonstance, *f.* circumstance, occasion.

circonstanciel, -elle, circumstantial, bothersome.

circonvolution, *f.* convolution.

circuit, *m.* circuit, circumference; — **amortisseur,** damping circuit; — **d'alimentation,** supply circuit; — **d'arrivée,** incoming circuit; — **de chauffage,** heating or filament circuit; — **dérivé,** derived or shunt circuit; — **inducteur,** — **primaire,** primary circuit; — **secondaire,** secondary circuit; — **utile,** output circuit; **hors** —, disconnected.

circulaire, *f.* circular (letter).

circulaire, circular.

circulairement, circularly, in a circle.

circulant, circulating, in circulation.

circularité, *f.* circularity.

circulation, *f.* circulation, traffic, street traffic, motion of protoplasm in cells; — **de la sève,** circulation of the sap.

circulatoire, circulatory.

circuler, to circulate, move round, revolve, travel, pass from hand to hand; **faire** —, to circulate, cause to move about.

circumnation, *f.* nutation, circumnutation.

circumscissile, *f.* circumscissile.

circumscription, *f.* the outline of an organ.

cire, *f.* wax, wax candle, beeswax; — **à cacheter,** sealing wax; — **à dorer,** — **des doreurs,** gilder's wax; — **blanche,** bleached or white wax; — **d'abeille(s),** beeswax; — **de cirier,** myrtle wax; — **de couture,** sewing wax; — **de myrica,** myrtle wax; — **d'Espagne,** Spanish (sealing) wax; — **de suint,** wool grease or wax; — **fossile,** fossil-wax, ceresin; — **minérale,** mineral wax, ozol cerite; — **vierge,** virgin wax, maidenwax.

cirer, to wax, black, polish, varnish, waterproof.

cireux, -euse, waxy.

cirier, *m.* wax myrtle (*Myrica cerifera*), wax chandler.

cirier, wax-producing.

cirières, *f.pl.* working bees.

ciron, *m.* mite, flesh worm.

cirque, *m.* circus, circular basin, cirque.

cirre, *m.* tendril, cirrus, tentacle.

cirré, cirrate, cerriferous.

cirreux, -euse, cirrous, barbellate.

cirrhifère, cirriferous, producing tendrils.

cirrhiforme, cirrhiform, resembling a tendril.

cirrhose, *f.* cirrhosis.

cirriflore, with flowers having tendrils.

cirripèdes, *m.pl.* Cirripedia.

cirse, *m.* horse thistle (Cirsium).

cirure, *f.* coating of wax, wax polish.

cis-, prefix cis-, (cisleithan).

cisaillage, *m.* shearing, cutting.

cisaille, *f.* clippings, cuttings, parings; *pl.* shears, nippers, wire cutter.

cisaillement, *m.* cutting, shearing, clipping, shearing (stress).

cisailler, to shear, clip, cut.

ciseau, *m.* chisel; — **à chaud,** hot chisel; — **à froid,** cold chisel.

ciseaux, *m.pl.* scissors, shears, clippers; — **à tondre,** sheep shears; — **de calfat,** calking iron; — **de jardinier,** pruning shears.

ciselage, *m.* grape thinning.

ciseler, to chase, engrave, chisel, carve, cut, shear, prune.

ciselure, *f.* chasing, embossing, engraving.

ciséron, *m.* chick-pea (*Cicer arietinum*).

cisoires, *f.pl.* bench shears.

cissampelos, *m.* cissus, bushrope.

cissoïde, *f.* cissoid (curve).

ciste, *m.* rockrose, cistus.

cistule, *f.* cistella, the apothecia of lichens.

citadelle, *f.* citadel.

citadin, *m.* citizen (of a city).

citation, *f.* citation, quotation.

cité, *f.* city.

citer, to cite, mention, quote, summon, name.

citerne, *f.* cistern, tank, reservoir, tanker; **wagon** —, tank car.

citerneau, *m.* small cistern.

citoyen, -enne, *m.&f.* citizen.

citraconique, citraconic (acid).

citragon, *m.* garden balm (*Melissa officinalis*).

citral, *m.* citral, geranial.

citrate, *m.* citrate; — **d'ammonique liquide,** solution of ammonium citrate; — **de potasse,** potassium citrate; — **de sodium,** — **de soude,** sodium citrate.

citraté, citrated.

citrène, *f.* citrene, d-limonene.

citrin, citrine, lemon-colored.

citrine, *f.* lemon oil, false topaz, citrine.

citrique, citric.

citron, *m.* lemon, lime, citron (*Citrus medica*).

citron, lemon-colored.

citronnade, *f.* lemonade.

citronnat, *m.* candied lemon peel.

citronné, like lemons, prepared with lemon.

citronnellal, *m.* citronellal.

citronnelle, *f.* citronella grass (*Cymbopogon nardus*), southernwood, citron water, Barbados water (liqueur).

citronnellol, *m.* citronellol.

citronnier, *m.* lemon tree (*Citrus limonia*).

citrouille, *f.* pumpkin, gourd.

cive, *f.* chive (*Allium schoenoprasum*).

civette, *f.* civet (perfume), civet cat, chive.

civière, *f.* (hand) barrow, stretcher, litter.

civil, *m.* civilian.

civil, civil, civilian, courteous.

civilisateur, *m.* civilizer.

civilisateur, -trice, civilizing.

civilisation, *f.* civilization.

civilisé, civilized.

civiliser, to civilize.

cladocarpe, *m.* pleurocarp, a club-shaped fruit.

cladode, *m.* cladophyll.

cladope, with branched stipe.

cladorrhize, with branched roots.

claie, *f.* screen, sieve, perforated wooden tray, hurdle, grid.

clair, *m.* light, clearing (in wood); **argent —,** ready money; **il fait —,** it is day (light); **lait —,** whey.

clair, bright, light, clear, thin, open, shining, transparent, evident, plain, unfertile (egg).

clair-bassin, *m.* buttercup, tall crowfoot.

clairçage, *m.* fining, clearing, purging (of sugar).

clairce, *f.* finings, fine liqueur, clearing sirup, clairce.

claircer, to clear, purge, decolorize.

claire, *f.* bone ash, clarifier, boiler (for sugar refining).

claire-étoffe, *f.* soft solder, an alloy of tin and lead.

clairement, clearly, plainly, distinctly.

claire-soudure, *f.* soft solder.

clairet, *m.* light red wine.

clairet, pale light (colored).

clairette, *f.* a sparkling white wine, lamb's-lettuce, corn salad (Valerianella), a disease of silkworms.

claire-voie, *f.* openwork, lattice, grating, skylight; **à —,** in openwork, open, loose-woven, thin (sowing).

clairière, *f.* glade, opening, gap blanc.

clairiéré, thinly-stocked, roomy, thinned out by age.

clair-obscur, *m.* light and shade, chiaroscuro.

clairsemé, thinly sown, scattered, sparse.

clairvoyant, shrewd, clear-sighted keen-witted, discerning.

clamander, to clamp.

clameau, *m.* clamp, cramp, (timber) dog, vise.

clameauder, to clamp.

clameur, *f.* clamor, outcry.

clamp, *m.* clamp, fish, forceps.

clandestine, *f.* broomrape (Orobanche).

clapet, *m.* clack valve, flap valve, rectifier, damper; **— à couronne,** cup (bell-shaped) valve.

clapier, *m.* rabbit hutch.

clapotement, *m.* splashing sound.

clapoter, to ripple, chop, plash (of sea).

claquement, *m.* chattering.

claquer, to clap, bang, clatter, smack, crack, burst.

claqueter, to clapper (stork), cackle, cluck (hen).

clarifiant, *m.* clarifying agent, clarifier, fining.

clarifiant, clarifying.

clarificateur, *m.* clarifier, fine.

clarification, *f.* clarification, purification.

clarifier, to clarify, purify.

clarifieur, *m.* clarifier.

clarté, *f.* light, clearness, transparency, brightness, limpidity, clarity.

classe, *f.* class, order, rank, category; **— d'âge,** age class; **— de fertilité,** quality class, grade of fertility; **— de hauteur,** height class.

classement, *m.* classification, classing.

classer, to classify, class.

classeur, *m.* filing cabinet, classifier, (ore) separator.

classification, *f.* classification, sorting, grading.

classifier, to classify, rate, arrange.

classique, classic, classical, regular, standard.

clastique, clastic, fragmentary.

clause, *f.* clause.

clausile, *m.* clausilus, macropodal embryo.

clausthalite, *f.* clausthalite.

clavaire, *f.* clavaria, club-top, mushroom

clavalier, *m.* prickly ash (*Zanthoxylum americanum*).

clavé, clavate, club-shaped.

claveau, *m.* vustulous eruption of sheep pox.

clavecin *n.* harpsichord.

Clavelée, *f.* sheep pox, variola, ovinia, scab-rot.

claveleux, -euse, pertaining to sheep pox.

clavelisation, *f.* inoculation against sheep pox.

clavetage, clavettage, *m.* keying, cottering.

claveter, clavetter, to key, wedge, fasten with cotters.

clavette, *f.* key, (bolt) pin, cotter; — de calage, — de serrage, cotter, tightening key.

claviculaire, clavicular.

clavicule, *f.* clavicle, collarbone.

clavier, *m.* keyboard.

clavifolié, with club-shaped leaves.

claviforme, clavate, club-shaped.

clayon, *m.* small screen, wicker tray (for draining cheeses), small hurdle, wattle enclosure.

clayonnage, *m.* wattlework.

clayonner, to groin, protect by groins.

clef, clé, *f.* key, plug (of cock), wrench, wedge, hook (of chain), keystone, crown, hitch, switch; — à béquille, spanner; — à dovilles, socket wrench; — à écrous, screw key, nut wrench, spanner; — à molette, monkey wrench; — anglaise, monkey wrench, coach wrench; — d'appel, call button; — de calage, adjusting, tightening key; — ouverte, open spanner.

cleistantherie, *f.* cleistanthery, the anthers of a cleistogamous flower remaining inside.

cleistocarpe, *m.* cleistocarp.

cleistogame, cleistogamous.

cleistogamie, *f.* cleistogamy.

cleistopétalie, *f.* cleistopetaly, permanent closing of corolla.

clématite, *f.* virgin's-bower (Clematis); — des bois, common clematis, traveler's joy (*Clematis vitalba*).

clenche, *f.* latch, catch, handle, pawl.

clerc, *m.* clerk, clergyman, scholar.

clichage, *m.* stereotyping.

cliché, *m.* (engraved) plate, block, cut, electro, stereotype (plate), (photographic) plate or negative, cliché; — à demi-teintes, half-tone cut, process block; —

d'impression, printing block; — de projection, lantern slide.

clicher, to stereotype.

client, *m.* client, customer, patient, purchaser.

clientèle, *f.* patronage, customers, consumers, custom, clientele, practice of medicine.

clignement, *m.* blinking, winking.

climacorhize, climacorhizal, surface of root not smooth.

climat, *m.* climate.

climatique, climatic.

clinandre, *m.* clinandrium.

clinanthe, *m.* clinanthium.

clin d'oeil, *m.* twinkling (of an eye).

clinide, *m.* clinidium, a filament in a pycnidium producing spores.

clinique, *f.* clinic.

clinique, clinic, clinical.

clinisme, *m.* clinism, inclination of the axis.

clinode, *m.* clinode, the conidiophore of certain fungi.

clinoïde, clinoid (apophysis).

clinomètre, *m.* clinometer.

clinorhombique, clinorhombic.

clinospore, *f.* stylospore, an exogenous spore.

clinostat, *m.* clinostat.

clinotropique, clinotropic, obliquely placed (organ).

clinquant, *m.* tinsel, Dutch gold, foil.

clinquant, tinseled, glittering.

cliquet, *m.* click, catch, (drill) brace, pawl, ratchet.

cliqueter, to rattle, clank, click.

cliquetis, *m.* rattling, clank(ing), jingle, jingling, knocking.

clissage, *m.* wickering (of bottles), setting (of fractured limb).

clisse, *f.* screen, cheese drainer, wicker covering (of bottle), splint.

clissé, wickered, wicker.

clisser, to wicker, cover with wicker, splint (limb).

clivable, cleavable (rock).

clivage, *m.* cleavage.

cliver, to cleave.

cloaque, *m.* cloaca.

cloche, *f.* bell jar, bell glass, bell, cover (of dish), hood, blister, bubble, bleb, covered glass, cylinder, dish, receiver, holder;

— **à cultures,** culture dish (with cover); — **à douille,** open-top bell jar; — **à vide,** vacuum bell jar; — **d'air,** air chamber; — **de bruyere,** heath bell; — **de soupage,** valve cage; — **de verre,** bell glass, receiver, bell jar; — **graduée,** graduated gas burette.

clocher, to cover with a bell jar.

clocher, *m.* steeple, belfry.

clochette, *f.* little bell jar, little bell, (any) small bellflower; — **des champs,** bindweed; — **d'hiver,** snowdrop (*Galanthus nivalis*).

cloison, *f.* partition, septum, wall, guide blade, cross wall, separating wall, partition wall, dissepiment.

cloisonnage, *m.* partition work, framework.

cloisonné, divided into compartments, cellular, cloisonné, valved, septate.

cloisonnement, *m.* partitioning, partition, septation.

cloisonner, to partition (off).

clone, *m.* clone.

clonique, clonic (spasm).

cloporte, *f.* sow bug, wood louse.

cloque, *f.* brown rust, blister, blight, exoascales; — **du pêcher,** peach-leaf curl.

clore, to close, enclose, shut.

clos, *m.* enclosure, close; — **de vigne,** vineyard.

clos, closed, enclosed, shut, tight, ended. **à la nuit** —, at nightfall.

closeau, *m.* closerie, *f.* small enclosure, small estate, cottage garden.

closier, *m.* cotter, crofter.

clostre, *f.* name for woody fibers.

clôture, *f.* enclosure, fence, fencing, wall, closing, conclusion, closure; — **en fil de fer,** wire fence.

clôturer, to close, terminate, shut in, fence.

clou, *m.* nail, spike, stud, rivet, (shoe) peg, boil, carbuncle; — **à ferrer,** horseshoe nail; — **à tête perdue,** brad; — **barbelé,** spike; — **de Biskra,** Biskra button, Oriental sore; — **de bouche,** tack; — **de girofle,** clove; — **d'épingle,** wire tack; — **de soufflet,** tack; — **en fonte,** cast nail.

clouage, clouement, *m.* nailing.

cloué, pinned.

clou-épingle, *m.* brad.

clouer, to nail, rivet, pin, nail down.

clouterie, *f.* nail (works) factory.

club, *m.* club.

clupéine, *f.* clupeine (a protamine).

cluse, *f.* a valley that crosses a mountain chain.

clypéolaire, clypeolate, shield-shaped.

clypéole, *f.* clypeole.

clypéolé, shield-shaped.

clystère, *m.* clyster, injection.

cm., *abbr.* (centimètre), centimeter.

cmc., *abbr.* (centimètre cube), cubic centimeter.

cmq., *abbr.* (centimètre carré), square centimeter.

cnique, *m.* Cnicus.

coadné, coadnate, connate.

coagglutiner, to cross agglutinate.

coagglutinine, *f.* coagglutinin, cross agglutinin.

coagulabilité, *f.* coagulability.

coagulable, coagulable.

coagulant, *m.* coagulant.

coagulant, coagulating.

coagulase, *f.* coagulase (an enzyme).

coagulateur, *m.* coagulator, coagulant.

coagulateur,-trice, coagulative, coagulatory.

coagulation, *f.* coagulation.

coaguler, to coagulate, congeal.

coagulum, *m.* coagulum, coagulant.

coaille, *f.* inferior wool (from the tail).

coalescence, *f.* coalescence.

coaliser, to unite, combine.

coalition, *f.* coalition, union, league.

coaltar, *m.* coal tar.

coaltarement, *m.* tarring.

coaltarer, to tar, treat with coal tar.

coaltarisation, *f.* tarring.

coaltariser, to tar, treat with coal tar.

coarcture, *f.* coarcture, crowding together.

coassant, croaking (frog).

coassocié, *m.* copartner, associate.

coauteur, *m.* joint author.

cobalt, *m.* cobalt; — **gris,** cobalt glance, cobaltite; — **d'outremer,** cobalt blue.

cobaltage, *m.* coating or plating with cobalt.

cobaltammine, *f.* cobaltammine.

cobalteux, -euse, cobaltous.

cobaltifère, cobaltiferous.

cobaltine, *f.* cobaltite, cobaltine, glance cobalt.

cobaltique, cobaltic.

cobaltisage, *m.* coating or plating with cobalt.

cobaltiser, to cover, plate with cobalt.

cobaye, *m.* guinea pig, cavy.

coboidine, *f.* linnaeite.

cobolt, *m.* grey powdered (metallic) arsenic.

cobre, *m.* cobra.

coca, *m.* coca.

cocaine, *f.* cocaine.

cocaïniser, to cocainize, inject cocaine.

cocalon, *m.* a silkworm cocoon of inferior value.

coccidés, *m.* coccidae.

Coccidiose, *f.* coccidiosis.

coccifère, cocciferous, bearing scarlet berries.

coccinelle, *f.* — à deux points, ladybird; — de l'érable, cottony maple-leaf scale.

coccobacille, *f.* coccobacillus.

coccobacilliare, coccobacillary.

coccocarpé, (plants) with scarlet fruit.

coccus, *m.* coccus; — prodigiosus, *Bacillus prodigiosus.*

coccygien, -enne, coccygeal.

coche, *f.* notch, nick, cut, score, slit, sow.

côché, fertile (egg).

cochelet, *m.* cockerel.

cochène, *m.* mountain ash (Sorbus).

cochenillage, *m.* cochineal dyeing or dye bath.

cochenille, *f.* cochineal (insect), scale insect, coccid.

cocheniller, to dye with cochineal.

cochenillier, *m.* cochineal fig, plant, nopal.

cocher, to score, jog, notch.

côcher, to tread (hen).

cochet, *m.* cockerel.

cochevis, *m.* crested lark.

cochinchine, *f.* Cochin China.

cochléaire, cochleate, cochlear, spoon-shaped.

cochléarifolié, with spoon-shaped leaves.

cochléariforme, spoon-shaped.

cochon, *m.* swine, hog, pig, pork, dross, slag; — d'Inde, guinea pig.

cochonner, to farrow.

coco, *m.* coconut, coco, licorice water; beurre de —, coconut oil; noix de —, coconut.

cocon, *m.* cocoon.

coconnage, *m.* formation of cocoons.

coconner, to cocoon.

cocose, *n.* a butter substitute from coconut oil.

cocotier, *m.* coco, coconut palm (*Cocos nucifera*).

cocotte, *f.* foot-and-mouth disease.

cocrête, *f.* (yellow) rattle (*Rhinanthus crista-galli*); — des prés, lousewort, red rattle (*Pedicularis palustris*).

coction, *f.* coction, boiling, digestion.

code, *m.* code, codex, pharmacopoeia; — forestier, *m.* forest code.

codéine, *f.* codeine.

codex, *m.* codes, pharmacopoeia.

codifier, to codify.

codiophylle, codiophyllous, with wooly pubescent leaves.

coecum *m.* see CAECUM.

coefficient, *m.* coefficient; — d'abaissement, coefficient of lowering; — d'ecoulement, coefficient of discharge; — de dilatation, coefficient of expansion; — de forme, form factor, factor of shape; — de rétrécissement, coefficient of contraction; — de rugosité, roughness factor; — de sûreté, — de sécurité, coefficient of safety, safety factor.

coeliaque, coeliac (artery).

coeloblaste, *m.* coeloblast, noncellular algae and fungi.

coenosphère, *f.* coenocentrum.

coenzyme, *f.* coenzyme.

coercible, coercible, compressible.

coercitif, -ive, coercive.

coesium, *m.* see CAESIUM.

coeur, *m.* heart, dorsal vessel, courage, soul, core, interior, middle part, heartwood, stomach; — du bois, heartwood, core; — mort, decaying heartwood; — rouge, red heartwood, brown oak; à contre-coeur, unwillingly; de bon (grœd) —,willingly; en —, heart-shaped, cordate; en — renversé, obcordate.

coexistence, *f.* coexistence.

coexister, to coexist.

coferment, *m.* coferment, coenzyme.

cofféine, *f.* see CAFÉINE.

coffrage, *m.* form, casing.

coffre, *m.* chest, box, coffer, coffer dam, seat (of coaches).

coffre-fort, *m.* strong box, safe.

coffrer, to encase.

coffret, *m.* small chest or box, muffle.

cognac, *m.* cognac, (Cognac) brandy.

cognasse, *f.* wild quince.

cognassier, *m.* quince, quince tree (*Cydonia oblonga*).

cognée, *f.* ax, woodcutter's ax, trimming ax.

cogner, to drive, knock, beat.

cognu, known (old form).

cohabitation, *f.* dimorphism.

cohérence, *f.* coherence.

cohérent, coherent, adherent.

cohérer, to cohere.

cohéreur, *m.* coherer.

cohésion, *f.* cohesion, unity, close relationship.

cohésionner, to cause (molecules) to cohere.

cohibition, *f.* restraint.

cohobateur, cohobating.

cohobation, *f.* cohobation, redistillation.

cohober, to cohobate, redistill.

cohorte, *f.* cohort, a group of orders.

coiffage, *m.* fuse cap.

coiffe, *f.* cap, cover, rootcap, pileorhiza, gall, calyptra, mesentery (of slaughtered animals), headdress, lining (of hat), hood; — **de fusée,** fuse cap; — **de racine,** rootcap; — **en caoutchouc,** rubber cap.

coiffer, to cap, cover, (the head), dress (the hair), intoxicate.

coiffure, *f.* headdress.

coin, *m.* corner, wedge, stamp, die, mark, quoin, angle, fillet, corner, cupboard, corner tooth (of horse).

coinçage, *m.* wedging, fastening with wedges.

coincement, *m.* choking up, wedging, jamming.

coincer, to wedge, jam, drive in (wedges), arrest, corner, stick, bind.

coïncidence, *f.* coincidence, confluence.

coïncider, to coincide, agree.

coing, *m.* quince.

coir, *m.* coir, coconut fiber.

coït, *m.* sexual intercourse, copulation, coitus.

coke, *m.* coke; — **de pétrole,** petroleum coke, oil coke.

cokéfaction, cokéification, *f.* coking.

cokéfier, to coke.

cokerie, *f.* coking plant.

coketier, *m.* one that makes or sells coke.

cokeur, *m.* coke burner, coke-oven stoker.

col, *m.* neck, collar, pass (of mountain), col, saddle, cervix (of bladder); — **de cygne,** swanneck, gooseneck; — **droit, wide** (straight) neck; à — **court,** short-necked.

cola, *m.* kola, Cola.

colature, *f.* filtration, straining, filtrate, strained liquid, colature.

colchicacée, *f.* a plant belonging to genus Colchicum.

colchicéine, *f.* colchiceine.

colchicine, *f.* colchicine.

colchique, *m.* meadow saffron (Colchicum).

colcotar, *m.* colcothar.

coléogène, *f.* coleogen, a ring-shaped group of cells.

coléogone, *f.* a tissue from which endoderm develops.

coléophylle, *m.* coleoptile.

coléoptère, *m.* coleopteron, beetle; *pl.* Coleoptera.

coléoptile, *m.* coleoptile, coleoptilum.

coléorhize, *f.* coleorhiza.

colère, *f.* anger, passion, wrath, rage.

colibacille, *f.* colon bacillus.

colibacillose, *f.* colon-bacillus infection.

colibri, *m.* humming bird.

colifichet, *m.* trinket, bird cake, small tripod, sizing machine; *pl.* rubbish.

colique, *f.* stomach-ache, colic.

colique, colic (artery).

colis, *m.* package, parcel, luggage, bale; **par — postal,** by parcel post.

collaborateur, -trice, *m.&f.* collaborator, fellow laborer, helpmate, contributor.

collaboration, *f.* collaboration, assistance, cooperation, contribution.

collaborer, to collaborate, contribute to, work jointly.

collage, *m.* gluing, sticking, fining, clarifying, size, sizing; — **des coupes,** fixing sections to slides.

collagène, *m.* collagen.

collagène, collagenous, collagenic.

collaire, plumes —s, neck feathers.

collapsus, *m.* collapse (of patient).

collatéral, collateral, standing side by side.

collationner, to compare, collate, check.

colle, *f.* paste, glue, size, sizing, cement, fining; — **à bouche,** lip glue, mouth glue; — **à chenilles,** insect paste; — **à gélatine,**

animal glue;—**au baquet,** parchment glue; **— blanche liquide,** liquid glue;—**d'amidon,** starch paste; **— de pâte,** (flour) paste; **— de peau,** skin (hide) glue; **— de poisson,** fish glue, isinglass; **— d'os,** bone glue; **— forte,** glue; **— végétale,** vegetable (glue) size.

collecteur, *m.* collector, collecting pipe, commutator, condenser; **— de poussier,** dust catcher.

collectif, -ive, collective.

collection, *f.* collection; **— de plantes sèches,** herbarium.

collectionner, to collect.

collectionneur, -euse, *m.&f.* collector.

collectivement, collectively.

collectivité, *m.* collectivity.

collectrice, *f.* collector.

collège, *m.* college, municipal high school.

collégien, *m.* collegian.

collégien, -enne, collegiate.

collègue, *m.* colleague.

collenchyme, *f.* collenchyma.

coller, to glue, paste, stick, adhere, cling, size, fine, clarify (wine), fasten with slip; **se —,** to stick, adhere, cake, be glued, sized, fined.

collerette, *f.* flange, collar, ring, involucre, annulus (mushroom); **— de la vierge,** stitchwort (Alsine).

collet, *m.* flange, collar, shoulder, neck, throat, crown (of anchor), collum, root neck.

colleur, *m.* gluer, paster, size, sizing.

colleuse, *f.* gluing machine, sizer.

collidine, *f.* collidine.

collier, *m.* collar, ring, necklace.

collifère, colliferous, bearing a collar.

colliger, to collect.

collimateur, *m.* collimator.

collimateur, collimating.

collimation, *f.* collimation.

colline, *f.* hill, elevation.

colliquatif, -ive, exhausting.

collision, *f.* collision.

collocyte, *f.* collenchyma cell.

collode, *m.* collodion.

collodié, collodionized.

collodion, *m.* collodion; **— cantharidé,** cantharidal (blistering) collodion; **— élastique,** flexible collodion; **— vésicant,** blistering (cantharidal) collodion.

collodionnage, *m.* collodion varnishing.

collodionné, collodionized.

collodionner, to collodionize.

colloïdal, *m.* colloidal substance.

colloïdal, -aux, colloid(al).

colloïde, *m.* colloid.

colloïde, colloid(al).

collotypie, *f.* collotypy, collotype.

colloxyline, *f.* colloxylin (variety of soluble guncotton).

collutoire, *m.* collutory, mouthwash.

collyre, *m.* collyrium, eyewash.

colmatage, *m.* warping (of land), colmatage.

colmater, to warp, raise land (by artificial inundation).

colocynthine, *f.* colocynthin (a bitter, purgative glucoside).

colombate, *m.* columbate, niobate.

colombe, *f.* dove.

Colombie, *f.* Colombia, Columbia.

colombier, *m.* dovecote, pigeon house; **format —,** colombier (size of paper, about 23 by 34 in.).

colombine, *f.* pigeon dung, columbine (*Aquilegia vulgaris*), columbin, capon's feather; **— panachée,** meadow rue.

colombite, *f.* columbite.

colombium, *m.* columbium, niobium.

colombo, *m.* calumba, colombo; **— d'Amérique,** American columbo (*Frasera caroliniensis*).

colomine, *f.* talcose clay.

colomnaire, columnar.

colomnifère, columnar.

colon, *m.* farmer, colonist.

côlon, *m.* large intestine, colon.

colonel, *m.* colonel.

colonial, -aux, colonial, of the colonies.

colonie, *f.* colony, settlement.

coloniser, to colonize.

colonne, *f.* column, pillar, post, barrel, upright; **— à plateaux,** rectifying column; **— à rectifier,** rectifying column; **— vertébrale,** spinal column.

colophane, *f.* colophony, rosin, common resin.

colophaner, to treat with rosin.

coloquinte, *f.* colocynth, bitter apple (*Citrullus coloncynthis*).

colorabilité élective, *f.* differential staining.

colorant, *m.* colorant, coloring matter, dye, dyestuff, stain; — **acridique,** acridine dye; — **azoaminé,** aminoazo dye; — **azoïque,** azo dye; — **bisazoïque,** disazo dye; — **ignifuge,** fireproof color(ing); — **oxazinique,** oxazine dye.

colorant, coloring, dyeing.

coloration, *f.* coloration, coloring.

coloré, colored, dark-red (wine), ruddy, florid (complexion), stained.

colorement, *m.* coloring, coloration.

colorer, to color, tinge, dye; **se** —, to be colored, take on color.

coloriage, *m.* coloring.

colorier, to color.

colorieur, *m.* colorer, (textile) printing roller, print roll.

colorifique, colorific.

colorigène, colorigenic, color-producing.

colorimètre, *m.* colorimeter.

colorimétrique, colorimetric.

coloris, *m.* coloring, hue, tint.

colorisation, *f.* coloration.

coloriste, *m.* colorer, colorist, painter.

colossal, -aux, colossal, gigantic, huge.

colosse, *m.* colossus, giant.

colostrum, *m.* colostrum, beestings.

coltar, colthar, see COALTAR.

coltarer, see COALTARER.

colubridés, *m.pl.* Colubridae.

columbium, *m.* see COLOMBIUM.

columelle, *f.* columella.

columnaire, columnar.

colza, *m.* colza, rape, rapeseed; huile de —, colza oil, rapeseed oil.

comanique, comanic (acid).

comateux, -euse, comatose.

combat, *m.* combat, contest, strife, struggle, fight, battle.

combattre, to fight, combat, contend, strive, oppose, struggle against, battle with.

combien, how, how much, how many, how long, how far.

combinabilité, *f.* combinableness.

combinable, combinable.

combinaison, *f.* combination, compound, scheme, contrivance; — **binaire,** binary compound.

combinateur, *m.* combiner, controller.

combinateur, -trice, combining.

combinatoire, combinatory, combinative.

combiné, *m.* combination, compound.

combiné, combined joint, concerted, united.

combiner, to combine, contrive, join.

combineur séquentiel, *m.* sequence switch.

comble, *m.* roof, roofing, height, summit, overmeasure, spillage, top, climax; — **à pannes,** purlin roof; — **d'intensité,** maximum intensity.

comblé, heaped up, quite full, piled up.

comblement, *m.* filling (up).

combler, to fill (in), fill up, overload, fulfill, heap (up), crown, make up.

comburable, combustible.

comburant, *m.* fuel, supporter of combustion.

comburant, combustive, burning, comburent.

comburé, burned.

comburer, to burn.

combustibilité, *f.* combustibility.

combustible, *m.* fuel, combustible, firing; — **aggloméré,** artificial (briquetted) fuel.

combustible, combustible, inflammable.

combustion, *f.* combustion, incineration; — **de la fumée,** smoke consumption; — **lente,** slow combustion; — **spontanée,** spontaneous combustion; — **vive,** quick combustion.

comédon, *m.* comedo, blackhead.

coménique, comenic (acid).

comestible, *m.* provisions, food, comestibles.

comestible, edible, eatable.

cométaire, cometary.

comète, *f.* comet.

comifère, comose, tufted.

comique, comic, comical, funny.

comité, *m.* committee, board.

comizophyte, stamens adnate to corolla.

commandant, *m.* commanding officer, commander.

commande, *f.* order (for goods), driving, drive, driving gear, control, lever, operation, lashing (of ropes), command; — **sur le volant,** flywheel drive; à — **directe,** direct-acting, direct drive; de —, ordered, pretended, feigned; **faire une** —, to give an order.

commandement, *m.* command, commandment.

commander, to command, order, drive, control, operate, govern; **se** —, to command oneself, work together.

commanditaire, *m.* dormant partner.

commandite, *f.* commandite; **société en —,** sleeping, limited partnership.

commanditer, to support (finance) as sleeping partner.

comme, as, as . . . as, like, as if, as much as, as to, how, since, because, as it were, in the matter of, in regard to, a kind of, so, almost, nearly, how.

commélinées, *f.pl.* Commelinaceae.

commémoratif, -ive, commemorative.

commémorer, to remember, commemorate.

commencement, *m.* beginning, commencement, origin, start; **au —,** at the outset, at first.

commencer, to begin, commence.

commensal, *m.* commensal.

commensalisme, *m.* symbiosis, commensalism.

commensurabilité, *f.* commensurability.

comment, how, why, in what manner, wherefore, what!

commentaire, *m.* commentary, comment, remark, memoir.

commenter, to comment on, annotate, criticize (adversely).

commerçant, *m.* merchant, trader.

commerçant, commercial, mercantile.

commerce, *m.* commerce, trade, business, intercourse, tradespeople; **— de bois,** timber trade; **— de détail,** retail trade; **— en gros,** wholesale trade; **— extérieur,** foreign trade; **— intérieur,** home trade; **faire le — de,** to deal in.

commercer, to trade, traffic, deal.

commercial, -aux, commercial, business

commercialement, commercially.

commère, *f.* mammy.

commettant, *m.* customer.

commettre, to commit, perpetrate, appoint, lay (rope).

comminuer, to comminute.

comminution, *f.* comminution.

commis, *m.* clerk, bookkeeper; **— voyageur,** commercial traveler.

commis, committed, appointed, laid (a rope).

commissaire, *m.* commissioner, commissary.

commission, *f.* commission, committee, board, message, errand, trust, commission business.

commissionnaire, *m.* commission agent, porter, messenger.

commissionner, to commission.

commissure, *f.* commissure, seam of coherence of two carpels; **— cérébrale supérieure,** anterior dorsal commissure.

commode, *f.* commode, chest of drawers.

commode, convenient, commodious, comfortable, handy, accommodating, fit.

commodément, conveniently, comfortably.

commodité, *f.* convenience, comfort, conveyance; *pl.* (public) water closets, accommodations.

commotion, *f.* shock, commotion, concussion.

commotionné, suffering from shell shock or concussion.

commotionner, to concuss, shock.

commun, *m.* common generality (of persons).

commun, common, ordinary, usual.

communal, -aux, common (pasture).

communauté, *f.* community.

commune, *f.* commune, community, town hall, common people, municipality.

communément, commonly, generally, usually.

communicant, communicating.

communicatif, -ive, communicative, infectious; **encre communicative,** copying ink.

communication, *f.* communication, intercourse, connection.

communiqué, *m.* communiqué, official news.

communiqué, communicated.

communiquer, to communicate, impart, share, confer; **se —,** to communicate, be communicative.

commutateur, *m.* switch, commutator, push button; **— à balais,** brush commutator; **— à manette,** lever switch; **— disjoncteur,** cutout, break-circuit; **— inverseur,** reversing switch; **— permutateur,** change-over (universal) switch; **— unipolaire,** single-pole switch.

commutatrice, *f.* rotary converter.

compacité, *f.* compactness, binding nature; **— (du sol),** consistency, binding nature (of soil).

compact, compact, dense, heavy, binding.

compagne, *f.* companion, helpmate.

compagnie, *f.* company, society, covey, herd.

compagnon, *m.* companion, workman, fellow, mate, partner; — **blanc,** white campion (*Lychnis alba*); — **rouge,** red campion (*Lychnis dioica*).

compagnonnage, *m.* trade union.

comparabilité, *f.* comparability.

comparable, comparable, to be compared.

comparablement, comparably.

comparaison, *f.* comparison, similitude; **à (en) — de,** in comparison with; **faire la — de,** to compare; **hors de —,** beyond comparison; **par —,** by comparison, comparatively; **sans —,** beyond comparison, without making any comparison.

comparaître, to appear.

comparateur, *m.* comparator.

comparateur, -trice, comparing.

comparatif, -ive, comparative.

comparativement, comparatively, by comparison.

comparé, compared, comparative; — **à,** as compared with.

comparer, to compare, liken; **se —,** to be compared.

compartiment, *m.* compartment, division, department, knot, chamber.

compartimentage, *m.* division into compartments, partitioning.

compartimenter, to divide into compartments.

comparu, appeared.

compas, *m.* compass, compasses; — **à coulisse,** slide calipers; — **à diviser,** spring (bow) divider, dividing tester; — **à jauge,** inside calipers; — **à pointes sèches,** dividers; — **de calibre,** calipers; — **de mesure,** dividers, compasses; — **d'épaisseur,** calipers; — **de route,** steering compass, sea compass; — **forestier,** caliper, diameter (tree) gauge; — **gyroscopique,** gyrocompass.

compassé, measured with compasses, stiff, formal, precise, regular, set.

compasser, to measure with compasses, proportion, gauge, regulate, measure, weigh.

compassier, *m.* compass, instrument maker.

compatibilité, *f.* compatibility.

compatible, compatible.

compatir, to be compatible, sympathize.

compatriote, *m.* compatriot, fellow countryman.

compendieusement, compendiously, concisely, briefly.

compendieux, -euse, compendious, summarized, abridged.

compendium, *m.* compendium.

compensateur, *m.* compensator.

compensateur, -trice, compensating.

compensatif, -ive, compensative.

compensation, *f.* compensation, amends, equalization.

compensatoire, compensatory.

compenser, to compensate, offset, counterbalance, equalize; **se —,** to compensate each other.

compère, *m.* daddy.

compétemment, competently.

compétence, *f.* competence, province, competition.

compétent, competent, suitable, proper.

compétiteur, -trice, *m.&f.* competitor, rival.

compétiteur, -trice, competing, rival.

compilateur, -trice, *m.&f.* compiler.

compilation, *f.* compilation.

compiler, to compile.

complaire, to please, humor, comply with; **se —,** to delight in, be pleased at.

complaisamment, obligingly, willingly.

complaisance, *f.* complaisance, complacency, kindness.

complaisant, complaisant, obliging, kind.

complaît, (he) pleases.

complant, *m.* planting, vineyard, plantation.

complanter, to plant (vines and trees together).

complective, (leaf) covered in the bud on sides and top.

complément, *m.* complement.

complémentaire, complementary, complemental.

complémentophile, complementophilic.

complet, *m.* complement.

complet, complete, total, full.

complètement, completely, entirely, thoroughly.

compléter, to complete, make up, perfect, finish off, fill up, complement, supplement.

complexe, complicated, complex.

complexion, *f.* constitution, sum of physical characters.

complexité, *f.* complexity.

complication, *f.* complication.

complimenter, to compliment.

complimenteur, -euse, complimentary.

compliqué, complicated, complex, elaborate, intricate.

compliquer, to complicate, make complex, render intricate; se —, to become complicated.

complu (complaire), complied with.

comportement, m. behavior, attitude, conduct.

comporter, to permit, allow, admit of, contain, involve, behave, comport oneself, act, go on, comport with.

composant, m. component, constituent.

composante, f. component.

composé, m. compound, composite; — benzénique, aromatic compound; — binaire, binary compound; — carburé, carbon compound; — chimique, chemical compound; — chloré, chlorine compound; — d'addition, addition compound or product; — de substitution, substitution product; — nitré, nitro compound; — phosphoré, phosphorus compound; — sulfoné, sulphonated compound; — ternaire, ternary compound.

composé, compound, composed, composite (flower), formal, complex, complicate.

composée, f. composite plant, composite; pl. Compositae.

composer, to compose, compound, make up, constitute, form, settle; se — (de), to be composed of, consist of.

compositeur, -trice m.&f. composer, compositor, setter.

compositif, -ive, composite.

composition, f. composition, composing, compounding, mixture, construction, constitution; — centésimale, percentage composition; — d'amorce, primer composition; — fulminante, fulminating (detonating) composition, priming; — fusante, rocket (fuse) composition; — lente, slow-burning composition; — vive, quick-burning (meal-powder) composition.

compost, m. compost, artificial manure.

composter, to compost, manure with compost.

compote, f. compote (of fruit), stewed fruit.

compound, compound.

compréhensibilité, f. comprehensibility, understanding.

compréhensible, comprehensible, distinct.

compréhensif, -ive, comprehensive.

compréhension, f. comprehension, understanding.

comprenant, comprising, including, containing, understanding, comprehending.

comprendre, to comprise, include, contain, comprehend, understand, realize; il n'y comprend rien, he understands nothing about (it).

compresseur, m. compressor; — à graisse, grease gun or cup; — d'air, air pump.

compressibilité, f. compressibility.

compressible, compressible; être —, to be elastic, spring.

compressif, -ive, compressive.

compression, f. compression.

comprimable, compressible.

comprimant, m. compressed tablet.

comprimant, compressed, restrained, compressing.

comprimé, m. compressed tablet.

comprimé, compressed, condensed.

comprimer, to compress, restrain, check, condense.

comprimeur, m. compressor.

comprirent (comprendre), (they) included, comprised, comprehended.

compris, (comprendre), comprised, enclosed, understood, included; non —, not included; y —, including.

compromettre, to compromise, injure, implicate, impair, endanger.

compromis, m. mutual agreement, compromise.

compromis, compromised.

comptabilité, f. bookkeeping, accounting, accountability, responsibility.

comptable, m. accountant, bookkeeper, responsible person; — forestier, forest accountant.

comptable, accountable, responsible.

comptage, m. counting, telling off, enumeration.

comptant, m. ready money, cash.

comptant, ready (money), in cash.

compte, m. account, count, calculation, regard, reckoning; — en définitive, final settlement; — rendu, report, statement; pl. proceedings of the French Academy; à —, on account; à ce compte-là, all the better because; au bout du —, after all, on the whole, considering everything; rendre — de, to explain, make clear; se rendre — de, to account for, take account of, realize; **tenir — de,** to take account of; **tout — fait,** taking everything into consideration.

compte-fils, *m.* thread counter, weaver's glass, linen prover.

compte-gouttes, *m.* dropper, dropping tube.

compter, to count, charge, number, comprise, take into account, consider, reckon, include, expect, calculate, look forward to, intend, count on, depend on, rely on; **se —,** to be counted, count oneself, be reckoned; **à — de,** reckoning from.

compte-tours, *m.* revolution indicator, moto meter.

compteur, *m.* counter, meter, register, indicator, recorder, reckoner, calculator; **— à eau,** water meter; **— à gaz,** gas meter; **— indicateur,** speedometer; **— universel,** calculating machine.

compteuse, *f.* teller.

comptoir, *m.* shop counter, branch bank, bank, warehouse.

compulsif, -ive, compelling, restraining, compulsory.

computation, *f.* computation.

computer, to compute.

comte, *m.* count.

comté, *m.* county, shire, earldom.

concassage, *m.* concassation; *f.* breaking, crushing, pounding.

concasser, to break, crush, grind, bray, pound, bruise.

concasseur, *m.* crusher, crushing mill, disintegrator.

concaulescence, *f.* the coalescence of axes.

concave, concave.

concavifolié, with concave leaves.

concavité, *f.* concavity.

concéder, to concede, grant, yield.

concentrable, capable of being concentrated.

concentrateur, *m.* concentrator.

concentration, *f.* concentration.

concentré, *m.* concentrate.

concentré, concentrated.

concentrer, to concentrate, condense, evaporate; **se —,** to be concentrated, center.

concentrique, concentric.

conceptacle, *m.* conceptacle.

conceptaculum, *m.* conceptacle.

conceptible, conceivable.

conception, *f.* conception, thought, notion, idea.

conceptionnel, -elle, conceptual.

conceptuel, -elle, conceptual.

concernant, concerning, relating to.

concerner, to concern, relate to, be concerned with, regard; **en ce qui concerne,** with regard to, as far as.

concert, *m.* concert, concord, unison.

concerté, concerted, united.

concerter, to contrive, arrange, plan, concert.

concession, *f.* concession, grant.

concessionnaire, *m.* concessionaire, grantee, patentee.

concevable, conceivable, imaginable.

concevoir, to conceive, imagine, understand, perceive, think out, express; **se —,** to be conceived; **cela se conçoit,** that may be conceived, that is natural.

conchiforme, shell-shaped.

conchoïdal, -aux, conchoid(al).

conchoïde, *f.* conchoid.

conchoïde, conchoid(al).

conchylien, -enne, conchylaceous, containing shells.

concierge, *m.&f.* doorkeeper, porter.

conciliation, *f.* conciliation, reconciliation.

conciliatoire, conciliatory (measure).

concilier, to reconcile, conciliate, gain, win.

concis, concise.

concision, *f.* conciseness, brevity.

concluant, decisive, conclusive.

conclure, to conclude, finish, infer, arrive at; **se —,** to be concluded.

conclusif, -ive, conclusive.

conclusion, *f.* conclusion, decision, close, in short, in conclusion.

concoction, *f.* concoction, digestion (of food).

conçoit (concevoir), (he) imagines; **ça se —,** that is natural.

concolore, of uniform color.

concombre, *m.* cucumber; **— de mer,** sea cucumber.

concomitant, concomitant, attendant.

concordance, *f.* concordance, agreement, accordance.

concordant, concordant, harmonious.

concorder, to agree, live in accord.

concourir, to concur, contribute, cooperate, compete, (of events) coincide, (of lines) converge, combine, unite.

concourme, *f.* curcumin (yellow dyestuff).

concours, *m.* concourse, gathering, concurrence, co-operation, aid, agreement, competition, (of lines) convergence, meeting; **hors (de) —,** beyond competition.

concréfier, to (render) concrete.

concrescence, *f.* concrescence, coalescence.

concret, concrete, solid.

concréter, to concrete, solidify, render concrete, congeal (blood).

concrétion, *f.* concretion, coagulation, concrescence, self-grafting; **— calculeuse,** urinary calculus; **—s calcaires,** chalkstones.

concrétionnaire, concrétionné, concretionary.

concrétionner (se), to form into concretions, cake together.

conçu (concevoir), conceived.

concubin, concubinary.

concurrement, concurrently, jointly, in competition with.

concurrence, *f.* concurrence, competition, opposition, amount.

concurrencer, to compete with.

concurrent, *m.* competitor, rival.

concurrent, competing, concurrent.

concurrentiel -elle, competing, rival.

concuter, to strike, encounter by concussion.

concuteur, *m.* striker.

condamner, to condemn, block up, shut up, blame, censure.

condensabilité, *f.* condensability.

condensable, condensable.

condensant, condensing.

condensateur, *m.* condenser; **— à plateaux,** plate condenser.

condensateur, condensing.

condensation, *f.* condensation, condensing; **— par l'extérieur, — par surface,** surface condensation; **— par mélange,** condensation by injection.

condenser, to condense.

condenseur, *m.* condenser; **— à (par) injecteur (à jet),** jet, injection condenser; **— à (par) surface,** surface condenser; **— tubulaire,** tubular, surface condenser; **sans —,** noncondensing.

condenseur, -euse, condensing.

condenseuse, *f.* condensing apparatus.

condiment, *m.* condiment, seasoning, spice.

condimentaire, condimenteux, -euse, condimental, *m.* condimentary.

condisciple, *m.* fellow student.

condit, *m.* candied fruit, confection, comfit.

condition, *f.* condition, service, state, terms, offer, circumstance(s), nature, quality; **— de fonctionnement,** working condition; **— naturelle,** working condition; **à — de,** providing, on condition of; **à — que,** on condition that, provided that.

conditionné, conditioned.

conditionnel, -elle, conditional

conditionnement, *m.* conditioning (of silk).

conditionner, to condition, put into good condition, season (wood).

conductance, *f.* conductance.

conducteur, -trice, *m.&f.* leader, conductor, manager, superintendent, guide, driver, wire; **— d'alimentation,** feeder, supply circuit.

conducteur, -trice, conducting (electricity or heat), transmitting, driving.

conductibilité, *f.* conductibility, conductivity.

conductible, conductible, conductive.

conduction, *f.* conduction.

conductivité, *f.* conductivity.

conduire, to conduct, convey, carry, drive, control, take, run, lead, bring, direct, go along with; **se —,** to conduct oneself, be conducted, behave (oneself), act; **— la machine,** to attend to the machine.

conduit, *m.* conduit, pipe, duct, channel, shoot, passage, culvert; **— à (de) gaz,** gas pipe; **— d'échappement,** exhaust pipe; **— de décharge,** delivery or waste pipe; **— de (la) vapeur,** steam pipe; **— de vent,** blast pipe.

conduite, *f.* conduit, pipe, main, hose, channel, conducting, conduction, control, direction, management, conduct, carrying conveyance, delivery, guidance, behavior, driving; **— d'amenée,** drain or waste pipe; **— d'amenée d'eau,** flume, wheel pit, penstock; **— d'aspiration,** suction tube or duct; **— de vent,** ventilation shaft, blast pipe; **— forcée,** pressure pipe, penstock.

conduplicable, susceptible to being folded lengthwise.

conduplicatif, -ive, conduplicate.

condyle, *m.* condyle.

cône, *m.* cone, strobile, conical roller; **— de broussin,** bur; **— d'éboulis, — de dé-**

jections, cone of detritus; — **de friction,** cone clutch; — **droit,** (up)right cone; — **tronqué,** truncated cone; **ajuster** —, to taper; **donner du — à,** to taper; **en** —, conical; **faire** —, to taper, make conical.

côné, coned, conical.

côner, to cone, taper.

conf., *abbr.* (conferez), compare.

confection, *f.* making, construction, manufacture, production ready-made clothes, confection.

confectionné, *m.* manufactured article.

confectionné, made, ready-made (clothes).

confectionnement, *m.* making (up).

confectionner, to make, make up, construct, manufacture, prepare; **se** —, to be made (up).

confectionneur, *m.* maker, manufacturer, clothier.

conférence, *f.* conference, discussion, lecture, comparison; **mâitre de** —, lecturer.

conférencier, -ère, *m.&f.* lecturer.

conférer, to confer, compare.

confertiflore, with flowers close together.

confertifolié, with leaves close together.

confervacé, resembling Conferva.

conferve, *f.* Conferva, confervoid alga, silkweed, hairweed.

confervoide, resembling Conferva.

confesser, to confess, acknowledge.

confession, *f.* confession, admission.

confiance, *f.* confidence, reliance, trust.

confiant, confiding, trusting, confident, reliant.

confidemment, confidentially, in confidence.

confidence, *f.* confidence.

confident, *m.* confidant.

confident, confidential.

confidentiel, -elle, confidential.

confidentiellement, confidentially.

confier, to confide, commit, entrust, extend; **se — à,** to confide (in), trust in, rely on, place confidence in.

configuration, *f.* configuration, shape.

configurer, to configure, shape.

confin, *m.* confine, limit, border.

confiner, to confine, restrain, border (on).

confins, *m.pl.* borders (of country).

confire, to preserve (fruits), candy (peel), pickle (in vinegar), soak (in late), dip, steep (skins), can.

confirmatif, -ive, confirmative, corroborative.

confirmation, *f.* confirmation.

confirmatoire, confirmatory.

confirmer, to confirm, sanction, ratify; **se** —, to be confirmed.

confisant, (confire), preserving.

confiserie, *f.* confectionery, confectioner's shop, preserving (of fruit, etc.).

confiseur, *m.* confectioner.

confisquer, to confiscate, forfeit.

confit, *m.* preserved meat, bran mash, maceration tub, vat; **mettre en** —, to bate (skins).

confit, preserved, pickled, steeped, bated.

confiture, *f.* preserve(s), jam, marmalade.

confiturerie, *f.* preserves, preserve making, jam factory.

confiturier, *m.* maker or seller of preserves.

conflit, *m.* conflict, strife, contention.

confluence, *f.* running together (of vesicles).

confluent, blended together, connate, confluent.

confluer, to flow together, meet, join.

confondre, to confound, mingle, blend, confuse, mix; **se** —, to be confounded, mingle, blend (into), be confused, be identical (with), coincide (with).

confondu, confounded, mingled, blended, amazed.

conformation, *f.* conformation, form, structure.

conforme, conformable, congruent.

conformé, formed, shaped.

conformément, conformably, according (to), in accordance (with).

conformer, to conform; **se — à,** conform oneself (to), comply with.

conformité, *f.* conformity, agreement.

confort, *m.* comfort; **pneus** —, balloon tires.

confortable, *m.* comfort, ease, convenience.

confortable, comfortable.

confortant, strengthening, corroborant.

confrère, *m.* colleague.

confrontation, *f.* comparison, confrontation, contrast.

confronter, to confront, compare, stand face to face, border (on).

confus, confused, indistinct, intricate, mixed.

confusément, confusedly, in confusion.

confusion, *f.* confusion, tumult.

congé, *m.* leave (of absence), (notice of) leaving, discharge, hollow, groove.

congédier, to discharge, dismiss.

congélabilité, *f.* congealableness.

congelable, congealable, freezable.

congelant, congealing, freezing.

congélation, *f.* congelation, congealment, freezing, cold storage, frostbite.

congeler, to congeal, freeze, coagulate.

congénère, *m.* congener.

congénère, congeneric, congenerous, of same species.

congénital, -aux, congenital, grown to anything, connate.

congestif, -ive, congestive, congested.

congestion, *f.* congestion.

congestionné, flushed, red (face).

congestionner, to congest.

conglobé, conglobate, congested.

conglomérat, *m.* conglomerate.

conglomération, *f.* conglomeration.

congloméré, conglomerated, conglomerate.

conglomérer, to conglomerate.

conglutinant, agglutinant, agglutinative.

conglutine, *f.* conglutin.

conglutiner, to conglutinate.

congratulation, *f.* congratulation.

congrégation, *f.* community.

congrégé, congregated.

congrès, *m.* congress, meet.

congru, congruous.

congruence, *f.* congruence, equivalence.

congruité, *f.* congruity.

conhydrine, *f.* conhydrine.

conicine, *f.* conicine, coniine, conine.

conicité, *f.* conicalness, conicity, taper (of bullet).

conidanges, *f.pl.* mother cells of conidia.

conidie, *f.* conidium.

conidiophore, *m.* conidiophore.

conifère, *m.* conifer, coniferous tree; *pl.* Coniferae.

conifère, coniferous.

coniférine, *f.* coniferin.

coniférylique, coniferyl.

coniflore, with cone-shaped flowers.

coniforme, coniform, conical.

conigène, borne on the cone of fir (tree).

coniien, conine, *f.* conine, coniine.

coniothéque, *f.* the loculus of an anther.

conique, conical, conic, tapering

conjecturalement, conjecturally, by guess.

conjecture, *f.* conjecture, guess.

conjecturer, to conjecture, guess.

conjoindre, to unite, conjoin, join.

conjoint, conjoined, joined, joint, conjunct, conjugate.

conjointement, (con)jointly.

conjoints, *m.pl.* futurs —, future couple.

conjoncteur, *m.* circuit closer, switch key.

conjoncteur-disjoncteur, *m.* self-closing circuit breaker.

conjonctif, -ive, conjunctive, connective.

conjonction, *f.* conjunction, union.

conjonctive, *f.* conjunctiva (of eye).

conjonctivite, *f.* conjunctivitis.

conjoncture, *f.* trade outlook.

conjugaison, *f.* conjugation, fusion of two gametes.

conjugal, -aux, conjugal, marital.

conjugation, *f.* copulation.

conjugué, conjugate, conjugated.

Conjuguées, *f.pl.* Conjugatae.

conjunctive fondamentale, *f.* cortical tissue of petiole.

connaissance, *f.* knowledge, acquirements, attainments, understanding, learning, acquaintance, consciousness.

connaissant, knowing, acquainted with, skilled in.

connaissement, *m.* bill of lading.

connaisseur, *m.* connoisseur, judge, expert, specialist.

connaître, to know, understand, be acquainted with, perceive, experience, be aware of; se —, to be versed (in), know each other, understand, be an expert in; faire —, to make known, bring to light, introduce, reveal; il se connaît en, he is a judge of; on lui connaît de grands faits, they attribute great things to him.

conné, connate (leaf), congenitally united.

connecter, to connect, switch.

connectif, -ive, connective.

connection, *f.* connection.

connexe, connected, related, associated.

connexion, *f.* connection, association, connexion; — directe, direct action.

connexité, *f.* connexity, relatedness, understood.

connivent, converging, connivent.

connu, m. known.

connu, known, well-known, understood; universellement —, world famous.

conoïde, conoid.

conque, f. concha (of the ear).

conquérir, to conquer, gain, subdue, win.

conquête, f. conquest.

conquiert, (conquérir), (he) conquers.

conquis, conquered.

conquit (conquérir), (he) conquered.

consacré, usual, in use.

consacrer, to devote, consecrate, sanction, appropriate.

consanguin, consanguineous, inbred (horse).

conscience, f. consciousness, conscience, conscientiousness.

consciencieusement, conscientiously.

consciencieux, -euse, conscientious.

conscient, conscious.

consécutif, -ive, consecutive.

consécutivement, consecutively.

conseil, m. counsel, council, board, resolution, course, advice.

conseiller, to counsel, advise; se —, to be advised, seek counsel (from).

conseiller, m. counselor, adviser, councilor.

conseilleur, m. adviser.

consensus, m. consensus, sympathy.

consentement, m. consent.

consentir, to consent (to), agree to.

conséquemment, consistently, consequently.

conséquence, f. consequence, effect, inference, importance, sequel; en —, in consequence, consequently, accordingly.

consequent, m. consequent, consequence; — d'un rapport, consequent of a ratio; par —, consequently, therefore.

conséquent, consistent, consequent, rational.

conservable, that can be preserved.

conservant le gram, Gram-positive.

conservateur, m. preserver, keeper, conservator, conservative.

conservateur, -trice, preserving, conserving, preservative, conservative.

conservatif, -ive, conservative, preservative.

conservation, f. preservation, conservation, forest circle, storage; — du gibier, game preserving, rearing of game.

conservatoire, conservatory.

conserve, f. conserve (a confection), preserved food, preserves, glass jar; pl. colored glasses; — alimentaire, preserved food, canned goods; — d'écorce d'orange, confection of orange peel; — de légumes, preserved vegetables; — de poissons, preserved (canned) fish; — de viande, preserved (canned) meat.

conserver, to preserve, keep, keep up, conserve, maintain, (fruits) be preserved, be kept, last, can, pot; — au sel, to preserve in brine, salt (down).

considérable, considerable, large, great, notable.

considérablement, considerably.

considération, f. consideration, reflection, motive, esteem, regard.

considéré, considered, esteemed, considerate.

considérer, to consider, take into consideration, look at, contemplate, esteem; se —, to be considered, consider oneself.

consignataire, m. consignee.

consignateur, -trice, m.&f. consignor, shipper.

consignation, f. consignment.

consigner, to deposit, consign, enter, record, keep in, refuse admittance to, give orders.

consistance, f. consistency, consistence, credit, firmness, density (of crop), standard density; — des tiges, density of crop; — du massif, degree of density or closeness.

consistant, consistent, firm, stable, holding together, solid; — en, consisting of.

consister, to consist of, be composed of.

consocié, congenital or gamophyllous, associated in consociation.

consolation, f. consolation.

console, f. bracket, wall fixture, extension arm.

consoler, to console, comfort.

consolidation, f. consolidation, strengthening.

consolider, to consolidate, fund, strengthen, heal, unite, anchor; se —, to consolidate, become solid, grow firm, heal.

consommable, consumable (food).

consommateur, m. consumer.

consommation, f. consumption, consummation.

consommé, consumed, consummate(d).

consommer, to consume, consummate, use up, complete, finish, eat.

consomptible, consumable.

consomptif, -ive, consumptive.

consomption, *f.* consumption, wasting, decline.

consonance, *f.* consonance, harmony.

consortium, *m.* syndicate, consortium, relations of lichen life.

consoude, *f.* comfrey (Symphytum); — **officinale,** common comfrey (*Symphytum officinale*); — **royale,** field larkspur, king's-consound (*Delphinium consolida*).

constamment, constantly, with constancy, all the time.

constance, *f.* constant, constancy, persistence, perseverance, stability, permanence.

constant, constant, certain, unchanging, firm, unquestionable.

constante, *f.* constant quantity; — **empirique,** empirical constant; — **lipémique,** "lipemic index."

constatation, *f.* verification, declaration, ascertaining, statement, authentication, verifying.

constater, to ascertain, verify, state, prove, establish, discover, aver, check up, find out.

constellation, *f.* constellation.

constellé, adorned, set, studded.

consterner, to astound, amaze, terrify.

constipant, constipating, binding.

constipation, *f.* costiveness, constipation.

constipé, constipated, costive.

constiper, to constipate.

constituant, *m.* constituent, ingredient.

constituant, constituent, component.

constituer, to constitute, put, place, fix, form, build, establish, make, make up; **se** —, to be formed, constituted.

constitutif, -ive, constituent, constitutive.

constitution, *f.* constitution, complexion; — **chimique,** chemical structure; — **du peuplement,** formation of a wood; — **du sol,** composition of the soil.

constrainte, *f.* compulsion.

constructeur, *m.* constructor, builder, maker, manufacturer.

construction, *f.* construction, building, structure, texture (of land), manufacture; — **au-dessus** du sol, — **en élévation,** superstructure work, high building.

construire, to construct, build, rear, erect, lay out.

consulat, *m.* consulate.

consultatif, -ive, consultative, consulting, advisory.

consulter, to consult, confer with, advise with; **se** —, to consider, think a matter over, consult one another, be consulted.

consumant, consuming (fire, fever).

consumer, to consume, waste; **se** —, to be consumed, waste away, waste one's fortune, time, energy.

contabescenz, *m.* contabescence, abortion of stamens.

contact, *m.* contact, touch; **au** — **de,** in contact with.

contagieux, -euse, contagious, infectious.

contagion, *f.* infection, contagion.

contamination, *f.* contamination.

contaminer, to contaminate, infect, pollute; **se** —, to be contaminated.

contemplateur, -trice, *m.&f.* contemplator.

contemplation, *f.* contemplation, meditation.

contempler, to contemplate, reflect, behold.

contemporain, contemporary, contemporaneous.

contenance, *f.* capacity, contents, extent, air, area, countenance, bearing.

contenant, *m.* container.

contenant, containing, restraining.

contendant, contending, rival.

contenir, to contain, restrain, include, comprise, keep in, hold in check; **se** —, to keep within bounds, forbear.

content, content, satisfied, pleased, glad.

contentement, *f.* satisfaction, content.

contenter, to content, please, give satisfaction, satisfy; **se** —, to content oneself, be satisfied (with), be contented with.

contentieux, -euse, disputed, contentious.

contention, *f.* contention, containing, application, exertion, intenseness.

contenu, *m.* contents, volume.

contenu, restrained, contained.

conter, to tell, relate.

conterie, *f.* rough Venetian glassware.

contestation, *f.* contestation, dispute, strife, contest.

conteste, *f.* dispute.

contester, to contest, dispute, deny.

contexture, *f.* texture.

contiennent (contenir), (they) include.

contient (contenir), (it) includes.

contigu, contiguous, adjacent, adjoining, bordering on.

continent, *m.* continent, mainland.

continental, -aux, continental.

contingent, *m.* quota, portions, share.

contint (contenir), (it) contained.

continu, *m.* continuum.

continu, continual, continuous, continued, (of an electric current) direct, uninterrupted, incessant; **en —,** continuously, by the continuous process, by direct current.

continuation, *f.* continuation, continuance.

continuel, -elle, continual, unceasing.

continuellement, continually.

continuer, to continue, proceed with; **se —,** to be continued, continue, extend.

continuité, *f.* continuity, continuance.

continûment, continuously, without a break.

contondant, bruising, blunt.

contorté, contorted, twisted or bent.

contortupliqué, contortuplicate, twisted back upon itself.

contour, *m.* contour, outline, circuit, circumference, girth; **— de forces,** polygon of forces.

contournement, *m.* convolution; **— en spirale,** torsion of fibers, twisted growth.

contourner, to trace the outline of, shape, pass round, turn around, wind around, twist, distort, deform, draw, entwine, outline; **se —,** to become bent.

contractant, *m.* contracting party, contractor.

contractant, contracting.

contracté, contracted, narrowed, shortened.

contracter, to contract, shrink, acquire, draw together, make a mutual agreement; **se —,** to be contracted, shrink, contract.

contractif, -ive, contractive (force).

contractile, contractile.

contraction, *f.* contraction, reduction of section.

contracture, *f.* contracture (of muscle).

contradiction, *f.* contradiction.

contradictoire, contradictory, inconsistent.

contradictoirement, contradictorily.

contraignant, constraining, compelling.

contraindre, to constrain, compel, force, make, oblige, pinch; **se —,** to refrain.

contraint, constrained, forced, cramped.

contrainte, *f.* constraint, restraint.

contraire, *m.* contrary, opposite; **au —,** on the contrary; **au — de,** contrary to, against.

contraire, contrary, injurious, adverse, opposite.

contrairement, contrarily, contrary; **— à,** unlike.

contrarier, to contradict, oppose, counteract, thwart, baffle.

contrariété, *f.* contrariety, annoyance, contradiction.

contrastant, contrasting.

contraste, *m.* contrast.

contraster, to contrast.

contrat, *m.* contract, deed; **— de louage,** lease, contract for service.

contravention, *f.* breach of forest law.

contrayerva, *f.* contrayerva (*Dorstenia contrayerva*).

contre, *m.* con, opposite side; **— à —,** side by side, in opposite direction; **le pour et le —,** pro and con; **par —,** per contra, on the other hand; **pour et —,** for and against; **tout —,** very close, ajar.

contre, against, close (by), near, close to, hard by, contrary to, in opposition, in exchange for.

contre-accélération, *f.* retardation, negative acceleration.

contre-allumage, *m.* backfire.

contre-balancer, to counterbalance; **se —,** to counterbalance each other, cancel out.

contrebande, *f.* contraband.

contre-bas, downward(s), (lower) down.

contre-charge, *f.* counterpoise.

contre-coeur, *m.* back (of a chimney); **à —,** reluctantly, unwillingly.

contre-coup, *m.* rebound, counterstroke, result.

contre-courant, *m.* countercurrent, back draft.

contredire, to contradict; **se —,** to give oneself away.

contredisant, contradicting, contradictory.

contredit, *m.* contradiction, objection, rejoinder; **sans —,** unquestionably.

contrée, *f.* country, region.

contre-échange, *m.* (mutual) exchange.

contre-écrou, *m.* set screw, lock nut, jam nut, check nut.

contre-électrometeur, counterelectromotive.

contre-émailler, to counterenamel, enamel back part.

contre-épreuve, *f.* counterproof, check test.

contre-essai, *m.* control experiment, check test.

contrefaçon, *f.* counterfeiting, counterfeit, imitate, infringement (of patent), forgery, pirated edition or copy.

contrefacteur, *m.* counterfeiter, forger, infringer, pirate.

contrefaction, *f.* counterfeiting, forgery.

contrefaire, to imitate, mimic, disguise, counterfeit, forge.

contrefait, counterfeit(ed), forged.

contre-feu, *m.* back plate (of hearth), counterfire, backfire.

contre-fiche, *f.* brace, strut, stay, truss.

contrefit (contrefaire), (he) forged.

contrefont (contrefaire), (they) imitate.

contreforce. *f.* counterforce.

contrefort, *m.* buttress, support, (mountain) spur, projection, outer hill.

contre-greffe, *f.* double grafting.

contre-hacher, to crosshatch.

contre-hachure, *f.* crosshatch(ing).

contre-haut, en —, higher up.

contremaître, -esse, *m.&f.* overseer, foreman, boss.

contremander, to countermand, cancel, revoke.

contremarque, *f.* countermark.

contremarquer, to bishop (horse's teeth).

contre-oxydeur, *m.* enameler.

contre-paroi, *f.* casing (of a furnace).

contre-partie, *f.* contrary.

contre-pente, *f.* reverse slope.

contre-peser, to counterbalance, counterpoise.

contre-pied, *m.* opposite course, wrong way; **à — de,** contrary to.

contre-plaque, *f.* reinforcing plate, back plate, wall plate, guard plate.

contre-plaqué, laminated, built up in layers; **— en trois épaisseurs,** three-ply (wood).

contrepoids, *m.* counterweight, counterpoise.

contre-poil, *m.* wrong way (of the hair); **à —,** against the grain, the wrong way.

contrepoison, *m.* antidote.

contre-porte, *f.* screen (storm) door.

contre-pression, *f.* counterpressure, back pressure.

contre-réforme, *f.* revision.

contre-saison, out of season.

contresens, *m.* wrong sense, wrong side (of fabrics), wrong way, wrong meaning, opposite.

contresigner, to countersign.

contretemps, *m.* mishap, mischance, hitch, wrong time, inconvenience; **à —,** inopportunely.

contre-tirage, *m.* back draft.

contre-vapeur, *f.* returning steam, back pressure.

contrevenir, to contravene, offend, transgress, infringe.

contrevent, *m.* shutter, (wind) brace, back draft.

contre-vis, *f.* check screw.

contribuant, *m.* contributor.

contribuant, contributing.

contribuer, to contribute, assist, tend, conduce.

contribution, *f.* contribution, tax, share.

contrôlable, that may be checked, verified.

contrôlage, *m.* controlling, checking.

contrôle, *m.* control, checking, testing, assaying, check, mark, stamp, inspection, register, roll, list, controller's office; **— de la charge,** load test, testing of load.

contrôlement, *m.* controlling, supervision.

contrôler, to control, check, calibrate, stamp, mark (as gold), prove, verify, observe, register.

contrôleur, -euse, *m.&f.* controller, checker, inspector, superintendent, censurer.

controuvé, forged, false, fabricated.

controuver, to invent, fabricate.

controverse, *f.* controversy, dispute.

controverser, to dispute, controvert, discuss, debate.

contus, contused, bruised.

contusion, *f.* contusion, bruise, lesion.

contusionner, to bruise, contuse.

convaincant, convincing.

convaincre, to convince, convict, persuade; **se —,** to be convinced.

convaincu, convinced.

convainquant, convincing, convicting.

convallaire, *f.* lily of the valley (*Convallaria majalis*).

convallarine, *f.* convallarin.

convariantes, *f.pl.* convariants, individuals of same age or generation, likely to vary.

convection, *f.* convection.

convenable, *m.* propriety.

convenable, suitable, expedient, becoming, fit, proper, convenient, well-bred; **peu —,** unseemly, improperly.

convenablement, suitably, properly, conveniently, fitly.

convenance, *f.* convenience, conformity, fitness, agreement, suitableness, propriety, proportion.

convenant, convenient, suitable, fit, proper.

convenir, to admit, acknowledge, agree, suit, behoove, grant, argue, be suitable, be expedient, be proper, be right, be advisable, be fit; il (lui) **convient,** it suits (him); il **convient des conditions,** he agrees on the conditions; il **en convient,** he admits (it).

convention, *f.* convention, contract, agreement.

conventionnel, -elle, conventional.

convenu, agreed, stipulated.

convergence, *f.* convergence, convergency.

convergent, converging, convergent, leaf veins running from base to apex in curved manner.

converger, to converge.

conversation, *f.* conversation.

converse, converse (proposition).

converser, to converse, change direction.

conversible, convertible.

conversion, *f.* conversion, converting, changing, transformation.

converti, *m.* convert.

converti, converted.

convertibilité, *f* convertibility.

convertible, convertible.

convertir, to convert, change, turn, transform.

convertissable, convertible.

convertissant, converting.

convertissement, *m.* conversion, converting.

convertisseur, *m.* converter, static transformer; **— à vapeur de mercure,** mercury-vapor rectifier; **— rotatif,** rotary converter.

convexe, convex.

convexité, *f.* convexity.

conveyeur, *m.* moving belt.

conviction, *f.* conviction.

convier, to invite, bid, beg, urge.

convint (convenir), (it) was suitable.

convoi, *m.* train, convoy, funeral; **— de marchandises,** freight train; **— de petite vitesse,** slow train; **— de voyageurs,** passenger train; **— direct,** express train.

convoiter, to covet, desire.

convoiteux, -euse, covetous.

convoitise, *f.* eagerness, greed, lust.

convoluter, to convolute.

convolutif, -ive, convolute.

convolvulacé, convolvulé, convolvulaceous.

convolvulacées, *f.pl.* Convolvulaceae.

convolvuline, *f.* convolvulin.

convolvulus, *m.* bindweed (Convolvulus).

convoquer, to convoke, call, summon, convene.

convulser, to convulse, frighten.

convulsif, -ive, convulsive.

convulsion, *f.* convulsion, spasm.

convulsionner, to convulse.

convulsivement, convulsively.

conyse, *f.* fleawort, fleabane.

coopératif, -ive, co-operative.

coopération, *f.* co-operation.

coopérer, to co-operate.

coordinal, -aux, co-ordinal, of the same natural order.

coordination des plants, *f.* arrangement of plants.

coordonnateur, -trice, co-ordinating.

coordonné, co-ordinated, co-ordinate.

coordonnée, *f.* co-ordinate; **—s polaires,** polar co-ordinates.

coordonner, to co-ordinate, arrange.

copahu, *m.* copaiba; **baume de —,** copaiba balsam.

copaïer, *m.* copaiba tree (Copaifera).

copal, *m.,* **copale,** *f.* copal(resin).

copaline, *f.* copalite, copaline.

copalme, *m.* copalm, sweet gum.

copayer, *m.* copaiba tree (Copaifera).

copeau, *m.* chip, shaving (of wood), splinter, cutting.

copeaux, *m.pl.* wood wool; **— d'alésage,** borings; **— de tour,** turnings.

copie, *f.* copy, imitation, transcript, illustration, image, cut, picture, diagram; **en —,** copied in duplicate.

copier, to copy, imitate.

copieusement, copiously, abundantly.

copieux, -euse, copious.

copiste, m. copier, copyist.

coprah, copra, copre, m. copra; huile de —, coconut oil.

coprécipitine, f. coprecipitin, group precipitin.

copropriété, f. joint property, joint ownership.

copulation, f. copulation, coupling, conjugation.

copulatives, f.pl. copulatives, dissepiments not readily separating axis or walls of pericarp.

copuler, to copulate, couple.

coq, m. cock, rooster, vane; — de bruyère, grouse, woodcock, capercaillie; — de bouleau, black game, black grouse; — nain, Bantam cock; maître —, ship's cook.

coque, f. shell, husk, coccus, cocoon, hull (of ships), body, kink (in rope); — d'oeuf, eggshell; — du Levant, Indian berry, cocculus indicus (Anamirta cocculus).

coquelicot, m. corn poppy, field poppy, red poppy (Papaver rhoeas).

coquelourde, f. pasqueflower (Anemone pulsatilla), mullein pink, rose campion (Lychnis coronaria).

coqueluche, f. whooping cough, pertussis, reed bunting.

coqueluchon, m. aconite, monkshood, wolfsbane (Aconitum napellus).

coqueret, m., coquerelle, f. winter cherry, ground cherry, alkekengi, (Physalis alkekengi).

coquetterie, f. coquetry, fastidiousness.

coquillage, m. shellfish, shell(s), chilling.

coquille, f. shell, peel, skin, demy, chill (mold), casing, housing, printer's error, grate; — de coussinet, bearing bush, brass, pillow, axle box.

coquiller, to cast in a chill mold.

coquilleux, -euse, shelly, conchitic.

coquillier, conchiferous, full of shells.

cor, m. (hunting) horn, corn (on toe), sitfast; — de chasse, hunting horn, bugle.

coraces, m.pl. crows.

coracoïde, coracoid.

corail, m. coral.

corailleux, -euse, corallien, -enne, coral-(line).

coralline, f. coralin.

coraux, m.pl. see CORAIL.

corbeau, m. raven, corbel, crow, bracket; — freux, rook.

corbeille, f. basket, flower bed; — protectrice, wire guard.

corbeillée, f. basketfull.

corbillon, m. small basket.

corbine, f. carrion crow.

cordage, m. rope, cord, cordage, measuring (of wood) by cords, roping (of bales); — goudronné, tarred rope.

cordat, m. coarse serge, coarse wrapping cloth.

corde, f. cord, twine, rope, (bow) string, line, thread, cord (vocal), ligament, chord (of segment), cord (wood), span; — à boyau, catgut; — à feu, fuse, slow match; — à violon, violin string; — de (à) piano, piano wire; — en fil de fer, wire rope; — en laine de bois, wood-wool rope.

cordé, corded, twisted, stringy, cordate, heart-shaped.

cordeau, m. line, string, cord, rope, match, fuse, measuring tape, planting cord, selvage; — détonant, detonating fuse; — porte-feu, fuse, match.

cordée, f. cord (of wood).

cordeler, to twist, twine (into rope).

cordelette, f. small cord or string.

cordelière, f. girdle.

cordeline, f. ferret, selvage, (glass) rod or tube.

cordelle, f. horse-line, towrope.

corder, to cord, twist, string, measure (by the cord), stack; se —, to become stringy, (vegetables) twist, be corded.

corderie, f. ropemaking, cordage factory, ropewalk, ropery.

cordial, -aux, cordial, stimulating, restorative.

cordialité, f. cordiality, joviality.

cordier, m. ropemaker or dealer.

cordier, cordage(-industry).

cordiérite, f. cordierite, iolite.

cordifolié, with heart-shaped leaves.

cordiforme, cordate, heart-shaped.

cordon, m. string, cord, strand, band, tape, ribbon, row, line, cordon, border, milled edge (of coin); — de cardinal, Persicaria; — de gazon, turf border; — libéro, ligneux, fibrovascular bundle; — medullaire, spinal cord; —s nerveux, funiculi of a nerve.

cordonner, to twist, plait, cord, braid.

cordonnerie, *f.* shoemaking or trade.

cordonnet, *m.* braid, small cord, silk twist, string, milling (on coin).

cordonnier, *m.* shoemaker.

cordouan, *m.* cordovan (leather).

corète, *f.* jute plant.

coriace, coriaceous, tough, leatherlike (meat).

coriacé, coriaceous, leatherlike.

coriaire, *f.* tanner's sumac (*Coriaria myrtifolia*).

coriandre, *f.* coriander (seed).

coriandrol, *m.* coriandrol, linalool.

coricide, *m.* corn cure.

corindon, *m.* corundum; — **granulaire,** emery, slime.

corindonique, corundum (bearing).

corme, *f.* serviceberry.

cormier, *m.* service tree (*Sorbus domestica*).

cormophytes, *f.pl.* cormophytes.

cormoran, *m.* cormorant.

cornaille, *f.* horn raspings.

cornaline, *f.* carnelian.

cornard, *m.* wheezing horse, glassworker's hook, stag beetle.

corne, *f.* horn, prong, turned-down corner (of book), hoof, antenna (of stag beetle), cornu, spur, cornel berry (Cornus); (*pl.*), cornua; — **d'abondance,** cornucopia, horn of plenty; — **de cerf,** hartshorn.

corné, corneous, horny, dog-eared.

cornéal, -aux, corneal.

cornée, *f.* cornea.

cornéenne, *f.* hornfels, aphanite.

corneille, *f.* crow, wood pimpernel, loosestrife (*Lysimachia nemorum*); — **mantelée,** hooded crow; — **noire,** carrion crow.

cornéliane, *f.* carnelian.

cornéole, *f.* woodwaxen, dyer's furze (*Genista tinctoria*).

corner, to sound a horn, ring, din, horn, turn down the corner (of book).

cornet, *m.* cornet, small horn, (screw) bag, dice box, turbinate bone.

cornetier, *m.* hornworker.

corniche, *f.* cornice, water caltrop (fruit).

cornichon, *m.* gherkin.

cornicule, *f.* cornicle.

corniculé, corniculate (petal).

corniculifère, corniculiferous, bearing horns or protuberances.

cornier, *m.* cornel (Cornus).

cornier, corner.

cornière, *f.* angle iron, corner iron.

corniolle, *f.* water chestnut, fruit of water caltrop (*Trapa natans*), scorpion senna (*Coronilla emerus*).

cornouille, *f.* cornel berry cornelian cherry (*Cornus mas*).

cornouiller, *m.* cornel tree (Cornus), red dogwood (*Cornus sanguinea*); — **mâle,** male dogwood, cornelian cherry (*Cornus mas*); — **sanguin,** red dogwood tree (*Cornus sanguinea*).

cornu, horned, horny, angular, absurd, cornute (leaf), spurred (wheel), rugged-hipped (horse).

cornue, *f.* retort, converter, water caltrop; — **à goudron,** tar (still), retort; — **à lessives,** a vessel for carrying lye.

corocore, *m.* caracore, Malay proa.

corollacé, corollaceous.

corollaire, *m.* corollary (tendril).

corollaire, corolline.

corolle, *f.* corolla; **appartenant à la —,** corolline; **ressemblant à la —,** corolline.

corollé, corollate.

corollifère, corolliferous.

corolliflore, *f.* plant with stamens on the corolla.

corolliforme, corolla-shaped.

corollin, corolline, corolla-like, petaloid.

corollique, found in the corolla.

coronaire, coronary (vein).

coroniforme, crown-shaped.

coronille, *f.* sicklewort(*Coronilla scorpioides*); — **bigarée,** purple coronilla.

coronope, *m.* swine's-cress, wart cress (Coronopus).

coronule, *f.* coronet, coronule.

corossol, *m.* custard apple (*Annona reticulata*).

corozo, *m.* vegetable ivory (from corozo palm).

corporation, *f.* guild, union.

corporel, -elle, corporeal, bodily.

corporifier, to corporify, materialize, solidify.

corps, *m.* substance, compound, (solid) body, frame, staff, corps; — **azoté,** nitrogenous substance, nitrogen compound; — **composé,** compound; — **creux,** hollow body; — **de la voie,** roadbed, body of a railway; — **desséchant,** dehydrating agent; — **dissout,** solute; — **étranger,** foreign (substance) body; — **gras,** fatty substance; — **isolant,** insulator; — **ligneux,** wood; — **médical,** medical profession; — **nourricier,** host; — **réducteur,** reducing agent, reducer; — **simple,** element.

corpusculaire, corpuscular.

corpuscule, *m.* corpuscle.

correct, correct, accurate.

correctement, correctly, accurately.

correcteur, -trice, *m.&f.* corrector, proofreader.

correcteur, correcting; **aimant —,** compensating magnet.

correctif, -ive, corrective, guard.

correction, *f.* correction, rectification, correctness; — **des torrents,** damming (fixation) of torrents; — **d'une rivière,** canalization of a river.

corrélatif, -ive, correlative.

corrélation, *f.* correlation, reciprocal relation, reciprocal influence in organisms.

corrélativement, correlatively.

correspondance, *f.* correspondence, communication, connection, intercourse, relation, harmony.

correspondant, *m.* correspondent, person speaking.

correspondant, corresponding, correspondent, homologous.

correspondre, to correspond, agree.

corridor, *m.* corridor, hall, passage.

corrigé, *m.* corrected copy.

corrigé, corrected.

corriger, to correct, rectify; **se —,** to be corrected, correct oneself, reform, amend.

corrigiole, *f.* strapwort (*Corrigiola littoralis*).

corroborant, *m.* corroborant, tonic.

corroborant, corroborative (proof), corroborant.

corroboratif, -ive, corroborative, confirmatory.

corroboration, *f.* strengthening.

corroborer, to corroborate, strengthen.

corrodant, *m.* corrosive.

corrodant, corroding, corrosive, erosive.

corrodé, paraissant comme —, erose, eroded.

corroder, to corrode, eat away, erode, wash out, attack; **se —,** to become corroded, be eroded.

corroi, *m.* currying (leather), claying, puddling, puddle (of clay).

corroierie, *f.* curriery.

corrompre, to corrupt, soften (leather, iron), deprive (wax) of plasticity, spoil, adulterate; **se —,** to become corrupted, tainted, softened, spoil.

corrompu, corrupt(ed), putrid, vitiated.

corrosif, -ive, corrosive, eating away caustic.

corrosion, *f.* corrosion, erosion, pitting, wearing away, rusting.

corroyage, *m.* currying, puddling, welding.

corroyer, to curry (leather), work, forge, weld (iron), puddle, pug, knead, dress down, plane (wood), roll, pass (iron) through rolling mill, rough-plane.

corroyère, *f.* tanner's sumac (*Coriaria myrtifolia*).

corroyeur, *m.* currier, leather dresser.

corrugation, *f.* corrugation, wrinkling.

corruptible, corruptible, perishable.

corruption, *f.* corruption, putrefaction, rottenness, decay, deterioration.

cors, *m.pl.* points, tines.

corsé, full (strong) bodied, thick, full-flavored.

corselet, *m.* corselet, prothorax.

corser, to give body, volume, flavor, strength to.

cortex, *m.* cortex, bark.

cortical, -aux, cortical.

cortication, *f.* cortication, the formation of cortex.

corticine, *f.* corticin.

cortifère, cortex-bearing.

cortiforme, barklike.

cortine, *f.* cortina, the filamentous annuli.

cortiqueux, -euse, corticous.

corubis, *m.* artificial corundum, alundum.

corvée, *f.* base service, statute labor.

corvette, *f.* corvette, sloop.

corydale, *m.* corydalis, birthwort (*Corydalis fabacea*).

corymbe, *m.* corymb, partial umbel.

corymbifère, corymbose, bearing corymbs.

corymbiforme, corymbose.

corymbuleux, -euse, arranged in small corymbs.

coryza gangreneuse, *f.* malignant catarrh (ox, sheep).

cosécante, f. cosecant.

cosensibiliser, to cross-sensitize, to co-sensitize.

cosinus, m. cosine.

cosmesthésie, f. cosmaesthesia, sensibility to external stimuli.

cosmétique, cosmetic.

cosmique, cosmic(al).

cosmogonie, f. cosmogony.

cosmopolite, f. cosmopolite, a plant of universal distribution.

cosse, f. pod, husk, hull, shell, cod, peel; — **en forme de coeur,** heart-shaped thimble.

cosseterie, f. cutting into chips or slices.

cossette, f. cossette, chip, slice.

cossettes, f.pl. dry chicory roots.

cosson, m. pea weevil, new shoot.

cossu, strong in the pod (beans).

cossus, m. — **gâte-bois,** goat moth; — **ronge-bois,** carpenter moth.

costule, f. veinlet, small rib.

costume, m. costume, uniform.

cotangente, f. cotangent.

cotarnine, f. cotarnine.

cote, f. quota, (indicating) dimensions, height from datum line, quotation, portion, share.

coté, classified, lettered, numbered, marked.

côte, f. rib, slope, shore, coast, seashore, costa, keel; — **à** —, side by side; — **de soie,** floss, cappadine.

côté, m. side, form, direction, part; — **de terre ferme,** landslide; — **du vent,** windward side; — **sous le vent,** lee side, sheltered from the wind; **à** —, on one side, near, sideways, by, close to; **à** — **de,** by the side of, beside, close to, near, at the side of; **de . . .** —, on . . . side; **de** —, on one side, sideways, awry; **de ce côté-ci** (-là), in this (that) direction; **de mon** —, for my part; **de tous** —s, in all directions; **du** —, on the side; **du** — **de,** with respect to, regarding; **d'un** —, on one side, on (the) one hand.

coteau, m. slope, hillside, hill, hillock.

côtelette, f. cutlet.

coter, to classify, number, letter, mark, quote, put down references.

coteux, m. calyx marked by outstanding veins.

côtier, coasting, pertaining to the coast, coastal.

côtière, f. sloping bed, malt floor, side (hearth) stone, (one of the parts of) mold (for lead pipes).

cotignac, m. quince marmalade.

cotignelle, f. quince wine.

cotin, cotinus, m. young fustic, fustet (Cotinus).

cotisation, f. clubbing together, assessment, share, quota.

cotiser, to assess; **se** —, to club together.

coton, m. cotton, down, cotton wool; — **azotique,** guncotton, nitrated cotton; — **brut,** raw cotton; — **courte soie,** short-staple(seed) cotton; — **de verre,** glass wool; — **en laine,** raw cotton, cotton wool; — **explosif,** nitrocellulose, collodion cotton; — **fulminant,** guncotton, fulminating cotton.

coton, — **hydrophile,** absorbent cotton (wool); — **longue soie,** long-staple cotton; — **minéral,** mineral wool, slag wool, mineral cotton; — **poudre,** m. guncotton; — **retors,** cotton thread; — **sauvage,** silkweed, swallowwort; **huile de** —, cottonseed oil.

cotonéastre, m. Cotoneaster.

cotonnade, f. cotton fabric, cotton goods.

cotonné, cottoned, cottony, downy, (of hair) woolly.

cotonner, to become nappy or downy. become fluffy; **se** —, to become nappy or downy, get mealy (of fruits).

cotonnerie, f. cotton plantation, cotton growing, cotton mill.

cotonneux, -euse, cottony, spongy, fleecy, woolly.

cotonnier, m. cotton worker, cotton plant.

cotonnier, cotton (industry).

cotonnière, f. cottonweed.

coton-poudre, m. guncotton.

côtoyer, to coast, (go) along, skirt, border on.

cotre, m. — **de la douane,** revenue cutter.

cotret, m. small fagot.

cotylédon, m. cotyledon, seed lobe, seed leaf.

cotylédonaire, cotyledonary, cotyledonous.

cotylédoné, cotyledonous, possessing seed lobes.

cotyléphore, with small capsules.

cotylifère, with slight excavation.

cotyloïde, f. cotyloid cell.

cou, m. neck; — **de cygne,** gooseneck.

couchage, *m.* couching, layering (of plants).

couchant, *m.* west, decline, setting sun, sunset.

couchant, laying, setting.

couchart, *m.* coucher (in papermaking).

couche, *f.* layer, bed, stratum, coat, coating, germinating bed, couch, parturition, covering, deposit, ring (annual), zone, bearing bush, brass, pillow, axle box; — **arable,** superficial soil, topsoil; — **cambiale,** cambium zone; — **cornée,** horny layer; — **de fumier,** hotbed; — **d'impression,** priming coat; — **membraneuse,** plasma membrane; — séparatrice, absciss layer; — **subereuse,** cuticle; à —s **épaisses,** coarse-zoned, broad-ringed, broad-zoned (wood); à —s minces, fine-zoned.

couché, lying, fallen, recumbent, lying down; — sur le sc[1], prostrate.

coucher, to lay (flat), lay down, lay on, put to bed, set, write down, insert, spread, coat (paper), couch, lie (down), sleep; se —, to retire, rest, go to bed, lie (down), set (of sun).

coucher, *m.* lying, setting (sun), bed, bedding, retiring; — **de soleil,** sunset.

couchette, *f.* bunk, berth.

coucheur, *m.* layer.

couchis, *m.* bed, layer, stratum, solepiece, wallpiece.

coucou, *m.* cuckoo, cuckoo clock, cowslip, white clover, shamrock, daffodil; — **gris,** cuckoo.

coude, *m.* elbow, knee, angle, bend, turn, bow, loop, crank, shoulder; — **d'aspiration,** suction bend.

coudé, bent, having an elbow, kneed, cranked.

couder, to bend (form of an elbow), knee, crank; se —, to form an elbow.

cou-de-cygne, *m.* gooseneck.

coudoyer (se), to elbow each other.

coudrage, *m.* soaking in tan liquor.

coudraie, *f.* hazel copse, hazel grove.

coudran, *m.* see GOUDRON.

coudranner, see GOUDRONNER.

coudre, to sew, stitch, tack on.

coudrée, *f.* arid land.

coudrement, *m.* soaking (skins) in tan liquor, steeping.

coudrer, to soak in tan liquor.

coudret, *m.* tanning vat.

coudrier, *m.* hazel, hazel (bush) tree, filbert tree (Corylus).

couenne, *f.* pigskin, rind (of bacon), pork rind, (diphtheric) membrane.

couette, *f.* socket, feather bed, sea gull.

coulage, *m.* pouring, running, flowing, leaking, casting, leakage, guttering (of candle), scalding; — à **noyau,** core casting; — **en coquille,** chill casting; — **en terre,** loam casting.

coulant, *m.* slide, slip, runner (of plant), stolon, sucker.

coulant, running, flowing, fluid, sliding, loose (ground), caving, fluent, smooth, accommodating.

coulé, *m.* cast, casting.

coulé, cast.

coulée, *f.* casting, tapping, tap, running off, discharge, channel from furnace to mold, molding, outflowing.

coulement, *m.* flow, flowing.

couler, to flow, leak, run (out), glide, slide, slip, creep, drop (of fruit), sink, run (through the mold), stream, draw, cast, pour, mold, found, tap, strain, cause to flow, run, pass (time), run lead into, scuttle; se —, to glide, slip, slide, ruin oneself.

couleur, *f.* color, coloring, dye, a variety of smalt, hue, tinge, paint; — à **cuve,** — **pour cuve,** vat dye; — à l'eau, water color; — à l'huile, oil color; — à **mordants,** mordant dye, dye requiring use of mordants; — azoïque, azo dye; — **bon teint,** — **grand teint,** fast color; — **claire,** light color; — composée, secondary color.

couleur, — **d'aniline,** aniline (color); — de chair, flesh color; — d'imprimerie, colored printing ink; — du **recuit,** tempering color; — **en détrempe,** distemper color; — **foncée,** dark color; — **grattée,** — **lavée,** pale color; — **matrice,** primary color; — **primitive,** primary color; — **solide,** fast (color) dye; — **vitrifiable,** vitrifiable (enamel) color; mettre en —, to color, stain.

couleur, *m.* caster, pourer, scalder.

couleuvre, *f.* snake.

couleuvrée, *f.* traveler's-joy (*Clematis vitalba*) white bryony (*Bryonia alba*), red bryony (*Bryonia dioica*); — **de Virginie,** Virginia snakeroot (*Aristolochia serpentaria*).

couleuvrine, *f.* bistort, snakeweed.

coulis, *m.* grout, liquid filling, thin mortar, molten metal, strained broth, (meat) jelly.

coulisse, *f.* groove. slot, slide, slotted link, sliding shutter, wood strip; à — sliding, with sliding shutter.

coulisseau, *m.* small groove, slide block, slider, guide block, guide, chute, shoot, link, ram.

coulissement, *m.* sliding, sliding motion.

coulisser, to slide, provide with guides.

couloir, *m.* strainer, colander, hopper, corridor, gangway, passage, passageway, lobby, runway, shoot, chute; — **d'égouttage,** drainage canal; — **de quille,** keel strake.

couloire, *f.* strainer, colander, filter.

coulomb, *m.* coulomb.

coulombmètre, *m.* coulomb meter, coulometer.

coulure, *f.* running (down), running out (from mold), washing away of pollen (by rain), abortion, failure of crop.

coumarine, *f.* coumarin.

coumarique, coumaric.

coumarone, *m.* coumarone, benzofuran.

coumarylique, coumarilic.

coumys, *m.* kumiss.

coup, *m.* stroke, blow, shock, shot, report, gust, hit, discharge (from gun), blast, sound, glass, swallow, draft (of liquor), move, throw, bruise, wound, bolt, thrust; **à —,** all at once; **à — de,** with blows from; **à — sûr,** surely, with certainty; **à tout —,** every time; — **de bâton,** blow, beating; — **de bélier,** water hammer.

coup, — **de chaleur,** heatstroke; — **de collier,** great effort, renewed effort; — **de dé(s),** cast of the dice, (fig.) venture; — **de feu,** shot, explosion, heating; — **de fusil,** shot; — **d'état,** decisive act; — **de main,** helping hand, bold action; — **de mine,** blast; — **de pioche à bourrer,** packing, tamping; — **de soleil,** sunstroke; — **de tête,** capricious decision; — **de tonnerre,** thunderclap; — **de vent,** gust of wind; — **d'oeil,** glance; — **sur —,** one after another, in rapid succession.

coup, après —, afterward, too late; **du —,** now at last, offhand; **du même —,** at the same time; **d'un seul —,** at one stroke, at one time; **du premier —,** at the first stroke; **encore un —,** once more; **pour le —,** this time, for once; **sur le —,** outright, on the post, instantly; **tout à —,** suddenly; **tout d'un —,** all at once, suddenly.

coupable, *m.* culprit.

coupable, guilty, culpable.

coupage, *m.* cutting, mixing, dilution (of alcohol), mixture, blend (of wines), green meat.

coupant, *m.* (sharp) edge.

coupant, cutting; **bord —,** cutting edge.

coupe, *f.* cutting, cut, section, felling, (cutting) area; — **à champagne,** champagne glass; — **blanche,** clear-cutting, clear-felling; — **claire,** open felling; — **d'amélioration,** improvement felling; — **d'éclaircie,** thinning; — **de fil,** cut with the grain; — **définitive,** final cutting.

coupe, — **d'extraction,** extraction of old trees; — **d'isolement,** very heavy thinning; — **en lisière,** shelterwood strip system; — **en long,** longitudinal section; — **en sève,** felling in the growing season; — **en travers,** cross section; — **horizontale,** horizontal section; — **transversale,** cross section; — **usée,** felled area.

coupé, cut, broken up, cut off, blunt, truncate(d), intersected.

coupe-air, *m.* trap (in pipe).

coupeau, *m.* cutting, chip.

coupe-bourgeon, *m.* common vine grub.

coupe-circuit, *m.* circuit breaker, cutout, fuse.

coupe-feu, *m.* fire belt, fire trace.

coupe-légumes, *m.* vegetable cutter.

coupellation, *f.* cupellation, assaying.

coupelle, *f.* cupel, pan, cup.

coupeller, to cupel, assay, test.

coupelleur, *m.* cupeler, assayer.

coupement, *m.* cutting, sawing.

coupe-navets, *m.* turnip cutter, pulper.

coupe-net, *m.* wire cutters, nippers.

coupe-paille, *m.* chaffcutter, chaffchopper.

couper, to cut, cut off, turn off, cut out, mix, blend (wine), dilute, interrupt, mow, intersect, chop down, dock, fell, stop, check, lop off, clip, trim (a hedge), snip off, geld (animal); **se —,** to crack (skin), cut oneself, part, split, cleave, cut each other, cross, intersect, wear into holes, contradict oneself, interfere.

couper; — **à blanc,** to clear-cut; — **à blanc étoc,** to cut clear, clean-cut, coppice; — **à travers,** to cut across the grain; — **en soulevent le copeau,** to cut across the grain; — **rez-tronc,** to cut level with the ground.

couperet, *m.* chopper, cleaver, blade, knife, enameler's file.

couperose, *f.* copperas, vitriol, acne rosacea; — **bleue,** blue vitriol, copper sulphate; — **verte,** green vitriol, copperas, ferrous sulphate.

coupe-tubes, coupe-tuyaux, *m.* tube (pipe) cutter.

coupeur, -euse, *m.&f.* cutter; **coupeuse de feuilles,** leaf cutter (bee).

coupeur, cutting.

couplage, *m.* coupling, connection.

couple, *f.* couple, pair; *m.* pair, couple, torque, timber, frame, cell (battery); **maître —,** midship frame.

couplement, *m.* coupling, connection.

coupler, to couple, join, connect, switch.

couplet, *m.* hinge, hinged joint, couplet.

coupoir, *m.* cutter.

coupole, *f.* cupola, dome.

coupon, *m.* coupon, part (of shares), remnant (of cloth), small felling, raft section.

coupure, *f.* cut, ditch, trench, opening, slit, incision, small paper money.

cour, *f.* court, yard.

courable, chaseable, warrantable.

courage, *m.* courage, valor, heart, temper.

courageux, -euse, courageous.

couramment, commonly, currently, readily.

courant, *m.* current, stream, course, course (of year), run, current month, inst.; **— à haute fréquence,** high-frequency current; **— biphasé,** two-phase (diphase) current; **— continu,** continuous (direct) current; **— de charge,** charging current; **— d'échauffement du filament,** filament current; **— de jusant,** ebb stream; **— de marée,** tidal stream; **— de retour,** return current.

courant, — dérivé, shunt current; **— discontinu,** intermittent current; **— inducteur,** primary (inducing) current; **— induit,** induced current; **— monophasé,** single-phase (monophase) current; **— parasite,** eddy current; **— redressé,** rectified current; **— tellurique, — terrestre,** earth current; **— vagabond,** stray current; **— vibré,** alternating current; **—s de Foucault,** eddy currents.

courant, current, present, general, common, lineal, running; **au —,** acquainted.

courbable, bendable.

courbage, *m.* bending, curving, deflection.

courbaril, *m.* courbaril, West Indian locust tree (*Hymenaea courbaril*).

courbarine, *f.* courbaril (resin), animé gum.

courbature, *f.* lameness, lumbago.

courbe, *f.* curve, turn, bend, knee, contour, bow, curvature; **— de niveau,** contour (line); **— d'agglutination,** curve of aggluti-

nation; **— heliocoïdale,** helical line or curve.

courbe, curved, bent.

courbé, curved, bent, bowed.

courbement, *m.* bending, curving.

courber, to bend, curve, bow, sag, stoop; **se —,** to bend, bow, be bent, stoop, curve.

courbes, *f.* figured wood, knees.

courbure, *f.* curvature, curve, bending, flexure.

coureur, *m.* runner, racer, rover, wanderer.

courge, *f.* gourd, squash; **— à la moelle,** marrow squash; **— citrouille,** pumpkin; **huile de —,** pumpkin-seed oil.

courir, to run, be current, run about, move, flow, pursue, run after, frequent, roam over, travel over, overrun, circulate.

courlieu, *m.* curlew; **petit —,** whimbrel.

courlis, *m.* common curlew; **— corlieu,** whimbrel.

couronnant, (leaves) terminating the stem or its divisions.

couronne, *f.* crown, ring, halo, coronet, corona, rim, (of wheel), hoop, wreath; **— des blés,** corn cockle, corn rose; **— royale,** field melilot.

couronné, stag-headed, top-dry.

couronnement, *m.* crowning, coronation, cap, coping, surrounding, top, stag-headedness, withering (of tree) from top; **— en éventail,** wing top, flat burner.

couronner, to crown, cap, wreathe; **se —,** to become stag-headed.

cou-rouge, *m.* redbreast.

courra (courir), (it) will run.

courrier, *m.* courier, messenger, mail, post, correspondence.

courroi, *m.* roller.

courroie, *f.* belt, strap, band, chain; **— de transmission,** driving belt, belt; **— sans fin,** endless belt.

courroucer, to anger, vex, irritate.

courroyer, to put on the roller.

cours, *m.* course, price, rate, market price, quotation, length, current, progress, stream; **— d'eau,** stream, watercourse; **— de chimie,** chemistry course; **— en madriers,** plank covering, timber planking; **au — de, en — de,** in the course of; **avoir —,** to be current; **faire un — d'histoire,** to lecture on history.

course, *f.* course, race, run, excursion, trip, cruise, errand (business), walk, ride, stroke (of piston), lift (of valve), throw (of eccentric crank), flight (of projectile), running, career; — ascendante, upstroke; — d'essai, test run; — de vitesse, race, sprint, dash.

coursier, *m.* (war) horse, millrace, guide channel, floatboard.

coursive, *f.* runway, catwalk.

court, *m.* (grass) court.

court, short, brief; à —s jours, for a short time, on short notice.

courtage, *m.* brokerage, commission

court-circuit, *m.* short circuit.

court-circuiter, to short-circuit.

courtement, shortly, briefly.

courte-queue, docked (of dog, horse), dock-tailed.

courtier, *m.* broker; — de change, jobber, dealer.

courtilière, *f.* mole cricket.

court-jointé, short-pasterned (of horse).

courtois, courteous.

coaru (courir), sought after, accrued.

cous, *m.* whetstone.

cousant (coudre), sewing.

cousent (coudre), (they) are sewing.

couseuse, *f.* sewer, stitcher.

cousin, *m* gnat, midge.

cousse, *f.* pod.

coussin, *n.* cushion, pillow, pad.

coussiner, to cushion, pad.

coussinet, *m.* pad, cushion, journal bearing, box (journal), pillow, pulvinus, pillow block, bush, bushing, brass, chair (rail), bilberry, cranberry, whortleberry; — à billes, ball bearing; — à fileter, — de filière, screw die; — de tête de bielle, connecting rod crank end bearing.

cousso, *m.* cusso, kousso, brayera (*Hagenia abyssinica*).

cousu (coudre), sewed.

coût, *m.* cost, price, expense.

coûtant, à (au) prix —, at cost price.

couteau, *m.* knife, knife-edge (of balance), cutter, whittle; — affilé, sharp knife; — de chasse, hunting knife; — émoussé, blunt knife; — rond, round-edged knife.

coutelet, *m.* small knife.

coutellerie, *f.* cutlery (trade and wares).

coutelure, *f.* cut (of parchment by knife).

coûter, to cost, be expensive, be painful.

coûteux, -euse, costly, expensive, dear.

coutil, *m.* ticking, drill, duck.

coutre, *m.* colter, plow colter, cutter.

coutrier, *m.* subsoil plow.

coutrière, *f.* colter clip.

coutume, *f.* custom, habit; de —, customary, as usual.

coutumier, accustomed, customary.

coutumièrement, customarily, usually.

couture, *f.* seam, sewing, stitching, cicatrix.

couvage, *m.* incubation, hatching (of eggs).

couvain, *m.* nest of insect eggs, brood comb (of bees), brood.

couvaison, *f.* brooding time.

couvé, addled, hatched.

couvée, *f.* sitting (eggs), hatching, brood, hatch.

couvent, *m.* convent, monastery.

couver, to brood (on), sit on eggs, incubate, hatch.

couvercle, *m.* cover, lid, cap, top (cover); — à anneau, ring(ed) cover; — en verre, glass cover.

couvert, *m.* cover, envelope, shelter, lodging, leaf canopy, foliage, leafage.

couvert, covered, obscure, secret, concealed, hid, overcast, deep-colored (wine), cloudy (weather), sheltered (country); à — épais, densely crowned; à — léger, thinly crowned.

couverte, *f.* glaze, covering, facing.

couverture, *f.* covering, cover, roof, roofing, blanket, coverlet, security; — d'humus, leaf mold; — du sol, soil covering, surface cover; — herbacée du sol, herbaceous soil covering; — morte, leaf litter; — morte du sol, ground litter, dead soil covering; — vivante, herbaceous soil covering.

couveuse, *f.* sitting hen, incubator.

couvoir, *m.* incubator.

couvre-amorce, *m.* primer cap.

couvre-objet, *m.* cover glass.

couvrir, to cover, face, screen, conceal, wrap up, defray expenses, shelter; se —, to cover oneself, be covered, be cloudy.

covelline, *f.* covellite, covelline.

crabe, *m.* crab.

crachat, *m.* sputum, spit, spittle; — de lune, star jelly, nostoc.

crachement, spitting, expectoration.

cracher, to spit, expectorate, sputter, prime foam, spark, throw out, splutter (pen).

cracque, *f.* fissure.

craie, *f.* chalk; — **de Briançon,** steatite, soapstone; — **lavée,** prepared chalk.

craindre, to fear, be afraid of, dread, stand in awe of, hesitate; **je crains qu'il ne vienne,** I fear he will come; **je crains qu'il ne vienne pas,** I fear he will not come; **je ne crains pas qu'il vienne,** I do not fear he will come.

craint, feared, dreaded.

crainte, *f.* fear, apprehension, dread, awe; **de — de,** for fear of; **de — que,** for fear that, lest; **sans —,** fearless.

craintif, -ive, timorous, timid, fearful.

cramoisi, crimson.

crampe, *f.* cramp, spasm; — **de poitrine,** angina pectoris.

crampon, *m.* cramp, cramp iron, clamp, staple, fulcrum, prop, rhizoid, aerial root, climbing iron, climbing spur, clinging root.

cramponner, to cramp, clamp; **se —,** to cling.

cramponnet, *m.* small cramp, staple (of lock).

cran, *m.* notch, catch, cag, nick, defect, flaw, horse-radish (*Armoracia rusticana*), whit, scurvy grass (*Cochlearia officinalis*), cran (of herrings); — **de Bretagne,** horse-radish (*Armoracia rusticana*).

cranage, *m.* notching.

crâne, *m.* skull, cranium.

crâne, bold, plucky, gallant.

crangon, *m.* shrimp.

crânien, -enne, cranial, relating to the skull.

crapaud, *m.* toad, mushroom anchor, low chair, contagious foot rot (sheep), canker of the hoof, clip.

crapaudine, *f.* bearing, bushing, socket, discharge valve, grating, strainer, step bearing, ironwort, contagious foot rot (sheep).

craquant, crackling.

craquelage, *m.* crackle manufacture.

craquelé, *m.* crackling.

craquelé, crackled.

craquelée, *f.* crackle (ware).

craqueler, to crackle, cover (china) with cracks.

craquelin, *m.* cracknel (biscuit), hard biscuit.

craquelure, *f.* cracking, small cracks (in paint).

craquement, *m.* cracking, crackling, creaking, crepitation.

craquer, to crack, crackle, creak.

craquetant, crackling.

craquètement, *m.* crackling, decrepitation.

craqueter, to crackle, decrepitate, chirp, clatter.

craquette, *f.* impurities removed from melted butter.

craqure, *f.* crack, fissure, split.

craspedodrome, *m.* lateral veins run from midrib to margin.

crassane, *f.* bergamot (pear).

crasse, *f.* crasses, dross, hammer slag, waste metal, scum (of metals), dirt, filth, clinker.

crassement, *m.* fouling.

crasser, to foul, clog, dirty.

crassicaule, thick-stemmed.

crassifolié, with thick leaves.

crassilobé, with voluminous lobes.

crassinervé, with projecting veins.

crassipétale, with thick petals.

crassule, *f.* Crassula, thick leaf.

cratère, *m.* crater, shell hole.

cratériflore, with cup-shaped flowers.

cratériforme, crater-shaped, crateriform.

cravant, *m.* brant (brent) goose, barnacle.

crave, *m.* chough, red-legged crow.

crayer, to chalk, mark with chalk.

crayer, *m.* vitrified coal ashes.

crayère, *f.* chalk pit.

crayeux, -euse, chalky, cretaceous.

crayon, *m.* crayon, pencil, carbon, sketch, outline, lead pencil, pencil (of caustic), chalk soil, marl; — **à dessiner,** drawing pencil; — **de carbone,** carbon rod, carbon; — **de charbon,** charcoal crayon, blue-black pencil; — **de couleur,** colored pencil; — **de graphite,** — **de mine de plomb,** lead pencil; — **d'une lampe à arc,** carbon (pencil) of arc lamp; — **électrique,** arc-light carbon.

crayonnage, *m.* drawing, pencil sketch.

crayonner, to draw, mark, sketch in pencil.

crayonneux, -euse, chalky, marly (soil).

créance, *f.* credit, credence, belief, claim, debt.

créancier, *m.* creditor, mortgagee.

créateur, -trice, *m.&f.* creator, establisher; **Le Créateur,** God.

créateur, -trice, creative, inventive.

créatine, *f.* creatine.

créatinine, *f.* creatinine.

création, *f.* creation, invention, establishment, production; — d'un peuplement, formation of a crop of trees.

créatrice, see CREATEUR.

créature, *f.* creature, being.

crébrisulce, with grooves close together.

crèche, *f.* crib, manger; — de pourtour, pile sheathing.

crédibilité, *f.* credibility.

crédit, *m.* credit, repute, influence.

créditer, to credit, trust.

crédule, credulous.

créer, to create, produce, invent, give rise to, establish; — un sous-étage, to underplant.

crémaillère, *f.* chimney hook, rack, toothed bar, rack bar, gear rack, arm sling.

crémant, creaming (wine).

crémation, *f.* cremation.

crème, *f.* cream, sirupy liqueur; — d'amandes, almond cream; — de chaux, cream of lime; — de glace, ice cream; — de menthe, peppermint liqueur; — de soufre, sublimed sulphur; — de tartre, cream of tartar; — fouettée, whipped cream; — froide, cold cream.

crément, *m.* the absorbed part of food, increment.

crémer, to cream, cremate.

crémerie, *f.* creamery, dairy.

crémeux, -euse, creamy.

crémocarpe, *m.* cremocarp.

crémomètre, *m.* creamometer.

crémone, *f.* casement bolt.

crémosperme, (seed) holding to placenta at top or middle.

crené, crenated, notched, milled.

créneau, *m.* battlement, embrasure.

crénelé, crenate, crenulate.

créneler, to indent, notch, cog, tooth.

crénelure, *f.* indentation, indenting, notching, creneling (of leaf).

créosol, *m.* creosol.

créosotage, *m.* creosoting.

créosotal, *m.* creosotal.

créosote, *f.* creosote.

créosoter, to creosote.

crêpage, *m.* dressing of crepe, crimping, crisping.

crêpe, *m.* crepe, crape, pancake.

crêper, to crimp, crisp, crape.

crépi, *m.* rough coat (of plaster), plastering.

crépine, *f.* strainer, fringe, caul, omentum

crépir, to grain, pebble (leather), roughcast, whitewash, plaster, rough-coat, crisp, crimp (hair).

crépissage, *m.* graining, rough coating, pebble dashing.

crépitant, crackling, crepitating, crepitant.

crépitation, *f.* crépitement, *m.* crackling, crepitation, sputtering.

crépiter, to crepitate, crackle, sputter.

crépon, *m.* crépon.

crépu, woolly, crisp, curled, frizzled, crispate, crinkled.

crépusculaire, crepuscular, dusky; lumière —, twilight.

crépusculais, *m.pl.* Crepuscularia, hawk moths.

crépuscule, *m.* twilight (either dawn or dusk).

crèque, *f.* bullace.

créquier, *m.* bullace (tree) (*Prunus insititia*).

crescent, *m.* lunated spot, lunule.

crésol, *m.* cresol.

cresserelle, *f.* kestrel.

cresson, *m.* cress; — alénois, garden peppergrass, garden cress (*Lepidium sativum*); — de fontaine, water cress (*Roripa nasturtium-aquaticum*); — des prés, cuckooflower, lady's-smock, meadow cress (*Cardamine pratensis*); — doré, chrysoplene, saxifrage; — du Mexique, Indian cress, garden nasturtium (*Tropaeolum majus*).

cressonette, *f.* cuckooflower, meadow cress.

cressonnière, *f.* water-cress pond.

crésyl, *m.* cresyl.

crésyle, *f.* cresyl (the radical).

crésyler, to disinfect (with cresyl).

crésylique, cresylic.

crésylol, *m.* cresol; — sodique dissous, sodium cresolate solution.

crétacé, cretaceous, chalky.

crête, *f.* crest, ridge, tuft, comb.

crêté, crested, tufted.

crételer, to cackle.

crétin, *m.* cretin.

cretons, *m.pl.* greaves, cracklings.

creusage, creusement, *m.* hollowing, grooving, digging, sinking, excavation.

creuse, hollow.

creuser, to dig, excavate, hollow (out) go thoroughly into, cut, delve; **se —,** to become (grow) hollow; **il se creuse (le cerveau),** he racks (his brain).

creuset, *m.* crucible, melting (pot), hearth; **— brasqué,** brasqued crucible; **— de porcelaine,** porcelain crucible; **— en argent,** silver crucible; **— en biscuit,** unglazed porcelain crucible; **— en charbon,** carbon crucible; **— en fer forgé,** wrought-iron crucible; **— en fonte,** cast-iron crucible; **— en grès,** stoneware crucible; **— en plombagine,** graphite crucible, black-lead crucible; **— en terre,** clay crucible.

creusiste, *m.* crucible maker.

creusure, *f.* hollowing, hollow.

creux, *m.* cavity, hollow, trough, hole, mold, pit (of stomach), cavity, depth, groove; **gravure en —,** intaglio.

creux, -euse, hollow, tubular, barren, empty, shallow, deep, unsubstantial (food).

crevasse, *f.* crevice, fissure, crack, crevasse, rift, flaw, burst.

crevassé, cracked.

crevasser, to crack, chink, split, chap; **se —,** to crack, crevice.

crève, *f.* death.

crève-chien, *m.* black nightshade (*Solanum nigrum*).

crever, to burst, break, crack, perish (of animals), puncture (tire); **se —,** to burst, stuff oneself with food; **— de faim,** to be starving; **— de rire,** to split with laughter, die.

crevette, *f.* shrimp, prawn; **— d'eau douce,** *f.* fresh-water shrimp.

cri, *m.* cry, creaking, rustling, grating, scream, shriek, squeak; **— de l'étain,** tin cry, crackling of tin.

criard, crying, discordant.

criblage, *m.* sifting, riddling (of coal), screening.

crible, *m.* sieve, screen, riddle; **— à secousse,** shaking sieve, swing sieve.

cribler, to sift, screen, riddle.

cribleur, *m.* sifter, screener, separator.

criblure, *f.* siftings, screenings.

cric, *m.* jack, lifting jack; **— à vis,** screw jack.

cricri, *m.* chirping, cricket.

criée, *f.* sale by auction.

crier, to cry, cry out, squeak, bark, creak, shout, scold, proclaim, offer (for sale); **— à,** to cry out.

crime, *m.* crime, guilt.

criminel, -elle, criminal, guilty.

crin, *m.* (coarse) hair (of mane and tail of horse), horsehair, bristles; **— végétal,** vegetable fiber.

crinière, *f.* mane; **— à longue,** crinite.

crinoïde, *m.* crinoid.

crinoïde, crinoidal.

criocère, *f.* Crioceris; **— du lis,** lily beetle.

criquet, *m.* locust, cricket.

criqûre, *f.* crack, flaw, fissure.

crise, *f.* crisis.

crispatif, -ive, irregularly folded, as though curled.

crispation, *f.* shriveling, puckering.

crisper, to shrivel, contract, clench, wrinkle, irritate, cockle; **se —,** to become wrinkled, shrivel, contract.

crispiflore, crispifloral, having curled flowers.

crispifolié, crispifolious.

crissement, *m.* grating, grinding, squeak.

crisser, to grate, grind.

crissure, *f.* ridge, crease, wrinkle.

crista, *f.* crista, a crest or terminal tuft.

cristal, -aux, *m.* crystal, crystal glass, flint glass; **— armé,** ferroglass, wire glass; **— de roche,** rock crystal; **— d'Islande,** Iceland spar, transparent calcite; **— taillé,** cut glass; **— violet,** crystal violet; **cristaux de soude,** soda crystals; **cristaux de Vénus,** crystallized, distilled verdigris; **cristaux hémièdres,** hemedrism, hemihedral forms; **cristaux réunis,** compound crystal; **en cristaux,** hooked.

cristallerie, *f.* crystal-glass cutting, crystal-glass factory.

cristallière, *f.* rock-crystal mine.

cristallifère, crystalliferous.

cristallin, *m.* crystalline lens.

cristallin, crystalline.

cristalline, *f.* crystallin, ice plant (*Mesembryanthemum crystallinum*).

cristallinité, *f.* crystallinity.

cristallisabilité, *f.* crystallizability.

cristallisable, crystallizable, glacial; **acide acétique —,** glacial acetic acid.

cristallisant, crystallizing.

cristallisation, *f.* crystallization, crystallizing; — fractionnée, fractional crystallization.

cristallisé, crystallized, crystalline.

cristalliser, to crystallize.

cristallisoir, *m.* crystallizing dish, pan, vessel, crystallizer, glass pot, Petri dish; — double, glass pot with lid, Petri dish.

cristallite, *f.* crystallite.

cristallitique, crystallitic.

cristallogénie, *f.* crystallogenesis, crystallogeny.

cristallogénique, crystallogenic.

cristallographe, *m.* crystallographer.

cristallographie, *f.* crystallography.

cristallographique, crystallographic.

cristalloïde, crystalloid.

cristallométrique, crystallometric.

cristaux, *m.pl.* see CRISTAL.

cristé, crested, tufted, cristate.

criste-marine, *f.* (sea) samphire (*Crithmum maritimum*).

critérium, critère, *m.* criterion.

critiquable, open to criticism.

critique, *f.* criticism, critique, censure, fault-finding; *m.* critic.

critique, critical.

critiquer, to criticize, censure, find fault.

croc, *m.* hook, catch, crunch (the sound), grapnel, canine tooth, tusk; — à sarcler, weed lifter.

crocéine, *f.* crocein.

crocher, to hook.

crochet, *m.* hook, small hook, clasp, catch, rabble, bend, turn, tusk, fang, crochet hook, bracket, crotchet, poker; — de serrurier, skeleton key.

crochetage, *m.* superficial hoeing, wounding the soil.

crochu, crooked, hooked.

crocique, croconic.

crocodile, *m.* crocodile.

crocoïse, *f.* crocoite, crocoisite, mineral lead chromate.

croconique, crocique, croconic.

crocus, *m.* crocus, saffron (*Crocus sativus*).

croire, to believe, think, trust; se —, to think oneself, depend (rely) on oneself, be believed.

croisade, *f.* crusade.

croisé, *m.* twill, crusader.

croisé, crossed.

croisée, *f.* sash (window), window, transept, crossing.

croisement, *m.* crossing, intersection, crossbreeding; — d'espèces, species hybridization.

croiser, to cross, cross out, lap (fold) over, twill, lay across, hybridize, interbreed, twist slightly, thwart, cruise; se —, to (inter)cross, be crossed, cross each other, intersect each other.

croisette, *f.* little cross, cross-stone, crosswort.

croiseur, *m.* cruiser, cross lode, cross vein; — aérien, large dirigible.

croisière, *f.* cruising.

croisillon, *m.* crossbar, crossarm, crosspiece.

croissance, *f.* growth, habit; — bien régulière, regular growth; — en spirale, torsion of fibers, twisted growth; — ondulée, wavy-fibered growth; à — hâtive, fast-growing.

croissant, *m.* crescent (malarial parasite), crescent-shaped object, sickle, roll, tread (of tire), top swage.

croissant, growing, increasing.

croit (croire), (he) believes.

croît (croître), (it) grows.

croître, to grow, increase, rise, augment, wax; — en massif, to grow under cover, to grow in a dense crop.

croix, *f.* cross, dagger, obelisk (printing); — ou pile, heads or tails; en —, crosswise.

cron, *m.* sandy soil, shell sand.

croquant, *m.* gristle.

croquant, crisp, crunching.

croquer, to crunch, sketch (roughly), be crisp.

croquet, *m.* hard biscuit (covered with almonds).

croquette, *f.* croquette, (corn) rattle.

croquis, *m.* sketch, scheme, design, survey, measure.

crosne, *m.* Chinese artichoke (*Stachys sieboldi*).

crosse, *f.* butt, buttstock, crosshead, bend, crook, crosier, circinate aestivation.

crossette, *f.* slip, cutting.

crot, *m.* box.

crotalaire, *f.* sunn, Bengal hemp.

crotale, *m.* crotalum, rattlesnake.

crotaphite, *m.* crotaphite, temporal fossa.

crotine, *f.* crotin.

croton, *m.* croton.

crotoné, containing croton oil.

crotonique, crotonic.

crotte, *f.* dung, dropping, dirt, filth, mire.

crotter, to dirty, soil, foul; **se —,** to get dirty.

crottin, *m.* (horse) dung, (sheep) droppings.

crotton, *m.* sugar greaves.

crouler, to fall, fall in, sink, crumble, ruin.

croupe, *f.* brow of a hill, crupper, buttocks, croup, hip; **— boiteuse,** half or false hip.

croupi, stagnant, foul.

croupion, *m.* rump, end of spine, tail base, coccyx.

croupir, to stagnate, grow foul, lie in filth, wallow.

croupissant, stagnating, stagnant.

croupon, *m.* butt (leather).

croustade, *f.* pie, pastry.

croûte, *f.* crust, scab, fissured part of bark, skin (of casting), caking, coating, scurf.

croûteux, -euse, crusted.

crown gall, *m.* crown gall.

crown-glass, *m.* crown glass.

croyable, credible, likely, believable.

croyance, *f.* belief, credence, faith, trust.

croyant, *m.* believer.

croyant, believing.

cru, *m.* growth, (particular) vineyard, fabrication.

cru, believed, thought, crude, raw, hard, rude, harsh, indecent, uncooked, indigestible, coarse; **eau — e,** hard water; **vin du —,** local wine.

crû (croître), grown.

cruauté, *f.* cruelty.

cruche, *f.* pitcher, jug.

cruchée, *f.* pitcherful, jugful.

cruchette, *f.* small jug.

cruchon, *m.* small jug, stone (hot-water) bottle.

crucifères, *f.pl.* Cruciferae, crucifers.

cruciforme, cross-shaped, four-petaled (corolla).

crucigère, bearing a cross.

crudité, *f.* crudity, crudeness, indigestibility, coarseness, hardness; *pl.* raw vegetables or fruit.

crue, *f.* growth, increase, rise, flood, freshet; **grande —,** spring tide, high water.

cruel, -elle, cruel, tiresome.

cruellement, cruelly, severely.

cruenté, bloody.

cruor, *m.* cruor.

crural, -aux, crural.

crurent (croire), (they) believed.

crûrent (croître), (they) grew.

crusocréatinine, *f.* crusocreatinine.

crustacés, *m.pl.* Crustacea, crustaceans, shellfish.

crut (croire), (he) believed.

crût (croître), (it) grew; (croire), he believed.

cryogène, *m.* cryogen, freezing mixture.

cryohydrate, *m.* cryohydrate.

cryolithe, cryolite, *f.* cryolite.

cryomètre, *m.* cryometer.

cryophore, *m.* cryophorus.

cryoplankton, *m.* glacial vegetation.

cryoscopie, *f.* cryoscopy.

cryothérapie, *f.* treatment (of skin) by extreme cold.

cryotropisme, *m.* cryotropism, movements influenced by cold.

crypte, *m.* crypt, front cavity of a stoma.

cryptogame, *f.* cryptogam.

cryptogame, cryptogamic, cryptogamous.

cryptogamie, *f.* cryptogamy (concealed fructification).

cryptogamique, cryptogamic.

cryptographique, cryptographic.

cryptohybride, *f.* a hybrid displaying unexpected characters.

crypton, *m.* krypton.

cryptomère, cryptomerous, with latent characters.

cryptopine, *f.* cryptopine.

cryptopore, *m.* cryptopore, stoma below the plane of epidermis.

cryptorrhynque, *m.* **— de l'aune,** willow weevil, willow borer.

ctenophytes, *f.pl.* vegetable parasites causing death of host plant.

cubage, *m.* cubic (capacity) contents, cubature, air space, finding the cubic contents, wood mensuration; **— des terrassements,** taking out quantities.

cubature, *f.* cubature.

cube, *m.* cube, volume, hexahedron, cubic number; **— d'un nombre,** cube, cubic number.

cube, cubic, cubical, cube.

cubèbe, *m.* cubeb.

cubébin, *m.*, **cubébine,** *f.* cubebin.

cuber, to cube, measure in cubic units, gauge.

cubicite, *f.* analcite.

cubilose, *f.* substance like gelose.

cubilot, *m.* cupola, cupola furnace, converter.

cubique, *f.* cubic (curve).

cubique, cubic, cubical, cube.

cubital, cubital (muscle).

cubitus, *m.* ulna (bone of forearm), cubitus.

cuboïde, cuboid.

cubo-octaèdre, *m.* cuboctahedron.

cucullé, hooded, hood-shaped, cucullate.

cucullifère, with appendages shaped like a horn.

cucullifolié, with hoodlike leaves.

cuculliforme, hooded, hood-shaped.

cucuméroide, resembling a cucumber.

cucumiforme, cucumiform, shaped like a cucumber.

cucurbifère, with gourd-shaped fruits.

cucurbitacé, cucurbitaceous.

cucurbitacées, *f.pl.* Cucurbitaceae.

cucurbite, *f.* cucurbit, boiler.

cueillage, *m.* gathering, picking.

cueillette, *f.* gathering, picking, harvest, plucking.

cueilleur, *m.* picker, gatherer.

cueillir, to gather, pick, pluck, coil (rope).

cuffat, cufat, *m.* cage, hoisting bucket.

cuiller, cuillère, *f.* spoon, ladle, auger bit, cutter; — à fondre, — de coulée, casting ladle; —e en corne, horn spoon; —e en verre, glass spoon.

cuillerée, *f.* spoonful, ladleful.

cuilleron, *m.* bowl (of a spoon), small spoon, alula, winglet (of dipter).

cuir, *m.* leather, hide, skin; — à oeuvre, shaft or upper leather; — à semelle(s), sole leather; — brut, rawhide; — chagrin, shagreen; — chamoisé, chamois (leather), shammy; — chromé, chrome(-tanned) leather; — ciré, — en cire, wax leather; — cordouan, cordwain, cordovan (leather); — corroyé, curried leather; — cru, rough leather, rawhide.

cuir, — de boeuf, ox (cow) hide, neat's leather; — de buffle, buff leather; — de montagne, asbestos, earth or mountain flax; — de Russie, Russia leather; —

embouti, cup leather; — en croûte, tanned skin; — en huile, oiled leather; — en suif, tallowed leather, waxed leather; — étiré, scraped leather; — factice, artificial leather, imitation leather; — fort, — gros, sole leather, crop hide or butt; — hongroyé, Hungarian leather.

cuir, — jusé, ooze leather; — laqué, — verni, patent (lacquered) leather; — lissé, smooth leather, grained leather; — maroquiné, morocco leather; — mégissé, tawed leather (hide); — mou, — molleterie, soft leather; — moulé, molded leather; — préparé, dressed leather; — rond, thick leather; — salé, salted hide; — tanné, tanned hide, tanned leather; — vert, green (raw) hide.

cuirasse, *f.* armor (plate), cuirass, breastplate, chest bandage.

cuirassé, *m.* armored vessel, battleship.

cuirassé, armored, armor-plated, armed, enclosed, covered, prepared.

cuirassement, *m.* armoring, armor, armor plating.

cuirasser, to armor, armor plate, supply with a cuirass, arm, prepare, enclose, protect.

cuire, to burn, fire, bake, kiln (bricks), boil (down), char (wood), cook, bake, roast, ripen, be cooked, bake, smart, hurt.

cuirer, to cover with leather.

cuisage, *m.* charring.

cuisant, burning, biting, smarting, sharp, suitable for cooking.

cuisent (cuire), they are burning.

cuiseur, *m.* burner, kilnman, fireman, boiler, cooker, kettle man, cookstove.

cuisine, *f.* kitchen, cooking, fare, cuisine, food, cookery, kitchenwork.

cuisiner, to cook.

cuisse, *f.* thigh, leg, hindquarter, drumstick, femur; à —s égales, isosceles, equally sided.

cuisson, *f.* burning, boiling, cooking, firing, smart, smarting, baking; — en dégourdi, biscuit baking (porcelain).

cuissot, *m.* haunch (of venison).

cuit, burned, boiled, cooked, done, ripe.

cuite, *f.* burning, baking, firing, boiling, cooking, batch, whey; — en grains, boiling (sugar) to grain.

cuivrage, *m.* coppering, copper plating; — galvanique, copper electroplating.

cuivre. *m.* copper; *pl.* brasses, brass or copper fittings, utensils; — bleu, azurite;

— **brut**, — **noir**, black or coarse copper; — **électrolytique**, electrolytic copper; — **gris**, tetrahedrite, gray copper ore; — **hydro-siliceux**, chrysocolla; — **jaune**, brass; — **laminé**, sheet or rolled copper.

cuivre, — **natif**, native copper; — **panaché**, bornite; — **phosphaté**, copper phosphate; — **pyriteux**, chalcopyrite, copper pyrites; — **rosette**, rosette (rose) copper; — **rouge**, pure copper, cuprite, red (ruby) copper ore; — **sulfaté**, copper sulphate; — **vierge**, virgin (native) copper.

cuivré, coppered, copper-colored.

cuivrer, to copper, coat, sheathe with copper, copperplate, make copper colored.

cuivreux, -**euse**, cuprous (oxide), copper, coppery, cupreous (ore).

cuivrique, cupric.

cujelier, *m.* wood lark.

cul, *m.* bottom, breech, stern (of ship), haunches, rump (of animal); — **brun**, brown-tail moth; — **de ver**, web cataract (horse's eye).

culasse, *f.* breech, culet, culasse, cylinder head.

culbute, *f.* somersault, fall.

culbuter, to turn over, somersault, tumble, fall, overturn, upset, overthrow, fail, collapse.

culbuteur, *m.* tripper, rocker, rocker arm, tumbler.

cul-de-lampe, *m.* bracket.

cul-de-poule, *m.* swelling, protuberance.

cul-de-sac, *m.* blind alley, short dead end.

culée, *f.* abutment, butt (of hide).

culicidés, *m.pl.* Culicidae, mosquitoes and gnats.

culinaire, culinary.

culmifère, *m.* cereal.

culmigène, growing on culm (of grasses).

culminant, culminating, climax.

culminer, to culminate.

culot, *m.* slag, residue, culot, mass (of unpoured metal), bottom, base, youngest member, nestling, last chick hatched, last animal born (of litter), centrifuged deposit.

culotte, *f.* Y tube, Y pipe, breeches, rump (of beef).

culpabilité, *f.* debt, guilt, fault.

culte, *m.* worship.

cultivable, arable, tillable.

cultivant, susceptible of cultivation.

cultivateur, -**trice**, *m.&f.* farmer, agriculturist, cultivator, grower, breeder.

cultivateur, -**trice**, agricultural, farming.

cultivation du sol, *f.* cultivation of the soil, husbandry.

cultivé, cultivated, grown.

cultiver, to cultivate, till, improve; **se** —, to be cultivated.

cultrifolié, with knife-shaped leaves.

cultural, -**aux**, cultural.

culture, *f.* culture, cultivation, tillage, cultivated land, crop, rearing, breeding; — **des bois**, silviculture; — **en gouttelette**, droplet culture; — **en strie**, streak culture.

cumène, *m.* cumene.

cumin, *m.* cumin (*Cuminum cyminum*); — **des prés**, caraway (*Carum carvi*); — **noir**, black cumin, fennelflower, black caraway (*Nigella sativa*).

cuminique, cumic, cuminic (acid).

cuminol, *m.* cuminole, cumaldehyde.

cumul, *m.* accumulation.

cumulatif, -**ive**, cumulative.

cumulé, accrued (interest).

cunéiforme, cuneate, wedge-shaped, cuneiform, triangular.

cuniculture, *f.* rabbit breeding.

cunila, *n.* Cunila.

cupide, greedy, covetous.

cupidité, *f.* cupidity, covetousness.

cupréine, *f.* cupreine.

cupressifolié, with cypress leaves.

cupressiforme, in the form of a cypress.

cupressoïde, cupressoid, with foliage like the cypress.

cuprifère, cupriferous.

cuprique, of or pertaining to copper, cupric.

cupro-ammoniaque, *f.* cuprammonia.

cupro-ammonique, cuprammonium.

cuproxyde, *m.* cupric oxide.

cupulaire, cupular, cup-shaped.

cupule, *f.* (small) cup, cupule, acorn cup, husk (of filberts).

cupulé, cupulate, cupuliferous, cup-bearing.

cupulifère, *m.* cupuliferous tree.

cupuliforme, cup-shaped.

curabilité, *f.* curability (of disease).

curable, curable.

curaçao, *m.* curaçao (liqueur).

curage, *m.* cleaning, clearing, water pepper, persicary.

curain, *m.* saline incrustation, muddy sediment of salt produced by boiling (in saltmaking).

curare, *m.* curare (poison).

curarine, *f.* curarine.

curariser, to administer curare to (animal).

curatier, *m.* tanner.

curatif, -ive, curative.

curculionide, *m.* weevil.

curcuma, *m.* curcuma, turmeric.

cure, *f.* cure, treatment.

curée, *f.* hounds' fees, quarry.

cure-feu, *m.* poker.

curement, *m.* cleaning, clearing.

cure-môle, *m.* dredger.

curer, to clear, clean out, cleanse, grass (flax).

curette, *f.* cleaner, scraper, spoon, curette.

cureur, *m.* cleaner, cleanser, dredger.

curiethérapie, *f.* radium therapy.

curieusement, curiously.

curieux, *m.* curious person, fact.

curieux, euse, curious, careful, interested (in), inquisitive.

curin, *m.* see CURAIN.

curiosité, *f.* curiosity, inquisitiveness.

curoir, *m.* plow stick.

curseur, *m.* slider, slide, index, runner (of Wheatstone bridge), slide rule.

cursif, -ive, running, cursory.

curtirostre, short-billed.

curure, *f.* dirt, mud, slime.

curvatif, -ive, (leaf) with a tendency to curl.

curvembryé, curvembryonic, with curved embryo.

curvicaule, having a curved stem.

curvicosté, curvicostate, with curved ribs or veins.

curvifolié, curvifoliate.

curvilatère, with curved side or sides.

curviligne, curvilinear, curvilineal.

curvinervé, curvinerved, having curved nerves.

curvipétale, curvipetal, (causes) tending to curve an organ.

curvisérié, curviserial, (orthostichies) slightly twisted spirally.

curvisète, with curved stems or stalks.

cuscamine, *f.* cuscamine.

cusconine, *f.* cusconine.

cuscute, *f.* dodder (Cuscuta).

cusparine, *f.* cusparine.

cuspide, *f.* cusp.

cuspidé, cuspidate, tipped with a cusp.

cuspidifolié, with cuspidate leaves.

cusson, *m.* bruchus.

cussonné, worm-eaten.

custode, *m.* custodian, keeper.

cutané, cutaneous; **maladie —e,** skin disease.

cuticole, subcutaneous (parasite).

cuticulaire, cuticular, of the cuticle.

cuticule, *f.* cuticle, epidermis.

cuticuleux, -euse, cuticular.

cutine, *f.* cutin, cutose.

cutinisation, *f.* cutinization.

cutiniser, to cutinize.

cuti-pronostic, *m.* prognosis based on skin tests.

cutiréaction, *f.* cutaneous test.

cutose, *f.* cutose, suberin.

cuvage, *m.* fermentation (of grape mash) in vats, vat room.

cuve, *f.* vat, tub, tank, copper, cistern, trough, jar, cell, tray, interior (of a blast furnace), shaft, stack, coop; — à la **couperose,** copperas vat; — à lavage, washing tray; — à lessiver, lixiviating tank; — de fermentation, fermenting tub, fermenter; — de précipitation, precipitation tank; — de trempage, steeping cistern; — d'indigo, indigo vat; — matière, mash tun, mashing vat.

cuveau, *m.* small vat, tub or tank.

cuvée, *f.* contents of a vat, tubful.

cuvelage, *m.* tubbing, lining, casing, timbering.

cuveler, to tub, line, case.

cuve-matière, *m.* mash vat, mash tun.

cuver, to ferment (in a vat).

cuvette, *f.* cup, cistern (of barometer), tray, dish (for developing), bulb (of thermometer), basin, sink, bowl, pan, basin-shaped valley, cove, swale; — à culture, culture dish; — à photographie, photographer's tray; — d'égouttage, drip cup; **en forme de —,** basin-shaped.

cuvier, *m.* tub.

C.V., *abbr.* (cheval-vapeur), horsepower.

cyanacétique, cyanacetic.

cyanamide, *f.* cyanamide; — calcique, calcium cyanamide.

cyanamine, *f.* cyanamide.

cyanate, *m.* cyanate.

cyane, *m.* cyanogen.

cyanhydrique, hydrocyanic, prussic.

cyanide, *m.* cyanide.

cyanine, *f.* cyanine.

cyanique, cyanic, tinge of blue.

cyaniser, to cyanize.

cyanite, *f.* cyanite.

cyanocarpe, with blue fruit.

cyanofer, *m.* ferrocyanogen.

cyanoferrate, *m.* ferrocyanide, ferrocyanate.

cyanoferre, *m.* ferrocyanogen.

cyanoferrique, ferrocyanic.

cyanoferrure, *m.* ferrocyanide.

cyanogène, containing cyanogen.

cyanogénèse, *f.* cyanogenesis.

cyanophile, cyanophilous.

cyanophycées, *f.pl.* blue-green algae (Cyanophyceae).

cyanophycine, *f.* the blue coloring matter of algae.

cyanose, *f.* cyanosis, cyanose.

cyanuration, *f.* cyanidation, cyaniding.

cyanure, *m.* cyanide; — d'argent, silver cyanide; — de potassium, potassium cyanide; — jaune, potassium ferrocyanide.

cyanuré, cyanided, cyanide, cyanide of.

cyanurer, to cyanide.

cyanurique, cyanuric.

cyathiforme, cup-shaped, cyathiform.

cyathium, *m.* cyathium.

cycadées, *f.pl.* Cycadaceae.

cycas, *m.* sago palm (*Cycas revoluta*).

cyclamen, *m.* cyclamen (*Cyclamen indicum*), sowbread (*Cyclamen europaeum*).

cyclamine, *f.* cyclamin.

cycle, *m.* cycle, period, one turn of a helix, a whorl in floral envelopes; — des transformations, cycle of development; — fermé, closed cycle.

cyclecar, *m.* light automobile.

cyclique, cyclic(al), foliar structures arranged in whorls.

cyclocéphale, cyclocephalous.

cycloheptane, *m.* cycloheptane.

cycloheptène, *m.* cycloheptene.

cyclohexane, *m.* cyclohexane.

cyclohexanol, *m.* cyclohexanol.

cyclohexanone, *f.* cyclohexanone.

cyclohexène, *m.* cyclohexene.

cyclohexylamine, *f.* cyclohexylamine.

cycloïdal, cycloidal.

cycloïde, *f.* cycloid.

cyclomètre, *m.* cyclometer, curve measurer.

cyclone, *m.* tornado, cyclone.

cyclopentadiène, *m.* cyclopentadiène.

cyclopentane, *m.* cyclopentane.

cyclopentanone, *f.* cyclopentanone.

cyclopentène, *m.* cyclopentene.

cyclophylle, with orbicular flowers.

cyclose, *f.* cyclosis, moving of protoplasm granulations to interior of cells.

cyclosperme, cyclospermous.

cycnoche, *m.* swanneck, swanflower.

cygne, *m.* swan; — sauvage, whooping swan.

cylindracé, somewhat cylindric.

cylindrage, *m.* calendering, rolling.

cylindre, *m.* cylinder, roller, roll, mangle, calender, arbor, spindle; — à chemise, jacketed cylinder; — à cingler, shingling roll; — à vapeur, steam cylinder; — broyeur, crusher roll, cracker; — compresseur, roller, press cylinder; — creux, hollow cylinder.

cylindre, — dégrossisseur, coarse-crushing roll, roughing cylinder, washing engine; — droit, right cylinder; — effilocheur, washing machine; — finisseur, finishing roll; — pressure, pressure cylinder; — raffineur, beating machine; un deux —s, a two-cylinder motor.

cylindré, rolled.

cylindrée, *f.* cylinderful, charge, cylinder capacity, piston displacement.

cylindrer, to calender, roll, cylinder.

cylindreur, *m.* calenderer, roller.

cylindriflore, with cylindrical flowers.

cylindrique, cylindric(al), round, elongated, (leaves) rolled.

cylindriquement, cylindrically.

cymbalaire, *f.* Kenilworth ivy, ivy-leaved toadflax (*Cymbalaria muralis*).

cymbiforme, cymbiforme, boat-shaped.

cyme, *f.* cyme, bunch, a flower cluster: en forme de —, cymose.

cymène, cvmol, *m.* cymene.

cymeux, -euse, cymose.

cynanque, *m.* cynanchum.

cynarhode, *f.* cynarrhodion, a fruit like the dog rose.

cynips, *m.* gall wasp.

cynorrhodon, *m.* hip (of rose).

cynosure, *f.* cockscomb grass, dog's-tail (*Cynosurus echinatus*).

cynurine, *f.* kynurine.

cyphelles, *f.pl.* cyphella.

cyprès, *m.* cypress (Cupressus); — **chauve,** swamp cypress.

cypripède, *m.* lady's-slipper (Cypripedium).

cypsela, *f.* cypsela.

cystalgie, *f.* pain in the bladder, cystalgia.

cystéine, *f.* cysteine.

cysticerque, *m.* bladder worm.

cystide, *f.* cystid.

cystine, *f.* cystine

cystique, cystic.

cystite, *f.* inflammation of bladder.

cystocarpe, *m.* cystocarp, a sporophore in red algae.

cystocèle, *f.* hernia of bladder.

cystoïde, cystoid, resembling a cyst.

cystolithe, *m.* cystolith, stone, calculus.

cystospore, *f.* cystospore, a carpospore.

cytase, *f.* cytase, an enzyme.

cytise, *m.* — **aubour,** laburnum, golden chain; — **des Alpes,** Scotch laburnum.

cytisine, *f.* cytisine.

cytoblaste, *m.* cytoblast.

cytode, *f.* cytode, a nonnucleated protoplasmic mass.

cytogénétique, cytogenetic.

cytohyaloplasma, *f.* hyaloplasma.

cytokinèse, *f.* cytokinesis, cell division by mitosis.

cytologie, *f.* cytology.

cytologique, cytologic(al).

cytolyse, *f.* cytolysis.

cytolytique, cytolytic.

cytoplasma, *m.* cytoplasm, protoplasm without the nucleus.

cytoplastine, *f.* cytoplastin, albuminoid substance of the protoplasm.

cytosarc, *m.* cellular protoplasm.

cytosine, *f.* cytosine.

cytotoxines, *f.pl.* cytotoxins, enzymelike productions.

cytotoxique, cytotoxic.

cytotropisme, *m.* cytotropism, cytolaxis.

D

d', contraction of de.

d'abord, at once, right away, at first.

dachshund, *m.* badger dog.

dactyle, *m.* orchard grass (*Dactylis glomerata*).

dactylifère, bearing dates.

dactylo, *m.&f.* typist.

dactylographe, *m.* typewriter; —, *f.* typist.

dactylographie, *f.* typewriting.

dactylotype, *f.* typewriter.

dactylotypie, *f.* typewriting.

dague, *f.* scraper, scraping knife, dagger, dirk, tusk, one-year-old stag's horn, brock, dag.

daguerréotyper, to daguerreotype.

daguerréotypie, *f.* daguerreotypy.

daguet, *m.* brocket, pricket.

dahlia, *m.* dahlia.

dahline, *f.* dahlin, inulin.

daigner, to deign, condescend, be pleased.

d'ailleurs, besides, moreover.

daim, *m.* buck, deer, fallow deer; — **adulte,** fallow buck three years old; — **male,** buck.

daine, *f.* doe.

daleau, *m.* scupper, escape hole for liquids, vent.

dallage, *m.* flagging, flagstone pavement.

dalle, *f.* slab, plate, flagstone, flag, gutter, spout, whetstone, slice (of fish), glass plate; — **rodée,** ground-glass plate.

daller, to flag, pave with flagstones.

dalot, *m.* escape hole (for liquids), vent.

dam., *abbr.* (décamètre), decameter.

damage, *m.* ramming, tamping.

damas, *m.* damask (linen, silk), damask steel, Damascus blade, damson (plum) (*Prunus domestica*).

damasquinage, *m.* damascening, damaskeening.

damasquiner, to inlay, damascene, damaskeen.

damasquinerie, *f.* damascening, damaskeening.

damasquineur, *m.* damascener, damaskeener.

damasquinure, *f.* damascening, damaskeening.

damassé, damask(ed).

damassin, *m.* light cotton damask.

dame, *f.* lady, dame, dam, rammer, queen (at cards); *pl.* checkers; — bonne, mountain spinach, garden orach (*Atriplex hortensis*).

dame-jeanne, *f.* demijohn, wicker bottle.

damer, to ram, tamp.

dammar, *m.* dammar (resin).

dammara, *m.* Dammara, Agathis; — austral, kauri pine (*Dammara australis*).

damoiselle, *f.* damsel, rammer, beetle.

dandy, *m.* dandy, fop.

danger, *m.* danger, risk, trouble, peril.

dangereux, -euse, dangerous, perilous.

danois, Danish.

dans, in, into, within, at, with, from.

danse, *f.* dance.

danser, to dance, prance, flicker.

dans-oeuvre, in the clear.

daphné lauréolé, *f.* spurge laurel (*Daphne laureola*).

daphnine, *f.* daphnin.

d'après, according to.

dard, *m.* pointed flame, tip of flame or blowpipe, dart, spindle, sting, (of flowers) pistil, fruit spur, javelin, stimulus, stinger.

darder, to dart, shoot with a dart, shoot at, spear.

dari, *m.* dari, durra, Indian millet (*Holcus sorghum*).

dart, — à fruit, fruit spurt.

dartre, *f.* tetter, skin affection, scab, scurf.

dartrose, *f.* dartrose.

dasyanthe, with flowers provided with hairs.

dasycarpe, with fluffy (downy) fruit.

dasyphylle, dasyphyllous, with woolly leaves.

dasystémone, with woolly stems.

dasystyle, with woolly style.

date, *f.* date, standing.

daté, dated.

dater, to date; à — de, beginning with.

datiscine, *f.* datiscin (dye).

datte, *f.* date.

dattier, *m.* date palm (*Phoenix dactylifera*).

datura, *m.* — stramonium, thorn apple (*Datura stramonium*).

daturine, *f.* daturine (hyoscyamine).

dauber, to steam.

dauphin, *m.* dolphin, gargoyle, mouth of gutter pipe, shoe, boot.

dauphinelle, *f.* larkspur (Delphinium).

d'autant, — mieux, so much the better.

davantage, more, longer, farther, any more.

de, of, from, in, during, because of, with, by, off, out of, upon, on, as, about, out, for, at, to; de ce que, because, although; de là, hence; pas de, no; rien de plus facile que de, nothing easier than.

de (partitive), some or any.

dé, *m.* die (*pl.*) dice, thimble, block, gage, stamp.

déalbation, *f.* dealbation, bleaching.

déambuler, to walk along, (per)ambulate.

déassimilation, *f.* production of extracellular bacterial substances.

déauration, *f.* gilding.

débâcle, *f.* breaking up (of ice), debacle, downfall, crash, panic.

débâcler, to open, clear, break up.

déballage, *m.* unpacking.

déballer, to unpack.

débander, to relax, unbend, unbind, slacken, disband.

débaptiser, to change the name of.

débarcadère, *m.* landing place.

débardage, *m.* clearance of felling area.

débarder, to clear the felling area.

débarquement, *m.* landing, debarkation, unloading.

débarquer, to land, disembark, unload.

débarrasser, to free, clear, clear up, clear away, rid, disencumber, disembarrass; se —, to free oneself, get rid (of), rid oneself of, extricate oneself from.

débarrer, to unbar, free from inequalities of dyeing.

débarreur, *m.* a mender of defects in dyeing (with brush).

débat (débattre), (he) disputes.

débat, *m.* debate, dispute, trial, discussion.

débâtir, to demolish, unbaste, untack.

débattable, debatable.

débattre, to debate, discuss, dispute; se —, to struggle, writhe, flounder, be debated.

débauche, *f.* debauch, dissipation; — **de couleurs,** riot of color.

débenzoler, to remove benzole or benzene from.

débenzoleur, *m.* benzole remover.

débet, *m.* debit.

débile, feeble, weak, sickly (plant).

débilement, weakly, feebly.

débilitant, debilitating, weakening.

débilité, *f.* debility, weakness, feebleness.

débit, *m.* deliver, flow, yield, supply, output, product, turnover, market, conversion, feed, debit, sale, retail shop, license to sell, cutting, sawing, discharge, delivery, utterance, strength, intensity, traffic, spreading out; — **d'eau,** flow of water; **de —, de bon —,** salable, in demand.

débitage, *m.* cutting up, sawing (up); — **des coupes,** cutting of sections.

débiter, to deliver, furnish, discharge, retail, cut up, saw up, debit, speak, utter; **se —,** to be delivered, sell, be sold, convert, saw up, be told.

débiteur, -trice, debtor.

débituminisation, *f.* debituminization.

débituminiser, to debituminize.

déblai, *m.* excavation, cutting, cut; *pl.* excavated material, rubbish.

déblayement, déblaiement, *m.* excavation, cutting away, clearing.

déblayer, to cut (away), clear (away), dig off, excavate.

debobineuse, *f.* rereeling machine, rewinder.

déboire, *m.* aftertaste, mortification, anguish, disappointment, distaste.

déboisement, *m.* clearing, deforestation, clearance.

déboiser, to deforest, clear of timber.

déboîtement, *m.* dislocation.

déboîter, to disconnect, dislocate, remove; **se —,** to come out of joint.

débonder, to unbung, open the vent of; **se —,** to burst forth, escape, run out, be emptied.

débondonner, to unbung.

débonnaireté, *f.* good humor.

débordement, *m.* overflow(ing), deflexion, flood, outburst, inundation.

déborder, to overflow, run over, project, overlap, burst forth; **se —,** to overflow, fall into, empty, burst forth.

débosseler, to straighten.

débouché, *m.* outlet, opening, issue, exit market, sale.

débouchement, *m.* unstopping, uncorking, outlet, setting (of fuse).

déboucher, to uncork, unstop, unstopper, clear, empty, open, unplug, set (a fuse) fall into, issue, pass out, debouch, emerge.

déboucler, to unbuckle, uncurl.

débouilli, *m.* boiling.

débouillir, to boil.

débouillissage, *m.* boiling.

débourbage, *m.* washing, trunking, sluicing, clearing.

débourber, to wash, clear (of mud or slime), trunk, sluice, clean (out).

débourbeur, *m.* clearing or washing cylinder, a worker that cleanses mud or slime from utensils, (ore) washer, cleanser.

débourrage, *m.* removal of wadding, untamping.

débourrement, *m.* opening of buds and appearance of green leaves.

débourrer, to scrape hair off (skins), remove the wadding or padding, untamp, depilate.

débourser, to disburse, expend.

debout, upright, standing, on one end, erect, up.

débraiser, to draw fire from.

débrayage, *m.* disconnecting, disconnection, release motion.

débrayer, to disconnect, disengage, throw out of gear, uncouple.

débrider, to stop, unbridle one's horse, incise, slit up.

débris, *m.* remains, debris, rubbish, waste, dirt, ruins, residue; — **de bois,** wood waste, chips of wood; — **de métal,** scrap; — **d'os,** bone meal.

débrouillement, *m.* analysis, discussion.

débrouiller, to untangle, unravel, clear up.

débroussaillement, *m.* clearance of scrub.

débrûler, to deoxidize.

débrutir, to rough-polish.

débucher, débusquer, to make a start (of deer), dislodge.

début, *m.* beginning, debut, first appearance, first play, lead, outset.

débutant, *m.* beginner.

débuter, to begin, start, lead, open, make a start.

déc(a) . . . , ten . . . dec(a) . . . .

deçà, (on) this side of, here, on this side; — **et delà,** here and there, on all sides; **en —,** on this side, short of; **par —,** on this side of.

décacanthe, having ten thorns or spines.

décacheter, to unseal, open.

décade, *f.* decade.

décadence, *f.* decadence, decline.

décaèdre, decahedral.

décagramme, *m.* decagram (10 grams).

décahydrate, *m.* decahydrate.

décaisser, to unpack, unbox, pay out, plant out.

décalage, *m.* unwedging, unkeying, difference of phase, divergence, displacement, brush lead; — **en arrière,** lag.

décalcification, *f.* decalcification, removal of calcareous matter.

décalcifier, to decalcify, delime.

décaler, to shift, space, displace; **ondes décalées,** waves out of phase.

décalitre, *m.* decaliter, ten liters.

décalquer, to transfer (a design), trace off.

décamètre, *m.* decameter, ten meters.

décamper, to decamp, move.

décandre, decandrian, decandrous, having ten stamens.

decandrie, *f.* decandria, hermaphrodites with ten stamens.

décantage, *m.* decanting, decantation.

décantateur, *m.* decanter.

décantation, *f.* decantation, decanting.

décanter, to decant.

décanteur, *m.* decanter.

décanteur, decanting.

décapage, décapement, *m.* scouring, cleaning, scraping.

décaper, to scour, clean, scrape, pickle, dip, cleanse surface, flux (metal before soldering), uncap.

décapétalé, decapetalous.

décapeur, *m.* scourer, pickler, cleaner.

décapiter, to cut head off, top (a tree).

décarbonater, to decarbonate.

décarbonisation, *f.* decarbonization.

décarboniser, to decarbonize.

décarburant, decarbonizing, decarburizing.

décarburation, *f.* decarbonization, decarburation.

décarburer, to decarbonize, decarburize.

décarbureur, *m.* purifier (worker).

décare, *m.* decare.

décastère, *m.* decastere (ten cubic meters).

décatir, to sponge, steam, ungloss.

décatissage, *m.* sponging, unglossing.

décatisseur, *m.* sponger.

décédé, deceased.

décèlement, *m.* disclosure, revealing, detection.

déceler, to reveal, disclose, make known, detect, betray, divulge.

décembre, *m.* December.

décemfide, decemfid, ten-cleft.

décennal, -aux, decennial, ten-year.

décentrage, *m.,* **décentration,** *f.,* **décentrement,** *m.,* decentration, decentering.

décentrer, to decenter.

déception, *f.* deception.

déceptivement, deceptively.

décerner, to award, assign, decree, grant, confer.

décès, *m.* decease, death, demise.

décevant, deceiving, disappointing.

décevoir, to deceive, disappoint.

déchaîner, to unchain, release, let loose.

déchalement, *m.* ebb, fall (of tide).

décharge, *f.* discharge, outlet, relieving, vent, rubbish heap, unloading, release.

déchargement, *m.* unloading, discharging.

déchargeoir, *m.* outlet, sluice, vent, exhaust pipe, waste pipe.

décharger, to discharge, release, remove the charge from, unload, empty; **se —,** to discharge, be discharged, unload or relieve oneself, fade out, empty.

déchargeur, *m.* discharger, unloader, (lightning) arrester.

décharner, to emaciate.

déchasseur, *m.* lifting device.

déchaumage, *m.* stubble plowing.

déchaumer, to plow stubble land.

déchaussement, *m.* (**par la gelée**), frost lifting.

déchausser, to take off footgear, lay bare; **se —,** to be lifted by frost.

déchéance, *f.* forfeiture, decay, decadence.

décherra (déchoir), (it) will decline.

déchet, *m.* loss, waste; *pl.* waste, refuse, diminution, refuse wood, offal timber; — **sur le poids,** weighing loss; —**s de coupes,** wood refuse, slashings.

déchiqueté, shredded, torn to pieces, laciniate.

déchiqueter, to cut up, cut in pieces, tear to pieces.

déchiqueteur, *m.* chipper, chopping machine.

déchiré, torn, lacerated.

déchirement, *m.* tearing, rending, upheaval.

déchirer, to tear, rend, lacerate, shatter, torture, slander.

déchirure, *f.* tear, rent, slit, rip, laceration, rupture.

déchoir, to fall, decay, lose, decline, fall off; — **d'un brevet,** to forfeit a patent.

déchu, fallen.

de-ci, here.

déciare, *m.* deciare (ten square meters).

décidé, decided, settled, determined.

décidément, decidedly, firmly, positively.

décidence, *f.* collapse, falling.

décider, to decide, determine, settle; **se —,** to be decided, decide; **se — à,** to make up one's mind; — **de,** to decide on, decide to.

décidu, deciduous.

décigrade, *m.* 0.1 grade.

décigramme, *m.* decigram (0.1 gram).

décilitre, *m.* deciliter (0.1 liter).

décimal, -aux, decimal.

décimale, *f.* decimal (fraction).

décime, *m.* decime (0.1 franc); *f.* tenth (part), tithe.

décimètre, *m.* decimeter (0.1 meter).

décintrement, *m.* removal of centering, center stocking.

décirage, *m.* removal of wax.

décirer, to remove wax from.

décisif, -ive, decisive, conclusive, peremptory.

décision, *f.* decision, determination.

décistère, *m.* decistere (0.1 cubic meter).

déclanche, *f.* disengaging gear, release device.

déclanchement, see DÉCLENCHEMENT.

déclancher, see DÉCLENCHER.

déclaration, *f.* declaration, notification, verdict.

déclarer, to declare, make known, assert, disclose; **se —,** to declare oneself, appear, occur, break out (of disease).

déclenchement, *m.* unloosing. release, throwing out of gear.

déclencher, to unlatch, unloose, release, disconnect, disengage, throw out of gear.

déclic, *m.* click, catch, trip, trigger, pawl, nippers, releasing mechanism.

déclin, *m.* decline, close, termination, waning.

déclinaison, *f.* declination, declension, variation, deflection.

décliné, declinate, bent or curved downward.

décliner, to decline, deviate, fall off, wane, refuse.

déclive, *f.* slope.

déclive, sloping, inclined.

décliver, to slope, be inclined.

déclivité, *f.* declivity, slope, gradient, grade.

déclore, to unclose, open.

décoagulation, *f.* liquefaction.

décoaltariser, to free from coal tar.

décocher, to shoot, discharge, let fly.

décocté, *m.* decoction.

décoction, *f.* decoction (act).

decoffrer, to remove the form.

décoiffage, *m.* uncapping (a fuse), uncorking.

décoiffer, to uncap, uncork.

déçoit (décevoir), (he) deceives.

décollage, décollement, *m.* separation, loosening, detachment (of adhesions), ungluing, coming unglued, taking off, cracking, plucking.

décoller, to separate, leave the ground, take off, detach, unstick, unglue, loosen, deglutinate, decollate, decapitate; **se —,** to come unglued, come off, work loose.

décolletage, *m.* removal of the tops of plants, screw cutting.

décolleter, to cut the tops of (root crops), cut (off), turn (screws).

décolorant, *m.* bleaching agent.

décolorant, decolorizing, bleaching.

décoloration, *f.* decolorization, fading, bleaching, decolorizing, discoloration.

décoloré, colorless, decolorized.

décolorer, to decolorize, discolor, change hue, bleach, take away the color.

décoloris, *m.* loss of color.

décombant, decumbent, reclining.

décombres, *m.pl.* rubbish, debris.

décombustion, *f.* deoxidation.

décomposable, decomposable.

décomposant, decomposing (agent).

décomposé, decomposed, decomposit, decompound.

décomposer, to decompose, alter, disintegrate, discompose, decompound, analyze, divide; **se —,** to decompose, change, be altered, break up, become decomposed; **se — à l'air,** to weather, disintegrate, effloresce.

décomposition, *f.* decomposition, alteration, disintegration; **— de forces,** resolution of forces.

décompresseur, *m.* relief cock.

décompte, *m.* deduction, discount, analysis (of account).

décompter, to deduct, calculate.

déconcert, *m.* disaccord.

déconfiture, *f.* discomfiture, collapse, failure, insolvency.

décongélation, *f.* thawing (out).

décongeler, to thaw, thaw out.

déconnecter, to disconnect, switch off, turn out.

déconnexion, *f.* release, releasing.

déconseiller, to counsel (advise) against.

déconsidérer, to discredit, bring disrepute.

déconstruire, to demolish, raze.

décor, *m.* decoration, scenery.

décorateur, *m.* decorator.

décoratif, -ive, decorative.

décoration, *f.* decoration, scenery.

décorder, to untwist (rope).

décordonnage, *m.* cleaning stampers (of powder mill).

décorer, to decorate, adorn.

décorner, to dehorn, dishorn (cattle).

décortication, *f.,* **décorticage,** *m.* decortication, disbarking, husking, barking, peeling.

décortiquer, to decorticate, bark, strip off the bark, husk, hull, pare, shell, peel, pulp, mill.

découdre, to unsew, unpick, unstitch, rip (off), tusk (boar).

découenner, to flay, skin (pig).

découler, to trickle, flow, drop, proceed, issue, spring, follow, flow from and with, agree with.

découpage, *m.* cutting.

découpe, *f.* shortening, cutting into small pieces.

découper, to cut, cut up, cut off, cut out, carve, cut into small pieces, punch,

shorten; **se —,** to be cut (up), stand out, show up.

découpeur, *m.* cutter, carver.

découpeuse, *f.* cutter, shearing machine, stamping machine.

découpler, to uncouple (hounds, horses), disengage, throw out of gear.

découpoir, *m.* cutter, cutting machine, shear, stamp.

découpure, *f.* cutting (out), cut, stamping, piece cut out, segment, denticulation.

décourager, to discourage; **se —,** to be discouraged, disheartened.

décourber, to straighten (out), unbend.

décousant (découdre), unpicking, ripping up.

décousu (découdre), ripped open.

découvert, *m.* overdraft, deficit.

découvert, discovered, uncovered, open, exposed; **à —,** uncovered, unprotected, exposed, openly, plainly, in the open air, unsecured.

découverte, *f.* discovery, detection.

découvreur, -euse, *m.&f.* discoverer.

découvrir, to discover, uncover, expose, disclose, recede (of the sea), denude, lay open, open, perceive; **se —,** to be discovered, come to light, expose oneself, uncover oneself.

décramponner, to unclamp, let go, loosen.

décrassage, *m.* cleaning, scouring.

décrasser, to clean, scour, wash, scrape.

décréditer, to discredit.

décrément, *m.* decrease, diminution.

décrépit, decrepit, dilapidated.

décrépitation, *f.* decrepitation.

décrépiter, to decrepitate.

décrépitude, *f.* decrepitude, decay.

décret, *m.* decree, order.

décreusage, décreusement, *m.* ungumming, scouring.

décreuser, to ungum, scour, strip, cleanse.

décrier, to decry, discredit, disparage.

décrire, to describe; **se —,** to be described.

décrivant, describing.

décrochage, *m.* unhooking, uncoupling.

décrocher, to unhook, take off, unfasten, lift, disengage, disconnect.

décroissance, *f.,* **décroissement,** *m.* decrease, diminution, abatement, decline, tapering.

décroissant, decreasing, tapering, diminishing.

décroît, *m.* decrease of livestock, last quarter (moon).

décroître, to decrease, diminish, decline.

décrotter, to clean, clean off, scrape, brush.

décrouir, to anneal.

décrouissage, *m.* annealing.

décroûtage, *m.* cleaning.

décroûter, to remove the crust (from), clean, uncrust, skin, rough-plane.

décru (decroître), diminished.

décruage, *m.* ungumming, scouring.

décrue, *f.* decrease, fall, decrement, retreat (glacier).

décruer, to ungum, scour, cleanse.

décrûment, décrusage, *m.* ungumming, scouring.

décruser, to ungum, scour.

déçu, deceived.

décuire, to thin (sirup); se —, to become thin.

décuivrer, to deprive of copper.

décuplateur, -euse, multiplying by ten.

décuple, tenfold.

décupler, to decuple, multiply by ten.

décurrent, decurrent, extending downward.

décurtation, *f.* fall of one-year-old healthy branches; — des rameaux, fall of twigs.

décussation, *f.* decussation, crossing.

décussé, decussate.

décuvage, *m.*, décuvaison, *f.* racking off, tunning.

décuver, to rack off, tun (wine).

dédaigner, to disdain, scorn.

dédaigneusement, disdainfully, scornfully.

dédaigneux, -euse, disdainful, scornful.

dédain, *m.* disdain, scorn.

dédale, *m.* labyrinth, maze.

dédaléen, inextricable, labyrinthine, perplexing.

dédaller, to remove the flags from, unpave.

dedans, *m.* inside, interior.

dedans, inside, within, in. — et dehors, within and without; au —, on the inside, within, at home; de —, from within; du —, from the interior; en —, within, (on the) inside, in, reserved (person); en — de, within, inside of; par —, inside, within.

dédicace, *f.* dedication.

dédicatoire, dedicatory.

dédier, to dedicate.

dédire (se), to retract.

dédit, *m.* forfeit, penalty, retraction, compensation, bond.

dédommagement, *m.* compensation, damages, remuneration.

dédommager, to indemnify, compensate.

dédorage, *m.*, dédorure, *f.* ungilding.

dédoré, tarnished.

dédorer, to ungild.

dédoublable, divisible into two, decomposable.

dédoublage, *m.* dilution, hydrolysis.

dédoublé, *m.* brandy (made by diluting alcohol with water).

dédoublé, decomposed, diluted.

dédoublement, *m.* reduplication, division (breaking) into two parts, double decomposition, resolution.

dédoubler, to divide into two, split, decompose, resolve, dilute, hydrolize, reduce one half, unline, unsheathe; se —, to divide (split) into two, decompose, be resolved, break into parts, diminish one half, be unlined, unfold.

déductif, -ive, deductive.

déduction, *f.* deduction, subtraction.

déductivement, deductively.

déduire, to deduce, deduct, recite, subtract, infer, write off.

dédurcir, to soften.

défaillance, *f.* extinction, failing, lapse, fainting, faintness, syncope, swoon, exhaustion, deliquescence, failure.

défaillant, faltering, weak, feeble, becoming extinct.

défaillir, to fail, become feeble, default, faint.

défaire, to unmake, undo, demolish destroy, untie, pull to pieces, disengage, ungear, take off, defeat, rid; se — de, to be undone, weaken, spoil, make (do) away with, part with, sell off, rid oneself, get rid (of).

défait, undone, loose, drawn.

défaite, *f.* disposal, defeat, lame excuse, evasion, overthrow.

défalquer, to deduct, subtract.

défarination, *f.* defarination, suppressed formation of starch.

défatiguer, to refresh, rest.

défausser, to straighten.

défaut, *m.* default, want, lack, absence, deficiency, imperfection, defect, flaw, fault, blemish; **à — de, au — de,** for want of; **faire —,** to fail to appear, be absent, be wanting; **sans —,** faultless.

défaveur, *f.* disfavor.

défavorable, unfavorable, disadvantageous, unsuitable.

défavorablement, unfavorably.

défécation, *f.* defecation, clarification.

défectible, imperfect, incomplete

défectif, -ive, defective.

défectueusement, defectively, faultily.

défectueux, -euse, defective, imperfect.

défectuosité, *f.* defect, imperfection, flaw.

défendeur, -eresse, *m.&f.* defendant.

défendre, to defend, back, protect, shelter, support, prohibit, forbid; **se —,** to defend oneself, forbear, deny, refrain.

défends, *m.pl.,* **bois en —,** closed wood.

défendu, forbidden, prohibited, defended.

défensable, open.

défense, *f.* defense, forbidding, protection, prohibition, guard (of tree), fender, tusk (of a boar), fang, razor.

défenseur, *m.* defender, supporter.

défensif, -ive, defensive, protective.

déféquer, to defecate, clarify, clear, purify.

défera (défaire), (he or it) will demolish, tear down.

déférer, to confer, bestow, denounce, inform against, accuse, defer.

déferrage, *m.* removal of iron, unshoeing (a horse).

déferrer, to remove the iron, deferrize, unshoe (a horse), disconcert.

défervescence, *f.* subsidence of fever.

défeuillaison, *f.* defoliation, leaf fall.

défeuillé, defoliate, leafless.

défeuiller, to defoliate, strip; **se —,** to cast off or shed leaves.

défi, *m.* defiance, challenge.

défiance, *f.* mistrust, distrust, diffidence.

défiant, distrustful, mistrustful.

défibrage, *m.* separation (removal) of fibers; **— transversal,** cross grinding; **— longitudinal,** long grinding.

défibrer, to disintegrate or remove the fibers of (sugar cane), grind (wood for pulping).

défibreur, *m.* disintegrator, fiber separator, pulp grinder, cane shredder.

défibrination, *f.* defibrination.

défibriner, to defibrinate, remove fibrin.

déficeler, to untie, undo, take the string off.

déficit, *m.* deficit, deficiency, shortage.

déficitaire, deficient; **année —,** lean year; **récoltes —s,** short crops.

défier, to challenge, dare, defy; **se —,** to mistrust, distrust, be suspicious, defy each other.

défigurer, to disfigure, distort, mar.

défilage, *m.* unthreading, raveling, breaking in, tearing up of rags.

défilé, *m.* defile, narrow pass, ravine, half stuff.

défiler, to unthread, ravel, untwist, separate (rags) into threads, break in (rags).

défileur, *m.* unthreader.

défileuse, *f.* unthreader, (paper) breaker.

défilochage, *m.* unthreading.

défilocher, to shred (rags).

défini, definite, defined, determined, cymose (in inflorescence).

définir, to define, decide, describe, explain: **se —,** to be defined, become clear.

définissable, definable.

définitif, -ive, definitive, final, positive; **en définitive,** definitively, finally, after all.

définition, *f.* definition; **— de la qualité,** quality classification, determination of quality.

définitivement, definitively, finally.

défit (défaire), (he) unmade, undid.

défixation, *f.* removal of enzyme from substance.

déflagrant, deflagrating.

déflagrateur, *m.* deflagrator, spark gap.

déflagration, *f.* deflagration, combustion.

déflagrer, to deflagrate.

déflation, *f.* deflation.

défléchir, to deflect, turn aside.

déflecteur, *m.* deflector, baffle plate.

déflegmation, *f.* dephlegmation, concentration (of alcohol).

déflegmer, to dephlegmate, rectify, concentrate.

défleuraison, *f.* blossom fall.

défleuri, confused.

défleurir, to shed the blossoms, nip flowers.

déflexion, *f.* deflection.

déflocheuse, *f.* rag-tearing machine.

défloraison, *f.* fall of the blossom.

défloré, deflorate, stale (news).

déflorer, to strip (plant) of its blooms.

défoliaison, *f.* exfoliation, defoliation.

défoliation, *f.* defoliation, leaf fall.

défoncer, to plow up, break up, stave in.

défont (défaire), (they) unmake or undo.

déforestaticn, *f.* deforestation, clearance.

déformateur, -trice, deforming, distorting.

déformation, *f.* deformation, strain, distortion.

déformé, deformed, distorted, misshapen.

déformer, to deform, put out of shape, twist, distort; **se —,** to become deformed, change form.

défournage, défournement, *m.* discharge, taking (bread) from an oven, drawing (of pottery) from kiln.

défourner, to draw (pottery) from kiln, (bread) from oven, discharge, empty.

défourneur, *m.* drawer, ovenman.

défourneur, -euse, discharging.

défourneuse, *f.* coke ram, coke-pushing machine.

défourni, *m.* refuse wood.

défraîchir, to fade, take off newness.

défrayer, to meet the expense of, defray, entertain.

défrichement, *m.* clearing, (act of) reclaiming, uprooting.

défricher, to clear, make tillable, reclaim, develop, clear up, uproot, grub up.

défroque, *f.* cast-off garment, casting off, cast.

defruiter, to strip of fruit.

dsfunt, defunct, deceased.

dégagé, disengaged, free, open.

dégagement, *m.* disengagement, liberation, isolation, setting free, release, discharge (of spores), emission, evolution, clearing, redemption, isolating, escape, formation, interrupting the leaf canopy; **— de la cime,** freeing of the crown; **tube à —,** exit tube.

dégager, to disengage, liberate, emit, evolve, release, free, set free, clear (a way), redeem, give out (heat), escape, generate, isolate; **se —,** to be disengaged, get away, be emitted, be cleared, get clear, be given off, disengage oneself, escape.

dégaler, to clean, comb (skins).

dégarnir, to dismantle, strip, untrim, unvarnish.

dégarnissage, *m.* degarnishing, clearing away, thinning.

dégât, *m.* damage, havoc, injury, waste; **—s du gibier,** injury by game; **—s du vent,** damage by storm, by wind; **—s par la gélee,** damage by frost.

dégauchir, to smooth, dress, surface, roughplane, polish.

dégazonnement, *m.* removal of sods.

dégazonner, to remove sods, cut turf.

dégel, *m.* thaw, thawing.

dégélatiner, to deprive of gelatin, degelatinize.

dégeler, to thaw, melt.

dégénération, *f.* degeneration.

dégénéré, degenerate.

dégénérer, to degenerate, decline.

dégénérescence, *f.* degeneration, running out, reversion; **— graisseuse,** fatty degeneration.

dégénérescent, degenerative.

dégermer, to degerminate, deprive of the germ, sprout (from malt).

dégîter, to dislodge (fox from hole).

dégivrage, *m.* cooling.

déglaçage, déglacement, *m.* removal of luster or ice.

déglacer, to thaw, melt the ice, deprive of luster.

dégluement, *m.* removal of glue.

dégluer, to remove the glue from, unglue, remove birdlime from (a bird).

déglutir, to swallow.

déglyceriner, to deglycerinize.

dégommage, *m.* ungumming.

dégommer, to ungum, scour, boil off (raw silk).

dégonflement, *m.* deflation, going down, reduction, subsiding (of swelling).

dégonfler, to deflate, let air out, reduce, bring down; **se —,** to be deflated, go down, subside.

dégor, *m.* discharge pipe or tube.

dégorgement, *m.* clearing, unstopping, scouring.

dégorgeoir, *m.* outlet, outflow, spout, a mill for scouring cloth, priming wire.

dégorger, to clear, unstop, scour, cleanse, purify, wash, remove sediment, hollow out, disgorge, discharge, flow out, overflow;

se —, to discharge, empty itself, overflow, become unstopped.

dégoudronnage, *m.* removal of tar.

dégoudronner, to remove (extract) tar, untar.

dégoudronneur, *m.* tar remover.

dégourdi, *m.* first baking, biscuit-baked (porcelain), baking.

dégourdi, slightly warm, in the state of biscuit, tepid, revived, lively, sharp, quick, shrewd.

dégourdir, to warm slightly, take the chill off of, give biscuit baking to (porcelain), remove numbness from, revive, sharpen, brighten; **se —,** to revive, grow warm, brighten up.

dégoût, *m.* disgust, repugnance, mortification.

dégoûtant, disgusting, sickening.

dégoûter, to disgust; **se —,** to grow disgusted.

dégouttement, *m.* dripping, dropping.

dégoutter, to drip, drop, trickle.

dégradation, *f.* degradation, deterioration, damage, diminution, graduation.

dégrader, to degrade, wear down, reduce, lower, tone down, damage, diminish, deteriorate; **se —,** to be degraded, lower oneself.

dégrafer, to unhook, unfasten.

dégraissage, *m.* scouring, degreasing.

dégraissant, *m.* scour, cleaner, degreaser

dégraissant, scouring, degreasing.

dégraisse, *m.* fat removed from meat.

dégraissé, with fat removed.

dégraisser, to scour, clean, take fat from (carcass), grease, skim off, scum, remove greasy stains, correct the greasy taste of (wine), treat (wine) for ropiness, bevel off, trim, impoverish (soil), dress down (wood), drain, emaciate.

dégraisseur, *m.* scourer, grease extractor.

dégraissis, *m.* grease removed by scouring.

dégraissoir, *m.* scraper, a machine for scouring.

dégras, *m.* dégras, refuse oil, grease pressed from chamois skins.

degré, *m.* degree, step, stair, grade, stage, point, extent; **— de chaleur,** degree of heat; **de — en —,** step by step, by degrees; **par —s,** by degrees, gradually.

dégression, *f.* formation of new species by activation of latent properties.

dégrever, to reduce, diminish, relieve of, free (from tax).

dégrossage, *m.* reduction, drawing down (wire).

dégrosser, to reduce, draw (wire).

dégrossir, to rough down, roughhew, give a rough preliminary dressing, rough-grind, concentrate (ores) roughly.

dégrossissage, *m.* roughing down, roughplaning, removal of gross impurities.

dégrossisseur, *m.* roughing roll, a workman or instrument that roughs down, glass grinder.

déguerpir, to scamper off, clear out.

déguisement, *m.* disguise, concealment.

déguiser, to disguise, conceal, hide.

dégustateur, -trice, *m.&f.* taster.

dégustation, *f.* tasting (of wine, tea), testing for flavor.

déguster, to taste, sample.

déharnacher, to unharness.

déhiscence, *f.* dehiscence; **— poricide,** poricidal dehiscence.

déhiscent, dehiscent.

dehors, *m.* outside, exterior, foreign countries; **au —,** out, outside, abroad, externally, without; **de —,** from outside, from without; **en —,** out, outward (on the) outside, frank, without; **en — de,** outside of, apart from, without, beyond; **par —,** outside of, without.

dehors, out, without, outside, out of doors, externally.

déhydracétique, dehydracetic, dehydroacetic.

déhydratation, *f.* dehydration.

déité, *f.* deity, god.

déjà, already, before, yet.

déjection, *f.* dejection, evacuation; *pl.* dejecta, excrement.

déjeter, to warp, buckle.

déjeuner, *m.* breakfast, lunch; **— de soleil,** an easily fading fabric.

delà, beyond, on the other side of, further on; **au —,** beyond, farther; **au — de,** beyond, further on; **de —,** from beyond; **en —,** further on, beyond; **en — de,** further on; **par —,** beyond, farther.

délabrement, *m.* dilapidation

délabrer, to shatter, ruin, dilapidate.

délai, *m.* delay, time allowed (for completion), period of time, time.

délaiement, *m.* see DÉLAYAGE.

délainer, to strip of wool, remove wool bandage (from tree).

délaissé, forsaken, abandoned.

délaisser, to abandon, desert, forsake, neglect.

délaitement, délaitage, *m.* drying of butter, removal of milk.

délaiter, to deprive of milk, work, dry (butter).

délaiteuse, *f.* butter-drying machine.

délardement, *m.* removal of fat.

délassant, refreshing.

délassement, *m.* repose, recreation.

délavage, *m.* dilution, washing out (of colors).

délaver, to dilute, wash out (colors).

délayage, délayement, *m.* mixing, mixture, dilution, thinned material.

délayant, *m.* diluent.

délayant, mixing.

délayer, to add water to a powdered material, temper, dilute.

délecter, to delight.

délégation, *f.* draft, check.

délégué, *m.* delegate.

délégué, delegated.

déléguer, to delegate.

délétère, deleterious, noxious, harmful.

délibération, *f.* deliberation, conference, consideration.

délibéré, deliberate, intentional, bold.

délibérément, deliberately, boldly.

délibérer, to deliberate, decide, consult.

délicat, delicate, dainty, tender, weak, nice, critical.

délicatement, delicately, daintily.

délicatesse, *f.* delicacy, daintiness, elegance.

délice, *m.* delight, pet, darling.

délicieux, -euse, delicious, delightful.

délié, slender, fine, thin, loose, clever.

délier, to untie, loose, loosen, unbind, liberate.

délimitation, *f.* demarcation, boundary settlement.

délimiter, to delimit, demarcate, limit, mark off, settle the limits.

délinquant, *m.* offender, delinquent, trespasser.

déliquescence, *f.* deliquescence.

déliquescent, deliquescent.

déliquium, *m.* deliquium, deliquescence.

délirant, delirious, raving.

délire, *m.* delirium.

délissage, *m.* shredding (of rags), rag cutting.

délisser, to sort rags and paper, shred rags

délisseuse, *f.* rag cutter.

délit, *m.* false bedding (of schistose stone), (stone bedded) against the grain, offense, offence, misdemeanor.

délitation, *f.* crumbling, weathering, disintegration.

déliter, to crumble, disintegrate, exfoliate, split, weather, cleave, slake (lime), effloresce.

délitescence, *f.* subsiding, delitescence, disintegration, efflorescence.

délitescent, efflorescent, capable of being disintegrated.

délivrance, *f.* deliverance, release, delivery (of the afterbirth).

délivre, *m.* afterbirth.

délivrer, to deliver, release, set free.

délogement, *f.* dislodgement.

déloger, to dislodge, oust, remove, quit.

delphinette, *f.* delphinium, larkspur.

delphinidés, *m.pl.* Delphinidae.

delphinine, *f.* delphinine.

delphinium, *m.* Delphinium.

delta, *m.* delta.

deltoïde, deltoid(al) (muscle, leaf, plate).

déluge, *m.* deluge, flood.

délustrer, to deprive of luster, sponge, steam.

délutage, *m.* unluting, opening of (gas retort).

déluter, to unlute, remove the lute from.

démaclage, *m.* stirring (of melted glass).

démacler, to stir (melted glass).

démagnétiser, to demagnetize.

demain, *m.* tomorrow; **de — en quinze,** a fortnight from tomorrow.

demande, *f.* request, petition, application, claim, demand, question, requirements.

demandé, wanted, in demand.

demander, to ask, desire, demand, beg, request, require, seek, need, ask for; **se —,** to ask oneself, wonder.

demandeur, -eresse, *m.&f.* asker, petitioner, plaintiff.

démangeaison, *f.* itching.

démanteler, to dismantle.

démarcation, *f.* demarcation.

démarche, *f.* step, walk, gait, bearing, measure, demeanor.

démariage, *m.* thinning (of plants), divorce.

démarrage, *m.* starting.

démarrer, to start, untie, get away.

démarreur, *m.* starter.

démasclage, *m.* removing the first cork layer.

démasquer, to open, unmask, show up.

démêlage, *m.* disentangling, combing out (of hair), teasing out (of wool), mashing, mash.

démêler, to disentangle, unravel, straighten out (affairs), recognize, discern, separate, distinguish; **se** —, to become disentangled, extricate oneself, be cleared up, stand out.

démembrement, *m.* dismemberment.

démembrer, to dismember, disjoint.

déménagement, *m.* removal, moving out.

démence, *f.* dementia, insanity.

démentir, to contradict, belie, deny; **se** —, to contradict oneself, be inconsistent, fall off, drop off, cease, (of building) give way.

démesuré, enormous, huge, immoderate, excessive, boundless, unmeasured.

démesurément, enormously, immeasurably, excessively, immoderately.

démettre, to dismiss, dislocate, put out of joint; **se** —, to resign, be dislocated.

demeurant, *m.* remainder, rest, survivor, residue.

demeurant, living; **au** —, all the same, after all, however.

demeure, *f.* delay, dwelling, residence, abode, home: **à** —, fixed permanent(ly), immovable; **en** —, in arrears.

demeurer, to live, dwell, reside, stay, remain, stop.

demi, half; **à** —, half, by half, semi-

demi-auget, *m.* half-bucket.

demi-bec, *m.* halfbeak.

demi-blanc, nearly white, half-bleached.

demi-brisé, half-broken.

demi-cercle, *m.* semicircle, half circle; — **gradué,** semicircular protractor.

demi-charge, *f.* half load.

demi-cheval, *m.* half horsepower.

demi-circonférence, *f.* semicircumference.

demi-circulaire, semicircular.

demi-civilisé, half-civilized.

demi-cristal, *m.* semicrystal ware.

demi-cylindre, *m.* half cylinder.

demi-diamètre, *m.* semidiameter, radius.

démieller, to remove wax from (honey).

demi-ellipse, *f.* semiellipse.

demi-ellipsoïde, *m.* semiellipsoid.

demi-ferme, *f.* half truss.

demi-ferme, semifirm.

demi-fin, *m.* twelve-carat gold.

demi-fin, half-fine.

demi-fleuron, *m.* individual flower in Compositae.

demi-fluide, semifluid.

demi-futaie, *f.* young high forest.

demi-gras, semibituminous.

demi-heure, *f.* half-hour.

demi-jour, *m.* twilight, dim light.

demi-journée, *f.* half day.

demi-ligne, *f.* half line.

demi-litre, *m.* half liter.

demi-longueur, *f.* half-length.

demi-lunes, *f.pl.* halfbalk, wainscot.

déminéralisation, *f.* demineralization, excessive elimination of mineral or inorganic salts.

demi-opale, *f.* semiopal.

demi-rond, half-round.

demi-ronde, *f.* half-round file.

démis (démettre), dislocated.

demi-sang, *m.* half-breed.

demi-siècle, *m.* half century.

demi-solide, semisolid.

demi-sphère, hemisphere.

démission, *f.* resignation.

demi-teinte, *f.* half tone, half tint.

demi-tige, *f.* nursery-grown deciduous tree.

demi-tour, *m.* half turn, half wheel, half circle.

demi-translucide, half-translucid.

démocratie, *f.* democracy.

démocratique, democratic.

démodé, old-fashioned, out of fashion, obsolete.

demoiselle, *f.* unmarried woman, dragonfly, beetle, rammer.

démolir, to demolish, destroy, pull down.

démonstrateur, *m.* demonstrator.

démonstratif, -ive, demonstrative, demonstrating.

démonstration, *f.* demonstration,

démonstrativement, convincingly, conclusively.

démontable, dismountable, capable of being taken apart or down, demountable, collapsible (boat).

démontage, m. dismounting, disassembling, decomposition.

démonter, to dismount, take to pieces, take apart, take down, knock down, disconnect, unmount (gems), disconcert; se —, to be disconcerted, be dismounted, dislocate, run down, go to pieces.

démontrabilité, f. demonstrability.

démontrable, demonstrable.

démontrer, to demonstrate, show, prove.

démoulage, m. removal (of pattern) from the mold, stripping (of casting), lifting (of porcelain).

démouler, to remove (pattern) from the mold, strip, lift.

démultiplicateur, m. reducing gear, gearing down (device).

démultiplier, to reduce (gear ratio), gear down.

dénaturant, m. denaturant.

dénaturant, denaturing.

dénaturateur, m. denaturer.

dénaturation, f. denaturation, denaturing, alteration.

dénaturé, denatured, altered, unnatural, perverted.

dénaturer, to denature, change the nature, alter, convert, refine (iron), pervert, change, distort, disfigure.

dendrite, f. dendrite.

dendritique, dendritic.

dendrographie, f. dendrology, dendrography.

dendroïde, dendroid, treelike in form.

dendrologique, dendrological.

dendromètre, m. dendrometer, hypsometer.

dendrométrie, f. dendrometry, tree measurement.

dénégation, f., déni, m. denial.

dénicher, to take (bird, eggs) out of the nest, rout out (animal), fly away.

dénier, to deny, refuse.

dénitrer, to denitrate, denitrify.

dénitrifiant, denitrifying.

dénitrificateur, m. denitrifying agent.

dénitrificateur, provoking denitrification.

dénitrification, f. denitrification.

dénitrifier, to denitrify (soil).

dénivellation, f. difference of level, contouring.

dénombrement, m. numbering, enumeration, counting, census, list, telling off.

dénombrer, to count, enumerate, number, tell off.

dénominateur, m. denominator.

dénomination, f. denomination, appellation, name.

dénommer, to name, denominate, call.

dénoncer, to denounce, declare.

dénonciation, f. denouncement, notice.

dénoter, to denote, describe.

dénouement, m. solution, conclusion, result.

dénoyauter, to pit, stone.

denrée, f. commodity, produce, product, foodstuff; —s alimentaires, provisions, food products, foodstuffs, comestibles.

dense, dense, compact, thick.

densiflore, with flowers crowded together.

densifolié, with numerous leaves crowded together.

densimètre, m. densimeter, hydrometer.

densimétrique, densimetric.

densité, f. density, denseness; — absolue, absolute density.

dent, f. tooth, cog, prong, tusk; — du gibier, browsing by game; —s d'une feuille, serrations of a leaf; en —s de scie, serrated.

dentaire, dental.

dent-de-chien, f. dogtooth violet (Erythronium denscanis).

dent-de-lion, f. dandelion (Taraxacum).

dent-de-loup, f. pin, bolt, catch, rafter nail, mortise bolt; pl. burnisher.

denté, toothed, cogged, denticulated, serrated, serrate, dentate; — en scie, serrate; roue —e, cogwheel.

dentelaire, f. leadwort (Plumbago).

dentelé, jagged, notched, dentate, toothed, denticulated, serrated, lacelike.

denteler, to notch, jag, indent.

dentelle, f. lace.

dentelure, f. denticulation, indentation, notching.

denter, to tooth, cog.

denticide, dispersing the teeth of pericarp.

denticulé, denticulate, indented.

dentier, m. set of false teeth, denture, an instrument for cutting soap.

dentifrice, *m.* dentifrice, tooth paste or powder.

dentifrice, tooth-cleaning.

dentine, *f.* dentine.

dentiste, *m.* dentist.

dentition, *f.* dentition, cutting of teeth.

denture, *f.* serrated edge, teeth, cogs, gearing, set of teeth.

dénudation, *f.* denudation, stripping bare.

denudé, denuded, glabrous.

dénuder, to denude.

dénué, devoid, destitute, void.

dénuer, to divest, deprive, strip.

dénutrition, *f.* denutrition, wasting.

déodore, *m.* deodar (*Cedrus deodara*).

dépaissance, *f.* grazing, feeding on, grazing ground.

dépaqueter, to unpack, open.

dépareillé, incomplete, unmatched, odd.

dépareiller, to break, spoil (a set or pair).

déparer, to strip of ornament, spoil, mar.

déparquement, *m.* unpenning of cattle, letting (sheep) out of fold.

départ, *m.* division, separation, sorting, parting (of metals), departure, exit, place of departure, discharge (of firearm), departure platform.

département, *m.* department.

départir, to separate, part, dispense, deal out, distribute, bestow, scatter; **se —,** to depart from, desist from, renounce.

dépasser, to pass or go beyond, exceed, surpass, draw out, unwind, excel, extend, overgrow, outgrow.

dépecer, to cut up, dismember.

dépêche *f.* dispatch.

dépêcher, to dispatch, hasten; **se —,** to hasten, be quick.

dépeindre, to describe, depict, paint, portray.

dépendance, *f.* dependence, appendage, subordination.

dépendant, dependent, subordinate.

dépendre, to depend, be subject to, belong, rest upon, take down.

dépens, *m.pl.* costs, expense, charges; **aux — de,** at the expense of.

dépense, *f.* expenditure, expense, outlay, consumption, discharge, outflow (of water), dispensary, pantry.

dépenser, to expend, spend, consume, squander; **se —,** to be spent, waste one's energy.

dépensier, unthrifty.

déperdition, *f.* loss, waste, dissipation, discharges, escape, leakage.

dépérir, to waste away, wither (away), spoil, perish, decline, languish, wilt, die, starve, decay, etiolate.

dépérissant, withering, wilted, dying, decaying.

dépérissement, *m.* decline, decay, perishing, withering, dying away.

dépérulation, *f.* deperulation, throwing off bud scales in leafing.

dépêtrer, to disengage, extricate.

depeupler, to depopulate; **se —,** to open out, become thinned.

déphlegmation, *f.* dephlegmation, concentration.

dephlogistiquer, to dephlogisticate, reduce inflammation.

déphosphoration, *f.* dephosphorization.

déphosphorer, to dephosphorize.

dépiauter, to skin (a rabbit).

dépigmentation, *f.* latency of color in flowers.

dépilage, *m.* unhairing, removal of pillars, taking off the fleece.

dépilatif, -ive, depilatory.

dépilation, *f.* unhairing, depilation.

dépilatoire, depilatory.

dépiler, to depilate, remove the hair, clear of pillars.

dépiquer, to tread out (corn), transplant (shoots).

dépister, to track down (game), discover, mislead, put (hounds) off scent.

dépit, *m.* spite, resentment, anger; **en — de,** in spite of.

déplaçable, displaceable, adjustable.

déplacé, out of place, displaced.

déplacement, *m.* displacement, movement, shifting.

déplacer, to displace, shift, move, remove; **se —,** to be displaced, change one's residence.

déplaire, to displease, offend, be displeasing; **se —,** not to thrive or flourish, to be displeased.

déplaisant, disagreeable, unpleasant.

déplaisir, *m.* dislike, aversion.

déplantage, *m.* unplanting, transplanting.

déplâtrage, *m.* removal of plastering, taking (limb) out of plaster.

déplétion, *f.* depletion.

déplier, to unfold, spread out.

déplombage, *m.* removal of lead (seals).

déplomber, to remove lead (seals) from, strip of lead.

déplorable, deplorable.

déployer, to display, deploy, unfurl, unfold, spread out, unroll.

déplu (déplaire), displeased.

déplumer, to pluck (chicken); **se —,** to molt.

dépolarisant, *m.* depolarizer.

dépolarisant, depolarizing.

dépolarisation, *f.* depolarization.

dépolariser, to depolarize.

dépoli, *m.* depolished surface, focusing screen.

dépoli, depolished, rough, unpolished, unground, ground, frosted (glass).

dépolir, to remove the polish of, grind, frost (glass); **se —,** to lose its polish, become dull.

dépolissage, *m.* depolishing, grinding, frosting.

dépommoir, *m.* probang (for cattle).

dépopulation, *f.* depopulation.

déposage, *m.* deposition, settling.

déposer, to deposit, lay or put down, depose, form a deposit, settle; **se —,** to be deposited, form a deposit, settle.

dépositaire, *m.&f.* depositary, trustee.

déposition, *f.* deposition, declaration.

dépôt, *m.* deposit, depot, deposition, depository, trust, warehouse, sediment, abscess, precipitate; **— marin,** marine deposit.

dépotage, dépotement, *m.* decanting, decantation, unpotting, bedding out (of plants).

dépoter, to decant, unpot (a plant), dump (night soil).

dépouille, *f.* skin, hide, crop (of fruit), slough, exuvia, spoil, booty, mortal remains, draw, draft.

dépouillement, *m.* skinning, stripping, spoliation, analysis (of report).

dépouiller, to skin, strip, deprive, plunder, despoil, rob, peel, shed, unclothe, lay bare, reap, gather (crops), abstract (accounts), examine (reports), taper, give clearance to; **se —,** to shed its leaves, strip oneself.

dépourvoir, to deprive, leave destitute, fail to provide.

dépourvu, deprived, unprovided (with), destitute, devoid; **— de branches,** clean-boled; **au —,** unawares, off guard.

dépoussiérage, *m.* dust separation, dust extraction.

dépraver, to deprave.

déprécier, to depreciate, undervalue.

déprenant, separating, dissolving.

déprendre, to separate, part; **se —,** to rid oneself, melt, run (of jelly).

dépressible, depressed (easily), compressed.

dépression, *f.* depression, lowering, reduced pressure, hollow, vacuum; **— du terrain,** depression of the ground.

déprimé, depressed, sunk down.

déprimer, to depress, press down.

dépris, separated, parted.

dépriser, to depreciate, underrate.

depuis, from, since, during, later, after, for; **— que,** since; **il est parti — une semaine,** he left a week ago.

dépulpage, *m.* reduction to pulp, pulping.

dépulper, to reduce to pulp, pulp.

dépulpeur, *m.* pulper, peeling drum.

dépurateur, -trice, dépuratif, -ive, depurative, purifier.

dépuration, *f.* depuration, clearing of liquid.

dépuratoire, depuratory, depurative.

dépurer, to depurate, cleanse.

député, *m.* deputy.

député, deputed.

déracinement, *m.* uprooting.

déraciner, to uproot, eradicate.

déraison, *f.* irrationality.

déraisonnable, unreasonable, unwise.

déraisonnement, *m.* folly.

dérangé, out of order, deranged.

dérangement, *m.* derangement, trouble, disturbance.

déranger, to derange, disarrange, disturb, unsettle, put out of order, displace, incommode; **se —,** to get out of order, trouble oneself, become deranged.

dérater, to spleen (dog).

dératisation, *f.* rat extermination.

dérayer, to cut the boundary furrow.

dérayure, *f.* water furrow.

déréglable, likely to get out of order.

déréglé, out of order, irregular, inordinate, disorderly.

déréglément, *m.* irregularity, disorder.

dérégler, to put out of order, disarrange, disorder; **se —,** to get out of order.

dérésinage, *m.* extraction of resin.

dérision, *f.* derision.

dérisoire, derisive, ridiculous.

dérivant, *m.* derivative.

dérivant, derived, originating.

dérivatif, -ive, revulsive, derivative.

dérivation, *f.* derivation, diversion, origin, shunt(ing), tapping of current.

dérive, *f.* deviation, drift.

dérivé, *m.* derivative.

dérivé, derived, secondary, derivative, product, shunt(ed) (current).

dérivée, *f.* derivative, differential quotient.

dériver, to be derived, drift, originate, spring from, proceed from, derive, divert, unrivet, shunt.

dériveter, to unrivet.

derle, *f.* kaolin, porcelain clay.

derma, *f.* derma, surface of an organ.

dermatite, *f.* dermatitis, inflammation of the skin.

dermatocalyptrogène, *f.* dermacalyptrogen, histogen producing root cap and root epidermis.

dermatocystide, *f.* dermatocyst, inflated hairs on surface of the sporophore.

dermatogène, *f.* dermatogen.

dermatomycose, *f.* mycosis of skin.

dermatose, *f.* dermatosis.

dermatosome, *f.* dermatosomes, granular bodies forming the cell wall.

derme, *m.* derma, dermis, true skin.

dermique, dermic, dermal.

dermoblaste, *f.* embryo with membranous cotyledons, dermoblastus.

dernier, -ère, last, utmost, final, recent, latter, farthest; **au — point,** intensely, extremely; **ce —,** the latter.

dernièrement, lately, recently, of late, lastly.

dérobé, hidden, concealed, shelled, stolen, secret; **à la —e,** secretly, on the sly; **pied —,** worn hoof; **récolte —,** catch (snatch) crop.

dérober, to denude, strip, skin, shell, peel, steal, rob, save, hide, conceal, shroud, screen; **se —,** to escape, steal away, slip away, be hidden, slink away, avoid.

dérochage, *m.* scouring, dipping, pickling.

dérocher, to scour, pickle, dip, remove rocks from, make (sheep) fall from rocks.

dérogation, *f.* digression, impairment.

dérogé, derogated.

dérompage, *m.* cutting (of rags).

dérompoir, *m.* rag-cutting machine.

dérompre, to cut (rags), break up land.

dérouillement, *m.* removal of rust.

dérouiller, to remove rust from, brush up.

déroulement, *m.* unrolling, unwinding, unfolding.

dérouler, to unroll, unfold, evolve, unwrap, spread out, display; **se —,** to be unfolded, extend.

déroutant, mystifying, bewildering.

dérouter, to confuse, baffle, mislead, disconcert.

derrière, *m.* back (part), hind part, rear, stern (of ship), backside, posterior, buttocks, rump.

derrière, behind, after; **par —,** behind, from behind, in the rear.

des (de les), of the, from the, some, any.

dès, from, starting from, at, since, by, as early as; **— avant,** (even) before; **— là,** from then, ever since, therefore; **— l'abord,** from the outset; **— lors,** from that time, ever since, consequently, therefore, hence, since then; **— lors que,** since, as, seeing that; **— que,** as soon as, when, as, since.

désabonner, to discontinue a subscription to a periodical.

désabuser, to disabuse, undeceive.

désaccord, *m.* disagreement, discord, variance, disaccord.

désaccoupler, to disconnect, disengage, throw out of gear.

désaccoutumer, to disaccustom.

désacidification, *f.* deacidification.

désacidifier, to deacidify, disacidulate.

désaciérer, to unsteel (blade).

désactiver, to render inactive.

désaération, *f.* deaeration, airing.

désaérer, to deaerate.

désaffleurer, to fail, push (tiles) out of level, be out of level, project.

désagencer, to disarrange, disorganize, throw (machine) out of gear.

désagrafer, to unhook, unfasten.

désagréable, disagreeable, uncomfortable, unpleasant.

désagréablement, disagreeably, unpleasantly.

désagrégation, *f.* disintegration, disaggregation, weathering.

désagrégeable, disintegrable.

désagrégeant, disintegrating.

désagréger, to disintegrate, disaggregate, weather (rock); se —, to disintegrate, weather.

désagrément, *m.* unpleasantness, disagreeableness.

désaigrir, to remove the sourness from.

désailement, *m.* severing the wings from the seed.

désailer, to sever the wings from the seed.

désaimantation, *f.* demagnetization.

désaimanter, to demagnetize.

désajuster, to throw out of adjustment, disarrange.

désallaiter, to wean.

désaltérer, to quench the thirst of, refresh, slake.

désamidase, *f.* deamidase, desamidase.

désamorçage, *m.* unpriming, exhausting (of arc), running down (dynamo).

désamorcer, to unprime (fuse or siphon).

désappointer, to disappoint.

désapprendre, to forget, unlearn.

désappris, forgotten, unlearned.

désapprouver, to refuse, reject.

désarçonner, to unhorse, unseat, throw.

désargentage, *m.*, **désargentation,** *f.* desilverization, desilvering.

désargenter, to desilverize, desilver.

désarmé, disarmed, unarmed.

désarmer, to disarm.

désarroi, *m.* disarray, disorder, confusion.

désarticuler, to disarticulate, disjoint, dislocate.

désassembler, to disassemble.

désassimilateur, -trice, disassimilating, catabolizing.

désassimilation, *f.* disassimilation, catabolism.

désassimiler, to disassimilate, catabolize.

désassociation, *f.* disassociation, dissociation.

désassocier, to disassociate, dissociate.

désassombrir, to lighten, brighten, cheer up.

désassorti, unmatched, odd, broken (set or collection).

désastre, *m.* disaster.

désastreusement, disastrously.

désastreux, -euse, disastrous.

désaturé, unsaturated, superheated (steam).

désavantage, *m.* disadvantage, drawback, loss.

désavantageusement, disadvantageously.

désavantageux, -euse, disadvantageous, unfavorable, detrimental.

désavouer, to disown, disclaim, deny.

désazotation, *f.* denitrification.

désazoter, to denitrify.

descellement, *m.* unsealing.

desceller, to unseal, unbed, unfasten, chip off rough parts of (glass).

descendance, *f.* descent, lineage, pedigree.

descendant, *m.* descendant, offspring.

descendant, descending.

descenderie, *f.* a descending passage or shaft.

descendre, to descend, go down, fall, take down, land, let down, come down, step down, prolapse, drop; il **descend à** (**un hotel**), he stops at (a hotel); il **descend de** (**la voiture**), he gets out of (the carriage).

descendu, descended.

descente, *f.* descent, slope, taking down, down pipe, fall, prolapse, falling, declivity, downstroke, rupture, hernia.

descripteur, *m.* describer.

descripteur, descriptive.

descriptif, -ive, descriptive.

description, *f.* description; — **du peuplement,** description of crop.

désemballer, to unpack.

désembrayage, see DÉBRAYAGE.

désembrayer, see DÉBRAYER.

désemparer, to leave, quit, abandon.

désemplir, to make less full, half empty; se —, to become or get less full.

désempoisonner, to free from poison.

désempoissonner, to unstock (pond).

désenchantement, *m.* disenchantment, disillusion.

désencroûter, to free from crust or scale (boiler).

désenflammer, to extinguish (a blaze).

désenfler, to deflate, reduce swelling.

désenflure, *f.* deflation, going down (of swelling).

désenfourner, (see défourner), to remove from oven or kiln.

désengager, to free from obligation, disengage.

désengrener, to disengage, throw out of gear.

désenrayer, to unlock, release, disconnect.

désensibiliser, to desensitize.

désentortiller, to disentangle, unravel.

désenvelopper, to open, undo.

déséquilibre, *m.* lack of balance, absence of equilibrium.

déséquilibré, unbalanced, out of equilibrium.

déséquilibrer, to throw out of (equilibrium) balance.

désert, *m.* desert, wilderness, waste.

désert, solitary, wild, waste, deserted, forsaken.

déserter, to desert, abandon.

désertique, desert.

désespérance, *f.* despair.

désespérant, discouraging, despairing.

désespéré, desperate, hopeless, sorry.

désespérément, desperately.

désespérer, to despair, give up all hope, drive to despair, dishearten.

désespoir, *m.* despair, hopelessness.

désétamage, *m.* detinning.

désétamer, to detin.

déshabiller, to undress, lay bare.

désherbage, *m.* cleaning (a field), weeding.

désherbant, weeding, eradicating weeds.

déshériter, to disinherit.

déshonnête, improper, immodest.

déshonneur, *m.* dishonor, disgrace.

déshonoré, topped, pollarded.

déshuiler, to remove the oil from.

déshydratant, *m.* dehydrating agent.

déshydratant, dehydrating.

déshydratation, *f.* dehydration.

déshydrater, to dehydrate; se —, to dehydrate, lose water.

déshydrogénation, *f.* dehydrogenation.

déshydrogéner, to dehydrogenize.

désidératum, *m.* desideratum, object, wish.

désignation, *f.* designation.

désigner, to designate, denote, appoint, point out, indicate.

désillusion, *f.* disillusion.

désincrustant, *m.* disincrustant, antiscale composition, scale preventive.

désincrustation, *f.* removal of incrustation or scale (from boiler).

désincruster, to scale (a boiler).

désinence, *f.* termination (of word).

désinfectant, *m.* disinfectant, fungicide.

désinfecter, to disinfect.

désinfecteur, *m.* disinfector.

désinfection, *f.* disinfection, fumigation.

désintégrateur, *m.* disintegrator.

désintégration, *f.* disintegration, weathering.

désintégrer, to disintegrate, crush.

désintéressé, disinterested, unbiased.

désintéressement, *m.* disinterestedness, impartiality.

désir, *m.* desire, wish, longing.

désirable, desirable.

désirer, to desire, wish, long for, wish for.

désireux, -euse, desirous, anxious, eager.

désistement, *m.* abandonment.

desman, *m.* desman; — musqué, muskrat.

desmidiés, *f.pl.* desmids, Desmidiaceae.

desmine, *f.* desmine, stilbite.

desmochondres, *m.pl.* microsomes, small granules.

desmotropie, *f.* desmotropy, desmotropism.

désobéir, to disobey.

désobstruant, désobstructif, -ive, deobstruent, aperient.

désodorer, to deodorize.

désodorisation, *f.* deodorization.

désodoriser, to deodorize.

désoeuvré, idle, unoccupied.

désoeuvrement, *m.* unemployment.

désoeuvrer, to render idle, separate (paper).

désolation, *f.* desolation, sorrow, vexation, grief.

désolé, desolate, sorry, disconsolate, heartbroken, afflicted.

désoperculé, deoperculate, shedding the operculum.

désoperculer, to uncap (honeycomb).

désordonné, disorderly, unruly, immoderate, unrestrained.

désordonner, to disorder, disturb.

désordre, *m.* disorder, irregularity, confusion, disturbance.

desorganisateur, *m.* disorganizer.

désorganisateur, -trice, disorganizing.

désorganisation, *f.* disorganization.

désorganiser, to disorganize; se —, to become disorganized.

désorientation, *f.* confusion.

désormais, henceforth, hereafter, from now on, henceforward.

désosser, to bone, remove the bones from.

désoufrage, *m.* desulphurization.

désoufrer, to desulphurize, desulphur.

désoxydant, *m.* deoxidizer.

désoxydant, deoxidizing.

désoxydation, *f.* deoxidation, deoxidization.

désoxyder, to deoxidize; se —, to be deoxidized.

désoxygénation, *f.* deoxygenation, deoxidation.

désoxygéner, to deoxygenate, deoxidize.

despotique, despotic, absolute.

despotisme, *m.* despotism.

despumation, *f.* despumation, scumming.

despumer, to despumate, skim, clarify (by removing scum).

desquamation, *f.* desquamation, peeling (of skin).

desquels, desquelles, of whom, of which, from whom, from which.

dessabler, to remove sand from.

dessaignage, *m.* freeing from blood (hides).

dessaigner, to free from blood, wash (hides or skins).

dessaisir, to get loose, release.

dessalage, *m.*, dessalaison, *f.*, dessalement, *m.* removal of salt (from fish, meat), soaking.

dessalé, freed from salt, wide-awake (person).

dessaler, to remove salt from, put (meat) in soak.

desséchant, drying, desiccating.

desséché, dried up, desiccated.

dessèchement, *m.* drying, withering, drainage, exsiccation, desiccation, insiccation.

dessécher, to dry, desiccate, dry up, drain (land), season (wood), wither, exsiccate, wither (plant).

dessécheur, *m.* drier.

dessein, *m.* design, intent, aim, plan, purpose, intention; à —, by design, intentionally, on purpose; à — de, with the intent of, in order to; à — que, to the end that, in order that; avec —, designedly; sans —, unintentionally, unknowingly.

desseller, to unsaddle.

desserrage, desserrement, *m.* release, loosening, thinning, slackening, inter-rupting (opening out) the leaf canopy; — du massif, opening out the crop.

desserrer, to loosen, thin, slacken, admit light, interrupt the leaf canopy; se —, to be loosened, work loose.

dessert, *m.* dessert.

desservir, to serve, supply, accommodate, remove, clear off, disserve, damage.

dessiccant, desiccating, desiccant.

dessiccateur, *m.* desiccator, dryer; — à vide, vacuum desiccator.

dessiccatif, -ive, desiccative.

dessiccation, *f.* desiccation, drying (up), seasoning.

dessin, *m.* design, drawing, sketch, outline, plan, pattern, figure.

dessiner, to draw, sketch, outline, design, delineate; se —, to appear, be visible, take form.

dessoler, — une terre, to change rotation (of crops).

dessouchement, *m.* uprooting, grubbing up.

dessoucheuse, *f.* stump puller.

dessoudage, *m.* unsoldering.

dessouder, to unsolder; se —, to come unsoldered.

dessoudure, *f.* unsoldering.

dessoufler, to deflate.

dessoufrage, *m.* desulphurization.

dessoufrer, to desulphurize.

dessous, *m.* underside, bottom, wrong side, lower part.

dessous, below, under, underneath, beneath; au —, underneath; au — de, beneath; de —, under, out from under; en —, underneath, face down, underhand.

dessuintage, *m.* scouring, steeping, removal of suint from wool.

dessuinter, to scour (wool), deprive of suint, steep.

dessus, *m.* upper side, upper part, top, right side (of cloth), cover (of book), back (of hand), upper hand.

dessus, uppermost, above, over, on, upon, beyond; au — de, above; ci —, above; de —, from off, from, upper, top; en —, on, upon, over; par —, over, on, upon.

destin, *m.* destiny, fate.

destinataire, *m.* addressee, recipient, consignee.

destination, *f.* destination, purpose.

destiné, destined, born.

déstinée, *f.* destiny, fate.

destiner, to destine, resolve, design, doom, intend.

destitué, destitute, devoid.

destructeur, -trice, destructive, ruinous, deadly.

destructibilité, *f.* destructibility.

destructible, destructible.

destructif, -ive, destructive.

destruction, *f.* destruction, overthrow.

désuintage, *m.* scouring.

désulfuration, *f.* desulphuration, desulphurization.

désulfurer, to desulphurize.

désunion, *f.* division.

désunir, to disunite, disjoin, take apart.

détachant, stain-removing.

détacher, to detach, (un)loose, untie, undo, free, remove, take down, take out stains, clean, loosen; **se —,** to come undone, be detached, stand out, break loose, break off, become loosened.

détail, *m.* detail, retail, particular, plotting (of details); **en —,** in detail, (by) retail.

détaillant, *m.* retailer.

détaillant, retail.

détaillé, detailed.

détailler, to detail, enumerate, retail.

détaler, to scamper away, hike off.

détanner, to detan, detannate, deprive of tannin.

détartrage, *m.* removal of tartar or scale, decarbonizing.

détartrer, to remove tartar or scale from, decarbonize, scale, fur, clean (casks or boilers), soften (water).

détartreur, *m.* detartarizer, scaling device, water softener.

détartrisation, *f.* removal of tartar or scale.

détecteur, *m.* detector.

détection, *f.* detection.

déteignant, decoloring.

déteindre, to decolor, lose color, fade, come off, bleach.

dételer, to unyoke, detach, uncouple, unharness, disengage, disconnect.

détendeur, *m.* pressure-reduction valve, pressure reducer.

détendre, to expand, cut off (steam), unbend, relax, release, slacken, loosen, take down; **se —,** to become slack, relax, expand (steam), unbend, loosen.

détenir, to hold, detain, keep back.

détente, *f.* expansion (of steam or gases), (steam) cutoff, detent, catch, trigger, power stroke, relaxation, loosening, brandishing, stop; **soupage de —,** expansion valve.

détenu, detained, withheld.

détergent, detergent.

déterger, to cleanse, purify.

détérioration, *f.* deterioration.

détériorer, to deteriorate, spoil, impair, wear out.

déterminable, determinable.

déterminant, *m.* separate material particles in the germ cells.

déterminant, determinative, decisive, determinant.

détermination, *f.* determination, resolution, determining, calculation.

déterminé, determined, determinate, caused, definite, fixed, given.

déterminer, to determine, cause, divert, fix, specify, occasion, bring about; **se —,** to determine, resolve, make up one's mind.

déterrement, *m.* excavation.

déterrer, to unearth, dig up, exhume, ferret out.

détersif, -ive, detersive, detergent, abstergent.

détestable, detestable.

détient (détenir), (he) holds.

détirer, to stretch, draw out.

détiser, to rake out (fire).

détitrer, to deprive; **— un alcool,** to bring an alcohol below proof.

détonant, *m.* explosive.

détonant, detonating, explosive.

détonateur, *m.* detonator, fog signal.

détonation, *f.* detonation, explosion.

détoner, to detonate, explode.

détonneler, to draw (wine) from the cask.

détordre, to untwist, unravel.

détors, untwisted.

détortiller, to untwist, disentangle.

détouper, to unstop, take tow out, clean off (thorns).

détour, *m.* turning, deviation, winding, detour, evasion.

détourné, diverted, secluded, indirect, roundabout, out of the way.

détournement, *m.* diversion.

détourner, to divert, deflect, turn off, turn aside, avert, deter, distract, dissuade, track, search (for game); **se —,** to deviate, turn away, turn aside.

détrapage, *m.* thinning stool shoots.

détraqué, out of order, deranged, broken-gaited (horse).

détraquer, to put out of order, rack to pieces, derange, divert; **se —,** to get out of order, become deranged, (health) break down, (digestion) get upset.

détrempe, *f.* distemper, annealing (of steel).

détremper, to mix, dissolve (solids), dilute, moisten, soak, slake (lime), anneal (steel), soften, enervate; **se —,** to be mixed, be thinned, slake, (of steel) lose its temper, anneal, soften; **champ détrempé,** sodden, soppy field.

détresse, *f.* distress, sorrow.

détrichage, *m.* sorting (of wool).

détricher, to sort (wool).

détriment, *m.* detriment, loss, detritus, harm, injury, prejudice.

detripler, to divide into three parts.

détritage, *m.* crushing (olives).

détriter, to crush (olives).

détrition, *f.* detrition, wearing away (teeth), disintegration.

détritique, detrital.

détritoir, *m.* crushing mill (for olives).

détritus, *m.* detritus, debris, roughage (of food), offal; **— végétaux,** *m.pl.* vegetable refuse, dead soil covering.

détroit, *m.* strait, pass, defile, inlet, outlet, sound.

détromper, to correct mistake.

détrôner, to dethrone.

détruire, to destroy, ruin.

détruisant, destroying, destructive.

dette, *f.* debt, obligation.

deuil, *m.* mourning.

deutérophloëme, *m.* secondary phloem or inner bark.

deutéroplasma, *f.* deuteroplasm, paraplasm.

deuteroxylème, *m.* secondary xylem, metaxylem.

deutocarboné, bicarbureted.

deutocérébron, *m.* deutocerebrum.

deutoplasma, *f.* metaplasm, protoplasm containing the formative or granular material.

deutoxyde, *m.* **— de mercure,** mercuric oxide; **— de plomb,** minium, red lead.

deutylène, *m.* ethylene.

deux, *m.* two, second.

deux, two, second, a few; **— à —,** every two, in pairs; **— fois,** twice; **tous —,** both; **tous les — jours,** every other day.

deuxième, *m.* second story.

deuxième, second.

deuxièmement, secondly, in the second place.

deux-points, *m.* colon (:).

devait (devoir), (he) was obliged to.

dévaler, to descend, lower, go down.

devancer, to precede, outstrip, outdo, overtake, anticipate, forestall, go before.

devancier, *m.* predecessor, forbear.

devant (devoir), owing.

devant, *m.* front, fore part, front part; **prendre les —s,** to forestall, get before.

devant, in front of, ahead, before.

dévastation par chenilles, *f.* damage by caterpillars.

dévaster, to devastate, waste.

développable, developable.

développante, *f.* involute, evolvent.

développateur, *m.* developer, developing agent.

développé, developed, evolved, full-grown, evolute.

développement, *m.* development, evolution, unfolding, growth; **— de la cime,** crown development.

développer, to develop, unwrap, undo, expand, explain, uncoil, unwind, open, unfold; **se —,** to spread out, develop, expand, extend, be unwrapped.

devenir, *m.* gradual growth.

devenir, to become, grow into, get, turn; **— sauvage,** to run wild; **qu'est-il devenu?** what has become of him?

déverdir, to lose green color.

déverdissage, *m.* loss of green color.

dévernir, to remove the varnish from.

dévernissage, *m.* removal (loss) of varnish.

dévers, *m.* inclination, slope, warping; **par —,** before, near; **par — soi,** for oneself.

dévers, toward, leaning.

déversement, *m.* pouring, outlet, emptying, inclination, warping.

déverser, to slope, incline, lean, warp, bend, flow, run, pour forth, fall into, empty, spread.

déversoir, *m.* overflow, spillway, lock, weir.

déviateur, -trice, causing deviation, deflecting.

déviation, *f.* deviation, deflection, turning aside, curvature (spine); — **du complément,** deviation of complement.

dévider, to unwind, wind, reel.

deviendra (devenir), (he) will become.

dévier, to deviate, swerve, turn aside, deflect; **se —,** to grow crooked, warp.

deviner, to divine, guess, think, conjecture, predict.

devint (devenir), (he) became.

devis, *m.* estimate; — **estimatif,** estimate of quantities and costs.

devise, *f.* catchword, device, motto, foreign bill, bill of exchange.

dévissage, *m.* unscrewing.

dévisser, to unscrew.

dévitaliser, to devitalize.

dévitrification, *f.* devitrification.

dévitrifier, to devitrify; **se —,** to become devitrified.

dévoiement, *m.* cant, tilt, inclination; — **de corps,** diarrhea.

dévoiler, to unveil, reveal, disclose.

devoir, to owe, should, ought, must, be obliged, be necessary, have to, be scheduled or destined to, be due to, be bound to, be compelled.

devoir, *m.* task, obligation, debit, duty.

dévolu, devolved, devolving, vested in.

dévoré, *m.* the devoured.

dévorer, to devour, swallow, consume, eat up.

dévotion, *f.* devotion.

dévoué, devoted.

dévouement, *m.* devotion, application.

dévouer, to devote, dedicate.

dévoyer, to turn aside, slope, cant, incline, loosen, purge.

devra (devoir), (he) will owe, be obliged to.

devrait (devoir), (he) should.

dextérité, *f.* dexterity, skill.

dextre, right-hand, right.

dextrement, skillfully.

dextrine, *f.* dextrin.

dextrogyre, dextrorotatory, dextrogyrate.

dextroracémique, dextrotartaric.

dextrorse, dextrorse, towards the right hand.

dextrorsum, dextrorse, clockwise, from left to right.

dextrose, *f.* dextrose.

dg., *abbr.* (décigramme), decigram.

diabase, *f.* diabase.

diabète, *m.* diabetes.

diabétique, diabetic.

diabétomètre, *m.* diabetometer, saccharometer.

diable, *m.* devil, deviling machine (for tearing rags), a barrow or truck, (twowheeled) trolley.

diacaustique, diacaustic.

diacétate, *m.* diacetate.

diacétyle, *m.* diacetyl.

diachylon, diachylum, *m.* diachylon, diachylum.

diachyme, *f.* mesophyll, interior parenchyma of a leaf.

diacode, *m.* diacodion, sirup of poppies.

diadelphe, diadelphous, with two groups of stamens.

diadelphie, *f.* Diadelphia.

diagénèse, *f.* diagenesis.

diagéotropique, diageotropic.

diagnose, *f.* diagnosis, diagnostic character.

diagnostic, *m.* diagnosis.

diagnostique, diagnostic.

diagnostiquer, to diagnose, diagnosticate.

diagomètre, *m.* diagometer.

diagonal, -aux, diagonal.

diagonale, *f.* diagonal, twill, cross-brace; **en —,** diagonally.

diagonalement, diagonally.

diagramme, *m.* diagram; — **des forces,** stress diagram of distribution.

diagraphe, *m.* diagraph, pantograph.

diagrède, *m.* scammony.

diakène, *f.* diakinesis.

dialdéhyde, *m.* dialdehyde.

dialecte, *m.* dialect.

diallage, *m.* diallage.

diallagique, diallagic.

dialogite, *f.* manganese spar, carbonate of manganese, rhodochrosite, dialogite.

dialycarpellé, apocarpous.

dialydesmie, *f.* dialydesmy, the separating of a stele into bundles.

dialypétale, polypetalous.

dialysable, dialyzable.

dialyse, *f.* dialysis.

dialysépale, dialysepalous, polysepalous.

dialyser, to dialyze.

dialyseur, *m.* dialyzer.

dialyseur, dialyzing.

dialystaminé, with separate stamens.

dialystélie, *f.* dialystely, steles in stem remain separate.

diamagnétique, diamagnetic.

diamagnétisme, *m.* diamagnetism.

diamagnétomètre, *m.* diamagnetometer.

diamant, *m.* diamond; — **à rabot,** — **de vitrier,** glazier's diamond; — **brut,** rough diamond; — **taillé,** cut diamond.

diamantaire, *m.* diamond cutter.

diamantaire, diamondlike, brilliant.

diamanté, glittering, glistening, sparkling.

diamantifère, diamantiferous, diamond-bearing, rich in diamonds.

diamantin, diamantine, diamondlike.

diamésogames, *m.pl.* plants pollinated by external agencies.

diamétral, diametral, diametric(al), diametrically opposite.

diamétralement, diametrically.

diamètre, *m.* diameter, bore. circle (of a propeller); — **à hauteur d'homme,** diameter breast high; — **à mi-hauteur,** mid-diameter; — **au gros bout,** diameter at the butt end; — **au petit bout,** diameter at the top end; — **extérieur,** outside diameter; — **intérieur,** inside diameter.

diamino-acide, *m.* diamino acid.

diamylique, diamyl.

diandre, diandrous, possessing two stamens.

diandrie, *f.* diandria, plants of two stamens.

Diane, *f.* Diana, silver.

diapause, *f.* suspended animation.

diapédèse, *f.* diapedesis.

diaphane, diaphanous, translucent, transparent.

diaphanéité, *f.* diaphaneity, translucency.

diaphorèse, *f.* diaphoresis, perspiration.

diaphorétique, diaphoretic, sudorific.

diaphragmant, using diaphragm.

diaphragmation, *f.* diaphragmation.

diaphragmatique, (pod) divided into two or more monosperm loculi.

diaphragme, *m.* diaphragm, membrane, a dividing membrane; — **du nez,** septum of the nose.

diaphragmé, provided with a diaphragm.

diaphyse, *f.* diaphysis, proliferation of the inflorescence.

diapositif, *m.* diapositive, transparent positive.

diapré, variegated, gaily colored.

diarche, *m.* diarch.

diarrhée, *f.* diarrhea, flux; — **verte,** green diarrhea.

diaspase, *f.* diaspasis, daughter nuclei in amitosis torn asunder.

diaspore, *m.* diaspore.

diastaltic, diastaltic.

diastase, *f.* diastase, diastasis, enzyme.

diastasique, diastatic, diastasic, enzymic.

diastèle, *f.* polystelic structure.

diaster, *m.* karyokinesis.

diastole, *f.* diastole, the slow dilation of a contractile vesicle.

diastomose, *f.* liver rot, (liver) fluke disease.

diathermane, diathermic, diathermous.

diathermanéité, *f.* diathermaneity.

diathermansie, *f.* diathermancy.

diatmese, *f.* diatmesis, division of daughter nuclei in amitosis.

diatomée, *f.* diatom.

diatomine, *f.* diatomine.

diatomique, diatomic.

diatomite, *f.* diatomite, tripoli.

diatrophe, heterotrophic, deriving nourishment from without.

diatropisme, *m.* diatropism.

diazine, *f.* diazine.

diazo, *m.* diazo compound.

diazobenzène, *m.* diazobenzene.

diazocomposé, *m.* diazo compound.

diazométhane, *m.* diazomethane.

diazotation, *f.* diazotization.

diazoture, *m.* dinitride.

dibenzyle, *m.* dibenzyl.

dicalcique, dicalcium, dicalcic.

dicarbonique, dicarboxylic, dicarbonic.

dicéluphe, with twofold cover.

dicéphale, dicephalous.

dicétone, *f.* diketone.

dichloré, dichloro.

dichogame, dichogamous.

dichogamie, *f.* dichogamy.

dichopétale, with bifid petals.

dichotome, dichotomous, bisected, forked.

dichotomie, *f.* dichotomy, bifurcation.

dichotomophylle, w th dichotomous leaves.

dichotrophe, *m.* heterotroph.

dichotypie, *f.* dichotypy, two different forms of the same stock.

dichroanthe, with flowers of two colors, bicolored.

dichroé, having two colors, bicolor, dichromatic.

dichroïde, dichroïque, dichroic.

dichroïsme, *m.* dichroism, fluorescence.

dichroïte, *m.* dichroite, iolite.

dichromatique, dichromatic.

dichrone, biennial.

diclésie, *f.* diclesium, an achene in a free perianth.

dicline, unisexual, diclinous, dioecious.

diclinie, *f.* dicliny, diclinism, male and female organs separate.

dicotylédoné, dicotyledonous.

dicotylédonées, *f.pl.* dicotyledons.

dicranocère, prong-horned (antelope).

dictame, *m.* dittany (Dictamnus); — blanc, white dittany, fraxinella (*Dictamnus albus*); — de Crête, Cretan dittany (*Origanum dictamnus*).

dictateur, *m* dictator.

dicter, to dictate.

dictionnaire, *m.* dictionary.

dictyocarpe, with reticulate fruits.

dictyodrome, dictyodromous, with reticulate venation.

dictyoide, resembling a net, reticulate.

dictyorrhize, with netlike roots.

dicycle, having two whorls.

dicyclique, biennial.

dicyme, *f.* dicyme, a cyme in which first axes again form cymes.

dicypellion, *m.* pinkwood (*Dicypellium caryophyllatum*).

didelphe, *m.* Didelphis, opossum.

didyme, *m.* didymium.

didyme, didymous, divided into two lobes.

didymère, anther with stunted connective.

didyname, didynamous, (flower) with four stamens in two pairs.

didynamie, *f.* didynamy.

diécique, dioecious.

dièdre, *m.* dihedron.

dièdre, dihedral.

diélectrique, dielectric, nonconducting.

diélectrolyse, *f.* dielectrolysis.

dientomophilie, *f.* dientomophily, individuals adapted for various insect pollination.

diérèse, *f.* diaeresis.

diérésile, (fruit) with numerous carpels in a circle.

diète, *f.* diet.

diététique, *f.* dietetics.

diététique, dietetic(al).

diéthylénique, diethylenic, diethylene.

dieu, *m.* god, God.

différé, deferred, postponed, put off.

différemment, differently.

différence, *f.* difference, distinction; — finie, fundamental difference; à la — de, unlike, contrary to, differently from; à la — que, with this difference that, except that.

différenciation, *f.* differentiation.

différencier, to differentiate, distinguish; se —, to be differentiated, differ.

différend, *m.* difference, dispute, disagreement.

différent, different, unlike, various, several

différentiation, *f.* differentiation.

différential, *m.* differential.

différentiel, -elle, differential.

différentielle, *f.* differential.

différentier, to differentiate.

différer, to defer, postpone, delay, hold over, differ, be unlike; se —, to be deferred, be put off, be intersected, be broken.

difficile, *m.* difficulty.

difficile, difficult, hard.

difficilement, with difficulty, difficultly, not easily.

difficulté, *f.* difficulty, difference, obstacle, objection; sans —, without difficulty, undoubtedly; sans plus de —, without further ado.

difficultueux, -euse, difficult.

diffluent, diffluent (tumor).

difforme, deformed, misshapen.

difformité, *f.* deformity, malformation.

diffracter, to diffract.

diffractif, -ive, diffractive.
diffraction, f. diffraction.
diffus, diffuse, prolix, diffused, ample, spreading.
diffuser, to diffuse.
diffuseur, m. diffuser, broadcaster, sprayer, atomizer.
diffusibilité, f. diffusibility.
diffusible, diffusible.
diffusif, -ive, diffusive.
diffusion, f. diffusion. extension. spread, broadcasting.
diflorigère, with two flowers, bifloral.
digame, digamous, two sexes in same cluster.
digamie, f. digamy, double fertilization.
digérable, digestible.
digérer, to digest, assimilate, put up with, mature (tumor); se —, to be digested.
digesteur, m. digester, extractor (siphon).
digestibilité, f. digestibility.
digestible, digestible.
digestif, -ive, digestive.
digestion, f. digestion.
digital, m. Clavaria.
digital, -aux, digital.
digitale, f. digitalis, purple foxglove (Digitalis purpurea).
digitaléine, f. digitalein, glucoside of the digitalis.
digitaline, f. digitalin.
digité, digitate(d), fingered.
digitifolié, digitate-leaved.
digitiforme, fingerlike.
digitinervé, digitinervate.
digitoxine, f. digitoxin.
digne, worthy, deserving, dignified, upright.
dignement, worthily, suitably, deservedly.
dignité, f. dignity, nobleness.
digresser, to digress.
digression, f. digression.
digue, f. dam, dike, barrier, jetty, pier.
diguer, to dam, dike.
digyne, digynous (flower), (ovary) with two separated carpels.
digynée, f. digynia, a gynaecium of two pistils.
dihybride, f. dihybrid.
dihydrate, m. dihydrate; — de térébenthène, terpin hydrate.

dihydrobenzène, m. cyclohexadiene, dihydrobenzene.
dika, m. dika, bread tree (Irvingia barteri).
dilatabilité, f. expansibility, dilatability, extensibility.
dilatable, expansible, dilatable.
dilatant, m. dilator.
dilatant, expanding, dilating, distending.
dilatation, f. expansion, dilation, dilatation.
dilater, to expand, dilate, distend.
dilation, f. delay, postponement.
dilemme, m. dilemma.
diligence, f. diligence, industry, promptness, haste.
diligent, diligent, prompt.
diligenter, to press, hasten, urge.
dilué, diluted, dilute.
diluer, to dilute; se —, to be diluted.
diluteur, diluting, diluent.
dilution, f. dilution.
diluvial, -aux, diluvien, -enne, diluvial, diluvian.
diluvium, m. diluvium.
dimanche, m. Sunday.
dîme, f. tithe, tenth, dime (ten cents).
dimension, f. dimension, size, measure, over-all length.
dimensionner, to proportion, rate.
dimère, m. dimer.
dimère, dimerous, having each whorl in two parts.
diméthylarsénique, dimethylarsenic, cacodylic.
diméthylbenzène, m. dimethylbenzene, xylene.
diméthylcétone, f. dimethyl ketone, acetone.
diméthylique, dimethyl.
diminuer, to diminish, lessen, decrease, lower.
diminution, f. diminution, lessening, decrease, running out.
dimorphe, dimorphous, dimorphic, having two forms.
dimorphisme, m., dimorphie, f. dimorphism.
dinanderie, f. copper and brass utensils.
dinde, f. hen turkey, turkey.
dindon, m. turkey.
dîner, to dine.
dîner, m. dinner.

dioctaèdre, dioctahedral.

diodange, *f.* diodange, sporangium of ferns.

diode, *f.* diode, reproductive body in vascular plants.

diodocarpe, *m.* sporocarp.

diodon, *m.* porcupine fish, globefish.

diodophytes, *f.pl.* diodophytes, vascular plants, prothallial plants.

dioecie, *f.* dioecia.

dioecique, dioecious, unisexual.

dioggot, diogot, *m.* daggett, birch-bark oil.

dioïque, dioecious.

dionée, *f.* catchfly (Dionaea).

dionine, *f.* dionin.

dioptase, *f.* dioptase.

dioptrique, *f.* dioptrics.

dioptrique, dioptric(al).

diorite, *f.* diorite.

dioritique, dioritic.

dioscorés, *f.pl.* Dioscoreaceae.

dioxindol, *m.* dioxindole.

dioxybutyrique, dihydroxybutyric.

dioxyquinone, *f.* dihydroxyquinone.

dioxytartrique, dihydroxytartaric, dioxytartaric.

dipentène, *m.* dipentene, cinene, inactive limoene.

dipeptide, *n.* dipeptide.

dipérianthé, (flower) with double perianth in two whorls.

dipétale, dipetalous.

diphasé, diphase, diphasic, two-phase.

diphénylacétylène, *m.* diphenylacetylene, tolane.

diphénylamine, *f.* diphenylamine.

diphénylcarbinol, *m.* diphenylcarbinol, benzohydrol.

diphénylcétone, *f.* diphenyl ketone, benzophenone.

diphényle, *m.* biphenyl, diphenyl.

diphénylglyoxal, *m.* benzil, dibenzoyl.

diphtérie, *f.* diphtheria; — aviaire, bird pox, sorehead, chicken pox.

diphtérique, diphtheritic, diphtheric.

diphyle, diphyletic, (hybrid) from two types of descent.

diphylle, diphyllous, having two leaves.

diphyte, *f.* parasite passing from one plant to another in its development.

diplanetique, diplanetic, with two kinds of motile spores.

diplécolobé, (embryo) with cotyledons bent like S.

diplégie, *f.* diplegia, bilateral paralysis.

diplobacille, *f.* diplobacillus.

diplocaryon, *m.* nucleus with 2n chromosomes.

diplocaule, (plant) with terminal bud unable to form sex organs.

diplocytes, *f.pl.* diplocytes, fungi with two conjugated nuclei.

diploèdre, *m.* diplohedron, diploid.

diploédrique, diplohedral.

diplogénèse, *f.* diplogenesis.

diploide, *f.* diploid.

diplombique, diplumbic.

diplôme, *m.* diploma.

diplômé, *m.* holder of a diploma, charter.

diplophase, *m.* diplophase.

diplospore, *m.* diplospore, spore with only one cell sister.

diplostémone, diplostemonous.

diplostique, diplostic, diarch.

diplostome, diplostemonous.

diplosymétrique, symmetrical in relation to two rectangular planes.

diplotégie, *f.* diplotegia, a state of inferior capsules.

diploxylée, diploxylic, (vascular bundles) in which centrifugal part of the wood is secondary.

dipode, biped, dipodous.

dipropylique, dipropyl.

dipsacé, dipsacaceous.

diptère, two-winged, dipterous.

diptères, *m.pl.* Diptera.

diptérocécidie, *f.* dipterocecidia, galls produced by dipterous flies.

dipyridyle, *m.* bipyridyl, dipyridyl, dipyridine.

dire, to say, tell, relate, call, name, speak, predict; se —, to be said, be told, be spoken, profess to be; c'est-à-dire, that is to say, that is; comme qui dirait, as you might say, as much, so to speak; en —, to reproach; il n'y a pas à —, there is no denying it; pour ainsi —, so to speak; pour tout —, in a word; vouloir —, to mean.

dire, *m.* statement, opinion, saying, account, report; à vrai —, truly, to tell the truth; **au — de**, in the words of, according to.

direct, direct, straight(forward).

directement, directly, immediately.

directeur, *m.* director, manager, superintendent.

directeur, -trice, directing, guiding.

directif, -ive, directive, guiding.

direction, *f.* direction, management, directorship, managership, steering gear.

directive, *f.* instructions, directions.

directrice, *f.* directrix, directress, manageress, guide-vane.

dirent (dire), (they) said.

dirigeable, dirigible.

diriger, to direct, guide, send, manage, conduct; **se —**, to go; **se — vers**, to migrate, sweep over.

disaccharide, *f.* disaccharide, disaccharose.

disamare, *m.* double samara.

disbroder, to wash (silk) after dyeing.

discernable, discernible, visible.

discerner, to discern, distinguish.

disciflore, with flowers on disc.

discigyne, with an ovary implanted on a disc.

disciple, *m.* disciple, follower.

discipline, *f.* discipline.

discoïde, discoïdal, -aux, discoid(al), flat and circular, disk-shaped.

discolore, discolored, bicolor(ed), of two different colors.

discontinu, discontinuous, intermittent, interrupted, irregular.

discontinuation, *f.* discontinuance, cessation.

discontinuer, to discontinue, stop; **se —**, to be discontinued.

discontinuité, *f.* discontinuity.

disconvenance, *f.* unfitness, disproportion, incongruity.

disconvenant, incongruous.

disconvenir, to be unsuitable, deny.

discordance, *f.* discordance, inconsistency.

discordant, discordant, inharmonious.

discorde, *f.* discord, strife.

discorder, to be discordant, (colors) clash.

discourir, to discourse, talk.

discours, *m.* discourse, speech, talk.

discrédit, *m.* discredit.

discret, discreet, prudent, sly.

discrètement, discreetly, modestly.

discrétion, *f.* discretion; **à —**, at discretion, at will.

discussion, *f.* discussion, argument, debate.

discutable, debatable, disputable, questionable.

discuter, to discuss, debate; **se —**, to be discussed.

disert, eloquent, fluent.

disette, *f.* dearth, scarcity, poverty, want, famine.

disexuel -elle, with two sexes.

disgracié, deformed, disfigured, ill-favored.

disgracieux, -euse, awkward, ungracious, uncomely, homely, ill-favored.

disjoindre, to disjoin, sever, separate; **se —**, to come apart, become disjoined.

disjoint, disjoined, disjointed, separated, disjunct.

disjoncteur, *m.* circuit breaker, cutout, disjunctor.

disjonctif, -ive, (stamens) attached under the disc.

disjonctiflore, with flowers separated from one another.

disjonction, *f.* disjunction, separation, varying degrees of separation in organs.

dislocation, *f.* dislocation, dismemberment, fault.

disloquer, to dislocate, put (limb) out of joint, dismember, break up.

disomose, *f.* gersdorffite, nickel glance.

disparaissant, disappearing, ephemeral, vanishing.

disparaître, to disappear, vanish.

disparate, dissimilar, unlike.

disparité, *f.* disparity.

disparition, *f.* disappearance.

disparu, missing, extinct, vanished.

dispendieux, -euse, expensive, costly.

dispensable, dispensable.

dispensaire, *m.* dispensary.

dispense, *f.* exemption.

dispenser, to dispense, do away (with), free (from), diffuse (light), excuse, exempt, dispense with; **se —**, to exempt oneself, be dispensed, avoid, dispense (with).

disperme, dispermous, two-seeded.

disperser, to disperse, scatter; **se —**, to be dispersed, scatter.

dispersif, -ive, dispersive.

dispersion, *f.* dispersion, scattering, leakage, dissipation.

disponibilité, *f.* availability.

disponible, available, disposable.

dispos, fit, well, active, agile, alert, nimble.

disposé, disposed.

disposer, to dispose, arrange, place, lay out, use, make use of, have at one's disposal, have for sale, dispose (of); **se —,** to be disposed or arranged, prepare, get (make) ready, prepare to.

dispositif, *m.* arrangement, device, preparation, apparatus, means, disposition, enacting terms; **— de transport,** carrying device, conveyor.

disposition, *f.* disposition, arrangement, disposal, order, draft, bill of exchange, service, plan, humor, mood; **— des feuilles,** leaf arrangement.

disproportion, *f.* disproportion.

disproportionné, disproportionate.

disproportionnel, -elle, disproportional.

disproportionnément, disproportionately.

disproportionner, to disproportion.

dispute, *f.* dispute, discussion, disputation.

disputer, to dispute, oppose, quarrel over, contend with.

disque, *m.* disk, disc, valve head.

disruptif, -ive, disruptive.

dissécable, dissectible.

dissecteur, *m.* dissector.

dissection, *f.* dissection.

dissemblable, dissimilar, different, unlike.

dissemblance, *f.* dissimilarity, unlikeness.

dissemblant, dissimilar, unlike.

disséminateur, -trice, disseminating, disseminator.

dissémination, *f.* dissemination, dispersion, sowing, spreading (of germs).

disséminé, disseminated.

disséminer, to disseminate, scatter, sow, spread.

dissentiment, *m.* disagreement.

dissentir, to dissent.

dissépale, with two sepals.

disséquer, to dissect.

dissertation, *f.* dissertation, treatise.

disserter, to discourse, debate, dispute.

dissidence, *f.* dissension.

dissimilarité, *f.* dissimilarity.

dissimuler, to conceal, hide, dissemble

dissipation, *f.* dissipation, wasting.

dissiper, to dissipate, waste, dispel, divert, scatter; **se —,** to dissipate, amuse oneself, vanish, disappear.

dissociable, dissociable.

dissociation, *f.* dissociation.

dissocier, to dissociate, disunite, resolve.

dissolubilité, *f.* solubility, dissolubility, dissolvability.

dissoluble, soluble, dissoluble, dissolvable.

dissoluté, *m.* solution.

dissolutif, -ive, dissolvent, solvent.

dissolution, *f.* solution, disintegration, dissolution.

dissolvant, *m.* solvent, dissolvent.

dissolvant, dissolving, dissolvent, solvent, resolvent.

dissonance, *f.* dissonance, discord.

dissoudre, to dissolve, liquefy, disintegrate, decompose, disperse, melt; **se —,** to be dissolved, go in solution.

dissous, dissolved.

dissoute, dissolved.

dissymétrique, asymmetric(al), unsymmetrical, dissymetrical.

distal, distal, terminal.

distance, *f.* distance, interval, span.

distannique, distannic.

distant, distant, remote, afar.

distendre, to distend, stretch, strain.

distendu, distended, trained.

distension, *f.* distention, overstretching.

disthène, *m.* disthene, cyanite.

distiche, distichous, two-ranked.

distillable, distillable.

distillat, *m.* distillate.

distillateur, *m.* distiller.

distillation, *f.* distillation; **— fractionnée,** fractional distillation; **— lente des carbons,** carbonization of coal; **— sèche,** dry distillation.

distillatoire, distilling, distillatory; **plante —,** pitcher plant, nepenthes.

distillatum, *m.* distillate.

distillé, *m.* product of distillation.

distiller, to distill.

distillerie, *f.* distillery, distilling.

distinct, distinct, clear, different, separate.

distinctement, distinctly, plainly.

distinctif, -ive, distinctive, characteristic.
distinction, *f.* distinction, eminence; **sans —,** indiscriminately.
distinctivement, distinctively.
distinguable, distinguishable.
distingué, distinguished, eminent.
distinguer, to distinguish, discern, discriminate; **se —,** to distinguish oneself, be distinguishable, noticeable.
distique, distichous, two-ranked.
distomatose, *f.* rot (in sheep).
distordre, to distort, twist (one's ankle).
distordu, distors, distorted.
distorsion, *f.* distortion, twisting, torsion.
distraction, *f.* distraction, diversion.
distraire, to distract, divert, separate; **se —,** to amuse oneself.
distrayant, diverting, entertaining.
distribuer, to distribute, issue, deal out, arrange, lay out.
distributeur, *m.* distributer, dispenser, (steam) valve or regulator, guide-vane apparatus; **— alimentaire,** perforated feed pipe.
distributif, -ive, distributive.
distribution, *f.* distribution, issue, delivery, division, valve gear (for steam), timing gear.
distributivement, distributively.
district, *m.* district, region, division.
distyle, distylous.
dit (dire), (he) says, (he) said, told, called, aforesaid, so-called.
ditartrique, ditartaric.
dithymol biiodé, dithymol diiodide, thymol iodide.
dito, ditto.
ditopogamie, *f.* ditopogamy, stamen and stigma are in the same flower.
ditrope, making two turns on itself.
diurèse, *f.* diuresis.
diurétique, diuretic.
diurnal, -aux, diurnal, daily.
diurne, diurnal, daily; **insectes —s,** diurnal insects; **oiseau —,** day bird.
divaguer, to wander, digress, ramble.
divalent, divalent, bivalent.
divariqué, spreading, divaricate, extremely divergent.
divellent, divellent, tearing apart.
divergence, *f.* divergence, divarication.

divergent, diverging, divergent.
diverger, to diverge.
diverginervé, with radiating main nerves.
divers, various, varied, different, diverse, divers, several.
diversement, diversely, differently, variously.
diversicolore, variegated.
diversifié, diversified.
diversifier, to diversify, vary, variegate.
diversiflore, with flowers of more than one kind.
diversion, *f.* diversion, change.
diversité, *f.* diversity, variety, difference.
divertir, to divert; **se —,** to be diverted, amuse oneself.
dividende, *m.* dividend.
divi-divi, *m.* divi-divi (the pods of *Caesalpinia coriaria*).
divin, divine.
divinateur, *m.* diviner, soothsayer.
divinateur, -trice, foreseeing, prophetic.
diviniser, to deify.
divinité, *f.* divinity, deity.
divisé, divided, finely divided.
divisément, separately.
diviser, to divide, parcel, portion out, divide finely, powder; **se —,** to divide, be divided, break up; **— en degrés, to** graduate; **— par catégories,** to sort.
diviseur, *m.* divider, divisor.
divisibilité, *f.* divisibility.
divisible, divisible.
division, *f.* division, separation, block; **— en degrés,** graduation.
divisionnaire, divisional, divisionary.
divorce, *m.* divorce.
dix, ten, tenth.
dix-huit, eighteen, eighteenth; **en —,** eighteenmo, in octodecimo.
dix-huitième, eighteenth.
dixième, tenth.
dix-millième, *m.* ten-thousandth.
dix-neuf, nineteen, nineteenth.
dix-neuvième, nineteenth.
dix-sept, seventeen, seventeenth.
dix-septième, seventeenth.
dixylé, dixylic, (petioles) **having two** vascular bundles.

dizaine, *f.* ten, about ten; **deux —s,** twenty, a score; **une — de,** ten, about ten, half a score.

dl., *abbr.* (décilitre), deciliter.

dm., *abbr.* (décimètre), decimeter.

docile, docile, submissive, manageable.

docilement, readily.

docimasie, *f.* docimasy, assaying.

docimasiste, *m.* assayer.

docimastique, docimastic(al), pertaining to assaying.

docte, learned.

docteur, *m.* doctor; **— honoris causa,** honorary doctor's degree, doctor's degree.

doctorat, *m.* doctorate, degree of doctor.

doctrinal, -aux, doctrinal.

doctrine, *f.* doctrine.

document, *m.* document.

documentaire, documentary.

documenter, to document, support on documentary evidence, by facts.

dodécadaire, dodecahedral.

dodécaèdre, *m.* dodecahedron.

dodécagynie, *f.* plants with twelve pistils.

dodécandrie, *f.* flowers with twelve (10 to 20) like stamens.

dogme, *m.* dogma.

dogue, *m.* bloodhound; **— anglais,** mastiff.

doigt, *m.* finger, toe, digit; **— annulaire,** ring finger; **— auriculaire, petit —,** little finger; **— majeur, — du milieu,** middle finger; **— du pied,** toe; **— indicateur,** index finger, forefinger; **— postérieur,** hind toe; **à deux —s de,** within a hair's breadth of, on the verge of.

doigtier, *m.* finger stall, finger protector, common foxglove (*Digitalis purpurea*).

doit (devoir), (he) owes, must.

doit, *m.* debtor, debit (side).

doive (devoir), (he) should, ought.

dolabriforme, dolabriform, hatchet-shaped.

dolage, *m.* smoothing, planing.

doler, to shave, smooth, pare, whiten, slick, chip with an adz.

dolérite, *f.* dolerite.

doléritique, doleritic.

dolichosis, *m.* dolichosis, retardation of growth in length.

dolichostylé, long-styled, dolichostylous.

doloire, *f.* adz, broad ax.

dolomie, dolomite, *f.* dolomite.

dolomitique, dolomitic.

domaine, *m.* domain, crown land, property, sphere, power, province, demesne.

domatie, *f.* domatium.

dôme, *m.* dome.

domesticité, *f.* domesticity, domesticated state.

domestique, *m.* servant, hand.

domestique, domestic, tame, domesticated, house.

domicile, *m.* abode, dwelling.

dominant, dominant, prevailing, predominant, dominating.

dominateur, -trice, dominant.

domination, *f.* domination, sway, supremacy, dominance.

dominé, suppressed, overtopped, dominated.

dominer, to dominate, predominate, overlook, command, outgrow, suppress, overtop, prevail, rule, cover, shelter.

dommage, *m.* damage, injury, harm, hurt; **c'est —,** it is a pity.

dommageable, detrimental, prejudicial, injurious.

dompter, to tame, subdue, quell.

dompte-venin, *m.* (white) swallowwort (*Cynanchum vincetoxicum*).

don, *m.* gift, present, endowment.

donc, therefore, then, now, accordingly, and so.

donnée, *f.* datum, known quantity, information, idea, theme, factor, notion, admission; *pl.* data.

donner, to give, attribute, ascribe, grant, allow, yield, strike, run, fall, fight, apply; **se —,** to be given, give oneself up, have; **— du cône,** to taper; **— du froid,** to slacken the fire; **— la chaude à,** to heat, glow; **— le vent,** to turn on the blast, blow in; **— lieu à,** to give rise to; **— naissance à,** to give rise to; **— prise à,** to give a footing, give reason for; **— raison à,** to agree with; **— sur,** to open upon, overlook; **— vent à,** to give vent to.

donneur, -euse, *m. & f.* giver, donor, guarantor.

dont, whose, of whom, from whom, with whom, of which, from which, with which, whereof, whence.

dorade, *f.* goldfish, golden carp.

doradille, *f.* **— des murailles,** wall rue (*Asplenium ruta-muraria*).

dorage, *m.* gilding, glazing (cake), browning (meat).

doré, gilded, gilt, gold-plated.

dorénavant, henceforth, hereafter.

dorer, to gild; **se —,** to be gilded.

doreur, *m.* gilder.

dorge, *m.* burl.

dorine, *f.* golden saxifrage (*Chrysosplenium americanum*).

dormant, *m.* sleeper, post, frame, casing.

dormant, dormant, inactive, dead, fixed, standing, stagnant.

dormir, to sleep, be dormant, be latent, be still, be stagnant.

dormitif, -ive, dormitive, soporific.

dorsal, dorsal.

dorsifère, dorsiferous, borne on the back.

dorsifixe, dorsifixus, fixed on the back.

dorsinastie, *f.* dorsinasty, epinasty.

dorsiventrale, dorsiventral.

dort (dormir), (he) sleeps, (it) is dormant.

dorure, *f.* gilding; **— à la pile, — galvanique, — à mordant,** pigment gilding; **— au mercure, — au sauté,** amalgam gilding; **— au trempé,** wet gilding.

doryphore, *m.* potato beetle.

dos, *m.* back, ridge, top, back mark; **en — d'ane,** saddle-backed.

dosable, determinable, measurable (ingredient).

dosage, *m.* determination, preparation, quantitative analysis, proportioning, mixture, proportion, titration, dosage, liqueuring; **au —,** properly proportioned.

dose, *f.* proportion, amount, dose, quantity, portion.

doser, to determine, prepare, titrate, mix or proportion properly, dose, liqueur.

doseur, *m.* determiner, regulator, doser.

dosimétrie, *f.* dosimetry.

dosimétrique, dosimetric.

dosseau, *m.* slab, sidepiece, outside plank.

dossier, *m.* back (of chair), file, papers.

doter, to endow, dower, furnish, supply.

dothiénentérie, *f.* typhoid fever.

douane, *f.* custom, customhouse.

douanier, *m.* customhouse officer.

douanier, of customs.

douara, *n.* pearl millet (*Pennisetum glaucum*).

doublage, *m.* doubling, folding in half, lining.

double, *m.* double, duplicate, counterpart, copy.

double, double, duplicate; **— voie,** double track; **à — effect,** double-effect, double-acting; **à — fond,** double-bottomed.

doublé, *m.* plated wares.

doublé, doubled.

double-fond, *m.* double bottom.

doublement, *m.* doubling, folding.

doublement, doubly.

double-pesée, *f.* double weighing.

doubler, to double, fold, line, sheath, coat with, supplement, sail around; **se —,** to be doubled.

doublet, *m.* doublet.

doubleur, *m.* doubler, maker of plated ware.

doublier, *m.* (calico) doubler.

doublon, *m.* two-year old (horse, ox, bull).

doublure, *f.* lining, sheathing, scaling, scale, fold.

douçâtre, sweetish.

douce, see DOUX.

douce-amère, *f.* bittersweet (*Solanum dulcamara*).

douceâtre, sweetish.

doucement, sweetly, softly, gently.

doucereux, *m.* sweetishness.

doucereux, -euse, sweetish, insipid.

doucette, *f.* an inferior soda ash, corn salad (*Valerianella locusta* var. *olitoria*).

douceur, *f.* sweetness, softness, delight, pleasure, mildness, pleasantness; **en —,** cautiously, gently, mildly.

douche, *f.* douche, shower bath.

douci, *m.* grinding down.

doucin, *m.* wild apple rootstock.

doucir, to grind down, polish.

doucissage, *m.* fine grinding, grinding down, setting (of tool).

doué, endowed (with), gifted.

douelle, *f.* stave.

douer, to endow, bestow upon, furnish.

douille, *f.* socket, tubular casing, sleeve, bushing, bush, case.

douillet, soft, downy, tender, delicate.

douillon, *m.* wool of inferior quality.

douleur, *f.* pain, grief, woe, sorrow, soreness, throe.

douloureux, -euse, painful, woeful, dolorous, sore, sorrowful.

doum, *m.* doom palm, doum palm (*Hyphaene thebaica*).

dourine, *f.* dourine (disease of horses).

dousil, *m.* spigot, spigot hole.

doute, *m.* doubt, uncertainty, distrust; hors de —, beyond doubt; sans —, without doubt, doubtless(ly), unquestionably, probably.

douter, to doubt, question; — de, to doubt, distrust; se —, to suspect, surmise; il s'en doute, he suspects (it).

douteux, -euse, doubtful, dubious, uncertain, questionable.

douve, *f.* stave, ditch, fluke, flukeworm. spearwort (Ranunculus).

Douvres, *n.* Dover.

doux, *m.* sweet, softness, gentleness.

doux, douce, sweet, soft, gentle, mild, smooth, pleasant, easy running, easy, new, unfermented, fresh.

douzaine, *f.* dozen, about twelve.

douze, twelve, twelfth.

douzième, twelfth.

douzil, *m.* spigot, spigot hole.

doyen, *m.* dean, senior, oldest member.

Dr, *abbr.* (Docteur), Doctor.

dracéna, *m.* dragon tree (*Dracaena draco*).

drachen-ballon, *m.* kite balloon.

drachme, *f.* dram, drachma.

dragage, *m.* dredging, dragging.

drage, *f.* brewer's refuse grains, draff.

dragée, *f.* bonbon, sugar almond, sweetmeat, sugarplum, sugar-coated pill, small shot; — de cheval, buckwheat.

drageoire, *f.* groove and fillet.

drageon, *m.* sucker, root sucker, (radical) shoot.

drageonnement, *m.* sending up suckers.

drageonner, to sprout from roots, send out suckers.

dragon, *m.* dragon, dragoon, flying lizard.

dragonnier, *m.* dragon tree (*Dracaena draco*).

drague, *f.* dredge, dredger, drag, brewer's refuse grains.

draguer, to dredge, drag.

dragueur, *m.* dredger.

drain, *m.* drain, drain tile, small culvert.

drainage, *m.* drainage, draining.

drainer, to drain.

drap, *m.* cloth, sheet (for bed); — feutre, felted cloth.

drapant, *m.* pressing board (paper), cloth weaver.

drapeau, *m.* flag.

draper, to make (a fabric) into finished cloth, drape, cover, censure.

draperie, *f.* cloth manufacture, factory, or trade, drapery, cloth.

drapier, *m.* draper, cloth merchant.

drastique, *m.* drastic, strong purgative.

drastique, drastic.

drayage, *m.* shaving, fleshing, buffing.

drayer, to shave, flesh, scrape, buff.

drayoire, *f.* fleshing iron, shaving knife.

drayure, *f.* shaving, fleshing (of hide).

drêche, *f.* spent wash, spent malt, brewer's grains.

dréger, to ripple.

drépanium, *m.* drepanium, a sickle-shaped cyme.

dressage, *m.* dressing, squaring, training (animal).

dressé, dressed, erect, with point erect.

dressement, *m.* dressing, straightening.

dresser, to set, set up, dress, align, work, arrange, erect, raise, stand up, straighten, level, train, draw up, break in; se —, to rise, stand erect or upright.

dressoir, *m.* sideboard, dresser, dressing or straightening board, rod.

drille, *f.* (seed-planting) drill, bit, borer.

drogue, *f.* drug.

droguer, to drug, adulterate, dope (horse), physic.

droguerie, *f.* drugs, drug shop.

droguiste, *m.&f.* druggist.

droit, *m.* right, law, custom, duty, toll, fee; — de douane, (customs) duty; — d'entrée, entrance fee, import duty; — de parcours, right-of-way; — de paturage, right of pasture; — devant, straight ahead; — d'indemnité, claims of compensation, claims of indemnific tion; à bon —, with good reason; au — de, at right angles with; de —, by right, rightfully; tous —s réservés, all rights reserved.

droit, right, straight, plumb, upright, direct, righteous, dextro, right-handed.

droite, *f.* right hand, right, right line, straight line.

droitement, rightly, justly

drôle, funny, droll.

dromadaire, *m.* dromedary.

dropax, *m.* dropax, a depilatory.

drosère, *f.* sundew, lustwort (Drosera); — à feuilles rondes, moorgrass.

drousser, to scribble (wool).

dru, thick, strong, crowded, dense, sturdy, lively, vigorous, full-fledged (birds), (grass) grown thickly.

drupacé, drupaceous, drupelike.

drupe, *m.* stone fruit.

drupéole, *m.* drupel(et).

drupéolé, resembling a small drupe.

drupifère, bearing drupes.

druse, *f.* druse.

du, (de le), of the, from the, by the, in the, some, any.

dû (devoir), due, ought, owing, owed, owned; **il aurait — le lui dire,** he ought to have told him.

dû, *m.* due.

dualisme, *m.* dualism.

dualiste, *m.* dualist.

dualiste, dualistic.

dualistique, dualistic.

dualité, *f.* duality.

dubitatif, -ive, doubtful.

duboisine, *f.* duboisine, hyoscamine.

duc, *m.* duke, horned owl; **grand —** eagle owl.

duché, *m.* duchy, dukedom.

ductile, ductile, pliant, malleable.

ductilité, *f.* ductility, docility.

due (dû, devoir), due, ought, owing.

duitage, *m.* weft, woof.

duite, *f.* weft thread, weft yarn.

dulcamarine, *f.* dulcamarin.

dulcifiant, *m.* dulcifier.

dulcifiant, sweetening, dulcifying.

dulcification, *f.* sweetening, dulcification.

dulcifier, to sweeten, dulcify, correct.

dulcine, *f.* dulcin, sucrol.

dulcinée, *f.* dulcinea, sweetheart.

dulcitane, *f.* dulcitan.

dulcite, *f.* dulcitol, dulcite, dulcose.

dûment, duly, in due form.

dumping, *m.* dumping.

dune, *f.* dune, down, sand dune.

duodécimal, -aux, duodecimal.

duodécimfide, (calyx) divided into twelve denticulations.

duodécimo, in the twelfth place.

duodénum, *m.* duodenum.

duplicata, *m.* duplicate.

duplication, *f.* duplication.

duplicature des fleurs, *f.* double blossom.

duplicité, *f.* duplicity, double-dealing.

duquel (de lequel), of which, from which, of whom, from whom.

dur, hard, harsh, tough, difficult, rough, stiff, callous, stoutly.

durabilité, *f.* durability, durableness.

durable, durable, lasting.

duralumine, *f.* duralumin.

duramen, *m.* duramen, heartwood.

duraminisation, *f.* duraminisation, change from sapwood to heartwood.

durant, during, for, lasting; **— que,** while.

durci, hardened.

durcir, to harden, indurate, chill; **sol durci par le soleil,** earth baked by the sun.

durcissage, *m.* hardening.

durcissement, *m.* hardening, setting.

durcisseur, *m.* hardener.

durcisseur, -euse, hardening.

durée, *f.* duration, time, continuance, durability, life; **— de travail,** working hours.

durelin, *m.* British oak (*Quercus robur*).

durement, hard, harshly.

dure-mère, *f.* dura mater.

durer, to last, endure, continue, remain.

dureté, *f.* hardness, callosity, toughness, hardiness, harshness; **— temporaire,** temporary hardness (water).

durillon, *m.* callosity, corn, callus.

dus (pl. of dû), due, owing.

dus (devoir), (I) owed.

dut (devoir), (he) owed.

duvet, *m.* down, (of cloth) nap, fluff, soft hair, tomentum.

duveté, duveteux, -euse, downy, pubescent.

dyke, *m.* dike.

dyname, dynamie, *m.* a unit of work, 1,000 kilogrammeters.

dynamique, *f.* dynamics.

dynamique, dynamic(al).

dynamitage, *m.* dynamiting.

dynamite, *f.* dynamite.
dynamiterie, *f.* dynamite factory.
dynamiteur, *m.* dynamite maker, dynamite handler.
dynamitière, *f.* dynamite magazine.
dynamo, *f.* dynamo, generator.
dynamogénèse, *f.* dynamogenesis, production of energy.
dynamomètre, *m.* dynamometer.
dynamophore, energy-producing (food).
dyne, *f.* dyne.
dysenterie, *f.* dysentery.
dysentérique, dysenteric.
dyslysine, *f.* dyslysin.

dysogéogène, (rock) disintegrating very slowly.
dysorexie, *f.* loss of appetite.
dyspepsie, *f.* indigestion.
dysphotique, dysphotic (vegetation).
dyspnée, *f.* dyspnea.
dysprosium, *m.* dysprosium.
dyssymétrie, *f.* asymmetry, dissymmetry.
dyssymétrique, asymmetrical, unsymmetrical.
dystropes, *f.pl.* (insects) not favoring fertilization.
dytique, *m.* dytiscus, water beetle.
dytique, dytiscid.

E

eau, eaux, *f.* water, watering place, (clear) liquid, filtrate, gloss, sweat, tears, liquor, amniotic fluid, aqueous extract, flood; — **à blanchir**, whitewash; — **acidulée**, carbonated water, acidulated (aerated) water; — **à dégraisser**, scouring water; — **amidonnée**, starch water; — **ardente**, ardent spirit, distilled liquor; — **à refroidir**, cooling water; — **aromatique**, aromatic water, an aromatic distillate; — **blanche**, bran mash, bran and water, lead water; — **boneuse**, muddy water; — **bromée**, bromine water.

eau, — **calcaire**, calcareous water, limewater; — **carbonique**, carbonated or acidulous water; — **chaude**, hot water; — **chlorée**, chlorine water; — **chlorurée**, chlorinated water; — **courante**, running water; — **crue**, hard water; — **d'alimentation**, feed water, drinking water; — **d'amandes amères**, bitter almond water; — **d'ammoniaque**, ammonia water; — **d'aneth**, dill water; — **d'arrosage**, rinsing or flushing water.

eau, — **de brome**, bromine water; — **de carrière**, quarry sap; — **de chaux**, limewater; — **de chlore**, chlorine water; — **de ciel**, rain water; — **de colle**, size, glue water; — **de Cologne**, Cologne water, eau de Cologne; — **de conduite**, main or city water; — **de cristallisation**, water of crystallization; — **de décharge**, waste water; — **de fenouil**, fennel water; — **de fleur d'oranger**, orange-blossom water; — **de fontaine**, well water, spring water.

eau, — **de Goulard**, Goulard water; — **d'Égypte**, black hair dye; — **de Javel**, Javelle water, bleaching liquid; — **de jour**, surface water, open water; — **de la**

reine de Hongrie, Hungary water; — **de lavage**, wash water; — **de malt**, malt extract, malt water; — **de marée**, tidewater; — **de menthe verte**, spearmint water; — **de mer**, sea water; — **de mine**, mine water.

eau, — **de naffe,** — **de naphe**, orange-flower water; — **dentifrice**, tooth wash; — **de pluie**, rain water; — **de puits**, well (pump) water; — **de rase**, oil of turpentine; — **de rose**, rose water; — **de savon**, soapsuds, soap water; — **de Seltz**, Seltzer water, carbonated water, soda water; — **de source**, spring water, well water; — **de toilette**, toilet water; — **de trempe**, tempering or hardening water; — **de vie**, brandy, whisky; — **de ville**, city (town) water, tap water.

eau, — **distillée**, distilled water; — **distillée de fleurs d'oranger**, orange-flower water; — **distillée de rose**, rose water; — **divine de Fernel**, yellow mercurial lotion; — **dormante**, stagnant water; — **douce**, fresh (not salt) water, soft water; — **dure**, hard water; — **ferrugineuse**, iron water, chalybeate water; — **fraiche**, fresh (not stale) water, cool water; — **froide**, cold water, cooling water; — **gazeuse**, carbonated water, charged water, aerated water; — **glacée**, ice water; — **gradée**, graduated brine; — **grasse**, fat water, first sour water.

eau, — **hépatique**, sulphurous hepatic water, sulphureted water; — **hydriodatée**, iodine water; — **incrustante**, hard water, muddy water; — **magnésienne**, solution of magnesium carbonate; — **magnesifère**, magnesian water, bitter salt water; — **martiale**, chalybeate water; — **mère**, mother liquor; — **météorique**, rain water, atmospheric water; — **minérale,** mineral water; — **oxygénée,** hydrogen peroxide, hydrogen dioxide, oxygenated water; — **parfumée,** scented (perfumed) water; — **pluviale,** rain water; — **potable,** drinking water; — **profande,** subterranean water.

eau, — **régale,** aqua regia, nitrohydrochloric acid; — **salée,** salt water, sea water, an aqueous solution of sodium chloride; — **saline,** saline water; — **saumâtre,** brackish water; — **seconde,** lye water, weak nitric acid; — **séléniteuse,** selenium water, a natural water containing calcium sulphate; — **souterraine,** underground water, subsoil water, ground water; — **sulfurée,** hepatic water; — **sulfureuse,** sulphur water.

eau, — **thermale,** thermal water, hot-spring water; — **tombée,** (depth of) rainfall; — **trouble,** turbid water, cloudy water; — **usée,** waste water; — **vanne acide,** acidulous drain water; — **vierges,** (raw) brine; — **vive,** spring water, running water; **cours d'—,** stream, watercourse; **faire —,** to leak; **source d'—,** stream, watercourse; **vapeur d'—,** aqueous vapor; **—x ammoniacales,** ammoniacal liquor; **—x de cuite,** boiling liquor; **—x d'égout,** sewage; **—x des villes,** refuse water from cities, sewage; **—x faibles,** — petites, a weak solution, weak brine or wash; **—x fortes,** a strong solution; **—x grasses,** fat water, first sour water, water containing grease; **—x résiduaires,** waste waters, residual liquor, sewage; **—x sures,** sour liquor; **—x vannes,** liquid manure.

eau-de-vie, *f.* brandy, whisky, aqua vitae; — **de cidre,** applejack; — **de grain,** whisky; — **de vin,** brandy, wine spirit.

eau-forte, *f.* aqua fortis, nitric acid.

eau-mère, *f.* mother liquor.

eaux, *f.pl.* see EAU.

eaux-mères, *f.pl.* mother liquors.

eaux-vannes, *f.pl.* liquid manure.

ébahissement, *m.* astonishment, amazement.

ébarbage, ébarbement, *m.* trimming, dressing.

ébarber, to trim, remove rough edges, dress, scrape, clean, clip, pare, chip, shave.

ébarbeur, -euse, *m.&f.* barley awner, fettler.

ébarboir, *m.* scraper, scraping iron.

ébarbure, *f.* trimming, outer leaves (of salad), burr.

ébardoir, *m.* scraper, scraping iron.

ébats, *m.pl.* pastime, sport, gambol.

ébattré(s'), to sport, gambol.

ébauchage, *m.* rough-shaping, blocking out.

ébauche, *f.* rough-shaping, sketch, rough sketch, rough draft.

ébauché, formed (of buds).

ébaucher, to rough-shape, outline, block out, roughcast, sketch.

ébauchoir, *m.* dotting needle.

ébène, *f.* ebony (wood).

ébéner, to ebonize.

ébénier, *m.* ebony (tree) (*Diospyros ebenum*).

ébéniste, *m.* cabinetmaker, cabinetworker.

éberthien, -enne, pertaining to the typhoid bacillus.

éberthiforme, of the typhoid bacillus group.

ébeurrer, to remove the butter from.

ébiseler, to chamfer, bevel.

éblouir, to dazzle, fascinate; **s'—,** to be dazzled.

éblouissant, dazzling, transplendent.

éblouissement, *m.* dazzling, dimness of vision.

ébonite, *f.* ebonite, vulcanite, hard rubber.

éborgnage, *m.* nipping off of buds.

éborgner, to disbud (fruit tree).

ébotter, to cut back (branches), pollard (tree).

ébouche, *f.* bud, anlage, primordium.

ébouillanter, to scald.

ébouillir, to boil down, boil away.

éboulement, *m.* falling in, crumbling, downthrow, caving in, collapse, landslide.

ébouler, to fall in, crumble, give way, cave in, bring down, wreck.

ébouleux, -euse, crumbling, sliding, slipping.

éboulis, *m.* fallen material, pile of rubbish.

ébourgeonnement, ébourgeonnage, *m.* disbudding.

ébourgeonner, to nip the buds off.

ébouriffer, to disorder.

ébourrage, *m.* unhairing (of skins).

ébourrer, to unhair (skins).

ébout (ébouillir), (it) is boiling.

ebracté, ebracteate.

ébranchement, *m.* lopping (off).

ébrancher, to lop (off).

ébranché, branchless, lopped (tree).

ébranlement, *m.* shaking, agitation, commotion, shock, concussion, perturbation.

ébranler, to shake, agitate, disturb, unsettle, undermine, convulse, affect; **s'—,** to shake, be shaken, get under way, be agitated.

ébréché, notched.

ébrécher, to notch, damage, nick, break.

ébriété, *f.* drunkenness, ebriety, alcoholism.

ebronder, to deoxidize (iron wire).

ébrouage, *m.* washing (of wool).

ébroudage, *m.* drawing (of wire).

ébroudeur, *m.* wire drawer (worker).

ébroudir, to draw (wire) through a fine gauge.

ébrouement, *m.* sneezing, snorting (horse).

ébrouer, to wash (wool), bran, hull.

ébrouissage, *m.* washing (of wool).

ébruiter, to make known, noise abroad, report.

ébrutage, *m.* rough cutting (of diamonds).

ébulliomètre, *m.* ebullioscope.

ébullition, *f.* boiling, ebullition.

éburge, *n.* iron shovel.

éburine, *f.* eburine, eburite.

éburnation, *f.* eburnation.

éburné, resembling ivory; **substance —e,** dentine.

éburnification, *f.* conversion of bone into ivorylike mass.

écachement, *m.* crushing.

écacher, to crush, squash, squeeze, bruise, flatten, laminate, squeeze moisture from (paper).

écaillage, *m.* scaling, peeling.

écaille, *f.* scale, flake, chip, shell; **— de bourgeon,** bud scale; **— de bulbe,** bulb scale; **— séminifère,** seed scale.

écaillement, *m.* scaling, peeling.

écailler, to scale, chip off, flake (off), shell, exfoliate; **s'—,** to scale off, peel off, chip off, flake off.

écaillette, *f.* small scale, shell, or flake.

écailleux, -euse, scaly, squamous, scutate, fissile.

écale, *f.* shell, hull, husk, cod, peel, skin, parings.

écaler, to shell, hull, husk, shuck; **s'—,** to fall out of pods.

écalyptré, deprived of calyptra.

écalyptrocarpe, *f.* fruit without calyptra.

écang, *m.* scutching blade.

écarlate, scarlet.

écarner, to take off or remove the corners, angles.

écart, *m.* variation (of temperature), deviation, difference, error, margin, scarf joint, stepping aside, digression, strain, swerving, discrepancy; **— absolu,** absolute error; **— d'épaule,** shoulder strain (of horse); **— moyen,** mean error; **à l'—,** aside, apart; **mettre à l'—,** to set aside; **se tenir à l'—,** to keep out of the way.

écarté, removed, distant, isolated.

écartement, *m.* removal, separation, spreading apart, distance (apart), spacing, gauge (of tracks), setting aside, planting distance.

écarter, to remove, disperse, scatter, spread apart, separate, turn aside, set aside, divert, depart from; **s'—,** to deviate, depart, diverge, open, turn aside, be moved away, swerve, err, disperse, be dispersed.

écarteur, *m.* retractor.

écatir, to luster (cloth by pressing).

écatissage, *m.* glossing (of cloth).

écaudé, ecaudate, tailless.

ecboline, *f.* ecboline, ergotinine, an alkaloid.

ecbolique, ecbolic.

ecchymose, *f.* ecchymosis.

ecgonine, *f.* ecgonine.

échafaud, *m.* scaffold, platform, staging.

échafaudage, *m.* scaffolding.

échalas, *m.* vine prop, vine stake, hop pole.

échalassement, échalassage, *m.* propping.

échalote, *f.* shallot (*Allium ascalonicum*).

échancré, sloped, emarginate, notched.

échancrer, to indent, notch, cut, groove, hollow out.

échancrure, *f.* groove, indentation, hollowing (out), notch(ing), sweeping, curving.

échandole, *f.* shingle.

échange, *m.* exchange, interchange; **en — de,** in exchange for, in return for.

échangeable, exchangeable.

échanger, to exchange, interchange, barter, renew, replace.

échantignole, *f.* bracket, purlin cleat, chime.

échantillon, *m.* sample, specimen, pattern, templet, gauge, mold, matrix, standard fir or pine plank.

échantillonnage, *m.* sampling, verifying, gauging.

échantillonner, to sample, gauge, test, blend, prepare patterns, verify, check.

échappade, *f.* escape, partition in a kiln, slip (of the burin).

échappée, *f.* escape (of cattle), space, interval, width of an opening.

échappement, *m.* escape, escapement, leakage, outlet, flue, exhaust.

échapper, to escape, avoid, run off, shun; **s'—,** to escape, get away, vanish, forget oneself, slip out, come undone; **— à, — de,** to escape (from), get away from, be lost to; **laisser —,** to set free, let off, let escape, let slip, overlook, pass over.

écharbot, *m.* water chestnut (*Trapa natans*).

écharde, *f.* splinter, prickle.

échardonnage, *m.* clearing (ground) of thistles, (wool) picking.

écharnement, écharnage, *m.* fleshing (of hides).

écharner, to flesh (hides).

écharnoir, *m.* fleshing knife.

écharpage, *m.* scribbling, carding (of wool).

écharpe, *f.* brace, tie, jib of a crane, scarf, sash, sling; **en —,** slantwise, aslant.

écharper, to card (wool), pass a rope (sling) around, slash.

échasses, *f.pl.* stilts.

échassier, *m.* stilt(bird), stilt walker.

échauboulure, *f.* pimple, pustule.

échaudage, *m.* scalding, whitewash(ing), incubation.

échaude, *f.* welding heat; shriveled by the sun.

échaudé, scalded; **blé —,** wheat shriveled by the sun.

échaudement, *m.* shriveling (of wheat) by the sun.

échauder, to scald, scour, whitewash, burn; **s'—,** to become hot, inflame.

échaudillon, *m.* piece of iron to be welded.

échaudoir, *m.* scalding vat, scalding house.

échaudure, *f.* scald.

échauffage, *m.* heating.

échauffaison, *f.* heat rash.

échauffe, *m.* sweating, sweating room.

échauffé, heated, rotten (wool), moldy, fusty, smutty (corn, hay).

échauffement, *m.* heating, overheating, constipation.

échauffer, to heat, overheat, warm, vex, inflame, excite. cause fermentation; **s'—,**

to heat, become (over)heated, get hot, grow warm, warm up, become warm, be heated, become excited, dry rot (of wood), become rotten (of yarn).

échauffeur, *m.* heat exchanger.

échauffure, *f.* heat rash.

échéance, *f.* maturity, falling due, expiration.

échéant (echoir, to fall to one's lot), falling due; **le cas —,** should the occasion arise, in case of need.

échec, *m.* check, failure, defeat, repulse, blow, loss.

échelle, *f.* scale, ladder, graduation; **— à poisson,** fish ladder; **— d'eau,** water gauge; **— de proportion, — proportionnelle,** diagonal scale; **— graduée,** graduated scale; **— mobile,** sliding scale; **— réduite,** reduced scale.

échelon, *m.* step, rung of a ladder, round, stirrup.

échelonne, ladder-form.

échelonnement, *m.* disposition, distribution at intervals or in a series.

échelonner, to arrange in a series, place at intervals, distribute, proportion.

échenal, écheneau, échenet, *m.* gutter, channel, gate channel (of mold).

écheniller, to destroy caterpillars.

échenilloir, *m.* tree pruner.

écherra (échoir), (it) will fall due.

échet (échoir), (it) is falling due.

écheveau, *m.* hank, skein.

échevette, *f.* small skein (in France 100 meters).

échicoter, to remove stumps of branches

échine, *f.* spine, back(bone), echinus.

échiné, muricate, prickly, echinate.

échinococcose, *f.* infection with echinococcus.

échinocoque, *m.* echinococcus.

échinoderme, *m.* echinoderm.

échinon, *m.* cheese mold.

échinope, échinops, *m.* globe thistle (Echinops).

échinulé, echinulate, with small prickles.

échiquier, *m.* checkerboard, exchequer; **en —,** checkered.

écho, *m.* echo; **faire —,** to echo.

échoir, to fall due, mature, expire. happen, lapse, chance, fall to.

échoppe, *f.* dotting needle.

échouer, to run around, strand, fail, wreck.

éohoyait (échoir), (it) was falling due.

échu (échoir), due.

écide, *m.* aecidium.

écidiocytes, *f.pl.* aeciospores.

écidiole, *f.* spermatium.

écidiospore, *f.* aeciospore, spore formed in aecidium (sporocarp).

écimer, to top, pollard.

éclabousser, to splash, spatter.

éclair, *m.* lightning, flash, fulguration, blick.

éclairage, *m.* lighting, illuminating, illumination, light; — **au gaz**, gas lighting; **gaz d'** —, illuminating gas.

éclairant, illuminating, lighting, light-giving.

éclairci, thinned out by age.

éclaircie, *f.* clear (spot), space, clearing, thinning, glade, enlightenment, rift; — **forte**, heavy thinning; — **naturelle**, gradual opening of the crop by age; — **par le bas**, thinning in the lower story; — **par le haut**, thinning in the upper story; **faire une** —, to thin.

éclaircir, to clear, clear up, clarify, thin, elucidate, enlighten, brighten, polish; **s'** —, to clear (up), become clear, become thin, become bright, open out the crop.

éclaircissage, **éclaircissement**, *m.* clearing up, explanation, opening out the crop.

éclaircissant, *m.* thinning, clarifying.

éclaire, *f.* celandine (*Chelidonium majus*).

éclairé, lighted, illuminated, well-informed, educated, enlightened, open.

éclairement, *m.* illumination, lighting, clearness.

éclairer, to light, illuminate, enlighten, reconnoiter, watch, throw light on, instruct, inform, light up, shine, lighten, flash; **s'** —, to be lighted, become enlightened, inform oneself.

éclaireur, *m.* scout.

éclampsie, *f.* eclampsia.

éclat, *m.* burst, bursting (ot noise), report, crash, fragment, splinter, shiver, chip, luster, brightness, brilliancy, flash, crack, shake (in wood), splendor, pomp, glory, glitter; *pl.* chips, chippings, clippings, shavings, — **de bois**, splinter; — **métallique**, metallic luster.

éclatant, bursting, brilliant, bright, sparkling, dazzling, shining, loud, shrill.

éclatement, *m.* bursting.

éclater, to burst, shiver, split, splinter, crack, thunder, explode, break out, divide, partition (roots of plant), shine, flash, glitter, spark.

éclateur, *m.* spark gap.

éclimètre, *m.* gradient recorder, gradometer.

éclipse, *f.* eclipse.

éclipser, to eclipse, obscure; **s'** —, to become eclipsed.

écliptique, *f.* ecliptic.

éclissage, *m.* fishing, scarf-jointing.

éclisse, *f.* splinter, splint, fish, fishplate, (wooden) wedge, split wood.

éclisser, to put (limb) in splints.

éclore, to open (out), bloom, dawn, appear, hatch out, unfold, come out, burst forth.

éclosion, *f.* dawning, appearance, germination (of spore), opening (flowers), blossoming, bloom, unfolding, hatching, hatch, blowing.

écluse, *f.* sluice, lock, floodgate, dam.

ecobuage, *m.* removal of sods, weeding, weed burning.

écobuer, to remove sods and burn the covering herbage, burn the soil covering, assort.

école, *f.* school, schoolhouse, blunder.

écolier, *m.* scholar, pupil, schoolboy.

écollage, *m.* fleshing (of hides).

écologie, *f.* ecology, biology.

écologique, ecological, biological.

écomoie, *m.* steward, bursar.

économe, economical.

économie, *f.* economy, organism, animal body; *pl.* savings, thrift; — **forestière**, forest economy; — **politique**, political economy, national economy.

économique, *f.* economics.

économique, economic(al), cheap.

économiquement, economically.

économiser, to economize, save.

économiste, *m.* economist.

écope, *f.* ladle, scoop.

écorçage, *m.* stripping, barking (of trees), peeling (of oranges), husking (rice).

écorce, *f.* bark, rind, peel, husk, (earth's) crust, shell, cortex, skin, surface; — **à tan**, tanbark, tanning bark; — **de bigarade**, bitter orange peel; — **de bourdaine**, frangula, alder buckthorn bark; — **de chêne**, oak bark; — **de citron**, lemon peel: — **de géoffrée**, cabbage-tree bark; — **de limon**, lemon peel; — de margousier, margosa bark.

écorce, — de racine de berbérides, berberis-root bark; — de ronce noire, blackberry bark; — de tan, tanbark; — de Winter, Winter's bark; — éleuthérienne, cascarilla, eleuthera bark, (*Croton eluteria*); — jeune et luisante, smooth bark, silver bark; — molle, soft bast, phloem without fibers; — rude et grossière, coarse bark; — subereuse, corky bark; — vive, phellogen.

écorcement, *m.* see ÉCORÇAGE.

écorcer, to strip, bark, peel, husk, ross.

écorceuse, *f.* barker, barking machine.

écorchement, *m.* excoriation, skinning, flaying.

écorcher, to skin, flay, peel off, taste rough, grate on, fleece, abrade, murder (a language).

écorchure, *f.* excoriation, abrasion.

écorcier, *m.* place where tanbark is kept.

écorner, to remove the horns of (animal), break off the corner of, chip, curtail, cut down (allowance), trim.

écornure, *f.* chip, fragment, corner broken-off.

écossais, -aise, *m.&f.* Scotch, Scot.

écossais, Scottish.

Écosse, *f.* Scotland.

écosser, to husk, hull, shell.

écot, *m.* stump, lopped tree, quota, share.

écotage, *m.* stemming (of tobacco).

écoté, lopped (tree or branch).

écoter, to stem (tobacco leaves, etc.).

écoulé, of last month, ult(imo).

écoulement, *m.* outflow, flowing (off), running (out), exudation (of sap), drainage, draining, discharge, vent, sale, disposal; jaugeage par —, graduation for delivery.

écouler, to flow out, run out, drain off, let out, pass, sell, disperse, elapse, pass on, empty, discharge, dispose of; s'—, to flow (away) out, run out, pass (away).

écourgeon, *m.* winter barley.

écourter, to shorten, curtail, dock, clip (ears).

écoussage, *m.* stain, dark spot.

écoutant, *m.* hearer, auditor.

écoutant, listening.

écouter, to listen (to), hear, hearken to, pay attention to.

écouteur, *m.* receiver, headphone.

écouvillon d'ouate sterile, *m.* swab.

écran, *m.* screen, baffle plate, baffle.

écrasage, crushing, squashing.

écrasé, crushed, flat, flattened.

écrasement, *m.* crushing, squashing, crush, overwhelming; — d'une chaudière, implosion of a boiler.

écraser, to crush, bruise, flatten, squash, squeeze, overwhelm, masticate (rubber).

écrasite, *f.* ecrasite.

écrémage, *m.* écrémaison, *f.* skimming, scumming, separating.

écrémé, skimmed, separated, skim (milk).

écrémer, to skim, cream, take off the cream, separate.

écrémette, *f.* skimmer, skimming ladle.

écrémeuse, *f.* separator, skimmer.

écrémoir, *m.* écrémoire, *f.* skimmer.

écrêtement, *m.* removal of (cocks') combs, knocking off the tops (of flowers, corn).

écrevisse, *f.* crawfish, crayfish, crab, lobster, lever-grip tongs, cancer.

écrier, to scour (iron wire); s'—, to cry out, exclaim.

écrin, *m.* case, casket.

écrire, to write, spell.

écrit, *m.* writing, pamphlet, work.

écrit, written.

écriteau, *m.* bill, placard.

écriture, *f.* entry, item, (hand-)writing; *pl.* accounts, books, writing.

écrivain, *m.* writer, clerk, author.

écriveur, *m.* writer.

écrou, *m.* screw nut, female screw; — aile, — à oreilles, thumb-nut screw, wing-nut screw, wing screw, fly nut; — d'arrêt, check nut; — de serrage, clamp nut, set nut.

écrouelles, *f.pl.* scrofula.

écrouellet, *m.* mole.

écrouir, to cold-hammer, hammer-harden.

écrouissement, écrouissage, *m.* cold-hammering, hammer-hardening.

écroulement, *m.* falling in or down, downfall.

écrouler, to crumble, fall in; il s'écroule, it collapses.

écroûtage, *m.* scarifying (of land).

écroûter, to remove the crust from, scarify (land).

écru, raw, natural-colored, unbleached, badly worked (iron).

ectasie, *f.* dilatation, distension, ectasis.

ectauxèse, *f.* ectauxesis, the growth of an organ outwards.

ectobaside, *f.* external basidium.

ectoblaste, *m.* ectoblast.

ectocline, *f.* pulpy receptacle.

ectocyste, *m.* ectocyst.

ectodermique, ectodermic, ectodermal.

ectoparasite *f.* ectoparasite, an exterior parasite.

ectophage, *m.* ectoparasite.

ectoplasme, *m.* ectoplasm, ectosarc.

ectospore, *f.* stylospore, a spore borne on a filament.

ectostroma, *f.* stains (infection) marked on leaves.

ectothèque, ectothecal, naked-spored.

ectotrophe, ectotrophic.

écu, *m.* shield, crown (five-franc piece), scutum; *pl.* money; — d'eau, pennywort; herbe aux écus, moneywort, creeping Jennie (*Lysimachia nummularia*).

écueil, *m.* reef, shelf, rock.

écuelle, *f.* ecuelle, bowl, saucer, porringer.

écumage, *m.* skimming, scumming.

écumant, foaming, frothing, scummy.

écume, *f.* foam, froth, lather, scum, dross, glut; — de fonte, kish; — de mer, meerschaum, sea foam.

écumer, to skim, scum, skim off, foam, froth.

écumeresse, *f.* large skimmer.

écumette, *f.* small skimmer.

écumeur, *m.* skimmer.

écumeux, -euse, foamy, frothy, scummy.

écumoire, *f.* skimmer, scummer.

écurage, *m.* cleaning, scouring.

écurement, *m.* water furrow, scouring.

écurer, to clean, scour.

écureuil, *m.* squirrel.

écurie, *f.* stable.

écusson, *m.* bud for grafting, shield (embryo of grasses), scutellum.

écussonage, *m.* shield budding.

écussoner, to bud.

écussonnoir, *m.* budding knife.

écuyer, *m.* equestrian, prop, stay (of tree).

eczéma, *m.* eczema.

édaphique, edaphic.

édaphophytes, *f.pl.* edaphophytes, plants with assimilation organs in air.

édelweiss, *m.* edelweiss.

édenté, edentate, toothless.

édestine, *f.* edestin, globulin constituent of wheat flour.

édicule, *m.* small edifice, shelter.

édification, *f.* building (up), edification, enlightenment.

édifice, *m.* building, edifice, structure, fabric (of society).

édifier, to build, build up, erect, edify, enlighten.

éditer, to publish, edit.

éditeur, -trice, *m.&f.* editor, publisher.

édition, *f.* edition, editing, publishing.

édredon, *m.* eider down.

éducateur, *m.* educator, breeder, raiser, grower; — d'abeilles, beekeeper.

éducation, *f.* education, raising, breeding, rearing, keeping, training, exhaustion (of steam); — des bois, rearing trees, tending trees; — du peuplement, tending of crop.

éducatrice, *f.* educator.

édulcorant, edulcorant, sweetening (substance).

édulcoration, *f.* sweetening.

édulcorer, to edulcorate, sweeten.

édule, edible, esculent.

effacement, *m.* turning, flattening, effacement, dimness.

effacer, to efface, obliterate, erase, blot out, disappear, — dans le lit du vent, to come up into the wind.

effaçure, *f.* obliteration, erasure, effacement.

effaner, to strip off the leaves.

effectif, -ive, effective, actual, real, efficacious; — du gibier, stock of game.

effectivement, effectively, sure enough, in effect, in reality, really, actually.

effectuer, to effect, work out, make, carry out; s' —, to be accomplished, take place, be effected.

effervescence, *f.* effervescence.

effervescent, effervescent.

effet, *m.* effect, action, purpose, result, bill, public funds; *pl.* goods, belongings; — brisant, rending (shattering) effect; — parasitaire, interference, disturbance; — utile, useful effect, useful output; —s publics, securities, stocks; à — retarde, delay action; à cet —, for this purpose; à double —, double acting; à l'— de, for the purpose of; à simple —, single acting; en —, in effect, in reality, indeed, in fact; sans —, ineffective.

effouillaison, *f.* leaf fall.

effeuillement, *m.* defoliation.

effeuiller, to strip off leaves.

efficace, efficacious, effective, effectual.

efficacement, efficaciously, effectively, efficiently.

efficacité, *f.* efficacy, effectiveness, efficiency.

efficient, efficient.

effilage, *m.* drawing out, raveling out.

effilé, drawn out, tapered, tapering, thinned, slim, slender, raveled, fringed, virgate; **tube —,** drawn-out tube.

effiler, to draw out, taper, thin, ravel (out), unravel; **s'—,** to be drawn out, taper, ravel out.

effilocher, to ravel out, cut, tear (rags), break (wool, cotton waste).

efflanqué, lean-flanked, raw-boned (animal), gaunt.

effleurement, effleurage, *m.* shaving, buffing, skimming, graze.

effleurer, to take off the surface, skim, scrape, shave, graze, brush, touch, touch on, strip of flowers, plow lightly; **s'—,** to be scraped.

effleurir, to effloresce, weather, disintegrate.

effloraison, *f.* flowering.

efflorescence, *f.* efflorescence, rash, eruption.

efflorescent, efflorescent.

effluent, effluent.

effluve, *m.* effluvium, emanation, electric (brush) discharge.

effondrement, *m.* subsoiling, trenching.

effondrer, to break in, break down, dig deep, plow deep, subsoil, draw (fowl), cut (fish); **s'—,** to give way, fall in, slump (prices).

effondrilles, *f.pl.* sediment, dregs.

efforcer, to strive, endeavor, try, attempt.

effort, *m.* effort, force, endeavor, stress, strain, rupture; **— de cisaillement,** shearing stress; **— de remorquage,** towing force; **— de rupture,** breaking stress; **— de tension,** tensile stress; **— de torsion,** torque; **— de traction,** tractive effort, pull; **sans —,** easily.

effraie, *f.* (dark-breasted) barn owl.

effravant, frightful, appalling, dreadful.

effraye, *f.* see EFFRAIE.

effrayer, to frighten, terrify; **s'—,** to take fright; **il s'effraie,** he becomes frightened.

effréné, unbridled, ungovernable.

effritement, *m.* exhaustion (of land), crumbling, disintegration, weathering.

effriter, to render friable, cause to crumble, disintegrate, exhaust (soil); **s'—,** to crumble, become exhausted.

effroi, *m.* fright.

effronté, shameless, bold.

effroyable, frightful, awful.

effroyablement, frightfully, tremendously.

effruiter, to strip fruit from tree.

effulguration, *f.* flashing, glowing.

effusion, *f.* effusion.

égagropile, *m.* wool ball.

égal, *m.* equal, equivalent; **à l'— de,** equal to, in comparison with, like; **cela m'est —,** it is all the same to me.

égal, equal, uniform, even, same, alike, like.

également, equally, likewise.

égaler, to equalize, equal, level, make even.

égalisation, *f.* equalization, leveling.

égaliser, to equalize, even, smooth, make level, demolish.

égaliseur, *m.* equalizer, regulator.

égalisoir, *m.* separating sieve.

égalitaire, equalizing, leveling, at par.

égalité, *f.* equality, evenness, regularity; **à — de . . . ,** where there is equality of . . . , with equal . . . .

égard, *m.* regard, respect, consideration; **à l'— de,** with respect to, with regard to, regarding, as to; **à cet —,** in this respect; **à tous —s** in all respects; **eu — à,** considering.

égaré, stray, strayed, erring, lost, mislaid, misled.

égarement, *m.* mislaying, wandering, error, wildness, mistake.

égarer, to lead astray, mislead, let wander, mislay, bewilder, derange; **s'—,** to go astray, be mislaid, lose one's way, err, lose oneself.

égaux, *m.pl.* see ÉGAL.

égayer, to cheer, gladden, lighten, enliven, prune, lop, rinse, water.

égermage, *m.* degerming.

égermer, to remove the germ of.

égide, *f.* protection, aegis.

églanduleux, -euse. eglandular.

églantier, *m.* wild brier, dog rose (*Rosa canina*); — odorant, sweetbrier (*Rosa eglanteria*).

églantine, *f.* dog brier, wild rose, sweetbrier, dog rose (*Rosa canina*).

église, *f.* church.

égoïsme, *m.* egotism, egoism.

égoïste, egotistic, selfish.

égorgement, *m.* sticking, cutting throat.

égout, *m.* sewer, drain, dripping, drip, gutter, channel, incline, slope, vent, sirup (water).

égouttage, égouttement, *m.* draining, dripping.

égoutter, to drain, strain, let drip; s'—, to drain, drip.

égoutteur, *m.* drainer, drain.

égouttoir, *m.* drainer, draining rack, plate rack, draining board.

égoutture, *f.* drainings, drippings.

égrainage, *m.* shelling (corn), picking (grapes), ginning (cotton).

égrappage, *m.* picking (grapes) from the bunch, cleaning (ore).

égrapper, to pick (grapes) from the bunch, clean (ore).

égrappeur, *m.* picker.

égrappoir, *m.* fruit-crushing and stalk-removing machine.

égratigner, to scratch, rake over (the soil).

égratignure, *f.* scratch.

égrenage. égrènement, *m.* shelling, picking, ginning.

égrener, to shell (grains), stone, core, dehusk, shell out, pick (grapes from the bunch), gin (cotton), tell (of beads); s'—, to fall, drop from, seed, crack (steel).

égrisage, *m.* grinding (glass or diamond).

égrisée, *f.*, égrisé, *m.* diamond powder, bort.

égriser, to grind (glass, marble).

égrotant. unhealthy.

égrugeage, *m.* seeding (grapes), pilling (flax), mealing (gunpowder).

égrugeoir, *m.* mortar, rubber, mealer, rippling bench.

égruger, to grind, meal, pulverize, seed.

égrugeur, -euse, grinding, pulverizing.

égrugeure, *f.* grindings.

égyptien, -enne, Egyptian.

eh bien, well, well then (Oh, well), eh, ah, oh.

eider, *m.* common eider.

eisodiale, eisodial, anterior (as outer pore of stomate).

éjecter, to eject.

éjecteur, *m.* ejector.

éjecteur, ejecting.

éjection, *f.* ejection.

éjointer, to clip wings of (duck).

élaborant, élaborateur, -trice, elaborative, elaborating.

élaboration, *f.* elaboration. production.

élaborer, to elaborate, work out, prepare, render assimilable.

élagage, *m.* pruning, lopping, branches lopped off; — natural, self-pruning.

élaguer, to prune, lop, trim.

élagueur, *m.* pruner.

élaïdine, *f.* elaidin.

élaïdique, elaidic.

élaile, *f.* ethyline.

élaïne, *f.* elain, olein.

élaïomètre, *m.* oleometer, elaeometer.

élaioplaste, *f.* elaioplast.

élaldéhyde, *m.* paraldehyde.

élan, *m.* spring, start, dash, jump, impulse, burst, flight, buoyance, ardor, glow (of enthusiasm), elk, moose-deer.

élancé, slender, slim, tall, lank.

élancer, to spring, rush, shoot (forth), dart, dash, fly. bound; s'—, shoot forth, spring, throw oneself.

élargir, to enlarge, widen, stretch, extend, spread, broaden.

élargissement, *m.* widening, enlargement, dilatation.

élarvement, *m.* removal of larvae.

élasticité, *f.* elasticity.

élastine, *f.* elastin.

élastique, elastic.

élatère, *m.* click beetle, elater (of liverwort)

élatéridé, *f.* click beetle.

élatérine, *f.* elaterin.

élatérite, *f.* elaterite.

élatérium, *m.* squirting (wild) cucumber (*Ecballium elaterium*).

élatéromètre, *m.* elatrometer.

élatine, *f.* waterwort (Elatine).

élavage, *m.* washing (out).

élaver, to wash (out) rags.

élcotropisme, *m.* elcotropism.

électif, -ive, elective, selective.

élection, *f.* election, choice, preference; **d'—,** chosen, preferred, best.

électivement, electively, by choice.

électivité, *f.* electivity, selectivity.

électre, *m.* electrum.

électricien, *m.* electrician.

électricisme, *m.* electric (system) phenomena (in general).

électricité, *f.* electricity.

électrification, *f.* electrification.

électrifier, to electrify.

électrique, electric, electrical.

électriquement, electrically.

électrisable, electrifiable.

électrisant, electrifying.

électrisation, *f.* electrification.

électriser, to electrify, charge with electricity, electrize; **s'—,** to become electrified.

électriseur, *m.* electrifier, electrotherapist.

électro-aimant, *m.* electromagnet.

électrochimie, *f.* electrochemistry.

électrochimique, electrochemical.

électrode, *f.* electrode.

électrodynamique, *f.* electrodynamics.

électrodynamique, electrodynamic.

électrogène, generating (electricity).

électrolysable, electrolyzable.

électrolysation, *f.* electrolyzation.

électrolyse, *f.* electrolysis.

électrolyser, to electrolyze.

électrolyte, *m.* electrolyte.

électrolytique, electrolytic.

électromagnétique, electromagnetic.

électromagnétisme, *m.* electromagnetism.

électrométallurgie, *f.* electrometallurgy.

électromètre, *m.* electrometer.

électromoteur, *m.* electromotor, electric motor.

électromoteur, -trice, electromotive.

électron, *m.* electron.

électronégatif, -ive, electronegative.

électronique, electronic.

électrophore, *m.* electrophorus.

électroscope, *m.* electroscope.

électroscopique, electroscopic.

électrosis, *m* electrosis, reaction of a plant to an electrical current.

électrosoudure, *f.* electric welding.

électrostatique, *f.* electrostatics.

électrostatique, electrostatic.

électrotactisme, *m.* electrotaxis.

électrotechnique, *f.* electrotechnics.

électrotechnique, electrotechnic(al).

électrothérapie, *f.* electrotherapy.

électrotrieuse, *f.* magnetic (ore) separator.

électrotropisme, *m.* electrotropism in plant functions.

électrotype, *m.* electrotype.

électrotypie, *f.* electrotyping, electrotype.

électrotypique, electrotypic.

électrum, *m.* electrum.

électuaire, *m.* electuary (a confection); **— de séné composé,** confection of senna (sirup and tamarind pulp).

élégamment, elegantly.

élégance, *f.* elegance.

élégant, elegant, nice.

élément, *m.* element, rudiment, part, ingredient, constituent, cell (of a storage battery); **— constitutif,** constituent part, constituent in a crop; **— de rechange,** reserve or spare part; **— minéral,** ash (mineral) constituent; **— nutritif,** foodstuff, nutrient, nourishing substance; **— crystallogène,** crystal-producing element; **—s de réduction,** end cells.

élémentaire, elementary, rudimental.

élémi, *m.* gum elemi.

éléolé, *m.* a preparation with oil as base.

éléoleucite, *f.* elaioplast, plastid-forming oil.

éléolithe, *m.* eleolite, elaeolite.

éléoplaste, *m.* elaioplast.

éléoptène, *m.* eleoptene, elaeoptene.

éléosaccharum, *m.* elaeosaccharum.

éléphant, *m.* elephant.

éleuthérien, -enne, eleuthera, cascarilla.

éleuthérogyne, (plant) with free ovary.

éleuthérogynie, *f.* plant with free ovary.

éleuthéropétale, eleutheropetalous, having free petals.

éleuthérosépale, eleutherosepalous, with distinct sepals.

élevage, *m.* raising, breeding, rearing (of animals), stock farm, ranch; **— avicole,** poultry husbandry.

élévateur, *m.* elevator, hoist; **— à godets, — à augets,** bucket elevator.

élévation, *f.* elevation, increase, rise, involution, raising, rising, altitude.

élévatoire, elevatory, elevating.

élève, *m.&f.* student, pupil, young animal, seedling; *f.* raising, breeding (animals and plants); — **bénévole,** special student.

élevé, high, lofty, raised, elevated; **moins** —, of lower growth.

élèvement, *m.* raising.

élever, to raise, step up, elevate, erect, bring up, educate, grow, breed, produce, rear, exalt, lift, increase; **s'** —, to rise, arise, be raised, be produced, run up, amount, ascend.

éleveur, *m.* (stock) breeder.

éleveuse, *f.* brooder, incubator.

élevure, *f.* pimple, pustule.

éligible, eligible.

élimé, threadbare.

éliminateur, -trice, eliminating, eliminative.

élimination, *f.* elimination.

éliminatoire, eliminatory.

éliminer, to eliminate, expel, remove.

élire, to elect, choose.

élixation, *f.* elixation, steeping, seething.

élixir, *m.* elixir; — **de propriété,** tincture of aloes and myrrh; — **fébrifuge d'Huxam,** Huxham's tincture of bark (compound tincture cinchona composita).

ellagique, ellagic.

elle, she, her, it.

elléboro, *m.* hellebore (Helleborus).

elléborine, *f.* helleborin, helleborine.

elle-même, herself, itself.

elles, they, them.

elles-mênes, themselves.

ellipse, *f.* ellipse, ellipsis.

ellipsoïdal, -aux, ellipsoidal, ellipsoid.

ellipsoïde, *m.* ellipsoid.

elliptique, elliptic(al), flat.

éloge, *m.* eulogy, praise, commendation; **faire l'** — **de,** to speak in praise of.

élogieux, -euse, eulogistic, laudatory.

éloigné, removed, far, far away, distant, remote, averse.

éloignement, *m.* removal, remoteness, distance, absence, aversion, neglect.

éloigner, to remove, defer, pull away, divert, turn aside, banish, avert, dismiss, estrange; **s'** —, to withdraw, go away, send away, depart, be distant, recede, get farther, be withdrawn, be different, be averse.

élongation, *f.* elongation, branches and roots grown in length.

éloquence, *f.* eloquence.

éloquent, eloquent.

élu (élire), elected, elect.

élucider, to elucidate.

éluder, to elude, evade.

élyme, *m.* lyme grass (Elymus).

élytre, *m.* elytrum (the horny outer wing of insects), wing covering, vagina; — **s lisses,** smooth wing cases; — **s ponctuées en séries,** striate-punctated elytrae; — **s ponctué-striées,** punctate-striate elytrae; — **s sillonnées,** furrowed wing cases; — **s striées,** channeled wing cases; — **s strié-ponctuées,** striate-punctated elytrae; — **s troquées,** truncated elytrae; — **s unies,** smooth wing eases.

élytricule, *m.* elytriculus, a floret in Compositae.

émacié, emaciated, very thin, wasted.

émail, *m.* enamel, glaze; — **de cobalt,** smalt.

émaillage, *m.* enameling, glazing.

émailler, to enamel, glaze, dot, fleck.

émaillerie, *f.* enameling.

émailleur, *m.* enameler.

émaillure, *f.* enameling, enamelwork, spot, patch.

émanatif, -ive, emanative.

émanation, *f.* emanation, efflux, offspring.

émancipateur, -trice, emancipating

émancipation, *f.* emancipation, liberation.

émaner, to emanate, issue forth, proceed.

émaraudin, emerald green.

émarginé, emarginate, notched.

émasculer, to castrate, geld.

émaux, *m.pl.* see ÉMAIL.

emballage, *m.* packing, packing case.

emballement, *m.* racing, excitement.

emballer, to pack, wrap up; **s'** —, to bolt run away, (of a reaction) become violent, race (engine).

emballeur, *m.* packer.

emballotter, to pack, bale.

embarcadère, *m.* loading place.

embarcation, *f.* small vessel, craft.

embargo, *m.* embargo.

embarillage, *m.* barreling (of goods).

embariller, to barrel.

embarquer, to ship, embark.

embarras, *m.* obstruction, encumbrance, hindrance, impediment, embarrassment; — **gastrique,** stomach trouble.

embarrassant, embarrassing, perplexing.

embarrasser, to obstruct, embarrass, trouble, inconvenience, puzzle; **s'— de,** to be embarrassed, trouble oneself about.

embase, *f.* base, shoulder, seating, flange.

embassure, *f.* walls (of a glass furnace).

embauchoir, *m.* last, boot tree.

embaumement, *m.* embalming.

embaumer, to embalm, perfume.

embecquer, to feed (bird), cram (fowl).

embellir, to embellish, beautify, improve in appearance.

embellissement, *m.* embellishing, improving, adornment.

emblaison, *m.* seed time.

emblavage, *m.* sowing.

emblée (d'), at the first onset, at once, with ease.

emblème, *m.* emblem, device.

emboire, to smear or coat with grease or wax; **s'—,** to soak in.

emboîtement, *m.* encasing, fitting, jointing, socket.

emboîter, to fit, interlock, fit into, nest, joint, brush, interlock, dovetail, mortise, clamp; **s'—,** to fit into each other, go together; **il emboite le pas,** he falls into step.

emboîture, *f.* clamp, collar, socket, joint, joining, box.

embonpoint, *m.* corpulence, stoutness.

embouche, *m.* rich pasture land.

emboucher, to fatten (animal).

embouchure, *f.* mouthpiece, mouth, opening, anastomosis.

embout, *m.* tip, ferrule, mouthpiece (of speaking tube).

embouteillage, *m.* bottling.

embouteiller, to bottle.

embouti, stamped, pressed, covered with metal.

emboutir, to stamp, press, shape, cover with metal.

emboutissage, *m.* stamping, sheathing with metal.

embranchement, *m.* branching (off), junction, branch, branch line, bifurcation.

embrancher, to join, connect; **s'—,** to join each other, branch, be joined, be branched, ramify; — **sur,** to branch off from.

embrasé, burning, on fire.

embrasement, *m.* burning, conflagration, fire.

embraser, to set fire to, fire, kindle, inflame; **s'—,** to catch fire, kindle.

embrassant, embracing, stem clasping.

embrassement, *m.* embracing.

embrasser, to embrace, comprise, include, surround, take in, clasp, kiss.

embrayage, *m.* engaging, clutch, coupling, throwing into gear.

embrayer, to engage, throw into gear, connect.

embrèvement, *m.* skew notch.

embrochage, *m.* insertion, series mounting.

embrocher, to insert, cut in, arrange in series, spit (piece of meat).

embrouillement, *m.* clouding, confusion.

embrouiller, to embroil, confuse, cloud, fog, obscure.

embrumé, misty, hazy.

embrun, *m.* spray, fog.

embrunir, to embrown, darken.

embryogénie, *f.* embryogeny.

embryologie, *f.* embryology.

embryon, *m.* embryo, germ, fetus.

embryonatus, embryonate, having an embryo.

embryoné, embryonate, having an embryo.

embryonifère, embryoniferous.

embryonnaire, embryonic, embryo.

embryotège, embryotega.

embu (emboire), smeared with grease.

embûche, *f.* ambush, snare.

embuvant (emboire), smearing.

émeraude, *f.* emerald bird of paradise.

éméraudine, *f.* emeraldine, rose chafer (*Cetonia aurata*).

émerge, above water.

émergé, emersed.

émergence, *f.* emergence, emergency, outgrowth from the surface like prickles.

émerger, to emerge.

émeri, *m.* emery; **à l'—,** coated with emery, ground with emery; **bouchon à l'—,** ground stopper; **flacon bouché à l'—,** ground stoppered bottle.

émeril, *m.* emery, slime.

émerillon, m. merlin, swivel.

émeriser, to coat with emery.

émérite, emeritus, skilled, experienced.

émersion, f. emergence, emersion.

émerveillable, astonishing.

émerveillé (de), astonished (at).

émerveiller, to astonish, amaze; s'—, to wonder; il s'en émerveille, he marvels at (it).

émet (émettre), (it) emits.

émétine, f. emetine.

émétique, m. emetic (producing vomiting), tartar emetic; — de bore, soluble cream of tartar.

émetteur, m. transmitter, sending apparatus, drawer; — à lampes, tube transmitter.

émetteur, -trice, sending, transmitting.

émettre, to emit, express, utter, give out, put in circulation, send out, transmit, put forth; — en téléphonie, radiocast; — en télégraphie, transmit by code.

émeut (émouvoir), (he or it) moved, stirred; il s'émeut, he is getting stirred up.

émier, to crumble.

émiettement, m. crumbling.

émietter, to crumble, break into pieces, pulverize.

émigration, f. emigration, migration.

émigrer, to emigrate, migrate.

émincer, to mince, slice up.

éminemment, eminently, to a high degree.

éminence, f. eminence, height; en —, par —, eminently.

éminent, eminent, high, superior.

émis (émettre), emitted, uttered.

émissif, -ive, emissive.

émission, f. emission, sending, transmission, broadcasting, issue; — à étincelles, spark transmission; — à arc, — par arc, arc transmission.

émit (émettre), (it) emitted.

emmagasinage, emmagasinement, m. storing, storage, accumulation; — frigorifique, cold storage.

emmagasiner, to store, accumulate, store up, stock.

emmailloter, to swaddle, swathe.

emmanché, having a handle, helved, attached, joined, begun.

emmanchement, m. providing with a handle, attaching, joint, beginning.

emmancher, to put a handle to, helve, attach, fit, joint, set out; s'—, to fit on, be begun, be done.

emmêler, to entangle, mix up.

emménagements, m.pl. accommodations, appointments.

emménagogue, emmenagogue.

emmener, to take away, carry away, convey, lead away.

emmeulage, m. stacking.

emmeuler, to stack.

emmieller, to sweeten with honey.

emmortaiser, to mortise.

emmouflement, m. putting in the muffle furnace.

emmoufler, to put in the muffle furnace.

émodine, f. emodin.

émoi, m. care, anxiety, flutter.

émollient, emollient.

émonctoire, m. emunctory.

émondage, m. tree pruning.

émonde, f. pollarding or lopping system, loppings.

émonder, to prune, trim, lop.

émondoir, m. pruning knife, pruning hook, trimming ax.

émotion, f. emotion, disturbance, agitation, feeling.

émotter, to remove sods.

émottoir, m. grub hoe.

émoucheter, to clean (flax).

émoudre, to grind, sharpen.

émoulage, m. grinding.

émoulant, m. grinding.

émoulu, ground, sharpened.

émoussé, hebetate, obtuse, dull, blunt.

émousser, to dull, blunt, make blunt, bevel, chamfer, remove the moss, free from foam; s'—, to become deadened, become blunt.

émousseur, m. foam remover.

émouvoir, to move, raise, stir up, excite, arouse; s'—, to be moved, be roused, be anxious, be troubled.

empailler, to pack with straw, stuff, covering (plants) with straw.

empailleur, m. taxidermist.

empanser, (of food) to blow(animal).

empaquetage, m. packing (up).

empaqueter, to pack (up), wrap up, do up, package.

emparer(s') de, to take up, seize, assimilate, take possession of, master.

emparquer, to fold (sheep).

empasme, *m.* empasm.

empasteler, to dye blue with woad.

empâtage, *m.* embedding, the process of making soap paste, saponification (of fat, oil), cramming (fowl), pasting, making of a paste.

empâtement, *m.* embedding, puffiness, fattening (fowl).

empâter, to embed, make pasty, lay on thickly, paste, smear, impaste (colors), fill with paste, turn thick, become pasty, clog (a file), stuff (fowls).

empâteur, *m.* (poultry) fattener.

empattement, *m.* foundation, footing, foot, base, node, articulation (of stem), root swelling.

empaumure, *f.* crown, palm (of glove).

empêché, prevented, hindered, puzzled, at a loss, constrained.

empêchement, *m.* hindrance, obstacle, obstruction.

empêcher, to prevent, hinder, impede, forebear, keep from, oppose; **s'—,** to refrain help; **il s'en empêche,** he keeps from (it).

empeigne, *f.* vamp, upper (of shoe).

empennage, *m.* fin, tail plane, empennage, feathers (of arrow).

empereur, *m.* emperor.

empesage, *m.* starching.

empeser, to starch, stiffen with starch.

empester, to infect, stink, taint, contaminate.

empêtre, *m.* Empetrum.

empêtrer, to entangle, embarrass, hobble (animal).

emphase, *f.* emphasis.

emphatique, emphatic(al).

emphysémateux, -euse, emphysematous, bladdery.

emphysème, *m.* emphysema.

emphytéose, *f.* hereditary tenure.

emphytéote, *m.* long-lease tenant.

empierrement, *m.* macadamizing, broken stone (for roads), road metal, macadam, ballasting, stone ballast.

empierrer, to macadamize, metal.

empiètement, *m.* encroachment, infringement.

empiéter, to encroach on, infringe on.

empilement, empilage, *m.* piling, piling up, stacking, stratification, checkerwork.

empiler, to pile, pile up, stack.

empire, *m.* empire, reign, rule, sway.

empirer, to render worse, grow worse.

empirie, *f.* empiricism, experimenting.

empirique, empiric(al).

empirisme, *m.* empiricism: — **medical,** quackery.

empiriste, *m.* empiricist.

emplacement, *m.* site, location, position, ground situation.

emplastique, emplastic, adhesive.

emplastration, *f.* budding.

emplâtre, *m.* plaster; — **adhésif,** adhesive plaster; — **contre les cors,** corn plaster; — **de cantharides,** cantharides (cerate) plaster; — **de litharge,** lead (litharge) plaster; — **de plomb,** diachylon, lead plaster; — **de poix cantharide,** cantharidal pitch plaster; — **de poix de Bourgogne,** Burgundy pitch plaster; — **de savon,** soap plaster; — **diachylon,** diachylon, lead plaster; — **resineux,** adhesive plaster; — **simple,** lead plaster; — **vésicatoire,** blistering plaster.

emplette, *f.* purchase.

empli, *m.* a container for heating beet sugar.

empli, filled, full.

emplir, to fill, fill up.

emplissage, *m.* filling (up); **jaugeage —,** graduation for contents.

emploi, *m.* use, employment, position, place, business.

emploiement, *m.* employment.

emplombage, *m.* leading, lining or coating with lead.

emplomber, to lead, coat with lead.

employé, *m.* clerk, employee.

employé, employed.

employer, to employ, use, spend; **s'—,** to be employed, occupy oneself.

employeur, -euse, *m.&f.* employer.

emplumé, feathered, quilled.

emplumer, to feather.

empoigner, to seize, grasp.

empois, *m.* starch paste, paste; — **d'amidon,** starch paste.

empoise, *f.* bearing.

empoisonnant, poisonous, poisoning.

empoisonnement, *m.* poisoning; — **de sang,** blood poisoning.

empoisonner, to poison, intoxicate.

empoisser, to pitch, cover with wax.

empoissonnement, *m.* stocking with fish.

emporté, fiery, runaway.

emporte-pièce, *m.* punch, puncher.

emporter, to carry away, carry off, carry along, remove, take away; **s'—,** to be carried off, fly into a passion, bolt, run to wood (tree); **l'— sur,** to prevail over, outweigh, surpass, exceed.

empourpré, purple, purpled, crimson.

empourprer, to purple, color purple, tinge with crimson.

empreignant, impressing, stamping.

empreindre, to impress, imprint, stamp, leave a print; **s'—,** to be imprinted, leave a print.

empreint, impressed, imprinted, stamped.

empreinte, *f.* impress(ion), (im)print, stamp, priming, first coat, mark, impress, ridge, mold; **— d'une fossile,** fossil remain.

empressement, *m.* eagerness, haste, zeal, alacrity.

empresser, to hasten, be eager; **s'—,** to be eager, hasten, be zealous.

emprésurage, *m.* addition of rennet.

emprimerie, *f.* tan(ning) vat.

emprise, *f.* land taken, encroachment, enterprise.

emprisonner, to imprison, hold tight.

emprunt, *m.* borrowing, loan; **d'—,** borrowed, assumed, artificial.

emprunté, borrowed, assumed, artificial, embarrassed, of constraint.

emprunter, to borrow; **il le lui emprunte,** he borrows it from him.

empuantir, to become putrid, infect.

empyrée, *m.* empyrean.

empyreumatique, empyreumatic.

empyreume, *m.* empyreuma.

ému (**émouvoir**), effected, moved, touched, excited.

émulation, *f.* emulation, rivalry, competition.

émule, *m.* rival, competitor.

émulsif, *m.* emulsive substance.

émulsif, -ive, emulsive.

émulsine, *f.* emulsin.

émulsion, *f.* emulsion; **— bacillaire,** tuberculin bacillary emulsion; **— d'essence de térébenthine,** emulsion of oil of turpentine; **-- d'huile de foie de morue,** emulsion of cod-liver oil; **— d'huile de ricin,** emulsion of castor oil.

émulsionnement, *m.* emulsification.

émulsionner, to emulsify, roll with emulsion, make soft or liquid.

en, (*prep.*), in, at, to, for, like, into, within, out of, in the form of, with, as, on, of, by, while; (*pron.*), some, any, of, from, by, or, with it, her, him, them; **— ce que,** from the fact that; **— sorte que,** so that.

énantioblasté, enantioblastic.

énantiomorphe, enantiomorphous, enantiomorphic.

énantiostyle, (styles) protruded right or left of axis with stamens opposite.

énation, *f.* enation.

encablure, *f.* cable's length.

encadrement, *m.* framing, frame.

encadrer, to frame, insert, surround.

encaissage, *m.* incasing, planting (shrubs) in tubs.

encaissant, incasing.

encaisse, *f.* cash on hand.

encaissement, *m.* incasing, embankinɲ (river), trenching (ground), recovery (of money due).

encaisser, to incase, envelop, enclose, collect, receive (money), ballast (a road), embank (river).

encan, *m.* auction.

encaquer, to barrel, pack (like sardines).

encarpium, *m.* encarpium, fruit stalk of fungi (sporophore).

en-cas, encas, *m.* article kept for emergency, large parasol.

encastelure, *f.* navicular disease, hoofbound.

encaster, to place in saggers.

encasteur, *m.* placer, sagger, stacker.

encastrement, *m.* fitting in, housing, recess, groove, bed.

encastrer, to fit in, set in, embed, fit into recess; **s'—,** to fit together.

encaustique, *f.* polishing wax, (bees)wax polish, encaustic (painting); **— à parquet,** floor wax.

encaver, to store (wine) in a cellar.

enceignant, enclosing.

enceindre, to surround, enclose.

enceint, surrounded.

enceinte, *f.* circumference, enclosure, (inclosed) space, chamber, hall, circuit.

enceinte, pregnant.

encens, *m.* incense.

encéphalique, encephalic.

encerclement, *m.* encircling.

enchaînement, *m.* chaining, linking, connection, series, concatenation.

enchaîner, to link, connect, arrange, chain (up), enchain, restrain, bind, captivate; s'—, to be linked, be connected.

enchantemen', *m.* enchantment, delight.

enchanter, to delight, enchant, charm, fascinate.

enchanteur, -eresse, enchanting, charming, delightful.

enchâssé, incased, surrounded.

enchâsser, to insert, mount, set (in), enshrine, encase, mortise, inbed.

enchâssure, *f.* insertion, housing.

enchaussenage, *m.* liming.

enchaussener, enchaussumer, to lime (hides).

enchaussenoir, enchaussumoir, *m.* lime pit, liming vat.

enchaux, *m.* lime liquor, limewash.

enchère, *f.* bid, bidding, auction.

enchérir, to raise the price of, bid for, become dearer; — sur, to outbid, improve on, surpass.

enchérissement, *m.* rise, increase (in price).

enchevaucher, to rabbet, overlap.

enchevêtré, entangled, confused.

enchevêtrement, *m.* entangling, entanglement, confusion; — des racines, matting of the roots.

enchevêtrer, to entangle, tangle, interlace, halter; s'—, to get entangled, or interlaced, become embarrassed.

enchyléma, énchylème, *f.* enchylema, fluid portion of cytoplasm.

encirer, to wax.

enclanchement, *m.* interlocking.

enclave, *f.* boundary, enclosure, recess.

enclaver, to enclose, fit, lock, let in, wedge, dovetail, house, key, inclose (land within other lands).

enclenchement, *m.* throwing into gear, interlocking, engaging, coupling, engaging gear.

enclencher, to throw into gear, engage, couple, interlock.

enclin, inclined, prone, ready.

encliquetage, *m.* ratchet mechanism; — à frottement, friction catch, strut action, pawl motion.

enclore, to enclose, fence in.

enclos, *m.* enclosure.

enclos, enclosed, included.

enclume, *f.* anvil.

enclumeau, enclumot, *m.* small anvil, stake anvil.

enclumette, *f.* small portable anvil.

encoche, *f.* notch, slot.

encochement, *m.* notching.

encocher, to notch.

encoffrer, to enclose, hoard, lock up.

encoignure, encognure, *f.* corner, angle, elbow.

encollage, *m.* sizing (of paper), size, gumming, gluing, shutting up.

encoller, to size, gum, shut up.

encolleur, *m.* sizer.

encollure, *f.* welding point.

encolure, *f.* neck and withers, neck; — de cygne, arched neck.

encombrant, encumbering, cumbersome, obstructing, bulky, taking up too much room.

encombre, *m.* obstacle, hindrance, accident.

encombrement, *m.* obstruction, encumbrance, congestion, space occupied, clearance.

encombrer, to obstruct, clog, encumber.

encontre (à l'), to the contrary, in opposition to; à l'— de, against, counter to, contrary to.

encore, still, even, yet, besides, also, again, once more, further, moreover, however, one more; — autant, as much again; — que, though, although; — une fois, once more; et —, even then; ou —, or else.

encourageant, encouraging.

encouragement, *m.* encouragement.

encourager, to encourage, stimulate.

encourir, to incur, run, hasten.

encouru, incurred.

encrage, *m.* inking.

encrassement, *m.* fouling, sooting (up), choking.

encrasser, to foul, make dirty, soil, grease, soot up, clog, choke; s'—, to become fouled, get dirty or choked.

encre, *f.* ink; — à copier, copying ink; — à écrire, writing ink; — à marquer, marking ink; — à tampon, ink for pads, marking ink; — copiante, copying ink; — d'imprimerie, — d'impression, printing ink; — dragon, — indélébile, indelible ink; — sympathique, sympathetic (invisible) ink.

encré, inked.

encrené, (iron) twice heated and hammered.

encrer, to ink.

encreur, inking (roller).

encrier, *m.* inkstand, ink block.

encrivore, ink-removing.

encroûtement, *m.* incrusting, incrustation, crust; — **calcique,** chalky deposit.

encroûter, to incrust, cake with mud, plaster; **s'—,** to become incrusted, sink into a rut.

encuivrage, *m.* (of firearms) metal fouling.

encuvage, *m.* putting into a vat.

encuvement, *m.* putting into a vat, depression, hollow.

encuver, to put into a vat or tub.

encyclie, *f.* one of a series of rings.

encyclopédie, *f.* encyclopedia.

endaubage, *m.* canned beef, preserved meat.

endauber, to preserve, tin, can (meat), stew.

endémicité, *m.* existence of endemic.

endémie, *f.* endemic.

endémique, endemic.

endentement, *m.* toothing, cogging.

endenter, to provide with teeth, tooth, cog, ratch, mesh, scarf, dovetail.

endhyménine, *f.* intine, innermost coat of pollen grain.

endigué, choked, suppressed, dammed.

endiguement, *m.* damming torrents, embankment, reclamation.

endiguer, to dam (up), dam in, embank, dike in, reclaim.

endive, *f.* broad-leaved chicory; — **frisée,** curled endive.

endobaside, *m.* endobasidium, an enclosed basidium.

endobiotique, endobiotic, like internal parasite.

endocarde, *m.* endocardium.

endocardite, *m.* endocarditis.

endocarpe, *m.* endocarp.

endocaryogamie, *f.* crossing two flowers of same individual.

endocellulaire, endocellular.

endochrôme, *m.* endochrome.

endocrâne, *m.* endocranium.

endocrine, endocrine (gland).

endocycle, *f.* endocycle, normal lignous region.

endoderme, *m.* endoderm.

endodermogènes, *f.pl.* endodermogens, vascular cryptogams.

endogamie, *f.* endogamy.

endogène, *m.* endogen.

endogène, endogenous.

endoglobulaire, in round cells.

endolithique, endolithic, lichens growing on stony substratum.

endommager, to damage, hurt.

endopachyte, *f.* secondary wood.

endophloeum, *m.* the inner bark, bast.

endophytes, *f.pl.* endophytes.

endoplasme, *m.* endoplasm.

endoplèvre, *f.* endopleura, tegmen.

endormi, asleep, sleeping, sluggish, dormant.

endormie, *f.* thorn apple.

endormir, to put to sleep, lull, deceive; **s'—,** to fall asleep.

endosmomètre, *m.* endosmometer.

endosmose, *f.* endosmosis.

endosmotique, endosmotic.

endosperme, *m.* endosperm.

endospermique, endospermic, (embryo) with an endosperm.

endospore, *m.* endospore.

endossé, endorsee, transferee.

endossement, *m.* endorsement, transference.

endosser, to endorse, put on, assume, take on, put into circulation.

endostome, *f.* endostome.

endothécium, *m.* endothecium.

endothélium, *m.* endothelium.

endothermique, endothermic.

endotoxine, *f.* endotoxin.

endotrophe, endotrophic (mycorhiza).

endoxyle, *f.* pith crown.

endozoique, endozoic, living inside an animal.

endroit, *m.* place, locality, side, aspect, point, spot, right side (of material), passage (in a book); **à l'— de,** with regard to; **par —,** here and there, in places.

enduire, to coat, cover, smear, overlay, plaster, dope, daub; — **de couleur,** to paint.

enduisant, adherent (matter).

enduiseuse, *f.* coating machine.

enduit, *m.* coat, coating, facing, lining, layer, wash, paint, (water) proofing, dope, glaze, glazing, plaster(ing); — **de noir,** blacking, black wash; — **pour planchers,** floor covering, flooring.

enduitage, *m.* impregnation, injection, steeping.

endurci, hardened.

endurcir, to harden, indurate, inure; **s'—,** to become hardened, callous, tough, harden, set.

endurcissement, *m.* hardening, hardness, induration.

endurer, to endure, bear, allow, permit.

énergétique, energetic, energizing (food).

énergide, *f.* energid.

énergie, *f.* energy, force; — **actuelle,** kinetic energy; — **cinétique,** kinetic energy; — **interne,** potential energy; — **thermique,** heat (thermal) energy; — **utilisable,** available energy.

énergique, energetic, powerful, forcible.

énergiquement, energetically.

énervé, enervous, enervate, destitute of veins or nerves.

énerver, to enervate, weaken.

enfaiter, to cover the ridge of (a roof) finish off.

enfance, *f.* infancy, childhood, puerility, childishness.

enfant, *m.&f.* child, infant, offspring; — **du diable,** stinkhorn (*Ithyphallus impudicus*).

enfantement, *m.* childbirth, delivery, labor.

enfanter, to bring forth, produce, bear, be delivered of.

enfer, *m.* hell, inferno; — **de Boyle,** Boyle's vessel for oxidation of mercury.

enfermer, to confine, enclose, surround, close, comprise, include, contain, shut up, shut in.

enfeuiller, to cover with leaves.

enfieller, to render (food) bitter, embitter, sour.

enfiévré, feverish.

enfiévrer(s'), to grow feverish, get excited.

enfilade, *f.* row, succession, series, sequence, suite.

enfiler, to thread, string, thread on, go along (a road), engage in.

enfilure, *f.* row, series, string, stringing.

enfin, in short, at last, finally, after all, at length.

enflammé, burning, ignited, on fire, inflamed, fiery (sun).

enflammer, to set fire to, ignite, kindle, cause to burn; **s'—,** to take fire, catch fire, inflame, irritate.

enflé, inflated, swollen, swelled, puffed up, turgid, tumid.

enflement, *m.* swelling, inflation.

enfler, to inflate, swell, puff out, distend, exaggerate; **s'—,** to swell (up), be inflated, (of waters) rise.

enfleurage, *m.* enfleurage (of fats and oils).

enfleurer, to impregnate (fats and oils) with perfume of flowers.

enflure, *f.* swelling, inflation, tumor, pride, turgidity.

enfoncé, broken, sunk, sunken, deep, smashed.

enfoncement, *m.* driving in, sinking, recess, alcove, hollow, depth, depression.

enfoncer, to sink, lower, plunge, drive in, drive down, thrust, break open, deepen, put bottom into (cask); **s'—,** to plunge, penetrate, sink down.

enfonçoir, *m.* mallet, beater.

enfonçure, *f.* depression, hollow, recess, cavity, bottom (of cask), cloven wood for coopers.

enformer, to shape, fashion.

enfoui, buried, sunken.

enfouir, to bury, plow in, hide in the ground.

enfouissement, *m.* hiding in the ground, burial.

enfourchement, *m.,* fork link, forked mortise.

enfourchure, *f.* fork, crotch (of tree).

enfournage, enfournement, *m.,* **enfournée,** *f.* charging a furnace, placing (bread) in oven.

enfourner, to put (pottery) in kiln, put (bread) in oven, start, begin; **s'—,** to start, enter (into); — **en échappade,** put into compartments in the kiln.

enfourneur, *-m.* ovenman, kilnman.

enfourneur, -euse, charging.

enfreignant, infringing.

enfreindre, to infringe, violate.

enfuir(s'), to leak, flee, escape by flight, run away, disappear.

enfumage, *m.* smoking out (of bees), blackening, heating.

enfumé, smoky.

enfumer, to smoke, fill with smoke, smoke out (animals), intoxicate; **s'—,** to be smoked.

enfutailler, enfûter, to cask, barrel (wine).

enfuyant, fleeing, leaking.

engagement, *m.* engagement, promise, agreement.

engager, to engage, insert, compel, fit, agree to, begin, invite, urge, lead, involve, induce, pawn; **s'—,** to be engaged in, engage oneself, begin, agree, bind oneself, get entangled, enter, be involved, enlist.

engainant, forming a sheath.

engainé, (stem) surrounded by sheaths.

engainer, to sheathe, ensheathe (stalk).

engallage, *m.* galling, weighting.

engaller, to gall, steep in galls.

engaver, to feed (nestling).

engazonner, to sow (ground) with grass seed.

engeance, *f.* breed, strain (poultry).

engendrement, *m.* production, breeding (of disease), generation (of heat).

engendrer, to produce, form, give rise to, set up, engender, occasion, generate, breed, sire, beget, cause; **s'—,** to be produced, be bred.

engerber, to sheaf, sheave, bind, pile up.

engin, *m.* engine, device, appliance, gear, hoist.

englober, to include, engulf, surround, unite, blend.

engloutir, to swallow up, engulf, absorb, devour.

engluanter, to englue.

engluement, engluage, *m.* liming of twigs, coating with glue.

engluer, to coat with glue, lime (twigs), stick, glue, smear with birdlime.

engobage, *m.* coating with engobe, slip painting.

engobe, *f.* engobe (slip or paste).

engober, to decorate, coat (pottery) with engobe (slip).

engommage, *m.* gumming, glazing, dressing.

engorgement, *m.* stopping up, clogging, entanglement, obstruction.

engorger, to stop up, choke (up), obstruct, block, clog, congest; **s'—,** to be stopped up, clogged.

engoué, obstructed, choked up.

engouement, *m.* infatuation, obstruction, choking up.

engouer, to choke up, obstruct.

engouffrer, to ingulf; **s'—,** to be ingulfed, rush into, swallowed up.

engoulevent, *m.* goatsucker, nightjar.

engourdi, torpid, benumbed, heavy.

engourdir, to numb, benumb, dull, deaden.

engourdissement, *m.* numbness, dullness, torpid state, torpor.

engrain, *m.* lesser spelt, one-grained wheat.

engrainer, to sheathe.

engrais, *m.* fertilizer, manure, dung, fattening feed or pasture; — **azoté,** nitrogenous fertilizer; — **chimique,** chemical fertilizer; — **composé,** mixed fertilizer; — **d'os,** bone fertilizer, bone manure; — **flamand,** liquid manure (from human excrement); — **potassique,** potash fertilizer; — **vert,** mulching (manuring with leaves or herbage).

engraissage, *m.* fertilizing, fattening (animals).

engraissé, fertilized, fattened, greasy, enriched, wet-beaten (paper).

engraissement, *m.* fertilizing, fattening, hydration.

engraisser, to manure, fertilize, fatten, cram, make fat, thrive, grease, make greasy; **s'—,** to grow fat.

engranger, to garner, get in (corn), stock (in a barn).

engrenage, *m.* gear, gearing, gear wheels, cog, stowage (of casks), complication; — **à vis sans fin,** worm gear; — **conique,** bevel gear; — **cylindrique,** — **droit,** spur gear; — **en biais,** bevel gear; —**s de distribution,** timing gears; **à** —, geared.

engrènement, *m.* engaging, coming into gear, contact, catch, feeding (threshing machine or animals).

engrener, to engage, gear, put into gear, be in gear, intercog, indent, feed with grain, begin; **s'—,** to gear, be in gear, catch, be geared, begin.

engrenure, *f.* gearing, catching, interlocking, toothing, cogging.

engrumeler, to clot, curdle, coagulate.

enhalide, *f.* formation of spermatophytes on soil in salt water.

enherbé, overrun with weeds or grass.

enherbement, *m.* becoming overgrown with grass.

enherber, to be overgrown with grass, form grass cover.

enhydre, enhydrous, containing water.

énieller, to rid (crop) of corn cockles.

énigmatique, enigmatic, puzzling.

énigme, *f.* enigma, riddle.

enivrant, intoxicating.

enivrement, *m.* intoxication.

enivrer, to intoxicate, elate.

enjolivement, *m.* embellishment, ornament.

enjoliver, to embellish, adorn.

enkysté, encysted.

enkystement, *m.* encystment, encystation.

enkyster(s'), to encyst, to become encysted.

enlacer, to lace, enlace, entwine, interlace.

enlaçure, *f.* — **des bois,** joining of timbers.

enlaidir, to make ugly, disfigure.

enlevage, *m.* discharge, removal.

enlèvement, *m.* removing, removal, carrying off, abduction, removal of litter; — **de la poussière,** dust extraction; — **de l'herbe,** grass cutting; — **de mottes gazonnées,** removal of sods.

enlever, to remove, take away, carry off, take off or out, discharge (calico), lift (up); **s'—,** to be removed, come out or off, rise, be lifted; — **à,** to take away from, remove from; — **la tourbe (de),** to dig (cut) peat.

enligner, to arrange.

enlisement, *m.* sinking in sand.

enluminer, to color, illuminate, flush, redden.

enluminure, *f.* illumination, coloring, colored print.

ennéagyne, enneagynous, having nine pistils.

ennéandre, enneandrous, with nine stamens.

ennéanthère, (plant) with nine stamens.

ennéasépale, with nine sepals.

ennemi, *m.* enemy, foe.

ennemi, hostile, inimical.

ennoblir, to ennoble.

ennui, *m.* ennui, tedium, care, annoyance, weariness.

ennuyant, tiresome, annoying, irksome, tedious.

ennuyer, to tire, weary, annoy, bother; **s'—,** to grow weary.

ennuyeux, -euse, tiresome, annoying.

énodé, énoué, enodal, without knots or nodes.

énoncé, *m.* statement, declaration, expression.

énoncé, stated, expressed.

énoncer, to state, enunciate, declare, express.

énonciation, *f.* statement, enunciation, expressing, delcaring.

énorme, enormous, huge.

énormément, enormously, immensely.

énormité, *f.* enormity, vastness, absurdity.

énouer, to burl (cloth).

enquérir(s'), to inquire.

enquête, *f.* inquiry, investigation, inquest.

enquiert (enquérir), (he) inquires.

enquis (enquérir), inquired.

enraciné, deep-rooted.

enracinement, *m.* root system, digging in (of sapling), taking root (of tree); **à — profond,** deep-rooted; **à — superficiel,** shallow-rooted.

enraciner, to root; **s'—,** to take root.

enragé, *m.* madman.

enragé, mad, raging, furious, infected with rabies.

enrayage, enrayement, *m.* checking, dragging, skidding, stopping, laying out (of furrows).

enrayer, to check, restrain, jam, stop, brake, lock, skid (as a wheel), spoke (a wheel), plow the first furrow, draw, lay out (furrows).

enrayure, *f.* groove, mortise, notch, furrow.

enregistrement, *m.* registration, registry, registering, recording, record.

enregistrer, to register, record, enroll.

enregistreur, *m.* register, recorder, recording device.

enregistreur, -euse, registering, recording.

enrichi, enriched, concentrated.

enrichir, to enrich; **s'—,** to be enriched, enrich oneself.

enrichissement, *m.* enrichment, enriching.

enrobage, *m.* coating, covering, embedding.

enrober, to cover, wrap, coat, cover with a protecting envelope, embed; — **des viandes,** to coat (cooked) meats with gelatin.

enrochage, *m.* caking, stone bedding, enrockment, riprap.

enrocher, to cake, harden, enrock, riprap.

enrouement, *m.* hoarseness.

enrouillé, rusted, rusty.

enrouiller, to rust.

enroulage, enroulement, *m.* winding, rolling (up); — **en dérivation,** shunt winding.

enrouler, to wind (cable), roll (up), wrap up; **s'—,** to be wound, be rolled, wind coil.

ensabler, to sand (up), cover with sand, run aground, ballast.

ensachage, *m.* bagging, sacking.

ensacher, to bag, sack, put into sacks.

ensafraner, to saffron, dye with saffron.

ensanglanter, to cover or stain with blood.

enseigne, *f.* sign, ensign, flag.

enseignement, *m.* instruction, teaching.

enseigner, to teach, instruct, show, inform.

enseigneur, *m.* teacher, instructor.

ensemble, *m.* ensemble, whole, entirety, aggregate, general effect, harmony, unity, group, series; **d'—,** general.

ensemble, together, at the same time.

ensemencement, *m.* sowing, seeding, scattering, inoculation (of media); **— naturel,** natural seeding, natural seedlings.

ensemencer, to sow, seed, scatter, inoculate (media), disseminate.

enserrer, to enclose, squeeze, contain, put (plant) in hothouse.

ensevelir, to bury, absorb.

ensifolié, ensiform-leaved.

ensiforme, ensiform, sword-shaped.

ensilage, *m.* storing in a silo, ensiling, ensilage; **— sec,** sweet silage.

ensiler, to store in a silo, ensile, ensilage.

ensilotage, *m.* ensilage.

ensiloter, to store in silo.

ensimage, *m.* greasing, oiling (of wool); **huile d'—,** textile oil.

ensimer, to oil, grease (wool).

ensoleillé, sunny, sun-kissed.

ensoleillement, *m.* sunning, insolation.

ensoufrage, ensoufrement, *m.* sulphuring, sulphuration.

ensoufrer, to sulphur, sulphurize, treat with sulphur.

ensoufroir, *m.* sulphuring chamber.

ensucrer, to sugar, sweeten.

ensuifer, to tallow, smear with tallow.

ensuit (ensuivre), (it) follows, results.

ensuite, then, next, afterward, after, later; **— de,** after.

ensuivre (s'), to follow, ensue, result.

entacher, to taint, blemish, contaminate; **— d'erreurs,** to vitiate.

entaille, *f.* (graduation) mark, cut, dentation, gash, groove, notch, nick, jag, slash; **— circulaire,** girdle notch; **— pour l'abatage,** felling notch; **— repère,** guide mark.

entaillé, notched.

entailler, to cut, groove, notch, gash, nick, jag.

entaillure, *f.* notch, mark.

entame, *f.* first cut, outside slice.

entamer, to begin on, start, open up, make the first cut, touch, cut (slightly), injure, encroach on, prevail on.

entamure, *f.* cut, incision, first piece, outside slice.

entartrage, *m.* incrustation, scale.

entartré, covered with scale.

entartrer, to become covered with scale.

entassement, *m.* heap, pile, heaping, crowding (of plants), piling (up), massing, accumulation.

entasser, to heap (up), pile, pile up, accumulate, huddle; **s'—,** to be heaped, crowd.

ente, *f.* graft, grafted shoot.

entendement, *m.* understanding.

entendoir, *m.* drying loft.

entendre, to hear, mean, understand; **s'—,** to be heard, be skillful in, be a judge of; **— à,** to consent to; **— dire,** to hear say; **il entend dire cela,** he hears that; **il s'entend avec lui,** he comes to an agreement with him; **ils s'entendent,** they agree.

entendu, heard, understood, capable, intelligent, skillful; **bien —,** of course; **c'est —,** all right.

enténébrement, *m.* darkness, darkening.

enténébrer, to wrap in darkness, benight.

entente, *f.* understanding, agreement, meaning, ability.

enter, to graft (tree).

entérite, *f.* enteritis, inflammation of the bowels.

entérocoque, *f.* enterococcus.

entérokinase, *f.* enterokinase (a ferment).

enterrage, *m.* planting.

enterrement, *m.* burial, interment.

enterrer, to bury, inter.

en-tête, *m.* heading (of letter).

entêté, stubborn, headstrong, affected with headache, intoxicated, conceited.

enthousiasme, *m.* enthusiasm.

enthousiasmer, to rouse to enthusiasm.

enthousiaste, *m.&f.* enthusiast.

enthousiaste, enthusiastic, entirely devoted.

entiché, unsound, worm-eaten, damaged, spotted (fruit).

entier, *m.* whole, entirety, total, integer, whole number, integral number.

entier, -ère, entire, whole, headstrong, integral (numbers), self-willed; **en —**

wholly, entirely, fully, in full; **tout** —, whole, entire, wholly.

entièrement, entirely, wholly.

entité, *f.* entity.

entodiscal, entodiscalis, inserted within a disc.

entoilé, cloth-covered, mounted on cloth.

entomogame, *m.* entomogamy, pollination of flowers by insects.

entomologie, *f.* entomology.

entomologiste, *m.* entomologist.

entomophage, entomophagous.

entomophiles, *f.* entomophilae.

entomorhize, growing on insects.

entonnage, entonnement, *m.*, **entonnaison,** *f.* barreling, putting in casks.

entonner, to barrel, put in casks, swallow, intone, sing.

entonnerie, *f.* cellar, place where casks are filled.

entonnoir, *m.* funnel, tunnel, infundibulum, crater; — **à boule,** bulbed funnel, separating funnel, funnel-shaped depression in gelatin slab; — **à brome,** bromine funnel; — **à filtration chaude,** funnel, hot-water funnel (for hot filtration); — **à longue douille,** long-stemmed funnel.

entonnoir, — **à robinet,** funnel with stopcock, dropping funnel; — **à séparation,** separating (separatory) funnel; — **à tamis,** sieve funnel; — **cannelé,** channeled (ribbed) funnel; — **en grès,** stoneware funnel; — **en verre,** glass funnel; — **percé de trous,** perforated funnel; **en** —, funnel-shaped, infundibular.

entophytogénèse, *f.* production of entophytes.

entoplasma, *m.* protoplasm.

entorse, *f.* strain, sprain, shock, wrench, twist.

entortiller, to twist, wrap (up), entangle; **s'**—, to twist, twine, coil, get entangled.

entour, *m.* environs, surroundings, associate; **à l'**— **de,** around, in the vicinity.

entourage, *m.* entourage, surroundings, things around, those around, case, casing, border, frame, mounting, setting.

entourer, to surround, encircle, enclose; **s'**—, to surround oneself.

entrailles, *f.pl.* entrails, intestines, bowels, guts, interior, center, womb, heart, core.

entrain, *m.* spirit, animation; — **de,** in the act of.

entraînable, easily carried away.

entraînement, *m.* carrying away or over, communicating gear, driving, pulling, enthusiasm, impulse, denudation, erosion; — **à la vapeur,** steam distillation; — **par l'eau,** denudation, erosion, washing away.

entraîner, to carry away, carry along, carry over, carry down, entail, draw, drag away, involve, win, captivate, train, keep going, drive, drag, draw along; **il s'entraîne,** he is training.

entraîneur, *m.* trainer, driving device.

entrave, *f.* hindrance, obstacle, impediment, shackle.

entraver, to hinder, impede, fetter, inhibit.

entre, between, among, in; — **eux,** — **elles,** together; — **nous,** between ourselves; **d'**—, of, from among.

entrebâiller, to open partly, set ajar.

entrechoquer(s'), to collide, clash, encounter one another.

entrecoupé, interrupted, intermittent, broken, incomplete.

entrecouper, to intersect, interrupt, intersperse.

entrecroisement, *m.* crossing, intercrossing, decussation.

entre-croiser, to intercross, intersect, interlace, interweave.

entre-détruire(s'), to destroy one another.

entre-deux, *m.* space between, interval, partition.

entre-deux, betwixt and between, so-so.

entrée, *f.* entree, entrance, entry, inlet, admission, coming in, access, ingress, opening, mouth.

entrefer, *m.* air gap, clearance.

entre-hivernage, *m.* winter plowing.

entrelacé, intertwined.

entrelacement, *m.* interlacing, network (of branches), interweaving, intertwining.

entrelacer, to interweave, interlace, intertwine (branches), twine.

entre-lame, *f.* fillet.

entre-luire, to glimmer, shine feebly.

entremêlé, intermingled.

entremêlement, *m.* intermingling, intermixture.

entremêler, to intermix, intermingle, mix (up).

entremise, *f.* intervention, interposition, mediation, medium.

entre-noeud, *m.* internode.

entreposage, *m.* storage, placing in bond.

entreposer, to store, warehouse, put in bond.

entreposeur, *m.* (bonded) warehouse keeper.

entrepositaire, *m.* bonder.

entrepôt, *m.* (bonded) warehouse, mart.

entreprenant, enterprising, venturesome.

entreprendre, to undertake, attempt, contract, attack, encroach.

entrepreneur, *m.* contractor, builder, employer, manufacturer, producer, grower.

entrepris, undertaken, ventured.

entreprise, *f.* undertaking, venture, project, enterprise, plant concern, contract, establishment, encroachment.

entrer, to enter, be admitted, take part in, bring in, come in, get in, put in, penetrate, insert, import; — **en ébullition,** to begin boiling; — **en fonctionnement,** to begin to work; — **en ligne de compte,** to take into account; **laisser** —, to admit.

entre-rail, *m.* space between the rails.

entre-temps, *m.* meantime, interval; **dans** —, meanwhile, in the meantime.

entretenir, to maintain, keep, keep up, talk with, converse with, entertain, carry on, sustain, keep in repair; **s'** —, to be kept (up), maintain oneself, support each other, talk, converse.

entretien, *m.* support, maintenance, upkeep, keeping in repair, care, supply, clothing, talk, conversation, dressing (of blisters); — **des bois,** tending of woods; — **du bétail,** cattle rearing, cattle breeding; — **du peuplement,** tending of crop; — **du sol,** preservation of fertility.

entretoise, *f.* crossbar, crosspiece, crossbrace, stay, tie, girder, rib.

entretoisement, *m.* counterbracing, mutual bracing, crosspiece, propping.

entretoiser, to brace, crossbrace.

entrevit (entrevoir), (he) suspected, foresaw.

entrevoir, to catch a glimpse of, foresee dimly, suspect.

entrevoyant, foreseeing.

entrevu, caught sight of.

entrevue, *f.* interview.

entropie, *f.* entropy.

entr'ouvert, open, half-open.

entr'ouvrir, to open, open partly, spread open; **s'** —, to half-open, set ajar, spread open.

énucléer, to pit, stone.

énumération, *f.* enumeration.

énumérer, to enumerate.

envahir, to invade, overrun, encroach upon.

envahissant, invasive, growing prolifically.

envahissement, *m.* invasion, encroachment; — **de mauvaises herbes,** *m.* becoming overgrown with weeds, invasion of weeds, becoming weedy; — **par la bruyère,** becoming overgrown with heather.

envahisseur, *m.* invader.

envasement, *m.* silting up, swampiness, bogginess.

envaser, to choke up with mud, silt up; **s'** —, to become muddy.

enveloppant, enveloping, enclosing.

enveloppe, *f.* envelope, bag, cover, covering, coat, casing, wrapping, case, exterior, jacket, appearance, tunic, integument involucral leaf; — **florale,** perigone, perianth.

enveloppement, *m.* enveloping, wrapping, packing, embryonic condition, sheath (of seed).

envelopper, to envelop, wrap (up), enclose, surround, cover, involve, include, enfold.

envenimé, infected, poisoned.

envergure, *f.* extent, breadth, spread (wings), span.

enverra (envoyer), (he) will send.

enverrage, *m.* vitrification, addition of cullet to the pots to facilitate fusion.

enverrer, to vitrify.

envers, *m.* wrong side, reverse, back side; **à l'** —, wrong side out, inside out, upside down.

envers, toward, to, in relation to.

envie, *f.* envy, longing, desire, wish; **avoir** — **de,** to feel like.

envier, to envy, long for, desire.

envieux, -euse, envious, desirous.

environ, about, nearly, thereabouts; **aux** —**s de,** near, around.

environment, surrounding.

environnant, amplexicaul (leaf).

environner, to surround, encompass, environ.

environs, *m.pl.* environs, vicinity, neighborhood.

envisager, to consider, regard, face, look at.

envoi, *m.* sending, shipment, parcel, consignment, envoy, despatch remittance.

envoie (envoyer), (he) is sending.

envoiler, to warp, bend (in tempering).

envoilure, *f.* curve (of scissors), bending.

envol, *m.* (birds) taking flight.

envoler(s'), to fly, fly away, fly off, take flight.

envoûté, archlike, arched.

envoûter, to vault, overarch.

envoyer, to send, send out, transmit; — chercher, to send for; — dans le monde, to broadcast.

envoyeur, -euse, *m.&f.* sender.

enzootie, *f.* enzootic disease.

enzymatique, enzymic, enzymatic.

enzyme, *f.* enzyme.

éocène, Eocene.

éolienne, *f.* wind motor, windmill.

éolipyle, éolipile, *m.* aeolipile, wind valve, ventilator.

éosine, *f.* eosin.

éosinophile, eosinophile.

éosinophilie, *f.* eosinophilia.

eosinophilique, eosinophile, eosinophilic.

eosinophilogene, eosinophilogenic.

épagneul, *m.* spaniel.

épaillage, *m.* removal of straw (from wool), stripping (sugar cane) of the lower leaves, cleaning (of gold).

épailler, to deprive (wool) of vegetable matter, strip (sugar cane) of the lower leaves, clean (gold).

épais, *m.* thickness, density.

épais, thick, dense.

épaisseur, *f.* thickness, density, depth; — du couvert, density of the cover or leaf canopy.

épaissir, to thicken.

épaississant, *m.* thickener.

épaississant, thickening.

épaississement, *m.* thickening.

épanchement, *m.* effusion, outpouring, overflowing.

épancher, to pour out, shed, effuse; s'—, to be effused, be extravasated, overflow, unbosom.

épandage, *m.* spreading, scattering (of manure), distribution (of water), irrigation; champ d'—, sewage farm.

épandre, to spread, scatter (manure), distribute, shed, strew.

épanouir, to open, expand, cause (flower) to bloom, bud, sprout, stool; s'—, to expand, be spread out, blow, brighten up open, bloom.

épanouissement, *m.* expansion, spread, opening (of flowers), cheerfulness, blowing, pole shoe.

éparapétale, parapetalous.

épargne, *f.* saving, economy, thrift.

épargner, to save, economize, spare.

éparpillement, *m.* scattering.

éparpiller, to scatter, strew.

épars, *m.* bar, crossbar.

épars, scattered, dispersed, thin.

éparvin, *m.* bone spavin.

épaté, flat, flattened, broken off, crippled.

épaule, *f.* shoulder, shoulder joint.

épaulement, *m.* epaulement, shoulder.

épauler, to support, protect, shoulder, sprain shoulder, bend at angles.

épeautre, *m.* spelt, German wheat (*Triticum aestivum* var. *spelta*).

épée, *f.* sword.

épeler, to spell.

épellation, *f.* spelling.

épenchyme, *f.* epenchyma, fibrovascular tissue.

épène, *f.* epenchyma.

épépiner, to take out the seeds, stone, core.

éperdu, dismayed, distracted, aghast.

éperdument, furiously, desperately.

éperon, *m.* spur, buttress, breakwater, wrinkle; — de chevalier, larkspur, consound (*Delphinium consolida*).

éperonné, calcarate, spurred.

éperonner, to spur, spur on.

éperonnière, *f.* larkspur (Delphinium).

épervier, *m.* sparrow hawk, sweep net.

épervière, *f.* hawkweed, rattlesnake weed (*Hieracium venosum*).

épharmonie, *f.* epharmony.

épharmosis, *f.* epharmosis, adaptation of plants under new conditions.

éphémère, *m.* spiderwort (Tradescantia), dayfly, ephemeron, green drake.

éphémère, ephemeral, short-lived.

éphydrogamique, (plant) fructifying on water.

épi, *m.* ear, head (of grain), awn, spike, tuft, jetty, dike.

épiage, *m.* épiation, *f.* earing, heading (of grain).

épiaire, *f.* grande —, hedge woundwort (*Stachys sylvatica*);— des marais, marsh woundwort (*Stachys palustris*).

épibasal, epibasal.

épiblaste, *m.* epiblast.

épiblastèse, *f.* growth due to development of reproductive corpuscles.

épiblème, *f.* epiblema.

épibole, *f.* epibole, epiboly.

épicarpe, *m.* epicarp, erocarp.

épicaule, growing on stem.

épice, *f.* spice, aromatic condiment.

épicé, spiced, pungent.

épicéa, *m.* spruce (Picea), Norway spruce (*Picea abies*); — **blanc,** white spruce (*Picea glauca*); — **piqué par le bostriche,** worm-eaten spruce.

épicer, to spice.

épicerie, *f.* groceries, spices, grocery shop, grocery business.

épichilium, *m.* epichilium.

épichlorhydrine, *f.* epichlorohydrin.

épicier, *m.* grocer.

épicline, epiclinal, seated upon the receptacle.

épicrâne, *m.* scalp.

épicurien, -enne, *m.&f.* epicurean.

épicycloïde, *f.* epicycloid.

épidémie, *f.* epidemic; — **hydrique,** waterborne epidemic.

épidémique, epidemic.

épidendre, to live on trunks of trees.

épiderme, *m.* epidermis, cuticle.

épidermique, epidermal.

épidote, *m.* epidote.

épié, spicate, disposed in a spike.

épierrer, to remove stones from.

épierreur, -euse, *m.&f.* stone remover.

épigastre, *m.* epigastrium.

épigé, epigeous.

épigène, epigenous.

épigénèse, *f.* epigenesis.

épigone, *m.* epigonium.

épiguanine, *f.* epiguanine.

épigyne, epigynous.

épigynie, *f.* epigyny.

épilation, *f.,* **épilage,** *m.* depilation, unhairing.

épilatoire, depilatory.

épilepsie, *f.* epilepsy, falling sickness.

épiler, to depilate, eradicate (hair), unhair.

épillet, *m.* spikelet.

épilobe, *m.* willow herb (*Epilobium angustifolium*).

épimaque, *m.* plumebird (a bird of paradise).

épimène, epimenous, the perianth being superior.

épimère, *m.* epimeron.

épinage, *m.* drawing off of the aqueous layer.

épinaie, *f.* brake, thicket, thornbush.

épinard, *m.* spinach (*Spinacia oleracea*); — **de Hollande,** round-leaved summer spinach.

épinastie, *f.* epinasty.

épinceter, épinceler, épincer, to burl (cloth).

épine, *f.* thorn, thorn tree, spine, a tube used to draw off the aqueous layer, a sharp point, prickle, backbone, barberry (*Berberis vulgaris*); — **blanche,** hawthorn, whitethorn (Crataegus); — **noire,** sloe, blackthorn (*Prunus spinosa*); — **vinette,** barberry (*Berberis vulgaris*); **petite —,** spinule.

épinéphrine, *f.* epinephrin, adrenalin.

épinette, *f.* fir (Abies), (chicken) coop; — **rouge,** tamarack (*Larix americana*).

épineux, -euse, thorny, prickly, spinose, muricate.

épine-vinette, *f.* barberry (*Berberis vulgaris*).

épingle, *f.* pin.

épingler, to pin, prick, clean out (with pin).

épinglette, *f.* priming needle, pricker.

épinière, spinal.

épinoche, *f.* stickleback.

épinycte, epinyctous, ephemeral, (flowers) beginning to open in the evening.

épiontologie, *f.* epiontology.

épipétal, epipetalous (borne upon the petals).

épipelté, epipeltate.

épiphlocodique, epiphloedic, epiphloedal

épiphlose, *f.* epidermis.

épiphragme, *m.* epiphragm.

épiphylle, epiphyllous, growing on leaves.

épiphyllosperme, bearing seed on leaflike organs.

épiphyte, *f.* epiphyte.

épiplasme, *m.* protoplasm, glycogen mass.

épiploon, *m.* epiploon, omentum.

épipode, *m.* epipodium.

épirrhize, epirrhizous.

épisarcine, *f.* episarcine.

épisépale, episepalous.

épispastique, epispastic.

épisperme, *m.* episperm, the outer covering of the seed.

épispermique, embryo deprived of endosperm, exalbuminous.

épisser, to splice.

épistaminal, growing on the stamens, as hairs.

épistaminé, gynandrous.

épister, to reduce to paste (by crushing).

épisternum, *m.* episternum.

épistome, *m.* epistome, nasus.

épistrophie, *f.* epistrophy.

épiteospore, *f.* uredospore, urediniospore.

épithalle, *m.* epithallus, the cortical layer of lichens.

épithalame, *m.* epithalamium, marriage ode.

épithecium, *m.* epithecium.

épithélial, -aux, epithelial.

épithélioide, epithelioid.

épithélioma, *m.* epithelioma (epithelial cancer).

épithélium, epithelium, cuticle.

épithème, *m.* epithem.

épithète, *f.* epithet.

épitrope, epitropous, anatropous (ovule).

épizoïque, epizoic.

épizootie, *f.* epizootic disease.

épizootique, epizootic.

épluchage, *m.* picking, peeling, paring, weeding, thinning, scrutiny.

éplucher, to pick, peel, clean, weed, scrutinize, examine.

épluchures, *f.pl.* pickings, leavings, parings, refuse.

épointer, to break the point of.

éponge, *f.* sponge, capelbow (horse); — **de platine,** platinum sponge.

épongeage, *m.* sponging.

éponger, to sponge.

époque, *f.* epoch, time, period, era, season; — **de la mue,** moulting time; — **de passage des oiseaux,** time of migrations; — **du rut,** rutting time; — **glaciaire,** ice age.

épouillage, delousing.

épouse, *f.* spouse, wife.

épousée, *f.* bride.

épouser, to espouse, marry, take the form of, follow.

époussetage, *m.* dusting.

épousseter, to dust.

époussetoux, -euse, dusting, dust-removing.

époussetoir, *m.* small dusting brush.

époussette, *f.* duster, dust brush, dustcloth.

épouti, *m.* burl, filth (in wool).

époutier, époutir, to burl (cloth), clean, pick.

épouvantable, frightful, dreadful.

épouvantail, *m.* scarecrow.

épouvante, *f.* terror, fright.

épouvanter, to frighten, scare, terrify, dismay.

époux, *m.* spouse, husband.

épreignant, pressing (out).

épreindre, to squeeze, press (out), express.

épreint, squeezed, pressed (out).

épreuve, *f.* test, trial, experiment, proof, testing, assay, essay; — **à chaud,** hot test; — **à froid,** cold test; — **à outrance,** resistance or breaking test; — **de cassure,** breaking or fracture test; — **de chaleur,** heat test; — **de pression,** pressure test; — **de traction,** tensile or tension test; **à l'—,** on trial; **à l'— de,** proof (against); **à l'— de l'eau,** waterproof; **à l'— du feu,** fireproof; **mettre à l'—,** to assay, try, prove, test.

éprouver, to test, try, prove, feel, experience, undergo.

éprouvette, *f.* eprouvette, test tube, (narrow upright) cylinder, (liquor) gauge, gas bottle, test bar, test piece, small spoon (for fluxes), probe; — **à dessécher les gaz,** drying cylinder, drying bottle; — **divisée,** — **graduée,** graduated cylinder.

épuisable, exhaustible.

épuisant, exhausting.

épuise, *f.* — **volante,** pumping station.

épuisé, exhausted, impoverished, spent, drained, worn out, out of print.

épuisement, *m.* exhaustion, draining, fainting, impoverishment of the soil, thinning out (of bacteria).

épuiser, to exhaust, extract, spend, drain, use up; **s'—,** to waste, wear out, become exhausted.

épulpeur, *m.* pulper (to extract beet juice).

épuratein, *m.* purifier, scrubber, refiner, filter, cleaner, knotter, jog knotter.

épurateur, purifying, refining, cleaning.

épuratif, -ive, purifying (process or apparatus).

épuration, *f.* purification, purifying, refining, scrubbing, filtering, cleaning, refinement; — bactérienne, bacteriological purification; — biologique, biological purification.

épuratoire, purifying.

épure, *f.* — des aires, leveling of surfaces.

épuré, purified, filtered, refined.

épurement, *m.* purification, refinement.

épurer, to purify, filter, refine, scrub, clarify, clean, cleanse; s'—, to be purified.

équanimité, *f.* evenness of temper.

équarrir, to square, flay, strip, cut up (carcass).

équarrissage, *m.* squaring (timber), cutting up (of carcass), knacker's trade; d'—, on the square.

équateur, *m.* equator.

équation, *f.* equation; — du premier degré, linear equation; — du second degré, quadratic equation; — du troisième degré (ordre), cubic equation.

équatorial, -aux, equatorial.

équerrage, *m.* bevel, squaring, beveling.

équerre, *f.* square, angle meter, right-angled brace, right-angled knee or elbow; à—, square.

équerrer, to square, bevel.

équiangle, equiangular.

équidés, *m.pl.* horses, Equidae.

équidistant, equidistant.

équienne, even-aged, of the same year.

équilatéral, -aux, équilatère, equilateral.

équilibrage, *m.* equilibration, balancing.

équilibre, *m.* equilibrium, balance; — chimique, chemical equilibrium; — de poids, counterbalance of weight; — dynamique, stability.

équilibrer, to equilibrate, balance, poise, place in equilibrium.

équimoléculaire, equimolecular.

équinoxe, *m.* equinox.

équinoxial, -aux, equinoctial.

équipage, *m.* equipment, outfit, gearing, equipage, crew; — de chasse, hunting equipment.

équipe, *f.* gang, crew, team, party, shift (of workmen); — de nuit, night shift.

équipé, equipped, ready for use.

équipement, *m.* equipment, outfit.

équiper, to equip, fit out; s'—, to equip oneself.

équipotentiel, -elle, equipotential.

équisétacé, equisetaceous.

équitable, equitable, just.

équitant, equitant, folded over, as if astride.

équité, *f.* equity, justice.

équivalant, equivalent.

équivalence, *f.* equivalence.

équivalent, *m.* return service.

équivalent, equivalent.

équivaleur, *f.* equivalence.

équivaloir, to be equivalent, equal (to).

équivaut (équivaloir), (it) is equal to.

équivoque, *f.* ambiguity (of expression), equivocation.

équivoque, equivocal, doubtful, ambiguous.

érable, *m.* maple (Acer); — à sucre, sugar maple, rock maple (*Acer saccharum*); — champêtre, common maple, field maple (*Acer campestre*); — faux platane, sycamore (*Acer pseudo-platanus*); — plane, Norway maple (*Acer platanoides*).

éradication, *f.* rooting out.

érafler, to scratch, graze.

éraflure, *f.* grazing wound.

erbine, *f.* erbia, erbium oxide.

erbium, *m.* erbium.

erbue, *f.* clay flux.

ère, *f.* era, period.

érecteur, *m.* erector (muscle).

érection, *f.* erection.

éreinter, to break back (of horse), exhaust.

éremacausie, *f.* full (gradual) decomposition.

éremophyte, *f.* eremophyte, desert plants.

érepsine, *f.* erepsin.

erg, *m.* erg (unit measure), sand-hill.

ergésie, *f.* ergesis, the ability of an organ to exhibit reaction.

ergostérine, *f.* ergosterol.

ergot, *m.* ergot, spur, lug, stub (on fruit tree); — de seigle, ergot of rye.

ergoté, ergoted, spurred (as rye), dew-clawed.

ergotine, *f.* ergotin(e).

ergotinine, *f.* ergotinine.

ergotique, ergotic.

ergotisme, *m.* ergotism, ergotic poisoning.

erianthe, *m.* erianthus.

érianthe, woolly-flowered.

éricacé, ericaceous, heathlike.

éricacées, f.pl. Ericaceae, heathers, heaths.

éricétin, ericetinous, growing on heaths.

ériciné, ericaceous.

éricoïde, ericoid, (leaves) like those of heaths.

ériger, to erect, establish.

érigéron, m. fleabane (Erigeron).

érinose, f. erinose.

ériocalicé, with hairy calyx.

ériocarpe, with hairy fruit.

ériocaule, with hairy stem.

érioclade, with hairy branches.

érion, m. wool, nap, down.

ériopétale, with hairy petals.

ériophylle, eriophyllous, woolly-leaved.

ériosperme, with hairy seed.

ériostachyé, with flowers on hairy head.

ériostémone, with hairy stem.

ériostyle, with hairy style.

érisma, f. erisma, rachis in grasses.

érismata, f. filament.

ermine, f. ermine.

erminette, f. adz, trimming ax.

ermite, m. hermit.

érodé, eroded.

éroder, to erode.

érosif, -ive, erosive.

érosion, f. erosion; — par lavage, erosion, denudation (by water).

érostré, erostrate, beakless.

errant, wandering, erring, errant, vagrant, mistaken.

erratique, erratic, irregular, wandering.

errer, to err, wander, be mistaken, stray.

erreur, f. error, mistake.

erroné, erroneous, mistaken.

erronément, erroneously, by mistake.

ers, m. tare, vetch, lentil.

ersatz, m. substitute.

éructation, f. eructation, belching.

érudit, m. scholar, learned man.

érugineux, -euse, aeruginous.

éruptif, -ive, eruptive.

éruption, f. eruption, rash.

érysipèle, m. erysipelas.

érysiphés, f.pl. powdery mildew (Erysiphaceae).

érythème, m. erythema.

érythrée, f. erythraea.

érythrine, f. erythrine, red cobalt.

érythrite, f. erythrite, cobalt bloom.

érythrocyte, f. erythrocyte.

érythrodextrine, f. erythrodextrin.

érythrophylle, f. erythrophyllin, red coloring of leaves.

érythrorhize, with red roots.

érythrosine, f. erythrosin.

érythrosperme, with red seed.

érythrostome, m. erythrostomum, an aggregate fruit.

es, (you) are.

és (en les), Maître ès arts, master of arts.

esc , abbr. (escompte), discount.

escalade, f. scaling.

escalader, to scale.

escalier, m. staircase, stairs, stairway.

escarbille, f. (coal) cinder, clinker.

escarbot, m. hister; — de la farine, meal beetle.

escarboucle, f. carbuncle.

escargot, m. snail, spiral, spiral staircase; en —, spiral.

escarner, to flesh (hides).

escarole, f. escarole, endive (Cichorium endivia).

escarpé, escarped, steep, cragged, abrupt, precipitous.

escarpement, m. escarpment, steep, crag.

eschare, f. scab, eschar.

escient, à bon —, knowingly.

esclavage, m. slavery.

esclave, m.&f. slave.

esclavon, -onne, Slavonian, Slavic.

escompte, m. discount, discounting; — hors banque, market rate (of discount).

escompter, to discount; s'—, to be discounted.

escope, f. see ÉCOPE.

escorage, m. — du lin, rolling after retting (flax).

escorter, to escort, convoy.

escouade, f. squad; — d'ouvriers, shift or squad of workmen.

escoupe, f. round shovel, scoop.

escourgeon, m. winter barley.

escrime, f. fencing.

escrimer(s'), to fight, scrimmage.

escroc, *m.* swindler.

escroquer, to swindle, cheat.

esculine, *f.* esculin.

ésérine, *f.* eserine, physostigmine.

espace, *m.* space, room; — clos, closed (or enclosed) space; — couvert, shaded area, sheltered space; — mort, dead space, waste space; — nuisible, noxious clearance, waste space; — vide, vacuum, open space.

espacé, roomy, thinly stocked.

espacement, *m.* spacing, interval; — des plantes, planting distance.

espacer, to space, space out, separate.

espagnol, *m.* Spanish (language).

espagnol, Spanish.

espalier, *m.* railing, trellis, espalier.

espar, *m.pl.* spar.

esparcette, *f.* sainfoin (onobrychis).

espèce, *f.* species, kind, sort; *pl.* specie, breed.

espérance, *f.* hope, expectation.

espérer, to hope, expect.

espinal, *m.* espinal.

esplanade, *f.* esplanade, open level space, terrace.

espoir, *m.* hope, expectation, expectancy.

esprit, *m.* spirit, mind, wit, soul, meaning, intellect; — de bois, wood spirit, methylated spirit; — de camphre, spirit of camphor; — de cannelle, spirit of cinnamon; — de nitre, spirit of niter; — de pétrole, petroleum spirit; — de raifort composé, compound spirit of horseradish; — de sel, spirit of salt, hydrochloric acid; — de sel ammoniac, spirit of sal ammoniac; — de vinaigre, glacial acetic acid; — d'orange composé, compound spirit of orange.

esprit-de-vin, *m.* spirits of wine, alcohol.

esquille, *f.* splinter (of bone).

esquilleux, -euse, splintery, comminuted.

Esquimau, *m.* Eskimo.

esquinancie, *f.* quinsy.

esquine, *f.* chinaroot, see SQUINE.

esquisse, *f.* sketch, outline, draft.

esquisser, to sketch, outline.

esquiver, to evade, avoid; s'—, to escape, give the slip.

essai, *m.* test, testing, analysis, assay, assaying, trial, sample, proof, attempt, experiment; — à blanc, blank test; — à chaud, hot test; — à froid, cold test; — d'agglutination, agglutination test; —

de pression, (air) pressure test; — des graines, germinating test; — d'essai, sample for analysis; — par voie humide, wet analysis, wet assay(ing); — par voie sèche, dry analysis, dry test, dry assay.

essaim, *m.* swarm, hive, flock.

essaimage, *m.* swarming.

essaimer, to swarm.

essangease, *m.* soaking (clothes in laundries).

essartement, *m.* clearing; — de protection, severance felling.

essarter, clear, assort.

essayage, *m.* testing.

essayé, analyzed, essayed.

essayer, to test, try, assay, analyze, taste, attempt, endeavor.

essayerie, *f.* assay office.

essayeur, *m.* assayer.

esse, *f.* the letter S, S-shaped object, wire gauge.

esséminer, to disseminate, scatter (seed).

essence, *f.* (essential) oil (of lemons, etc.), oil of turpentine, essence, substance, any volatile liquid, benzine, petrol, gasoline, species, variety (of tree), nature, spirit, natural quality, forest species; — d'abri, nurse, shelter tree, shelter wood; — d'absinthe, oil of wormwood; — d'aiguilles de pin, pine-needle oil, pine oil; — d'amandes amères, essence or oil of bitter almonds; — d'aneth, oil of dill; — d'automobile, gasoline, petrol.

essence, — de bigarade, oil of bitter-orange peel; — de bouleau, — de bétula, oil of sweet birch; — de cajeput, oil of cajuput; — de carvi, oil of caraway; — de citron, oil of lemon; — de copahu, oil of copaiba; — d'écorce de bouleau, birch-bark oil; — de couverture, soil-protecting tree; — de fenouil, oil of fennel.

essence, — de genièvre, oil of juniper; — de lavande, oil of lavender; — de lumière, light demander, light-bearing tree; — de malt, extract of malt; — de menthe poivrée, oil of peppermint; — de menthe verte, oil of spearmint; — de mirbane, oil (or essence) of mirbane, nitrobenzene; — de moutarde, mustard oil; — de muscade, nutmeg oil, oil of myristica.

essence, — de pin sylvestre, pine-needle oil, pine oil; — de Portugal, sweet-orange oil; — de pouliot américaine, oil of (American) pennyroyal; — de romarin, oil of rosemary; — **de santal,** oil of sandalwood; — **de térébenthine,** oil of turpentine; — **d'eucalyptus,** oil of eucalyptus; — **d'ombre,** shade bearer, shade-bearing tree; — **feuil:ue,** broad-leaved tree; — **minérale,** gasoline, petrol; —**s de couverture,** oil-protecting species.

essencifier, to essentificate.

essentiel, -elle, essential, indispensable, ethereal, volatile.

essentiellement, essentially, mainly.

esser, to gauge (wire).

essieu, *m.* axle, axletree, axis, spindle.

essor, *m.* flight, progress, advance, swing, development, free play, soaring.

essorage, *m.* drying, seasoning, removal by suction.

essorer, to dry (air, wringing, or centrifugal dryer), season, drain.

essoreuse, *f.* drying machine, centrifuge, centrifugal drier, centrifugal filter.

essouchement, *m* grubbing out stumps, uprooting.

essoucher, to grub out stumps, root up.

essuie-main(s), *m.* (hand) towel.

essuie-plume, *m.* penwiper.

essuyage, *m.* wiping.

essuyant, wiping, drying, enduring.

essuyer, to wipe, wipe off, dry, sweep away, endure, suffer, stand.

est, (he) is.

est, *m.* east; à l'—, eastward.

estacade, *f.* stockade, platform, breakwater.

estagnon, *m.* container (made of copper or tin), oil can.

estaminet, *m.* café, bar.

estampage, *m.* stamping, branding.

estampe, *f.* stamp, print.

estamper, to stamp, impress.

estampeuse, *f.* stamping machine.

estampillage, *m.* stamping.

estampille, *f.* stamp, mark, trademark.

estampiller, to stamp, mark.

estampilleuse, *f.* stamping machine.

estampoir, *m.* stamp, punch.

est-ce-que, (introducing a question), **est-ce-qu'il est là?** is he there?

esthésie, *f.* esthesis.

estimation, *f.* estimation, estimate, valuation.

estime, *f.* estimation, esteem, dead reckoning, respect.

estimé, valued.

estimer, to estimate, value, esteem, rate, regard, think, deem; **s'**—, to consider oneself.

estival, -aux, summer, estival (plant).

estivation, *f.* estivation.

estoc, *m.* stock, trunk (of tree), tuck (a sword), estoc.

estomac, *m.* stomach, maw, gizzard, paunch.

estompé, blunted, dull, softened.

estragol, *n.* estragole.

estragon, *m.* tarragon (*Artemisia dracunculus*).

estran, *m.* tidal flat.

estraquelle, *f.* ladle.

estrasse, *f.* floss silk.

estrigue, *m.* annealing oven, leer.

estropier, to cripple, mutilate, mangle.

estuaire, *m.* estuary, strand.

esturgeon, *m.* sturgeon.

ésule, *f.* spurge.

et, and; et . . . et, both . . . and.

établage, *m.* stalling (cattle).

étable, *f.* stable.

établi, *m.* workbench.

établi, established.

établir, to establish, fix, set, place, build, make, erect, make certain, set up, settle, prove; **s'**—, to establish oneself, be established.

établissement, *m.* establishment, settlement, setting up, business house, building, establishing, colony, works, company, plant, installation; — **de plans,** surveying; — **d'un projet,** plan verification.

étage, *m.* story, floor, layer, row, tier, degree, step, level, stage, grade, range; — **inférieur,** under tier, lower story; — **supérieur,** upper tier, upper story; **au premier** —, on the second floor; **de premier** —, initial stratum.

étagé, growing in stories.

étager, to arrange in rows or tiers.

étagère, *f.* shelf, set of shelves; — **en bois,** wooden stand.

étai, *m.* stay, prop.

étaim, *m.* fine carded wool.

étain, *m.* tin, pewter; — **battu,** tin foil; — **de glace,** bismuth; — **en feuilles,** sheet tin; — en grains, grain tin; — en grenaille, granulated tin; — en verges, bar tin.

étairion, f. etaerio, aggregate fruit.

étal, m butcher's stall.

étalage, m. display, bosh (of furnace).

étale, slack, spreading, spread out.

étalement, m. — du sang, making a blood film.

étaler, to spread, spread out, display, expose, show off; s'—, to spread (out), open, smear.

étalon, m. standard, stallion.

étalon, standard, normal; pile —, standard cell.

étalonnage, étalonnement, m. standardization, calibration, gauging, stamping, marking.

étalonner, to standardize, calibrate. gauge, stamp, mark (weights), (of stallion) to serve.

étamage, m. tinning, tin plating, tinned work; — au zinc, galvanizing, zincking; — des glaces, silvering of mirrors or glass.

étamer, to tin, tin-plate, silver, zinc, galvanize.

étamine, f. bolting cloth, tammy, stamen; — pétaliste, petaloid stamen; à trois —s, triandrous, triandrian.

étampage, m. stamping, swaging.

étampe, f. stamp, punch, swage, die.

étamper, to stamp, punch, swage.

étamper, f. nail hole.

étamure, f. coating of tin, tinning metal.

étanche, water and air tight, tight; — à la vapeur, steamtight; — à l'eau, — d'eau, watertight; à —, tight.

étanchéité, f. tightness (air, water).

étancher, to stop, check the flow of, stanch, make tight, calk, quench (thirst).

étançon, m. prop, stay, stud, shore, frame, standard.

étandard, m. standard.

étang, m. pond, pool; — à frai, — d'éclosion, breeding pond; — d'accroissement, — d'alevinage, pond for fry; — d'hivernage, wintering pond.

étant, being, standing.

étape, f. stage, halting place, stop.

état, m. state, condition, calling, profession, trade, status, list, report, account, government, location; — clair, open stand, open crop; — clairière, roomy stand, glade;

— critique, critical state; — d'assiette des coupes, plan of cuttings; — de massif, close crop, dense stand; — d'isolement, isolated state or stand; — gaseux, gaseous state; — naissant, nascent state.

état, à l'— brut, unwrought; à l'— de, in the position of; à l'— libre, — de liberté, in the free state; en l'—, this being the case; en — de, in a condition to; faire — de, to depend, count on, intend, value; faire — que, to think, believe; hors d'—, out of repair.

étatisation, f. nationalization.

état-major, m. staff, staff office.

Etats-Unis, m.pl. United States.

étau, m. vise; — à main, hand vise.

étaux, m.pl. see ÉTAL and ÉTON. butcher shops or stalls, vises.

étayer, to prop, stay, support, shore up, hold up, strengthen.

etc., abbr. (et caetera), and so forth.

été (être), been.

été, m. summer.

éteignant, quenching.

éteindre, to extinguish, put out, quench, slake (lime), cancel, deaden, dim, soften, allay; s'—, to be extinguished, go out, be quenched, become extinct, die down, be dimmed.

éteint, extinguished, slaked, out, dull faint, dying, extinct, dead.

étendage, m. diluting, stretching, flattening, steaming, drying yard, drying rack or room, clothes lines.

étendard, m. standard, banner, flag.

étendoir, m. drying loft.

étendre, to dilute, stretch, stretch out, spread, spread out, extend, enlarge, flatten; s'—, to be diluted, sprawl, spread, stretch oneself, extend, reach, expand, dwell (upon).

étendu, dilute, diluted, stretched, extended, extensive, long, broad.

étendue, f. extension, reach, extent, compass, length, expanse, space, extensiveness.

éternel, -elle, eternal, everlasting.

éternellement, eternally, forever, forevermore.

éternité, f. eternity.

éternuement, éternûment, m. sneezing sneeze.

éternuer, to sneeze.

êtes (you), are.

étêté, pollarded.

étêtement, *m.* pollarding, topping.

étêter, to remove the head of, top (trees), pollard.

éthal, *m.* ethal, cetyl alcohol.

éthalique, cetyl (alcohol).

éthanal, *m.* ethanal, acetaldehyde.

éthane, *m.* ethane.

éthanol, *m.* ethanol, ethyl alcohol.

éther, *m.* ether, ester; — acétique, acetic ester (acetic ether); — acide, ester, acid ester, hydrogen ester; — alcoolerise, ether and alcohol (1:1); — amylazoteux, — amylnitreux, amyl nitrite; — azoteux, nitrous ester, nitrous ether, ethyl nitrite; — azotique, nitric ester, ethyl nitrate; — chlorhydrique, hydrochloric ester, ethyl chloride; — de pétrole, petroleum ether; — éthylique, ethyl ether, ethyl ester.

éther, — hydriodique, ethyl iodide; — hydrochlorique, ethyl chloride; — méthylique, methyl(ic) ether, methyl ester; — neutre, neutral ester, neutral ether; — nitreux, nitrous ester or ether, ethyl nitrite; — nitrique, nitric ester, ethyl nitrate; — officinal, (purified) ether; — ordinaire, ordinary ether; — sel, ethereal salt, ester; — sulfuré, thio ether, sulphur ether; — sulfurique, sulphuric ether.

éthéré, ethereal, volatile.

éthérificateur, *m.* etherifying apparatus.

éthérification, *f.* esterification, etherification.

éthérifier, to esterify, etherify.

éthérique, ethereal, ether, pertaining to ether.

éthérisation, *f.* etherification, esterification, etherization.

éthériser, to etherify, esterify, etherize.

éthérolique, having other as the excipient.

éther-oxyde, *m.* ether, organic oxide.

éther-sel, *m.* ethereal salt, ester.

éthionique, ethionic.

éthiops, *m.* ethiops; — martial, black iron oxide; — minérale, black mercuric sulphide.

ethmoïde, *m.* ethmoid.

ethnographie, *f.* ethnography.

éthologique, ethological.

éthoxyle, *m.* ethoxyl.

éthylamine, *f.* ethylamine.

éthylbenzine, éthylbenzène, *f.* ethylbenzene.

éthyle, *m.* ethyl.

éthylé, ethyl, ethylated.

éthylène. *m.* ethylene.

éthylénique, ethylene, ethylenic.

éthyler, ethylate.

éthylidénique, ethylidene.

éthylique, ethyl, ethylic.

étiage, *m.* low-water mark.

étincelant, sparkling, sparking, flashing, brilliant.

étinceler, to sparkle, flash, glitter.

étincelle, *f.* spark, flash, scintillation; — électrique, electric spark.

étincellement, *m.* sparkling, glittering, scintillation.

étiolé, etiolated, blanched, whitened.

étiolement, *m.* etiolation, whitening, declining.

étioler, to etiolate, become pale, be whitened, blanch, wilt.

étioline, *f.* etiolin.

étiologie, *f.* etiology.

étiologique, etiological.

étique, emaciated, hectic.

étiqueter, to label, docket, ticket.

étiquette, *f.* etiquette. label, formality.

étirable, ductile, elastic, drawable.

étirage, *m.* drawing (out), stretching.

étiré, drawn, long, drawn out.

étirer, to draw, draw out, roll, stretch (out), lengthen, mill; s'—, to be drawn out, be stretched, stretch oneself.

étisie, *f.* emaciation, wasting, decline.

étoffe, *f.* fabric, cloth, stuff, material, a tawing liquor (of alum and salt).

étoffé, ample, stuffed, padded, stout, stocky.

étoile, *f.* star, radial (star-shaped) crack; — de mer, starfish; à la belle —, in the open air.

étoilé, starry, stellate, star-shaped.

étoiler, to crack (radially), star.

étonnamment, astonishingly.

étonnant, astonishing.

étonnement, *m.* astonishment, wonder; surprise, amazement.

étonner, to astonish, amaze, astound. crack, rend, fissure; s'—, to be astonished, wonder.

étouffant, suffocating, stifling.

étouffé, suppressed, choked, dammed.

étouffée, *f.* stewing, stew.

étouffement, *m.* suffocation, stifling, smothering.

étouffer, to choke, suffocate, stifle, smother, muffle, suppress; s'—, to be suffocating, be suffocated, choke.

étouffoir, *m.* charcoal extinguisher, damper.

étoupage, *m.* packing, stuffing.

étoupe, *f.* tow, oakum, packing, stuffing, waste (of cotton).

étoupement, *m.* stopping, calking, packing.

étouper, to stop (crevices) with tow or oakum, pack, stuff, calk, chinse.

étouperie, *f.* tow packing cloth.

étoupille, *f.* primer, fuse, quick match; — à friction, friction fuse or primer; — à percussion, percussion fuse; — de sûreté, safety fuse.

étoupiller, to fuse, provide with quick match.

étourderie, *f.* inadvertency, carelessness, blunder.

étourdi, thoughtless, heedless, stunned, numb, slightly warm, parboiled, allayed.

étourdiment, heedlessly, inconsiderately.

étourdir, to stun, benumb, allay (pain), take chill (off water), parboil.

étourdissement, *m.* dizziness, stunning, astonishment.

étourneau, *m.* starling.

étrange, strange, odd, queer, foreign.

étrangement, strangely, oddly.

étranger, *m.* foreigner, stranger, strangeness, foreign parts; à l'—, abroad, in foreign lands.

étranger, -ère, foreign, strange, extraneous, unknown.

étrangeté, *f.* strangeness, peculiarity.

étranglement, *m.* constriction, narrowing, choking, choke, throttling, strangling, condensing.

étrangler, to constrict, narrow, choke, throttle, strangle, condense.

étrangleur, *m.* throttle valve.

étrangloir, *m.* choking frame.

étranguillon, *m.* quinsy, strangles.

être, to be, exist, have; en —, to be, be at, be of it.

(il) **est**, ça y —, that's it, it's done now; c'est-à-dire, that is; c'— à faire, it is to be done; c'— ça, that's right; c'— égal, that makes no difference; c'— là, that is; c'— qu'il a tort, that fact is that he is wrong; c'— son affaire, that suits him; ce m'— égal, it is all the same to me; est-ce que, (introducing a question), est-ce qu'il — là? is he there?

(il) **est**, there is, there are; il — allé, he went; il — d'accord avec moi, he agrees with me; il — en train de lire, he is busy reading; il — hors de lui, he is beside me; il — venu, he came.

(il) **est**, il en — ainsi, such is the case; il en — de même, the same is the case; il serait pour, he would be capable of; il y —, he understands; il y — pour quelque chose, he has a hand in it; n'est-ce pas? isn't (he, she, etc.)?; on — deux, we are two; qu'ils soient . . . ou, either . . . or; quoi qu'il en soit, however that may be; soit, so be it, that is, let there be given; soit . . . soit, either . . . or.

être, *m.* being, existence, creature, tree trunk; — vivant, living being.

étrécir, to narrow, shrink; s'—, to become narrower, shrink, contract.

étrécissement, *m.* narrowing, shrinking, shrinkage, contraction.

étreignant, embracing, tying.

étreindelle, *f.* a press bag of horsehair.

étreindre, to bind, tie, tie up, clasp, press.

étreint, bound.

étreinte, *f.* binding, tie, embrace, constraint, clasp, pressure, clasping, hugging.

étrésillon, *m.* prop, stay, shore, brace.

étrésillonnement, *m.* propping.

étrier, *m.* stirrup, support, strap, brace, hoop, band, clamp.

étriper, to gut (fish), draw (chicken).

étroit, narrow, tight, slender, strait, strict, close, intimate.

étroitement, narrowly, tightly, closely, straitly.

étroitesse, *f.* narrowness.

étude, *f.* study, research, designing, office, rough plan, design; à l'—, under investigation.

étudiant, *m.* student.

étudier, to study, design.

étui, *m.* case, box, cover, envelope, sheath.

étuvage, *m.* stoving, drying (sugar), baking (mold), fomenting, seasoning (wool), sweating (leather).

étuve, *f.* oven, closet, bath, drying room, drying oven, stove, sweating room, (hot-air or vapor) bath, incubator; — à air chaud, hot-air oven; — à cultures, incubator; — à dessécher dans le vide, vacuum drying oven; — à eau, hot-water oven; — à vapeur, steam (heated) oven: — froide, ice closet, refrigerator; sécher à l'—, to dry in the oven.

étuvement, *m.* stoving, drying.

étuver, to stove, heat, dry (in an oven), sweat, bake, disinfect, foment, stew, steam.

eu (avoir), had.

eucalyptus, eucalypte, *m.* blue gum tree (Eucalyptus).

eucalyptol, *m.* eucalyptole, cineole.

eucharis, *m.* eucharis.

euchlorine, *f.* euchlorine (mixture of chlorine and chlorine dioxide).

euclase, *f.* euclase.

euclidien, -enne, Euclidean.

eudiomètre, *m.* eudiometer.

eudiométrique, eudiometric(al).

euéphémère, euephemerous, (flower) opening and closing within 24 hours.

eufrais, *f.* eyebright, see EUPHRAISE.

eugénine, *f.* eugenin.

eugénol, *m.* eugenol.

eugéogène, eugeogenous, disintegrating easily in air.

euglobuline, *f.* euglobulin.

eupatoire, *f.* eupatory, hemp agrimony (*Eupatorium cannabinum*).

euphorbe, *f.* euphorbia, spurge (Euphorbia).

euphorbiacées, *f.pl.* Euphorbiaceae.

euphotique, euphotic, (hydrophytes) receiving abundant light.

euphotométrique, euphotometric,leaves turn to light.

euphraise, euphrasie, *f.* euphrasy, eyebright (*Euphrasia officinalis*).

euphylle, *f.* euphylla, true leaves.

euphytoide, *f.* euphytoid, a plant parasite.

eupione, *f.* eupione.

eurent (avoir), (they) had.

eurhodine, *f.* eurhodine.

eurhodol, *m.* eurhodol.

Europe, *f.* Europe.

européen, *m.* European.

européen, -enne, European.

europhène, *m.* europhen.

europium, *m.* europium.

euryhaline, euryhaline.

euryphote, euryphotic, adapted to light of varying intensity.

eusporangié, eusporangiate.

eut (avoir), (he) had.

eutectique, eutectic.

eutexie, *f.* eutexia.

eutrope, eutropous.

eux, they, them.

eux-mêmes, themselves.

évacuant, evacuant.

évacuation, *f.* evacuation, emptying, shipping, exhaust, eduction; tuyautage d'—, evacuation pipe.

évacuer, to evacuate, empty, remove, ship; s'—, to be evacuated, empty.

évader, to evade, escape.

évaginulé, deprived of sheath.

évaluable, evaluable, appraisable, estimable.

évaluation, *f.* valuation, estimation, estimate, assessment.

évaluer, to evaluate, value, estimate, rate, calculate.

évalve, evalvis, destitute of valves.

évanescent, evanescent.

évanouir, to vanish, disappear, fade, faint; faire —, to clear, eliminate.

évanouissement, *m.* disappearance, elimination, fainting, swoon, senselessness.

évaporable, evaporable.

évaporateur, *m.* evaporator.

évaporatif, -ive, evaporative.

évaporation, *f.* evaporation.

évaporatoire, evaporating (process).

évaporé, evaporated, flighty, irresponsible.

évaporer, to evaporate, vent; s'—, to evaporate, find vent, become heedless, volatilize.

évaporeuse, *f.* evaporator.

évaporisation, *f.* evaporation.

évaporomètre, *m.* evaporometer, atmometer.

évasé, bell-mouthed, widened, enlarged, flaring, funnel-shaped, splayed.

évasement, *m.* widening, enlargement, spreading, splay, width (of the opening).

évaser, to widen (opening), enlarge, spread flare; s'—, to widen, become flaring.

évasion, *f.* escape, flight, evasion.

évasure, *f.* widening, (flaring) opening.

évêché, *m.* bishopric.

éveil, *m.* warning, alarm, awakening, hint.

éveiller, to wake, awake, be aroused, arouse, quicken, call forth; s'—, to wake up, awake.

éveilleur, *m.* awakener, starter.

événement, *m.* event, occurrence, emergency, issue, consequence.

évent, *m.* open air, mustiness, flatness, aperture, air hole, vent, escape hole, flash hole, cavity, crack, spout (of whale).

éventable, alterable in air, easy to ventilate.

éventage, *m.* airing, exposure.

éventail, *m.* fan, batswing (burner), rhipidium, fan-shaped tree, (enameler's) screen; **d'—, en forme d'—,** fan-shaped, flabellate; **en —,** fan-shaped.

éventer, to air, expose to the air, damage, spoil by exposure to air, make (wine) musty or flat, discover, divulge, scent, get wind of, find out, fan; **s'—,** to be spoiled by exposure to air (food), (of wine) become flat.

éventrer, to rip open, disembowel, eviscerate, gut.

éventualité, *f.* eventuality, contingency.

éventuel, -elle, possible, contingent, eventual.

éventuellement, possibly, contingently, eventually.

éventure, *f.* crack, flaw.

évêque, *m.* bishop.

éverdumer, to bleach, blanch (green juices).

évertuer(s'), to do one's utmost.

évidage, *m.* hollowing (out), scooping out.

évidé, hollowed (out), hollow, grooved.

évidement, *m.* hollow, groove, cavity, recess, aperture, grooving.

évidemment, evidently, obviously, certainly.

évidence, *f.* evidence, obviousness, clearness; **de toute —,** fully, evident; **mettre en —,** to make evident, obvious.

évident, evident, obvious, plain, clear.

évider, to hollow (out), scoop (out), groove

évidure, *f.* hollow, cavity, groove.

évier, *m.* sink, gutter stone, sinkstone.

éviscérer, to eviscerate.

évitable, avoidable.

évitement, *m.* avoiding, avoidance, shunting, siding, turnout.

éviter, to avoid, prevent, shun; **— à,** to spare.

évolué, evolved, developed.

évoluer, to evolve, perform evolutions, develop.

évolutif, -ive, evolutive, evolutional.

évolution, *f.* evolution.

évoquer, to evoke, call to mind, call forth.

exact, exact, accurate, precise, strict, punctual.

exactement, exactly, accurately.

exactitude, *f.* exactness, exactitude, accuracy, precision, correctness.

exagération, *f.* exaggeration.

exagéré, extreme.

exagérément, too much.

exagérer, to exaggerate, make too large, increase, magnify.

exalbuminé, exalbumineux, -euse, exalbuminous.

exalgine, *f.* exalgin.

exaltation, *f.* exaltation, (over-)excitement.

exalter, to exalt, excite, extol, magnify.

examen, *m.* examination; **— sur pied,** field inspection.

examinateur, -trice, *m.&f.* examiner.

examiner, to examine, inquire into, assay, inspect.

exanthématique, exanthematic.

exanthème, *m.* exanthema.

excavateur, *m.* excavator, digging machine.

excavation, *f.* excavation, excavating.

excaver, to excavate, hollow out.

excédant, excess, surplus, tiresome.

excédent, *m.* excess, surplus; **— de puissance,** margin of power.

excéder, to exceed, go beyond, tire, wear out.

excellemment, excellently.

excellence, *f.* excellence, excellency; **par —,** preeminent(ly), par excellence.

excellent, excellent, to the highest degree.

exceller, to excel.

excentrer, to throw off center, decenter, make eccentric.

excentricité, *f.* eccentricity.

excentrique, eccentric, one-sided, abaxial.

excentriquement, eccentrically.

excepté, except, excepting, save, but, excepted; **— que,** except that, unless.

excepter, to except, make an exception of, exclude.

exception, *f.* exception; **à l'— de,** with the exception of; **par —,** by way of exception; **sauf —,** with certain exceptions.

exceptionnel, -elle, exceptional.

exceptionnellement, exceptionally, in exceptional cases.

excès, *m.* excess, exaggeration, immoderation; **— de maturité,** overripeness; **à l'—,** **jusqu'à l'—,** to excess, immoderately.

excessif, -ive, excessive, extreme, extravagant, immoderate.

excessive, *f.* excess.

excessivement, excessively, extremely, immoderately.

excipient, *m.* excipient, vehicle, menstruum.

excise, *f.* excise.

exciser, to dissect out.

excision, *f.* excision.

excitant, exciting, excitant, stimulating.

excitateur, *m.* exciter, discharger, excitator.

excitateur, -trice, exciting.

excitatif, -ive, excitative.

excitation, *f.* excitation, excitement, stimulus.

exciter, to excite, provoke, incite, irritate, arouse, stimulate.

exclure, to exclude.

exclusif, -ive, exclusive.

exclusion, *f.* exclusion, cutoff; **à l'— de,** exclusive of, excepting.

exclusivement, exclusively.

exclut, excluded.

excoriation, *f.* excoriation.

excorier, to excoriate.

excortiquer, to excorticate, decorticate.

excrément, *m.* excrement.

excrémenteux, -euse, excrémentiel, -elle, excrémentitiel, -elle, excremental, excrementitious.

excreta, *m.pl.* excreta.

excréter, to excrete.

excréteur, -trice, excretive, excretory.

excrétion, *f.* excretion.

excrétoire, excretive, excretory.

excroissance, *f.* excrescence, outgrowth; **— du trône,** tree wart, rind gall.

excurrent, excurrent.

excursion, *f.* excursion, trip, ramble, (piston) stroke.

excursionner, to take an excursion.

excusablement, excusably.

excuser, to excuse.

exécutable, executable, feasible, practicable.

exécuter, to execute, do, carry out, carry on, make, construct, perform; **s'—,** to be performed, be accomplished, be executed.

exécuteur, -trice, exécutif, -ive, executive.

exécution, *f.* execution, performance, doing, carrying out.

exemplaire, *m.* copy, model, pattern, sample.

exemplaire, exemplary.

exemple, *m.* example, copy, pattern; **à l'— de,** in imitation of; **par —,** for example, for instance, of course, indeed.

exempt, free, exempt; **— d'alcool,** nonalcoholic.

exempter, to exempt.

exercer, to exercise, perform, use, practice, exert, train; **s'—,** to be exercised, exerted.

exercice, *m.* exercise, practice, work, financial (fiscal) year.

exérèse, *f.* cutting away (diseased matter).

exert, exserted (stamens).

exfoliation, *f.* exfoliation, peeling off.

exfolié, sloughed off.

exfolier, to exfoliate, cast or shed leaves.

exhalaison, *m.* exhalation, evaporation, effluvium, breath; *pl.* fumes.

exhalation, *f.* exhalation, exhaling, evaporation.

exhaler, to exhale, emit, send forth, breathe out; **s'—,** to be exhaled, evaporate.

exhaussement, *m.* raising, height, rising elevation.

exhausser, to raise, erect, increase height.

exhausteur, *m.* exhauster.

exhaustion, *f.* exhaustion, removal by suction.

exhéréder, to disinherit.

exhiber, to exhibit, show.

exhibition, *f.* exhibition.

exhilarant, exhilarant; **gaz —,** laughing gas.

exhyménine, *f.* extine, the outer coat of a pollen grain.

exhumer, to unearth, bring to light.

exigeant, exacting, accommodating, fastidious; **peu —,** nonexacting, unexacting.

exigence, *f.* need, exigency, exactingness, unreasonableness, requirement, demand, exigence, claim.

exiger, to exact, demand, require, necessitate, need; **s'—,** to be required.

exigibilité, *f.* expiration, maturity.

exigu, tiny, small, narrow, slender, scanty.

exine, *f.* extine.

existant, existent, existing, extant, on hand, in force.

existence, *f.* existence, being, life, stock on hand.

exister, to exist, live, be.

exitèle, *f.* valentinite.

exocarpe, *m.* epicarp, exocarp, the outer layer of a pericarp.

exocet, *m.* flying fish.

exochomophyte, *f.* exochomophyte, surface-rooting and mat-forming plants.

exoderme, *m.* exoderm.

exodermis, *m.* exodermis.

exoénergétique, energy-liberating (reaction).

exogamie, *f.* exogamy.

exogène, exogenous, exogenetic; **plante —,** exogen.

exogyne, exogenous.

exomphale, *f.* umbilical hernia, exomphalos.

exonde, projecting out of the waves, above the water.

exophtalmie, *f.* exophthalmos.

exorbitamment, exorbitantly, excessively.

exorrhize, exorhizal.

exosmose, *f.* exosmosis.

exosmotique, exosmotic.

exospore, *m.* exospore.

exostome, *m.* exostome.

exostose, *f.* exostosis.

exothermique, exothermic.

exotique, exotic, foreign; **plantes —s,** *f.pl.* exotics.

exotiquement, exotically.

exotoxines, *f.pl.* extracellular toxins.

exotrophie, *f.* exotrophy, development of lateral shoots.

exotropie, *f.* exotropy.

expansibilité, *f.* expansibility, dilatability.

expansible, expansible.

expansif, -ive, expansive.

expansion, *f.* expansion, dilation.

expansivité, *f.* expansiveness.

expectation, *f.* expectation.

expectative, *f.* hopes, expectation.

expectorant, expectorant.

expectoration, *f.* expectoration, sputum.

expédient, expedient.

expédier, to send, forward, ship, despatch, expedite, draw up contract, send off.

expéditeur, *m.* sender, shipper.

expéditeur, -trice, forwarding.

expéditif, -ive, expeditious, prompt.

expédition, *f.* expedition, despatch, sending, forwarding, shipment, shipping, copy, duplicate.

expéditionnaire, *m.* sender, shipper, forwarding agent, copying clerk.

expéditionnaire, shipping, forwarding, copying, expeditionary.

expéditivement, expeditiously, speedily.

expérience, *f.* experiment, test, experience, experimentation, trial.

expérimental, -aux, experimental.

expérimentalement, experimentally.

expérimentateur, -trice, *m.&f.* experimenter.

expérimentation, *f.* experimentation, series of experiments.

expérimenté, experienced, efficient, skilled, tried, tested.

expérimenter, to experience, try, test, experiment, make experiment of, try out.

expert, *m.* expert, connoisseur, specialist.

expert, expert, skilled.

expertement, expertly, ably.

expertise, *f.* expert appraisement, valuation, survey, expert's report.

expertiser, to examine, appraise, value, survey.

expiration, *f.* expiration.

expirer, to expire, exhale, die.

explicable, explicable, explainable.

explicatif, -ive, explanatory.

explication, *f.* explanation.

explicité, *f.* explicitness.

explicite, explicit, plain.

explicitement, explicitly, plainly.

expliquer, to explain, interpret; **s'—,** to be explained, understand, come to an explanation.

exploit, *m.* exploit, achievement.

exploitabilité, *f.* exploitability; **— absolue,** economic exploitability, economic maturity.

exploitable, mature, ripe for the ax, exploitable.

exploitant, *m.* exploiter, operator, grower, cultivator, contractor, producer.

exploitant, exploiting.

exploitation, *f.* exploitation, operation, working, operating, business, cultivation, farming, utilization, produce, yield; **— abusive,** destructive working; **— à menu taillis,** short rotation coppice system;

— à outrance, destructive working; **— à rendement soutenu,** sustained working; **— de bois,** forest utilization; **— de prairies,** grass farming, pasture farming; **— des coupes,** management of cuttings; **— des forêts,** forest management.

exploitation, — des mines, mining; **— en menu taillis,** short rotation coppice system; **— en tailles de branches,** lopping system; **— en têtards,** pollarding system; **— par coupe unique,** clear-cutting system; **— par groupes,** shelterwood group system, gap felling; **— rendement fixe annuel,** exploitation for a sustained annual yield.

exploiter, to exploit, work, operate, farm, manage, cut, fell.

exploiteur, m. exploiter, operator, manager, farmer, grower.

explorateur, m. explorer.

explorateur, -trice, exploring.

exploration, f. exploration; **— du sol,** soil examination.

explorer, to explore, examine.

exploser, to explode, burst.

exploseur, m. exploder.

explosible, explosive.

explosif, -ive, explosive; **— à la nitroglycérine,** nitroglycerin explosive; **— de sûreté,** safety explosive.

explosion, f. explosion, outburst; **faire —,** to explode, burst, crack, fracture, rupture.

expoliation, f. removal of dead parts (of tree).

exponentiel, -elle, exponential.

exportateur, -trice, m.&f. exporter.

exportation, f. exportation, export.

exporter, to export.

exporteur, m. exporter.

exposant, m. exponent, exhibitor, petitioner, index.

exposant, exposing, explaining.

exposé, m. exposition, statement, account, outline.

exposé, exposed, explained.

exposer, to expose, explain, expound, state, set forth, exhibit; **s'— à,** expose oneself, be liable, be exposed.

expositif, -ive, explanatory, expository, descriptive.

exposition, f. exposition, exposure, statement, account, world's fair, aspect, explanation.

exprès, m. messenger, express train.

exprès, express, expressly, on purpose.

expressément, expressly, explicitly, purposely.

expressif, -ive, expressive.

expression, f. expression, form, phrase; **— complex,** complex number; **— fractionnaire,** improper fraction, composed fraction.

exprimable, expressible.

exprimer, to express, squeeze out, utter; **s'—,** to express oneself, be expressed.

exproprier, to expropriate, dispossess.

expuer, to spit out.

expulser, to expel.

expulseur, -trice, expulsif, -ive, expulsive.

expulsion, f. expulsion, evacuation.

expurgade, f. removal of obnoxious trees.

exquis, exquisite, rare, choice.

exquisement, exquisitely.

exsangue, exsanguin, exsanguine, anemic.

exsert, exserted, protruding out of.

exsiccateur, m. desiccator.

exsiccation, f. desiccation, drying.

exsudant, exuding, sudorific.

exsudat, m. exudate; **— peritoneal,** peritoneal exudate; **— pleural,** pleural exudate.

exsudation, f. exudation, exudate.

exsuder, to exude.

extase, f. ecstasy, rapture, trance.

extasier, to ravish, make ecstatic.

extenseur, m. extensor (muscle), shock absorber.

extensibilité, f. extensibility, dilatability.

extensif, -ive, extensive, tensile, expansive.

extension, f. extension, extent, prolongation, enlargement.

exténuer, to extenuate, weaken, diminish.

extérieur, m. exterior, outside, surface, foreign thing; **à l'—,** abroad, outside.

extérieur, exterior, external, outer, outside, outward, foreign.

extérieurement, on the outside, externally, outwardly.

extérioration, f. exterioration.

exterminateur, m. exterminator.

exterminateur, -trice, exterminating.

exterminer, to exterminate, eradicate.

externe, external, outer, exterior.

extincteur, m. extinguisher.

extincteur, -trice, extinguishing.

extinction, *f.* extinction, slaking (of lime), suppression, loss, dying out, going out, termination.

extinguible, extinguishable, quenchable.

extirpateur, *m.* extirpator, scuffler, weeding machine.

extirpation, *f.* extirpation, eradication.

extirper, to extirpate, eradicate.

extra, extra.

extra-axillaire, extra-axillary.

extra-calciné, supercalcined.

extracellulaire, extracellular.

extra-courant, *m.* extra current, self-induced current.

extracrescent, growing out.

extracteur, *m.* extractor, exhauster.

extractif, -ive, extractive.

extraction, *f.* extraction, discharge, drawing, removal; — de la tourbe, turf cutting; — des graines, seed extraction, seed husking; — des plantes, transplanting, lifting; — des souches, grubbing up stumps; — les graines (des cônes), to husk.

extrados, *m.* extrados, top face.

extra-fin, extra (super) fine.

extrafolié, extrafoliaceous, borne on the outside of the leaves.

extraire, to extract, draw, cut out, take out, husk, lift, grub up, wash, leach (out), lixiviate.

extrait, *m.* extract, abstract; — de bois colorant, extract of dyewood; — de colchique acétique, acetic extract of colchicum; — de fiel de boeuf, purified oxgall; — de garance, madder extract; — de levure, yeast extract; — de malt, malt extract; — de teinture, extract of dyewood; — de viande, meat extract; — éthéré, ethereal (ether) extract; — sec, dry matter; — tannant, tanning (tannic) extract.

extraordinaire, extraordinary, remarkable.

extraordinairement, extraordinarily.

extrapolation, *f.* extrapolation.

extra-terrestre, extraterrestrial, beyond the earth.

extrayant, extracting.

extrême, extreme, excessive, extremity; à l'—, to an extreme.

extrêmement, extremely, utterly.

extrémité, *f.* extremity, end, utmost degree, top end.

extrorse, extrovorse, extrorse.

exubérant, exuberant, copious, rank.

F

f., *abbr.* franc.

fabagelle, *f.,* **fabago,** *m.* bean caper.

fable, *f.* fable.

fabricant, *m.* manufacturer, maker.

fabrication, *f.* manufacture, manufacturing, making, make, baking, brewing, boiling (soap); — d'articles en osier, basketry; — en grande série, — en masse, quantity production; — journalière, daily output.

fabrique, *f.* manufacture, making, factory, manufactory, works, mill, plant, make, fabric.

fabriqué, manufactured.

fabriquer, to make, manufacture, fabricate; se —, to be made.

fabuleux, -euse, fabulous.

façade, *f.* façade, front.

face, *f.* face, front, aspect, surface, appearance, (cutting) edge, ventral side; — à, facing, opposite to; — interne, ventral side; — supérieure, ventral side; en — de, opposite, in front of. over against; faire — à, to be opposite, face.

facette, *f.* facet, little face, bevel.

facetter, to facet, cut facets upon.

fâcher, to get angry, be offended.

fâcheusement, sadly, awkwardly.

fâcheux, -euse, troublesome, sad, unfavorable, provoking, annoying, unpleasant, disagreeable, grievous, vexatious; il est —, it is a pity.

facial, -aux, facial.

facies, *m.* facies, aspect, appearance.

facile, facile, easy, ready, quick.

facilement, easily, readily.

facilité, *f.* facility, ease, fluency, readiness.

faciliter, to facilitate, make easy.

façon, *f.* making, fashion, manner, look, mode, way, make, workmanship, appearance, care, attention, dressing, ceremony, culture, laborer's wages; à la — de, after the manner of; de — à, de — que, so tha' so as to, in such a way as, with the result that; de — (indirecte), in an (indirect) way; de cette —, in this way.

façonnement, façonnage, *m.*, **façonnerie,** *f.* fashioning, rough conversion.

façonner, to fashion, shape, form, model, shape, make, stack, make up, work, finish (off), mold, form, polish, dress, figure, accustom, convert; **se —,** to accustom oneself.

fac-similaire, in facsimile.

fac-similé, *m.* facsimile.

fac-similer, to reproduce in facsimile.

factage, *m.* delivery, transport, porterage.

facteur, *m.* factor, manufacturer, clerk, carrier, porter, postman; **— d'amplitude,** crest factor; **— de charge,** load factor; **—de conversion,** reducing factor; **— d'empilage,** reducing factor (for converting stacked contents into solid measure); **— de puissance,** power factor; **— de sécurité,** factor of safety.

factice, artificial, factitious, factious, imitatión; **—s de caoutchouc,** rubber substitutes.

facticement, artificially, factitiously.

faction, *f.* faction, sentry.

factorielle, *f.* factorial.

facture, *f.* invoice, bill, style, composition.

facturer, to invoice, bill.

facturier, *m.* invoice clerk, invoice book.

facule, *f.* facula.

facultatif, -ive, facultative, optional.

facultativement, optionally, at discretion.

faculté, *f.* faculty, ability, power, quality, college, right, virtue, property, adaptability; **— de rejeter,** power of reproduction from stool, reproductive faculty; **— germinative,** germinating power.

fadasse, very insipid, very dull, washed out.

fade, insipid, flat, tasteless, washed out (color).

fadement, insipidly.

fadet, somewhat insipid.

fadeur, *f.* insipidity, sickliness, flatness dullness.

fagot, *m.* fagot, bundle of sticks, bavin.

fagotage, *m.* fagot binding.

fagotaille, *f.* brushwood revetment.

fagoue, *f.* sweetbread, pancreas.

fahlunite, *f.* fahlunite.

faible, *m.* weakness, failing, weak point, weak side, weak person.

faible, weak, feeble, slight, faint, deficient, light, low, small.

faiblement, weakly, feebly, slightly, faintly.

faiblesse, *f.* weakness, feebleness, faintness, lowness, lightness, deficiency, defect, debility.

faiblir, to weaken, become weak, lose strength, relax, slacken, fail.

faïence, *f.* faïence, earthenware, delftware, crockeryware.

faïencerie, *f.* earthenware or crockery factory or business.

faïencier, *m.* maker or dealer of earthenware and pottery.

faillant, failing, about to end.

faille, *f.* coarse-grained silk material, grosgrain, fissure, rent, fault, dike; **— des couches,** dislocation of strata.

faillé, faulted.

failli, failed, bankrupt; **tendon —,** close tendon.

faillibilité, *f.* fallibility.

faillible, fallible.

faillir, to fail, err, be on the point (of), come near, almost to do.

faillite, *f.* failure, bankruptcy.

faim, *f.* hunger; **avoir —,** to be hungry.

faîne, *f.* beech mast, beechnut, beech tree (Fagus); **huile de —,** beechnut oil.

fainéant, lazy, indolent.

fainéanter, to loaf.

fainéantise, *f.* laziness.

fainée, *f.* beech mast.

faire, to make, do, cause, perform, produce, form, study, accustom, take (a ride or a walk), charge (for goods), matter, signify, travel, go, be; **se —,** to be made, take place, be done, be used to, make oneself, be formed, become; **se — à,** to become accustomed to; **— abstraction de,** to take away; **— aller,** to move, work; **— attention,** to be careful; **— autorité,** to be authoritative; **— bien,** to do well, do right.

faire, — cas, to prize; **— communiquer,** to connect; **— défaut,** to be lacking; **— de la radiotéléphonie,** to talk by radio; **— du mal,** to inflict injuries; **— face à,** to offset; **— fonctionner,** to run; **faire —,** to have made, cause to be made, have done; **ne — que,** inf., only; **ne — que,** to make only, do only, do nothing but; **ne — que de venir,** to have just come; **— la cuisine,** to cook; **— la première coupe,** to make a first felling; **— le tour de,** to go around; **— le vide,** to evacuate, create a vacuum; **— part de,** to inform, acquaint; **— partie de,** to form a part of; **— peur,** to scare.

faire, — **que,** to cause that, see that; — **retour,** to return; — **ses classes,** to study, learn; — **sortir,** to bring out, develop; — **un cours,** to give a course; — **voir,** to show; — **état de (cet homme),** to esteem (that man); — **grand cas de (vous),** to think highly of (you); — **le compte de,** to settle with; — **le coup,** to do the deed; — **preuve de cela,** to prove that; — **tout son possible,** to do one's best; **en** — **autant,** to do the same; **(en)** — **la remarque,** to notice (it).

faire, (le) — **valoir,** to bring out (its) merits; **(le)** — **venir,** to send for (him); **(lui)** — **bon acceuil,** to welcome (him); **(lui)** — **de la peine,** to hurt (him); **(lui)** — **dire,** to send word to (him); **(lui)** — **horreur,** to horrify (him); **(lui)** — **la part,** to give (him) his share; **(lui)** — **part de la nouvelle,** to inform (him) of the news; **(lui)** — **savoir,** to inform (him).

faire, il se fait à sa tâche, he is becoming accustomed to his task; **ça se fait,** it is done; **cela fait défaut,** that is lacking; **cela ne fait rien,** that makes no difference; **comment se fait-il que . . . ?,** how does it happen that . . . ?; **il fait beau temps,** the weather is fine; **il fait chaud,** it is warm; **il fait froid,** it is cold; **il fait jour,** it is light; **qu'est-ce que cela vous fait?,** what's that to you?

faire, m. making, doing, execution.

faisable, feasible, practicable.

faisan, m. pheasant.

faisceau, m. bundle, group, number, fascicle, cluster, fasciculus, sheaf, tuft, bunch; — **du liber,** phloem strand; — **fibro-vasculaire,** fibrovascular bundle; — **libéro-ligneux,** fibrovascular bundle; — **lumineux,** pencil of light, beam (parallel, convergent, or divergent).

faiseur, -euse, m.œf. maker, doer, author, publisher.

faisselle, f. cheese drainer.

fait, m. fact, deed, act, event, matter, affair, point, question, share; **au** — come to the point; **dans le** —, in fact; **de** —, in fact, indeed; **de ce** —, from this fact, hence; **du** — **de,** because of; **du** — **que,** from the fact that.

fait, en —, as a matter of fact; **en** — **de,** in point of, in the matter of, as to; **en** — **(il le demande),** as a matter of fact (he asks for it); **être** — **de,** to become, suit; **il est de** — **que . . . ,** it is a fact that . . . ; **par le** —, in fact, by the fact; **tout** —, ready-made; **tout à** —, altogether, entirely.

fait, made, done, qualified, fit, finished, mature, fully developed, fixed.

faitage, m. ridge, purlin, roof truss.

faîte, m. top, summit, ridge, apex, watershed.

faîtière, f. ridge tile, skylight.

faîtière, ridge.

faix, m. weight, burden, load.

falaise, f. cliff, steep slope.

falciforme, falcated, falciform, sickle-shaped.

fallacieusement, fallaciously.

fallacieux, -euse, fallacious, deceptive.

falloir, to be wanting, be lacking, be necessary, be needing, need, be requisite, must, ought, should; **s'en** —, to be wanting, be nearly, be on the point; **il lui fallait venir,** he should have come; **il faudra qu'il vienne,** he will have to come; **il a fallu,** it was necessary; **il a fallu un livre,** a book was needed.

falloir, comme il faut, as should be, respectably, well-bred, proper; **de bien s'en faut,** by a good deal; **il lui faut quelque chose,** he needs something; **il me faut un tube à essai,** I need a test tube; **il me le faut,** I must have it; **il s'en faut,** far from it; **il s'en faut de beaucoup que,** it is far from being true that; **peu s'en faut,** very nearly.

falourde, f. large fagot of stakes.

falque, falciform, sicklelike.

falsificateur, -trice, m.&f. adulterator, falsifier.

falsification, f. falsification, forging, adulteration, counterfeiting, sophistication.

falsifié, adulterated.

falsifier, to falsify, adulterate, counterfeit, forge, sophisticate, debase.

falsinerve, falsinervis, (nerves of cellular tissue) without fibrovascular bundles.

falun, m. shell marl.

falunage, m. manuring with marl.

faluner, to manure with shell marl.

falunière, f. shell-marl pit.

fameux, -euse, famous, renowned.

familiariser, to familiarize, accustom to; **se** —, to become familiar, familiarize oneself, accustom oneself to.

familiarité, f. familiarity.

familier, -ère, familiar, intimate, free.

familièrement, familiarly, freely.

famille, f. family; **en** —, at home.

famine, _f._ famine.

fanage, _m._ tedding, tossing (of hay), foliage, leaves (for litter), fading, wilting.

fanal, _m._ lantern.

fane, _f._ grass drying for hay, dry leaf, fallen leaves, floral envelopes, involucre (of buttercup); —**s de navets,** turnip tops.

fané, faded, withered, flaccid.

faner, to fade, wilt, toss, ted (hay); **se —,** to fade, wilt, wither.

faneur, _m._ haymaker.

faneuse, _f._ tedder, flap over, rotary haymaker.

fange, _f._ mud, mire, dirt.

fangeux, -euse, muddy, miry, dirty.

fanon, _m._ dewlap (of ox), fetlock, whalebone.

fantaisie, _f._ fancy, imagination, whim, floss silk (thread).

fantasmagorie, _f._ phantasmagoria.

fantasque, fantastic, fanciful.

fantastique, fantastic, fanciful, chimerical.

faon, _m._ fawn; — **chevrette,** roe fawn; — **de biche,** fawn of red deer.

farcin, _m._ farcy, glanders.

farcineux, -euse, affected with farcy.

farcir, to stuff (poultry).

fard, _m._ paint (face), rouge, make-up.

fardage, _m._ sample package.

fardeau, _m._ burden, load, weight.

farder, to paint (face), disguise, weigh heavily, settle, sink.

fardier, _m._ two wheels for removing logs.

farigoule, _f._ thyme (Thymus).

farinacé, farinaceous, flourlike.

farine, _f._ flour, meal, farina; — **d'avoine,** oatmeal; — **de blé,** wheat flour; — **de bois,** sawdust; — **de forage,** bore dust; — **de froment,** wheaten flour; — **de lin,** linseed meal; — **de maïs,** corn meal, Indian maize flour.

farine, — **de pois,** pea meal; — **de seigle,** rye flour; — **de viande,** meat meal; — **d'orge,** barley flour; — **fossile,** infusorial earth, kieselguhr, diatomite, fossil meal; — **lactée,** malted milk; **ver de —,** meal worm.

fariner, to flour, dust, dredge with flour, become flourlike, get mealy.

farineux, _m._ starchy food.

farineux, -euse, floury, mealy, farinose, farinaceous.

farinose, _f._ farinose.

faro, _m._ Brussels beer.

faroba, _m._ locust tree, varnish tree.

farouche, _m._ crimson clover, carnation clover (_Trifolium incarnatum_).

farouche, wild, fierce, savage, ferocious,

fasciation, _f._ fasciation.

fasciculaire, growing in bunches, fascicular, scopiform.

fascicule, _m._ fascicle, small bundle, instalment (of a publication), cluster; — **de feuilles,** leaf bundle.

fasciculé, tasseled, fascicular.

fascie, _f._ bar, fascia, a crossband, as of color.

fascié, fasciate(d) (stems), banded, striped.

fascine, _f._ fascine, fagot.

fasciner, to fascinate, enchant, captivate.

fastidieux, -euse, tedious, tiresome.

fastigié, fastigiate.

fastueux, -euse, pompous, magnificent, gaudy.

fatal, fatal, inevitable, deadly.

fatalement, inevitably, mortally, fatally.

fatalisme, _m._ fatalism.

fatigant, fatiguing, wearisome, tedious, tiresome, irksome.

fatigue, _f._ fatigue, toil, hardship, weariness, stress, strain.

fatigué, fatigued, tired, overworked.

fatiguer, to fatigue, tire, weary, strain, impair, overwork, wear; **se —,** to be fatigued, grow tired, be impaired, tire oneself, tire.

fau, _m._ beech, beech tree (Fagus).

faubourg, _m._ suburb.

faucard, _m._ reed mower.

faucher, to mow, cut (down).

fauchet, _m._ hay rake.

faucheur, _m._ mower, reaper.

faucheuse, _f._ mower (machine).

faucille, _f._ sickle, reaping hook.

fauciller, to mow with sickle.

faucon, _m._ falcon.

faude, _m._ place where charcoal is burned, place where ores are calcined.

fauder, to fold (double) lengthwise, mark center line.

faudra (falloir), (it) will be necessary, (he will fail.

faulde, _m._ place to burn charcoal.

faulx, *f.* scythe, falx, sickle-shaped fold (of brain).

faune, *f.* fauna.

fausse, see FAUX.

faussé, inaccurate, vitiated.

fausse-chenille, *f.* false caterpillar, hymenopterous larva.

faussement, *m.* bending, warping.

faussement, falsely, erroneously, wrongly.

fausse-ombelle, *f.* cyme.

fausser, to force, bend, warp, spring, affect adversely, strain, distort, violate, falsify, break (one's word); **se —,** to be bent, be warped, bend, warp, strain.

fausset, *m.* spigot, peg, pin, falsetto.

fausseté, *f.* falseness, falsity, falsehood, insincerity, untruth.

fausse-teigne, *f.* larger wax moth.

faut (falloir), (it) is necessary.

faut (faillir), (he) is failing.

faute, *f.* fault, lack, want, mistake, error, defect; **— de,** for want of, through not; **à— de,** in default of; **faire—,** to fail, be wanting; **sans—,** without fail.

fauteuil, *m.* armchair.

fautif, -ive, faulty, erroneous.

fautivement, erroneously, wrongly.

fautre, *m.* felt.

fauve, *m.* fawn color, rust.

fauve, fallow, fawn-colored, tawny; **bête —,** fallow deer.

fauvette, *f.* hedge sparrow; **— des jardins,** garden warbler.

faux, *m.* imitation, forgery, falsehood, scythe.

faux, fausse, false, imitation, counterfeit, mock, wrong, faithless, bogus, erroneous, falsely, erroneously; **— argent,** mica; **— baumier,** false balm tree; **— cumin,** nutmeg flower (*Nigella sativa);* **— fenouil,** deadly carrot (*Thapsia garganica);* **— fruit,** spurious fruit; **— jour,** false light; **— teint,** fugitive (color or dye); **à —,** falsely, wrongly, in vain, out of perpendicular; **fausse côte,** false rib, floating rib; **fausse oronge,** fly agaric, fly amanita (*Amanita muscaria*).

faux-bourdon, *m.* faux-bourdon, drone.

faux-fond, *m.* false or movable bottom.

faux-frais, *m.pl.* incidental expenses.

faux-fruit, *m.* spurious fruit.

faux-fuyant, *m.* bypath, shift, evasion.

faux-jour, *m.* false light.

faux-roseau, *m.* canary grass (*Phalaris canariensis*).

favelle, *f.* favella, a conceptacle of certain red algae.

favelotte, *f.* garden bean, broad bean.

favéolé, favose, faveolate, honeycombed, alveolate.

faveur, *f.* favor, vogue, kindness, cover, help; **à la — de,** by help of, by means of; **en — de,** in behalf of, in consideration of.

faviforme, faviform.

favique, pertaining to favus.

favorable, favorable, auspicious, propitious, suitable.

favorablement, favorably.

favori, -ite, favorite.

favoriser, to favor, assist, befriend.

favus, *m.* favus; **— des poules,** variety of favus affecting fowls.

fayard, *m.* beech, beech tree (Fagus).

fayence, fayencerie, *f.* see FAÏENCE.

fébrifuge, febrifuge.

fébrile, febrile.

fécal, -aux, fecal.

fécaloïde, fecal.

féces, *f.pl.* feces, stool.

fécond, fecund, fertile, fruitful, rich, prolific, genial, abundant, germinable.

fécondant, fecundating, fertilizing, impregnating, genial.

fécondateur, -trice, fecundating.

fécondation, *f.* fecundation, impregnation, fertilization, pollination, fructification.

féconder, to fecundate, fertilize, impregnate, pollinate, render prolific, make fruitful.

fécondité, *f.* fecundity, fertility, fruitfulness, prolificness, wealth.

fécule, *f.* fecula, starch, potato starch, green matter of plants; **— de blé,** wheat starch; **— de maïs,** cornstarch, maize starch; **— de pommes de terre,** potato starch.

féculence, *f.* feculence, thickness, turbidity and starchiness (of solution).

féculent, *m.* starchy substance, starchy vegetable.

féculent, feculent, thick, turbid, starchy, containing starch.

féculer, to reduce (potatoes) to starch or fecula.

féculerie, *f.* starch factory, fecula manufacture.

féculeux, -euse, starchy, amylaceous, containing starch.

féculier, m. starch (maker) manufacturer.

feignant, pretending.

feindre, to feign, simulate, scruple, pretend.

feint, feigned, assumed, imitation, counterfeit.

feldspath, m. feldspar; — **argiliforme,** kaolin; — **nacré,** moonstone; — **opalin,** labradorite; — **vert,** amazon stone, amazonite.

feldspathique, feldspathic.

fêle, f. blowpipe.

fêler, to crack.

félin, m. cat (family).

félin, feline.

felle, f. see FÊLE.

fêlure, f. crack, split, fissure, fracture.

f.é.m., abbr. (force électromotrice), electromotive force.

femelle, female.

femme, f. woman, wife.

fémur, m. femur, thigh bone.

fenaison, f. hay harvest.

fenchène, m. fenchene.

fenchone, f. fenchone.

fenchyle, m. fenchyl.

fendage, m. splitting, cleaving.

fenderie, f. slitting, cutting, slitting machine, slitting mill.

fendeur, m. splitter, woodcutter, (diamond) cleaver.

fendille, f. little crack, flaw, cleft, fissure, crevice.

fendillement, m. cracking, splitting, chinking.

fendiller, to crack, fissure, split, chap.

fendre, to split, cleave, hew, rend, slit, crack, rip up, part, fissure, chink, cut; **se** —, to be split, part, be rent, crack, gape, chink, chap.

fendu, split, cleft, cloven, fissile.

fenestré, fenestrate.

fenêtre, f. window, opening, aperture, port; — **à guillotine,** sash window.

fenêtrer, to perforate, provide with windows.

fenil, m. hay barn, hayloft.

fenouil, m. fennel (*Foeniculum vulgare*); — **amer,** bitter fennel, wild fennel, horse fennel (*Seseli hippomarathrum*); — **bâtard,** dill (*Anethum graveolens*); — **d'eau,** water fennel (*Oenanthe phellandrium*); — **de mer,**

— **marin,** sea fennel, samphire (*Crithmum maritimum*); — **officinal,** sweet fennel (*Foeniculum dulce*); — **puant,** dill (*Anethum graveolens*); — **sauvage,** hemlock.

fenouillette, f. fennel water, fennel brandy.

fente, f. anther, slit, slot, split, crack, flow, chink, cleft, fissure, crevice, cut, gap, rent, groove; — **de coeur,** — **de forêt,** heartshake; — **de dessiccation,** seasoning crack; — **d'insolation,** sun crack, sun blister; — **de retrait,** crack due to warping.

fenugrec, m. fenugreek (*Trigonella feonumgraecum*).

féodal, -aux, feudal.

féodalité, f. feudalism.

fer, m. iron, tool, horseshoe, sword, steel; *pl.* fetters; — **à cheval,** horseshoe; — **aciéreux,** steely iron, fine-grained iron; — **affiné,** refined iron, malleable iron; — **à grain(s) gros,** coarse-grained iron; — **aigre,** cold-short iron; — **à loupe,** bloom iron; — **à souder,** soldering iron; — **à T,** T iron: — **au bois,** — **au charbon de bois,** charcoal iron.

fer, — **battu,** wrought iron, hammered iron; — **blanc,** white cast iron, white iron, tinned iron, tin, tin plate; — **brûlé,** burnt iron, overburnt iron; — **brut,** puddle(d) bar; — **carbonaté,** iron carbonate; — **carburé,** iron carbide; — **cassant,** brittle iron, short iron; — **cassant à chaud,** hot-short iron, red-short iron; — **cassant à froid,** cold-short iron; — **chromé,** chrome iron ore, chrome iron, chromite; — **corroyé,** wrought iron; — **coulé,** cast iron, pig iron; — **cru,** crude iron.

fer, — **de ferraille,** scrap iron; — **de fonte,** cast iron; — **de forge,** wrought iron; — **de lance,** lance head; — **de riblons,** scrap iron; — **écroui,** hammer-hardened iron; — **en bandes,** hoop iron; — **en feuilles,** sheet iron; — **en gueuse,** pig iron; — **en lames,** sheet iron; — **en limaille,** iron filings; — **en loupes,** bloom iron; — **en rubans,** hoop iron; — **en saumon,** pig iron; — **en tournure,** iron turnings.

fer, — **étiré,** rolled iron, drawn iron; — **fendu,** slit iron; — **fin,** refined iron; — **fin au bois,** charcoal iron; — **fondu,** cast iron; — **fondu à air chaud,** hot-blast iron; — **fondu à air froid,** cold-blast iron; — **forgé,** wrought iron; — **galvanisé,** galvanized iron; — **laminé,** rolled iron, laminated iron; — **limoneux,** limonite; — **magnétique,** magnetite, magnetic iron (ore); — **martelé,** hammered iron; — **martial,** copperas; — **métallique,** metallic iron; — **météorique,** meteoric iron; — **mou,** soft iron.

fer, — natif, native iron; — **nerveux,** fibrous iron; — **noir,** black sheet iron; — **oligiste,** hematite, oligist iron; — **oölithique,** oölitic iron ore; — **oxydé,** iron oxide, oxide of iron; — **oxydé rouge,** red iron ore; — **oxydulé,** magnetite; — **plat,** hoop iron; — **puddlé,** puddled iron; — **raffiné,** refined iron; — **soudé,** weld iron; — **spathique,** spathic iron; — **tenace,** tough iron; — **zingué,** galvanized iron.

fera (faire), (he) will make or do.

féramine, *f.* iron pyrites.

fer-à-nerf, *m.* fibrous iron.

fer-blanc, *m.* tin plate, tin.

ferblanterie, *f.* tinware, tinwork.

ferblantier, *m.* tinsmith.

féret, *m.* ferret, hard kernel, core (in stone).

ferlet, *m.* peel.

fermage, *m* rent.

fermail, *m.* clasp.

fermant, closing.

ferme, *f.* farm, farmhouse, homestead, lease, girder, truss, rib, frame, roof, rent, set price.

ferme, firm, strong, stout, resolute, steady, closed, binding.

fermé, closed, close, tight.

ferme-circuit, *m.* circuit closer.

fermement, firmly, stoutly, resolutely.

ferment, *m.* ferment, enzyme, yeast; — **figuré,** — **organizé,** organized ferment; — **nitreux,** microorganisms of nitrosation, agents of oxidation of ammonia to nitrous acid; — **nitrique,** microorganisms of nitratation, agents of oxidation of nitrous acid to nitric acid; — **peptonisant,** peptonizing enzyme.

fermentable, fermentable.

fermentaire, fermentatif, -ive, fermentative, fermenting.

fermentation, *f.* fermentation, ferment; — **alcoolique,** alcoholic fermentation; — **gallique,** fermentation of tannin; — **panaire,** fermentation of bread dough, leavening; — **sucrée,** sugar reaction.

fermenter, to ferment.

fermentescibilité, *f.* fermentability.

fermentescible, fermentable.

fermer, to close, shut, fasten, bind, bolt, pin, lock, seal, stop (up); **se —,** to close (up), shut (up), be closed, be enclosed.

fermeté, *f.* firmness, solidity, steadfastness, stability.

fermeture, *f.* closing, sealing, fastening, shutter, blind, closure, cutoff (steam), (ordnance) fermeture, shutting; — **de la chasse,** closed season (for game); — **du massif,** closing of leaf canopy, dense leaf canopy.

fermier, *m.* farmer, lessee, tenant.

fermière, *f.* farmer's wife.

fermoir, *m.* clasp, catch, snap, chisel.

féroce, ferocious, fierce, wild.

ferraille, *f.* old iron, scrap iron.

ferrate, *m.* ferrate.

ferre, *f.* fine iron grindings, bottle pincers.

ferré, chalybeate, fitted with iron, iron, metaled, macadamized (roads), shod (horses), versed (in), harsh, hard, nailed, spiked, iron-pointed.

ferrement, *m.* ironwork, iron tool.

ferrer, to fit or mount with iron, bind with iron, iron, smooth, metal, macadamize (a road), shoe (a horse).

ferrerie, *f.* ironwork, iron trade.

ferret, *m.* hard kernel in stones, glass ferret, tag, tab (of shoe lace); — **d'Espagne,** red iron ore, hematite.

ferreux, ferrous.

ferrico-, ferric (in double salts).

ferricyanhydrique, ferricyanic.

ferricyanure, *m.* ferricyanide.

ferrifère, ferriferous, iron-bearing.

ferrique, ferric.

ferrite, *n.* ferrite.

ferro-chrome, *m.* ferrochrome, ferrochromium.

forrocyanhydrique, ferrocyanic.

ferrocyanique, ferrocyanic.

ferrocyanure, *m.* ferrocyanide.

ferronerie, *f.* ironworks, iron foundry, hardware.

ferroprussiate, *m.* ferroprussiate.

ferro-silicium, *m.* ferrosilicon.

ferrotier, *m.* glassworker.

ferro-tungstène, *m.* ferrotungsten.

ferrotypie, *f.* ferrotype.

ferroviaire, *m.* railroad.

ferrugineux, *m.* chalybeate.

ferrugineux, -euse, ferruginous, chalybeate.

ferruginosité, *f.* ferruginous quality, ferrugination.

ferrure, *f.* ironwork, shoeing, shoes (of horses), strap.

fertile, fertile, fruitful, germinable.

fertilement, fertilely, fruitfully.

fertilisable, fertilizable.

fertilisant, fertilizing.

fertilisation, *f.* fertilization, enrichment.

fertiliser, to fertilize, manure.

fertilité, *f.* fertility, richness (of soil).

férule, *f.* giant fennel (*Ferula communis*).

férulique, ferulic.

fesse, *f.* buttock, breech.

festin, *m.* feast, banquet.

feston, *m.* festoon.

festonner, to festoon.

fête, *f.* fete, feast, festival, birthday, holiday.

fétichisme, *m.* fetichism.

fétide, fetid, sickening, rank, foul, offensive.

fétidité, *f.* fetidness, foulness.

fétu, *m.* straw, bit, whit, stalk.

fétuque, *f.* fescue (Festuca).

fétus, *m.* see FOETUS.

feu, *m.* fire, flame, combustion, cautery, light, fireplace, household, home, heat, hearth; — **d'affinerie,** fining forge;—, ?**'artifice,** firework, fireworks; — **d'éclairage,** light; — **de réduction,** reducing flame; — **de signaux,** signal light; — **d'oxydation,** oxidizing flame; — **follet,** ignis fatuus, will-o'-the-wisp, jack-o'-lantern; — **grisou,** firedamp; — **nu,** naked flame, free flame; **à — nu,** with the naked or free flame, with direct flame; **en —,** on fire; **mettre à (en) —,** to fire, set on fire; **mettre hors de —, mettre hors —,** to put out of blast; **mettre le — à,** to set on fire.

feu, late, deceased.

feuillade, *f.* foliaceous expansion.

feuillage, *m.* foliage, frond.

feuillagé, (of plant) in leaf.

feuillaison, *f.* foliation, leafing.

feuille, *f.* leaf, foil, sheet, plate, scale, list, waybill, newspaper, carp fry; — **anglaise,** sheet rubber; — **caduque,** deciduous leaf; — **charmue,** water-storing leaf; — **composée pennée,** compound pinnatifid leaf; — **d'argent,** silver foil; — **de cuivre,** copper foil; — **de fer,** sheet of iron; — **de ferblanc,** sheet of tin (plate).

feuille, — **de papier,** sheet of paper; — **de plomb,** lead foil; — **d'étain,** tinfoil; — **d'or,** gold foil; — **florale,** bract; — **laminée,** rolled sheet; — **primordiale,** cotyledon; — **séminale,** cotyledon, seed leaf;

— **volante,** loose sheet, fly sheet, pamphlet; — **primordiales,** leaves developing after the cotyledons; **à —s persistantes,** evergreen; **à —s raides et coriaces,** stiffleaved.

feuillé, foliate, leafy, leaf, leafed, foliaceous.

feuille-morte, dead-leaf colored.

feuiller (se), to sprout, leaf.

feuillet, *m.* leaf (of a book), layer, thin plate, gill (of fungi), plate, folio, sheet, omasum, paunch; — **magnétique,** magnetic shell; — **vasculaire,** parablast.

feuilletage, *m.* lamination.

feuilleté, foliated, laminated.

feuilleter, to turn the leaves (of), peruse, leaf over, fan, laminate.

feuilletis, *m.* edge (of cut stone). cleavage in slate.

feuilleton, *m.* fly sheet, scientific article.

feuillette, *f.* small leaf, leaflet, cask (30.8 gallons).

feuillir, to put forth leaves, leaf, leave.

feuillu, *m.* broad-leaved tree.

feuillu, leafy, foliose, (tree) with caducous leaves.

feuillure, *f.* groove, rabbet, slit, recess.

feurre, *m.* straw.

feutrable, capable of being felted.

feutrage, *m.* felting, matting, tomentum, felt covering; — **d'herbes,** sward, grassy turf.

feutrant, felting.

feutre, *m.* felt, padding, packing, tomentum; — **d'herbe,** sward, grassy turf; — **pour papeteries,** felt for paper mills.

feutré, felted, felt, padded, tomentose.

feutrament, *m.* felting.

feutrer, to felt, pad, pack, cover with felt.

fève, *f.* bean, berry, chrysalis, lampas; — **de cacao,** cacao bean (*Theobroma cacao*); — **de Calabar,** Calabar bean (*Physostigma venenosum*); — **de haricot,** kidney bean (*Phaseolus vulgaris*); — **de Saint-Ignace,** St.-Ignatius's-bean (*Strychnos ignatii*); — **des marais,** Windsor bean, broad bean (*Vicia faba*); — **de soja,** soja bean, soy bean (*Glycine max*); — **pichurim,** pichurim bean (*Nectandra pichurim*); — **tonka,** tonka bean, seed of *Dipteryx odorata*.

féverole, *f.* (dried) bean, horse bean, field bean (*Vicia faba*).

févier, *m.* honey locust tree (*Gleditsic triacanthos*).

février, *m.* February.

fiançailles, *f.pl.* betrothal, engagement.

fibre, *f.* fiber, grain (of wood), string; — **d'écorce,** fiber of phloem, cortex, pericycle; — **de remplacement,** substitute fiber; — **libérienne,** bast fiber, phloem fiber; — **ligneuse,** wood fiber, grain of wood; — **primaire,** midrib, primary rib; — **soumise à la traction,** — **tendue,** fiber in tension, stretched fiber; — **vulcanisée,** vulcanized fiber; **à** — **fine,** fine-grained; **à** — **grossière,** coarse-grained; **à** — **tordues,** upset; **à** — **torse,** cross-grained, twisted-grained; — **de bois,** wood fiber, grain.

fibré, fibered, fibrous.

fibreux, -euse, fibrous, stringy, grained (of wood); **racine fibreuse,** fibrous root.

fibreuse de bois, wood cellulose.

fibrillaire, fibrillar, fibrillary.

fibrille, *f.* fibril, fibrilla, rootlet.

fibrilleux, -euse, fibrillose.

fibrillifère, with fibrils.

fibrine, *f.* fibrin.

fibrineux, -euse, fibrinous.

fibrin-ferment, *m.* fibrin ferment, thrombin.

fibrinogène, *m.* fibrinogen.

fibrinolyse, *f.* fibrinolysis.

fibrocartilagineux, -euse, fibrocartilaginous.

fibrocellulaire, fibrocellular.

fibroïde, fibroid (tumor).

fibroïne, *f.* fibroin.

fibrome, *m.* fibrous tumor.

fibrovasculaire, fibrovascular, composed of vessels and fibers.

fic, *m.* wart (of horse).

ficaire, *f.* ficaria, lesser celandine, pilewort (*Ficaria verna*).

ficeler, to tie, tie up (with string).

ficelle, *f.* string, twine, packthread, trick.

fiche, *f.* peg, pin, bolt, plug, key, stake, slip (of paper), ticket, memorandum, tag.

ficher, to stick in, fix, fasten, drive in, insert, point (wall); **il lui fiche la paix,** he lets him alone.

ficiforme, like a fig.

ficoïde, *f.* fig marigold (Mesembryanthemum); — **glaciale,** ice plant (*Mesembryanthemum crystallinum*).

fictif, -ive, fictitious, imaginary.

fidèle, *m.* follower.

fidèle, faithful, loyal, true, accurate.

fidèlement, faithfully, loyally, trustfully.

fidélité, *f.* fidelity, faithfulness, loyalty, truth, accuracy.

fiel, *m.* gall, bile; — **de boeuf,** oxgall; — **de terre,** earthgall, lesser centaury (*Centaurium umbellatum*), fumitory.

fielleux, -euse, like gall, bitter.

fiente, *f.* dung, excrement.

fienter, to dung, manure.

fier (se), to trust, confide in, rely upon.

fier, fière, proud, fierce, great, haughty.

fièrement, proudly.

fierté, *f.* pride.

fièvre, *f.* fever; — **aphtheuse,** aphthous fever, foot-and-mouth disease; — **charbonneuse,** anthrax, splenic fever.

fiévreux, -euse, *m.&f.* fever patient.

fiévreux, -euse, feverish, restless.

figement, *m.* congealment, congealing, solidifying, clotting.

figer, to coagulate, congeal, fix, solidify.

figue, *f.* fig; — **de Barbarie,** Barbary fig, prickly pear (*Opuntia compressa*).

figuerie, *f.* fig garden.

figuier, *m.* fig tree (Ficus).

figuratif, -ive, figurative, emblematic.

figuration, *f.* representation.

figure, *f.* figure, form, shape, face, countenance, illustration.

figuré, *m.* representation, image, cut, diagram.

figuré, figurative.

figurer, to figure, appear, represent; **se** —, to be represented or figured, imagine, picture oneself.

fil, *m.* wire, thread, string, yarn, filament, fiber, edge (of sharp tools), grain (of wood), flaw, vein (in stones), stream (water), inoculating needle; — **à lier,** binding twine; — **à plomb,** plumb line; — **conducteur,** conducting wire, conductor; — **couvert,** insulated wire; — **d'acier,** steel wire; — **d'aller,** lead wire; — **d'amorce,** priming wire; — **d'araignée,** spider line, cross wire; — **d'archal,** brass wire; — **de bois,** grain of wood.

fil, — **de clavecin,** piano wire; — **de cocon,** untwisted silk fiber; — **de laiton,** brass wire; — **d'emballage,** pack thread, wrapping twine; — **de soie,** silk thread; — **écru,** raw yarn; — **en bois,** match splint; — **fusible,** fuse wire, — **laminé,** rolled wire; — **métallique,** (stitching) wire; — **nu,** uninsulated wire; — **recouvert,** insulated wire, covered wire; — **retors,** twine, double yarn; — **thermique,** heating wire; **sans** —, wireless.

filage, *m.* spinning, Filago.

filaire, *f.* threadworm.

filament, *m.* filament, thread form; — **de charbon,** carbon filament; — **de mais,** corn silk.

filamenteux, -euse, filamentous, stringy, filiform.

filandre, *f.* thread, string, filament, a defect in glass, gossamer.

filandreux, -euse, stringy, tough (meat), thready, streaked (marble).

filant, ropy (of liquids).

filardeux, -euse, (of stones) veiny, streaky, veined (marble).

filasse, *f.* tow, bast fiber, unspun fibers, oakum, stringy meat; — **de montagne,** asbestos, earth or mountain flax.

filateur, *m.* spinner.

filature, *f.* spinning, spinning mill.

file, *f.* file, row, rank; — **de rails,** line of rails, set of tracks; **à la** —, in a single file.

filé, *m.* thread.

filé, spun.

filer, to spin, slip, pay out, wiredraw, draw, run, twist (tobacco), spin out, pour out (oil), conduct, shoot by, steal by, file along, slip away, flow (in a viscous stream), be ropy, rope, spin, (of stars) shoot, (of time) fly, (of lamps) smoke.

filet, *m.* thread, band of fibers, fiber filament, thread (of screw), fillet, stream (of liquid), net, netting, dash, stroke, runner (of strawberry), setting rule, snare, stamen filament, string, frenum; — **à droite,** right-handed thread (of screws); — **carré,** square or flat thread; — **de vis,** screw thread; — **en houppe,** wart of hairs; — **extérieur,** outside or male screw; — **femelle,** inside or female thread.

filetage, *m.* wiredrawing, threading (of screws), thread, pitch.

fileté, threaded.

fileter, to thread, draw (wire), cut a thread, wiredraw, screw (bolt).

filiation, *f.* filiation, descendency.

filicoïde, filiform, thread-shaped.

filicule, suspended by a filament.

filière, *f.* drawplate, screw plate, wire gauge, tap.

filifolié, with filiform leaves.

filiforme, filiform, threadlike.

filigrane, *m.* filigree, (paper) watermark — **ombré,** embossment.

filigrané, filigree(d).

filin, *m.* rope, cordage.

filipendule, *f.* dropwort (*Filipendula hexapetala*).

filipendulé, filipendulous, with tubercles at end of filiform roots.

fille, *f.* daughter, girl, maid, nun; **la — avant la mère,** liver leaf (Hepatica).

filoche, *f.* sheave.

filon, *m.* vein, lode, seam.

filon-couche, *n.* sill, vein-bearing rock.

filonien, -enne, pertaining to veins; **gîte** —, vein deposit.

filoselle, *f.* filoselle, floss silk.

filosité, *m.* spindle sprout.

fils, *m.* son, boy; *pl.* of FIL.

filtrable, filterable.

filtrage, *m.* filtering, filtration, straining.

filtrant, filtering, filter, filterable; **virus** —, filterable virus.

filtrat, *m.* filtrate.

filtrateur, *m.* filterer.

filtrateur, filtering.

filtration, *f.* filtration, filtering.

filtratum, *m.* filtrate.

filtre, *m.* filter, strainer; — **à aspiration,** suction filter; — **à bougie,** candle filter, cylinder filter; — **à l'amiante,** asbestos filter; — **à plis,** plaited filter, folded filter; — **à pression,** pressure filter; — **à vide,** vacuum filter; — **dégrossisseur,** preliminary filter; — **en papier,** paper filter; — **plissé,** plaited filter, folded filter.

filtre-presse, *m.* filter press.

filtrer, to filter, percolate, filtrate; **se** —, to be filtered, filter; — **dans le vide,** to filter with suction.

filtreur, *m.* filterer, strainer.

fimbrié, fimbriate, fringed.

fimbrille, *f.* fimbrilla, a diminutive fringe.

fin, *f.* end, close, termination, aim, purpose, conclusion; — **de l'année,** annual balance of account; — **de l'automne,** late autumn, fall; — **de l'été,** late summer; **à la** —, at last, in the end, after all; **en — de,** at the end of; **en — de compte,** in the end, after all; **sans** —, endless.

fin, *m.* fine metal, fine coal, coal dust, fine point, fine linen.

fin, fine, small, shrewd, sharp, thin, slender, delicate, acute.

finage, *m.* refining (metal).

final, final, last, conclusive.

finale, *f.* ending, termination; *m.* finale, conclusion.

finalement, finally, lastly, ultimately.

financiel, -elle, financial.

financier, *m.* financier.

financier, -ère, financial.

financièrement, financially.

fine, see FIN.

finement, finely, artfully, shrewdly.

fine-métal, *m.* white pig.

finerie, *f.* finery, refinery.

fines, *f.pl.* slack or small coal.

finesse, *f.* finesse, fineness, slenderness, sharpness, keenness, acuteness, nicety, delicacy, cunning, subtlety, ingenuity, shrewdness.

fin-fin, extra fine.

fini, *m.* finish.

fini, finished, finite, complete, ended, over.

finir, to finish, end, complete, be over, cease, terminate; **se —,** to end, finish, be at an end, cease, be finished, be completed; **à n'en plus —,** endless; **il finit de s'habiller,** he finishes dressing; **il finit par s'ennuyer,** he finally becomes bored; **il en finit avec lui,** he is getting rid of him.

finissage, *m.* finishing.

finisseur, -euse, *m.&f.* finisher.

finisseur, -euse, finishing, final.

finlandais, Finnish, Finlander.

fiole, *f.* flask, bottle, phial, vial; **— à gouttes,** dropping bottle; **— à jet,** washing (wash) bottle; **— à vide,** filtering (suction) flask; **— ballon,** round flask, balloon flask; **— ballon à fond plat,** flat-bottomed flask; **— conique pour filtrer,** filter flask; **— de mesure,** graduated flask; **— étalon,** standard (normal) flask; **— Gayon,** conical flask with ground cap; **— jaugée,** graduated (volumetric) flask.

fiord, *m.* fiord, inlet.

fiorite, *f.* fiorite.

firent (faire), (they) made or did.

firmament, *m.* firmament, heavens.

firme, *f.* firm.

fisc, *m.* fisc, state treasury, exchequer.

fiscal, -aux, fiscal.

fissent (faire), (they) might make or do.

fissile, fissile, cleavable.

fissilité, *f.* fissility, fissibility, cleavability.

fissipares, *m.pl.* Fissipara.

fissipare, fissiparous.

fissuration, *f.* fissuration; cracking.

fissure, *f.* fissure, cleft, rent, crack; **— due à la poussée,** crack due to expansion.

fissuré, fissured.

fissurer, to fissure, split, crack.

fistulaire, fistular, tubelike.

fistule, *f.* fistula.

fistuleux, -euse, fistulous, fistular.

fit (faire), (he) made or did.

fixage, *m.* fixing.

fixateur, *m.* fixing solution, fixative, fixer, fixator, fastener, binding material or agent, agglutinant, binder; **— chimique,** chemical fixing agent.

fixation, *f.* fixation, setting, putting into place, fastening; **— de la possibilité,** fixation of annual yield, determination of capability; **— du complément,** fixation of complement; **— par contenance** fixation of annual yield by area; **— par volume,** method of determining yield by volume.

fixe, *m.* fixed substance, fixed star, fixed salary, settled weather, set fair.

fixe, fixed, stationary, regular, firm, set.

fixé, fixed.

fixement, fixedly.

fixer, to fix, fasten, determine, settle, make solid, look steadfastly at, establish, clamp; **se —,** to fix, become fixed, settle down, be fastened, adhere.

fixité, *f.* fixity, stability, fixedness.

fjord, *m.* see FIORD.

flabellé, flabellate, fan-shaped.

flabellifolié, with fan-shaped leaves.

flabelliforme, flabelliform, fan-shaped.

flache, *f.* flaw, crack, depression, wane, rough edge, dull edge, inequality, blaze (on tree).

flacherie, *f.* flaccidity (of silkworm).

flacheux, -euse, flawy, waney, dull-edged.

flacon, *m.* bottle, vessel, flask; **— absorbeur,** absorption bottle; **— à capsule à vis,** bottle with screw cap; **— à densité,** specific-gravity bottle; **— à lait,** milk bottle; **— à pied,** a bottle with a base; **— à pression,** pressure bottle; **— à réactifs,** reagent bottle; **— à robinet,** bottle with stopcock; **— à tare,** weighing bottle; **— bouche à l'émeri,** glass-stoppered bottle; — **compte-gouttes,** dropping bottle; — **laveur,** wash bottle; — **pour réactifs,** reagent bottle; — **tubulé,** tubulated bottle.

flagellaire, flagellate.

flagellateur, *m.* scourger.

flagellation, *f.* flagellation, scourging.

flagelle, *f.* flagellum.

flagellé, flagellate.

flageller, to scourge, whip.

flagellés, *m.pl.* flagellata.

flagellifère, provided with runners.

flagelliforme, flagelliform.

flagellose, *f.* flagellosis.

flagellum, *m.* flagellum.

fiageoler, to wobble.

flageolet, *m.* (small) kidney bean (*Phaseolus vulgaris*).

flagrant, flagrant, decided, notorious.

fiair, *m.* smell, scent, sense of smell (game).

fiairer, to smell, smell out, scent.

flamand, *m.* Flemish (language).

flamand, Flemish.

flamant, *m.* flamingo.

flambage, *m.* flaming, singeing (hair).

flambard, *m.* blazing (piece of) coal.

flambé, flamed.

flambeau, *m.* torch, candle, candlestick, light.

flambée, *f.* blazing fire.

flamber, to flame, pass through fire, blaze, singe, fire, sterilize (in flame), char, skin dry (mold), yield.

flambeur, *m.* singer.

flamboiement, *m.* flaming, blazing.

flamboyer, to blaze, flame, flash.

flambure, *f.* (dyeing) spot (not evenly colored).

flamme, *f.* flame, fire, passion, streamer, pennant; — **du coup de feu,** thin or blowpipe flame; — **éclairante,** luminous (lighting) flame; — **nue,** naked (free) flame; — **oxydante,** oxidizing (oxidation) flame; — **réductrice,** reducing flame; —**s perdues,** waste heat.

flammé, flamelike, wavy.

flammèche, *f.* spark of burning matter, flake.

flammette, *f.* small flame.

flan, *m.* baked-custard tart, disk, plate, planchet, blank, flong.

flanc, *m.* flank, side, bosom; *pl.* womb.

Flandre, *f.* Flanders.

flanelle, *f.* flannel.

flâner, to lounge, loaf, saunter.

flanquer, to flank, throw.

flaque, *f.* puddle, pool, bog, slough.

fiasque, *f.* flask (for mercury).

flasque, *m.* side plate, cheek plate, support; — **de chaine,** chain link.

flasque, flabby, flaccid, soft, weak.

flatter, to fatter, caress, soothe, delight, humor.

flatteur, -euse, flattering, caressing.

flatueux, -euse, flatulent, windy (food).

flatulence, *f.* flatulence.

flatuosité, *f.* flatulence, flatus.

flavaniline, *f.* flavaniline.

flavine, *f.* flavin.

flavone, *f.* flavone.

fléau, *m.* beam, arm (of balance), bar, flail, scourge, plague.

flèche, *f.* arrow, pointer (of balance), stem (of tree), shoot, beam (of plow), top, head (horse), deflection, bend, sag, height, rise (of an arch), spire, lance head; — **à treillis,** lattice jib; — **d'eau,** arrowhead (Sagittaria); — **de la cote,** arrowhead.

fléchi, flexed.

fléchière, *f.* arrowhead (Sagittaria).

fléchir, to bend, flex, bow, move, yield, give way, fall (of prices).

fléchissement, *m.* bending, yielding, falling, sagging, deflection, height (of arch).

fléchisseur, *m.* flexor.

flegmatique, phlegmatic, lymphatic.

flegme, *m.* phlegm.

fléole, *f.* timothy, cat's-tail grass (*Phleum pratense*).

flétri, flaccid.

flétrir, to wither, fade, brand, dry up, blast, blight.

flétrissement, *m.* wilt disease.

flétrissure, *f.* withering, fading, wilting, brand, stigma, marcescence.

fletter, to smooth, dress.

fleur, *f.* flower, surface, blossoming, blossom, bloom (on fruits), hair side, grain side (of skin), mold (on wine), mildew specks, inner bark, prime flour, level; — **d'antimoine,** flower of antimony; — **de cinabre,** a powdery form of cinnabar; — **de farine,** fine (superfine) flour; — **de lis,** fleurde-lis, flower-de-luce, iris; — **de terre,** fumitory; — **du vendredi saint,** wood anemone.

fleur, — en languette, ray flower; — **en tube,** disk flower; **—s unisexuées,** unisexual flowers; **—s d'arnique,** arnica flowers, arnica; **—s de benjoin,** flowers of benzoin; **—(s) de carthame,** safflower (*Carthamus tinctorius*); **—(s) de cobalt,** cobalt bloom; **—(s) de soufre,** flowers of sulphur; **—s de zinc,** flowers of zinc, zinc bloom; **à — de,** (almost on a) level with, even with, flush with, close to.

fleurage, *m.* crystals, (an opaqueness) produced on glass by hydrofluoric acid, grits, groats, pollard, bran, husks.

fleuraison, *f.* flowering time, efflorescence.

fleurant, fragrant, perfumed.

fleurée, *f.* foam (in dyeing).

fleurer, to smell, be fragrant.

fleuret, *m.* floss silk, drill, bit; — **de laine,** choice wool.

fleurette, *f.* floweret.

fleuri, flowery, florid, blooming, blossoming.

fleurir, to flower, bloom, blossom, flourish.

fleuriste, *m&f.* florist, floriculturist.

fleuron, *m.* flower, ornament, floret, rosette.

fleuronné, having florets, floreted.

fleuve, *m.* river, stream.

flexibilité, *f.* flexibility.

flexible, flexible, pliant, pliable.

flexion, *f.* flexion, bending, sagging, inflection, inflexion, flexure, deflection.

flexueux, -euse, flexuous, flexuose, sinuous.

flexure, *f.* flexure, bending, fold.

flint, *m.* silex.

flint-glass, *m.* flint glass.

flocon, *m.* flock, floccule, flake, tuft.

floconnement, *m.* flocculation, forming into flakes, floccules.

floconner, to flocculate.

floconneux, -euse, flocculent, flocky, flaky, fleecy, floccose.

floculant, flocculating.

floculation. *f.* flocculation.

floculer, to flocculate.

floculeux, -euse, flocculent, somewhat floccose.

floquet, *m.* red campion (*Lychnis dioica*).

floraison, *m.* flowering, efflorescence, blooming, blossoming, bloom.

floral, -aux, floral, flower.

floran, *m.* beating trough (paper).

flore, *f.* flora, plant life; — **intestinale,** intestinal flora.

florentin, Florentine.

floribond, floribund, abounding in flowers.

floriculture, *f.* flower growing.

florifère, floriferous, flower-bearing.

florification, *f.* florification, the process of flowering.

floriform, flower-shaped.

florissait (fleurir), (it) bloomed.

florissant, flourishing.

floristique, *m.* floristics, study of plant species.

florule, *f.* little flora, local flora.

floscule, *f.* flosculum, a floret.

flosculeux, -euse, flosculous.

floss, *m.* floss.

flot, *m.* wave, billow, flood, ocean, stream, flow, flood tide, floating, float; — **et jusant,** ebb and flood.

flôtre, *m.* felt.

flottable, floatable.

flottage, *m.* floating, driving (timber), drift, rafting; — **à bûches perdues,** drifting, floating; — **en trains,** rafting.

flottant, floating, flowing, wavering, undecided, evasive.

flotte, *f.* float, buoy, fleet.

flotté, bois —, driftwood.

flottement, *m.* undulation, floating, wavering.

flotter, to float, swim at the surface, wave, flutter, drift, waver, fluctuate.

flotteur, *m.* float, raftsman, water gauge; — **immergé,** loaded float.

flouve, *f.* sweet vernal grass (*Anthoxanthum odoratum*).

fluate, *m.* fluate.

fluaté, chaux —e, fluorite, fluor spar.

fluctuant, fluctuating.

fluctuation, *f.* fluctuation.

fluctuer, to fluctuate, waver.

fluctueux, -euse, fluctuating.

fluer, to flow, run, bleed.

fluet, slender, thin.

fluide, *m.* fluid, liquid, gas.

fluide, fluid, liquid.

fluidement, fluidly, like a fluid.

fluidifiant, fluidifying.

fluidificateur, -trice, fluidifying.

fluidification, *f.* fluidification.

fluidifié, melted, liquefied.

fluidifier, to fluidify, render fluid.

fluidique, fluidic, fluid.

fluidité, *f.* fluidity, fluidness, fluency.

fluoborate, *m.* fluoborate.

fluor, *m.* fluorine, fluorite, fluor.

fluoré, containing fluorine.

fluorène, *m.* fluorene.

fluorescéine, *f.* fluorescein.

fluorescence, *f.* fluorescence; **entrer en —,** to fluoresce.

fluorescent, fluorescent.

fluorescigène, fluorescent.

fluorhydrate, *m.* hydrofluoride.

fluorhydrique, hydrofluoric

fluorifique, containing fluorine.

fluorine, *f.* fluorite, fluor spar, mineral calcium fluoride.

fluorure, *m.* fluoride.

fluosilicate, *m.* fluosilicate.

flûte, *f.* flute.

flûté, fluted.

flûteau, *m.* water plantain (Alisma).

fluvial, -aux, voie —e, waterway.

flux, *m.* flux, flow, tide, flood; **— blanc,** white flux; **— de dispersion,** leadage flux, stray flux; **— de sang,** dysentery.

fluxion, *f.* fluxion, inflammation, **— lunatique,** moon blindness, nyctalopia.

focal, -aux, focal.

foetal, -aux, fetal.

foetus, *m.* fetus, embryo.

foi, *f.* faith, proof, belief, fidelity, trust; **de bonne —,** honestly; **ma —,** upon my word.

foible, *m.* foible, weakness.

foie, *m.* liver, hepar; **— de morue,** cod liver; **— de soufre,** liver of sulphur, hepar, hepar sulphuris, potassa sulphurata.

foin, *m.* hay, grass; **— de Parnasse,** white buttercup, grass-of-Parnassus (Parnassia).

foire, *f.* diarrhea, market (place).

fois, *f.* time, repetition; **à la —,** at the same time, together; **de — à autre,** from time to time; **deux —,** twice; **encore une —,** once more; **tout à la —,** at once, at the same time; **une —,** once; **une — que,** when once, once as; **une bonne —,** once for all.

foison, *f.* abundance; **à —,** abundantly.

foisonnant, abundant, extensive, swelling.

foisonner, to swell, expand, increase, abound.

fol, folle, crazy, foolish; see **FOU.**

folâtre, frolicsome, sportive.

folâtrer, to sport, gambol.

foliacé, foliaceous, like a leaf, leafy.

foliaire, foliar.

foliaison, *f.* leafing, foliation.

foliation, *f.* foliation, leafing.

foliatus, foliate, leaved.

folie, *f.* madness, insanity, folly, frenzy, passion.

folié, foliate(d), leafy.

foliifère, foliipare, foliiferous, leaf-bearing.

foliole, *f.* foliole, leaflet; **composé de —s,** folilate, having leaflets.

foliolé, foliate, foliolate, having leaflets.

folle, crazy, foolish; see **FOL.**

follement, madly, foolishly.

follet, -ette, wanton, sportive, lively; **feu —,** ignis fatuus, will-o'-the-wisp, jack-o'-lantern.

folliculaire, follicular.

follicule, *f.* follicle, locule; **— ovarienne,** follicle sac.

folliculeux, -euse, folliculous.

fomentation, *f.* fomentation.

fomenter, to foment, stir up.

fonçage, *m.* bottoming (a cask), deepening, sinking of a shaft.

fonçailles, *f.pl.* head (of cask).

foncé, deep, dark (colors), deepened, dusky.

fonceau, *m.* a table or slab on which glass pots are molded.

foncement, *m.* deepening, bottoming.

foncer, to deepen, darken (a color), drive, bore, dig, sink, lower, bottom (casks).

fonceuse, *f.* coating machine, paper-staining machine.

foncièrement, fundamentally, at (the) bottom, thoroughly.

fonction, *f.* function, office, work, working, running; **en — de,** as a function of; **faire — de,** to act as, serve as.

fonctionnaire, *m.* official, functionary.

fonctionnel, -elle, functional, working, characteristic.

fonctionnement, *m.* functioning, operation, action, working, operating, running; — **à vide,** running light, idle motion.

fonctionner, to function, work, operate, act; — **en moteur,** to run as a motor.

fond, *m.* bottom, foundation, base, ground, head, background, ground coat, back, depth; — **de course,** bottom of the stroke, lower dead center; — **de galets,** — **de gravier,** gravel bottom, flinty ground; — **noir,** dark ground (illumination); **à —,** thoroughly; **à — plat,** flat-bottomed; **à — rond,** round-bottomed; **au —,** at bottom, at heart, after all, in the background; **de — en comble,** thoroughly, entirely; **du —,** in the background; **faire — sur,** to rely, depend on.

fondage, *m.* founding, smelting.

fondamental, -aux, fundamental, basic; **couleur —e,** primary color.

fondamentalement, fundamentally.

fondant, *m.* flux, dissolvent.

fondant, melting.

fondateur, -trice, *m.&f.* founder.

fondation, *f.* foundation, endowment; — **sur pieux,** — **sur pilotis,** pile foundation.

fondé, *m.* authorized person; — **de pouvoir,** attorney; **être —,** to be well founded, be justified.

fondé, founded, well-founded, justified.

fondement, *m.* foundation, reliance, basis, reality, substructure, base, fundament, basement.

fonder, to found, base, lay the foundation, build, consolidate, endow, establish; **se —,** to be founded, rely upon.

fonderie, *f.* foundry, smelting, founding; — **de bronze,** bronze (brass) foundry; — **de cuivre jaune,** — **de laiton,** brass foundry, — **de fer,** iron foundry, iron-works.

fondeur, *m.* founder, smelter, caster.

fondoir, *m.* melting house; — **de suif,** tallow-melting house.

fondre, to melt, smelt, fuse, found, cast, dissolve, disperse, resolve, blend, sink, fall (upon), thaw, pounce, rush (upon), burst; **se —,** to melt (down), fuse, blend, dissolve.

fondrier, *m.* projecting wall of a furnace (salt).

fondrilles, *f.pl.* sediment; see EFFONDRILLES.

fonds, *m.* fund, capital, ready money, stock, field, landed property, ground, land, soil, business; *pl.* of FOND; — **de**

roulement, working capital; — **et superficie,** *f.* soil and crop; — **social,** capital.

fondu, melted, fused, cast, molten, split.

fonger, to blot (paper), soak up ink.

fongicide, *f.* fungicide, anticryptogam.

fongicide, fungus-destroying.

fongiforme, fungiform, mushroom-shaped.

fongine, *f.* fungin.

fongique, fungic.

fongivore, feeding on fungi.

fongoïde, fungoid.

fongosité, *f.* fungosity.

fongueux, -euse, fungous.

fongus, *m.* fungus.

font (faire), (they) make or do.

fontaine, *f.* fountain, spring, well, cistern.

fonte, *f.* cast iron, pig iron, pig, melting, fusion, melt, founding, casting, smelting, blending (of wools), (of type) font; — **à l'air froid,** cold-blast pig; — **à noyau,** cored (hollow) casting; — **au bois,** — **au charbon de bois,** charcoal (blast) iron, charcoal pig; — **au coke,** coke pig; — **blanche,** white cast iron, white pig; — **blanchie,** fine iron, chilled work.

fonte, — **crue,** pig iron; — **d'acier,** cast steel; — **d'affinage,** forge pig; — **de bois,** charcoal iron; — **de fer,** cast iron; — **de moulage,** foundry iron, malleable pig iron; — **des tulipes,** smoulder of tulips; — **dure,** hard (cast) iron; — **électrique,** electric smelting; — **en gueuse(s),** pig iron; — **en saumon(s),** pig iron.

fonte, — **épurée,** washed metal, refined cast iron; — **graphiteuse,** black pig iron, graphitic pig; — **grise,** gray cast iron, gray pig; — **malléable,** malleable cast iron; — **moulée,** cast iron(ware); — **raffinée,** refined cast iron, refined pig; — **spiegel,** spiegeleisen, mirror iron; — **truitée,** mottled pig, mottled cast iron; **en —,** cast-iron.

forage, *m.* boring, drilling, borehole.

foraminulé, foraminulose, pierced by little holes.

forçage, *m.* forcing.

force, *f.* force, power, strength, energy; — **ascensionelle,** lift, lifting power; — **centrifuge,** centrifugal force; — **centripète,** centripetal force; — **de cheval,** horsepower; — **déviatrice,** deflecting force; — **du courant,** amperage, current intensity; — **freinante,** braking force, brake horse-power.

force, — **majeure,** act of God; — **morte,** vis mortua, dead force; — **nécessaire,** want of power, power consumption; — **vive**, living force, driving force, kinetic; energy; —**s utiles**, useful power; **à** —, extremely, much, hard; **à** — **de**, by the strength of, by dint of, with; **à toute** —, absolutely, at any cost; **de** —, by force, forcibly; **de toutes ses** —**s**, with all his might.

force, much, many.

forcé, compulsory, forced.

forcement, *m.* forcing.

forcément, forcibly, necessarily, by force.

forcer, to force, break open, break through, compel, oblige; — **une décimale**, to approximate a decimal.

forces, *f.pl.* shears.

forcettes, *f.pl.* small shears.

forcière, *f.* breeding pool.

forer, to bore, drill.

forerie, *f.* boring, drilling, boring machine, boring house.

foresterie, *f.* forestry, forest management.

forestier, *m.* forester.

forestier, **-ére**, forest, sylvan, forestal.

foret, *m.* drill, borer, boring tool, gimlet, auger.

forêt, *f.* forest, wood; — **domaniale**, state forest; — **en défends**, protection forest; — **feuillue**, deciduous forest; — **jardinée**, selection forest; — **particulière**, private forest; — **provenant de semis**, seedling forest; — **résineuse**, coniferous forest; — **vierge**, virgin (primeval) forest, jungle.

foreur de bouchon, *m.* cork borer.

foreuse, *f.* boring (drilling) machine.

forfait, *m.* contract, crime.

forfait, forfeited.

forficule, *f.* earwig.

forge, *f.* forge, smithy; — **à l'anglaise**, rolling mill; — **Catalane**, Catalan forge; — **d'affinerie**, refinery, refining hearth.

forgé, forged.

forgeable, forgeable, malleable.

forgeage, **forgement**, *m.* forging.

forger, to forge, work, invent; **se** —, to fancy, create.

forgerie, *f.* forging, smithing, forgery.

forgeron, *m.* smith, blacksmith, click or snapping beetle.

formaldéhyde, *m.* formaldehyde.

formanilide, *f.* formanilide.

format, *m.* form and size; — **normal**, standard size.

formateur, **-trice**, **formatif**, **-ive**, formative, forming, creative.

formation, *f.* formation, forming; — **du massif**, completion of the leaf canopy; — **houillère**, carboniferous formation; —**s secondaires**, substitute associations.

formative, formative, giving form, plastic.

forme, *f.* form, mold, shape, figure, habit, frame, make, last; *pl.* manners; **sous (la)** — **de**, in the form of.

formel, **-elle**, formal.

formellement, formally.

formène, *m.* methane.

forménique, methane, pertaining to methane.

former, to form, make, shape, mold, frame, fashion, train, ᴗe —, to form, be formed, assume a shape; — **massif**, to form a complete leaf canopy; **il se forme**, he is developing, there is developed.

formiamide, *f.* formamide.

formiate, *m.* formate, formiate; — **d'ammoniaque**, ammonium formate.

formidable, formidable, terrible, tremendous, enormous.

formique, formic.

formol, *m.* formol, formaldehyde, formaldehyde solution.

formolé, treated with formaldehyde or formol.

formoler, to treat with formol or formaldehyde.

formulable, capable of being formulated.

formulaire, *m.* formulary.

formulation, *f.* formulation.

formule, *f.* formula, rule; — **brute**, empirical formula; — **chimique**, chemical formula; — **de constitution**, constitutional (structural) formula.

formuler, to formulate, state, express (in a formula); **se** —, to be formulated, be stated, have the formula.

formyle, *m.* formyl.

fort, *m.* strong part, strong person, thick part, main point, strength, depth, height, fort, lair, stock of game.

fort, strong, high, violent, loud, great, large, powerful, thick, hard, difficult, stout, violent, severe, skillful, heavy (soil), fast, greatly, extremely, very, very much, exceedingly, strongly; — **peu soluble**, very slightly soluble.

fortement, strongly, extremely, very much, tightly, forcibly, hard, much, vigorously, greatly.

forteresse, *f.* fortress.

forticule, *f.* issue, cautery.

fortifiant, *m.* tonic.

fortifiant, strengthening, fortifying.

fortifier, to strengthen, fortify; **se —,** to grow strong.

fortuit, fortuitous, accidental.

fortuitement, fortuitously, by chance.

fortune, *f.* fortune, chance, luck; **de —,** by chance, accidentally.

fortuné, fortunate.

forure, *f.* drilled hole, bore.

fosse, *f.* pit, hole, depression, cavity, ditch, socket, trench, grave, fossa; **— à fumier et à purin,** cesspit, midden pit; **— d'aisances,** cesspool; **— nasale,** nasal fossa; **— septique,** septic tank, anaerobic tank; **—s orbitaires,** orbital cavities, orbits.

fossé, *m.* ditch, drain, trench, moat; **— d'adduction,** feeder (drain); **— d'arrêt,** trap trench (for insects or mice).

fossette, *f.* dimple, dent, little pit.

fossile, *m.* fossil.

fossile, fossil.

fossoyeur, *m.* sexton or burying beetle, gravedigger.

fossoyeur, -euse fossorial, burrowing (insect).

fou, fol, folle, crazy, mad, insane, unsteady, loose, idle, wild, foolish, extravagant, enormous, wanton.

foudre, *f.* lightning, thunderbolt, thunder; **chute de —,** thunderbolt; **coup de —,** clap of thunder.

foudre, *m.* tun, vat, hogshead, large cask.

foudrier, *m.* cooper, hooper.

foudroiement, *m.* striking down with lightning.

foudroyer, to strike by lightning, destroy, blast, thunder.

fouet, *m.* whip, whipcord, birch rod, lash, whipping, flagellum, tearing of ligament, tearing of muscle fiber, a workman who anneals bottles.

fouetté, streaked.

fouetter, to whip, lash, beat, pelt, streak, plaster, roughcast.

fougère, *f.* fern, brake; **— mâle,** male fern (Dryopteris).

fougueux, -euse, fiery, ardent.

fouille, *f.* excavation, digging, cavity, pit, trench.

fouille-merde, *m.* dung beetle.

fouiller, to excavate, dig, search, burrow into (ground), explore.

fouilleuse, *f.* subsoil plow, trenching plow.

fouillis, *m.* medley, confusion.

fouine, *f.* stone or beech marten, long pitchfork.

fouir, to dig, sink a well, burrow, excavate.

foulage, *m.* pressing, treading (of grapes), kneading.

foularder, to full, mill.

foule, *f.* multitude, great number, crowd, fulling, treading, lot; **en —,** in crowds.

foulée, *f.* pile (of skins), tread (of step), foiling, spur, footprint.

foulement, *m.* pressing, treading (of grapes).

fouler, to press, crush, squeeze, press (grapes), full or mill (cloth), tread, tramp, trample, oppress, sprain, strain, wrench.

foulerie, *f.* fulling mill.

fouleur, *m.* fuller, millman, wine presser.

fouloir, *m.* wine press, treading vat, beater, rammer, fulling mill; **— à trémie,** hopper crusher.

foulon, *m.* fuller; **terre à —,** fuller's earth.

foulonner, to full, mill (cloth).

foulonnier, *m.* fuller, millman.

foulque, *f.* coot.

foulure, *f.* fulling, milling, treading, sprain, wrench, footprint, track.

four, *m.* furnace, kiln, oven, stove; **— à acier,** steel furnace; **— à boulanger,** baker's oven; **— à briques,** brickkiln; **— à cémenter,** cementation furnace; **— à chaux,** limekiln; **— à coke,** coke oven; **— à creusets,** crucible furnace; **— à cuire,** annealing oven, baking oven; **— à cuve,** shaft furnace.

four, — à émailler, enameling furnace; **— à fondre,** melting furnace; **— à gaz,** gas furnace; **— à incinérer,** incinerator; **— à manche,** small shaft furnace, cupola; **— à moufle,** muffle furnace; **— à puddler,** puddling furnace; **— à réchauffer,** reheating furnace, heating (warming) furnace; **— à recuire,** annealing furnace; **— à réverbère,** reverberatory furnace; **— à ruche,** beehive oven.

four, — à sole, open-hearth furnace; **— à soufre,** sulphur furnace; **— continu,** continuous (working) furnace or kiln; **— d'affinage,** refining furnace; **— de fusion,** melting furnace; **— de galère,** gallery furnace; — **électrique,** electric furnace; — **liégeois,** Liége or Belgian furnace; — **Pasteur,** Pasteur sterilizer; — **rotatif,** rotary kiln, rotary furnace; — **soufflé,** blast furnace, furnace under draught.

fourbir, to furbish, clean.

fourbure, f. laminitis, founder.

fourche, f. fork, frame, branchwood; — **à foin,** pitchfork, hayfork; **tube à —,** Y tube.

fourché, forked, furcate, branched, dichotomous.

fourchée, f. forkful (of hay).

fourcher, to fork, branch.

fourchet, m. two-pronged fork, foot rot.

fourchette, f. (small) fork, table fork, horny frog (horse's hoof), belt guide (machinery).

fourchon, m. prong (of fork), tine.

fourchu, forked, branching, furcate, dichotomous; **pied —,** cloven hoof.

fourchure, f. fork (in road), forking, branching, furcation, bifurcation.

fourgon, m. poker, stirrer, wagon, car, van.

fourme, f. cheese.

fourmi, f. ant, emmet; — **lion,** m. ant lion; **—s blanches,** termites.

fourmilier, m. anteater, ant thrush.

fourmilière, f. anthill, ant heap, formicary, spongy horn, pumiced hoof.

fourmillement, m. crawling, crowding, swarming.

fourmiller, to swarm, teem, abound, tingle.

fournaise, f. furnace.

fourneau, m. furnace, kiln, oven, stove, (mine) chamber, pothole; — **à charbon,** charcoal pit; — **à gaz,** gas stove; — **à incinération,** muffle furnace; — **à sécher,** drying stove, drying kiln; — **à vent,** blast furnace, air or draught furnace; — **de calcinage,** calcining furnace, calciner; — **de cuisine,** kitchen stove, cooking range; — **de grillage,** roasting furnace; — **de mine,** blasthole; — **d'essai,** assay furnace; — **générateur,** generator; **haut —,** blast furnace.

fournée, f. ovenful, batch, charge (in furnace), baking.

fournette, f. small reverberatory furnace.

fourni, furnished, provided, thick (hair, forest).

fournilles, f.pl. firewood, brushwood.

fournir, to furnish, supply, stock, provide, deliver, pay (for).

fournissement, m. furnishing, contribution in shares (to a company).

fournisseur, m. furnisher, supplier, caterer, purveyor, seller.

fourniture, f. furnishing, provision, supply, supplies, seasoning (of dish), delivery, consignment.

fourrage, m. forage, fodder, bulk, foraging.

fourrager, to forage, search, pillage, ravage, rummage.

fourragère, fit for forage or fodder.

fourré, m. thicket, covert.

fourré, thrust, bushy, thickly wooded (country), furry, fur-lined.

fourreau, m. cover, covering, case, casing, sheath, scabbard.

fourrer, to line with fur, cram, stuff, thrust, put, line with bushes, serve (rope).

fourreur, m. furrier.

fourrier, m. quartermaster.

fourrure, f. fur, furring, wood lining.

fourvoyer, to mislead; **se —,** to lose oneself, err, be wrong.

fouteau, foyard, m. beech, beech tree (*Fagus sylvatica*).

fovéole, f. foveola, small pit.

fovéolé, foveolate, pitted.

fovilla, f. fovilla, the contents of the pollen grain.

foyer, m. foyer, hearth, home, center, fire, source of light, seat, focus, furnace, firebox, lobby, focus (of lens); — **à grille,** grate hearth; — **à réchauffer,** reheating hearth; — **d'affinage,** refining hearth; — **d'infection,** center of infection; — **réel,** real focus.

fracas, m. crash, noise, din, tumult.

fracasser, to smash, shatter, break.

fraction, f. fraction, portion, part, section; — **composée,** improper fraction, compound fraction.

fractionnaire, fractional.

fractionné, fractional, fractionated; **distillation —e,** fractional distillation.

fractionnement, m. fractionation, fractioning.

fractionner, to fractionate, fraction.

fracture, f. fracture, breaking, break, broken bone, rupture.

fracturer, to fracture, break, force (lock).

fragile, fragile, brittle, breakable, frail.

fragilité, f. fragility, brittleness, frailness, weakness.

fragment, *m.* fragment, piece.

fragmentaire, fragmentary.

fragmentation, *f.* fragmentation.

fragmenter, to divide into fragments.

fragon, *m.* butcher's-broom (*Ruscus aculeatus*).

fragrance, *f.* fragrance.

fragrant, fragrant.

frai, *m.* spawning, spawn, small fry, hard roe;— de gɪenouille, frog spawn.

fraîche, see FRAIS.

fraîchement, freshly, recently, coolly.

fraîcheur, *m.* coolness, chilliness, freshness, light air.

fraîchir, to get cooler, (of wind) freshen.

frais, *m.* coolness, freshness, fresh air, cool place.

frais, *m.pl.* expenses, cost, costs, outlay, charges;— d'abatage, felling expenses; — d'exploitation, operating costs, working expenses; — scolaires, tuition; à — communs, at joint expense.

frais, fraîche, fresh, new, recent, cool, youthful, recently.

fraisage, *m.* milling.

fraise, *f.* strawberry, cutter, milling tool, reamer, countersink, crow (of calf and lamb), ruffle, frill, mesentery.

fraiser, to fraise, bore, ream, countersink, mill, plait, ruffle.

fraiseuse, *f.* milling machine.

fraisier, *m.* strawberry plant (Fragaria).

fraisil, *m.* coal cinders, mixture of charcoal dust and earth.

framboésie, *f.* frambesia.

framboise, *f.* raspberry (Rubus).

framboisé, raspberry-flavored, having a raspberrylike surface.

framboiser, to flavor with raspberry.

framboisier, *m.* raspberry bush (Rubus).

franc, *m.* franc.

franc, franche, free, frank, pure, unadulterated, true, real, full, candid, clear, complete, whole, quite, frankly, freely, entirely, fully, completely; de — pied, sprung from seed.

français, *m.* French (language); cap. Frenchman.

français, French.

franc-alleu, *m.* allodium, interest in lands.

France, *f.* France.

franche, see FRANC.

franchement, frankly, openly, readily, purely, free.

franchir, to pass over, cross, get over, exceed, surmount, clear.

franchise, *f.* exemption, immunity, frankness, freedom.

franchissement, *m.* clearing, passing, crossing.

franco, free (of charge), carriage free, postpaid.

frange, *f.* fringe.

frangé, fimbriate, fringed, frayed.

franger, to fringe.

frangibilité, *f.* frangibility.

frangible, frangible, breakable, fragile.

franguline, *f.* frangulin.

frappage, *m.* striking, stamping, impression.

frappant, striking, impressive.

frappé, struck, stamped, forcible, iced, (of cloth) strong, close; être — de, to be affected with.

frappement, *m.* striking.

frapper, to strike, be affected, hit, beat, strike down, strike (attention), stamp, press, stand out, impress, ice (liquids), affect.

frappeur, *m.* striker, tapper, beater, knocker, puncher, stamper.

frappeur, -euse, striking, beating.

fraude, *f.* fraud.

frauder, to adulterate, defraud, cheat, smuggle.

frauduleusement, fraudulently.

frauduleux, -euse, fraudulent.

frayer, to fray, rub (of a stag), graze, wear away, spawn, open a way, associate.

frayère, *f.* spawning place.

frayeur, *f.* terror, fright.

frayoir, *m.* mark of rubbing (by stags).

frayure, f. rubbing.

frein, *m.* brake, check, curb, bit, bridle, ligament; — à air comprimé, air brake.

freindre, to shrink, break to pieces.

freiner, to brake, check, restrain.

freinte, *f.* shrinkage, loss.

frelatage, frelatement, *m.,* **frelaterie, frelatation,** *f.* adulteration, sophistication.

frelater, to adulterate, sophisticate.

frelateur, *m.* adulterator, sophisticator.

frêle, frail, fragile, faint, weak, feeble.

freloche, *f.* butterfly net.

frelon, *m.* hornet.

freluquet, *m.* dandy, whippersnapper.

frémir, to quiver, vibrate, tremble, shake, quake, shudder, (of water) simmer, boil, rustle, roar.

frémissement, *m.* rustling, agitation, vibration, simmering, shuddering, trembling, fremitus.

frêne, *m.* ash (tree or timber) (Fraxinus); — **à fleurs,** flowering ash, manna ash (*Fraxinus ornus*); — **épineux,** prickly ash (Zanthoxylum); — **pleureur,** — **parasol,** weeping-ash (*Fraxinus excelsior pendula*).

fréquemment, frequently, often.

fréquence, *f.* frequency; **basse** —, low frequency, audio frequency; **haute** —, high frequency, radio frequency.

fréquent, frequent.

fréquentation, *f.* attendance, frequenting.

fréquenter, to attend (classes) to frequent, keep company with, resort (to), associate (with).

frequin, *m.* a cask for sugar.

frère, *m.* brother.

fresque, *f.* fresco.

fret, *m.* freight, chartering, load, cargo; **faux** —, dead freight.

fréter, to freight, load, charter.

fretin, *m.* small fry.

frette, *f.* hoop, ring, band, tire.

fretté, hooped, iron-banded.

fretter, to hoop, flange.

freux, *m.* rook; **colonie de** —, rookery.

friabilité, *f.* friability.

friable, friable, easily pulverized.

friche, *f.* fallow, fallow land, waste; **en** —, fallow, uncultivated, untilled.

friction, *f.* friction, rubbing.

frictionner, to rub, chafe.

friganoptosie, *f.* fall of branches.

frigide, frigid.

frigidité, *f.* frigidity, coldness.

frigo, *m.&f.* chilled meat.

frigorifère, *m.* cold chamber, refrigerator.

frigorification, *f.* refrigeration, chilling (of meat).

frigorifier, to produce cold, freeze (meat), refrigerate.

frigorifique, refrigerating, freezing, cold-producing, frigorific.

frigorigène, *m.* cold-producing substance.

frigoule, *m.* thyme.

frimas, *m.* hoarfrost, rime.

frimousse, *f.* grimace, wry face.

fringotter, fringoter, to chirp (of chaffinch).

friquet, *m.* tree sparrow.

frire, to fry.

frise, *f.* border.

Frise, *f.* Friesland.

frisé, undulate, wavy.

friser, to curl, crisp, touch, skim, graze, be near.

frisson, *m.* shivering, shudder, thrill, shiver, rigor.

frissonner, to shiver, shudder, quiver, shake.

frit, fried.

fritons, *m.pl.* cracklings.

frittage, *m.* fritting, calcination.

fritte, *f.* frit, fritting, roasting.

fritter, to frit, fuse partially, roast.

fritteur, *m.* fritter.

fritteux, -euse, fritty, of the nature of a frit.

frittier, *m.* fritter.

friture, *f.* frying, fried food, frying (oil).

froid, *m.* cold, coldness, chill, frigidity; **à** —, cold, in the cold; **avoir** —, to be cold.

froid, cold, indifferent, frigid, lifeless.

froidement, coldly, frigidly.

froideur, *f.* coldness, chill, frigidity.

froidure, *f.* cold, coldness, cold weather, frostbite.

froidureux, -euse, cold, chilly.

froissé, offended, hurt.

froissement, *m.* rumpling, rubbing, ruffling.

froisser, to rub, rumple, bruise, ruffle, trample, offend, hurt.

froissure, *f.* (slight) bruise, rumple, wrinkle,

frôlement, *m.* grazing, rustling, brushing. contact.

frôler, to graze, brush, touch lightly, pass over.

fromage, *m.* cheese, crucible stand; — **à la crème,** cream cheese; — **à pâte ferme,** — **dure,** — **résistante,** hard cheese; — **à pâte molle,** soft cheese; — **blanc,** cottage cheese; — **boursouflé,** cheese with cavities resulting from abnormal fermentation; — **fermenté,** fermented cheese; — **frais,** fresh cheese — **gras,** rich cheese, cream cheese; — **maigre,** skim-milk cheese; — **mou, soft** cheese.

fromager, *m.* cheese mold, cheese maker, silk-cotton tree.

fromagerie, *f.* cheese factory, cheese dairy, cheese room.

fromageux, -euse, cheesy.

froment, *m.* wheat (Triticum); — de Turquie, des Indes, d'Espagne, Indian corn, maize (*Zea mays*); — épeautre, spelt (*Triticum sativum* var. *spelta*); — touselle, beardless wheat; **faux** —, rye grass (*Lolium perenne*).

fromentacé, frumentaceous.

fromentacée, *f.* frumentaceous plant.

fromental, *m.* oat grass, rye grass (*Lolium perenne*).

fromenteux, -euse, wheat-producing, rich in wheat.

frondaison, *f.* foliage, leaves, leafing time.

fronde, *f.* sling, splint (for fractured jaw), frond (of ferns), foliage.

frondicole, growing on leaves.

frondiculé, ramose, ramous, branching, having many branches.

frondule, *f.* little foliage.

front, *m.* brow, forehead, front, face, head; — de taille, face of workings; de —, abreast, in front, simultaneously.

frontal, *m.* frontal bone.

frontière, *f.* frontier, boundary, border.

frontispice, *m.* frontispiece.

frottage, *m.* rubbing.

frottant, rubbing, sliding.

trottement, *m.* friction, rubbing; — de glissement, sliding friction; — de roulement, rolling friction; à —, barely, but just.

frotter, to rub, grind, scour, polish, associate, meddle (with).

frotteur, *m.* rubbing contact, brush, rubber, floor polisher, friction piece.

frottis, *m.* polish, thin coat, wash, smear preparation.

frottoir, *m.* rubber, polisher.

frou-frou, froufrou, *m.* froufrou, rustling, humming bird.

fructifère, fructiferous, fruit-bearing.

fructification, *f.* fructification.

fructifier, to fructify, be fruitful, sporulate.

fructiflore, (the) entire flower (is) an ovary.

fructose, *f.* fructose, levulose.

fructuaire, relating to fruit.

fructueusement, fruitfully, profitably.

fructueux, -euse, fruitful, profitable.

frugal, -aux, nonexacting.

frugalité, *f.* frugality, soberness, nonexactingness.

frugivore, frugivorous, fruit-eating.

fruit, *m.* fruit, product, profit; — agrégé, collective or anthocarpus fruit; — à noyau, stone fruit, drupe; — à pépins, pome, fruit with pips, pomaceous fruit, fruit with soft kernels; — comestible, edible fruit; — composé, compound fruit, etaerio, syncarp; — déhiscent, schizocarp; — sec, seed vessel; —s de fenouil, fennel fruit.

fruité, fruity, of fruit.

fruiter, to fruit, bear fruit.

fruitier, *m.* fruit tree, fruiterer, fruit room.

fruitier, -ère, fruit, fruit-bearing.

frumentacé, frumentaceous, cereal.

frustrer, to disappoint, deprive, frustrate, baffle.

frustule, *f.* crust, corrosion, frustula.

frutescent, frutescent, shrubby, fruticose.

fruticuleux, -euse, frutescent, becoming shrubby.

fruticulus, *m.* fruticulus, a small shrub.

frutiqueux, -euse, frutescent, shrubby, fruticose.

fucacé, fucaceous.

fuchsia, *m.* Fuchsia.

fuchsine, *f.* fuchsin; — acide, fuchsin (acid); — phéniquée, carbol fuchsin.

fuchsiné, containing fuchsin.

fucoïde, fucaceous.

fucose, *n.* fucose.

fugace, fugitive, fleeting, transient, fugacious.

fugacité, *f.* fugitiveness, evanescence, transiency.

fugitif, -ive, fugitive, transient, fleeting.

fui, escaped, fled.

fuir, to flee, fly, escape, leak, bleed, (of colors) fade, retreat, recede, evade, shun, avoid.

fuite, *f.* flight, evasion, leak, leakage, escape, fading, avoidance.

fulcracé, fulcraceous, of or belonging to the fulcra.

fulcre, *f.* fulcrum; *pl.* fulcra.

fulcré, fulcrate, having fulcra.

fulgide, flashing, refulgent.

fulguration, *f.* fulguration, flashing.

fulgurite, *m.* fulgurite.

fuligineux, -euse, sooty, smoky, fuliginous, dusky.

fuliginosité, *f.* fuliginosity, smokiness, sootiness, soot.

fuligo, *m.* soot.

fuligule-milouin, *m.* pochard.

fullomanie, *f.* disease of plant, too much foliage, no flowers or fruit.

fulmicoton, *m.* guncotton.

fulminant, fulminating.

fulminate, *m.* fulminate.

fulmination, *f.* fulmination, detonation.

fulminer, to detonate, fulminate, rage.

fulminique, fulminic.

fulminurique, fulminuric.

fumade, *f.* dunging

fumage, *m.* fuming, smoking, manuring.

fumaison, *m.* manuring, dunging.

fumant, fuming, smoking, steaming, reeking.

fumarate, *m.* fumarate.

fumarine, *f.* fumarine, protopine.

fumarique, fumaric.

fumarolle, *f.* fumarole.

fumé, smoked, smoke-cured, smoky, fumed, manured.

fumée, *f.* fume, smoke, steam, vapor, flue gas, reek; **sans —,** smokeless.

fumées, *f.pl.* droppings, dung of deer, fumets.

fumer, to fume, smoke, steam, reek, manure.

fumerolle, *f.* fumarole, smoke hole.

fumeron, *m.* half-kilned (smoky) charcoal.

fumet, *m.* smell, aroma, odor, bouquet (of wine).

fumeterre, *m.* fumitory (*Fumaria officinalis*).

fumets, *m.pl.* fumets, red-deer dung.

fumeux, -euse, smoky, (of wine) fumy, heady.

fumier, *m.* manure, dung, dunghill, fertilizer; **— de ferme,** farm manure.

fumigateur, *m.* fumigator.

fumigation, *f.* fumigation.

fumigatoire, fumigating.

fumigène, smoke-producing, fumigenic.

fumiger, to fumigate.

fumiste, *m.* builder of chimneys, stove setter.

fumisterie, *f.* building of chimneys, fireplaces.

fumivore, *m.* smoke consumer.

fumivore, smoke-consuming.

fumure, *f.* manuring, fertilizing, manure, fertilizer.

funèbre, funereal, mournful.

funérailles, *f.pl.* funeral rites, obsequies.

funéraire, funereal.

funeste, fatal, disastrous, sinister, baneful, direful.

fungicide, *f.* fungicide.

fungicide, fungicidal.

fungine, *f.* fungin.

fungus, *m.* fungus; **— de la moisissure,** mold fungus.

funiculaire, *m.* funicular curve, catenary, funicular (or cable) railway.

funiculaire, funicular.

funicule, *m.* funiculus, flagellum.

funifère, with appendages.

funiliforme, funiform, ropelike.

fur, *m.* **au (à) — et à mesure,** in proportion as.

furculaire, furcular.

furent (être), (they) were.

furet, *m.* ferret.

furetage, *m.* selection felling in coppice.

fureur, *f.* fury, rage, passion.

furfuracé, furfuraceous, branny, scurfy.

furfurane, *m.* furan, furfuran.

furfuration, *f.* furfuration.

furfurine, *f.* furfurine.

furfurol, *m.* furfurole, furfural.

furieux, -euse, furious, mad, violent, wild, tremendous, extraordinary.

furoncle, *m.* furuncle, boil.

furonculose, *f.* furunculosis.

furtif, -ive, furtive, stealthy.

fusain, *m.* spindle tree, prick timber (*Euonymus evropaea*), charcoal pencil, charcoal crayon, charcoal sketch.

fusant, fusing.

fuseau, *m.* spindle, pin, axle.

fusée, *f.* fuse, fuze, rocket, fusee, sky rocket, axle arm, barrel, journal, spindle, spindleful; **— à durée,** time fuse; **— à effet retardé,** fuse with retarded action; **— à temps,** time fuse; **— d'éclairage;** light rocket; **— de sûreté,** safety fuse; **— fusante,** time fuse, burning fuse; **— instantanée,** quick match, high-sensitive fuse; **— lumineuse,** light rocket,

— percutante, percussion fuse; **— réglée,** time fuse; **— retardée,** delay-action fuse.

fusée-signal, *f.* signal rocket.

fuselage, *m.* frame, fuselage.

fuselé, spindle-shaped, fusiform.

fuser, to fuse, melt, crackle, deflagrate, fizz, burn slowly, burn without exploding, slake, (of colors) spread, run.

fusibilité, *f.* fusibility.

fusible, fusible.

fusiforme, spindle-shaped.

fusil, *m.* (small) gun, rifle, musket, tinderbox, steel (for striking flint); **— de chasse,** gun, fowling piece.

fusion, *f.* fusion, melting, blending, liquefaction; **— à la soufflerie,** sealing with the blast; **en —,** melted, molten.

fusionnement, *m.* fusion, blending.

fusionner, to fuse, amalgamate, unite, blend.

fussent (être), (they) might be.

fustet, fustel, *m.* fustet, Venetian sumac (*Cotinus coggygria*).

fustoc, *m.* fustic (*Chlorophora tinctoria*).

fut (être), (he) was.

fût, *m.* cask, barrel, shaft, trunk, bole, stem, pole, post, rib, stock, frame, handle, brace; **à — cylindrique,** cylindrical, non-tapering; **à — décroissant,** tapering; **à — dénudé,** clear-boled; **à — droit,** straight-boled; **à — irrégulier,** crooked; **à — soutenu,** full-boled. cylindrical.

futaie, *f.* timber forest, high forest, high wood, seedling forest; **— de haute,** full-grown; **— jardinée,** selection forest.

futaille, *f.* cask, barrel, casks, barrels, stave.

futaillerie, *f.* cask wood.

futailleur, *m.* cooper, hooper. barrelmaker.

futée, *f.* a cement (composed of glue and sawdust).

futile, futile.

futur, future.

fuyant, escaping, fleeing, fading, leaky, (of colors) fugitive, tapering, receding, streamline.

fuyard, *m.* fugitive, shy, timid.

G

g., *abbr.* (gramme), gram.

g.+, *abbr.* Gram-positive.

g.—, *abbr.* Gram-negative.

g. à gtte., *abbr.* (goutte à goutte), drop by drop.

gabarit, *m.* maximum structure, gauge, caliber.

gâchage, *m.* mixing (mortar), rinsing (linen).

gâche, *f.* staple, trowel, spatula.

gâcher, to mix, wet, temper (mortar), slake (lime), bungle, waste, botch, spoil, rinse (linen), harrow (corn); **— clair, — lâche,** to mix thin; **— serré,** to temper hard, mix stiff.

gâchis, *m.* wet mortar, mud, slush, mess.

gadelle, *f.* red currant.

gadellier, *m.* red currant bush (Ribes).

gadolinium, *m.* gadolinium.

gadoue, *f.* night soil.

gaffe, *f.* boat hook, gaff.

gages, *m.pl.* wages.

gagnage, *m.* pasture ground.

gagner, to gain, get, obtain, overcome, earn, win, arrive at, extend, reach; **— de proche en proche,** spread through.

gai, gay, spirited, lively, bright, loose, free, easy, having play.

gaïac, *m.* guaiacum, lignum vitae, pockwood (*Guaiacum officinale*).

gaïacol, *m.* guaiacol.

gaiement, gaily, blithely, merrily.

gailleterie, *f.* coal in small lumps, small coal, beans and nuts.

gailleteux, -euse, lumpy (coal).

gain, *m.* gain, profit.

gaine, *f.* sheath, case, shaft, flue, priming tube, scabbard, ocrea, lorica; **— foliaire,** sheath, spathe; **— ovigère,** egg tube, ovariole; **— scléreuse,** mestome sheath, leaf sheath.

gainer, to sheath, incase.

gainerie, *f.* making or selling of sheaths or cases.

gainier, *m.* sheath maker, Judas tree (*Cercis siliquastrum*).

gainule, *f.* vaginula, a small sheath.

gaize, *f.* gaize, extremely hard strata.

gala, *f.* milk consistency.

galactagogue, galactagogue, increasing secretion of milk.

galactane, *m.* galactan.

galactate, *m.* lactate.

galactine, *f.* galactine, galactan of the seeds.

galactique, lactic, galactic.

galactite, *m.* milkstone.

galactocyte, *f.* latex tube.

galactogène, lactiferous.

galactose, *f.* galactose, galactosis.

galactozyme, *m.* galactozyme.

galanga, *m.* galanga, galingale (*Alpinia galanga*).

galantine, *f.* snowdrop (*Galanthus nivalis*).

galazyme, *m.* a fermented mixture made from milk (of asses and mares).

galbanifère, producing galbanum.

galbanum, *m.* galbanum.

galbe, *m.* entasis, curved lines, contour, swell.

galbule, *m.* galbulus (cone of cypress).

gale, *f.* scabies, itch, scale, scurf, mange, scab, scum; — **commune,** common scab; — **noir,** potato wart, black scab of potato; — **poudreuse,** powdery scab, corky scab; — **verruqueuse,** powdery scab.

galé, *m.* sweet gale, Dutch myrtle (*Myrica gale*).

galéga, *m.* goat's rue (*Galega officinalis*).

galéiforme, galeate, hood-shaped.

galène, *f.* galena.

galénique, galenic(al).

galéopsis, *m.* hemp nettle (*Galeopsis tetrahit*), stinking nettle.

galère, *f.* crucible (gallery) furnace, galley; *pl.* compulsory labor.

galerie, *f.* gallery (beetles), shelf, platform, shaft, level, draft, gallery, drift, adit; — **oblique,** stellate gallery; — **principale,** principal or mother gallery; — **rayonnante,** stellate gallery.

galéruque, *f.* — **de l'aune,** alder leaf beetle.

galet, *m.* pebble, boulder, gravel, shingle, pulley, disk, roller; — **de roulement,** live (friction) roller.

galette, *f.* flat cake, press cake, slab, disk, ship biscuit.

galeux, -euse, scabby, itchy, mangy, scurfy, covered with gall.

Galicie, *f.* Galicia (Poland).

galicien, -enne, Galician.

galiope, *f.* hemp nettle (Galeopsis).

galipot, *m.* galipot, solidified resin, rosin.

galipoter, to coat with galipot.

gallate, *m.* gallate.

galle, *f.* gall, oak apple, nutgall, gallnut, gallfly, knot, lump; — **de chêne,** oak gall; **noix de —,** nutgall, gallnut.

galléine, *f.* gallein.

gallérie, *f.* bee moth.

Galles, *f.* Wales.

gallicole, *f.* gall wasp.

gallifère, bearing oak apples, yielding gallium.

gallinacé, *f.* obsidian, volcanic glass.

galliote, *f.* herb bennet (*Geum urbanum*).

gallique, gallic, Gallic.

gallium, *m.* gallium.

gallois, *m.* Welshman.

gallois, Welsh.

gallon, *m.* gallon, a kind of nutgall (on some oaks).

gallotannique, gallotannic.

galop, *m.* gallop.

galoper, to gallop.

galvanique, galvanic.

galvanisation, *f.* galvanization, galvanizing.

galvanisé, galvanized.

galvaniser, to galvanize, (electro-)plate, zinc.

galvano, *m.* electrotype, electro.

galvanocautère, *m.* galvanocautery.

galvanomètre, *m.* galvanometer.

galvanoplastie, *f.* galvanoplastics, electroplating.

galvanoplastique, galvanoplastic.

galvanotactisme, *m.* galvanotaxis, electrotaxis.

galvanotyper, to electrotype.

gambier, *m.* gambier, pale catechu.

gambir, *m.* gambier, yellow catechu.

gamelle, *f.* bowl, mess kettle.

gamétange, *f.* gametangium.

gamète, *m.* gamete.

gamétisation, *f.* change of cell into a gamete.

gamétocyte, *f.* gametocyte.

gamétophyte, *f.* gametophyte, the gameteforming phase.

gamin, *m.* gamin, boy, apprentice.

gamme, *f.* gamut, scale, range.

gamodesmie, *f.* gamodesmy, the vascular bundles are fused together.

gamogénèse, *f.* gamogenesis.

gamopétale, gamopetalous.

gamosépale, gamosepalous.

gamostèle, *f.* gamostele, the fusion of several steles.

gamostélie, *f.* gamostely, a polystelic stem.

gamostyle, with fused styles.

gamotrope, gamotropic, position of flowers when expanded.

gamotropisme, *m.* gamotropism, movements of nutation.

ganache, *f.* jaw, mandible, chin.

Gand, *m.* Ghent.

gangliforme, gangliform.

ganglion, *m.* ganglion, nerve knot; — cérébroide, brain, cerebral ganglion.

ganglionné, with piliferous nodes.

gangrène, *f.* gangrene, slough, canker.

gangreneux, gangréneux, -euse, gangrenous, cankerous.

gangue, *f.* gangue, gang, veinstone, covering membrane.

ganil, *m.* ganil (a brittle limestone).

ganse, *f.* braid, loop, cord, eye, ear.

gant, *m.* glove, gauntlet.

gantelée, *f.* foxglove (Digitalis).

gantelet, *m.* gauntlet.

ganterie, *f.* glovemaking, glove trade.

garançage, *m.* madder dyeing.

garance, *f.* madder (*Rubia tinctorum*).

garance, madder (dye).

garancer, to dye with madder.

garanceux, *m.* a coloring matter extracted from previous madder dyeing.

garancine, *f.* garancine.

garant, *m.&f.* voucher, warrant, guarantor, guaranty, surety.

garantie, *f.* guarantee, security, warranty, voucher.

garantir, to guarantee, warrant, protect, shelter, preserve, ensure; se —, to protect oneself.

garçon, *m.* boy, apprentice, journeyman, waiter, lad, fellow, man, bachelor.

garde, *f.* guard, care, keeping, watch, protection, custody; *m.* guard, keeper, warden, adjutant, forest guard; — chasse, gamekeeper; — caisse, *m.* case protector,

savechest; — général, ranger; — pêche, river guard; — vente, guard of a felling area; en —, on guard; sur ses —s, on the alert.

garde-corps, *m.* railing, guardrail.

garde-feu, *m.* fireguard, fender.

garde-fou, *m.* railing.

garde-fourneau, *m.* furnaceman.

garde-meuble, *m.* lumber room.

gardénia, *m.*, gardénie, *f.* gardenia.

garder, to keep, guard, beware, preserve, protect; se —, to keep, take care not to, remain unspoiled, beware (of), keep (from); il se garde de, he guards against; il se garde d'être vu, he takes care not to be seen.

garderie, *f.* beat, domain.

garde-robe, *f.* wardrobe, water-closet, southernwood (*Artemisia abrotanum*), feces, stool.

garde-temps, *m.* chronometer.

garde-vue, *m.* eye shade, lamp shade.

gardon, *m.* roach.

gare, *f.* (railway) station, switch, shunt; — de bifurcation, junction station; — de triage, station where trains are made up.

garenne, *f.* warren, preserve.

garer, to keep out of the way.

gargariser, to gargle.

gargarisme, *m.* gargle, throat wash, gargarism.

gargouillement, *m.* gurgling, splashing sound.

garnir, to furnish, supply, provide, fit, fill, line, face, trim, mount, garnish, decorate, garret (joints), stuff (chair), pack (piston).

garnissage, *m.* furnishing, filling, trimming.

garniture, *f.* filling, lining, facing, mounting, set, fitting(s), trimming(s), furnishing, furniture, border, outfit, decoration, packing, stuffing, bearing box.

garou, *m.* spurge flax (*Daphne gnidium*), mezereon (*Daphne mezereum*), spurge olive.

garouille, *f.* kermes oak (*Quercus coccifera*).

garrigue, *f.* waste land, heath.

garrot, *m.* withers, shoulders, tourniauet, tongue.

garus, *m.* elixir of Garus.

gaspillage, *m.* squandering, wasting.

gaspiller, to squander, waste.

gastérothalame, *f.* closed sporangia sunken in thallus.

gastrique, gastric.

gastrite, *f.* gastritis, inflammation of the stomach.

gastronomique, gastronomic(al).

gastrula, *f.* gastrula.

gâté, *m.* spoiled part.

gâté, decayed, spoiled, damaged, tainted, unsound, deteriorated.

gâteau, *m.* cake, comb.

gâte-bois, *m.* goat moth.

gâter, to spoil, waste, damage, corrupt; **se —,** to spoil, be spoiled, decay, deteriorate.

gattilier, *m.* agnus castus, chaste tree (*Vitex agnus astus*).

gattine, *f.* gattine, disease of silkworms.

gauche, left, left-handed, levorotatory, levo, awkward.

gauchement, awkwardly, clumsily.

gaucherie, *f.* awkwardness, clumsiness.

gauchi, warped (by the wind).

gauchir, to warp, shrink out of true.

gauchissement, *m.* warping.

gaudage, *m.* dyeing with weld.

gaude, *f.* weld, dyer's rocket, dyer's-weed (*Reseda luteola*).

gauder, to dye with weld.

gaudronné, repand.

gaufrage, *m.* stamping.

gaufre, *f.* honeycomb, waffle.

gaufrer, to emboss, goffer, crimp, plait, stamp.

gaule, *f.* pole, stick, rod, staff, fishing rod, switch, sapling.

gauler, to beat (fruit tree).

gaulis, *m.* crop of saplings.

gaulthérie, *f.* Gaultheria; — **couchée,** wintergreen (*Gaultheria procumbens*).

gavage, *m.* cramming, stuffing.

gave, *f.* crop (of birds).

gaver, to cram.

gayac, *m.* see GAÏAC.

gayacol, see GAÏACOL.

gaz, *m.* gas; — **acide carbonique,** carbonic-acid gas, carbon dioxide; — **acide chlorhydrique,** hydrochloric acid; — **azote.** nitrogen gas; — **brûlés,** exhaust gases; — **carbonique,** carbon dioxide, carbonic-acid gas; — **d'échappement,** exhaust gases;

— **d'éclairage,** illuminating gas; — **de combustion,** combustion gas; — **de gueulard,** waste gas, blast-furnace gas; — **de houille,** coal gas; — **des marais,** marsh gas.

gaz, — **détonant,** detonating gas; — **de ville,** city gas, town gas; — **mixte,** semi-water gas, producer gas; — **naturel,** natural gas; — **nobles,** inert gases, rare gases; — **parfait,** perfect gas, ideal gas; — **pauvre,** producer gas; — **perdu,** waste gas; — **riche,** oil gas; — **suffocant,** asphyxiating gas; — **sulfhydrique,** hydrogen sulphide; — **sulfureux,** sulphur dioxide.

gazage, *m.* gassing.

gaze, *f.* gauze; — **métallique,** wire gauze, wire netting; — **oxygénée,** antiseptic gauze.

gazé, *m.* black-veined white butterfly.

gazéifiable, gasifiable, which can be volatilized.

gazéificateur, *m.* gasifier, gas producer, flare lamp.

gazéification, *f.* gasification, carbonating; — **du charbon,** coking process, coal distillation.

gazéifier, to gasify, volatilize, carbonate, charge (as water).

gazéiforme, gasiform, gaseous.

gazéité, *f.* gaseous nature or state.

gazer, to cover with gauze, veil, gas (troops), roast, reduce with gaseous fuel.

gazette, *f.* gazette, newspaper, sagger.

gazeux, -euse, gaseous, of gas, foamy.

gazier, *m.* gauze maker, gas, gasman.

gazier, -ère, pertaining to gas.

gazifère, gas-producing.

gazofacteur, *m.* gasifier.

gazogène, *m.* gas generator, gas producer.

gazogène, gasogene, carbonating.

gazolène, *m.* gazoline, *f.* gasoline.

gazolyte, gasifiable.

gazomètre, *m.* gasometer, gasholder.

gazomoteur, *m.* gas engine, gas motor.

gazon, *m.* grass, turf, sod, sward, lawn; — **d'Olympe,** thrift; — **mousse,** mossy saxifrage (*Saxifraga hypnoides*); **couvert d'un — naissant,** grassy, grass-covered.

gazonnage, *m.* sodding.

gazonné, covered with turf.

gazonnement, *m.* turfing.

gazonner, to turf. sod, make a lawn to.

gazouillant, warbling, chirping, babbling, purling.

geai, *m.* jay.

géant, *m.* giant.

géant, gigantic, huge.

géantisme, *m.* giantism.

géastre, *m.* geaster.

gécine cendré, grey-headed green wood-pecker.

géine, *f.* gein (humic acid).

geitonocarpie, *f.* production of fruit as outcome of geitonogamy.

geitonogamie, *f.* geitonogamy, fertilization among flowers of same branch.

gel, *m.* gel, frost, freezing.

gelable, congealable.

gélatine, *f.* gelatin, jelly; — **détonante,** explosive gelatin.

gélatiné, gelatinized, gelatined.

gélatineux, -euse, gelatinous, jellylike.

gélatinifiable, gelatinizable.

gélatinifier, to gelatinate, convert into gelatin.

gélatiniforme, gelatiniform, gelatinoid.

gélatinisation, *f.* gelatinization, gelatinizing.

gélatinisé, containing gelatin.

gélatiniser, to gelatinize, coat with gelatin.

gélatino-bromure, *f.* gelatinobromide.

gélatino-chlorure, *f.* gelatinochloride.

gélatose, *f.* gelatose.

gèle, *m.* gel.

gelé, frozen, frostbitten.

gelée, *f.* jelly, frost, freeze, freezing; — amenant le déchaussement, soil-lifting frost; — **automnale,** autumnal frost, early frost; — **blanche,** hoarfrost, rime frost; — **printanière,** spring frost, late frost; — **végétale,** vegetable jelly, pectin.

geler, to freeze, congeal, solidify.

gélif, -ive, congealing easily, frost-cracked (tree, stone).

gélification, *f.* gelification, becoming gelatinous.

gélifier, to gelatinize, gelatinate, jellify.

gélignite, *f.* gelignite.

gelinotte, *f.* hazel grouse.

gélis, frost-cracked, frost-cleft.

gélivure, *f.* frost-crack.

gélose, *f.* gelose, agar.

gélosé, containing agar.

gelsémine, *f.* gelsemine, gelsemium.

Gémeaux, *m.pl.* Gemini, the Twins.

gémelliflore, with flowers arranged in pairs.

gémination, *f.* gemination.

géminé, double, geminate, twin.

géminiflore, with flowers arranged in pairs.

gémir, to groan, moan.

gémissement, *m.* moaning, groaning.

gemmage, *m.* incising, tapping (of trees for resin).

gemmaire, pertaining to buds.

gemmation, *f.* budding, gemmation.

gemme, *f.* gem, pine resin, turpentine, leaf bud, resin.

gemme, precious, gemmeous, glittering.

gemmer, to incise, tap pine trees for resin, bleed, bud, germinate; — **à mort,** to tap to death; — **à vie,** to tap without killing.

gemmifère, gemmiferous, bud-bearing, gemmate.

gemmiforme, bud-shaped, gemmiform.

gemmipare, gemmiparous, producing buds.

gemmulation, *f.* gemmulation, gemmule formation.

gemmule, *f.* embryo, plumule, gemmule.

gênant, annoying, in the way, troublesome.

gencive, *f.* gum.

gêne, *f.* trouble, inconvenience, constraint, uneasiness.

gêné, constrained, uneasy.

généagenèse, *f.* geneagenesis, parthenogenesis.

généalogie, *f.* genealogy, pedigree.

génépi, génipi, *m.* wormwood (Artemisia).

gêner, to impede, hinder, restrict, obstruct, embarrass, annoy, trouble; **se —,** to put oneself to inconvenience; **il se gêne,** he is ill at ease.

général, -aux, general, universal, common; **en —,** in general, as a rule.

généralement, generally, in general.

généralisateur, -trice, generalizing.

généralisation, *f.* generalization.

généraliser, to generalize; **se —,** to become generalized.

généralité, *f.* generality.

générateur, *m.* generator, boiler, gas producer; — **à vapeur,** — **de vapeur,** steam generator, boiler.

générateur, -trice, generating, generative, producing.

génératif, -ive, generative.

génération, *f.* generation, production, procreation, propagation, breeding.

génératrice, *f.* generatrix, generator; — **gasogène,** gas generator or producer.

générer, to generate, produce.

généreusement, generously.

généreux, -euse, generous, fertile (soil).

générique, generic.

génériquement, generically.

générosité, *f.* generosity.

genèse, *f.* genesis, origin.

génésie, *f.* generation.

génésique, genetic.

genestrolle, *f.* woodwaxen, dyer's green-weed (*Genista tinctoria*).

genêt, *m.* broom (Cytisus), whin, genista; — **à balai(s),** common broom (*Cytisus scoparius*); — **des teinturiers,** dyer's genista, dyer's-broom, woodwaxen (*Genista tinctoria*); — **épineux,** furze, gorse, whin (*Ulex europaeus*); — **poilu,** hairy greenwood (*Genista pilosa*).

génétique, genetic.

genette, *f.* whin (Genista), narcissus.

Genève, *f.* Geneva.

genévrette, *f.* a drink made from juniper berries.

genévrier, *m.* juniper tree (*Juniperus communis*); — **sabine,** common savin, savin tree (*Juniperus sabina*).

génial, -aux, of a genius, geniuslike.

géniculé, geniculate, kneed.

géniculiflore, with sessile flowers.

génie, *m.* genius, intellect, disposition, spirit, engineering, engineers.

genièvre, *m.* juniper tree (*Juniperus communis*), juniper berry, gin.

genièvrerie, *f.* gin distillery.

génisse, *f.* heifer.

genista, *m.* genista.

génital, -aux, genital.

géniteur, *m.* sire, generator.

génito-urinaire, genitourinary.

genou, *m.* knee, elbow, joint, ball-and-socket joint; *pl.* lap; **à genoux,** kneeling.

genouillé, geniculate, abruptly bent.

genouillère, *f.* kneecap, kneepiece, joint, connection, elbow joint, toggle joint.

genre, *m.* kind, family, sort, type, genus, style, way, mode, manner, fashion, taste,

gender, genre (painting); — **d'exploitation,** method of operation.

gens, *m.pl.* people, folks, men, nations, servants.

gent, *f.* tribe, race, brood.

gentiane, *f.* gentian.

gentianine, *f.* gentianin, gentisin.

gentianose, *f.* gentianose.

gentil, -ille, pretty, elegant, genteel gracious.

gentilhomme, *m.* nobleman, gentleman.

génuine, genuine.

génuinité, *f.* genuineness.

géoblaste, *f.* geoblast.

géocarpe, géocline, geocarpic, (fruit) maturing underground.

géode, *f.* geode, druse.

géodèse, *m.* geodesist.

géodésie, *f.* geodesy.

géodésique, geodetic.

géodique, geodic.

géodynamique, *f.* geodynamics.

géoécien, -enne, growing on the ground.

géoffrée, *f.* Geoffraea.

géogène, geogenous, growing on the ground

géogénie, *f.* geogeny.

géognosie, *f.* geognosy.

géographe, *m.* geographer.

géographie, *f.* geography.

géographique, geographic(al).

géographiquement, geographically.

géologie, *f.* geology.

géologique, geologic(al).

géologue, *m.* geologist.

géomètre, *m.* geometrician, surveyor; — **du pin,** pine looper moth.

géométrie, *f.* geometry.

géométrique, geometric(al).

géomorphogénie, *f.* geomorphogeny.

géonyctitropique, geonyctitropic, (sleep movements) requiring gravity.

géoperception, *f.* geoaesthesia, capacity to respond to gravity.

géophile, *f.* geophyte, producing perennial underground buds.

géophilie, *f.* geophily, terrestrial species.

géophysique, *f.* geophysics.

géophyte, producing perennial underground buds.

géoplagiotropisme, *m.* geoplagiotropism, growing obliquely to the ground.

géostrophisme, *m.* geotortism, torsion due to gravitation.

géotonus, *m.* geotonus, normal position to gravity.

géotortisme, *m.* geotortism, torsion by gravitation.

géotrophie, *f.* geotrophy, growth due to gravity.

géotropisme, *m.* geotropism.

gérance, *f.* management, managership.

géraniacées, *f.pl.* Geraniaceae.

géranial, *m.* geranial, citral.

géraniol, *m.* geraniol.

géranium, *m.* geranium.

gérant, *m.* manager, director.

gérant, managing, directing.

gerbe, *f.* sheaf, spout, jet, small heap.

gerber, to stack, pile, sheaf, bind in sheaves.

gerbeur, *m.* binder.

gerbier, *m.* stack (of corn), barn.

gerbiforme, sheaflike.

gerbillon, *m.* small sheaf.

gerce, *f.* clothes moth.

gercement, *m.* — **du bois,** cracking of the wood.

gercer, to crack, chap.

gerçure, *f.* crack, fissure, rent, split, shake, cleft, chink, chap.

gerçuré, cracked, fissured, shaky (wood).

gérer, to manage, carry on.

gergelin, *m.* oil of sesame.

germandrée, *f.* germander.

germanique, Germanic, German.

germaniser, to Germanize.

germanium, *m.* germanium.

germant, germinating.

germe, *m.* germ, sprout, seed, bud, shoot, ovary, embryo, microorganism; — **invisible,** ultramicroscopic organism; —**s bacteriens,** bacteria.

germer, to germinate, sprout, spring up.

germification, *f.* transformation into a germ.

germinal, -aux, germinal.

germinateur, *m.* germinating apparatus.

germination, *f.* germination.

germoir, *m.* hotbed, seedbed, malt floor, germinating apparatus.

gérofle, *f.* clove, see GIROFLE.

gerzeau, *m.* corn cockle (*Agrostemma githago*).

gésier, *m.* gizzard.

gésine, *f.* lying-in.

gésir, to lie, rest, be.

gesse, *f.* pea, chick-pea (*Cicer arietinum*), vetch, vetchling (Lathyrus); — **cultivée,** chickling vetch (*Lathyrus sativus*); — sylvestre, wood vetch, (*Vicia sylvatica*), wild lathyrus, everlasting pea (*Lathyrus latifolius*).

gestation, *f.* gestation, pregnancy.

geste, *m.* gesture.

gesticuler, to gesticulate.

gestion, *f.* management; **agent de —,** manager, superintendent.

gestionnaire, *m.* manager.

gestionnaire, managerial.

geyser, *m.* geyser.

gibbeux, -euse, gibbous, gibbose, more convex, protuberant.

gibbifère, (corolla) with expanded gorge.

gibbiflore, with gibbous petals.

gibbosité, *f.* gibbosity.

gibelet, *m.* gimlet.

gibier, *m.* game; — **à plumes,** wildfowl, feathered game; — **à poil,** furred game, ground game; — **comestible,** edible game; — **d'eau,** waterfowl; — **émigrant,** migratory game; — **sédentaire,** indigenous (standing) game; **gros —,** big game; **menu —,** small game.

gible, *m.* set of bricks (in the kiln).

giboulée, *f.* shower of rain.

giboyeux, -euse, abounding in game.

giclée, *f.* spray, squirt.

gicler, to squirt, spurt, spray.

gicleur, *m.* atomizer, nozzle, sprayer.

gigantesque, gigantic, huge.

gigartin, resembling small grape seed.

gigue, *f.* haunch (of venison).

gillénie, *f.* Gillenia, Indian physic.

gillon, *m.* mistletoe (*Viscum album*).

gingembre, *m.* ginger.

gingeole, *f.* jujube.

gingibrine, *f.* powdered ginger.

gingival, -aux, gingival.

ginguet, -ette, poor, weak, of little value.

ginseng, *m.* ginseng.

giobertite, *f.* magnesite.

gipon, *m.* wool cloth used for tallowing hides.

girandole, *f.* girandole, chandelier, cluster; — **d'eau,** water horsetail.

girasol, *m.* girasol, fire opal, sunflower, heliotrope.

giration, *f.* gyration.

giratoire, gyratory.

giraumont, giraumon, *m.* pumpkin.

girofle, *m.* clove.

giroflée, *f.* gillyflower stock, wallflower (*Cheiranthus cheiri*); **cannelle —,** clove bark.

giroflier, *m.* clove tree (*Eugenia aromatica*).

girolle, *f.* chanterelle.

giron, *m.* lap, presence.

girouette, *f.* wind vane, weather flag, weathercock.

gisant, lying; **bois —,** felled timber.

gisement, *m.* deposit, bed, layer, vein, stratum, bearing; — **en amas,** massive deposit; — **en couche,** stratified (bedded) deposit; — **en filons,** vein (lode) deposit.

gisent (gésir), (they) are lying.

gît (gésir), (he) is lying.

gîte, *m.* deposit, bed, lodging, lair (of deer), shelter, haunt, crops, lower millstone, leg of beef, covert, form.

githago, *m.* corn cockle (*Agrostemma githago*).

givre, *m.* white (hoar) frost, rime.

givré, frosty, frosted.

glabre, glabrous, smooth, denuded.

glabrescent, glabrescent, glabrate, almost glabrous.

glabréité, with glabrous (smooth) surface.

glabriscule, almost glabrous.

glabrisme, *m.* glabrism, the smoothness of normally hairy parts.

glabritie, *f.* glabrous surface.

glaçage, *m.* icing, glazing, glaze.

glaçant, freezing, icy.

glace, *f.* ice, plate glass, mirror, glass, lens, glass window, flaw (in diamond), ice cream, sugar icing; — **de fond,** ground ice; — **fondante,** melting ice.

glacé, iced, icy, frozen, chilly, icy-cold, biting.

glacer, to ice, freeze, chill, glaze, calender.

glacerie, *f.* plate-glass works, ice-cream trade.

glaceur, *m.* glazer.

glaceux, -euse, flawed (gems), flawy, icy.

glaciaire, glacial.

glacial, glacial, icy, frigid, freezing, cold.

glaciale, *f.* ice plant (*Mesembryanthemum crystallinum*).

glacier, *m.* glacier, plate-glass maker, maker of ices, dealer in ices.

glacière, *f.* icehouse, icebox, ice chest, refrigerator, ice-cream freezer.

glacis, *m.* slope, sloping surface, glacis, glaze.

glaçon, *m.* piece or block of ice, icicle.

glaçure, *f.* glaze, glazing.

gladié, ensiform, gladiate.

glaïeul, *m.* corn flag, garden gladiolus (Gladiolus); — **des marais,** swamp iris (*Iris pseudacorus*); — **puant,** fetid iris (*Iris fetidissima*).

glairage, *m.* glairing.

glaire, *f.* glair, white of egg, mucus, phlegm, flaw (in diamond).

glairer, to glair, coat with glair.

glaireux, -euse, glairy, glaireous, viscous, slimy.

glairine, *f.* glairin.

glaisage, *m.* coffering (shaft), claying (borehole).

glaise, *f.* clay, loam, vault of a glass furnace; **terre —,** clay soil, loamy soil, potter's earth, clay.

glaiser, to clay, treat, dress or line with clay, coffer.

glaiseux, -euse, clayey, loamy.

glaisière, *f.* clay pit.

gland, *m.* acorn, mast, tassel, glans; — **de terre,** earthnut, tuberous rooted pea.

glandage, *m.* — **secondaire,** after-pannage.

glandaire, glandular.

glande, *f.* gland, kernel; — **collétérique,** — **sébifique,** colleterial or sebific glands; — **sébacée,** sebaceous gland; — **surrenale,** suprarenal (adrenal) gland; —**s calcaires,** salt glands.

glandée, *f.* oak mast, pannage; — **abondante,** full mast; — **et fainée,** high mast, beech or acorn mast; — **peu abondante,** — **rare,** quarter mast.

glandifère, acorn-bearing, glandiferous.

glandulaire, glanduleux, -euse, glandular, glandulous.

glane, *f.* bunch, gleanings.

glaner, to glean.

glapir, to yelp, yap.

glarimètre, *m.* glarimeter.

glaubérite, *f.* glauberite.

glaucescence, *f.* glaucescence.

glaucescent, glaucescent, becoming seagreen.

glaucière, glaucienne, *f.* glaucium.

glaucome, *m.* glaucoma.

glauconie, *f.* glauconite.

glaucophylle, with glaucous leaves.

glauque, glaucous, bluish green, sea-green.

glaux, *m.* sea milkwort (*Glaux maritima*).

glèbe, *f.* clod (of earth), sod, soil.

glène, *f.* socket, glene.

glénoïdal, -aux, glenoid.

gleton, *m.* cleavers, goose grass.

glette, *f.* litharge.

gliadine, *f.* gliadin.

glissade, *f.* glissade, slipping, slide; — de terre, landslide.

glissant, sliding, slippery.

glissement, *m.* sliding, slipping, gliding, landslip, landslide.

glisser, to slide, slip, glide, creep, run; se —, to glide, creep, steal.

glisseur, *m.* slider, slide block, glider.

glissière, *f.* guide, groove, slide bar, slide, guide rod.

glissoir, *m.* sliding block, slide, timber slide, flume.

glissoire, *f.* slider, sliding block, slide, shoot, guide.

global, -aux, total, aggregate, gross, lump, summary.

globe, *m.* globe, ball, earth, sphere, orb, glass cover.

globeux, -euse, globose, globular.

globiforme, globular, globe-shaped.

globigerines, *f.pl.* Globigerinae; boueå —, globigerina ooze.

globine, *f.* globin.

globoïde, *f.* globoid.

globulaire, *f.* globe daisy, globularia.

globulaire, globular, spherical.

globule, *m.* globule, drop (of water), corpuscle; — blanc, white corpuscle, leucocyte; — pyoïde, pus cell; — rouge, red corpuscle.

globuleux, -euse, globulous, globular.

globulicide, destroying blood corpuscles.

globulifolié, with globulous leaves.

globulin, *m.* leucocyte.

globuline, *f.* globulin.

globulinurie, *f.* globulinuria.

globulolyse, *f.* globulolysis, hemocytolysis.

globulose, *f.* globulose.

glochide, *f.* glochidium.

glochidé, glochidial.

gloiocéphale, (fungus) with glabrous and viscous hood.

gloire, *f.* glory, boast, honor, fame, splendor, fixed sun, aureole, nimbus.

gloirifier, to glorify, praise; se —, to glory in.

glomérule, *f.* glomerule, glomerulus, small mass, flower cluster, tuft.

glomérulé, glomerulate.

glomérulifère, glomeruliferous, bearing clusters.

glonoïne, *f.* glonoin.

glorieux, -euse, glorious, splendid.

glossaire, *m.* glossary, vocabulary.

glossines, *f.pl.* tsetse flies.

glossologie, *f.* glossology.

glossopétale, with ligulate petals.

glotte, *f.* glottis.

glouglou, *m.* gurgle, gurgling.

gloume, *f.* see GLUME.

glouteron, *m.* burdock (Arctium), common burdock, clotbur (*Arctium lappa*).

glover, *m.* Glover tower.

glu, *f.* birdlime, glue; — marine, marine glue.

gluant, glutinous, sticky, gummy, viscous.

gluau, *m.* limed twig.

glucine, *f.* glucina, beryllia.

glucinium, *m.* glucinium, beryllium.

glucomètre, *m.* glucometer.

gluconique, gluconic.

glucosane, *f.* glucosan.

glucose, *f.* glucose, grape sugar.

glucosé, containing glucose.

glucoser, to treat with glucose.

glucoside, *m.* glucoside.

glucosique, producing glucose.

glucosurie, *f.* see GLYCOSURIE.

gluer, to smear with birdlime, lime (twigs), make sticky.

glumacé, glumaceous, chaffy.

glume, *f.* husk, glume, chaff.

glumé, glumose.

glumelle, *f.* palea of grasses, glumella.

glumelléen, coming from a palea.

glumellule, *f.* glumellule, palea, lodicule.

glutamine, *f.* glutamine.

glutamique, glutamic.

glutarique, glutaric.

gluten, *m.* gluten, glue.

glutenase, glutenase.

glutinatoire, agglutinating.

glutine, *f.* glutin, tree gum.

glutineux, -euse, glutinous.

glycémie, *f.* glycemia.

glycéré, glycérat, *m.* glycerite, glycerole, glycerin (solution of a medicinal substance in glycerin); — d'amidon, glycerite (glycerin) of starch; — de phénol, glycerite of phenol; — de tannin, glycerite of tannic acid; — d'hydrastis (du Canada), glycerite of hydrastis.

glycéride, *f.* glyceride.

glycérie flottante, *f.* manna grass (Glyceria).

glycérine, *f.* glycerin, glycerol; — phénique, glycerite of phenol.

glycériné, glycerinated.

glycériner, to glycerinate, glycerinize, treat with glycerin.

glycérolé, *m.* glycerite, glycerin.

glycéryle, *m.* glyceryl.

glycicarpe, with sweet fruits.

glycide, *m.* glycide.

glycin, *m.* glycin (a photographic developer).

glycine, *f.* glycine, beryllia, glucina, Glycine, wistaria (licorice vetch).

glycinium, *m.* see GLUCINIUM.

glycocholique, glycocholic.

glycocolle, *m.* glycocoll, glycine.

glycogénase, *f.* glycogenase.

glycogene, *m.* glycogen.

glycogène, glycogenous, glycogenic.

glycogénèse, glycogénie, *f.* glycogenesis, glycogeny.

glycogénique, glycogenic.

glycol, *m.* glycol.

glycolyse, *f.* glycolysis.

glycolytique, glycolytic.

glycoprotéide, *n.* glucoprotein, glycoprotein.

glycosamine, *f.* glycosamine.

glycose, *f.* see GLUCOSE.

glycoside, *m.* see GLUCOSIDE.

glycosurie, *f.* glycosuria, diabetes mellitus.

glycosurique, glycosuric.

glycuronique, glucuronic.

glyoxaline, *f.* glyoxaline, imidazole.

glyoxylique, glyoxylic.

gneiss, *m.* gneiss.

gneissique, gneissic.

gnesiogamie, *f.* gnesiogamy, fertilization between different individuals of same species.

go, tout de —, easily, immediately.

gobelet, *m.* goblet, cup, rocket case; — à cuire, beaker; —s, pennywort.

gobeleterie, *f.* hollow-glass works.

gobe-mouches, *m.* flycatcher, flytrap, sundew.

gober, to swallow, gulp down.

gobetis, *m.* rough cast, pointing (wall).

goder, to pucker, bag.

godet, *m.* cup, bowl, shallow dish, calyx, cup (of acorn), cell, bucket, bagging, puckering; — à huile, oil cup; — favique, scutula, of favers.

goéland, *m.* gull; — argenté, herring gull; — rieur, black-headed gull.

goémon, *m.* seaweed.

goitre, *m.* goiter, struma.

golfe, *m.* gulf, sinus.

gommage, *m.* gumming.

gomme, *f.* gum, gumma, rubber, gum (disease in fruit trees); — adragante, gum tragacanth; — arabique, gum acacia, gum arabic; — d'acajou, cashew gum; — élastique, gum elastic, caoutchouc, (India) rubber; — explosive, explosive gelatin; — laque, lac, shellac; — molle, crude turpentine.

gommé, gummed, gum, gummy, containing gum.

gomme-grattoir, eraser, rubber.

gomme-gutte, *f.* gamboge, gum guttae.

gomme-laque, *f.* gum-lac.

gommelaquer, to shellac, treat with gum-lac.

gommeline, *f.* gommelin, dextrin.

gommement, *m.* gumming.

gommer, to gum, mix with gum, erase, rub out.

gomme-résine, *f.* gum resin.

gommeux, -euse, gummy, sticky, gummous, gummiferous, gum-yielding (trees), gummatous (tumor).

gommier, *m.* gum tree (Eucalyptus).

gommifère, gummiferous, gum-yielding.

gommose, *f.* gummose.

gonade, *f.* gonad.

gond, *m.* hinge pin.

gonder, to hinge.

gondole, *f.* eye bath, gondola.

gondoler, to buckle, warp, blister (of paint).

gonflage, *m.* inflation, swelling.

gonflé, inflated, distended, swollen, swelled.

gonflement, *m.* swelling, distention, inflation, enlargement.

gonfler, to swell (out), distend, inflate, pump up; **se** —, to swell, be swollen.

gongyle, *m.* propagulum, gonidium in lichens.

gongylode, resembling a round head.

gonidema, *f.* microgonidium.

gonidie, *m.* gonidium.

gonie, *f.* ascospores and basidiospores.

gonimique, gonimic, relating to gonidia.

gonimoblaste, *m.* gonimoblast.

gonimolobes, *f.pl.* gonimolobes, the terminal tufts of gonimoblasts.

goniomètre, *m.* goniometer.

goniomyce, with angular forms.

goniosperme, with angular seed.

goniothèque, *f.* goniotheca.

gonocoque, *m.* gonococcus.

gonocyste, *f.* microgonidium.

gonohyphema, *f.* microgonidium.

gonomères, *f.pl.* idiomeres.

gonoophyte, with angular fruit.

gonophore, *f.* gonophore.

gonorrbée, *f.* gonorrhea.

gonosperme, with angular seed.

gonotoconte, *f.* gonotokont, zeugite, fungus spore from fusion of two nuclei.

gorge, *f.* throat, orifice, mouth, neck, gullet, groove, furrow, hollow, recess, gorge, defile.

gorge-de-pigeon, dove-colored shot (silk).

gorgée, *f.* mouthful, draught.

gorger, to fill, gorge, stuff.

gorgonzola, *m.* Gorgonzola (cheese).

gosier, *m.* throat, gullet, pharynx, fauces.

gosiller, to be carried over.

gossypin, gossypine, cottony, flocculent.

gossypiphore, cotton producing.

gothique, Gothic.

gouache, *f.* water colors.

goudron, *m.* tar; — **à calfater,** navy pitch; — **de bois,** wood tar; — **de houille,** coal tar; — **de lignite,** lignite tar; — **de tourbe,** peat tar; — **minéral,** mineral tar, asphalt; — **végétal,** vegetable tar.

goudronnage, *m.* tarring, pitching.

goudronner, to tar, pitch.

goudronnerie, *f.* tar works, tar shed.

goudronneux, -euse, tarry.

gouet, *m.* arum, cuckoopint, (*Arum maculatum*); — **à trois feuilles,** jack-in-the-pulpit, Indian turnip (*Arisaema triphyllum*); — **serpentaire,** common dragon (*Dracunculus vulgaris*).

gouffre, *m.* gulf, abyss, vortex.

gouffré, corrugated, ribbed, chamfered, fluted, grooved.

gouger, to gouge, scoop out.

goujon, *m.* gudgeon, dowel, pin, peg.

goujure, *f.* indentation notch.

goulot, *m.* neck (of a bottle), spout.

goulotte, *f.* gullet, channel.

goupille, *f.* pin, peg, key, splint, gudgeon.

goupillé, split.

goupiller, to pin, dowel, key, peg.

goupillon, *m.* test-tube brush (for lamp, bottle, or gum), sprinkler, Aspergillus.

goupillonner, to cleanse with a brush.

gourbet, *m.* beach grass, marram grass, (*Ammophila arenaria*).

gourd, swollen, benumbed.

gourde, *f.* gourd, flask, water bottle, sac.

goure, *f.* adulterated drug.

gourer, to adulterate (drugs).

goureur, *m.* adulterator.

gourmand, *m.* water sprout, epicormic branch; **branches gourmandes,** suckers.

gourmander, to prune.

gourme, *f.* strangles.

gourmette, *f.* curb chain.

gournable, *f.* treenail.

gousse, *f.* pod, legume, husk; — **d'ail,** clove of garlic; — **de vanille,** vanilla bean.

gousset, *m.* armpit, gusset, bracket, brace.

goût, *m.* taste, flavor, bouquet (of wine), relish, liking.

goutellete, *f.* see GOUTTELETTE.

goûter, to taste, relish, enjoy, like, smell; se —, to be tasted, be liked, like each other.

goûter, *m.* lunch.

goutte, *f.* drop, minim, dram, bead, blister, gout; — à —, drop by drop, dropwise; (en) — pendante, hanging drop; (en) — suspendue, hanging drop.

goutte-de-sang, *f.* pheasant's-eye (*Adonis aestivalis*).

gouttelette, *f.* droplet, little drop, globule.

goutter, to drop, drip.

goutteux, -euse, gouty (person).

gouttière, *f.* gutter, eaves, spout, groove, furrow, cradle, splint, trough.

gouvernable, governable.

gouvernail, *m.* rudder, helm; roue du —, steering wheel.

gouvernateur, *m.* regulator, governor.

gouverne, *f.* guidance, guide, rudder, control.

gouvernement, *m.* government, direction, steering, management.

gouverner, to govern, rule, direct, steer, manage.

gouverneur, *m.* governor, ruler.

gouvion, *m.* large bolt, pin.

goyave, *f.* guava (fruit).

goyavier, *m.* guava(tree) (*Psidium guajava*).

grabeau, *m.* fragment, impurities (from drugs).

grabelage, *m.* sorting, cleansing.

grabeler, to cull (the refuse from).

grâce, *f.* grace, thanks, pardon, mercy, favor, charm; — à, thanks to, owing to; de —, please; de bonne —, willingly.

gracieusement, graciously, kindly.

gracieux, -euse, graceful, gracious, kind.

gracilifolié, with slender leaves.

gracilipède, with filiform stipe.

gradatif, -ive, gradational.

gradation, *f.* gradation; — des âges, age gradation.

grade, *m.* grade, rank, degree.

gradient, *m.* — de la température, drop in temperature.

gradin, *m.* step, tier, shelf, bench.

graduation, *f.* graduation, calibration; — d'échelle, scale division.

gradué, *m.* graduate.

gradué, graduated, progressive.

graduel, -elle, gradual.

graduellement, gradually, little by little.

graduer, to graduate, calibrate.

graille, *f.* crow.

grain, *m.* grain, texture, berry, particle, atom, bead, pellet, bush, bushing, seed, stone, pustule, shower, squall; — de raisin, grape; — d'orge, barleycorn (Hordeum); — truité, mottled grain; —s chromatiques, chromidia, chromatin granules; à —s fins, fine-grained; à —s serrés, close-grained; à gros —s, coarse-grained.

grainage, *m.* granulation.

graine, *f.* seed, seed grain, berry, eggs (of silkworms); — ailée, winged seed; — à vers, wormseed; — d'ambrette, amber seed, musk seed; — d'anis, aniseed; — d'aspic, canary seed; — d'Avignon, Avignon berry, French berry; — d'écarlate, kermes, kermes berry; — de girofle, cardamom; — de lin, flaxseed, linseed; — de paradis,, grain of paradise; — de perroquet, saffron seed; — de Perse, Persian berry.

graine, — de Turquie, Indian corn, maize (*Zea mays*); — de vers à soie, silkworm eggs; — forestière, forest seed, seed of forest trees; — légère, winged seed; — lourde, heavy seed; — musquée, musk seed, amber seed; — oléagineuse, oil seed; — vaine, unfruitful seed; —s de coton, cottonseed (residue); —s des Moluques, Molucca grains, tilley seed; —s de Tilly, tilley (seed); — tinctoriale, kermes, kermes berry; monter en —, to go to seed.

grainelé, granulated.

graineler, to grain (paper); see GRENELER.

grainer, to granulate; see GRENER.

graineterie, *f.* seed trade or shop.

grainetier, *m.* seed dealer.

grainier, *m.* seedsman, collection of seeds.

grainure, *f.* crystalline fracture.

graissage, *m.* greasing, lubricating, grease, oiling, becoming ropy; — à barbotage, splash lubrication; — sous pression, forced lubrication.

graisse, *f.* fat, grease, ropiness (of wine), lubricant; — alimentaire, grease for food, edible fat; — balsamique, — benzoinée, benzoinated lard; — de laine, wool grease, lanolin; — de ménage, cooking fat, oleomargarine; — de mouton, mutton suet; — de porc, lard; — minérale, mineral fat; — végétale, vegetable fat; goutte de —, fat globule.

graisser, to grease, lubricate, smear, oil, clog (a file), (of wine) become ropy.

graisseur, *m.* lubricator, greaser, oiler, oil box, oilhole, wetness, hydration (of paper); — **à godet,** grease cup; — **mécanique,** self-acting lubricator.

graisseur, -euse, lubricating, greasing.

graisseux, -euse, greasy, fatty, ropy.

gram, *m.* Gram solution, Gram stain, Gram's method.

gramen, *m.* lawn grass.

graminée, *f.* grass, gramineous plant, species of grass.

graminiforme, gramineous.

graminoïde, gramineous.

graminopétale, with lineate petals.

grammaire, *f.* grammar.

grammatite, *f.* grammatite, tremolite.

gramme, *m.* gram.

gramme-degré, small calorie, gram calorie.

grand, *m.* adult, grown up, great man, grandee, greatness, grandeur.

grand, great, large, big, open, wide, high, tall, full, main, grand; — **air,** open air; **en —,** on a large scale, full-size; —**e masse,** main body; —**e mauve,** wild mallow (*Malva sylvestris*); —**e vitesse,** great velocity, high-speed; —**s blés,** wheat and rye; —**s déplacements,** low-power work with microscope; **tout — ouvert,** wide open.

grand-duc, *m.* grand duke, eagle owl.

Grande-Bretagne, *f.* Great Britain.

grandelet, biggish.

grandement, greatly, grandly, vastly, very much.

grandeur, *f.* grandeur, greatness, largeness, size, quantity, magnitude, length, height, stature.

grandiose, grand, sublime.

grandir, to grow, increase, grow large, magnify, make greater, cause to grow; **se —,** to become tall or greater.

grandissement, *m.* growth, enlargement, magnification.

grand-paon, *m.* great peacock moth.

grand'peine, *f.* **à —,** with a great deal of trouble.

graner, to hatch.

grange, *f.* barn.

granifère, graniferous.

granit, *m.* granite.

granitaire, granitelike, granitic.

granitelle, *f.* granitell.

graniteux, -euse, granitique, granitic.

granitoïde, granitoid.

granula, *f.* granula, granule.

granulage, *m.* granulation.

granulaire, granular.

granulateur, *m.* granulator.

granulation, *f.* granulation.

granulatoire, *m.* granulator.

granule, *m.* granule.

granulé, granulated.

granuler, to granulate, grain, hatch.

granuleuse, *f.* granule cell.

granuleux, -euse, granular, granulated, miliary.

granulicaule, with granular stem.

granuliforme, granuliform, granular.

granulochromidie, *f.* chromidium, gonidium.

granulomes, *f.pl.* granulomata.

granulose, *f.* granulose.

graphique, *m.* diagram, graph, drawing.

graphique, graphic, graphical.

graphiquement, graphically.

graphite, *m.* graphite, black lead, plumbago.

grappe, *f.* cluster, bunch, raceme, mass, grapeshot, candylamata, string; — **composée,** compound raceme; — **de raisin,** bunch of grapes, grapeshot; — **d'ombellie,** umbellike raceme; — **pendante,** catkin.

grapperie, *f.* grape growing.

grappillage, *m.* gleaning (grapes).

grappillon, *m.* small (bunch) cluster.

grappin, *m.* grapple, hook.

gras, *m.* fat, flesh, meat, calf (of leg).

gras, fat, fatty, greasy, unctuous, plump, thick, ropy, oily (wine), soft (metals), loamy, clayey (soil), heavy, rich, fertile, moist, wet (paper), slippery, boldface (print); **série —se,** fatty series.

grassement, plentifully.

grasset, *m.* stifle joint.

grateron, *m.* goose grass, catchweed, cleavers (*Galium aparine*), common burdock (*Arctium lappa*), sweet woodruff (*Asperula odorata*).

gratification, *f.* gratuity, allowance, bonus.

gratifier, to gratify, ascribe, confer.

gratiole, *f.* Gratiola; — **officinale, hedge hyssop** (*Gratiola officinalis*).

gratioline, *f.* gratiolin.

gratis, gratis, free (of charge).

gratitude, *f.* gratitude.

grattage, *m.* scraping, erasure, teaseling.

gratte, *f.* scraper.

gratte-boësse, *f.* wire (scratch) brush.

gratte-ciel, *m.* skyscraper.

gratte-cul, *m.* hip (berry of wild rose).

grattelle, *f.* itching.

grattement, *m.* act of scratching.

gratter, to scrape, scratch, rub, erase, tease, nap, flatter.

gratteron gaillet, *m.* cleavers, goose grass.

grattoir, *m.* scraper, rake, eraser.

gratuit, gratuitous, free.

gratuitement, gratuitously, free, for nothing.

grauwacke, *f.* graywacke.

gravats, *m.pl.* screenings (of plaster), rubbish.

grave, heavy, grave, deep (sounds), serious, sedate.

gravé, engraved.

graveler, to gravel.

graveleux, -euse, gravelly, gritty.

gravelle, *f.* gravel, crude tartar crystals.

gravement, gravely, seriously.

graver, to cut, engrave, carve, etch, imprint, impress; **se —,** to be engraved, be etched, split, crack.

graveur, *m.* engraver.

gravide, gravid, pregnant.

gravier, *m.* gravel, grit.

gravière, *f.* gravel pit, ringed plover, mixed sowing of vetch and lentils.

gravillon, *m.* fine gravel.

gravimètre, *m.* gravimeter.

gravimétrique, gravimetric.

gravir, to climb, clamber, ascend, mount.

gravitation, *f.* gravitation.

gravité, *f.* gravity, seriousness, weight, importance; **— spécifique,** specific gravity.

graviter, to gravitate, revolve.

gravois, *m.pl.* screenings (of plaster), rubbish.

gravure, *f.* engraving, print, etching.

grayer, *m.* verderer.

grayerie, *f.* forest jurisdiction.

gré, *m.* will, inclination, liking, wish; **bon —, mal —,** whether one will or not; **de —, à —,** by (mutual) agreement; **en**

—, in good part, liking for; **savoir bon —,** to be pleased; **savoir mauvais —,** to be displeased.

grèbe, *m.* **— castagné,** little grebe, dabchick; **— huppé,** great crested grebe.

grec, grecque, Greek.

Grèce, *f.* Greece.

grecque, *f.* fret, fretwork.

greffage, *m.* grafting.

greffe, *f.* graft, grafting, record office (of court); **— à cheval,** saddle graft; **— écusson,** shield budding; **— en anneau,** ring grafting; **— en couronne,** crown grafting; **— en fente,** cleft grafting; **— en flûte,** flute budding; **— en placage,** veneer grafting; **— mixte,** graft between plants not closely related; **— par approche,** inarching; **— sur racine,** root grafting.

greffer, to graft; **se —,** to be grafted.

greffoir, *m.* grafting or budding knife.

greffon, *m.* graft, scion.

grégaire, gregarious.

grégarine, *f.* Gregarina.

grège, raw (silk).

grêle, *f.* hail.

grêle, slender, thin, shrill, scanty, small; **intestin —,** small intestine.

grêlé, damaged by hail.

grêler, to hail, grail.

grêlet, rather slender or thin.

grêleux, -euse, seamed, unequal.

grêlon, *m.* hailstone.

gremille, *f.* ruff.

grenade, *f.* grenade, pomegranate; **— à fusil,** rifle grenade; **— à main,** hand grenade.

grenadier, *m.* grenadier; pomegranate (tree) (*Punice granatum*).

grenadin, sirup —, sirup of pomegranate.

grenadine, *f.* grenadine (sirup or silk); sirup of pomegranate.

grenage, *m.* granulation.

grenaille, *f.* refuse grain, tailings, small shot, grain, granulated metal; **— de charbon,** carbon granules; **— de plomb,** small shot, lead shot; **en —,** granulated.

grenaillement, *m.* granulation.

grenailler, to granulate.

grenaison, *f.* seeding, corning.

grenat, *m.* garnet.

grenatique, of or containing garnets.

greneler, to grain (leather, paper).

grener, to seed, run to seed, grain (salt), corn, granulate, stipple, shred (wax), (of silkworms) lay eggs.

grèneter, to grain (leather).

grèneterie, *f.* seed trade; see GRAINETERIE.

grènetier, *m.* seed dealer; see GRAINETIER.

grenétine, *f.* grenetine (a white gelatin).

grènetis, *m.* mill, milling.

grenette, *f.* little grain, Avignon berry.

grenier, *m.* granary, attic, garret; en —, in bulk.

grenoir, *m.* granulator, granulating machine, corning sieve, corning house.

grenouille, *f.* frog.

grenouillet, *m.* Solomon's seal (Polyganatum).

grenouillette, *f.* buttercup (Ranunculus), frogbit (*Hydrocharis morsus-ranae*), ranula, frogtongue.

grenu, *m.* grain, granular structure, granularity.

grenu, grainy, granular, granulous.

grès, *m.* sandstone, grit(stone), paving stone, stoneware; — à meule(s), millstone grit; — bigarré, variegated sandstone, bunter sandstone; — cérame, stoneware; — dur, grit(stone).

grésil, *m.* broken or pounded glass, sleet, hail.

grésillement, *m.* shriveling, chirping (of crickets), pattering (sound like sleet).

grésiller, to shrivel (up), scorch, trim (edges), crackle, sputter, sleet.

gresserie, *f.* sandstone quarry, stoneware.

grève, *f.* sandbank, strand, mortar sand, strike (of workers); faire —, to be on strike; se mettre en —, to go on strike.

grevé, burdened with rights.

grever, to burden, encumber.

gréviste, *m.* striker.

grièche, sharp, sour.

grief, *m.* grievance.

grieu, *m.* firedamp, marsh gas.

grièvement, seriously, severely.

griffage, *m.* seizing, marking, notching, scratching, blazing.

griffe, *f.* claw, talon, paw, clamp, catch, prong, hook, scratch, stamped signature, (signature) stamp, root(let), tendril, bark blazer, scratcher; —s de girofle, clove stems.

griffer, to scratch, claw, seize with claws, grip, blaze, mark, notch, lash.

griffette, *f.* clawlet.

grignard, *m.* hard stoneware, hard sandstone.

grignons, *m.pl.* olive(-oil cake), olive husks, crusts.

grignoter, to nibble.

gril, *m.* grid (iron), grill.

grillage, *m.* roasting, network; — en tas, roasting in heaps.

grillager, to lattice, grate, fence, rail off.

grille, *f.* grille, grill, grate, grating, grid, rack, furnace, roasting furnace, calciner, fire grate, lattice, railing; — à analyse(s), combustion furnace; — à mouvement va-et-vient, shaking grate; — à secousses, shaking grate; — de foyer, fire grate; — en escalier, step grate.

griller, to broil, grill, toast, roast, calcine, singe, scorch, burn.

grilloir, *m.* roaster, singeing machine.

grimpant, climbing.

grimpante, *f.* climbing plant, climber.

grimper, to climb, creep.

grimpereau, *m.* tree creeper.

grincer, to grind, creak, grate, gnash.

grindélie, *f.* gum plant (Grindelia).

gringalet, *m.* weakling, stripling.

griotte, *f.* a sour cherry, morel, morello, mahaleb cherry, a spotted marble.

griottier, *m.* sour cherry tree; — du nord, morello.

grippage, *m.* seizing, abrasion, friction.

grippe, *f.* influenza, equine distemper, pinkeye.

grippeler, to shrink, seize.

grippement, *m.* seizing, gripping.

gripper, to seize, grip, bind, catch, snatch, contract, shrink, crinkle (up), grind, run hot; se —, to shrink, shrivel.

grippure, *f.* abrasion, worn part.

gris, *m.* gray, grey; — acier, steel gray; — cendré, ash gray; — clair, light gray; — d'acier, steel gray; — de fer, iron gray, iron liquor; — foncé, dark gray.

gris, gray, pale (wine), tipsy.

grisaille, *f.* grizzled color, gray, mixed white and black, pepper-and-salt (cloth), grey poplar.

grisard, *m.* white (or gray) poplar wood, badger, hard sandstone.

grisard, dark gray, grayish brown.

grisâtre, grayish.

grise, *f.* red spider.

griser, to gray, make gray, paint, tint with gray, intoxicate; **se** —, become intoxicated.

grisette, *f.* white rot, whitethroat.

grisou, *m.* firedamp, marsh gas.

grisoumètre, *m.* firedamp detector.

grisouteux, -euse, containing firedamp.

grive, *f.* thrush, throstle; — **chanteuse,** song thrush; — **de gin,** — **draine,** missel thrush; — **des vignes,** — **mauvis,** redwing; — **litorne,** fieldfare; — **mauvis,** redwing; — **musicienne,** song thrush.

Groenland, *m.* Greenland.

grogner, to grunt, growl, grumble.

groisil, *m.* cullet, broken glass.

gronder, to scold, growl, roar, grumble.

gros, *m.* bulk, mass, body, main part, — **bois,** heavy timber; — **bout,** butt end; **à** — **grain,** coarse-grained; **en** —, wholesale, in general.

gros, grosse, big, large, much, great, coarse, bulky, loud, full, high, gross, thick, heavy, dark, deep (colors), rich; **grosse,** pregnant.

gros-bec, *m.* hawfinch, grosbeak; — **linotte,** linnet; — **tarin,** siskin; — **vulgaire,** hawfinch.

groseille, *f.* currant; — **à maquereau,** — **verte,** gooseberry (Grossularia).

groseiller à grappes, currant bush (Ribes).

groseillier, *m.* (red) currant bush, currant; — **épineaux,** — **à maquereau,** gooseberry bush, gooseberry.

grosse, *f.* gross, large piece.

grosse, see GROS.

grossement, in general, as a whole, roughly speaking.

grosserie, *f.* wholesale trade.

grossesse, *f.* pregnancy.

grosseur, *f.* size, swelling, growth, tumor, bulk, bigness, thickness, girth; — **de boîtes,** can size.

grossier, *m.* coarse material, wholesaler.

grossier, -ère, coarse, rough, rude, rank, gross, plain.

grossièrement, coarsely, roughly, rudely, grossly, clumsily.

grossièreté, *f.* coarseness, grossness, roughness, bluntness, clumsiness.

grossification, *f.* grossification.

grossir, to grow large or larger, augment, exaggerate, increase, swell, magnify, enlarge; **se** —, to grow larger, be magnified.

grossissant, magnifying, enlarging, amplifying.

grossissement, *m.* increasing, swelling, magnification, magnifying, magnifying power.

grossulaire, grossular.

grotesque, grotesque, odd, ridiculous.

grotte, *f.* grotto.

grouillement, *m.* swarming, crawling.

grouiller, to swarm.

groupe, *m.* group, set, unit, clump (of trees), cluster, formation; — **des noeuds adhérents,** knot cluster; **par** —**s,** in groups.

groupement, *m.* grouping, group, gathering; — **fonctionnel,** characteristic group.

grouper, to group, connect up, join up, couple; **se** —, to group, bunch, gather.

gruau, *m.* meal, coarse ground wheat flour, grits, groats, gruel, small crane; — **de sarrasin,** buckwheat grits; — **d'orge,** pearl barley.

grue, *f.* crane; — **à vapeur,** steam crane; — **de chargement,** loading hoist.

gruer, to grind, reduce to meal.

grugeon, *m.* piece of sugar.

gruger, to crunch.

grume, *f.* bark, log.

grumeau, *m.* clot, flock, curd (of milk), lump, particle.

grumeleux, -euse, clotted, curdled, clotty, coagulated, lumpy.

grumelure, *f.* cavity (in casting).

gruyère, *m.* Gruyere cheese.

guaco, *m.* guaco.

guanase, *f.* guanase.

guaner, to manure with guano.

guanidine, *f.* guanidine.

guanier, -ère, of or yielding guano, guaniferous.

guanine, *f.* guanine.

guano, *m.* guano; — **de poisson(s),** fish (guano) manure; — **phosphate,** phosphatic guano.

guayac, *m.* pockwood, lignum vitae (Guaiacum).

gué, *m.* ford.

guède, *f.* woad, pastel.

guéder, to dye with or steep in woad, cram, stuff.

guéderon, *m.* woad dyer.

guenille, *f.* rag, tatter; *pl.* old clothes.

guêpe, *f.* wasp.

guêpier, *m.* wasp's nest; — vulgaire, bee eater.

guère, ne . . . —, scarcely, hardly, but little, not much.

guéret, *m.* fallow (plowed) land, furrow.

guérir, to cure, heal; se —, to recover, heal, be cured (of).

guérison, *f.* cure, recovery, healing (of wound).

guérissable, curable.

guérite, *f.* cabin, shelter, sentry box, a lookout turret, (locomotive) cab, (telephone) booth.

guérit-tout, *m.* great wild valerian, cat's valerian, allheal.

guerre, *f.* war, warfare, strife, contention.

guerrier, *m.* warrior.

guerrier, -ère, warlike, martial.

guet, *m.* watch, lookout.

guêtre, *f.* gaiter, legging.

guetter, to watch, lie in wait for, watch for.

gueulard, *m.* throat, mouth (of blast furnace), mouth (of sewer).

gueule, *f.* mouth, muzzle (of gun), chops, jaws, throat; en —, labiate.

gueule-de-lion, gueule-de-loup, *f.* snapdragon (*Antirrhinum majus*).

gueuse, *f.* pig (iron); — des mères, sow (iron).

gueuset, *m.* small pig.

gui, *m.* mistletoe (*Viscum album*).

guichet, *m.* wicket, shutter, little window.

guidage, *m.* guiding, alignment.

guide, *m.* guide, guide rod, slide bar, leader, guidebook, guides; *f.* rein.

guider, to guide, conduct, direct, lead, steer.

guifette, *f.* — noire, black tern.

guignier, *m.* heart cherry (tree), sweet cherry, bigarreau; — à fruit jaune, yellow gean, yellow bigarreau.

guildive, *f.* rum.

guillage, *m.* fermenting, brewing.

guillemet, *m.* quotation mark, quote.

guillemeter, to put in quotation marks.

guiller, to ferment, brew.

guilleret, sprightly, merry, chipper.

guilloire, *f.* fermentation vat.

guillotine, *f.* guillotine.

guimauve, *f.* marshmallow (*Althaea officinalis*).

guindeau, *m.* windlass.

guinder, to raise, hoist.

guinée, *f.* guinea, Guinea.

guipage, *m.* wrapping, winding, taping, insulation.

guiper, to wind, wrap (tape).

guirlande, *f.* garland, wreath.

guise, *f.* guise, manner, way, fashion, wise, humor; à ma —, as I please; en — de, by way of, instead of.

gummifère, gummiferous, gum-bearing.

gurgu, gurjum, *m.* gurjun (*Dipterocarpus alatus*).

gustation, *f.* taste.

gustative, flavor, flavoring; sensation —, sensation of taste.

gutta-percha, *f.* gutta-percha.

guttation, *f.* guttation.

gutte, *f.* gamboge, gum guttae.

guttifère, guttiferous.

guttural, -aux, guttural.

gymnanthe, gymnanthous, naked-flowered.

gymnase, *m.* gymnasium.

gymnocarpe, gymnocarpous, naked-fruited.

gymnocarpien, -enne, with naked fruit.

gymnocéphale, with naked flowers.

gymnocidie, *m.* gymnocidium.

gymnogyne, gymnogynous, with naked ovary.

gymnomonospermée, *f.* plant with single naked fruit.

gymnorhize, with naked roots.

gymnos, naked.

gymnosperme, *f.* gymnosperm.

gymnosperme, gymnospermous.

gymnospermie, *f.* gymnospermy, bearing naked fruit.

gymnostome, gymnostomous, peristome of mosses destitute of teeth.

gymnostyle, when style is naked.

gynandre, gynandrous.

gynécée, *f.* gynecium.

gynécée, gynoeceum, *m.* gynoecium, pistils collectively.

gynizus, *m.* gynixus, gynizus, stigma (in orchids).

gynobase, *f.* gynobase.

gynobasé, with a gynobase.

gynobasique, gynobasic.

gynodiécie, *m.* gynodioecism, with female and hermaphrodite flowers.

gynodyname, *m.* gynodynamus, female element is preponderant.

gynoeceum, *m.* gynoeceum, all carpels of a flower.

gynomonécie, *f.* gynomoneocism.

gynophore, *m.* gynophore.

gynophorien, -enne, born on a gynophore.

gynophoroïde, resembling a gynophore.

gynostège, *m.* gynostegium.

gynostème, *m.* gynostemium, pistil and stamens in orchids.

gypaete barbu, bearded vulture.

gypse, *m.* gypsum, plaster of Paris; — cunéiforme, — en fer de lance, — spathique, sparry or specular gypsum.

gypseux, -euse, gypseous.

gypsifère, gypsiferous.

gyratoire, gyratory.

gyroid, spiroid.

gyrome, *f.* gyroma, buttonlike shield of Gyrophora.

gyronea, *f.* annulus of ferns.

gyroscopique, gyroscopic.

gytja, *f.* humus produced under lake waters.

H

habile, capable, skillful, clever, able, competent.

habilement, ably, cleverly, skillfully.

habileté, *f.* ability, skill, cleverness.

habillage, *m.* dressing, preparing.

habillement, *m.* clothing.

habiller, to dress, clothe, fit, equip, cover, trim, prune (plants or roots), become.

habit, *m.* clothes, garment, habit, dress, (dress) coat.

habitable, inhabitable.

habitant, *m.* inhabitant, resident, denizen, inmate.

habitat, *m.* habitat (of plant or animal).

habitation, *f.* habitation, dwelling, haunt, abode, habitat, settlement, plantation.

habiter, to inhabit, live in, frequent, dwell in, reside.

habitude, *f.* habit, custom; **comme d'—,** as usual; **d'—,** ordinarily.

habitué, accustomed, used.

habituel, -elle, habitual, usual, customary.

habituellement, habitually, usually, customarily.

habituer, to habituate, accustom, inure; **s'— à,** to accustom oneself to.

hache, *f.* ax, hatchet; — d'abatage, felling ax; — de fendage, cleaving ax.

hache-fourrage, *m.* chaff cutter.

hacher, to chop, hack, cut (in pieces), hash, shred, mince, hatch, mangle.

hachette, *f.* hatchet; — à marteau, hammer head hatchet.

hachisch, hachich, *m.* hashish, Indian hemp.

hachoir, *m.* chopper, chopping board, chopping knife.

hachure, *f.* hatching, shading; **faire des —s,** to hatch.

hadrocentrique, hadrocentric.

hadrocyte, *f.* pitted element.

hadromal, *m.* hadromal.

hadromase, *m.* hadromase.

haem-, see HEM-.

haie, *f.* hedge, fence, hedgerow.

haine, *f.* hatred, hate, aversion, enmity.

haïr, to hate, detest, abhor.

haïssable, detestable.

halbran, *m.* young wild drake or duck, flapper.

halbrené, broken-feathered.

hale, *m.* towline.

hâle, *m.* heat of sun, sunburn, hot wind.

hâlé, tanned, sunburnt.

haleine, *f.* breath; **en —,** in good condition, in practice.

haler, to haul, pull, tow.

hâler, to burn, brown, tan (from sun).

haleter, to pant, puff and blow.

halimètre, *m.* see HALOMÈTRE.

hall, *m.* hall, room.

halle, *f.* market (place).

hallier, *m.* thicket, brushwood, net.

halo, *m.* halo, areola, halation, ring, corona.

halochimie, *f.* the chemistry of salts.

halogène, *m.* halogen.

halogène, halogen, halogenous.

halogéné, halogenated, halogen.

halographie, *f.* halography, description of salts.

haloïde, haloid, halide.

haloir, *m.* hemp kiln, drying room.

halologie, *f.* a treatise on salts.

halologique, pertaining to the treatise on salts.

halomètre, *m.* halometer, salt gauge.

halométrie, *f.* halometry, determination of salts.

halophile, halophilous.

halophytes, *f.pl.* halophytes.

haloplankton, *m.* haloplankton, the floating vegetation of salt-water.

halot, *m.* rabbit hole.

halotechnie, *f.* the technique of preparing salts.

halte, *f.* halt, stop.

haltère, *m.* dumbbell; *pl.* poisers (of Diptera).

halurgie, *f.* technique of preparing salts.

halurgique, relating to saltmaking.

hambourg, *m.* small cask, keg.

hameau, *m.* hamlet.

hameçon, *m.* fishhook.

hameçonné, hooked, barbed (spike).

hamigère, with hook-shaped hair.

hampe, *f.* staff, shank, handle, scape, stem, stalk, thin flank (of beef), shaft, trunk.

hamuleux, -euse, with small hooked hairs.

hanche, *f.* hip, haunch, leg, coxa.

hanebane, *f.* henbane (*Hyoscyamus niger*).

hangar, *m.* (open) shed, hangar.

hanneton, *m.* May bug, June bug, cockchafer.

hannetonner, to clear (trees) of cockchafers.

hanter, to haunt, frequent.

hapaxanthique, hapaxanthous, manocarpic.

haplobiotique, hapaxanthous, monocarpic.

haplocaryon, *m.* nucleus with *n* chromosomes.

haplocaule, haplocaulous, with unbranched stem.

haplochlamydé, monochlamydeous, having a single perianth.

haploïde, *f.* haploid, gametophyte.

haplopétale, haplopetalous.

haplophase, *m.* haploid, gametophyte.

happant, adhering, adhesive.

happelourde, *f.* imitation jewel.

happement, *m.* adhering, clinging.

happer, to snap up, nab, snatch, seize, catch (insects), adhere, stick.

haptophore, *f.* haptophore.

haptotropisme, *m.* haptotropism, tropism induced by stimulus.

haras, *m.* stud, stud farm.

harasse, *f.* crate (for glass).

harasser, to tire out, exhaust.

harde, *f.* herd.

hardé, soft-shelled (egg).

harderie, *f.* iron sulphate.

hardi, bold, daring, hardy, intrepid.

hardiesse, *f.* boldness, audacity, hardiness, daring.

hardiment, boldly, daringly.

hareng, *m.* herring.

haricot, *m.* bean, kidney bean (*Phaseolus vulgaris*); — **de Lima,** Lima bean (*Phaseolus limensis*).

harle, *m.* merganser.

harmaline, *f.* harmaline.

harmonica, *m.* harmonica.

harmonie, *f.* harmony, keeping.

harmonier, to harmonize.

harmonieusement, harmoniously.

harmonieux, -euse, harmonious.

harmonique, harmonic.

harmoniquement, harmonically.

harmoniser, to harmonize.

harmotome, *m.* harmotome.

harnacher, to harness.

harnais, *m.* harness, equipment, armor, trappings, gear, gearing.

harper, *m.* springhalt.

harponner, to harpoon.

hart, *f.* withe.

harveyage, *m.* Harveyizing.

harveyiser, to Harveyize.

hasard, *m.* chance, accident, hazard, risk; **à tout —,** at all events, at all hazards; **au —,** at random; **au — de,** at the risk of; **par —,** by chance, accidentally.

hasarder, to hazard, risk, venture; **il se hasarde,** he takes a chance.

hasardeux, -euse, hazardous, risky, unsafe.

haschisch, *m.* Indian hemp; see HACHISCH.

hase, *f.* she-hare, doe-hare.

hasté, hastate, lanceolate.

hastiforme, lanciform, lanceolate, hastate, hastiform, spear-shaped.

hatchetine, *f.* hatchettite, hatchettine.

hâte, *f.* haste, hurry, hastiness; **à la —, en —,** in haste, in a hurry, hastily.

hâter, to hasten, hurry, force, expedite, make haste

hâtif, -ive, premature, early, precocious, hasty, early ripe.

hâtiveau, *m.* early fruit or vegetable.

hâtivement, hastily, early.

hauban, *m.* stay wire, wind brace.

haubanage, *m.* bracing, stays.

hausse, *f.* rise, advance, lift, prop, support, rear sight (of rifle).

hausse-col, *m.* gorget.

hausser, to raise, rise; **se —,** to rise, be raised, clear (weather).

haut, *m.* height, top, summit.

haut, high, deep, upper, bright (colors), tall, loud, lofty, great, exalted, highly, above, loudly, aloud, aloft; **— de course,** top of the stroke, upper dead center; **— fourneau,** blast furnace; **—e pression,** high pressure; **au — de,** at the top of; **de — en bas,** downward, from above, from top to bottom; **d'en —,** upper, from above; **en —,** above, upstairs; **en — de,** at the top of; **par —,** upward; **par en —,** by the upper way, up above; **plus —,** above; **tout —,** loudly, aloud.

haute-futaie, *f.* crop of trees in high forest.

hautement, highly, loudly, aloud, boldly.

haute-tige, *f.* high-grown transplant.

hauteur, *f.* height, altitude, depth, elevation, eminence, haughtiness; **— d'eau,** watermark; **— de ceinture, — d'homme,** chest or breast high; **— de chute relative,** relative fall; **— de la dent,** depth of tooth; **— de la vitesse,** height of velocity, height due to velocity; **— de pluie,** depth of rainfall; **— du tenon,** depth of dovetail; **— relative,** height above the ground; **— totale,** total height; **— utile,** efficient or useful height; **à la —,** up to the mark; **à la — de,** on a level with.

haut-fourneau, *m.* blast furnace.

haut-parleur, *m.* loudspeaker.

haut-perchis, *m.* crop of high poles.

haüyne, *f.* hauynite, hauyne.

havane, *m.* Havana cigar or tobacco; **La Havane,** Havana.

havane, light chestnut.

haveron, *m.* wild oats (*Avena fatua*).

havre, *m.* harbor, haven, port.

hebdomadaire, weekly.

hebdomadairement, weekly, once a week.

hébéanthe, with hairy corolla.

hébécarpe, hebecarpus, (fruit) with downy pubescence.

hébéclade, (branches) with downy pubescence.

hébégyne, with pubescent pistils.

hébépétale, with pubescent petals.

hébépode, with pubescent stipe.

hébéter, to stupefy, deaden.

hébétude, *f.* hebetude, dullness, dazed condition.

hébraïque, Hebrew, Hebraic.

hébreu, Hebrew, Jewish, Jew.

hécatophylle, hecatophyllous, with 50 pairs of leaflets.

hécistotherme, *m.* hekistotherm, plant thriving with minimum of heat.

hect., *abbr.* (hectolitre), hectoliter.

hectare, *m.* hectare (10,000 square meters, 2.471 acres).

hectique, hectic (fever).

hectogramme, hecto, *m.* hectogram (100 grams).

hectolitre, hecto, *m.* hectoliter (100 liters).

hectostère, *m.* hectostere (100 cubic meters).

hédéoma, *m.* hedeoma, American pennyroyal (*Hedeoma pulegioides*).

hédéra, *f.* ivy (*Hedera helix*).

hédéracé, hederaceous.

hédéré, hederaceous, resembling ivy.

hédonal, *m.* hedonal.

hédérifolié, with ivy leaves.

hékistothermique, hekistothermic (plant) requiring little heat and light.

hélas, alas.

hélianthe, *m.* sunflower (*Helianthus annuus*).

hélianthème, *m.* rockrose (Helianthemum).

hélianthine, *f.* helianthin, methyl orange.

hélice, *f.* helix, screw, propeller, airscrew; **— en bois,** wooden airscrew, propeller.

hélichrysine, *f.* helichrysin, a yellow coloring.

hélicine, *f.* helicin.

hélicisme, *m.* helicism, torsion at advanced period.

hélicoïdal, -aux, helicoidal.

hélicoïde, helicoid, helical, spiral, helicoidal, (of turbine) downward flow, axial flow.

héliocentrique, heliocentric.

héliochromie, *f.* heliochromy, color photography.

héliofuge, fleeing the sun.

héliographie, *f.* heliography.

héliogravure, *f.* heliogravure.

héliolithe, *f.* oligoclase, sunstone.

héliophile, heliophile, heliophilous.

héliophobe, heliophobic.

héliophyte, heliophyte, (plants) adapted to full sunlight.

hélioplastie, *f.* a process of photoengraving.

hélioscope, *m.* helioscope.

héliotactisme, *m.* heliotaxis.

héliotrope, *m.* turnsole (*Heliotropium peruvianum*).

héliotropine, *f.* heliotropin, peperonal.

héliotropism, *m.* heliotropism.

héliotypie, *f.* heliotypy.

hélium, *m.* helium.

hellébore, *m.* hellebore, Christmas rose (*Helleborus niger*).

helminthe, *m.* helminth, intestinal worm.

helminthique, helminthic.

helminthothèque, with cylindrical seeds, worm-shaped, vermiform.

hélobié, growing in marshes.

hélophytes, *m.* swamp plants.

helvétique, Swiss.

hem, dry cough.

hématéine, *f.* hematein.

hématie, *f.* red blood corpuscle, erythrocyte; —s lavées de mouton, washed sheep's corpuscles.

hématimètre, *m.* hematimeter.

hématine, *f.* hematin.

hématique, hematic.

hématite, *f.* hematite; — brune, brown hematite, limonite; — rouge, red hematite.

hématite, hematitic, blood-red.

hématocarpe, (fruit) speckled red.

hématocrité, *m.* hematocrit.

hématode, marked with red spots.

hématogène, *m.* hematogen.

hématologie, *f.* hematology.

hématolyse, *f.* hematolysis, hemolysis.

hématope, with deep-red stipe.

hématophage, blood-sucking (insect).

hématophylle, with blood-red leaves.

hématosperme, with seed of blood-red color.

hématoxyle, *m.* logwood tree (*Haematoxylon campechianum*).

hématoxyline, *f.* hematoxylin; — ferrique, iron hematoxylin.

hématurie, *f.* hematuria.

hémi-, hemi-, half-.

hémicarpe, *m.* half-fruit, half of naturally dividing fruit.

hémicellulose, *m.* hemicellulose.

hémichryse, partly gilded.

hémicleistogamie, *f.* hemicleistogamy, flowers open slightly.

hémicyclique, hemicyclic.

hémicylindrique, hemicylindrical.

hémièdre, hemihedral.

hémiédrie, *f.* hemihedrism, hemihedry.

hémiédrique, hemihedral.

hémihédrie, *f.* hemihedrism, hemihedral forms.

hémigamotrope, hemigamotropous, (flowers) which open and shut imperfectly.

hémigoniaire, hemigonaris, sex organs changed into petals.

hémigymnocarpe, hemigymnocarpous, (spores) maturing in closed receptacles.

hémiologame, *f.* calyx with male, female, and hermaphroditic flower.

hémine, *f.* hemin.

hémiparasite, *m.* hemiparasite, facultative saprophyte.

hémipermeable, semipermeable.

hémiprismatique, hemiprismatic.

hémiptère, *m.* hemipterous insect, bug; *pl.*, Hemiptera.

hémisaprophyte, *m.* hemisaprophyte, facultative parasite.

hémisphère, *m.* hemisphere.

hémisphérique, hemispheric(al).

hémisyngynique, half-adherent.

hémitétracotylie, *m.* hemitetracotyledon, one cotyledon is apparently divided into three.

hémitricotylie, *f.* partial division of one cotyledon.

hémitrope, hemitrope, hemitropic.

hémitropie, *f.* hemitropism, twin crystallization; axe d'—, twinning axis.

hémochromogène, *m.* hemochromogen.

hémoglobine, *f.* hemoglobin.

hémoglobinomètre, *m.* hemoglobinometer.

hémoglobinurie, *f.* hemoglobinuria.

hémolysant, hemolyzing.

hémolyse, *f.* hematolysis, hemolysis.

hémolysine, *f.* hemolysin.

hémolytique, hematolytic, hemolytic.

hémomètre, *m.* hemometer.

hémoptysie, *f.* hemoptysis, spitting of blood.

hémopyrrol, *m.* hemopyrrole.

hémorragie, hémorrhagie, *f.* hemorrhage, bleeding.

hémorragique, hémorrhagique, hemorrhagic.

hémorroïde, hémorrhoïde, *f.* hemorrhoid, pile.

hémostase, *f.* hemostasia.

hémostatique, hemostatic.

hénequen, *m.* henequen, sisal hemp (*Agave sisalina*).

henné, *m.* henna (*Lawsonia inermis*).

hennir, to neigh.

hépar, *m.* hepar.

hépatique, *f.* hepatica, liverwort; — blanche, white buttercup, grass-of-Parnassus.

hépatique, hepatic.

hépatisation, *f.* hepatization.

hépatiser, to become hepatized.

hépatite, *f.* hepatite, hepatitis.

heptagyne, (plant) with seven pistils.

heptagynie, *f.* heptagynia.

heptandrie, *f.* heptandria, plants with seven stamens.

heptane, *m.* heptane.

heptanthéré, with seven stamens.

heptapétale, heptapetalous, having seven petals.

heptaphylle, heptaphyllous.

heptarine, heptandrous, possessing seven stamens.

heptarrène, with seven stamens.

heptasépale, possessing seven sepals.

heptastémone, with seven stamens.

heptose, *f.* heptose.

heptylène, *m.* heptylene.

heptylique, heptyl.

héraut, *m.* herald.

herbacé, herbaceous, gramineous.

herbage, *m.* herbage, grass (-land), pasture, green vegetables.

herbager, to put (animals) out to grass.

herbe, *f.* herb, grass, weed; — à faulon, soapwort (*Saponaria officinalis*); — à jaunir, yellowweed, dyer's rocket (*Reseda luteola*), woodwaxen (*Genista tinctoria*); — à l'ail, garlic mustard (*Alliaria officinalis*), English treacle; — à loup, wolfsbane, monkshood (Aconitum); — au lait, milkwort (*Polygala vulgaris*); — au scorbut, scurvy grass (*Cochlearia officinalis*); — au soleil, sunflower (*Helianthus annuus*); — au verre, saltwort (*Salsola kali*).

herbe, — aux abeilles, meadowsweet (*Spiraea salicifolia*); — aux anes, evening primrose, tree primrose (*Oenothera biennis*); — aux chantres, hedge mustard (*Sisymbrium officinale*); — aux chats, catnip, catmint (*Nepeta cataria*), great wild valerian, cat's valerian, allheal; — aux goutteux, goutweed, ashweed (*Aegopodium podagraria*); — aux massues, club moss (Lycopodium); — aux tanneurs, tanner's sumac (*Rhus coriaria*); — aux puces, fleabane (*Pulicaria dysenterica*); — aux teinturiers, woodwaxen, dyer's greenweed (*Genista tinctoria*); — d'amour, mignonette (*Reseda odorata*); — des sorciers, thorn apple (*Datura stramonium*).

herbe, — empoisonnée, alkekengi (*Physalis alkekengi*), belladonna, deadly nightshade (*Atropa belladonna*); — jaune, yellowweed, dyer's rocket (*Reseda luteola*); — marine, seaweed; — mauvaise, weed; — parfaite, boneset (*Eupatorium perfoliatum*); — vivante, sensitive plant; couvert de mauvaises —s, envahi de mauvaises —s, weedy, infested with weeds; en —, green, unripe, in the blade, prospective, in embryo.

herber, to grass, bleach on the grass, croft.

herbeux, -euse, grassy.

herbier, *m.* herbarium.

herbification, *f.* everything with relation to plants.

herbivore, *m.* herbivore.

herbivore, herbivorous, grass-eating.

herboriser, to botanize.

herbue, *f.* pasture land, clay flux, loam.

hercogamie, *f.* hercogamy.

Hercule, *m.* Hercules.

herculéen, -enne, Herculean.

hère, *m.* fawn of red deer.

héréditaire, hereditary.

hérédité, *f.* heredity, inheritance, succession.

hérissé, bristling, bristly, spiky, covered, shaggy, rough, prickly (stem), hirsute.

hérisser, to bristle, ruffle up (feathers), cover, surround, cast (wall); **se —,** to bristle, stand on end (hair), become rough.

hérisson, *m.* hedgehog, brush, cogwheel, spur wheel, toothed cylinder.

hérissonne, *f.* woolly bear (caterpillar).

hérissonné, (surface) strewn with stiff thorns.

héritage, *m.* heritage, inheritance.

hériter, to inherit.

héritier, *m.* heir.

hermaphrodite, hermaphroditic, perfect (of flowers), hermaphrodite, androgynous.

herméticité, *f.* airtightness, imperviousness.

hermétique, hermetic(al), airtight.

hermétiquement, hermetically, tightly.

hermétisme, *m.* Hermetism (alchemy).

hermétiste, *m.* follower of the Hermetic philosophy.

hermine, *f.* stoat, ermine.

herniaire, *f.* rupturewort (*Achyranthes polygonoides*).

herniaire, hernial.

hernie, *f.* rupture, hernia.

herniole, *f.* rupturewort (*Achyranthes polygonoides*).

héroïque, heroic.

héron, *m.* heron.

héros, *m.* hero.

herpisme, *f.* herpism, creeping by variously shaped pseudopodia.

hersage, *m.* harrowing.

herse, *f.* harrow; — **accouplée,** jointed harrow; — **faite de branches,** bush harrow.

herser, to harrow.

hertzotropisme, *m.* hertzotropism, movement due to Hertzian waves.

hésitation, *f.* hesitation, hesitancy.

hésiter, to hesitate, waver.

hespéridie, *m.* hesperidium.

hespéridine, *f.* hesperidin.

hétairionnaire, (compound fruits) from ovaries with one style.

hétéroblastique, heteroblastic.

hétéracanthe, covered with thorns.

hétérandre, heterandrous, with two sets of stamens.

hétéranthe, with different flowers.

héteranthérie, *f.* heteranthery, flowers with distinct stamens.

hétérauxèse, *f.* heterauxesis.

hétéraxone, (diatome) with unequal transverse axes.

hétéroalbumose, *m.* heteroalbumose.

hétérocarpe, heterocarpous.

hétérocarpie, *f.* heterocarpy, having two kinds of fruit.

hétérocarpien, -enne, heterocarpian, heterocarpous.

hétérocarpique, heterocarpian.

hétérochlamydé, heterochlamydeous.

hétérocline, heterocline, heterocephalous.

hétéroclite, heteroclite, anomalous in formation.

hétérocyclique, heterocyclic.

hétérocyste, heterocystous.

hétérodichogamie, *f.* heterodichogamy, dichogamy.

hétérodiodé, heterosporous.

hétéroecie, *m.* heteroecium, a metoecious parasite.

hétérogame, heterogamous.

hétérogamètes, *f.pl.* heterogametes.

hétérogamie, *f.* heterogamy, alternation of generations.

hétérogène, heterogeneous, unlike.

hétérogénéité, *f.* heterogeneity, heterogeneousness.

hétérogénèse, *f.* heterogenesis.

hétérogénie, *f.* heterogeny, heterogenesis.

hétérogénisme, *m.* heterogenesis.

hétéroïde, heteroideus, diversified in form.

hétérolobe, heterolobous, having unequal lobes.

hétéromalle, heteromallus, spreading in all directions.

hétéromère, heteromerous.

hétéroméricarpie, *f.* heteromericarpy, heterocarpy occurring between parts of the same fruit.

hétéromérie, *f.* heteromeri.

hétéromésogamie, *f.* heteromesogamy, individuals vary in fertilization.

hétéromorphe, heteromorphous.

hétéromorphisme, *m.*, heteromorphism.

hétéromorphose, *m.* heteromorphosis.

hétéropétale, heteropetalous, with dissimilar petals.

hétérophylle, heterophyllous, having leaves of different forms.

hétérophyllie, *f.* heterophylly.

hétéroplasie, *f.* heteroplasia.

hétéroplastides, *m.pl.* heteroplastids, differing cells performing different functions.

hétéroprotéose, *m.* heteroproteose.

hétéroprothallie, *f.* heteroprothally, production of unisexual prothallia.

hétéros, dissimilar.

hétérosporé, heterosporous.

hétérosporie, *f.* heterospory.

hétérostémone, heterostemonous, with unlike stamens.

hétérostylé, heterostylic, with unlike styles.

hétérostylie, *f.* heterostyly.

hétérotaxie, *f.* heterotaxy, heterotaxis.

hétérothermique, heterothermic, porous silicious (soil) absorbing and losing warmth.

hétérotome, heterotomous.

hétérotrique, heterotrichous, having two types of cilia.

hétérotrope, heterotropal, heterotropous, amphitropous.

hétérotrophe, *m.* heterotroph.

hétérotrophique, heterotrophic.

hétérotype, heterotypic.

hétérovalve, (fruits) with unlike valves.

hétéroxanthine, *f.* heteroxanthine.

hétéroxène, heteroxenous.

hétéroxénie, *m.* heteroccium, a metoecious parasite.

hétérozygote, *m.* heterozygote.

hêtre, *m.* beech (*Fagus sylvatica*); — à crête de coq, cut-leaved beech; — parasol, weeping beech (*Fagus sylvatica pendula*); — pourpre, copper beech (*Fagus sylvatica purpurea*).

heuchère, *f.* Heuchera; — d'Amérique, alumroot (*Heuchera americana*).

heudécaphylle, having eleven leaves.

heule, *m.* resin oozing from pistachio.

heure, *f.* hour, time, o'clock; à cette —, at this time, now; à l'—, on time, per hour, an hour; à l'— actuelle, at the present time, today, nowadays; à l'— qu'il est, by this time, nowadays; à la bonne — fine!; à toute —, at any time; de bonne —, early; pour l'—, for the present; sur l'—, at once, immediately.

heureusement, happily, successfully, luckily, fortunately.

heureux, -euse, happy, fortunate, felicitous, glad, good, successful, lucky.

heurt, *m.* shock, blow, knock, mark, bruise, crown (of bridge).

heurter, to hit, strike (against), knock (against), run against, jar, offend, jostle, bump, collide with, shock; se —, to knock, clash, hit oneself, strike, collide.

heurtoir, *m.* door knocker, catch, stop, buffer, sill (of lock gate).

hévé, *m.* hevea, Pará rubber tree (*Hevea brasiliensis*).

hévée, *f.* seringa (Hevea).

hexacanthe, *m.* hexacanth.

hexachlorure, *m.* hexachloride.

hexacoque, formed of six shells.

hexaèdre, *m.* hexahedron.

hexaèdre, hexahedral.

hexaédrique, hexahedral.

hexafluorure, *m.* hexafluoride.

hexagonal, -aux, hexagonal.

hexagone, *m.* hexagon.

hexagyne, hexagynian.

hexahydraté, hexahydrated.

hexahydrure, *m.* hexahydride.

hexalépide, hexalepidus, six-scaled.

hexaméthylène, *m.* hexamethylene.

hexandre, hexandrous, with six stamens.

hexandrie, *f.* hexandry, with six stamens.

hexanitré, hexanitrated.

hexanthéré, hexastemonous, hexandrous, six-stamened.

hexanthérie, *f.* plant with six united stamens.

hexaphylle, hexaphyllous, six-leaved.

hexapode, *m.* hexapod.

hexaptère, hexapterous, six-winged.

hexarrhène, hexastemonous, six-stamened.

hexasépale, hexasepalous, with six sepals.

hexastémone, hexastemonous, six-stamened.

hexatomique, hexatomic.

hexavalent, hexavalent.

hexite, *f.* hexite.

hexobiose, *f.* sucrose.

hexone, *f.* hexonic base

hexonique, hexonic, hexone.

hexosane, *m.* hexosan.

hexose, *f.* hexose.

hexyle, *m.* hexyl.

hexylène, *m.* hexylene.

hexylique, hexyl, hexylic.

hiatus, *m.* hiatus, space, opening.

hibernacle, *f.* hibernacle, hibernaculum, winter quarters.

hibernal, -aux, occurring in winter.

hibou, *m.* owl; — **brachyôte,** short-eared owl.

hideur, *f.* hideousness.

hideux, -euse, dreadful, hideous, shocking.

hidrotique, hidrotic, sudorific.

hie, *f.* ram, beetle, pile driver.

hièble, *f.* dwarf elder (*Sambucus ebulus*).

hiémal, wintry, winter.

hiémalis, hiemal, relating to winter.

hier, to ram, drive, creak.

hier, yesterday.

hiérarchie, *f.* hierarchy.

hilarant, exhilarating, laughing; **gaz —,** laughing gas.

hile, *m.* hilum.

hilifère, hiliferous, bearing a hilum or a scar.

hilus, *m.* hilum.

hippobosque, *m.* horse tick.

hippocrépiforme, horseshoe-shaped, hippocrepiform.

hippurate, *m.* hippurate.

hippurique, hippuric.

hirondelle, *f.* swallow; — **de fenêtre,** martin; — **de mer,** common tern; — **de mer caujek,** Sandwich tern; — **de rivage,** sand martin.

hirsutie, *f.* hirsuties, hairiness.

hirtiflore, with hairy flowers.

hirudine, *f.* hirudin.

hispide, hairy, bristly, hispid.

hispidulé, somewhat hispid.

hispiduleux, -euse, hispidulous, minutely hispid.

hisser, to raise, hoist, haul up.

histéranthé, (flowers) appearing before the leaves.

histidine, *f.* histidine.

histochimie, *f.* histochemistry.

histochimique, histochemic(al).

histogène, histogenetic, histogenic.

histogénèse, *f.* histogenesis.

histogénie, *f.* histogeny.

histogénique, histogenetic, histogenic.

histoire, *f.* history, story.

histologie, *f.* histology.

histologique, histologic(al).

histologiste, *m.* histologist.

histolyse, *f.* histolysis.

histomères, *f.pl.* chromidia, extranuclear particles of chromatin.

histone, *f.* histone.

historien, *m.* historian.

historique, *m.* history (facts).

historique, historic, historical, of history.

historiquement, historically.

hiver, *m.* winter.

hivernage, *m.* hibernation, wintering (of cattle).

hivernal, -aux, winter, wintry, hibernal.

hiverner, to hibernate, winter.

hl., *abbr.* (hectolitre), hectoliter(s).

hobereau, *m.* hobby (a small falcon).

hoche, *f.* notch, dent, nick.

hochepot, *m.* ragout of beef and vegetables.

hochequeue, *m.* wagtail; — **jaune,** blue-headed wagtail.

hocher, to shake, notch, nick, dent.

hollandais, *m.* Dutch (language), Dutchman.

hollandais, Dutch.

hollande, *f.* Dutch porcelain, delftware, holland (cambric), Holland; *m.* Dutch (Edam) cheese.

hollander, *m.* hollander, pulper.

holmium, *m.* holmium.

holoblastique, holoblastic.

holocauste, *m.* holocaust, burnt offering.

holocristallin, holocrystalline.

holoèdre, *m.* holohedron.

holoèdre, holohedral.

holoédrie, *f.* holohedrism.

holoédrique, holohedral.

holoparasite, *m.* holoparasite.

holopétale, with whole petals.

holophyte, *f.* holophyte.

holoplankton, *m.* holoplankton.

holosaprophyte, *f.* holosaprophyte.

holosericeus, velutinous, velvety.

holostée, *f.* holosteum.

holothurie, *f.* holotnurian.

homalogoné, composed of flat articulations.

homalophylle, with flat leaves.

homalotropisme, *m.* homalotropism, diatropism.

homard, *m.* lobster.

homéopathie, *f.* homeopathy.

homéoplasie, *f.* homeoplasia.

homéotropie, *f.* homeotropy.

homicide, homicidal, murderous.

hommage, *m.* homage, respects.

hommasse, mannish, masculine, manlike.

homme, *m.* man, husband; — de bien, worthy man; — de coeur, man of spirit; — de métier, specialist, expert.

homoblastique, homoblastic.

homocarpe, homocarpous, having similar fruit.

homochlamydé, homochlamydeous.

homocline, homoclinous, (flower) showing phenomenon of autogamy.

homoclinie, *f.* autogamy.

homodrome, homodromous.

homodromie, *f.* homodromy.

homoeoplasie, *f.* homeoplasy, hyperplasia.

homofocal, -aux, homofocal.

homogame, homogamous.

homogamie, *f.* homogamy.

homogène, homogeneous.

homogénéocarpe, with homogeneous fruits.

homogénéisateur, *m.* homogenizer, viscolizer.

homogénéiser, to make homogenous, fix (milk).

homogénéité, *f.* homogeneity, uniformity.

homogènement, homogeneously.

homogentisique, homogentinisique, homogentisic.

homohétérostylie, *f.* homoheterostyly, with similar and dissimilar styles in the same species.

homoide, with same form as its tegument.

homoiochlamydé, uniform (perianth).

homoiodipérianthé, calyx and corolla with same number of divisions.

homologable, capable of being homologized.

homologie, *f.* homology.

homologique, homologic(al), homologous.

homologue, *m.* homologue.

homologue, homologous.

homologuer, to confirm, endorse, homologize, homologate.

homomorphe, homomorphic, uniform in shape.

homonéméen, -enne, composed of homogeneous filaments.

homonyme, correspondent, homologous.

homopétale, homopetalous.

homophylle, with similar leaves.

homoplastide, *m.* homoplastid, organism derived from similar cells.

homoptère, homopterous.

homopyrrol, *m.* homopyrrole.

homoquinine, *f.* homoquinine.

homosporé, homosporous, similar-seeded.

homothermique, homothermic, homoiothermal.

homotrope, homotropous.

homotype, homotypic, homologous.

homovalve, with similar valves.

homozygote, *m.* homozygote.

hongre, castrated, gelded.

Hongrie, *f.* Hungary.

hongroierie, *f.* Hungarian-leather manufacture.

hongrois, Hungarian.

hongroyer, to dress (leather) by the Hungarian method.

hongroyeur, *m.* tanner of Hungary leather.

honnête, honest, honorable, proper, civil, polite, decent, reasonable, moderate.

honnêtement, honestly.

honnêteté, *f.* honesty, propriety, decency.

honneur, *m.* honor, credit; en —, in favor.

honorable, honorable, fitting, creditable.

honorablement, honorably.

honoraire, *m.* honorarium; *pl.* fees.

honoraire, honorary.

honorer, to honor, respect, be an honor to, be a credit to.

honte, *f.* shame, disgrace; avoir —, to be ashamed.

honteusement, shamefully.

honteux, -euse, shameful, ashamed, disgraceful, timid.

hôpital, *m.* hospital.

hoquet, *m.* hiccough.

horaire, *m.* timetable.

horaire, hourly, horary, pertaining to the time of day.

horde, *f.* horde.

hordéacé, hordéiforme, hordeaceous.

hordéine, *f.* hordein.

horizon, *m.* horizon.

horizontal, -aux, horizontal.

horizontalité, *f.* horizontal position.

horloge, *f.* clock; — **de la mort,** deathwatch (beetle).

horlogerie, *f.* clock and watchmaking.

hormis, except, excepting, but, save; — **que,** except that, save that.

hormonal, -aux, of or pertaining to hormones.

hormone, *f.* hormone.

hormogonie, *m.* hormogonium.

hormospore, *m.* microgonidium.

hormotila, *m.* hormogone.

hornblende, *f.* hornblende, amphibole.

horreur, *f.* horror, awe.

horrible, horrible, dreadful, awful.

horripilation, *f.* horripilation, goose flesh.

hors, outside, out of, except, beyond, save, out, but; — **de,** out of, outside of, away from, beyond.

hors-d'oeuvre, *m.* relish, appetizer, side dish.

hortensia, *m.* hydrangea.

horticole, horticultural.

horticulteur, *m.* horticulturist, gardener.

hospitalier, -ere, hospitable.

hostile, hostile, adverse.

hostilité, *f.* hostility, enmity.

hôte, *m.* host, guest.

hôtel, *m.* hotel, mansion, town house, building; — **de ville,** city hall, town hall.

hotte, *f.* hood, basket, hod, scuttle; — **à bon tirage,** hood with a good draft.

hottée, *f.* basketful, hodful.

hotter, to carry in a basket.

houage, *m.* hoeing.

houblon, *m.* hop (*Humulus lupulus*); **cuillir le** —, to pick the hops.

houblonnage, *m.* addition of hops.

houblonner, to hop, flavor with hops.

houblonnier, *m.* hop raiser.

houblonnier, -ère, hop-growing, hop.

houblonnière, *f.* hop field, hop garden.

houe, *f.* hoe, mattock; — **à cheval,** horse hoe, cultivator.

houement, *m.* hoeing.

houer, to hoe.

hougne, *f.* soft grass.

houille, *f.* coal, pit coal, anthracite coal; — **à coke,** coking coal; — **à courte flamme,** short-flame coal; — **à longue flamme,** long-flame coal, sintering coal; — **anthraciteuse,** (semi)anthracite coal; — **blanche,** water power, white coal; — **bleue,** wind power, tide power; — **brune,** brown coal, lignite; — **carbonisée,** coke, carbonized coal; — **crue,** rough coal.

houille, — **demi-grasse,** semibituminous coal; — **grasse,** fat coal; — **luisante,** glance coal; — **maigre,** lean coal, nonbituminous coal, semianthracite; — **maréchale,** forge coal; — **sèche,** dry coal, sintering coal; — **terreuse,** earthy (brown) coal, lignite; — **tout-venante,** rough coal, run of mine; — **verte,** stream power.

houiller, coal, carboniferous, coal-producing.

houillère, *f.* coal mine, colliery.

houilleur, *m.* coal miner.

houilleux, -euse, coal, coal-bearing.

houillifier, se —, to carbonize.

houillite, *f.* anthracite.

houle, *f.* swell, surge, sea roller.

houlette, *f.* shepherd's crook, spatula-shaped instrument, garden trowel, hand ladle.

houlque, *f.* — **laineuse,** meadow soft grass, velvet grass (*Holcus lanatus*), Yorkshire grass; — **molle,** cock's tail, creeping soft grass (*Holcus mellis*); — **sucrée,** sugar cane (*Saccharum officinarum*).

houppe, *f.* tuft, top-knot, crest.

houppé, tufted.

houpper, to trim, tuft, comb (wool).

houppette, *f.* small tuft.

houppier, *m.* summit.

hourdis, *m.* hollow brickwork, pugging.

hourettes, *f.pl.* fagots or small twigs for charcoal.

houssage, *m.* dusting, sweeping.

housse, *f.* housing, cover, saddle cloth, rough form.

housser, to dust, sweep, cover up.

houssoir, *m.* duster, whisk, feather brush.

houx, *m.* holly, holly tree (*Ilex aquifolium*).

hoyau, *m.* prong hoe, mattock.

hublot, *m.* scuttle, sidelight, porthole.

huche, *f.* bin, hopper, chest, kneading trough, tub.

huilage, *m.* oiling, lubrication.

huile, *f.* oil; — **à brûler,** burning oil, lamp oil; — **à gaz,** gas oil; — **à graisser,** lubricating oil; — **à immersion,** immersion oil; — **alimentaire,** edible oil; — **à mécanisme,** — **à mouvement,** machine oil; — **animale,** animal oil, bone oil; — **brute,** crude oil, raw light oil; — **comestible**, edible oil; — **d'aniline**, aniline oil; — **d'arachide(s)**, peanut (earth-nut) oil; — **de baleine**, whale oil; — **de bois**, wood-tar oil, wood oil.

huile, — **de cèdre**, cedar (wood) oil; — **de coco**, coconut oil; — **de colza**, colza oil; — **de coprah**, copra oil; — **de coton**, cottonseed oil; — **de créosote**, creosote oil; — **de croton**, croton oil; — **de foie de morue**, cod-liver oil; — **de goudron**, tar oil; — **de graissage**, lubricating oil; — **de houille**, coal-tar oil; — **de lard**, lard oil.

huile, — **de lin**, linseed oil; — **de lin crue**, raw linseed oil; — **de lin cuite**, boiled linseed oil; — **de machine**, (sewing) machine oil; — **de morue**, cod-liver oil; — **de naphte**, mineral sperm oil; — **de navette**, rape oil, colza oil; — **de noix**, walnut oil; — **de palme**, palm oil; — **de paraffine**, paraffin oil; — **de pétrole**, petroleum; — **de pied(s) de boeuf**, neat's-foot oil; — **de poisson**, fish oil.

huile, — **de rabette**, rübsen (seed) oil; — **de résine**, resin oil; — **de ricin**, castor oil; — **de schiste**, schist oil; — **de sésame**, sesame oil; — **de succin**, amber oil; — **de suif**, tallow oil, olein; — **de table**, salad oil; — **de térébenthine**, oil of turpentine; — **de vaseline**, liquid paraffin, vaseline oil; — **de vitriol**, oil of vitriol; — **du styrax**, storax oil; — **empyreumatique**, fusel oil; — **essentielle**, essential oil.

huile, — **fixe**, fixed oil; — **lampante**, lamp oil, kerosene; — **légère**, light oil; — **lourde**, heavy oil; — **lubrifiante**, lubricating oil; — **minérale**, mineral (sperm) oil, naphtha, petroleum; — **moyenne**, medium oil; — **non-siccative**, nondrying oil; — **pour turc rouge**, Turkey-red oil; — **siccative**, drying oil; — **végétale**, vegetable oil, plant oil; — **verte**, green oil; — **vierge**, virgin oil; — **volatile**, volatile oil.

huilement, *m.* oiling.

huiler, to oil, lubricate, exude oil.

huilerie, *f.* oil works, oil mill, oil shop.

huileux, -euse, oily.

huilier, *m.* oiler, oilcan, oil merchant, oil cruet, oil stand, castor.

huis clos, closed door, private.

huissier, *m.* usher.

huit, eight, eighth.

huitaine, *f.* about eight, eight days, week, eight, eight-day clock.

huit-cors, *m.* stag of eight points.

huitième, eighth.

huître, *f.* oyster.

huîtrier-pie, *m.* oyster catcher.

hulotte, *f.* common wood owl; — **chothuant**, tawny owl.

humage, *m.* inhalation.

humain, *m.* human being, man.

humain, human, humane.

humainement, humanly, humanely.

humanitaire, humanitarian.

humanité, *f.* humanity, mankind.

humant, inhaling, breathing.

humate, *m.* humate.

humble, humble, meek, lowly, low.

humblement, humbly, meekly.

humectage, *m.* moistening, dampening.

humectant, humectant, moistening, diluent.

humectation, *f.* humectation, moistening, infiltration, dampening, wetting.

humecter, to moisten, dampen, wet, infiltrate; **s'** —, to become moist or damp, be moistened.

humer, to suck up, suck in, inhale, breathe.

huméral, -aux, humeral (ligament).

humérus, *m.* humerus.

humescent, becoming humid, slightly damp.

humeur, *f.* humor, moisture, fluid, temper, mood, ill humor, sucker, inhaler, sniffer; — **aqueuse**, aqueous humor.

humeux, -euse, of or pertaining to humus.

humicole, living in humus-rich soil.

humide, humid, moist, damp, wet.

humidement, damply, humidly, moistly.

humidier, to dampen, mo sten, humidify.

humidification, *f.* dampening, moistening.

humidifier, to dampen, moisten, humidify.

humidifuge, repelling moisture.

humidité, *f.* humidity, moisture, dampness; — **du sol**, ground humidity.

humiliant, humiliating.

humilier, to humiliate, humble.

humique, humic.

humus, *m.* humus, vegetable mold; — **acide**, raw humus, sour humus; — **brut**, sour humus; — **charbonneux**, bituminous humus; — **de bruyère**, heather humus; — **doux**, mild humus.

huppe, *f.* tuft, topknot, hoopoe; — **puput**, hoopoe.

hurler, to howl, yell, cry out.

hutte, *f.* hut, cottage.

hyacinthe, *f.* hyacinth.
hyacinthine, *f.* hyacinth.
hyalin, hyaline.
hyalite, *f.* hyalite, hyalitis.
hyalogène, *m.* hyalogen.
hyaloïde, hyaloïdien, hyaloid.
hyalome, *f.* substance between meshes of linin.
hyaloplasma, *f.* hyaloplasma.
hyalosperme, hyalospermous, having transparent seeds.
hyalotechnie, *f.* glassmaking, technique of glass.
hyalotechnique, pertaining to glassmaking.
hyalurgie, *f.* art of making glass.
hyalurgique, pertaining to glassmaking.
hybridation, *f.* hybridization.
hybride, *m.* hybrid, mongrel, bastard.
hybrider, to hybridize, cross.
hybridisme, *m.* hybridism.
hybridité, *f.* hybrid character, hybridity.
hybridogamie, *f.* hybridogamy, hybrids between different species.
hydantoïne, *f.* hydantoin.
hydarthrose, *f.* hydarthrosis.
hydatode, *f.* hydathode.
hydracide, *m.* hydracid, halogen acid.
hydracrylique, hydracrylic.
hydragogue, hydragogue.
hydralcool, *m.* alcohol diluted with water.
hydralcoolique, dilute alcoholic.
hydramide, *m.* hydramide.
hydrargyre, *m.* hydrargyrum, mercury.
hydrargyride, pertaining to mercury.
hydrargyrie, *f.* hydrargyria, mercurialism.
hydrargyrique, hydrargyric, mercurial.
hydrargyrisme, *m.* hydrargyrism, mercurialism.
hydrargyrure, *m.* amalgam, alloy with mercury.
hydraste, *m.* Hydrastis; — du Canada, goldenseal (*Hydrastis canadensis*).
hydrastine, *f.* hydrastine.
hydrastinine, *f.* hydrastinine.
hydratable, capable of being hydrated.
hydratation, *f.* hydration.
hydrate, *m.* hydrate, hydroxide; — de carbone, carbohydrate; — de chaux, hydrated lime, calcium hydroxide; — de potasse, caustic potash, potassium hydrox-

ide; — de soude, caustic soda, sodium hydroxide.
hydraté, hydrated, hydrous, hydroxide of; baryte —e, barium hydrate.
hydrater, to hydrate; s'—, to become hydrated, hydrate, deliquesce.
hydratique, hydrate.
hydratropique, hydratropic.
hydraulicien, *m.* hydraulic engineer, hydraulician.
hydraulique, *f.* hydraulics.
hydraulique, hydraulic.
hydrazine, *f.* hydrazine.
hydrazobenzène, *m.* hydrazobenzene.
hydrazoïque, hydrazoic.
hydrazone, *f.* hydrazone.
hydre, *f.* hydra.
hydrémie, *f.* hydremia.
hydrer, to hydrogenize, hydrogenate.
hydriodate, *m.* hydriodide.
hydriodique, hydriodic.
hydrique, aqueous, of water, water, waterborne.
hydrobromate, *m.* hydrobromide.
hydrobromique, hydrobromic.
hydrocarbonate, *m.* hydrocarbonate.
hydrocarbone, *m.* hydrocarbon.
hydrocarboné, hydrocarbon, hydrocarbonic.
hydrocarbure, *m.* hydrocarbon.
hydrocarpie, *f.* hydrocarpy, maturation of fruit in water.
hydrocaule, hydrocaulus.
hydrocèle, *f.* hydrocoele.
hydrocellulose, *f.* hydrocellulose.
hydrocéphalie, *f.* hydrocephalus.
hydrocérame, *m.* porous pottery for cooling water.
hydrocéramique, hydroceramic.
hydrochlorate, *m.* hydrochloride.
hydrochlorique, hydrochloric.
hydrochore, dispersed by water.
hydrocinnamique, hydrocinnamic.
hydrocleistogamie, *f.* flowers closed because submersed.
hydroconion, *f.* shower bath.
hydrocyanate, *m.* hydrocyanide.
hydrocyanique, hydrocyanic.
hydrocystes, *f.pl.* hydrocysts.

hydrocyte, *f.* element inside the cambial zone.

hydrodynamique, *f.* hydrodynamics.

hydro-électrique, hydroelectric.

hydro-extracteur, *m.* hydroextractor, centrifugal drying machine.

hydrofluate, *m.* hydrofluoride.

hydrofuge, repelling water, waterproof, watertight, rainproof.

hydrogame, *m.* hydrogam, cryptogam.

hydrogel, *m.* hydrogel.

hydrogénable, hydrogenizable.

hydrogénant, hydrogenizing, hydrogenating.

hydrogénase, *f.* hydrogenase.

hydrogénation, *f.* hydrogenization, hydrogenation.

hydrogène, *m.* hydrogen; — **arsénié,** arseniureted hydrogen, arsine; — **carboné,** hydrocarbon; — **phosphoré,** phosphureted hydrogen, phosphine; — **sulfuré,** sulphureted hydrogen, hydrogen sulphide.

hydrogéné, hydrogenated.

hydrogéner, to hydrogenize, hydrogenate; **s'—,** to be hydrogenized, be hydrogenated.

hydrogénique, hydrogen.

hydrogéniser, to hydrogenize.

hydrogénure, *m.* hydride.

hydrogéologie, *f.* hydrogeology.

hydrogère, carrying water.

hydrogiton, living in vicinity of water.

hydrographe, *m.* hydrographer.

hydrographe, hydrographic.

hydrographie, *f.* hydrography.

hydrographique, hydrographic(al).

hydroïde, *m.* hydroid, a tracheid.

hydroiodique, hydriodic.

hydrolase, *f.* hydrolyzing diastase.

hydrolat, *m.* water.

hydrologie, *f.* hydrology.

hydrolycyte, *f.* vacuole.

hydrolysable, hydrolyzable.

hydrolysation, *f.* hydrolyzation, hydrolysis.

hydrolyse, *f.* hydrolysis.

hydrolyser, to hydrolyze.

hydrome, *f.* hydrome, water-conducting tissue.

hydromégathermique, hydromegathermic.

hydromel, *m.* hydromel.

hydromètre, *m.* hydrometer.

hydrométrie, *f.* hydrometry, water measurement.

hydrominéral, **aux,** relating to mineral waters.

hydromorphose, *m.* morphosis due to habitat in water.

hydronaphtol, *m.* hydronaphthol.

hydrophane, *f.* hydrophane.

hydrophile, absorbent (cotton wool), hydrophilous.

hydrophthorique, hydrofluoric.

hydrophytes, *f.pl.* hydrophytes.

hydropique, *m.* dropsical subject.

hydropique, dropsical, hydropic.

hydropisie, *f.* dropsy.

hydroplastie, *f.* hydroplasty.

hydropneumatique, hydropneumatic.

hydropyridique, hydropyridine.

hydrorrhée, *f.* hydrorrhea.

hydrosilicate, *m.* hydrous silicate.

hydrosol, *m.* hydrosol.

hydrosphère, *f.* hydrosphere.

hydrospore, *m.* spore disseminated by water.

hydrostatique, *f.* hydrostatics.

hydrostatique, hydrostatic(al).

hydrosulfate, *m.* hydrosulphide.

hydrosulfite, *m.* hydrosulphite.

hydrosulfure, *m.* hydrosulphide.

hydrosulfureux, **-cuse,** hydrosulphurous.

hydrosulfurique, hydrosulphuric.

hydrotimétrique, hydrotimetric.

hydrotrophie, *f.* trophy due to reaction of water.

hydrotropisme, *m.* hydrotropism.

hydroxyde, *m.* hydroxide.

hydroxydé, combined with hydrogen and oxygen, hydroxide of.

hydroxylamine, *f.* hydroxylamine.

hydroxyle, *m.* hydroxyl.

hydruration, *f.* hydrogenization, hydrogenation.

hydrure, *m.* hydride; — **d'éthyle,** ethyl hydride, ethane.

hyène, *f.* hyena; — **rayée,** striped hyena.

hyétomètre, *m.* gauge, pluviometer.

hygiène, *f.* hygiene.

hygiénique, hygienic.

hygiéniquement, hygienically.

hygrine, *f.* hygrine.

hygrique, hygric.

hygromètre, *m.* hygrometer.

hygrométricité, *m.* hydronasty, hygroscopic property.

hygrométrique, hygrometric(al), hygroscopic.

hygrophile, hygrophilous.

hygroplasma, *m.* hygroplasm.

hygroscopicité, *m.* hydronasty.

hygroscopie, *f.* hygroscopy.

hygroscopique, hygroscopic(al).

hylémyie du blé, *m.* gout fly.

hylésine, *f.* bast beetle; — crénelé, great black ash bark beetle; — piniperde, pine beetle.

hylobe, *m.,* — du mélèze, larch weevil; — du pin, pine weevil.

hylophyte, *f.* hylophyte, a forest plant.

hyménium, *m.* hymenium.

hyménocarpe, with membranous fruit.

hyménode, hymenode, having a membranous texture.

hyménomycète, *m.* hymenomycete, pileate fungus.

hyménophore, *m.* hymenophore.

hyménoptère, *m.* hymenopterous insect.

hyménorhize, with membranous roots.

hymique, humic.

hyocholalique, hyocholic.

hyoïde, *m.* hyoid bone.

hyoscine, *f.* hyoscine.

hyoscyamine, *f.* hyoscyamine.

hypanthe, *f.* lower part of calyx, hypanthium.

hypanthode, *m.* hypanthodium, a syconium.

hyperbole, *f.* hyperbole, hyperbola, exaggeration.

hyperbolique, hyperbolic(al).

hyperboloïde, hyperboloid.

hyperchimère, *f.* hyperchimera, a graft hybrid.

hyperchlorhydrie, *f.* hyperchlorhydria.

hyperchlorique, perchloric, hyperchloric.

hypercondriaque, hypochondriacal.

hyperémie, *f.* hyperemia, congestion.

hyperglycémie, *f.* hyperglycemia, glycohemia.

hyperhydrique, hyperhydric, overflow for water in tissues.

hypériodique, hyperiodic, periodic.

hypermétrope, hypermetropic, far-sighted.

hypermétropie, *f.* hypermetropy, far-sightedness.

hyperoxyde, *m.* hyperoxide, peroxide.

hyperplasie, *f.* hyperplasia.

hypersensibilité, *f.* hypersensibility.

hypertonique, hypertonic.

hypertrophie, *f.* hypertrophy.

hypertrophié, hypertrophied.

hypertrophytes, *f.pl.* hypertrophytes, parasitic fungi causing hypertrophy in the tissues.

hypèthre, *m.* skylight.

hyphe, *f.* hypha (molds).

hyphéar, *m.* hyphear, a kind of mistletoe.

hyphema, *m.* microgonidium.

hyphomycètes, *f.pl.* Hyphomycetes.

hyphosporé, in form of filaments.

hyphydrogamiques, *f.pl.* hydrogams, cryptogams.

hypnal, *m.* hypnal.

hypnée, *f.* feather moss (Hypnum).

hypnocyste, *f.* hypnospore, a resting spore.

hypnophile, growing among moss.

hypnospore, *f.* hypnospore, a resting spore.

hypoazoteux, -euse, hyponitrous.

hypoazotique, hyponitric.

hypoazotite, *m.* hyponitrite.

hypobasal, -aux, hypobasal.

hypoblaste, *m.* hypoblast.

hypocalicie, *f.* plant with calyx under the ovary.

hypocarpe, (enlarged peduncle) under the ovary.

hypocarpogé, hypocarpogean, hypogeous.

hypochilium, *m.* hypochilium.

hypochloreux, -euse, hypochlorous.

hypochlorhydrie, *f.* hypochlorhydria.

hypochlorique, hypochloric.

hypochlorite, *m.* hypochlorite.

hypocratériforme, hypocrateriform, hypocraterimorphous.

hypoderme, *m.* hypoderma, hypodermis, hypoderm; — du boeuf, ox warble fly, cattle grub.

hypodermique, hypodermic.

hypodicarpé, with two low ovaries.

hypogé, hypogeous.

hypogène, hypogenous, produced beneath.

hypoglycémie, *f.* hypoglycemia.

hypogyne, hypogynic, hypogenous.

hypogynie, *f.* hypogyny.

hypolampre, glossy below.

hyponastie, *f.* hyponasty.

hyponastique, hyponastic.

hyponomeute, *m.* ermine moth.

hypopelté, hypopeltate, having base of limb on inferior face.

hypopétalé, hypopetalous, having an hypogynous corolla.

hypophléode, hypophloeodal, living beneath the bark within the periderm.

hypophloeodique, hypophloeodic.

hypophosphate, *m.* hypophosphate.

hypophosphite, *m.* hypophosphite; — de chaux, calcium hypophosphite; — de potasse, potash hypophosphite; — de soude, sodium hypophosphite.

hypophosphoreux, -euse, hypophosphorous.

hypophosphorique, hypophosphoric.

hypophylle, *m.* hypophyllium, a scalelike leaf below a cladophyll.

hypophyllocarpe, with fruit borne on under side.

hypophyse, *f.* hypophysis.

hypophyte, hypophyte.

hypoplasie, *f.* hypoplasy, incomplete development.

hypopode, *m.* hypopodium.

hypoptéré, hypopteratus, (wings) produced from below.

hypoptères, *f.pl.* hypopteries, a wing growing from below.

hypostase, *f.* hypostasis, the suspensor of an embryo.

hypostomium, *m.* hypostomium, cells forming the lower portion of the stomium.

hyposulfate, *m.* hyposulphate, dithionate.

hyposulfite, *m.* hyposulphite, thiosulphate; — de potasse, hyposulphite of potash.

hyposulfureux, -euse, hyposulphurous, thiosulphuric.

hyposulfurique, hyposulphuric, dithionic.

hypoténuse, *f.* hypotenuse.

hypothalle, *m.* hypothallus.

hypothecium, *m.* hypothecium.

hypothèque, *f.* mortgage; — en premier rang, first (prime) mortgage.

hypothéquer, to mortgage, hypothecate.

hypothèse, *f.* hypothesis, supposition.

hypothétiquement, hypothetically.

hypothionique, dithionic.

hypotonique, hypotonic.

hypotrophe, hypotropous.

hypoxanthine, *f.* hypoxanthine.

hypoxyde, *m.* suboxide.

hypoxylie, *f.* hyponasty.

hypsomètre, *m.* hypsometer, dendrometer.

hypsophylla, *f.* hypsophyll, bract.

hysope, *f.* hyssop.

hystérésis, *f.* hysteresis.

hystérie, *f.* hysteria.

hystérophore, having a volva.

hysteroplasma, *f.* hysteroplasma, the more fluid part of protoplasm.

hystrelle, *f.* hystrella, a simple pistil.

I

ïambe, *m.* iambus, iamb.

iatrochimie, *f.* iatrochemistry.

iatrochimique, iatrochemical.

ibéride, *f.* candytuft (Iberis).

ibérien, -enne, Iberian.

ibis, *m.* ibis.

ichnanthe, with striated flowers.

ichneumon, *m.* ichneumon fly.

ichneumonide, *m.* ichneumon fly, ichneumon wasp.

ichor, *m.* ichor.

ichtyocolle, *f.* isinglass, fish glue.

ichtyol, *m.* ichthyol.

ici, here, now, hither; d'ici, henceforth; d'ici là, until then; d'ici longtemps, for a long time; d'ici peu, before long; par ici, this way, through here.

ici-bas, here below.

ici-même, in this very place, in this work.

icmadophile, loving damp places.

iconogène, *m.* eikonogen.

icosaèdre, icosahedral.

icosandrie, *f.* Icosandria.

ictère, *m.* icterus, jaundice; — épidemique, infectious jaundice.

ictus, *m.* sudden attack, stroke.

idéal, -aux, ideal.

idéaliser, to idealize.

idéaliste, *m.* idealist.

idéaliste, idealistic.

idéalité, idealism.

idée, *f.* idea, thought, notion, bit, small amount, sketch, outline.

identification, *f.* identification.

identifier, to identify.

identique, identical, the same.

identiquement, identically.

identité, *f.* identity, sameness.

idioblaste, *m.* idioblast, a biophore, an ultimate cell unit.

idiome, *m.* idiom, language.

idiomères, *f.pl.* idiomeres, structures evolved during the resting stage.

idioplasma, *m.* idioplasm.

idiothalame, idiothalamous, having different coloration than the thallus.

idiotique, idiotic, idiomatic.

idiotisme, *m.* idiom, idiocy, imbecility.

idite, *f.* iditol.

idocrase, *f.* idocrase, vesuvianite.

idolâtre, *m.&f.* idolater.

idolâtre, idolatrous.

idolâtrie, *f.* idolatry.

idole, *f.* idol.

idrialite, *f.* idrialite.

idryle, *m.* idryl, fluoranthene.

idylle, *f.* idyl.

Iéna, *n.* Jena (glass).

if, *m.* yew tree (Taxus), draining rack (for bottles).

igname, *f.* yam, Indian potato.

igné, igneous, fiery.

ignescent, ignescent.

ignicolore, of a bright metallic red, flame-colored.

ignifère, igniferous.

ignifugation, *f.* fireproofing.

ignifuge, *m.* fireproof, fire-extinguishing material.

ignifuge, rendering noninflammable, fire-resisting.

ignifuger, to fireproof.

ignition, *f.* ignition; en ——, ignited.

ignitubulaire, chaudière ——, smoke-tube boiler, fire-tube boiler.

ignorance, *f.* ignorance.

ignorant, ignorant, unacquainted with.

ignoré, unknown.

ignorer, to not know, be ignorant of, not to know, be unaware of.

il, he, it, there; il n'y a pas de, there is no; il y a, there is, there are; il y a dix ans, ten years ago.

île, *f.* island, isle.

iléon, *m.* ileum.

ilex, *m.* ilex, holm oak (*Quercus ilex*), holly (Ilex).

ilicine, *f.* ilicin.

ilion, *m.* ilium.

illégal, -aux, illegal, unlawful.

illégalement, illegally.

illégitime, illegitimate.

illimitable, illimitable, boundless.

illimité, unlimited, limitless.

illipé, *m.* illipe.

illisible, illegible.

illisiblement, illegibly.

illogique, illogical.

illogiquement, illogically.

illuminant, illuminating.

illumination, *f.* illumination.

illuminer, to illuminate, illumine, enlighten.

illusion, *f.* illusion, delusion.

illusionner, to deceive, delude.

illusoire, illusory, illusive.

illustration, *f.* illustration, rendering, illustrious, celebrity.

illustre, *m.* illustrious man.

illustre, illustrious, eminent, renowned, famous.

illustrer, to illustrate, render illustrious.

îlot, *m.* small island, islet.

ils, they.

im., *abbr.* (immobile), nonmotile.

image, *f.* image, picture, likeness, resemblance, diagram, copy; —— réelle, real image; —— renversée, inverted image.

imagerie, *f.* chromo (color) printing, picture trade.

imaginable, imaginable.

imaginaire, imaginary.

imagination, *f.* imagination.

imaginer, to imagine, conceive, contrive, think out; s'——, to imagine, believe, fancy, picture to oneself.

imago, *f.* imago.

imbattable, invincible.

imberbe, beardless, without barbels.

imbiber, to soak, imbibe, impregnate; **s'—**, to be imbued, soak, imbibe, penetrate, saturate.

imbibition, *f.* imbibition, imbuement, absorption.

imboire, to moisten, wet; **s'—**, to be imbued.

imbricatif, -ive, imbricative.

imbrice, imbricate.

imbrifuge, rainproof, waterproof.

imbriqué, imbricate.

imbrûlable, unburnable, incombustible.

imbu (imboire), imbued.

imbuvable, undrinkable, not fit to drink.

imbuvant (imboire), soaking, steeping.

imidazol, *m.* imidazole.

imide, *m.* imide.

imidé, imido, imino-(compound).

imidogène, *m.* imidogen.

iminé, imino.

imitateur, -trice, *m.&f.* imitator.

imitateur, -trice, imitative, imitating.

imitatif, -ive, imitative.

imitation, *f.* imitation.

imiter, to imitate.

immaculé, immaculate, pure.

immangeable, uneatable.

immaniable, unmanageable, unwieldy.

immanquablement, infallibly, without fail.

immatériel, -elle, immaterial, intangible.

immaturité, *f.* immaturity.

immédiat, immediate, instant, at hand; **analyse—e**, qualitative (proximate) analysis.

immédiatement, immediately, instantly, at once.

immémorial, -aux, immemorial.

immense, immense, vast, important, great.

immensément, immensely, hugely.

immensité, *f.* immensity, vastness, infinity.

immergé, immersed.

immerger, to immerse, immerge, season under water.

immersif, -ive, immersible.

immersion, *f.* immersion, dipping.

immesurable, immeasurable.

immesure, unmeasured.

immeuble, *m.* real (landed) estate, house, real property.

immeuble, immovable, fixed, real (estate).

imminent, imminent, impending.

immiscibilité, *f.* immiscibility.

immiscible, immiscible.

immobile, immovable, motionless, non-motile.

immobilisation, *f.* rendering immovable.

immobiliser, to immobilize, make inactive, render immovable, fasten, fix, stop.

immobilité, *f.* immobility, immovability.

immodéré, immoderate, excessive.

immodérément, immoderately.

immonde, impure, unclean.

immondice, *f.* impurity, dirt, filth.

immoral, -aux, immoral, dishonest.

immortaliser, to immortalize.

immortalité, *f.* immortality.

immortel, -el e, immortal.

immortelle, *f.* immortelle, everlasting, cudweed, sandy everlasting, life everlasting, (Antennaria).

immotif, -ive, (which) does not change place.

immotiflores, *f.pl.* immotiflorae, wind-pollinated plants with fixed flowers.

immuable, immutable, unchangeable.

immun, immune.

immunisant, *m.* protective serum.

immunisant, immune, immunizing.

immunisation, *f.* immunization.

immuniser, to immunize.

immunité, *f.* immunity.

impact, *m.* impact, shock.

impair, odd, uneven (of numbers), azygous.

impalpable, impalpable, intangible.

imparfait, imperfect, incomplete, inexact.

imparfaitement, imperfectly.

imparinervé, nerves (veins) in odd numbers.

imparipenné, imparipinnate, unequally pinnate.

impartialité, *f.* impartiality.

impartible, indivisible.

impassible, impassible, unaffected.

impastation, *f.* impastation.

impatience, *f.* impatience.

impatient, impatient.

impatiente, *f.* touch-me-not (Impatiens), yellow balsam, quick-in-the-hand (*Impatiens pallida*).

impatientant, annoying, provoking.

impatienter (s'), to become impatient.

impayable, invaluable, ridiculous, priceless.

impayé, unpaid.

impédance, *f.* impedance, apparent resistance.

impénétrabilité, *f.* impenetrability.

impénétrable, impenetrable, impervious.

impératif, -ive, imperative.

impérativement, imperatively.

impératoire, *f.* imperatoria, masterwort.

imperceptibilité, *f.* imperceptibility.

imperceptible, imperceptible.

imperfection, *f.* imperfect on.

impérieusement, imperiously.

impérieux, -euse, imperious, domineering, imperative.

impérissable, imperishable, undying.

impermanence, *f.* impermanence, impermanency.

impermanent, not permanent, impermanent.

imperméabilisation, *f.* waterproofing.

imperméabilité, *f.* impermeability, tightness.

imperméable, *m.* waterproof fabric, raincoat.

imperméable, impermeable, impervious, waterproof; — à l'air, airtight; — à l'eau, waterproof.

imperméant, impermeable.

impermutable, unexchangeable.

impersonnel, -elle, impersonal.

imperturbable, imperturbable.

impétiolaire, with sessile leaves.

impétueux, -euse, impetuous, wild, headlong.

impitoyable, pitiless, merciless, unrelenting.

impitoyablement, pitilessly, mercilessly, unmercifully.

implanter(s'), to become implanted, lodged, fixed, take root.

implicite, implicit, inferred, implied.

impliquer, to imply, implicate, involve, infer.

impluviosité, *f.* want of rain.

impollué, impolluted.

impondérabilité, *f.* imponderability.

impondérable, imponderable.

importable, importable.

importance, *f.* importance, size, weight, moment.

important, *m.* important point, main thing, consequential person.

important, important, essential, large.

importateur, -trice, *m.&f.* importer.

importateur, -trice, importing.

importation, *f.* importation, import.

importer, to be important, be of importance, matter, be of moment, import; s'—, to be imported; **n'importe,** no matter; **n'importe comment,** any way (at all); **n'importe où,** anywhere; **n'importe peu,** it matters little; **n'importe quand,** at any time; **n'importe quel,** any; **qu'importe?** what does it matter?

importun, troublesome.

imposant, imposing, stately, majestic, commanding.

imposer, to impose, require, make necessary, levy (a tax), tax, overawe, enjoin, obtrude; s'—, to become required, strike; il **s'impose,** he is obtrusive.

impossibilité, *f.* impossibility; **être dans l'— de, se trouver dans l'— de,** to be unable to.

impossible, impossible.

imposte, *f.* abutment, butment.

impôt, *m.* tax, duty, impost; — **foncier,** land tax.

impotence, *f.* impotence, lameness.

impraticabilité, *f.* impracticability.

impraticable, impracticable, unfeasible.

imprégnation, *f.* impregnation, injection, steeping.

imprégner, to impregnate, fertilize, saturate, imbibe, inject.

impressif, -ive, impressive.

impression, *f.* impression, impress, edition, (textile) printing, effect, (painting) priming, mark, brand, ground.

impressionnant, impressive, impressing, sensational.

impressionné, marked by depressions.

impressionner, to impress, affect, expose (to), make an impression on, produce an image.

imprévoyance, *f.* want of foresight.

imprévu, *m.* unforeseen (unexpected) event

imprévu, unforeseen, unexpected.

imprimé, *m.* printed paper or book; *pl.* printed matter.

imprimer, to print, imprint, implant, impart, impress, communicate, prime.

imprimerie, *f.* printing, printing office, press, typography.

imprimeur, *m.* printer.

imprimeuse, *f.* small printing machine.

imprimure, *f.* priming, grounding.

improbabilité, *f.* improbability.

improbable, improbable, unlikely.

improbablement, improbably.

improductif, -ive, unproductive.

improductivité, *f.* unproductiveness.

impromptu, impromptu, without warning.

improportionnel, -elle, without any common ratio.

impropre, incorrect, wrong, improper, unfit, unsuited, inaccurate, unclean, unsuitable; — au service, unserviceable, unfit for service.

improviser, to improvise.

improviste (à l'), unexpectedly, unawares, suddenly.

imprudence, *f.* imprudence.

impubère, impubic, impuberal.

impuissance, *f.* impotence, inability, disability.

impuissant, impotent, powerless.

impulser, to impel.

impulsif, -ive, impulsive.

impulsion, *f.* impulse, impulsion, impetus.

impulvérisé, unpulverized.

impunément, with impunity.

impur, impure, filthy, unchaste.

impureté, *f.* impurity, contamination.

impurification, *f.* contamination.

impurifié, unpurified.

imputer, to impute, attribute, ascribe, charge.

imputrescibilisation, *f.* prevention of decay.

imputrescibilité, *f.* imputrescibility.

imputrescible, imputrescible, incorruptible.

inabordable, inaccessible, unapproachable.

inacceptable, unacceptable.

inaccessible, inaccessible.

inaccommodable, irreconcilable.

inaccompagné, unaccompanied.

inaccompli, unaccomplished, unfinished.

inaccordable, irreconcilable.

inaccoutumé, unaccustomed, unusual.

inachevé, unfinished.

inactif, -ive, inactive, sluggish.

inaction, *f.* inaction.

inactivement, *m.* inactively.

inactivité, *f.* inactivity, inertness.

inadéquat, inadequate.

inadhérence, *f.* lacking adherence.

inadhérent, unadhesive, not adherent.

inalliable, incompatible, incapable of being alloyed.

inallié, unalloyed.

inaltérabilité, *f.* inalterability.

inaltérable, inalterable, unalterable, unchangeable.

inaltéré, unaltered, unchanged.

inanalysé, unanalyzed.

inangulé, inangulate, devoid of angles.

inanimé, inanimate, lifeless, unconscious.

inanition, *f.* inanition, exhaustion.

inanthéré, inantherate, having no anther pistillate.

inanthérifère, without anthers.

inaperçu, unobserved, unnoticed.

inappendicé, inappendiculate, without appendages.

inappendiculé, inappendiculate, without appendages.

inappétence, *f.* inappetence, want of appetite.

inapplicabilité, *f.* inapplicability.

inapplicable, inapplicable, unsuitable.

inappliqué, unapplied, unattentive.

inappréciable, inappreciable.

inappréciablement, inappreciably.

inappréciation, *f.* inappreciation.

inapprêté, unfinished, undressed, uncooked, unprepared.

inapte, inapt, unfit, unsuited.

inaptitude, *f.* inaptitude, unfitness.

inaqueux, -euse, not aqueous.

inarticulé, inarticulate.

inassimilable, that is not assimilable.

inattaquable, unattackable, unassailable, untouched; — aux acides, acidproof, acid-resisting, noncorrosive; — par l'acide sulfurique, resisting sulphuric acid.

inattendu, unexpected.

inauguration, *f.* inauguration.

inaugurer, to inaugurate.

inautorisé, unauthorized.

incalcinable, incapable of being calcined.

incalculable, incalculable.

incandescence, f. incandescence.

incandescent, incandescent, white-hot.

incanescence, m. incanescence, becoming gray.

incapable, incapable, unfit, unable.

incapacité, f. incapacity, incompetency.

incarnadin, m. pink, flesh color.

incarnadin, incarnadine, rosy, pale-red.

incarnat, m. carnation (Dianthus caryophyllus).

incarnat, rosy, flesh-colored, pink.

incassable, unbreakable.

incendiaire, incendiary.

incendie, m. fire, conflagration, burning, (volcanic) eruption; — des cimes d'une forêt, crown fire.

incendier, to set fire to, burn down.

incération, f. mixing or covering with wax, inceration.

incérer, to mix or cover with wax, incerate.

incertain, uncertain, questionable.

incertitude, f. uncertainty, incertitude, doubt.

incessamment, incessantly, unceasingly, continually, immediately.

incessant, incessant, continuous, constant, unceasing.

incidemment, incidentally.

incidence, f. incidence, casualty.

incident, incident, falling upon.

incidentel, -elle, incidental.

incinérateur, m. incinerator.

incinération, f. incineration.

incinérer, to incinerate, burn to ashes, cremate, ash.

incisé, incised

inciser, to incise.

incisif, -ive, incisive, incisory.

incision, f. incision.

incisives, dents —, f.pl. incisors, nippers.

incitation, f. stimulus, instigation.

inciter, to incite.

incivilisé, uncivilized.

inclément, inclement, severe.

inclinable, inclinable.

inclinaison, f. inclination, dip, slope, gradient, angle; — d'un toit, descent of a roof.

inclination, f. inclination, attachment, affection.

incliné, inclined, bent, abrupt, slanting, sloping.

incliner, to incline, bend, dip, tilt, slant, overhang; s'—, to incline, slope, lean, bow, dip (of strata); il s'incline, he bows, yields.

inclus, enclosed.

inclusif, -ive, inclusive.

inclusion, f. inclusion, insertion, enclosure, encysted state, imbedding; — à la paraffine, imbedding in solid paraffin.

inclusivement, inclusive(ly).

incoagulable, uncoagulable, incoagulable.

incoercibilité, f. incoercibility.

incoercible, incoercible.

incohérence, f. incoherence, incoherency.

incohérent, incoherent, rambling.

incohésion, f. incohesion, incoherence.

incoloration, f. colorlessness.

incolore, colorless, uncolored, without color, faulty in color.

incombant, incumbent (anther).

incomber, to be incumbent (on).

incomburant, not supporting combustion.

incombustibilité, f. incombustibility.

incombustible, incombustible, fireproof, refractory.

incommensurabilité, f. incommensurability.

incommensurable, incommensurable.

incommodant, unpleasant, disagreeable, annoying.

incommode, inconvenient, incommodious, troublesome, bothersome.

incommoder, to inconvenience, incommode, trouble, disturb, hinder, disagree.

incommodité, f. inconvenience, awkwardness.

incomparable, incomparable, unequalled.

incomparablement, incomparably.

incompatibilité, f. incompatibility.

incompatible, incompatible, inconsistent.

incomplet, imperfect, unsaturated.

incomplètement, incompletely, imperfectly

incompréhensible, incomprehensible.

incompressibilité, f. incompressibility.

incompressible, incompressible.

incomprimé, uncompressed.

inconcevable, inconceivable.

inconciliabilité, *f.* irreconcilability.

inconciliable, irreconcilable.

inconducteur, nonconducting.

inconfiance, *f.* lack of trust.

inconformité, *f.* unconformity.

incongelable, uncongealable, nonfreezing.

incongelé, uncongealed, unfrozen.

incongru, incongruous, improper.

incongruité, *f.* incongruity.

incongrûment, incongruously.

inconnu, unknown.

inconnue, *f.* unknown (quantity).

inconsciemment, unconsciously.

inconscience, *f.* unconsciousness.

inconscient, unconscious.

inconséquence, *f.* inconsistency.

inconséquent, inconsistent, inconsequent, inconsiderate.

inconstance, *f.* inconstancy, unsteadiness.

inconstant, inconstant, variable, unsteady.

incontaminé, uncontaminated.

incontestable, incontestable, indisputable, unquestionable, undeniable.

incontestablement, incontestably, unquestionably, undeniably.

incontinu, not continuous, discontinuous.

inconvenable, not fitting.

inconvénient, *m.* inconvenience, loss, harm, disadvantage.

inconverti, unconverted, unregenerate.

inconvertible, inconvertible, unconvertible.

incorporer, to incorporate, embody mix.

incorrect, incorrect.

incorrection, *f.* incorrectness, inaccuracy, error.

incorrigé, uncorrected.

incrassant, nutritive, incrassative.

incrassé, incrassate, made stout, thickened.

incrément, *m.* increment.

increscent, increscent, growing.

incristallisable, uncrystallizable.

incroyable, incredible.

incroyablement, incredibly.

incrustation, *f.* incrustation, crust, coating, scale.

incruster, to encrust, inlay

incubation, *f.* incubation, brooding, sitting.

incube, incubous.

incuit, uncooked, underdone, unburned, unboiled.

inculquement, *m.* impressing, engraving.

inculquer, to inculcate.

inculte, uncultivated. waste, rude, not planted.

incultivable, unculturable.

incurie, *f.* carelessness.

incurvé, incurvated, curved.

incurver, to incurve, curve inward.

incurvifolié, (leaves) bent inward.

indamine, *f.* indamine.

inde, *m.* indigo (blue).

Inde, *f.* India; bois d'Inde, logwood; Indes orientales, East Indies.

indébrouillable, inextricable, indistinct, undetermined, indefinite.

indécis, undecided, doubtful, indeterminate.

indécision, *f.* indecision, uncertainty.

indécomposable, indecomposable, irresolvable.

indécomposé, undecomposed.

indécouvrable, indiscoverable.

indécrit, undescribed.

indédoublable, undecomposable, unresolvable, unconvertible.

indéfini, indefinite, unlimited.

indéfiniment, indefinitely.

indéfinissable, indefinable, unaccountable.

indéformable, inseparable, undetachable; — à la chaleur, heat-resisting, stable under heat.

indéfrichable, uncultivable, waste (land).

indéfriché, uncleared, uncultivated.

indéhiscent, indehiscent.

indélébile, indelible.

indemniser, to indemnify, compensate.

indemnité, *f.* indemnity, compensation.

indémonstrable, undemonstrable, improvable.

indéniable, undeniable.

indenter, to indent.

indépendamment, independently.

indépendance, *f.* independence.

indépendant, independent.

indéréglable, foolproof, not likely to get out of order.

indescriptible, indescribable.

indescriptiblement, indescribably.

indésirable, undesirable.

indétermination, f. indetermination, irresoluteness.

indéterminé, indeterminate, indefinite.

index, m. index, index finger, forefinger.

indican, m. indican.

indicanurie, f. indicanuria.

indicateur, m. indicator, gauge, index finger, forefinger, directory, guide; — de (la) vitesse, speed indicator; — de niveau d'eau, water gauge; — de pente, gradient recorder; — de pression, pressure gauge; — de tirage, draft gauge, draft indicator; — de vide, vacuum gauge.

indicateur, -trice, indicatory, indicating.

indicatif, -ive, indicative, indicatory.

indication, f. indication, sign, mark, instruction.

indicatrice, f. indicatrix.

indice, m. sign, mark, indication, index, factor, value, number, of a constant, point reached on scale; — de réfraction, index of refraction; — des blessures, hit mark, wound (by shot).

indicible, inexpressible, unspeakable.

indien, m. Indian.

indien, -enne, Indian.

indienne, f. printed calico, cotton print.

indiennerie, f. printed cotton-goods industry, calico printing.

indienneur, m. calico printer.

indifféremment, indifferently.

indifférence, f. indifference, inertness, neutrality (of salt).

indifférent, indifferent, unconcerned, neutral (solution).

indigence, f. poverty.

indigène, indigenous, native, aborigine.

indigent, indigent, poor.

indigéré, undigested.

indigeste, indigestible, undigested.

indigestible, indigestible.

indigne, unworthy.

indignement, unworthily.

indigner, s'—, to be indignant; il s'indigne, he is becoming indignant.

indigo, m. indigo; — blanc, indigo white; — bleu, indigo blue.

indigoterie, f. indigo (plantation) factory.

indigotier, m. indigo plant (Indigofera tinctoria), indigo manufacturer.

indigotine, f. indigotin.

indigotique, indigotic.

indiquer, to indicate, point out, sketch, show; s'—, to be indicated.

indirect, indirect.

indirubine, f. indirubin.

indiscret, indiscreet, imprudent.

indiscrétion, f. indiscretion.

indiscutable, indisputable, unquestionable.

indiscuté, unquestioned, beyond a doubt.

indispensable, indispensable.

indisposer, to upset, set against.

indisputablement, indisputably.

indissolubilité, f. insolubility.

indissoluble, undissolvable, indissoluble.

indissolublement, indissolubly.

indissous, undissolved.

indistinct, indistinct, hazy, confused.

indistinctement, indistinctly, without distinction.

indistinguible, indistinguishable.

indium, m. indium.

individu, m. individual, person.

individualisation, f. individualization.

individualiser, to individualize.

individualité, f. individuality.

individuel, -elle, individual.

individuellement, individually.

indivis, undivided (of ownership).

indivisé, undivided.

indivisibilité, f. indivisibility.

indivisible, indivisible.

in-dix-huit, decimo-octavo, (in) eighteenmo.

indochinois, Indo-Chinese.

indogène, m. indogen.

indol, m. indole.

indolence, f. indolence, idleness, apathy.

indolent, indolent, lazy, sluggish.

indolore, painless.

indomptable, indomitable, untamable, uncontrollable.

indompté, untamed, wild, indomitable.

indophénine, f. indophenin.

indophénol, m. indophenol.

indosable, indeterminable.

indosé, undetermined.

in-douze, duodecimo, (in) twelvemo.

indoxyle, *m.* indoxyl.

indoxylique, indoxyl(ic).

indu, undue, not due.

indubitable, indubitable, undoubted.

inductance, *f.* inductance.

inducteur, *m.* inductor, field, induction coil, field magnet.

inducteur, -trice, inducing, primary.

inductif, -ive, inductive.

inductile, not ductile.

inductilité, *f.* inductility.

induction, *f.* induction, inducement; — propre, self-induction.

induire, to induce, infer, conclude.

induisant (induire), inferring.

induit (induire), induced, inferred, secondary.

induit, *m.* induced (secondary) circuit, armature (of a dynamo), rotor; — en anneau, ring-wound armature.

indulgence, *f.* indulgence, forbearance.

induline, *f.* induline.

indûment, unduly.

induplicative, induplicative, edges bent abruptly or rolled inward.

indupliqué, induplicate (with edges folded inward).

induration, *f.* induration, hardened tissue.

induré, indurated.

indurer, to indurate, harden.

industrie, *f.* industry, business, manufacture, trade, occupation, activity, calling, skill; — frigorifique, refrigeration industry; — tinctoriale, dyeing industry; — vinicole, wine-growing industry.

industriel, *m.* manufacturer.

industriel, -elle, industrial, manufacturing.

industrieusement, skillfully, industriously.

industrieux, -euse, skillful, ingenious, dexterous.

induvial, -aux, induvial, calyx persistent, covering the fruit.

induvies, *f.pl.* induviae, scale leaves, any persistent portion of perianth.

inébranlable, immovable, unshakable, firm.

inébriant, intoxicating.

inédit, unpublished.

ineffaçable, indelible.

ineffectif, -ive, ineffective.

inefficace, ineffective, ineffectual, inefficacious.

inégal, -aux, uneven, unequal, unlike, scabby, scurfy.

inégalement, unequally, unevenly.

inégaliser, to render unequal or uneven.

inégalité, *f.* inequality, unevenness.

inélastique, inelastic.

inéligible, ineligible.

inéluctable, irresistible, ineluctable.

ineptie, *f.* ineptitude, inaptness.

inépuisable, inexhaustible.

inépuisé, unexhausted.

inépuré, unpurified, unrefined.

inéqualiflore, with unequal flowers.

inéqualifolié, with unequal leaves.

inéquienne, uneven-aged.

inéquilatéral, inequilateral, having two sides unequal.

inéquilobé, inequilobate, with lobes of unequal size.

inéquitable, inequitable, unfair.

inéquivalvé, inequivalve, having two unequal valves.

inerme, inerm, inermous, without spines or prickles.

inerte, inert, sluggish, lifeless.

inertie, *f.* inertia, inertness, sluggishness; moment d'—, moment of inertia.

inespéré, unhoped, unexpected.

inespérément, unexpectedly.

inestimable, inestimable, priceless.

inétirable, undrawable, not ductile.

inévitabilité, *f.* inevitableness.

inévitable, inevitable.

inexact, inexact, inaccurate, incorrect, not punctual.

inexactement, inaccurately, impunctually.

inexactitude, *f.* inaccuracy, inexactness.

inexcusablement, inexcusably.

inexistant, nonexistent.

inexistence, *f.* nonexistence.

inexorable, inexorable, unrelenting.

inexorablement, inexorably.

inexpérience, *f.* inexperience.

inexpérimenté, inexperienced, unskilled, untried.

inexplicable, inexplicable, unexplainable.

inexpliqué, unexplained.

inexploitable, nonexploitable, uncultivable (land).

inexploité, unexploited, undeveloped, unworked, untilled.

inexploré, unexplored.

inexplosible, nonexplosive.

inexprimable, inexpressible, unutterable.

inexprimé, unexpressed.

inextensible, inextensible.

inextinguible, inextinguishable.

inextricable, inextricable.

infaillibilité, *f.* infallibility.

infaillible, infallible, unerring, sure, unfailing.

infaisable, unfeasible, impracticable.

infâme, infamous, foul, filthy, squalid.

infantile, infantile.

infatigable, indefatigable, unwearied, untiring.

infavorable, unfavorable.

infécond, infecund, infertile, sterile, barren, unfruitful.

infécondité, *f.* infertility, infecundity, barren, fruitlessness.

infect, foul, infectious, infected.

infecté, infected, contaminated.

infecter, to infect.

infectieux, -euse, infectious.

infection, *f.* infection, infestation, contagion; — **aigue,** acute infection; — **purulente,** pyemia.

inféoder, to make subservient.

infère, lower, inferior, of lower growth.

inférence, *f.* inference, conclusion.

inférer, to infer.

inférieur, inferior, lower, less (than), at the bottom.

inférieurement, below, inferiorly.

infériorité, *f.* inferiority.

infermenté, unfermented.

infermentescible, unfermentable.

infernal, -aux, infernal.

infertile, infertile, sterile.

infester, to infest, overrun.

infidèle, unfaithful, unreliable, untrustworthy.

infiltration, *f.* infiltration, percolation.

infiltrer, to infiltrate, seep in, spread, infilter, percolate, trickle through; — **dans le terrain,** to sink into a porous stratum.

infime, lowest, mean, very low, very small.

infini, infinite (quantity), numberless, infinity; **à l'—,** infinitely, ad infinitum, to infinity.

infiniment, infinitely, far, extremely, indefinitely without end; — **petit,** *m.* infinitesimal, minute particle; **les —s petits,** *m.pl.* infinitesimal creatures.

infinité, *f.* infinity, very large number.

infinitésimal, -aux, infinitésime, infinitesimal.

infirmation, *f.* nullification, weakening.

infirmer, to nullify, invalidate, weaken, cancel (letter).

infirmité, *f.* infirmity.

inflammabilité, *f.* inflammability.

inflammable, inflammable.

inflammateur, *m.* igniter, ignition apparatus, igniting charge.

inflammation, *f.* inflammation, ignition, burning, firing; — **spontanée,** spontaneous combustion.

inflation, *f.* inflation.

infléchi, inflexed (stamen).

infléchir, to inflect, bend, turn, deflect.

inflexibilité, *f.* inflexibility, unbendingness.

inflexible, inflexible, unyielding, unbending, rigid.

inflexion, *f.* inflection.

infliger, to inflict.

inflorescence, *f.* inflorescence.

influence, *f.* influence.

influencer, to influence.

influent, influential, influencing.

influenza, *f.* influenza.

influer, to have influence, influence, sway.

in-folio, folio.

infondibuliforme, infundibular.

informateur, *m.* informer.

information, *f.* information, inquiry, investigation.

informe, formless, unformed, shapeless, informal, irregular.

informé, *m.* inquiry.

informé, informed.

informer, to inform, acquaint; **s'—,** to inquire, ask.

informulé, unformulated.

infortune, *f.* misfortune, adversity.

infortuné, *m.* unfortunate, unhappy person.

infortuné, unfortunate.

infra-axillaire, infra-axillary, situated below the axil.

infranchissable, unsurmountable, impassable.

infra-rouge, infrared.

infrastructure, *f.* substructure.

infréquemment, infrequently, seldom.

infréquence, *f.* infrequency.

infréquent, infrequent, rare, uncommon.

infriable, not friable.

infructueusement, unprofitably, without success.

infructueux, -euse, unfruitful, unprofitable, fruitless, unavailing, unsuccessful.

infrutescence, *f.* collective or anthocarpus fruit.

infundibuliforme, funnel-shaped, infundibular.

infus, intuitive, native, innate.

infusé, *m.* infusion.

infusé, infused.

infuser, to instill, infuse; s'—, to be infused.

infusibilité, *f.* infusibility.

infusible, infusible.

infusion, *f.* infusion.

infusoire, infusorial.

infusoires, *m.pl.* Infusoria.

infusum, *m.* infusion.

ingaranti, not guaranteed, unwarranted.

ingelif, -ive, not rent by frost, not frost cleft, not freezing easily.

ingénier (s'), to do one's best, strive.

ingénieur, *m.* engineer; — chimiste, chemical engineer; — des ponts et chaussées, civil engineer; — géographe, topographical engineer, cartographer; — hydrographe, hydrographer.

ingénieusement, ingeniously, cleverly.

ingénieux, -euse, ingenious, clever.

ingéniosité, *f.* ingeniousness, ingenuity.

ingénu, ingenuous, simple.

ingénuité, *f.* ingenuousness.

ingérer, to ingest, introduce into the stomach; s'—, to interfere.

ingestion, *f.* ingestion.

ingrain, *m.* spelt.

ingrat, ungrateful, unproductive, unpleasing, unpromising, barren.

ingrédient, *m.* ingredient, constituent.

inguérissable, incurable.

ingurgiter, to swallow, gulp down.

inhabile, unskillful, inapt, not qualified, incapable.

inhabilement, unskillfully, awkwardly.

inhabileté, *f.* want of skill, inexpertness.

inhabitable, uninhabitable.

inhabitué, unaccustomed.

inhalateur, *m.* inhaler.

inhalateur, -trice, inhaling.

inhérent, inherent.

inhibiteur, -trice, inhibitory, inhibiting, inhibitive.

inhibition, *f.* inhibition.

inhibitoire, inhibitory.

inhospitalier, -ère, inhospitable, unfriendly.

inhumain, inhuman, cruel, savage.

inhumectation, *f.* lack of moisture, dryness.

inhumecte, unmoistened.

inimaginable, unimaginable, unthinkable.

inimitié, *f.* enmity, hostility.

ininflammable, noninflammable.

inintelligible, unintelligible.

inintentionellement, unintentionally.

ininterrompu, uninterrupted, unbroken, continuous.

initial, -aux, initial.

initiale, *f.* initial, first meristematic cell.

initiateur, *m.* initiator.

initiative, *f.* initiative.

initier, to initiate.

injectable, injectable.

injecté, injected, congested, inflamed, impregnated, creosoted (wood).

injecter, to inject, impregnate; s'—, to be injected, be flushed, become congested.

injecteur, *m.* injector.

injecteur, -trice, injecting.

injection, *f.* injection, impregnation, congestion (of tissue).

injudicieusement, injudiciously.

injure, *f.* injury, insult, wrong, abuse.

injurieux, -euse, injurious.

injuste, unjust, unfair.

injustement, unjustly, wrongly.

injustice, *f.* injustice, unfairness.

injustifiable, unjustifiable.

inlacé, like a catkin.

inlassablement, untiringly.

inlide, like a catkin.

irné, innate, inborn, inbred.

ir nervation, f. innervation.

innervé, nerveless (leaf).

innocence, f. innocence.

innocent, innocent, harmless.

innocuité, f. innocuousness, harmlessness.

innombrable, innumerable, countless.

innombrablement, without number.

innovation, f. innovation.

innover, to innovate, introduce

innucellé, (plant) whose ovules lack nucellus and integuments.

innumérable, innumerable.

inobscurci, unobscured.

inobservable, not observable, hardly perceptable.

inobservé, unobserved, unnoticed.

inobstrué, unobstructed.

inocarpe, with fibrous fruit.

inoccupé, unoccupied, unemployed, vacant.

in-octavo, octavo, 8vo.

inoculable, inoculable.

inoculation, f. inoculation.

inoculer, to inoculate, infect.

inodore, odorless, inodorous.

inoffensif, -ive, inoffensive, harmless, innocuous, out of action.

inofficiel, -elle, unofficial.

inondation, f. inundation, flood.

inonder, to inundate, overspread, deluge, flood.

inophylle, inophyllous, with reticulate or threadlike veins.

inopiné, unexpected, sudden, unforeseen.

inopportun, inopportune, ill-timed.

inorganique, inorganic.

inosite, f inosite, inositol.

inositurie, inosurie, f. inosituria, inosuria.

inoubliable, unforgettable.

inouï, unheard of.

inovulé, lacking ovules.

inoxydabil.té, f. unoxidizability.

inoxydable, unoxidizable, inoxidable, which cannot be oxidized, rust-resisting.

inquart, m. inquartation, f. quartation.

in-quarto, quarto.

inquiet, unquiet, restless, uneasy, anxious.

inquiéter, to worry; s'—, to concern oneself, care about, bother about, worry.

inquiétude, f. anxiety, worry, disquietude, uneasiness.

insaisissable, imperceptible, unseizable.

insalubre, insalubrious, unhealthy.

insalubrité, f. unhealthfulness.

insanité, f. insanity.

insapide, tasteless, insipid.

insaponifiable, unsaponifiable.

insatiable, insatiable.

insatisfait, unsatisfied.

insaturable, insaturable, not saturable.

insaturé, unsaturated.

insciemment, unknowingly, unconsciously.

inscriptible, inscribable.

inscription, f. inscription, matriculation, registration.

inscrire, to inscribe, write down, set down, register, enter; s'—, to be inscribed, be recorded, register.

inscrit, inscribed.

insécable, indivisible.

insecte, m. insect; — nuisible, destructive insect; — rongeur, leaf-chewing insect, —s cuticoles, biting insects; —s piqueurs, biting insects.

insecticide, m. insecticide.

insectifuge, f. insectifuge.

insectivore, insectivorous.

insectivores, m.pl. Insectivora.

insécurité, f. insecurity.

in-seize, sixteenmo, 16 mo.

inséminer, to inseminate, fertilize artificially.

insensé, insensate, insane, foolish, senseless.

insensibilisateur, m. anesthetic, anesthetizing apparatus.

insensibilisateur, -trice, producing insensibility.

insensibiliser, to insensibilize. anesthetize.

insensibilité, f. insensibility.

insensible, insensible, insensitive, unfeeling, imperceptible.

insensiblement, insensibly, imperceptibly.

insensitif, -ive, insensitive.

inséparable, inseparable.

insérable, insertable.

insérant, m. advertiser.

insérer, to insert, put in circuit, put in, jab, notch.

insertion, *f.* insertion; — **de l'antenne,** antennary socket.

insexe, asexual.

insidieux, -euse, insidious.

insigne, *m.* mark.

insigne, remarkable, noted, signal.

insignifiance, *f.* insignificance, unimportance.

insignifiant, insignificant, unimportant.

insinuer (s'), to insinuate, creep into, worm in.

insipide, insipid, tasteless.

insipidité, *f.* insipidity, tastelessness.

insistance, *f.* insistence, entreaty, urging, perseverance.

insister, to insist, persist, urge, stress.

insolation, *f.* insolation, exposure to the sun, sun blister.

insoler, to expose to the sun.

insolide, not solid, unsound.

insolidité, *f.* insolidity, unsoundness.

insolite, unusual, unwonted.

insolitement, unusually.

insolubiliser, to render insoluble.

insolubilité, *f.* insolubility.

insoluble, insoluble.

insolution, *f.* nonsolution.

insolvabilité, *f.* insolvency.

insolvable, insolvent.

insomnie, *f.* insomnia.

insondable, unfathomable.

insonore, not sonorous, soundproof.

insouciance, *f.* carelessness, neglect, heedlessness.

insoucieux, -euse, neglectful.

insoudable, unweldable.

insoupçonné, unsuspected.

insoutenable, untenable, indefensible, insupportable.

inspecter, to inspect, survey.

inspecteur, *m.* inspector, deputy conservator, assayer.

inspection, *f.* inspection, survey, examination.

inspirant, inspiring.

inspiration, *f.* inspiration, inhaling.

inspirer, to inspire, inhale, instill; **il s'inspire de,** he draws inspiration trom.

inspissation, *f.* thickening.

inspisser, to concentrate, inspissate.

instabilité, *f.* instability, unstableness.

instable, unstable, unsteady.

installateur, *m.* installer.

installation, *f.* installation, plant, set; — **de chauffage,** heating plant; — **de cuisson,** cooking unit; — **de force,** power plant.

installer, to install; **il s'installe,** he sets up.

instaminé, instaminate, not bearing stamens.

instamment, urgently, insistently, earnestly.

instance, *f.* instance, solicitation, insistence, entreaty.

instant, *m.* instant, moment; **à l'—,** instantly, immediately, a moment ago; **à tout —,** in a moment.

instantané, *m.* snapshot, instantaneous photograph.

instantané, instantaneous.

instantanéité, *f.* instantaneousness.

instantanément, instantaneously, instantly.

instar, à l'— de, after the fashion, manner of, like.

instaurer, to establish, set up, found.

instiguer, to incite, prompt.

instiller, to instill.

instinct, *m.* instinct.

instinctif, -ive, instinctive.

instinctivement, instinctively.

instituer, to institute, establish.

institut, *m.* institute, institution.

instituteur, *m.* institutor, founder, teacher.

institution, *f.* institution.

instructeur, *m.* instructor, teacher.

instructif, -ive, instructive.

instruction, *f.* instruction, investigation, inquiry, direction; — **d'entretien,** service instruction, instruction for employment.

instruire, to instruct, teach, investigate, train, inform, educate.

instruit, instructed, learned, educated, well-read, aware.

instrument, *m.* instrument, utensil, apparatus, implement, tool; — **d'arpentage,** surveying instrument; — **de menage,** household utensil; —**s de bord,** tools and repair material.

instrumental, -aux, instrumental.

insu, *m.* ignorance, unknown; **à l'— de,** unknown to; **à mon —,** unknown to me.

insuccès, *m.* failure, want of success.

insuffisamment, insufficiently.

insuffisance, *f.* insufficiency, inadequacy, incapacity.

insuffisant, insufficient, unsufficing, incompetent, inadequate.

insufflation, *f.* insufflation, inflation, blowing up, blast, breath.

insuffler, to insufflate, breathe, blow up, inflate.

insulaire, *m.&f.* islander.

insupportable, unbearable, insufferable.

insurmontable, insurmountable.

intact, intact, whole, unblemished.

intaille, *f.* intaglio.

intarissable, inexhaustible.

intégrable, integrable.

intégral, -aux, integral, complete, entire.

intégrale, *f.* integral.

intégralement, integrally, entirely, in its entirety, wholly.

intégrant, integral, integrant.

intégration, *f.* integration.

intégrer, to integrate.

intégriflore, with entire flowers.

intégrifolié, integrifolious, with entire leaves.

intégrité, *f.* integrity, entireness.

intégument, *m.* integument, bud covering.

integumenta floralia, *f.pl.* floral envelopes.

intellect, *m.* intellect, understanding.

intellectuel, -elle, intellectual.

intelligemment, intelligently.

intelligence, *f.* intelligence, intellect, understanding, collusion.

intelligent, intelligent, learned.

intelligible, intelligible, intellectual, clear.

intempérant, intemperate.

intempéré, unrestrained, immoderate.

intempérie, *f.* inclemency, bad weather.

intempestif, -ive, untimely, inopportune, unseasonable.

intenable, untenable.

intendance, *f.* intendancy, direction, management.

intendant, *m.* intendant, director, manager, steward.

intense, intense, severe, heavy (traffic).

intensif, -ive, intensive.

intensifier, to intensify.

intensité, *f.* intensity, intenseness; — du courant, current (intensity) strength; — d'une bougie, candle power.

intensivement, intensively, intensely, extremely.

intenter, to bring, enter (an action).

intention, *f.* intention, purpose; à l'— de, in honor of, for the sake of.

intentionnel, -elle, intentional.

interalliés, *m.pl.* interallies.

interatomique, interatomic.

intercalaire, intercalary, growth not apical, between apex and base.

intercalation, *f.* intercalation, interpolation, insertion.

intercaler, to intercalate, put in, sandwich, insert, interpose.

intercéder, to intercede, plead.

intercellulaire, intercellular.

intercepter, to intercept, cut off, shut off, stop.

interchangeabilité, *f.* interchangeability.

interchangeable, interchangeable.

interdiction, *f.* interdiction, prohibition, suspension.

interdire, to interdict, forbid, prohibit, suspend, disconcert, dumfound, bewilder.

interdit, *m.* interdict, prohibition.

intéressant, interesting.

intéresser, to interest, affect, injure, concern, involve; s'—, to become interested.

intérêt, *m.* interest, concern, advantage, share.

interférence, *f.* interference.

interférent, interfering.

interférer, to interfere.

interfoliacé, interfoliaceous, interpetiolar (between the petioles).

interfolier, to interleave.

interfolium, *m.* internode.

interieur, *m.* interior, inside, home, home life; — de la cellule, cell cavity; — des terres, inland; à l'—, inside, internally.

intérieur, interior, internal, inner, inward, inside, homemade, domestic.

intérieurement, internally, inwardly, interiorly, inside.

interjeter, to interject.

interligne, *m.* space between lines; *f.* lead (in printing).

interligner, to interline, lead.

intermède, *m.* intermediate, interlude.

intermédiaire, *m.* intermediary, medium, agent, middleman, mediator.

intermédiaire, intermediate, intermediary, intervening; **par l'— de,** by means of.

intermédiat, intermediate.

interminable, interminable, endless, unending.

intermittent, intermittent, discontinuous.

international, -aux, international.

interne, *m.* interne.

interne, internal, inward, inside.

internodial, internodal.

interoculaire, interocular.

interpeller, to summon, challenge.

interpétiolaire, interpetiolar (between the petioles), interfoliaceous.

interpolaire, interpolar.

interpolateur, *m.* interpolator.

interpolateur, -trice, interpolating.

interpolation, *f.* interpolation.

interpoler, to interpolate.

interposé, interposed.

interposer, to interpose, put in; **s'—,** to be interposed, intervene.

interpositif, -ive, interposed.

interposition, *f.* interposition.

interprétation, *f.* interpretation, construction.

interprète, *m.&f.* interpreter.

interpréter, to interpret.

interrane, underground, subterranean.

interrogation, *f.* interrogation, question.

interroger, to interrogate, question, consult.

interrompre, to interrupt, impede, break, cut off; **— et retablir,** to make and break.

interrompu, interrupted, broken (of a line), incomplete.

interrupteur, *m.* interrupter, circuit breaker, switch.

interrupteur, -trice, interrupting.

interruption, *f.* interruption; **— du massif,** interruption of the leaf canopy.

interrupti-penné, interruptedly pinnate.

intersection, *f.* intersection.

interstellaire, interstellar.

interstice, *m.* interstice, empty space.

interstitiel, -elle, interstitial.

interstrié, interstrial, interstriated.

interurbain, interurban.

intervalle, *m.* interval, space, clearance.

intervalvaire, with partition between valves of pericarp.

intervenir, to intervene, interpose, occur, come in, come into play, interfere.

intervention, *f.* intervention, interposal, agency.

interversion, *f.* inversion.

interverti, inverted, invert, interverted.

intervertir, to invert, transpose, interchange.

intervertissement, *m.* inverting, inversion.

intervient, (he) intervenes.

intestin, *m.* intestine, gut, bowel, rectum; **— moyen,** mid-intestine; **— postérieur,** hind intestine; **— terminal,** small intestine, ileum.

intestinal, -aux, intestinal.

intigé, acaulous, stemless.

intime, intimate, close, internal, inmost.

intimement, intimately, closely.

intimer, to give notice to.

intimité, *f.* intimacy, close connection, inmost part.

intine, *f.* intine, the innermost coat.

intitulé, *m.* title.

intituler, to entitle, call.

intolérable, intolerable, unbearable.

intolérance, *f.* intolerance.

intonation, *f.* intonation, pitch (of sounds).

intoxicant, poisoning, poisonous, toxic.

intoxication, *f.* poisoning, intoxication.

intoxiquer, to poison.

intracellulaire, intracellular.

intracérébral, intracerebral.

intrados, *m.* intrados, bottom face.

intraduisible, untranslatable.

intrafolié, intrafoliaceous.

intraglobulaire, contained in a round cell.

intraire, intrarious.

intraitable, intractable, beyond treatment.

intramarginal, intramarginal, placed within the margin.

intraméabilité, *m.* intrameability, protoplasm permits substances to pass into its vacuoles.

intramoléculaire, intramolecular.

intransférable, nontransferable.

intransmissible, nontransmissible.

intransmuable, nontransmutable.

intransparence, *f.* nontransparence, opacity.

intransparent, nontransparent.

intransportable, untransportable.

intraperitoneal, intraperitoneal.

intravasculaire, intravascular.

intraveineux, -euse, intravenous.

intrépide, intrepid, fearless.

intrinsèque, intrinsic(al).

introduction, *f.* introduction, putting in admission, induction.

introductoire, introductory.

introduire, to introduce, put in, bring in; s'—, to be introduced, get in, find one's way in, introduce oneself.

introfléchi, bent inside from without.

introrse, introrse, facing inward, towards the axis.

introublé, undisturbed.

introuvable, undiscoverable, not to be found.

intrus, *m.* intruder.

intrusion, *f.* intrusion.

intuitif, -ive, intuitive.

intuition, *f.* intuition.

intuitivement, intuitively.

intumescence, *f.* intumescence, swelling.

intussusception, *f.* intussusception.

inulase, *f.* inulase.

inule, *f.* inula, elecampane.

inulénine, *f.* inulenin, a subordinate constituent of insulin.

inuline, *f.* inulin.

inusable, impossible to wear out or use up, durable.

inusé, unused, unworn.

inusité, not in use, unusual, unused, obsolete.

inutile, useless, unnecessary, needless, vain.

inutilement, uselessly, vainly, to no purpose, needlessly.

inutilisable, unutilizable, useless, unserviceable.

inutilisé, unutilized.

inutiliser, to render useless.

inutilité, *f.* inutility, uselessness, useless thing.

invagination, *f.* invagination.

invaginé, invaginate(d).

invalide, invalid.

invalidement, invalidly.

invalider, to invalidate.

invalidité, *f.* infirmity, invalidity.

invar, *m.* invar.

invariabilité, *f.* invariability, invariableness.

invariable, invariable, unchangeable.

invariablement, invariably, constantly.

invariant, *m.* invariant, nonvariant.

invasion, *f.* invasion, inroad, attack.

inveiné, without veins or nerves.

invendable, unsalable, unmarketable.

invendu, unsold.

inventaire, *m.* inventory; — d'un peuplement, valuation, survey.

inventer, to invent, find out, contrive.

inventeur, *m.* inventor, discoverer.

inventif, -ive, inventive.

invention, *f.* invention, discovery.

inventorier, to inventory, take stock.

invérifiable, unverifiable.

invérifié, unverified.

inverse, *m.* contrary, reverse, reciprocal.

inverse, inverse, inverted, opposite, contrary.

inversement, inversely, contrariwise.

inverser, to reverse.

inverseur, *m.* inverter, reversing device.

inversion, *f.* inversion, reversal.

invertase, *f.* invertase, invert sugar.

invertébré, *m.* invertebrate.

inverti, reversed, sexually perverted; sucre —, invert sugar.

invertine, *f.* invertin, invertase.

invertir, to reverse, invert.

investigateur, *m.* investigator.

investigateur, -trice, investigating, inquiring.

investigation, *f.* investigation, research.

invincible, invincible, unconquerable.

inviscant, coating, daubing.

inviscation, *f.* inviscation with saliva, coating with a viscous material, smearing daubing.

invisibilité, *f.* invisibility.

invisible, invisible.

invisibles, *m.pl.* ultramicroscopic organisms.

invisquer, to inviscate with saliva, coat with viscous material, smear, daub.

invitant, inviting.

invitation, *f.* invitation.

inviter, to invite, bid, attract.
invocation, f. invocation, prayer.
involontaire, involuntary.
involontairement, involuntarily.
involucelle, m. involucel, partial involucre.
involucral, -aux, involucral.
involucre, m. involucre.
involucré, involucrate.
involucrum, m. involucrum, involucre.
involuté, involute(d).
involutif, -ive, involutive.
involution, f. involution, degeneration.
involvant, enveloping and protecting.
invoquer, to invoke, appeal to, call upon.
invraisemblable, improbable, unlikely.
invraisemblance, f. improbability, unlikelihood.
iodate, m. iodate.
iode, m. iodine.
iodé, iodized, coated or impregnated with iodine.
ioder, to iodize, combine or impregnate with iodine.
iodeux, iodous.
iodhydrate, m. hydriodide.
iodhydrique, hydriodic.
iodide, m. iodide.
iodifère, iodiferous, containing iodine.
iodique, iodic.
iodisme, m. iodism.
iodoamidonné, starch iodide.
iodoforme, m. iodoform.
iodoformé, iodoformized, iodoform.
iodol, m. iodol.
ioduration, f. iodination, iodization, iodation.
iodure, m. iodide; — d'alcoyle, alkyl iodide; — d'argent, silver iodide; — de méthyle, methyl iodide; — de potassium, potassium iodide; — d'éthyle, ethyl iodide; — mercurique, mercuric iodide.
ioduré, combined with iodine, iodinated, iodized, containing iodine, iodated, containing an iodide.
iodurer, to iodize, iodinate, iodate (with iodide).
iodydrate, m. hydriodide.
iolite, f. iolite.
ion, m. ion; —s hydrogènes, hydrogen ions, hydrions.

ionisation, f. ionization.
ioniser, to ionize.
ionium, m. ionium.
ionone, f. ionone.
ipéca, m. ipecac.
ipécacuana, m. ipecacuanha.
ira (aller), (he) will go.
irait (aller), (he) would go.
iridé, like an iris.
iridées, f.pl. Iridaceae.
iridescent, iridescent.
iridié, containing iridium.
iridine, f. iridin.
iridique, iridic, iridian, iridal.
iridium, m. iridium.
iridosmine, f. iridosmine.
iris, m. iris, rainbow, flag, or sword lily (Iris); — de Florence, Florentine iris (Iris florentina); racine d'—, orrisroot.
irisation, f. irisation, iridescence.
irisé, iridescent, irisated, irised, having colors like the rainbow, changeable.
iriser, to make iridescent; s'—, to become iridescent.
irlandais, m. Irish, Irishman.
irlandais, Irish.
irone, f. irone.
ironique, ironical.
irradiation, f. irradiation.
irradier, to radiate, irradiate, emit rays.
irrationnel, -elle, irrational.
irréalisable, unrealizable, unattainable, impracticable, that cannot be done.
irrecevable, inadmissible.
irréconciliabilité, f. irreconcilability.
irréconciliable, irreconcilable.
irrécouvrable, irrecoverable.
irrécupérable, irreparable, irrecoverable.
irrécusable, unexceptionable, unchallengeable, indisputable.
irréductibilité, f. irreducibleness.
irréductible, irreducible.
irréduit, unreduced.
irréel, -elle, unreal.
irréfléchi, thoughtless, heedless.
irréfutable, irrefutable, indisputable.
irrégularité, f. irregularity.
irrégulier, -ère, irregular, anomalous.
irrégulièrement, irregularly.

irrémédiable, irremediable, not to be remedied, helpless, irretrievable.

irréparablement, irreparably.

irréprochable, irreproachable.

irrésistible, irresistible.

irrésolu, irresolute, unsolved.

irrésoluble, irresolvable, unsolvable.

irrespirable, irrespirable.

irréussite, f. failure, lack of success.

irréversible, irreversible.

irrévocable, irrevocable.

irrévocablement, irrevocably.

irrigable, that can be irrigated.

irrigation, f. irrigation, watering.

irriguer, to irrigate, water.

irritabilité, f. irritability.

irritable, f. irritable.

irritant, irritant, irritating.

irritation, f. irritation.

irriter, to irritate, provoke, enrage, excite.

irrorateur, m. spray, diffuser.

irruption, f. irruption, inroad.

isadelphe, isadelphous.

isandré, isandrous.

isanthère, with equal anthers.

isatine, f. isatin.

isatinique, isatinic.

isatropique, isatropic.

ischium, m. ischium.

ischurétique, ischuretic.

isentrope, f. adiabatic curve.

iséthionique, isethionic.

isinglass, m. isinglass; — végétal, agar.

islandais, Iceland, Icelandic.

isoapiol, m. isoapiole.

isobutylique, isobutyl.

isocèle, isosceles, equally sided.

isochromatique, isochromatic.

isochrone, isochronique, isochronous, isochronic, isochron l.

isocline, f. isoclinic line, isoclinal.

isocyanate, m. isocyanate.

isocyanique, isocyanic.

isocyanurique, isocyanuric.

isocyclique, isocyclic, eucyclic (flower) with isomerous whorls.

isodimorphe, isodimorphous, isodimorphic.

isodiodé, isosparous.

isodiodie, f. isodiody (producing diodes giving rise to unisexual prothallia).

isodique, centripetal.

isodulcite, f. isodulcite.

isodyname, isodynamous, equally developed

isodynamique, isodynamic.

isoète, m. quillwort (Isoetes).

isogamète, f. isogamete.

isogamie, isogamous.

isogène, isogenous, having the same origin.

isogone, isogonic, isogonal, (hybrids) combining parental characters in equal degree.

isogyne, isogynous, having the pistils similar.

isolable, isolable.

isolant, m. insulating substance, insulator.

isolant, isolating, insulating.

isolatéral, -aux, isolateral, bilateral, arranged on opposite sides.

isolateur, m. insulator.

isolateur, -trice, isolating, insulating.

isolation, f. isolation, insulation.

isolé, isolated, insulated, detached, alone, straggling, scattered.

isolément, m. isolation, insulation; état d'—, isolated state.

isolément, separately, individually, isolatedly.

isoler, to isolate, separate, insulate, detach; s'—, to be isolated or insulated, isolate oneself, separate.

isoleucine, f. isoleucine.

isolichénine, f. isolichenin.

isologie, f. isology.

isologue, m. isologue.

isologue, isologous.

isoloir, m. insulator, insulating stool.

isomaltose, f. isomaltose.

isomère, m. isomer.

isomère, isomerous, isomeric.

isomérie, f. isomerism.

isomérique, isomeric.

isomériser, to isomerize.

isomérisme, m. isomerism.

isométrique, isometric(al).

isomorphe, isomorphous, isomorphic.

isomorphie, f., isomorphisme, m. isomorphism.

isophyllie, f. isophylly, leaves alike in shape and size.

isopropylique, isopropyl.

isoprothallie, *f.* isoprothally, prothallia similar in sexual character.

isoscèle, see ISOCÈLE.

isospare, isosparous.

isosporé, isosporous, homosporous, having one kind of spore.

isostémone, isostemonous.

isostémonopétale, with equal number of stamens and petals.

isostique, isostic, (mother root) with more than two xylem bundles.

isotherme, *f.* isotherm.

isotherme, isothermal.

isothermique, isothermic.

isotonie, *f.* isotonicity.

isotonique, isotonic.

isotope, *m.* isotope.

isotopie, *f.* isotopy.

isotrope, isotropic.

isotrophyte, *f.* isotrophyte, a parasitic fungus.

isotropie, *f.* isotropy, isotropism.

issu, issued, issuing, descended, born or sprung from, descending from, proceeding (from).

issue, *f.* issue, outlet, egress, passage, offal, refuse, garbage, waste; à l'— de, on leaving, at the end of; —s de blé, *m.pl.* sharps, middlings by-products.

isthme, *m.* isthmus.

itaconique, itaconic.

italien, *m.* Italian.

italien, -enne, Italian.

italique, italic (type).

italiqué, italicized.

iterative, iterative.

itinéraire, *m.* itinerary, route.

iule, *m.* catkin, inlus, galley worm.

ivoire, *m.* ivory; d'—, ivory white.

ivoirine, *f.* ivorine.

ivraie, *f.* darnel, rye grass (*Lolium perenne*).

ivresse, *f.* intoxication, ecstasy.

ixia, *f.* Ixia.

J

jable, *m.* croze, chime, chimb (of cask), croze, groove (of cask stave).

jabler, to notch, groove.

jaborandi, *m.* jaborandi.

jabot, *m.* crop (of a bird).

jacapucayo, *m.* sapucaia (Lecithis), Brazil-nut tree (*Bertholletia excelsa*); **noix de —,** Brazil nut, monkey-pot.

jacée, *f.* knapweed (*Centaurea jacea*); **— des blés,** cornflower, bachelor's-button, bluebottle (*Centaurea cyanus*); **petite —,** wild pansy (*Viola tricolor*).

jachère, *f.* fallow (land).

jachérer, to fallow, plow up.

jacinthe, *f.* hyacinth (Hyacinthus).

jacobée, *f.* tansy ragwort (*Senecio jacobaea*).

jacquerotte, *f.* earthnut pea, tuberous pea, heath pea (*Lathyrus tuberosus*).

jade, *m.* jade.

jadien, -enne, containing jade.

jadis, of old, once, formerly, of yore.

jaguar, *m.* jaguar.

jaïet, *m.* see JAIS.

jaillir, to spout, gush (out), spurt, jump, **burst out,** (of light) flash, (of spark) fly;

faire —, to pass, strike, produce (sparks), spout, flash.

jaillissant, spouting, gushing out.

jaillissement, *m.* spouting, gushing, spurting.

jais, *m.* jet.

jalap, *m.* jalap.

jalapine, *f.* jalapin.

jalapique, jalapic.

jallot, *m.* vessel for molding tallow.

jalon, *m.* stake, pole, guidepost, beacon, tracing picket.

jalonnement, *m.* staking out, beaconing, enclosure, demarcation.

jalonner, to stake out, mark out, plot.

jalot, *m.* a vessel for molding tallow.

jalouse, jealous.

jalousé, latticed, screened (window).

jalousement, jealously.

jalouser, to envy, be jealous of.

jalousie, *f.* jalousie, jealousy, envy, grating. Venetian blind, sweet William (*Dianthus barbatus*); **— des jardins,** rose campion (*Lychnis coronaria*).

jaloux, -ouse, m.&f. jealous person or creature.

jaloux, -ouse, jealous, envious.

jamais, ever, (with ne expressed or understood) never; à —, forever; à grand —, no, never; à tout —, for ever and ever; pour —, forever

jambage, m. jamb, foundation wall, downstroke (of written letter).

jambe, f. leg, shank, tibia, gaskin.

jambière, f. legging, gaiter, sleeve.

jambon, m. ham; — des jardiniers, evening primrose.

jante, f. rim (of wheel), face, felly, curb (horse).

janvier, m. January.

Japon, m. Japan.

japon, m. sapanwood, Japanese porcelain.

japonais, Japanese.

japonique, japonic (acid).

japonner, to fire a second time so as to give appearance of Japanese porcelain.

jarde, f. curb, bog spavin.

jardin, m. garden; — alpin, rock garden; — potager, vegetable garden.

jardinage, m. gardening, garden produce, garden plot, horticulture, flaw (in gem), single-tree method.

jardinatoire, f. selection felling.

jardiner, to plant after thinning, garden.

jardineux, -euse, (of gems) cloudy, having dark spots.

jardinier, m. gardener; — de la forêt, pine beetle.

jardinier, garden.

jardinière, f. flower stand, pruning saw, ground beetle.

jargon, m. jargon, a variety of zircon from Ceylon.

jarosse, jarousse, f. vetch.

jarre, f. jar.

jarret, m. bend of the knee, elbow, hock, ham, hough, hamstring.

jars, m. gander.

jasmin, m. jasmine, jessamine (Jasminum); — de Virginie, trumpet creeper (Campis radicans); — en arbre, syringa, mock orange (Philadelphus coronarius); — sauvage, false jasmine, yellow jasmine (Gelsemium sempervirens).

jasminoïde, resembling jasmine.

jaspage, m. marbling (paper).

jaspe, m. jasper; — noir, siliceous schist, flinty slate; — sanguin, bloodstone.

jaspé, jasperlike in blending of colors.

jasper, to marble, mottle, jasperize.

jaspure, f. marbling, mottling.

jatte, f. bowl, pan, basin.

jattée, f. bowlful.

jauge, f. gauge, standard, calibration mark; — de vapeur, steam gauge; — du vide, vacuum gauge.

jaugeage, m. gauging, calibration, graduation, tonnage (of ship); — par écoulement, calibration (graduation) for delivery; — par emplissage, calibration (graduation) for contents.

jauger, to gauge, measure, calibrate.

jaugeur, m. gauger, calibrator.

jaugeur, gauging, measuring, calibrating.

jaunâtre, yellowish.

jaune, yellow; — citron, lemon yellow; — de cadmium, cadmium yellow; — de chrome, chrome yellow; — de zinc, zinc yellow; — d'ocre, yellow ocher; — d'oeuf, yellow of egg, egg yolk; — foncé, dark yellow; — minéral, mineral yellow; — pâle, pale yellow; — solide, fast yellow; — terne, tawny.

jaune-citron, m. lemon yellow.

jaune-paille, m. straw yellow.

jaunet, m. bachelor's-button, buttercup (Ranunculus acris); — d'eau, yellow water lily (Nymphozanthus advena).

jaunet, -ette, yellowish.

jaunir, to yellow, color or turn yellow.

jaunissage, m. yellowing, ripening.

jaunissant, yellowing, ripening (of grain).

jaunisse, f. jaundice, icterus, yellows.

jaunissement, m. yellowing, turning yellow, ripening (of grain).

javanais, Javanese.

javelage, m. drying (of corn in the fields).

javel, m., javelle, f. eau de —, Javelle water, eau de Javelle, solution containing potassium hypochlorite.

javeline, f. javelin, spear.

javelle, f. small heap (of sea salt), handful, swath, loose sheaf, bundle, windrow, fagot.

javeller, to windrow.

javellisation, f. sterilizing, javellization.

jayet, m. jet; see JAIS.

je, I.

jécoral, -aux, pertaining to liver, hepatic.

jérose, *f.* rose of Jericho (*Anastatica hierochuntica*).

jervine, *f.* jervine.

jésus, *m.* a size of paper, superroyal, long royal.

jet, *m.* jet, throw, cast, stream, gush, spouting, spurt, flash, sprout, shoot, casting, pouring, shot, outline, sketch; — **à la mer,** jettison, throwing overboard; — **à moule,** runner, channel, jet; — **d'eau,** jet of water; — **de flamme,** jet of flame, blowpipe flame; — **de fossé,** earth thrown up; — **de lumière,** flash of light; — **de pierre,** stone's throw; — **de sable,** sand blast; — **de vapeur,** steam jet.

jetage, *m.* throwing, running of mucus from nostrils.

jeté, cast, thrown.

jeter, to throw, cast, hurl, toss, shed, sprinkle, strew, put, place, emit, shoot forth, discharge, sprout, take root, run, throw down, utter, lay, calculate; **se —,** to be thrown, throw oneself, cast oneself, fall, fling, empty; **il se jette dans,** he goes in for.

jeu, *m.* play, clearance, looseness, backlash, action, working, set, game, gambling, throw, hand, stake, trick, sparkle (of gems); — **de mots,** pun; — **de puids,** set of weights; — **d'orgues,** organ stop, stop key; — **du joint,** gap at joint; — **inutile,** backlash, lost motion; — **perdu,** lost motion; **en —,** in gear; **mettre en —,** to set in action, concern, stake.

jeudi, *m.* Thursday.

jeun (à), fasting, sober, without food.

jeune, *m.* young person, young (animal).

jeune, young, youthful, green, younger, junior; — **brin,** young seedling; — **futaie,** young high forest.

jeûne, *m.* fast, fasting.

jeunement, recently, youthfully.

jeûner, to fast.

jeunesse, *f.* youth.

jeûneur, *m.* faster.

joaillerie, *f.* jeweler's trade, jewelery.

joaillier, *m.* jeweler.

joie, *f.* joy, mirth, glee.

joignant, joining, adjoining, adjacent.

joindre, to join, unite, adjoin, meet, assemble, fit, add: **se —,** to join, be joined, couple. unite, meet; — **à chaud,** to weld.

joint, *m.* joint, seam; — **à boulet,** ball-and-socket joint; — **en about,** butt joint; — **que,** — **à ce que,** beside which, added to which.

joint, joined, united, jointed.

jointage, *m.* jointing, jointing the boards.

jointée, *f.* double handful.

jointer, to joint; — **les douves,** to joint staves.

jointif, -ive, joined, touching, contiguous.

jointoyer, to point (wall).

jointure, *f.* joint, articulation.

joli, pretty, nice, handsome, fine, good.

joliment, well, finely, very, much, prettily, rather nicely.

jonc, *m.* cane, reed, rush; — **à balais,** reed; — **des chaisiers,** bulrush (*Scirpus lacustris*); — **d'Inde,** rattan (Calamus); — **fleuri** flowering rush (*Butomus umbellatus*); — **odorant,** lemon grass, sweet calamus (*Andropogon schoenanthus*), sweet rush (*Cymbopogon schoenanthus*); **canne de —,** rattan (walking stick).

joncacées, *f.pl.* Juncaceae.

jonchaie, *f.* reed field.

jonché, strewed (soft white).

jonchée, *f.* (soft white) cheese (made in rush basket), strewed material, strewing (branches and flowers).

joncher, to strew, cover, scatter, litter.

jonction, *f.* junction, joining, joint, connection.

jongler, to juggle.

jonquille, *f.* jonquil, pale yellow.

joubarbe, *f.* houseleek (*Sempervivum tectorum*), stonecrop (Sedum); — **âcre,** common stonecrop (*Sedum acre*).

joue, *f.* cheek, side, jaw, flange.

jouer, to play, come into play, have clearance, fit loosely, work loose, work, act, be discharged, operate, start, spring, warp; — **de,** to use, work, imitate, baffle, deceive, feign, stake, make game of; **se —,** to amuse oneself; **il joue à se faire tuer,** he runs the risk of being killed.

jouet, *m.* clamp, fishplate, toy, plaything; — **s en bois,** wooden toys.

joueur, *m.* player.

joug, *m.* yoke, crossbeam, crosshead, subjection.

jouir, to enjoy, be in possession of, possess-

jouissance, *f.* enjoyment, right to interest, possession, use.

joule, *m.* joule.

jour, *m.* day, daylight, light, aperture; space, clearance, way, daytime, life, **à —,** open, openly, openwork; **au petit —,** at dawn; **en plein —,** in open daylight; **grand —,** broad daylight; **mettre au —,** to bring to light, publish; **huit —s,** a week; **quinze —s,** fortnight; **tous les —s,** every day; **un beau —,** some fine day.

journal, *m.* (news)paper, journal, diary, daybook.

journalier, *m.* day laborer.

journalier, -ère, daily, diurnal.

journalisme, *m.* journalism.

journaux, *m.pl.* see JOURNAL.

journée, *f.* day, daytime, day's journey, day's work, day's wage(s); **toute la —,** all day long.

journellement, daily, every day, day by day.

joûtir, to ripen (fruit under glass).

joyau, *m.* jewel.

joyeusement, cheerfully, joyously.

joyeux, -euse, joyful, merry, **joyous,** cheerful.

juchoir, *m.* roost, roosting place.

judiciaire, *f.* judgment.

judiciaire, judicial.

judiciairement, judicially.

judicieusement, judiciously, discreetly.

judicieux, -euse, judicious, wise, prudent.

jugal, *m.* cheek bone.

juge, *m.* judge, justice.

jugement, *m.* judgment, opinion, verdict.

juger, to judge, believe, deem.

juglandine, *f.* juglandin.

juglon, *m.* juglone.

jugulaire, jugular.

juif, *m.* Jew, Hebrew.

juif, -ive, Jewish.

juillet, *m.* July.

juin, *m.* June.

jujube, *m.* jujube (the paste).

jujube, *f.* jujube (the fruit).

jujubier, *m.* jujube tree (Zizyphus).

julacé, amentaceous, amentiform, catkinlike.

jule, *m.* catkin.

julep, *m.* julep.

jumeau, *m.* gemellus muscle.

jumeau, -elle, twin, double.

jumelé, double, twin.

jumelles, *f.pl.* binoculars, cheeks, sidepieces, gemels; — **de campagne,** field glasses.

jument, *f.* mare.

jumenteux, -euse, thick, turbid (of urine).

jungle, *f.* jungle.

Jura, *m.* Jura (range).

jurassique, Jurassic (system); — **supérieur,** Upper Jurassic, Upper Oölite.

juré, *m.* juryman, juror.

juré, sworn.

jurer, to swear, vow, assert, curse, clash, jar, squeak.

juridiction, *f.* jurisdiction.

juridique, juridical, judicial.

jurisprudence, *f.* jurisprudence, law.

jus, *m.* juice, sap, gravy; — **de raisin,** grape juice; — **de tannée,** tan liquor, ooze.

jusant, *m.* ebb, ebb tide.

jusée, *f.* tan(ning) liquor.

jusque, to, till, until, up to, as far as, even; **jusqu'à ce que,** till, until; **jusqu'à la position,** even the position; **jusque-là,** so far, till then, hitherto, at that point; **jusqu'en,** until; **jusqu'ici,** thus far, up to the present, until now; **jusqu'où,** how far.

jus uiame, *f.* henbane (Hyoscyamus); — **noire,** black henbane (*Hyoscyamus niger*).

juste, just, fair, exact, correct, right, tight, close-fitting, well, tightly, exactly, precisely; **au —,** just, exactly, precisely.

justement, just, exactly, justly, precisely, rightly.

justesse, *f.* accuracy, exactness, correctness, justness, precision.

justice, *f.* justice, righteousness, right.

justifiable, justifiable.

justifiant, justifying.

justifier, to justify, vindicate, prove.

jute, *m.* jute.

juter, to exude juice, be juicy.

juteux, -euse, juicy.

juxtaposé, juxtaposed, in juxtaposition.

juxtaposer, to juxtapose, place side by side.

juxtaposition, *f.* juxtaposition.

K

kaieput, *m.* see CAJEPUT.

kaïnite, *f.* kainite.

kaki, *m.* Japanese persimmon, Chinese date plum (*Diospyros kaki*).

kaki, *m.* khaki.

kakodyle, *m.* see CACODYLE.

kakodylique, see CACODYLIQUE.

kaléidoscopique, kaleidoscopic.

kali, *m.* kali, glasswort, prickly saltwort (*Salsola kali*), potash.

kalidie, *f.* cystocarp.

kalium, *m.* potassium.

kalmie, *f.* kalmia; — à larges feuilles, mountain laurel (*Kalmia latifolia*).

kamala, *f.* kamala, rottlera.

kanten, *m.* kanten, agar-agar.

kaolin, *m.* kaolin, porcelain clay, China clay; — lithomarge, porcelain earth or clay; lait de —, slip.

kaoliniser, to kaolinize.

kaoutchouc, *m.* see CAOUTCHOUC.

kapok, *m.* kapok.

karabé, *m.* amber.

karabique, succinic.

karat, *m.* carat.

karatas, *m.* karatas, silk grass.

karyocinèse, *f.* nuclear division, mitosis; see CARYOCINÈSE.

karyocinétique, karyokinetic.

karyogamète, *f.* karyogamete, nucleus of a gamete.

karyogamie, *f.* karyogamy, the union of gametonuclei.

karyoïdes, *m.pl.* karyoids, minute spherical bodies attached to chlorophyll plate.

karyokinèse, *f.* see KARYOCINÈSE.

karyoplasma, *m.* nucleoplasm, nuclear protoplasm.

karyosome, *m.* karyosome.

kassu, *m.* kassu, betel-nut extract from *Areca catechu*.

katablast, *m.* katablast, a shoot from an underground stock.

kataplasie, *f.* functional degeneration.

katatonique, catatonic, tending to decrease a stimulus.

kava, *f.* kava (shrub or drink) (*Piper methysticum*).

kéfir, *f.* kefir.

kéramide, *m.* ceramidium, a carpostome.

kéraphyllocele, *f.* keratoma.

kératenchyme, *m.* ceratenchyma, sieve tubes becoming horny.

kératine, *f.* keratin.

kératiniser, to keratinize, become horny.

kératite, *f.* keratosis, inflammation of the cornea.

kératogène, keratogenous.

kératoïde, keratoid, horny.

kératoplastique, hardening the skin, keratoplastic.

kératose, *f.* keratosis.

kermès, kermes; — minéral, kermes mineral, red antimony; chêne —, kermes oak (*Quercus coccifera*).

kérosène, *m.* kerosene.

kérosolène, *m.* rhigolene.

ketmie, *f.* hibiscus, rose mallow (Hibiscus); — comestible, gumbo.

khôl, *m.* kohl.

kieselguhr, kieselgur, *m.* kieselguhr, diatomaceous (infusorial) earth, diatomite.

kilo, *m.* kilo (kilogram).

kilogramme, *m.* kilogram (1,000 grams).

kilogrammètre, *m.* kilogram-meter.

kilomètre, *m.* kilometer.

kilométrique, kilometric(al).

kilowatt, *m.* kilowatt.

kilowatt-heure, *m.* kilowatt-hour.

kina, *m.* see QUINQUINA.

kinase, *f.* kinase.

kinate, *m.* see QUINATE.

kinesis, *m.* kinesis.

kinine, *f.* see QUININE.

kinique, see QUINIQUE.

kino, *m.* kino.

kinoplasma, *m.* kinoplasm, active elements of protoplasm.

kinovine, *f.* see QUINOVINE.

kiosque, *m.* shed, small building.

kirsch, kirschwasser, *m.* kirsch, kirschwasser.

kiwi, *m.* kiwi, apteryx.

km., *abbr.* (kilomètre), kilometer.

knoppern, *m.* gall, gallnut, from *Cynips quercus calicis*.

koa, *m.* koa (*Acacia koa*).

kohol, *m.* see KOHL.

kola, *m.* kola, Cola.

kolatier, *m.* cola tree (*Cola acuminata*).

koumis, *m.* kumiss.

kousso, *m.* cusso, brayera; see COUSSO.

krypton, *m.* krypton.

kummel, *m.* kümmel (the liqueur).

kupfernickel, *m.* kupfernickel, niccolite.

kw.-an, *abbr.* (kilowatt-an), kilowatt year.

kwas, *m.* kvass (Russian beer).

kyaniser, to cyanize; see CYANISER.

kyste, *m.* cyst.

kystique, kysteux, -euse, cystic.

L

l', contraction of le or la, the.

la, fem. sing. the; *pron.*, her, she, it; **de —, de l', des,** some, any; — **voilà,** there she is.

là, there, then; **cà et là,** here and there; **ce livre-là,** that book; **ces livres-là,** those books; **cette plume-là,** that pen; **de là,** thence, from there, whence, from then; **dés là,** at once, from that time on; **là où,** where; **par là,** that way.

lab, *m.* rennet, lab, labenzyme, lab ferment.

là-bas, down there, over yonder, below.

labdanum, *m.* labdanum.

label, *m.* label, mark.

labelle, *m.* labellum.

labenzyme, *m.* the enzyme of rennet, rennin.

labeur, *m.* labor, toil, cultivation, work; **cheval de —,** plow horse.

lab ferment, *m.* the enzyme of rennet.

labial, -aux, labial.

labiatiflore, labiatiflorous, corolla divided into two liplike portions.

labié, labiate, lipped.

labiées, *f.pl.* labiates (Labiatae).

labile, labile, easily disturbed.

labitome, *f.* cutting forceps.

labium, *m.* labellum.

laboratoire, *m.* laboratory, hearth, chamber of furnace; — **de recherches chimiques,** chemical (research) laboratory.

laborieusement, laboriously, painfully.

laborieux, -euse, laborious, industrious, painful, diligent.

laboriosité, *f.* laboriousness.

labour, *m.* tillage, plowing, tilled (plowed) land, dressing, field labor; — **à bras,** hand tillage; **terres de —,** plowed land.

labourable, arable, tillable.

labourage, *m.* tillage, plowing; — **à plat,** paring.

labourer, to till, plow, turn, graze, drudge.

laboureur, *m.* tiller, plowman.

laboureuse à vapeur, *f.* steam plow.

labradorite, *f.*, labrador, *m.* labradorite.

labre supérieure, *f.* labrum.

labyrinthe, *m.* labyrinth, maze.

labyrinthique, labyrinthine, labyrinthian.

lac, *m.* lake; — **salant,** — **salé,** salt lake.

laçage, *m.* lacing.

laccase, *f.* laccase.

laccifère, lac-bearing, lac-producing.

laccolithe, *m.* laccolith.

lac-dye, *m.* lac dye.

lacer, to lace (up), tie up.

lacération, *f.* laceration.

lacéré, lacerated, torn, laciniate.

lacérer, to lacerate, tear (up).

laceron, *m.* sow thistle (Sonchus).

lacertiforme, lacertilian, lacertian, lizardlike.

lacet, *m.* lace, loop, zigzag, hairpin bend, winding, noose, bird snare, springe, running or sliding knot; — **de la locomotive,** irregular oscillating motion of the engine; — **de routes,** road network; — **de soulier,** shoelace; — **d'une fiche simple,** bolt of a hinge; — **en cuir,** leather lace.

laceur, *m.* net knitter or maker.

lâche, *m.* coward.

lâche, loose, slack, lax, mean, base, cowardly.

lâchefer, *m.* tap bar, frit iron.

lâchement, *m.* loosening, freeing (of animal).

lâchement, loosely, slackly, feebly, sluggishly.

lâcher, to release, loosen, loose, relax, slack off, slacken, empty, open, drop, let go, fire off, discharge, let off, let out, let slip; **se —,** to loosen, become slack, slip, get loose, let out, go off (of gun).

lacinie, *f.* lacinia, slash (in leaf).

lacinié, laciniate(d), slashed (petals).

lacinifolié, laciniate-leaved.

laciniure, *f.* jag, dent, cut, tear, rent.

lacinule, *f.* lacinula, a small lacinia.

lacinulé, lacinulate, finely laciniate.

lacis, *m.* network, plexus; — **de racines,** matted roots.

lac-laque, *f.* lac lake.

lacmoide, *m.* lacmoid.

laconique, laconic.

là-contre, to the contrary, against that.

lacryma-christi, *m.* Lachryma Christi (an Italian wine).

lacrymal, *m.* lachrymal bone.

lacrymal, -aux, lachrymal.

lacrymogène, *m.* lachrymator, tear gas.

lacrymogène, tear (-exciting) -producing.

lacrymule, *f.* small tear, small drop.

lactacidase, *m.* diastase existing in alcoholic zymase.

lactaire, lactary, lacteal.

lactalbumine, *f.* lactalbumin, milk albumin.

lactame, *n.* lactam.

lactamide, *f.* lactamide.

lactamique, of or pertaining to a lactam, lactamic.

lactase, *f.* lactase.

lactate, *m.* lactate; — **d'ammoniaque,** ammonium lactate; — **d'antimoine,** antimony lactate; — **de fer,** lactate of iron, ferrous lactate.

lactation, *f.* lactation.

lacté, milky, lacteous, milk, lacteal, lactescent; **fiévre —e,** milk fever; **La Voie —e,** the Milky Way.

lactéiforme, resembling milk.

lactéine, lactéoline, *f.* lactein, lactolin, condensed milk.

lactescence, *f.* lactescence, milkiness.

lactescent, lactescent, milky, resembling milk.

lacticémie, *f.* lacticemia, presence of lactic acid in the blood.

lactide, *f.* lactide.

lactifère, lactiferous, milk-bearing (gland, plant).

lactiflore, lactiflorous, having milk-white flowers.

lactine, *f.* see LACTOSE.

lactique, lactic.

lactobacilline, *f.* culture of lactic-acid bacteria.

lacto-densimètre, *m.* lactodensimeter, milk gauge.

lactoline, *f.* see LACTÉINE.

lactomètre, *m.* lactometer

lactone, *f.* lactone.

lactophénol, *m.* medium to observe algae.

lactoscope, *m.* lactoscope.

lactose, *f.* lactose, milk sugar, lactine.

lactosé, containing lactose.

lactosérum, *m.* milk serum.

lactucarium, *m.* lactucarium inspissated juice of *Lactuca virosa.*

lactucé, resembling lettuce.

lacune, *f.* lacuna, gap, blank, cavity, depression; — **aérifère,** air space.

lacuneux, -euse, lacunose, full of lacunas or gaps.

lacunosité, *f.* porosity.

lacustre, *m.* lake-dweller, lacustrian.

lacustre, lacustrine, lacustral, lacustrian.

ladanifère, bearing labdanum.

ladanum, *m.* see LABDANUM.

là-dedans, in there, inside, within.

là-dehors, out there, outside, without.

là-dessous, under there, under that, underneath.

là-dessus, on that, about that, thereupon.

là-devant, on ahead, yonder.

ladite, fem. of LEDIT.

ladre, measly, measled, leprous.

ladrerie, *f.* leprosy, measles (of pigs), measliness, bladder-worm disease.

lageniforme, lageniform, flask-shaped.

lagopède d'Écosse, *m.* red grouse.

lagopode, lagopodous, possessing hairy or feathery feet.

lagre, *m.* a sheet of glass.

lagune, *f.* lagoon.

là-haut, up there, above.

laiche, *f.* sedge, sedge grass, reed (Carex); — **des sables,** sand sedge, sand carex (*Carex arenaria*).

laiciser, to turn into state property.

laid, ugly, bad, unbecoming, ill-looking, uncomely.

laideur, *f.* ugliness.

laie, *f.* sow (wild), ride, service path, slate band, slate; — **principale,** main ride, major ride; — **sommière,** major ride.

lainage, *m.* fl ece, teaseling, raising woolen goods; — **du drap,** cloth raising or teaseling.

laine, *f.* wool, woolen cloth; — **à broder,** (colored) embroidering wool; — **à tricoter** knitting wool — **beige,** natural (colored) wool, not dyed wool; — **d agneau,** lamb's wool; — **de bois,** wood wool, wood fiber; — **de Cachemire** cashmere wool; — **de court soie,** short-stapled wool; — **de laitier,** slag wool, mineral (wool) cotton; — **de lama,** llama yarn; — **de longue soie,** long-stapled wool.

laine, — **de mouton,** sheep wool; — **de nettoyage,** cotton waste for cleaning; — **de peigne,** worsted yarn, worsted; — **de pin(s),** vegetable wool, forest wool; — **de scorie(s),** slag wool, mineral wool; — **de verre,** glass wool; — **en cordes,** wood wool in cords; — **en suint,** greasy wool; — **en (de) toison,** fleece wool; — **mère,** finest wool, back or sp ne wool; *pl.* prime locks; — **précipitée,** pushed wool.

lainer, to teasel, tease, raise, nap, cord.

laine-renaissance, *f.* wool recovery.

lainerie, *f.* manufacture of woolens, woolen goods, woolen shop, teaseling machine.

laineur, *m.* teaseler.

laineux, -euse, woolly, lanuginous, lanate, tomentose, velvet.

lainier, -ère, wool, pertaining to wool.

laïque, *m.* lay, layman.

laïque, lay, laical.

lais, *m.* standard tree, accretion, standard, alluvium, silt.

laisser, to leave, let, allow, omit, let go, forsake; — **de côté,** to leave or lay aside, allow oneself; — **là,** to let alone, abandon; **il le laisse faire,** he leis him alone; **il se laisse faire,** he offers no resistance; **ne — pas de,** to cease to, stop, leave off; **ne — pas de faire quelque chose,** to do something nevertheless, not to fail to do something.

laisser-aller, *m.* unconstraint.

laisser-faire, *m.* noninterference.

laisser-passer, *m.* pass, permit.

lait, *m.* milk, milk diet; — **artificiel,** artificial milk, remade milk; — **battu,** buttermilk, fumitory (Fumaria); — **caillé,** curdled milk, coagulated milk; — **clair,** whey; — **complet,** unskimmed or rich milk; — **concentré,** condensed milk, evaporated milk; — **condensé,** condensed milk· — **d'amande(s),** milk of almond(s); —

d'amiante, asbestos milk, asbestos suspension; — **de beurre,** buttermilk; — **de chaux,** milk of lime, limewash, whitewash; — **de couleuvre,** cypress spurge; — **de femme,** human milk, mother's milk.

lait, — **de jument,** mare's milk; — **de malt,** malt milk; — **desséché,** dried milk; — **écrémé,** skimmed milk; — **en conserve,** milk preserve; — **en poudre,** milk powder, powdered milk; — **entier,** whole milk; — **grumeleux,** — **mammitique,** garget milk; — **pasteurisé,** pasteurized milk; — **Sainte-Marie,** milk thistle (*Carduus marianus*); — **tourné,** curdy milk; — **végétal,** latex; — **visqueux,** ropy milk.

laitage, *m.* milk diet, milk food.

laitance, *f.* (cement) laitance, milt, grout; — **de poisson,** fish roe.

laité, soft roed , milt.

laiterie, *f.* dairy, dairying, dairy room, dairy farm; — **à vapeur,** steam dairy.

laiterol, *m.* floss-hole plate, tap hole.

laiteron, *m.* sow thistle, hare's lettuce, hogweed (Sonchus).

laiteux, -euse, milky, milk, lacteal, lacteous.

laitier, *m.* milkman, dealer, dairyman, milkwort (Polygala) slag, scoria, cinder, dross; — **commun,** milkwort; — **granulé,** granulated blast-furnace slag.

laitier, -ère, milk.

laitière, *f.* milkmaid, milk cart.

laiton, *m.* brass; — **bruni,** old brass; — **de fonte,** cast brass; — **en feuilles,** sheet brass; — **massif,** solid brass.

laitonnage, *m.* brassing.

laitonner, to brass plate, (cover with) brass.

laitue, *f.* lettuce; — **à couper,** loose-leaved lettuce, cutting lettuce; — **frisée,** curled lettuce; — **pommée,** cabbage lettuce, heading lettuce; — **romaine,** romaine, Cos lettuce.

laize, *f.* width, breadth (of cloth).

lama, *m.* llama, lama; — **alpaga,** alpaca; — **vigogne,** vicuña.

lamanage, *m.* pilotage.

lamarckisme, *m.* Lamarckism.

lambeau, *m.* shred, scrap, rag, bit, fragment, ribbon.

lambourde, *f.* dwarf shoot. spur shoot, groin of a groined ceiling.

lambris, *m.* paneling, wainscoting, inlaying, canopy.

lambruche, *f.* fox grape.

lame, *f.* foil, thin plate, sheet, jet, pane, film, blade (of knife), lamina, scale, strip, slice, tinsel, slide, wave, band, billow, diaphragm, surge, core bar, shell bar; — **criblée,** sieve plate; — **d'eau,** sheet of water, water space; — **de fond,** ground swell; — **de platine,** platinum foil; — **de sang étalé,** blood-film preparation; — **de scie,** saw blade; — **de verre,** microscope slide, sheet of glass; — **ordinaire,** microscope slide; — **porte-objet,** slide.

lamé, laminated, laminate.

lamellaire, lamellar.

lamellation, *f.* lamellar arrangement, condition of being lamellated.

lamelle, *f.* lamella, lamina, leaflet, scale, small disk, glass cover; — **couvre-objet,** cover glass; — **moyenne,** middle lamella.

lamellé, lamellate(d), foliated, fissile, flaky, scaly.

lamelleux, -euse, lamellar, lamellate, laminate, scaly.

lamellibranche, lamellibranchiate.

lamellifère, lamelliferous.

lamelliforme, lamelliform.

lamentable, woeful, grievous.

lamentablement, lamentably.

lamenter, to lament, bewail, mourn; **il se lamente,** he complains.

lamette, *f.* small plate, strip, lamella, small leaf, small blade, clasp.

lamier, *m.* dead nettle (Lamium); — **taché,** spotted dead nettle.

laminage, *m.* rolling, flattening, lamination.

laminaire, *f.* laminaria.

laminaire, laminated, laminar, flaky, scaly.

laminale, laminal.

laminarine, *f.* laminarin, a polysaccharide.

lamine, *f.* lamina, thin plate.

laminé, laminate(d), rolled, calendered, flattened.

laminer, to roll (metal), flatten, laminate, calender, plate glaze.

laminerie, *f.* rolling mill.

lamineur, *m.* roller, mill hand, calenderer.

lamineur, rolling.

lamineux, -euse, laminose, laminate. **tissu —,** cellular tissue.

laminiforme, laminiform.

laminoir, *m.* flatting mill, roller, roll. **rolling mill.**

lampant, *m.* kerosene.

lampant, (oil) suitable for burning in a lamp, purified, illuminating, refined, clear.

lampas, *m.* lampas.

lampe, *f.* lamp, audion, tube, oil lamp; — **à deux électrodes,** two-element vacuum tube, diode; — **à essence,** benzine (gasoline) lamp; — **à gaz,** gas (lamp) burner; — **à huile,** oil lamp, paraffin lamp; — **ampoule,** electric-light bulb; — **à pétrole,** kerosene lamp, coal-oil lamp; — **à vide,** vacuum tube; — **de poche,** pocket lamp, flashlight; — **de sûreté,** safety lamp; — **émettrice,** power tube; — **témoin,** indicator lamp; — **triode,** three-element vacuum tube, triode; **à la —,** by lamplight.

lampe-heure, *f.* lamp-hour.

lamper, to become luminous, phosphorescent.

lamperon, *m.* wick holder, bowl (of lamp).

lampette, *f.* corn cockle (Agrostemma githago), ragged robin (Lychnis floscuculi).

lampisterie, *f.* lamp manufacture.

lampourde, *f.* burweed (Xanthium)

lamproie, *f.* lamprey.

lamprophyllé, with glossy and luminous leaves.

lampsane, *f.* nipplewort (Lapsana communis).

lampyre, *m.* lampyris, firefly, glowworm.

lançage, *m.* launching.

lance, *f.* spear, lance, nozzle, pole, staff; **en fer de —,** lanceolate; **en forme de fer de —,** lanceolate, lance-shaped.

lance-bombes, *m.* trench mortar.

lancée, *f.* attack of pain; *pl.* shooting pains.

lance-flammes, *m.* flame projector.

lancement, *m.* launching, throwing, flinging, emission, starting, cranking.

lancéolaire, lanceolate.

lancéolé, lanceolate, spear-shaped.

lancer, to launch, throw, dart, emit, cast, issue, send out, start, crank, hurl, fling, release (pigeons), slide; **se —,** to spring, rush, start, dash forward.

lancette, *f.* lancet.

lande, *f.* heath, moor; *pl.* sandy tracts.

langage, *m.* language, speech, tongue.

langes, *m.pl.* swaddling clothes, fetal envelopes, felts.

langoureux, -euse, languishing.

langouste, *f.* spiny lobster.

langue, *f.* tongue, language, pointer; — **de bois,** wooden tongue; — **de cerf,**

hart's-tongue (*Phyllitis scolopendrium*); — de chien, hound's-tongue (*Cynoglossum officinale*); — de serpent, adder's-tongue (Ophioglossum).

langue-de-boeuf, *f.* wood rotting of the oak.

languette, *f.* small tongue (of wood, metal), pointer (of balance), small strip (of tin foil), ligula.

langueur, *f.* languor, languidness.

languir, to languish, pine, long, drag, linger, droop.

languissamment, languidly, languishingly, droopingly.

languissant, languishing, languid, lingering.

laniaire, laniary (teeth).

lanière, *f.* strap, thong, (narrow) strip, lash.

lanifère, laniferous, wool-bearing.

laniflore, laniflorous.

lanigère, lanigerous, wool-bearing.

lanoline, *f.* lanolin.

lantanier, *m.* lantana.

lanterne, *f.* lantern, window, glass case, core barrel; — vitrée, glass gauge.

lanthane, *m.* lanthanum.

lanugineux, -euse, lanate, woolly.

lanure, *f.* lunure, cup shake.

lapaxic, *m.* evacuation.

laper, to lap (water, milk).

lapereau, *m.* young rabbit.

lapidaire, *m.* lapidary.

lapidaire, lapidary.

lapin, *m.* rabbit, coney; — de garenne, wild rabbit; — de hanneton, cockchafer grub; — du Brésil, guinea pig.

lapine, *f.* doe-rabbit, doe.

lapiner, to litter.

lapinière, *f.* rabbit hutch.

lapis, lapis-lazuli, *m.* lapis lazuli.

lapon, *m.* Lappic, Laplander.

lapon, Lappic.

Laponie, *f.* Lapland.

laps, *m.* lapse.

lapsus, *m.* slip of the tongue.

laque, *f.* lake, lac, gum-lac; *m.* lacquer; — à modèle, pattern varnish; — carminée, carmine lac; — colorée, drop lac, varnish color; — de bois rouge, redwood lake; — de boule, fine or round lac; — de garance, madder lake; — de nitro-cellulose, pyroxyline lacquer; — de première couche, filler; — en bâton, stick-lac; — en écailles, — en feuilles, shellac; — en grains, seed-

lac, grained lac; — en masses, lump lacquer; — en plaques, shellac; gomme —, gum lake.

laque, lac.

laqué, lacquered, japanned, laked.

laquebleu, *m.* litmus.

laquelle, who, whom, which, that, which one.

laquer, to lacquer, lake, varnish.

laqueur, *m.* japanner, varnisher.

laqueux, -euse, of or pertaining to lac, lake (tint).

lard, *m.* bacon, lard, fat; — de baleine, blubber; gros —, — gras, fat bacon; petit —, — maigre, lean streaky bacon; pierre de —, soapstone, steatite.

lardacé, lardaceous, amyloid.

larder, to interlard, lard, sprinkle, stab, pierce, stick, jab.

lardeux, -euse, lardaceous, lardy, fat.

lardite, *f.* soapstone, steatite.

larêche, *f.* larch.

large, breadth, width, open (high) sea; au —, off, away, well off, comfortably, spaciously; au long et au —, far and wide; de —, in width, wide; en —, broadways, broadwise; en long et en —, up and down, to and fro.

large, broad, wide, large, great, ample, lax loose, liberal, generous, largely.

largement, wide(ly), largely, broadly, amply, abundantly, fully.

larget, *m.* sheet billet, flat bar.

largeur, *f.* breadth, width, wideness, span, gauge, beam, broadness, amplitude; — au fort, extreme breadth; — dans oeuvre, width in the clear; — de l'installation, over-all width; — de voie, gauge of way, width between the rails; — du bassin, width of pelvis; — du jour, width of opening; — en couronne, breadth at the top; — en fond, width at the bottom; — extrême, breadth over all; — utile, working width.

largine, *f.* largin, silver albuminate.

largue, loose, slack, started, free, large.

larguer, to let go, slacken, loose (rope), release, let off (steam), cast loose or adrift; — l'écoute du vent, to start the weather sheet; — un câble, to loosen or yield a cable; — une voile, to unfurl a sail.

larix, *m.* larix, larch (*Larix decidua*).

larmaire, lacrimiform.

larme, *f.* tear, drop.

larmeux, -euse, in tears, in drops.

larmier, *m.* larmier, dripstone, corona, inner canthus; — **bombé et réglé,** vaulted head; — **d'un mur de clôture,** coping of enclosure wall.

larmoiement, *m.* lachrymation, watering (weeping) of the eyes, epiphora.

larmoyer, to water, weep, shed tears.

larvaire, larval.

larve, *f.* larva, grub, caterpillar; — **apode,** maggot; — **d'élatéride,** wireworm; — **du hanneton,** larva of cockchafer.

larvicide, larvicide.

larvicole, larvicolous (parasite).

larviforme, larva-shaped.

larvipare, larviparous.

laryngé, laryngeal.

larynx, *m.* larynx.

las, *m.* mow (of barn).

las, weary, tired.

lascif, -ive, lascivious.

laser, *m.* laser, laserwort (Laserpitium).

lasianthe, lasianthous, woolly-flowered.

lasiocampe, — **du pin,** *f.* pine moth, egger.

lasiocarpe, lasiocarpous, pubescent-fruited.

lasiocéphale, with flowers in woolly capitulum.

lasioglotte, with woolly legumes.

lasiope, with woolly foot.

lasiosperme, with pubescent fruit.

lasiostachyé, with flowers in pubescent spike.

lasser, to lasso, weary, tire; **se —,** to become fatigued, grow tired, be wearied.

lassitude, *f.* sagging, lassitude, weariness, weakness, tiredness.

lasting, *m.* lasting.

latanier, *m.* macow tree (Latania).

latemment, latently.

latence, *f.* temps de —, latent period of induction.

latent, latent, hidden, concealed.

latéraire, *f.* cross rafter, trimmer of rafters.

latéral, -aux, lateral, on the side; **chaine —e,** side chain; **tube —,** side neck.

latéralement, laterally, sideways.

latériflore, laterifloral.

laterinervé, laterinerved, with lateral veins.

latérite, *m.* laterite.

latex, *m.* latex.

laticifère, laticiferous, latex-bearing.

laticifères, *m.pl.* milk vessels, laticiferous ducts, cells, or vessels.

latiflore, with broad flowers.

latifolié, latifoliate, broad-leaved.

latin, *m* Latin.

latin, Latin, lateen (sail).

latisepté, latiseptate, with broad partitions.

latitude, *f.* latitude.

latrine, *f.* latrine, privy, water closet, lavatory.

latte, *f.* lath, batten; — **à mesurer,** measuring staff; — **à semis,** seed lath.

lattis, *m.* lathwork, latticed partition.

laudanine, *f.* laudanine.

laudanisé, containing laudanum.

laudanosine, *f.* laudanosine.

laudanum, *m.* laudanum; — **de Sydenham,** wine of opium.

laumonite, *f.* laumontite, laumonite.

lauracées, *f.pl.* Lauraceae.

lauréat, *m.* laureate, prize winner.

lauréole, *f.* daphne, laurel; — **femelle,** mezereon (*Daphne mezereum*); — **male,** spurge laurel (*Daphne laureola*).

laurier, *m.* laurel, bay tree, sweetbay (*Laurus nobilis*); — **benzoin,** spice-bush (*Benzoin aestivale*); — **commun,** — **noble,** true laurel, bay laurel, sweet bay (*Laurus nobilis*); — **des marais,** rosebay, great laurel (*Rhododendron maximum*); — **de St. Antoine,** rosebay, blooming sally, fireweed (*Epilobium angustifolium*).

laurier-cerise, *m.* cherry laurel (*Prunus laurocerasus*).

laurier-rose, *m.* oleander, rosebay (*Nerium oleander*).

laurier-sauce, *m.* true laurel (*Laurus nobilis*).

laurier-tin, *m.* laurustine, (*Viburnum tinus*).

laurier-tulipier, *m.* great-flowered magnolia (*Magnolia grandiflora*).

laurine, *f.* laurin.

laurinées, *f.pl.* Lauraceae.

laurique, lauric (acid).

laurose, *m.* oleander.

lauryle, *m.* lauryl.

lavable, washable.

lavabo, *m.* washstand, lavabo, lavatory.

lavage, *m.* washing, spilled liquid, slop, wash, injection, lixiviation, extraction; — **à dos,**

fleece washing, irrigation, lavage; — au crible, riddling, jigger washing; — de filets, washing of nettings; — de laine, wool washing; — fractionné, fractional washing; — par voie humide, wet washing; — par voie sèche, dry washing.

lavande, *f.* lavender (Lavandula); — aspic, aspic, French lavender (*Lavandula spica*); — commune, garden lavender (*Lavandula spica*).

lavandière, *f.* laundress, washerwoman, washing machine, (gray) wagtail.

lavanèse, *f.* goat's-rue.

lavange, *f.* avalanche.

lavaret, *m.* pollan, powan.

lavasse, *f.* siliceous stone, downpour, watery soup or wine, dishwater.

lavatère, *f.* Lavatera; — en arbre, tree mallow (Lavatera).

lave, *f.* lava; — coulante, flowing lava; — lithoïde, stony lava.

lavé, washed, thin, faint.

lavée, *f.* amount (of wool) washed at one time.

lavement, *m.* washing, enema, injection.

laver, to wash, wash out, cleanse, lave, rinse, buddle (ores), potch, poach; — à fond, to wash thoroughly; — le câble, to clean or scrub the cable; — les drêches, to sparge the draff; — les minerais, to wash ores.

laverie, *f.* lavatory, washhouse.

laveur, *m.* washer, washing machine, purifier, gas washing bottle; — à boulets, ball washer; — de chiffons, rag washer; — de drap, cloth washer.

laveur, -euse, washing.

laveuse, *f.* see LAVEUR; — de linge, washerwoman, laundress.

lavique, lavatic, lavic.

lavis, *m.* wash, washing, tinting, coloring, wash design.

lavoir, *m.* washer, washing machine, washhouse, laundry, cleaning rod, buddle.

lavure, *f.* washings, dishwater, swill, slop, bloody supersecretions, dross, sweepings; — de chair, bloody slime.

laxatif, -ive, laxative, aperient.

laxiflore, lax, loose flowering.

laxité, *f.* laxity, slackness, relaxed state.

layer, to blaze (trail).

layette, *f.* layette, small drawer, box, baby linen.

layon, *m.* minor ride, cross ride, service path.

lazulite, *f.* lazulite, blue spar.

le, the, him, he, it.

lé, *m.* breadth, width (of cloth), towing path; — du drap, breadth of the cloth.

lécanorine, *f.* lecanorin (lecanoric acid).

léchage, *m.* licking.

lécher, to lick, touch, finish off, surround.

lécithine, *f.* lecithin.

leçon, *f.* lesson, lecture, reading.

lecteur, *m.* reader.

lecture, *f.* reading, perusal; — brute, rough reading; — de l'échelle, reading of scale; — des indications de l'aiguille, pointer reading.

lecus, *m.* lecus, corm, a bulblike fleshy stem or base of stem.

ledit, the said, the above, the aforesaid.

lédon, *m.*, lède, *f.* marsh andromeda, Labrador tea, wild rosemary (*Ledum palustre*).

légal, -aux, legal, lawful, legitimate.

légalement, legally, lawfully.

légalisé, legalized.

légaliser, to authenticate, legalize, certify.

légalité, *f.* legality, lawfulness.

légendaire, legendary.

légende, *f.* legend, story.

léger, *m.* lightness.

léger, -ère, light, quick, slight, feeble, slender, nimble, frivolous, easy; à la légère, lightly.

légèrement, lightly, slightly, nimbly, imprudently.

légèreté, *f.* lightness, slightness, weakness.

légiférer, to legislate.

légion, *f.* legion, countless number.

législateur, *m.* legislator.

législateur, -trice, legislative.

législatif, -ive, legislative.

législation, *f.* legislation.

légiste, *m.* legist.

légitimation, *f.* legitimation, proof of identity.

légitime, *f.* legal share, rightfulness, legitimacy.

légitime, legitimate, lawful, right.

légitimement, legitimately, lawfully.

légitimer, to justify, warrant.

léguer, to bequeath, leave by will.

légume, *m.* legume, vegetable; — **en conserve,** canned vegetable; — **tuberculifère,** tuber vegetable; — **vert,** leaved vegetable.

légumine, *f.* legumin, vegetable casein.

légumineuse, *f.* legume, pod; *pl.* Leguminosae, leguminous plants.

légumineux, -euse, leguminous.

léguminode, composed of several legumes.

lehm, *m.* loam, loess.

léianthère, with glossy anthers.

léiocome, *m.* leiocome, dextrin.

léiope, with glossy feet.

léiopile, with glabrous and glossy cap.

léioplaque, forming very glossy discs.

léiostachyé, with glossy spikes.

leishmania, *f.* Leishmania.

lemnien, -enne, Lemnian.

lendemain, *m.* next day, morrow, day after; **le — de,** the day after.

lenitif, -ive, lenitive, laxative, emollient, palliative.

lent, slow.

lentement, slowly.

lenteur, *f.* slowness; **avec —,** slowly.

lenticelle, *f.* lenticel.

lenticulaire, lenticulé, lenticular, lensshaped, lentiform.

lenticule, *f.* duckweed (Lemna).

lentifère, with lentils.

lentigère, with lentils.

lentigo, *m.* freckle, lentigo.

lentille, *f.* lens, lentil, bob (of pendulum); — **avant,** field lens; — **collectrice,** condensing lens; — **convergente,** converging lens; — **divergente,** diverging lens; —**s collées,** cemented lenses.

lentisque, *m.* mastic (tree), lentiscus (*Pistacia lentiscus*).

léopard, *m.* leopard.

lépicène, *f.* lepicena, the glume in grasses.

lépidanthé, having scaly appearance.

lépidine, *f.* lepidine.

lépidoïde, cupressoid, with foliage like the cypress.

lépidolite, *m.* lepidolite, lithia mica.

lépidophyton, *m.* *Aspergillus lepidophyton.*

lépidoptères, *m.pl.* Lepidoptera.

lépisme, *m.* lepisma, silverfish.

lèpre, *f.* leprosy.

lépreux, -euse, *m.&f.* leper.

lépreux, -euse, leprous.

leproïde, leprosylike.

leptacanthe, with thin thorns.

leptanthe, with small flowers.

leptocarpe, with long, slender fruit.

leptocaule, with slender stem.

leptocentrique, leptocentric.

leptomiase, *f.* leptomiasis.

leptopétale, with narrow petals.

leptophylle, leptophyllous, slender-leaved.

leptorhize, with slender roots.

leptosépale, with narrow and linear divisions of the calyx.

leptosperme, with very small seeds.

leptosporangié, leptosporangiate.

leptostachyé, with slender spikes.

leptostyle, with filiform style.

lequel, who, whom, which, which one, that.

les, the, them, they.

lesdites, lesdits, *pl.* the above, aforesaid.

léser, to injure.

lésion, *f.* injury, damage, lesion, mutilation.

lesquels, lesquelles, who, whom, which (ones).

lessivage, *m.,* **lessivation,** *f.* leaching, washing, lixiviation; — **forcé,** overhead digestion.

lessive, *f.* lixivium, lye, washing, solution buck(ing), wash; — **bouillante,** boiling not caustic solution; — **caustique,** caustic alkali solution, solution of potassium hydroxide; — **de soude,** solution of sodium hydroxide; — **normal de soude,** normal sodium hydroxide solution, sodium hydroxide; — **residuaire,** residual lye, waste lye; — **soudique,** caustic soda, sodium hydroxide.

lessiver, to lixiviate, leach, extract, wash, soak in lye, scour, buck.

lessiveur, *m.* lixiviator, leacher, rag boiler, digester, kier; — **de chiffons,** rag boiler; — **de paille,** straw boiler.

lessiveur, -euse, lixiviating, leaching.

lessiveuse, *f.* an apparatus for lixiviating, rag boiler, washing machine, washer, washerwoman, laundress.

lest, *m.* ballast.

leste, light, brisk, nimble, active, smart.

lestement, briskly.

lester, to ballast.

léthalité, *f.* lethality.

léthargie, *f.* lethargy.

lette, *f.* water hollow between the dunes.

lettrage, *m.* lettering.

lettre, *f.* letter, type, character; *m.* man of letters, scholar; — **bloquée,** turned letter; — **chargée,** registered (insured) letter; — **de créance,** credentials; — **de gage,** mortgage bond, debenture; — **d'envoi,** covering letter; — **de propagande,** publicity letter; — **de voiture,** waybill, bill of lading.

lettre, — **explicative,** letter of reference; — **hypothécaire,** mortgage (deed); — **minuscule,** small letter, minuscula; — **recommandée,** registered letter; —**s égyptiennes,** block types; —**s marchées,** broken types; —**s tombées,** broken types; **au pied de la** —, literally.

leucacanthe, with white thorns.

leucanthe, leucanthous, white-flowered.

leucanthème, white-flowered.

leucanthéré, with white anthers.

leucéine, *f.* leucine.

leucémie, *f.* leukemia, leucocythemia.

leucémique, leukemic.

leucine, *f.* leucine.

leucique, leucic.

leucite, *f.* leucite, sclerotitis, plastid.

leucocéphale, with white flowers united in head.

leucocidine, *f.* leucocidin, toxic products causing degeneration of leucocytes.

leucocytaire, leucocytic.

leucocyte, *m.* leucocyte, white blood corpuscle.

leucocythémie, *f.* leucocythemia, leukemia.

leucocytolyse, *f.* leucocytolysis.

leucocytorrhée, *f.* bowel evacuation rich in leucocytes.

leucocytose, *f.* leucocytosis.

leucocytosique, exciting leucocytosis.

leucodérivé, *m.* leuco derivative.

leucolyse, *f.* leucolysis.

leucomaïne, *f.* leucomaine.

leuconique, leuconic.

leucopénie, *f.* leucopenia.

leucopile, with white hood (cap).

leucoplaste, *m.* leucoplast.

leucopyre, with white fruit.

leucopyrite, *f.* leucopyrite.

leucorhize, with white roots.

leucorrhée, *f.* leucorrhea, the whites.

leucosine, *f.* vegetable protein with albumen and globulin.

leucosperme, with white seeds.

leucotoxine, *f.* leucotoxin.

leucoxanthe, marked white and yellow.

leucoxyle, with white wood.

leur, to them, them, their, theirs; **le** —, **la** —, **les** —**s,** theirs.

leurrer, to lure, entice.

levage, *m.* raising, hoisting, lifting.

levageur, *m.* fitter.

levain, *m.* leaven, barm, yeast, ferment; — **en poudre,** baking powder; — **pur,** starter, pure culture.

levant, *m.* east, Levant.

levant, rising, eastern, orient.

levé, *m.* surveying, survey.

levé, raised, erect, up, risen, lifted up.

levée, *f.* raising, lifting, removal, collecting, gathering (crops), sprouting (of corn), stroke, lift, surveying, plotting, embankment, dike, dam, levy; — **de terre,** earth bank, embankment.

lève-gazon, *m.* sod lifter.

lever, to raise (up), lift, hoist, set up, remove, take up, collect, gather (crops), survey, plot, sketch, draw up, rise, get up, start up, spring up, shoot, sprout, come up, part, pick, germinate; **se** —, to rise, be raised, clear up; — **sur lessive,** to separate (soap) from the lye.

lever, *m.* rising, surveying, plotting, plan, sketch; — **du soleil,** sunrise.

lève-roue, *m.* screw jack.

lève-soupape d'échappement, *m.* exhaustvalve lifter.

leveur, *m.* taker-off, layer.

levier, *m.* lever, arm, crowbar; — **à main de pompe,** pump handle; — **coudé,** crank lever, bent lever; — **de réglage,** adjustment lever.

levifolié, with glossy leaves.

léviga eur, *m.* levigator.

lévigation, *f.* levigation, levigating, trituration.

léviger, to levigate.

lévogyre, levorotatory, levogyre, levogyrate.

levoir, *m.* lay stool.

lévosine, *f.* carbohydrate found in seeds of barley, wheat, and rye.

levrander, to annoy, harass.

levraut, *m.* leveret, young hare.

lèvre, *f.* lip, edge, border, rim, labium, labia, lower lip; — **supérieure,** labrum, upper lip.

levrette, *f.* harrier bitch, greyhound bitch.

lévrier, *m.* greyhound.

lévuline, *f.* levulin.

lévulinique, levulinic.

lévulosane, *m.* levulosan.

lévulose, *f.* levulose, fructose, (levo)glucose.

levulosé, containing levulose, fructose.

levure, *f.* yeast, barm, leaven, ferment; — **basse,** bottom yeast, low yeast; — **de dépot,** bottom yeast; — **de vin,** wine lees; — **durable,** permanent yeast, zymin; — **ferment,** yeast enzyme; — **haute,** top yeast, high yeast; — **lactique,** lactic ferment, bacillus causing souring of milk; — **pressée,** compressed yeast; — **sèche,** pressed yeast.

levurier, *m.* maker or seller of yeast.

lexicographe, *m.* lexicographer.

lexique, *m.* lexicon, dictionary, glossary.

lézard, *m.* lizard, swift.

lézarde, *f.* split, crevice, crack, chink.

lézarder, to crack, chink, idle, bask in sun.

liage, *m.* tying, binding.

liais, *m.* hard limestone, heddle bar.

liaison, *f.* binding, joining, connection, bond, union, linkage, tie, mortar, relation, liaison.

liane, *f.* liana, tropical creeper; *pl.* climbing plants; — **à serpents,** snakeweed.

liant, *m.* pliancy, pliability, binder; — **pour couleurs,** color agglutinant.

liant, pliant, springy, easily worked, flexible, elastic, tenacious, tough, malleable, supple, gentle, mild, compliant.

liard, *m.* black poplar (*Populus nigra*).

lias, *m.* Lias.

liasique, liassique, Liassic.

liasse, *f.* bundle, package, file, cord, string.

libelle, *m.* libel, air bubble (in liquid inclusion).

libellé, *m.* wording, terms used.

libellé, drawn up, worded.

libeller, to draw up, word, sign and date.

libellule, *f.* dragonfly.

liber, *m.* liber, bast, inner bark, bastrope, phloem; — **mou,** liber, soft bast.

libéral, -aux, liberal.

libéralement, liberally, generously.

libéralité, *f.* liberality, generosity.

libérateur, -trice, liberating.

libération, *f.* liberation, discharge, release, deliverance.

libéré, liberated, discharged.

libérer, to liberate, discharge, free, release.

libérien, -enne, pertaining to bast or phloem.

liberté, *f.* liberty, freedom, free play, clearance, windage; **en** —, free; **mettre en** —, to set free, liberate; **mise en** —, setting free.

libidibi, *n.* divi-divi (*Caesalpinia coriaria*).

libraire, *m.* bookseller.

libraire-éditeur, *m.* publisher.

librairie, *f.* bookstore, book trade, library.

libration, *f.* libration.

libre, free, open, clear, inside (of dimensions), exempt, unconfined, unstamped, out of gear, disengaged; **à l'air** —, **au** —, in the open air, in free air.

libre-échange, *m.* free trade.

librement, freely, spontaneously.

libroplaste, *m.* libroplast, elaioplast.

lice, *f.* lists, field, list, rail, warp (of tapestry).

licence, *f.* license, leave, permit, licentiate's degree.

licencié, *m.* licensee, licentiate (intervening degree between bachelor and doctor).

licet, *m.* permission, permit, leave.

lichen, *m.* lichen; — **de renne,** reindeer moss (*Cladonia rangiferina*); — **des rochers,** archil (*Roccella tinctoria*); — **tartareux,** tartarean moss (*Lecanora tartarea*).

licheneux, -euse, lichenous.

lichénicole, lichenicolous, dwelling in or on lichens.

lichénine, *f.* lichenin.

lichénique, lichenic.

licite, licit, lawful, permissible.

licou, licol, *m.* halter, head collar.

lie, *f.* lees, dregs, grounds, yeast, barm; *pl.* lees; — **de vin,** wine lees, purple color; — **des vins,** argol.

lié, bound, tied.

liebig, *m.* beef extract.

liège, *m.* cork, periderm; — **femelle,** inferior (female) layer of cork; — **mâle,** first (male) layer of cork; — **mère,** phellogen.

liégeux, -euse, corky, corklike, suberous.

lien, *m.* bond, tie, relation, connection, band, strap, ligature, bonds.

lier, to bind, tie, join, connect, form a connection, fasten, link, knit, enter into, engage in, thicken (sauce); **se —,** to. be bound, be tied, become intimate, thicken, bind; — **en bottes,** to bundle, sheave.

lierre, *m.* ivy (*Hedera helix*); — **terrestre,** ground ivy (*Nepeta hederacea*).

iies, *f.pl.* lees.

liesse, *f.* mirth, gaiety.

lieu, *m.* place, turn, estate, circle, rank, locus, reason, spot, occasion, cause, room; *pl.* premises; — **d'expédition,** shipping station; — **exposé à la gelée,** frost locality; — **peuplé de roseaux,** patch of reeds; **au — de,** instead of; **au — que,** whereas; **avoir —,** to take place; **donner —,** to occasion, cause; **en aucun —,** nowhere; **en premier —,** in the first place; **en tout —,** everywhere; **il y a — de,** it is well to, there is (every) reason for.

lieue, *f.* league, mile.

lieur, *m.* binding attachment.

lieuse, *f.* reaper binder.

lieutenant, *m.* lieutenant.

lièvre, *m.* hare.

ligament, *m.* ligament.

ligature, *f.* ligature, band, binding, lashing, tie, joint; — **sur la gorge,** neck-groove binding.

ligaturer, to tie, splice, bind, fasten, ligature.

ligne, *f.* line, order, rank, path, range (of mountains); *pl.* streaks; — **à plomb,** plumb line, sounding line; — **conductrice,** power line; — **courbe,** curved line; — **d'eau,** water line; — **de chemin de fer,** railway line; — **de contact,** contact conductor, trolley wire; — **de courant,** stream line; — **de démarcation,** boundary line.

ligne, — **de faîte,** crest line, ridge line, pass; — **de flottaison,** length at the water line; — **de mire,** line of sight; — **de plants,** planting row; — **de raccordement,** junction line; — **de sang,** blood line; — **de semis,** line of seedlings; — **de tir,** line of fire, fire zone; — **de visée,** line of sight; — **enfoncées,** channels.

ligne, — **fémorale,** coxal line; — **garde-feu** safety strip, safety belt; — **infléchie,** inflected line; — **libre,** disengaged line; — **neutre,** dead line; — **pointillée,** — **ponctuée,** dotted line; **hors de —,** out of line, out of the way; **hors —,** beyond comparison, extraordinary; **mettre en — de compte,** to take into account.

ligné, lineate, lined.

lignée, *f.* line, lineage.

ligner, to draw lines, rule.

ligneux, *m.* lignin.

ligneux, -euse, ligneous, woody.

lignification, *f.* lignification.

lignifié, lignified, encrusted with a ligneous substance.

lignifier, to lignify.

lignine, *f.* lignin.

lignite, *m.* lignite, brown coal; — **aggloméré,** brown-coal briquette; — **brut,** crude or raw lignite.

lignocérique, lignoceric.

lignone, *f.* lignin, wood cellulose, lignone.

lignose, *f.* lignose.

lignosité, *f.* woodiness, ligneous quality.

ligroine, *f.* ligroine.

ligue, *f.* league.

ligule, *f.* ligule, ligula.

ligulé, ligulate, tongue-shaped.

ligulifère, having ligules.

lilacine, *f.* lilacin, syringin.

lilas, lilac.

liliacé, liliaceous.

liliacées, *f.pl.* Liliaceae.

lilial, lilylike, lily-white.

liligère, bearing lilies.

limace, *f.* slug, Archimedean screw, water screw.

limaçon, *m.* snail, cochlea, limaçon, great wheel, snail wheel, Archimedean screw.

limage, *m.* filing.

limaille, *f.* filings, file dust; — **de fer,** iron filings; — **de forage,** iron borings.

limailleux, -euse, kishy.

limande, *f.* rule, straightedge, ruler, level.

limas, *m.* slug, Archimedean screw.

limature, *f.* filings.

limbaire, resembling the limb of a petal or of a leaf.

limbe, *m.* limb, rim, border, edge, limbus, graduated arc, lamina, leaf blade.

limbifère, limbate.

limbique, marginal, limbic.

lime, *f.* file, lime (the fruit); — **à archet,** bow file, riffler; — **à arronder,** cabinet or round-off file; — **à bouter,** sharp file; — **à ébarber,** planchet file; — **à queue de rat,** rattail file; — **batarde,** bastard file; — **carrée,** square file; — **demi-douce,** second-cut file; — **demi-ronde,** half-round file; — **douce,** smooth file; — **grosse,** coarse file; — **plate,** flat file.

lime-bois, *m.* wood-eater, timber beetle, shipworm.

limer, to file, polish; — **en travers,** to file across.

limette, *f.* sweet lime.

limettier, *m.* plant or tree of the genus Citrus.

limeur, *m.* filer, filing machine.

limeur, filing.

limier, *m.* bloodhound.

limitatif, -ive, limiting, restrictive.

limitation, *f.* limitation.

limitativement, in a limited manner.

limite, *f.* limit, boundary, border; — **d'allongement,** yield point or ratio; — **d'écoulement,** yield point or ratio; — **de flexion,** limit in bending; — **de la végétation arborescente,** limit of aborescent vegetation, timber line.

limité, limited.

limiter, to limit, confine, bound, restrain.

limiteur de tension, *m.* potentiometer, voltage divider.

limnoplankton, *m.* limnoplankton, the plankton of fresh waters.

limon, *m.* mud, slime, shaft, thill, citron, lemon, sediment, ooze, silt, clay; — **extérieur,** outer string, wall string.

limonade, *f.* lemonade; — **sèche,** lemonade powder.

limonage, *m.* enriching soil with silt, silting up.

limone, *f.* limonin.

limonène, *m.* limonene.

limoner, to clean, scale (fish).

limoneux, -euse, muddy, slimy, turbid, growing in mud.

limonier, *m.* lemon tree (*Citrus limonia*).

limonine, *f.* limonin.

limonite, *f.* limonite, bog-iron ore.

limoniteux, -euse, limonitic.

limousine, *f.* limousine.

limpide, limpid, clear, transparent.

limpidité, *f.* limpidity, limpidness, clearness, transparency.

linure, *f.* filings.

lin, *m.* flax, linen, linseed; — **de rouissage,** red or steeped flax; — **en bois,** flax straw; — **en paille,** flax straw; — **froid,** late flax;

— **peigné,** hackled flax; — **purgatif,** purging flax; — **roui,** retted flax; — **sauvage,** toadflax, common flax; — **teillé,** scutched flax; **huile de** —, linseed oil.

linacé, linaceous.

linacées, *f.pl.* Linaceae.

linaigrette, *f.* cotton grass (Eriophorum).

linaire, *f.* toadflax, flaxseed (Linaria).

linaloé, *f.* linaloa.

linalol, *m.* linaloöl.

linamarine, *f.* linamarin.

linceul, *m.* shroud.

lincrusta, *m.* lincrusta.

linéaire, linear.

linéal, -aux, lineal (descent).

linéarilobé, (leaves) divided into linear lobes.

linette, *f.* linseed, flaxseed.

linge, *m.* cloth, calico, rag, linen, clothes; — **de dessous,** underwear; — **sale,** dirty linen; — **uni,** plain linen.

linger, *m.* draper.

lingerie, *f.* linen trade, linen, ready-made linen shop.

lingot, *m.* ingot, slug, bullion, cube.

lingotière, *f.* ingot mold.

lingotiforme, ingot-shaped.

lingual, -aux, lingual, tonguelike.

linguet, *m.* pawl, catch.

linguiforme, linguiform, tongue-shaped.

linguiste, *m.&f.* linguist.

linier, *m.* flax worker.

linier, -ère, flax, of flax.

linière, *f.* flax field, flax plot.

liniment, *m.* liniment; — **calcaire,** lime liniment; — **crotoné,** liniment of croton oil; — **savonneux camphré,** soap liniment.

linine, *f.* linin.

linition, *f.* coating, anointing.

Linné, *m.* Linnaeus.

linnéite, *f.* linnaeite.

linographe, *m.* linographer.

linographie, *f.* linography.

linoléate de plomb, *m.* lead linoleate.

linoléique, linoleic.

linoléum, *m.* linoleum.

linome, *m.* linom, linim.

linon, *m.* lawn.

linophanie, *f.* linophany.

linotte, *f.* linnet; — **à bec jaune,** twite.

linoxine, *f.* linoxyn.

lint, *m.* lint.

linteau, *m.* lintel, transom, cross timber, intertie.

linters, *m.pl.* linters (fleecy fiber).

lion, *m.* lion.

liorhize, liorhizal, pertaining to Liorhizae.

liparolé, *m.* ointment, unguent.

lipase, *f.* lipase.

lipaséidine, *f.* lipaseidin, the fat-splitting enzyme in castor-bean seeds.

lipasique, lipolytic, lipase.

lipémie, *f.* lipemia.

lipide, *f.* lipide.

lipochrome, *m.* lipochrome.

lipogénèse, *f.* lipogenesis.

lipoïde, lipoid.

lipolyse, *f.* lipolysis.

lipolytique, lipolytic.

lipomateux, -euse, lipomatous.

lipomatose, *f.* lipomatosis, fatty degeneration.

lipome, *m.* lipoma, adipoma.

lipurie, *f.* lipuria.

liquater, to liquate.

liquation, *f.* eliquation, liquation.

liquéfaction, *f.* liquefaction.

liquéfiable, liquefiable.

liquéfié, liquefied.

liquéfier, to liquefy.

liquescence, *f.* liquescence.

liqueur, *f.* liqueur, liquor, fluid, liquid, solution; — **aqueuse,** aqueous solution; — **d'acétate de fer,** solution of ferric acetate; — **d'ammoniaque,** ammonia water; — **de Fehling,** Fehling solution; — **des cailloux,** liquor of flints, water-glass solution; — **de virage,** indicator liquid

liqueur, — **d'Hoffman,** spirit of ether, anodyne liquor of Hoffman; — **épreuve,** testing solution; — **iodo-iodurée,** iodine-iodide solution; — **surnageante,** supernatant liquid; — **témoin,** indicator liquid; — **titrée,** standardized solution; — **type,** standard solution; — **vésicant,** blistering liquid.

liquidambar, *m.* liquidambar.

liquidateur, *m.* liquidator.

liquidation, *f.* liquidation, treatment of soap to remove impurities; — **après inventaire,**

stock-taking sale; — **de fin d'année,** yearly settlement; — **générale,** clearance sale.

liquide, *m.* liquid, fluid, liquor; — **conservateur,** preserving liquid; — **d'essorage,** liquor discharged by hydroextraction; — **excitateur,** exciting liquid; — **témoin,** test liquid.

liquide, watery.

liquider, to liquidate; **se** —, to pay off, settle.

liquidité, *f.* liquidity, fluidity.

liquoreux, -euse, liqueurlike (wine), (wine) still, sweet and soft.

liquoriste, *m.* wine and spirit maker or seller.

lire, to read; **se** —, to be read; — **une épreuve,** to correct, read the clean proof.

lire, *f.* lira (an Italian coin).

lirelle, *f.* lirella, a linear apothecium of lichens.

lirelleux, -euse, lirellous.

lirelliforme, lirelliform, shaped like a lirella.

lis, *m.* lily; — **d'eau, des étangs,** water lily (*Nymphaea alba);* — **des marais,** sweet flag, sweet calamus, sweet rush (*Acorus calamus);* — **des teinturiers,** dyer's rocket, dyer's-weed, yellowweed (*Reseda luteola);* — **des valées,** lily of the valley (*Convallaria majalis).*

lisage, *m.* agitation of hanks of silk, reading in.

lisait (lire), (he) was reading.

lisant, reading, agitating.

lise, *f.* quicksand.

lisent (lire), (they) are reading, agitating.

liser, to agitate in the vat.

liséré, *m.* border, edge, binding.

lisérer, to border, edge, trim, bind.

liseron, *m.* bindweed, small morning-glory (Convolvulus); — **des champs,** bindweed; — **grand,** bearbine.

lisette, *f.* common vine grub.

liseur au semple, *m.* card cutter.

liseuse, *f.* book mark.

lisibilité, *f.* legibility.

lisible, legible, easy to read.

lisiblement, legibly.

lisière, *f.* selvage, list, border, edge, strip, verge, skirt; — **de la toile à voiles,** selvage of canvas; — **du bois,** edge of wood, skirt.

lissage, *m.* smoothing, black wash.

lisse, *m.* smoothness, warp (of cloth), string, cord, rail, region of the root extending from the rootcap to the piliferous zone.

lisse, *f.* head, head beam, ridge beam; — à double courbure, harpin; — de bastinage, rail, breast rail; — de garde-corps, main rail, rough tree rail; — de hourdi, main or wing transom; — de l'arrière, stern rail; — de recouvrement, rib flange; — des oeuvres mortes, top timber line; — du fort, bilge harpin; — plane, ribband.

lisse, smooth, sleek, glossy, glabrous.

lissé, *m.* smoothness, polish, gloss, glaze.

lissé, smoothed, smooth, glossed, glabrous.

lissée, *f.* glazing.

lisser, to smooth, gloss, polish, sleek, glaze, calender, burnish, preen (feathers).

lisserons, *m.pl.* shafts.

lisseur, *m.* smoother, glazer.

lisseuse, *f.* smoothing machine.

lissoir, *m.* smoother, glosser, smoothing tool, polishing iron, finishing house.

lissoir, smoothing, glossing.

lissure, *f.* smoothing, glazing, polishing, glossiness, glaze.

liste, *f.* list, catalogue, blaze; — de pièces, specifications; — de prix, catalogue, price list.

listeau, *m.* ledge, listel, fillet.

lit, *m.* bed, layer, stratum, channel, course; — bactérien, bacteria bed (sewage disposal); — de fusion, mixture bed; — de gueuse, sow channel; — du canal, canal bottom; — du vent, wind's eye, teeth of the wind, rhumb; — majeur, flood bed, alluvial plain.

liteau, *m.* band, stripe.

litée, *f.* litter (of animals).

liter, to place in layers, cover the selvage of (fabrics).

literie, *f.* bedding.

litharge, *f.* litharge, lead oxide.

lithargé, lithargyré, containing litharge.

lithiase, lithiasie, *f.* lithiasis.

lithine, *f.* lithia, lithium oxide.

lithiné, containing lithia.

lithinifère, containing lithia or lithium.

lithionite, *f.* lepidolite, lithia mica, lithionite.

lithique, lithic; acide —, lithic acid, uric acid.

lithium, *m.* lithium.

lithochromie, *f.* lithochromy.

lithocolle, *f.* lapidary cement.

lithofellique, lithofellinique, lithofellic, lithofellinic.

lithofracteur, *m.* lithofracteur.

lithographe, *m.* lithographer.

lithographie, *f.* lithography, lithograph.

lithographier, to lithograph.

lithographique, lithographic(al).

lith ïde, lithoid, lithoidal.

lithologie, *f.* lithology, petrography.

lithologique, lithological.

lithomarge, *f.* porcelain earth or clay.

lithophanie, *f.* lithophany.

lithophile, lithophilous, saxicolous, dwelling on rocks.

lithophytes, *m.pl.* Lithophyta, lithophytes, plants growing on rocks.

lithosphère, *f.* lithosphere, the solid part of the earth.

litière, *f.* litter; — de feuilles mortes, litter of dead leaves or needles; — de tourbe, peat litter or moss; — faite d'aiguilles, needle litter; — faite de branche, branch litter.

litige, *m.* legal dispute, litigation.

litigieux, -euse, disputable, contested.

litre, *m.* liter.

litreuse pour fûts, *f.* cask-gauging apparatus.

littéraire, literary.

littéral, -aux, literal.

littéralement, literally.

littérateur, *m.* man of letters.

littérature, *f.* literature.

littoral, *m.* littoral, coast, seashore, shore.

litzendraht, *m.* braided wire, high-frequency cable.

livarot, *m.* Livarot cheese.

livèche, *f.* lovage, sea parsley (*Levisticum officinale*).

livide, *m.* livid color.

livide, livid.

lividité, *f.* lividness, lividity.

livrable, deliverable, to be delivered.

livraison, *f.* delivery (of goods), part, supply, purveyance, number, issue, fascicle, consignment.

livrancier, *m.* deliverer, shipper, contractor, seller.

livre, *f.* pound (half kilogram), pound (sterling).

livre, *m.* book; — **de bureau,** office book — **de factures,** invoice book; — **de généalogie,** register, studbook, herdbook, pedigree; — **de magasin,** store book, stock book.

livrée, *f.* livery, appearance.

livrer, to deliver, yield (up), abandon, hand over, furnish, give up, afford, leave, supply, provide with; **se** — (à), to devote oneself, trust, surrender, give oneself up to, deliver up.

livret, *m.* small book, booklet, catalogue, (local) timetable.

livreuse, *f.* delivery van.

lixiviateur, *m.* lixiviator.

lixiviation, *f.* lixiviation, leaching.

lixiviel, -elle, lixivial.

lixivier, to lixiviate, leach.

llanos, *m.pl.* llanos, a special type of savanna forming vast plains.

lobe, *m.* lobe; — **olfactif,** olfactory lobe; — **protocérébral,** protocerebral lobe.

lobé, lobed, lobate.

lobéliacées, *f.pl.* Lobeliaceae.

lobélie, *f.* lobelia, cardinal flower (*Lobelia cardinalis*); — **enflée,** Indian tobacco (*Lobelia inflata*).

lobéline, *f.* lobeline alkaloid of *Lobelia inflata*, lobelin (resin).

lobule, *f.* lobule, lobelet.

lobulé, lobuleux, lobulate, lobulous.

local, *m.* place, locality, premises, room, habitation.

local, -aux, local.

localement, locally.

localisation, *f.* localization.

localiser, to localize, locate; **se** —, to become localized.

localité, *f.* locality, place.

locataire, *m.&f.* tenant, renter, hirer, lessee.

locatif, -ive, pertaining to renting.

location, *f.* hiring, hire, renting, rent.

loch, *m.* log (of ship).

lochage, *m.* loosening.

loche, *f.,* — **franche,** common loach; — **de rivière,** groundling.

locher, to shake, shake loose, loosen.

locomobile, *f.* — **à moteur,** petrol, oil, and gas engine; — **à vapeur,** steam engine.

locomoteur, *m.* locomotor.

locomoteur, -trice, of locomotion.

locomotive, *f.* locomotive, engine.

locomotivité, *f.* locomotivity.

loculaire, loculate, locular.

loculamenteux, loculamentous.

loculé, loculate (d)

loculicidé loculicide.

locus *m.* spot, place, locus.

locuste, *f.* locust, grasshopper; — **des marais,** swamp locust, water locust

loden, *m.* shag, coarse cloth.

loess, *m.* loess.

loevis, smooth, glossy.

lof, *m.* weather side, luff.

lofer, to luff, spring a luff.

logarithme, *m.* logarithm.

logarithmique, *f.* logarithmic curve.

logarithmique, logarithmic(al).

logatome, *m.* articulation.

loge, *f.* lodge, cell, booth, box, loculus, anther lobe, cabin, hut, cavity, shed.

logé, lodged, put up (of wine) in casks.

logement, *m.* lodging, bed, groove, accommodation, hole, seat, dwelling, apartment, quarters, room.

loger, to lodge, live, quarter, put (up), pack, keep, place, set, harbor, let in.

logette, *f.* pollen chamber, locellus.

logique, *f.* logic.

logique, logical.

logiquement, logically.

logis, *m.* home, house, dwelling.

logne, *f.* perch, tongue.

loi, *f.* law, standard (coin), rule; — **de douane,** customs law; — **de la chasse,** game law, game act; — **des brevets,** patent law; — **fiscale,** tax law; — **forestière,** forest law; — **naturelle,** natural law.

loin, — **de,** far from, far, remote, afar, distant, long; **au** —, at a distance, afar, far off; **bien** —, very far; **de** —, from afar, at a distance, long ago; **de** — **en** —, at long intervals, here and there.

lointain, *m.* distance, background.

lointain, far, distant, remote.

loir, *m.* dormouse.

loisible, allowable, permissible, optional.

loisir, *m.* leisure, spare time; **à** —, leisurely

lomatocarpe, (fruit) surrounded by a thickened border.

lomatorhizé, with bordered roots.

lombago, *m.* lumbago.

lombaire, lumbar, of the loins.

lombric, *m.* earthworm.

lomentacé, lomentaceous, bearing or resembling loments.

lomentum, *m.* loment, a legume or pod constricted between the seeds.

lonchophylle, with very long, linear, and lanceolate leaves.

londonien, -enne, London, of London.

Londres, *f.* London.

long, *m.* length.

long, longue, long, slow, tedious; à la longue, in the long run, in the course of time; au —, at length, at great length; au — de, along; au — et au large, far and wide; de —, lengthwise; de — en large, en — et en large, up and down, to and fro; en —, lengthwise; le — de, along; tout le —, all along.

longailles, *f. pl.* staves, staff wood.

longer, to go along, run along, follow, line, walk along, skirt.

longeron, *m.* stringer, sill mainbeam, slider.

longévif, -ive, long-lived.

longévité, *f.* longevity.

longicaude, longicaudate, long-tailed.

longicaule, longicauline, long-stemmed.

longicorne, longicorn.

longifolié, long-leaved.

longilobé, divided into elongate lobes.

longimane, longimanous, long-handed.

longipède, long-footed.

longipédonculé, with long peduncle.

longipenne, longipennate, long-winged.

longirostre, longirostral, long-billed.

longiscape, with very long stems (stalks).

longisiliqueux, with long siliques.

longispinuleux, with long thorns.

longistaminé, longistaminate, (flowers) with long stamens.

longistyle, with very long styles.

longitude, *f.* longitude; — géodésique, terrestrial longitude.

longitudinal, -aux, longitudinal.

longitudinalement, longitudinally, lengthwise.

longophylle, with very long leaves.

longrine, *f.* longitudinal beam or girder.

longtemps, a long time, long, a long while;

aussi — que, so long as; depuis —, for a long time.

longtron, *m.* oval hole.

longue, *f.* long time; à la —, in the long run.

longuement, a long time, a great while, long, a long while.

longuerine, *f.* girder.

longuet, a little long, rather long, linear form.

longueur, *f.* length, slowness, distance; — d'onde, wave length; en —, lengthwise, at length.

longue-vue, *f.* field glass, spyglass, telescope.

looch, *m.* soothing emulsion, lincture, lohoch, electuary.

lophante, resembling an egret or pappus.

lophine, *f.* lophine.

lophogone, with fringed angles like a crest.

lophotrique, lophotrichous.

lophyre du pin, *m.* pine sawfly.

lopin, *m.* bit, piece, patch, morsel, bloom, slab.

loque, *f.* rag, tatter, shred, bittersweet (*Solanum dulcamara*); — pourriture, foul brood.

loquet, *m.* latch, catch.

loqueteau, *m.* snap, catch, falling latch.

loquette, *f.* carding roll, roll, scrap (of bread).

lorétine, *f.* loretin.

lorgnette, *f.* binocular; — de spectacle, opera glass.

lorgnon, *m.* eyeglass, eyeglasses.

lorifolié, with elongated leaves similar to straps.

loriot, *m.* golden oriole.

lorique, *f.* testa, outer layer of ovule.

lors, then, at the time; — de, at the time of, at the moment of, during, in; — . . . que, when; — même que, although, even though, when even; depuis —, since that time; dès —, from that time, since then, therefore, hence; dès — que, since, seeing that; pour —, in that case, so, then.

lorsque, when, in case that.

losange, *m.* lozenge, rhomb, diamond.

lot, *m.* lot, prize (in lottery), portion, share, parcel.

loterie, *f.* lottery.

lotier, *m.* bird's-foot trefoil (Lotus); — des marais, marsh bird's-foot trefoil.

lotiforme, lotiform, having the form of a lotus.

lotion, *f.* lotion, washing.

lotionner, to wash, bathe, sponge.

lotir, to sort out, grade, divide into lots, allot, assign, parcel out, sample (ores).

lotissage, lotissement, *m.* sampling, dividing, sorting.

loto, *m.* lotto.

lotte, *f.* hollow between dunes; — **commune,** burbot, eelpout.

lotus, *m.* lotus (tree).

louable, praiseworthy, commendable, laudable.

louage, *m.* leasing, renting, letting, hiring.

louange, *f.* praise, commendation.

louanger, to praise, commend.

louche, *m.* turbidity, cloudiness; *f.* ladle, shovel, reamer, broach.

louche, ambiguous turbid, cloudy, squinting, squint-eyed, suspicious.

louchement, *m.* becoming turbid, clouding, squinting.

loucher, to be cross-eyed.

louchet, *m.* (narrow) spade.

louchette, *f.* glasses (for strabismus).

louchir, to become turbid or cloudy.

louchissement, *m.* becoming turbid, clouding.

louer, to rent, hire, let, praise.

loueur, *m.* one who rents or lets out.

louis, *m.* louis (a 20-franc piece).

loup, *m.* wolf, wool breaker, willow, devil, defect, flaw, error, mask, facepiece (of gas mask), crowbar, drawer, salamander, bear; — **batteur,** beating opener.

loupe, *f.* magnifying glass, reading glass, lens, eyeglass, sebaceous tumor, hump, bloom, wen, knobby growth, wart, figured wood, burr; — **à main,** reading glass; — **microscope,** magnifying glass.

lourd, heavy, slow, dull, turbid, grave, awkward, expensive, towering, sultry, close, weighty, burdensome, inactive, lifeless, stagnant.

lourdement, heavily, grossly.

lourdeur, *f.* heaviness, dullness.

loutre, *f.* otter, minx; — **vulgaire,** common otter.

louve, *f.* she-wolf, sling, cran iron, grapnel, dog, ram.

louvetage de laines, *m.* wool breaking.

louveter, to card.

louvoyer, to veer, put or beat about, cruise.

lover un cordage, to coil a rope.

loxanthère, with oblique anthers.

loxodromie, *f.* loxodromics.

loyal, -aux, sincere, honest, loyal, legal, lawful.

loyalement, loyally, fairly, honestly, faithfully.

loyauté, *f.* honesty, uprightness, loyalty.

loyer, *m.* rent, wages, pay, reward.

lu, read.

lubie, *f.* whim, caprice.

lubrifaction, *f.* lubrication.

lubrifiage, *m.* lubrication.

lubrifiant, *m.* lubricant.

lubrifiant, lubricating.

lubrificateur, *m.* lubricator.

lubrificateur, lubricating.

lubrification, *f.* lubrication, oiling, greasing.

lubrifier, to lubricate, oil, grease.

lubrifieur, *m.* lubricator, oil cup.

lucane cerf-volant, *m.* stag beetle.

lucarne, *f.* eyelet hole, lower window, dormer window, roof trap door.

lucet, *m.* whortleberry (*Vaccinium myrtillus*), sound boarding.

lucide, lucid, clear, shining.

lucidement, lucidly, clearly.

lucidité, *f.* lucidity, clearness.

luciferase, *f.* luciferase.

luciférine, *f.* luciferin.

lucifuge, lucifugous.

lucilie, *f.* greenbottle, blowfly.

lucinocte, (corolla) open at night and closed during the day.

luciole, *f.* firefly, lightning bug, glowworm.

lucratif, -ive, lucrative.

lues, *f.* lues, syphilis.

luétine, *f.* luetin.

luette, *f.* uvula, epiglottis.

lueur, *f.* glimmer, gleam, flash, (faint) light.

luge, *f.* sled.

lugubre, mournful, lugubrious.

lui, him, he, it, to him, to her, to it, her; c'est —, it is he.

lui-même, himself, itself.

luire, to shine, gleam, glitter, glimmer, glisten, sparkle.

luisance, *f.* shining, glass, luster, gleam.

luisant, *m.* gloss, sheen, luster.

luisant, shining, glossy, glittering, glowing, glimmering; ver —, glowworm.

luisante, *f.* bright star.

luisard, *m.* micaceous hematite.

lumachelle, *f.* lumachelle limestone.

lumbago, *m.* lumbago.

lumen, *m.* lumen.

lumière, *f.* light, opening, aperture, orifice, mouth, priming hole (of gun), port, slot, slit, sight (hole); *pl.* knowledge, insight; — à gaz, gaslight; — à incandescence, incandescent light; — à vapeur, steam port; — d'échappement, exhaust port; — du jour, daylight; — froide, cold light, luminescence; — incident, incident light; — polarisée, polarized light; — solaire, sunlight.

lumignon, *m.* candle end, lampwick, tallow dip, lantern.

luminaire, *m.* lighting, luminary, star, sun.

luminescence, *f.* luminescence.

luminescent, luminescent.

luminet, *m.* eyebright.

lumineusement, luminously, brightly.

lumineux, -euse, luminous, of light, bright, reflecting light.

luminifère, luminiferous.

luminosité, *f.* luminosity, luminousness, sheen.

lunaire, *f.* satinpod, moonwort (*Lunaria biennis*).

lunaire, lunar.

lunaison, *f.* lunation.

lundi, *m.* Monday.

lune, *f.* moon, caprice, whim, (alchemy) luna, silver; — cornée, silver chloride; — d'eau, white water lily (*Nymphaea alba*).

lunetier, *m.* spectacle maker.

lunette, *f.* telescope, spyglass, field glass, spectacles, peephole, sight glass, eye glass, lunette, cab window (locomotive); — à viseur, sighting telescope.

lunifère, bearing lunate markings.

luniforme, crescent-shaped, lunate.

lunulé, lunulate.

lunure, *f* lunure, internal sapwood.

lupin, *m.* lupine (Lupinus).

lupinelle, *f.* crimson or scarlet clover (*Trifolium incarnatum*), sainfoin (*Onobrychis vicaefolia*).

lupinine, *f.* lupinin (glucoside), lupinine (alkaloid).

lupulin, *m.* lupulin.

lupuline, *f.* lupulin, black medic (*Medicago lupulina*), yellow trefoil.

lupus, *m.* lupus.

lurent (lire), (they) read (past tense).

lusol, *m.* crude benzene.

lustrage, *m.* glossing, lustering, shininess.

lustre, *m.* luster, chandelier, brilliancy, splendor.

lustré, lustered, glossy.

lustrer, to luster, gloss, glaze, polish, give a gloss to, calender, dress.

lustroir, *m.* polisher.

lut, *m.* lute, luting.

lutation, *f.* luting.

lutécium, *m.* lutecium.

lutéine, *f.* lutein.

lutéoline, *f.* luteolin.

luter, to lute, seal.

lutidine, *f.* lutidine.

lutidique, lutidinic.

lutte, *f.* struggle, contest, strife, fighting.

lutter, to strive, struggle, wrestle, fight, control (parasites).

lutteur, *m.* wrestler, striver.

luxation, *f.* luxation, dislocation.

luxe, *m.* luxury, richness.

luxuriant, luxuriant.

luzerne, *f.* lucerne, alfalfa, purple medick, snail clover (*Medicago sativa*).

lycée, *m.* lycée, lyceum, secondary school, high school.

lycéen, *m.* lycée pupil.

lychnide, *f.*, lychnid, *m.* lychnis, rose campion (*Lychnis coronaria*).

lyciet, *m.* boxthorn, teaplant (*Lycium barbarum*).

lycope, *m.* bugleweed (Lycopus).

lycopode, *m.* lycopodium, club moss, staghorn moss (Lycopodium).

lycopodiacées, *f.pl.* Lycopodiaceae.

lycose, *f.* lycosa, wolf spider, tarantula.

lyddite, *f.* lyddite.

lymphatique, lymphatic, clear, pellucid.

lymphe, *f.* lymph, sap.

lymphocyte, m. lymphocyte.
lymphocytose, f. lymphocytosis.
lynx, m. lynx.
lyré, lyrate, lyriform, lyre-shaped.
lys, m. lily; fleur de —, iris, flag.
lysatine, f. lysatine.
lysé, having undergone lysis.
lysidine, f. lysidine.

lysigène, lysigenous.
lysimaque, m. loosestrife (Lysimachia vulgaris).
lysimètre, m. lysimeter.
lysine, f. lysine (an amino acid), lysin (antibody causing cell dissolution).
lysis, f. lysis.
lysol, m. lysol.

M

M., abbr. (Monsieur), Mr., Sir; (Majesté), Majesty; (Mars), March.

m., abbr. (mètre), meter; (minute), minute; (mon), my; (midi), south; (mobile), mobile; (mort), died; (masculin), masculine.

ma, (fem. of MON), my.

macadam, m. macadam, macadam road.

macadamisage, macadamisation, f. macadamizing, macadamization.

macadamiser, to macadamize.

macareux, m. sea parrot, puffin.

macaron, m. macaroon.

macaroni, m. macaroni.

macédoine, f. medley.

macératé, m. (liquid) product of maceration.

macérateur, m. macerater.

macérateur, -trice, macerating.

macération, f. maceration, macerating.

macératum, m. product of maceration.

macéré, m. macerated product.

macéré, macerated.

macérer, to macerate, steep, soak.

machaon, m. swallowtail butterfly.

mâche, f. — commune, corn salad, lambs'-lettuce (Valerianella locusta).

mâché, chewed, worn, fretted, galled, pulped.

mâche-bouchon, m. cork press or squeezer.

mâchefer, m. clinker, ashes, scoria, slag, dross.

mâchement, m. chewing, mastication.

mâcher, to chew, masticate, disintegrate (paper) to form pulp, cut roughly, champ (fodder), mince.

mâchicatoire, m. masticatory.

machinal, -aux, mechanical, automatic.

machinalement, mechanically.

machine, f. machine, engine, dynamo, mechanism, machinery; — à abattre les arbres, tree-felling machine; — à arroser, sprinkling machine; — à battre, threshing machine; — à boutons, pulp strainer, knotter; — à calandrer, calender; — à decortiquer, peeling machine, sheller; — à déroder des tronches d'arbres, tree-stump assorting machine.

machine, — à detente, expansion engine; — à écrire, typewriter; — à extirper des tronches d'arbres, tree-stump grubber; — à foncer, grounding or paper-staining machine; — à force centrifuge, centrifugal machine; — à huile minérale, oil engine: — à imprégner, impregnating machine; — à laver, washing machine, washer; — alimentaire, feeding engine; — à lisser, smoothing or satining machine.

machine, — alternative, alternating-current dynamo; — à nettoyer les semences, seed grain cleaning machine; — à pâte de bois, wood pulp machine; — à pétrole, gasoline engine; — à remplir, filler, bottling machine; — à satiner, glazing machine, calenders; — à traire, milking machine, mechanical milker.

machine, — à triturer, triturating apparatus; — à vapeur, steam engine; — à vent, blast engine; — fixe, stationary engine; — marine, marine engine; — motrice, motor; — pour la preparation de la pâte de papier, pulp-preparing machine; — rotative, rotary engine; — soufflante, blowing engine, blast engine.

machine-outil, f. machine tool.

machinerie, f. machinery, machine construction, engine room.

machiniste, m. machinist, engineer.

mâchoire, f. jaw, clamp, maxilla, jawbone, mandible, dolt, blockhead.

mâchurer, to soil, smear daub, blur, mark, smudge.

macis, m. mace.

macle, *f.* wide-meshed net, twin crystal, macle, chiastolite, druse, water caltrop (Trapa).

maclé, macled. (of glass) stirred, (of crystals) twinned, of the nature of crossed twins, hemitrope.

macler, to mix, stir (glass), twin, crystallize in crossed twins.

maclurine, *f.* maclurin.

maçon, *m.* mason.

maçonnage, *m.* masonry.

maçonner, to build, lay, cement, do mason's work, line with masonry.

maçonnerie, *f.* masonry, stonework, brickwork; — **brute en blocage,** rubblework.

macquage, *m.* breaking (of hemp, flax).

macracanthe, with long and strong thorns.

macranthe, large-flowered.

macradène, with large glands.

macre, *f.* water chestnut, water caltrop (Trapa natans).

macreuse, *f.* scoter, black duck.

macrobotryte, with large clusters.

macrocalicé, with large calyx.

macrocéphale, macrocephalic, large-headed.

macrocère, with a very long spur like a horn.

macroconidie, *m.* macroconidium, a large asexual spore or conidium.

macrocyste, *m.* macrocyst, ascogone, archicarp.

macrodiode, *f.* macrospore.

macrogamète, *f.* macrogamete.

macrokyste, *m.* macrocyst.

macronucléus, *m.* macronucleus.

macrophage, *m.* macrophage, large mononuclear leucocytes.

macroplankton, *m.* macroplankton, pleuston plants.

macropode, macropodous, with an enlarged or elongated hypocotyl.

macroprothalle, *f.* macroprothallium.

macroptère, macropterous, with unusually large fins or wings.

macropycnide, *m.* macrospore.

macroscléréide, *m.* macrosclereids, long stone cells with blunt ends.

macroscopique, macroscopic.

macrosperme, with large seeds.

macrosporange, *m.* macrosporangium, megasporangium.

macrospore, *f.* macrospore.

macrosporophylle, *f.* macrosporophyll, carpel, megasporophyll.

macrostème, with long projecting stamens.

macrostémone, with long projecting stamens.

macrostyle, macrostylous, long-styled.

macroure, macrurous, spiked, long-tailed.

macrozoospore, *f.* macrozoospore.

maculage, *m.,* **maculation,** *f.* spotting, maculation.

maculature, *f.* mackled sheets, waste power.

macule, *f.* macule, stain, spot, macula, cording quire, millwrapper.

maculé, spotted, blotched, maculate.

maculer, to spot, stain, soil, maculate.

madame, *f.* madame, madam, Mrs.

madéfaction, *f.* moistening, dampening, wetting.

madéfier, to moisten, wet.

mademoiselle, *f.* young lady, Miss.

madère, *m.* Madeira (wine).

madrage, *m.* mottling or marbling (of soap).

madré, mottled, marbled, spotted, speckled, figured, wavy, curled.

madréporique, madreporic.

madrier, *m.* plank, thick board, joist, timber, beam, sleeper.

madrure, *f.* vein, spot, mottle, mark, speckle.

magasin, *m.* shop, store, storehouse, warehouse, magazine; — **à couvain,** brood chamber, brood apartment; **en —,** in stock.

magasinage, *m.* storage, storing.

magasinier, *m.* warehouseman, storekeeper.

magdaléon, *m.* roll, stick, magdaleon.

mage, *m.* magian; *pl.* Magi.

magenta, *m.* magenta (color).

magie, *f.* magic; — **blanche,** white magic; — **noire,** black art.

magique, magic, magical.

magistère, *m.* magistery.

magistral, *m.* magistral, roasted copper and iron pyrites.

magistral, -aux, *m.* magisterial, masterly, authoritative, magistral.

magma, *m.* magma.

magnanage, *m.,* **magnanerie,** *f.* cocoonery, silkworm breeding.

magnanime, magnanimous, high-minded.

magnanimement, magnanimously.

magnau, *m.* silkworm.

magnésie, *f.* magnesia.

magnésié, containing magnesia.

magnesien, *m.* magnesium derivative, magnesian limestone.

magnésien, -enne, magnesian, pertaining to magnesia or magnesium.

magnésique, magnesic, magnesium.

magnésite, *f.* sepiolite, meerschaum.

magnésium, *m.* magnesium.

magnétique, magnetic.

magnétiquement, magnetically.

magnétisant, magnetizing.

magnétisation, *f.* magnetization.

magnétiser, to magnetize, mesmerize, hypnotize.

magnétisme, *m.* magnetism.

magnétite, *f.* magnetite.

magnéto, *f.* magneto.

magnification, *f.* magnification.

magnificence, *f.* magnificence, splendor.

magnifier, to magnify.

magnifique, magnificent, splendid, grand, gorgeous.

magnolier, *m.* magnolia tree, magnolia, beaver tree, swampwood, beaverwood.

magnum, *m.* magnum (two-liter bottle).

mai, *m.* May.

maie, *f.* trough (for kneading), bread bin.

maigre, *m.* lean meat, lean.

maigre, lean, thin, poor, meager, scanty, slender, raw, unfertile, poor, close burning, hard, light-faced.

maigrement, meagerly, poorly.

maigreur, *f.* leanness, thinness, emaciation.

maigrir, to grow thin or lean.

mail, *m.* mallet, maul.

maillage, *m.* breaking (of flax).

maille, *f.* mesh, opening, link, stitch, spot, speck, aperture, mallet, leucoma, bud (of vine, melon), pith rays, quarter grain; — de réseau, sieve mesh.

maillechort, *m.* maillechort, German silver.

maillet, *m.* mallet, maul, hammerhead (shark).

mailloche, *f.* beetle, large maul or mallet.

maillon, *m.* link, stitch, shackle. noose.

maillot, *m.* swaddling clothes.

maillure, *f.* medullary spot, pith fleck.

main, *f.* hand, paw, scoop, handle, handful, (of paper) quire or 25 sheets, (of silk) skein, feel (of cloth), hook; — d'oeuvre, labor charges, labor; à la —, by hand, ready, on hand, handmade, by guess; à pleines —s, by the handful; de — de maître, in a masterly manner; la haute —, the greatest authority; sous —, underhand, secretly; sous la —, at hand.

main-brune, *f.* a coarse gray paper.

main-d'oeuvre, *f.* manual labor, cost of labor, workmanship.

maine, *f.* handful.

main-forte, *f.* assistance, help, aid.

mainmise, *f.* freeing, seizure, annexation.

maint, many a, many.

maintenant, now, at present, maintaining, preserving; — que, now that.

maintenir, to keep, maintain, keep up, preserve, sustain, support, hold fast; se —, to keep, remain, last, be maintained, keep up, hold out, maintain oneself, hold one's own; il se maintient, he holds his own.

maintien, *m.* maintenance, keeping (up), upkeep, preservation, attitude, bearing, carriage; — du massif, maintenance of density (in a crop).

maïolique, *f.* see MAJOLIQUE.

mairain, *m.* wood for cooperage, stem, bole.

maire, *m.* mayor.

mais, but, why, more, indeed; — non, why no; — si! why yes!

maïs, *m.* maize, Indian corn (Zea mays); — sucre, sweet corn; huille de —, corn oil.

maison, *f.* house, home, family, firm, habitation; — de commerce, business house; — d'habitation, dwelling; — de santé, private hospital, sanitarium; — de ville, — commune, town hall; — forestière, forest house; à la —, at home.

maisonnée, *f.* the household.

maisonnette, *f.* cottage; — hygromètre à boyau, weather box.

maitre, *m.* master, owner, proprietor, teacher, instructor, head, chief, master (workman); — d'école, schoolmaster; — ès arts, master of arts; — verrier, master glassmaker.

maîtresse, *f.* mistress, chief, main, principal, superior.

maîtrise, *f.* mastery, mastership.

maîtriser, to master, control, govern, subdue.

majesté, *f.* majesty, grandeur. sublimity.

majestueux, -euse, majestic, sublime.

majeur, *m.* middle (second) finger.

majeur, greater, major, great, of age, most.

majolique, *f.* majolica.

majoration, *f.* additional charge, increase (in price), increased allowance, excess.

majorer, to overvalue, increase (the cost or price), add an allowance to, make additional charge.

majorité, *f.* majority.

majuscule, capital (letters).

makis, *m.* thicket, bush, shrub.

mal, maux, *m.* evil, trouble, wrong, injury, misfortune, disease, harm, ache, ill; — **chimique,** necrosis of the lower jaw, phosphorus necrosis; — **de taupe,** poll evil; — **du garrot,** saddle gall.

mal, bad, ill, badly, poorly, wrong, amiss; **au plus —,** very ill, past recovery.

malacanthe, with flowers in head soft to touch.

malachite, *f.* malachite.

malacoderme, malacodermous.

malacogame, malacophilous, pollinated by snails.

malacophile, malacophyllous.

malade, *m.&f.* sick person, invalid.

malade, ill, sick, diseased, poorly.

maladie, *f.* malady, disease, illness, complaint, sickness, disorder, mania, passion; — **contagieuse,** epidemic, contagious disease, epizooty; — **cutanée,** skin disease; — **de coeur,** heart disease; — **de l'écorce de l'épicéa,** Norway-spruce bark disease; — **de l'encre du chataignier,** root rot of the chestnut; — **des aiguilles du sapin,** white-spruce leaf spot; — **des chiens,** distemper; — **des taches,** leaf spot.

maladie, — du bois, disease of wood or trees; — **du chou,** stem rot; — **du jaune,** yellow disease; — **du renflement,** rickets; — **du rond,** disease caused by *Armillaria mellea* and other fungi; — **du rouge,** needle-shedding disease; — **naviculaire,** navicular disease; — **ronde,** seedling root rot; — **traumatique,** disease caused by wound; — **vermineuse,** vermination.

maladif, -ive, sickly, unhealthy.

maladresse, *f.* awkwardness, clumsiness, blunder.

maladroit, awkward, clumsy, unskillful, unhandy.

maladroitement, awkwardly, unskillfully, maladroitly, bunglesomely.

malaga, *m.* Malaga (wine).

malaguette, *f.* melegueta pepper, grains of Paradise, guinea grains (*Amomum melegueta*)

malaire, malar, zygomatic.

malais, malai, Malay, Malayan.

malaise, *m.* discomfort, uneasiness.

malaisé, difficult, arduous, hard.

malaisément, with difficulty, unwillingly.

Malaisie, *f.* Malaysia, Malay Archipelago.

malambo, *m.* malambo (bark of *Croton malambo*).

malandre, *f.* rotten knag, knot, malanders.

malandreux, -euse, knaggy, snaggy, malandrous.

malaria, *f.* malaria, marsh fever.

malate, *m.* malate.

malavisé, unwise, imprudent, rash.

malaxage, *m.,* **malaxation,** *f.* malaxation, kneading (dough), malaxing, working (butter).

malaxer, to malax, malaxate, kneading (dough), mixing (cement), massage, work (butter).

malaxeur, *m.* malaxator, worker, cement mixer, (for clay) pugmill, butterworker.

mâle, *m.* male, cock, buck, dog, bull, courage.

mâle, male, manly.

maléique, maleic.

malentendu, *m.* misunderstanding.

malfaisant, injurious, harmful, malicious, mischievous, noxious, unhealthy, malevolent.

malfil, *m.* press bags of coarse woolen cloth (for fatty acids).

malformation, *f.* malformation.

malgré, in spite of, notwithstanding, though; — **que vous en ayez,** in spite of all you may say or do.

malhabile, unskillful, awkward, clumsy.

malherbe, *f.* mezereum (*Daphne mezereum*), European leadwort (*Plumbago europaea*).

malheur, *m.* misfortune, unhappiness, illluck; **par —,** unfortunately, unhappily.

malheureusement, unfortunately, unhappily, unluckily.

malheureux, -euse, *m.&f.* destitute person, unfortunate wretch, poor creature.

malheureux, -euse, unfortunate, unhappy, unlucky, wretched, poor.

malhonnête, dishonest.

malhonnêtement, dishonestly.

malice, *f.* malice, wickedness, mischief, spite, maliciousness.

malicieusement, maliciously, viciously.

malicieux, -euse, malicious, vicious.

maliforme, maliform, shaped like an apple.

maligne, malignant, wicked.

malin, malignant, malicious, cunning.

malique, malic.

malle, *f.* trunk, mail.

malléabiliser, to malleableize.

malléabilité, *f.* malleability, pliability.

malléable, malleable.

malléer, to beat, hammer out (metal).

malléine, *f.* mallein.

malodorant, having a bad odor.

malon, *m.* brick.

malonique, malonic.

malpighien, -enne, Malpighian.

malpropre, dirty, unclean, squalid, slovenly.

malpropreté, *f.* dirtiness, uncleanliness, indecency.

malsain, unhealthy, unwholesome, pernicious.

malt, *m.* malt; — **á'avoine,** rolled oats, oatmeal; — **pelleté,** turned malt; — **touraillé,** kiln-dried malt.

maltage, *m.* malting.

maltais, Maltese.

maltase, *f.* maltase.

malter, to malt.

malterie, *f.* malthouse.

malteur, *m.* maltster, maltman.

malthe, *f.* maltha, mineral tar.

maltose, *f.* maltose.

maltraiter, to maltreat, abuse.

malvacées, *f.pl.* Malvaceae.

malvenant, unthrifty, badly grown, badly shaped.

malvoisie, *f.* malmsey, malvasia.

mamelle, *f.* breast, udder, bag, mamma, teat, nipple; **glande** —, mammary gland.

mamelon, *m.* mammilla, nipple, teat, hummock, papilla, mamelon, gudgeon, rounded hill, protuberance.

mamelonné, mammillary, mammillate(d), rounded, warted, knobbed.

mamillaire, mammillary.

mamillé, mamilleux, -euse, mammillate, with teat-shaped processes, papillate.

mammaire, mammary (gland).

mammifère, *m.* mammal, mammifer, breast bearer.

mammifère mammalian, mammiferous.

mammiforme, mammiform, mammillary.

mammite, *f.* mammitis; — **contagieuse,** *f.* infectious mastitis.

manche, *m.* handle, holder, haft.

manche, *f.* sleeve, hose, channel, tube, shaft, pipe; — **à air,** ventilating pipe; — **du sternum,** manubrium; **la Manche,** the (English) Channel.

mancheron, *m.* plow handle, stilt.

manchette, *f.* cuff, sleeve of wire netting; — **annulaire,** annulus.

manchon, *m.* sleeve, collar, flange, jacket, disk, casting, housing, shell, socket, mantle, muff, clutch, coupling (head), cylinder; — **d'accouplement,** clutch, coupling; — **élastique,** flexible coupling; — **pour éclairage,** incandescent mantle.

manchonnier, *m.* cylinder blower.

manchot, *m.* penguin.

mancienne, *f.* wayfaring tree.

mancône, *m.* sassy (poisonous) bark, mancona bark (of *Erythrophloeum guineënse*).

mandarine, *f.* mandarin, tangerine.

mandarinier, *m.* mandarin tree (*Citrus nobilis deliciosa*).

mandat, *m.* mandate, commission, order, draft, warrant, money order, mandator, customer; — **de poste,** post-office money order.

mandataire, *m.* proxy, representative, attorney.

mandchon, Manchu, Manchurian.

mandélique, mandelic.

mander, to send for, instruct, summon.

mandibule, *f.* mandible, beak, jaw.

mandragore, *f.* mandrake (Mandragora); — **officinale,** mandrake (*Mandragora officinarum*).

mandrin, *m.* mandrel, former, punch, swage, chuck, reel-up drum.

mandriner, to work or form upon a mandrel, chuck (work on lathe), punch, gauge, drift (holes), swage (iron), expand.

manducation, *f.* mastication, act of eating.

manège, *m.* adroit conduct, horse gear.

mânes, *m.pl.* shades, manes.

maneton, *m.* crankpin.

manette, f. handle, hand lever, trowel.

manganate, m. manganate.

mangane, m. manganese.

manganèse, m. manganese; — carbonaté, manganese spar, rhodochrosite, carbonate of manganese; — oxydé, pyrolusite.

manganésé, containing manganese.

manganésien, -enne, containing manganese.

manganésifère, manganiferous.

manganésique, manganic.

manganeux, manganous.

manganique, manganic.

manganite, f. manganite.

manganium, m. manganese.

mangeable, eatable, edible.

mangeaille, f. feed (for fowls, domestic animals), victuals, (soft) food.

manger, to eat, consume, wear away, damage, ruin, corrode, destroy, feed; se —, to be eaten, be consumed.

manger, m. food, eating.

mangeur, m. eater.

mangle, f. mangrove fruit.

manglier, m. mangrove (Rhizophora mangle).

mangold, m. beet.

mangrove, f. mangrove.

mangue, f. mango.

manguier, m. mango tree (Mangifera indica).

mani, m. resin (from the Guiana candlewood tree).

maniable, manageable, easy to handle or control, handy, workable.

maniage, m. handling, kneading (of clay).

manicome, f. lunatic asylum.

manie, f. mania.

maniement, m. feeling, handling, conduct, rustle (of silk).

manier, to handle, manipulate, feel, manage.

manier, m. feel, handling, touch.

manière, f. manner, way, method, sort, kind; — de fonctionner, working manner, procedure; — de voir, opinion, point of view; à la — de, in the manner of; de — à, so as, so as to, in such a way as; de — que, de — à ce que, in such a way that, so that; de . . . —, in . . . way, manner; par — de, by way of.

manifestation, f. manifestation.

manifeste, manifest, evident, obvious, plain.

manifestement, manifestly.

manifester, to manifest, display, show, give evidence of, make known; se —, to manifest oneself, be made manifest.

manigaux, m. pl. lever (of bellows).

maniguette, f. see MALAGUETTE.

manihot, m. see MANIOC.

manilla, m. Manila hemp, Manila cigar, manilla.

manille, f. shackle, connecting link, Manila rope, manila.

maniment, m. feeling, handling, conduct.

manioc, m. manioc, cassava (Manihot).

manipulateur, m. manipulator, (telegraph) key.

manipulation, f. manipulation, handling.

manipule, m. handful.

manipuler, to manipulate, handle.

manipuleur, -euse, m.&f. manipulator.

manivelle, f. crank, crank handle, winch.

mannane, m. mannan, a hemicellulose.

manne, f. manna, basket, hamper; — en sorte, common manna.

mannequin, m. mannikin, wicker crate, basket, hamper, phantom.

mannipare, producing manna.

mannitane, f. mannitan.

mannite, f. mannite, mannitol, manna sugar.

mannitique, mannitic.

mannoheptite, f. mannoheptite.

mannonique, mannonic.

mannose, f. mannose.

manoeuvre, m. workman, (unskilled) laborer.

manoeuvre, f. maneuver, working, managing, operation, maneuvering, intrigue, control, handling, evolution (of a ship), rope, tackle.

manoeuvrer, to work, operate, manipulate, manage, control, handle, maneuver.

manomètre, m. manometer, pressure gauge; — du vide, vacuum gauge; — pour pneu, tire gauge.

manométrique, manometric(al).

manouvrier, m. workman, day laborer.

manquant, m. absentee, missing person.

manquant, missing, absent, wanting.

manque, m. want, lack, deficiency, shortage; de — de, for want of.

manqué, missed, unsuccessful, abortive.

manquement, *m.* lack, want, failure, omission.

manquer, to be lacking, be wanting, be short, fail, miss, want, be deficient, be in need of; **se —,** to be missed; **— de,** to lack, want, escape, miss; **ne pas — de faire quelque chose,** not to neglect to do something; **s'en — de,** to be far from it; **il ne manque plus que cela,** that is the last straw; **il manque de le faire,** he fails to do it.

mansarde, *f.* curb roof, attic, garret.

mansienne, *f.* wayfaring tree.

manteau, *m.* mantle, cloak, mask, shell, casing, outer mold, hood, mantelpiece; **— de dame,** lady's-mantle (*Alchemilla vulgaris*); **— royal,** columbine (*Aquilegia vulgaris*); **recouvert d'un — de poils courts et doux,** downy.

mantelet, *m.* mantlet.

mantidés, *m.pl.* Mantidae, carnivorous insects.

mantisse, *f.* mantissa.

manubrium, *m.* manubrium.

manucodiates, *f.pl.* Manucodiatae, Paradisea.

manuel, -elle, manual, made by hand.

manuellement, manually, with the hand.

manufacture, *f.* (manu)factory, (iron) mill, manufacture.

manufacturer, to manufacture.

manufacturier, *m.* manufacturer.

manufacturier, manufacturing.

manuluve, *f.* hand bath.

manuscrit, *m.* manuscript.

manutention, *f.* management, administration, handling (of goods), storehouse, bakery.

manutentionnaire, *m.* manager, storekeeper.

manutentionner, to handle (goods), prepare, work up (tobacco), bake (bread).

manzanilla, *m.* manzanilla (dry and brown sherry).

mappe, *f.* plan, map.

mappemonde, *f.* map of the world.

maquereau, *m.* mackerel.

maquette, *f.* dummy (of book).

maquiller, to make up, paint (the face).

maquis, *m.* thicket, bush, shrub.

maraîchage, *m.* truck farming, fruit farming, nursery.

maraîcher, *m.* market gardener, greengrocer.

maraîcher, market.

marais, *m.* marsh, morass, bog, swamp, moor, fen, market garden, truck farm; **— salant,** saltern, salt garden, saline, salt marsh; **— tourbeux,** peat bog, peat moor.

marante, *f.* arrowroot (Maranta).

marasca, *m.,* **marasque,** *f.* marasca (*Prunus cerasus marasca*).

marasquin, *m.* maraschino (liqueur).

marbre, *m.* marble, marbling, slab, stone, block, dressing plate, imposing table, cast-iron surface plate.

marbré, marbled, mottled, variegated, veined.

marbrer, to marble, mottle, vein, marver.

marbrière, *f.* marble quarry.

marbrure, *f.* marbling, mottling.

marc, *m.* marc, murc, residue, residuum, pomace, grape husks and seeds; **— de baies,** berry pomace, berry baggings; **— de vendange,** marc of grapes; **eau de vie de —,** white brandy.

marcassite, *f.* marcasite, white iron pyrites.

marcassin, *m.* young wild boar.

marceau, *m.* sallow, goat willow (*Salix caprea*).

marceline, *f.* marceline (fabric).

marcescent, marcescent, subpersistent, faded, withering, fading.

marchand, *m.* merchant, dealer; **— de gros,** wholesale dealer; **— de volaille,** poulterer; **— grainier,** seed dealer.

marchand, commercial, mercantile, market, salable, marketable.

marchandage, *m.* bargaining.

marchander, to bargain, haggle.

marchandise, *f.* merchandise, goods, wares, freight; **— disponible,** spot goods; **—s de rebut,** rejections.

marche, *f.* course, procedure, advance, progress, running, sailing, walk, walking, gait, travel, march, step, stair, motion, rate, speed, pedal, track, hoofprint; **— à vide,** no load; **en —,** while running, in motion; **mettre en —,** to start, set going.

marché, *m.* market, price, bargain, sale, purchase, contract; **— indigène,** domestic market, home market; **— mondial,** world market; **à bon —,** cheap, cheaply; **bon —,** cheapness; **faire bon — de,** to hold cheap; **par-dessus le —,** into the bargain.

marchepied, *m.* step, footboard, stepping-stone, footpath.

marcher, to go, progress, run, travel, advance, move on, work, tread, walk, march, behave; **faire —,** to make go, tread.

marcher, *m.* walking, walk.

marcheur, marcheux, *m.* treader, walker.

marchoir, marcheux, *m.* kneading trough, kneading place, kneader or treader.

marciume, *m.* soft rot.

marcottage, *m.* layering; **— par cépée,** cuttings of root.

marcotte, *f.* layer, runner.

marcotter, to make layers, layer.

mardi, *m.* Tuesday.

mare, *f.* pool, pond, fen.

marécage, *m.* marsh, swamp, bog, moor, morass.

marécageux, -euse, marshy, swampy, boggy.

maréchal, *m.* marshal; **— ferrant,** blacksmith, farrier.

maréchalerie, *f.* blacksmith shop, farriery, blacksmith work.

marée, *f.* tide, fresh (sea) fish; **— descendante,** falling tide, ebb tide; **— haute,** high water, high tide; **— montante,** rising tide, flood tide; **odeur de —,** odor of fresh fish.

marémoteur, tide-driven (machine), tidal.

marennine, *f.* marennin, a light blue pigment.

margarate, *m.* margarate.

margarine, *f.* margarine (artificial butter), margarin (glyceryl margarate).

margarinerie, *f.* margarine manufacture.

margarinier, *m.* margarine maker.

margarique, margaric (acid).

margarite *f.* margarite.

marge, *f.* margin, border, edge, (spare) time.

margeoir, *m.* a plate to cover both openings of (glass) furnace.

marger, to close or stopper (glass furnace).

marginal, -aux, marginal.

marginé, edged, marginate.

marginer, to margin, annotate.

margotin, *m.* small fagot.

margousier, *m.* margosa, China tree, bead tree (*Melia azedarach*).

marguerite, *f.* marguerite, daisy (*Chrysanthemum frutescens*).

mari, *m.* husband.

mariage, *m.* marriage.

marié, *m.,* bridegroom; *pl.* married couple.

mariée, *f.* bride.

marier, to marry, unite; **se —,** to marry, blend, go with, harmonize.

marin, *m.* seaman, mariner, sailor.

marin, marine, maritime, sea.

marinade, *f.* marinade, pickle, brine, pickled meat.

marine, *f.* navy, marine.

mariné, pickled, sea-damaged.

mariner, to pickle, sour, silage.

maringouin, *m.* mosquito, sand fly.

maritime, maritime, sea.

marjolaine, *f.* marjoram (Origanum); **— commune,** sweet marjoram (*Marjorana hortensis*); **— sauvage,** wild marjoram (*Origanum vulgare*).

mark, *m.* (German) mark.

marmelade, *f.* marmalade, jam, (mixed) fruit sauce, compote; **— de pommes,** apple sauce.

marmenteau, *m.* reserved tree.

marmite, *f.* pot, pan, kettle, boiler; **— de Papin,** Papin digester; **— émaillée,** enameled saucepan; **— norvégienne,** cooking box.

marmitée, *f.* potful, kettleful.

marmoré, marbled, mottled, variegated, veined.

marmoriforme, marblelike.

marmorisation, *f.* marmarization, marbleization.

marmoriser, to marmarize.

marmotte, *f.* marmot.

marmotter, to mutter, mumble.

marnage, *m.* marling (of soil), liming, chalking.

marne, *f.* marl; **— cendrée,** earthy marl; **— schisteuse,** slaty marl, marl slate; **— à foulon,** fuller's earth.

marner, to marl.

marneux, -euse, marly, marlaceous.

marnière, *f.* marlpit.

maro, *m.* maro (Papican apron).

marocain, *m.* Moroccan.

marocain, Moroccan.

maronage, *m.* right to building timber.

maroquin, *m.* morocco, morocco leather.

maroquinage, *m.* morocco tanning or dressing.

maroquiné, morocco-tanned, morocco-dressed.

maroquiner, to morocco tan, morocco dress, give a morocco finish.

maroquinerie, *f.* morocco tanning or dressing, morocco factory, morocco article.

maroquinier, *m.* morocco tanner or dresser.

marouflage, *m.* glued-on lining (as of canvas).

maroufle, *f.* strong glue (for remounting pictures).

maroufler, to line (with canvas), remount (painted picture) on new foundation.

maroute, *f.* mayweed, dog fennel, stinking camomile (*Anthemis cotula*).

marquage, *m.* marking.

marquant, striking, conspicuous, prominent.

marque, *f.* mark, brand, make, proof, trademark, sign, stamp, marker, score, tally; — **d'eau,** water gauge, watermark; — **de fabrique,** trade-mark; — **de subdivision,** graduation on scale, scale line.

marqué, marked, evident, spotted.

marquer, to mark, blaze, spot, brand, scratch, stamp, note, sign, indicate, denote, assign; **se —,** to be noted, be distinguished, appear; — **d'un blanchis,** to blaze, mark; — **en délivrance,** to mark trees for felling.

marqueter, to mark with spots, inlay.

marqueterie, *f.* marquetry, inlaid work, patchwork.

marquette, *f.* a cake of virgin wax.

marqueur, -euse, *m.&f.* marker, stamper, brander.

marquise, *f.* type of rocket, awning, canopy, shelter.

marron, maroon, runaway, run wild, unlicensed.

marron, *m.* (large edible) chestnut, Spanish chestnut (*Castanea sativa*), chestnut (color), maroon, firecracker, marker, tag; — **d'Inde,** horse chestnut (*Aesculus hippocastanum*), buckeye.

marronnier, *m.* Spanish chestnut tree; **faux —,** — **d'Inde,** horse chestnut tree (*Aesculus hippocastanum*), buckeye.

marrube, *m.* horehound (Marrubium), madwort, bugleweed; — **blanc,** — **commun,** white horehound, common horehound (*Marrubium vulgare*).

mars, *m.* March, Mars (iron).

marsala, *m.* Marsala (wine).

marseillais, of Marseilles.

marsouin, *m.* porpoise.

marsupiflore, with flowers resembling pouches.

marsupium, *m.* marsupium.

marte commune, *f.* pine marten.

marteau, *m.* hammer, striker, knocker, clapper; — **à queue,** tilt hammer; — **à vapeur,** steam hammer; — **compteur,** die hammer, numbering hammer; — **forestier,** range hammer; — **pioche,** a combined hammer and pick, pickax; — **plantoir,** planting hatchet.

marteau-foulon, *m.* heavy hammer for softening hides.

marteau-pilon, *m.* power hammer, drop hammer.

martelage, *m.* hammering, marking trees for felling; — **au martinet,** tilting.

marteler, to hammer, hammer out, torment, worry, mark for felling.

martelet, *m.* small hammer.

martial, -aux, martial, of iron, warlike, ferruginous.

martiner, to hammer, tilt.

martinet, *m.* tilt hammer, martin (the bird), swift.

martin-pêcheur, *m.* kingfisher.

martin roselin, rose-colored starling.

martre, *f.* marten; — **commune,** pine marten.

marum, *m.* cat thyme (*Teucrium marum*).

marute, *f.* mayweed; see MAROUTE.

mascaron, *m.* mask (an ornamented piece).

masculin, *m.* masculine, male, mannish.

mash, *m.* (bran)mash.

masque, *m.* mask, disguise, blind, pretense, screen, cover; — **d'abatage,** killing mask; — **respirateur,** gas mask, respirator.

masquer, to mask, hide, disguise, screen, cover, conceal.

massacrer, to massacre.

masse, *f.* mass, weight, fund, capital (stock), volume, (varying) quantity, estate, share, block (of stone), heap, sledge hammer, maul, beetle, stake (in games), ground, earth; — **active,** assets; — **de fibres,** bundle of fibers; — **d'équilibrage,** counterpoise, counterweight; — **passive,** liabilities; **en —,** in a mass, at large, in a body, together; **par —,** by the lump.

massé, *m.* sponge, bloom.

masseau, *m.* shingled bloom.

masselet, *m.* small bloom.

masselotte, *f.* deadhead (of casting), (feed-)head, sprue, runner, weight, plunger.

masser, to mass, massage; **se** —, to form masses or in crowds.

masseur, -euse, *m.&f.* masseur.

massiau, *m.* shingled bloom.

massicot, *m.* massicot, yellow lead, paper cutter, lead oxide, litharge.

massif, *m.* solid mass, main part, shell, clump of shrubs, substructure, pier, wall, tower, group, block, mass of masonry, mass of mountains, wood, crop of trees, close crop or stand; — **complet,** — **plein,** complete crop or stand; — **d'abri,** shelterwood; — **interrompu,** incomplete crop or stand; **croître en** —, to grow under cover.

massif, -ive, massive, solid, bulky.

massivage, *m.* tamping, packing.

massivement, massively, heavily.

massiver, to tamp, pack, beat, mortar.

massiveté, *f.* massiveness.

massoque, *f.* slab (half-bloom).

massoquette, *f.* small bloom.

massoquin, *m.* slab (half-bloom).

massoy, *n.* massoy, massoy bark.

massue, *f.* club, knob; — **feuilleté,** lamellate club; **en forme de** —, club-shaped, clavate.

mastic, *m.* mastic, cement, compound, putty; — **à greffer,** grafting wax; — **asphaltique,** asphaltic cement; — **de fer,** rust cement.

masticage, *m.* cementing, puttying, filling, stopping.

mastication, *f.* mastication, chewing.

masticatoire, masticatory.

masticine, *f.* masticin.

mastigophores, *f.pl.* Mastigophora.

mastiquer, to cement, fill in, putty, masticate, manducate, chew.

mastite, *f.* mastitis.

mastoïde, mastoid.

mat, *m.* mat (dull) surface, dull finish.

mat, mat, dead, dull, unpolished, not bright, heavy (dough), frosted.

mât, *m.* mast, pole, tower; — **d'amarrage,** mooring mast.

maté, *m.* maté, Paraguay tea (*Ilex paraguayensis*).

matelas, *m.* layer, mattress, lining, backing, cushion, padding; — **à air,** air cushion; — **à eau,** water bed.

matelasser, to pad, stuff, cushion.

matelassure, *f.* padding, stuffing, lining, backing.

matelot, *m.* sailor, seaman.

mater, to mat, dull, render mat, deaden, caulk, hammer, flatten, beat down, burr, work (dough), make heavy.

matérialiser, to materialize.

matériaux, *m.pl.* materials, material; — **bruts,** raw material(s).

matériel, *m.* material, equipment, stock, stores, plant, implements (of farm); — **brut,** raw or starting material; — **d'emballage,** pack(ag)ing material; — **normal,** normal growing stock; — **pour expéditions,** shipping material; — **roulant,** rolling stock.

matériel, -elle, material, of matter, corporeal, coarse.

matériellement, materially, sensually, coarsely.

maternel, -elle, maternal, motherly.

materniser, to imitate human milk, imitate one's mother.

maternité, *f.* maternity, motherhood, lying-in hospital.

mathématicien, *m.* mathematician.

mathématique, *f.(pl.)* mathematics.

mathématique, mathematical.

mathématiquement, mathematically.

mathématiquer, to do mathematics.

matico, *m.* matico (*Piper angustifolium*).

matière, *f.* matter, substance, subject, cause, material, pus, purulent matter, feces, excreta; — **agglutinante,** binding material, fixing agent, adhesive; — **blanchissante,** bleaching agent, bleaching compound; — **brute,** raw material; — **caséeuse,** caseous matter; — **colorante,** coloring matter, dye, pigment.

matière, — de remplacement, substitute; — **dulcifiante,** sweetening; — **fibreuse,** paper fiber; — **nutritive,** food material; — **organique,** organic matter; — **première,** raw material; — **résiduaire,** residual product, by-product; — **sucrée,** carbohydrate, saccharine matter, —**s d'argent,** silver bullion; **en** — **de,** in point of, in matters of, with respect to; **tables des** —**s,** table of contents.

matin, *m.* morning, forenoon; **le** — **de,** **bonne heure,** early in the morning.

mâtin, *m.* watchdog, house dog.

matinal, -aux, early, morning.

mâtiné, mongrel, cross-bred.

matinée, *f.* matinee, morning, forenoon, the morning period.

matir, to mat, dull, render mat, deaden.

matité, *f.* dullness.

matoir, *m.* stamp, riveting hammer, matting tool, calking chisel.

matou, *m.* male cat, tomcat.

matras, *m.* matrass, bolthead; — **d'essayeur,** assayer's matrass (glass tube); — **jaugé,** volumetric flask; — **Pasteur,** culture flask with ground cap.

matricaire, *f.* feverfew (*Chrysanthemum parthenium*), wild camomile (*Matricaria chamomilla*).

matrice, *f.* matrix (of rocks), womb, uterus, standard weight or measure.

matrice, primary (colors), mother, parent.

matriculer, to register, enroll.

matrone, *f.* matron, midwife.

mattage, *m.* matting, dulling.

matte, *f.* matte, mat, coarse metal; — **blanche,** white metal; — **crue,** crude (rough) mat; — **de cuivre,** copper mat.

matteau, *m.* bundle (of hanks), hank (of raw silk).

matter, to render mat, dull, deaden.

mattoir, *m.* matting tool.

maturatif, *m.* maturant, maturative.

maturatif, -ive, maturative.

maturation, *f.* maturing, ripening, maturation, refining; — **du fromage,** cheese ripening, curing.

maturer, to mature, ripen, refining.

maturité, *f.* maturity, ripeness.

maudire, to curse.

maudit, cursed, abominable, horrible.

mauerparenchym, *m.* muriform parenchyma.

mauresque, Moorish.

mausolée, *m.* mausoleum.

mauvais, bad, injurious, ill-natured, ill, badly, wrong; **faire —,** to be hard, be disagreeable (weather); **trouver —,** to dislike.

mauve, *f.* mauve, mallow (Malva), sea gull; — **sauvage,** wild mallow (*Malva sylvestris*).

mauvéine, *f.* mauveine.

mauvis, *m.* redwing.

maux, *m.pl.* see MAL.

maxillaire, maxillary.

maxima, maximum.

maxime, *f.* maxim, axiom.

maximum, *m.* maximum, acme; **au —,** at (to) the maximum.

maye, *f.* trough (for kneading).

mayonnaise, *f.* mayonnaise.

mazéage, mazage, *m.* refining of pig (cast) iron.

mazée, *f.* refined metal.

mazér, to refine (pig iron).

mazerie, *f.* refinery.

mazout, *m.* mazut, fuel oil.

me, me, to me.

méandriforme, sinuous, sinuate.

méat, *m.* meatus; — **au intervalle,** intercellular space; — **intercellulaire,** small triangular intercellular space.

mécanicien, *m.* mechanic, machinist, engineer.

mécanique, *f.* mechanics, mechanism.

mécanique, mechanical.

mécaniquement, mechanically.

mécaniser, to mechanize, make a machine.

mécanisme, *m.* mechanism.

mécery, *f.* various kinds of white, black, or reddish opium.

méchage, *m.* matching, fumigating (of casks).

méchamment, wickedly.

méchanceté, *f.* wickedness.

mechanomorphose, *f.* mechanomorphosis, mechanical changes in structure.

méchant, bad, wicked, unkind, mean, wretched, naughty.

mèche, *f.* wick, match, touch, fuse, tinder, core, heart, bit, drill, auger, lock, curl, ringlet, rove.

mécher, to match, fumigate (casks).

méchoacan, *m.* mechoacan.

mécompte, *m.* miscount, error, mistake, misconception, disappointment.

méconate, *m.* meconate.

méconine, *f.* meconin.

méconinique, meconinic.

méconique, meconic.

méconium, *m.* meconium.

méconnaissable, unrecognizable.

méconnaissance, *f.* misjudgment, misconstruction.

méconnaitre, to disregard, repudiate, slight, disown, misunderstand, not know, not recognize, mistake, undervalue, not know again, not appreciate.

méconnu, unrecognized, unappreciated, misunderstood.

mécontent, discontented, dissatisfied.

mécontentement, *m.* dissatisfaction, displeasure.

mécontenter, to displease, dissatisfy.

médaille, *f.* medal, medallion.

médaille, *m.* medalist.

médaillé, decorated, medaled.

médaillon, *m.* medallion, locket.

médecin, *m.* physician, doctor.

médecine, *f.* medicine, physic.

médian, median, central, mesial.

médiastin, *f.* mediastinum, a very thin transverse partition.

médical, -aux, medical, medicinal (of plants).

médicament, *m.* medicament, medicine.

médicamentaire, medicinal.

médicamentation, *f.* treatment with medicine.

médicamenter, to treat with medicine, medicate.

médicamenteux, -euse, medicinal.

médicastre, *m.* quack.

médicateur, -trice, medicinal, remedial, curative.

médication, *f.* medication.

médicinal, -aux, medicinal.

médicinalement, medicinally.

médicinier, *m.* physic nut (*Jatropha curcas*).

médiéval, -aux, medieval.

médifixe, medifixed, fixed at the middle.

mediintestin, *m.* mid-intestine.

médiocre, mediocre, moderate, ordinary.

médiocrement, indifferently, moderately.

médire, to slander.

méditatif, -ive, meditative, thoughtful.

méditation, *f.* meditation, thought.

méditer, to meditate, contemplate, plan, think over.

méditerrané, mediterranean, interior.

Méditerranée, *f.* Mediterranean (Sea).

méditerranéen, -enne, of the Mediterranean.

médium, *m.* medium, compromise.

médius, *m.* middle finger, medius, second finger.

médivalve, medivalvis, fixed on the middle of the valves.

médoc, *m.* Medoc (claret).

médullaire, medullary, medullar, pithy.

médulle, *f.* medulla, pith.

méfiant, mistrustful, suspicious.

méfier, to be suspicious, mistrust, distrust; **il se méfie de lui,** he distrusts him.

mégalocarpe, with large fruit.

mégalomère, with greatly divided calyx.

mégalorhize, with large coarse roots.

mégalosperme, with coarse fruit.

méganthe, with large flowers.

mégaphylle, megaphyllous, with large leaves or leaflike expansions.

mégaplankton, *m.* macroplankton, pleuston plants.

mégarde, *f.* inadvertence; **par —,** inadvertently.

mégaspore, *f.* macrospore.

mégasporophylle, *f.* megasporophyll, carpel, sporophyll with megaspores.

mégastachyé, with large spikes.

mégatherme, *m.* megatherm, a tropical plant.

mégie, *f.* tawing (leather).

mégir, mégisser, to taw, dress (light skins).

mégis, *m.* alum steep, tawing paste.

mégis, tawed.

mégisserie, *f.* tawing, dressing.

mégissier, *m.* tawer.

mégistothermique, megistothermic, requiring high uniform temperature.

mehlfrucht, *f.* cereal.

meilleur, better, best; **faire —,** to be better.

meiophylle, *f.* meiophylly, suppression of one or more leaves in a whorl.

meiostémone, meiostemonous, with fewer stamens than petals.

méiotaxie, *f.* meiotaxy, suppression of entire whorls.

méjuger, to misjudge.

mélaleuque, black and white.

mélalome, with dark-colored border.

mélam, *m.* melam.

mélamine, *f.* melamine.

mélampyre, *m.* cowwheat (Melampyrum); **— des prés,** yellow cowwheat (*Melampyrum pratense*).

mélampyrite, *f.* melampyrite, dulcite.

mélananthère, with dark-colored anthers.

mélancolique, melancholy, gloomy.

mélange, *m.* mixture, mixing, crossing (breeds), blending, mingling, medley, jumble; — **de graines,** seed mixture; — **détonant,** explosive mixture; — **par pieds isolés,** mixture by isolated plants; — **réfrigérant,** freezing mixture.

melangé, mixed (breed), motley; **peuplement** —, mixed crop.

mélangeoir, *m.* mixer, mixing machine.

mélanger, to mix, blend, mingle, potch, poach.

mélangeur, *m.* mixer.

mélanine, *f.* melanin.

mélanique, melanotic, melanosed.

mélanite, *f.* melanite.

mélanosperme, melanospermous, having dark-colored seeds or spores.

mélanoxyle, with dark-colored wood.

mélanthère, with dark-colored anthers.

mélaphyre, *m.* melaphyre.

mélasperme, melanospermous, having dark-colored seeds or spores.

mélasse, *f.* molasses, treacle.

mélassigène, melassigenic.

mélassique, molasses, melassic (acid)

melastome, *f.* melastoma.

mêlé, mixed.

mêlée, *f.* melee, fray, scuffle.

mêler, to mix, mingle, blend, mix up, jumble; **se** —, to mix, be mixed, take part.

mélèze, *m.* larch tree (*Larix decidua*); — **du Japon,** Japanese larch (*Larix kaempferi*).

mélézitose, *f.* melezitose.

méliacées, *f.pl.* Meliaceae.

mélianthe, *m.* honeyflower.

mélibiose, *f.* melibiose.

mélicoque, *m.* honeyberry (*Melicocca bijuga*).

méligèthe, *m.* pollen beetle.

mélilot, *m.* melilot, sweet clover (Melilotus); — **blanc,** sweet clover, Bokhara clover (*Melilotus alba*); — **officinal,** yellow melilot (*Melilotus officinalis*).

mélilotique, melilotic.

mélinet, *m.* honeywort (Cerinthe).

mélinite, *f.* melinite.

mélisse, *f.* balm, balm mint (Melissa); — **officinale,** lemon balm, meliss balm (*Melissa officinalis*).

mélissique, melissic, melissylic.

mélissophylle, with leaves similar to those of melissa.

mélissique, melissic, melissylic.

mélitococcie, *f.* Malta fever.

mélitococcique, of Malta fever.

mélitococciques, *pl.* Malta-fever patients.

mélitose, *f.* melitose.

mélitriose, *f.* melitriose, raffinose, a sugar occurring in beets and germinating cereals.

mélitte, *f.* melittis.

mélittophiles, melittophilae, sweet-scented flowers pollinated by large bees.

mellate, *m.* mellitate.

melléolé, *m.* electuary.

melléolique, a medicine composed of honey and a powder.

mellique, mellitic.

mellite, *m.* mellite, honey preparation; — **de borax,** borax honey; — **de vinaigre,** oxymel.

mellithate, *m.* mellitate.

mellithe, *m.* mellite.

mellitique, **mellithique,** mellitic.

mellon, *m.* mellon.

mellophanique, meliophanic.

mélodieux, melodious.

méloé, *m.* oil beetle (Meloe).

melon, *m.* melon; — **d'eau,** watermelon.

melone, *f.* canteloupe.

mélongène, *f.* eggplant.

membrane, *f.* membrane, film; — **cellulaire,** cellular membrane; — **du corps vitré,** hyaloid membrane; — **muqueuse,** mucous membrane; — **ondulante,** undulating membrane; — **séreuse,** serous membrane; — **vibrante,** diaphragm.

membrané, **membraneux,** -euse, membranous, webbed.

membre, *m.* member, limb; — **postérieur,** hind limb.

membrure, *f.* frame, framework, limbs.

même, same, very, self, even, also, more, likewise.

même, *m.* same thing, same; **à** — (**la bouteille**), from (the bottle) itself; **à** — **de,** able to; **à peu près de** —, much the same; **au** — **temps,** at the same time; **au** — **temps que,** at the same time with; **dans le**

— **sens que,** directly as; **dans le — temps,** at once.

même, de —, the same, likewise; **de — que,** in the same manner as, so as, the same as, just as; **de — que . . . de — . . . , just as . . . so . . . ,** the same as . . . **so . . . ; en — que, as well as; en — temps,** at the same time; **par là —, par cela —,** for that very reason; **peu à — de,** unable to; **tout de —,** all the same, for all that; **un —,** one and the same.

mémoire, *f.* memory, remembrance.

mémoire, *m.* memoir, memorandum, report, bill, article, account, dissertation, memorial.

mémorable, memorable.

mémorandum, *m.* memorandum, notebook.

mémorial, *m.* memoirs, notebook, daybook.

menaçant, threatening.

menace, *f.* menace, threat.

menacer, to threaten, menace.

ménage, *m.* housekeeping, household, family, home, household goods, management, thrift, saving.

ménagé, careful.

ménagement, *m.* care, caution, consideration, circumspection, exploitation.

ménager, to manage, arrange, make, contrive, cut, dispose, take care of, make the most of, be careful of, spare, save, husband, obtain, secure; **se —,** to take care of oneself, spare oneself, be careful, get on well, be cared for.

ménager, *m.* housekeeper, economizer.

ménager, -ère, household, waste (water), careful, thrift, saving.

ménagère, *f.* housewife, housekeeper.

mendélien, Mendelian.

mendéliser, to Mendelise.

mendélisme, *m.* Mendelism.

mendiant, *m.* beggar.

mendier, to beg, implore.

mène, *f.* lye tank.

mener, to lead, conduct, take, carry, direct, guide, carry on, draw (a line), treat; **— à bien,** to carry out; **— à bonne fin,** to go through with, carry out; **il mène un grand train,** he lives high.

meneur, *m.* leader, driver.

méningé, meningeal.

méningite, *f.* meningitis.

méningocoque, *m.* meningococcus.

méniscoïde, meniscoid, crescent-shaped.

ménispermacées, *f.pl.* Menispermaceae.

ménisperme, *m.* menisperm.

ménisque, *m.* meniscus; — (**convergent**), convexo-concave lens.

mensonge, *m.* lie.

mensonger, -ère, untrue, deceitful, false.

menstrue, *m.* menstruum, solvent.

menstrues, *f.pl.* menses.

mensuel, -elle, monthly.

mensuellement, monthly, each month.

mensurabilité, *f.* measurability, mensurability.

mensurable, measurable, mensurable.

mensurateur, -trice, measuring.

mensuration, *f.* mensuration, measuring.

mental, -aux, mental.

mentalement, mentally.

menteur, -euse, *m.&f.* liar.

menteur, -euse, lying, false, deceitful.

menthe, *f.* mint, a liqueur flavored with mint; **— à épi,** spearmint (*Mentha spicata*); **— à feuilles rondes,** round-leaved mint (*Mentha rotundifolia*); — **anglaise,** peppermint (*Mentha piperita*); — **aquatique,** water mint (*Mentha aquatica*); — **crépue,** curled mint (*Mentha crispa*).

menthe, — de chat, catnip, catmint (*Nepeta cataria*); — **de cheval,** horsemint (*Mentha longifolia*); — **poivrée,** peppermint (*Mentha piperita*); — **pouliot,** pennyroyal (*Mentha pulegium*); — **sauvage,** horsemint (*Mentha longifolia*); — **verte,** spearmint (*Mentha spicata*).

menthe-coq, *f.* costmary, alecost (*Chrysanthemum balsamita*).

menthène, *m.* menthene.

menthol, *m.* menthol.

mentholé, mentholique, mentholated, containing menthol.

menthone, *f.* menthone.

mention, *f.* mention.

mentionner, to mention, speak of.

mentir, to lie, deceive, tell a falsehood, fib.

menton, *m.* chin.

mentonnet, *m.* catch, tappet, cam, wiper, lug.

mentonnière, *f.* chinpiece, muffle plate, mentum.

menu, *m.* small fragments, smalls, fines, detailed account, bill of fare, menu.

menu, small, fine, slender, minute, thin, minor; — **bois d'oeuvre,** small timber;

— **bois mort,** dead fallen branches, dry fallen wood; — **branchage,** twigs; — **charbon,** small coal; — **plomb,** small shot, bird shot; — **taillis,** brushwood; —**s produits,** minor produce; **menues houilles,** small coal, slack.

menuchon, *m.* scarlet pimpernel (*Anagallis arvensis*).

menuisage, *m.* cutting down (wood) to required size.

menuise, *f.* twigs, sticks.

menuiserie, *f.* joinery, carpentry, carpenter's shop, small objects of wood and metal.

ményanthe, *m.* buck bean (*Menyanthes trifoliata*).

méphitique, mephitic, noxious.

méphitiser, to render poisonous or mephitic, vitiate.

méphitisme, *m.* vitiation, mephitism.

méplat, flat, half-flat.

méprendre (se), to be mistaken, mistake.

mépris, *m.* contempt, scorn; **au (en) — de,** in defiance of, without regard to; **avec —,** scornfully.

méprise, *f.* mistake, oversight.

mépriser, to scorn, despise, condemn.

mer, *f.* sea, ocean, tide; **sur —,** at sea.

mercantile, commercial.

mercaptal, *m.* mercaptal.

mercaptan, *m.* mercaptan.

mercenaire, *m.* hireling, mercenary.

mercenaire, mercenary.

mercerisage, *m.* mercerizing, mercerization.

merceriser, to mercerize.

merci, *m.* thanks.

merci, *f.* mercy.

mercredi, *m.* Wednesday.

mercure, *m.* mercury; *cap.* Mercury; — **doux,** calomel, mercurous chloride; — **fulminant,** fulminating mercury; — **précipité blanc,** white precipitate; — **sulfuré,** cinnabar, vermilion.

mercureux, mercurous.

mercuriale, *f.* mercury (Mercurialis), price list, schedule; — **annuelle,** garden mercury; — **vicace,** dog's mercury (*Mercurialis perennis*).

mercurialiser, to mercurialize.

mercuriaux, *m.pl.* mercurials.

mercuriel, -elle, mercurial, (of) mercury.

mercurifère, containing mercury.

mercurique, mercuric (salt).

merde, *f.* excrement, dung; — **d'oie,** a greenish yellow color; — **du diable,** asafetida.

mère, *f.* mother, matrix, mold, source, dam.

mère, fine, pure, mother; — **laine,** mother wool; — **poule,** mother hen, clucker; **vin de la — goutte,** wine from first pressing, unpressed must; **eau —,** mother liquor.

mérenchyme, *f.* merenchyma, spherical cellular tissue.

méricarpe, *m.* mericarp.

méridien, -enne, meridian; — **protoplasmique,** spindle fiber.

méridienne, *f.* meridian, meridian line, compass plant.

méridional, -aux, *m.* southerner.

méridional, -aux, meridional, southern, meridian.

mérinos, *m.* merino (sheep).

mériphyte, *m.* meriphyte, vascular tissue of the leaf.

merise, *f.* wild cherry (*Prunus avium*), mazzard; — **de Virginie,** chokecherry (*Prunus virginiana*).

merisier, *m.* wild cherry tree (*Prunus avium*), mazzard.

mérisme, *m.* merism, a repetition of homologous parts.

méristèle, *f.* meristele, a leaf bundle.

méristème, *m.* meristem.

méristémone, with branched or ramified stamens.

mérite, *m.* merit, worth, person of merit.

mériter, to merit, deserve, require, be worthy; **se —,** to be merited.

mérithalle, *m.* internode.

méritoire, meritorious, deserving.

merlan, *m.* whiting; — **bleu,** mackerel.

merle, *m.* blackbird; — **à gorge noire,** black-throated thrush; — **à plastron,** ring ouzel; — **grive,** song thrush; — **mauvis,** redwing; — **noir,** blackbird, ouzel.

merleau, *m.* young blackbird.

merlin, *m.* cleaving ax.

merluche, *f.*, **merlus,** *m.* stockfish, dried codfish, hake.

merrain, *m.* stavewood, staves, main stem; — **douvain,** cooper's wood.

mérule, *m.* wood fungus, dry rot.

merveille, *f.* wonder, marvel; **à —,** wonderfully, admirably.

merveilleusement, wonderfully, marvelously.

merveilleux, *m.* marvelous (part), marvelousness.

merveilleux, -euse, wonderful, marvelous.

mes, *pl.* my.

mésaconique, mesaconic.

mésange, *f.* titmouse; — **charbonnière,** great titmouse *(Parus major);* — **des saules,** willow titmouse; — **huppée,** crested titmouse *(Parus cristatus);* — **lugubre,** somber titmouse; — **noire,** coal titmouse *(Parus ater);* — **nonnette,** marsh titmouse *(Parus palustris).*

mésarche, mesarch.

mésarque, mesarch.

mésaventure, *f.* mishap, accident.

mescal, *m.* mescal.

mésembryanthème, *m.* fig marigold (Mesembryanthemum).

mésentère, *m.* mesentery.

mésenterin, with irregular undulations on the surface.

mésentéron, mid-intestine.

mésestimer, to underrate, underestimate, undervalue.

mésitine, *f.* mesitine, mesitite.

mésitylène, *m.* mesitylene.

mésitylénique, mesitylenic.

mésoblaste, *m.* mesoblast, mesoderm.

mésoblastesis, *m.* microgonidium, a small gonidium.

mésocarpe, *m.* mesocarp.

mésochilium, *m.* mesochilium, the center of the labellum in certain orchids.

mésocotyl, *f.* mesocotyl, an interpolated node in the seedling of grasses.

mésoderme, *m.* mesoderm.

mésophylle, *m.* mesophyllum.

mésophyte, *f.* mesophyte.

mésopode, *m.* mesopodium.

mésosperme, *m.* mesosperm, middle coat of a seed, the sarcoderm.

mésospore, *f.* mesospore.

mésostylé, mesostylic, medium-styled.

mésotartrique, mesotartaric.

mésotherme, mesothermal, requiring a moderate degree of heat.

mésothermique, mesothermic, dwelling in the temperate zone.

mésothorax, *m.* mesothorax.

mésothorium, *m.* mesothorium.

mésozoïque, Mesozoic.

mesquin, mean, poor, paltry, petty, stingy.

mesquinerie, *f.* punctiliousness.

message, *m.* message.

messager, *m.* messenger, carrier.

messagerie, *f.* carrying trade, rapid transport, conveyance of goods, shipment, coach office.

messieurs, *m.pl.* see MONSIEUR.

mestome, *m.* mestome.

mesurable, measurable.

mesurage, *m.* measuring, measurement surveying.

mesure, *f.* measure, measurement, measuring, proportion, dimension, limit, determination, bound; **à —,** in proportion, successively, accordingly; **à — de,** in proportion to, according to; **à — que,** in proportion as, according as, as; **au fur et à —,** gradually, in proportion as; **dans la — de,** to the extent of, as much as; **dans la — du possible,** as much as possible.

mesure, en — de, in a position to; **être en — de,** to be able to; **— à pied,** a measure with foot or base; **— cylindrique,** measuring cylinder; **— de capacité,** stacked measure; **— de retrait,** measure of contraction or shrinkage; **— en ruban, tape measure;** **— unitaire,** unit measure; **outre —,** unmeasurably, beyond measure; **sans —,** immeasurably, without measure.

mesurer, to measure, weigh, consider, compare; **se —,** to be measured, measure oneself, one's strength; **— au niveau,** **— avec le niveau,** to level, take the level; **— de superficie,** square measure, superficial measure; **— le contour,** to take the girth (of a tree).

mesureur, *m.* measurer; **— de la pâte,** pulp meter.

mésuser, to misuse.

met (mettre), (he) puts, sets, places.

métabiose, *f.* metabiosis.

métabisulfite, *m.* metabisulphite, pyrosulphite.

métabolique, metabolic.

métabolisme, *m.* metabolism.

métacarpe, *m.* metacarpus.

métacellulose, *f.* metacellulose, fungus cellulose.

métachromasie, *f.* metachromatism.

métachromatine, *f.* metachromatin.

métachromatique, metachromatic.

métagénèse, *f.* metagenesis, alternation of generations.

métagynie, *f.* metagyny, protandry.

métairie, *f.* farm.

métakinése, *f.* metakinesis.

métal, *m.* metal, bullion; — **alcalino-terreux,** alkaline-earth metal; — **antifriction,** white or babbitt metal, pewter, antifriction metal; — **appreté,** dressed metal; — **brut,** raw metal, coarse metal; — **commun,** base or ignoble metal; — **d'allumage,** inflammable metal; — **de cloche,** bell metal; — **de fonte,** cast metal; — **déployé,** expanded metal; — **de rouissage,** retting process.

métal, — **doux,** soft metal; — **étiré,** drawn metal; — **fraisé,** milled metal; — **ignoble,** base or ignoble metal; — **natif,** native metal; — **ondulé,** corrugated metal; — **pour repoussage,** metal for embossing; — **précieux,** precious metal, noble metal; — **repoussé,** embossed metal; — **scié,** sawn metal; — **soudé,** wrought iron; — **vierge,** virgin metal, native ore.

métalbumine, *f.* metalbumin, pseudomucin.

métaldéhyde, *f.* metaldehyde.

métalléité, *f.* metallicity, metalleity.

métallescence, *f.* metallic luster.

métallescent, of metallic luster.

métallifère, metalliferous, metal-bearing.

métalliforme, metalliform, metallike.

métallin, metalline, metallic.

métallique, metallic, metal.

métalliquement, metallically, in metal, through metal, by means of metal.

métallisation, *f.* **métallisage,** *m.,* metallization, metallizing, plating, bronzing.

métalliser, to metallize, convert into metal.

métallochimie, *f.* the chemistry of metals.

métallochimique, relating to the chemistry of metals.

métallochromie, *f.* metallochromy.

métallogénie, *f.* metallogeny, treating origin of metalliferous deposits.

métallographe, *m.* metallographist.

métallographie, *f.* metallography.

métallographique, metallographic.

métalloïde, *m.* metalloid, nonmetal.

métalloïde, metalloid(al).

métalloïdique, metalloid(al).

métallurgie, *f.* metallurgy.

métallurgique, metallurgic(al).

métallurgiste, *m.* metallurgist.

métamère, *m.* metamere, somite.

métamère, metameric.

métamérie, *f.* metamerism.

métamorphique, metamorphic.

métamorphisme, *m.* metamorphism.

métamorphose, *f.* metamorphosis, transformation.

métamorphoser, to metamorphose, transform; **se** —, to change completely.

métandrie, *f.* metandry, protogyny.

métanthèse, *f.* metanthesis, retarded floral development.

métaphase, *m.* metaphase.

métaphloème, *f.* metaphloem.

métaphosphate, *m.* metaphosphate.

métaphosphorique, metaphosphoric.

métaphylle, *f.* metaphylla, the mature leaf.

métaphysicien, *m.* metaphysician.

métaphysique, *f.* metaphysics.

métaphysique, metaphysical.

métaphysiquement, metaphysically.

métaphyte, *n.* metaphyte.

métaplasie, *f.* metaplasia.

métaplasma, *f.* metaplasm.

métargon, *m.* metargon.

métastannique, metastannic.

métathorax, *m.* metathorax.

métatonique, metatonic, (stimulus) reversing action.

métatype, *f.* mutation of type.

métaux, *m.pl.* see METAL.

métaxénie, *f.* metaxeny.

métaxylème, *f.* metaxylem.

métazoaires, *m.pl.* Metazoa.

méteil, *m.* mixed crops of wheat and rye.

météore, *m.* meteor.

météorique, meteoric.

météorite, *m.* meteorite.

météoritique, meteoritic.

météorologie, *f.* meteorology.

météorologique, meteorologic(al).

météorologiste, *m.* meteorologist.

méthane, *m.* methane.

méthanique, methane, pertaining to methane series.

méthanol, *m.* methanol, wood alcohol.

méthanomètre, *m.* methanometer.

méthémoglobine, *f.* methemoglobin.

méthionique, methionic.

méthode, *f.* method, system, process, way, elementary method; — **d'aménagement,** method of determining the regular yield; — **de fonctionnement,** working manner, procedure; — **des taux indicateurs,** indicating method; — **pondérale,** gravimetric method.

méthodique, methodical.

méthodiquement, methodically.

méthodiste, methodizing.

méthol, *m.* methyl alcohol.

méthoxyle, *m.* methoxyl.

méthylal, *m.* methylal.

méthylamine, *f.* methylamine.

méthylate, *m.* methylate.

méthyle, *m.* methyl.

méthyle-éthyle-cétone, *f.* methyl ethyl ketone, butanone.

méthylène, *m.* methylene; — **de bois,** wood spirit, wood (methyl) alcohol.

méthylénique, methylene, of methylene.

méthylique, methyl(ic), of methyl.

méthyliser, to methylate.

méthylorange, *m.* methyl orange.

méthylure, *m.* methide.

méthylviolet, *m.* methyl violet.

méticuleux,-euse, meticulous, (over)scrupulous, fastidious, overnice.

méticulosité, *f.* meticulousness, overcarefulness, punctiliousness.

métier, *m.* trade, profession, handicraft, occupation, loom, frame; — **à bras,** — **à la main,** hand loom; — **à filer,** spinning frame; — **à tisser,** loom; — **secondaire,** supplementary business, side-line business.

métière, *f.* evaporating basin (for salt).

métis, *m.* cross.

métis, hybrid, half-breed, mongrel, redshort.

métissage, *m.* crossing, crossbreeding.

métisser, to crossbreed, cross; **se** —, to cross.

métol, *m.* metol.

métrage, *m.* measuring, measurement, (metric) length.

mètre, *m.* meter (39.37 inches), folding pocket rule; — **anglais,** yard; — **carré,** square meter (1.196 sq. yd.); — **cube,** cubic meter, solid meter (35.1 cu. ft.).

métré, *m.* measurement (in meters).

mètre-kilogramme, *m.* kilogrammeter.

métrer, to measure in meters.

métrique, metric(al).

métrogonidie, *m.* microgonidium, a small gonidium.

métrologie, *f.* metrology.

métrologique, metrological.

métrophotographique, photogrammetric

métropole, *f.* metropolis, mother country, capital.

métropolitain, metropolitan, home, archiepiscopal, Paris subway.

mets, *m.* (article of prepared) food, dish, mess, meal.

mettre, to put, place, set, put on, lay, take (of length of time), put down, devote to, reduce, invest (funds), use, employ, make, cause to consist; **se** —, to go, get, put or place oneself, stand, sit, begin, dress, be set, be worn; **se** — **à,** to begin, set about, occupy oneself with; — **à l'abri de,** to protect from; — **à l'épreuve,** to put to the test, make a trial of; — **à mort,** to kill; — **à profit,** to take advantage of, exploit; — **au courant,** to post; — **au jour,** to bring to light, publish; — **au point,** to set up, finish, perfect, adjust.

mettre, — **bas,** to yearn, bring forth; — **de côté,** to put by; — **en action,** to start; — **en balance,** to compare; — **en cave,** to store in a cellar; — **en commande,** to order; — **en défends,** to enclose, fence, hedge, close; — **en dérivation,** to shunt, put in shunt; — **en honneur,** to bring into repute; — **en jauge,** to heel in, put into a trench (plants in a nursery); — **en liberté,** to set free, liberate; — **en marche,** to start.

mettre, — **en oeuvre,** to employ, set going, use, cut grooves (for resin); — **en pâture,** to pasture; — **en place,** to set up; — **en réserve,** to keep, store away; — **en route,** to start; — **en valeur,** to sell, realize, turn something into money; — **hors,** to put out, expend; — **sur pied,** to set up, build; — **un anneau,** to ring, band; **mettez que,** suppose that; **se** — **d'accord,** to agree.

méture, *f.* corn bread.

meuble, *m.* piece of furniture, device; *pl.* furniture, personal estate, chattel.

meuble, movable, personal (property) easily worked, porous, loose (soil); **terre** — light soil, running soil, loose ground.

meubler, to furnish (a room), stock (farm), store.

meule, *f.* millstone, grindstone, polishing wheel, stack, charcoal kiln, rick (of hay), heap, pile, hotbed, clamp, manure heap, pearled burr, burr (of antler); — à aiguiser, grindstone; — à charbon, charcoal pile, charcoal kiln; — couchée, horizontal pile, lying pile; — debout, vertical pile; — de foin, haystack, hayrick; — d'émeri, emery wheel.

meuler, to mill.

meulier, pertaining to millstones, for grinding.

meulière, *f.* millstone, burrstone, molar tooth, millstone quarry, millstone grit.

méum, *m.* baldmoney, spicknel (*Meum athamanticum*).

meunerie, *f.* (flour) milling, milling trade.

meunier, *m.* miller, miller's-thumb, downy mildew of lettuce, cockroach.

meurt (mourir), (he) is dying.

meurtre, *m.* murder.

meurtri, bruised.

meurtrier, -ere, murderous, deadly.

meurtrir, to bruise.

meurtrissure, *f.* bruise, contusion.

meut (mouvoir), (he) is moving.

meute, *f.* pack of hounds.

meuvent (mouvoir), (they) are moving.

mexicain, Mexican.

Mexique, *m.* Mexico.

mézéréon, *m.* mezereon (*Daphne mezereum*), mezereum.

mi-, half, semi-, mid-.

miaou, *m.* miaou, mew.

mica, *m.* mica.

micacé, micaceous.

micaschiste, *m.* mica schist.

micellaire, micellar.

micelle, *f.* micella.

mi-chemin (à), half-way.

micocoulier, *m.* nettle tree (*Celtis australis*).

mi-collé, half-sized.

micracanthe, presenting small spines.

micranthe, micranthé, with small flowers.

microaérophile, microaerophilic, anaerobic.

microbase, *f.* microbasis, a variety of the carcerule.

microbe, *m.* microbe, microorganism, germ; — filtrant, filterable virus.

microbicide, *m.* microbicide.

microbien, -enne, microbial, microbic, bacterial.

microbiologie, *f.* microbiology.

microbique, microbic.

microcarpe, microcarpous.

microcéphale, microcephalic.

microchimie, *f.* microchemistry.

micrococcus, microcoque, *m.* micrococcus.

microconidie, *m.* microconidium, pycnoconidium.

microcosmique, microcosmic.

microcytase, *f.* microcytase.

microdiode, *f.* microdiode, microspore.

microdonte, microdont, (calyx) with small teeth.

microgamète, *f.* microgamete.

microgamétocyte, *m.* microgametocyte.

microgonidie, *f.* microgonidium, a small gonidium.

micrographe, *m.* micrographer, microscopist.

micrographie, *f.* micrography.

microkyste, *f.* microcyst.

microlithe, *m.* microlith.

micrologie, *f.* micrology.

micromelittophiles, *f.pl.* micromelittophilae, flowers pollinated by hymenoptera.

micromètre, *m.* micrometer; — objectif, stage micrometer.

micrométrie, *f.* micrometry.

micrométrique, micrometric(al).

micromillimètre, *m.* micromillimeter, micron.

micromyophiles, *f.pl.* micromyophilae, flowers pollinated by small diptera.

micron, *m.* micron, micromillimeter.

micronucléus, *m.* micronucleus.

micro-organisme, *m.* microorganism.

microphage, *f.* microphage.

microphone, *m.* microphone; — à liquides, liquid microphone.

microphonique, microphonic.

microphotographie, *f.* photomicrography, photomicrograph.

microphyte, *f.* microphyte.

microprothalle, *m.* microprothallus, reduced prothallus.

micropycnide, *m.* microspore.

micropyle, *m.* micropyle.

micro-réactions, *f.pl.* microreactions.

microsclérote, *f.* microsclerote, a modified sclerotium.

microscope, *m.* microscope.

microscopie, *f.* microscopy.

microscopique, microscopic(al).

microscopiste, *m.* microscopist.

microsomes, *m.pl.* microsomes, small granules embedded in the protoplasm.

microspectroscope, *m.* microspectroscope.

microsporange, *m.* microsporangium.

microspore, *m.* microspore.

microsporon, *m.* microspore.

microsporophylle, microsporophyll.

microstachyé, with flowers arranged in small spikes.

microstémone, with small stamens.

microstyle, microstylous, having short styles.

microsublimation, *f.* microsublimation.

microtherme, *f.* microtherm.

microthermique, microthermic.

microtome, *m.* microtome.

microzoaire, *m.* microzoan.

microzoospore, *f.* microzoospore.

microzyma, *f.* microzyme, bacterial ferment.

midi, *m.* noon, twelve o'clock, midday, south; *cap.* southern France.

mi-distance, à —, half-way.

mi-doux, half-soft.

mi-dur, semihard, half-hard.

mie, *f.* crumb, the soft part of bread.

miel, *m.* honey; — **boraté,** borax honey; — **despumé,** clarified honey; — **en rayon,** comb honey; — **en sections,** box honey; — **extrait,** extracted honey; — **rosat,** honey of rose; — **vierge,** virgin honey.

mielat, miellat, *m.* honeydew.

miellé, honeyed, sweetened with honey, honey-colored.

miellée, *f.* honeydew, honey flow.

mielleux, -euse, honeyed, like honey, luscious.

miellure, *f.* honeydew.

mien, mienne, my, mine, of mine, my own.

miette, *f.* crumb. morsel, little bit.

mieux, better, rather, best, more; **aimer —,** to prefer; **au —,** at best, as well as possible, in the best possible way; **au — de,** in the interest of; **bien —,** even more, rather;

de son —, the best he can; **du — qu'on peut,** the best one can; **le —,** the best.

mi-fruit, *m.* half-share of (farm) produce.

mignon, tiny, dainty, delicate.

mignonnette, *f.* mignonette (Reseda), coarse ground pepper, wild chicory, clover (*Trifolium arvense*), fine pebbles, mignonette lace.

mignotise des génevois, thyme (*Thymus vulgaris*).

migraine, *f.* migraine, sick headache. megrim.

migrateur, -trice, migrating, migratory.

migration, *f.* migration, roading.

migratoire, migratory.

migrer, to migrate, pass.

mijoter, to simmer, stew slowly.

mil, *m.* millet.

mil, one thousand.

mi-laine, *m.* half-wool fabric, half-wool.

milan, *m.* kite (the bird), glede.

milanais, Milanese.

mildiou, mildew, *m.* mildew, brown rot.

mildiousé, mildewed.

miliaire, miliary.

milieu, *m.* milieu, medium, atmosphere, environment, region, middle portion, middle, midst, intermediate course, center, middle point; — **nutritif,** nutrient medium; — **réducteur,** reducing medium; **au — de,** in the midst of.

militaire, *m.* soldier, military man.

militaire, military.

militer, to militate, contend.

mille, *m.* a thousand, mile, league.

mille, thousand, a thousand.

mille-feuille, *f.* milfoil, yarrow, nosebleed (*Achillea millefolium*); — **aquatique,** water violet (*Hottonia inflata*).

millefleur, with very numerous flowers.

millénaire, *m.* millennium.

mille-pattes, mille-pieds, *m.pl.* centipede, millipede, thousand legs.

mille-pertuis, *m.* St.-John's-wort (Hypericum); — **velu,** Aaron's-beard.

millet, *m.* millet grass (*Panicum miliaceum*); — **de Hongrie,** German millet (*Setaria italica*); — **épars,** millet grass (*Panicum miliaceum*); — **long,** canary grass (*Phalaris canariensis*); **grand —,** sorghum (*Holcus sorghum*).

milliampèremètre. *m.* milliammeter.

milliard, *m.* billion, thousand millions.

millième, thousandth.

millier, *m.* thousand.

milligramme, *m.* milligram.

millilitre, *m.* milliliter.

millimètre, *m.* millimeter.

million, *m.* million.

millionième, millionth (part).

millionnaire, *m.* millionaire.

millistère, *m.* millistere, liter, cubic decimeter.

milouin, *m.* pochard.

mimétisme, *m.* mimetism.

mimique, *f.* mimicry.

mimique, mimical.

mimose, *f.* sensitive plant (*Mimosa pudica*).

mimule, *m.* monkey flower (Mimulus).

mince, thin, slender, slight, shallow, insignificant; — filet, filar strand of protoplasm.

mincer, to mince, cut fine.

minceur, *f.* thinness, slenderness, scantiness.

mine, *f.* mine, ore, graphite, pit, lead, blast, shot, appearance, look, aspect, mien, source, countenance; — à ciel ouvert, surface mine, strip mine; — anglaise, minium; — brute, raw ore; — de fer, iron mine, iron ore; — de houille, colliery, coal mine, coalpit; — de(s) marais, bog iron ore, swamp ore, limonite; — de plomb, lead (mine), graphite, black lead pencil; — douce, zinc blende, sphalerite; — en rapport, producing mine; — grillée, roasted or burnt ore.

mine-orange, *f.* orange lead, orange minium.

miner, to mine, undermine, destroy, consume.

minerai, *m.* ore; — brut, raw ore; — d'argent, silver ore; — de fer, iron ore; — de fer à fleur de terre, meadow, swamp, or lake ore; — d'étain, tin ore.

minéral, *m.* mineral; — d'Hermès, mercury.

minéral, -aux, mineral; chimie —e, inorganic chemistry; source —e, mineral spring, spa.

minéralisable, mineralizable.

minéralisation, *f.* mineralization.

minéraliser, to mineralize.

minéralogie, *f.* mineralogy.

minéralogique, mineralogical.

minéralogiste, *m.* mineralogist.

minéraux, *m.pl.* see MINERAL.

minerie, *f.* rock-salt mine.

minette, *f.* minette (iron ore), black trefoil, black medic, nonesuch (*Medicago lupulina*), yellow clover, hop clover (*Trifolium procumbens*); — dorée, black medic, yellow trefoil, (*Medicago lupulina*).

mineur, *m.* miner, minor.

mineur, minor, mining, burrowing (insect).

miniature, *f.* miniature.

minier, -ère, mining, pertaining to mines.

minière, *f.* ore-bearing material, surface mine.

minima, *f.* minimum.

minimant, at a minimum.

minime, unimportant, extremely small, smallest, tiniest.

minimer, minimiser, to minimize.

minimum, *m.* minimum; au —, at the minimum, to a minimum.

minimum, minimum.

ministère, *m.* ministry, office; — du commerce, board of trade.

ministre, *m.* minister; premier —, prime minister.

minium, *m.* minium, red lead; — de fer, red ocher.

minoratif, *m.* laxative.

minoratif, -ive, laxative.

minorité, *f.* minority.

minoterie, *f.* flour mill, flour milling.

minuit, *m.* midnight.

minuscule, *f.* small letter, lower-case letter.

minuscule, small, tiny, minute.

minute, *f.* minute, rough draft, moment, instant; à la —, per minute, a minute.

minutieusement, minutely, scrupulously.

minutieux, -euse, minute. thorough, particular, trifling, detailed.

miocène, Miocene.

mi-parti, equally divided, halved, half . . . and half.

mi-partir, to divide into halves.

mi-partition, *f.* division into halves.

mirabelle, *f.* mirabelle (plum).

miracle, *m.* miracle; fait à —, admirably well done.

miraculeusement, miraculously.

miraculeux, -euse, miraculous, marveious.

mirage, *m.* mirage, looming.

mirbane, *f.* mirbane; essence de —, oil of mirbane, nitrobenzene.

mire, *f.* sight, sighting, aiming, foresight, leveling rod, picket, rod, boar's tusk.

mire, *m.* doctor, apothecary.

mirent (mettre), (they) placed or put.

mire-oeuf, mire-oeufs, *m.* candling apparatus, egg tester.

mirepoix, *m.* stock sauce (of meat and vegetables).

mirer, to look at, aim at, examine against the light, candle (eggs).

mirette, *f.* Venus's-looking-glass (Specularia).

miroir, *m.* mirror, looking glass, speculum, blaze (on trees to be felled), medullary (pith) rays, quarter grain; — ardent, burning glass; — de Vénus, Venus's-looking-glass (*Specularia speculum-veneris*); — frontal, head mirror; — plan, plane mirror; — réflecteur, reflector; oeufs au —, eggs fried in butter.

miroirs, *m.pl.* pith rays.

miroitant, flashing, sparkling, glittering, specular.

miroitement, *m.* flashing, glitter(ing).

miroiter, to glitter, shine, reflect light, flash, polish.

miroiterie, *f.* mirror manufacture, factory, or trade.

miroitier, *m.* cutter, maker, or seller of mirrors.

mis (mettre), (I or you) placed or put.

miscellannées, *m.pl.* miscellanies.

miscibilité, *f.* miscibility.

miscible, miscible, that can be mixed.

mise, *f.* putting, placing, circulation (of money), outlay, share (capital), dress, stake, bid; — à prix, valuation, estimate, price fixing; — au point, completion, carrying out, focalization, adjustment, improvement, perfection of a method, fine adjustment; — de feu, lighting, ignition, firing (furnace); — de feu chimique, chemical ignition; — de fonds, putting up funds, expenditure; — en coupe, first felling; — en défends, closing.

mise, — en état du sol, preparation of soil; — en feu, firing (furnace); — en liberté, setting free, liberation; — en marche, starting; — en moule, molding (setting) the curd; — en oeuvre, putting into play; — en pratique, putting into use; — en service, putting into service; — en solution, dissolving; — en valeur, utilization; — hors, disbursement of money, putting out (a furnace); être de —, to be admissible, fashionable, presentable.

misérable, miserable, wretched, mean.

misérablement, miserably, wretchedly.

misère, *f.* misery, destitution, trifle, want, distress.

miséricorde, *f.* mercy.

mi-soie, *m.* half-silk (fabric).

mispickel, *m.* mispickel, arsenopyrite.

missbildung, *f.* monstrosity.

missent (mettre), (they) might place or put, (they) placed.

mission, *f.* mission.

missionnaire, *m.* missionary.

mistelle, *f.* wine made of unfermented grape juice blended with alcohol.

mit (mettre), (he) placed.

mitaine, *f.* mitten, mitt.

mitapsis, *f.* synapsis.

mite, *f.* mite, moth worm, clothes moth; — du fromage, cheese mite.

mité, moth-eaten.

mitelle, *f.* mitella, miterwort (Mitella).

mithridate, *m.* mithridate, antidote.

mitigeant, mitigating, modifying.

mitiger, to mitigate, soften, relax; se —, to be mitigated.

mitochondrie, *f.* mitochondria.

mi-toile, *m.* half-linen (fabric).

mitome, *m.* mitome.

miton, *m.* crumb (of the loaf).

mitonner, to simmer, prepare quietly.

mitose, *f.* mitosis.

mitoyen, intermediate, middle, partition.

mitra, *f.* mitra.

mitraille, *f.* scrap metal, scrap, grapeshot; — cuivre, scrap (old) copper; — d'étain, scrap (old) tin; — de fer, scrap iron.

mitrailleuse, *f.* mitrailleuse, machine gun.

mitre, *f.* miter, cowl (for chimney), thick paving block.

mittelband, *m.* connective (of anther).

mi-vent, *m.* half-grown fruit tree (in exposed position).

mi-vitesse, *f.* half speed.

mixie, *f.* mixie, fusion of two similar nuclei.

mixote, *m.* mixote, artificial mingling of spore material.

mixotrophe, mixotrophic.

mixte, mixed.

mixtion, *f.*, mixtionnage, *m.* mixture; mixtion, gold size, mordant, a mixture of tallow and oil for covering the etched parts (engraving).

mixtionner, to compound (drug).

mixture, *f.* mixture; — **frigorific,** freezing mixture.

mm., *abbr.* (millimètre), millimeter.

Mn, *abbr.* (manganèse), manganese.

mnémotechnique, mnemotechnic.

mnésique, mnesic.

moabitique, Moabitic.

mobile, *m.* moving body, motive power, mover, spring motive.

mobile, mobile, movable, moving, variable, pivoting.

mobilier, *m.* furniture.

mobilier, -ère, movable, personal.

mobilisation, *f.* mobilization.

mobiliser, to mobilize.

mobilité, *f.* mobility, motility (of bacteria).

modalité, *f.* modality, circumstance, details.

mode, *m.* mode, method, manner, means.

mode, *f.* mode, fancy, fashion, way, millinery; — **de traitement,** mode (method) of treatment; — **de vente,** mode of sale; — **d'exploitation,** method of felling; **à la** —, in the fashion, after the style, fashionable.

modelage, *m.* modeling.

modelé, *m.* hill shading, representation of hill features (on map).

modèle, *m.* model, pattern, type.

modeler, to model, shape, make a model or pattern.

modeleur, *m.* modeler.

modérateur, *m.* moderator, regulator, governor, damper.

modérateur, -trice, governing, restraining, regulating.

modération, *f.* moderation, restraint, reduction.

modéré, moderate(d), reduced, temperate.

modérément, moderately.

modérer, to moderate, reduce, restrain, diminish.

moderne, modern.

moderniser, to modernize.

modeste, modest, humble.

modestement, modestly.

modestie, *f.* modesty, coyness.

modicité, *f.* smallness, moderateness.

modifiable, modifiable, alterable.

modifiant, modifying

modificateur, *m.* modifier.

modificateur, -trice, modifying.

modification, *f.* modification.

modifier, to modify; **se** —, to change, alter, be modified.

modique, small, reasonable, moderate.

modiquement, in a small way, at a low price.

modulateur, *m.* modulating.

modulation, *f.* modulation.

module, *m.* module, modulus; — **d'allongement,** coefficient of elongation.

moduler, to modulate.

moelle, *f.* marrow, pith, medulla; — **allongée,** medulla oblongata; — **d'os,** bone marrow; — **épinière,** spinal cord.

moelleux, *m.* mellowness, softness, ease, grace, soft stuff.

moelleux, -euse, marrowy, pithy, mellow, velvety, soft, easy.

moellon, *m.* ashlar, rough stone, sandstone for grinding mirrors; — **brut,** rubble.

moeurs, *f.pl.* manners, habits, ways, customs, morals.

mofétisé, mephitic, noxious.

mofette, *f.* noxious gas, chokedamp.

moi, *m.* ego, self.

moi, me, to me, I.

moie, *f.* stack, mow (of hay).

moignon, *m.* stump.

moi-même, myself.

moindre, less(er), smaller, least, slightest; —**s carrés,** least squares; **le** —, the least, slightest.

moindrement, less; **le** —, the least.

moine, *m.* monk, blister (on iron), stag without antlers.

moineau, *m.* sparrow; — **domestique,** house sparrow.

moins, least, minus sign, dash.

moins, less, not so, least, minus, except; — **de,** less than, — **élevé,** inferior, of lower growth; **à** — **de,** for less than, unless, without, barring; **à** — **que,** (with **ne** and the subj.) unless, if not; **au** —, at least, above all, however; **de** —, less, short, missing, too little; **de** — **en** —, less and less, gradually less; **du** —, at least, at all events, still, nevertheless.

moins, en —, minus, less; **en** — **de,** in less than; **le** —, least; **non** — **que . . . ,** as well as . . . , quite as much as . . . ; **pas** —, no less; **pas le** —, not in the least, by no means; **pour le** —, at least; **rien** —

que, anything but, nothing less than; tout au —, at the very least.

moins-value, *f.* depreciation, diminution in value, reduced value or revenue.

moirage, *m.* watering (of fabrics).

moire, *f.* moire, watered fabric, watering, honeysuckle (*Lonicera caprifolium*).

moiré, *m.* moiré, watering, water, watered effect; — métallique, crystal tinplate.

moiré, watered, moiré (silk), chatoyant.

moirer, to moiré (fabrics), water, mottle.

moirure, *f.* watering, watered effect, moiré.

mois, *m.* month, month's pay; *pl.* menses, periods.

moise, *f.* tie, brace, binding piece.

moisi, *m.* moldiness, mildew, mold.

moisi, mildewy, moldy, molded, musty, foxed, foxy.

moisir, to mildew, make moldy, mold, become moldy; se —, to grow moldy.

moisissure, *f.* mold, mildew, moldiness, mustiness; — du piston, aspergillus.

moissine, *f.* long stem of vine (with grape bunch attached).

moisson, *f.* harvest (of cereals), crop.

moissonner, to harvest, reap, gather, mow.

moissonneur, *m.* reaper, harvester.

moissonneuse, *f.* corn reaper.

moite, moist, damp, humid.

moiteur, *f.* moistness, perspiration, dampness, humidity.

moitié, half; à —, half, at half; de —, by half.

moitir, to moisten, make damp.

moka, *m.* Mocha (coffee).

mol, see MOU.

molaire, *f.* grinder, cheek tooth, molar.

molaire, molar.

molarite, *f.* millstone, grit.

molasse, *f.* see MOLLASSE.

môle, *m.* horst, ridge fault, heaved block, sunfish.

moléculaire, molecular.

molécularisation, *f.* conversion into molecules.

moléculariser, to arrange in molecules.

molécule, *f.* molecule, particle.

molécule-gramme, *f.* gram molecule.

molendinacé, molendinaceous, furnished with winglike expansions.

molène, *f.* mullein (Verbascum); — blattaire, moth mullein (*Verbascum blattaria*); — médicinale, great (common) mullein (*Verbascum thapsus*).

molester, to molest, trouble.

molet, *m.* pincers, nippers.

moleter, to mill, polish, knurl.

molets, *m.pl.* boggy, swampy ground.

moletage, molettage, *m.* milling, polishing.

molette, *f.* serrated or embossed roller or wheel, small pestle, muller, milling wheel, cutting wheel (for glass), shepherd's-purse, windgall.

moleter, moletter, to mill, polish.

moliant, pliable, soft.

mollasse, *f.* molasse, sandstone.

mollasse, soft, flabby, lacking in body, slow.

molle, *f.* bundle of split osiers.

molle, see MOU.

mollement, softly, gently, feebly, weakly, effeminately, gracefully.

mollesse, *f.* softness, slackness, mellowness, flabbiness, mildness, weakness, sluggishness, effeminacy, ease and luxury.

mollet, *m.* calf (of leg).

mollet, -ette, moderately soft, tender.

molleter, to mill, polish.

molleterie, *f.* light sole leather.

mollir, to soften, become soft, faint. (of fruit) become mellow, slacken, ease off, sag, weaken, yield.

molliuscule, somewhat mellow.

mollusque, *m.* mollusk, shellfish.

moly, *m.* wild garlic (*Allium vineale*).

molybdate, *m.* molybdate.

molybdaté, molybdate of.

molybdène, *m.* molybdenum.

molybdénite, *f.* molybdenite.

molybdeux, molybdous.

molybdine, *f.* molybdite, molybdic ocher.

molybdique, molybdic (acid).

molybdurane, *m.* uranium molybdate.

moment, *m.* moment, momentum, time, instant; — de rotation, moment of torsion; — sollicitant, applied moment; au — de, just as, at the moment of; au — où, at the time when; dans le —, in a moment, instantly; dans le — où (que), at the moment when, as soon as; du — que, as soon as, seeing that, since; en ce —, at present, just now; pour le —, at the moment, in the meantime; **par —**, at times, at intervals.

momentané, momentary.

momentanément, momentarily.

momie, *f.* mummy.

momification, *f.* mummification, desiccation of tissues.

momitié, mummified.

momifier, to mummify.

mon, my.

monacrorhize, monacrorhize, (plants) with roots from a single mother cell.

monade, *f.* monad.

monadelphe, monadelphous

monamide, *f.* monamide.

monamine, *f.* monamine.

monandre, monandrous.

monandrie, *f.* monandria.

monangique, monangic, (sporangium) enclosed by a hoodlike indusium.

monanthe, monanthous, one-flowered.

monanthème, with solitary flowers.

monanthère, having one anther.

monaptère, with a single wing.

monarchie, *f.* monarchy.

monarde, *f.* Monarda.

monarque, *m.* monarch.

monaxile, haplocaulous.

monaxone, monaxon.

monazite, *f.* monazite.

monbin, *m.* hog plum (Spondias).

monceau, *m.* heap, mound, pile.

mondation, *f.* cleaning, blanching, peeling.

monde, *m.* world, people, persons, society, company, crew, servants, universe; **au —** in the world; **du — sali,** of saline earth; **le plus vite du —**, as quickly as possible; **tout le —**, everybody, everyone.

mondé, in pods; **orge —**, hulled barley, barley water.

monder, to clean, cleanse, peel, hull, blanch (almonds), stone (raisins).

mondial, -aux, world, world-wide.

mondificatif, -ive, mundificant, cleansing.

mondifier, cleansing (wound).

monergique, monergic.

monétaire, monetary, money.

monétiser, to monetize, mint.

mongolien, -enne, mongolique, Mongolian.

moniliforme, necklace-shaped, moniliform.

moniteur, *m.* monitor, adviser.

monnaie, *f.* money, currency, mint, coin.

monnayage, *m.* minting, coining, mintage.

monnayer, to coin, mint.

monnayère, *f.* pennycress (*Thlaspi arvense*), moneywort (*Lysimachia nummularia*).

monoatomique, monatomic.

monobase, monobasal, monobasic.

monobasique, monobasic.

monobromure, *m.* monobromide.

monocarbonique, monocarboxylic.

monocaréné, with a single carene.

monocarpe, monocarpic.

monocarpien, -enne, monocarpous.

monocarpique, monocarpic, biennial (plants).

monocéphale, monocephalous.

monochlamydé, monochlamydeous.

monochlorhydrique, éther —, ethyl chloride.

monochromatique, monochromatic, single-colored.

monochrome, *m.* monochrome.

monochrome, monochrome, monochromic, single-colored.

monoclinal, -aux, monoclinal.

monocline, monoclinous.

monoclinie, *f.* monocliny, stamens and pistils in the same flower.

monoclinique, monoclinic.

monocotyle, monocotyledonous.

monocotylédone, monocotyledonous.

monocotylédonées, *f.pl.* monocotyledons.

monoculaire, monocular.

monocyclique, monocyclic, annual (plants).

monocylindrique, one-cylinder.

monodyname, monodynamous, with one stamen longer than the others.

monoecie, monoecious.

monogame, monogamous.

monogamie, *f.* monogamy.

monogénèse, *f.* monogenesis.

monogonie, *f.* monogony, asexual reproduction.

monographe, *m.* monographist, monograph writer.

monographie, *f.* monograph.

monogyne, monogynous.

monohybride, *m.* monohybrid.

monohydrate, *m.* monohydrate.

monohydraté, monohydrated.

monoicodimorphie, *f.* cleistogamy.

monoïque, monoecious.

monolepsis, *m.* monolepsis, false hybridism.

monomagnésique, monomagnesium.

monôme, *m.* monomial, single term.

monomer, monocarpous.

monomère, monomerous.

monométallique, monometallic.

monomorphe, monomorphous, (all flowers) of one form only.

mononucléaire, *f.* mononuclear leucocyte.

mononucléaire, mononuclear.

monopérianthe, single perianth.

monopérigyne, with perigynous stamens on monocotyledons.

monopétale, monopetalous.

monophasé, single-phase.

monophylétique, monophyletic.

monophylle, monophyllous.

monophyte, holding only one species, (parasite) developing on only one species.

monoplan, monoplane.

monoplanétique, monoecious.

monoplaste, *m.* monoplast.

monopode, *m.* monopodium.

monopode, monopodous.

monopodial, monopodial.

monopole, *m.* monopoly.

monopoler, to monopolize, have a monopoly.

monopolisation, *f.* monopolizing.

monopoliser, to monopolize.

monopyrène, monopyrenous, having a single stone or kernel.

monorchide, monorchid, with a single tubercle.

monosaccharide, *f.* monosaccharide; *pl.* monosaccharides, monosaccharoses.

monosépale, monosepalous, unisepalous.

monosoc, *m.* single plow.

monosperme, monospermous, one-seeded.

monospermie, *f.* monospermy.

monospermique, monospermous.

monosporange, *m.* monosporangium.

monospore, monosporous, having a single spore.

monostachyé, monostachyous, with only one spike.

monostéarine, *f.* monostearin.

monostélie, *f.* monostely.

monostigmaté, monostigmatous, with one stigma only.

monosyllabique, monosyllabic.

monosymphytogyne, with adherent ovary.

monothalamé, monothalamic, single-chambered.

monothèle, with one ovary.

monotone, monotonous.

monotonement, monotonously.

monotonie, *f.* monotony.

monotrique, monotrichous.

monotrope, *m.* Monotropa; — suce-pin, bird's-nest, pinesap, false beechdrops.

monotype, monotypic(al).

monovalent, monovalent, univalent.

monsieur, *m.* sir, master, Mr., gentleman.

monstre, *m.* monster.

monstrueux, -euse, monstrous, enormous.

monstruosité, *f.* monstrosity.

mont, *m.* mount, mountain.

montacide, *m.* monte-acide, acid elevator.

montage, *m.* rising up, boiling up (liquids), increase, carrying up, lifting, hoisting, mounting, assembling, setting up, connecting, wiring; — au baume, mounting in balsam; — en tandem, cascade connection.

montagnard, *m.* mountaineer.

montagne, *f.* mountain.

montagneux, -euse, mountainous.

montant, *m.* upright, standard, post, stem (of plants), total, amount, high flavor, arm, strut.

montant, rising, ascending, high.

monte, *f.* covering, mount, breeding season.

monté, lifted, raised, mounted, erected, equipped, fitted, strongly colored.

monte-acide, *m.* monte-acide, acid pump, acid elevator.

monte-charge, *m.* freight elevator, goods-lift, hoist.

montée, *f.* rise, rising, ascent, gradient, uphill pull, climb, step, height.

monte-jus, *m.* monte-jus, an apparatus for pumping liquids.

monter, to rise, ascend, mount, go up, equip, supply, increase, embark, lift (up), raise, hoist, carry up, take up, set up (of apparatus), pilot, erect, assemble, man (a vessel), connect up, establish, get up, deepen, flush up, heighten (colors) wind (timepieces); se —, to rise, amount, supply oneself, stock up, ride, be raised, become excited, irritated; — à **graine**, to go to seed; — **sa garde**, to be on guard.

monteur, *m.* setter, mounter, erecter, producer, fitter (of pipes).

monticole, monticolous, inhabiting mountainous regions.

monticule, *m.* hillock, mound.

montmartrite, *m.* montmartrite, gypsum containing calcium carbonate.

montrable, presentable, capable of being shown.

montre, *f.* watch, display, show, show (case) window, show place, sample (of goods); — à **repos**, stop watch; — **marine**, chronometer.

montrer, to show, exhibit, display, point out, teach, set forth, manifest, look; se —, to show oneself, appear.

montueux, -euse, mountainous, hilly.

monture, *f.* mount, mounting, setting up, fitting, frame, handle, socket, attachment, equipment.

monument, *m.* monument, building.

moquer(se), to make fun of, to ridicule.

morailles, *f.pl.* pincers, nippers, tongs.

moraillon, *m.* hasp, clasp.

moraine, *f.* moraine, dead wool, skin wool, erratic, block, boulders; — **frontale**, terminal moraine; — **médiane**, medial moraine; — **profonde**, ground moraine.

moral, *m.* morale, (state of) mind, moral nature.

moral, moral, mental, intellectual.

morale, *f.* morals, ethics, moral science, reprimand, morality.

moralement, morally.

morbide, morbid.

morceau, *m.* piece, fragment, part, bit, morsel.

morceler, to divide, parcel out, cut up.

morcellement, *m.* division, breaking up.

mordache, *f.* clamp, jaw, clip, tongs.

mordacité, *f.* mordacity, corrosiveness, causticity.

mordançage, *m.* mordanting, mordant action.

mordancer, to mordant.

mordant, *m.* mordant, corrosiveness, causticity, keenness, cutting power, holder, clamp, gold size.

mordant, mordant, biting, cutting, corrosive, piercing, keen, pungent, caustic.

mordicant, biting, corrosive, acrid.

mordiller, to nibble.

mordoré, reddish brown, bronze (color).

mordorure, *f.* bronze finish (on leather).

mordre, to bite, corrode, bite at, nibble, catch, nip, cut, penetrate, effect, attack, act, take effect, work, take hold, engage, carp (at), encroach, reflect (on).

mords, *m.* jaw, bit.

mordu, premorse.

moré, *m.* mulberry wine.

morelle, *f.* nightshade (Solanum); — **furieuse**, belladonna, deadly nightshade (*Atropa belladonna*); — **grimpante**, bittersweet (*Solanum dulcamara*); — **noire**, black (common) nightshade (*Solanum nigrum*).

moresque, Moorish; see **MAURESQUE**.

moret, *m.* whortleberry (*Vaccinium myrtillus*).

morfil, *m.* wire edge (on tools), raw ivory.

morfondu, infertile (silkworm eggs).

morfondure, *f.* cold, panzoötic catarrhal fever or influenza (horse, dog).

morgeline, *f.* chickweed.

moricandie, *f.* — **des champs**, field moricandie.

morille, *f.* morel (*Morchella esculenta*).

morin, *m.* morin.

morindine, *f.* morindin.

morindone, *f.* morindone.

morne, sad, dull, gloomy, mournful. dismal.

morphesthésie, *f.* morphaesthesia, tendency to definite relations of symmetry.

morphine, *f.* morphine.

morphique, morphine, of morphine.

morphogène, producing a modification of the form.

morphogenie, *f.* morphogeny.

morpholine, *f.* morpholine.

morphologie, *f.* morphology.

morphologique, morphologic(al).

morphologiquement, morphologically.

morphose, *f.* morphosis.

morphosis, *m.* morphosis.

morruol, morrhuol, *m.* morrhuol.

mors, *m.* jaw (of vise), end of the blowpipe (glass), bridle bit, mouth bit.

morsure, *f.* bite, biting.

mort, *m.* dead person, dead body; *f.* death; — **au chien,** meadow saffron (*Colchicum autumnale*); — **aux mouches,** fly poison; — **aux poules,** henbane (*Hyoscyamus niger*); — **aux rats,** ratsbane, arsenic; — **aux vaches,** celery-leaved crowfoot (*Ranunculus sceleratus*); — **en cime,** stagheadedness; **à** —, to death, mortally.

mort, dead, dormant, faded, lifeless, still, stagnant.

mortaise, *f.* mortise, hole.

mortaiser, to mortise, slot.

mortalité, *f.* mortality.

mort-chien, *m.* meadow saffron (*Colchicum autumnale*).

morte, dead.

morte-eau, *f.* dead water, slack water, neap (tide).

mortel, -elle, mortal, deadly, fatal.

mortellement, mortally, fatally.

morte-saison, *f.* off season, dead or slack season.

mort-gage, *m.* mortgage.

mortier, *m.* mortar, mixture; — **à chaux et à sable,** sand-lime mortar; — **à prise lente,** slowly hardening mortar; — **de chaux,** lime mortar; — **de limon,** loam or clay mortar; — **de tranchée** trench mortar; — **du compass,** compass box; — **en agate,** agate mortar; — **en verre,** glass mortar; — **liquide,** grout, thin mortar; — **rayé,** rifled mortar.

mortifié, mortified, dead, chagrined.

mortifier, to mortify, hang (game), make (meat) tender.

mortine, *f.* leaves of the myrtle (*Coriaria myrtifolia*).

mort-né, stillborn.

morts-bois, *m.pl.* useless shrubs, forest weeds, worthless scrub.

mort-terrain, *m.* dead ground (without ore).

morue, *f.* cod, codfish; — **rouge,** red bacterial pigmentation of salted codfish; — **salée,** — **sèche,** salt cod; **huile de foie de** —, cod-liver oil.

morve, *f.* glanders, nasal mucus, rot (of plants), farcy.

morvé, *m.* mistletoe.

morver, to get the rot.

morveux, -euse, glandered, of glanders.

mosaïque, *f.* mosaic.

Moscou, *f.* Moscow.

moscouade, moscovade, *f.* muscovado, raw sugar.

mot, *m.* word, saying, motto, answer, witticism; — **à** —, word for word; **au bas** —, at the lowest estimate; — **d'ordre,** countersign.

moteur, *m.* motor, engine, mover, author, motive power; — **à air carburé,** internal-combustion engine; — **à air chaud,** hot-air engine, hot-air motor; — **à alcool,** spirit motor; — **à deux temps,** two (stroke)-cycle engine, two-stroke motor; — **à essence,** gasoline engine, petrol engine; — **à explosion(s),** internal-combustion engine; — **en dérivation,** shunt motor, shunt-wound motor; — **thermique,** heat engine, heat motor; — **triphasé,** three-phase motor.

moteur, -trice, motive, propelling, moving, driving.

motif, *m.* motive, reason, incentive, cause, ground.

motif, determining.

motiver, to motivate, be the motive of, justify, warrant, allege, be the cause of.

motocharrue, *f.* motor plow.

motoculteur, *m.* — **à fraise,** rotary cultivator; — **à fraises rotatives,** ground mill.

motocyclette, *f.* motorcycle.

moto-propulseur, *m.* motor and propeller.

mototracteur, *m.* tractor.

motrice, *f.* motor, motor cell (in grass leaf).

motrice, driving, moving.

motricité, *f.* motility, motivity.

motte, *f.* mound, lump, clod, mass, ball (on roots), hillock; — **à brûler,** sod of peat, briquette (of spent tan); — **de beurre,** pat of butter; a small butter; — **de gazon,** grassy turf, sod, matting of grass; — **de pot,** pot ball; — **de terre,** root ball, ball of earth, clod; — **gazonnée,** sod.

motton, *m.* lump (in porridge).

mou, *m.* slack (of rope), lights, lungs (of animals), soft stuff.

mou, soft, flabby, mellow, weak, sultry (weather), close, slack, feeble, lax, spineless, sluggish, inactive, lifeless, flat (wine), lazy, spiritless.

mouche, *f.* fly, spot, chin tuft, bull's-eye; — **à viande,** blowfly, flesh fly; — **de la betterave,** beet-leaf miner, beet fly; — **des cerises,** cherry fruit fly; — **domestique,** common fly; — **du chou.** cabbage fly.

moucher, to wipe, blow (nose), square or trim the end of, snuff (a candle).

moucheron, *m.* gnat, midge, small fly.

mouchet, *m.* hedge sparrow.

moucheté, mottled.

moucheter, to spot, speckle.

mouchette, *f.* outer fillet (of drip molding), residue from sifted plaster, nose ring for pigs.

moucheture, *f.* spot, speckle, speck, slight scarification.

mouchoir, *m.* handkerchief.

mouchure, *f.* nasal mucus, snuff (of candle), waste end (of board).

moudre, to grind, mill.

mouette, *f.* gull.

moufette, *f.* skunk.

moufle, *f.* mitten, mufflers (gloves), tackle, system of pulleys, tie(bar), clamp.

moufle, *m.* muffle.

moufler, to put in a muffle, tie(walls).

mouflon, *m.* mouflon, wild sheep.

mouillade, *f.* wetting, moistening.

mouillage, *m.* wetting, moistening, anchorage; — **du sol,** moistening of the soil.

mouillé, wetted, moist, wet, watery.

mouillement, *m.* wetting.

mouiller, to wet, water, moisten, steep, soak, cast (anchor), anchor, moor, put down.

mouilleur, *m.* wetter, sprinkler.

mouilleuse, *f.* spray damper.

mouilleux, -euse, wet (ground).

mouilloir, *m.* water pot for moistening fingers.

mouillure, *f.* wetting, damping, moisture.

moulage, *m.* grinding, milling, molding, casting, millwork; — **au plâtre,** plaster cast; — **creux,** — **à noyau,** hollow casting; — **en argile,** — **en terre,** loam molding.

moulait (moudre or mouler), (he) was grinding or casting.

moulant, grinding, milling, molding.

moule, *f.* mussel.

moule, *m.* mold, cast, form, matrix; — **à beurre,** butter mold; — **à fonte,** casting mold; — **à fromage,** cheese mold, chessel, cheese hoop; — **à paraffine,** trough for embedding in paraffin; — **de gueuse,** channel, pig; — **de travail,** working mold.

moulé, *m.* block letters, print.

moulé, molded, cast.

moulée, *f.* swarf, grindings.

moulent (moudre or mouler), (they) are grinding or casting.

mouler, to mold, cast, form, print, shape; **se —,** to be molded, cast, model oneself.

moulerie, *f.* casting works, foundry.

mouleur, *m.* molder, caster.

mouleuse, *f.* printer (butter).

moulin, *m.* mill; — **à avoine,** oat mill; — **à baryte,** heavy spar mill; — **à beurre,** churn; — **à blé,** corn mill; — **à caillé,** curd mill; — **à canne,** cane mill, sugar mill; — **à craie,** chalk mill; — **à cylindre(s),** cylinder or roller mill.

moulin, — **à eau,** water mill; — **à farine,** flour mill; — **à marée,** tide mill; — **à meules,** mill with millstones; — **à sucre,** sugar mill; — **à vapeur,** steam mill; — **à vent,** windmill, wind turbine; — **à vis,** screw mill.

mouliné, worm-eaten.

mouliner, to throw (silk), grind, eat into (wood).

moulinet, *m.* winch, reel, turnstile, twirl, turn, whirl, windlass, paddle(wheel).

moulu, ground, milled, powdered.

moulure, *f.* molding (ornamental).

mouospora, *f.* yeast, infecting Daphnia.

mour, *m.* muzzle, nose.

mourant, dying, fading, moribund, faint, pale.

mouret, *m.* see MORET.

mourir, to die, die out, be spent, stop, perish; **se —,** to be dying, die out, expire; — **par la gelée,** to freeze to death; **faire —,** to put to death, cause to die, kill, worry.

mourmane, *m.* coast at Murmansk.

mouron, *m.* — **d'eau,** water pimpernel, brookweed (*Samolus valerandi*); — **des oiseaux,** chickweed (*Stellaria media*); — **rouge,** scarlet pimpernel (*Anagallis arvensis*).

mourut (mourir), (he) died.

moussache, *f.* manioc flour, arrowroot, cassava starch, tapioca meal.

moussant, frothing, foaming, foamy, frothy.

mousse, *f.* froth, foam, lather, whipped cream, sponge, moss; — **de platine,** platinum sponge; — **des rennes,** reindeer moss (*Cladonia rangiferina*); — **foliacée,** foliaceous moss; — **jaune,** biting stonecrop, wall pepper (*Sedum acre*); — **marine perlée,** carrageen (*Chondrus crispus*); —**s feuillées,** frondiferous mosses.

mousse, blunt, dull.

mousseline, *f.* muslin.

mousseline, mousseline.

mousser, to froth, foam, lather, scum, (of wine) sparkle, effervesce, fizz.

mousseron, *m.* edible mushroom.

mousseux, -euse, mossy, moss, spongy, foamy, foaming, frothy, lathering.

moussoir, *m.* potcher roll.

mousson, *f.* monsoon.

moussu, mossy, moss-grown.

moustique, *m.* mosquito, gnat, sand fly.

moût, *m.* must (of grapes), wort (of beer), unfermented wine or fruit juice, mash; — **sucré,** sweetwort.

moutarde, *f.* mustard; — **blanche,** white mustard (*Brassica alba*); — **des champs,** charlock, wild mustard (*Brassica arvensis*); — **des moines,** horse-radish (*Amoracia rusticana*); — **noire,** black (brown) mustard (*Brassica nigra*); **huile de —,** mustard-seed oil.

moutardin, moutardon, *m.* charlock, wild mustard (*Brassica arvensis*), white mustard (*Brassica alba*).

mouton, *m.* sheep, mutton, sheep leather, ram, rammer, monkey, (pile) driver, beetle, drop hammer.

moutonneux, -euse, cloudy.

mouture, *f.* grinding, milling (corn), fee for grinding, multure, a mixture of wheat, rye, and barley.

mouvant, moving, actuating, motive, unstable, changeable, quick (sand).

mouve-chaux, *m.* stirrer, stirring pole.

mouvement, *m.* motion, movement, works (of machinery), impulse, agitation, commotion, variation, fluctuation, change, activity, traffic, animation, life; — **à secousses,** shaking motion; — **croissant,** accelerated motion; — **de roulis,** rocking (rolling) motion; — **de terrain,** undulation, variation; — **de translation,** transfer or continuous motion; — **de va-et-vient,** reciprocal motion, motion to and fro, alternating motion.

mouvement, — **en avant,** forward motion; — **rapide,** coarse motion; — **retardé,** decreasing motion; — **rotatif,** rotary motion, rotation; — **sautillant,** skipping motion; — **tangage,** pitching motion; — **tourbillonnaire,** vortical motion; **mettre en —,** to set in motion, start; **se mettre en —,** to start, stir, move.

mouveron, *m.* a wooden stirrer.

mouvoir, to move, stir up, excite, prompt, impel; **se —,** to be moved, move, move about; **faire —,** to cause to move, move, set going.

moyau, *m.* beam cover (of a press).

moyen, *m.* mean, means, middlings, abilities, reasons, way, medium, method, manner; **au — de,** by means of; **le moyen âge,** the Middle Ages.

moyen, -enne, middle, mean, average, medium, intermediate, median, mediocre, middling.

moyenageux, -euse, medieval, of the Middle Ages.

moyennant, on (a certain) condition, by means of; — **que,** provided that.

moyenne, *f.* mean, average, medium, middlings; **en —,** on an average.

moyennement, moderately, fairly, on an average, middlingly.

moyette, *f.* stook, shock.

moyeu, *m.* hub, nave, (egg) yolk, (fruit) stone, preserved plum, core (in casting).

mu (mouvoir), moved.

muable, changeable, mutable.

mucate, *m.* mucate.

mucédine, *f.* mucedin.

mucédinée, *f.* mold.

mucilage, *m.* mucilage.

mucilagineux, -euse, mucilaginous, slimy viscous.

mucine, *f.* mucin.

mucinoïde, *n.* mucoid.

mucique, mucic.

mucoïde, *m.* mucoid.

mucoïde, mucoid.

muconique, muconic.

Mucorinées, *f.pl.* Mucoraceae.

mucosité, *f.* mucus, mucosity, slime.

mucron, *m.* mucro, terminal point.

mucroné, mucronated, pointed.

mucronifolié, with mucronate leaves.

mucronulé, mucronulate.

mucus, *m.* mucus.

mue (mouvoir), moved.

mue, *f.* coop, cage, molting, molting season, casting, skin, molt.

muer, to molt, shed hair.

muet, -ette, mute, silent, dumb, speechless.

mufle, *m.* muffle, muzzle, snout; — **de veau,** snapdragon (*Antirrhinum majus*).

mufleau, *m.* snapdragon (*Antirrhinum majus*).

muflier, *m.* snapdragon (Antirrhinum).

mugir, to roar, bellow. low, moo.

mugissement, m. roaring, bellowing, lowing.

muguet, m. lily of the valley (*Convallaria majalis*), thrush (human mouth).

muid, m. hogshead.

muire, m. salt water, mother liquor, brine (in the salines).

mulasse, f. young mule.

mulâtre, m. mulatto.

mule, f. she-mule.

mulet, m. mule.

muletier, m. muleteer, mule driver.

mulot, m. field mouse, wood mouse.

multangulaire, multangulé, multangular.

multibulbeux, -euse, producing many bulbs.

multicapsulaire, multicapsulate.

multicaule, multicauline, many-stemmed.

multicellulaire, multicellular.

multicolore, multicolor, gaily colored, variegated.

multicoque, multicoccous.

multicorne, with appendices resembling a horn.

multicuspidé, multicuspid(ate).

multidigité, multidigitate, many-fingered.

multiembryonné, (seed) with several embryos.

multifère, multiferous, often bearing, fruitful.

multifide, multifid(ous), many-cleft.

multiflore, multifloral, many-flowered.

multifoliolé, multifoliolate, with many leaflets.

multiforé, pierced by holes.

multiforme, multiform, many-sided.

multifurcation, f. polytomy.

multijugué, (a pinnate leaf) with more than five pairs of leaflets.

multilatéral, -aux, multilateral.

multilatère, multilateral, many-sided.

multilobé, multilobular.

multiloculaire, multilocular, many-celled.

multinervé, multinervate.

multinervulé, with numerous small nerves (veins).

multinoueux, -euse, with numerous nodes.

multiovule, multiovulate, with many ovules.

multipare, multiparous.

multipartit, multipartite, many times divided.

multipétalé, with many petals

multiple, multiple, manifold.

multipliant, multiplying.

multiplicande, m. multiplicand.

multiplicateur, m. multiplier.

multiplicateur, -trice, multiplying, multiplicative.

multiplicatif, -ive, multiplicative.

multiplication, f. multiplication.

multiplicité, f. multiplicity.

multiplié, double (flower).

multiplier, to multiply; se —, to be multiplied, multiply, reproduce, increase.

multiplinervé, with multiple veins.

multipolaire, multipolar.

multirotatoire, multirotatory.

multisilique, multisiliquous, having many siliques.

multitérébénique, polyterpene.

multitude, f. multitude, great crowd.

multivalve, multivalve, having many valves.

muni de, provided with; — de stomates, stomatiferous.

municipal, -aux, municipal, city.

munir, to supply, furnish, provide with; se —, to provide oneself, be supplied.

munition, f. munition, ammunition, provisions, stores, supplies, provisioning.

munjeestine, f. munjistin.

munjeet, m. munjeet, Indian madder (*Rubia cordifolia*).

munster, m. Münster cheese.

muqueuse, f. mucous membrane.

muqueux, -euse, mucous.

mur, m. wall; — de refend, partition, wall; — en briques, brick wall; — orbe, dead or blind wall.

mûr, ripe, mature, mellow, exploitable, shabby, worn out (clothes); impartaitement —, unripe, immature.

murage, m. walling (up), masonry.

mûraie, f. mulberry plantation.

muraille, f. wall, side (of ship); couleur de —, gray (color).

muraillement, m. walling.

mural, -aux, growing on walls.

mûral, -aux, mulberry; calculs muraux, mulberry calculus.

mûre, f. mulberry (the fruit); — de haie, blackberry (Rubus); — de ronce, blackberry (Rubus); — **sauvage,** blackberry, bramble (Rubus).

mûre, ripe, mature.

mureau, *m.* tuyère wall.

mûrement, maturely.

murent (mouvoir or murer), (they) moved or wall (up).

murer, to wall in, stop up, block (up).

muret, *m.* low wall.

murexane, *m.* murexan, uramil.

murexide, *f.* murexide.

murexoïne, *f.* murexoin.

murger, *m.* cone of detritus.

mûri, ripened, ripe, matured.

muriate, *m.* muriate, chloride, hydrochloride; — **d'ammoniaque,** ammonium chloride.

muriaté, muriated.

muriatique, muriatic, hydrochloric (acid).

muriculé, muriculate.

muride, *m.* bromine, muride.

mûrier, *m.* mulberry tree (Morus); — **à papier,** paper mulberry (*Broussonetia papyrifera*); — **sauvage,** bramble (Rubus).

muriqué, muricate.

mûrir, to ripen, mature, develop, maturate.

mûrissant, ripening, maturing.

mûrisseur, *m.* ripener.

murmure, *m.* murmur, grumbling, prattling, babbling.

murmurer, to murmur.

mûron, *m.* blackberry, wild raspberry (Rubus).

murrayine, *f.* murrayin (extracted from *Murraya exotica*).

mus (mouvoir), moved.

musaraigne, *f.* shrew.

musc, *m.* musk, musk deer.

muscade, *f.* nutmeg.

muscadelle, *f.* muscatel, muscadine, muscat.

muscadet, *m.* muscatel (wine).

muscadier, *m.* nutmeg tree (Myristica).

muscari, *m.* grape hyacinth (Muscari).

muscarine, *f.* muscarine.

muscat, muscat, muscatel (wine and grape), muskpear.

muscidés, *m.pl.* Muscidae, flesh flies.

muscinées, *f.pl.* Musci, mosses.

muscle, *m.* muscle; — **fléchisseur,** flexor.

muscologie, *f.* muscology.

muscovite, *f.* muscovite, white mica.

musculaire, muscular.

musculeux, -euse, muscular, musculous, brawny.

musculine, *f.* musculin.

museau, *m.* muzzle, nose, snout, face (of the negro).

musée, *m.* museum.

museler, to muzzle, gag.

muser, to muse.

musette, *f.* nose bag, air bell, paper defect.

muséum, *m.* museum (of natural history).

musical, aux, musical.

musif, see MUSSIF.

musique, *f.* music, band, sediment, settlings.

musqué, musk, musky, musked, scented with or like musk, (of language) affected.

musquer, to musk, perfume with musk.

mussif, mosaic; or —, mosaic gold, disulphide of tin.

musulman, Moslem, Mussulman.

mut (mouvoir), (it) moved.

mutabilité, *f.* mutability.

mutage, *m.* mutage (of wine).

mutation, *f.* mutation, change, transfer.

muté, muted, transferred.

muter, to mute, check (the fermentation).

muteuse, *f.* muting apparatus.

mutilation, *f.* mutilation, maiming.

mutiler, to mutilate, maim.

mutiner (se), to mutiny.

mutique, mutic, pointless, blunt, muticous, awnless, without granules or thorns.

mutualisme, *m.* mutualism.

mutuel, -elle, mutual, reciprocal.

mutuellement, mutually, each other, reciprocally.

mycélieu, mycelial.

mycélium, *m.* mycelium.

mycétocarpe, *m.* spongy fruit.

mycétologie, *f.* mycology.

mycétome, mycetoma; — **à grains noir,** black Maduromycosis.

mycoderme, *m.* mycoderma.

mycodermique, mycodermic, mycodermatoid.

mycodomatie, *f.* domatia due to a fungus.

mycologie, *f.* mycology.

mycologique, mycological.

mycoplasma, *f.* mycoplasma.

mycoprotéine, *f.* mycoprotein.

mycorrhizes, *m.pl.* mycorhiza.

mycose, *f.* mycose, trehalose, mycosis.

mycosique, mycotic.

mydriase, *f.* mydriasis.

mydriatique, *m.* mydriatic.

mydriatique, mydriatic.

myéline, *f.* myelin.

mylabre, *m.* a member of Mylabris.

myoblaste, *m.* myoblast.

myocaillot, *m.* muscle coagulum.

myocardite, *m.* myocardite.

myope, myopic, near-sighted.

myophiles, *f.pl.* plants pollinized by Diptera.

myoplasma, *m.* muscle plasma.

myose, *f.*, **myosis,** *m.* myosis.

myosérum, *m.* muscle serum.

myosine, *f.* myosin.

myosinogène, *m.* myosinogen.

myosotis, *m.* forget-me-not (Myosotis); — **palustre,** — **des marais,** forget-me-not (*Myosotis palustris*).

myosure, *f.* mousetail (Myosurus).

myotique, myotic.

myrcène, *m.* myrcene.

myriacanthe, with numerous spines.

myriade, *f.* myriad, immense number.

myriagramme, *m.* myriagram (10 kilograms).

myrianthe, with numerous flowers.

myrica, *m.* wax myrtle, candleberry myrtle (Myrica); — **cérifère,** wax myrtle (*Myrica cerifera*); — **galé,** sweet gale, bog myrtle (*Myrica gale*); **cire de —,** myrtle wax.

myricacées, *d.pl.* Myricaceae.

myricine, *f.* myricin.

myricique, myricyl.

myriophylle, myriad-leaved, tenuifoliate, pinnatifid.

myrique, *m.* bayberry, candleberry myrtle (*Myrica cerifera*).

myristica, nutmeg.

myristicacées, myristicées, *f.pl.* Myristicaceae.

myristicine, *f.* myristicin.

myristicinique, myristicic, myristicinic.

myristicol, *m.* myristicol.

myristine, *f.* myristin.

myristique, myristic.

myristone, *f.* myristone.

myrmécodrome, (plant) with cavities used by ants.

myrmécophilie, *f.* association of plants with ants.

myrmécoxène, myrmecoxenous, supplying both food and shelter.

myrobalan, myrobolan, *m.* myrobalan (plum).

myrobalanier, *m.* myrobalan tree (Terminalia).

myronique, myronic.

myrosine, *f.* myrosin.

myroxyle, *m.* member of Myroxylon.

myrrhe, *f.* myrrh (the gum resin).

myrrhé, perfumed with myrrh.

myrrhide, *f.*, **myrrhis,** *m.* myrrhis.

myrrhine, *f.* myrrhin.

myrrhique, myrrhic.

myrrhol, *m.* myrrhol.

myrtacées, *f.pl.* Myrtaceae.

myrte, *m.* myrtle (Myrtus).

myrtifolié, with myrtle leaves.

myrtille, *f.* bilberry, blueberry, whortleberry (*Vaccinium myrtillus*); — **rouge,** mountain cranberry, red whortleberry (*Vaccinium vitis-idaea*).

myrtoïde, myrtoid.

mystère, *m.* mystery.

mystérieusement, mysteriously.

mystérieux, -euse, mysterious.

mysticisme, *m.* mysticism.

mystifier, to mystify.

mystique, mystic(al).

mythologie, *f.* mythology.

myure, myurus, long and tapering like a mouse's tail.

myxobactéries, *f.pl.* myxobacteria, bacteria forming colonies united by gelatinous covering.

myxogastères, *f.pl.* myxogasters.

myxome, *m.* myxoma.

myxomycètes, *f.pl.* Myxomycetes.

myxophytes, *f.pl.* Myxomycetes.

myxospore, *m.* myxospore. a spore formed in the sporangia.

N

n, à la n-ième puissance, to the nth power.

n', contraction of ne.

nacarat, nacarat, orange red, nacarat (grape).

nacelle, *f.* boat, wherry, (weighing) scoop, car, gondola (of balloon), boat for combustion furnace; — à fond plat, flat-bottomed boat; — en verre pour pesées, glass weighing scoop; — motrice, power car.

nacre, *f.* nacre, mother-of-pearl.

nacré, nacreous, like mother-of-pearl.

nacrer, to give a pearly luster to.

naevus, *m.* nevus, birthmark, mole.

nagana, *f.* nagana (equines).

nage, *f.* rowing, swimming, rowlock.

nagement, *m.* swimming (of fishes).

nageoire, *f.* fin (of fish), cork, bladder, float, couchstool.

nager, to swim, row, float.

nageur, *m.* swimmer.

nageur, -euse, swimming.

naguère, naguères, a short time ago, but lately, recently, but now.

naïf, -ïve, naïve, artless, original, natural, simple, unaffected.

nain, dwarfish, dwarf; oeuf —, yolkless egg.

nainisme, *m.* nanism.

naissance, *f.* source, root (of tongue or plant), origin, rise, birth, springing line, descent, extraction, springer; donner — à, to give rise to, produce; prendre —, to originate, arise, be formed, appear, spring up.

naissant, nascent, newly born, growing, incipient, dawning (day), beginning, rising, faint, pale (color).

naître, to be born, spring up, dawn, come up, grow, originate, arise, proceed; faire —, to give birth to, give rise to, arouse, occasion, cause to grow, raise, breed, produce.

naïvement, naïvely, candidly.

naïveté, *f.* naïveté, artlessness, simplicity.

naniser, to dwarf (plant).

nanisme, dwarfishness.

nantir, to secure (a creditor), provide.

nantissement, *m.* collateral security.

napacé, napiform, turnip-shaped.

napel, *m.* monkshood, aconite (*Aconitum napellus*).

napelline, *f.* napelline.

naphtadil, naphtagil, *m.* ozocerite, neftgil.

naphtalène, *m.* naphthalene.

naphtaline, *f.* naphthalene; — blanche en boules, moth balls.

naphte, *m.* naphtha (mineral oil); — de huile, coal-tar naphtha; — de pétrole, petroleum naphtha.

naphtène, *m.* naphthene.

naphténique, naphthene.

naphtionique, naphthionic (acid).

naphtofurane, *m.* naphthofuran.

naphtoïque, naphthoic.

naphtol, *m.* naphthol.

naphtoler, to naphtholize, treat or impregnate with naphthol.

naphtolsulfoné, naphtholsulphonic.

naphtoquinone, *f.* naphthoquinone.

naphtylamine, *f.* naphthylamine.

naphtyle, *m.* naphthyl.

naphtylène, *m.* naphthylene.

naphtylique, naphthylic, naphthyl.

naphtylol, *m.* naphthol.

napiforme, napiform, turnip-shaped.

napoléon, *m.* napoleon (20-franc piece).

napolitain, Neapolitan, of Naples.

nappage, *m.* table linen.

nappe, *f.* nappe, sheet (ice, fire, lead, water), surface, level (water), bed, stratum, deposit (of petroleum), sheet, tablecloth, layer, cloth; — aquifère, subsoil water; — souterraine, subsoil water, underground water.

naquit (naître), (it) was born.

nar, *m.* spikenard; see NARD.

narcéine, *f.* narceine.

narcisse, *m.* narcissus; — à bouquets, polyanthus daffodil, French daffodil (*Narcissus tazetta*); — des poetes, poet's daffodil, poet's narcissus, pheasant's-eye (*Narcissus poeticus*); — des prés, — trompette, (common) daffodil (*Narcissus pseudonarcissus*).

narcissiflore, with narcissus flowers.

narcose, *f.* narcosis.

narcotico-âcre, narcotico-acrid.

narcotine, *f.* narcotine.

narcotinique, narcotinic.

narcotique, narcotic.

narcotiser, to narcotize.

narcotisme, *m.* narcotism.

nard, *m.* nard (the ointment), nard, matweed, matgrass (*Nardus stricta*); — de montagne, setwall (*Valeriana celtica*); — indien, nard (the ointment), spikenard (*Nardostachys jatamansi*), Indian beard grass (*Andropogon nardus*); — sauvage, asarabacca (*Asarum europaeum*).

narée, *f.* drowsiness, dullness.

narguer, to defy, set at defiance.

narine, *f.* nostril.

narrateur, *m.* narrator.

narré, *m.* account, story, narrative.

narrer, to narrate, relate.

nasal, -aux, nasal.

nase, *m.* beaked carp, nose carp.

naseau, *m.* nostril (of an animal).

nasière, *f.* nose ring.

nasitor, nasitort, *m.* garden cress, garden peppergrass (*Lepidium sativum*).

nasse, *f.* bow net, weir basket, (lobster) pot.

nasturce, *m.* water cress (Rorippa); — officinal, (common) water cress (*Rorippa nasturtium-aquaticum*).

natal, natal, native (of things).

natation, *f.* swimming.

natatoire, natatory, swimming.

nates, *m.pl.* nates, buttocks.

natif, -ive, native. natural, pure.

nation, *f.* nation, people.

national, -aux, national.

nationalement, nationally.

nationaliser, to nationalize.

nationalité, *f.* nationality.

nativité, *f.* birth.

natrium, *m.* natrium, sodium.

natron, natrum, *m.* native soda, natron.

natte, *f.* mat, matting, plait, braid, plaiting; — de Russie, Russian mat, bast mat.

natter, to plait, braid, cover with mats.

naturalisation, *f.* naturalization.

naturalisé, naturalized, of foreign origin but reproducing itself as a native.

naturaliser, to naturalize.

naturaliste, *m.* naturalist.

naturaliste, naturalistic.

nature, *f.* nature, kind; — des propriétaires, status of proprietor; — du sol, condition of the soil; de — à, of a nature to.

nature, inherent, plain.

naturel, *m.* nature, native, constitution, disposition, temper, naturalness; — du sol, soil condition; au —, naturally, according to life or nature, cooked simply.

naturel, -elle, natural, native.

naturellement, naturally, of course, plainly.

naucifere, nuciferous, with fruits resembling small nuts.

naucus, *m.* naucus, certain cruciferous fruits which have no valves.

naufrage, *m.* shipwreck, wreck, ruin.

naufragé, shipwrecked, wrecked, castaway.

nauséabond, nauseous, foul, sickening.

nausée, *f.* nausea, seasickness.

nauséeux, -euse, nauseous, nauseating causing nausea.

nautile, *m.* nautilus, nautilus shell.

nautique, nautical.

naval, naval.

navarin, *m.* mutton stew.

navet, *m.* turnip (*Brassica rapa*), root, rape; — hatif, early garden turnip.

navette, *f.* rape, colza, turnip rape, rapeseed, colza oil, pig (of lead), shuttle, shift (of workers), netting needle; — en bois, wooden shuttle; huile de —, rape(-seed) oil.

naviculaire, navicular, boat-shaped.

navigabilité, *f.* seaworthiness.

navigateur, *m.* navigator.

navigateur, seafaring.

navigation, *f.* navigation, navigating.

naviguer, to navigate, sail.

navire, *m.* ship, vessel, liner, ocean-going ship; — à vapeur, steamer, steamship; — à voiles, sailing vessel; — de charge, cargo vessel, freighter.

ne, accompanied by pas with the verb in between (ne le faites pas, do not do it), not, no; ne . . . aucun, none; ne . . . guère, scarcely, hardly; ne . . . guère que, scarcely but; ne . . . jamais, never; ne . . . ni . . . ni . . . , neither . . . nor . . . ; ne . . . nul, none, none; ne . . . nulle part, nowhere; ne . . . nullement, not at all, none; ne . . . pas, not, no.

ne, ne . . . pas encore, not yet; ne . . . pas que, not only; ne . . . personne, no one, nobody; **ne . . . plus,** no more, no longer; **ne . . . plus du tout,** not at all; **ne . . . plus que,** only; **ne . . . plus rien,** nothing more; **ne . . . point,** not at all; **ne . . . que,** only, but; **ne . . . rien,** nothing; **ne . . . rien que,** nothing but.

né, born, well-born; — **à terme,** fully matured.

néanmoins, nevertheless, however, notwithstanding.

néant, *m.* nothing, nothingness, naught; **mettre à —,** to dismiss, annul.

nébulaire, nebular.

nébuleuse, *f.* nebula.

nébuleux, -euse, nebulous, cloudy, turbid.

nébulosité, *f.* nebulosity.

nécessaire, *m.* necessary, necessaries, case, work basket, outfit.

nécessaire, necessary, requisite, needful, indispensable.

nécessairement, necessarily, needs, inevitably.

nécessité, *f.* necessity, need; **de —,** necessarily, of necessity; **par —,** through necessity.

nécessiter, to necessitate, compel, require, force.

nécrobiose, *f.* necrobiosis.

nécrogène, necrogenous.

nécrologie, *f.* necrology, obituary.

nécromancie, *f.* necromancy.

nécromancien, necromant, *m.* necromancer.

nécrophage, necrophagous.

nécrophore, *m.* carrion beetle.

nécrose, *f.* necrosis, canker (in wood); — **du bois,** wood canker.

nécroser, to necrose, mortify, cause necrosis, canker.

nectaire, *m.* nectary, honey tube.

nectar, *m.* nectar.

nectaré, nectareous, nectariferous.

nectarifère, nectariferous.

nectarilyme, *f.* nectarilyma, any appendages to a nectary.

nectarine, *f.* nectarine, thin-skinned freestone peach.

nectarostigmate, *m.* an indication of a nectariferous gland.

nectarothèque, *f.* nectarotheca, portion of a flower surrounding a nectariferous pore.

néerlandais, *m.* Dutchman, Dutch.

néerlandais, Dutch.

néfaste, of evil omen, unlucky.

nèfle, *f.* medlar (fruit).

néflier, *m.* medlar (tree) (*Mespilus germanica*).

neftgil, *m.* ozocerite; see NAPHTADIL.

négatif, -ive, negative.

négation, *f.* negation, denial, negative (word).

négative, *f.* negative (proposition).

négativement, negatively.

négativité, *f.* negativeness, negativity, negativism.

négligé, neglected, unheeded, careless, negligent.

négligeable, negligible, that can be neglected.

négligemment, carelessly, negligently, casually.

négligence, *f.* negligence, neglect, want of care.

négligent, negligent, careless, neglectful.

négliger, to neglect, disregard, slight, ignore; **se —,** to be careless, be negligent.

négoce, *m.* trade, business, traffic, commerce.

négociable, negotiable.

négociant, *m.* merchant, trader.

négociateur, *m.* negotiator.

négociation, *f.* negotiation.

négocier, to negotiate, trade, do business.

nègre, black, negro.

negundo, *m.* negundo, ash-leaved maple, box elder (*Acer negundo*).

neige, *f.* snow; — **fraîche,** new snow.

neiger, to snow.

neigeux, -euse, snowy, snow-white.

néisme, *f.* neism, the origin of an organ on a given place.

nélombo, nelumbo, *m.* Egyptian lotus (*Nymphaea lotus*), (East) Indian lotus (*Nelumbo nucifera*).

némate, *f.* — **de l'épicéa,** spruce saw fly; **grande** — **du mélèze,** larch saw fly.

nématoblaste, *m.* nematoblast.

nématocyste, *m.* nematocyst.

nématode, *m.* threadworm.

nématophyte, filamentous (plant).

ne-me-touchez-pas, *m.* touch-me-not, yellow balsam, quick-in-the-hand (*Impatiens pallida*).

némoblaste, with filiform embryo.

né-mort, stillborn.

ne m'oubliez pas, m. forget-me-not (Myosotis).

nénuphar, m. water lily, pond lily, waterbells (Nymphaea); — jaune, yellow water lily (Nuphar luteum).

néodyme, m. neodymium.

néoformation, f. neoformation.

néogene, neogenic.

néon, m. neon.

néonate, neonatal, pertaining to the newborn.

néophron percnoptère, m. Egyptian vulture.

néophyte, f. neophyte, a newly introduced plant.

néoplastique, pertaining to new growth.

néoytterbium, m. ytterbium.

Népâl, Népaul, m. Nepal.

nèpe, f. water scorpion.

népenthès, m. pitcher plant (Nepenthes).

népérien, -enne, Napierian.

néphélémètre, m. nephelometer.

néphéline, f. nephelite, nepheline.

néphélinite, f. nephelinite.

néphélion, m. nebula.

néphéloïde, cloudy, cloudlike.

néphrectomie, f. nephrectomy.

néphrétique, f. nephrite, jade, nephritis; m. nephritic, kidney remedy.

néphrétique, nephritic.

néphrite, f. nephrite, nephritis, inflammation of the kidneys, renal disease.

néphrosta, f. nephrosta, sporangia of Lycopodium.

nerf, m. nerve, fiber, sinew, ligament, tendon, thew; — de boeuf, pizzle.

néritique, neritic.

néritte, f. rosebay, spiked willow herb.

néroli, m. neroli, orange-flower essential oil; — bigarade, oil of bitter-orange flowers; essence de —, oil of neroli, neroli.

néroplankton, m. neroplankton, neritic plankton.

nerprun, m. buckthorn (Rhamnus); — purgatif, purging buckthorn (Rhamnus cathartica); huile de —, buckthorn oil.

nervale, (tendril or cirrus) completely developed in leaf extension.

nervation, f. venation, neuration.

nerve, f. vein.

nervé, veined, nerved, costate, nervate.

nerveux, -euse, nervous, fibrous, sinewy, strong, hard, sensible.

nervifolié, ribbed (leaf).

nervosisme, f. nervousness.

nervule, f. vein, nervule.

nervure, f. vein, nerve, rib (of leaf), web, fin, feather, flange, fillet, nervation, neuration, groove; — médiane, midrib, middle vein; à —s anastomosées, nettedveined.

nervuré, ribbed, veined, flanged, grooved.

nesslérisation, f. Nesslerization.

nesslériser, to Nesslerize, test with Nessler reagent.

net, net, clean, pure, clear, evident, definite, distinct, sharp, sharply defined, neat, unmixed, frank, fair, plain, open, flatly, entirely, on the spot, clearly; mettre au —, to make a fair copy of.

nettement, neatly, clearly, distinctly, entirely, plainly, sharply, frankly, flatly.

netteté, f. cleanness, clearness, neatness, distinctness, clarity, plainness; — de l'image, definition of image, superb definition.

nettoiement, m. cleaning, cleansing, thinning (a thicket); faire un —, to clean, thin a thicket.

nettoyage, m. cleaning, purifying.

nettoyer, to clean, scour, sweep, swab, wipe, wash out, screen (corn), gin (cotton), thin a thicket.

nettoyeur, m. cleaner.

nettoyeuse, f. cleaning machine.

neuf, neuve, new, nine, ninth, fresh; à — anew, afresh, as good as new.

neuf-trous, m. nine-hole course.

neurine, f. neurine.

neurone, m. neuron.

neurotrique, with hairy veins.

neutralement, neutrally.

neutralisant, m. neutralizing body.

neutralisation, f. neutralization.

neutraliser, to neutralize; se —, to be neutralized, neutralize each other, become neutral.

neutralité, f. neutrality.

neutre, neutral, neuter; fleur —, asexual flower.

neutriflore, neutriflorus, having neuter ray florets.

neutrophilique, neutrophilic.

neuve, see NEUF.

neuvième, ninth.

neveu, *m.* nephew; *pl.* descendants, posterity.

névralgie, *f.* neuralgia.

névrine, *f.* neurine.

névrite, *f.* neuritis.

névroptère, *m.* neuropterous insect.

newtonien, -enne, Newtonian.

nez, *m.* nose, smell, scent, beak, promontory.

ni, nor, or; **ni . . . ni,** neither . . . nor.

niable, deniable.

niaouli, *f.* cajuput (*Melaleuca leucadendron*).

niccolique, nickel, of nickel.

niche, *f.* niche, kennel, nook.

nichée, *f.* nest(ful) (of birds), brood.

nicher(se), to nestle.

nichet, *m.* nest egg.

nicheur, nest-building, nesting.

nickel, *m.* nickel.

nickelage, *m.* nickeling, nickel-plating, nickelage.

nickeler, to nickel, nickel-plate, nickelize.

nickeleux, -euse, nickelous.

nickelglanz, *m.* nickel glance, gersdorffite.

nickélifère, nickeliferous, nickel-bearing.

nickéline, *f.* niccolite, kupfernickel.

nickélique, nickelic.

nickélisage, *m.* nickel-plating.

nickéliser, to nickel-plate.

nickelure, *f.* nickeling, nickel-plating.

nicol, *m.* Nicol (prism).

nicotéine, *f.* nicoteine.

nicotianine, *f.* oil of tobacco, nicotianin.

nicotine, *f.* nicotine.

nicotinisme, *m.* nicotinism.

nicotique, nicotinic, nicotic.

nicotiser, to nicotinize, nicotize.

nictation, *f.* winking, nictation.

nid, *m.* nest, berth, place; **— à couvée,** brood nest; **— d'abeilles,** honeycomb.

nidification, *f.* nest building.

nidifier, to build a nest.

nidoreux, -euse, nidosus, nidorosus, having a foul smell.

nid-trappe, *f.* trapnest.

nie (nier), (he) denies.

nielle, *f.* niello, smut, blight, rusi, corn cockle (*Agrostemma githago*); **— des blés,**

— des champs, corn cockle, corn companion (*Agrostemma githago*).

nier, to deny; **se —,** to be denied.

nigelle, *f.* fennelflower (Nigella).

nigrine, *f.* nigrine, a ferruginous variety of rutile.

nigripède, with dark-colored foot.

nigrisperme, with dark-colored seed.

nigrite, *f.* nigrite.

nigrosine, *f.* nigrosine.

Nil, *m.* Nile.

nille, *f.* tendril, loose (tool) handle.

niobate, *m.* columbate, niobate.

niobé, *n.* Niobeoil.

niobium, *m.* columbium, niobium.

niqueter, to nick.

nitidiflore, with brilliant flowers.

nitidifolié, with shiny leaves.

niton, *m.* niton.

nitramidine, *f.* nitramidine.

nitrant, nitrating.

nitratation, *f.* nitration.

nitrate, *m.* nitrate; **— d'ammoniaque,** ammonium nitrate; **— de baryte,** barium nitrate; **— de chaux,** nitrate of lime, calcium nitrate; **— de plomb,** lead nitrate; **— de potasse,** potassium nitrate.

nitrater, to nitrate.

nitratier, -ière, pertaining to nitrates.

nitration, *f.* nitration.

nitre, *m.* niter, saltpeter, potassium nitrate; **— cubique,** cubic niter, sodium nitrate; **— lunaire,** silver nitrate.

nitré, nitrated, nitro.

nitrer, to nitrate.

nitreux, -euse, nitrous.

nitrière, *f.* niter bed, saltpeter bed, niter works.

nitrificateur, *m.* nitrifier.

nitrificateur, -trice, nitrifying, causing nitrification.

nitrification, *f.* nitrification.

nitrifier, to nitrify; **se —,** to become nitrous, be converted into niter, nitrate.

nitrile, *m.* nitrile.

nitrique, nitric.

nitrite, *m.* nitrite; **— de soude,** sodium nitrite.

nitrobacter, *m.* Nitrobacter.

nitrobenzène, *m.,* **nitrobenzine,** *f.* nitrobenzene.

nitrocalcite. *m.* nitrocalcite, nitrate of lime.

nitrocellulose, *f.* nitrocellulose, cellulose nitrate, collodion cotton.

nitrocoton, *m.* guncotton.

nitroforme, *m.* nitroform, trinitromethane.

nitro-gélatine, *f.* nitrogelatin, gelatin dynamite.

nitrogène, *m.* nitrogen.

nitrogéné, nitrogenous.

nitrogéner, to nitrogenize, azotize.

nitroglycérine, *f.* nitroglycerin.

nitroleum, *m.* nitroleum, nitroglycerin.

nitrolique, nitrolic.

nitrométhane, *m.* nitromethane.

nitromètre, *m.* nitrometer.

nitromonade, *f.* Nitrobacter.

nitromuriatique, nitromuriatic, nitrohydrochloric.

nitronaphtaline, *f.* nitronaphthalene.

nitrophénol, *m.* nitrophenol.

nitrophile, nitrophilous, thriving in nitrogenous soils.

nitrophytes, *f.pl.* nitrophytes, potashloving plants.

nitroprussiate, *m.* nitroprussiate, nitroprusside.

nitrosaccharose, *f.* nitrosaccharose.

nitrosamine, *f.* nitrosamine.

nitrosé, converted into a nitroso compound or a nitrite.

nitrosité, *f.* nitrous quality or state.

nitrosochlorure, *m.* nitrosochloride.

nitrosococcus, *m.* Nitrosomonas.

nitrosophénol, *m.* nitrosophenol.

nitrosubstitué, nitrosubstituted.

nitrosulfate, *m.* nitrosulphate.

nitrosyle, *m.* nitrosyl.

nitrosylsulfurique, nitrosylsulphurie.

nitrotoluène, nitrotoluol, *m.* nitrotoluene.

nitrure, *m.* nitride.

nitryle, *m.* nitryl.

niveau, *m.* level, gauge, horizontal line, leveling instrument; — **à bulle,** spirit level, water level; — **à bulle d'air,** air, spirit level; — **d'eau,** water level, water line; — **de la mer,** (mean) sea level; **au —, de —,** on a level, level, horizontal; **au — de,** on a level with; **mettre au —, mettre de —** to level, bring to a level.

niveler, to level, level down, grade; **se —,** to become level.

niveleur, *m.* leveler.

niveleur, -euse, leveling.

nivellement, *m.* leveling, surveying.

nivéole, *f.* snowdrop (*Galanthus nivalis*).

N.O., *abbr.* (nord-ouest), northwest.

nobélite, *f.* dynamite.

noble, noble, great, generous.

noble-épine, *f.* hawthorn (*Crataegus oxyacantha*).

noblement, nobly.

noblesse, *f.* nobility, nobleness.

noce, *f.* wedding, marriage, wedding festivities.

nochère, *f.* gutter, (leaded) skylight.

nocif, -ive, noxious, injurious, harmful, toxic, pathogenic.

nocivité, *f.* noxiousness, harmfulness.

noctuelle, *f.* night moth, noctua, owl moth; — **piniperde,** pine beauty.

nocturne, *m.* nocturne.

nocturne, nocturnal, in the night, nightflowering.

nocturnement, by night, nocturnally.

nocuité, *f.* nocuousness, harmfulness, injuriousness.

node, *f.* knob, swelling.

nodosité, *f.* node; — **radiculaire,** nodule of root.

nodule, *m.* nodule, small node.

noduleux, -euse, nodular, noduled, nodulous.

nodus, *m.* node.

Noël, *m.* Christmas.

noeud, *m.* knot, node, joint, knarl, snag, knag, knob, knuckle, bond, tie; — **adhérent,** intergrown knot; — **de bois mort,** horny knot, loose knot; — **délié** — **détaché,** loose knot; — **pourri,** decayed knot.

noie (noyer), is drowning.

noir, *m.* black, (animal) black, charcoal, bull's-eye, bruise, brown rust, black mold (on plants); — **animal,** animal charcoal, bone black; — **de bougie,** candle black; — **de charbon,** coal black, carbon black; — **de Chine,** India ink; — **de fumée,** lampblack; — **de liège,** cork black; — **de schiste,** slate black: — **de vigne,** vine black.

noir, — **d'imprimerie,** printer's ink; — **d'ivoire,** ivory black; — **d'os,** bone black.

spodium, animal charcoal; — **foncé,** deep
black; — **outremer,** ultramarine black;
— **pour fonderie,** founder's black, powdered charcoal or coal; — **végétal,** vegetable black; **mettre au —,** to blacken;
mettre dans le —, to hit the mark.

noir, black, dark, dirty.

noirâtre, blackish, dark.

noirceur, *f.* blackness, baseness.

noircir, to blacken, grow or become black,
turn black, darken, black wash, stain.

noircissement, *m.* blackening, dyeing in
black, blackwashing.

noircisseur, *m.* dyer in black, blackener.

noircissure, *f.* black spot, smudge.

noisetier, *m.* hazel tree (*Corylus avellana*).

noisette, *f.* hazelnut, filbert, hazel color.

noix, *f.* nut, walnut, kernel, clamp (for
fastening a rod to an iron stand), plug (of
cock), tumbler (of gunlocks), cap (of knee),
sprocket (of chainwheel), drum (of
capstan or winch), semicircular groove,
cone (of mill); — **d'acajou,** cashew nut;
— **d'Alep,** Aleppo gall; — **d'arec,** betel
nut, areca nut.

noix, — de banda, nutmeg; — **de broche,**
spindle wharve; — **de coco,** coconut,
cocoanut; — **de cola,** kola nut; — **de
galle,** gallnut, nutgall, gall, oak gall; —
de vomique, nux vomica; — **d'Inde,** coconut; — **muscade,** nutmeg; — **palmiste,**
areca nut; **huile de —,** walnut oil.

nom, *m.* name, noun, fame; — **de guerre,**
assumed name; — **du genre,** genus name;
— **social,** firm name; **au — de,** in the
name of, for the sake of; **de —,** by name,
in name, nominally.

nomade, *m.* nomad.

nomade, nomadic.

nombrable, numerable, countable.

nombre, *m.* number, quantity; — **atomique,**
atomic number; — **carré,** square number;
— **de,** a number of, many; — **entier,**
integer, whole number; — **impair,** odd
number; — **inférieur,** minority, smaller
number; — **pair,** even number; — **rompu,**
fraction, broken number; — **rond,** round
number.

nombre, — sourd, irrational or surd number; — **suffisant,** sufficient number, quo-
rum; **au — de,** among, in the number of;
dans le —, among, in the number; **du —
de,** one of, among, in the number of; **en —,**
many, in numbers; **faire —,** to count;
mettre au — de, to count; **sans —,**
numberless, countless, innumerable.

nombrer, to number, reckon, count.

nombreux, -euse, numerous; **peu —,** few
in number.

nombril, *m.* navel, hilum, eye (of fruit); —
de Vénus, navelwort (*Cotyledon umbilicus*).

nomenclateur, *m.* nomenclator, classified
list, vocabulary.

nomenclature, *f.* nomenclature, catalogue,
list.

nomenclaturer, to name methodically,
compile a list.

nominal, -aux, nominal.

nominalement, nominally.

nominativement, by name.

nommément, namely, particularly, to wit.

nommer, to name, call, mention, nominate,
elect, appoint; **se —,** to be called, be
named, state one's name.

nomophylla, *f.* collection of plant leaves.

non, no, not, un-, in-, non-; — **pas,** no; —
plus, neither, either, no longer, no more;
— **plus que,** not (no) more than; — (**pas)
que,** not that; — **seulement,** not only; **ni
moi — plus,** nor I either.

non-activité, *f.* nonactivity.

non-adhérence, *f.* nonadherence.

nonane, *m.* nonane.

nonantième, ninetieth.

nonchalamment, carelessly.

nonchalance, *f.* carelessness.

nonchalant, careless.

non-conducteur, *m.* nonconductor, nonconducting.

non-cultivé, uncultivated.

non-désagrégé, unweathered.

non-disponibilité, *f.* nonavailability.

non-disponible, nonavailable, unavailable.

non-dosé, *m.* undetermined quantity.

non-dosé, undetermined.

non-dressé, unbroken.

non-égrappage, *m.* not picking from the
cluster.

non-existence, *f.* nonexistence.

non-isolé, uninsulated.

nonne, *f.* nun moth.

nonnius, *m.* nonius, vernier.

nonobstant, notwithstanding, nevertheless.

nonoïque, nonoic.

nonopètale, having nine petals.

nonose, *f.* nonose.

non-ouvré, unworked, unwrought, uncut.

non-réussite, *f.* failure, abortiveness.

non-succès, *m.* failure.

non-tanin, *m.* nontannin, nontan.

non-toxicité, *f.* nontoxicity, nonpoisonousness.

non-transparent, opaque.

nonuple, ninefold.

nonuple-effet, *m.* nonuple effect (apparatus).

non-usage, *m.* disuse, nonusage.

non-valeur, *f.* worthlessness, loss, deficiency, bad debt, unproductivity.

nonyle, *m.* nonyl.

nonylène, *m.* nonylene.

nonylique, nonylic (acid), nonyl, nonoic.

nopal, *m.* nopal, cochineal cactus or fig (*Nopalea cochinellifera*).

noper, to burl (cloth).

norberte, *f.* small black plum.

nord, north, northern.

nord-est, northeast (wind), northeastern.

nord-ouest, northwest (wind), northwestern.

nord-sud, north-south, north by south.

normal, -aux, normal, usual, regular, standard; — décime, decinormal.

normale, *f.* normal, perpendicular.

normalement, normally.

normalisation, *f.* standardization.

normand, Norman, evasive, crafty, shrewd.

Normandie, *f.* Normandy.

norme, *f.* norm, standard.

noropianique, noropianic (acid).

Norvège, *f.* Norway.

norvégien, -enne, Norwegian.

nos, our.

nostoccacées, *f.pl.* Nostocaceae, family of blue-green algae.

nota, *m.* note (marginal), footnote.

notabilité, *f.* notability.

notable, notable, eminent, considerable.

notablement, notably, appreciably, much, considerably.

notaire, *m.* notary.

notamment, especially, more particularly, specially, for example.

notation, *f.* notation.

note, *f.* note, mark, account, memorandum, grade, bill, invoice.

noter, to note, observe, notice, mark; se —, to be noted.

nothogamie, *f.* nothogamy, heteromorphic xenogamy.

notice, *f.* notice, account.

notification, *f.* notification.

notifier, to notify.

notion, *f.* notion, information, apprehension, idea, conception.

notocorde, notochorde, *f.* notochord, chorda dorsalis.

notoire, well-known, notorious.

notoirement, notoriously.

notorhize, notorhizal, (radicle) on back of cotyledons.

notoriété, *f.* notoriety, notoriousness, fame.

notre, our.

nôtre, ours, our own; *pl.* ours, our friends, people, our party; le —, la —, les —s, ours, our own.

nouage, *m.* knotting, joining. rickets, rachitis.

noue, *f.* marshy meadow.

noué, tied, rickety; fruit —, fertilized fruit.

nouer, to knot, tie, tie up, knit, form, engage in, set (of fruits).

nouet, *m.* bag of seasoning herbs, bag containing substance to be steeped.

noueux, -euse, knotty, gnarled, closed by nodes or joints.

nougat, *m.* nougat, nut-oil cake.

nouilles, *f.pl.* noodles; — aux oeufs, egg noodles.

nouillettes, *f.pl.* fine (small) noodles.

nourri, nourished, fed, full, thick, strong, rich, copious, hardened.

nourrice, *f.* nurse, nurse bee.

nourricerie, *f.* stock farm, silkworm farm.

nourricier, -ère, nourishing, nutritive, nutrient.

nourrir, to nurse, nourish, feed, fatten, rear, foster, forage, fodder, bring up, raise, produce, cherish; se —, to feed, live, keep oneself, thrive, improve oneself.

nourrissage, *m.* cattle feeding, nourishing, rearing.

nourrissant, nourishing, satisfying, substantial, nutritive, nutritious.

nourrissement, *m.* feeding; — spéculatif, stimulative feeding.

nourrisseur, *m.* feeder, dairyman, stock-raiser.

nourrisson, *m.* nursing infant, foster child.

nourriture, *f.* nourishment, nutriment, food, sustenance, feed, feeding, rearing, fattening (animals), nursing, tawing paste; — **sèche,** dry food, dry hay.

nous, we, us, to us, each other.

nous-mêmes, ourselves.

nouure, *f.* knotting, setting (of fruits), rickets.

nouveau, new, fresh, recent; **à** —, anew, again; **de** —, again, anew.

nouveau-né, *m.* new-born child.

nouveau-né, new-born.

nouveauté, *f.* novelty, newness, change, innovation, new fashion, new invention, new book, fancy goods.

nouveau-venu, *m.* **nouvelle-venue,** *f.* newcomer.

nouvel, -elle, new.

nouvelle, *f.* news, novel.

Nouvelle-Écosse, *f.* Nova Scotia.

nouvellement, newly, recently, lately.

Nouvelle-Orléans, *f.* New Orleans.

novaculite, *f.* novaculite.

novale, *f.* newly broken up (land).

novale, new, newly cleared.

novateur, *m.* pioneer, innovator.

novembre, *m.* November.

novemdigité, novemdigitate, nine-fingered.

novemfolié, with nine leaflets.

novemlobe, novemlobus, nine-lobed.

novemloculaire, with nine locules.

novemnervé, novemnervious, nine-nerved.

novice, *m.* novice, beginner.

novice, inexperienced.

noyau, *m.* nucleus, core (of a ring or mold), stone, kernel (of fruit), nut, stem, shank (of bolt), plug, key (of cock), hub (of wheel), heart; — **benzénique,** benzene nucleus; — **central,** bull's-eye, bullion, knob; — **d'induit,** armature core; — **en tole,** sheet iron core; — **germinal,** gamete nucleus; — **pentagonal,** five-membered nucleus.

noyé, drowned, immersed, underwater, flooded (carburetor), countersunk, blurred (outline).

noyer, to drown, flood, sink, immerse, bed in cement, drive (nail) in flush, bury, put under water, be drowned, countersink, blend, mix (colors) overslake (lime); **se** — to drown oneself.

noyer, *m.* walnut (wood or tree) (Juglans); — **beurre,** butternut tree, white walnut tree (*Juglans cinerea*), oil nut; — **blanc,** shagbark hickory (*Carya ovata*); — **commun,** English walnut, common European walnut (*Juglans regia*), walnut tree; — **gris,** white walnut, butternut (*Juglans cinerea*); — (**blanc**) **d'Amérique,** hickory, shellbark (*Carya ovata*); — **noir,** black walnut (*Juglans nigra*).

nu, bare, exposed, naked, nude, plain, denuded, destitute; **à** —, bare, naked, exposed.

nuage, *m.* cloud.

nuageux, -euse, cloudy, overcast, hazy.

nuance, *f.* shade, hue, tint, gradation, nuance, slight difference.

nuancement, *m* shading, blending, variation.

nuancer, to shade, gradate, blend, tint, vary, variegate; **se** —, to be shaded, blended, variegated, or tinted.

nubile, nubile, marriageable.

nucamentacé, nucamentaceous.

nucelle, *f.* nucellus.

nucifère, nut-bearing.

nucine, *f.* nucin, juglone.

nucléal, -aux, nucléaire, nuclear.

nucléase, *f.* nuclease.

nucléé, nucleated, nucleate (cell).

nucléifère, containing a nucleus.

nucléiforme, nucleiform, shaped like a nucleus.

nucléine, *f.* nuclein.

nucléique, nucleic.

nucléo-albumine, *f.* nucleoalbumin.

nucléobranches, *m.pl.* Nucleobranchiata.

nucléochondre, *f.* seeds of chromatin.

nucléohistone, *f.* nucleohistone.

nucléolaire, nucleolar.

nucléole, *m.* nucleolus.

nucléolé, nucleolated.

nucléoplasma, *f.* plasma of the nucleus.

nucléoprotéide, *m.* nucleoprotein.

nucléus, *m.* nucleus.

nucode, *f.* fruit composed of distinct nuts.

nuculaine, *f.* nuculane, nuculanium.

nucule, *f.* nutlet, achene, part of a schizocarp.

nuculeux, -euse, containing nucules.

nudicaule, nudicaulous, naked-stemmed.

nudiflore, nudiflorous.

nudifolié, with bare, glossy leaves.

nudipède, with leafless peduncle.

nudité, *f.* bareness, nudity, nakedness.

nue, *f.* cloud.

nue, (*fem.*) see NU.

nuée, *f.* (large) cloud, swarm, shower.

nui, injured, harmed.

nuire, to do harm, be harmful (to), be injurious, prejudice, hurt, wrong; — à, to injure, harm, be prejudicial to, interfere with.

nuisibilité, *f.* harmfulness, injuriousness.

nuisible, hurtful, harmful, injurious, noxious, detrimental, pernicious.

nuisiblement, injuriously, harmfully.

nuisit (nuire), (he) harmed, was injurious to.

nuit, *f.* night, darkness.

nul, nulle, no, not one, not any, none, nobody, null, void, of no value, unsounded, silent (letter), zero.

nullement, not at all, by no means, in no wise.

nullifier, to nullify.

nullité, *f.* nullity, invalidity, incapacity, nonentity.

nûment, nakedly, plainly, frankly.

numéraire, *m.* specie, coin.

numéraire, legal (value).

numéral, -aux, numeral.

numérateur, *m.* numerator; — à secteur, counting disk.

numération, *f.* numeration, numbering, notation, counting, count.

numérique, numerical.

numériquement, numerically.

numéro, *m.* number, size; — d'ordre. order (serial) number.

numérotage, *m.* numbering.

numéroter, to number.

nummulaire, *f.* moneywort (*Lysimachia nummularia*).

nuphar, *m.* yellow water lily (*Nuphar advenum*).

nuptial, -aux, nuptial, bridal.

nuque, *f.* nape of the neck, nucha.

nutation, *f.* nutation.

nutrescibilité, *f.* nutritive value.

nutrescible, nutritive, nutrient.

nutricier, -ère, nourishing, nutritious, nutrient.

nutricisme, *m.* nutricism.

nutriment, *m.* nutriment.

nutrimentaire, nutrimental, nutritive.

nutritif, -ive, nutritive, nutritious, nutrient, nourishing.

nutrition, *f.* nutrition, feeding, alimentation, nourishing.

nutritivité, *f.* nutritiveness, nutritiousness, food value.

nutrose, *f.* nutrose.

nychthémère, having ephemeral existence.

nyctaginées, *f.pl.* Nyctaginaceae.

nyctitropique, *m.* nyctitropism.

nyctitropisme, *m.* nyctitropism.

nymphe, *f.* nymph, nympha, pupa, chrysalis; — proprement dites, free or exarate pupa.

nymphié, provided with an ovarian nectary.

nymphion, *f.* ovarian nectary.

nymphose, *f.* pupation.

O

O., *abbr.* (ouest), west.

ô, O, oh.

obclavé, obovate.

obclaviforme, obclavate.

obcomprimé, obcompressed, flattened vertically or anteriorly.

obconique, obconic.

obcordé, obcordate.

obcordiforme, obcordate.

obcrénelé, (border) cut into salient angles.

obdiplostémone, obdiplostemonous, outer stamens opposite the petals.

obduction, *f.* post-mortem examination.

obéir, to obey, be obedient, yield (to force); — à, to obey, answer, respond to.

obéissance, *f.* obedience, elasticity (of copper).

obéissant, obedient, pliant, supple.

obélisque, *m.* obelisk.

obier, *m.* snowball tree, guelder-rose (*Viburnum opulus*).

obimbriqué, imbricated, against the grain.

obituaire, obituary.

objecter, to object, object to, oppose, reproach with.

objectif, *m.* objective (lens), object glass, aim, end; — **à sec,** dry objective; — **à immersion,** immersion objective.

objection, *f.* objection.

objectivement, objectively.

objet, *m.* object, article, thing, purpose, subject, aim, purport, view, — **courant,** easily salable article; — **de rechange,** spare part; —**s en bois,** woodenware, articles in wood.

oblancéolé, oblanceolate.

obligataire, *m.* bondholder.

obligation, *f.* obligation, duty, bond, debenture; — **de plombe,** sealing duty.

obligatoire, obligatory, compulsory.

obligé, obliged, necessary, indispensable.

obligeamment, obligingly.

obligeance, *f.* kindness, obligingness.

obligeant, obliging, kind.

obliger, to oblige, constrain, compel, bind.

obligulé, obligulate, (ligulate florets) extended on the inner side of the capitulum.

obliguliflore, obliguliflorous, florets obligulate.

obliguliforme, obligulate.

obliquangle, oblique-angled.

oblique, *m.* oblique muscle; *f.* oblique line.

oblique, oblique, slanting; **en** —, obliquely, slantingly.

obliquement, obliquely.

obliquer, to slant.

obliquité, *f.* obliquity, obliqueness.

oblitération, *f.* obliteration, canceling, obstruction, stopping.

oblitéré, obliterated, suppressed.

oblitérer, to obliterate, cancel, deface, stop, obstruct; **s'**—, to disappear, be obliterated.

oblong, -gue, oblong.

oblongifolié, with oblong leaves.

obnubilé, overcast.

oboval, obové, obovulé, obovate.

obovatifolié, with oboval (obovate) leaves.

obovoïde, obovoid, ovoid.

obpyramidal, obpyramidal.

obscur, dark (of colors), obscure.

obscurateur, *m.* darkener, obscurer.

obscurateur, -trice, darkening, obscuring.

obscurcir, to obscure, darken, cloud, dim, fog; **s'**—, to become obscure, grow dim or dark.

obscurcissement, *m.* darkening, obscuring, dimness.

obscurément, obscurely, darkly, confusedly, dimly.

obscurité, *f.* darkness, obscurity.

obséder, to beset, worry, besiege, haunt, obsess.

observable, observable.

observateur, *m.* observer.

observateur, -trice, observing, observant.

observation, *f.* observation, observance, remark, seeing.

observatoire, *m.* observatory.

observer, to observe, watch, notice, perform; **s'**—, to be observed, be circumspect, be on one's guard.

obsidiane, obsidienne, *f.* obsidian.

obsolescence, *f.* obsolescence, atrophy.

obsonine, *f.* opsonin.

obsonique, opsonic.

obstacle, *m.* obstacle, hindrance; **faire** —, to form an obstruction, hinder.

obstétrique, *f.* obstetrics.

obstination, *f.* obstinacy, pertinacity, stubbornness.

obstiné, obstinate, stubborn, self-willed, persistent.

obstinément, stubbornly, obstinately.

obstiner(s'), to be obstinate, insist on.

obstipation, *f.* constipation.

obstructeur, *m.* obstructer.

obstructif, -ive, obstructive.

obstruction, *f.* obstruction, stopping up, stoppage.

obstrué, obstructed, stopped.

obstruer, to obstruct, close, stop up, hinder; **s'**—, to become obstructed, stopped up.

obsubulé, obsubulatous, very narrow, pointed at base and widening toward the apex.

obsutural, obsuturalis, applied to suture of pericarp.

obtenir, to get, obtain, secure; **s'**—, to be obtained.

obtention, *f.* obtaining, obtainment.

obtenu, obtained.

obtient (obtenir), (he) obtains.

obtint (obtenir), (he) obtained.

obtondant, blunting.

obturateur, *m.* obturator, closing device, check, cutoff, valve, stopper, plug, cap, valve shutter, stop valve, stopcock, gas check (of gun), shutter (of camera).

obturateur, -trice, obturating, stopping, closing.

obturation, *f.* obturation, stopping, closing, obliteration, occlusion, shutting off.

obturbiné, obturbinatus, reverse topshaped.

obturer, to stop, seal, close, shut, obturate.

obtus, obtuse, dull, blunt.

obtusangle, obtuse-angled.

obtusé, obtuse, with rounded tip.

obtusifide, divided into numerous obtuse segments.

obtusiflore, with obtuse petals.

obtusilobé, obtusilobous.

obtusion, *f.* obtusion, dullness, obtuseness.

obus, *m.* shell, projectile, bomb; — à balles, shrapnel; — à étoile, star shell; — à gaz, gas shell, chemical projectile; — éclairant, light shell, star shell; — fusant, time shell; — lacrymogène, tear shell; — rayé, rifled projectile.

obuser, to shell.

obusier, *m.* howitzer.

obvallé, obvallate.

obvers, *m.* obverse.

obvier (à), to obviate, prevent, avoid.

obvoluté, obvolute.

obvolutifolié, with grooved leaves.

occasion, *f.* occasion, opportunity, cause, bargain; à l'— de, at the occasion of; d'—, accidentally, secondhand.

occasionnel, -elle, occasional.

occasionnellement, occasionally.

occasionner, to occasion, cause, give rise to.

occident, *m.* west, Occident, the West; d'—, western, west.

occidental, -aux, western, occidental; les occidentaux, the Occidentals, westerners.

occipito-basilaire, occipitobasilar.

occiput, *m.* occiput, back of the head.

occis, slain, killed.

occlure, to occlude.

occlus, occluded.

occlusion, *f.* occlusion, shutting up, cutoff, obstruction.

occulte, occult, hidden, secret.

occultement, occultly, secretly.

occupant, *m.* occupant.

occupant, occupying.

occupation, *f.* occupation, pursuit, employment, business, work.

occupé, occupied, engaged, busy.

occuper, to occupy, busy, employ; s'— de, to be occupied with, pay attention to, be busied with, concern oneself with, occupy oneself.

occurrence, *f.* occurrence, event; dans l'—, as it happens, in case of emergency; en l'—, under the circumstances.

occurrent, occurring, happening.

océan, *m.* ocean; L'Océan, the Atlantic; grand Océan, Pacific Ocean.

océanique, oceanic, ocean.

océanographe, *m.* oceanographer.

océanographie, *f.* oceanography.

ocelle, *m.* simple eye, ocellus, primary optic spot.

ocellus, *m.* ocellus, epidermal cell of leaf sensitive to light.

ochracé, ocracé, ocherous.

ochranthe, with pale yellow flowers.

ochrea, *f.* ocrea, a tubular sheathlike expansion at base of petiole.

ochrocarpe, ochrocarpous.

ochypétale, with broad petals.

ocimophylle, with leaves similar to those of basil.

ocre, *f.* ocher; — jaune, yellow ocher; — rouge, red ocher.

ocreux, -euse, ochreous, ocherlike.

octaèdre, *m.* octahedron.

octaèdre, octahedral.

octaédrique, octahedral.

octandrie, *f.* octandria.

octane, *m.* octane; *f.* octan (fever).

octante, eighty, fourscore.

octave, *f.* octave.

octobre, *m.* October.

octocarbure, *m.* octacarbide.

octofide, with eight denticulations.

octoflore, with eight flowers.

octogonal, -aux, octagonal.

octogone, *m.* octagon.

octogone, octagonal.

octohydrure, *m.* octahydride, octohydride.

octolobé, with eight lobes.

octoloculaire, with eight denticulations, eight-celled fruit or pericarp.

octoné, octamerous, arranged in eights.

octonervé, with eight veins.

octonique, octonic.

octopétale, octopetalous.

octopodes, *m.pl.* Octopoda.

octosépale, octosepalous, with eight sepals.

octosperme, octospermous, eight-seeded.

octostémone, octostemonous, with eight fertile stamens.

octovalent, octovalent.

octovalve, with eight valves.

octroi, *m.* octroi, town dues, city toll on goods to be consumed within a town, concession, grant.

octroiement, *m.* granting, conceding.

octroyer, to grant, accord, concede.

octuple, octuple, eightfold, eight times.

octyle, *m.* octyl.

octylène, *m.* octylene.

octylique, octylic.

oculaire, *m.* ocular, eyepiece; — à réticule, crossline eyepiece; — chercheur, finder eyepiece; — compensateur, compensating eyepiece; — micrométrique, micrometer eyepiece; témoin —, eyewitness.

oculaire, ocular, of the eye.

odeur, *f.* odor, smell, fragrance, scent; — de fleurs, fragrance; — de rance, rancid odor; à son —, by its odor.

odieux, *m.* odium, odiousness.

odieux, -euse, odious, hateful.

odomètre, *m.* odometer, distance meter, range finder.

odontalgie, *f.* toothache.

odontorhize, with roots similar to small teeth set in one another.

odorabilité, capability of being smelled.

odorable, capable of being smelled.

odorant, odorous, odoriferous, odorant, fragrant, sweet smelling.

odorat, *m.* smell, smelling, sense of smell.

odoratif, -ive, pertaining to smell.

odoration, *f.* smelling, odoration.

odorer, to smell, smell of, be fragrant.

odoriférant, odoriferous, fragrant.

odorifère, odorous.

odorifique, odoriferous, odorific, sweet smelling, fragrant.

oecogénie, *f.* plant ecology.

oecologie, *f.* ecology, biology.

oedémateux, -euse, edematous, oedematous.

oedématope, with swollen foot.

oedème, *m.* edema; — malin, malignant edema.

oeil, (*pl.* yeux) *m.* eye, hole, opening, bud, aperture, loop, luster, glass, sheen (of silk or pearls), bubble, sight, look, attention, notice, face (of type), daisy (*Chrysanthemum leucanthemum*); — à facettes, compound or faceted eye; — à feuilles, leaf bud; — composé, compound or faceted eye; — de chat, cat's-eye; — de perdrix, bird's eye, corn (between toes); — dormant, dormant bud; — latent, resting bud, dormant bud; — terminal, terminal bud; à l'—, by (with) the eye; à l'— nu, with the naked eye; aux yeux de, in the opinion of; coup d'—, glance.

oeil-de-boeuf, *m.* moonflower, bull's-eye, ox-eye daisy (*Chrysanthemum leucanthemum*).

oeil-de-chat, *m.* cat's-eye, nicker nut.

oeillade, *f.* look, glance, ogling.

oeillère, *f.* horse blinder, eyeflap, eyetooth, canine tooth.

oeillet, *m.* eyelet, eye, little hole, pink, carnation (Dianthus), loop of wire; — de poète, sweet William (*Dianthus barbatus*); — des prés, ragged robin (*Lychnis flosculi*).

oeilleton, *m.* eyehole, peephole, sucker, offset, layer.

oeilletonnage, *m.* layering.

oeilletonner, to layer (plant), remove buds from (plants).

oeillette, *f.* oil poppy, opium poppy (*Papaver somniferum*); huile d'—, poppy-seed oil.

oenanthal, *m.* oenanthal.

oenanthe, *m.* water dropwort (Oenanthe).

oenanthine, *f.* oenanthin.

oenanthique, oenanthic.

oenanthol, *m.* oenanthole, oenanthaldehyde.

oenanthylate, *m.* oenanthylate.

oenanthylique, oenanthylic.

oenobaromètre, *m.* oenobarometer.

oenolature, *f.,* oenolé, *m.* medicated wine.

oenolique, oenolic, pertaining to wine, having wine as excipient.

oenologie, *f.* oenology.

oenologique, oenological.

oenologiste, oenologue, *m.* oenologist.

oenolotif, *m.* medicated wine.

oenolotif, -ive, containing wine.

oenomel, *m.* oenomel, honey wine.

oenomètre, *m.* oenometer.

oenothère, *m.* evening primrose (Oenothera).

oenoxydase, *f.* oenoxydase.

oesophage, *m.* esophagus, gullet.

oestre, *m.* warble fly, botfly.

oesypien, growing on sheep droppings.

oeuf, *m.* egg, ovum; — **à couver,** hatching egg; — **à la chaux,** preserved egg; — **comestible,** eatable egg; — **couvé,** rotten egg, bad egg; — **de frigorifique,** cold-stored egg; — **de poule,** hen's egg; — **frais,** fresh egg, new-laid egg; — **hardé,** wind egg; **en** — **renversé,** obovate.

oeuvre, *f.* work, piece of work, act, deed, action, setting (of stone); *m.* work, performance, production, structure; **dans** —, inside measurement, in the clear; **hors d'** —, out of alignment, outside the main work, accessory, unmounted, not set; **mettre à l'** —, to set to work, commence operations; **mettre en** —, to use, avail oneself of, put in hand, employ, set (stones).

oeuvrer, to work.

offensant, offensive, obnoxious.

offense, *f.* offense.

offenser, to offend; **s'** —, to be offended.

offenseur, *m.* offender.

offensif, -ive, offensive, pathogenic.

offensivement, offensively, by attacking.

offert (offrir), offered.

office, *m.* office, function, duty, charge; *f.* pantry; **d'** —, officially; **faire l'** — **de,** to serve as, act as.

officiel, -elle, official.

officiellement, officially.

officier, *m.* officer, official; — **de santé,** health officer.

officieux, -euse, officious, serviceable, obliging, kind, semiofficial.

officinal, -aux, officinal, medicinal.

officine, *f.* dispensary, pharmacy, apothecary's shop and laboratory.

offrant, *m.* bidder.

offrant, bidding, offering.

offre, *f.* offer, proposal, tender, supply, quotation.

offrir, to offer, present, propose, afford, bid; **s'** —, to offer oneself, offer, be offered, presented, appear, turn up.

offusquer, to obscure, obfuscate, dazzle, offend.

ognon, *m.* onion, bulb; — **patate,** potato onion; — **rond,** pear-shaped onion.

oïdiose, *f.* disease due to Oïdium.

oïdium, *m.* vine mildew (Oïdium).

oie, *f.* goose; — **cendrée,** graylag (goose); — **rieuse,** white-fronted goose; — **sauvage,** wild goose, bean goose; — **vulgaire,** bean goose.

oignant, oiling, anointing.

oignon, *m.* onion, bulb, bulbous root, bulbil, offset, bunion — **à fleurs,** flower bulb; — **rouge foncé,** blood-red onion.

oignonière, *f.* onion bed.

oindre, to rub with oil, anoint.

oing, *m.* grease, lard.

oint (oindre), (he) anoints.

oint, anointed.

oiseau, *m.* bird, fowl, hod (of bricklayer); — **aquatique,** waterfowl; — **chanteur,** songbird; — **de leurre,** decoy duck; — **de passage,** bird of passage; — **de proie,** bird of prey; — **émigrant,** migratory bird; — **mouche,** humming bird; — **sédentaire,** sedentary bird.

oiseler, to go bird catching.

oiseleur, *m.* birdcatcher.

oiselier, *m.* bird fancier.

oiselle, *f.* hen bird.

oiseux, -euse, oisif, -ive, idle, doing nothing.

oisiveté, *f.* idleness, leisure.

oléacées, *f.pl.* Oleaceae.

oléagineux, *m.* oleaginous substance.

oléagineux, -euse, oleaginous, oily, oil-yielding.

oléandre, *m.* oleander (*Nerium oleander*).

oléastre, *m.* oleaster (*Elaeagnus angusti-folia*).

oléate, *m.* oleate.

oléfine, *f.* olefin.

oléicole, pertaining to olive culture.

oléiculteur, *m.* olive grower.

oléiculture, *f.* culture of the olive.

oléifère, oil-producing.

oléifiant, olefiant, producing or forming an oil.

oléifolié, with leaves similar to those of olive tree.

oléiforme, oily, of the consistency of oil.

oléine, *f.* olein.

oléinées, *f.pl.* Oleaceae.

oléique, oleic.

oléographie, *f.* oleography, oleograph.

oléomargarine, *f.* oleomargarine.

oléomètre, *m.* oleometer.

oléorésine, *f.* oleoresin; — de capsique, oleoresin of capsicum; — de gingembre, oleoresin of ginger; — de poivre noir, oleoresin of (black) pepper.

oléorésineux, -euse, oleoresinous.

oléosaccharure, oléosaccharat, *m.* oleosaccharum.

oléracé, oleraceous.

oléule, *f.* essence.

oléum, *m.* oleum.

olfactif, -ive, olfactory, olfactive.

olfaction, *f.* olfaction, (sense of) smell, smelling.

oliban, *m.* olibanum, frankincense.

olibène, *m.* olibene.

olide, *m.* lactone.

oligacanthe, oligacanthous, bearing few spines.

oliganthère, oligandrous, bearing few stamens.

oligiste, *m.* oligist, hematite.

oligiste, oligist, oligistic.

oligocarpe, oligocarpous, having few carpels.

oligocène, Oligocene.

oligocéphale, oligocephalous, with few capitula.

oligocérate, with few buds similar to horns (warts).

oligoclase, *f.* oligoclase, sunstone.

oligomère, oligomerous, one or more whorls with fewer members.

oligosperme, oligospermous, with few seeds.

oligostémone, with few stamens.

oligotrique, slightly hairy.

olivacé, olivaceous, olive-green.

olivaie, *f.* olive plantation.

olivaire, olivary, olive-shaped.

olivaison, *f.* olive season, olive crop.

olivâtre, inclined to olive in color, olive-hued.

olive, *f.* olive.

olive, olive, olive-green, olive-shaped (button or knob).

olivenite, *f.* olivenite.

oliverie, *f.* olive-oil works.

olivète, *f.* oil poppy.

olivier, *m.* olive (tree or wood) (Olea).

olivine, *f.* olivine.

ollaire, pierre —, *f.* steatite, potstone.

olocarpe, with fruit remaining whole.

olonervié, (leaves) with longitudinal veins.

olopétalaire, olopetalarious, transformed into petals.

oloptère, with whole wings.

ombelle, *f.* umbel.

ombellé, umbellate, having inflorescence in umbels.

ombellifère, *f.* umbellifer.

ombellifère, umbelliferous.

ombellifères, *f.pl.* Umbelliferae, umbelliferous plants.

ombelliférone, *f.* umbelliferone.

ombelliforme, umbelliform.

ombellique, umbellic.

ombellule, *f.* umbellule, umbellet, partial umbel, compound umbel.

ombellulifère, umbelliferous, bearing umbels.

ombilic, *m.* umbilicus, hilum, navel.

ombiliqué, umbilicate.

ombraculifère, umbraculiferous.

ombraculiforme, shaped like an umbrella.

ombrage, *m.* shade, umbrage, offense, mistrust, speck, leucoma, nebula; supportant l' —, shade-enduring.

ombrager, to shade, overshadow, screen.

ombrant, shading.

ombraticole, living in shade.

ombre, *f.* shadow, shade, umber, spirit; — brûlée, burnt umber; — du vent, beside, side away from wind; — naturel, raw umber; — portée, cast shadow, central shadow; à l' — de, under the shade (protection) of; faire —, to cast a shadow, put someone in the shade; sous l' —, sous —, under pretense.

ombre, *m.* umber, grayling; — chevalier, char; — commun, grayling.

ombré, shaded, striped (fur).

ombrellaire, like an umbrella.

ombrelle, *f.* umbel.

ombrellé, with an appendage like an umbrella.

ombrelliforme, like an umbrella.

ombrer, to shade.

ombreux, -euse, shady.

ombrométrie, *f.* rain measurement, pluviometry.

ombrophobe, heliophilous, adapted to full exposure to the sun.

omet, (he) is omitting.

omettre, to omit, neglect.

omis, omitted.

omission, *f.* omission.

omnibus, *m.* local or accommodation train.

omniscient, omniscient, all-knowing.

omnitige, (buds) having a like tendency to rise up.

omnivore, omnivorous.

omoplate, *f.* scapula, shoulder plate.

omopléphyte, with stamens united by their filaments.

omphalode, *m.* omphalode, scar at hilum of seed.

on, one, people, they, we, somebody; si l'on . . . , if one . . . (l' for sake of euphony).

onagraire, onagre, *m.* evening primrose, tree primrose (Oenothera).

once, *f.* ounce.

onciné, uncinate, hooked.

oncle, *m.* uncle.

onction, *f.* unction, oiling, anointing, inunction.

onctueux, -euse, unctuous, greasy, oily.

onctuosité, *f.* unctuousness, oiliness.

ondatra, *m.* muskrat.

onde, *f.* undulation, billow, wave, corrugation; —s amorties, damped waves; —s entretenues, undamped or continuous waves.

ondé, wavy, waved, wavelike, grained (wood), watered (silk), streaked (glass).

ondée, *f.* shower, downpour.

ondoiement, *m.* undulation, waving, wavy motion.

ondoyant, undulating, waving, swaying.

ondoyer, to undulate, wave, ripple.

ondulant, undulating, waving, undulant (fever).

ondulation, *f.* undulation, waving, vibration, piping, ribbing; *pl.* twists (of Spironema); — des fibres, wavy grain, curl.

ondulatoire, undulatory.

ondulé, undulated, wavy, corrugated, figured, undulating, mottled.

onduler, to undulate, ripple, wave, flutter, corrugate.

onduleux, -euse, undulating, wavy, billowy, undulous.

ondulifolié, with undulated leaves.

onéreux, -euse, burdensome, onerous, difficult.

ongle, *m.* nail, claw, hoof.

onglé, armed with claws or talons.

onglet, *m.* unguis, claw (of petal), miter, miter joint, notch, groove, nail cut, flat graver, ungula; — d'un pétale, claw.

onglette, *f.* claw.

ongletté, unguiculate.

onglon, *m.* ergot.

onguent *m.* ointment, unguent, salve, liniment; — antidartreux, dartre ointment; — citrin, citrine ointment; — mercuriel double, mercurial ointment, blue ointment.

onguiculé, unguiculate.

ongulé, nail-shaped, hoofed.

onobryché, recalling a sainfoin.

ont (avoir), (they) have.

ontogénèse, *f.* ontogenesis.

ontogénie, *f.* ontogeny.

onychomycose, *f.* onychomycosis.

onyx, *m.* nail (finger or toe), onyx.

onze, eleven, eleventh.

onzième, eleventh.

oocyste, *f.* sporocyst.

oogamie, *f.* oögamy.

oogone, *m.* oögonium.

oolithe, *m.* oölite.

oolithique, *f.* oölitic.

oomycètes, *f.* Oömycetes.

oophoridie, *m.* oöphoridium, the megasporangium in certain plants.

oosphère, *f.* oösphere.

oosporange, *m.* oösporangium, sacs or sporangia which produce oospores.

oospores, *f.pl.* Oösporeae, oöspores.

oothèque, *f.* oötheca, theca or sporangium of ferns.

opacifiant, *m.* opaquing agent

opacifiant, rendering opaque.

opacifier, to render opaque, opaque; s'— to become opaque.

opacimètre, *m.* plate tester.

opacité, *f.* opacity, opaqueness, denseness.

opalage, *m.* stirring under of the sugar crystals.

opale, *f.* opal, crust of sugar crystals.

opaler, to stir under the crust (of sugar crystals).

opalescence, *f.* opalescence.

opalescent, opalescent.

opalin, *f.* opaline, milk glass.

opalin, opaline.

opalisant, opalizing, opalescent.

opalisé, opalescent.

opaliser, to opalize.

opaque, opaque.

opérant, operating.

opérateur, *m.* operator, motor.

opératif, -ive, operative, active.

opération, *f.* operation, working, process, phenomenon; —s agricoles, field work.

opératoire, operative; **mode —,** modus operandi, method of preparation.

opercule, *m.* cover, cap, lid, operculum, diaphragm.

operculé, opercular, operculate, furnished with a lid.

opérer, to operate, work, act, effect, do, carry out, accomplish, operate upon; **s'—,** to be operated, be wrought, be effected, take place, perform.

ophidien, -enne, ophidian, snakelike.

ophiosperme, (embryo) resembling a small snake.

ophistodiale, *f.* opening of th. stomate.

ophite, *m.* ophite, ophicalcite, serpentine-bearing marble.

ophrys, *f.* Ophrys, arachnites; — **abeille,** bee orchis.

ophtalmie, *f.* ophthalmia.

opiacé, containing opium.

opiacer, to add opium to, opiate.

opianate, *m.* opianate.

opianique, opianic.

opiat, *m.* opiate, narcotic, electuary, tooth paste.

opiner, to give or state one's opinion, opine.

opiniâtre, obstinate, self-opinionated, stubborn.

opiniâtrément, stubbornly.

opiniâtrer (s'), to be obstinate.

opiniâtreté, *f.* obstinacy, stubbornness

opinion, *f.* opinion.

opium, *m.* opium.

oplacium, *m.* an expansion in the form of a cornet.

opobalsamum, *m.* opobalsam, balm of Gilead (*Commiphora meccanensis*).

opodeldoch, *m.* opodeldoc, soap liniment.

opolé, opolite, *m.* juice, sap (of plants).

oponce, *m.* prickly pear (Opuntia).

opopanax, *m.* opopanax, Hercules' allheal (*Opopanax chironium*).

opothérapie, *f.* opotherapy, organotherapy.

opothérapique, organotherapeutic.

opportun, opportune, convenient.

opportunément, opportunely, seasonably.

opportunité, *f.* opportunity, expediency, advisability.

opposant, *m.* opponent.

opposant, opposing.

opposé, *m.* opposite direction, side.

opposé, opposed, opposite, contrary.

opposer, to oppose, compare (with), put (against), go against, object (to); **s'—,** to be opposed, be contrary to, be against, object, stop, hinder.

oppositaire, *f.* an opposite condition.

opposite, *m.* opposite, contrary.

oppositiflore, oppositiflorous, having opposite peduncles.

oppositifolié, oppositifolious, with opposite leaves.

opposition, *f.* opposition, antithesis, resistance.

oppositipenné, opposite-pinnate.

oppresser, to oppress.

oppressif, -ive, oppressive.

oppression, *f.* oppression, crushing.

oppressivement, oppressively.

opprimer, to oppress.

opsonine, *f.* opsonin.

opsonique, opsonic.

opsonisant, opsonizing.

opsigonie, *f.* opsigony, production and development of proventitious buds.

opter, to choose.

opticien, *m.* optician.

optima, *m.pl.* optima, best.

optimiste, optimistic.

optimum, *m.* optimum, the best.

option, *f.* option.

optique, *f.* optics; d'—, optical.

optique, optic, optical.

optiquement, optically.

opulence, *f.* opulence, wealth.

opulent, opulent, wealthy.

or, now, but.

or, *m.* gold; — **affiné,** refined gold; — **au titre,** standard gold; — **couleur,** gold size; — **de coupelle,** refined gold; — **de départ,** parting gold; — **en feuille(s),** gold leaf; — **faux,** false gold; — **fulminant,** fulminating gold; — **haché,** rugged gilding; — **imité,** burnish gold; — **musif,** mosaic gold; — **potable,** auric chloride; — **vierge,** virgin gold, native gold; **d'**—, of gold, golden.

orage, *m.* storm.

orageux, -euse, stormy, tempestuous.

oraison, *f.* oration, speech.

oral, -aux, oral, verbal.

oralement, orally.

orange, *f.* orange (fruit) (Citrus); *m.* orange (color); — **amère,** bitter orange (*Citrus aurantium*); — **douce,** sweet orange (*Citrus sinensis*).

orange, orange, orange-colored.

orangé, orange, orange-colored.

orangelette, orangette, *f.* an orange fruit scarcely as large as a cherry.

oranger, *m.* orange tree (*Citrus aurantium*), orange seller.

orang-outan, orang-outang, *m.* orangutan.

orateur, *m.* orator, speaker.

oratoire, oratorical.

orbe, *m.* orb, sphere, orbit.

orbiculaire, *m.* orbicular muscle, sphincter.

orbiculaire, orbicular, round flowered.

orbiculé, orbicular, round flowered.

orbital, -aux, orbital.

orbite, *f.* orbit, socket (of the leg).

orcanette, orcanète, *f.* orcanet, alkanet, dyer's bugloss (*Anchusa officinalis*).

orcéine, *f.* orcein.

orchidé, with several lobes deeply divided.

orchidées, *f.pl.* Orchidaceae, the orchids.

orchis, *m.* testicle, orchis; — **à deux feuilles,** butterfly orchis; — **tacheté,** purple orchis, spotted orchis.

orcine, *f.* orcinol, orcin.

ordinaire, *m.* ordinary practice, ordinary.

ordinaire, ordinary, common, usual; **à l'**—, as usual; **d'**—, ordinarily, usually.

ordinairement, ordinarily, commonly, usually.

ordonnance, *f.* ordering, order, ordinance, disposition, prescription.

ordonnancer, to sanction, authorize, pass for payment.

ordonné, ordered, prescribed.

ordonnée, *f.* ordinate.

ordonner, to order, direct, command, regulate, ordain, arrange, prescribe.

ordre, *m.* order, discipline, command, nature, size, class, instruction; — **de marche,** running order; — **des coupes,** succession of fellings; **d'**—, by order.

ordure, *f.* ordure, excrement, dung, dirt, filth, dust, refuse, rubbish.

ore, *f.* tuyère plate.

orée du bois, *f.* skirt (of forest).

oreille, *f.* ear, lug, handle, lobe, attachment, projection, corner, moldboard (of plow); — **pendante,** lop ear; —**s d'ane,** — **de vache,** common comfrey (*Symphytum officinale*).

oreiller, *m.* pillow.

oreillette, *f.* auricle (grass), auricle (heart).

oreillon, *m.* ear flap (of cap), cutting, tragus (of bat); *pl.* mumps.

ores, d'— et déjà, now and henceforth, right now.

orfèvre, *m.* goldsmith.

orfévré, orfévri, wrought by the goldsmith.

orfèvrerie, *f.* goldsmith's art and work, articles of gold and silver.

orfraie, *f.* osprey, sea hawk.

organdi, *m.* organdie, book muslin.

organe, *m.* organ, means, medium, part, piece, member; — **auditif,** — **chordotonal,** auditory sensilla, auditory or chordotonal organ; — **buccal,** mouth part; — **d'accouplement,** aedeagus, penis, male intromittent organ; — **de manoeuvre,** control device; — **du tact,** tactile sensilla; — **foliacé,** foliaceous organ, leaf organ; — **répétiteur,** repeater; —**s de l'odorat,** olfactory sensilla; —**s du goût,** gustatory sensilla; —**s générateurs,** reproductive organs; —**s génitaux femelles,** female reproductive organs.

organique, organic.

organiquement, organically.

organisable, organizable.

organisant, organizing.

organisateur, *m.* organizer.

organisateur, -trice, organizing.

organisation, *f.* organization, management.

organisé, organized.

organiser, to organize; **s'**—, to become organized.

organisme, *m.* organism, organized body.

organoleptique, organoleptic.

organo-métallique, organometallic.

organsin, *m.* organzine (a silk thread), thrown silk.

organsiner, to throw silk, twist twice, organzine.

orge, *f.* barley; — **carrée,** four-rowed barley, bigg; — de **brasserie,** brewer's barley; — **mondée,** peeled barley; — **perlé,** pearl barley.

orgelet, *m.* sty (on eye).

orgue, *m.* organ.

orgueil, *m.* pride, arrogance, fulcrum.

orgueilleusement, proudly.

orgueilleux, -euse, proud, haughty, arrogant, vain.

orgyie, *f.* pale tussock moth.

oribus, *m.* resin torch.

orient, *m.* east, orient, Orient, the East, beginning, rise.

oriental, -aux, Oriental, eastern, east.

orientation, *f.* orientation, orienting, direction.

Orientaux, *m.pl.* the Orientals, easterners.

orienter, to orient, direct, guide, turn, get one's bearings set, turn toward the East; **s'—,** to discover where one is.

orifice, *m.* orifice, mouth, aperture, hole, opening, port.

origan, *m.* marjoram (Origanum); — **des marais,** hemp agrimony, water hemp (*Eupatorium cannabinum*).

originaire, native, dating from, original, originally from, originary.

originairement, originally.

original, -aux, original, queer, primitive, odd; **d'—,** first hand, from the original source; **en —,** in the original; **sur —,** from the original.

originalement, originally.

originalité, *f.* originality.

origine, *f.* origin, beginning, source, derivation, provenance; **dans l'—,** originally, in the beginning; **dès l'—,** from the very beginning, from the outset; **livre d'—s,** studbook.

originel, -elle, original, primitive.

originellement, originally, from the beginning.

orillon, *m.* ear, lug, handle, projection, moldboard (of plow).

oripeau, *m.* tinsel, foil.

orlean, *m.* orlean, annatto (dye for coloring butter).

ormaie, *f.* elm grove.

orme, *m.* elm (tree or wood) (Ulmus); — **à large feuilles,** — **blanc,** — de **montagne,** Scotch elm, wych-elm, mountain elm (*Ulmus glabra*); — **à petites feuilles,** — **rouge,** common elm (*Ulmus glabra*); — **à trois feuilles,** hop tree (*Ptelea trifoliata*); — **champêtre,** common or English elm (*Ulmus campestris*); — **fauve,** slippery elm (*Ulmus fulva*); — **subéreux,** rock elm, cork elm (*Ulmus racemosa*); — **tortillard,** dwarf elm (*Ulmus glabra minor*).

ormeau, *m.* young elm.

ormille, *f.* elm sapling, (hedge) row of young elms.

orne, *m.* manna (flowering) ash (*Fraxinus ornus*).

orné, ornamented, ornate.

ornement, *m.* ornament, decoration.

ornemental, -aux, ornamental, decorative.

ornementation, *f.* ornamentation, ornamenting.

ornementer, to ornament.

orner, to ornament, adorn, embellish, decorate.

ornière, *f.* rut, track, groove.

ornithine, *f.* ornithine.

ornithogames, *f.pl.* ornithophilae, plants pollinized by birds.

ornithologie, *f.* ornithology.

ornithophiles, *f.pl.* ornithophilae.

ornithoptère, *m.* ornithopter.

ornithurique, ornithuric.

orobanche, *f.* broomrape (*Orobanche*); — de **Virginie,** beechdrops (*Epiphegus virginiana*); — **rameuse du chanvre,** broomrape of hemp.

orobe, *f.* earthnutpea, flat pea (*Lathyrus sylvestris*).

orogénie, *f.* orogeny.

orogénique, orogenic.

orographie, *f.* orography.

orographique, orographic(al).

orographiquement, orographically.

oronge, *f.* amanita, orange agaric (*Amanita caesarea*).

orpaillage, *m.* gold washing.

orpailleur, *m.* gold washer.

orphelin, *m.* orphan.

orphelin, orphan, queenless.

orpiment, *m.* orpiment, yellow arsenic.

orpimenter, to mix or treat with orpiment.

orpin, *m.* orpine, stonecrop (*Sedum telephium*), orpiment.

orseille, *f.* archil, dyer's moss (*Roccella tinctoria*); — **de terre,** cudbear.

orsellinique, orsellinic.

orsellique, orsellic, lecanoric.

ort, gross.

orteil, *m.* toe, great toe.

orthacanthe, with straight spines.

orthochromatique, orthochromatic.

orthochromatisme, *m.* orthochromatism.

orthoclade, orthocladous, straight-branched.

orthoclase, *m.* orthoclase.

orthodérivé, *m.* ortho derivative.

orthogénèse, *f.* orthogenesis.

orthogonal, -aux, orthogonal.

orthographe, *f.* orthography, spelling.

orthographier, to spell.

orthophyte, *f.* plant, the gametophyte.

orthoptère, orthopterous; *m.pl.* Orthoptera.

orthorhombique, orthorhombic, trimetric.

orthose, *m.* orthoclase, orthose.

orthosérie, *f.* ortho series.

orthosperme, orthospermous, with straight seeds.

orthotrope, orthotropous.

orthovanadique, orthovanadic.

ortie, *f.* nettle; — **blanche,** white dead nettle (*Larium album*); — **brûlante,** — **grièche,** stinging nettle (*Urtica dioica*); — **de Chine,** China grass, ramie (*Boehmeria nivea*); — **de mer,** sea anemone; — **morte,** dead nettle (Lamium); — **puante,** hedge woundwort, hedge nettle (*Stachys sylvatica*); — **rouge,** red dead nettle (*Lamium purpureum*); — **textile,** green-leaved China grass (Boehmeria).

ortié, fièvre —, *f.* nettle rash, urticaria

ortol, *m.* ortol.

orvale, *f.* clary (*Salvia sclarea*).

orvet, *m.* slowworm, blindworm.

os, *m.* bone; — **de seiche,** — **de sèche,** cuttlefish bone; — **dissous,** dissolved bone (fertilizer); — **pulvérisé,** bone meal.

osazone, *f.* osazone.

oscillant, oscillating, (anther) mobile on a filament.

oscillariées, *f.pl.* Oscillatoriaceae.

oscillation, *f.* oscillation, vibration.

oscillatoire, oscillatory (movement).

oscille, *f.* sorrel (Rumex).

osciller, to oscillate, fluctuate, vibrate.

oscillogramme, *m.* oscillogram.

oscillographe, *m.* oscillograph.

oscinie, *f.* — **ravageuse,** frit fly.

osculaire, oscular (muscle).

osculateur, -trice, osculatory.

öse, *m.* rod with platinum wire sealed in.

oseille, *f.* common sorrel, dock (*Rumex acetosa*); **sel d'—,** salts of sorrel, acid potassium oxalate.

oser, to dare, venture.

oseraie, *f.* willow culture, osiery.

osier, *m.* osier, wicker, basket willow (*Salix viminalis*).

osmazome, *m.* osmazome.

osmiamique, osmiamic.

osmiate, *m.* osmate, osmiate.

osmié, containing osmium.

osmieux, osmious, osmous.

osmique, osmic.

osmiridium, *m.* iridosmium.

osmium, *m.* osmium.

osmiure, *m.* — **d'iridium,** iridosmium, osmiridium.

osmogène, *m.* osmogene, osmotic apparatus.

osmomètre, *m.* osmometer.

osmonde, *f.* osmund, flowering fern (Osmunda); — **royal,** royal fern (*Osmunda regalis*).

osmose, *f.* osmosis, osmose.

osmoser, to subject to osmosis, obtain by osmosis.

osmotique, osmotic.

osmotropisme, *m.* osmotropism, tropic stimulus due to osmotic action.

osone, *f.* osone.

osotétrazine, *f.* osotetrazine.

osotriazol. *m.* osotriazole, 1,2,5-triazole.

ossature, *f.* skeleton.

osséine, *f.* ossein.

osselet, *m.* knucklebone (sheep), ossicle, small bone.

ossements, *m.pl.* bones.

osseux, -euse, osseous, bony.

ossiculaire, ossicular, like ossicles.

ossicule, *m.* ossiculus, pyrene of a fruit.

ossiculé, provided with ossicula.

ossification, *f.* ossification, bone formation.

ossifier, to ossify.

ossuaire, *m.* ossuary, charnel house, heap of bones.

ostariphyte, having pulpy and drupaceous fruit.

ostéine, *f.* ostein, ossein.

ostensible, ostensible, open, above board.

ostensiblement, ostensibly, openly.

ostéoblaste, *f.* osteoblast.

ostéocolle, *f.* osteocolla, bone glue.

ostéogenèse, *f.* osteogenesis, formation of bony tissue.

ostéomyélite, *f.* osteomyelitis.

ostiole, *f.* stomatal opening.

ostiolé, ostiolate, furnished with an opening or mouth.

ostracés, *m.pl.* Ostracea, ostraceans.

ostraciser, to banish.

ostréicole, pertaining to oyster culture.

ostréiculture, *f.* oyster breeding.

ôté, removed, except, barring, with the exception of.

ôter, to remove, take away, take off, deprive, relieve, displace, except; **s'—,** to move away, remove, get away, deprive oneself (of).

otite, *f.* otitis; **— moyenne,** otitis media, tympanitis.

ou, or, either, or else; **ou . . . ou,** either . . . or; **ou bien,** or else, or otherwise, either; **ou bien . . . ou bien,** either . . . or else; **ou le bien . . . ou le mal,** either good or bad.

où, where, to what, to which, when, at, in, in which, at which, from or with which; **d'où,** whence, from where, from which, whither, wherein; **d'où vient que . . . ,** how is it that . . . ; **jusqu'où,** how far; **là où,** (there) where; **où que,** wherever. wheresoever; **par où,** by what way, by which, how; **partout où,** wherever; **vers où,** to which, whereto.

ouabaïne, *f.* ouabain.

ouate, *f.* wadding, cotton wool; **herbe à —,** silkweed, swallowwort, milkweed (*Asclepias syriaca*).

ouater, to wad, pad, line with wadding, pack.

ouateux, -euse, woolly, fleecy.

ouatier, *m.* silkweed.

oubli, *m.* forgetting, forgetfulness, neglect, oblivion, oversight.

oubliable, likely to be forgotten.

oublie, *f.* wafer, (ice cream) cone.

oublier, to forget, omit; **s'—,** to forge oneself.

oublieux, -euse, oblivious, forgetful.

ouest, *m.* west, West, western.

oui, yes.

ouï, heard.

ouï-dire, *m.* hearsay.

ouïe, *f.* hearing, audition, gill (of fish); **à l'— de la cognée,** within earshot.

ouillage, *m.* filling up (a cask with wine).

ouiller, to ullage, fill up (a cask by adding wine).

ouïr, to hear.

ouragan, *m.* hurricane.

ourdir, to warp, plait, twist, hatch, weave, concoct.

ourler, to hem.

ourlet, *m.* hem, (of metals) lap joint, seam, edge (of crater), border.

ours, *m.* bear.

ourse, *f.* (she) bear.

outarde, *f.* **grande —, — barbue,** great bustard; **— petite, — canepetière,** little bustard.

outil, *m.* tool, implement, instrument, utensil; **— à bras,** hand implements; **— forestier,** forest tool, forest implement; **— tranchant,** cutting tool, edged tool.

outillage, *m.* set of tools, equipment, outfit, implements, plant, tool outfit.

outillé, furnished with tools.

outiller, to equip, fit out, furnish with tools.

outrage, *m.* outrage, gross insult.

outrageant, outrageous, reproachful.

outrager, to outrage.

outrance, *f.* extreme, excess; **à —,** to excess, to the finish, to the uttermost.

outrancier, *m.* extremist.

outrancier, -ère, excessive, extravagant.

outre, *f.* leather bottle, skin.

outre, beyond, besides, beside, in addition to, further on; **— que,** beyond that, besides that, not only . . . but; **d'— en —,** through and through, clear through; **en —,** in addition, besides, furthermore, moreover.

outré, exaggerated, overdone, excessive, exasperated, extravagant.

outrecuidant, presumptuous, outlandish.

outre-Manche, on the other side of the Channel, in England.

outremer, *m.* ultramarine; — **lapis,** ultramarine from lapis lazuli.

outremer, ultramarine.

outre-mer, overseas, beyond the seas.

outre-monts, beyond the mountains, ultramontane; **d'—,** transmontane.

outre-passe, *f.* extra cuttings.

outrepasser, to go beyond, exceed.

outrer, to overdo, carry to excess, overstrain, exaggerate, provoke beyond measure.

outre-Rhin, beyond the Rhine.

ouvert, open, candid, quick, frank, ready, patent; **fleur —e,** flower in bloom.

ouvertement, openly, frankly.

ouverture, *f.* opening, aperture, overture, orifice, hole, chasm, mouth, gap, break, width, span, opportunity, readiness; — **à la scorie,** slag hole; — **de la chasse,** opening of the hunting season; — **de la chauffe,** stokehole, fire hole; — **de la pêche,** beginning of fishing season; — **de moule,** pouring funnel; — **de visite,** manhole; — **du service,** starting of work; — **relative,** aperture ratio; **à large —,** widenecked, widemouthed.

ouvrable, workable, working, tractable.

ouvrage, *m.* work, production, book, performance, piece of work, workmanship, hearth (of furnace), construction, structure; — **à forfait,** job work or piecework; — **à la mécanique,** machine work; — **de cordier,** ropemaker's or roper's work; — **de couture,** needlework; — **de fabrique,** factory product.

ouvrage, — **de l'émailleur,** enameling; — **de raclerie,** carved work; — **de soudure,** welding; — **en fer-blanc,** tinwork; — **en fonte,** cast, casting; — **martelé,** raised or hammered work; — **noir,** blacksmithing, blacksmith work; — **réticulé,** diamond work, net masonry, reticulated work or bond.

ouvragé, worked, wrought.

ouvrager, to work, figure.

ouvraison chimique, chemical working of wood.

ouvrant (ouvrir or ouvrer), opening or working.

ouvre, (he) is opening or working.

ouvré, worked, wrought, figured.

ouvreau, *m.* working (draft) hole, peep hole, air vent.

ouvrer, to work, make, convert (timber).

ouvrés, *m.pl.* wrought wood.

ouvreur, *m.* opener, vatman, dipper.

ouvrier, *m.* workman, worker, artisan, laborer, mechanic; — **à la journée,** casual laborer; — **compagnon,** fellow, journeyman, mate; — **domicilié,** resident or local worker; — **exercé,** skilled workman; — **fixé,** regular workman; — **fondeur,** founder, smelter; — **papetier,** papermaker; — **tiseur,** stoker; — **verrier,** glass blower; **simple —,** unskilled worker.

ouvrir, to open, unlock, cut open, cut through, unfold, disclose, begin; **s'—,** to dehisce, open, unbosom (unburden) oneself; — **le massif,** to interrupt the leaf canopy.

ouvroir, *m.* workshop, workroom.

ovaire, *m.* ovary.

ovalaire, oval (foramen).

ovalbumine, *f.* ovalbumin.

ovale, oval.

ovalifolié, with oval leaves.

ovarien, -enne, ovarian.

ovarifère, with an ovary.

ovarique, ovarian.

ovatifolié, with oval leaves.

ove, *m.* ovum, ovolo.

ové, egg-shaped, ovoid, ovate.

ovelle, *m.* ovellum, the carpel at the time of florescence.

ovidés, *m.pl.* Ovidae.

oviducte, *m.* oviduct.

oviforme, egg-shaped, oviform.

ovigère, having oviform fruit.

ovine, ovine, pertaining to sheep.

ovipare, oviparous.

ovipositeur, *m.* ovipositor (of insects)

ovisac, *m.* ovisac.

oviscapte, *m.* ovipositor.

ovocystes, *f.pl.* mother cells of embryo sac.

ovogenèse, *f.* ovogenesis, oogenesis.

ovoïdal, -aux, ovoid, egg-shaped.

ovoïde, *m.* ovoid.

ovoïde, ovoid, egg-shaped.

ovomucoïde, *m.* ovomucoid.

ovovitelline, *f.* ovovitellin, vitellin.

ovulaire, ovular.

ovulation, *f.* ovulation.

ovule, *m.* ovule; **renfermant des —s,** ovuliferous.

ovulé, ovulate, possessing ovules.

Oxa, *f.* Oajaca (Oaxaca).

oxacétique, hydroxyacetic, glycolic.

oxacétylurée, *f.* hydantoic acid.

oxacide, *m.* oxacid, oxygen acid, hydroxyacid.

oxalacétique, oxalacetic.

oxalantine, *f.* oxalantine, leucoturic acid.

oxalate, *m.* oxalate; — de fer, oxalate of iron, ferrous oxalate.

oxalaté, combined with an oxalate.

oxalide, *f.* wood sorrel (Oxalis).

oxalidées, *f.pl.* Oxalidaceae.

oxalimide, *f.* oxalimide.

oxalique, oxalic.

oxalurate, *m.* oxalurate.

oxalurie, *f.* oxaluria.

oxalurique, oxaluric.

oxalyle, *m.* oxalyl.

oxamide, *f.* oxamide.

oxanilide, *f.* oxanilide.

oxanilique, oxanilic.

oxazine, *f.* oxazine.

oxazol, *m.* oxazole.

oxéolat, oxéolé, *m.* acetolatum.

oxhydrile, hydroxylated, hydroxyl (group).

oxhydrilé, *m.* hydroxyl (group).

oxhydrique, oxyhydrogen.

oxhydryle, *m.* hydroxyl.

oxide, *m.* oxide.

oxime, *m.* oxime.

oxindol, *m.* oxindole.

oxyacétique, hydroxyacetic, glycolic.

oxyacétylénique, oxyacetylene.

oxyacide, *m.* hydroxy acid.

oxyadène, with pointed glands.

oxyazoté, combined with oxygen and nitrogen.

oxybenzoate, *m.* hydroxybenzoate.

oxybenzoïque, hydroxybenzoic.

oxybractée, with sharp bracts.

oxybromure, *m.* oxybromide.

oxycarboné, combined with oxygen and carbon.

oxycarbonisme, *m.* carbon monoxide poisoning.

oxycarpe, oxycarpous, (fruit) sharp-pointed.

oxycétone, *f.* hydroxyketone.

oxychlorique, oxychloric.

oxychlorure, *m.* oxychloride.

oxyclade, having sharp branches.

oxycyanure, *m.* oxycyanide.

oxydabilité, *f.* oxidizability, oxidability.

oxydable, oxidizable, oxidable.

oxydant, *m.* oxidizer, oxidizing agent.

oxydant, oxidizing.

oxydase, *f.* oxidase.

oxydation, *f.* oxidizing, oxidation.

oxyde, *m.* oxide; — alcoolique, alkyl oxide, ether; — d'argent, silver oxide; — de calcium, calcium oxide, quicklime; — de carbone, oxide of carbon, carbon monoxide; — de cuivre, copper oxide, cupric oxide; — de plomb fondu, litharge; — d'éthyle, ethyl oxide (ethyl ether); — jaune de plomb, yellow lead, wulfenite, mineral lead molybdate.

oxyde, — magnétique, magnetic oxide of iron; — mercurique jaune, yellow mercuric oxide; — nitreux, nitrous oxide, nitrogen monoxide; — noir de manganèse, black oxide of manganese, manganese dioxide; — platinique, platinum oxide; — puce de plomb, lead peroxide; — rouge de plomb, — salin de plomb, red lead, minium; — salin, mixed oxide.

oxyder, to oxidize; s'—, to become oxidized.

oxydeur, *m.* oxidizer.

oxydule, *m.* lower oxide; — de cuivre, cuprous oxide.

oxydulé, in the form of a lower oxide, oxidulated.

oxygénable, oxygenizable, oxidizable.

oxygénant, oxygenating, oxidizing.

oxygénation, *f.* oxygenation, oxidation.

oxygène, *m.* oxygen; — dissous, dissolved oxygen.

oxygéné, oxygenated; eau —e, peroxide of hydrogen.

oxygéner, to oxygenate, oxidize.

oxygénifère, carrying oxygen.

oxygénomorphose, *m.* morphosis due to oxygen in the air.

oxygras, hydroxyfatty.

oxyhémoglobine, *f.* oxyhemoglobin.

oxyhydrile, *m.* hydroxyl, hydroxyl group.

oxyhydrique, oxyhydrogen.

oxyhydrogène, oxyhydrogen.

oxyiodure, *m.* oxyiodide.

oxyioduré, combined with oxygen and iodine.

oxylophyte, *f.* a plant formation of dry acid soils.

oxyme, *m.* oxime.

oxymel, *m.* oxymel.

oxypétale, oxypetalous, having sharp-pointed petals.

oxyphile, oxyphile.

oxyphylle, oxyphyllous, having sharp-pointed leaves.

oxypyridine, *f.* hydroxypyridine.

oxyrhinique, *m.* oxyrhynchus.

oxysaccharum, *m.* oxysaccharum.

oxysel, *m.* oxysalt, salt of an oxacid.

oxysperme, with acuminate fruits.

oxysulfure, *m.* oxysulphide.

ozène, *m.* ozena.

ozocérite, ozokérite, *f.* ozocerite.

ozonateur, -trice, ozonizer.

ozone, *m.* ozone.

ozoner, to ozonize.

ozoneur, *m.* ozonizer.

ozonique, ozonic.

ozonisation, *f.* ozonization.

ozoniser, to ozonize.

ozoniseur, *m.* ozonizer.

ozonomètre, *m.* ozonometer.

ozonométrie, *f.* ozonometry.

ozotypie, *f.* ozotype.

P

P., *abbr.* (pour); p. 100 (pour cent), per cent.

pacage, *m.* grazing ground; — en forêt, forest pasture.

pacager, to graze on, feed on.

pacfung, *m.* nickel silver, paktong.

pachycarpe, pachycarpous, having a thick pericarp.

pachyderme, pachydermous, thick-skinned.

pachynème, with very thick wood.

pachypode, with thick podium.

pachyrhize, with thick roots.

pacifier, to pacify, appease; se —, to calm down.

pacifique, pacific, peaceable.

pacifiquement, peacefully, pacifically, quietly.

packfond, packfong, *m.* see PACFUNG.

padelin, *m.* glass crucible.

pagaie, *f.* paddle.

pagayer, *n.* to paddle.

page, *f.* page (of book); *m.* page (boy).

paginer, to page, paginate.

pagodite, *f.* pagodite, agalmatolite.

pagure, *m.* hermit crab.

paiement, *m.* payment.

païen, -enne, pagan, heathen.

paillage, *m.* mulching.

paillasse, *f.* straw mattress, (laboratory) bench, flask stand, masonry support, foundation.

paillasson, *m.* mat; — en roseau, reed mat.

paillassonnage, *m.* covering (of seedlings) with (straw) matting.

paille, *f.* straw, haulm, culm, scale (of iron), blister, flaw, chip; — d'avoine, chaff; — de bois, wood fiber, wood wool; — de céréales, straw of corn; — de fer, hammer scale, forge scale; — de froment, wheat straw; — de lin, flax straw; — de maïs, cornstalks; — de seigle, rye straw; — d'orge, barley straw; — fourrageuse, forage or fodder straw; — hachée, chopped straw chaff; menue —, chaff.

paille, straw-colored.

paillé, *m.* stable litter.

paillé, straw-colored, flawy, scaly.

pailler, to mulch, cover or protect with straw.

pailler, *m.* farmyard, straw stack, dunghill, straw loft.

paillet, *m.* pale red wine, mat, straw mat.

paillet, pale red.

pailleté, spangled, pailletted, lamellar, laminar.

paillette, *f.* paillette, lamella, leaflet, lamina, shining particle of gold dust, flake (of mica), spangle, (forge) scale, flaw (in gem), palea, pale, patellula, chaff.

pailleux, -euse, straw, strawy (manure), flawy (iron, glass).

paillis, *m.* mulch.

paillon, *m.* scale, plate, foil, large spangle, link, wisp of straw, bundle of straw, straw case (for bottle); — de soudure, leaf solder, grain of solder.

pailloner, to cover with sheet metal or tin.

paillot, *m.* door mat, straw mattress, bed.

pail-mel, *m.* feed made of straw and molasses.

pain, *m.* bread, loaf, wafer, cake (of wax), blow, punch, lump, pig (of metal), block (of soap); — à cacheter, (sealing) wafer, — azyme, unleavened bread, azyme bread; — bis, brown bread, Graham bread; whole-rye bread; — biscuité, bread baked longer than usual; — blanc, white bread, whole-wheat bread; — complet, wholemeal bread; — de beurre, pat of butter; — de cire, cake of wax, wax wafer.

pain, — de coucou, wood sorrel, shamrock (*Oxalis acetosella*); — de froment, wheat bread; — de grenouilles, water plantain; (Alisma); — de munition, ration bread; — d'épice, gingerbread; — de pourceau, sowbread (*Cyclamen europaeum*); — de sarrisin, buckwheat bread; — de savon, cake of soap; — de seigle, rye bread; — de sucre, sugar loaf; — d'olives, oil cake.

pain, — grillé, toast; — levé, raised bread, leavened bread; — mollet, unleavened bread; — noir, black bread, rye bread; — rassis, stale bread; — salé, salt lick; — viennois, Vienna bread; arbre de —, breadfruit tree (*Artocarpus communis*); en —, in the form of a cake; fruit à —, breadfruit; petit —, (French) roll.

pair, m. par, peer; au —, at par; de —, on a par, on an equal footing; hors de —, unrivaled, peerless; sans —, without a peer, peerless.

pair, equal, even, like; nombre —, even number.

paire, f. pair, couple, team; — de meules, couple of millstones, stone mill.

pairement, (of numbers) evenly.

paisible, peaceful, peaceable, calm, quiet.

paisiblement, peacefully, calmly.

paisseau, m. vine stake, vine prop, hop pole.

paisson, f. pasture (in forests); — pleine, full mast.

paître, to feed on, graze, pasture, browse.

paix, f. peace, quiet, rest.

palacé, palaceous, spade-shaped.

palais, m. palate, palace, court (of justice), law.

palan, m. tackle.

palatin, m. palate bone.

palatum, m. palate.

pale, f. pale, stake, pile, bung, vane, blade, sluice gate, hatch.

pâle, pale, wan.

paléa, f. palea.

paléacé, paleaceous, chaffy.

paléiforme, lamellar.

paléobotanique, f. paleobotany.

paléole, f. glumelle, palea of grasses.

paléolifère, paleoliferous, bearing paleae.

paléontologie, f. paleontology.

paléontologique, paleontological.

paléontologiste, m. paleontologist.

paléozoïque, Paleozoic.

paléozoologie, f. paleozoology.

paletot, m. overcoat, greatcoat.

palette, f. paddle, paddle board, spatula, pallet, palette, drill plate.

palétuvier, m. mangrove (Rhizophora), button tree (Conocarpus).

pâleur, paleness, pallor, pallidness.

palier, m. landing (of stairs), floor, level stretch, stage, degree, bearing; — à billes, ball bearing; — graisseur, self-oiling bearing.

palingénésie, f. regeneration.

palingénétique, palingenetic.

pâlir, to turn pale, grow dim, blanch, make pale.

palisage, m. nailing up (plants).

palissade, f. stockade, palisade, railing, trellis fence.

palissader, to stockade, fence with palisades.

palisson, m. stretching machine (for leather).

palissoner, to stretch (leather).

palladeux, -euse, palladious.

palladique, palladic.

palladium, m. palladium.

palliatif, -ive, palliative.

pallidiflore, with pale flowers.

pallium, m. cerebral cortex.

pallophotophone, m. sound film.

palma-christi, m. castor-oil plant, palma Christi (*Ricinus communis*).

palmaire, of the breadth of the palm (about three inches).

palmarosa, f. palmarosa (*Cymbopogon martini*), ginger grass (*Cymbopogon flexuosus*).

palmatifide, palmatifid, cut in a palmate fashion.

palmatiflore, with palmate flowers.

palmatifolié, with palmate leaves.

palmatiforme, palmatiform, (venation) arranged in a palmate manner.

palmatilobé, palmatilobate, palmately lobed.

palmatinervé, palmatinerved, palmately nerved.

palmatipartit, palmatipartite, palmately cut nearly to the base.

palmatisèqué, palmatisect, palmately cut.

palme, *f.* palm, palm tree; *m.* palm, hand's-breadth.

palmé, palmate, palmated, webbed, digitate.

palmer, *m.* micrometer gauge.

palmette, *f.* palmetto.

palmier, *m.* palm tree, palm; — à chanvre, hemp palm; — à huile, — épineux, oil palm (*Elaeis guineensis*); — nain, dwarf fan palm (*Chamaerops humilis*); — sagou, sago palm (Cycas).

palmifère, palmiferous, producing palms.

palmifide, with segments arranged like fingers.

palmifolié, with palmate leaves.

palmiforme, palmiform.

palmilobé, palmately lobed.

palmipartite, palmately parted.

palminervé, palminerved, palmately nerved.

palmiste, *m.* cabbage palm, palm trees, palmetto; huile de —, palm kernel, palm-nut oil; ver —, palm grub, worm, weevil.

palmite, *m.* palm marrow, palmite.

palmitine, *f.* palmitin.

palmitique, palmitic.

palombe, *f.* ring dove, wood pigeon.

palon, *m.* wooden spatula.

pâlot, rather pale, wan.

palpable, palpable, evident.

palpablement, palpably.

palpe, *f.* feeler; — maxillaire, maxillary palpus.

palper, to feel, finger, handle.

palpitant, palpitating, panting.

palpiter, to palpitate, throb.

paludéen, *m.* malarial patient.

paludéen, -enne, marshy, malarial, paludal.

paludeux, -euse, paludose, growing in marshy places.

paludicole, paludose.

paludier, *m.* worker in a salt garden.

paludique, paludic, malarial.

paludisme, *m.* paludism, malaria.

palus, *m.* marsh, alluvial plain.

palustre, marsh, swampy, palustral, paludal, paludous.

pâmoison, *f.* faint, swoon.

pampas, *f.pl.* pampas; herbe des —, pampas grass (*Cortaderia argentea*).

pampe, *f.* blade, leaf (of corn).

pamplemousse, *m.* shaddock, grapefruit, pomelo (*Citrus maxima*).

pampre, *m.* vine branch, tendril, shoot, vine, vine leaves.

pan, *m.* face, facet, lappet, side, surface, section, panel, flap, piece of wall.

panache, *m.* plume, tuft, panache, crest, stripe, variegation.

panaché, variegated, plumed, mixed, streaked, of various colors.

panacher, to variegate, plume, impart colors to (flowers).

panachure, *f.* variegation (on flowers), stripe of color.

panage, *m.* pannage.

panaire, panary, pertaining to making of bread.

panais, *m.* parsnip.

panama, *m.* Panama hat, soapbark, quillai bark (from *Quillaja saponaria*).

panaposporie, *f.* panapospory.

panard, toe wide, splay-footed, base narrow, cow-hocked.

panaris, *m.* foul, whitlow, paronychia, felon.

pancalier, *m.* savoy cabbage.

pancarte, *f.* placard, bill, (printed) sheet, (show) card, folder.

panchromatique, panchromatie.

pancréas, *m.* pancreas.

pancréatine, *f.* pancreatin.

pancréatique, pancreatic.

pandanus, *m.* screw pine (Pandanus).

pandoré, (leaves) covered with devastating plant lice.

pandure, fiddle-shaped.

pandurifolié, with panduriform leaves.

panduriforme, fiddle-shaped.

pané, containing bread.

panerée, *f.* basketful.

pangène, *m.* corpuscle carrier of hereditary properties in sex nuclei.

pangénèse, *m.* pangenesis.

panic, *m.* panic grass (Panicum), (Echinochloa); — d'Italie, Italian millet (*Setaria italica*); — pied-de-coq, barn (yard) grass (*Echinochloa crus-galli*).

panicaut, *m.* Eryngium; — maritime, sea holly (*Eryngium maritimum*).

panicule, *f.* panicle.

paniculé, paniculate.

paniculiforme, paniculiform, panicle-shaped.

panier, *m.* basket, pannier, beehive, willow.

panifiable, suitable for bread making; **farine —,** flour for bread, "strong" wheat flour.

panification, *f.* panification, breadmaking.

panifier, to convert into bread.

panique, *f.* panic, sudden fright.

panne, *f.* (hog's) fat, plush, panne, truss, peen (of hammer), pantile, purlin (of roof), engine trouble; **avoir une —, rester en —,** to have engine trouble, stop, break down; **mettre en —,** to bring to, heave to.

panneau, *m.* snare, trap, panel, pane (of glass), face (of stone), board, glass frame (for plants), riffle.

panneauter, to bring (seedlings) under glass.

panneresse, *f.* stretcher.

panniforme, panniform, having the appearance or texture of felt.

panorama, *m.* panorama.

panorpe, *f.* scorpion fly.

panphotométrique, panphotometric, (leaves) adapting their position to direct and diffused light.

panse, *f.* belly, paunch.

pansement, *m.* dressing (of wounds), grooming (of a horse).

panser, to dress (wounds), groom (a horse).

pansu, pot-bellied, pursy.

pantage, *m.* bundling (of hanks).

pantalon, *m.* trousers; **— d'eau,** emergency water-ballast bag.

pante, *f.* bundle (of hanks).

panteler, to pant, gasp.

pantenne, pantène, *f.* wicker tray.

panter, to bundle (hanks).

pantographe, *m.* pantograph.

pantoufle, *f.* slipper.

panulé, brown like the crust of bread.

paon, *m.* peacock, peafowl; **— de jour,** peacock butterfly.

paonner, to strut, peacock, preen oneself.

papaine, *f.* papain.

papavéracé, papaveraceous, resembling the poppy.

papavéracées, *f.pl.* Papaveraceae.

papavérine, *f.* papaverine.

papavérique, papaveric.

papaye, *f.* papaya, papaw (the fruit).

papayer, *m.* papaya tree (*Carica papaya*).

pape, *m.* pope.

paperasse, *f.* official papers, waste paper.

papeterie, *f.* papermaking, paper mill, paper factory, paper trade, stationery business, writing case.

papetier, *m.* papermaker, paper seller, stationer, stationer's shop.

papetier, -ère, papermaking, stationery.

papier, *m.* paper; **— à calquer,** tracing paper; **— à copier,** carbon paper, copying paper; **— à dessin, — à dessiner,** drawing paper; **— à écrire,** writing paper; **— à épruive,** proof paper; **— à filtrer,** filter paper; **— à filtrer durci,** hard filter paper; **— à journaux,** newspaper stock, news; **— à la cuve,** old paper, wastepaper, refuse; **— à la main,** handmade paper; **— albuminé,** albumenized paper.

papier, — à lettre, letter paper; **— à poncer,** tracing paper, pounce paper; **— à réactif,** test paper; **— argenté,** silver paper; **— au cyano-fer, — au ferro-prussiate,** blueprint paper; **— autocopiste,** copying paper; **— autotypie,** half-tone paper; **— blanc,** white paper, blank paper, first form; **— bleu de tournesol,** blue litmus paper; **— brouillard,** blotting paper; **— bulle,** scratch paper, scribbling paper; **— buvard,** blotting paper.

papier, — carbone, carbon paper; **— chimique,** carbon paper, chemically prepared paper; **— ciré,** wax paper; **— collé,** sized paper; **— couché,** coated paper, stained paper; **— court,** short-term paper, short bill (of exchange); **— d'amiante,** asbestos paper; **— d'argent,** tin foil; **— de bois,** paper made from wood pulp; **— de chanvre,** hemp paper; **— de coton,** cotton paper; **— de curcuma,** turmeric paper.

papier, — de fantaisie, fancy paper, colored paper; **— d'emballage,** wrapping paper, packing paper; **— d'émeri,** emery paper; **— de paille,** straw paper; **— de poste,** letter paper; **— d'épreuve,** test paper; **— de rebut,** waste paper; **— de riz,** rice paper; **— de soie,** silk paper, a fine white tissue paper; **— d'étain,** tin foil; **— de tenture,** wallpaper, paper hangings; **— de tournesol,** litmus paper; **— de verre,** glass paper, sandpaper.

papier, — doré, gilt paper, gold paper; **— durci,** hard paper, compressed paper; **— écolier,** exercise or scribbling paper; **— écu,** office paper; **— emporétique,** filter(ing) paper; **— entoilé,** cloth-mounted (drawing) paper; **— faïence,** pottery tissue; **— filigrane,** watermarked (filigreed) paper; — **filtre**, filter paper, blotting paper; — **gaufré**, embossed paper; — **glacé**, glazed paper; — **gommé**, gummed paper; — **goudronné**, tar paper, tarred paper.

papier, — **héliographique**, blueprint paper; — **huilé**, oil paper, transparent paper; — **hygiénique**, toilet paper; — **irisé**, iridescent paper; — **isolateur**, insulating paper; — **joseph (de soie)**, filter paper, light-gray tissue paper; — **lissé**, glazed paper; — **mâché**, papier-mâché; — **madré**, grained paper; — **marbré**, marbled paper; — **maroquiné**, morocco paper; — **marqué**, stamped paper; — **mécanique**, machine-printed wallpaper; — **mince**, thin paper.

papier, — **nacré**, nacreous, mother-of-pearl paper; — **peigne**, grained paper, stained paper; — **peint**, wallpaper, paper hangings; — **pelucheux**, fleecy paper; — **pelure**, onionskin, foreign-post paper; — **photographique**, photographic paper; — **ponce**, pumice-stone paper; — **poudré**, gunpowder, powder paper; — **quadrillé**, cross-section paper, millimeter paper; — **rayé**, striped paper, linear paper; — **(à) réactif**, test paper; — **réglé**, ruled paper; — **rouge de tournesol**, red litmus paper.

papier, — **(à) sable**, sandpaper; — **sans colle**, printing paper; — **satiné**, satined paper, plated paper; — **savonné**, soap paper; — **sensible**, sensitive paper, sensitized paper; — **timbré**, stamped paper; — **toile**, tracing cloth; — **tournesol**, litmus paper; — **tue-mouches**, flypaper; — **valeur**, security, bond, scrip; — **végétal**, tracing paper; — **velouté**, flock paper; — **volant**, loose sheet of paper.

papier-carton, *m.* paperboard, (fine) cardboard.

papier-cuir, *m.* leather paper.

papier-goudron, *m.* tar paper.

papier-monnaie, *m.* paper money.

papier-parchemin, *m.* parchment paper.

papier-tenture, *n.* wallpaper, paper hangings.

papilionacé, papilionaceous.

papilionacée, *f.* leguminous plant.

papillaire, papillary, papillate (gland).

papille, *f.* papilla.

papillé, papillate.

papilleux, -euse, papillose, verrucose, warty.

papillifère, papilliferous.

papilliforme, nipple-shaped.

papillon, *m.* butterfly, insert, leaflet, extra sheet, fly bill, butterfly valve, throttle, damper, butterfly nut; — **blanc du chou**, cabbage (white) butterfly; — **de nuit**, moth, miller; — **nocturne**, moth.

papilloter, to dazzle, be dazzling, glitter, blink.

pappe, *f.* pappus.

pappeux, -euse, papposy, downy.

pappifère, pappiferous.

pappiforme, pappiform, resembling pappus.

pappophore, with a tuft.

paprika, *m.* paprika.

papule, *f.* papule, pimple, papula.

papuleux, -euse, papulose, papillose.

papulifère, papuliferous, bearing pustules.

papuliforme, like a papula.

papyracé, papery, chartaceous.

papyrus, *m.* papyrus.

pâque, *f.* passover.

paquebot, *m.* packet, mail boat, steamship.

pâquerette, *f.* English daisy (*Bellis perennis*), selfheal (*Prunella vulgaris*).

Pâques, Pâque, *m.* Easter.

paquet, *m.* parcel, packet, package, bundle, mail, mailpacket, mail boat, pile, fagot (metal).

paquetage, *m.* packing, pack, piling, fagoting.

paqueter, to pack, fagot, pile.

paqueteur, -euse, *m.&f.* packer.

par, through, by, about, on, upon, per, out of, into, in, from, at, for, with, during; — **entre**, between; — **ici (là)**, this (that) way; — **(un temps pareil)**, in (such weather): **de** —, by the order of, in; **par-ci par-là**, hither and thither, now and then.

parabanique, parabanic (acid).

parablaste, *m.* parablast.

parabole, *f.* parable, parabola.

parabolique, parabolic(al).

paraboliquement, parabolically.

paraboliser, to parabolize, make parabolic.

paraboloïde, *m.* paraboloid.

paracarpe, *m.* paracarpium, aborted ovary.

paracaséine, *f.* paracasein.

paracentrique, paracentric.

parachèvement, *m.* finishing, completion, perfecting.

parachever, to finish, perfect, complete in full.

parachloré, parachloro.

paracholérique, choleralike.

paracorolle, *f.* paracorolla.

paracrésol, *m.* paracresol, p-cresol.

paracyanogène, *m.* paracyanogen.

paracyanure, *m.* paracyanide.

paracyste, *m.* paracyst, the antheridium of Pyronema.

parade, *f.* parade, show.

paradérivé, *m.* para derivative.

paradis, *m.* paradise.

paradisier, *m.* bird of paradise.

paradoxal, -aux, paradoxical.

paradoxalement, paradoxically.

paradoxe, *m.* paradox.

parafe, *m.* signature.

paraffènes, *m.pl.* paraffin series.

paraffinage, *m.* paraffining.

paraffine, *f.* paraffin, paraffin wax; — fondue, melted paraffin; — liquide, liquid (petrolatum) paraffin.

paraffiner, to paraffin.

paraffinique, paraffin.

paraffinoïde, paraffinoid, pertaining to a group of scents such as those of rose, lime, and elder.

parafoudre, *m.* lightning conductor, lightning arrester.

parage, *m.* lineage, descent, degree, district, place, parts, dressing, sizing, trimming.

paraglosse, *f.* paraglossa.

paragraphe, *m.* paragraph, section mark.

parahéliotropisme, *m.* paraheliotropism.

paraison, *f.* ball (of molten glass).

paraisonner, to blow (glass).

paraisonnier, *m.* blower of fine table glassware.

paraissant, appearing.

paraître, to appear, shine, be visible, be published, be conspicuous, seem, look; chercher à —, to show off; faire —, to publish, bring out (a book), show, display, cause to appear; laisser —, to show, allow to be seen; à ce qu'il paraît, apparently.

paraître, *m.* appearance.

parajour, *m.* shade (for a lamp).

paralactique, paralactic.

paraldéhyde, *m.* paraldehyde.

paralinine, *f.* paralinin.

parallactique, parallactic(al).

parallaxe, *f.* parallax.

parallèle, *m.* parallel, comparison, (of a turbine) parallel flow, downward flow, parallel ruler, parallel line.

parallèle, parallel.

parallèlement, in a parallel manner, parallel(ly).

parallélépipède, *m.* parallelepiped.

parallélépipédique, parallelepipedal.

parallélinervé, parallelinervate, (leaves) with veins or nerves parallel.

parallélique, parallel.

paralléliser, to parallelize, make (lines) parallel, compare.

parallélisme, *m.* parallelism.

paralléliveiné, parallel-veined, straight-veined.

parallélodrome, parallelinervate.

parallélogrammatique, parallelogrammatic-(al).

parallélogramme, *m.* parallelogram; — de Wheatstone, Wheatstone bridge.

paralysant, paralysateur, -trice, paralyzing.

paralyser, to paralyze, annul, palsy, inhibit (secretions); se —, to become paralysed.

paralysie, *f.* paralysis, palsy, lameness.

paralytique, paralytic.

paramagnétique, paramagnetic.

paramagnétisme, *m.* paramagnetism.

paramécic, *f.* paramecium.

paramètre, *m.* parameter.

paramitome, *m.* paramitome the ground substance of protoplasm.

paramos, *m.* paramos, extensive fell fields in South America.

paramoustique, *m.* mosquito-preventive essence.

paramylon, *m.* paramylum, a mucilaginous substance.

paranastie, *f.* paranasty, continued growth lengthwise of lateral parts.

parangon, *m.* paragon, comparison, flawless gem.

parangon, flawless.

parant (parer), dressing, trimming, ornamental.

paranthèse, *f.* parenthesis.

paranucléine, *f.* paranuclein.

paranucléique, paranucleic.

paranymphion, *m.* kind of nectary.

parapet, *m.* parapet, balustrade.

parapétale, *m.* parapetalum, any appendage to a corolla.

parapétalifère, having parapetals.

parapétaloïde, parapetaloid, bearing a parapetalum.

parapétalostémone, with stamens carried on the parapetals.

paraphylle, *f.* foliaceous expansion.

paraphyllie, *f.* paraphyllia, foliaceous expansions.

paraphyse, *f.* paraphysis, sterile filaments.

paraplasme, *f.* paraplasma, metaplasma.

parapluie, *m.* umbrella, hood (of chimney).

pararosaniline, *f.* pararosaniline.

parasérie, *f.* para series.

parasitaire, parasitic(al).

parasite, *m.* parasite.

parasite, parasitic(al).

parasité, parasitized.

parasiter, to parasite.

parasiticide, parasiticide.

parasitique, parasitic(al).

parasitisme, *m.* parasitism.

parasitologie, *f.* parasitology.

parasol (en), weeping, hanging umbellate.

parasoleil, *m.* lens hood, sunshade.

parastades, *f.pl.* parastades, cellular filaments, coronal rays.

parastème, *f.* parastamen, an abortive stamen.

parastique, *f.* parastichy, secondary spiral.

parasymbiose, *f.* parasymbiosis, peculiar parasitism of a fungus on a lichen.

paratagma, *f.* mass of micellae.

paratartrique, paratartaric, racemic.

parathecium, *m.* parathecium, circumscribing walls of the lichen thecium.

paratonnerre, *m.* lightning rod, lightning arrester.

paratrachial, paratracheal, applied to wood elements about the vessels.

paravent, *m.* draft screen, folding screen.

paraxanthine, *f.* paraxanthine.

par-brise, *m.* windshield (front and side).

parc, *m.* park, enclosure, depot, pen, fold; — à gibier, game preserve.

parcage, *m.* enclosing, penning, folding (of sheep), manuring of ground by cattle.

parcellaire, *m.* division into compartments.

parcelle, *f.* small fragment, particle, plot (of land), parcel, portion, lump. compartment; — de fouêt, wood lot.

parceller, to parcel, portion out, subdivide.

parce que, *m.* wherefore, cause, reason.

parce que, because, since, as, inasmuch as.

parchemin, *m.* parchment; — végétal, papier —, vegetable parchment, parchment paper.

parcheminé, parchmentlike, parchmentized, dried.

parcheminer, to parchmentize.

parcheminerie, *f.* parchment making or trade.

parchemineux, -euse, parchmentlike.

parcheminier, -ère, *m.&f.* parchmentmaker.

par-ci, here, hither, now; par-ci par-là, here and there, now and then.

parcimonie, *f.* parsimony, stinginess, sparingness.

parcimonieux, -euse, parsimonious, stingy, niggardly.

parcourir, to go over, go through, run over, look over, run through, examine, survey, travel over, traverse, peruse.

parcours, *m.* course, distance covered, run, trip, journey.

par-dedans, within.

par-dessous, under, beneath, underneath.

pardessus, *m.* overcoat.

par-dessus, above, over, in addition to.

par-devant, in front, before.

pardon, *m.* forgiveness, pardon.

pardonnable, pardonable, excusable.

pardonner, to pardon, excuse, forgive.

paré (parer), dressed, trimmed.

pare-éclats, *m.* splinterproof shield or guard.

pare-étincelles, *m.* spark arrester, spark catcher.

parégorique, paregoric.

pareil, *m.* like, equal, match; sans —, peerless, matchless, without equal, exceptional.

pareil, -eille, like, alike, similar, equal, identical, same, like that, such.

pareille, *f.* the like, same, equal; sans —, matchless, without equal.

pareillement, in like manner, also, too, similarly, likewise, equally.

parelle, *f.* patience dock (*Rumex patientia*), parella (lichen) (*Leconora parella*); — des marais, water dock (*Rumex aquaticus*); — sauvage, curled dock (*Rumex crispus*).

parement, *m.* adorning, dressing, ornament, adornment, decoration, face, facing (of wall), cuff (of sleeve), curb, curbstone, one stick (in a fagot).

parencéphale, *m.* cerebellum.

parenchymal, -aux, parenchymal.

parenchymateux, -euse, parenchymatous.

parenchyme, *m.* parenchyma; — libérien, liber, soft bast; — ligneux, wood parenchyma, woody tissue, prosenchyma; — phloème, phloem parenchyma; — radial, medullary rays, pith rays.

parent, *m.* relation, parent, kinsman, ancestor.

parent, related, relative.

parenté, *f.* relationship, kinship, relatives, parentage.

parenthèse, *f.* parenthesis, digression, bracket.

parer, to prepare, dress, trim, finish, pare, adorn, embellish, ward off, avert, guard against, avoid, shelter, provide, protect, clear; se —, to adorn oneself, be adorned, be dressed, ripen, mature, make a show, protect oneself.

parère, *m.* arbitration.

pare-soleil, *m.* sunshade, visor.

paresse, *f.* laziness, idleness.

paresseux, -euse, idle, lazy, sluggish, slow, indolent, slothful.

pare-torpilles, *m.* torpedo net.

pareur, *m.* finisher, dresser, trimmer.

pareuse, *f.* sizing machine.

parfaire, to finish, perfect, complete, fill out, make up; se —, to be completed.

parfaisant, completing.

parfait, *m.* perfection; — au café, ice cream with coffee flavor.

parfait, perfect, faultless, complete, thorough, perfected, finished.

parfaitement, perfectly, completely, exactly.

parfera (parfaire), (he) will complete.

parfit (parfaire), (he) completed.

parfois, sometimes, now and then, at times, occasionally.

parfondre, to fuse thoroughly and uniformly.

parfont (parfaire), (they) are completing.

parfum, *m.* perfume, odor, fragrance, agreeable odor, smell, scent.

parfumer, to scent, perfume, fumigate.

parfumerie, *f.* perfumery.

parfumeur, *m.* perfumer.

pari, *m.* bet, wager.

paria, *m.* pariah, outcast.

pariade, *f.* pairing, mating, copulation (of birds and insects).

parier, to bet, wager.

pariétaire, *f.* pellitory (Parietaria); — officinale, wall pellitory (*Parietaria officinalis*).

pariétal, -aux, parietal.

parinervé, with two like veins marginally located.

paripenné, paripinnate, abruptly pinnate, with an equal number of leaflets.

parisien, -enne, Parisian, Paris.

parité, *f.* parity, equality, equivalence, evenness (of numbers).

pari-valence, *f.* even valence.

parjure, *m.* perjury, perjurer.

parjure, perjured.

parlant, speaking, expressive, talkative, lifelike.

parlement, *m.* parliament.

parlementaire, parliamentary.

parlementer, to parley.

parler, *m.* speaking, speech.

parler, to speak, talk, converse; se —, to be spoken (of), speak to each other; sans — de, to say nothing of, not to mention.

parleur, *m.* sounder (of telephone), speaker.

parménie, *f.* bear's-foot, stinking hellebore (*Helleborus foetidus*).

parmentiera, *f.* candle tree (*Parmentiera cerifera*).

parmentière, *f.* potato.

parmesan, *m.* Parmesan cheese.

parmesan, Parmesan, of Parma.

parmi, among, amid, amongst, amidst. in the midst of.

parnassie, *f.* grass of Parnassus (Parnassia): — des marais, white buttercup.

paroi, *f.* wall, side, partition, border tree. coat (seeds), facing (of masonry), lining, shell (of boiler); — cellulaire, cell wall, cellular membrane; — de derrière, tailboard; — de l'estomac, coat(ing) of the stomach; — **extérieure,** exterior surface; **à — épaisse,** thick-walled.

paroir, *m.* paring knife, scraper.

parole, *f.* word, speech, parole, tone of voice, eloquence, oratory, trust, promise, sentiment; — **d'honneur,** upon my word.

parotidien, -enne, parotid.

parou, *m.* dressing (of cloth).

parquer, to pen (cattle), pen up, park (car), manure (by cattle).

parquet, *m.* parquet, hardwood floor, inlaid floor, platform, felling area, back of a mirror frame, small enclosure, (hen) run.

parrain, *m.* sponsor.

parsemer, to strew, sprinkle, spread, spangle, scatter, disseminate, stud, spread.

part (partir), (he) departs.

part, *m.* birth, childbirth, newborn, parturition fetus.

part, *f.* part, portion, share, information, concern, source, interest, side, place; **à —,** apart, aside, separately, except, alone, odd, by oneself; **à — cela,** except for that; **autre —,** elsewhere; **d'autre —,** on the other side, then again, on the other hand, furthermore, besides, in another connection; **de — en —,** through and through, from one side to the other, right through; **de — et d'autre,** on both sides, here and there, on all sides, reciprocally.

part, de la — de, on the part of, in the name of, from; **des deux —,** on both sides; **de toute —, de toutes —s,** on all sides, in all quarters, in every direction; **d'une —,** on the one hand, ⌄n one side; **faire — de,** to inform, communicate; **faire la — de,** to make allowance for; **nulle —,** nowhere; **pour une grande —,** to a great extent; **quelque —,** somewhere.

partage, *m.* division, partition, portion, share, lot, distribution, watershed.

partagé, divided, reciprocal, supplied, endowed.

partageable, divisible.

partager, to divide, share, partake (of), supply, endow, parcel, participate in; **se —,** to be divided, part.

partant, consequently, therefore, starting, departing.

parténement, *m.* basin for concentration of brine.

parterre, *m.* parterre, flower garden, garden plot; — **de la coupe,** felling area.

parthénocarpie, *f.* parthenocarpy.

parthénogénèse, *f.* parthenogenesis.

parthénogonidie, *m.* parthenogonidium, zooid of a protozoan colony, with asexual reproduction.

parthénospore, *m.* parthenospore, a spore produced without fertilization.

parti, *m.* party, advantage, use, cause, resolution, treatment, determination, decision, side, part, means, way, contract, measure, profit.

parti, departed, partite (leaf); — **pris,** prejudice; **sans — pris,** without taking sides; **tirer — de,** to derive advantage from.

partial, -aux, partial.

partialement, partially.

partialité, *f.* partiality.

participant, *m.* participant.

participant, participating.

participation, *f.* participation, sharing.

participe, *m.* participle.

participer (à), to participate in, share in, take part in; — **de,** to partake of, participate of.

particulaire, elementary, atomic, of particles, particulate.

particulariser, to particularize.

particularité, *f.* peculiarity, particularity, particular; *pl.* particulars.

particule, *f.* particle.

particulier, *m.* particular, private life, individual.

particulier, -ère, particular, separate, odd, special, peculiar, individual, singular; **en —,** in particular, particularly, privately.

particulièrement, particularly, in particular, specially, peculiarly, especially.

partie, *f.* part, party, project, plan, portion, match, specialty, lot (of goods), entry (in bookkeeping), game, deal, ingredient, constituent; — **aérienne de la plante,** aerial part of the plant; — **d'about,** end (part) piece; — **de bois en massif,** dense crop; — **de milieu,** middle (part) piece; **—s naturelles,** genitals; **en —,** partly, in part; **en grande —,** to a great extent; **en majeure —,** for the most part; **faire — de,** to form part of, be a part of.

partiel, -elle, partial.

partiellement, partially, in part, in parts, piecemeal.

partir, *m.* start, departure; **à — de,** beginning with or from, since the time of.

partir, to start, set out, depart, move, leave, proceed, dart (off), shoot (gun), go off, go away, come off, part; **faire —,** to make go, run, send off, dispatch, fire (gun), remove (stain), set off (fireworks), bring (food) to boil.

partisan, *m.* partisan.

partition, *f.* partition.

partout, everywhere, in every direction; **— où . . . ,** wherever **. . . ; de —,** everywhere, all over, on all sides; **en tout et —,** at all times and in all places.

parturition, *f.* parturition, childbirth.

paru (paraître), appeared, published (books).

parure, *f.* attire, adornment, ornament, dress, finery, set (of gems), match, parings (of leather), trimmings (of meat).

parurent (paraître), (they) appeared.

parut (paraître), (he) appeared.

parvenir, to arrive, attain, come, succeed, reach, get; **— à,** to arrive at, reach, succeed in.

parvenu, *m.* parvenu, self-made person, upstart.

parvenu, arrived, fortunate, successful.

parviflore, parviflorous, having smaller flowers than in its congeners.

parvifolié, parvifolious, with small leaves.

parvint (parvenir), (he) succeeded.

parvoline, *f.* parvoline.

pas, *m.* step, pace, passage, threshold, way, pass, strait, conduct, footstep, progress, thread channel (of screw), pitch (of screw), precedence; **— à —,** step by step; **— d'âne,** coltsfoot (*Tussilago farfara*); **— de clerc,** blunder, false step; **— d enroulement,** winding pitch; **— maître,** past master; **de ce —,** directly, this instant.

pas, not any, none; **— de,** no; **— du tout,** not at all; **— fécondé,** barren, uncovered; **— mal,** quite a lot; **— ne,** not, no, not any, none; **— un,** not one, none, not a, no; **ne . . . pas,** not, no; **presque —,** scarcely any.

passable, passable, tolerable, pretty good.

passablement, passably, tolerably.

passage, *m.* passing, passage, transit, transition, road, way.

passager, *m.* traveler, passenger.

passager, -ère, transient, transitory, fugitive, temporary, passing.

passagèrement, transiently, for a short time, temporarily.

passant, *m.* passer-by, passenger!

passant, passing, public, crowded; **en —,** casually.

passe, *f.* pass, situation, passing, channel.

passé, *m.* past, time past.

passé, beyond, after, past, passed, former, bygone, last, gone, over, faded, out of fashion; **— cette date,** after this date; **beauté —e,** faded beauty.

passée, *f.* time or season of migration, deer path.

passe-fleur, *f.* pasque flower (*Pulsatilla vulgaris*).

passe-lait, *m.* milk strainer.

passementerie, *f.* passementerie, trimmings, making or selling of passementerie.

passe-méteil, *m.* maslin, a grain mixture containing $\frac{2}{3}$ wheat, $\frac{1}{3}$ rye.

passe-partout, *m.* master key, pit saw, compass saw, crosscut saw, passport, frame.

passe-perle, *m.* extremely fine wire.

passe-pierre, sea-fennel, samphire, (*Chritmum maritinum*).

passeport, *m.* passport.

passer, to pass, pass away, pass by, pass along, get, go, come, pass through, proceed, expire, die, surpass, omit, (of colors) fade, last, be considered, strain, sift, pass over, excuse, cross, satisfy, go beyond, exceed, excel, survive, slip on, outlast, put on, allow, grant, forgive, spend (time), enter into (a contract), sell; **ça se passe tous les jours,** that happens every day; **il passe pour tel,** he is considered as such.

passer, — au tamis, to sift, screen; **— chez,** to call on, look in; **— de rouleau,** to roll the seed; **— en revue,** to review; **— par,** to pass by, pass through, go through; **— rapidement,** to migrate, sweep over; **en — par,** to submit to, put up with it; **faire —,** to cause to pass, carry over, bring over, pass round, hand around, remove, get rid of, cure (defects); **laisser —,** to let pass, let go; **se —,** to pass (away), wear (away), decay, fade, happen, spend, go, proceed, be content (with), dispense (with), come to pass, take place; **se — de,** to do without; **s'en —,** to do without.

passerage, *f.* pepperwort, peppergrass (Lepidium); **— cultivé,** garden cress (*Lepidium sativum*).

passereau, *m.* sparrow.

passerelle, *f.* footbridge, bridge, plank bridge.

passerie, *f.* plumping liquor (of hides).

passerose, *f.* hollyhock (*Althaea rosea*).

passe-temps, *m.* pastime.

passe-velours, *m.pl.* cockscomb (Celosia).

passe-violet, *m.* violet (blue) color.

passible, passible, sensible, liable.

passif, *m.* liabilities, debt.

passif, -ive, passive.

passiflore, *f.* passion flower (Passiflora).

Passiflorées, *f.pl.* Passifloraceae.

passion, *f.* passion, love, last suffering, strong emotion.

passionnaire, *f.* passion flower (Passiflora).

passionnément, passionately, fondly.

passionner (se), to become extremely fond (of).

passivement, passively.

passiveté, passivité, *f.* passivity, passiveness.

passoire, *f.* strainer, colander.

pastel, *m.* pastel, woad, crayon, pastel drawing.

pastenade, *f.* parsnip.

pastèque, *f.* watermelon.

pasteur, *m.* pastor, minister, shepherd.

pasteur, pastoral.

pasteurellose, *f.* pasteurellosis.

pasteurien, -enne, of Pasteur, Pasteurian.

pasteurisateur, *m.* pasteurizer, heater.

pasteurisation, *f.* pasteurization.

pasteuriser, to pasteurize, sterilize.

pastillage, *m.* ornamentation.

pastille, *f.* pastille, lozenge, troche, tablet, pellet, pill, rubber patch; — **à brûler,** fumigating (pastille) candle; — **du sérail,** aromatic pastille.

pastilleur, -euse, *m.&f.* a worker or machine making pastilles.

pastorien, -enne, of Pasteur, Pasteurian.

patate, *f.* sweet potato (*Ipomoea batatas*).

patchouli, *m.* patchouli (*Pogostemon patchouli*).

pâte, *f.* paste, dough, pulp, stuff, clay, mill cake, sort, kind, (printer's) pie; — **à autocopier,** copying paste; — **alimentaire,** edible paste; — **à papier,** paper pulp; — **à potage,** edible paste; — **de bois,** wood pulp, wood flour, cellulose; — **de chiffons,** rag pulp; — **demi-chimique,** semichemical (paper) pulp.

pâte, — **dentifrice,** tooth paste; — **de paille,** paper pulp from straw; — **de papier,** papier-mâché; — **de remplage,** filling compound or paste; — **d'Italie,** Italian paste,

edible paste; — **du fromage,** body and texture of cheese; — **effilochée,** half or first stuff; — **mécanique,** mechanical pulp, wood pulp, cellulose; — **molle,** soft paste; — **pour chaussures,** shoe polish; — **raffinée,** stuff.

pâté, *m.* meat pie, patty, pasty, blot (of ink), block (of houses), clump (of trees).

pâtée, *f.* mash; — **d'élevage,** chicken mash; — **de ponte,** laying mash; — **sèche,** dry mash.

patelin, *m.* (small) assay crucible.

patelliforme, patelliform, shaped like a small disk, circular and rimmed.

patellule, *f.* patellula, diminutive patella.

patent, patent, obvious, evident.

patentable, subject to a license.

patente, *f.* license, patent; — **de santé,** bill of health; — **nette,** clean bill (of health).

patenté, *m.* licensee.

patenté, licensed, established, recognized, patented.

patenter, to license.

patère, *f.* curtain holder, hat peg, clothes peg.

paternel, -elle, paternal, fatherly.

pâteux, -euse, pasty, doughy, viscous, sticky, viscid, cloudy, thick, clammy, muddy.

pathétiquement, pathetically.

pathogène, *f.* pathogen.

pathogène, pathogenic.

pathogénie, *f.* pathogenesis.

pathogénique, pathogenic.

pathologie, *f.* pathology; — **végetale,** diseases of plants, phytopathology.

pathologique, pathological.

pathologiquement, pathologically.

pathologiste, *m.* pathologist.

patiemment, patiently.

patience, *f.* patience, forbearance, patience dock (*Rumex patientia*).

patient, patient, enduring.

patienter, to be patient, take patience.

patin, *m.* support, sill block, base, plate, foot, flange, shoe, sole piece, slide block, skate.

patine, *f.* patina.

patiner, to skate, skid, slip, patinate.

patio, *m.* patio, inner court, paved floor.

pâtir, to suffer.

pâtis, *m.* grazing ground, pasture.

pâtissage, *m.* pastry making.

pâtisser, to make (into) pastry, work up (piecrust).

pâtisserie, *f.* pastry, pastry making, pastry (shop) business.

pâtissier, -ère, *m.&f.* pastry cook.

pâtisson, *m.* squash (melon).

pâton, *m.* lump of dough, lump, ball (of paper).

patouille, *f.* ore separator.

patouillet, *m.* mixing or washing apparatus.

patouilleux, -euse, muddy, splashy, sloppy.

pâtre, *m.* shepherd, herdsman.

patrie, *f.* country, native land, fatherland.

patrimoine, *m.* patrimony, inheritance.

patriote, *m.* patriot.

patriote, patriotic.

patron, *m.* patron, master, employer, pattern, model, template, stencil.

patronage, *m.* patronage, stenciling.

patronner, to patronize, protect, support, stencil, cut out on a pattern.

patrouiller, to patrol, dabble, paddle.

patte, *f.* paw, foot, claw, leg (of insects), foot (of a glass), clip, clamp, hook, flange, flap, tongue, tab, strap, catch, root, root swelling, butt swelling, tarsus; — abdominale, proleg; — de coq, cockspur; — de la racine, root swelling, butt swelling; — d'oie, goosefoot (Chenopodium); — d'ours, club moss (Lycopodium); — fausse, proleg; — membraneuse, proleg; —s ambulatoires, gradient or walking legs; —s natatoires, swimming legs; —s ravisseuses, seizing legs; —s saltatoires, hopping or leaping legs.

patte-d'oie, *f.* crossroads.

pattinsonage, *m.* pattinsonizing, Pattinson's desilvering process.

pattu, large-pawed, feather-legged.

pâturage, *m.* pasturage, pasture (land), grazing.

pâturant, pasturing, grazing, herbivorous.

pâture, *f.* food, feed, fodder, forage, pasture, pasturage.

pâturer, to pasture, graze, feed.

pâturin, *m.* meadow grass (Poa); — commun, bird grass, rough-stalked meadow grass (*Poa trivialis*); — des prés, smooth meadow grass, Kentucky bluegrass, June grass (*Poa pratensis*); — maritime, sea spear grass (*Puccinellia maritima*).

paturon, *m.* pastern.

paucicellulaire, composed of a small number of cells.

paucidenté, with few teeth.

pauciflore, pauciflorous, few-flowered.

paucijugué, paucijugate, with only a few pairs of leaflets.

paucinervé, slightly veined.

paucirugueux, -euse, slightly rugose (wrinkled).

pauciséminé, oligospermous, few-seeded.

paucité, *f.* paucity.

paume, *f.* palm (of the hand), tennis, tennis court.

paumure, *f.* crown (of stag's head).

paupière, *f.* eyelid, eyelash.

pause, *f.* pause, stop, rest, interval, suspension.

pauser, to pause.

pauvre, *m.* poor material, poor person.

pauvre, poor, needy, indigent, wretched, miserable, mean, barren.

pauvrement, poorly.

pauvreté, *f.* poverty, dearth, poorness.

pavage, *m.* paving, pavement; — en béton, concrete paving.

pavé, *m.* paving stone, paving block, paving, pavement, paved (road) street, highway; — aggloméré, artificial stone pavement, concrete paving block; — en bois, wood paving; — en brique(s), brick paving; — en pierre, stone paving.

pavé, paved.

pavement, *m.* paving, (ornamental) pavement.

paver, to pave.

pavie, *f.* clingstone (peach).

pavier, *m.* Pavia.

pavillon, *m.* pavilion, opening, mouth (of funnel), horn, the mouthpiece (of telephone), colors, flag, summerhouse, building, wing (of a building).

pavot, *m.* poppy (Papaver); — Argémone, rough or prickly poppy (*Argemone platyceras*); — cornu, horn(ed) poppy (*Glaucium flavum*); — rouge, corn poppy (*Papaver rhoeas*); — somnifère, opium poppy, maw seed (*Papaver somniferum*); huile de —, poppy-seed oil.

payable, due, payable.

payant, *m.* payer.

pavant, paying.

paye, *f.* pay. salary.

payement, *m.* payment.

payer, to pay, pay for, give proof; **se —,** to pay oneself, be paid, be satisfied; **s'en —,** to have a good time.

pays, *m.* country, home, land.

paysage, *m.* landscape, region.

paysagiste, *m.* landscape painter.

paysan, *m.* peasant, countryman.

paysan, -anne, rural, rustic, peasant.

paysanne, *f.* countrywoman, farmer's wife.

paysannerie, *f.* peasantry.

Pays-Bas, *m.pl.* Netherlands, Low Countries.

péage, *m.* toll, charge, tollhouse.

peau, *f.* skin, hide, peel, husk, coating, film; **— à saucisses,** sausage skin; **— brute,** rawhide, raw skin; **— chamoisée,** chamoised skin, chamois; **— de blaireau,** badger's skin; **— de buffle,** buff, buffalo hide or skin; **— de chamois,** chamois (leather); **— de cheval,** horsehide; **— de chevreau,** kid leather; **— de chien marin,** sealskin; **— de daim,** buckskin.

peau, — de lapin, rabbitskin; **— de mouton,** sheepskin; **— de raisin,** grape skin; **— de vache,** cowhide; **— de vélin,** vellum; **— passée en mégie,** tawed leather; **— pleine,** (unsplit) hide; **— préparée,** dressed hide; **— refendue,** split skin, split hide; **— tannée,** tanned (skin) hide; **— verte,** green hide, pelt, fresh hide.

peaucier, cutaneous.

peausserie, *f.* skin dressing, trade in skins or hides.

peaussier, *m.* dealer in skins, pelts.

peaux, *f.pl.* see PEAU.

pébrine, *f.* pébrine.

peccant, morbid.

pechblende, *f.* pitchblende, uraninite.

pêche, *f.* peach, fishing, fishery; **— à la ligne,** angling; **— amande,** *f.* peach-almond hybrid.

péché, *m.* sin.

pécher, to sin.

pêcher, to draw out, fish (out), pick up, get; **— à la ligne,** to angle.

pêcher, *m.* peach tree (*Prunus persica*).

pêcherie, *f.* fishery, fishing.

pêcheur, *m.* fisher, fisherman.

pêcheur, fishing.

pêchurane, *f.* pitchblende, uraninite.

pécore, *f.* creature, animal.

pectase, *f.* pectase.

pectate, *m.* pectate.

pecteux, -euse, pectous, pectinous.

pectinase, *m.* pectinase.

pectine, *f.* pectin, vegetable jelly, pectose.

pectiné, pectinate(d).

pectineux, -euse, pectinous, pectous.

pectique, pectic.

pectoral, -aux, pectoral.

pectose, *f.* pectose.

pécuniaire, pecuniary.

pédale, *f.* pedal, treadle.

pédalé, pedate (leaf).

pédalier, *m.* pedal board.

pédaliforme, in form of a pedal.

pédalinervé, (petiole) divided at tip into two very divergent veins.

pédate, pedate.

pédatiforme, in form of a pedal.

pédatilobé, pedatilobate, pedatilobed, palmate, with supplementary lobes at the base.

pédatinervé, (petiole) divided at tip into two very divergent veins.

pédé, pedate.

pédiaire, pedal.

pédicelle, *m.* pedicel, pedicle, flower stalk.

pédicellé, pedicellate.

pédicellulé, with very thin pedicel.

pédicular, *f.* lousewort, wood betony (*Pedicularis canadensis*).

pédicular, pediculous.

pédicule, *m.* pedicel, stalk.

pédiculé, pediculate.

pédiculus, *m.* louse.

pédicure, *m &f.* chiropodist.

pédigrée, *f.* pedigree.

pédilé, pedilatous, furnished with a pedilus.

pédogamie, *f.* union of protoplasts of neighboring cells.

pédonculaire, peduncular.

pédoncule, *m.* peduncle, fruit stalk, floral shoot, stem.

pédonculé, petiolate, pedunculate.

pédonculé, pedunculate(d), stalked.

pegmatite, *f.* pegmatite.

peignage, *m.* combing, carding (of wool).

peignant (peigner. peindre), combing, painting.

peigne, *n.*. comb, card, gill, chaser, hackle; — **d'égrugeoir,** ripple, rippling comb.

peignée, *f.* cardful.

peignent (peigner, peindre), they are combing, painting.

peigner, to comb, card, hackle, chase, thrash, drub.

peignerie, *f.* combing works.

peigneuse, *f.* combing machine.

peille, *f.* rags.

peinchebec, *m.* pinchbeck (alloy of copper and zinc).

peindre, to paint, portray, describe, depict, dye, delineate; **se** —, to be portrayed, depicted; — **à la chaux,** to whitewash; — **à l'aquarelle,** to paint in water color; — **à l'huile,** to paint in oil.

peine, *f.* pain, trouble, difficulty, punishment, affliction, grief; **à** —, scarcely; hardly; **à grand'** —, with great difficulty; **avec** —, with difficulty, with regret; **avoir** —, to have difficulty in; **sans** —, without difficulty, easily, willingly.

peiné, grieved, labored.

peiner, to labor, toil, pain, grieve, vex.

peint, painted, colored, (he) is painting.

peintre, *m.* painter; — **en bâtiments,** — **décorateur,** house painter.

peintre-verrier, *m.* glass painter.

peintre-vitrier, *m.* house painter and glazier.

peinture, *f.* painting, coating, paint, picture, color; — **à la chaux.** whitewashing, lime paint or color; — **à (la) gouache,** — **à l'aquarelle,** water-color painting; — **à l'huile,** oil painting, oil paint; — **d'apprêt,** priming, glass staining; — **en détrempe,** distemper painting.

peinturer, to paint, coat, bedaub.

pelage, *m.* unhairing, coat, wool, fur, covering, skin.

pélagique, pelagic, inhabiting the open ocean.

pelain, *m.* lime pit; see PELIN.

pelainage, *m.* liming.

pelaineur, *m.* limer.

pelan, *m.* bark.

pelanage, *m.* unhairing, depilation; — **à la chaux,** liming.

pelaner, to unhair, depilate.

pélargonique, pelargonic.

pelé, peeled, bald, naked, bare.

pêle-mêle, pell-mell, in confusion, confusion, confusedly.

pêle-mêler, to jumble up, mix confusedly.

peler, to peel (skins, hides), depilate, strip, skin, pare, ross, turf off, unhair, bark (trees), clean, scald (carcass), scrape (a crucible), strip off (the hair), make bald; **se** —, to peel off, grow bare, come off.

pèlerin, *m.* pilgrim.

pèlerinage, *m.* pilgrimage.

pèleriner, to go on a pilgrimage.

pélican, *m.* pelican.

pelin, *m.* lime pit.

pellage, *m.* shoveling.

pellagre, *f.* pellagra, Italian leprosy.

pelle, *f.* shovel; — **à puiser,** long-handled shovel.

pellée, *f.* shovelful, spadeful.

peller, to shovel.

pellerée, *f.* shovelful, spadeful.

pelletage, *m.* shoveling.

pelletée, *f.* shovelful, spadeful.

pelleter, to shovel.

pelleterie, *f.* fur, skins, fur making, fur trade.

pelletier, *m.* furrier.

pelletierine, *f.* pelletierine.

pelliculage, *m.* stripping (of plate or film).

pelliculaire, pellicular; **effet** —, skin effect.

pellicule, *f.* pellicle, thin skin, film, crust, filmy skin; — **d'air,** layer or stratum of air, air space.

pelliculé, covered with a pellicle or film, pelliculate.

pelliculer, to strip (a plate or film).

pellucide, pellucid, limpid.

pellucidité, *f.* pellucidness, pellucidity.

pélogène, pelogenous, applied to those rocks which yield a clayey detritus and the plants which thrive thereon.

pélophile, pelophilous, pelogenous.

pélorie, *f.* peloria, condition of abnormal regularity.

pélorié, pélorisé, offering abnormal regularity.

pelotage, *m.* winding (of skeins of wool) into a ball.

pelote, *f.* ball, wad, pellet, card (of cotton), pad, star (on horse).

peloter, to make into a ball, roll up into a ball, curl up, plot (soap).

peloteur, *m.* winder (of thread).

peloton, *m.* ball, lump, cluster, group, troop, platoon, glomerule.

pelotonner, to wind; se —, to be wound or rolled up, curl up, roll up, huddle up, gather, cluster (of bees).

pelouse, *f.* lawn, greensward, grass plot.

pelté, peltate, shield-shaped, bucklershaped.

peltidé, peltideous, orbicular or bucklershaped.

peltiforme, peltiform.

peltinervé, peltinerved, with ribs arranged as in a peltate leaf.

peluche, *f.* plush.

peluché, shaggy, hairy (flower).

pelure, *f.* peel, skin, rind, peeling, paring, fruit scale, pelt, wool; — d'oignon, purplish-brown color, onion skin, a thin writing paper, onionskin; papier — (d'oignon), foreign note paper, pelure paper.

pelvien, -enne, pelvic.

pelvis, *m.* pelvis.

pemmican, *m.* pemmican.

pénalité, *f.* penalty, fine.

penaud, sheepish.

penchant, *m.* inclination, slope, bent, verge, decline, liking, penchant, propensity, descent, gradient, fall, declivity.

penchant, inclining, shelving, sloping.

pencher, to incline, lean, bend, tilt, tip, slope; se —, to bend (over), lean, stoop, be inclined; **il se penche,** accidents (often) happen.

pendaison, *f.* hanging, suspension.

pendant, *m.* pendant, counterpart.

pendant, hanging, pendant, pending; fruits —s par (les) racines, growing (standing) crops; oreilles —es, lop ears.

pendant, pending, during, for; — que, while, whilst.

pendiller, to dangle, hang loose, swing.

pendoir, *m.* hook or cord (for hanging meat).

pendre, to hang (up), suspend, depend, be pending; se —, to hang on, cling to.

pendu, hung.

pendule, *f.* clock.

pendule, *m.* pendulum, governor; — à boules, ball governor.

penduliflore, with pendant flowers.

pendulifolié, with pendant leaves.

pêne, *m.* bolt; *f.* pitch mop, tar brush.

pénéanthées, *f.pl.* impoverished flowers.

pénéplaine, *f.* peneplain.

pénétrabilité, *f.* penetrability.

pénétrable, penetrable.

pénétrant, penetrating, piercing, keen.

pénétration, *f.* penetration, entering, insight; — du bois, impregnation of wood, wood steeping; — du sol par l'humidité, moistening of the soil.

pénétrer, to penetrate, touch deeply, impress, reach, enter, pervade, pierce, get into; se —, to penetrate each other, combine, mix, become impregnated with; se — de, to learn, impress upon one's mind.

pénible, painful, hard, laborious, distressing.

péniblement, painfully, laboriously.

péniche, *f.* barge.

pénicillé, brushlike, aspergilliform.

pénicilliforme, penicilliform, shaped like an artist's pencil.

péninsule, *f.* peninsula.

pénis, *m.* aedeagus, penis, male intromittent organ.

pénitence, *f.* penitence, penance.

pénitent, penitent, repentant.

pennage, *m.* plumage.

pennaticisé, pinnaticissus.

pennatifide, pinnatifid.

pennatifolié, with pinnatifid leaves.

pennatilobé, pinnately lobed.

pennatipartite, pinnatipartite, pinnately parted.

pennatistipulé, with pinnatifid stipules.

penne, *f.* feather, pinion feather, warp end, pinna.

penné, pinnate, pinnatifid.

pennifide, pinnatifid.

pennifolié, with pinnatipartite leaves.

penniforme, pinniform.

penniglumé, with plumous glumes.

pennilobé, pinnately lobed.

penninervé, pinniveined.

penniveiné, pinniveined.

pénombre, *f.* penumbra, dim light, partial shadow.

pensant, thinking.

pensée, *f.* thought, mind, intention, meaning, idea, opinion, conception, outline, pansy (*Viola tricolor*), heartsease.

penser, to think, think of, imagine, picture, be in great danger, reflect, opine; il pense à le lire, he intends to read it.

penseur, m. thinker.

penseur, -euse, thoughtful, reflective.

pension, f. pension, allowance, board (and lodging), boarding house, pension, boarding school (fees).

pensionnaire, m.&f. pensioner, boarder.

pensionner, to pension.

pensivement, thoughtfully.

penstémone, pentandrous, with five stamens.

pentacamare, with five follicles.

pentacarboné, pentacarbon, containing five carbon atoms.

pentacarpe, with fruit formed of five carpels.

pentachlorure, m. pentachloride.

pentachotome, divided in five parts.

pentacoque, pentacoccous, with five cocci.

pentacycle, describing five turns.

pentacyclique, pentacyclic.

pentadactyle, pentadactyl.

pentadelphe, pentadelphous, with five fraternities or bundles of stamens.

pentaèdre, m. pentahedron.

pentagonal, -aux, pentagonal.

pentagone, m. pentagon.

pentagone, pentagonal.

pentagyne, pentagynous.

pentakène, (fruit) divided into five lobes of maturity.

pentamère, pentamerous.

pentaméthylénique, pentamethylene.

pentandrie, f. Pentandria.

pentanthe, with five flowers.

pentapétale, pentapetalous.

pentarche, pentarch, (roots) having five xylem strands.

pentasépale, pentasepalous, having five sepals.

pentasperme, pentaspermous, five-seeded.

pentasulfure, m. pentasulphide.

pentathionique, pentathionic.

pentatomique, pentatomic.

pentavalent, pentavalent, quinquivalent.

pente, f. slope, inclination, incline, gradient, declivity, pitch, fall, bent, propensity, tendency, decline, downgrade; — abrupte, precipitous; — ascendante, ascent, up-

grade, acclivity; — assez rapide, moderate slope; — descendante, descent, downgrade, declivity; — douce, gentle slope; — rapide, steep slope; aller en —, to slope, incline.

penthiophène, m. penthiophene.

pentite, f. pentite, pentitol.

pentosane, m. pentosan.

pentose, m. pentose.

penture, f. hinge (strap).

pénurie, f. scarcity, dearth, shortage, lack, poverty, penury.

péonine, f. paeonin.

péparifère, pepo-bearing.

péperin, peperino, m. peperino.

pépier, to chirp, peep.

pépin, m. pip, small seed (of fruit), stone (of fruit), pith, kernel.

pépinière, f. nursery (for trees); — fixe, permanent nursery; — volante, temporary nursery.

pépiniériste, m. nurseryman.

pépite, f. nugget.

péplide, f. Peplis.

pepo, m. pepo, peponidium, a gourd fruit.

pépon, peponide, m. peponidium, a gourd fruit.

pepsine, f. pepsin.

pepsinogène, m. pepsinogen.

peptide, m. peptide.

peptogène, m. peptogen.

peptolytique, peptolytic.

peptonate, m. peptonate.

peptone, f. peptone.

peptoné, peptonized, peptone, containing peptone.

peptonification, f. conversion into peptone.

peptonifier, to convert into peptone.

peptonisant, peptonizing.

peptonisation, f. peptonization.

peptonisé, peptonized, containing peptone.

peptoniser, to peptonize.

peptonurie, f. peptonuria.

peptotoxine, f. peptotoxine.

peranosis, f. peranosis, change in the permeability of protoplasm.

péraphylle, f. paraphyllium, stipule.

pérat, m. lump coal, patent fuel.

perborate, m. — du soude, sodium perborate;

perçage, m. piercing, boring, drilling.

perçant, *m.* piercer, borer.

perçant, piercing, penetrating, sharp, keen.

perce, *f.* drill, punch, borer, piercing tool, hole; **mettre en —**, to tap, broach a cask.

percé, pierced, bored.

perce-bois, *m.* wood borer.

perce-bouchon, *m.* cork borer.

percée, *f.* opening, taphole, opening (in a wood), glade.

perce-feuille, *f.* hare's-ear (*Bupleurum rotundifolium*).

perce-fournaise, *m.* tapping bar.

perce-oreille, *m.* earwig.

percement, *m.* piercing, opening, digging, perforation, opening (in forest or wall).

perce-muraille, *f.* wall pellitory (Parietaria).

perce-pierre, *f.* parsley piert (*Aphanes arvensis*), saxifrage (Saxifraga).

percepteur, *m.* perceiver, collector of taxes.

perceptibilité, *f.* perceptibility.

perceptible, perceptible.

perception, *f.* perception, collection of taxes.

percer, to pierce, bore, drill, punch, lance, tap, broach, cut (holes), perforate, penetrate, open (doors, streets), break through, appear, leak out, peep out, tunnel out, drive (a tunnel), wet thoroughly, become known; **se —**, to be pierced, be bored.

percerette, *f.* drill, cork borer.

perceur, *m.* piercer, driller, borer; **— de liège**, cork borer.

perceuse, *f.* boring (drilling) machine.

percevable, perceivable, collectible.

percevant, perceiving, collecting.

percevoir, to perceive, hear, collect, gather (taxes).

perchage, *m.* poling.

perche, *f.* pole, rod, perch, stake; **— à houblon**, hop pole; **— argentée**, calico bass; **— noire**, small-mouthed black bass; **— truite**, large-mouthed black bass.

percher, to pole, perch, roost.

perchis, *m.* pole wood.

perchlorate, *m.* perchlorate.

perchloré, perchloro.

perchlorique, perchloric.

perchlorure, *m* perchloride, bichloride.

perchoir, *m.* perch, roost.

perchromique, perchromic.

perclus, crippled, stiff-jointed.

percoir, *m.* punch, punch drill, borer, broach, awl; **— carrée**, square rod.

perçoit (percevoir), (he) perceives, collects.

percolateur, *m.* percolator.

perçu (percevoir), perceived, collected.

percussion, *f.* percussion.

percutant, percussive, percussion, striking.

percuter, to strike, percuss, sound (chest) by percussion.

percuteur, *m.* striker, hammer, firing pin, plunger.

perdant, *m.* loser.

perdant, losing.

perditance, *f.* leakance (leakage conductance).

perdre, to lose, waste, spoil, fail, fall, ebb, ruin, undo; **se —**, to lose oneself, be lost, lose one's way, stray, disappear, blend (colors), be damaged; **— de vue**, to lose sight of; **— sa couleur**, to change color, become discolored.

perdreau, *m.* young partridge.

perdrix, *f.* partridge; **— grecque**, Greek partridge; **— grise**, common gray partridge; **— rouge**, red-legged French partridge.

perdu, lost, ruined, done for, random, leisure, idle, remote, out of the way, invisible, flush.

perdurable, very durable, lasting forever.

perdurer, to be very durable, last long.

père, *m.* father, sire.

péreirine, *f.* pereirine.

perembryon, *f.* perembryum, that part of monocotyledonous embryo investing the plumule and radicle, not externally distinguishable.

péremptoire, peremptory.

perennant, perennate, lasting the whole year through.

pérenne, *f.* small nettle.

pérenne, perennial, lasting, durable, persistant.

pérennité, *f.* perpetuity, perenniality.

péréquation, *f.* equalization.

perfection, *f.* perfection; **à la —**, **en —**, **dans la —**, to perfection, perfectly.

perfectionné, perfected, improved.

perfectionnement, *m.* improvement, perfecting, improving, perfection.

perfectionner, to improve, perfect; **se —**, to improve oneself, be improved, perfected.

perfectionneur, *m* perfecter, improver.

perfeuillé, perfoliate, when a stem apparently passes through a leaf.

perfide, treacherous, false.

perfidement, perfidiously, treacherously, basely.

perfidie, *f.* perfidy, treachery.

perfolié, perfoliate.

perforage, *m.* perforation, boring.

perforant, perforating.

perforateur, *m.* drill, punch, borer, perforator.

perforateur, -trice, perforating.

perforatif, -ive, perforative.

perforation, *f.* perforation, perforating, hole, puncture.

perforatrice, *f.* drill, borer.

perforer, to perforate, pierce, punch, bore, drill, puncture.

perhydrure, *m.* perhydride.

périandrique, periandricous, used of a nectary when it is ranged around the stamens.

périanthe, *m.* perianth (sepals and petals).

perianthé, with a perianth.

périanthien, coming from a perianth.

périblème, *m.* periblem, a layer of nascent cortex beneath the epidermis.

péricambium, *m.* pericambium, the pericycle.

péricarde, *m.* pericardium.

péricardique, pericardial.

péricardite, *m.* pericardite.

péricarpe, *m.* pericarp, seed vessel.

péricarpial, -aux, pericarpial.

péricarpique, pericarpic.

péricentrique, pericentral, round or near the center.

périchèse, *m.* perichaetium.

périchet, *m.* perichaetium, the involucre around the base of the seta in mosses.

périclade, *m.* pericladium, the sheathing base of a leaf surrounding the supporting branch.

péricladium, *m.* pericladium.

périclase, *f.* periclase, periclasite.

périclinanthe, (involucral bracts) tending in vertical direction.

péricline, *m.* periclinium, the involucre of the capitulum in Compositae.

périclines, *m.pl.* periclines, periclinal valves.

péricliniforme, in form of periclinium.

périclinoïde, resembling periclinium.

péricliter, to be in danger.

péricorollé, with perigynous corolla.

péricycle, *m.* pericycle, the external layer of the stele.

péricyclogène, (vascular plant, endogenous organs) borne in the pericycle of mother organ.

péridérivé, *m.* peri derivative.

périderme, *m.* periderm.

péridesme, *m.* peridesm.

péridie, *m.* peridium.

péridiniens, *f.pl.* organisms of the plankton.

péridinine, *f.* peridinin, a coloring matter.

péridiscal, -aux, around the disc.

péridium, *m.* peridium.

péridot, *m.* peridot, yellow-green chrysolite.

péridotite, *m.* peridotite.

périer, *m.* tapping bar.

périgée, *m.* perigee.

périgone, *m.* perigone.

périgoniaires, perigoniarious, (flowers) doubled from multiplication of perigone.

périgyne, perigynous.

périhélie, *m.* perihelion.

péril, *m.* peril, danger, risk; **mettre en —,** to imperil, jeopardize.

périlleusement, perilously, dangerously.

périlleux, -euse, perilous, dangerous.

périmer, to lapse.

périméristème, *f.* perimeristem, meristem giving rise to secondary formations.

périmètre, *m.* perimeter, area; **— du couvert,** shaded area.

périnée, *m.* perineum.

perintègre, perinteger, quite entire.

période, *f.* period, cycle, revolution, epoch; *m.* pitch, degree, height, point, stage; **— d'attente,** transition period, regulation period; **— d'égalisation,** transition period; **— de l'accouplement,** flight time, pairing time; **— de prohibition,** closed season; **— d'essaimage,** flight time, pairing time; **— d'une seule chiffre,** simple or single repetend.

periodé, periodo-.

périodicité, *f.* periodicity, rhythm, frequency.

périodique, *m.* periodical.

périodique, periodic(al), intermittent.

périodiquement, periodically.

périople, *m.* periople, coronary frog band.

périoste, *m.* periosteum.

péripétie, *f.* vicissitude, turn of fortune, event, catastrophe.

périphérie, *f.* periphery, circumference.

périphérique, peripheral, on the circumference.

périphoranthe, *m.* periphoranthium, involucre of Compositae.

périphore, *m.* fleshy support to ovary with corolla and stamens attached.

périphorique, adhering to the periphore.

périphylles, *f.pl.* periphylls, hypogynous scales or lodicules of grasses.

périphyses, *m.* periphyses in certain fungi.

périplasme, *m.* periplasm.

périplaste, *f.* periplast.

périplocé, winding around plants.

péripneumonie des bovidés, *f.* cattle pleuropneumonia.

périptéré, fruits and seeds with a winglike membranous border.

périr, to perish, die, lapse, spoil.

périsperme, *m.* perisperm.

périspermique, perispermic.

périsporangium, *m.* perisporangium, an indusium.

périspore, *f.* perispore, spore covering.

périssable, perishable.

péristachyon, *m.* peristachyum, glume of grasses.

péristaltique, peristaltic, vermicular.

péristaminé, (apetalous flower) with perigynous stamens.

péristème, *m.* perianth.

péristome, *m.* peristomium.

péristylique, situated around the style.

périthèce, *f.* perithecium.

périthecium, *m.* perithecium.

péritoine, *m.* peritoneum.

péritonéal, -aux, peritoneal.

péritonite, *f.* peritonitis.

péritrique, peritrichous, (whole surface) beset with cilia.

péritrope, peritropal, peritropous.

périxyle, perixylic, exarch, mesarch.

perlasse, *f.* pearl ash.

perle, *f.* bead, pearl; — **de borax,** borax bead; — **de verre,** glass bead.

perlé, pearled, pearly, beaded, beady, thick (liquids), (of sugar) twice-boiled; **orge —,** pearl barley.

perler, to pearl, adorn, bead, twice boil (sugar), execute to perfection, form beads.

perlier, -ère, pearl-bearing, producing pearls.

perlière, *f.* cudweed (Helichrysum).

perlite, *f.* perlite.

perlitique, perlitic.

permanence, *f.* permanence, permanency, lastingness; **en —,** permanent(ly).

permanent, permanent, standing, constant, enduring.

permanganate, *m.* permanganate.

permanganique, permanganic.

perméabilité, *f.* permeability, leakage, perviousness.

perméable, permeable, pervious.

perméant, (fluid) permeable.

permettre, to permit, allow, let; **se —,** to allow oneself, indulge.

permis, *m.* permit, license; — **d'exploiter,** felling permit; — **de chasse,** shooting license, game license; — **de pêche,** fishing permit; — **de vidange,** permit to remove wood.

permis, permissible, permitted, lawful.

permission, *f.* permission, leave.

permissioner, to authorize, license.

permit (permettre), (he) permitted.

permixtion, *f.* intimate mixture.

permutabilité, *f.* interchangeability, permutability.

permutable, interchangeable, permutable.

permutation, *f.* exchange, interchange, permutation.

permuter, to exchange, permute, interchange.

permutite, *f.* permutite.

pernette, *f.* support (for a sugar mold).

pernicieusement, perniciously.

pernicieux, -euse, pernicious, injurious.

pernion, *m.* chilblain.

pernitrate, *m.* pernitrate.

perocidion, *m.* perichaetium.

Péronosporées, *f.pl.* Peronosporaceae

pérot, *m.* sapling.

Pérou, *m.* Peru.

peroxydase, *f.* peroxidase.

peroxydation, *f.* peroxidation.

peroxyde, *m.* peroxide, — **d'azote,** nitrogen peroxide; — **de plomb,** lead dioxide, lead peroxide; — **d'hydrogène,** hydrogen peroxide.

peroxydé, peroxidized.

peroxyder, to peroxidize.

perpendiculaire, perpendicular.

perpendiculairement, perpendicular(ly), at right angles.

perpétuation, *f.* perpetuation, perpetuance.

perpétuel, -elle, perpetual, everlasting.

perpétuellement, perpetually.

perpétuer, to perpetuate; **se —,** to be perpetuated.

perpétuité, *f.* perpetuity; **à —,** endlessly, forever, for life, perpetually; **en —,** for all time.

perplexe, perplexed, perplexing, embarrassed.

perrier, *m.* quarryman, tapping bar.

perrière, *f.* quarry, slate quarry.

perron, *m.* steps, platform.

perroquet, *m.* parrot.

perrotine, *f.* perrotine (calico printing).

perruche, *f.* hen parrot.

perruthénique, perruthenic.

pers, sea-green, blue-green, gray, greenish blue.

persan, Persian.

perse, *f.* chintz; *cap.* Persia.

persécuter, to persecute.

perséite, *f.* perseite, perseitol.

persel, *m.* persalt.

persévérance, *f.* perseverance.

persévérant, persevering.

persévérer, to persevere.

persicaire, *f.* persicary, lady's-thumb (*Polygonum persicaria*).

persicifolié, with peach leaves.

persicot, *m.* persico, cordial flavored with peach kernels.

persiflage, *m.* derision.

persil, *m.* parsley (*Petroselinum hortense*); — **à grosse racine,** turnip-rooted parsley (*Petroselinum hortense radicosum*); — **des marais,** marsh parsley (*Peucedanum palustre*).

persillé, spotted with green, (cheese) covered with blue mold.

persique, Persian.

persistance, *f.* persistence, persistency.

persistant, persistent, lasting, perennial, indeciduous, durable.

persister, to persist, persevere.

personnalité, *f.* personality, individuality, egoism.

personne, *f.* person; *m.* anyone, anybody; *pl.* people, (with verb) no one, none, nobody.

personné, masked.

personnée, *f.* personate plant.

personnel, *m.* personnel, staff; — **conducteur,** crew.

personnel, -elle, personal, selfish.

personnellement, personally.

perspectif, -ive, perspective.

perspective, *f.* perspective, view.

perspicacité, *f.* perspicacity.

perspiratoire, perspiratory.

persuader, to persuade, convince; **se —,** to convince oneself.

persuasif, -ive, persuasive.

persuasion, *f.* persuasion.

persulfate, *m.* persulphate.

persulfure, *m.* persulphide.

persulfuré, persulphide of.

persulfurique, persulphuric.

perte, *f.* loss, waste, destruction; — **sèche,** sheer loss, absolute loss; **à —,** at a loss; **à — de vue,** out of sight, farther and farther away; **en pure —,** as a total or dead loss.

pertérébrant, lancinating, shooting, grinding (pain).

pertinemment, pertinently.

pertuis, *m.* hole, opening, taphole, drain, strait, narrows.

perturbateur, *m.* disturber.

perturbateur, -trice, disturbing.

perturbation, *f.* perturbation, disturbance, restlessness.

perturber, to disturb, perturb.

pérule, *f.* perula, a leaf-bud scale.

pervenche, *f.* periwinkle (Vinca).

pervers, perverse, wicked.

perversion, *f.* perversion, warp.

pervertir, to pervert.

pervertissement, *m.* perverting.

pesable, weighable.

pesage, *m.* weighing.

pesamment, heavily, weightily.

pesant, *m.* weight.

pesant, heavy, weighty, having weight, ponderous, slow, dull.

pesanteur, *f.* gravity, heaviness, weight, dullness, sluggishness; — spécifique, specific gravity.

pèse, *m.* densimeter, hydrometer.

pèse-acide, *m.* acid hydrometer, acidimeter.

pèse-alcali, *m.* alkali hydrometer, alkalimeter.

pèse-alcool, *m.* alcoholometer.

pèse-colle, *m.* glue hydrometer.

pesée, *f.* weighing, amount weighed at one time, weight, force, leverage, effort, trust; **methode par** —, gravimetric method.

pèse-esprit, *m.* spirit hydrometer.

pèse-éther, *m.* hydrometer for ether.

pése-filtre, *m.* weighing bottle.

pèse-gouttes, *m.* dropper, drop counter.

pèse-lait, *m.* milk hydrometer, lactometer.

pèse-lettre, *m.* postal (letter) scales.

pèse-liqueur, *m.* alcoholometer.

pèse-moût, *m.* must hydrometer, saccharometer.

pèse-papier, *m.* paperweight.

peser, to weigh, ponder, dwell upon, be heavy, have weight, bear down upon, lay stress on; se —, se faire —, to weigh in, be weighed.

pèse-sel, *m.* salt gauge, salinometer.

pèse-sucre, *m.* must hydrometer.

pesette, *f.* assay scales, a small precision balance.

pèse-tube, *m.* weighing-tube support.

peseur, *m.* weigher.

pèse-vin, *m.* wine hydrometer, oenometer.

peson, *m.* balance, scales; — à ressort, spring (balance) scales.

pesse, *f.* mare's-tale (*Hippuris vulgaris*).

pessière, *f.* spruce crop.

peste, *f.* pest, plague, bubonic plague, attack, torment; — bovine, cattle plague; — des eaux, water thyme (*Philotria canadensis*); — des oiseaux, chicken typhus, bird pox (fowls).

pesteux, -euse, pestiferous, of plague.

pétale, *m.* petal.

pétalipare, petalodeus, having petals.

pétaliser, to change into petals, petalize.

pétalite, *f.* petalite.

pétalodé, (double flower) through abnormal transformation of other organs into petals.

pétaloïde, petaloid.

pétalomanie, *f.* petalomania, abnormal multiplication of petals.

pétalostémone, with stamens inserted on the corolla.

pétard, *m.* petard, shot, blast, cartridge, firecracker, fog signal.

pétarder, to blast, blow up.

pétasite, *m.* butterbur (*Petasites officinalis*).

péter, to burst, explode, crackle, pop, crack, break wind.

péterolle, *f.* (fire)cracker.

pétillant, crackling, semisparkling, lively, sprightly.

pétillement, *m.* crackling, sparkling, bubbling.

pétiller, to crackle, fizz, bubble, sparkle, be eager.

pétiolacé, petiolacecus.

pétiolaire, petiolar.

pétiole, *m.* petiole, leafstalk.

pétiolé, petiolate, pedunculate, stalked.

pétiolule, *m.* petiolule, partial petiole.

pétiolulé, petiolulate.

pétioluleux, -euse, with long petiolule.

petit, *m.* little, little one, young (animals), cub, youngling, babe; *m.pl.* the young.

petit, small, little, short, weak, mean, low, petty, slender, thin; — à —, little by little; — autel, flue bridge; — bleu, pneumatic-tube letter; — bois, wood lot, grove, spinney; — bout, top end; — cardamome, cardamom (*Elettaria cardamomum*); — chêne, wall germander (*Teucrium chamaedrys*); — lait tournesolé, litmus whey; — nard, wild sarsaparilla (*Aralia nudicaulis*).

petit, — nom, given name, Christian name; — pois, garden pea, shelling pea; — ravin, gully; —e chasse, small-game shooting; —e eau, weak (solution) liquor; —e éclaire, figwort (*Scropularia marylandica*), pilewort, lesser celandine (*Ficaria verne*); —e oseille, wood sorrel, shamrock; —e vérole, variola, smallpox; en —, on a small scale, in miniature; un — peu, a very little, a little; —s blès, oats and barley; — scolyte, lesser elm-bark beetle.

petit-cheval, *m.* donkey engine.

petit-deux, *m.* sugar loaf (cone) weighing a kilogram.

petitement, little, but little, poorly, meanly, in small quantities.

petite-oie, *f.* giblets.

petitesse, *f.* smallness, littleness, diminutiveness, slenderness, pettiness, meanness.

petit-fils, *m.* grandson, grandchild.

petit-grain, *m.* small immature orange; essence de —, orange-leaf oil, petitgrain oil.

petit-gris, *m.* miniver, squirrel (fur).

pétition, *f.* petition, request; faire une — de principe, to beg the question, petitio principii.

pétitionner, to petition.

petit-lait, *m.* whey, serum of milk.

petit-poivre, *m.* agnus castus, chaste tree (*Vitex agnus-castus*).

pétré, stony.

pétreau, *m.* sucker.

pétri, kneaded, molded.

pétrifiant, petrifying, petrifactive.

pétrification, *f.* petrifaction, calcification.

pétrifier, to petrify, encrust with lime.

pétrin, *m.* kneading trough; — mécanique, kneading machine.

pétrir, to knead, shape, mold, form.

pétrissable, capable of being kneaded or molded.

pétrissage, pétrissement, *m.* kneading, molding, massage.

pétrisseur, *m.* kneader.

pétrisseuse, *f.* kneading machine.

pétrocincle bleu, *m.* blue rock thrush.

pétrographie, *f.* petrography, lithology.

pétrographique, petrographic(al).

pétrolage, *m.* kindling with paraffin or petroleum.

pétrole, *m.* petroleum, mineral oil, gasoline, kerosene; — brut, crude petroleum, crude oil; — d'éclairage, kerosene, illuminating oil; — lampant, kerosene, illuminating oil; — ordinaire, kerosene, illuminating oil.

pétrolène, *m.* petrolene.

pétroler, to kindle or set fire to with petroleum or kerosene, oil (pools) against mosquitoes.

pétrolerie, *f.* petroleum works, oil refinery.

pétrolien, -enne, of petroleum.

pétrolier, *m.* tank (oil) steamer.

pétrolier, -ère, petroleum, of petroleum.

pétrolifère, petroliferous, oil-bearing.

pétrologie, *f.* petrology.

pétrophytes, *f.pl.* petrophyta rock plants.

pétrosilex, *m.* felsite.

pétrosiliceux, -euse, felsitic.

pétulant, petulant, pert, lively.

pétunia, *f.* petunia.

pétunsé, pétunzé, *m.* petuntse, China stone.

peu, *m.* little, bit, a little, few, a few.

peu, little, few, not very, not much, not, not long; — à —, little by little, gradually; — après, soon after, not long after; — au point, little or none; — boisé, sparsely wooded; — de chose, very little, not much, small matter; — d'entre eux, few of them; — exigeant, unexacting, not fastidious.

peu, — importe, no matter; — profond, shallow; à —, little by little, gradually; à— près, almost, nearly, about, pretty nearly; avant —, before long; dans —, soon, presently, shortly; depuis —, recently, lately; d'ici —, shortly, before long; en — de temps, shortly, in a little while.

peu, encore un —, a little more, a few more; homme de —, lowborn man; ni — ni point, none at all; pour — que, if only, if ever, if . . . only a little; pour un —, very nearly; quelque —, somewhat, rather; si — que, however little; si — que rien, the least little bit, a very little.

peucédan, *m.* Peucedanum; — officinal, hog's-fennel (*Peucedanum officinale*).

peuplade, *f.* tribe, people, horde.

peuple, *m.* people, nation, common people.

peuple, vulgar, common.

peuplé, populated, peopled, stocked.

peuplement, *m.* crop, stand, wood, standing crop, stock, forest crop, forest soil, growth; — accessoire, secondary stand, dominated crop; — d'arbres de futaie, high forest, formed tree; — d'abri, shelterwood, nurse crop, parent stand; — dominant, superior stand, dominant crop; — mélangé, mixed crop or stand; — principal, dominant crop; — pur, pure crop or stand; — serré, dense crop, crowded crop; — type, indicating wood, index forest.

peuplement-type, *m.* index forest.

peupler, to people, stock, inhabit, populate, multiply, breed.

peuplier, *m.* poplar (tree and wood) (Populus); — bâtard, — grisaille, gray poplar (*Populus canescens*); — baumier, balsam poplar (*Populus tacamahaca*); — blanc, white poplar, silver poplar (*Populus alba*); — noir, black poplar (*Populus nigra*); — tremble, aspen (*Populus tremula*).

peur, *f.* fear, fright; avoir — de, to be afraid of; de — de, for fear of; de — que, for fear that, lest; **faire — à,** to frighten; **prendre —,** to take fright.

peureux, -euse, fearful.

peut (pouvoir), (he) can, is able; **il n'en — plus,** he cannot stand it any longer; **il n'y — rien,** he can do nothing; **il se —,** it is possible.

peut-être, perhaps, possibly, maybe.

peuvent (pouvoir), (they) can.

pézise, *f.* cup fungus (Peziza).

phaeophylle, *f.* phaeophyll, the coloring matter of brown algae.

phaenologie, *m.* phenology, recording the periodical phenomena of plants.

phaeoplaste, *m.* phaeoplast, chromatophores of Fucoideae.

phagédène, *f.* phagedena, rodent ulcer.

phagédénique, phagedenic.

phagocytaire, phagocytic.

phagocyte, *m.* phagocyte.

phagocyter, to phagocytose.

phagocytisme, *m.* phagocytic action.

phagocytose, *f.* phagocytosis.

phagolyse, phagocytolyse, *f.* phagolysis, phagocytolysis.

phalange, *f.* phalanx.

phalaris, *m.* canary grass (Phalaris).

phalène, *f.* moth, phalaena; **— du pin,** pine looper moth.

phalère, *f.* phalera (moth).

phalloide, phalloid.

phanéranthe, phaneranthous, with manifest flowers.

phanéranthère, phanerantherous, (anthers) protruding beyond the perianth.

phanérocotylèdone, with apparent cotyledons.

phanérogame, *f.* phanerogam, flowering plant.

phanérogame, phanerogamic, phanerogamous.

phanéropore, phaneroporous, (stomata) in the same plane as the epidermis.

phanérostémone, with very distinct stamens.

phare, *m.* lighthouse, beacon, headlight.

pharmaceutique, *f.* pharmaceutics.

pharmaceutique, pharmaceutic(al).

pharmacie, *f.* pharmacy, chemist's and druggist's shop, dispensary, medicine chest.

pharmacien, *m.* pharmacist, chemist, druggist, apothecary.

pharmacodynamique, *f.* pharmocodynamics.

pharmacognosie, *f.* pharmacognosy.

pharmacologie, *f.* pharmacology.

pharmacologique, pharmacologic(al).

pharmacologiste, pharmacologue, *m.* pharmacologist.

pharmacopée, *f.* pharmacopoeia.

pharmacosidérite, *f.* pharmacosiderite, cube ore.

pharmacothèque, *f.* medicine chest.

pharyngien, -enne, pharyngeal.

pharynx, *m.* pharynx.

phase, *f.* phase, stage, aspect, period.

phaséole, *f.* haricot bean (Phaseolus).

phaséoline, *f.* phaseolin.

phaséolunatine, *f.* phaseolunatin.

phasme, *m.* phasma.

phasmidé, *m.* phasmid, specter insect.

phégoptère, growing at the foot of beeches.

phellandre, *f.* water fennel (*Oenanthe phellandrium*).

phellandrène, *m.* phellandrene.

phelloderme, *m.* cork, phelloderm.

phellogène, *m.* phellogen.

phelloide, *m.* phelloid, nonsuberized layers in the phellem.

phellonique, *f.* phellonic acid.

phellosine, *f.* agglomerated cork.

phénacétine, *f.* phenacetin.

phénacite, phénakite, *f.* phenacite.

phénanthrène, *m.* phenanthrene.

phénate, *m.* phenolate, phenate; **— de soude,** sodium phenolate.

phénazine, *f.* phenazine.

phénéthol, *m.* phenetole.

phénétidine, *f.* phenetidine.

phénétol, *m.* phenetole.

phénicine, *f.* indigo purple.

phénique, pertaining to phenol; **acide —,** carbolic acid, phenol.

phéniqué, phenolized, carbolized, carbolic.

phéniquer, to treat or impregnate with phenol, carbolize, carbolate.

phénocarpe, with very apparent fruit.

phénogame, *m.* phanerogam.

phénogamie, *f.* phanerogamy.

phénol, *m.* phenol, carbolic acid.

phénolique, phenolic, pertaining to or containing phenol.

phénologie, *f.* phenology.

phénol-phtaléine, *f.* phenolphthalein.
phénol-sodique, sodium carbolate.
phénol-sulfoné, phenolsulphonic.
phénoménal, -aux, phenomenal, amazing.
phénoménalement, phenomenally.
phénomène, *m.* phenomenon.
phénoménologie, *f.* phenomenology.
phénostémone, with stamens longer than the calyx.
phénotype, *m.* phenotype.
phénylacétamide, *f.* phenylacetamide.
phénylacétique, phenylacetic.
phénylalanine, *f.* phenylalanine.
phénylamide, *f.* phenylamide.
phénylamine, *f.* phenylamine, aniline.
phényle, *m.* phenyl.
phénylène, *m.* phenylene.
phényléthane, *m.* phenylethane.
phénylglvcérine, *f.* phenylglycerol.
phénylglycocolle, *m.* phenylglycocoll.
phénylglycol, *m.* phenylglycol.
phénylglycolique, phenylglycolic.
phénylhydrazine, *f.* phenylhydrazine.
phénylique, phenyl, phenylic.
phénylméthane, *m.* phenylmethane, toluene.
phénylurée, *f.* phenylurea.
phéolépide, with brown scales.
phéophtalme, with brown oscillating stains.
phéopode, with brown stem.
philologue, *m.* philologist.
philosophale, pierre —, philosopher's stone.
philosophe, *m.* philosopher.
philosophe, philosophic(al).
philosopher, to philosophize.
philosuphie, *f.* philosophy.
philosophique, philosophic(al).
phlébite, *f.* phlebitis.
phlébodermé, with veined tegument.
phlébophore, with veins.
phlegme, *m.* phlegm, coolness.
phléogonimique, growing on cortex.
phléoécien, -enne, growing on cortex.
phlobaphène, *m.* phlobaphene.
phloème, *m.* phloem, the inner bark, liber.
phloeoterme, *m.* endoderm.
phlogistication, *f.* phlogistication.
phlogistique, *m.* phlogiston.
phlogistique, phlogistic, inflammatory.

phlogistiquer, to phlogisticate.
phlogopappe, with egret color of fire.
phloorrhétine, *f.* phloretin.
phloridzine, phloorrhizine, *f.* phlorizin.
phloroglucine, *f.* phloroglucin, phloroglucinol.
phloxine, *f.* phloxin.
phlyctène, *f.* phlyctena, pimple, blister, vesicle.
phobisme, *m.* phobism, repulsion of plants.
phocénine, *f.* phocenin.
phoenicé, phoeniceous, scarlet, red with a little yellow added.
phoenicine, *f.* phenicin, phoenicine.
phoenicopyre, with fruits red and pyriform.
phoque, *m.* seal.
phoranthe, *m.* phoranthium.
phormium, phormion, *m.* New Zealand flax (*Phormium tenax*).
phosgène, *m.* phosgene, carbonyl chloride.
phosphame, *n.* phospham.
phosphatage, *m.* fertilizing with phosphates, treating grapes with phosphates.
phosphate, *m.* phosphate; — d'ammoniaque, ammonium phosphate; — de chaux, phosphate of lime, calcium phosphate; — de chaux hydraté, precipitated calcium phosphate; — de fer, iron phosphate; — de potasse, potassium phosphate; — de soude, sodium phosphate.
phosphaté, phosphatic, phosphated, containing phosphate.
phosphatide, *n.* phosphatide.
phosphatique, phosphatic.
phosphaturie, *f.* phosphaturia.
phosphène, *m.* phosphene, firefly.
phosphine, *f.* phosphine.
phosphite, *m.* phosphite.
phosphoglycérate, *m.* phosphoglycerate, glycerophosphate.
phosphomolybdique, phosphomolybdic.
phosphoprotéides, *m.* nucleoproteids.
phosphore, *m.* phosphorus, phosphor; — amorphe, amorphous phosphorus; — blanc, white phosphorus; — rouge, red phosphorus.
phosphoré, phosphorous, containing phosphorus.
phosphorer, to phosphorate, phosphorize, treat or combine with phosphorus.
phosphorescence, *f.* phosphorescence.
phosphorescent, phosphorescent.

phosphoreux, -euse, phosphorous.

phosphoride, *m.* phosphide.

phosphorique, phosphoric, phosphorus.

phosphoriser, to phosphorize, phosphorate.

phosphorisme, *m.* phosphorism.

phosphorite, *f.* phosphorite, fibrous apatite, phosphate rock.

phosphoritique, phosphoritic.

phosphorogénique, phosphorogenic.

phosphoroscope, *m.* phosphoroscope.

phosphure, *m.* phosphide.

phosphuré, containing phosphide, phosphoreted.

photauxisme, *m.* influence of total radiation on speed of growth of stem.

photoblaste, *m.* photoblast, shoot adapted to live in light and air.

photocéramique, *f.* photoceramics.

photochimie, *f.* photochemistry.

photochimique, photochemical.

photochromie, *f.* photochromy, color photography.

photochromogravure, *f.* photographic color printing.

photocleistogamie, *f.* photocleistogamy, pseudocleistogamy due to lack of light.

photocopie, *f.* photographic reproduction, proof or print.

photo-électrique, photoelectric(al).

photoépinastie, *f.* photoepinasty, epinasty induced by the action of light.

photogène, *m.* photogen.

photogène, photogenic.

photogénèse, photogénie, *f.* production of light, photogenesis.

photogénétique, photogenic.

photogénique, actinic, photogenic.

photoglyptie, *f.* woodburytype, photogravure.

photogramme, *m.* photographic print.

photogrammétrie, *f.* photogrammetry.

photographe, *m.* photographer.

photographie, *f.* photography, photograph, picture.

photographier, to photograph.

photographique, photographic.

photograveur, *m.* photoengraver, process engraver.

photogravure, *f.* photoengraving; — **en relief,** photoengraving in relief, heliography.

photokinèse, *f.* photokinesis, movement induced by light.

photolithographie, *f.* photolithography.

photomécanique, photomechanical.

photomètre, *m.* photometer; — à **éclats,** flicker photometer.

photométrie, *f.* photometry.

photométrique, photometric(al).

photomicrographie, *f.* photomicrography.

photomicrographique, photomicrographic.

photopeinture, *f.* coloring of photographs.

photophile, heliophilous, adapted to full exposure to the sun.

photophore, *m.* photophore.

photoplastographie, *f.* photoplastography, woodburytype.

photopoudre, *f.* flashlight powder.

photosphère, *f.* photosphere.

photosynthèse, *f.* photosynthesis.

phototactique, phototactic.

phototactisme, *m.* phototactism, phototaxis.

phototaxie, *f.* phototaxis, phototaxy.

photothérapie, *f.* phototherapy.

photothérapique, phototherapeutic.

phototopographie, *f.* phototopography.

phototopographique, phototopographic.

phototortisme, *m.* tortion due to action of light.

phototrophie, *f.* phototrophy, trophy due to action of light.

phototropisme, *m.* phototropism.

phototypie, *f.* phototypy, phototype.

phototypographie, *f.* phototypography, halftone reproduction.

photozincographie, *f.* photozincography.

phragme, *m.* phragma, spurious dissepiment in fruits.

phragmigère, phragmiger, divided by partitions.

phragmite, *m.* reed; — **des joncs,** sedge warbler.

phragmoplaste, *m.* phragmoplast, connecting spindle between two nuclei in the same cell.

phrase, *f.* phrase, sentence.

phrénésie, *f.* frenzy, delirium.

phtalate, *m.* phthalate.

pntaléine, *f.* phthalein; — **du phénol,** phenolphthalein.

phtalimide, *f.* phthalimide.

phtaline, *f.* phthaline.

phtalique, phthalic.

phtalyle, *m.* phthalyl.

phtanite, *f.* phthanite, chert.

phtartique, deleterious, deadly, deleterion.

phtiriasis, *f.* lousiness, pediculosis.

phtisie, *f.* phthisis.

phycoblastème, *m.* microgonidium, a small gonidium.

phycochrome, *m.* phycochrome.

phycocyanine, *m.* phycocyanin, blue coloring matter in algae.

phycoèrythrine, *m.* phycoerythrin, red pigment of Floridean algae.

phycologie, *m.* phycology, department of botany that relates to algae.

phycophèine, *m.* phycophaein.

phycostème, *m.* nectary.

phycoxanthine, *m.* phycoxanthin, diatomin, yellowish-brown pigment of algae.

phyle, *m.* phylum.

phyllamphore, (leaf) having an amphore.

phyllastrophyte, with verticillate leaves.

phylle, *f.* sepal.

phyllidium, *m.* phyllidium, homologue of the leaf in the gametophyte.

phylloblaste, *m.* phylloblastus.

phyllocéphale, with leafy capitulum.

phylloclade, *f.* phylloclade, a green flattened or rounded stem functioning as a leaf.

phylloclade, *m.* celery-topped pine, celery pine (Phyllocladus).

phyllocyanine, *f.* phyllocyanin.

phyllode, *m.* phyllodium.

phyllodé, resembling a leaf.

phylloderme, phyllodermé, (fruit-bearing membrane) folded in form of leaf.

phyllodial, (cup) of an ascideous phyllode.

phyllodie, *f.* phyllody, metamorphosis of floral organs into leaves.

phyllodiné, phyllodineous, with phyllodes in place of leaves.

phyllogénie, *f.* phylogenesis.

phyllogonie, *f.* theory of leaf production.

phylloïde, leaflike.

phylloïdé, phyllolichéné, affecting the form of foliaceous expansions.

phyllolobé, with foliaceous cotyledons.

phyllomanie, *f.* phyllomania, abnormal production of leaves.

phyllome, *m.* phyllome.

phyllomorphie, *f.* phyllody, metamorphosis of floral organs into leaves.

phyllophage, phyllophagous.

phyllophile, growing among leaves.

phyllopode, *m.* phyllopode.

phyllopode, phyllopodous.

phyllopodium, *m.* phyllopodium.

phylloporphyrine, *f.* phylloporphyrin.

phyllotaonine, *f.* phyllotaonin, product of chlorophyll, dull green in tint.

phyllotaxie, *f.* phyllotaxy.

phylloxanthine, *f.* phylloxanthin; — de la vigne, vine fretter.

phylloxéra, *m.* Phylloxera.

phylloxéré, attacked by the Phylloxera.

phylloxérien, -enne, phylloxérique, pertaining to Phylloxera.

phylule, *f.* phyllula.

phylogénèse, *f.* phylogenesis, phylogeny.

phylogénie, *f.* phylogeny.

phylotaxie, *f.* phyllotaxis, posture of the leaves.

phylum, *m.* branching.

physaline, *f.* physalin.

physalite, *f.* physalite, coarse intumescent topaz.

physicien, *m.* physicist, natural philosopher.

physico-chimie, *f.* physical chemistry.

physico-chimique, physicochemical.

physico-mathématique, physicomathematical.

physico-mécanique, physicomechanical.

physiographe, *m.* physiographer.

physiographie, *f.* physiography.

physiologie, *f.* physiology.

physiologique, physiologic(al).

physiologiquement, physiologically.

physiologiste, physiologue, *m.* physiologist.

physionomie, *f.* physiognomy, countenance, looks.

physionomique, physiognomical.

physique, *f.* physics, natural philosophy; *m.* physique (of person).

physique, physical.

physiquement, physically.

physocarpe, with enlarged or vesiculose fruits.

physodes, *f.pl.* physodes, vesicles in algae containing assimilation products.

physostigma, *m.* Calabar bean.

phytamie, *f.* growth of plants.

phytine, *f.* phytin, phosporous compound derived from seeds.

phytobiologie, *f.* plant biology.

phytocécidie, *f.* cecidium, galls produced by fungi or insects.

phytochimie, *f.* phytochemistry, plant chemistry.

phytodomatie, *f.* phytodomatia, shelters in which other plants live.

phytogamie, *f.* phytogamy, cross-fertilization of flowers.

phytogène, phytogenic, phytogenous.

phytogénésie, *f.* phytogenesis, origin and development of the plant.

phytogéographie, *f.* phytogeography, science of plant distribution.

phytognomie, *f.* phytognomy.

phytographie, *f.* descriptive botany.

phytol, *m.* phytol.

phytolaccine, *f.* phytolaccin.

phytolaque, *f.* common pokeweed, redweed, (*Phytolacca decandra*).

phytologie, *f.* phytology, botany.

phytome, *m.* phytoma, vegetative substance of all plants.

phytomélane, *f.* phytomelane, a black structureless layer in the pericarp of Compositae.

phyton, *m.* phyton.

phytonymphie, *f.* first appearance of the flower.

phytopaléontologie, *f.* plant paleontology, paleobotany.

phytopathogène, phytopathogenic.

phytopathologie, *f.* phytopathology, diseases of plants.

phytophages, *m.pl.* Phytophaga (plant eater).

phytoplankton, *m.* phytoplankton, floating pelagic plant organisms.

phytopte, *m.* rust mite; — du poirier, pearleaf blister mite, rust mite.

phytosterine, *f.* phytosterin, phytosterol.

phytotocées, *f.pl.* modifications of pistil becoming a fruit.

phytotomie, *f.* phytotomy, plant anatomy or histology.

pian, *m.* frambesia, yaws.

pianiste, *m.* pianist.

piano, *m.* piano.

piastre, *f.* piaster.

pic, *m.* pick, pickax, poker, (mountain) peak, woodpecker; — à feu, poker; — épeiche, great spotted woodpecker; — épeichette, lesser spotted woodpecker; — vert, green woodpecker; à —, perpendicularly, straight up.

picadil, *m.* hearth glass, highly viscous glass.

picadon, *m.* container in which soda is broken up (for soap).

picamare, *f.* picamar.

picéine, *f.* picein.

picène, *m.* picene.

piciforme, piceous, viscous.

pickeler, to pickle.

picnomètre, *m.* pycnometer.

picoïde, *f.* woodpecker; — tridactyle, three-toed woodpecker.

picoline, *f.* picoline.

picolique, picolinic.

picotement, *m.* pricking, tingling, itching, smarting.

picoter, to prick, cause to tingle, smart, tingle, sting, peck (at fruit).

picramique, picramic (acid).

picrate, *m.* picrate.

picraté, picrated, containing picrate or picric acid.

picride, *f.* oxtongue (*Picris echioides*).

picrique, picric.

picrocarmin, *m.* picrocarmine.

picrol, *m.* picrol.

picromel, *m.* picromel.

picrotine, *f.* picrotin.

picrotoxine, *f.* picrotoxin.

picvert, *m.* see PIVERT.

pie, *f.* magpie.

pie, piebald; — blanc, white and black; — noir, black and white; — rouge, red and white.

pièce, *f.* piece, part, room, apartment, each, head (of cattle), cask, barrel (of wine), fragment, bit, document, tip, casting, bloom, work, lobe; — à —, bit by bit, piecemeal; — de blé, cornfield; — de bois, wood in the round; — (de) coulée, casting; — de frottement, friction piece, galling leather; — de longueur, long log; — déplaçable, slide part; — de rechange, spare part.

pièce, — de terre, strip, plot, piece of land; — dorsale, galea exterior; — héritée, heirloom; — intercalaire, intermediate or connecting piece; — intermaxillaire, lacinea interior; — manquée, spoiled casting; — moulée, casting; — plate, flat part; — profilée, shaped piece; à la —, singly; mettre en —s, to break or tear into bits or fragments.

pied, *m.* foot, base, leg (of furniture), seed piece, hoof, stem, stalk, plant, slope, foothold, footing, track, footprint, heel, metrical foot, ground tripod; — à coulisse, sliding caliper; — à —, step by step; — bornier, boundary tree; — de boeuf, neat's foot.

pied, — de chèvre, goutweed, ashweed (*Aegopodium podagraria*); — de coq, cocksfoot, orchard grass (*Dactylis glomerata*); — de griffon, bear's-foot, setterwort, fetid hellebore (*Helleborus foetidus*); — de lièvre, rabbit-foot clover (*Trifolium arvense*); — de lion, lion's-foot, edelweiss (*Leontopodium alpinum*); — de Madura, Madura foot; — de travers, wry hoof; — d'oiseau, serradella, bird's-foot (*Ornithopus sativus*).

pied, — encastelé, contracted hoof; — palmé, webfoot, palmated foot; —s fouisseux, burrowing legs; à —, on foot, walking; de — ferme, firmly, resolutely; de plain —, on the same level, floor; sur —, on root, afoot, up and about, well, standing (crops), (beef) on the hoof.

pied-d'alouette, *m.* larkspur (Delphinium).

pied-de-chat, *m.* cat's paw, everlasting cudweed (*Gnaphalium dioicum*).

pied-de-chèvre, *m.* crowbar, claw bar, spike drawer.

pied-droit, *m.* side wall, pier, pillar.

piédestal, *m.* pedestal.

pied-paroi, *f.* border tree.

piège, *m.* trap, snare.

pie-grièche, *f.* shrike, butcherbird; — commun, magpie.

pie-mère, *f.* pia mater.

piéride, *f.* butterfly; — gazée, black-veined white butterfly.

pierraille, *f.* broken stones, rubble, gravel.

pierre, *f.* stone, calculus, gem, boulder; — à aiguiser, whetstone, grindstone, oilstone, rubber, hone; — à bâtir, building stone; — à briquets, flint; — à brunir, burnishing stone; — à cautère, potassium hydroxide, caustic stick; — à chaperon, coping stone, cover plate; — à chaux, limestone; — à feu, flint, firestone; — à filtrer, filtering stone; — à fusil, gunflint.

pierre, — à paver, paving stone; — à plâtre, gypsum, plaster stone; — à repasser, hone, oilstone; — calcaire, limestone; —

d'achoppement, stumbling block; — d'aimant, loadstone; — de lard, soapstone, steatite; — de lune, moonstone, adularia; — de mine, ore-bearing rock, ore; — de savon, soapstone; — de taille, freestone ashlar, hewn stone, cut stone; — de touche, touchstone.

pierre, — de vin, tartar, wine stone; — factice, — fausse, imitation (artificial) gem or stone; — fine, precious stone, jewel; — fondamentale, foundation stone; — infernale, infernal stone, lunar caustic; — meulière, millstone; — ponce, pumice stone, pumice; — précieuse, precious stone, jewel; — réfractaire, firestone, fireproof stone; —s roulées, boulder or rubble stones, pebblestone; —s stratifiées, stratified stones or rocks.

pierreries, *f.pl.* precious stones, gems.

pierrette, *f.* small stone, pebble.

pierreux, -euse, stony, of stone, saxatile.

piétin, *m.* (contagious) foot rot (of sheep).

piétinement, *m.* tramping, stamping.

piétiner, to tramp, trample, stamp, tread, crush with the feet.

piéton, *m.* foot passenger, pedestrian.

pieu, *m.* stake, pile, post, mast, picket.

pieusement, piously.

pieux, -euse, pious, devout.

piézochimique, piezochemical.

piézoélectricité, *f.* piezoelectricity.

piézoélectrique, piezoelectric.

piézomètre, *m.* piezometer.

piézotropisme, *m.* piezotropism, movement by compression acting as stimulus.

pigamon, *m.* meadow rue (Thalictrum).

pigeon, *m.* pigeon, dove, thick builder's plaster, trowelful, hard lump, nodule (in lime); — ramier, ringdove, wood pigeon.

pigeonnier, *m.* dovecote, pigeon house.

pigment, *m.* pigment; — biliaire, bile pigment.

pigmentaire, pigmentary, pigmented.

pigmentation, *f.* pigmentation.

pigmenté, pigmented.

pigmenteux, -euse, pigmentary.

pigmentogène, pigment-forming.

pignadas, *f.pl.* pignadas, crop of maritime pines.

pignon, *m.* pignon, gable, pinion, gear, pine nut; — d'Inde, physic nut, croton seed.

pilage, *m.* pounding, crushing, grinding.

pile, *f.* pile, heap, reverse, tail (of coins), stack (of wood), pillar, cell, battery, beating trough, stamping trough (for sugar), bed (of powder mill), pulping machine, beater, pier (of bridge), mole, vat; — **à ballon,** cell with a balloon; — **à charbon,** carbon cell; — **à immersion,** plunge (immersion) battery; — **à liquide,** fluid (cell) battery; — **au bichromate,** bichromate cell; — **à un seul liquide,** single-fluid cell.

pile, — **de bois,** woodpile, stack of wood; — **défileuse,** washing engine, washer; — **de polarisation,** storage battery, polarization cell; — **étalon,** normal cell; — **sèche,** dry cell, dry battery; — **secondaire,** secondary battery, storage battery, accumulator; — **thermo-électrique,** thermoelectric pile, thermopile, thermoelectric battery; **croix ou —,** head or tail.

pilé, crushed, ground.

pileaté, pileate, having the form of a cap.

piléiforme, pileiform, pileus-shaped.

piléolé, provided with a pileola.

piléorhize, pilorhize, *f.* pileorhiza, rootcap.

piler, to pound, crush, bruise, grind, bray, powder, pestle, beat.

pilet, *m.* pintail, pintailed duck.

pileur, *m.* pounder, grinder.

pileux, -euse, pilose, pilous, hairy.

pilier, *m.* pillar, shaft, pier, column, post.

pilifère, piliferous.

pillard, *m.* pillager, plunderer.

piller, to plunder, pillage.

pilocarpe, with fruit covered with hair.

pilocarpidine, *f.* pilocarpidine.

pilocarpine, *f.* pilocarpine.

pilon, *m.* pestle, tamper, beater, stamp, crusher, rammer, (shingling) hammer.

pilonnage, *m.* ramming, tamping.

pilonner, piloner, to ram, tamp, pound, pulp, beat, stamp.

piloselle, *f.* mouse-ear hawkweed (*Hieracium pilosella*).

pilosisme, *m.* pilosism, abnormal hairiness in plants.

pilosiuscule, pilosiusculous, slightly hairy.

pilot, *m.* pile (stake), heap of salt.

pilotage, *m.* piloting, pile driving.

pilote, *m.* pilot.

piloter, to pilot, guide, drive piles into.

pilotis, *m.* pilework, piling, set of piles.

pilulaire, *f.* pillwort.

pilulaire, pilular, like pills.

pilule, *f.* pill; — **de mercure,** —**s bleues,** —**s mercurielles simples,** mass of mercury, blue mass, mercury pill.

pilulier, *m.* pill machine.

pilulifère, with fruits united in single globular mass.

piluliflore, with flowers united in rounded capitula.

pimarique, pimaric (acid).

pimélique, pimelic (acid).

piment, *m.* pepper, red pepper, pimento, capsicum; — **annuel,** Spanish pepper (*Capsicum frutescens*); — **de la Jamaique,** — **des Anglais,** — **poivre,** allspice, pimento (*Pimenta officinalis*); — **rouge,** red pepper.

pimenta, *f.* Pimenta.

pimprenelle, *f.* burnet (Sanguisorba); — **aquatique,** water pimpernel, brookweed (*Samolus valerandi*).

pin, *m.* pine (tree and wood) (Pinus), Scotch pine (*Pinus sylvestris*); — **à résine,** — **de poix,** pitch pine, resin pine; — **à trochets,** rigid pine (*Pinus rigida*); — **blanc,** — Weymouth, white pine, Weymouth pine (*Pinus strobus*); — **cembro,** — **pinier,** cembra pine, Swiss pine, stone pine (*Pinus cembra*); — **chétif,** dwarf pine; — **Corse,** Corsican pine (*Pinus nigra* var. *calabrica*).

pin, — **d'Alep,** Aleppo pine (*Pinus halepensis*); — **de Banks,** Banks' pine, jack pine (*Pinus banksiana*); — **de mature,** — **rouge,** Norway pine, red pine (*Pinus resinosa*); — **des Landes,** — **maritime,** maritime pine, cluster pine (*Pinus pinaster*); — **jaune,** yellow pine; — **noir,** black pine, Austrian pine (*Pinus nigra*); — **sylvestre,** (common) pine, Scotch pine, Scotch fir (*Pinus sylvestris*).

pinacle, *m.* pinnacle.

pinacoline, *f.* pinacolin.

pinacone, *f.* pinacone, pinacol.

pinasse, *m.* pine.

pinastre, *m.* maritime pine, cluster pine (*Pinus pinaster*).

pinçade, *f.* pinching.

pinçage, *m.* pinching off, nipping off, nip.

pince, *f.* pincers, pliers, nippers, claw, tongs, forceps, tweezers, clasp, clamp, pinchcock, clip, crowbar, grip, hold, incisor, front tooth; — **à bourre,** cotton-plug forceps; — **à cordages,** rope catch; — **à coupelles,** cupel forceps; — **à crampons,** spike tongs; — **à creusets,** crucible tongs; — **à dissection,** forceps.

pince, — à mâchoires, pincers, pliers; — à ressort, spring pinchcock, spring clamp; — à vis, screw pinchcock, screw clamp; — coupante, cutting pliers, nippers, cutter; — en bois, wooden clamp; — plate, flat-nosed pliers, flat-bit tongs; — pour éprouvettes, test-tube holder; — pour porte-objets, slide forceps; — ronde, roundnosed pliers.

pinceau, m. brush, painter's pencil, hair pencil, fasciculus, bundle.

pincée, f. pinch (of salt).

pincement, m. pinching, nipping, feeling of constriction.

pincer, to pinch, nip, hold fast, grip, squeeze, catch, nip off, top (plant).

pincette, f. light pinch, nip, (small) pincers, nippers, tweezers, pliers, tongs.

pinchbeck, m. pinchbeck (alloy of copper and zinc).

pinçon, m. clip, mark, bruise.

pinçure, f. crease, wrinkle, pinching.

pinéal, -aux, pineal.

pineau, m. a black grape or wine (of Burgundy).

pinède, f. pine land.

pinène, m. pinene.

pineraie, f. pine wood, stand of pines.

pingouin, m. penguin; — macroptère, razorbill.

pinifolié, like pine leaves.

pinipicrine, f. pinipicrin.

pinique, pinic.

pinite, f. pinitol, pinite.

pinnatifide, pinnatifid.

pinnatilobé, pinnatilobed.

pinnatiséqué, pinnatisect.

pinné, pinnate.

pinninervé, pinninerved, pinnately nerved.

pinnule, f. sight vane, pinnule.

pinoïde, acicular, slender or needle-shaped.

pinson, m. finch, chaffinch.

pinta, f. pinta.

piochage, m. digging, breaking (of ground); — superficiel, scratching or wounding the soil; — des montagnes, brambling.

pioche, f. pickax, mattock, hoe.

piochement, m. digging.

piocher, to dig, peck, pick, hoe.

pionnier, m. pioneer.

pipe, f. pipe, tube, pipette, large cask.

pipe-de-tabac, Dutchman's-pipe (Aristolochia sipho).

pipeau, m. bird call.

piper, to lure by whistling.

pipéracées, f.pl. Piperaceae.

pipérazine, f. piperazine.

pipéridine, f. piperidine.

pipéridone, f. piperidone.

pipérine, f. piperine.

pipérique, piperic.

pipéroïde, m. piperoid; — de gingembre, piperoid of ginger.

pipéronylique, piperonylic.

pipette, f. pipette, glass tube; — à boule, bulb pipette; — compte-gouttes, dropping pipette, dropper; — divisée, graduated pipette; — effilée, pipette with drawn-out point; — étalon, standard pipette, normal pipette; — jaugée, calibrated pipette.

pipette-mélangeur, m. mixing pipette.

pipit, m. pipit; — des arbres, tree pipet; — des prés, meadow pipit; — rousseline, tawny pipit.

piptospore, m. basidiospore disseminating by falling on soil.

piquage, m. streak.

piquant, m. prickle, thorn sting, prick, spine, piquancy, sharpness, pungency, (of lye) causticity.

piquant, prickling, stinging, piquant, sharp, tart, biting, prickly, pungent, strong.

piqué, m. piqué, quilting.

piqué, pricked, worm-eaten, larded (meat), punky, dozy, beginning to rot, sour (wine), decayed.

pique-feu, m. poker.

piquer, to prick, sting, stab, prickle, spur (horse), goad (oxen), bite, sour (wine), puncture, pierce, poke (fire), scale off (boiler), dig, stick, insert, thrust in, gnaw, eat into, spot, pit, mold, quilt, lard, pique, excite, stimulate, stir; se —, to prick, sting oneself, turn sour, be strong, get excited, pride oneself (on), become worm-eaten, moth-eaten, become spotted with rust, dust, mold.

piquet, m. peg, stake, pin, picket, tapering cane, lay stool.

piqueter, to stake out, mark out, spot, dot.

piquette, f. piquette, wine of poor quality.

piqueur, m. overseer, stitcher, pricker, biting insect.

piqûre, f. prick, pricking, puncture, sting, pitting, hypodermic injection, pit, worm hole, moth hole, spot, speck, perforation, hole, quilting, stab culture; — **profonde,** deep stab culture.

piraterie, *f.* piracy.

pire, *m.* worst.

pire, worse, worst.

piriforme, pear-shaped, pyriform.

pirogue, *f.* canoe.

pirole, *f.* wintergreen (Pyrola).

piroplasme, *m.* piroplasma.

pis, *m.* worst, dug, udder, teat.

pis, worse, worst; **aller de mal en —,** to go from bad to worse; **de — en —,** worse and worse; **et qui — est . . . ,** and what is worse. . . .

pis-aller, *m.* last resource, make shift.

pisciculteur, *m.* pisciculturist.

pisciculture, *f.* pisciculture.

pisé, *m.* lining material, lining puddled clay.

pisifère, having peas.

pisolithe, *f.* pisolite.

pisolithique, pisolitic.

pissasphalte, *m.* pissasphalt, maltha.

pissat, *m.* urine (of animals).

pissée, *f.* slag duct, slag channel, urine.

pissement, *m.* urination.

pissenlit, *m.* dandelion (*Taraxacum officinale*).

pissette, *f.* washing bottle, wash bottle.

pisseux, -euse, of or resembling urine.

pissode, *m.* small pine weevil.

pissote, *f.* escape pipe, waste pipe (of wood).

pistache, *f.* pistachio (nut).

— **de terre,** peanut, earthnut.

pistachier, *m.* pistachio tree (*Pistacia vera*).

piste, *f.* track, trail, trace, footprint.

pistil, *m.* pistil, gynoecium.

pistillaire, (vessels) going from stigma to ovary.

pistillé, pistillate.

pistillidie, *f.* pistillidium, archegonium, organ analogous to pistil.

pistillidium, *m.* pistillidium, archegonium.

pistillifère, pistilliferous, pistillate.

pistilliforme, in form of a pistil.

pistillipare, (flower organs) changed into pistils.

pistillistémone, (plant) male organ rests on female organ.

pistolet, *m.* pistol.

piston, *m.* piston.

pitchpin, *m.* pitch pine, (long-leaf) yellow pine (*Pinus palustris*).

pite, *f.* pita, agave, American aloe (*Agave americana*); *m.* pita, pita fiber, hemp, or flax.

piteusement, piteously, sadly.

piteux, -euse, pitiable, woeful.

pitié, *f.* pity, compassion.

piton, *m.* eyebolt, screw ring, screw eye, peak.

pitoyable, pitiful, pitiable, lamentable.

pitoyablement, pitifully, woefully.

pittacal, *m.* pittacal.

pitte, *f.* see PITE.

pittoresque, picturesque, pictorial, quaint.

pittoresquement, picturesquely, quaintly.

pituitaire, pituitary (gland).

pituite, *f.* pituite, phlegm, mucus, gastric catarrh.

pivert, *m.* green woodpecker.

pivoine, *f.* peony (Paeonia); *m.* bullfinch.

pivot, *m.* pivot, pin, bearing, axis, stud, taproot, main root.

pivotant, pivoting, taprooted.

pivoter, to pivot, turn (on a pivot), have a taproot, form taproots.

placage, *m.* plating (of metals), patchwork; **bois de —,** veneering wood.

plaçage, *m.* placing, selling, allotment.

placard, *m.* cupboard, poster, bill, placard, galley proof.

place, *f.* place, seat, room, position, ground, public square, locality, spot, market, merchants, figure, digit; — **à charbon,** charring place, kiln site; **à la — de,** instead; **à quatre —s,** four passenger; — **d'essai,** sample plot; **faire — à,** to make a place, make room, give way; **la deux —s,** light roadster; **sur —,** on the spot.

placement, *m.* placing, investment, sale.

placenta, *m.* placenta.

placentaire, placental.

placentarien, -enne, giving rise to placenta.

placentation, *f.* placentation.

placentature, *f.* placentation.

placente, *f.* placenta.

placentifère, placentiferous, bearing placentae.

placentiforme, placentiform, quoit-shaped.

placentoïde, *f.* placentoid, organ occurring in the anthers of certain dicotyledons.

placer, to place, put, set, invest, sell, dispose, plant; **se —,** to be placed, place oneself, be put, obtain a position, be invested, sell (goods).

placoplaste, *f.* placoplast.

plafond, *m.* ceiling, crown, top, bottom (of a canal).

plage, *f.* beach, strand, shore, region.

plagièdre, plagihedral.

plagioclase, *m.* plagioclase.

plagiodrome, plagiodromous, with tertiary leaf veins at right angles to the secondary veins.

plagionite, *f.* plagionite.

plagiopode, with oblique or arched peduncle.

plagiotrope, plagiotropic, the direction of growth oblique or horizontal.

plaider, to plead, argue, sue, litigate.

plaidoyer, *m.* address to the court, speech for the defense.

plaie, *f.* wound, sore, plague, hurt, injury, scar, ulcer.

plaignant, *m.* plaintiff, prosecutor, complainant.

plaignant (plaindre), complaining.

plain, level, flat, even, plane, plain, open, flush.

plaindre, to pity, complain of, grudge; **se —,** to complain, moan.

plaine, *f.* plain; **— d'alluvion,** flood plain; **— élevée,** plateau, elevated plain.

plainer, to unhair (leather with lime).

plain-pied, *m.* flat, suite of rooms on one floor; **de —,** on the same level, smoothly, on same floor or course.

plainte, *f.* complaint, plaint, moan, groan.

plaintif, -ive, plaintive, doleful.

plaire, to be pleasing, be agreeable, please; **se —,** to be pleased (with), be happy, take pleasure (in), delight in; **— à,** to please, take pleasure in; **s'il vous plaît,** (if you) please.

plaisamment, pleasantly, amusingly.

plaisant, *m.* fun, joker.

plaisant, pleasing, pleasant, ludicrous. funny, agreeable, humorous.

plaisanter, to joke.

plaisent (plaire), (they) are pleasing.

plaisianthé, diclinous (plant), with flowers having calyx and corolla.

plaisir, *m.* pleasure, delight.

plaît (plaire), (he) is pleasing; **cela me —,** I am pleased, I like it.

plamage, *m.* unhairing skins with lime.

plamer, to unhair skins with lime.

plan, *m.* plane, plane (flat) surface, plateplain, plan, scheme, map, plot, design, ground, (surface) project, drawing; — **d'aménagement,** working plan; — **de construction,** working drawing; — **de mise au point,** focusing plane; — **des courbes de niveau,** contour map; — **d'études,** curriculum.

plan, — **du foyer,** focal plane;— **général d'exploitation,** working plan, report; — **incliné,** inclined plane, incline;— **inférieur,** lower wing; — **parcellaire,** management map; — **spécial d'exploitation,** detailed scheme of working; **au second —** upstage, in the background; **premier —** foreground.

plan, plane, plain, smooth, level.

planage, *m.* smoothing, planing.

planche, *f.* board, plank, sheet (of metal), plate, block (for printing), cut, plate, bed, border, land (between water furrows), flower bed, nursery lines; — **alignée,** squared board; — **épaisse,** plank.

planchéier, to floor (a room), board, plank.

plancher, *m.* floor, flooring, platform, ceiling.

planchette, *f.* (small) board, plane table.

plançon, *m.* big slip, sapling, twig; — **marine,** ship timber.

plan-concave, plano-concave.

plan-convexe, plano-convex.

plancton, *m.* plankton.

plane, *f.* drawing knife, paring knife, turning chisel.

plane, *m.* Norway maple (*Acer platanoides*); **faux —,** sycamore maple (*Acer pseudoplatanus*).

plané, *m.* rolled gold, gold casing.

planer, to make smooth, plane, planish (metals), shave (wood), soar, float, hover, look down, glide.

planétaire, planetary.

planète, *f.* planet.

planeur, *m.* glider.

planimétrie, *f.* planimetry.

planimétrique, planimetric.

planisiliqué, with flat silique.

planitique, inhabiting the plains.

plankton, *f.* see PLANCTON.

planogamète, *f.* planogamete.

plant, *m.* young plant, set, twig, plant. plantation, grove, bed, planting; — **d'un**

an, yearling plant; — **en motte,** ball plant; — **pris en forêt,** wild seedling, wildling; — **récépé,** truncated plant; — **repiqué,** transplant; — **sauvage,** plant taken from woods, wildling.

plantage, *m.* planting, plantation.

plantain, *m.* plantain (Plantago); — **d'eau,** water plantain (Alisma).

plantaire, plantar.

plantanier, *m.* plantain (of banana type).

plantard, *m.* sapling, slip, twig.

plantation, *f.* plantation, planting, raising (of plants), breeding (under glass); — **ceinture,** belt planting; —**coordonnée,** regular planting; — **d'arbres,** protective planting; — **en butte,** mound planting; — **en carré,** square planting; — **en fente,** planting in notches, notching.

plantation, — **en lignes,** planting in lines; — **en mottes,** ball planting; — **en quinconce,** planting in quincunx; — **en sillons,** furrow planting; — **par pieds isolés,** planting by single plants; — **par touffes,** bush planting, multiple planting; — **sur buttes,** planting on mounds; — **sur les dunes,** protection of dunes by planting.

plante, *f.* plant, sole (of the foot); — **adventice,** casual plant; — **alimentaire,** food plant; — **alpine,** rock plant, alpine plant; — **aquatique,** hydrophyte, aquatic plant; — **à tige grimpante,** climber; — **à tige volubile,** creeper, creeping plant; — **à vrille,** ranking plant; — **caractéristique du sol,** plant restricted to certain soils; — **condimentaire,** potherbs.

plante, — **de suspension,** hanging plant; — **exigeante,** exacting plant; — **fourragère,** forage plant; — **grimpante,** climbing plant, climber, creeper; — **herbacée,** herbaceous plant, herb; — **horticole,** garden plant; — **ligneuse,** woody plant; — **naturalisée,** prepared plant; — **nuisible,** destructive weed, noxious weed; — **ombrophyle,** shade-loving plant.

plante, — **peu exigeante,** nonexacting plant; — **potagère,** vegetable, potherb, kitchen plant; — **préférente du sol,** plant partial to certain soil; — **rampante,** trailing plant; — **sarclée,** set free by weeding; — **sarmenteuse,** climbing plant, liana; — **sociale,** gregarious plant, social plant; — **sousligneuse,** undershrub; — **vivace,** perennial plant, perennial.

planté, planted, situated, placed, erect, standing.

planter, to plant, set (flowers), set up, place, drive.

planteur, *m.* planter.

plantoir, *m.* planting peg, dibble.

plantulation, *f.* germination.

plantule, *f.* plantlet, seedling, sprout.

plantureux, -euse, abundant, copious, fertile, rich.

planure, *f.* shavings.

plaque, *f.* (bed) plate, pane, slab, table, diaphragm, veneer, anode, strip, layer, badge, patch, spot, flat block, plate culture; — **à fente,** slot plate; — **chauffante,** heating plate, hot plate; — **criblée,** sieve plate; — **de culture,** plate culture; — **de fond,** bottom plate; — **de gazon,** sod.

plaque, — **en porcelaine,** porcelain plate; — **équatoriale,** equatorial plate; — **filtrante,** filter plate; — **opaline,** milk-glass plate; — **pelliculaire,** film; — **photographique,** photographic plate; — **réglementaire,** number plate, license plate; — **sensible,** sensitive plate; — **souple,** film.

plaqué, *m.* plated goods, plated metal, veneer, electroplate.

plaqué, plated.

plaqueminier, *m.* ebony tree (*Diospyros lotus*); — de **Virginie,** persimmon tree, (*Diospyros virginiana*).

plaquer, to plate, lay on (plaster), lay down (sod), veneer, line, face.

plaquette, *f.* small plate, thin book.

plasma, *m.* plasma, plasm; — **sanguin,** blood plasma.

plasmode, *m.* plasmode, plasmodium.

plasmodesmes, *f.pl.* plasmodesma, connecting threads of protoplasm.

plasmodies, *f.pl.* colonies of nuclei united by a single plasma.

plasmogonie, *f.* formation of plasmodium at beginning of sexuality.

plasmolyse, *f.* plasmolysis.

plasmosome, *f.* plasmosome.

plasmospore, *f.* spore reduced to nucleus surrounded by protoplasm.

plasome, *f.* plasome, a living element of protoplasm.

plaste, *m.* leucoplast, specialized colorless protoplasmic granule.

plasticité, *f.* plasticity.

plastide, *f.* chromatophore.

plastidule, *f.* plastidule, a primordial cell of protoplasm.

plastine, *f.* plastin.

plastique, *f.* art of modeling.

plastique, plastic.

plastochondrie, plastoconte, *f.* mitochondria.

plastogamie, *f.* plastogamy.

plastogénie, *f.* plastogeny, cytoplastic elements undergoing a reorganization.

plastron, *m.* plastron, breastplate.

plat, *m.* flat, flat part (of hand), plate, sheet, dish, pan, scale (of balance); à —, flatwise, flat.

plat, flat, (of vessels) gold or silver, dead, still, stagnant, plane, level, even.

platane, *m.* plane tree, (Platanus); — d'Amérique, American plane, buttonwood, water beech, sycamore (*Platanus occidentalis*).

plateau, *m.* pan, scale (of balance), basin (of scales), disk, plate, slab, board, table, tray, plank, plateau, mesa, tableland, platform, shoal; — de cylindre, cylinder cover.

plate-bande, *f.* border, grass border, nursery bed.

plate-forme, *f.* platform, rail truck, tram, roadbed (of railway).

platelage, *m.* floor (of a bridge) planking.

platement, flatly, flat, plainly.

platinage, *m.* platinizing, platinization, platinum plating.

platine, *f.* plate, platen, stage (of microscope), flat part, lock (of a gun), slab of leveling stand; — à chariot, mechanical stage; — à dissection, dissecting table; — chauffante, heating stage.

platine, *m.* platinum, platina; éponge de mousse —, platinum sponge; — en éponge, platinum sponge; — en fil, platinum wire; — en lame, — laminé, platinum foil; — en livret, platinum leaf (in books); — spongieux, spongy platinum, platinum sponge; noir de —, platinum black.

platiné, platinized.

platiner, to platinize, plate with platinum.

platineux, -euse, platinous.

platinifère, platiniferous.

platinique, platinic.

platiniser, to platinize.

platinocyanure, *m.* platinocyanide.

platinoïde, *m.* platinoid.

platinotypie, *f.* platinotype.

platinotypique, platinotype.

platinure, *f.* platinizing, platinization.

platisilique, with compressed and broad siliques.

plâtrage, *m.* plastering (wall).

plâtras, *m.* old plaster, rubbish.

plâtre, *m.* plaster, plaster of Paris, plasterwork, plaster cast, (face) paint; — aluné, Keene's cement, marble cement; — cru, unburned gypsum.

plâtré, plastered.

plâtre-ciment, *m.* hydraulic lime.

plâtrer, to plaster (over), dress (soil) with gypsum, clear (wine) with gypsum.

plâtreux, -euse, gypseous, chalky (water or soil).

plâtrière, *f.* gypsum quarry, plaster kiln.

platycarpe, platycarpic, platycarpous, broad-fruited.

platycéphale, with flattened cap (hood).

platyglossate, broad-tongued.

platylobé, platylobate, broad-lobed.

platylome, with broad border.

platyneure, with broad veins.

platypède, with stem expanded at the base.

platypétale, platypetalous.

platypode, with broad peduncles.

platystémon, *m.* California poppy (Eschscholtzia).

plausible, plausible.

playon, *m.* reaper blade.

plécolépide, with scales fused at the base.

plectenchyme, *f.* plectenchyma, a tissue of woven hyphae.

plectobasidié, *f.* basidiomycete in which basidia develop in islets.

pléiade, *f.* pleiad.

plein, *m.* plenum, full part, fullness, solid part, full stroke, middle; en —, full, in the middle, at the full, fully, entirely.

plein, full, replete, solid, copious, fully, whole, entire, close-grained (wood), open, (of the air, etc.) filled, the middle of, dense, close, full (with young), pregnant; de — gré, at one's own free will; de — saut, all at once, suddenly; en — jour, in broad daylight; en — moisson, in the middle of harvest; en — vent, in the open air.

pleinement, fully, entirely.

pleiocarpe, (bulbs) producing several branches in succession.

pleiochasium, *m.* pleiochasium, axis of a cyme producing more than two branches.

pleiocycle, pleiocyclic, perennial.

pleiocyclique, hardy, perennial.

pleiomère, *f.* pleiomery, having more whorls than normal.

pleiontismus, *m.* simultaneous production of protandrous and protogynous individuals.

pleiopétale, pleiopetalous, with flowers double.

pleiophylle, pleiophyllous, with numerous leaves from the same point.

pleiophyllie, *f.* pleiophylly.

pleioxe, (parasite) able to live on several plants.

plénière, plenary, full, complete.

plenitude, *f.* fullness, plenitude.

plenum, *m.* plenum.

pléochroïsme, *m.* pleochroism.

pléogamie, *f.* pleogamy, methods of pollination varying in respect of time.

pléomorphe, pleomorphic.

pléomorphisme, *m.* pleomorphism.

pléone, *m.* pleon, aggregate of molecules smaller than a micella.

pléospore, *m.* spore accompanied by numerous sister cells.

pléostémone, with several stamens.

plérome, *f.* plerome.

pléthore, *f.* plethora, superabundance.

pleural, -aux, pleural.

pleurer, to weep, weep for, mourn, bewail, (of eyes) water, (of vines) bleed, start and drip.

pleurésie, *f.* pleurisy.

pleurétique, pleuritic.

pleuridie, *f.* lateral expansion of the labellum.

pleurocarpe, pleurocarpous.

pleurodiscal, -aux, pleurodiscous.

pleurogyne, pleurogynus, (a tubercular elevation) rising close to the ovary.

pleurogynique, inserted on the body of the free ovary.

pleuronervé, with a lateral vein.

pleuroplaste, pleuroplastic, (leaf) in which the central portion first attains permanency.

pleuropneumonie bovine, *f.* pleuropneumonia of cattle.

pleurorrhize, pleurorhizal, (embryo) with its radicle against one edge of the cotyledons.

pleurospermé, (plant) having ovary with parietal placentation.

pleurospore, *f.* pleurospore, spore formed at the sides of a basidium.

pleurothallé, pleurothalline.

pleurotribe, pleurotribal, pleurotribe.

pleurotriche, (zoospore) with lateral flagella.

pleurotrope, pleurotropous, flattened on the sides.

pleurs, *m.pl.* tears.

pleuston, *m.* pleuston, plants that float.

pleuvoir, to rain.

plèvre, *f.* pleura.

plexéoblaste, plexeoblastus, (cotyledons) rising aboveground.

pli, *m.* fold, plait, crease, wrinkle, cover, envelope, letter, paper, bend (of the arm), bent, undulation (of ground), thin spot in wall of pollen grain.

pliable, pliable, flexible.

pliage, *m.* folding, creasing, bending.

pliant, *m.* folding chair, campstool.

pliant, pliant, pliable, docile, folding, collapsible.

plicatile, plicate.

plié, folded, plicate, rimose, camptotropal, orthotropal (ovule).

plier, to fold, bend, bow down, submit, plait, double, curve, sag, yield; se —, to be folded, obey, yield, adapt oneself, be bent, bend.

plieur, -euse, *m.&f.* folder.

plinger, to dip, plunge (wick of candle).

pliocène, Pliocene.

pliotron, *m.* pliotron (a type of vacuum tube).

plissage, *m.* plaiting, folding.

plissé, plaited, kilted, plicate.

plissement, *m.* plaiting, corrugation, vernation, estivation, folding, fold, warping (of strata).

plisser, to plait, crease, corrugate, fold, warp; se —, to be plaited, crease, wrinkle.

plissure, *f.* plaiting, pleating.

ploc, *m.* hair (from cow or dog), waste wool, sheathing (calking) felt.

plocage, *m.* carding (wool).

plococarpe. *m.* plococarpium, fruit composed of follicles ranged around an axis.

ploie (ployer), (it) is bending, folding.

plomb, *m.* lead, shot, plummet, plumb line, lead seal, poisonous gas, hydrogen sulphide; — **antimonié,** lead containing antimony; — **blanc,** white lead; — **brulé,** lead ashes, litharge; — **carbonaté,** carbonate of lead, white lead, cerusite; — **de chasse,** shot, small shot; — **de sonde,** plummet, sounding lead.

plomb, — **de sûreté,** safety plug; — **d'oeuvre,** raw lead, lead goods, work lead; — **doux,** refined lead; — **durci,** hard(ened) lead; — **en saumon,** pig lead; — **filé,** lead wire; — **fusible,** (lead or safety) fuse, — **laminé,** sheet lead, rolled lead; **à —;** plumb, vertical(ly), upright.

plombage, *m.* leading, plumbing, sealing (with lead), lead work, lead glazing, rolling down (ground).

plombagine, *f.* black lead, plumbago, graphite.

plombaginé, similar to plumbago or leadwort.

plombate, *m.* plumbate.

plombé, leaded, leaden, livid, leady, lead covered.

plomber, to lead, cover with lead, weight, plumb, sound, seal (with lead), roll (ground); **se —,** to take on a leaden hue.

plomberie, *f.* lead work, plumbing, lead works, plumber's shop.

plombeur, *m.* sealer, (heavily weighted) roller.

plombeux, plumbous, of lead.

plombier, *m.* plumber.

plombier, pertaining to lead or lead working.

plombifère, plumbiferous, lead-bearing.

plombique, plumbic, of lead.

plongé, immersed, dipped, submarine (plant).

plongeant, plunging, downward, descending, falling freely, in straight lines.

plongée, *f.* plunge, dive, dip, slope, incline.

plongement, *m.* plunging, dip.

plongeon, *m.* dabchick, didapper, diver.

plonger, to plunge, submerge, descend, immerse, dip (candles), dive; **se —,** to plunge, immerse oneself, dive.

plongeur, *m.* plunger, dipper, diver, diving apparatus, diving bird, dishwasher.

plongeur, -euse, plunging, diving.

ploque, *f.* hair (from cow or dog), waste wool.

plot, *m.* block, plug; *pl.* logwood.

ployable, flexible.

ployer, to bend, bow, yield, fold (up); **se —,** to bend, give way, yield, be folded.

ployure, *f.* fold, bend.

plu (plaire, pleuvoir), pleased, rained.

pluche, *f.* plush.

pluie, *f.* rain, shower; — **battante,** — **diluvienne,** — **en hallebardes,** downpour, pelting rain.

plumage, *m.* plumage, feathers.

plumassier, *m.* feather merchant.

plumbaginé, similar to plumbago.

plume, *f.* pen, quill, feather, feathers, plume; — **et poil,** fur and feather-

plumer, to plume, pluck.

plumet, *m.* switch.

plumeux, -euse, feathery, plumose, plumous.

plumule, *f.* plumule.

plupart, *f.* most part, most, majority, most people, greater part; **la — du temps,** most of the time, in most cases, generally; **pour la —,** for the most part, mostly.

pluralité, *f.* plurality, majority, multiplicity, plural (in grammar).

plurent (plaire), (they) pleased.

pluricarpellaire, formed of several carpels.

pluricolore, of many colors, variegated.

pluriel, -elle, plural.

pluriflore, many-flowered.

pluriloculaire, multilocular.

plurinucléé, with several nuclei.

pluriovulé, with several ovules.

plurisérié, composed of several series.

plurivalent, multivalent, polyvalent.

plurivalve, plurivalve.

plurivariant, multivariant.

plus, *m.* more, most, maximum.

plus, more, most, longer, plus, moreover, besides, in addition, also (with negative) no more; — . . . —, the more . . . the more; — **de,** no more; — **grande hauteur,** greatest height; — **haut,** higher, above; — **loin,** farther, farther on; — **offrant,** highest bidder.

plus, — **ou moins,** more or less; — **rien,** nothing more; — **tôt,** sooner; **au —,** at (the) most, at best, at the most; **c'est tout ce qu'il y a de —,** simple, nothing is simpler; **de —,** besides, in addition, moreover; furthermore; **de — en —,** more and more; **en —,** besides; **en — de,** in addition to.

plus, le (la) —, the most; **ne . . . —,** no more, no longer; **ni — ni moins,** neither more nor less, exactly; **non —,** either, neither; qui —, qui moins, some more, some less; sans —, without more, without further; tout au —, at the very most.

plusianthé, diclinous, (plants) with flowers having calyx and corolla.

plusieurs, several, many; à — reprises, several times, repeatedly.

plus-value, *f.* increase, increase in value, appreciation, premium, surplus value or revenue.

plut (plaire, pleuvoir), (it) pleased, (it) rained.

plutonien, -enne, plutonique, plutonic, igneous.

plutôt, rather, sooner; — que, rather than.

pluvial, -aux, of rain, pluvial.

pluvier, *m.* plover; — doré, golden plover.

pluvieux, -euse, rainy, wet.

pluviographe, *m.* self-recording rain gauge.

pluviomètre, *m.* gauge, pluviometer.

plyrontophyte, (plant) with stamens borne from inner wall of calyx.

pneu, *m.* pneumatic tire, rubber tire; gros — (à basse pression), balloon tire.

pneumathode, *m.* respiratory root, original lenticels.

pneumatique, *m.* (pneumatic) tire; *f.* pneumatics.

pneumatique, pneumatic, air.

pneumatochimie, *f.* the chemistry of gases.

pneumatochymifère, carrying air and liquid simultaneously.

pneumatophore, *m.* pneumatophore.

pneumobacille, *m.* pneumobacillus.

pneumonanthe, (flower) resembling a blister full of air.

pneumonie, *f.* pneumonia.

pneumonique, *m.* pneumonia patient.

pneumonique, pneumonic.

poacé, resembling meadow grass.

poche, *f.* pocket, bag, pouch, sack, case, crop (of bird), sac, ladle, bordered pit; — à huile, oil pan; — à resine, resin gall; — copulatrice, bursa copulatrix.

pocher, to poach (eggs), get baggy (trousers).

pochette, *f.* small pocket, pocket kit, pouch.

pochon, *m.* ladle.

poculiforme, poculiform, shaped like a goblet.

podagraire, *f.* goatweed goutwort (*Aegopodium podagraria*).

podetium, *m.* podetium.

podium, *m.* podium, a footstalk

podocarpe, *m.* yellowwood.

podocarpe, podocarpous.

podogyne, *m.* podogynium, a gynophore.

podogynique, podogynicus.

podomètre, *m.* measuring wheel, pedometer.

podophylle, *m.* May apple (*Podophyllum peltatum*).

podophyllin, podophylline, *f.* podophyllin.

podoptère, podopterous, having winged peduncles.

podosperme, *m.* funiculus.

podsol, *m.* bleached sand.

poêle, *m.* stove, pall; *f.* shallow (frying) pan.

poêlon, *m.* small saucepan with handle, casserole.

poésie, *f.* poetry, poesy.

poète, *m.* poet.

poétique, poetic(al).

poids, *m.* weight, gravity, load, burden, importance; — atomique, atomic weight; — brut, gross weight; — curseur, sliding weight; — des graines, seed weight; — mobile, sliding weight; — moléculaire, molecular weight; — mort, dead weight, dead load; — net, net weight; — public, weighing house; — spécifique, specific gravity, specific weight; à — d'or, at an extremely high price.

poignarder, to stab, stick, jab with a stinger.

poignée, *f.* handful, handle, grip, hilt, stock, holder; — de mains, handshake; à —, by handfuls.

poignent (poindre), (they) are stinging.

poignet, *m.* wrist, wristband, cuff.

poil, *m.* hair, nap, bristle, fur, pile, coat (of animals), down, pubescence, hair of animals; — absorbant, root hair; — de chameau, camel's hair; — de chèvre, goat's hair; — de laitier, cinder or slag wool; — de lièvre, hare hair; — de scorie, slag wool; — follet, down hair; — grossier, coarse hair; — radiculaire, root hair; garni de —s cours, setulose; pourvu de —s en ligne, pourvu de —s en touffe, bearded.

poileux, -euse, hairy, shaggy.

poilu, hairy, bristly, pilose.

poinçon, *m.* point, awl, bodkin, pricker, piercer, pouch, stamp, hallmark, die, punch, puncheon, large cask.

poinçonnage, poinçonnement, *m.* punching, stamping, marking, pricking.

poinçonner, to prick, bore, punch, stamp, mark.

poinçonneuse, *f.* punching machine, punch.

poindre, to sting, appear, dawn, break.

poing, *m.* fist, hand; **coup de —,** cuff, punch.

point, *m.* point, dot, period, place, pole, shooting pain, speck, spot, bubble, mark, position, stitch, (with **ne** expressed or understood), not at all, no, none, not, not any; **— d'appui,** support, fulcrum, center of motion, point of support, shoulder, abutment; **— d'ébullition,** boiling point; **— de congélation,** freezing point; **— de départ,** starting point, initial point; **— de fuite,** vanishing point, accidental or visual point; **— de fusion,** melting point.

point, — de jonction, point of junction, junction; **— de mouvement,** center of motion; **— de passage,** control point; **— de repère,** reference point, datum point, bench mark, guide mark, base, starting point, station, **— de rosée,** dew point; **— de tangence,** point of contact.

point, — de vaporisation, evaporating point; **— de vue,** point of view, focus, standpoint, point of sight.

point, — d'inflammation, flash point; **— d'inflexion,** point of inflexion, turning point; **— d'insertion,** point of attachment; **— du tout,** not at all; **— fixe,** fixed point, fulcrum; **— idéal, — imaginaire,** mathematical point; **— mort,** dead point, dead center; **— perdu,** point of inflexion, turning point; **— vertical,** zenith.

point, à —, opportunely, on the dot, just in time; **à ce —,** to that extent; **au — de,** at the point; **au — de vue de,** from the point of view of; **au — du jour,** daybreak; **de — en —,** in every detail, precisely, exactly; **mettre au —,** to put to a focus, focus, perfect, effect, adjust; **ne . . . —,** not, not at all, no, (expressed or understood); **sur le — de,** on the verge of, about to.

pointage, *m.* pointing, aiming, checking.

pointe, *f.* point, head, witticism, top (of a tree), apex (of the heart), peak, horn, (of crescent) tip, nail, tack, pin, bit, dash, touch, break (of day); **— arrière,** the point in the rear, the rear tapering to a point; **— de prosternum,** apex of the prosternal process; **en —,** tapering in a point, to a point, pointed; **recouvert de —s dures et courtes,** muricate, prickly, echinate.

pointeau, *m.* center punch, needle (of a needle valve).

pointement, *m.* appearance, outcrop (of ores), check (mark), sprouting (of plant).

pointer, to prick, pierce, stick, stab, check (off), register, aim, point (a telescope), appear, sprout, spring up, rise, soar, rear (of horse).

pointillage, pointillement, *m.* dotting, stippling.

pointillé, *m.* dotted line, stippling, dots.

pointillé, dotted, spotted, stippled.

pointiller, to dot, stipple, cavil, split hairs, annoy, nag.

pointilleux, -euse, particular.

pointu, pointed, sharp.

poire, *f.* pear, (pear-shaped) bulb; **— à couteau,** dessert pear; **— à cuire,** stewing pear, cooking pear; **— (soufflante) en caoutchouc,** rubber (bulb) bottle; **en —,** pear-shaped, pyriform.

poiré, *m.* perry, fermented pear juice.

poireau, *m.* leek (*Allium porum*), top onion, pearl onion, wart.

poirée, *f.* Swiss chard, sea-kale beet, silver-beet (*Beta vulgaris* var. *cicla*).

poirier, *m.* pear (tree or wood) (Pyrus).

pois, *m.* pea, pellet; **— à cautères,** issue pea, **— cassé,** split pea; **— chiche,** chick-pea; Egyptian pea (*Cicer arietinum*); **— de senteur,** *m.* sweet pea (*Lathyrus odoratus*), lady pea; **— des champs,** field pea (*Pisum arvense*); **— d'iris,** orrisroot pellet, irisroot pea; **— fulminant,** pellet primer, detonating ball, toy torpedo; **— velus,** cowhage (*Stizolobium pruriens*).

poison, *m.* poison, venom; **— du coeur,** cardiac poison.

poissement, *m.* pitching, coating with pitch.

poisser, to pitch, coat with pitch, make sticky, soil.

poisseux, -euse, sticky, pitchlike, coated with pitch.

poisson, *m.* fish; **— blanc,** white fish; **— d'eau de mer, — de mer,** sea fish; **— d'eau douce,** fresh-water fish; **— laité,** soft-roed fish; **— ouivé,** hard roed fish; **— rapace,** voracious fish; **— rouge,** golden carp, goldfish.

poissoneux, -euse, abounding with fish.

poitrine, *f.* chest, breast, lungs, breast cut, bosom, brisket.

poivre, *m.* pepper; **— de Cayenne,** cayenne pepper, red pepper (*Capsicum frutescens*);

— des murailles, wall pepper, biting yellow stonecrop (*Sedum acre*); — noir, black pepper (*Piper nigrum*).

poivré, peppery, pungent, caustic.

poivrer, to pepper, season, infect with disease.

poivrette, *f.* black caraway (*Nigella sativa*).

poivrier, *m.* pepper plant (Piper), pepper box.

poix, *f.* pitch; — blanche, white pitch; — de Bourgogne, Burgundy pitch, galipot, white resin; — de cordonnier, shoemaker's pitch, common black pitch; — noire, black pitch, common pitch; — sèche, hard (stone) pitch; — terrestre, mineral pitch, bitumen; — végétale, wood pitch.

poix-résine, *f.* rosin.

polachaine, polakène, *f.* polachena, fruit like a cremocarp but with five carpels.

polaire, polar, north pole.

polarimètre, *m.* polarimeter.

polarisant, polarizing.

polarisateur, *m.* polarizer.

polarisateur, -trice, polarizing.

polarisation, *f.* polarization, polarizing; — rotatoire, rotatory polarization.

polariscope, *m.* polariscope.

polarisé, polarized.

polariser, to polarize.

polariseur, *m.* polarizer.

polarite, *m.* a water purifier.

polarité, *f.* poiarity; —s multiples, many branches.

polder, *m.* polder, sunk meadow.

poldérien, -enne, pertaining to the polder.

pôle, *m.* pole; —s de même nom, similar (like) poles; —s de nom contraire, opposite (unlike) poles.

polémique, *f.* polemic.

polémique, polemic(al).

polémonie, *f.* Polemonium.

poli, *m.* polish, gloss.

poli, polished, bright, polite, refined.

polianite, *f.* polianite.

police, *f.* policy, police, civil administration, rules of order, bill, list; — d'assurance, insurance policy.

policer, to polish, civilize.

poliment, *m.* polishing, polish.

poliment, politely.

poliomyélite, *f.* poliomyelitis.

polioplasma, *m.* polioplasm, circulating portion of the cytoplasm.

polir, to polish, burnish, civilize.

polissable, polishable.

polissage, polissement, *m.* polishing, polish, smoothing.

polisseur, *m.* polisher.

polissoir, *m.* polisher, burnisher, polishing tool.

polissure, *f.* polishing, polish.

politique, *f.* policy, politics; *m.* politician.

politique, political, politic.

politiquement, politically, shrewdly.

pollachigène, polycarpic, with numerous carpels.

pollakanthique, polycarpic.

pollaplostémonopétale, having stamens within a multiple number of petals.

pollen, *m.* pollen.

polligère, having a farinaceous dust resembling pollen.

pollinaire, pollinarious, pertaining to fine flour.

pollinaires, *f.pl.* cystids projecting on the border of blades of some fungi.

pollination, *f.* pollination.

pollineux, -euse, farinose, pulverulent.

pollinide, *m.* pollinidium.

pollinie, *m.* pollinium.

pollinifère, polliniferous.

pollinique, pollinicous, composed of pollen.

pollinisation, *f.* pollination; — directe, autogamy, self pollination; — par le vent, wind pollination.

polliniser, to pollinize, pollinate.

pollinium, *m.* pollinium, a pollen mass.

pollinode, *m.* pollinodium.

pollinodie, *f.* male sex organ in Ascomycetes.

polluer, to pollute, defile.

pollution, *f.* pollution, contamination.

polonifère, containing polonium.

polonium, *m.* polonium.

polyadelphe, polyadelphous.

polyadène, polyadenous, with many glands.

polyakène, *m.* schizocarp with many carpels.

polyamine, *f.* polyamine.

polyandre, polyandrous.

polyandrie, *f.* polyandria.

polyanthe, polyanthème, many-flowered.

polyarche, polyarch, stele with many protoxylem groups.

polyatomique, polyatomic.

polyazoté, containing more than one nitrogen atom.

polybasique, polybasic.

polybasite, f. polybasite.

polycalathidé, having several calathides.

polycarpe, polycarpic, fruiting many times, indefinitely.

polycarpellé, polycarpellary.

polycarpien, -enne, polycarpic, fruiting many times, indefinitely.

polycarpique, polycarpic.

polycarpium, m. fruit with several carpels dry and fleshy.

polycellulaire, polycellular.

polycéphale, polycephalous, bearing many capitula.

polycérate, with fruits resembling a bundle of small horns.

polychloré, polychloro.

polychorionide, f. polychorionide, etaerio, aggregate fruit.

polychroique, polychroic, pleochroic, having many colors.

polychroïsme, m. polychroism, pleochroism.

polychrome, polychrome, polychromatic.

polychromie, f. colored photography.

polycilié, polyciliate, having numerous cilia.

polycladie, f. polyclady, a supernumerary development of branches and leaves.

polyclone, with a much divided branch.

polycopie, f. manifolding or stenciling process, (duplicated) copy.

polycotylédoné, polycotyledonous.

polydactyle, polydactylous.

polyderme, m. polyderm, tissue composed of endodermal and parenchymatous cells.

polyécique, (moss) with autoecious and dioecious individuals.

polyédral, polyhedral.

polyèdre, m. polyhedron.

polyèdre, polyhedral.

polyédrique, polyhedral, polyhedric.

polyembryonie, f. polyembryony, production of more than a single embryo in an ovule.

polyergique, polyergic.

polyflore, having many flowers.

polygale, m. milkwort (Polygala); — de Virginie, senega snakeroot, snakeroot (Polygala senega).

polygame, polygamous.

polygamie, f. polygamy.

polygénie, f. polygeny.

polygonal, -aux, polygonal, many-angled.

polygonaté, polygonatus, (stem) having many nodes.

polygone, m. polygon; — funiculaire, funicular or link polygon.

polygone, polygonal.

polygyne, polygynous, having many distinct styles.

polyhalite, f. polyhalite.

polymère, m. polymer.

polymère, polymeric.

polymérie, f. polymerism.

polymérique, polymeric.

polymérisation, f. polymerization.

polymériser, to polymerize.

polymérisme, m. polymerism.

polymorphe, polymorphous, pleomorphic.

polymorphisme, m., polymorphie, f. polymorphism, polymorphy, pleomorphism.

polynitrophiles, (bacteria) requiring nitrogen for normal growth.

polynôme, m. polynomial, multinomial.

polynucléaire, polynuclear.

polyoéicie, m. polyoecism, (plants) with flowers differing in sex.

polyol, m. polyhydric alcohol.

polyolal, m. aldose.

polype, m. polyp.

polypeptide, f. polypeptide.

polypérianthé, with several perianths.

polypétale, polypetalous.

polyphasé, polyphase.

polyphore, m. polyphore, torus with many pistils.

polyphragme, with several septa.

polyphyle, (hybrid) derived from several ancestral types.

polyphylétique, polyphyletic.

polyphylle, many-leaved.

polyphyllie, f. polyphylly.

polypier, m. polyp, coral.

polypode, m. polypodium.

polypode, polypod.

polypore, polyporous.

polyporéen, -enne, similar to a polyporus.

polyrhize, polyrhizal, having numerous rootlets.

polysaccharide, *m.* polysaccharide.

polysamare, with several samaras.

polysarcie, *f.* obesity.

polysépale, polyphyllous, polysepalous.

polysèque, polysecus, having an aggregate fruit.

polysète, with numerous hairs in form of silk.

polysiphoné, polysiphonous.

polysoc, *m.* multiple plow.

polysperme, many-seeded.

polyspermie, *f.* polyspermy, many-seededness.

polystachyé, polystachous, having many spikes.

polystélie, *f.* polystely, several steles, each with its own pericycle and endoderm.

polystélique, polystelic.

polystémone, polystemonous, having many stamens.

polystigmé, with numerous stigmata.

polystique, polystichous, (organs) borne in many series.

polysulfure, *m.* polysulphide.

polysulfuré, polysulphide of.

polytechnique, polytechnic.

polytérébénique, polyterpene.

polythalamé, (grasses) having flowers of different sexes on same base.

polythélé, polytheleus, (flower) containing several distinct ovaries.

polytome, polytomous.

polytomie, *f.* polytomy.

polytric, *m.* haircap moss (Polytrichum).

polytrique, polytrichous, having many hairs.

polyvalent, polyvalent.

pomacées, *f.pl.* pomaceous plants.

poméridien, -enne, pomeridianus, blossoming in the afternoon.

pommade, *f.* pomade, pomatum, ointment, salve; — **de goudron,** tar ointment; — **de noix de galle,** nutgall ointment; — **d'iode,** iodine ointment.

pomme, *f.* apple, pome, head (of cabbage, lettuce), knob, ball, rose (of sprinkler), pommel; — **d'amour,** tomato; — **d'arrosoir,** sprinkling rose, rosehead; — **de mât,** truck; — **de pin,** pine cone, pine-

apple; — **de reinette,** pippin; — **de terre,** potato; — **de terre glycerinée,** glycerinated potato; — **de terre sucrée,** sweet potato (*Ipomoea batatas*); — **d'étrier,** diamond knot; — **épineuse,** thorn apple (*Datura stramonium*); — **sauvage,** crab apple, wild apple.

pommé, grown to a round head, cabbaged.

pommeau, *m.* pommel.

pommelé, dappled, spotted.

pommelle, *f.* grating, strainer (over pipe).

pommeraie, *f.* apple orchard.

pommette, *f.* cheekbone.

pommier, *m.* apple tree (Malus); — **sauvage,** wild apple tree.

pommique, pomaceous, malic.

pompe, *f.* pump, syringe, pomp; — **à air,** air pump; — **à huile,** oil pump; — **à incendie,** fire engine, fire pump; — **à jet,** jet pump; — **alimentaire,** feed pump; — **à mercure,** mercury (air) pump.

pompe, — **à piston,** plunger pump; — **à pneumatique,** tire pump; — **aspirante,** suction pump, aspiring pump; — **à vide,** vacuum pump; — **centrifuge,** centrifugal pump, turbine pump; — **d'alimentation,** feed pump; — **foulante,** compressor, force pump; — **rotative,** rotary pump.

pomper, to pump, suck in, suck up.

pompeusement, pompously.

pompeux, -euse, stately, pompous.

pompholyx, *m.* pompholyx, flowers of zinc.

pompier, *m.* fireman, pumper, pump maker.

ponçage, *m.* pumicing, pouncing.

ponce, *f.* pumice, pumice stone, pounce, (thick black) stencil ink; **pierre** —, pumice stone.

ponceau, *m.* ponceau, corn poppy (*Papaver rhoeas*), poppy, poppy color, scarlet, flaming red, small bridge, culvert.

poncer, to pumice, rub or polish with pumice, pounce.

ponceux, -euse, pumiceous.

ponction, *f.* puncture, tapping.

ponctionner, to tap, puncture, prick.

ponctuage, *m.* spottiness.

ponctuation, *f.* punctuation, dotting, dot, pit; — **auréolée,** bordered pit.

ponctué, punctuated, dotted, pitted.

ponctuel, -elle, punctual.

ponctuer, to punctuate, dot.

pondérabilité, *f.* ponderability.

pondérable, ponderable, weighable.

pondéral, -aux, ponderal, gravimetric, pertaining to weight, by weight.

pondéralement, by weight, gravimetrically.

pondération, *f.* balance, equilibrium, poise.

pondéré, well-balanced, cool, level-headed.

pondérer, to balance, poise.

pondéreux, -euse, ponderous.

pondeur, -euse, (egg-) laying (moth, bird).

pondeuse, *f.* layer.

pondoir, *m.* laying nest, ovipositor.

pondre, to lay (eggs), produce.

pont, *m.* bridge, deck; — **levis,** drawbridge.

pont-boîte, *m.* transmission gear case.

ponte, *f.* laying (of eggs); — **des oeufs,** oviposition.

pontil, *m.* pontil, punty, pontee.

poophyte, *f.* poophyte, a plant inhabiting meadows.

populage, *m.* globeflower (Trollius), loosestrife (Lysimachia), marsh marigold (Caltha); — **des marais,** marsh marigold (*Caltha palustris*).

populaire, *m.* populace.

populaire, popular.

populairement, popularly.

populariser, to popularize, make popular.

population, *f.* population.

populeux, -euse, populous.

populine, *f.* populin.

porc, *m.* hog, pig, swine, pork, slag.

porcelaine, *f.* porcelain, china, cowrie (shell); — **de ménage,** household china; — **dure,** hard porcelain; — **mate,** unglazed porcelain; — **montée,** metal porcelain; — **tendre,** soft porcelain.

porcelainier, *m.* porcelain maker.

porcelainier, -ère, of, pertaining to porcelain.

porcelané, porcelanique, porcelaneous, porcelanic, glazed (like porcelain).

porc-épic, *m.* porcupine.

porche, *m.* porch.

porcherie, *f.* pigsty, pig farm.

porcin, porcine.

pore, *m.* pore, pit.

poreux, -euse, porous.

poricide, poricidal.

porogamie, *f.* porogamy.

porophylle, porophyllous, (leaves) with numerous transparent spots.

porosité, *f.* porosity, porousness, looseness.

porphyre, *m.* porphyry, slab of porphyry (for triturating drugs).

porphyré, resembling porphyry, porphyry.

porphyrique, porphyritic.

porphyrisation, *f.* grinding, trituration.

porphyriser, to grind, triturate (on slab).

porreau, *m.* leek, wart, top onion, pearl onion (*Allium porrum*).

port, *m.* carrying, wearing, carriage, bearing, postage, tonnage, burden (of ship), port, harbor, air, growth, habit; — **de relâche,** port of call; — **de stationnement, home** port, dock.

portabilité, *f.* portability.

portable, portable, wearable.

portage, *m.* conveyance, portage, carriage, bearing.

portant, *m.* support, handle, prop, keeper, armature (of magnet).

portant, bearing, carrying.

portatif, -ive, portable, easily carried, hand, pocket.

porte, *f.* door, gate, entrance; — **à coulisse,** sliding door, sliding sluice or valve; — **à guillotine,** drop door; — **de chauffe,** fire (box) door; — **du foyer,** fire door; — **graine,** mother tree, seed-bearing tree; — **vitrée,** glass door.

porte, portal, prone, inclined.

porté, *m.* wear, appearance.

porté, borne, prone, inclined, disposed.

porte-aiguillon, aculeate, furnished with a sting.

porte-allumettes, *m.* match (box) case.

porte-amorce, *m.* fuse cup, primer holder.

porte-ampoule, *m.* lamp (bulb) holder.

porte-balai, *m.* broomstick, brush holder.

porte-bougie, *m.* candlestick.

porte-bouquet, *m.* flower vase.

porte-charbon, *m.* carbon holder.

porte-cigare, *m.* cigar holder.

portée, *f.* litter, brood, gestation period, range, reach, ability, power, scope, capacity, reach of the hand, extent, distance, span, importance, significance, discharge, delivery, bearing (of beam), bearing surface; à (la) — **de,** within the range (reach) of, ready, convenient.

porte-feu, *m.* flame passage.

porte-feu, fire-transmitting.

portefeuille, *m.* portfolio, pocketbook, letter case; **en —,** in manuscript, unpublished.

porte-grain, *m.* disseminating agent, mother tree, seed-bearing tree.

porte-graines, seed-bearing.

porte-greffe, *m.* stock (trees), rootstock.

porte-lame, *m.* slide support.

porte-loupe, *m.* lens holder, magnifying-glass stand.

porte-manchon, *m.* mantle holder (in burner).

porte-mèche, *m.* wick (bit) holder.

porte-objectif, *m.* nosepiece (microscope).

porte-objet, *m.* object bearer, microscopic slide, holder, stage (of microscope); — **avec cavité,** hollow-ground microscope slide.

porte-outil, *m.* tool holder.

porte-plume, *m.* penholder.

porte-pneu, *m.* tire carrier.

porter, to bear, support, sustain, hold, carry, transport, convey, take, bring, wear, turn, produce, yield, be pregnant, go with young (of animals), direct, lead, put, give, deliver, influence, induce, dispose, incline, state, elect, reach, stand, pronounce upon; **se** —, to be borne, betake oneself, be supported, proceed, be inclined, go, move, tend, be (with regard to health), present oneself; — **coup,** take effect, tell, strike a blow; **il lui porte envie,** he envies him.

porte-tubes, *m.* tube holder.

porteur, *m.* carrier, bearer, holder, porter.

porteur, -euse, supporting.

porte-vent, *m.* mouth tube, blast pipe, tuyère.

porte-verre, *m.* glass holder, lens holder.

portier, *m.* porter, doorkeeper.

portillon, *m.* small door or gate.

portion, *f.* portion, part, share, ration.

portique, *m.* portico.

portland, *m.* Portland cement.

porto, *m.* port wine, port.

portrait, *m.* portrait, likeness.

posage, *m.* placing, laying.

pose, *f.* placing, laying, setting (up), putting up, stationing, (time) exposure, seating, attitude, posture, pose, shift (of workmen).

poser, to place, lay, put, set, set up, state, ask, pose, set down, lay down, rest, lie, suppose, grant; **se** —, to settle, land, pose; **cela posé,** that being granted.

positif, -ive, positive, certain.

position, *f.* position, status, situation, posture.

positive, *f.* positive; — **sur verre,** positive plate.

positivement, positively.

positivisme, *m.* positivism.

posologie, *f.* posology.

posséder, to possess, own, have, dominate, be master of.

possesseur, *m.* possessor, owner, proprietor.

possession, *f.* possession; — **de terres,** territorial property, landed property.

possibilité, *f.* possibility, capability, annual yield; — **globale,** total yield capacity; — **par contenance,** capability by area; — **par pieds d'arbre,** by single trees.

possible, *m.* possible, utmost; **autant que** —, as soon as possible.

possible, possible.

possiblement, possibly.

postal, -aux, postal.

postament, *m.* postament.

poste, *f.* post, mail, post office; *m.* post, station, position, place, appointment, shift, turn of duty, set, apparatus, operator's tower; — **á étincelles,** spark station; — **de commande,** operator's tower; — **d'équipage,** crew's quarters; — **émetteur à lampes,** vacuum-tube transmitting station.

postel, *m.* teasel.

poster, to post, station, place (men).

postérieur, posterior, hind, later, subsequent.

postérieurement, subsequently.

postérité, *f.* posterity.

postface, *f.* postscript, notice at the end of a book.

postfloration, *f.* postfloration.

posthume, posthumous.

postiche, superadded, artificial, false, provisional.

postintestin, *m.* hind intestine.

postphyllome, *m.* postphyllome, leaves.

posttrophophylle, *f.* posttrophophyll, sporophyll.

posttrophosporophylle, *f.* posttrophosporophyll, leaf serving nutrition and reproduction, sporophyll.

postulant, *m.* applicant, candidate.

posture, *f.* posture, attitude, position.

pot, *m.* pot, pitcher, jug, can, jar, crock; — **à colle,** glue pot; — **à eau,** water pitcher; — **à fleurs,** flower pot, garden pot; — **au beurre,** figwort, pilewort, lesser celandine (*Ficaria verne*); — **de fusion,** crucible, (s)melting foot.

potabilité, *f.* potability, drinkableness.

potable, drinkable, drinking.

potage, *m.* soup.

potager, *m.* kitchen garden, vegetable garden, kitchen stove, dinner pail.

potager, -ère, kitchen, culinary, garden, edible.

potamot, *m.* pondweed (Potamogeton).

potasse, *f.* potash, (specif.) caustic potash, potassium hydroxide, potassium carbonate; — **à la chaux,** caustic potash, potassium hydroxide; — **caustique,** caustic potash, potassium hydroxide; — **de suint,** potash from suint; — **nitratée,** potassium nitrate, saltpeter; — **perlasse,** pearlash(es), potassium carbonate; — **vitriolée,** potassium sulphate.

potassé, containing potash, combined with potassium.

potasser, to work at, study, grind.

potasserie, *f.* potash factory.

potasside, *m.* potassium or any of its compounds.

potassique, potassium, containing potassium, potassic.

potassium, *m.* potassium.

pot-au-feu, *m.* soup pot, beef broth.

pot-de-vin, *m.* gratuity, bonus.

poteau, *m.* post, pole, stake, prop; — **d'éclairage,** mast, pole, post, standard.

poteau-frontière, boundary post, landmark.

potée, *f.* potful, putty, luting loam, jeweller's red, red-ocher solution; — **d'émeri,** emery dust, emery slime, flour emery; — **de montagne,** rottenstone, ground pumice; — **d'étain,** putty powder.

potence, *f.* crutch, gallows, gibbet, support, arm, crosspiece, bracket, tube carrier, crane; **en — T,** T-shaped.

potentiel, *m.* potentiality.

potentiel, -elle, potential.

potentiellement, potentially.

potentille, *f.* cinquefoil, five-finger (Potentilla); — **ansérine,** silverweed, goose grass (*Potentilla anserina*).

potentiomètre, *m.* potentiometer, voltage divider.

poterie, *f.* pottery, vessels, earthenware, baked clay; — **de grès,** stoneware; — **d'étain,** tinware. pewter dishes or vessels;

— **de terre,** earthenware, (coarse) pottery, crockery; — **émaillée,** glazed (enameled) pottery; — **mate,** unglazed pottery; — **vernisée,** glazed pottery; **terre à —,** potter's clay.

potiche, *f.* potiche, flask, vase, Chinese or Japanese porcelain, test cut in timber.

potier, *m.* potter; — **d'étain,** tinman, pewterer.

potin, *m.* pinchbeck, brass, cock metal, pot metal.

potion, *f.* potion, draft, mixture.

potiron, *m.* pumpkin; **huile de —,** pumpkinseed oil.

pou, *m.* louse; — **des rameaux du sapin,** fir-tree louse; — **des serres,** white mealy bug; — **du mélèze,** larch aphis.

pouce, *m.* thumb, great toe, inch; — **carré,** square inch.

poucettes, *f.pl.* manacles, handcuffs.

poud, *m.* pood (Russian measure = 36 lb.).

poudingue, *m.* conglomerate, pudding stone.

poudrage, *m.* dusting.

poudre, *f.* dust, powder, explosive, pounce; — **à lessiver,** washing powder; — **à lever,** baking powder; — **à nettoyer,** cleaning powder; — **à polir,** polishing powder; — **à tremper,** tempering powder, cementing powder; — **de bois,** wood dust; — **de charbon,** coal dust, pounded charcoal; — **de chasse,** sporting powder, shooting powder; — **de coton,** guncotton; — **de lait,** milk powder, powdered milk, dried milk; — **de liège,** cork dust, cork powder.

poudre, — **dentifrice,** tooth powder; — **de plomb,** dust shot, lead powder; — **de riz,** rice powder, face powder; — **de savon,** soap powder; — **des épices,** aromatic powder; — **d'os,** bone meal, bone dust; — **en grains,** grained powder; — **fulminante,** fulminating powder, detonating powder, flash (light) powder; — **insecticide,** insect powder, insecticide; — **lissée,** glazed powder; — **météorifuge,** antiflatulent powder (for cattle); **en —,** powdered.

poudre-coton, *m.* guncotton.

poudre-éclair, *f.* flash powder.

poudrement, *m.* powdering.

poudrer, to powder, dust, pounce.

poudrerie, *f.* powder (mill), powder works.

poudrette, *f.* fine dust, dried and powdered night soil (for manure).

poudreux, -euse, dusty, powdery.

poudrier, *m.* worker in a powder mill.

poudroyer, to dust, be dusty, cover with dust.

pouillot, *m.* willow warbler; — **siffleur,** wood warbler; — **véloce,** chiffchaff.

poulailler, *m.* hen house, poultry house.

poulain, *m.* colt, young horse, foal.

poularde, *f.* (table) fowl.

poule, *f.* hen, fowl; — **couveuse,** sitting hen; — **d'eau,** moor hen; — **d'ornement,** ornamental fowl, fancy fowl; — **grasse,** goosefoot (*Chenopodium album*); — **naine,** bantam; **acier** —, blister steel.

poulet, *m.* chicken, fowl.

poulette, *f.* pullet.

poulevrin, *m.* priming powder.

pouliche, *f.* filly (foal).

poulie, *f.* pulley, block, trochlea; — **à gorge(s),** grooved pulley; — **folle,** loose pulley, idle pulley.

pouliner, to foal.

pouliot, *m.* pennyroyal (*Mentha pulegium*); — **américain,** American pennyroyal (*Hedeoma pulegioides*).

poulpe, *m.* octopus, poulp.

pouls, *m.* pulse; — **fréquent,** rapid pulse.

poumon, *m.* lung; **à pleins** —s, with deep breaths.

poupée, *f.* doll, manikin, puppet, finger bandage, poppet, headstock, crown graft.

pour, *m.* pro (and con).

pour, for, per, to, in order to, on account of, instead of, in order to, towards; — **ainsi dire,** so to speak; — **autant que,** in so far as; — **cent,** per cent; — **ce que,** because of; — **peu que,** if only, ever so little; — **que,** in order that, that, so that, for . . . to; **pour . . . que,** however, although.

pourboire, *m.* tip.

pourceau, *m.* hog, swine, pig.

pour-cent, *m.* per cent.

pourcentage, *m.* percentage; — **en volume,** percentage by volume.

pourparler, *m.* parley, conference.

pourpier, *m.* purslane.

pourpre, purple, scarlet, crimson.

pourpré, pourpreux, -euse, pourprin, p''rple or crimson, purplish.

pourquoi, why, wherefore.

pourra (pouvoir), (he) will be able.

pourri, *m.* rotten or decayed part.

pourri, rotten, rotted, putrid, spoiled, putrefied, fragile, brittle.

pourridié, *m.* — **du chêne,** oak disease on seedlings; — **des racine,** honey fungus, palisade fungus, shoestring rot.

pourrir, to rot, decay, putrefy, spoil, age, mature, ripen, cause to decay.

pourrissable, liable to rot, putrefiable.

pourrissage, *m.* rotting, aging, steeping.

pourrissant, rotting, putrefying.

pourrissoir, *m.* steeping vat.

pourriture, *f.* rot, rottenness, decay, putrefaction, rotten part, gangrene, spoilage; — **complète,** trunk rot; — **de l'aubier,** bluing, blue sap, sap rot; — **du bois du chêne,** white rot of oak; — **du coeur,** heart rot; — **du trèfle,** wilt of clover; — **grise,** gray mold; — **rose,** pink rot; — **rouge,** red rot; — **sèche,** dry rot; — **sous l'écorce,** white rot.

poursuite, *f.* pursuit, suit.

poursuivre, to pursue, chase, prosecute, continue, sue, proceed, follow.

pourtant, however, nevertheless, yet.

pourtour, *m.* circumference, environs, periphery, circuit, compass; — **du limbe,** leaf margin.

pourverra, (he) will provide.

pourvit, (he) provided.

pourvoi, *m.* appeal.

pourvoir, to provide (for), see (to), attend (to), furnish, endow, supply; **se** —, to provide oneself (with), appeal.

pourvu, provided (with); — **que,** provided (that), so long as, if only.

pouset, *m.* a red color obtained from cochineal.

pousse, *f.* growth, shoot, sprout, sprouting, scion, spray, ropiness of wine; — **de la 2ᵉ sève,** Lammas shoot, second shoot; — **principale,** leading shoot; — **terminale,** leading shoot, top shoot, leader.

poussée, *f.* thrust, pressure, push, sprouting, growth; — **de vent,** wind pressure.

pousser, to push, put forth, drive, thrust, shove, grow, sprout, shoot forth, shoot out, send forth, extend, carry, pursue, show, utter, stir up, actuate, urge, force; **se** —, to be pushed, elbow one's way, push oneself forward; — **des rejetons,** to sprout; — **des tiges,** to trail, sprout tendrils.

poussier, *m.* coal dust, screenings, gunpowder dust; — **de charbon,** charcoal dust; — **de minerai,** pulverized ore, ore slime.

poussière, *f.* dust, fine spray, powder, pollen; — de **charbon,** charcoal dust; — de **tourbe,** peat dust.

poussiéreux, -euse, dusty.

poussif, -ive, broken-winded, pursy.

poussin, *m.* chick, chicken; — à **rôtir,** roasting chicken.

poussoir, *m.* pusher, push button, valve lifter, driver, punch.

poussolane, *f.* see POUZZOLANE.

poutrage, *m.* framework; — **creux,** keel tunnel.

poutre, *f.* beam, rafter, girder, balk; — à **âme jumelée,** twin girder; — à **âme pleine,** solid girder; — de **rive,** shore girder; — en **treillis,** lattice girder.

poutrelle, *f.* small beam, joist.

pouture, *f.* stable fattening.

pouvoir, to be able, can, may be allowed; se —, to be possible, can be done; il n'en **peut plus,** he cannot stand it any longer; il n'y **peut rien,** he can do nothing; il se **peut,** it may be, it is possible; il se **pourrait,** it might be.

pouvoir, *m.* power, authority, might, capacity; — **adhérent,** adhesive power or force; — **agglutinant,** binding power; — **calorifique,** calorific power, heating power; — **conducteur,** conducting power, conductivity, conductibility; — des **pointes,** action of points, point effect; — **éclairant,** illuminating power; — **résolvant,** resolving power (of lens), capacity for showing delicate structure; — **rotatoire,** rotatory power.

pouzzolane, *f.* pozzuolana, pozzolana; — en **pierre,** trass.

practicable, possible, practicable, feasible.

prairial, plante —e, meadow plant.

prairie, *f.* meadow, prairie; — **tremblante,** trembling bog.

pralin, *m.* a mixture of earth and water for fertilizing and coating roots.

pralinage, *m.* treatment with "pralin."

praline, *f.* praline, burnt almond, crisp almond.

praliner, to brown almonds, make like pralines, treat with "pralin."

prame, *f.* praam, pontoon, barge.

prase, *f.* prase.

praséodyme, *m.* praseodymium.

praséolite, *f.* praseolite.

prasin, light green.

praticabilité, *f.* practicability, feasibility.

praticable, practicable, passable, feasible.

praticien, *m.* practitioner, artisan.

praticien, -enne, practicing, practical.

pratique, *f.* practice, use, method, exercise, experience, usage, routine, general use, customer.

pratique, practical, useful, experienced.

pratiquement, practically, in practice.

pratiquer, to practice, employ, use, do, perform, conduct, construct, make, contrive, cut, form, frequent, obtain; se —, to be practiced, be usual, be customary.

pré, *m.* meadow, pasture.

préachat, *m.* prepayment.

préacheter, to prepay.

préalable, *m.* preliminary.

préalable, previous, preliminary; au —, to begin with, previously, first of all.

préalablement, previously, first.

préambule, *m.* preface, introduction.

préaviser, to forewarn.

pré-bois, *m.* forest open to pasture, grazing forest.

précaire, precarious, delicate.

précairement, precariously.

précambrien, -enne, Pre-Cambrian.

précarité, *f.* precariousness, uncertainty.

précaution, *f.* precaution, prophylaxis.

précautionner, to warn, caution; se —, to guard against, provide.

précédemment, previously, already, before, above.

précédence, *f.* precedence, priority.

précédent, *m.* precedent.

précédent, preceding, precedent, former, previous.

précéder, to precede, have or take precedence over.

prêcher, to preach.

précieusement, very carefully, preciously, affectedly.

précieux, -euse, precious, valuable, costly, useful, affected.

précipice, *m.* precipice.

précipitable, precipitable.

précipitamment, precipitately, hurriedly.

précipitant, *m.* precipitant.

précipitant, precipitating.

précipitation, *f.* precipitation, acceleration (of the pulse), rainfall, precipitin reaction.

précipité, *m.* precipitate, precipitation, deposit; — **pulvérulent,** pulverulent precipitation; — **rouge,** red precipitate, red mercuric oxide.

précipité, precipitated, hasty, precipitous, steep.

précipiter, to precipitate, hurl, hurry; **se —,** to be precipitated, precipitate, precipitate oneself, spring forth, dash headlong, rush.

précipitine, *f.* precipitin; **—s de groupe,** group precipitins, cross-precipitins, co-precipitins.

précis, *m.* abstract, summary, epitome.

précis, precise, exact, definite, just.

précisément, precisely, exactly, just.

préciser, to state precisely, specify, define, determine exactly, state exactly.

précision, *f.* precision, preciseness, exactness; *pl.* data.

précité, previously cited, aforesaid, above-mentioned.

précoce, precocious, early, rapid (for methods).

précocité, *f.* precocity, prematureness, early maturity.

précomptage, *m.* deduction from the fixed annual yield.

préconcevoir, to preconceive.

préconçu, preconceived.

préconiser, to recommend, commend, praise, extol.

précurseur, *m.* precursor, harbinger, forerunner.

précurseur, precursory, premonitory.

prédécesseur, *m.* predecessor.

prédestiner, to predestinate.

prédiction, *f.* prediction, prophecy.

prédire, to predict, foretell.

prédisposer, to predispose.

prédominance, *f.* predominance.

prédominant, predominating, predominant, prevailing.

prédominer, to predominate.

prééminence, *f.* pre-eminence.

prééminent, pre-eminent.

préenlèvement, *m.* sampling.

préexistant, pre-existing, pre-existent.

préexister, to pre-exist, exist before.

préface, *f.* preface.

préférable, preferable.

préférablement, preferably.

préférant, particular (to a soil).

préférence, *f.* preference, distinction, mark of preference, choice; **de —,** preferably, rather.

préférer, to prefer, choose.

préfet, *m.* prefect, administrator of a French department.

préfeuille, *f.* prophyll.

préfixe, *m.* prefix.

préfloraison, *f.* prefloration, estivation.

préfoliaison, *f.* prefoliation, vernation, leafing.

préfoliation, *f.* prefoliation, estivation.

prégnant, pregnant, with young (animal).

préhensile, prehensile, adapted for catching hold.

prehnite, *f.* prehnite.

prehnitène, *m.* prehnitene.

prehnitique, prehnitic.

préintestin, *m.* fore intestine.

préjudice, *f.* prejudice.

préjudiciable, prejudicial, injurious, detrimental, harmful.

préjudiciel, -elle, preliminary, interlocutory.

préjudicier, to be prejudicial or detrimental.

préjugé, *m.* precedent, prejudice, presumption.

préjuger, to prejudge, foresee, predict.

prélart, prélat, *m.* tarpaulin.

prèle, prêle, *f.* horsetail (Equisetum,; — **des champs,** field horsetail, bottle brush, cat's-tail (*Equisetum arvense*).

prélèvement, *m.* deduction in advance, setting apart, sample, retained portion.

prélever, to retain, save out, remove, draw out, take (from), select, take off first, deduct (previously).

préliminaire, preliminary.

préliminairement, preliminarily, first.

prélude, *m.* prelude.

préluder, to prelude.

prématuré, premature.

prématurément, prematurely.

prémédité, premeditated.

préméditer, to premeditate.

prémices, *f.pl.* first fruits; — **du bétail,** firstlings of cattle.

premier, *m.* first, former, chief, leader.

premier, -ère, first, raw (materials), prime, former, early, foremost; — **choix,** first choice, best grade; — **en date,** first in the order of time; — **étage,** first (floor) story (above the ground floor); — **jet.**

first casting, first cast, rough sketch; **en** —, first, directly; **du — coup,** at the first attempt, directly, at once.

premierement, first, firstly, in the first place.

prémolaire, *f.* fore molar, fore grinder.

prémunir, to caution, forewarn; **se** —, to provide, be provided.

prémutation, *f.* plant entering mutation with new characteristics already latent.

prendre, to take, catch, seize, assume, make, get, receive, purchase, take on, adopt, solidify, cake, congeal, cement, freeze, curdle, set, take root, take hold, have effect, deduct; — **à,** to take from; — **à bail,** to lease, rent; — **appue,** to rest; — **congé de,** to take leave of; — **en recette,** to credit on account; — **feu,** to take fire, catch fire, ignite.

prendre, — **garde,** to take care, take heed; — **garde à ne pas,** to take care not to; — **naissance,** to originate, arise, appear, take rise; — **place,** to take place, occur, fit; — **racine,** to take root; — **retrait,** to shrink; **se** —, to be taken, take each other, be caught, solidify, congeal, curdle, coagulate, cake, set, set about, begin; **s'en — à,** to blame; **s'y** —, to go about it.

prendre, cela se prend, it is infectious; **il le prend dans,** he takes it from; **il le lui prend,** he takes it from him; **il prend fin,** it is coming to an end; **il prend par là,** he is going in that direction; **il prend parti pour,** he sides with; **il prend son parti,** he makes up his mind; **il se prend à,** he begins to; **il s'y prend pour,** he is going about it to; **qu'est-ce qui te prend?** what is the matter with you?

preneur, *m.* taker, captor, catcher, buyer, purchaser, lessee.

prénom, *m.* first name, Christian name.

prenons (prendre), let us take.

préoccupation, *f.* preoccupation, worry.

préoccuper, to preoccupy, engross, bother, be greatly occupied; **se — (de),** to preoccupy oneself (with), give attention (to), notice, be preoccupied (with), busy oneself.

préparant, preparing.

préparateur, -trice, *m.&f.* preparer (of specimens), mixer, maker, manufacturer; — **de laboratoire,** laboratory (lecture) assistant.

préparatifs, *m.pl.* preparations.

préparation, *f.* preparation, dressing, slide; — **du sol,** preparation of the soil; — **par impression,** impression preparation.

préparatoire, preparatory, preparative.

préparer, to prepare, make ready, dress (leather), cultivate, work; **se** —, to be prepared, prepare oneself, prepare, get ready; — **le sol,** to till, prepare the soil.

prépareur, *m.* preparer, mixer, demonstrator.

prépondérance, *f.* preponderance.

prépondérant, preponderant, prevailing, overpowering.

préposé, *m.* officer, agent, forest guard.

préposer (à), to set, place (over), put in charge (of).

prérogative, *f.* prerogative.

près, near, nearly, by, close to, close by, hard by; — **de,** near, close to, nearly, almost, about; **à** —, within, except for; **à peu** —, nearly, almost, practically, pretty nearly; **de** —, near, closely.

présage, *m.* omen, presage, foreboding.

présager, to presage, show, portend.

presbyte, presbyopic, affected with presbyopia.

presbytie, *f.* presbyopia, diminished power of accommodation for near objects.

prescription, *f.* prescription, limitation, superannuation, regulation; *pl.* specifications.

prescrire, to prescribe, enjoin.

prescutum, *m.* protergum, prescutum.

préséance, *f.* precedence, priority.

présence, *f.* presence, occurrence; **en — de,** in the presence of.

présent, *m.* gift, present.

présent, present; **à** —, at present, now.

présentation, *f.* presentation, appearance.

présentement, now, at present, presently.

présenter, to present, offer, show; **se** —, to appear, present oneself, be presented, occur.

préservateur. -trice, preservative, preserving.

préservatif, -ive, preservative, prophylactic.

préservation, *f.* preservation.

préserver, to preserve, keep, save.

présidence, *f.* presidency.

président, *m.* president.

présider, to preside, direct.

présomption, *f.* presumption.

présomptivement, presumptively.

presque, almost, nearly, scarcely, hardly, all but; — **jamais,** hardly ever, almost never; — **personne,** hardly anyone.

presqu'île, *f.* peninsula.

pressage, *m.* pressing.

pressant, pressing, urgent.

presse, *f.* (printing) press, clamp, crowd, haste, urgency, rush, clingstone peach, pressure, congestion; — **à bras,** hand press; — **à cintrer,** bending or shaping press; — **à foin,** hay press; — **à fromage,** cheese press; — **à huile,** oil press; — **à imprimer,** printing press.

presse, — **à jus,** press for extracting juices; — **à main,** (screw) clamp, hold fast; — **à paille,** straw compressor; — **à serrer,** clamp, holdfast; — **à tourbe,** peat press; — **à vis,** screw press, fly press; — **roulante,** employed press; **sous** —, in press.

pressé, pressed, crowded, in a hurry, in haste, very busy, urgent.

presse-artère, *m.* artery forceps.

pressée, *f.* pressing, pressure, amount (of juice) expressed, cider-pressful (of apples).

pressentiment, *m.* presentiment, misgiving.

pressentir, to have a presentiment of, forebode.

presse-papiers, *m.* paperweight.

presser, to press, crowd, compress, be urgent, be severe, urge, squeeze; **se** —, to be in a hurry, hasten, press, hurry.

pressette, *f.* small press.

presseur, *m.* presser, pressman.

presseur, -euse, pressing, pressure.

pression, *f.* pressure, compression, head, tension; — **du couvert,** suppression by vertical shade; — **effective,** pressure above atmospheric; — **latérale,** by lateral shade; — **osmotique,** osmotic pressure; **en** —, under pressure; **haute** —, high pressure; **moyenne** —, mean pressure.

pressis, *m.* expressed juice (from meat and vegetables).

pressoir, *m.* wine press, pressroom, press house.

pressurage, *m.* pressing, pressure, wine from second pressing.

pressurer, to press, squeeze, express.

prestance, *f.* portliness, bearing.

preste, quick, sharp.

prestement, quickly, nimbly.

prestesse, *f.* agility.

présumer, to presume, suppose.

présupposer, to presuppose.

présure, *f.* rennet, curds, rennet extract.

présurer, to curdle with rennet.

présurier, *m.* rennet maker.

prêt, *m.* loan, pay, exchange.

prêt, ready.

prétendre, to pretend, claim, lay claim to, maintain, aspire, require, lay, allege, mean, intend, assert, hope.

prétendu, pretended, alleged, so-called, supposed, false, sham.

prête-nom, *m.* person lending his name, figurehead, dummy.

prétention, *f.* pretension, claim.

prêter, to lend, loan, adapt, attribute, take (an oath), swear, give, stretch (cloth), give rise (to); **se** —, to be lent, lend oneself, favor, yield, lay itself open to, adapt oneself, lend each other.

prêteur, -euse, *m.&f.* lender.

prêteur, -euse, lending.

prétexte, *m.* pretext, pretense, show.

prétexter, to pretend.

prêtre, *m.* priest.

preuve, *f.* proof, evidence, reason, testing, argument; **en** —, as a proof; **faire** — **de,** to prove.

prévaloir, to prevail, get the advantage; **se** —, to take advantage (of), avail oneself (of).

prévalu, prevailed.

prévenant, kind, considerate, pleasing.

prévenir, to prejudice, predispose, inform, notify, prevent, warn, forestall, anticipate, precede, prepossess.

préventif, -ive, preventive, prophylactic.

prévenu, *m.* defendant.

prévenu, informed, prejudiced, biased, accused, charged.

prévision, *f.* anticipation, prevision, foresight, conjecture.

prévit (prévoir), (he) foresaw.

prévoir, to foresee, anticipate, provide (for).

prévoyance, *f.* foresight.

prévu (prévoir), foreseen, provided for.

prier, to pray, beseech, beg; **je vous en prie,** please.

prière, *f.* prayer, request, entreaty.

primaire, primary, primeval.

primauté, *f.* primacy, priority, preeminence.

prime, *f.* premium, bounty, subsidy, prism.

prime, prime, first; **de** — **abord,** to begin with, at first; **de** — **face,** at first sight; **de** — **saut,** in the first impulse, at the first attempt, at once.

primé, awarded with a prize, prize.

primer, to lead. take the lead, excel, surpass, take precedence of, come before, prime, award a prize to (cattle), precede, dig, harrow, dress (ground) for the first time.

primerolle, *f.* primrose, cowslip.

primerose, *f.* primrose (Primula), hollyhock (*Althaea rosea*).

primeur, *f.* freshness, newness, early season, early product, bloom, first of the season (flowers, fruits).

primevère, *f.* primrose, cowslip (*Primula veris*).

primine, *f.* first integument.

primitif, -ive, original, primitive, first, former, primeval, primary (colors).

primitivement, primitively, originally, at first.

primorde, *m.* primordium, an organ in its earliest condition.

primordial, primordial.

primulacées, *f.pl.* Primulaceae.

primuline, *f.* primuline.

principal, *m.* principal, principal thing, chief thing.

principal, -aux, principal, chief, main.

principalement, principally, chiefly.

principe, *m.* principle, element, source; — **actif,** agent; — **amer,** bitter principle; — **gras,** fatty principle; — **immédiat,** constituent, immediate principle; — **minéral,** ash constituent; — **nutritif,** nourishing substance; **en —,** as a principle, as a rule.

printanier, -ère, vernal, spring, early.

printemps, *m.* spring (the season).

priori (à), a priori.

priorité, *f.* priority.

prirent (prendre), (they) took.

pris (prendre), taken, caught, solidified, curdled.

prisable, estimable, valuable.

prise, *f.* hold, grasp, grip, solidification, congealing, setting, coagulation, taking, engagement, purchase, intake (of air), supply (of water), steam valve, steam cock, holding, clinch, dose, capture, prize, quarrel, fighting, felting, matting, meshing, pitching.

prise, — **cohésive,** cohesion, cohesive attraction; — **d'eau,** forebay, inlet; — **de courant,** current supply, current collection; — **d'essai,** sample; — **de vues,** picture taking; — **directe,** direct transmission; **aux —s avec,** struggling with; **en —,** in gear, geared.

priser, to appraise, value, snuff, prize.

prismatique, prismatic(al).

prismatocarpe, with prismatic fruit.

prisme, *m.* prism.

prismoïde, prismoidal.

prison, *f.* prison.

prisonnier, -ère, *m.&f.* prisoner.

prit (prendre), (he) took.

privation, *f.* privation, deprivation.

privé, *m.* private, privy.

privé, private, deprived, tame.

priver, to deprive (of), domesticate, tame; **se —,** to do without.

privilège, *m.* privilege.

privilégié, privileged, favored.

privilégier, to grant a privilege, privilege.

prix, *m.* price, reward, prize, cost, value, worth; — **courant,** market price, current price, rate; — **coûtant,** cost of production, first cost; — **d'achat,** purchase price; — **de détail,** consumer's price; — **de marché,** market price; — **de revient,** cost price, cost, net cost, producer price; — **de vente,** selling price; **à tout —,** at any price; **au — de,** at the price of, in comparison with.

prix-courant, *m.* price list.

proanthèse, *f.* proanthesis, flowering in advance of normal period.

probabilité, *f.* probability, likelihood.

probable, probable, likely.

probablement, probably, likely.

probaside, *f.* probasid, organ intermediate between a basidium and a sporophore.

probe, upright, honest.

probité, *f.* integrity, honesty.

problématique, problematic(al).

problème, *m.* problem.

proboscide, *f.* proboscis.

proboscidé, proboscideous, having a large terminal horn.

Proboscidiens, *m.pl.* Proboscidea.

procambiale, provascular, primary, primeval.

procambium, *m.* procambium, provascular strand.

procarpe, *m.* procarp.

procédé, *m.* process, way, method, procedure, proceeding, behavior, conduct; — **à chaud,** hot process; — **cultural,** cultural system; — **de contact,** contact process; — **des meules,** process of charring wood in heaps; — **expéditif**, quick (rapid) process; — **par contact**, contact process; — **par coupe unique**, clear-cutting system; — **par coupes successives**, successive-regeneration felling system.

procéder, to proceed, go about, go ahead (with), carry on, come from.

procédure, *f.* procedure, proceedings.

procès, *m.* process, trial, method, suit.

processionaire, *f.* — **du chêne**, oak procession moth; — **du pin**, pine procession moth.

processus, *m.* process, progress, course; — **suppuratif**, suppurative process.

procès-verbal, *m.* report, proceedings, minutes.

prochain, *m.* neighbor.

prochain, close, soon, next, early, near, nearest, coming, approaching.

prochainement, presently, soon, shortly.

proche, *m.* relative.

proche, near (to), next, nigh; **de — en —**, gradually, from place to place, from one place to another; **tout —**, close at hand.

proclamer, to proclaim, announce.

proclivité, *f.* slope, incline.

procuration, *f.* proxy, power of attorney.

procurer, to procure, obtain, provide, get; **se —**, to be procured.

prodiastase, *f.* proenzyme, zymogen.

prodigalement, prodigally, lavishly.

prodigalité, *f.* prodigality, lavishness.

prodige, *m.* prodigy, wonder.

prodigieusement, prodigiously, wonderfully.

prodigieux, -euse, prodigious, wondrous.

prodigue, prodigal, lavish, unsparing.

prodiguer, to lavish, waste, be prodigal of.

prodrome, *m.* preamble, introduction, prodrome.

producteur, -trice, *m.&f.* producer.

producteur, -trice, productive, producing, generating.

productible, producible.

productif, -ive, productive, fruitful.

production, *f.* production, output, yield, product, growth, berth, prolongation, manufacture, extraction; — **du bois**, production of wood; — **en matière**, volumetric yield, yield in material.

productivité, *f.* productivity, productiveness.

produire, to produce, bear, yield, introduce, bring forth; **se —**, to occur, show oneself, be produced, take place.

produisant, producing, introducing.

produit, *m.* product, produce, proceeds, yield; — **abrasif**, grinding material; — **accessoire**, accessory product, by-product; — **accidentels**, incidental produce, accident yield; — **alimentaire(s)**, food product, victuals, provisions, food; — **antirouille**, rust inhibitor, rust-proofing compositions; — **brut**, raw (product) material, gross proceeds.

produit, — **d'éclaircie**, thinning material; — **de coupure**, split product; — **de déchet**, waste product; — **de dédoublement**, product of disunion or separation; — **de désagrégation**, product of disintegration; — **de la coupe**, outturn of a felling, yield from felling; — **de laiterie**, dairy industry; — **de l'assimilation**, product of metabolism.

produit, — **en argent**, revenue; — **en matière**, volumetric yield; — **intermédiaire**, intermediate product, intermediate product of thinnings; —**s accidentels**, incidental products; —**s extraordinaires**, extra fellings; —**s ligneux**, —**s en nature de bois**, wood; —**s principaux**, principal produce; —**s secondaires**, by-products, residual products.

produit, produced, introduced.

proembryon, *m.* proembryo.

proéminence, *f.* prominence.

proéminent, prominent, protuberant.

proenzyme, *f.* proenzyme, zymogen.

profanation, *f.* profanation.

profaner, to profane.

proferment, proenzyme, zymogen.

professer, to profess, practice, teach, lecture, be a professor; — **un cours**, to give a course.

professeur, *m.* professor, teacher.

profession, *f.* profession, occupation, trade, calling, declaration.

professionnel, -elle, professional.

profil, *m.* profile, section, cross section; — **en travers**, cross profile.

profit, *m.* profit, advantage, gain, benefit.

profitable, profitable.

profiter, to profit, gain, derive profit, be profitable, take advantage (of), thrive, avail oneself, avail of.

profond, *m.* depth.

profond, deep, profound, deep-seated, complete, heavy, sound (of sleep); **peu —,** slight, shallow.

profondément, deeply, profoundly, deep, soundly.

profondeur, *f.* depth, profundity, thickness; **— de foyer,** depth of focus.

profus, profuse.

profusément, profusely.

profusion, *f.* profusion, abundance; **à —,** in profusion.

progamète, *f.* progamete, a cell dividing to form gametes, mother cell.

progéniture, *f.* offspring, progeny.

programme, *m.* program.

progrès, *m.* progress, improvement, advancement.

progresser, to progress, improve.

progressif, -ive, progressive.

progression, *f.* progression, progress.

progressivement, progressively, by degrees, gradually.

progressives, *f.pl.* root branches elongating and ramifying in the ground.

progressivité, *f.* progressiveness.

prohiber, to prohibit.

prohibitif, -ive, prohibitive.

prohibition, *f.* prohibition; **— de pâturage,** closure against grazing.

proie, *f.* prey, spoil, plunder; **en — à,** suffering from.

projecteur, *m.* projector, searchlight, headlight.

projectif, -ive, projective.

projectile, *m.* projectile, shot, missile.

projection, *f.* projection, scattering, projecting, splashing, casting.

projecture, *m.* projectura, a longitudinal projection on stems where leaf originates.

projet, *m.* project, design, plan; **— d'amenagement,** working plan.

projeter, to project, throw out, cast, forward, plan, purpose, design, throw; **se —,** to put out, project, be projected, stand out.

prokaryogamète, *f.* prokaryogamete, nucleus of a primary progamete.

prolepsis, *f.* prolepsis.

prolifération, prolification, *f.* proliferation.

prolifère, proliferous, prolific, fruitful.

proliférer, to proliferate.

prolifique, prolific.

proligère, proligerous.

proline, *f.* proline.

prolongation, *f.* prolongation, extension.

prolongé, prolonged, protracted.

prolongement, *m.* prolongation, extension, lengthening.

prolonger, to prolong, continue, extend, lengthen, produce; **se —,** to be prolonged, extend, continue.

promenade, *f.* drive, riding for pleasure, walk, promenade.

promener, to lead, direct, guide, flutter, display, lead about; **se —,** to go for a walk, promenade, walk, go.

promeneur, *m.* walker, pedestrian, promenader.

promesse, *f.* promise.

promettre, to promise.

promeut (promouvoir), (he) is promoting.

promis, promised.

promission, *f.* promise.

promit, (he) promised.

promontoire, *m.* promontory, headland, prominence, mountain spar, foothill.

promoteur, -trice, *m.&f.* promoter.

promoteur, -trice, promoting.

promouvoir, to promote.

prompt, prompt, rapid, ready, quick.

promptement, promptly, quickly, soon, readily.

promptitude, *f.* promptness, readiness, speediness, promptitude.

promu (promouvoir), promoted.

promulguer, to promulgate.

promycélium, *m.* promycelium.

prononcé, decided, pronounced.

prononcer, to pronounce, utter, speak, mark distinctly, deliver, bring out; **se —,** to pass judgment.

prononciation, *f.* pronunciation, pronouncement.

pronostic, *m.* prognostic, sign, prognosis, forecast.

pronostiquer, to prognosticate, predict, foretell.

pronotum, *m.* pronotum.

pronucléus, *m.* pronucleus.

propagande, *f.* propaganda, advertisement.

propagateur, -trice, propagating, spreading.

propagation, *f.* propagation, propagating, diffusion, spread.

propager, to propagate, extend, carry on; **se —,** to be propagated, spread, diffused, procreate, reproduce.

propagine, *f.* propagine, a bulblet.

propagule, *m.* bud for propagating, offset.

propane, *m.* propane.

propanoïque, propionic, propanoic.

propargylique, propiolic, propargyl(ic).

propension, *f.* propensity, tendency.

propényle, *m.* propenyl.

propeptone, *f.* propeptone.

propeptonique, propeptonic.

prophase, *f.* prophase.

prophète, *m.* prophet.

prophétiser, to prophesy.

prophylactique, prophylactic(al), preventive.

prophylaxie, *f.* prophylaxis, prevention.

prophylle, *m.* prophyll.

prophylloïde, prophylloid, like prophylla.

prophyse, *f.* sterile organs of mosses.

propice, propitious, favorable.

propiolique, propiolic.

propionate, *m.* propionate.

propione, *f.* propione.

propionique, propionic.

propolis, *f.* propolis, bee glue.

proportion, *f.* proportion, ratio, percentage; **à —, en —,** in proportion, proportionately; **à — de, en — de,** in proportion to; **à — que,** in proportion as.

proportionnalité, *f.* proportionality.

proportionné, proportioned, proportionate, suited.

proportionnel, -elle, proportional; **proportionelle moyenne,** mean proportional.

proportionnellement, proportionally, in proportion to.

proportionnément, proportionately, in due proportion.

proportionner, to proportion, gauge, fit; **se —,** to be proportioned, suit.

propos, *m.* discourse, talk, gossip, remark, purpose, pertinency; **à —,** apropos, seasonably, opportunely, by the way; **à — de,** apropos of, with regard to, about, concerning, **à ce —,** in this connection; **à quel —?** what is it about? **hors de —,** ill-timed, inopportune(ly).

proposer, to propose, suggest, propound; **se —,** to offer oneself, propose, plan, intend.

proposition, *f.* proposition, proposal, theorem.

propre, *m.* property, nature, characteristic.

propre, proper, own, one's own, clear, right, fit, self, clean, same, very, suitable, adapted, belonging to, natural, fitted, neat.

proprement, properly, cleanly, neatly, strictly; **— dit,** properly, so-called.

propreté, *f.* cleanness, cleanliness, tidiness.

propriétaire, *m.* proprietor, owner.

propriétaire, proprietary.

propriété, *f.* property, ownership, peculiarity quality, virtue, propriety.

propugnacule, *m.* brow antler.

propulseur, *m.* propeller.

propulseur, propelling, impulsive, propulsive.

propulsif, -ive, propulsive, propelling, driving.

propulsion, *f.* propulsion, propelling.

propylamine, *f.* propylamine.

propyle, *m.* propyl.

propylène, *m.* propylene.

propylénique, of propylene.

propylique, propyl, propylic.

prorata, *m.* proportional part, share.

prosaïque, prosaic.

proscolle, *f.* proscolla, viscid gland on the stigma of an orchid.

proscription, *f.* rejection.

proscrire, to proscribe, outlaw, exile, prohibit, reject.

prosécrétine, *f.* prosecretin.

prosenchyme, *m.* prosenchyma.

prosenthèse, *f.* prosenthesis, (whorled flowers) having gaps between two successive whorls.

prospecter, to prospect.

prospère, prosperous, favorable.

prospèrement, prosperously.

prospérer, to prosper, thrive.

prospérité, *f.* prosperity.

prosporie, *f.* prospory, precocious development of spores.

prostate, *f.* prostate, prostate gland.

prosternum, *m.* prosternum.

prosthétique, prosthetic.

prostré, prostrate(d), exhausted.

protagon, *m.* protagon.

protamine, *f.* protamine.

protaminique, protamine.

protandre, protandrous, (anthers) maturing before pistils.

protandrie, *f.* protandry.

protandrique, protandrous, exhibiting protandry.

protane, *m.* methane.

protargol, *m.* protargol.

protéase, *f.* protease, enzyme capable of acting upon proteid substances.

protecteur, *m.* protector.

protecteur, -trice, protecting, protective, fostering.

protection, *f.* protection, defense, shelter; — de la santé, health protection, hygiene.

protectrice, *f.* protectress, patroness.

protéger, to protect, defend.

protéide, *f.* proteid, protein.

protéiforme, jelly-forming, infinitely variable.

protéine, *f.* protein, albumen; — conjuguée, conjugated protein; — végétal, vegetable protein.

protéiné, proteaceous, relating to the family Proteaceae.

protéique, *m.* protein substance.

protéique, protein, pertaining to protein.

protéites, *f.pl.* complex albuminoid substances.

protéisme, *m.* proteism, changing shape.

protéoalbumoses, *m.pl.* original albumoses coming from albumins.

protéoïde, *n.* albuminoid, scleroprotein.

protéolyse, *f.* proteolysis.

protéolytique, proteolytic.

protéoplastides, *f.pl.* aleurone grains.

protéose, *f.* proteose.

protérandrie, *f.* proterandry.

protéranthé, proteranthous, flowering preceding leafing, hysteranthous.

protérogynie, *f.* proterogyny.

protérogynique, protogynous.

protestation, *f.* protestation, protest.

protester, to protest.

protêt, *m.* protest.

prothalle, *m.* prothallus, prothallium.

prothorax, *m.* prothorax.

prothrombine, *f.* prothrombin.

protiodure, *m.* protoiodide.

protistes, *m.pl.* protista.

protobaside, *m.* protobasidium.

protoblaste, *m.* protoblast.

protobromure, *m.* protobromide.

protocarbure, *m.* protocarburet.

protocatéchique, protocatechuic.

protocérébron, *m.* protocerebrum.

protocharde, *f.* protoxylem group.

protochlorure, *m.* protochloride; — de cuivre, protochloride of copper, cupric chloride; — de mercure, mercurous chloride; — d'étain, stannous chloride, protochloride of tin.

protochloruré, protochloride of.

protochordes, *m.pl.* primary vascular bundle.

protochromosome, *m.* protochromosome, chromatophile granulations uniting into two chromosomes at end of prophase.

protococcus, protocoque, *m.* protococcus.

protocole, *m.* record, minutes, protocol.

protocorme, *m.* protocorm.

protocyanure, *m.* protocyanide.

protoderme, *m.* protoderm.

protogamétophyte, *m.* part of plant containing the progametes.

protogynie, *f.* protogyny.

protogynique, protogynous.

protohadrome, *m.* protohadrome, protoxylem.

protokorme, *m.* tubercular substance constituting the proembryo.

protomycélium, *m.* protomycelium, plasmic mass formed between cells of parasitic fungi.

protone, *f.* protone.

protonéma, *f.* protonema.

protopectine, *f.* protopectin.

protophosphure, *m.* protophosphide, protophosphuret.

protophylle, *f.* protophyll, a cotyledon or primordial leaf.

protophyte, *f.* protophyte.

protoplasma, protoplasme, *m.* protoplasm.

protoplasmique, protoplasmic.

protoplaste, *m.* protoplast.

protoprotéose, *f.* protoproteose.

protorganie, *f.* spontaneous generation.

protosulfate, *m.* protosulphate; — de fer, ferrous sulphate.

protosulfure, *m.* protosulphide; — de fer, iron sulphide.

protothalle, *f.* protonema.

prototype, *m.* prototype, original.

protoxyde, *m.* protoxide, monoxide; — d'azote, nitrous oxide; — de plomb, lead monoxide.

protoxyder, to protoxidize.

protoxylème, *f.* protoxylem, the first elements of wood in a vascular bundle.

protozoaire, *m.* protozoan; *pl.* Protozoa.

protrophe, protrophic.

protrusion, *f.* protrusion.

protubérance, *f.* projection, protuberance.

proue, *f.* prow, stem.

prouesse, *f.* prowess, valor, feat.

proustite, *f.* proustite.

prouvable, provable.

prouver, to prove; se —, to be proved.

provenance, *f.* source, origin, production.

provenant, proceeding, coming, arising; — de, from, proceeding from.

provende, *f.* provisions, victuals, fodder.

provenir, to proceed (from), arise, come (from), arise (from).

proventifs, *m.pl.* adventitious buds developing without apparent cause.

proverbe, *m.* proverb; faire passer en —, to make or become proverbial.

proverbialement, proverbially.

provignement, *m.* layering (of vines).

provigner, to layer (vines), increase.

provin, *m.* layer (of a vine).

province, *f.* province.

provint (provenir), (it) proceeded (from).

provision, *f.* provision, stock, store; par —, provisionally.

provisoire, provisionnel, -elle, provisional, temporary.

provisoirement, provisionally.

provoquer, to provoke, cause, challenge, induce.

proximal, -aux, proximal, (part) nearest the axis.

proximité, *f.* proximity, nearness; à — de, near to.

prudemment, prudently, carefully.

prudence, *f.* prudence, discretion, caution.

prudent, prudent, discreet, cautious.

pruine, *f.* bloom (of fruits).

pruiné, covered with bloom, velvety.

pruineux, -euse, pruinose. frosted.

prune, *f.* plum.

pruneau, *m.* prune, dried plum.

prunée, *f.* plum sauce.

prunelaie, *f.* plum orchard.

prunelle, *f.* sloe, wild plum (*Prunus spinosa*), pupil (of the eye), prunella (the fabric), Prunella.

prunellier, *m.* sloetree, blackthorn (*Prunus spinosa*), bullace tree (*Prunus insititia*).

prunier, *m.* plum (tree) (Prunus); — épineux, blackthorn (*Prunus spinosa*); — damas, damson tree (*Prunus insitia*), plum tree; — myrabolan, cherry plum (*Prunus cerasifera*); — tardif, black cherry.

prussiate, *m.* prussiate, cyanide; — jaune, — jaune de potasse, potassium ferrocyanide; — rouge, — rouge de potasse, potassium ferricyanide.

prussien, -enne, Prussian.

prussique, prussic (hydrocyanic).

psammite, *m.* psammite, sandstone.

psammogène, psammogenous, producing a sandy soil.

psammophile, psammophilous.

psammophyte, *f.* psammophyte, a sand plant.

psautier, *m.* psalterium (stomach of ruminant).

pseudaxis, *m.* pseudaxis, sympodium.

pseudobulbe, *m.* pseudobulb, a thickened and bulblike internode in orchids.

pseudobulbille, *f.* pseudobulbil.

pseudocarpe, *m.* pseudocarp.

pseudocarpien, -enne, pseudocarpous.

pseudocéphalodie, *f.* pseudocephalodium, association with an alga produced on the protothallus.

pseudo-cils, *m.pl.* pseudocilia, protoplasmic filaments of certain green algae.

pseudo-cleistogamie, *f.* pseudocleistogamy, flowers are cleistogamous only under certain external conditions.

pseudo-élatères, *f.pl.* pseudoelaters, sterile cells in spore capsules of Anthoceros.

pseudo-éphémère, *f.* pseudoephemer, flower remaining expanded a little over a day and then closing.

pseudo-fruit, *m.* pseudo fruit.

pseudo-hybride, *f.* chimera.

pseudomorphe, pseudomorphous.

pseudomorphique, pseudomorphic.

pseudomorphisme, *m.* pseudomorphism.

pseudomucine, *f.* pseudomucin.

pseudopodes, *m.pl.* pseudopodia.

pseudo-pore, *m.* pseudopore, thickened rings without perforations.

pseudo-verticille, *m.* false whorl.

psiloglotte, with elongated fruits without hairs.

psilomélane, *m.* psilomelane, hydrous manganese oxide.

psilostachyé, bare-spiked.

psocide, *f.* psocid.

psychologie, *f.* psychology.

psychophiles, *f.pl.* psychophilae, plants pollinated by diurnal Lepidoptera.

psychotrine, *f.* psychotrine.

psychroklinie, *f.* psychrokliny, bending caused by freezing.

psychrométrie, *f.* psychrometry, hygrometry.

psychrophytes, *f.pl.* psychrophytes, alpine plants.

psyllides, *f.pl.* lerp insects.

psylomie, *f.* — de la carotte, carrot fly.

ptarmique, *f.* sneezewort (*Achillea ptarmica*).

ptéricoque, with winged capsules.

ptérigène, borne on ferns.

ptérigyne, pterigynous, wing-seeded.

ptérocarpe, pterocarpous, wing-fruited.

ptérocaule, pterocaulous, wing-stemmed.

ptérocéphalé, pterocephalous, with tufted seeds.

ptérodie, *f.* fruit surrounded by membranous wing.

ptéroide, pteroid, like a fern.

ptéropodé, pteropodous, wing-footed.

ptérostyle, with style expanded in form of wing.

ptérygope, pterygopous, having the peduncle winged.

ptine, *m.* Ptinus, bookworm.

ptygmature, with folded peduncles.

ptomaine, *f.* ptomaine.

ptyaline, *f.* ptyalin, salivary ferment.

pu (pouvoir), (he has) been able.

puant, stinking, foul, offensive.

puanteur, *f.* stink, stench, foul smell.

pubère, puber.

puberté, *f.* puberty, pubescence.

pubérulent, puberulent, puberulous.

pubescence, *f.* pubescence, hairiness of plants.

pubescent, pubescent.

pubiflore, with velvety calyx and corolla.

public, -ique, public.

publicateur, *m.* publisher.

publication, *f.* publication.

publicité, *f.* publicity, advertising.

publier, to publish.

publiquement, publicly.

puce, *f.* flea; — de terre, flea beetle; — pénétrante, chigoe, jigger.

puce, puce, dark-brown.

pucelage, *m.* periwinkle (*Vinca minor*).

puceron, *m.* aphis, plant louse, green fly; — de l'orme, elm woolly aphis; — lanigère, woolly aphis.

puddlage, *m.* puddling; — au gaz, gas puddling.

puddler, to puddle.

puddleur, *m.* puddler.

puer, to stink, smell.

puéril, puerile, childish.

puerpéral, -aux, puerperal.

pugioniforme, in the form of a dagger.

puis (pouvoir), (I) can.

puis, then, next, afterward; et —, and then. moreover, and besides.

puisage, *m.* drawing (of water).

puisard, *m.* cesspool, trap, draining well, sump.

puiselle, *f.* ladle.

puisement, *m.* drawing.

puiser, to draw (water), draw off, take, draw up, derive, imbibe; — dans, draw from, take (from).

puisoir, *m.* ladle.

puisque, since, as, because.

puissamment, powerfully, exceedingly, extremely.

puissance, *f.* power, thickness, force, horse-power, sway, authority, effectiveness; — calorifique, calorific power, heating power; — de cheval, — en chevaux, horse-power; — effective, effective horsepower, real power; — en bougies, candle power; — motrice, motive power.

puissant, *m.* powerful person.

puissant, powerful, great, mighty, potent, large, rich, puissant.

puits, *m.* well, shaft, pit, water tank; — à pompe, pump well; — artésien, artesian well; — de sondage, bored well; — foré, drive well, tube well; — phréatique, shallow well.

pulégone, *f.* pulegone.

pulicaire, *f.* fleabane (Pulicaria).

pullé, pullus, nearly black, ternate brown.

pulligère, with projecting pustules.

pullulation, *f.* multiplication, rankness.

pulluler, to multiply rapidly, pullulate, swarm, multiply, grow prolifically, grow rankly.

pulmonaire, *f.* lungwort (*Pulmonaria officinalis*).

pulmonaire, pulmonary.

pulpation, *f.* pulping.

pulpe, *f.* pulp, tissue, marrow.

pulper, to pulp, reduce to pulp.

pulpeux, -euse, pulpy, pulpous.

pulque, *m.* pulque (fermented drink).

pulsateur, *m.* pulsator.

pulsateur, -trice, pulsating, pulsatory.

pulsatif, -ive, pulsating.

pulsatille, *f.* pasqueflower (*Anemone pulsatilla*).

pulsation, *f.* pulsation, beating.

pulsatoire, pulsatory, pulsating.

pulsomètre, *m.* pulsometer.

pulvérin, *m.* mealed gunpowder, fine spray.

pulvérisable, pulverizable, (liquid) that may be sprayed or atomized.

pulvérisateur, *m.* atomizer, sprayer, pulverizer, spraying machine.

pulvérisateur, pulverizing, spraying, atomizing.

pulvérisation, *f.* pulverization, atomizing, spraying.

pulvériser, to pulverize, grind, reduce (to powder), dust, atomize, spray, meal (gunpowder), powder.

pulvériseur, *m.* pulverizer, grinder, disk harrow.

pulvérulence, *f.* pulverulence, dustiness.

pulvérulent, pulverulent, pulverized, powdery, powdered, farinose.

pulvinar, *m.* pulvinar, prominence on back of thalamus, pertaining to pulvinus.

pulviné, pulvinate(d).

pumicin, *m.* palm oil.

pumiqueux, -euse, pumiceous, pumicose.

punais, affected with ozena.

punaise, *f.* bug, thumbtack; — **des lits,** bedbugs.

punaisie, *f.* ozena.

punir, to punish, chastise.

punition, *f.* punishment.

pupation, *f.* pupation.

pupe, *f.* pupa, chrysalis; — **emmaillotée,** obtect pupa; — **incomplète,** free pupa; — **resserrée,** coarctate pupa.

pupillaire, pupillary (membrane).

pupille, *m.&f.* pupil.

pur, pure, free (from), clear, genuine, mere, downright, unmixed; **en pure perte,** uselessly, to no purpose; **pure de semence,** true from seed.

purée, *f.* puree, mash, thick soup, mixed fruit sauce.

purement, purely, merely.

purent (pouvoir), (they) were able, could.

pureté, *f.* purity, pureness, clearness, cleanness.

purette, *f.* a black ferruginous sand.

purgatif, -ive, purgative, purge, cathartic.

purgation, *f.* purging, purgation, purge, purgative.

purge, *f.* purging, purge, purgative, cleansing, disinfection, cleaning.

purgeoir, *m.* purifying or filtering tank.

purger, to purge, cleanse, purify, clean, clear, blow off, free.

purgeur, *m.* purger, cleanser, purifier, blow-off gear.

purgeur, purging, cleansing.

purifiant, purifying.

purificateur, *m.* purifier.

purificateur, -trice, purifying.

purification, *f.* purification, purifying, cleansing.

purifier, to purify, cleanse, refine; **se —,** be purified, become pure.

purin, *m.* liquid manure.

purine, *f.* purine.

purique, pertaining to purine.

puron, *m.* sweet whey.

purpurate, *m.* purpurate.

purpurin, *f.* purpurin, madder purple.

purpurin, purplish.

purpurique, purpuric, purple.

purpurope, with purple stalk.

pur-sang, thoroughbred, full-blood(ed).

purulent, purulent.

pus, *m.* pus, matter; — **bleu,** blue (and green) pus.

pusilliflore, with small flowers.

pussent (pouvoir), (they) might be able, (they) could.

pustule, *f.* pustule; — maligne, malignant pustule.

pustuleux, -euse, pustulous, pustular.

pusule, *f.* pusula, contractile vesicle.

put (pouvoir), (he) could.

putier, *m.* bird cherry.

putois, *m.* polecat, fitchet.

putréfactif, -ive, putrefactive.

putréfaction, *f.* decay, decomposition, rotting, putrefaction, moldering; — humide, wet rot.

putréfait, decayed, rotten, putrefied.

putréfier, to putrefy, rot, decompose, molder.

putrescent, putrescent.

putrescible, putrescible.

putride, putrid, tainted.

putridité, *f.* putridity.

pycnide, *f.* pycnidium, cavity containing gonidia.

pycnomètre, *m.* pyknometer.

pycnostachyé, pycnostachous, in compact spikes.

pyémie, *f.* pyemia.

pyine, *f.* pyin.

pylône, *m.* tower.

pylorique, pyloric.

pyocyanine, *f.* pyocyanin.

pyocyanique, *m. Bacillus pyocyaneus.*

pyocyanique, pertaining to *Bacillus pyocyaneus.*

pyocyanisé, infected with *Bacillus pyocyaneus.*

pyoémie, *f.* pyemia.

pyogène, pyogenic.

pyohémie, *f.* pyemia.

pyoïde, pertaining to pus.

pyoxanthose, *f.* pyoxanthose.

pyracanthe, pyracanthus, with yellow spines.

pyrale, *f.* pyralis; — des pommes, codling moth; — des pousses, pine-shoot tortrix; — tordeuse, leaf roller, tortrix moth.

pyramidal, -aux, pyramidal.

pyramide, *f.* pyramid.

pyramidon, *m.* pyramidone.

pyrane, *m.* pyran.

pyrazine, *f.* pyrazine.

pyrazol, *m.* pyrazole.

pyrazoline, *f.* pyrazoline.

pyrazolique, pyrazole, of pyrazole.

pyrazolone, *f.* pyrazolone.

pyrénacé, containing nutlets in a fleshy pericarp.

pyrène, *m.* pyrene.

pyrénine, *f.* pyrenin, constituent of the body of the nucleus.

pyrénoïde, *f.* pyrenoid.

pyrénomycete, *m.* pyrenomycete.

pyrèthre, *m.* pyrethrum (Chrysanthemum); — officinal, — vrai, pellitory, pellitory of Spain, bertram (*Anacyclus pyrethrum*).

pyridine, *f.* pyridine.

pyridique, of pyridine, pyridine (series).

pyridone, *f.* pyridone.

pyriforme, pyriform.

pyrimidine, *f.* pyrimidine.

pyrite, *f.* pyrites, iron (pyrites) pyrite.

pyriteux, -euse, pyritic(al).

pyritifère, pyritiferous.

pyroacétique, pyroacetic.

pyrocatéchine, *f.* pyrocatechol, pyrocatechin.

pyroélectricité, *f.* pyroelectricity.

pyroélectrique, pyroelectric(al).

pyrogallate, *m.* pyrogallate.

pyrogallique, pyrogallic; acide —, pyrogallic acid.

pyrogallol, *m.* pyrogallol, pyrogallic acid.

pyrogène, pyrogéné, pyrogenous, igneous.

pyrogénèse, *f.* pyrogenesis.

pyrogénésique, pyrogénétique, pyrogenetic.

pyrognostique, pyrognostic.

pyrole, *f.* false wintergreen, shinleaf (Pyrola).

pyroligneux, pyroligneous.

pyrolignite, *m.* pyrolignite.

pyrolusite, *f.* pyrolusite, mineral manganese dioxide.

pyromètre, *m.* pyrometer.

pyrométrie, *f.* pyrometry.

pyrométrique, pyrometric, pyrometrical.

pyromorphite, *f.* pyromorphite.

pyrone, *f.* pyrone.

pyronine, *f.* pyronine.

pyrope, *m.* pyrope.

pyrophore, *m.* pyrophorus.

pyrophorique, pyrophoric.
pyrophosphate, *m.* pyrophosphate.
pyrotartrique, pyrotartaric.
pyrotechnicien, *m.* pyrotechnist.
pyrotechnie, *f.* pyrotechnics, pyrotechny.
pyrotechnique, pyrotechnic(al).
pyroxyle, pyroxile, *m.* pyroxylin.

pyroxylé, pyroxylin, nitrocellulose.
pyroxyline, *f.* pyroxylin.
pyroxylique, pyroxylic, pyroligneous.
pyrroline, *f.* pyrroline.
pyruvique, pyruvic.
pyurie, *f.* pyuria.
pyxide, pyxidie, pyxidian, *m.* pyxidium.

Q

quadrangle, *m.* quadrangle.
quadrangulaire, quadrangulé, quadrangular.
quadrant, *m.* quadrant.
quadratique, quadratic, tetragonal.
quadrature, *f.* quadrature, squaring.
quadrialié, four-winged.
quadricanaliculé, furnished with four canals.
quadri-caréné, furnished with four carina.
quadrichloré, tetrachloro.
quadricoque, with four capsules.
quadricorné, with four appendages in the form of horns.
quadricotyledoné, quadricotyledonous, with four cotyledons.
quadricyle, *m.* quadricycle.
quadridenté, quadridentate.
quadridigité, quadridigitatus, divided into four divisions.
quadridigitipenné, with four pinnate digitate divisions.
quadriennal, -aux, quadrennial.
quadriérémé, with four eremi.
quadrifarié, quadrifarious, in four ranks, as leaves.
quadrifide, quadrifid, four-cleft (calyx).
quadriflorigère, with four flowers.
quadrifolié, quadrifoliate.
quadrifoliolé, quadrifoliolate, with four subordinate leaflets.
quadrilatéral, -aux, quadrilateral.
quadrilatère, *m.* quadrilateral, figure, square.
quadrillage, *m.* cross ruling, squaring, grid, checkering.
quadrillé, *m.* cross ruling, checkering.
quadrillé, cross ruled, checkered, ruled in squares.
quadriller, to cross rule, rule in squares, checker.
quadrilobé, quadrilobate.

quadriloculaire, quadrilocular, having four cells.
quadrine, with four subordinate leaflets.
quadripartit, quadripartite, four cleft nearly to the base.
quadriné-penné, with leaves composed of four leaflets.
quadripétale, with four petals.
quadriphylle, with four leaves.
quadrisoc, *m.* four-furrow plow.
quadrivalence, *f.* quadrivalence, tetravalent.
quadrivalvulé, with four small valves.
quadroxalate, *m.* quadroxalate, tetroxalate.
quadroxyde, *m.* quadroxide, tetroxide.
quadrupède, *m.* quadruped.
quadruple, *m.* quadruple.
quadruple, quadruple, fourfold.
quadrupler, to quadruple.
quai, *m.* quay, wharf, (rail) platform, pier, cross dike.
qualifiable, that can be called or characterized.
qualificatif, -ive, qualifying, qualitative.
qualification, *f.* title, qualification, designation.
qualifié, qualified.
qualifier, to qualify, call, name, style.
qualitatif, -ive, qualitative.
qualitativement, qualitatively.
qualité, *f.* quality, property, title, rank, capacity; — des semences, seed property; — du sol, quality of locality or soil; — marine, seaworthiness; en — de, in the capacity of.
quand, when, whenever, though, although, even though, even if; — même, even if, in any case, although, even though, despite all; depuis —, since when, how long since; jusqu'à —, how long, since when.
quant à, as for, as to, with respect to respecting, concerning.

quanta, *m.pl.* quanta.

quantième, *m.* day of the month.

quantitatif, -ive, quantitative.

quantitativement, quantitatively.

quantité, *f.* quantity, amount; — de pluie, amount of rainfall; — infiniment petite, differential, infinitesimal quantity; en —, in quantity, in bulk, in parallel.

quantum, *m.* quantum.

quarantaine, *f.* (about) forty, (some) forty, twoscore, Lent, quarantine, age of forty.

quarante, *m.* forty.

quarante, forty, fortieth; — cinq, forty-five.

quarantième, *m.* fortieth.

quarantième, fortieth.

quarre, *f.* groove (for resin), resin blaze, streak.

quarré, square.

quart, *m.* quarter, fourth, cup (holding a quarter of a liter), (nautical) watch; — de cercle, quadrant of a circle; — de réserve, quarter (of mean annual yield) in reserve; — de rond, quarter round.

quartanier, *m.* wild boar four years old.

quartation, *f.* quartation.

quartaut, *m.* a cask (holding quarter of a hogshead).

quarteron, *m.* twenty five, book of 25 leaves of gold or silver foil.

quartier, *m.* quarter, fourth part, district, lodging.

quartile, *f.* quartile.

quartine, *f.* quartine, a fourth integument of some ovules.

quartz, *m.* quartz, quartz crystal; — piézo-électrique, piezoelectric condenser.

quartzeux, -euse, quartzose, quartzous.

quartzifère, quartziferous.

quartzique, quartz, of quartz.

quartzite, *f.* quartzite.

quasi, almost.

quasi-labours, *m.pl.* cultivating.

quasi-radié, quasiradiatus, slightly radiant.

quassia, quassier, *m.*, quassie, *f.* quassia.

quassine, *f.* quassin.

quaternaire, Quaternary.

quaterné, quaternate.

quaternifolié, with quaternate leaves.

quatorze, *m.* fourteenth.

quatorze, fourteen, fourteenth.

quatorzième, *m.* fourteenth.

quatorzième, fourteenth.

quatre, *m.* fourth.

quatre, four, fourth; — à —, in great haste; à —, on all fours.

quatre-temps, *m.* four-cycle engine.

quatre-vingt, eighty, fourscore.

quatre-vingt-dix, ninety.

quatre-vingtième, *m.* eightieth.

quatre-vingtième, eightieth.

quatre-vingt-onze, ninety-one.

quatre-vingts, eighty, eightieth.

quatre-vingt-un, eighty-one.

quatrième, *m.* fourth, fourth floor; *f.* fourth class.

quatrième, fourth, quarter.

quatrillion, *m.* thousand trillions, quadrillion (10^{15}).

quayage, *m.* quayage, wharfage.

que, whom, which, that, what, when, how, why, how much, how many, namely, so that, let, may, if, whether, lest, for fear that, yet, although, since, as far as, unless, until, than, but, only, as, because; — de, how many; — ne vient-il? why doesn't he come? ne . . . —, only, but; qu'est-ce que, qu'est-ce qui, what?

quebracho, *m.* quebracho.

quel (quelle), what, which; — que, whatever; — que soit, whatever (whoever) may be; tel —, each one, such as it is.

quelconque, any, whatever, whatsoever; un . . . —, a given . . . , any, like any other.

quelque, some, any, whatsoever, a few, about, however; *pl.* a few, several; — chose, something; — part, somewhere; — peu, a little, somewhat; — . . . que, however, whatsoever, howsoever.

quelquefois, sometimes, now and then, occasionally.

quelques-uns, *m.pl.* some, some people, some, few.

quelqu'un, someone, somebody, anyone, anybody.

quenouille, *f.* quenouille-trained fruit tree.

querciné, similar to oak.

quercitannique, quercitannic.

quercite, *f.* quercite, quercitol.

quercitrine, *f.* quercitrin.

quercitron, *m.* quercitron oak (*Quercus velutina*), yellow oak, black oak.

querelle, *f.* quarrel.

quereller, to quarrel with.

querir, aller —, to go and fetch; envoyer —, to send for; venir —, to come for.

qu'est-ce que, what?

qu'est-ce qui, what?

question, f. question.

questionnaire, m. examination questions, quiz.

questionner, to question.

quetche, f. large plum.

quête, f. quest, search, (church) collection, tracking.

quêter, to track, seek.

quetsche, f. large plum, prune.

quetsche-wasser, m. plum brandy.

queue, f. tail, handle, end, queue, cue, stalk, valve stem, tail end; — de la grappe, grape stalk; — de renard des champs, mousetail, slender foxtail (grass) (Alopecurus myosuroides); en — d'aronde, dovetailed.

queue-d'aronde, f. arrowhead.

queue-de-cheval, f. horsetail (Equisetum).

queue-de-lion, f. motherwort (Leonotis leonurus).

queue-de-loup, f. field cowwheat (Melampyrum arvense).

queue-de-morue, f. flat brush.

queue-de-rat, f. rattail file, small wax taper.

queue-de-renard, f. foxtail (grass).

queurse, queurce, f. slater.

queurser, to slate.

queux, f. hone, whetstone.

qui, who, whom, which, that, he who, whoever, what, whomever; — que, whoever, whosoever, whomever, whatever; — que ce soit, whoever it may be, anyone; qui est-ce que, whom; qui est-ce qui, who.

quiconque, whoever, whichever, whomsoever, whomever, whosoever.

quiètement, quietly.

quillaja, quillaya, m. Quillaja; — savonneux, soapbark tree (Quillaja saponaria).

quille, f. skittle, iron wedge, ninepin, support, string piece, keel.

quina, m. see QUINQUINA.

quinaldinique, quinaldic, quinaldinic.

quinamine, f. quinamine.

quinate, m. quinate.

quincaille, f. hardware, piece of hardware.

quincaillerie, f. hardware, hardware business.

quincaillier, m. hardware merchant.

quinconcial, -aux, quincuncial.

quiné, quinate, growing together in fives.

quinhydrone, f. quinhydrone.

quinicine, f. quinicine.

quinidine, f. quinidine.

quiniflore, with flowers arranged in fives.

quinine, f. quinine.

quinique, quinic.

quinite, n. quinite, quinitol.

quinizarine, f. quinizarin.

quinoa, m. quinoa.

quinoïdine, f. quinoidine.

quinoléine, f. quinoline.

quinoléique, of quinoline, quinolinic.

quinone, f. quinone.

quinonique, of quinone, quinoid, quinonoid.

quinovine, f. quinovin.

quinoxaline, f. quinoxaline.

quinoyle, m. quinoyl.

quinquédenté, quinquedentate, with five teeth.

quinquééremé, with five eremi.

quinquéfarié, quinquefarious, five-ranked.

quinquéfide, quinquefid, five-cleft.

quinquéfoliolé, quinquefoliolate, with five leaflets.

quinquennal, quinquennial.

quinquépartit, quinquepartite, deeply divided into five parts.

quinquina, m. cinchona (tree and bark), Peruvian bark; — jaune, cinchona, yellow cinchona bark.

quintal, m. quintal (50 kg.), approx. hundredweight.

quintaux, m.pl. see QUINTAL.

quintefeuille, f. cinquefoil.

quintessence, f. quintessence.

quintessenciel, -elle, quintessential.

quintessencier, to extract the quintessence from, refine.

quintine, f. an embryonic sac.

quintuplation, f. quintupling.

quintuple, quintuple.

quintupler, to quintuple, increase fivefold.

quintuplinervé, quintuplinerved, five-veined.

quinzaine, f. (about) fifteen, fortnight

quinze, m. fifteenth.

quinze, fifteen, fifteenth.

quinzième, *m.* fifteenth.

quinzième, fifteenth.

quittance, *f.* receipt.

quittancer, to receipt.

quitte, free, out of debt, quit, rid, discharged.

quitter, to quit, leave, abandon, give up, release, renounce, free, leave off; **se** —, to be left; **ils se quittent,** they part.

quoi, what, which, that, wherewithal, that which; — **que, quoi . . . que,** whatever, however, whatsoever; — **qu'il en soit,** however that may be, be that as it may; **de** —, wherewith.

quoique, though, although, even if.

quote-part, *f.* share, quota.

quotidien, -enne, daily.

quotidiennement, daily, every day.

quotient, *m.* quotient.

quotité, *f.* quota, share.

quotter, (of gearing) to engage, catch.

R

rabais, *m.* reduction, allowance, discount, rebate, depreciation, diminution, deduction.

rabaissement, *m.* lowering, reduction, depreciation.

rabaisser, to lower, reduce, depreciate, abate, lessen.

rabane, *f.* matting; — **de raphia,** grass mat.

rabat (rabattre), (he) is lowering.

rabat, *m.* discount, reduction.

rabattage, *m.* lowering (prices), cutting back (of shoots).

rabattement, *m.* rabattement (of triangle), folding back.

rabatteur, *m.* beater.

rabattre, to lower, diminish, deduct, lessen, put down, bring down, press down, shut down, turn down, turn under, flatten, beat out, smooth, soften, beat up the game, head off (game), suppress, humble; **se** —, to be lowered, come down, fall back, turn off.

rabette, *f.* rape, rapeseed, kohlrabi.

rabique, rabic, rabietic, rabid (virus, animal).

râble, *m.* back (of hare), skimmer (for molten glass), fire hook, rabble, strike.

râbler, to stir, rabble.

rabonnir, to improve, better.

rabot, *m.* plane, scraper, cutter, stirrer, polisher; — **d'ébouage,** road scraper.

raboter, to plane, smooth, polish, dress, file down.

raboteuse, *f.* planing machine, planer.

raboteux, -euse, rough, uneven, unpolished, knotty.

rabougri, stunted, crippled, dwarfed.

rabougrir, to stunt, be stunted.

raccommodage, *m.* mending.

raccommodement, *m.* reconciliation.

raccommoder, to mend, repair, reconcile, right, correct; **ils se raccommodent,** they are making up a difference.

raccord, raccordement, *m.* joining, junction, linking, seam, connection, adjustment, reconciliation, coincidence.

raccorder, to join (up), unite, level, smooth, adjust, reconcile; **se** —, to join with each other, be joined, united.

raccourci, *m.* abridgment, shortening, epitome, summary, short cut.

raccourci, shortened, short, abridged; **en** —, foreshortened.

raccourcir, to shorten, abridge, abbreviate, contract; **se** —, to grow shorter, shorten, shrink.

raccourcissement, *m.* shortening, reducing (in length), contraction.

raccours, *m.* shrinking, shrinkage.

raccrocher, to hook again.

race, *f.* race, family, stock, variety, breed; — **indigène,** native breed, indigenous breed.

racème, *m.* raceme.

racémeux, -euse, racémiforme, racemose.

racémique, racemic.

racémuleux, -euse, in a small raceme.

rachat, *m.* commutation, buying back; — **des servitudes,** purchase of servitudes; — **par contrainte,** compulsory commutation.

rache, *f.* dregs (of tar).

rachéole, *f.* rachilla, very short axis of the spikelet of grasses.

racheter, to redeem, buy, buy back, commute, buy out.

racheux, râcheux, -euse, knotty (wood).

rachevage, rachèvement, *m.* finishing.

rachever, to finish.

rachis, *m.* rachis, midrib, spike, spine.

rachitique, rachitic, imperfect, weak, blighted, stunted, knotty.

racinage, *m.* (edible) roots, walnut dye, root crops, esculent roots.

racinal, *m.* beam, sill, sleeper.

racine, *f.* root; — adven*t*i.e, adventitious root; — **aerienne,** aerial root; — **alimentaire,** edible root; — **binaire,** diarch root; — **capillaire,** fibrous root, rootlet, root fibril; — **carrée,** square root; — **cubique,** cube root — **d'actée à grappes,** cimicifuga, black cohosh (*Cimicifuga racemosa);* — **d'arnique,** arnica rhizome; — **de Saint Christophe,** common baneberry, herb Christopher (*Actaea rubra*).

racine, — **de sumbul,** sumbul, muskroot (root of *Ferula sumbul);* — **douce,** licorice root; — **fibreuse,** fibrous root; — **petite,** rootlet radicle; — **pivotante,** taproot, main root; — **principale,** main root, taproot; — saponnaire, soap root; — **suceuse,** sucking root; — **superficielle,** shallow root; — **tracante,** superficial root; — **vivace,** perennial root.

raciner, to take root, root, dye (yellow) with walnut.

rack, *m.* arrack, rack (liquor).

raclage, *m.* scraping.

racle, *f.* guard board.

racler, to scrape (off), skive, wipe.

raclette, *f.* (small) scraper, hoe.

racloir, *m.* scraper, flax-dresser's knife.

raclure, *f.,* **raclon,** *m.* scrapings, dust.

raconter, to relate, recount, narrate, tell.

racornir, to harden (as horn), shrivel.

racornissement, *m.* hardening, hardness, shriveling.

rade, *f.* roadstead, road (of harbors).

radeau, *m.* raft, wood raft.

rader, to dress, strike.

radiable, radiable, liable to be struck off (from list).

radiaire, radiate(d).

radial, -aux, radial.

radialement, radially.

radiance, *f.* radiance, radiation.

radiant, radiant.

radiateur, *m.* radiator, condenser, cooler; — **à serpentins,** radiator coil.

radiateur, radiating.

radiatiflore, with radiated flowers.

radiatiforme, radiatiform, in the form of a ray.

radiation, *f.* radiation, irradiation, striking out.

radical, *m.* radical; — composé, compound radical; — des racines, radical or root sign.

radical, -aux, radical, fundamental, drastic.

radicalement, radically.

radicant, rooting, radicant.

radication, *f.* radication, the root system of a plant.

radicé, with long roots.

radicellation, *f.* radicellation, root system.

radicelle, *f.* rootlet, radicle.

radiciflore, radiciflorous, flowering apparently from the root.

radiciforme, in the form of a root.

radiculode, *f.* radiculoda, the apex of the radicle in grasses.

radicivore, rhizophagous, root-eating.

radicule, *f.* radicle, rootlet.

radiculeux, -euse, radiculose, bearing rootlets.

radié, radiated, radiate, stellate, radial, rayed.

radier, to radiate, beam, erase, strike off.

radieux, -euse, radiant.

radifère, radiferous, radium-bearing.

radioactif, -ive, radioactive.

radioactivité, *f.* radioactivity.

radiodiffusion, radicémission, *f.* broadcasting.

radiogonométrie, *f.* radiogoniometry.

radiographie, *f.* radiography, radiograph.

radiographier, to radiograph.

radiographique, radiographic(al).

radiolaire, *m.* radiolarian.

radiologie, *f.* radiology.

radiomètre, *m.* radiometer.

radiométrie, *f.* radiometry.

radiophonie, *f.* radiophony, broadcasting.

radiosarque, *f.* tubercular root.

radioscopie, *f.* radioscopy.

radioscopique, radioscopic(al), fluoroscopic.

radiotélégraphie, *f.* wireless telegraphy.

radiotélégraphique, wireless.

radiotélégraphiste, *m.* wireless telegrapher, radio expert.

radiotéléphonie, *f.* wireless telephony, radio.

radiothérapie, *f.* radiotherapy.

radiothorium, *m.* radiothorium.

radis, *m.* radish (*Raphanus sativus*); — de cheval, horse-radish (*Amoracia rusticana*); — oléifère, Chinese radish.

radium, *m.* radium.

radiumthérapeute, *m.* radium therapeutist.

radius, *m.* radius (bone).

radouber, to repair, mend, refit (vessels), work over.

radoucir, to calm, soften, render mild, smooth, anneal; se —, to soften, grow mild.

rafale, *f.* sudden squall, gust; — de grêle, hailstorm.

raffermir, to harden, make firm(er), strengthen, improve; se —, to grow stronger.

raffermissement, *m.* hardening, making firmer, strengthening, steadying (of prices), improvement.

raffinade, *f.* refined sugar.

raffinage, *m.* refining.

raffinase, *f.* raffinase.

raffiné, *m.* stuff.

raffiné, refined, delicate, nice, ripened, purified.

raffiner, to refine, beat, mill, finish.

raffinerie, *f.* sugar refinery.

raffineur, *m.* refiner, perfecting engine.

raffineur, -euse, refining.

raffineuse, *f.* refining (beating) engine.

raffinose, *f.* raffinose, a sugar occurring in beet and germinating cereals.

rafle, *f.* stalk (of grapes), cob (of maize), clean sweep, carrying off.

rafler, to carry off, sweep off, round up.

rafleux, -euse, rough, uneven.

rafraîchir, to cool, refresh, renew, touch up, renovate, revive, freshen (color), trim, stir up (mortar), prune (tree roots), sharpen, refine, recultivate (soil); se —, to refresh oneself, cool, become cool, rest.

rafraîchissant, *m.* refrigerant, laxative.

rafraîchissant, cooling, refreshing.

rafraîchissement, *m.* cooling, refreshment.

rafraîchisseur, rafraîchissoir, *m.* cooler, refrigerator, crystallizer.

rage, *f.* rage, madness, rabies, hydrophobia; — canine, madness, rabies.

ragot, *m.* solitary wild boar.

raid, *m.* expedition, journey.

raide, stiff, rigid, tight, taut, precipitous, steep, inflexible, quickly, swiftly, promptly, completely.

raideur, *f.* stiffness, rigidity.

raidir, to stiffen, make rigid, tighten, make taut, stretch; se —, to stiffen, grow stiff, bear up, harden.

raidissement, *m.* stiffening.

raie, *f.* line, stroke, scratch, stripe, streak, mark, furrow, ridge, groove, ray, band, zone.

raifort, *m.* radish (*Raphanus sativus*), horseradish (*Armoracia rusticana*); — cultivé, summer - and- winter radish; — sauvage, horse-radish (*Armoracia rusticana*).

rail, *m.* rail, trackway.

rainer, to groove, flute, score.

rainette, *f.* tree frog, rennet, pippin (apple).

rainure, *f.* groove, channel, furrow, slot, keyway.

raiponce, *f.* rampion (*Campanula rapunculus*).

raisin, *m.* grape, grapes; — d'Amérique, pokeweed, Virginia poke (*Phytolacca americana*); — de Corinth, (dried) currants; — de cuve, wine grape; — de loup, black nightshade (*Solanum nigrum*); — d'ours, bearberry (*Arctostaphylos uva-ursi*); — sec, dried grape, raisin; grain de —, grape; grappe de —, bunch of grapes; pepin de —, grape seed, grapestone.

raison, *f.* reason, ratio, (firm) name, rate, ground, cause, intellect, satisfaction, reparation, claim; — de plus, all the more reason; — d'être, reason for existence; — directe, direct ratio; — géométrique, geometrical ratio; — inverse, inverse ratio; — sociale, firm name, trade name.

raison, à —, rightly; à — de, at the rate of; à plus forte —, with greater reason, all the more; avoir —, to be right; avoir — de, to get the better of; en — de, by reason of, on account of, in ratio to, because of, at the rate of; être de —, imaginary being; plus (moins) que de —, more (less) than right.

raisonnable, reasonable, rational, right, moderate.

raisonnablement, reasonably, rationally, moderately.

raisonné, reasoned, rational, analytical, methodical, systematic.

raisonnement, *m.* reasoning, arguing, argument.

raisonner, to reason, argue, consider, study, discuss; — **sur,** to argue about, discuss.

rajeunir, to rejuvenate, restore, make young.

rajouter, to add (more) again.

rajustement, *m.* readjustment, putting in order.

rajuster, to readjust, repair, reconcile.

râle, *m.* rail; — **d'eau,** water rail (*Rallus aquaticus*); — **de genêt** — **des prés,** land rail, crake, corncrake.

ralentir, to slacken, slow up, retard, make slower, abate, mitigate; **se —,** to slacken, slow up, relax, abate.

ralentissement, *m.* slackening, slowing up, abatement.

ralentisseur, *m.* slackener, retarder.

ralliement, *m.* rallying.

rallier, to rally, rejoin, win the approval of.

rallonge, *f.* extension piece, back.

rallongement, *m.* lengthening, extension.

rallonger, to lengthen, make longer, extend; **se —,** to lengthen, be extended.

rallumer, to relight, revive, rekindle.

rallumeur, *m.* relighter, a pilot burner.

ramaire, rameous, ramal.

ramas, *m.* heap, pile, collection.

ramassage, *m.* collecting.

ramassé, compact, collected, clustered, thick-set, stocky, cobby.

ramasser, to gather, collect, pick up, take up; **se —,** to collect, gather, gather oneself, be collected, be gathered, double up, roll itself up.

ramasseur, ramassoir, *m.* collector, gatherer.

ramastre, *m.* ramastrum, a secondary petiole of compound leaves.

rame, *f.* ream (of paper), paddle, oar, (bean) stick, prop, pole; *pl.* brush.

ramé, branched (of peas), propped.

raméal, -aux, rameal, pertaining to a branch.

rameau, *m.* branch, small branch, shoot, twig, bough, limb, hypha, branchlet, (mineral) vein; — **bifurqué,** forked branch; — **désarticulé,** twig broken off (by squirrel).

ramée, *f.* green boughs, branches, arbor.

ramelle, *f.* subdivision of a secondary petiole of a pinnate leaf.

ramender, to manure again, redye (cloth).

ramener, to bring back, draw back, turn, restore, reduce, revive.

ramentacé, ramentaceous, possessing ramenta (scales of epidermis).

ramentum, *m.* ramentum, thin chaffy scales of the epidermis.

rameur, *m.* rower, oarsman.

rameux, -euse, branching, branched, ramose, ramous.

ramie, *f.* ramie, China grass (*Boehemeria nivea*).

ramier, *m.* ringdove, fagot wood.

ramier, ring.

ramification, *f.* ramification, (small) branch, branching.

ramifié, ramifying, spreading, ramified.

ramifier, to branch, ramify, branch out.

ramiflore, ramiflorous, flowering on the branches.

ramiforme, ramiform, shaped like a branch.

ramille, *f.* twig, shoot; *pl.* fagot wood; **menues —s,** small spray wood.

ramique, with numerous branches.

ramoitir, to remoisten.

ramollir, to soften.

ramollisable, softening.

ramollissement, *m.* softening.

rampant, *m.* slope, crawling, sloping part.

rampant, sloping, creeping, crawling.

rampe, *f.* flight of stairs, slope, rise, incline, inclined plane, ramp, footlights, rail, stairs, baluster, upgrade; — **à éclipse,** a row of burners; — **à gaz,** row of gas burners.

ramper, to climb, slope, incline, creep, crawl, wind.

ramulaire, ramular, pertaining to a branchlet.

ramule, *f.* spray, sprig, twig.

ramulifiore, ramiflorous, flowering on the branches.

ramure, *f.* branching, branches, antler, boughs.

ramuscule, *m.* branchlet, twig.

rance, *m.* rancidness, rancidity.

rance, rancid.

rancidité, *f.* rancidity, rancidness.

rancir, to become or grow rancid.

rancissement, *m.* becoming rancid, souring.

rancissure, *f.* rancidness, rancidity, rustiness.

rang, *m.* rank, row, line (of trees), tier, range, string of (onions), order, ridge; au — de, among.

rangé, ranged, steady, arranged in rows.

rangée, *f.* row, range, line.

ranger, to range, arrange, class, draw up, rank, set in order, bring back, put away, subdue; se —, to range oneself, make way, concur, step aside, line up; **il se range,** he settles down.

ranimer, to revive, restore, stir up (a fire).

rapacé, (root) in form of turnip.

râpage, *m.* rasping, grating, grinding (of wood).

rapatrier, to bring back to one's country.

râpe, *f.* rasp, grater, stalk (of grapes).

râpé, *m.* wine leavings (collected and re-bottled), rape wine.

râpé, grated, rasped, worn out, threadbare.

râper, to rasp, grate, grind (wood for pulp), wear threadbare.

râperie, *f.* pulp mill.

rapetisser, to make smaller, shorten, lessen, reduce, shrink, diminish, minimize, belittle; se —, to diminish, become shorter, shrink.

râpette, *f.* cleavers, catchweed (Gallium), German madwort (*Asperugo procumbens*).

raphané, similar to radish.

raphé, *m.* raphe.

raphia, *m.* raphia (fiber).

raphides, *f.pl.* raphides, needle-shaped crystals in the cells of plants.

rapide, *m.* rapid (in river), express (train).

rapide, rapid, swift, steep, fast, fleet.

rapidement, rapidly, swiftly, quickly, fleetly.

rapidité, *f.* rapidity, velocity.

rapiécer, to patch, piece.

rapine, *f.* rapine, plunder, pillage.

rapointir, to repoint, resharpen.

rappel, *m.* call, recall, return.

rappeler, to recall, call back, bring back, repeal; se —, to remember, recollect, recall.

rapport, *m.* relation, ratio, report, return, connection, outturn, yield, production, produce, agreement, relationship, conformity, correlation, accord, proceeds, likeness, proportion; — **du marché,** market report; — **entre les tensions,** ratio of stresses; **en — avec,** suitable to; **par — à,** with regard to, with respect to in propor-

tion to, with reference to, as related to; **sous ce —,** in this, that respect; **sous le — de,** with respect to, with regard to.

rapporté, inlaid, attacned, built up, not in one piece with.

rapporter, to refer, relate, report, quote, cite, bring back, retrieve, fetch and carry, take back, bring in connection with, yield, produce, bring in, revoke, rescind, repeal, add, insert, transport, turn, direct, aim at, protract plot, ascribe, attribute, compare; se —, to agree, relate, allude, refer, tally, rely (on) be referred, bear; **il s'en rapporte à,** he leaves the matter to.

rapporteur, *m.* reporter, recorder, protractor.

rapprêter, to dress again.

rapproché, brought nearer, close, near, near at hand, closely related.

rapprochement, *m.* bringing together, drawing nearer, reconciliation, comparison.

rapprocher, to bring nearer, bring together. compare, associate, reconcile, approach again, condense; se —, to be brought nearer, come near, reconcile, come together, draw nearer, draw closer together. approach, be allied (to), connect with.

rapt, *m.* rape, abduction.

râpure, *f.* rasping(s).

raquette, *f.* cactus, Indian pear, (thick, fleshy) point (of nopal).

rare, rare, thin, unusual, scarce, scant, infrequent.

raréfactibilité, *f.* capability of being rarefied.

raréfactif, -ive, rarefactive.

raréfaction, *f.* rarefaction, depletion.

raréfiable, rarefiable.

raréfiant, rarefying, rarefactive.

raréfier, to rarefy; se —, to be rarefied, become scarce.

rarement, rarely, seldom, hardly ever.

rarescence, *f.* becoming rarefied.

rarescent, becoming rarefied.

rarescibilité, *f.* capability of being rarefied.

rarescible, rarefiable.

rareté, *f.* rarity, tenuity, rareness, scarcity, scarceness.

raricosté, with several ribs.

rariétoilé, with few stars.

rariflore, with flowers far apart, sparsely flowered.

rariplissé, with folds far apart, scarcely pleated.

rarisillonné, hollowed by a small number of furrows.

rarissime, extremely rare.

ras, *m.* level.

ras, flat, level, bare, blank, short-napped, short-haired, close-cropped, shaven, close, shorn, cut short, smooth; **à —, au —,** even; **à — de, au — de,** level with, flush with, even with; **table —e,** clean sweep.

rasade, *f.* brimful glass.

rasage, *m.* shaving.

rase, *f.* turpentine, oil of turpentine, rosin, rosin oil

raser, to shave, graze, pull down, level, raze.

rasoir, *m.* razor, (microtome) knife.

rassasier, to satisfy, satiate, surfeit.

rassemblement, *m.* assembling, collecting, collection, gathering, assemblage, crowd.

rassembler, to collect, gather, assemble, reassemble, bring together, collect again.

rasseoir, to reseat, replace, calm.

rassis, *m.* settled state (of liquid); **prendre son —,** to settle (liquid) spirits.

rassis, settled, calm, staid, stale (bread), replaced.

rassortiment, *m.* reassortment, restocking.

rassortir, to reassort, rematch.

rassurant, reassuring.

rassurer, to strengthen, enhearten, secure, tranquilize, reassure.

rat, *m.* rat; **— à bourse, gopher; — d'Amérique,** guinea pig; **— de cave,** thin wax taper; **— musquée,** muskrat; **— surmulot,** brown rat.

ratafia, *m.* ratafia (liqueur).

ratanhia, *m.* rhatany (*Krameria triandra*).

ratatiné, shriveled, shrunken.

ratatiner, to shrivel (up), dry up, crinkle up.

rate, *f.* spleen, milt, female rat.

raté, *m.* misfire.

raté, miscarried, ineffectual, missed, failed.

râteau, *m.* rake.

râtelage, *m.* raking; **— de la litière,** raking of litter.

râtelée, *f.* rakeful (of hay).

râteler, to rake (up).

râtelier, *m.* rack, set of teeth, creel.

rater, to miss fire, fail, miscarry.

ratifier, to ratify, confirm.

ration, *f.* ration, allowance, portion, share, diet.

rationnel, -elle, rational, reasonable.

rationnellement, rationally, reasonably.

rationner, to put on rations.

ratissage, *m.* raking, scraping (of vegetables).

ratisser, to scrape, rake, hoe.

ratissoire, *f.* rake, hoe, scraper.

ratissure, *f.* rakings, scrapings.

raton, *m.* little rat, cheese cake.

rattachage, rattachement, *m.* connection, fastening.

rattacher, to fasten, tie again, connect (with), attach (to), reattach, refasten, join, tie (together); **se —,** to be fastened, be connected, be attached, be linked, resemble.

rattrapage, *m.* recovering, taking up.

rattraper, to recover, overtake, catch up, take up; **se —,** to seize hold (of), be recovered, make up.

rature, *f.* erasure, scrapings.

raturer, to erase, scrape.

raucité, *f.* hoarseness.

rauque, hoarse, harsh, raucous.

ravage, *m.* ravages, havoc, mischief.

ravager, to ravage, devastate, spoil.

ravale, *f.* mold board, leveler, levellator.

ravalement, *m.* cutting back to the ground, repointing.

ravaler, to rough-cast, rough-coat, clean, dress, redress, resurface, scrape, level (soil), swallow, retract, disparage, cut back to the ground, trim.

rave, *f.* rape, radish, beet, turnip (*Brassica rapa*).

ravi, entranced, ravished, delighted.

ravière, *f.* turnip field.

ravin, *m.* ravine, gulch, hollow road.

ravine, *f.* mountain torrent, mountain stream.

ravinement, *m.* hollowing out.

raviner, to wash out, gully, form a ravine.

ravir, to ravish, delight, carry off, tear away, enrapture; **à —,** delightfully, admirably.

raviser, to change one's mind.

ravison, *m.* ravison, Black Sea rape (*Brassica napus*).

ravissant, delightful.

ravisseur, *m.* ravisher.

ravitailler, to revictual, supply.

ravivage, *m.* reviving.

raviver, to revive, brighten up, freshen.

ravoir, to have again, get back again.

rayage, m. scratching, scoring.

rayé, streaked, crossed out, striped, rifled.

rayer, to scratch, stripe, streak, cross out, rifle, erase, rule, strike off.

ray-grass, m. rye grass (Lolium perenne).

rayon, m. ray, beam, radius, shelf, spoke, region, drill, furrow, rayon, artificial silk; — calorifique, heat ray; — de lumière, light ray; — de miel, honeycomb; — lumineux, light ray; — médullaire, medullary ray, pith ray; — visuel, line of sight, visual ray; —s X, X rays, Roentgen rays.

rayonnage, m. furrowing, drilling, shelving, shelves.

rayonné, radiant, radiating, shining.

rayonné, radiated, radiate.

rayonnement, m. radiation, radiance.

rayonner, to radiate, beam, shine, irradiate, drill, furrow, fit (room) with shelves.

rayure, f. stripe, streak, scratch, score, rifling, ruling, erasure, groove, furrow.

raze, f. oil of turpentine.

réabonnement, m. renewal of subscription.

réabsorber, to reabsorb.

réabsorption, f. reabsorption, reabsorbing.

réacquérir, to recover.

réactibilité, f. reactivity.

réactif, m. reagent, test.

réactif, -ive, reactive.

réaction, f. reaction, back pressure; — d'induit, armature reaction.

réactionnel, -elle, reactive, reacting, pertaining to reaction.

réactiver, to reactivate.

réadmettre, to readmit.

réaffirmer, to reaffirm.

réagglomérer, to reagglomerate, reform.

réagibilité, f. reactiveness.

réagir, to react, show a reaction.

réagissant, reacting.

réaimanter, to remagnetize.

réajuster, to readjust.

réalgar, m. realgar, red arsenic.

réalisable, realizable, that can be made real, feasible, attainable.

réalisateur, m. perfecter, inventor.

réalisation, f. realization, accomplishment, making.

réaliser, to realize, bring about, effect, carry out, sell, make real, accomplish; se —, to be realized; il se réalize, it becomes real.

réalité, f. reality; en —, really.

réapparaître, to reappear.

réapparition, f. reappearance.

réargenter, to resilver.

réassortir, to reassort.

réassurer, to reinsure.

rebaisser, to lower again.

rebâtir, to rebuild, reconstruct.

rebattre, to beat, hammer again, say over and over again.

rebattu, beaten again, trite, hackneyed.

rebelle, m.&f. rebel, revolter.

rebelle, obstinate, refractory, rebellious, rebel, disobedient.

rebeller, to rebel, revolt.

reblanchir, to become white again, rebleach.

rèble, m. cleavers, goose grass (Galium aparine).

reboisement, m. reafforesting.

reboiser, to reforest.

rebond, m. rebound.

rebondi, plump, round.

rebondir, to rebound.

rebord, m. edge, border, rim, flange, ledge. hem, brim; à —, with raised edge, flanged,

reboucher, to stop up again, cork again, stopper again.

rebouillir, to boil again.

rebours, m. cross-grained (wood), wrong way (of the grain or nap), contrary, reverse, wrong side; à —, au —, against the hair, nap, or grain, the wrong way, backward.

rebout (rebouillir), (it) is boiling again.

rebrouiller, to mix again, stir up again.

rebroussé, retrorse.

rebroussement, m. retrogression (in response to overstimulus).

rebrousser, to turn back, go back, retrogress, retrace one's way.

rebroyer, to regrind.

rebrûler, to reburn, distill again, soften (glass).

rebrunir, to brown again, reburnish.

rebut, m. repulse, repulsion, waste, refuse, rejected material; bois de —s, refuse wood; de —, refuse, waste; mettre au —, to discard, reject.

rebutant, repulsive.

rebuter, to reject, refuse, discourage, repulse, rebuff, dishearten.

recalciner, to recalcine.

récalcitrant, stubborn, recalcitrant.

recalculer, to recalculate.

recaler, to reset, relevel, readjust, smooth.

récalescence, *f.* recalescence.

récapituler, to recapitulate, sum up.

recarbonisation, *f.* recarbonization.

recarboniser, to recarbonize, recarburize.

recasser, to break again.

recassis, *m.* broken-up stubble land.

recaulescence, *f.* recaulescence, adnation of leaves or petioles to the stem.

recèlement, *m.* concealing.

recéler, to conceal, contain.

récemment, recently.

recense, *f.* restamping (of gold and silver).

recensement, *m.* census, counting, inventory.

récent, recent, new, late, distinct, fresh.

recepage, *m.* cutting back.

receper, to cut back, shorten, truncate, clip.

récépissé, *m.* receipt.

réceptacle, *m.* receptacle; — **fractifère,** fruit body, sporophore; — **séminal,** spermatheca.

récepteur, *m.* receiver, receptor; — **à lampes,** audion receiver.

récepteur, -trice, receiving.

réception, *f.* receiving, receipt, reception; — **à l'oreille,** receiving by ear.

réceptivité, *f.* receptivity.

récessif, -ive, recessive.

recette, *f.* receipt, recipe, acceptance, collection, formula, prescription.

recevable, receivable.

receveur, *m.* receiver, collector.

recevoir, to hold, receive, accept, get, take, approve, entertain.

rechange, *m.* replacement, change, spare part, duplicate; **de** —, extra, spare, duplicate.

réchapper, to escape, recover, rescue, save, deliver.

recharge, *f.* recharging.

rechargement, *m.* recharging, reloading.

recharger, to recharge, reload, renew, repair, reballast.

réchaud, *m.* small portable stove, hot plate, hotbed (for plants), mulch.

réchauffage, *m.* reheating, warming up.

réchauffement, *m.* reheating, new manure.

réchauffer, to reheat, warm over, heat again, heat, rewarm, warm up, rekindle; **se** —, to be warmed, heated, get hot.

réchauffeur, *m.* heater, warmer, forewarmer. feed-water heater, preheater.

réchauffoir, *m.* heater, plate warmer, hot plate.

rêche, harsh, rough.

recherche, *f.* research, investigation, pursuit, inquiry, examination, seeking, search, prospecting, effort, affectation, studied elegance; **à la** — **de,** in search of.

recherché, sought after, wanted, in request, in (great) demand, studied, affected, far-fetched, choice.

rechercher, to search for, search again, seek after, seek for, test for, inquire into, investigate, look for.

rechercheur, *m.* researcher, research worker, seeker, inquirer.

rechinser, to wash (wool) carefully.

rechute, *f.* relapse.

récidive, *f.* repetition, relapse, recurrence.

récif, *m.* reef.

récipé, *m.* recipe.

récipient, *m.* receiver, container, recipient, vessel, cistern, cylinder, receptacle.

réciprocité, *f.* reciprocity.

réciproque, *f.* like, same thing, reciprocal, converse.

réciproque, reciprocal, inverse (ratio), mutual, reciprocating, reversible.

réciproquement, reciprocally, conversely, mutually, vice versa.

recirer, to wax again, rewax.

récit, *m.* account, narration, recital, relation.

réciter, to relate, recite, tell.

réclamant, claiming; — **de la lumière,** light-demanding, intolerant of shade; — **de l'ombre,** shade-demanding.

réclamation, *f.* complaint, objection, claim, demand, reclamation.

réclame, *f.* advertising.

réclamer, to demand, beg, call for, claim (back), reclaim, object, seek eagerly, protest.

récliné, reclinate.

recluse, *f.* recluse.

recoin, m. corner.

reçoit (recevoir), (he) is receiving.

récolement, m. check counting, revision of the produce, control of fellings; — périodique, revision of working plans.

récoler, to revise.

recoller, to reglue.

recolorer, to recolor, counterstain; se —, to become colored again.

récolte, f. collection, harvest, crop, harvesting, yield; — de bois, wood harvest; — du lait, milk production; — manquée, bad crop, failure of crops.

récolter, to harvest, gather in, collect.

récolteur, m. collector, harvester.

récolteur, -trice, gathering, harvesting.

recommandable, to be recommended, recommendable, estimable.

recommandation, f. recommendation, registration (of letter), esteem, advice.

recommander, to recommend, register (a letter), instruct, advise, request.

recommencer, to recommence, begin again, rebegin.

récompense, f. reward, recompense, requital.

récompenser, to reward.

réconciler, to reconcile; se —, to be reconciled.

réconfortant, m. tonic.

réconfortant, strengthening, tonic, comforting.

réconforter, to strengthen, refresh, comfort, cheer.

reconnaissable, recognizable.

reconnaissance, f. recognition, verification, exploration, acknowledgment, gratitude, reward, reconnaissance.

reconnaissant, grateful.

reconnaître, to recognize, know, examine, verify, admit, identify, discover, locate, explore, acknowledge, concede, return; se —, to be recognized, get one's bearings, reflect, make out.

reconnu, recognized.

reconquérir, to reconquer, regain.

reconstituer, to reconstruct, reconstitute, put together again.

reconstitution, f. reconstitution.

reconstruire, to reconstruct.

reconvertir, to reconvert.

recoquillement, m. curling up, shriveling.

recoquiller, to curl up, turn up, shrivel.

record, m. record; — mondial, world's record.

recouchage, m. laying down again.

recouler, to recast, scrape (hides) to remove lime.

recoupage, m. recutting, cutting again.

recoupe, f. clippings, chippings, cuttings, scraps, fragments, blended (raw) spirit, second flour, second cutting (of hay), second crop.

recouper, to recut, cut again, blend (wines), cut across.

recoupette, f. very coarse (third) flour.

recourbé, bent, curved, recurved.

recourber, to bend back, bend (again), bend round, recurve; se —, to bend, be bent, curl.

recourbure, f. bend, bent state.

recourir, to have recourse, turn to, resort to, run again.

recours, m. recourse, resort, resource.

recourt (recourir), (he) is resorting to.

recouru (recourir), (he has) resorted to.

recouvert, covered, covered again, recovered.

recouvrement, m. recovery, covering, cover, hood, overlapping, lap, occlusion (of wounds); à —, by overlapping, lapping.

recouvrer, to recover, regain, collect (taxes).

recouvrir, to cover (over), mask, overlay, cap, coat, cover again, recover; se —, to become covered, cloud over.

récréer, to enliven, refresh.

récrément, m. recrement, dross.

recrépir, to regrain, rough-cast anew.

récrire, to rewrite, write an answer.

recrobiller, to curl up, shrivel.

recroiser, to recross, cross again.

recroître, to grow again, get longer, spring up again.

recroqueviller, to curl up, shrivel up.

récrouir, to reheat, anneal.

recrû, m. regrowth, second growth, natural reproduction.

recru, tired out, worn out.

recrutement, m. recruiting.

recruter, to recruit.

rectal, -aux, rectal.

rectangle, m. rectangle.

rectangle, rectangular.

rectangulaire, rectangular.

rectembryé, with embryo having straight radicle.

recteur, *m.* rector.

recteur, -trice, directing.

rectifiable, rectifiable.

rectificateur, *m.* rectifier.

rectificateur, -trice, rectifying.

rectificatif, -ive, recti.ying.

rectification, *f.* rectification, rectifying, redistillation

rectifier, to rectify, mend, righten; se —, to be rectified.

rectiligne, rectilinear.

rectinervé, rectinerved, straight-veined, parallel-veined.

rectitude, *f.* straightness, rectitude, uprightness.

recto, *m.* right side, right-hand page, recto.

rectrices, *f.pl.* rectrices, tail feathers, rudder feathers of the tail.

rectum, *m.* rectum.

reçu, *m.* receipt.

reçu (recevoir), received.

recueil, *m.* collection, compilation.

recueillement, *m.* collecting, gathering, collectedness.

recueilli, collected, gathered.

recueillir, to collect, gather, harvest, compile, receive; se —, to be collected, collect oneself, be picked up, collect one's thoughts.

recuire, to reheat, reburn, anneal (glass).

recuisant, annealing, reheating.

recuisseur, *m.* annealer.

recuisson, *m.* recooking, annealing, reheating.

recuit, *m.* annealing, tempering, reheating, reburning, reboiling.

recuit, annealed, cooked or baked again.

recuite, *f.* annealing of metals or glass, reheating of liquids.

recul, *m.* recoil, starting back, falling back. recession.

reculade, *f.* retreat, falling back, backward movement.

reculé, distant, remote.

reculer, to move back, set back, put back, recede, retreat, shrink, withdraw, recoil, extend, delay, postpone, defer; se —, to draw back, be extended.

reculons (à), backwards.

récupérable, recoverable.

récupérateur, *m.* recuperator, regenerator, hot-blast stove.

récupération, *f.* recuperation, recovery, regeneration.

récupérer, to recover, recuperate, regenerate.

reçurent (recevoir), (they) received.

recurer, to scour, clean (pots).

récurrence, *f.* recurrence.

recurvé, recurved, reflexed, bent backward.

récurvifolié, with leaves inflected at the end.

récuser, to challenge, object to.

reçut (recevoir), (he) received.

rédacteur, *m.* editor, writer.

rédaction, *f.* editing, editorial staff, drawing up, wording (a bill).

redan, *m.* recess, notch, projection, check.

redécouvrir, to rediscover.

redent, *m.* check, recess.

redescendre, to descend again, redescend, bring or carry down again.

redevable, *m.* debtor.

redevable, indebted, beholden.

redevenir, to become again, rebecome; — inculte, to run wild.

rédiger, to edit, draw up, write down, word.

redire, to repeat, find fault with, criticize, blame.

redissolution, *f.* redissolving, re-solution.

redissoudre, to redissolve.

redistillation, *f.* redistillation.

redistiller, to redistill.

redistribuer, to redistribute, distribute again.

redonner, to give again, give back, restore, return, begin again.

redorer, to regild.

redoublé, redoubled, doubled, double.

redoubler, to redouble, increase.

redoutable, redoubtable, dreadful, formidable.

redouter, to dread, fear.

redressage, redressement, *m.* straightening, rectification, resetting.

redresser, to straighten (up), reset, re-erect, put right, rectify, redress, correct, adjust, dress again, reform.

redresseur, *m.* rectifier, straightener, commutator.

redresseur, -euse, straightening.

redû, *m.* balance due.

réductase, *f.* reductase.

réducteur, *m.* reducer, reductor, reducing agent.

réducteur, -trice, reducing.

réductibilité, *f.* reducibility, reducibleness.

réductible, reducible.

réductif, -ive, reducing.

réduction, *f.* reduction.

réduire, to reduce, diminish, subjugate, lessen, oblige; **se —,** to be reduced, amount (to).

réduisant, reducing.

réduit, *m.* retreat, small house, lodgings, redoubt.

réduit, reduced, small, low.

réduplicatif, -ive, reduplicative, (edges) valvate and reflected.

réédifier, to rebuild.

réel, -elle, real, actual, true.

réellement, really, in reality, actually.

réensemencement naturel, *m.* natural reproduction or regeneration, self-sown seed.

réensemencer, to resow, reseed.

refaire, to make again, remake, redo, make over, repair, do again, mend, recover, revive, begin again; **se —,** to recover, be made again, reestablish oneself.

refaucher, to mow the aftermath.

réfection, *f.* refection, repair, reconstruction, meal, food, making over, recovery.

refend, *m.* partition.

refendre, to split, rip, divide lengthwise, slit.

refente, *f.* splitting, ripping.

refera (refaire), (he) will remake.

référence, *f.* reference.

référer, to refer, ascribe, attribute; **se —,** to refer, leave the matter (to).

refermer, to close again, close, shut again, reclose, shut up.

refiltrer, to refilter, filter again.

refit (refaire), (he) remade.

réfléchi, recurved.

réfléchi, reflected, deliberate.

refléchir, to bend again.

réfléchir, to reflect, reverberate, think (of).

réfléchissement, *m.* reflection, reflecting.

réflecteur, *m.* reflector, lamp shade.

réflecteur, reflecting.

reflet, *m.* reflection, reflex.

refléter, to reflect, reflect light.

réflexe, reflex.

réflexible, reflexible, capable of being reflected.

réflexion, *f.* reflection, reflexion, meditation, thought.

refluer, to flow back, fall back, ebb (of tide).

reflux, *m.* reflux, flowing back, ebb.

refondre, to remelt, recast, refound, remodel, reform, melt again.

refont (refaire), (they) are remaking.

refonte, *f.* recasting, remelting, remodeling.

réforme, *f.* reform, reformation, reforming, reshaping.

reformer, to form again, reshape, re-form.

réformer, to reform, correct, mend, discharge.

refouillement, *m.* hollowing out, hollow.

refouiller, to search anew, hollow out, carve, cut deeper, dig again.

refoulement, *m.* pressing back, compressing, driving back, delivery, discharge, backflow, suppression, turning back; **— des eaux,** rising of water.

refouler, to press back, drive back, repel, force (back), ram, compress, drive, upset, jump up, tamp, deliver, stem (a current), flow back, check, restrain, suppress, press again, remill.

refouleur, *m.* compressor, force pump, driver.

refournir, to refurnish, supply again.

réfractaire, refractory, rebellious, resistant.

réfracté, curved, retracted.

réfracter, to refract; **se —,** to be refracted.

réfracteur, *m.* refractor.

réfracteur, -trice, refracting.

réfractif, -ive, refractive.

réfraction, *f.* refraction, bending.

réfractoire, refractional.

réfractomètre, *m.* refractometer.

refrain, *m.* refrain.

réfranger, to refract.

réfrangibilité, *f.* refrangibility.

réfrangible, refrangible.

refrapper, to strike again.

refréner, to restrain, control, check.

réfrigérant, *m.* condenser, cooler, refrigerator, refrigerant, radiator; **— à ailettes,**

gilled or ribbed cooler, radiator; — **à air,
air** condenser; — **à boules, bulb** (condenser) cooler; — **à reflux,** — **ascendant,**
reflux condenser, back-flow cooler; —
descendant, ordinary condenser, inclined
condenser.

réfrigérant, cooling, refrigerating, freezing,
refrigerant.

réfrigérateur, *m.* refrigerator.

réfrigératif, -ive, refrigerant.

réfrigération, *f.* refrigeration, cooling,
chilling.

réfrigérer, to refrigerate, cool, chill.

réfringence, *f.* refractivity, refringency.

réfringent, refringent, refractive, refracting.

refroidir, to cool, chill, grow cold, get cold,
slacken, refrigerate, cool down.

refroidissement, *m.* cooling, chilling, coldness.

refroidisseur, *m.* cooler, radiator.

refroidisseur, -euse, cooling.

refroidissoir, *m.* cooler.

refuge, *m.* refuge, shelter.

réfugier (se), to take refuge, find shelter.

refus, *m.* refusal, refuse, denial, thing
refused; **battre à —,** to drive to excess,
to drive till progress ceases.

refuser, to refuse, deny, withhold, reject,
fail to work; **se —,** to be refused.

réfutation, *f.* refutation, disproof.

réfuter, to refute, confute.

regagnage, *m.* recovery, regaining.

regagner, to regain, recover.

regain, *m.* aftermath, eatage, aftergrowth.

régale, eau —, aqua regia.

régaler, to level, spread (earth) evenly,
entertain, regale, feast.

regard, *m.* regard, look, gaze, glance,
attention, opening, gauge, manhole, peephole, gully hole; — **vitre,** glass gauge;
au — de, in comparison with, with
regard to, opposite; **en —,** opposite,
facing.

regarder, to look at, regard, look, behold,
consider, concern, face, point toward, be
directed, pay attention to; **cela me
regarde,** that concerns me. **il y regarde
de près,** he is very particular.

regarnir, to refurnish, replant blanks, complete the crop by planting, repair.

regarnis, *m.pl.* replanted blanks.

regarnissage, *m.* completing the crop by
planting, afterculture, replanting.

regel, *m.,* **regélation,** *f.* freezing again,
renewed frost.

regeler, to freeze again.

régénérateur, *m.* regenerator.

régénérateur, -trice, regenerating, regenerative.

régénération, *f.* regeneration, reclaiming,
restocking; — **par coupes d'abri,** regeneration under a shelter wood.

régénérer, to regenerate, restock.

regermable, capable of sprouting again.

regetom, *m.* shoot, sprout, offset, scion.

régie, *f.* administration, excise, excise
office, management, regime.

régime, *m.* normal operation, regular
running, regulation, working conditions,
management, performance, government,
regime, regimen, system, rule(s), regulations, control, rate of charging or discharging, normal rate of speed, cluster,
raceme, treatment.

régime, — **carné,** meat diet; — **de basses
pression,** — **de dépression,** area of low
pressure; — **de fortes pluies,** rainy area;
— **de la futaie,** high or seedling forest
system; — **de pluies rares,** dry zone,
area with little rain; — **du taillis,** coppice
system; — **du taillis simple,** simple coppice method; — **du taillis sous futaie,**
coppice-with-standards system; — **pastoral,**
forestry combined with pasture; — **végétarien,** vegetarian diet; **en — permanent,**
in a steady state, at constant load.

région, *f.* region, locality, territory, part;
— **alizée,** region of the trade winds; —
basse, lowland; — **pluvieuse,** rainy area.

régir, to govern, rule, manage, direct.

registre, *m.* register, regulator, record,
damper.

registrer, to register.

réglable, regulable, variable, adjustable.

réglage, *m.* regulation, adjustment, ruling,
control, regulating.

règle, *f.* rule, ruler, guide bar, rod, strip,
model, pattern, example, order, pole,
perch; *pl.* menses; — **à calcul,** slide rule
or scale; — **conjointe,** chain rule, conjoined rule of three; — **de réclame,**
advertising rule; — **de retraite,** contraction
rule; — **des phases,** phase rule; — **de
trois,** rule of three, rule of proportion;
— **divisée,** scale, measure, rule; — **inclinée,**
sine rule; — **réduite,** reduced scale; **en —,**
in good order; **en — générale,** as a general
rule.

réglé, regulated, regular, ruled, fixed, steady.

règlement, *m.* regulation, settlement, adjustment, rule, order; — **de chasse,** hunting regulation; — **de normalisation,** specifications; — **de pâturage,** pasture regulation; — **d'exploitation,** working plan; — **spécial des coupes,** felling plan.

réglementaire, regular, regulatory, prescribed, regulation.

réglementation, *f.* formation or drawing up of regulations.

réglementer, to regulate, make regulations.

régler, to regulate, adjust, set, assign, control, order, manage, fix, decide, determine, settle, rule (paper); **se —,** to be regulated, regulate one's life.

réglette, *f.* small rule or scale, reglet; — **à curseur,** sliding rule.

régleur, *m.* regulator.

réglisse, *f.* licorice.

reglisser, to slide (slip) again.

réglure, *f.* ruling of paper.

régnant, reigning, prevailing, prevalent.

règne, *m.* reign, kingdom; — **végétal,** vegetable kingdom.

régner, to reign, rule, abound, prevail, extend, run.

regommer, to regum, gum again, retread.

regonflement, *m.* refilling, reinflation, swelling.

regonfler, to refill, pump up, swell.

regorgement, *m.* overflowing, abounding, overflow.

regorger, to overflow, run over, abound, regurgitate, vomit, disgorge.

regoûter, to taste again.

regratter, to scrape again, rub down.

regreffage. *m.* regrafting.

régressif, -ive, regressive, throw back.

régression, *f.* regression, retrogression.

regret, *m.* regret, grief, sorrow.

regrettable, regrettable.

regrettablement, regrettably.

regretter, to regret.

regriller, to roast again.

regros, *m.* tanning bark.

régulariflore, regulariflorous, head of Compositae contains tubular florets.

régularisation, *f.* regulation, regularization.

régulariser, to regularize, regulate, make regular, straighten out.

régularité, *f.* regularity, steadiness; — **du fût,** straightness of stem.

régulateur, *m.* regulator, governor; — **à force centrifuge,** Watt's governor; — **de chaleur,** thermoregulator.

régulateur, -trice, regulating, regulative.

régulation, *f.* regulation.

régule, *m.* regulus, antifriction metal.

régulier, - ère, regular, straight-boled, isomerical.

régulièrement, regularly.

régulin, reguline.

régurgiter, to regurgitate (food).

réhabiliter, to rehabilitate, reestablish.

rehaussé, heightened, set off.

rehaussement, *m.* raising, setting off.

rehausser, to raise, make higher, increase, enhance, heighten, intensify, set off, touch up.

rehaut, *m.* retouch, appreciation.

rehumecter, to remoisten, dampen again.

réhydrater, to hydrate again,

réimporter, to reimport.

réimpression, *f.* reprinting, reprint.

réimprimer, to reprint.

rein, *m.* kidney; *pl.* loins, back, reins.

reinaise, reniform.

reine, *f.* queen; — **des prés,** meadowsweet (*Filipendula ulmaria*).

reine-Claude, *f.* greengage (plum).

reine-marguerite, *f.* China aster (*Callistephus chinensis*).

reinette, *f.* pippin (apple), russet; — **d'or,** golden rennet.

réinsérer, to reinsert.

réintégration, *f.* reinstatement, restoration.

réintégrer, to reinstate, restore, reintegrate.

réitérer, to reiterate, repeat.

rejaillir, to rebound, gush (out), spurt out, be reflected, spring up, arise.

rejaillissement, *m.* rebounding, reflection, spouting, splashing.

rejauger, to gauge or calibrate again.

rejaunir, to yellow again.

réjection, *f.* rejection.

rejet, *m.* rejection, throwing off, thrown-out material, shoot (of tree), offshoot, sprout, escape pipe, second dipping, cast, afterswarm; — **de souche,** stump shoot, stool shoot; — **de tige,** stem shoot, epicormic shoot.

rejetable, rejectable.

rejeter, to reject, throw out, throw back, cast off, sprout, shoot, throw up, throw away, transfer; **se** —, to be rejected, fall back (on), dodge; — **de souche,** to sprout from the stool.

rejeton, *m.* shoot, sucker (of plant), off-spring, descendant, sprout, scion, spray, sprig.

rejettement, *m.* rejection, throwing back.

rejoindre, to rejoin, reunite, overtake; **se** —, to meet again.

réjouir, to rejoice, delight, entertain, amuse; **se** —, to rejoice.

réjouissance, *f.* rejoicing.

réjouissant, jovial, gladsome, amusing.

relâchant, *m.* laxative.

relâchant, relaxing.

relâche, *m.* relaxation, rest, respite, discontinuance; *f.* call.

relâché, relaxed, slack, loose, lax.

relâchement, *m.* relaxing, looseness, slackening, abatement, laxity, laxness.

relâcher, to relax, loosen, slacken, ease, abate, divert, touch at, let go, release, give up, yield, put into port; **se** —, to be relaxed, slacken, get loose, abate, moderate, grow milder.

relais, relai, *m.* relay, sandbank, sand flats.

relais-moteur, *m.* relay motor.

relaisser, to leave again.

relancer, to throw again, cast again.

relargage, *m.* salting out (soap), graining.

rélargir, to widen.

rélargissement, *m.* widening.

relarguer, to salt out (soap), grain.

relater, to relate, state.

relateur, *m.* relater.

relatif, -ive, relative, relating.

relation, *f.* relation, connection, account.

relativement, relatively; — **à,** with regard to, touching.

relativité, *f.* relativity.

relaver, to rewash, wash again.

relaxer, to relax, release.

relayer, to relieve (any one), change; **se** —, to relieve one another, relay.

reléguer, to relegate, banish.

relent, *m.* musty or moldy smell or taste.

rêler (se), to be fissured, become cracked or split.

relevage, *m.* raising, lifting.

relevant, dependent, depending.

relevé, *m.* raising, abstract, summary, statement.

relevé, raised, erect, spicy, pungent, exalted, high, noble.

relèvement, *m.* raising, setting up again, summary, account, increase, bearing, position, rising, uplift.

relever, to raise (up), lift up, take up, pick up again, gather up, draw up, raise again, restore, rebuild, relieve, enhance, heighten, release, remark on, survey, call attention to, note, notice, record, depend (on), recover (from), take the bearings of, exalt, extol, set off; **se** —, to rise again, recover, revive, be raised.

releveur, *m.* levator muscle.

relief, *m.* relief, contour, outline; *pl.* scraps, leavings.

relien, *m.* coarse gunpowder.

relier, to join, unite, bind, connect, bind again; **se** — **à,** to connect with, be bound up with.

religieusement, religiously, scrupulously, punctually.

religieux, -euse, religious, devout, punctual.

religion, *f.* religion.

relimer, to file again, polish up.

reliquat, *m.* remainder, residue, balance.

reliquataire, *m.* debtor.

reliquataire, remaining, left.

relique, *f.* relic, stable plant formation due to past climatic factors.

reliquefier, to reliquefy, liquefy again.

relire, to reread, read again, read over.

reliure, *f.* binding, bookbinding.

relu (relire), reread, read again.

reluire, to shine, glitter, glisten, gleam.

reluisant, shining, glittering, glossy.

relut (relire), (he) read over or again.

remaillage, *m.* graining (leather), mending.

remailler, to shave, grain, mend.

rémailler, to reenamel, enamel again.

remalaxer, to remalaxate, resoften (by kneading or rubbing).

remanent, remaining, residual, remanent.

remanents, *m.pl.* felling refuse, offal timber.

remaniement, remaniment, *m.* alteration, relaying, change, rehandling, repairing, modification.

remanier, to repair, rehandle, alter, modify, change, work over, revise.

remarquable, remarkable, noteworthy.

remarquablement, remarkably.

remarque, *f.* remark, notice, observation.

remarquer, to remark, observe, notice, re-mark.

remballer, to repack, pack again.

remblai, *m.* filling (up), fill, packing, embanking, mound, embankment, pugging.

remblayage, *m.* filling up.

remblayer, to fill (up), embank.

remboîter, to reassemble, fit together again, rebind (a book), set (bone).

rembouger, to fill up again.

rembourrer, to stuff, pad, line (of nests).

remboursement, *m.* reimbursement, repayment; contre —, C.O.D.

rembourser, to reimburse, repay, refund.

rembruni, dark, brown, gloomy.

rembrunir (se), to get darker.

remède, *m.* remedy.

remédier, to remedy, make amends.

remembrement, *m.* reallotment, reallocation.

remémorer, to recall to mind.

remener, to lead or take back.

remerciement, remerciment, *m.* thanks, acknowledgment.

remercier, to thank, decline, dismiss, discharge.

remesurage, *m.* remeasurement.

remesurer, to remeasure, measure again.

remettre, to put back, put again, set back, replace, recall to mind, return, remit (money), reput, deposit, hand over, deliver, give up, resign, entrust, refer, postpone, delay, restore, calm, reconcile, put on again, put off; — en marche, to start again; — en service, to put back into service.

remettre (se), to go back, return, defer, recall, begin again, start again, resume, be postponed, commit oneself (to), entrust with, rely (on), recover, be reconciled, compose one's self; il se remet, he is getting well; il se remet à travailler, he starts to work again.

rémige, wing.

rémige, *f.* wing feather, flight feather; *pl.* remiges.

réminiscence, *f.* reminiscence.

remis (remettre), (he) put back, returned.

remise, *f.* putting back, delivery restoration, remittance, discount, rebate, allowance, delay, low cover, wagon shed.

remit (remettre), (he) set back again, returned.

remmailler, to mend.

remmancher, to put a new handle on.

remodeler, to remodel.

remonder, to reclean.

remontage, *m.* ascending, putting together, reassembling.

remontant, everblooming, perpetual.

remonter, to ascend, remount, mount, go back to, go up again, rise again, go back, reascend, raise, take up again, set up again, reassemble, restock, wind a clock, revive; se —, to provide oneself with a fresh supply, be reassembled, be wound up, revive, recover one's spirits.

remontrer, to show, remonstrate, point out, show again, demonstrate.

remorque, *f.* towing, tow, pulling; se mettre à la —, to follow in the wake.

remorquer, to pull, draw.

remotifolié, with leaves removed from one another.

remotis (à), aside.

remou, *m.* air eddy, whirl, vortex.

remoudre, to regrind, grind again.

rémoudre, to resharpen, sharpen again.

remouiller, to wet again, remoisten.

remouillure, *f.* renewal of yeast.

remoulage, *m.* regrinding, remilling, remolding, bran.

remoulant, regrinding, remolding.

rémoulant, resharpening.

remouler, to remold, recast, mold anew.

remoulu, reground.

rémoulu, resharpened.

rempart, *m.* rampart, bulwark.

remplaçable, replaceable.

remplacement, *m.* replacement, replacing, substitution, reinvestment.

remplacer, to replace, take the place of, substitute, reinvest (funds); se —, to be replaced, replace each other, get up a new supply.

remplage, *m.* filling (up), backing.

rempli, filled, full, replete.

remplir, to fill (up), fill out, fill again, fulfil, cram, furnish, complete, carry out, make up, take up, occupy, accomplish.

stock, supply, repay, refill; se —, to fill, be filled, become full, repay.

remplissage, *m.* filling (up), filling in.

remplisseur, *m.* filler.

remplisseur, -euse, filling.

remporter, to carry away, take away, bear off, carry back, take back, gain, win.

rempoter, to repot.

remuable, movable.

remuage, *m.* moving, stirring, digging, shaking.

remuant, moving, restless, turbulent, stirring.

remuement, *m.* moving, stirring, removing, removal, disturbance, stir.

remuer, to move, agitate, stir, stir up, rouse, dig up, recast; se —, to bestir oneself.

remûment, *m.* moving, stirring, disturbance.

rémunérateur, -trice, remunerative, profitable, paying.

rémunération, *f.* remuneration, payment.

rémunérer, to reward, remunerate, pay.

remunir, to refurnish, supply again.

renaissance, *f.* revival, renaissance, renewal, new birth.

renaissant, renascent, reviving, springing up again.

renaître, to revive, return, be restored, be born again, spring up again, grow again, reappear.

rénal, renal.

renaquit (renaître), (it) revived.

renard, *m.* fox, leak, fissure, opening, gap, bloom, loop.

renarde, *f.* vixen.

rencaisser, to put (plants) in boxes again, rebox (orange trees).

renchérir, to advance, increase in price.

renchérissement, *m.* rise (advance) in price.

rencontre, *f.* meeting, encounter, occurrence, conjunction, occasion, coincidence.

rencontrer, to find, meet (with), encounter, hit on, light upon, happen on; se —, to meet, be found, be met with, run into, agree, tally.

rendement, *m.* yield, efficiency, outturn, returns, output, produce, performance, product, profit; — **brut,** gross yield, gross proceeds, gross revenue; — **effectif,** labor efficiency, output; — **en argent,** revenue; — **en nature,** yield in material; — **en matière,** yield in material, volumetric

yield; — **soutenu,** sustained yield; — **utile,** useful output, useful power.

rendez-vous, *m.* rendezvous, meeting place,

rendre, to return, render, yield, make. produce, restore, repay, give, carry, deliver, emit, throw out, give up, go, lead, surrender; se —, to be returned, return, go, lead, betake oneself, run, give up, surrender, submit, be exhausted, make oneself, give oneself up, become, proceed; — **compte de,** to give an account of; se — **compte de,** to find out, become aware of, ascertain.

rendre, *m* repayment, return; — **brut,** gross proceeds, gross revenue; — **en argent,** revenue.

rendu, *m.* return, returned article.

rendu, returned, arrived, exhausted, done up.

renduire, to coat (cover) again.

rendurcir, to harden, make harder.

rendurcissement, *m.* hardening.

rené (renaître), revived, born again.

renettoyer, to clean again, reclean.

renfermé, *m.* musty smell.

renfermement, *m.* enclosure, shutting up.

renfermer, to contain, include, comprise, comprehend, close in, confine, shut up (again); se —, to be contained, included, or confined, confine oneself.

renflammer, to rekindle; se —, to take fire again.

renflé, swelled, swollen, rounded, inflated, bladdery.

renflement, *m.* swelling, bulging, enlargement, bulge, swell, shoulder, boss, elevation, struma.

renfler, to swell (out), inflate.

renfoncement, *m.* hollow, recess, depression, cavity.

renforçage, *m.* strengthening, reenforcing.

renforçateur, *m.* intensifier, magnifier.

renforcé, strengthened, strong, substantial, intensified.

renforcement, *m.* strengthening, reenforcing.

renforcer, to strengthen, reenforce, intensify, magnify, increase; se —, to be strengthened, grow stronger or more pronounced.

renforceur, strengthening.

renfort, *m.* reenforcement, supply, strengthening piece.

rengorgement, *m.* showing off.

rengréner, to recoin, reengage, feed (hopper).

renier, to deny, disown.

reniflard, *m.* air valve, snifting valve, air intake.

renifler, to snuffle, sniff, snort.

réniforme, reniform, kidney-shaped.

rénipustulé, with kidney-shaped pustules.

reniveler, to level again.

renne, *m.* reindeer.

renom, *m.* renown, fame.

renommé, renowned, celebrated, famous.

renommée, *f.* renown, fame, reputation, report.

renommer, to name again, re-elect, make famous, celebrate.

renoncement, *m.* renouncing.

renoncer, to abandon, renounce, give up, forego, desist (from), waive.

renonciation, *f.* abandonment.

renonculacées, *f.pl.* Ranunculaceae.

renoncule, *f.* crowfoot, buttercup (Ranunculus); — **des champs,** corn crowfoot (*Ranunculus arvensis*).

renouée, *f.* polygonum, knotgrass, knotweed (*Polygonum aviculare*); — **liseron,** black bindweed (*Polygonum convolvulus*).

renouer, to renew, tie again, resume.

renouveau, *m.* revival, renascence.

renouveler, to renew, revive, resuscitate, recall to mind; **se** —, to be renewed, recur, happen again.

renouvellement, *m.* renewal, renewing, revival.

rénovation, *f.* renovation.

renseignement, *m.* (piece of) information, (piece of) intelligence, indication, hint, data.

renseigner, to inform, show, instruct, give information to, teach again; **se** —, to get information, find out, inquire.

rentamer, to cut again, resume, start anew.

rente, *f.* revenue, interest, rent, income, annuity, allowance, pension; — **foncière,** ground rent.

rentier, -ère, *m.&f.* person with independent income.

rentonner, to barrel again.

rentrainer, to carry away again.

rentrant, re-entering, re-entrant.

rentrée, *f.* re-entrance, re-entering, return home, re-opening, receipt, getting in, gathering.

rentrer, to re-enter, enter again, go back, return, take in, get in, resume, reopen, gather, come in, bring in.

rentr'ouvrir, to half-open again.

renverra (renvoyer), (he) will return.

renversable, invertible, reversible, overturnable.

renversé, thrown down, inverted, inversely.

renversement, *m.* inversion, reversal, reversing, overturning, overthrow, subversion, destruction.

renverser, to invert, turn upside down, overturn, spill, upset, overthrow, throw down, ruin, astound, reverse, amaze, fall over.

renvoi, *m.* return, sending back, dismissal discharge, reference, adjournment, gear, gearing, eructation, belch; — **de mouvement,** transmission.

renvoie (renvoyer), (he) is returning, may return.

renvoyer, to send again, send back, throw back, return, reflect (light), reverberate, dismiss, discharge, refer, postpone, adjourn, resend.

réoccuper, to reoccupy.

réorganiser, to reorganize.

réoxydation, *f.* reoxidation.

réoxyder, to reoxidize.

repaire, *m.* den, lair.

repaître, to feed.

répandage, *m.* scattering.

répandre, to spread, diffuse, scatter, become common, send forth, shed, strew, spread out, emit, give out, pour out, exhale, effuse, spill, utter, propagate; **se** —, to be spread, diffused, spread, be spilt, be distributed, extend.

répandu, spread, widely distributed, common, scattered, split, diffused, generally known, in vogue.

réparage, *m.* repairing.

reparaître, to reappear.

réparateur, *m.* repairer.

réparateur, -trice, repairing, reparative, restoring, refreshing.

réparation, *f.* repairing, repair, amends, reparation, distribution.

réparer, to repair, mend, overhaul, retrieve, restore, rectify, make up for.

repartir, to start again, set off again, go off again, reply, retort.

répartir, to distribute, share out, divide, apportion.

répartiteur, *m.* distributer, divider.

répartition, *f.* distribution, repartition, division.

reparu (reparaître), reappeared.

repas, *m.* meal, repast.

repassage, *m.* sharpening, grinding, ironing, recrossing.

repasser, to repass, pass again, return, look over, pass back, review, sharpen, whet, treat again, temper again, iron, iron out, go over again, say over again.

repayer, to repay.

repeindre, to repaint, paint again.

repentir (se), to repent.

repérage, *m.* marking, adjusting, fixing.

repercer, to pierce, perforate again.

répercussion, *f.* repercussion, reverberation, rebound.

répercuter, to reflect, rebound, react, check.

repère, *m.* mark, reference mark, adjusting mark, bench mark, guide mark; **point de —,** reference point, guiding point, clue, essential fact.

repérer, to mark, fix by guide marks, adjust, register, locate, watch closely.

répertoire, *m.* index, table, catalogue, register, repertory.

repeser, to reweigh, weigh again.

répéter, to repeat, rehearse, reflect (images); **il se répète,** he says the same thing over again.

répétiteur, -trice, repeating.

répétition, *f.* repetition, repeating.

repétrir, to knead again.

repeuplement, *f.* restocking, regrowth, regeneration; **— naturel,** second growth.

repiler, to pound again.

repiquage, repiquement, *m.* transplanting, lining out (in a nursery).

repiquer, to transplant, line out, prick out, repair (a road); **— des glands,** to dibble acorns; **— un meule,** to edge a millstone.

répit, *m.* respite.

replacement, *m.* replacing.

replacer, to replace, reinvest, put back.

replanter, to replant.

repli, *m.* fold, plait, winding, crease, sinuosity, coil.

réplicative, -ive, replicate, doubled down, upper part coming against the lower.

repliement, *m.* folding, bending back.

replier, to fold (up), fold again, bend (up), turn up, tuck in, coil up, bend back; **se —,** to fold up, fold one's self, coil up, bend back, wind, meander, fall back, retire.

réplique, *f.* reply.

répliquer, to reply, retort, regain, answer.

replonger, to plunge, dip, immerse again.

replum, *m.* replum, a framelike placenta.

repolir, to repolish, reburnish.

repolissage, *m.* repolishing, rubbing up, reburnishing.

repondre, to lay (egg) again.

répondre, to answer, respond, reply, agree, correspond, satisfy, fill, lead (to); **il en répond,** he is responsible for it; **il répond d'elle,** he is responsible for her.

répons, *m.* response.

réponse, *f.* answer, reply, response.

reporter, to carry back, carry, bring back, transport, place, put, carry forward, refer; **se —,** to be carried back, go back, refer.

repos, *m.* rest, repose, tranquillity, pause; **au —,** at rest; **en —,** at rest.

reposée, *f.* lair.

reposer, to rest, stand, settle (liquids), lay again, refresh, lie, repose, replace, put again, set again, calm, relieve; **se —,** to rest, settle (liquids) again.

reposoir, *m.* washing trough, settling vat.

repoudrer, to repowder, powder again.

repous, *m.* mortar or cement.

repoussant, repelling, repulsive, repellent.

repousser, to repel, push back, repulse, oppose, thrust back, drive back, reject, be repellent to, emboss, chase, throw out (branches) again, sprout again, grow again, deepen, recoil (a gun), resist (a spring); **se —,** to repel each other.

repoussoir, *m.* drift (bolt), starter, rammer, driver.

repouster, to shake, agitate (gunpowder). break up lumps.

réprécipitation, *f.* reprecipitation.

réprécipiter, to reprecipitate.

reprendre, to take up, start again, revive, dissolve, absorb, take again, recapture, take back, retake, take root again, get again, thrive, strike, get back, resume, reply, recover, heal up, repair, reprove, congeal again; **— ses droits,** to come into its own; **se —,** to recover oneself, be taken up, congeal again, heal, start again.

représentant, *m.* representative.

représentatif, -ive, representative.

représentation, *f.* representation, display, show.

représenter, to represent, describe, depict, present again, remind of; **se —,** to reappear, occur, be represented, fancy, present itself again, imagine, picture to oneself.

répression, *f.* repression.

réprimant, repressive.

réprimer, to repress, put down.

repris (reprendre), taken up again.

reprise, *f.* taking up again, repetition, return, resumption, recovery, resuming time, reviving (of plants); **à plusieurs —s,** several times, again and again, repeatedly.

reprit (reprendre), (he) took again.

reproche, *m.* reproach.

reprocher, to reproach, find fault (with), blame for.

reproducteur, -trice, reproductive, reproducing.

reproductibilité, *f.* reproducibility, reproductive power.

reproductible, reproducible.

reproductif, -ive, reproductive.

reproduction, *f.* reproduction, procreation, regeneration; **— agame,** agamous reproduction; **— asexuée,** asexual reproduction; **— dans la famille,** inbreeding; **— fidèle,** breeding true to type.

reproduire, to reproduce, breed (animals), republish; **se —,** to be reproduced, reproduce itself, reappear, occur again.

reptile, *m.* reptile.

repu (repaître), full, satieted.

républicain, *m.* republican.

republier, to republish.

république, *f.* republic, commonwealth.

répudier, to repudiate, renounce.

répugnant, repugnant, contrary.

répugner, to be repugnant, be reluctant.

répulsif, -ive, repulsive, repelling.

répulsion, *f.* repulsion.

repurger, to purge again.

réputation, *f.* reputation, repute.

réputé, reputed, well-known.

réputer, to repute, consider, reckon, deem, think.

requérir, to request, require, demand, claim, summon.

requerra (requérir), (he) will request.

requête, *f.* request, petition.

requiert (requerir), (he) is requesting.

requis, *m.* requisite.

requis (requerir), required, necessary, requisite, requested, proper.

resceller, to seal again.

réseau, *m.* network web, net, line, netting, system (of lines or wires), (diffraction) grating, screen (of wires), plexus, second stomach, reticulum; **— de chemins,** network of roads; **— de distribution,** distributing system, distributing network; **— déformable,** unstable frame; **— ferrés,** railroads, railroad systems; **— fluvial,** river system.

resécher, to dry again.

réséda, *m.* mignonette, sweet reseda (Reseda); **— gaude,** dyer's-weed (*Reseda luteola*), yellowweed; **— odorant,** garden mignonette (*Reseda odorata*).

reservage, *m.* reserving.

réserve, *f.* reservation, reserve, reserve supply, caution, claim, resist; **— à assiette fixe,** definitely located reserve; **— à assiette mobile,** shifting reserve; **à la — de,** with the exception of; **à la — que,** except that; **en —,** in reserve, aside, spare, in store.

réserver, to reserve, save up, lay by, keep, leave standing, store; **se —,** to save oneself, hold back, wait, reserve.

réservoir, *m.* reservoir, tank, fish pond, well, cistern, chamber, container, holder; **— à gaz,** gas holder, gasometer; **— alimentaire,** feed tank; **— de barrage,** storage basin.

résider, to reside, live, dwell, consist, lie.

résidu, *m.* residue, residuum, remainder, settlement, lees, settling, waste product; **— de la décomposition,** residual soil; **— sec,** dry matter.

résiduaire, residual, remaining.

résiduel, -elle, residual.

résigner, to resign, submit.

résillier, to cancel, annul, terminate.

résinage, *m.* extraction of turpentine, resin tapping, resin boxing.

résine, *f.* resin, rosin, colophany; **— commune,** common resin, liquid pitch; **— de gaiac,** guaiacum resin, guaiacum; **— de terebenthine,** rosin; **— elastique,** rubber, caoutchouc; **— jaune,** yellow resin; **— vierge,** crude turpentine.

résiner, to tap for resin, bleed, extract resin, treat with resin; **— à mort,** to tap so as to kill the tree; — **à vie,** to tap without killing the tree.

résineux, *m.* conifer, coniferous tree.

résineux, -euse, resinous, resiny, rosinous.

résinier, *m.* resin collector, resin tapper.

résinifère, resiniferous.

résinification, *f.* resinification, resinifying.

résinifier, to resinify.

résiniforme, resiniform.

résinigomme, *f.* gum resin.

résinigommeux, -euse, gum-resinous.

résino-amer, *m.* aloe.

résinoïde, resinoid.

résinose, *m.* disease of conifers.

résinosis, *f.* resinosis, abnormal production of resin in lacunae.

résistance, *f.* resistance, opposition, strength; — **à la flexion,** transverse strength, bending or deflective strength; — **à la pression,** compressive strength; — **à la tension,** tensile strength; — **à la traction,** tensile strength, resistance to tension, absolute tenacity.

résistance, — **à l'oscillation,** dynamic strength; — **au cisaillement,** shearing strength; — **continue de flexion,** bending stress durability; — **de mise en marche,** starting resistance; — **d'isolement,** insulation resistance; l:mite de —, yield point.

résistant, resisting, resistant, tough, unyielding, hardy, firm, hard; — **à la gelée,** frost hardy; — **au vent,** stormproof, wind firm.

résister, to resist, oppose, withstand, support, endure.

résolu, solved, resolute, determined, fixed.

résoluble, decided on, settled, capable of being solved.

résolument, resolutely.

résolutif, -ive, resolvent, discutient.

résolution, *f.* (re)solution, determination, resolve, reduction, conversion, canceling.

résolvant, *m.* resolvent.

résolvant, solving.

résolvent (résoudre), (they) are solving, reducing.

resomptif, -ive, restorative.

résonance, *f.* resonance, tuning; — **du courant,** current resonance.

résonner, to resound, reverberate, be resonant.

résorber, to resorb, reabsorb.

résorcine, *f.* resorcinol, resorcin.

résorption, *f.* resorption, reabsorption.

résoudre, to solve, resolve, settle, reduce, convert, dissolve, cancel, induce, persuade, decide upon; **se** —, to be solved, decide, separate (into), make up one's mind, resolve.

résous (résoudre), resolved, dissolved, converted.

résout (résoudre), (he) is resolving, solving, reducing.

respect, *m.* respect, regard, reverence.

respectable, respectable.

respecter, to respect, reverence, honor, have regard for.

respectif, -ive, respective, each, corresponding.

respectivement, respectively.

respirabilité, *f.* respirability.

respirable, respirable.

respirateur, *m.* respirator.

respirateur, -trice, respiratory.

respiration, *f.* respiration, breathing, breath.

respiratoire, respiratory.

respirer, to breathe, respire, inhale, long for.

resplendir, to shine brightly, gleam.

resplendissant, resplendent, effulgent.

responsabilité, *f.* responsibility, liability.

responsable, responsible.

ressac, *m.* surf, recoil of waves, undertow.

ressaisir, to seize again, reseize.·

ressasser, to sift again, rebolt (flour), examine again, scrutinize, repeat over and over.

ressaut, *m.* projection, shoulder, rise, jog, lug.

ressayer, to try again.

ressemblance, *f.* resemblance, likeness, similarity.

ressemblant, similar, alike, like, resembling.

ressembler, to be like, look like, take after, be alike, resemble; **se** —, to be alike, resemble each other.

ressense, *f.* restamping.

ressentir, to feel, experience, feel again, resent; **se** —, to be felt, feel the effects (of), have a liking for; **il s'en ressentit,** he feels the effects of it.

resserrement, *m.* contraction, tightening (of money), constipation.

resserrer, to contract, narrow, confine, restrain, compress, construct, condense, tie, rebind. bind together, tighten, draw tighter, shorten, put away, put back, hem in, constipate; **se** —, to be contracted, contract, shrink, become tighter, get colder.

resservir, to serve again, be used again.

ressort, *m.* elasticity, springiness, spring, province, jurisdiction, means, energy; — **à boudin,** — **à hélice,** helical spring, spiral spring; — **d'attache,** spring catch; — **de rappel,** return spring.

ressortir, to come, go out again, stand out, be evident, bring out, spring, show off, follow, result, be under the jurisdiction (of), be set off; **faire** —, to bring out to view, set forth. emphasize.

ressouder, to resolder, reweld, reunite.

ressoudure, *f.* resoldering, rewelding, joining together again.

ressource, *f.* resource, resources, means.

ressouvenir, to recollect, remember again.

ressouvenir, *m.* reminiscence.

ressuage, *m.* sweating, eliquation.

ressuer, to sweat, be liquated, sweat again.

ressui, *m.* sweating.

ressuiement, *m.* drying.

ressuivre, to follow again.

ressusciter, to resuscitate, revive, come to life, restore.

ressuyage, *m.* drying.

ressuyé, dried.

ressuyer, to dry, wipe (dry) again.

restant, *m.* remainder, rest, residue.

restant, remaining, left, residual.

restaurant, *m.* restaurant, restorative.

restaurant, restoring, restorative.

restauration, *f.* restoration; — **des vides,** replanting, afterculture.

restaurer, to restore.

reste, *m.* residue, remainder, rest, offal, garbage, remain(s), waste, rubbish, tailings, leavings, refuse, relic; **au** —, **du** —, besides, moreover, also, however, furthermore; **de** —, to spare, more than enough, left.

rester, to remain, stay, continue, be left, stop; **en** —, to leave off, stop, discontinue; **il en reste là,** he stops there; **il reste à voir,** it remains to be seen; **il reste en place,** he keeps still; **il reste un livre,** one book is left.

restibile, perennial.

restituer, to restore, give back.

restituteur, *m.* restorer.

restitution, *f.* restoration, restitution.

restreindre, to restrict, restrain, limit.

restreint, restricted, limited.

restriction, *f.* restriction, limitation, reservation.

restringent, astringent, restringent, styptic.

résultant, resulting, resultant.

résultante, *f.* resultant, resultant force.

résultat, *m.* result, consequence.

résulter, to result, be the consequence, follow, ensue, appear; **il en résulte que,** the result is that.

résumé, *m.* résumé, summary, summing up, recapitulation; **au** —, **en** —, to sum up, on the whole, in short.

résumer, to sum up, recapitulate, summarize, repeat.

résupiné, resupinate, upside down.

résurrection, *f.* resurrection.

rétablir, to reestablish, restore, put back; **se** —, to be restored.

rétablissement, *m.* reestablishment, restoration, recovery.

retaille, *f.* cutting, clipping, shred, paring.

retaillement, *m.* cutting again, repruning.

retailler, to cut again, recut, resharpen (pencil), dig up, prune again.

retaler, to spread again.

rétamage, *m.* retinning, replating.

rétamer, to retin, resilver, replate.

retanner, to tan again, retan.

retard, *m.* delay, lateness, slowness, hindrance, lag, retardment; **en** —, late, behind time, slow; **être en** —, to be late.

retardataire, lagging behind, late, backward.

retardateur, -trice, retarding, retardative.

retardation, *f.* retardation, decrease of speed.

retarder, to retard, delay, defer, set back, put off, be slow, he behind.

retasser, to develop cavities, pipe.

retassure, *f.* pipe (the cavity), shrinkage hole.

retâter, to feel (touch) again.

reteindre, to redye, color.

réteindre, to re-extinguish.

retenant, retaining, holding back.

retendre, to restretch, stretch again.

rétendre, to dilute again.

retène, *m.* retene.

retenir, to retain, hold, keep, hold back, hold on, restrain, withhold, hold up, sustain, get back, get again, remember; se —, to catch hold, cling, stop, refrain, be retained.

retenter, to try again, reattempt.

rétenteur, -trice, retaining, restraining.

rétentif, -ive, retentive.

rétention, *f*. retention, reservation.

retentir, to resound, echo, ring, re-echo.

retentissant, resounding.

retentissement, *m*. resounding, echoing, re-echoing, resonance, reverberation, renown, publicity, sound, effect, noise.

retenu, retained, reserved, prudent, discreet.

retenue, *f*. withholding, deduction, stoppage, docking, reservoir, dam, guy rope, reserve, moderation, discretion, fixture, retention; — des eaux, rising of water.

réticence, *f*. reserve, reticence.

réticulaire, reticular, reticulate, netted.

réticule, *m*. reticle, reticule, cross wires, cross hairs.

réticulé, reticulate(d), netted.

réticulum, *m*. reticulum.

retient (retenir), (he) is retaining.

rétine, *f*. retina.

rétinervé, retinerved, net-veined.

rétinien, -enne, retinal, of the retina.

retint (retenir), (he) retained, held back.

retiré, secluded.

retirement, *m*. withdrawing, withdrawal, contraction, crawling.

retirer, to withdraw, remove, pull out, draw back, take back, retire, get, obtain, extract, derive, take in, contract, draw out, take out; se —, to retire, withdraw, be withdrawn, leave, forsake, contract, shrink, subside.

retirure, *f*. hollow due to shrinkage.

rétisité, *f*. restiveness.

retombant, prostrate.

retombée, *f*. springer.

retomber, to fall again, fall back, fall down again, relapse, attribute.

retordre, to wring again, twist (again).

retors, *m*. twist, second twist.

retors, twisted, wrung, crafty, bent.

retorte, *f*. retort.

retouche, *f*. alteration.

retoucher, to retouch, dress, finish.

retour, *m*, return. comeback, turn, winding, angle, corner, recurrence, backing up, change, repetition, wane, decline (of life); — à l'état inculte, running wild, becoming a waste; au —, on returning; en —, in return, declining; en — d'équerre, at right angles; en — de, in return for; sans —, forever.

retourné, inverted, bent back.

retourner, to turn, turn over, turn upside down, invert, dig up, examine carefully, return, turn up, affect, move, go back; se —, to turn (over) round, manage, be turned; — à l'état sauvage, to run wild, run to bush; — le sol, to turn or trench the soil; il s'en retourne à, he returns to.

retracer, to retrace, relate.

rétracter, to contract, retract.

retrait, *m*. contraction, shrinking, shrinkage, withdrawal; prendre —, to shrink.

retrait, contracted, shrunken, warped.

retraite, *f*. retreat, retirement, retiring pension, redraft, pension, contraction, shrinkage, refuge, shelter.

retraiter, to treat (handle) again, retreat, retire, pension.

retranchement, *m*. subtraction, retrenchment, curtailment.

retrancher, to subtract, deduct, cut off, retrench, abridge, curtail, take away, leave out, stop.

retravailler, to work over again, finish.

retraverser, to retraverse.

rétréci, narrow, shrunk, contracted.

rétrécir, to narrow, contract, shrink, straighten, shorten.

rétrécissement, *m*. narrowing, contraction, shrinking, narrowness, stricture.

retreindre, to hammer out.

retremper, to soak again, temper again, reinvigorate.

rétribuer, to remunerate, pay.

rétribution, *f*. remuneration, reward, pay, salary.

rétroactivement, retroactively.

rétrogradation, *f*. retrogradation, retrogression, reversion.

rétrograde, retrograde, backward.

rétrograder, to retrogress, retrograde, go backward, revert.

rétrogression, *f*. retrogression, reversion towards simpler organization.

retrousser, to turn up, tuck up, hold up, bunch up.

retrouver, to find, find again, meet with, recover, meet again; **se —,** to be found, find one's way, be met with; **ils se retrouvent à,** they meet again at.

rets, *m.* net.

rétudier, to study again.

rétus, retuse.

réunion, *f.* union, reunion, assembly, junction, connection, meeting.

réunir, to reunite, unite, join, assemble, collect, gather together, call together; **se —,** to reunite, be collected, unite, gather, join together, assemble.

réussi, successful.

réussir, to succeed, be successful, result, issue, turn out, prosper, thrive, end, carry out well; **tout lui réussit,** he succeeds in everything.

réussite, *f.* issue, result, success.

revanche, *f.* return, revenge, requital; **en —,** in return, on the other hand, in compensation.

revaporisation, *f.* revaporization.

revaporiser, to revaporize.

rêve, *m.* dream.

revêche, harsh, rough, difficult to work, cross-grained, short, brittle.

revécu, revived.

réveil, *m.* waking, awakening, reveille.

réveille-matin, *m.* alarm clock, spurge, wartweed.

réveiller, to wake, awaken, quicken, rouse, revive; **se —,** to wake up, awake, revive, rouse up.

révélateur, *m.* detector, revealer, discloser, indicator, developer.

révélateur, -trice, revealing, disclosing.

révéler, to reveal, disclose, develop, detect; **se —,** to reveal oneself, be revealed, manifest itself, behave.

revenant, *m.* ghost, apparition.

revenant, returning, pleasing.

revendication, *f.* claim, demand.

revendiquer, to claim, demand.

revenir, to return, come back, come again, recur, recover, amount (to), result, arise, yield, redound, be due to, be suited (to), get over, be pleasing (to), come again; **— à,** to please, suit, agree with, return to, succeed in; **au même,** to amount to the same thing; **— sur,** to return to, retract, fall back on; **faire —,** to brown, let down the temper, restore, **il vous revient très bon marché,** it costs you very little.

revenu, *m.* revenue, income, rental, financial return, tempering; **bien —,** welcome back, welcome again.

revenue, *f.* return, new growth, young wood.

rêver, to dream, fancy, picture, muse, dream of, long for.

réverbération, *f.* reverberation, reflection.

réverbératoire, reverberatory.

réverbère, *m.* reflector, street lamp, reverberator; **à —,** reverberatory.

réverbérer, to reverberate, reflect.

reverdir, to grow, turn green again, revive, soak and soften (hides).

reverdissage, *m.* cleaning and softening (hides).

reverdissement, *m.* growing green again, new life.

rêverie, *f.* revery, guesswork, roving.

reverra (revoir), (he) will see again.

revers, *m.* reverse, back, wrong side, other side.

reverser, to pour out again, pour back, transfer, carry.

réversibilité, *f.* reversibility.

réversible, reversible.

réversion, *f.* reversion, a change backward.

revêtement, *m.* facing, casing, cover, covering, coating, sheathing, lining, revetment.

revêtir, to clothe, dress, face, coat, case, cover, line, sheathe, cover over, put on, assume, fit, equip; **— des planches,** to campsheet, campshed; **se —,** to put on, clothe oneself, array oneself, dress, assume.

revêtu, faced; **— d'une écorce,** covered with bark.

revidage, *m.* re-emptying, enlarging (a hole), reboring.

revider, to re-empty, enlarge (a hole), rebore.

revient (revenir), (he) is returning; **il — de,** he gets over; **il en — à,** he comes back to.

revient, *m.* cost; **prix de —,** cost price.

revint (revenir), (he) returned.

revirer, to veer, turn around.

reviser, réviser, to revise, inspect, examine.

revision, révision, *f.* revision, inspection, examination.

revisser, to screw again, screw up, tighten.

revit (revoir, reviser), (he) saw again, (he) is reviving.

revivificateur, -trice, revivifying, reviving.

revivification, *f.* revivification, reviving.

revivifier, to revivify, revive.

revivre, to revive, live again; **faire —,** to revive, bring to life again.

revoir, to revise, see again, inspect, examine again, oversee, resee; **au —,** good-by till later.

révolte, *f.* revolt, mutiny.

révoluté, révolutif, -ive, revolute.

révolutifolié, with revolute leaves.

révolution, *f.* revolution, rotation; **— de conversion,** temporary rotation for conversion.

révolutionnaire, revolutionary.

révolutionnairement, revolutionarily.

revolver, *m.* revolver, revolving nosepiece (of microscope).

révoquer, to revoke, recall; **— en doute,** to call in question.

revu (revoir), revised.

revue, *f.* review, survey, inspection, journal, examination, seeing again, magazine; **passer en —,** to review.

révulsif, -ive, revulsive, counterirritant.

rez, even with, level with; **à — de,** on a level with.

rez-de-chaussée, *m.* ground level, ground floor.

rez-terre, flush with the ground, at ground level.

rez-tronc, close to the stem.

rhabillage, *m.* repair, mending, overhaul, dressing again.

rhabiller, to repair, mend, reclothe, dress again.

rhacis, *f.* spinal column.

rhamné, similar to buckthorn, rhamnaceous.

rhamnétine, *f.* rhamnetin.

rhamnose, *f.* rhamnose.

rhaphanus, *m.* wild mustard (Raphanus).

rhaptocarpe, with striated fruit.

rhénan, Rhenish, Rhine.

rhéotropisme, *m.* rheotropism.

rhéostat, *m.* rheostat.

rheum, *f.* rhubarb (Rheum).

rhexigène, rhexigenous, (origin of tissues) formed by mechanical rupture.

rhexolyte, rhexolytic, (gemmae) detached by the rupture of a cell.

rhigolène, *m.* rhigolene.

Rhin, *m.* Rhine.

rhinchite, *f.* **— du bouleau,** birch-tree weevil.

rhinosclérome, *f.* rhinoscleroma.

rhipidium, *m.* rhipidium, a fan-shaped cyme.

rhizanthé, rhizanthous, root-flowered.

rhizelle, *f.* rhizel, base of the root.

rhizidium, *m.* rhizidium, rhizoid in the oophore condition.

rhizine, *f.* rhizina, root hairs of mosses.

rhizoblaste, *m.* rhizoblastus, an embryo that emits roots.

rhizoctone, *m.* violet root rot.

rhizodé, rootlike.

rhizoderme, *m.* rhizodermis, outermost cortical layer.

rhizoïde, *m.* rhizoid, a hair (frequently branched).

rhizomatique, relating to rhizome.

rhizome, *m.* rhizome, rhizoma, rootstock.

rhizomorphe, *m.* rhizomorph, a rootlike branched strand of mycelial hyphae.

rhizophore, *m.* rhizophore.

rhizophylle, rhizophyllous, (roots) proceeding from the leaves.

rhizophyse, *f.* rhizophysis, an expansion of the radicle.

rhizopode, furnished with roots at the foot.

rhizopodes, *m.pl.* Rhizopoda.

rhizosperme, with fruits near the root.

rhizotaxie, *f.* rhizotaxy, the system of arrangement of the roots.

rhodacanthe, with reddish spines.

rhodamine, *f.* rhodamine.

rhodique, rhodic.

rhodium, *m.* rhodium.

rhodizonique, rhodizonic.

rhodoleuque, rhodoleucous, reddish-white.

rhodosperme, with rose-colored seed.

rhodospermine, *f.* rhodospermin, crystallized phycoerythrin.

rhombe, *m.* rhomb, rhombus, rhombohedron.

rhombe, rhombic.

rhombé, rhombeous, rhombic, shaped like a rhomb.

rhombifolié, rhombifolious, rhomboidal-leaved.

rhombiforme, rhombiform, rhomb-shaped.

rhombique, rhombic.

rhomboèdre, *m.* rhombohedron.

rhomboédrique, rhombohedral.

rhomboïdal, -aux, rhomboidal.

rhomboïde, rhomboid.

rhubarbe, *f.* rhubarb (Rheum); — des **paysans,** — des **pauvres,** meadow rue (Thalictrum).

rhum, *m.* rum.

rhumatique, rheumatic.

rhumatisant, *m.* rheumatic patient.

rhumatismal, -aux, rheumatic.

rhumatisme, *m.* rheumatism.

rhume, *m.* cold; **gros** —, bad, violent cold.

rhus, *m.* sumac.

rhythme, *m.* rhythm.

rhytidome, *m.* rhytidome, rough or dead bark.

ri (rire), (he) laughed.

riant (rire), smiling, cheerful, laughing, pleasant.

ribésié, similar to currant.

ribésoïde, similar to currant.

riblon, *m.* scrap (iron or steel).

ribonique, ribonic.

ribose, *n.* ribose.

riche, *m.* rich material, rich person.

riche, rich, wealthy, abundant.

richement, richly.

richesse, *f.* richness, riches, wealth, abundance, stock, fertility (of soil); — **en volaille,** stock of poultry.

ricin, *m.* castor-oil plant (*Ricinus communis*), ricinus, castor bean; **huile de** —, castor oil.

ricine, *f.* ricin.

riciné, impregnated with castor oil.

ricinine, *f.* ricinine.

ricinoléique, ricinolique, ricinoleic.

ricocher, to rebound, ricochet.

ricochet, *m.* ricochet, series, succession.

ride, *f.* wrinkle, ridge, crease, fold, ripple, ripple mark, lanyard.

ridé, wrinkled, rugose, shriveled, corrugated, fluted.

rideau, *m.* curtain, screen, shelter belt, windbreak.

rider, to wrinkle, shrivel, ripple, ruffle, corrugate, flute, rib.

ridicule, *m.* ridiculousness, ridicule.

ridicule, ridiculous.

ridiculement, *m.* ridiculously.

rien, *m.* nothing, trifle; — **à faire,** nothing to be done; — **d'autre,** nothing else; — **du tout,** nothing at all; — **qu'à la voir.** merely by seeing her; — **que,** nothing but, only, merely, just; **de** —, don't mention it; **en** —, not at all, somewhat; **en moins de** —, in less than no time; **il n'en est** —, nothing of the kind; **ne** ... —, nothing.

rien, nothing, anything, something.

riflard, *m.* coarse file (for metals), paring chisel, jack plane.

rifler, to plane, pare, scratch, file, rasp.

rifloir, *m.* curved file, riffler.

rigide, rigid, severe, strict, harsh, tense.

rigidement, rigidly, strictly.

rigidifolié, with stiff leaves.

rigidité, *f.* rigidity, rigidness, tenseness.

rigolage, *m.* trenching.

rigole, *f.* channel, groove, runner, trough, (water) furrow, gutter, drain, ditch, (small) trench, rill, little furrow.

rigoler, to trench, plow deep, furrow, trench plow.

rigoleur, *m.* drill hoe.

rigoleuse, *f.* trenching plow.

rigoureusement, rigorously, strictly, harshly.

rigoureux, -euse, rigorous, severe, strict, sharp, accurate.

rigueur, *f.* rigor, severity, strictness; **à la** —, at the worst, if need be, rigorously, strictly speaking, in case of necessity.

rime, *f.* rima, cleft.

rimosipède, with slit stalk.

rinçage, rincement, *m.* rinsing.

rince-bouteilles, *m.* bottle rinser.

rincer, to rinse, drench.

ringard, *m.* poker, fire iron, tapping iron.

ringarder, to stir, poke (a fire).

ringent, ringent, wide open, gaping.

ringentiflore, ringentiflorous, with ringent corollas.

ringentiforme, ringentiform, apparently gaping.

ripaille, *f.* feasting; **faire** —, to feast on junket.

riper, to shift, scrape, slip, slide.

riposte, *f.* parry and thrust, pat answer.

riquet, *m.* cricket.

riquette, *f.* scrap iron.

rire, to laugh, mock, smile, jest.

rire, *m.* laughing, laughter.

ris, *m.* sweetbread, laughter, reef (in a sail).

risque, *m.* risk, hazard.

risquer, to risk, hazard.

rit (rire), (he) is laughing, (he) laughed.

rite, *m.* ceremony, rite.

rituel, -elle, ritualistic, *m.* ritual.

rivage, *m.* riveting, shore, bank.

rival, -aux, rival, competitor.

rivaliser, to compete, vie, rival.

rivalité, *f.* rivalry.

rive, *f.* shore, border, edge, bank (of a river).

rivement, *m.* riveting, clinching.

river, to rivet, clinch (nails).

riverain, *m.* borderer, coast dweller, neighbor.

riverain, bordering, shore dwelling.

rivet, *m.* rivet, clinch, clincher, riveting nail.

rivetage, *m.* riveting.

riveter, to rivet.

riveur, *m.*, riveter.

riveuse, *f.* riveter (machine).

rivicole, inhabiting the banks of rivers.

rivière, *f.* river, stream; La Rivière, Riviera.

rivure, *f.* riveting, rivet(ed) joint, rivet head, pin joint.

rixe, *f.* tiff, squabble.

riz, *m.* rice, paddy; huile de —, rice oil.

rizerie, *f.* rice mill.

rizière, *f.* rice field, rice plantation.

rob, *m.* rob.

robe, *f.* robe, gown, dress, coat (of animals), skin, husk, outer leaf, wrapper, envelope.

rober, to bark (madder), cover, wrap (cigar).

robinet, *m.* stopcock, cock, plug (of cock), faucet, spigot, tap, valve; — à deux voies, two-way stopcock; — à gaz, gas cock; — à goutte, drop tap or cock; — à pointeau, needle valve; — à quatre voies, four-way cock; — à trois voies, three-way cock; — à vapeur, steam cock; — à vis, screw cock, screw faucet.

robinet, — d'arret, stopcock; — d'eau, water cock; — de décharge, discharge cock, purge cock; — de purge, purging cock, blow valve, blowoff cock; — d'essai, try cock, pet cock; — de sûreté, safety cock, safety tap; — de vidange, drain cock, discharge valve; — mélangeur, mixing cock; — souffleur, blast cock.

robinetterie, *f.* manufacture of cocks and valves.

robinier, *m.* robinia, false acacia, locust tree (*Robinia pseudacacia*).

roborant, strengthening, tonic.

roburite, *f.* roburite.

robuste, strong, robust, stout, light-bearing.

robustesse, *f.* strength, stoutness, sturdiness.

robusticité, *f.* robustness.

roc, *m.* rock.

rocaillage, *m.* rockwork.

rocaille, *f.* rockwork.

rocailleux, -euse, stony, rocky.

roccella, roccelle, *f.* Roccella.

roccelline, *f.* roccellin, roccelline.

rocellique, roccellic.

rochage, *m.* frothing, foaming.

roche, *f.* rock, boulder, crude borax; — à sel, rock salt; — de fond, bedrock; — vive, solid rock, ore in sight.

rocher, to flux, froth, foam, vegetate (silver), spit, sprout.

rocher, *m.* rock, petrosal (bone).

rochet, *m.* ratchet, pawl, click, bobbin.

rocheux, -euse, rocky, stony; montagnes Rocheuses, Rocky Mountains.

rocou, *m.* annatto, roucou.

rocouer, to dye with annatto.

rocouyer, *m.* annatto tree, achiote (*Bixa orellana*).

rodage, *m.* grinding, polishing, wearing.

rodé, ground to fit.

roder, to grind, rub, polish, grind in (glass stopper).

rôder, to roam, rove, prowl, work freely, have play.

rodoir, *m.* grinder, polisher, lap, tanning vat.

Roentgen, Roentgen; rayons —, Roentgen rays.

rognage, *m.* clipping, trimming, paring.

rogne, *f.* scab (of plants), mange, itch, part of bark.

rognement, *m.* clipping, paring.

rogner, to clip, trim, pare, cut (down) off, curtail, reduce, shave.

rognoir, *m.* paring tool, clipper, trimmer, scraper.

rognon, *m.* kidney (of animals), nodule, kidney stone; — de silex, flint nodule.

rognure, *f.* cutting, clipping, trimming, shaving, paring, scrap.

roi, *m.* king.

roide, stiff, rigid, tight.

roideur, *f.* stiffness, rigidity.

roitelet, *m.* goldcrest, wren; — triple bandeau, firecrest, fire-crested wren.

rôle, *m.* roll, list, role, part played, stack of wood.

rollier, *m.* roller (bird).

romain, Roman.

romaine, *f.* steelyard.

roman, *m.* romance, novel.

roman, romance, romanic.

romarin, *m.* rosemary (*Rosmarinus officinalis*); — sauvage, marsh tea, wild rosemary (*Ledum palustre*).

rompre, to break, break up, snap off, break off, rend, break with, shatter, rupture, tear. refract (light), deaden, blend (colors), disrupt, interrupt; se —, to be broken, break up, break off; il se rompt, it breaks,

rompu, broken, experienced, used, snapped.

ronce, *f.* bramble, brier (Rubus); — noire, blackberry (*Rubus fruticosus*); —s artificielles, barbed wire.

roncier, *m.* bramble bush, blackberry bush.

ronciné, *m.* runcinate.

rond, *m.* round, ring, circle, disk, washer, wheel, bar iron; en —, round, roundly, circularly, in a ring.

rond, round, rotund, turnip-rooted, circular.

ronde, *f.* round, rounds; à la —, round, around, roundabout.

rondelle, *f.* (small) round, disk, ring, circlet, washer, round piece; — à ados, saddle plate; — de caoutchouc, rubber washer; — de joint, gasket, packing ring; — en cuir, leather washer; — en liège, cork ring or washer.

rondement, roundly, briskly, promptly, frankly.

rondeur, *f.* roundness, curve, rotundity.

rondin, *m.* thick stick, (round) billet, small log, roll; — piège, trap billet.

rondot, *m.* milk pan.

ronflement, *m.* roaring, rumbling.

ronfler, to roar, rumble, snore, boom, hum.

rongé, damaged (by insects), erose, eroded.

rongeage, *m.* discharging.

rongeant, *m.* dischagre.

rongeant, corroding, rodent (ulcer), gnawing.

rongement, *m.* corroding, gnawing.

ronger, to corrode, pit, eat away, etch, discharge, gnaw, wear, nibble, bite, consume.

rongeur, *m.* rodent; — du chêne, oak bark beetle.

rongeur, corroding, gnawing.

roquefort, *m.* Roquefort cheese.

roquet, *m.* rocket.

roquette, *f.* rocket (Hesperis).

roridé, roridus, dewy, covered with particles that resemble dewdrops.

rorqual, *m.* rorqual, finback; — rostré, beaked whale.

rosace, *f.* rosette, rosework, rose design.

rosacées, *f.pl.* Rosaceae.

rosage, *m.* oleander, rhododendron.

rosaniline, *f.* rosaniline.

rosat, rose, of roses; géranium —, rose geranium (*Pelargonium graveolens*).

rose, *m.&f.* rose; — bengale, rose bengale; — grimpante, climbing rose; — mousseuse, mossy rose; — rouillée, sweetbrier (*Rosa eglanteria*); — tremière, hollyhock (*Althaea rosea*).

rose, rose, rosy, pink.

rosé, rose, rosy, pink, pinkish, roseous.

roseau, *m.* reed, reed grass.

rosée, *f.* dew; — du soleil, sundew (Drosera)

roselé, rosulate, collected into a rosette.

roser, to rose, make pink.

roseraie, *f.* rosery, rose garden.

rosette, *f.* rosette, rose, disk, washer, rosette copper, red ink, red chalk, skin plow, jointer, skin coulter; — des saules, bushy top, rosette in willows.

rosier, *m.* rosebush, rose; — demi-tige, half standard rose; — des chiens, dog rose (*Rosa canina*); — haute tige, standard rose; — nain, dwarf rose; — remontant, perpetual flowering rose.

rosinduline, *f.* rosinduline.

rosolique, rosolic.

rossignol, *m.* nightingale.

rossolis, *m.* rosolio (liquor), sundew (Drosera).

rostellé, rostrate, rostellate.

rostellum, *m.* rostellum, a small beak.

rostre, *m.* rostrum, beak.

rostré, rostrate(d), beaked.

rosulaire, rosular. collected into a rosette.

rot, *m.* eructation, belch.

rotacé, rotate, wheel-shaped.

rotang, *m.* rattan.

rotateur, -trice, rotating, rotatory.

rotatif, -ive, rotary, rotative.

rotation, *f.* rotation, revolution; — des coupes, felling rotation.

rotatoire, rotatory.

rot-brun, *m.* brown rot.

rotengle, *m.* redeye.

rotin, *m.* rattan, calamus.

rôtir, to roast, bake, broil.

rôtissage, *m.* roasting.

rotondité, *f.* rotundity.

rotor, *m.* rotor.

rotule, *m.* kneecap, patella, swivel joint, ball-and-socket joint.

rotundifolié, rotundifolious, round-leaved.

rouable, *m.* hook, fire rake.

rouage, *m.* wheels, wheelwork, mechanism, wheel, toothed wheel, gear wheel.

rouan, *m.* roan.

roucou, *m.* see ROCOU.

roue, *f.* wheel, coil (of rope), runner; — à aubes, paddle wheel; — à cames, cam wheel, cogwheel; — à gorge, grooved friction wheel; — à rochet, ratchet wheel; — à sillon, furrow wheel; — avant, front wheel; — à vis, worm wheel.

roue, — de charrue, plow wheel; — de commande, driving wheel; — de derrière, hind wheel; — dentée, toothed wheel, cogwheel, gear wheel, gear; — de rechange, spare wheel; — en dessus, overshot wheel; — hydraulique, water wheel; — volante, flywheel, regulator; bateau à —s, side-wheel steamer.

rouelle, *f.* round slice, round (of beef).

rouet, *m.* sheave, pulley, spinning wheel.

rouge, *m.* red, blush, red heat, rouge; — à farder, rouge; — anglais, English red, colcothor; — blanc, white heat; — cramoisi, crimson-red; — de chair, flesh-colored; — de sang, blood red; — de toluylène, toluylene red; — d'indigo, indigo red; — du pin, pine-leaf cast, pine-twig blight; — groseille, currant red.

rouge, — naissant, nascent red, beginning red heat; — outremer, ultramarine red; — rose, rose-red (heat); — sang, blood red; — solide, fast red; — sombre, dark red, blood-red heat; — turc, Turkey red; — vif, bright red, bright red heat; —

vineux, wine red; au —, at red heat; au — blanc, at white heat.

rouge, red, red hot.

rougeâtre, reddish.

rouge-gorge, *m.* redbreast, robin redbreast.

rougeole, *f.* measles, cowwheat (*Melampyrum*).

rougeoleux, -euse, measles patient.

rouge-queue, *m.* redstart; — noire, black redstart.

rouget, *m.* — du porc, red fever, swine erysipelas.

rougeur, *f.* redness, flush, blush, glow.

rougir, to redden, tinge with red, turn red, heat red-hot, blush, make red; — à blanc, — en blanc, to heat white-hot; faire —, to heat (a metal) red-hot, color, blush.

roui, *m.* retting, steeping, staleness, musty odor.

roui, retted.

rouillage, *m.* rusting.

rouille, *f.* rust, blight, mildew, leaf spot, blister, sun scald; — aciculaire du pin, pine (leaf) needle rust; — blanche, white blister, white rust; — brune, yellow rust, orange leaf rust; — courneuse du pin, pine branch twist; — couronnée de l'avoine, crown rust of oats; — de cuivre, green coating on copper, verdigris; — des blés, rust of cereals; — jaune, yellow rust, stripe rust; — noire, black rust, stem rust; — vesiculaire du sapin, spruce needle rust.

rouillé, rusted, rusty, mildewed (plant).

rouiller, to rust, get rusted, mildew, blight; se —, to become mildewed.

rouilleux, -euse, rusty, rust-colored.

rouillure, *f.* rustiness, rust, blight.

rouir, to ret, steep; — sur pré, to dewret.

rouissage, *m.* retting.

rouissoir, *m.* retting (pit) place.

roulage, *m.* rolling, carriage, cartage (of goods), road traffic.

roulant, rolling, moving, running, easy (smooth) running.

roulé, cup-shaken, ring-shaken.

rouleau, *m.* roller, roll, cylinder, roller bearing, reel, land roller.

roulement, *m.* rolling, roll, running, working, rolling mechanism; — à billes, ball bearing.

rouler, to roll, roll up, revolve, coil, carry along, turn, run (on wheels), ride, circulate, be plentiful, rotate, wander; se —, to be rolled, roll over and over.

roulette, *f.* roller, caster, small wheel, roll, runner, cycloid, trochoid.

rouleur, *m.* roller, workman who rolls (barrels).

rouleur, -euse, rolling.

roulier, *m.* carter, wagoner, carrier, trucker.

roulier, -ère, carrying.

roulis, *m.* rolling.

rouloir, *m.* roller, roll, cylinder.

roulon, *m.* round, rung.

roulure, *f.* rolling, rolled edge, roll, (of wood) shake, cupshake.

roumain, Roumanian.

roure, see ROUVRE.

roussâtre, russet, russety, reddish, ruddy.

rousse, see ROUX.

rousserolle, *f.* reed warbler; — turdoide, great reed warbler; — verderolle, marsh warbler.

rousseur, *f.* russet color or quality.

roussi, *m.* scorching, blight, scorched odor, Russia leather.

roussi, browned, reddened, scorched.

roussiller, to scorch.

roussir, to turn russet or brown, redden, scorch, singe (linen).

roussissage, *m.* dyeing russet.

roussissement, *m.* turning brown or russet.

route, *f.* road, path, track, route, way, course, highway; **en —,** on the way; **en —,** forward; let us go!

routier, -ère, road, of roads.

routine, *f.* routine.

routinier, -ère, routine, hide bound.

routoir, *m.* retting hole or pond.

rouverin, rouverain, red-short, hot-short.

rouvert (rouvrir), reopened.

rouvre, *m.* British oak (*Quercus robur*), holm oak (*Quercus ilex*).

rouvrir, to reopen.

roux, -rousse, russet, reddish, red (hair), auburn, brownish.

royal, -aux, royal, kingly, regal.

royalement, royally.

royaume, *m.* kingdom, realm.

royauté, *f.* royalty.

ru, *m.* channel, brook, rill, watercourse.

rubace, rubacelle, *f.* rubicelle, orange-red gem spinel, red topaz.

ruban, *m.* ribbon, band, strip, strap, (measuring) tape.

rubané, ribboned, striped.

rubaner, to ribbon, tape.

rubasse, see RUBACE.

rubéfaction, *f.* rubefaction.

rubéfiant, rubefacient.

rubéfier, to redden.

rubéole, *f.* rubella, German measles.

rubescent, rubescent, turning red.

rubiacées, *f.pl.* Rubiaceae.

rubiaciné, similar to the madder.

rubicarpe, with red fruit.

rubidium, *m.* rubidium.

rubiforme, in the form of raspberry.

rubigineux, -euse, rubiginous, rusty, rust-colored, brownish-red.

rubine, *f.* ruby, rubin, fuchsin; — d'arsenic, ruby of arsenic.

rubioïde, similar to the madder.

rubis, *m.* ruby, jewel; — balais, balas ruby, spinel ruby; — spinelle, spinel ruby.

rubricaule, with red stem.

rubriflore, with red flowers.

rubrique, *f.* red chalk, red ocher, ruddle, rubric, method, title, trick.

rubro-maculé, with reddish spots.

rubro-marginé, with red borders.

ruche, *f.* hive, beehive, swarm.

ruchée, *f.* hiveful.

rucher, *m.* apiary.

rude, rough, harsh, hard, severe, rugged, uncouth, rude, rigid, uneven, formidable.

rudement, roughly, harshly, severely.

rudéral, -aux, ruderal, growing in waste places, or among rubbish.

rudesse, *f.* roughness, severity, harshness.

rudiment, *m.* rudiment.

rudimentaire, rudimentary, rudimental.

rue, *f.* street, rue; — de chèvre, — officinale, goat's-rue (*Galega officinalis*); — odorante, common rue (*Ruta graveolens*).

rueller, to mold up (vines).

ruer, to fling, hurl, kick; se —, to rush upon.

ruficarpe, with sumac fruit.

rufigallique, rufigallic.

rufigallol, *m.* rufigallol.

rugifolié, with rugous leaves.

rugir, to roar, bellow.

rugissement, *m.* roar, roaring, bellowing.

rugosité, *f.* rugosity, ruggedness, roughness, wrinkle, corrugation.

rugueux, -euse, wrinkled, rough, rugose.

ruine, *f.* ruin, decay, destruction; *pl.* ruins.

ruiner, to ruin, destroy; **se** —, to be ruined, go to ruin, ruin oneself.

ruineux, -euse, ruinous.

ruisseau, *m.* stream, brook, gutter, small stream, rivulet, streamlet.

ruisselant, streaming, running.

ruisseler, to stream, trickle, drip, irrigate, stream down, run, flow.

ruissellement, *m.* streaming, dripping through, leaking, leaching.

rumb, rhumb, *m.* rhumb, point.

rumeur, *f.* murmur, din, clamor, rumor, noise.

rumex, *m.* Rumex (docks and sorrels).

ruminant, *m.* ruminator, ruminant.

ruminé, ruminate, looking as though chewed.

ruminer, to ruminate.

ruolz, *m.* electroplated ware.

rupestre, rupestrine, rupicolous, growing on rocks.

rupicole, rupicolous, dwelling among rocks.

ruptile, ruptile, dehiscing in an irregular manner.

ruptinervé, ruptinerved, having ribs of straight-ribbed leaf interrupted.

rupture, *f.* rupture, breaking, break, bursting.

ruscine, similar to butcher's broom.

ruse, *f.* artifice, trick, cunning, ruse.

rusque, *f.* bark of the kermes oak.

russe, Russian.

Russie, *f.* Russia.

rusticité, *f.* hardiness.

rustique, simple, plain, hardy.

rut, *m.* rut.

ruthénique, ruthenic.

ruthénium, *m.* ruthenium.

rutilance, *f.* redness, glow(ing).

rutilant, glowing red, rutilant.

rutile, *m.* rutile.

rutiler, to shine, glow.

rutique, capric.

rythmer, to furnish time for, cadence.

S

s' (contraction of se), oneself, himself, etc.

sa, his, her, its, one's.

sabadilline, *f.* sabadilline.

sabine, *f.* savin (*Juniperus sabina*).

sable, *m.* sand. gravel, urinary sand, grit, sable; — **aurifère,** auriferous (gold-bearing) sand; — **blanchi,** bleached sand; — **de corindon,** emery; — **de fer,** iron filings; — **de mer,** sea sand; — **de moulage,** molding sand; — **de rivière,** river gravel, alluvial sand; — **flottant,** quicksand; — **mouvant,** shifting sand, drift sand; — **sec,** dry sand, parting sand.

sablé, sanded, graveled, speckled, molded or cast in sand.

sabler, to sand, gravel, grind, sandblast, cast in a sand mold, sable.

sablerie, *f.* place where sand molds are made.

sablet, *m.* Agaricus arenarius.

sableur, *m.* maker of sand molds.

sableuse, *f.* sandblast.

sableux, -euse, sandy, arenaceous, gritty, sabulous.

sablier, *m.* sandglass, sandbox, sandtable, sandbox tree (*Hura crepitans*).

sablière, *f.* sand pit, gravel pit.

sabline, *f.* sandwort (Arenaria).

sablon, *m.* very fine sand.

sablonner, to scour with sand, sand.

sablonneux, -euse, sandy, gritty.

sablonnière, *f.* sand pit, sandbox.

sabord, *m.* port, porthole.

sabot, *m.* wooden shoe, sabot, clog, shoe, socket, hoof (of horse), whipping top, lady's slipper.

sac, *m.* sack, bag, pouch. sac; — **à gaz, gas** bag, air cushion; — **à sable,** sandbag; — **de nuit,** — **de voyage,** traveling bag. wallet; — **embryonnaire,** embryo sac

saccade, *f.* jerk, start, jolt.

saccadé, jerky, abrupt, irregular.

saccager, to sack, pillage, ransack.

saccharate, *m.* saccharate.

sacchareux, -euse, saccharine.

saccharide, *m.* saccharide.

saccharifère, containing or yielding sugar, sacchariferous.

saccharifiable, saccharifiable.

saccharifiant, saccharifying.

saccharificateur, *m.* saccharifier.

saccharification *f.* saccharification.

saccharifier, to saccharify, convert into (impregnate with) sugar.

saccharimètre, *m.* saccharimeter.

saccharin, saccharine.

saccharine. *f.* saccharin.

sacchariné, saccharinous, sugary.

saccharique, saccharic.

saccharoïde, saccharoid(al).

saccharolé, *m.* a preparation containing sugar.

saccharomètre, *m.* saccharimeter.

saccharomyces, *m.pl.* Saccharomyces, yeast fungus, yeast.

saccharone, *f.* saccharone.

saccharophore, carrying sugar.

saccharose, *f.* saccharose, sucrose, cane sugar.

sacellus, *m.* sacellus, a one-seeded indehiscent pericarp.

sachant (savior), knowing.

sache (savoir), that (I or he) may know; **pas que je —,** not as far as I know.

sachet, *m.* small bag, sachet.

sacré, sacred, holy, sacral.

sacrifier, to sacrifice.

sacrum, *m.* sacrum.

safran, *m.* saffron, crocus (*Crocus sativus*); **— bâtard,** safflower, false saffron (*Carthamus tinctorius*).

safrané, saffron-colored, yellow, flavored with saffron.

safraner, to color or flavor with saffron.

safranière, *f.* saffron plantation.

safranine, *f.* safranine.

safranum, *m.* safflower, false saffron (*Carthamus tinctorius*).

safre, *m.* zaffer.

safrol, *m.* safrole.

sagace, sagacious, shrewd.

sagacité, *f.* sagacity, acuteness, shrewdness.

sagapénum, *m.* sagapenum.

sagard, *m.* master sawyer.

sage, *m.* wise man, sage.

sage, wise, modest, good, quiet, docile, gentle, sage, prudent.

sagement, wisely, prudently, soberly, sensibly.

sagesse, *f.* wisdom, modesty, sobriety, goodness.

sagette, *f.* arrow, arrowhead.

sagine, *f.* pearlweed (Sagina).

sagittaire, *f.* arrowhead (Sagittaria), adder's-tongue (Ophioglossum).

sagitté, sagittate, arrow-shaped (leaf).

sagou, *m.* sago.

sagoutier, sagouier, *m.* sago palm (Metroxylon), sago tree, coontie (Zamia).

saignée, *f.* blood letting, bleeding, trench

saignement, *m.* bleeding; **— de nez,** epistaxis, nosebleed.

saigner, to bleed, let blood, drain, tap.

saigneux, -euse, bloody.

saillant, leaping, projecting, salient, striking, remarkable.

saille, *f.* projection, ledge, jutting out, spurt, gush, flash, cheek, lug, flange, prominence, start, fit, sally, covering (by male); **faire —,** to project, protrude.

saillir, to project, stand forth, gush out, protrude, stand out.

sain, sound, sane, healthy, healthful, wholesome.

sainbois, *m.* spurge flax (*Daphne gnidium*), mezereum.

saindoux, *m.* lard.

sainegrain, *m.* fenugreek (*Trigonella foenumgraecum*).

sainement, soundly, wholesomely, sanely.

sainfoin, *m.* sainfoin (*Onobrychis viciaefolia*).

saint, *m.* saint.

saint, holy.

saisie, *f.* seizure.

saisir, to seize, lay hold, take hold of, catch hold, grasp, perceive, discern, understand, put in possession, attach, distrain, comprehend; **se —,** to lay hold of, possess oneself of, seize, take.

saisissable, seizable, attachable, capable of being seized.

saisissant, seizing, startling, striking, thrilling, piercing, nipping (cold), surprising, impressive, forceful.

saisissement, *m.* seizure, seizing, chill, sudden shock.

saison, *f.* season, proper time; **— d'abatage,** felling season; **— de la chasse,** shooting season; — des amours (en général), breeding season; — du frai, spawning time; — pour fauves, rutting season; — pour oiseaux, pairing season.

sait (savoir), (he) knows; il — compter, he can count; il ne — que faire, he doesn't know what to do; il en — bien long, he knows a thing or two.

saké, saki, m. sake, rice beer.

salade, f. salad; — d'aubes, breakage of vanes (of turbine).

salage, m. salting.

salaire, m. pay, wages, salary, reward, hire; — à la pièce, job wage, piecework rate.

salaison, m. salting, salt meat; pl. salt provisions; — humide, wet pickling of meat.

salange, m. (salt) season.

salanque, f. salt marsh (for making sea salt).

salant, salt, saline; marais —, salt marsh, saline.

salarier, to pay, pay wages to, reward.

sale, dirty, dull (color), soiled, foul, filthy, grayish.

salé, m. salt pork.

salé, salt, salted, salty, briny, piquant, keen.

salement, dirtily.

salep, m. salep

saler, to salt, season, charge exorbitantly, pickle.

saleté, f. dirtiness, filth, dirt, beastliness.

saleur, m. salter.

salicaire, f. loosestrife (Lythrum); — commune, purple loosestrife, willow herb (Lythrum salicaria).

salicifolié, resembling willow leaves.

salicine, f. salicin.

salicinées, f.pl. Salicaceae.

salicole, salt-producing, salt.

salicorne, salicor, f. glasswort (Salicornia), saltwort (Salsola).

salicotte, f. saltwort (Salsola).

saliculture, f. salt production.

salicylate, m. salicylate; — d'ammoniaque, ammonium salicylate; — d'amyle, amyl salicylate.

salicyle, m. salicyl.

salicyler, to salicylate, treat with salicylic acid.

salicyleux, -euse, salicylous.

salicylique, salicylic.

salicylol, m. salicylal, salicylic aldehyde.

salières, f.pl. hollows over the eyes (in horses).

salifère, saliferous, containing salt.

salifiable, salifiable.

salification, f. salification.

salifié, (acid or metal) combined in a salt.

salifier, to salify.

saligénine, f. saligenin, saligenol.

salignon, m. a lump (cake) of salt, saltcat.

saligot, m. water chestnut (Trapa natans).

salin, m. saline, salin (crude potash), saltworks, salt (garden) marsh.

salin, saline, salty, briny.

salinage, m. saltworks, concentration of the brine, salin (crude potash).

saline, f. saltworks, salt pit, salt (meat) fish.

salinier, m. salt dealer, saltmaker.

salinier, -ère, relating to salt production.

salinité, f. saltness, salinity.

salinomètre, m. salinometer.

salipyrine, f. salipyrin.

salir, to dirty, soil, make dirty, foul, stain, defile, pollute; se —, to become soiled or dirty, soil.

salitre, m. Chile saltpeter, sodium nitrate.

salivaire, salivary.

salivation, f. salivation.

salive, f. saliva, spittle.

saliver, to salivate.

salle, f. hall, room, ward, large room; — d'attente, waiting room; — d'audience, session hall, court; — de lecture, reading room; — de travail, workroom, laboratory.

salmiac, m. sal ammoniac, ammonium chloride.

salol, m. salol, phenol salicylate.

salon, m. drawing room.

salpêtrage, m. formation of saltpeter.

salpêtre, m. saltpeter, niter, nitrate of potash; — brut, crude saltpeter; — de chaux, nitrate of lime; — de Chili, Chile saltpeter, sodium nitrate; — de houssage, wall saltpeter, calcium nitrate; — de potasse, potash saltpeter, potassium nitrate; — de soude, sodium nitrate, Chile saltpeter.

salpêtrer, to saltpeter, cover or treat with saltpeter.

salpêtrerie, f. saltpeter (niter) works.

salpêtreux, -euse, saltpetrous.

salpêtrier, m. saltpeter maker.

salpêtrière, *f.* saltpeter works, niter works.

salpêtrisation, *f.* covering or treating with saltpeter.

salpiglossis, *m.* Salpiglossis.

salplicat, *m.* gold lacquer.

salse, *f.* salse, mud volcano.

salsepareille, *f.* sarsapari la.

salsifis, *m.* salsify, purple goat's-beard (*Tragopogon porrifolius*).

saltatoire, saltatory.

salubre, salubrious, healthy, healthful.

salubrité, *f.* salubrity, health, healthfulness.

saluer, to salute, greet, hail.

salure, *f.* saltness, saltiness.

salut, *m.* safety, salutation salute, greeting.

salutaire, salutary, wholesome, beneficial.

salviaire, salvié, similar to sage.

samalie, *f.* bird of paradise.

samare, *f.* samara, winged seed.

samarskite, *f.* samarskite.

sambucé, similar to elder.

sambuciné, resembling elder.

samedi, *m.* Saturday.

samole, *m.* brookweed, water pimpernel (*Samolus floribundus*).

sanction, *f.* sanction, penalty.

sanctionner, to sanction, approve, ratify, confirm.

sanctuaire, *m.* sanctuary.

sandal, *m.* see SANTAL.

sandaraque, *f.* sandarac, realgar (the resin).

sander, *m.* zander.

sang, *m.* blood, race, parentage, kinship; — **artériel,** arterial blood; — **chaud,** warm blood; — **veineux,** venous blood.

sang-dragon, sang-de-dragon, *m.* dragon's blood (*Calamus draco*), bloodwort, bloody dock (*Rumex sanguineus*).

sanglant, bloody, bleeding, sanguinary, cruel, cutting.

sangle, *f.* strap, band, girth, surcingle, saddle strap.

sangler, to strap, girth, bind.

sanglier, *m.* wild boar; — **mâle,** wild boar.

sanglot, *m.* sob, sobbing.

sangsue, *f.* leech.

sanguin, sanguine, blood, bloody, blood-colored.

sanguinaire, *f.* bloodroot (*Sanguinaria canadensis*).

sanguinaire, sanguinary.

sanguine, *f.* bloodstone, red hematite, red chalk.

sanguinelle, *f.* dogwood.

sanitaire, sanitary, healthy.

sans, without, but for, were it not for; — **façon,** without ceremony; — **que,** without: **cela va — dire,** (it is a matter) of course.

sansevière, *f.* bowstring hemp (Sansevieria).

sans-façon, *m.* unceremoniousness.

santal, *m.* sandalwood; — **blanc,** white sandalwood (*Santalum album*).

santaline, santaléine, *f.* santalin.

santé, *f.* health.

santonine, *f.* santonin, santonica.

santonique, santonic.

sanve, *f.* charlock, wild mustard (*Brassica arvensis*).

sapa, *m.* inspissated (thickened) fruit juice.

sapan, *m.* sapanwood.

sape, *f.* sap, sapping, undercut(ting), undermining.

saper, to sap, undermine, cut corn.

saperde, *f.* — **du peuplier,** poplar longicorn poplar borer.

saphène, *f.* saphenous vein, saphena.

saphir, *m.* sapphire.

saphirin, sapphirine.

sapide, savory, sapid.

sapidité, *f.* savoriness, sapidity, savor.

sapin, *m.* fir tree, spruce, pine; — **argenté,** fir tree, silver fir (*Abies alba*); — **baumier,** balsam fir (*Abies balsamea*); — **blanc,** — **de Riga,** white spruce (*Picea glauca*); — **épicéa,** Norway spruce (*Picea abies*); — **noir,** black spruce (*Picea mariana*); — **pectiné,** fir tree, silver fir (*Abies alba*); — **rouge,** red deal, Scotch pine (*Pinus sylvestris*), spruce.

sapindacé, sapindé, sapindaceous.

sapindus, *m.* soapberry (Sapindus).

sapinette, *f.* spruce, hemlock spruce (*Tsuga canadensis*), spruce beer; — **blanche,** white spruce (*Picea glauca*); — **noire,** black spruce (*Picea mariana*).

sapinière, *f.* fir forest.

saponacé, saponaceous, soapy.

saponaire, *f.* soapwort (Saponaria); — **officinale,** soapwort (*Saponaria officinalis*).

saponase, *f.* lipase.

saponé, *m.* a medicated soap.

saponifiable, saponifiable.

saponification, *f.* saponification.

saponifié, saponified.

saponifier, to saponify.

saponine, *f.* saponin.

saponite, *f.* saponite.

saponule, *f.* saponule.

sapotacées, *f.pl.* Sapotaceae.

sapote, sapotille, *f.* sapodilla (fruit), naseberry.

sapotier, **sapotillier,** *m.* sapodilla tree (*Sapota achras*).

saprémie, *f.* sapremia, septicemia.

saprogène, saprophytic.

sapromyophile, sapromyiophilous, (plants) pollinated by carrion flies.

saprophyte, *f.* saprophyte, saprophytic fungus.

saprophyte, saprophytic.

saproplankton, *f.* saproplankton, foul-water plankton.

sapucaia, *m.* sapucaia tree (Lecythis).

sarc, *m.* cellular protoplasm.

sarcelle, *f.* teal.

sarcine, *f.* sarcine, hypoxanthine, sarcina.

sarclage, *m.* weeding.

sarcler, to weed, weed out.

sarcloir, *m.* dutch hoe.

sarclure, *f.* weeds.

sarcobase, *f.* sarcobasis, a carcerule.

sarcocarpe, *m.* sarcocarp (fruit), fruit flesh.

sarcocarpe, with fleshy fruit.

sarcocarpien, -enne, with fleshy fruit

sarcocolle, *f.* sarcocolla (the exudate).

sarcocollier, *m.* Sarcocolla.

sarcode, *f.* sarcode, protoplasm.

sarcoderme, *m.* sarcoderm, a fleshy layer in seed coats.

sarcodines, *f.pl.* Sarcodina.

sarcodiphyte, (plant) with succulent or fleshy fruits.

sarcodique, pertaining to Sarcodina.

sarcolactique, sarcolactic.

sarcolobé, with thick and fleshy lobes.

sarcomateux, -euse, sarcomatous.

sarcome, *m.* sarcoma.

sarcose, *f.* sarcosis.

sarcosine, *f.* sarcosine.

sarcosperme, with fleshy seeds.

sarde, *f.* sard, Sardinian.

sardoine, *f.* sardonyx.

sarelle, *m.* cowwheat (*Melampyrum silvaticum*).

sariette, *f.* sawwort (*Serratula tinctoria*).

sarigue, *f.* opossum.

sarment, *m.* vine shoot, tendril, sarmentum, runner, cirrhus; —s de vignes, vine (shoots) twigs.

sarmenteux, -euse, climbing.

sarmentide, *m.* sarmentidium, a group of cymes or spikes arranged centrifugally.

sarque, *m.* cellular protoplasm.

sarracénie, *f.* pitcher plant (Sarracenia).

sarrasin, *m.* buckwheat (Fagopyrum), dross, waste, refuse; — de **Tartarie,** tartarian buckwheat (*Fagopyrum tataricum*); blé —, buckwheat.

sarrasin, Saracenic.

sarriette, *f.* savory, (Satureia), pepperwort, peppergrass (Lepidium); — **annuelle,** summer or annual savory; — **commune,** — **des jardins,** summer savory, bean tressel (*Satureia hortensis*); — de **montagne,** winter savory (*Satureia montana*).

sart, *m.* sea wrack (Fucus); see ZOSTÈRE.

sartage, *m.* combination of coppice with field crop.

sarter, — **au feu courant,** to burn the soil covering.

sas, *m.* sieve, bolt, screen, lock chamber.

sassafras, *m.* sassafras.

sassement, *m.* sifting, bolting.

sassenage, *m.* Sassenage cheese.

sasser, to sift, bolt, screen.

sasset, *m.* little sieve.

sasseur, *m.* sifter, bolter.

sassoline, *f.* sassolite, sassolin, mineral boric acid.

satiété, *f.* satiety.

satin, *m.* satin.

satinage, *m.* satining, glazing.

satiné, *m.* satiny appearance, gloss.

satiné, satin, satiny, glossy.

satiner, to satin, give a glossy surface to, look like satin, glaze.

satisfaction, *f.* satisfaction.

satisfaire, to satisfy, content, render satisfaction; se —, to be satisfied; il satisfait à ses devoirs, he discharges his duties properly.

satisfaisant, satisfying, satisfactory, pleasing.

satisfait, satisfied, content, pleased.

saturabilité, *f.* saturability.

saturable, saturable.

saturant, saturating, saturant.

saturateur, *m.* saturator.

saturation, *f.* satu ation.

saturé, saturated, surfeit; **non —,** unsaturated.

saturer, to saturate; **se —,** to be or become saturated.

satureur, *m.* saturator, one who saturates.

saturne, *m.* Saturn, lead; **extrait de —,** Goulard's extract.

saturnin, saturnine, of lead.

saturnisme, *m.* saturnism, plumbism, lead poisoning.

sauce, *f.* sauce.

saucer, to dip, sop, soak, steep, souse, drench.

saucisse, *f.* sausage.

saucisson. *m.* (large) sausage, powder hose, saucisson, aerial torpedo.

sauf, sauve, safe, unharmed, sound, unscathed, save, but, except.

sauge, *f.* sage (Salvia); **— des jardins,** garden sage (*Salvia officinalis*); **— des prés,** meadow sage (*Salvia pratensis*).

saugé, *m* (variety of) lilac.

saugé, containing sage.

saulaie, *f.* osiery, willow culture.

saule, *m.* willow, withy, osier, sallow (Salix); **— amandier,** almond-leaved willow (*Salix amygdaloides*); **— blanc,** white willow (*Salix alba*); **— cendré,** gray willow (*Salix cinerea*); **— étêté,** pollard willow; **— fragile,** crack willow, brittle willow (*Salix fragilis*); **— marsault, — marceau,** sallow, goat willow (*Salix caprea*); **— osier,** common osier (*Salix viminalis*); **— pentandrique,** sweet gale (*Myrica gale*), sweet osier; **— pleureur,** weeping willow (*Salix babylonica*); **— rampant,** creeping willow (*Salix repens*); **— viminal,** common osier (*Salix viminalis*).

saumâtre, somewhat salty, briny, brackish, saltish.

saumon, *m.* salmon, pig (of lead), ingot, head.

saumurage, *m.* pickling (in brine), brining (of meat).

saumure, *f.* pickling brine; **— vierge,** fresh brine (not yet ammoniated).

saumuré, pickled (in brine), brined (anchovies).

saunage, *m.,* **saunaison,** *f.* saltmaking, salt trade.

sauner, to make salt, deposit salt.

saunerie, *f.* saltworks, saltern, salt market.

saunier, *m.* saltmaker, salt merchant.

saupoudrage, *m.* powdering, sprinkling, dusting, salting, dredging.

saupoudrer, to sprinkle with salt, sprinkle, powder, dust.

saur, smoked and salted (herrings).

saura (savoir), (he) will know.

saurait (savoir), (he) would know.

saure, sorrel.

saurer, to cure, bloat, kipper (herrings).

saurien, -enne, saurian.

saurin, *m.* smoked herring, bloater.

sauris, *m.* herring brine.

saussaie, *f.* see SAULAIE.

saussurite, *f.* saussurite.

saut, *m.* leap, jump, spring, bound, fall; **— du vent,** shift of wind; **en — de mouton,** leapfrog, overhead.

sautage, *m.* explosion, blowing up, exploding.

sauter, to jump, leap, skip, vault, spring, bound, explode, blow up, fall, fail, blow out, burst, have out, omit, fly off, rush (at), tumble, cover (mare); **faire —,** to blow up, blast, explode, burst, make jump, get rid of, turn.

sauterelle, *f.* grasshopper, locust.

sauternes, *m.* sauterne, Sauternes wine.

sauteur, *m.* jumping, leaping, tupping ram.

sautiller, to hop, skip, trip, jump.

sauvage, *m.* savage, unsociable person, wild man.

sauvage, wild, bitter, rude, savage, shy, untamed.

sauvageon, *m.* wildling, wild stock, tree grown from seed.

sauvagerie, *f.* wildness, savagery, shyness, unsociability.

sauvagin, *m.* fishy taste or smell.

sauvagin, fishy (taste), marshy.

sauvagine, *f.* wild fowl, waterfowl, marsh fowl.

sauve, see SAUF.

sauvegarde, *f.* safeguard.

sauvegarder, to safeguard, protect.

sauvement, *m.* salvage.

sauver, to save, preserve, rescue, spare, conceal; **se —,** to be saved, escape, flee, indemnify oneself, make up (for).

sauvetage, *m.* lifesaving, rescue, salvage.

sauve-vie, *f.* wall rue (*Asplenium rutamuraria*).

sauvillot, *m.* privet (*Ligustrum vulgare*).

savait (savoir), (he) knew.

savamment, learnedly, knowingly.

savane, *f.* savannah.

savant, *m.* learned person, scholar, scientist, savant.

savant, learned, well-informed, scholarly, erudite, skillful, clever.

savent (savoir), (they) know.

saveur, *f.* savor, flavor, taste, relish.

savinier, *m.* savin (*Juniperus sabina*).

savoir, *m.* knowledge, learning, scholarship.

savoir, to know, know how, be able, understand, can, be aware of, be informed of; **— gré,** to be grateful, be pleased; — **(mauvais) gré à,** to ake it (un)kindly of; **à —,** to wit, namely, that is to say; **faire —,** to inform tell, notify; **se —,** to be known; **(un) je ne sais quel (charme),** (an) indescribable (charm); **(un) je ne sais quoi de (beau),** something indescribably (beautiful).

savoir-faire, *m.* ability, skill, judgment, wits; **il a du —,** he knows how to get along.

savoir-vivre, *m.* good breeding.

savon, *m.* soap; — **à barbe,** shaving soap; — **à base de potasse, — de potasse,** potash soap; — **à base de soude, — de soude,** soda soap; — **à foulon,** full'ng soap, fuller's soap; — **au sable,** sand soap; — **blanc,** white soap; — **calcaire,** lime soap, lime, liniment; — **de goudron,** tar soap.

savon, — de résine, rosin soap, resin soap; — **de suif,** tallow soap; — **de toilette,** toilet soap, fancy soap, scented soap; — **dur,** hard soap; — **en feuilles,** soap in the form of leaves; — **en poudr ,** soap powder; — **léger,** light soap; — **marbré,** marbled soap, mottled soap; — **minéral,** mountain soap, rock soap, unctuous halloysite.

savon, — mou, soft soap, brown soap; — **parfumé,** scented soap, perfumed soap; — **ponce,** pumice soap; — **pour bain,** bath soap; — **pour la barbe,** shaving soap; — **rapé,** planed soap; — **résineux,** rosin soap, resin soap, resinous soap; — **silicaté,** silicate soap, silicated soap; — **vert,** green soap.

savonnage, *m.* soaping.

savonner, to soap, wash in soap, lather.

savonnerie, *f.* soapmaking, soap factory.

savonnette, *f.* cake or ball of toilet soap, shaving brush.

savonneux, -euse, soapy.

savonnier, *m.* soapmaker, soap boiler, soapberry tree (Sapindus).

savonnier, -ère, of soap.

savonnière, *f.* soapwort (*Saponaria officinalis*).

savonule, *f.* combination of an e‑sence.

savourer, to relish, taste, savor, enjoy.

savoureux, -euse, savory, tasteful, pleasant to taste.

saxatile, growing on rocks.

sax , *m.* Saxon porcelain, Dresden china, Saxony.

saxifrage, *f.* saxifrage, stonebreak; **— sarmenteuse,** beefsteak saxifrage, strawberry geranium (*Saxifraga sarmentosa*).

saxon, -onne, Saxon.

scaber, scaber, rough scurvy.

scabieuse, *f.* scabious (Scabiosa).

scabre, scabrous, rough, unequal, scabby, scurfy, scaly.

scabreux, -euse, rough, difficult, perilous, dangerous.

scabride, scabrid, scabridous, somewhat rough.

scabriflore, scabrous (calyx).

scabrifolié, rough-leaved.

scabrisète, scabrous (peduncle).

scabriuscule, slightly rough.

scalariforme, scalariform, ladder-shaped.

scalène, scalenous.

scalpel, *m.* scalpel, surgeon's knife.

scammonée, *f.* scammony (*Convolvulus scammonia*), scammony (gum resin).

scandinave, Scandinavian.

scandium, *m.* scandium

scape, *f.* scape, leafless stem.

scapiflore, scapiflorous, (flowers) scapeborne.

scapiforme, scapiform, resembling a scape.

scapigère, scapigerous, scape-bearing.

scarabée, *m* (scarabaeid) beetle.

scarabéidé, *m.* scarabaeid.

scarieux, -euse, scarious.

scarificateur, *m.* scarifier.

scarification, *f.* scarification.

scarifier, to scarify, wound (soil).

scarlatine, *f.* scarlatina, scarlet fever.

scarlatineux, -euse, scarlatinal.

scatol, *m.* skatole.

sceau, *m.* seal; — de la Vierge, — de Notre-Dame, black bryony (*Tamus communis*); — de Salomon, Solomon's seal (*Polyganatum*); — d'or, goldenseal (*Hydrastis canadensis*).

scellage, *m.* sealing.

scellé, *m.* seal.

scellé, sealed.

scellement, *m.* sealing.

sceller, to seal, cement, bed, cramp in.

scelleur, *m.* sealer.

scénario, *m.* scenario.

scène, *f.* scene, stage, theater.

scepticisme, *m.* skepticism.

sceptique, *m.* skeptic.

sceptique, skeptical, skeptic.

sceptiquement, skeptically.

schabraque, *f.* housing, saddle cloth.

scheelite, *f.* scheelite.

scheidage, *m.* hand sorting (of ore).

schelling, *m.* shilling.

schéma, *m.* scheme, schema, diagram, outline, plan.

schématique, diagrammatic, schematic.

schématiquement, diagrammatically, schematically.

schématiser, to schematize, diagrammatize.

schilling, *m.* shilling.

schismatoptéride, (capsules) opening through a crack.

schistacé, schistaceous, slate-colored, deeptoned gray.

schiste, *m.* schist, shale, slate; — amphibolique, hornblende slate or schist; — ardoisier, slate; — argileux argillaceous schist; — calcaire, calcareous schist; — micacé, micaceous schist, mica slate; — talqueux, talc schist.

schisteux, -euse, schistose, schistous.

schistocarpe, (fruits) opening while splitting.

schizocarpe, *m.* schizocarp.

schizocarpium, *m.* schizocarp.

schizogène, schizogenous, fission-formed.

schizogonie, *f.* schizogony.

schizolysigène, schizolysigenous, arising from splitting or tearing of the tissues.

schizomycètes, *m.pl.* Schizomycetes.

schizostelie, *f.* schizostely.

schlamm, *m.* slime, tailings.

schlich, *m.* slimes, crushed ore.

schlot, *m.* saline incrustation.

schloter, to remove the first deposit from (the brine).

scholie, *m.* scholium.

sciadophylle, *f.* shade leaf.

sciage, *m.* sawing, cutting, sawn timber hand massage.

scie, *f.* saw, sawfish; — à arc, bow saw; — à châssis, frame saw, span saw; — à deux mains, two-handed saw; — à main, handsaw; — à métaux, metal circular saw; — à ruban, band saw; — circulaire, circular saw, disk saw; — d'arbre, pruning saw; — de jardinier, pruning saw; — de long, pit saw, cross-cutting saw; — en archet, bow saw; — sans fin, band saw, belt saw, ribbon saw.

sciemment, knowingly, wittingly, consciously.

science, *f.* science, knowledge.

scientifique, scientific.

scier, to saw, cut (grain); — le bois de long, to saw timber lengthwise.

scierie, *f.* sawmill, power saw.

scieur, *m.* sawyer; — de blé, cutter, reaper; de long, pit sawyer.

scille, *f.* squill (Scilla), wild hyacinth; — maritime, officinal squill, sea onion (*Urginea scilla*).

scinder, to divide, break down, split up, decompose.

scintiller, to scintillate, sparkle.

scion, *m.* scion, resting shoot.

sciophile, ombrophilous, rain-loving, sciophilous, shade-loving.

sciophyte, *f.* sciophyte, shade plant.

scirpe, *m.* club rush, bulrush (*Scirpus lacustris*).

scission, *f.* scission, division, cleavage, splitting, separation, schism, segmentation.

scissipare, fissiparous, reproduced by segmentation.

scissiparité, *f.* scissiparity, schizogenesis.

scissure, *f.* fissure, cleavage.

sciure, *f.* sawdust, saw cuttings; — de bois, sawdust.

sclarée, *f.* clary (*Salvia sclarea*).

scléranthe, *m.* knawel (*Scleranthus annuus*).

scléréide, *f.* sclereid, a sclerotic or stone cell.

sclérenchyme, *m.* sclerenchyma.

scléreux, -euse, sclerous.

sclérifié, lignified, converted into wood.

sclérites, *f.pl.* mechanical cells.

sclérocytes, *f.pl.* mechanical cells.

scléroïde, *f.* sclerite, large. thick-walled ideoblasts.

sclérophylle, stiff-leaved.

scléropode, with rigid peduncles.

scléroprotéine, *f.* scleroprotein.

sclérose, *f.* sclerosis.

sclérote, *f.* sclerote, sclerotium.

sclérotique, sclerotic.

scobiculé, scobiculatous, in fine grains like sawdust.

scobiforme, scobiform, having the appearance of sawdust.

scolaire, school, educational, scholastic, academic.

scolex, *m.* head (of tapeworm), scolex.

scolie, *f.* scolia.

scoliidés, *m.* Scoliidae.

scolitydes, *m.* bark beetles.

scolopendre, *f.* centipede, hart's-tongue fern (*Phyllitis scolopendrium*).

scolyte, *m.* bark beetle, cambium beetle, shot-hole borer; — **ruguleux,** fruit-tree bark beetle.

scombre, *m.* mackerel.

scombrine, *f.* scombrine.

scopolamine, *f.* scopolamine.

scopospore, *f.* ascospore, spore produced in an ascus.

scorbut, *m.* scurvy.

scorbutique, scorbutic.

scoriacé, scoriaceous.

scorie, *f.* slag, scoria, volcanic cinders, dross, cinder; — **crue,** raw slag, poor slag; — **de déphosphoration,** basic slag, Thomas slag, phosphatic slag; — **de fer,** iron slag, iron cinder; — **pauvre,** poor slag; — **riche,** rich (fining) slag.

scorification, *f.* scorification.

scorificatoire, *m.* scorifier.

scorifier, to scorify; **se** —, to be scorified, slag.

scorpioïde, scorpioid.

scorpion, *m.* scorpion; — **net,** scorpionet, little scorpion.

scorpione, *f.* scorpion grass, forget-me-not (Palustris), female scorpion.

scorpiure, *f.* caterpillar.

scorsonère, *f.* viper's-grass, black salsify (*Scorzonera hispanica*).

scototropisme, *m.* scototropism, turning toward darkness.

scrobicule, *f.* depression, pit of the stomach.

scrobiculé, scrobiculate, marked by minute depressions, pitted.

scrofulaire, *f.* figwort (Scrophularia).

scrofule, *f.* scrofula, struma.

scrofuleux, -euse, scrofulous, strumous.

scrophularinées, *f.pl.* Scrophulariaceae.

scrotiforme, scrotiform, pouch-shaped.

scrotum, *m.* scrotum.

scrubber, *m.* gas washer, scrubber.

scrupule, *m.* scruple, scrupulousness, qualm, moral anxiety.

scrupuleusement, scrupulously.

scrupuleux, -euse, scrupulous.

scruter, to scrutinize, examine closely.

scrutin, *m.* poll, ballot, vote, balloting.

sculptage, *m.* carving, sculpturing.

sculpter, to carve, sculpture.

sculpteur, *m.* sculptor.

sculpture, *f.* sculpture.

scutellaire, *f.* scutellaria, skullcap.

scutellaire, scutellar.

scutelle, *f.* scutellum, bony plate.

scutelliforme, scutellate, platter-shaped, shield-shaped, buckler-shaped.

scutelloïde, scutellate.

scutellum, *m.* metatergum, scutellum.

scutifolié, scutifoliate.

scutum, *m.* mesotergum, scutum.

scyphoïde, in the form of a cup.

scyphule, *m.* scyphulus, colesule of Hepaticae.

scyphuliforme, in the form of scyphulus.

se (s'), oneself, himself, herself, itself, themselves, to oneself, each, one another, to each other.

séance, *f.* seat, sitting, session, meeting, performance, séance.

séant, sitting, becoming, fitting, proper.

seau, *m.* pail, bucket, pailful, scoop; — **à lait,** milk pail.

sébacé, sebaceous.

sébacique, sebacic.

sébate, *m.* sebacate, sebate.

sébifère, sebiferous.

sébile, *f.* wooden (bowl) vessel.

séborrhée, *f.* seborrhea.

sébum, *m.* sebum, tallow.

sec, *m.* dryness, manger food, dry food; à —, to dryness, dry, free of moisture, dried up, without water; mettre à —, to drain.

sec, sèche, dry, dried (fruit), spare, lean, barren, meager, seasoned, matured, brittle, short, hard, sharp, harsh.

sécante, *f.* secant.

sécateur, *m.* pruning shears, cutter, osteotome, cutting secateur.

séchage, *m.* drying (of hay), seasoning (of wood); — à l'air, air drying; — au four, kiln drying.

sèche, *f.* cuttlefish.

sèche, see SEC.

séché, dried; — à l'air, air-dry; — à l'étuve, kiln dried.

séchée, *f.* drying, duration of drying.

sèchement, dryly, sharply, coldly.

sécher, to dry, dry up, season (wood), drain, wither, cure (malt), heat up, warm; se —, to dry oneself, become dry, dry up, wither.

sécheresse, *f.* dryness, leanness, harshness, coldness, dry season, drought.

sécherie, *f.* drying (room) shed, seed kiln, drying, drought, aridity, seed-husking establishment.

sécheron, *m.* dry meadow.

sécheur, *m.*, **sécheuse,** *f.* drier

séchoir, *m.* drying (place) room, drying apparatus, drier, drying shed or house.

second, *m.* second, second floor, second class; en —, in second place, second.

second, second.

secondaire, secondary, minor, subsequent.

secondairement, secondarily.

seconde, *f.* second of latitude or time, second class, second proof; à la —, per second, a second.

secondement, secondly.

seconder, to second, assist, support, help.

secondinaire, relating to inner tegument or afterbirth.

secondine, *f.* inner tegument, afterbirth.

secouage, secouement, secoûment, *m.* shaking, shake-up.

secouer, to shake, shake down (fruit), help; se —, to be shaken, bestir oneself, rouse oneself.

secoueur, *m.* shaker, form breaker; — de graines, seed riddle.

secourir, to succor, relieve, help, aid.

secours, *m.* help, succor, relief, attendance, aid, assistance.

secousse, *f.* shake, shaking, shock, jolt, concussion, jerk, blow.

secret, *m.* secret, privacy, secrecy, secretness; en —, in secret, privately, secretly.

secret, secret, hidden, recondite, discreet.

secrétaire, *m.* secretary, clerk, writing desk.

secrètement, secretly, privately.

sécréter, to secrete, carrot (pelts).

sécréteur, -trice, -euse, secreting, secretory.

sécrétine, *f.* secretin.

sécrétion, *f.* secretion; — lactée, lactation, milk secretion.

sécrétoire, secretory.

secte, *f.* sect.

secteur, *m.* sector, segment, local supply circuit, distribution main, sector, quadrant; — d'alimentation, feed (city) circuiı; — denté, notched segment, toothed (segment).

section, *f.* section, cross section, profile, working section; — longitudinale, longitudinal section; — transversale, cross section, transverse section.

sectionnel, -elle, sectional.

sectionnement, *m.* division, sectioning.

sectionner, to divide into sections, section, divide, amputate.

séculaire, secular, once in a century, of the ages, permanent, century old.

secundiflore, secund.

secundo, secondly.

sécuriforme, axe-shaped.

sécurigère, having a hatchetlike organ.

sécurité, *f.* security, secureness, safety.

sédatif, -ive, sedative.

sédation, *f.* mitigation, alleviation.

sédentaire, *m.* forest clerk.

sédentaire, sedentary.

sédiment, *m.* sediment, settlings, deposit precipitation.

sédimentaire, sedimentary.

sédimentation, *f.* sedimentation.

Sedlitz, *m.* Seidlitz.

sédon, *m.* stonecrop, orpine (Sedum).

séducteur, *m.* seducer.

séducteur -trice, seductive, enticing.

séduire, to seduce, attract, allure, delude, suborn.

séduisant, seductive, attractive.

ségétal, -aux, segetalis.

seglin, m. cheat grass, chess, brome grass (Bromus).

segment, m. segment; — d'une ligne, section, finite line.

segmentaire, segmental, segmentary.

segmenté, segmented.

ségrégatif, -ive, segregative.

ségrégation, f. segregation.

seiche, f. cuttlefish.

ʋeigle, m. rye (plant and grain) (Secale cereale); — de printemps, spring rye; — ergoté, spurred rye, ergot of rye.

seigne, f. great dodder (Cuscuta europaea); — petite cuscute, clover dodder, lesser dodder (Cuscuta eppithymum).

seigneur, m. lord.

seille, f. (wooden) pail, bucket.

seime, f. sand crack.

sein, m. midst, bosom, breast, heart, interior, gravid womb, lap; au — de, in the midst of, in.

seine, f. seine, dragnet.

seing, m. signature.

seirolytique, seirolytic, separation of hereditary characters.

seiospore, f. stylospore, spore borne on a filament.

séismique, seismic.

seizaine, f. (approximately) sixteen.

seize, sixteen, sixteenth.

seizième, sixteenth.

séjour, m. stay, sojourn, abode, standing, habitation.

séjourner, to stay, remain, stand, tarry, sojourn, make a stay, dwell, reside.

sel, m. salt; — admirable, sal mirabile, Glauber's salt, sodium sulphate; — amer, bitter salt, Epsom salt, magnesium sulphate; — ammoniac, sal ammoniac, ammonium chloride; — basique, basic salt, subsalt; — calcaire, lime salt, calcium salt; — commun, common salt, sodium chloride, kitchen salt; — d'argent, silver salt; — de cuisine, common salt, kitchen (table) salt; — de fer, iron salt.

sel, — de Glauber, Glauber's salt, sodium sulphate; — de nitre, saltpeter; — de phosphore, microcosmic salt; — de plomb, lead salt; — d'Epsom, Epsom salt, magnesium sulphate; — de roche, rock salt, mineral salt, halite; — de saline, common salt; — de Saturne, sugar of lead, plumbic acetate; — de soude, sodium salt, soda

salt; — d'oseille, salt of sorrel, (acid) potassium oxalate.

sel, — essoré, dried salt; — excitateur, exciting salt; — fin, fine salt; — fin-fin, very fine salt, powdered salt; — gemme, rock salt, halite; — inorganique, metallic salt, inorganic compound; — marin, common salt, sea salt; — minéral, mineral salt, inorganic salt; — neutre, neutral salt; — sédatif, sedative salt, boric acid; — volatil, smelling salt.

sélecteur, -trice, selective.

sélectif, -ive, selective.

sélection, f. selection; — des semences, seed amelioration, seed selection: — naturelle, natural selection.

sélectionner, to select, pick out.

sélénhydrique, hydroselenic.

séléniate, m. selenate, seleniate.

sélénié, seleniureted, combined with selenium.

sélénieux, -euse, selenious, selenous.

sélénifère, seleniferous.

sélénique, selenic.

sélénite, f. selenite, gypsum.

séléniteux, -euse, selenitic.

sélénium, m. selenium.

séléniure, m. selenide.

selle, f. seat, bench, stool, bowel evacuation, excrement, sella, slacken, saddle.

seller, to saddle (up); se —, to harden at the surface (of land).

sellerie, f. saddlery.

sellette, f. rack, seat. saddle, bedplate; — à traire, milking stool.

sellier, m. saddler.

selon, according to, pursuant to; — moi, in my opinion; — que, according as, as, according to; c'est —, that depends, that is as may be.

Seltz, Selters, m. Selter(s); eau de —, Seltzer water.

seltzogène, m. seltzogene, gazogene.

semage, m. sowing.

semaille, f. sowing, seed, seed time; — d'essai, experimental sowing.

semailler, to sow.

semaine, f. week, week's work, week's pay.

semaison, f. sowing time, self-seeding.

semblable, m.&f. fellow, like, equal, fellow creature.

semblable, like, alike, similar, such.

semblablement, similarly, likewise, also, too.

semblant, *m.* appearance, seeming, semblance.

sembler, to appear, seem, look.

semé, sown, sowed.

semelle, *f.* sole, foot, shoe, sill, sleeper, bedplate, dormer, bottom, step, tread; — d'ancrage, stay block; — en corton, cardboard sole; — en cuir, leather sole; — en feutre, felt sole, felt pad; — en liège, cork sole; — en paille, straw sole; — inférieure, lower flange.

semence, *f.* seed, seeds, sowing, seed pearl, semen, sperm, small nail, tack; — de carvi, caraway seeds; — de colza, rapeseed; — de graminées, grass seed; — de lin, linseed, flaxseed; -- de mauvaises herbes, weed seeds; — de taèfle, clover seed; — d'origine, pedigree seed; —s potagères, vegetable seed.

semenceau, *m.* seed-turnip, seed piece (potato).

semencier, *m.* seed-bearing tree.

semen-contra, *m.,* **semencine,** *f.* santonica, Levant wormseed, semen contra, semen cinae (*Artemisia pauciflora*).

semer, to sow, spread, strew, scatter, sprinkle, disseminate.

semestre, *m.* half year, six months, six-months' income, pay, duty, service or leave.

semestre, of six-months duration.

semestriel, -elle, semiannual, for six months.

semeur, *m.* disseminator.

semeuse, *f.* sowing drill; — à bras, hand sowing machine.

semi-anatrope, half-inverted.

semi-annuel, -elle, semiannual.

semi-bifide, half-cleft.

semi-biloculaire, half-bilocular.

semi-capsule, *f.* semicapsula, cupule.

semi-circulaire, semicircular, half round.

semi-cordé, semicordate, heart-shaped on one side only.

semi-digyne, semidigynus, (two carpels) cohering near the base only.

semidique, semidine.

semi-double, semidouble.

semi-flosculeux, -euse, semiflosculous.

semi-fluide, semifluid.

semi-hebdomadaire, semiweekly.

semi-lianes, *f.pl.* semilianes, scrambling plants in hedges and margins of forests.

semi-liquide, semiliquid.

semi-mensuel, -elle, semimonthly.

semimonopétalé, (plant) with corolla not freely gamopetalous.

séminal, -aux, seminal, spermatic.

séminase, *f.* seminase.

sémination, *f.* semination.

séminifère, seminiferous.

séminule, *f.* seminule, spore.

séminulifère, seminuliferous, bearing seminules.

semi-orbicule, semiorbicular, half-round or hemispherical.

semi-palmé, semipalmate, half-palmate.

semi-perméable, semipermeable.

semi-pluriloculaire, with half number of cells.

semi-radié, with half rays.

semis, *m.* seedling, sowing, seed bed, seed plot; — à la volée, broadcast sowing; — de graines légères, seedlings from light seeds; — de graines lourdes, seedlings from heavy seeds; — en plein, broadcast sowing; — en rigoles, sowing in trenches; — en sillons, sowing in furrows.

semis, — naissant, seedling growth; — par bandes, sowing in strips; — par places, sowing in patches; — par potets, sowing in holes, dibbling seed; — par trous, sowing in holes, dibbling; — préexistant, advance growth; — sur banquettes, sowing on ridges; — sur buttes, sowing on mounds.

semi-sphérique, hemispheric, hemispherical.

semi-staminaire, semistaminate, when part of the stamens are changed into petals.

semi-symphoyostémone, semisymphiostemonis, some stamens cohering, others remaining free.

semoir, *m.* sowing drill.

semoule, *f.* semolina, middlings.

séné, *m.* senna (*Cassia senna*); — américain, wild senna (*Cassia marilandica*); — indigène, bladder senna (*Colutea arborescens*).

seneçon, *m.* groundsel (Senecio).

senegrain, senegré, *m.* see SAINEGRAIN.

sénescence, *f.* senescence.

sénestre, rolling up in spirals from right to left and from top to bottom.

sénevé, *m.* mustard, black mustard (*Brassica nigra*), charlock (*Brassica arvensis*).

sénevol, *m.* mustard oil.

senousse, *f.* pigweed.

sens, *m.* sense, direction, way, signification, import, opinion, judgment, meaning, sensation; — **dessus dessous,** upside down; — **devant derrière,** back to front, hind part before; — **direct,** positive direction; — **rétrograde,** negative direction; **de** — **contraire,** in a contrary direction; **en** — **inverse,** in the reverse order, in a contrary direct on; **en** — **inverse des aiguilles d'une montre,** counterclockwise; **en tout** —, in every direction.

sensation, *f.* sensation, impression, feeling.

sensationnel, -elle, sensational.

sensé, sensible, judicious, intelligent.

sensément, sensibly, judiciously.

sensibilisateur, *m.* sensitizer.

sensibilisateur, -trice, sensitizing.

sensibilisation, *f.* sensitization.

sensibiliser, to sensitize, render sensitive

sensibilisinogène, *f.* sensibilisinogen.

sensibilisme, *m.* sensibilism.

sensibilité, *f.* sensitiveness, sensibility, feeling, tenderness.

sensible, sensitive, sensible, perceptible, appreciable, noticeable, obvious; — **au froid,** frost tender.

sensiblement, sensitively, sensibly, perceptibly, practically, obviously, appreciably.

sensitif, -ive, sensitive.

sensitive, *f.* sensitive plant.

sentence, *f.* sentence, decree, decision, maxim.

senteur, *f.* odor, scent, smell, perfume, fragrance.

sentier, *m.* path.

sentiment, *m.* feeling, perception, sensation, sense, consciousness, sentiment, opinion, concern; **au** —, by guess, by judgment, at will.

sentinelle, *f.* sentinel, sentry, guard(ing).

sentir, to feel, be conscious, sensible of, perceive, smell (odor), taste of; **se** —, to be felt, feel, be conscious of; **il se sent mieux,** he feels better.

seoir, to sit, fit, be becoming, suit, become.

sep, *m.* sole shoe, colter.

sépale, *m.* sepal.

sépalodie, *f.* sepalody, the metamorphosis of petals into sepals.

séparabilité, *f.* separability.

séparable, separable.

séparage, *m.* separation, sorting.

séparant, separating.

séparateur, *m.* separating funnel, separator.

séparateur, -trice, separating, separatory, separative.

séparatif, -ive, separative.

séparation, *f.* separation, partition, parting, elimination, segregation, precipitation.

séparatoire, *m.* separating funnel.

séparé, separated, separate, distinct.

séparément, separately.

séparer, to separate, sort, divide, sever, split, sunder, distinguish; **se** —, to be separated, separate, divide, branch off, separate down, be deposited, settle.

sépia, *f.* sepia, cuttlefish.

sept, seven, seventh.

septantaine, *f.* about seventy.

septemangulé, with seven angles.

septembre, *m.* September.

septemdigité, with seven digitations.

septemfoliolé, septemfoliolate having seven leaflets.

septemlobé, having seven lobes.

septemnervé, having seven veins.

septené, septenate, having parts in sevens.

septentrional, -aux, north, northern, septentrional.

septicémie, *f.* septicemia.

septicémique, septicemic.

septicide, septicidal.

septicité, *f.* septicity, septic quality.

septième, seventh.

septifère, septiferous, bearing a partition or dissepiment.

septiforme, septiform, having the appearance of a dissepiment.

septifrage, septifragal.

septile, septile, of or belonging to dissepiments.

septique, septic.

septivalent, heptavalent.

septoeil, *m.* lamprey.

septulifère, with a dissepiment.

septum, *m.* septum.

sépulcre, *m.* burial place, tomb, sepulcher.

sépulture, *f.* burial, vault, sepulture.

séquelle, *f.* series, gang, set, crew.

sera (être), (he) will be.

sérancer, to hackle. hatchel.

serbe, Serbian.

serein, *m.* serein, evening dew.

serein, serene, clear, calm, placid.

sereinage, *m.* dew retting.

sérénité, *f.* calmness.

séreux, -euse, serous.

serfouette, *f.* hoe.

serfouissage, *m.* hoeing.

sergent, *m.* cramp, screw frame, sergeant, ground beetle.

sériaire, serial.

séricéeux, sericeous, silky.

séricicole, sericultural, silkworm.

sériciculture, *f.* sericulture, silkworm raising.

séricifolié, with silky leaves.

séricine, *f.* sericin.

séricule, *m.* regent bird.

série, *f.* series, working section; — aromatique, aromatic series; — d'exploitation, working section; — de coupes, felling series; — grasse, fatty series; en —, in series.

sériel, -elle, serial.

sérieusement, seriously.

sérieux, *m.* seriousness; au —, seriously, in earnest.

sérieux, -euse, serious, earnest, weighty, grave.

serin, *m.*, canary.

sérine, *f.* serine, canary.

seringa, seringat, *m.* — des jardins, mock orange (Philadelphus).

seringue, *f.* syringe.

seringuer, to syringe, spray, squirt.

sérique, of serum, serous.

serment, *m.* oath.

séro-albumine, *f.* serum albumin, seralbumin.

séro-globuline, *f.* serum globulin, seroglobulin.

séro-physiologie, *f.* serum physiology.

séro-pronostic, *m.* seroprognosis.

sérosité, *f.* serosity, serum.

sérothérapie, *f.* serotherapy, serum therapy.

serpe, *f.* billhook, hedging bill, pruning knife.

serpent, *m.* serpent, snake.

serpentaire, *f.* serpentaria; — de Virginie, Virginia snakeroot (*Aristolochia serpentaria*).

serpente, *f.* silver paper, fine tissue paper.

serpenté, winding, serpentine.

serpenteau, *m.* serpent, young snake.

serpenter, to wind, curve, meander.

serpentère, *f.* bistort, snakeroot (*Polygonum bistorta*).

serpentin, *m.* worm, coil; — à vapeur, steam coil; — réchauffeur, reheating coil.

serpentin, serpentine.

serpentineux, -euse, serpentinous.

serpette, *f.* pruning knife.

serpolet, *m.* wild thyme, mother-of-thyme (*Thymus serpyllum*).

serradelle, *f.* serradella (*Ornithopus sativus*).

serrage, *m.* tightening, screwing tight, tightness, shrinkage, tension, pressure.

serratifolié, serratifolious, having serrate leaves.

serratile, like a saw, serrate.

serratule, *f.* sawwort (Serratula).

serrature, *f.* serrature, toothing of a serrate leaf.

serre, *f.* greenhouse, forcing house, propagating house, pressure, pressing; — chaude, hothouse, warmhouse; — froide, greenhouse, coolhouse.

serré, tightened, hard, cautiously, tight, serrate, tense, close, pressed closely, compact, crowded, together, dense.

serre-bouchon, *m.* swing-wire stopple.

serre-écrous, *m.* nut setter.

serre-étoupe, *m.* stuffing box.

serre-fil, serre-fils, *m.* binding screw, binding post, set screw.

serre-joint, serre-joints, *m.* clamp, cramp frame.

serrement, *m.* tightening, squeezing, pressing.

serrer, to tighten, fasten, clinch, clamp, screw down, close, shut, crowd, squeeze, press, press close, wedge in, clasp, lock, pinch, force, condense, put away, lock up, keep close to, lie close; se —, to be tightened, press close to each other, contract, crowd.

serrette, *f.* sawwort (*Serratula tinctoria*).

serricorne, serricorn.

serrière, *f.* plug, stopple.

serrure, *f.* lock.

serrurerie, *f.* locksmithing, ironwork.

serrurier, *m.* locksmith, ironworker.

sert (servir), (he) is serving; il s'en —, he makes use of (it).

sertir, to set (a stone), encase, frame.

sertule, *m.* sertulum, simple umbel.

sertulé, umbellate.

sertulifère, umbelliferous.

sérum, *m.* serum; — **sanguin,** blood serum.

sérumalbumine, *f.* serum albumin, seralbumin.

sérumthérapie, *f.* see SÉROTHÉRAPIE.

servante, *f.* servant, prop, support, bench vise.

serve, *f.* drinking pond (for cattle).

service, *m.* service, duty, use, attendance, branch, department, course, set; **qu'y a-t-il pour votre** —? what can I do for you?

serviette, *f.* napkin, towel.

servir, to serve, help, assist, cover (more), work, operate, pay interest or rent; **se** — (**de**), to make use of, use, employ, be served, help, be of use, be of service to, be used as.

serviteur, *m.* servant.

servitude (de), *f.* dependence (on).

servomoteur, *m.* auxiliary motor, servomotor.

ses, *pl.* his, her, its, one's.

sésame, *m.* sesame (Sesamum).

sésamoïde, sesamoid (bone).

séséli, *m.* meadow saxifrage (Seseli).

sesquichlorure, *m.* — **d'or,** auric chloride, trichloride of gold.

sesquifluorure, *m.* sesquifluoride.

sesquiiodure, *m.* sesquiiodide.

sesquioxyde, *m.* sesquioxide; — **de manganèse,** manganese oxide, sesquioxide of manganese.

sesquisel, *m.* sesquisalt.

sesquisulfure, *m.* sesquisulphide.

sessile, sessile (leaf), setting.

session, *f.* session, sitting.

sétacé, setaceous, bristled, bristlelike.

sétaire, *f.* Setaria.

séteux, -euse, having bristles.

sétifère, setose.

sétiflore, having bristlelike petals.

sétiforme, bristle-shaped.

sétigère, bristled, bristly.

sétipède, with bristle-shaped peduncle.

seuil, *m.* threshold, sill, divide, ground sill, ridge (of watersheds).

seul, *m.* one, one alone, only one.

seul, alone, mere, only, single, sole, lonely; **à lui** —, **à elle** —**e,** alone, by itself.

seulement, only, merely, solely, even, but; **non (pas)** —, not even, not only.

sève, *f.* sap, juice, vigor, life, strength; — **descendante,** descending sap; — **montante,** ascending sap, flow of sap.

sévère, severe, strict.

sévèrement, severely, strictly.

sévérité, *f.* severity.

séveux, -euse, sappy, full of sap.

sévir, to be severe, treat rigorously, rage, prevail.

sevrage, *m.* weaning.

sevrer, to wean, separate, ablactate, deprive.

sèvres, *m.* Sèvres (porcelain).

sexe, *m.* sex.

sexérémé, with fruit composed of six eremi.

sexfacié, (leaves) forming six series along the stem.

sexfide, sex(i)fid.

sexifère, sexiferous.

sexflore, bearing six flowers.

sexjugué, (pinnate leaf) with six pairs of leaflets.

sextuple, sixfold

sexualité, *f.* sexuality.

sexué, sexual.

sexuel, -elle, sexual.

seyait (seoir). (it) was becoming.

seyant, becoming, fitting.

seybertite, *m.* seybertite.

shellac, *m.* shellac.

shérardie, *f.* Sherardia.

shérardisat'on, *f.* sherardizing.

shérardiser, to sherardize.

shunt, *m.* shunt.

shunter, to shunt, to put in shunt.

si, if, whether, so, however, so much, yes; **si bien que,** with the result that; **si fait,** yes, indeed; **s'il en est ainsi,** if so; **si non,** if not; **si . . . que,** so . . . that; **si . . . ne,** unless.

sialagogue, *m.* sialagogue.

sialagogue, sialagogic.

sialozymase, *f.* ptyalin.

siamois, Siamese.

Sibérie, *f.* Siberia.

sibérien, -enne, Siberian.

siccatif, *m.* drying substance, dryer, siccative.

siccatif, -ive, drying, siccative; **huile —,** drying oil.

siccité, *f.* dryness.

sicilien, -enne, Sicilian.

sidéral, -aux, sidereal.

sidérique, pertaining to iron, sidereal.

sidérite, *f.* siderite, native ferrous carbonate, indigo-blue quartz.

sidérochrome, *m.* siderolite.

sidérose, *f.* siderite, siderosis.

sidérotechnie, *f* siderotechny.

sidéroxyle, sidéroxylon, *m.* ironwood (Sideroxylon).

sidérurgie, *f.* siderurgy.

sidérurgique, siderurgical.

siècle, *m.* century, period, age, world.

sied (seoir), (it) is fitting.

siège, *m.* seat, chair, stool, center bench, box, siege, habitat, occurrence, breech, buttock, distribution; **— à ressorts,** spring seat; **— de campagne,** camp chair; **— de soupape,** valve seat; **— du cocher,** driver's seat; **— du pointeau,** needle-valve seat; **— pliant,** folding chair, camp chair; **— réglable,** sliding seat; **— social,** headquarters, registered office.

siéger, to sit, reside, lie, be, have head office.

sien, sienne, his, hers, its one's own; **le —,** his own, her own, one's own; **les siens,** one's kindred or friends.

sieste, *f.* siesta.

sifflement, *m.* whistling, whizzing, hissing.

siffler, to whistle, whizz, hiss.

sifflet, *m.* whistle, hiss, spot, bevel.

sigillé, sigillate(d) (stem, pottery); **terre —e,** Lemnian earth.

sigmoïde, sigmoid.

signal, *m.* sign signal; **— avancé,** distant signal; **signaux horaires,** time signals.

signalé, signal, remarkable, conspicuous.

signalement, *m.* description (of a man).

signaler, to report, notify, call attention to, point out, give the description of, describe, signal, give a signal.

signaleur, *m.* signalman.

signalisation, *f.* signaling.

signataire, *m.* subscriber, signatory, signer.

signataire, signatory, signing.

signature, *f.* signing, signature.

signaux, *m.pl.* see SIGNAL.

signe, *m.* sign, mark, symptom, token, constellation; **— inverse,** opposite sign; **—s lactiques,** lacteal marks.

signer, to sign, stamp, mark.

signifiant, significant.

significatif, -ive, significant, significative.

signification, *f.* signification, notification, meaning, import, significance.

significativement, significantly.

signifier, to signify, mean, notify.

sil, *m.* ocher.

silence, *m.* silence, stillness; **sous —,** in silence.

silencieusement, silently, still.

silencieux, -euse, silent, silently, still.

silésien, -enne, Silesian.

silex, *m.* flint, silex, silica; **— corné,** hornstone, chert; **— xyloïde,** petrified wood, silicified wood.

silhouette, *f.* silhouette, profile, outline.

silicate, *m.* silicate; **— d'alumine,** silicate of alumina, aluminum silicate; **— de chaux,** calcium silicate, silicate of lime; **— de magnésie,** silicate of magnesia, magnesium silicate; **— de potasse,** potassium silicate, waterglass, soluble glass.

silicaté, siliceous, silicate; **matières —es,** silicates

silicatisation, *f.* silicating, silicification.

silicatiser, to silicate, silicify.

silice, *f.* silica, flint, silicious earth.

silicé, containing silica.

siliceux, -euse, siliceous, silicious.

silicicole, silicicolous.

silicié, silicon, containing silicon.

silicifère, siliciferous.

silicification, *f.* silicification.

silicifier, to silicify.

silicifuge, silicifugous.

silicique, silicic.

silicium, *m.* silicon, silicium.

silicule, *f.* silicle, silicula.

siliculiforme, in the form of a silicula.

silique, *f.* silique, pod.

sillon, *m.* furrow, groove, ridge, wrinkle, streak, wake, trail, train; **— antennale,** antennal scrobe.

sillonné, furrowed.

sillonner, to furrow, plow, wrinkle, groove streak.

silo, *m.* silo.

silotage, *m.* ensilage.

silphe, *m.* carrion beetle.

silure, *m.* sheatfish, catfish.

silurien, -enne, Silurian.

silvine, *f.* sylvite, sylvine, native potassium chloride.

simaruba, simarouba, *m.* Simarouba.

simarubacées, *f.pl.* Simaroubaceae.

similaire, similar, like.

similargent, *m.* imitation silver.

similarité, *f.* similarity.

simili, *f.* half tone; bijoux en —, imitation jewelry.

simili, imitation, artificial.

similiflore, similiflorous, with flowers all alike.

similigravure, *f.* process engraving, half tone.

similimarbre, *m.* imitation marble.

similiser, to mercerize, give a silk finish to.

similisymétrie, *f.* similisymmetry, two halves of a diatom valve are similar.

similitude, *f.* similitude, similarity, likeness, comparison.

similor, *m.* imitation gold.

simple, simple, plain, elementary, single, mere, ordinary, common, private (soldier), smooth, even, unarmed; à — effet, single-acting; la vue —, the naked eye.

simplement, simply, merely, plainly.

simples, *m.pl.* simples, medicinal herbs.

simplicicaule, with simple stalk.

simplicifolié, with simple leaves.

simplicité, *f.* simplicity, simpleness, plainness.

simplificateur, -trice, *m.&f.* simplifier.

simplificateur, -trice, simplifying.

simplification, *f.* simplification.

simplifier, to simplify; se —, to be or become simple(r) or simplified.

simulacre, *m.* simulacrum, image, imitation, semblance, shadow, feint, appearance, show, phantom.

simulation, *f.* simulation, malingering, feigning.

simulé, fictitious, false.

simuler, to simulate, pretend, feign.

simultané, simultaneous.

simultanéité, *f.* simultaneousness.

simultanément, simultaneously, at the same time.

sinalbine, *f.* sinalbin, mustard oil from *Brassica alba.*

sinapine, *f.* sinapine, mustard oil from *Brassica nigra.*

sinapique, sinapic (acid).

sinapisé, containing mustard.

sinapiser, to mix with mustard, sinapize.

sinapisme, *m.* sinapism, mustard plaster; — en feuille, mustard paper.

sinapoline, *f.* sinapoline.

sincère, sincere.

sincèrement, sincerely.

sincérité, *f.* sincerity, sincereness.

singe, *m.* monkey, ape, hoist, windlass, crab.

singularité, *f.* singularity, oddness.

singulier, -ère, singular, peculiar, odd.

singulièrement, singularly, peculiarly, oddly.

sinigrine, *f.* sinigrin.

sinistre, *m.* disaster, loss.

sinistre, sinister.

sinistrorse, sinistrorse, turned to the left.

sinistrorsum, sinistral, counterclockwise, from right to left.

sinon, otherwise, or else, if not, unless, except, save.

sinué, sinuous, strongly wavy, sinuate.

sinueusement, sinuously, windingly.

sinueux, -euse, sinuous, winding, meandering.

sinuosité, *f.* sinuosity, turn, bend.

sinus, *m.* sinus, concavity, sine.

sinusoïdal, -aux, sine-shaped, sinusoidal.

siotropisme, *m.* siotropism, stimulus by shaking.

siphoïde, siphonal, siphoniform.

siphon, *m.* siphon.

siphonal, -aux, siphonal.

siphoné, containing several energids in a membrane.

siphoner, to siphon.

siphonnement, *m.* siphoning.

siphonner, to siphon.

siphonogamie, *f.* siphonogamy.

sippage, *m.* Danish process of leather dressing.

sirex, *m.* — **géant,** yellow wood wasp, giant sirex; — **spectre,** steel-blue wood wasp.

siroco, *m.* sirocco (hot wind).

sirop, *m.* sirup, juice; — **aromatique,** aromatic sirup; — **composé,** compound sirup; — **couvert,** clayed sirup, treacle; — **d'amidon,** starch sirup; — **de baies,** concentrated berry juice; — **de betterave,** beet sirup; — **de capillaire,** sirup prepared from maidenhair fern; — **d'écorce de cerisier,** sirup of wild cherry.

sirop, — **de fécule,** glucose, potato-starch sirup; — **de framboises,** raspberry juice; — **de glycose,** sirup of glucose; — **de suc de citron,** (**limon**), sirup of lemon; — **de sucre,** sugar sirup, beetroot molasses; — **simple,** simple sirup, sirup; — **sudorifique,** compound sirup of sarsaparilla; — **vert,** green sirup, greens.

sirupeux, -euse, sirupy.

sis, seated, situated.

siserin, *m.* — **boréal,** mealy redpoll.

site, *m.* site, habitat.

sitone, *m.* — **rayé,** pea weevil.

sitôt, so soon, as soon, as soon as, once; — **que,** as soon as.

situation, *f.* situation, state, position, location, report, account; — **du marché,** market conditions; — **exposée à la gelée,** frost locality, frost hole; — **non abritée,** unsheltered situation, exposed site.

situé, situated, located.

situer, to place, situate, locate.

six, six, sixth.

sixième, sixth.

sizain, sixain, *m.* half dozen, package of six.

slave, Slavic, Slav.

smalt, *m.* smalt.

smaltine, *f.* smaltite, smaltine.

smaragdite, *f.* smaragdite.

smectique, smegmatic, detersive; **argile —,** fuller's earth.

smectite, *f.* smectite, greenish halloysite.

smegma, *m.* smegma.

smithsonite, *f.* smithsonite, native zinc carbonate.

sobole, *f.* bulblet, shoot.

sobolifère, soboliferous, bearing vigorous shoots.

sobre, sober, temperate, abstemious, moderate, sparing.

sobrement, soberly.

soc, *m.* plowshare.

social, *m.* firm, company.

social, -aux, social, gregarious.

sociétaire, *m.* member (of corporate body), associate.

société, *f.* society, community, association, company, partnership, house; — **anonyme,** joint-stock company; — **commerciale,** commercial company; — **d'assurances,** insurance company; — **de capitaux,** stock company; — **en nom collectif,** partnership, private company; — **fiduciare,** trust company; — **par actions,** joint-stock company.

socle, *m.* base, bottom, pedestal, stand, footing, foundation.

socotrin, Sokotrine (aloes).

soda, *m.* soda water.

sodalite, *f.* sodalite.

sodé, sodium, combined with sodium.

sodique, sodium, of sodium, containing soda.

sodium, *m.* sodium.

soeur, *f.* sister.

sofa, *m.* sofa; — **de repos,** couch.

soi, self, oneself, itself; **chez —,** at home, in one's own house.

soi-disant, so-called, would-be, self-styled, alleged, supposedly.

soie, *f.* silk, bristle (of hog), hair, seta, pin (of crank); — **artificielle,** artificial silk, imitation silk, rayon; — **cuite,** boiled silk, scoured silk; — **de porc,** hog bristle, bristle; — **écrue,** — **grège,** raw silk; — **moulinée,** — **ouvrée,** thrown silk; **plante à —,** silkweed, swallowwort.

soient (être), (they) may be, are.

soierie, *f.* silk fabric, silk goods, silk mill, silk trade.

soif, *f.* thirst.

soigné, well-finished, carefully done, cared for, attended to, careful.

soigner, to care for, take care of, look after, give attention to.

soigneur, *m.* caretaker, attendant.

soigneusement, carefully, with care.

soigneux, -euse, careful.

soi-même, oneself, itself, self; **en —,** in(to) oneself.

soin, *m.* care, attention, trouble, heed; **—s culturaux,** tending; **avoir —,** to take care.

soir. *m.* evening, afternoon.

soirée, *f.* evening, evening party.

soit (être), be, be it so, let . . . be, either, or.

soit . . . **soit,** either . . . or, so be it, all right.

soixantaine, *f.* sixty, about sixty, threescore.

soixante, sixty, sixtieth; — **et onze,** seventy-one.

soixante-cinq, sixty-five.

soixante-dix, seventy.

soixante-dixième, seventieth.

soixante-douze, seventy-two.

soixante-huit, sixty-eight.

soixante-quinze, seventy-five.

soixante-treize, seventy-three.

soixantième, sixtieth.

soja, *m.* soy bean, soja bean, soja, soya (*Glycine hispida*); **huile de** —, soy bean oil.

sol, *m.* soil, ground, floor, bottom, sol, earth; — **argileux,** clayey soil; — **calcaire,** calcareous soil; — **crayeux,** chalky soil; — **dur,** stiff soil; — **fangeux,** fen, bog, marsh, swamp; — **frais,** fresh soil; — **lehmeux,** loamy soil; — **maigre,** meager soil; — **marécageux,** marshy soil.

sol, — **marneux,** marly soil; — **mouilleux,** wet soil; — **neutre,** mild soil; — **pierreux,** stony soil; — **profond,** deep soil; — **sablonneux,** sandy soil; — **sec,** arid soil; — **superficiel,** shallow soil; — **tourbeux,** peaty soil, peat ground; — **vierge,** virgin soil.

solaire, solar, of the sun.

solamire, *f.* sieve (bolting) cloth.

solanacées, solanées, *f.pl.* Solanaceae.

solandres, *f.pl.* sallenders, malanders.

solanée, *f.* nightshade (Solanum).

solanine, *f.* solanine.

solanum, *m.* nightshade (Solanum).

solariser, to solarize.

soldat, *m.* soldier; — **des bois,** soldier(ant).

soldat, soldierly.

solde, *m.* balance, surplus stock, job lot; *f.* pay.

solder, to pay, settle, sell off.

sole, *f.* sole, horny sole, bottom, bed, sleeper, sill, hearth (of furnace), ground plate, plot, break, field.

soleil, *m.* sun, sunburn, sunflower (Helianthus); **au** — **levant,** at sunrise.

soleilleux, -euse, sunny.

soléine, *f.* a resin spirit.

solennel, -elle, solemn.

solennité, *f.* solemnity.

solénoïde, *m.* solenoid.

solfatare, *f.* solfatara.

solidaire, integral, solidary, united as one, in one piece.

solidarité, *f.* solidarity, community, joint responsibility, union of interests and responsibilities.

solide, *m.* solid.

solide, solid, lasting, fast, strong, firm.

solidement, solidly, tightly.

solidificateur, *m.* solidifier.

solidificateur, -trice, solidifying.

solidification, *f.* solidification.

solidifier, to solidify; **se** —, to become solid, congeal.

solidité, *f.* solidity, strength, firmness.

solitaire, *m.* recluse, hermit, solitary boar.

solitaire, solitary, lone.

solitairement, solitarily, lonesomely.

solitude, *f.* solitude, desert.

solive, *f.* joist, beam; — **ancienne,** one-half cubic meter; — **nouvelle,** one-tenth cubic meter.

sollicitation, *f.* solicitation, entreaty.

solliciter, to solicit, induce, influence, attract, act on, canvass, urge.

solséquial, -aux, (plant) following the apparent movement of the sun.

solstice, *m.* solstice.

solubiliser, to render soluble.

solubilité, *f.* solubility, solubleness.

soluble, soluble.

soluté, *m.* solution; — **de chaux,** lime water; — **de pancréatine,** the pancreatic solution.

solutif, -ive, capable of dissolving, laxative.

solution, *f.* solution, break of continuity, fault; — **au titre,** standard solution; — **de continuité,** breach of continuity; — **empirique,** standard solution; — **normale,** normal solution, standard solution; — **tartrique,** solution of tartar; — **type,** standard(ized) solution; **mettre en** —, to dissolve.

solutol, *m.* solutol.

solvabilité, *f.* solvency.

solvant, *m.* solvent.

somatophytes, *f.pl.* somatophytes.

somatotropisme, *m.* somatotropism, directive influence of the substratum on the growth of an organism.

sombre, dark, dull, gloomy, sad, somber, melancholy.

sombre, *m.* darkness.

sombrer, to founder, sink, go down.

sommaire, *m.* summary, synopsis, abstract.

sommaire, summary, brief, concise.

sommairement, summarily, briefly.

sommation, *f.* summons, summation, request, demand.

somme, *f.* sum, amount, epitome, load, burden, total, sand bar; — **globale,** lump sum; — **toute,** on the whole; **en** —, in short.

somme, *m.* short sleep, nap.

sommeil, *m.* sleep.

sommer, to summon, sum up, add.

sommes (être), (we) are.

sommet, *m.* summit, vertex, height, peak, apex, top, crown.

sommier, *m.* pack animal, support, bed, crossbeam, bearer, girder, mattress, register, cash book; — **du contrôle,** control register, control book.

sommité, *f.* top, summit, tip, flowering apex, prominent person.

somnifère, soporific, hypnotic, narcotic.

somnolence, *f.* drowsiness, somnolence.

somptueux, -euse, sumptuous.

son, *m.* bran, sound.

son, his, her, its, one's.

sondage, *m.* boring, sounding.

sonde, *f.* borer, boring machine, drill, driller, taster (for cheese), grain (sampler), sound, probe, sounding line, catheter, (sounding) lead; — **à beurre,** trier, grader; — **trayeuse,** milking tube.

sonder, to bore, sample (soil), taste (cheese), explore, sound, probe, examine, test, try, fathom.

sondeur, *m.* sounder, probe.

sondeuse, *f.* boring machine.

songe, *m.* dream.

songer, to dream; — **à,** to think of, reflect upon, consider, muse.

sonnailler, *m.* bellwether.

sonnant, striking, sounding, ringing, sonorous.

sonner, to sound, strike, ring; **faire** —, to sound, ring.

sonnerie, *f.* ringing, set of bells, call, electric bell.

sonnette, *f.* small bell, pile driver, rattle (of rattlesnake); **mouton de** —, monkey.

sonore, sonorous, loud sounding, resounding.

sonorité, *f.* sonorousness, sonority.

sons, *m.pl.* bran.

sophistication, *f.* sophistication, adulteration.

sophistiquer, to sophisticate, adulterate.

sopor, *m.* morbid sleep, sopor.

soporatif, -ive soporifère, soporifique. soporific, hypnotic.

sorbe, *f.* sorb, sorb apple, service berry, rowan, dogberry.

sorbet, *m.* sherbet.

sorbier, *m.* sorb, service tree; — **domestique,** service tree (*Sorbus domestica*); — **des oiseaux,** — **des oiseleurs,** mountain ash, rowan tree (*Sorbus aucuparia*).

sorbine, *f.* sorbin, sorbose.

sorbique, sorbic.

sorbite, *f.* sorbite.

Sorbonne, *f.* Sorbonne.

sorcellerie, *f.* sorcery.

sorcier, *m.* sorcerer.

sordide, sordid, dirty.

sore, *m.* sorus.

sorédie, *m.* soredium.

sorgho, sorgo, *m.* sorghum (*Holcus sorghum*); — **sucré,** sweet sorghum (*Holcus sorghum* var. *saccharatus*); — **commun,** broomcorn (*Holcus sorghum* var. *technicus*).

sorne, *f.* slag, dross, scoria.

sorose, *f.* sorose, a fleshy multiple fruit.

sort, *m.* lot, fate, chance, destiny, charm.

sorte, *f.* way, manner, sort, kind, quality class; **de** — **que, en** — **que,** so that, so as, with the result that; **de telle** —, in such a way, so much; **en quelque** —, in some way, in a measure.

sorteur, *m.* sorter.

sortie, *f.* going out, coming out, departure, outlet, exist, mouth, issue, leaving, leave, retirement, export, exportation, excursion, sortie, sally, outburst; — **du laitier,** slag hole, cinder notch.

sortir, to go out, come out, emerge, get out, be out, issue, come forth, go beyond, depart, leave, retire, proceed, come, spring, ensue, project, stand out, take out, bring out; **au** — **de,** at the end of, on coming out of; **faire** —, to bring out, take out, call out, turn out, emphasize.

sortir, *m.* leaving, departure.

sot, *m.* fool, dolt.

sot, sotte, silly, stupid, foolish.

sottise, *f.* stupidity, foolishness, folly.

soubassement, *m.* basement, subfoundation.

soubresaut, *m.* sudden start, bound, shock, jerk, bumping.

soubresauter, to leap, start, jolt, bump.

souche, *f.* stump, stub, stock, stem, trunk, rootstock, rhizome, ancestor, predecessor, progenitor, stool.

souchet, *m.* — **commun,** shoveler duck, galingale (Cyperus).

souchetage, *m.* counting the stumps.

souchon, *m.* small stump, short thick iron bar.

souci, *m.* care, anxiety, concern, marigold; — **d'eau,** marsh marigold (*Caltha palustris*)

soucier (se), to care, concern oneself, mind.

soucieux, -euse, anxious.

soucoupe, *f.* saucer, fruit stand.

soudable, weldable, capable of being soldered.

soudage, *m.* soldering, welding, joining, brazing.

soudain, sudden, unexpected, at once, suddenly.

soudainement, suddenly, unexpectedly.

soudaineté, *f.* suddenness.

soude, *f.* soda, soda ash, sodium carbonate, saltwort (Salsola), glasswort (Salicornia), sodium hydroxide; — **à la chaux,** caustic soda; — **artificielle,** artificial soda; — **boratée,** (native) borax, borate of sodium; — **brute,** crude soda, ball soda.

soude, — **carbonatée,** sodium carbonate; — **caustique,** caustic soda, sodium hydroxide; — **caustique liquide,** solution of sodium hydroxide, caustic soda solution; — **nitratée,** sodium nitrate; — **tartarisée,** potassium sodium tartrate.

soudé, united.

souder, to weld, solder, fuse, burn, braze, suture, unite, join; **se** —, to be welded, weld, unite, melt together, run together, grow together, self-graft, coalesce, consolidate; — **à l'autogène,** to solder autogenically; — **à recouvrement,** to lap weld.

soudeur, *m.* welder.

soudier, *m.* soda maker.

soudier, -ère, soda, of soda.

soudière, *f.* soda factory, soda works.

soudoir, *m.* soldering iron.

soudure, *f.* soldering, solder, welding, union, weld, soldered joint, suture, splice, fusing, burning, uniting, coalescence, self-grafting; — **à la résine,** rosin soldering; — **au plomb,** lead solder;

soudure, — **autogène,** autogenic soldering, lead burning; — **de cuivre,** hard solder, copper solder; — **de laiton,** brass solder, spelter solder; — **électrique,** electric welding; — **matée,** calked seam, calk weld; — **tendre,** soft (solder) soldering; **sans** —, without weld, seamless.

souffert (souffrir), suffered, undergone.

soufflage, *m.* glass blowing, blast.

soufflant, blowing.

soufflante, *f.* blower, blowing engine.

soufflard, *m.* air (mud) volcano, volcanic steam jet.

souffle, *m.* breath, breathing, exhalation, whiff, blowing, blast, murmur.

soufflement, *m.* blowing.

souffler, to blow, breathe, puff, pant, blow up, inflate, blow off, blow out, utter (a sound), sheathe (a ship).

soufflerie, *f.* bellows, blowing engine, blower, blast, blast lamp.

soufflet, *m.* bellows, blower, blowing machine, fan, fanner, slap.

souffleur, *m.* blower; — **de verre,** glass blower.

soufflure, *f.* blister, air hole, blowhole, flaw.

souffrance, *f.* suffering, suspense, pain.

souffrant, suffering.

souffrir, to suffer, endure, bear, allow, let.

soufrage, *m.* sulphuring.

soufre, *m.* sulphur, brimstone; — **doré d'antimoine,** red antimony sulphide; — **en bâtons,** stick sulphur, roll sulphur; — **en canons,** roll sulphur; — **en fleur(s),** flowers of sulphur; — **mou,** plastic sulphur; — **sélenifère,** selensulphur; — **végétal,** vegetable sulphur, lycopodium powder, witch meal; — **vierge,** — **vif,** virgin sulphur, native sulphur.

soufrer, to sulphur, spray or fumigate with sulphur, stum (wine).

soufreuse, *f.* sulphur sprayer.

soufrière, *f.* sulphur mine, sulphur pit.

soufroir, *m.* sulphuring stove or chamber.

souhait, *m.* wish, desire; **à** —, according to one's wishes, to one's liking.

souhaitable, desirable.

souhaiter, to wish, wish for, hope, long for, desire.

souillarde, *f.* sinkhole, sink stove, lye tub, lye vat.

souille, *f.* wallowing place, mud hole.

souillé, contaminated, soiled.

souiller, to contaminate, soil, tarnish, stain, taint.

souillure, *f.* contamination, stain, blot, spot, taint, impurity.

soûl, surfeited, gorged, satiated, full.

soulageant, easing.

soulagement, *m.* relief, alleviation, aid, comfort.

soulager, to relieve, alleviate, aid, ease, comfort, lighten.

soulèvement, *m.* rising, upheaval, uprising, swelling.

soulever, to raise, lift (up), break up, excite, rouse; **se —,** to rise, rise up, raise oneself, be raised, swell, work (of ferments).

soulier, *m.* shoe.

soulignement, *m.* underlining, stressing (of word).

souligner, to underline, emphasize.

soumettre, to subdue, subject, refer, submit, subjugate, force, overcome, present; **se —,** to be subjected, submit, yield, abide, consent; **— au régime forestier,** to subject to the control of the forest department.

soumis, submissive, obedient, subjected, subject.

soumission, *f.* submission, submissiveness, obedience, tender, bond.

soumissionnaire, *m.* bidder, tendering party.

soupape, *f.* valve, strike valve, discharge valve; **— à champignon,** mushroom valve, lift valve; **— commandée,** mechanically controlled valve; **— d'arrêt,** check valve, stop valve; **— de sûreté,** safety valve; **— de surpression,** safety valve; **à —,** valved, with valve(s).

soupçon, *m.* suspicion, distrust, conjecture, surmise, touch, dash.

soupçonnable, suspicious, susceptible.

soupçonner, to suspect, surmise.

soupçonneux, -euse, suspicious.

soupe, *f.* soup, meal.

souper, *m.* supper.

soupeser, to try the weight of, heft.

soupière, *f.* soup tureen.

soupirail, *m.* air hole, vent.

soupirer, to sigh.

souple, supple, pliable, pliant, flexible.

souplesse, *f.* flexibility, facility, pliability, suppleness.

source, *f.* source, spring (of water), fountain, rise, well, wellhead.

sourcil, *m.* eyebrow, brow.

sourcilier, -ère, superciliary (muscle).

sourd, deaf, dull, muffled, cloudy (gems), secret, hollow, dark.

sourdement, low, with a dull hollow sound, secretly, mysteriously.

sourdre, to spring (up), well (up), result, arise.

sourire, to smile.

sourire, *m.* smile.

souris, *f.* mouse; *m.* smile.

sous, under, beneath, below, in, at, underneath, sub.

sous-acétate, *m.* subacetate; **— de plomb,** lead subacetate.

sousalaire, subalar, underwing.

sous-arbrisseau, *m.* undershrub.

sous-azotate, *m.* subnitrate.

sous-bois, *m.* underwood, undergrowth.

sous-carbonate, *m.* subcarbonate, basic carbonate.

sous-chlorure, *m.* subchloride.

sous-classe, *f.* subdivision, subclass.

sous-cortical, -aux, subcortical.

sous-couche, *f.* substratum, underlying layer.

souscripteur, *m.* subscriber.

souscription, *f.* subscription.

souscrire, to subscribe, sign.

sous-cutané, subcutaneous.

sous-diviser, to subdivide.

sous-égaliser, to pass through a sieve.

sous-épidermique, subepidermal.

sous-espèce, *f.* subspecies.

sous-étage, *m.* lower story, underwood.

sous-genre, *m.* subgenus.

sous-groupe, sous-groupement, *m.* subgroup.

sous-jacent, subjacent, underlying, located below.

sous-lacustre, *f.* sublacustrine.

sous-location, *f.* subleasing, subletting.

sous-marin, submarine, submersed, undersea.

sous-maxillaire, submaxillary.

sous-multiple, submultiple.

sous-muriate, *m.* subchloride.

sous-nitrate, *m.* subnitrate, basic nitrate.

sous-officier, *m.* petty officer.

sous-oxyde, *m.* suboxide.

sous-parcelle, *f.* subcompartment.

sous-préfet, *m.* subprefect.

sous-pression, *f.* pressure from below.

sous-produit, *m.* by-product.

sous-région, *f.* subregion.

sous-sel, *m.* subsalt, basic salt.

sous-sol, *m.* subsoil, substratum, basement, mantle rock.

sous-station, *f.* substation.

sous-sulfure, *m.* subsulphide.

sous-tangente, *f.* subtangent.

sous-tendre, to subtend.

sous-titre, *m.* subtitle, subhead, caption.

soustraction, *f.* removal, withdrawal, abstraction, subtraction.

soustraire, to remove, subtract, withdraw, save, shelter, protect, defalcate; **se — (à),** to avoid, remove from, escape.

sous-traitant, *m.* subcontractor.

soustrate, *m.* substrate.

sous-ventrière, *f.* girth, belly band.

sous-vêtement, *m.* undergarment.

soute, *f.* storeroom, bunker, locker, storage can or tank, magazine.

soutenir, to sustain, support, uphold, keep up, maintain, prop up, bear, stand, endure, resist; **se —,** to support oneself, be sustained, be firm, last, continue, hold, persist, succeed, keep up, resist, stand up.

soutenu, sustained, supported, continued.

souterrain, *m.* underground passage, cave, vault, dungeon, subway; **— jumeau,** twin tube.

souterrain, underground, subterranean; **en —,** underground.

soutien. *m.* support, maintenance.

soutient (soutenir), (he) is supporting.

soutirage, *m.* drawing off, racking, clarifying (wine).

soutirer, to draw off (liquors), rack off, tap.

soutrage, *m.* clearing the soil covering.

souvenir (se), to remember, recollect, occur to the mind; **autant qu'il m'en souvient,** as far as I can remember.

souvenir, *m.* remembrance, recollection, souvenir, memory, memorial, memorandum book.

souvent, often, frequently.

souverain, sovereign, supreme.

souverainement, *f.* sovereignly, supremely.

souveraineté, *f.* sovereignty, supremacy.

souvient (souvenir), (he) recollects; **il m'en —,** I remember.

soya, *m.* see SOJA.

soyeuse, *f.* silkweed, milkweed, silky (fowl).

soyeux, -euse, silky, silken, covered with hairs as fine and soft as silk.

sozoïodol, *m.* sozoiodol.

spacieusement, spaciously.

spacieux, -euse, spacious, roomy, large.

spadelle, *f.* stirring tool.

spadice, *m.* spadix.

spadicé, spadiceous.

spadiciflore, spàdiciné, spadicifloral.

spagirie, *f.* alchemy.

spagirique, spagyric(al), alchemical.

spagirisme, *m.* alchemy, iatrochemistry

spagiriste, *m.* spagyrist.

spalt, *m.* asphalt, fluor spar, compact bitumen.

spananthe, spananthous, having few flowers.

sparadrap, *m.* sparadrap, adhesive plaster; **— de capsique,** capsicum plaster.

sparsiflore, sparsiflorous.

sparsifolié, sparsifolious.

sparsioplaste, *f.* elaioplast variable in position and number.

sparte, spart, *m.* esparto grass (*Lygeum spartum*).

spartéine, *f.* sparteine.

sparterie, *f.* wood plaiting, basketwork.

spasme, *m.* spasm.

spasmodique, spasmodic.

spath, *m.* spar, calcite; **— adamantin,** common corundum; **— calcaire,** calc-spar, calcite; **— fluor,** fluor spar, fluorite, native calcium fluoride; **— pesant, — lourd,** heavy spar, barite.

spathacé, spathaceous, spathe-bearing, **of** the nature of a spathe.

spathe, *f.* spathe, shuck (of maize).

spathelle, *f.* glumelle, palea of grasses.

spathellule, *f.* spathellula, palea of grasses.

spathiforme, spathose.

spathique, spathic, sparry.

spatial, -aux, spatial.

spatialiser, to spatialize, locate in space.

spatulage, *m.* stirring.

spatule, *f.* spatula, spattle, beater, slicer, stirrer, paddle, spoonbill; — **à encre,** ink slab; — **en bois,** wooden spatula or stirrer; — **en corne,** horn spatula; — **en platine,** platinum spatula; — **en verre,** glass spatula; — **racloir,** scrap spattle.

spatulé, hatcheled, spatulate, spoon-shaped, worked (flax).

spatuler, to stir with spatula or paddle, hatchel (flax).

spécial, -aux, special, peculiar, professional.

spécialement, specially, especially.

spécialiser, to specify, specialize.

spécialiste, *m.* specialist, expert.

spécialiste, specializing.

spécialité, *f.* speciality, specialty, special function.

spécieux, -euse, specious, plausible.

spécification, *f.* specification.

spécificité, *f.* specificity, specificness.

spécifier, to specify, state definitely.

spécifique, specific.

spécifiquement, specifically.

spécimen, *m.* specimen, sample.

spectacle, *m.* spectacle, sight, show, play.

spectateur, -trice, *m.&f.* spectator.

spectral, -aux, spectral.

spectre, *m.* specter, spectrum; — **d'émission,** emission spectrum; — **d'étincelles,** spectrum of sparks; — **magnétique,** magnetic tracing, magnetic curves; — **solaire,** solar spectrum; — **synoptique,** general spectrum.

spectrographe, *m.* spectrograph.

spectromètre, *m.* spectrometer.

spectrophotomètre, *m.* spectrophotometer.

spectroscope, *m.* spectroscope; — **à main,** hand spectroscope.

spectroscopie, *f.* spectroscopy.

spectroscopique, spectroscopic(al).

spectroscopiste, *m.* spectroscopist.

spéculaire, specular.

spéculatif, -ive, speculative.

spéculation, *f.* speculation.

spéculer, to speculate, ponder.

speiss, *m.* speiss.

speltre, *m.* spelter.

spergule, *f.* spurry, spergula.

spermaceti, *m.* spermaceti.

spermapode, *m.* spermapodium, a branched gynophore in Umbelliferae.

spermapodophore, *m.* spermapodophorum, spermapodium.

spermatange, *m.* spermatangium.

spermatide, *f.* spermatid.

spermatie, *m.* pycnoconidium, stylospore.

spermatine, *f.* spermatin.

spermatique, spermatic.

spermatocyste, *f.* spermatocyst, mother cell of antheridia.

spermatozoaire, *m.* spermatozoon.

spermatozoïde, *m.* spermatozoid.

sperme, *m.* sperm, semen.

spermidé, spermideous, producing seed.

spermine, *f.* spermine.

spermique, spermic, relating to a seed.

spermoderme, *m.* testa.

spermogonie, *f.* spermogone.

spermophore, *m.* spermophyte.

spermotoxine, *f.* spermotoxin.

sphacèle, *m.* gangrene.

sphaeroplaste, *f.* chromidium.

sphaigne, *m.* peat moss, bog sphagnum, peat reek.

sphanidophyte, *f.* plant whose fruit is crowned by teeth.

sphégide, *f.* fossorial wasp.

sphégifère, (plant) whose flower resembles a wasp.

sphène, *m.* sphene, titanite.

sphénoïdal, -aux, sphenoid.

sphère, *f.* sphere, globe.

sphéricarpe, with spherical fruit.

sphéricité, *f.* sphericity.

sphérique, spherical, spheric.

sphériquement, spherically.

sphéroblaste, *m.* spheroblast.

sphérocarpe, with round fruit.

sphéroïdal, -aux, spheroidal.

sphéroïde, *m.* spheroid.

sphéroïdique, spheroidic.

sphérosidérite, *m.* spherosiderite, concretionary siderite.

sphérosperme, with round seeds.

sphinx, *m.* hawk moth; — **du pin, pine** hawk moth; — **tête de mort,** death's-head hawk moth.

sphygmisme, *m.* sphygmism, the formation of contractile vacuoles through some stimulus.

spiauter, *m.* zinc, spelter.

spic, *m.* spike lavender (*Lavandula spica*).

spicanard, *m.* spikenard.

spiciforme, spikelike.

spicigère, spicigerous, bearing flower spikes.

spiculaire, spicular, dartlike.

spicule, *m.* spicule.

spiculé, spicate.

spiculifère, (plant) with flowers arranged in spikelets.

spiegel, *m.* spiegeleisen, spiegel.

spigélie, *f.* worm grass (Spigelia); — du Maryland, pinkroot, Carolina pink, (*Spigelia marilandica*).

spile, *m.* spilus, the hilum in grasses.

spilhémigone, (plant) with umbel marked in center by a round spot.

spilogoné, (plant) with umbel marked on circumference with angular spots.

spinelle, *m.* spinel.

spinelle, *f.* spinule.

spinellé, spinulose.

spinelleux -euse, spinellosous.

spinescence, *f.* spinescence.

spinescent, spinescent.

spinifère, thorny, spiny.

spinifolié, spinifolious, having spiny leaves.

spiniforme, spiniform.

spininervé, with denticulate veins.

spinocarpe, spinocarpous.

spinule, *f* spinule, a diminutive spine.

spinuleux, -euse, spinulose.

spinuliflore, (calyx) provided with spinules.

spiracle, *m.* spiracle, stigma.

spiral, -aux, spiral.

spirale, *f.* spiral, helix.

spiralé, made spiral, spiral-formed.

spire, *f.* spire, turn, winding.

spirée, *f.* Spiraea.

spirème, *f.* spireme.

spirille, *m.* spirillum.

spiritualiser, to spiritualize.

spirituel, -elle, spiritual, intellectual, intelligent.

spiritueux, *m.* spirituous liquor.

spiritueux, -euse, spirituous.

spirituosité, *f.* spirituousness, spirituosity.

spirobactéries, *f.pl.* spirobacteria.

spirochète, *m.* spirochete.

spirochétose, *f.* spirochetosis, disease of fowls due to Spirocheta.

spiroïdal, -aux, spiroïde, spiroid.

spirolobé, spirolobous, with the cotyledons spirally rolled.

spirostyle, with style twisted in a spiral.

spitzkasten, *m.* funnel box, pointed box.

splendeur, *f.* splendor, brightness.

splendide, splendid, magnificent.

splénique, splenic (artery).

spode, *f.* spodium.

spondyle, *m.* vertebra.

spongieux, -euse, spongy, porous.

spongine, *f.* spongin.

spongiole, *f.* spongiole, root end, radicle.

spongiosité, *f.* sponginess.

spontané, spontaneous, voluntary, run wild, self-sown, growing wild.

spontanéité, *f.* spontaneity, spontaneousness.

spontanément, spontaneously.

sporadique, sporadic.

sporange, *m.* sporangium.

spore, *f.* spore; — de conservation, resting spore.

sporidies, *f.pl.* conidia.

sporifère, spore bearing.

sporocarpe, *m.* sporocarp.

sporocyste, *m.* sporocyst, unicellular structure producing asexual spores.

sporoderme, *m.* sporoderm, integument of a spore.

sporogénèse, *f.* sporogenesis, origin and development of seeds or spores.

sporogone, *m.* sporogone, sporogonium.

sporogonie, *f.* sporogony.

sporophore, *m.* sporophore.

sporophylle, *f.* sporophyll.

sporophyte, *f.* sporophyte.

sporotrichose, *f.* sporotrichosis.

sporozaires, *m.pl.* Sporozoa.

sporozoïte, *f.* zygotoblast.

sport, *m.* sport, variation starting from a bud or seed.

sporulation, *f.* sporulation.

sporule, *f.* sporula, sporule, a small spore.

sporulé, sporulated.

sporuler, to sporulate.

spume, *f.* froth.

spumeux, -euse, spumous, spumy, foamy, frothy.

sputation, *f.* expectoration.

squale, *m.* dogfish.

squame, *f.* scale.

squamellifère, squamelliferous, scale-bearing.

squamelliforme, squamelliform, shaped like a scale.

squamellule, *f.* squamellula, scalelike appendage in tube of certain corollas.

squameux, -euse, squamous, scaly.

squamiforme, squamiform, scalelike.

squamule, *m.* squamule, the hypogynous scale of grasses.

squamuliforme, squamuliform, resembling a small scale.

squarreux, -euse, squarrose.

squelette, *m.* skeleton, carcass.

squine, *f.* chinaroot (*Smilax china*).

stabile, stabile, fixed.

stabilisateur, *m.* stabilizer, steadying surface, elevator.

stabilisateur, -trice, stabilizing.

stabilisation, *f.* stabilization.

stabiliser, to stabilize, establish firmly; **se —,** to become stabilized.

stabilité, *f.* stability, solidity.

stabiloplastes, *m.pl.* stabiloplasts, elaioplasts that are fixed in number and position.

stable, stable, steadfast, fixed.

stablement, stably, firmly, steadily, fixedly.

stabulation, *f.* keeping (cattle) in sheds.

stachys, *m.* — tubéreux, Chinese or Japanese artichoke (*Stachys sieboldi*).

stade, *m.* stage, stadium, period, stade, phase.

stage, *m.* period of probation, course; **faire un —,** to take a course.

stagiaire, *m.* forest probationer.

stagnant, stagnant, standing.

stagner, to stagnate, stand.

stalactite, *f.* stalactite.

stalactitique, stalactitic.

stalagmite, *f.* stalagmite.

stalagmomètre, *m.* stactometer, stalagmometer.

stalle, *f.* stall.

staminaire, (double flower) with supernumerary petals derived from the transformation of the stamens.

staminé, staminate.

stamineux, -euse, staminous.

staminifère, staminate.

staminiforme, shaped like a stamen.

staminipare, (flower) with organs taking the form of stamens.

staminode, *f.* staminodium.

staminule, *m.* rudimentary stamen.

stampe, *f.* country rock between veins.

stannage, *m.* tin mordanting.

stannate, *m.* stannate.

stanneux, -euse, stannous.

stannifère, stanniferous, tin-bearing.

stannine, *f.* stannite.

stannique, stannic (acid).

stapélie, *f.* Stapelia.

staphisaigre, *f.* stavesacre (*Delphinium staphisagria*).

staphylier, *m.* bladdernut tree (Staphylea).

staphylin, *m.* cocktail (beetle), rove beetle.

staphylococcie, *f.* staphylococcal infection.

staphylococcique, staphylococcal.

staphylocoque, *m.* staphylococcus.

staphylolysine, *f.* staphylolysin.

staphylotoxine, *f.* staphylococcus toxin.

stathouder, *m.* stadtholder.

statice, *m.* sea lavender (Limonium).

staticité, *f.* static condition.

statif, *m.* stand, framework (of microscope).

station, *f.* station, plant, stand, position, stop, stay, locality, standing; — **émission radiophonique,** broadcasting station; — **d'expériences,** experimental station; — **d'expériences agronomiques,** agricultural experiment station; — **terminus,** terminal.

stationnaire, stationary.

stationnement, *m.* stopping, standing, halt.

stationner, to stop, stand, station.

statique, *f.* statics.

statique, static(al).

statistique, *f.* statistics, statistic.

statistique, statistical.

statolithes, *m.pl.* statoliths, starch grains causing curvature by their weight.

stator, *m.* stator.

statosperme, statospermous, orthotropous.

statuaire, statuary.

statuette, *f.* statuette, small statue.

staude, *m.* shrub.

staudenartig, frutescent, shrublike.

staurolithe, *f.* staurolite.

staurophylle, staurophyllous, cruciate.

stéarate, *m.* stearate.

stéarine, *f.* stearin.

stéariner, to coat with stearin.

stéarinerie, *f.* stearin factory.

stéarique, stearic, stearin.

stéarone, *f.* stearone.

stéaroptène, *f.* stearoptene.

stéatite, *f.* steatite.

stéatiteux, -euse, steatitic.

stegmata, *f. pl.* stegmata, tabular cells containing a mass of silica.

stégocarpe, stegocarpic, stegocarpous.

stégomyie, *f.* stegomyia, yellow-fever mosquito.

stèle, *f.* stele.

stellaire, *f.* starwort (Stellaria).

stellaire, stellar.

stellinervé, star-ribbed (leaf).

stemmate, *m.* simple eye, ocellus.

sténographe, *m.* stenographer.

sténohaline, *f.* stenohaline.

sténopétale, stenopetalous, narrow-petaled.

sténophylle, narrow-leaved (plant).

sténostachyé, thin-spiked.

stéphocarpe, with rounded fruit.

steppe, *m.* steppe, plain.

stercoraire, *m.* dung beetle.

stercoraire, stercoraceous, relating to dung.

stère, *m.* stere, cubic meter, raummeter, stacked cubic measure; **bois de —,** cord wood.

stéréide, *m.* stereid, lignified cell from the stereome.

stéréoautographe, *m.* stereoautograph.

stéréoautographie, *f.* stereoautography.

stéréochimie, *f.* stereochemistry.

stéréoisomère, *m.* stereoisomer.

stéréoisomère, stereoisomeric.

stéréoisomérie, *f.* stereoisomerism.

stéréome, *m.* stereome.

stéréométrie, *f.* stereometry.

stéréophylle, with firm leaves.

stéréoplasma, *m.* stereoplasm, solid part of protoplasm.

stéréoscope, *m.* stereoscope.

stéréoscopique, stereoscopic.

stéréotactisme, *m.* thigmotaxis, result of mechanical stimulus.

stéréotélémètre, *m.* stereotelemeter.

stéréotropisme, *m.* stereotropism, a definite direction toward the substratum.

stéréotypage, *m.* stereotyping.

stéréotype, stereotype(d).

stéréotyper, to stereotype.

stéréotypie, *f.* stereotypy, stereotyping.

sterigmates, *f.pl.* sterigmata.

stérile, *m.* sterile material.

stérile, sterile, barren, infertile, fruitless.

stérilement, sterilely, unprofitably.

stérilisateur, *m.* sterilizer.

stérilisation, *f.* sterilization, disinfection.

stériliser, to sterilize.

stérilité, *f.* sterility, barrenness, infertility.

sternellum, *m.* metasternum, sternellum.

sternite, *m.* sternite.

sternum, *m.* breastbone, sternum.

sternutatif, -ive, sternutatoire, sternutatory, sternutative.

sthénique, sthenic.

stibial, -aux, stibial, antimonial.

stibié, stibiated, impregnated with antimony, containing antimony.

stibieux, -euse, stibious, antimonious.

stibine, *f.* stibnite, stibine.

stichidie, *m.* stichidium.

stichocarpe, stichocarpous, (fruit) disposed along a spiral line.

stictopétale, stictopetalous, (petals) covered with glandular points.

stigmataire, marked by sunken points.

stigmate, *m.* stigma, brand, mark, scar, cicatrix; *pl.* spiracles, stigmata.

stigmatiforme, stigmatiform, shaped like a stigma.

stigmatiphore, stigmatophorous, (style of Compositae) bearing stigmas.

stigmatique, (linear stigma) extended along certain styles.

stigmatiser, to stigmatize.

stigmatoïdé, stigmatic.

stigmatophore, stigmatophorous.

stigmatostège, *m.* tegument of the stigma.

stigmatostémone, with stamens implanted on the stigma.

stigmatule, *m.* organ using stigma.

stigmule, *f.* stigmula, a division of a stigma.

stilbène, *m.* stilbene.

stilbite, *f.* stilbite.

stil-de-grain, *m.* stil de grain, yellow lake.

stillant, dripping, dropping.

stillation, *f.* dropping, dripping, oozing, stillation.

stimulant, *m.* stimulant, stimulus, irritant.

stimulant, stimulating, stimulative.

stimulation, *f.* stimulation, stimulating, excitation.

stimule, *m.* stimulus.

stimuler, to stimulate, arouse.

stimuleux, -euse, stimulose, covered with stinging hairs.

stimulines, *f.pl.* stimulins.

stipe, *m.* stem of monocotyledons, culm, caudex; *pl.* stipes, footstalk.

stipelle, *f.* stipel.

stipe-tenace, *f.* esparto grass (*Lygeum spartum*), alfa grass (*Stipa tenacissima*).

stipifère, stipiferous, bearing small flower stalks.

stipiforme, stipiform.

stipité, stipitate, having a stipe or special stalk.

stipulacé, stipulaceous, belonging to a stipule, with large stipules.

stipulaire, stipular, having stipules.

stipule, *f.* stipule; **sans —s,** erstipulate.

stipulé, stipulate, having stipules.

stipuléen, -enne, stipulaneous, resulting from the transformation of a stipule.

stipuler, to stipulate.

stirpé, deep-rooted.

stock, *m.* stock (of goods).

stockage, *m.* keeping in stock.

stockfish, *m.* stockfish.

stoechiogénie, *f.* stoichiogeny.

stoechiologie, *f.* stoichiology.

stoechiométrie, *f.* stoichiometry.

stolon, *m.* sucker, runner, stolon.

stolonifère, stoloniferous.

stomacal, -aux, stomachic, gastric, of the stomach.

stomachique, stomachic.

stomate, *m.* stoma; *pl.* stomata; **— aquifère,** water pore.

stomoxes, *f.pl.* biting flies of the genus Stomoxys.

stoppage, *m.* stopping, fine darning.

stopper, to stop.

stoquer, to stoke, tend (a fire).

stoqueur, *m.* poker, stoker.

storax, *m.* storax.

stovaïne, *f.* stovaine.

stragule, *m.* stragulum, palea of grasses.

stramoine, *f.*, stramonium, *m.* stramonium, stramony, thorn apple (*Datura stramonium*).

strass, stras, *m.* strass, paste (jewelry).

strasse, *f.* floss silk, waste silk, coarse packing paper.

strate, *f.* stratum, layer.

stratégie, *f.* strategy.

stratégique, strategic.

stratification, *f.* stratification.

stratifié, stratified.

stratifier, to stratify, be stratified.

stratigraphe, *m.* stratigrapher.

stratigraphique, stratigraphic(al).

stratum, *m.* layer, stratum.

streptobacille, *m.* streptobacillus.

streptocarpe, streptocarpous, (fruit) marked spirally.

streptococcie, *f.* streptococcal infection.

streptocolysine, *f.* hemolysin of a streptococcus.

streptocoque, *m.* streptococcus.

streptothrices, *f.pl.* Streptotricheae

striation, *f.* striation.

strict, strict.

strictement, strictly.

striduleux, -euse, stridulous, noisy, piercing, creaking.

strie, *f.* stria, streak, stroke, furrow.

strié, striated, striate, striped, marked by parallel lines, channeled.

strier, to striate.

stries, *f.pl.* striae; **— enfoncées,** channels.

striguleux, -euse, grooved, reticulated.

striiflore, striated (corolla).

striphuocalicé, with very hairy calyx.

striure, *f.* striation, stria.

strobilacé, strobilaceous, relating to a cone.

strobile, *m.* strobile, strobila, cone.

stroma, *m.* stroma.

strombulifère, strombuliferous, (fruit) spirally twisted.

strombuliforme, strombuliform, (fruit) spirally twisted.

strontiane, *f.* strontia.

strontianite, *f.* strontianite, native strontium carbonate.

strontique, strontium, of strontium.

strontium, *m.* strontium.

strophante, stophantus, *m.* Strophanthus.

strophantine, *f.* strophanthin.

strophiole, *f.* strophiole, caruncle.

strophisme, *m.* strophism, tendency to twist in response to external stimulus.

structural, -aux, structural.

structure, *f.* structure; — **granuleuse,** crumb structure.

strychnée, *f.* plant of the genus Strychnos.

strychnine, *f.* strychnine, strychnia.

strychnique, strychnic.

strychniser, to strychninize, treat patient with strychnine.

strychnisme, *m.* strychninism.

strychnos, *m.* Strychnos; — **bois de couleuvre,** snakewood (*Strychnos colubrina*); — **vomiquier,** nux vomica (*Strychnos nux-vomica*).

stuc, *m.* stucco.

stucage, *m.* stuccoing.

stucatine, *f.* fine stucco.

studieusement, studiously.

studieux, -euse, studious.

stupéfactif, *m.* stupefactive.

stupéfactif, -ive, stupefactive.

stupéfait, stupefied, amazed.

stupéfiant, *m.* stupefacient, narcotic.

stupéfiant, stupefying, stupefacient.

stupéfier, to stupefy.

stupeur, *f.* stupor.

stupeux, -euse, stupeous, woolly.

stupide, stupid, dull.

stupidement, stupidly.

stupidité, *f.* stupidity, hebetude

stuquer, to stucco.

sturine, *f.* sturine.

stygien, -enne, stygious, (plants) growing in foul waters.

stylaire, stylar.

style, *m.* style, manner.

stylé, stylate, trained.

stylet, *m.* stylet, small style, probe.

styleux, -euse, stylose, having styles of remarkable length or persistence.

styliforme, style-shaped.

stylisque, *f.* columella.

stylopode, *m.* stylopod.

stylospore, *f.* stylospore, spore borne on a filament.

styphinique, styphnic.

stypticine, *f.* stypticin.

stypticité, stypticity, astringency.

styptique, *m.* styptic.

styptique, styptic, astringent.

styracées, *f.pl.* Styraceae.

styrax, *m.* storax (Styrax); — **officinale,** *Styrax officinalis.*

Styrie, *f.* Styria.

styrolène, *m.* styrene, styrolene.

styrolénique, of styrene, styrolene.

su, *m.* knowledge.

su (savoir), known.

suage, *m.* sweating, oozing, swage, paying, tallowing, paying stuff, tallow.

suager, to swage, pay, tallow.

suaire, *m.* shroud, pall.

suant, sweating, sweaty, welding.

suave, sweet, pleasant, suave, agreeable, soft.

suavement, sweetly, pleasantly, agreeably, softly.

suavité, *f.* sweetness, pleasantness, suavity, softness.

subacaule, subcaulous, with the stem hardly apparent.

subapiculaire, subapicularous, (stem) prolonged beyond an inflorescence without branch or leaf.

subauriculé, with small appendages like auricles.

subbiflore, with uniflorous or biflorous peduncles.

subbilobé, apparently divided into two lobes.

subbipennatifide, with nearly pinnatifid leaves.

subcaulescent, subcaulescent, with a very short stem.

subclaviforme, nearly club-shaped.

subcoalescent, almost closing.

subconoïde, nearly conoid, subconoid.

subconvoluté, subconvolute, partially convolute.

subcortical, -aux, subcortical.

subcutine, *f.* subcutin.

subdécurrent, with almost decurrent leaves.

subdenté, subdentate, imperfectly dentate.

subdéprimé, slightly depressed.

subdichotome, almost regularly dichotomous.

subdigité, almost digitate.

subdiviser, to subdivide; **se —,** to be subdivided.

subdivision, *f.* subdivision.

subérate, *m.* suberate.

subération, *f.* becoming suberized, suberification.

subéreux, -euse, suberose, suberous, corky.

subérification, *f.* suberization.

subérine, *f.* suberin.

subérique, suberic, suberyl.

subérisation, *f.* suberization.

subérone, *f.* suberone, cycloheptanone.

subérosité, *f.* corky character.

subéryle, *m.* suberyl.

subérylique, suberyl, of suberyl.

subfasciculé, nearly like a bundle.

subfoliacé, almost leaflike.

subfrondescent, almost like a frond.

subglabre, almost glabrous.

subhyménium, *m.* hyopthecium.

subintégrifolié, with nearly entire leaves.

subir, to undergo, experience, suffer, sustain; **se —,** to be undergone.

subit, sudden, unexpected.

subitement, suddenly.

subjuguer, to subjugate, master.

sublimable, sublimable.

sublimation, *f.* sublimation.

sublimatoire, *m.* sublimatory.

sublimatoire, sublimatory.

sublime, sublime, lofty, elevated, high.

sublimé, *m.* sublimate; **— corrosif,** corrosive sublimate, mercuric chloride.

sublimé, sublimed, sublimated.

sublimer, to sublimate, sublime; **se —,** to be sublimed, sublime.

sublyré, with leaves almost lyrate.

submergé, submerged, plunged.

submerger, to submerge.

submersible, submersed, submersible.

submersion, *f.* submersion; **— naturelle,** basin irrigation; **— simple,** check system, basin method.

subordination, *f.* subordination.

subordonné, *m.* subordinate.

subordonné, subordinate(d).

subordonner, to subordinate.

subpédiculé, with pedicel hardly visible.

subpétiolé, subpetiolate, under the petioles.

subpétioliforme, almost like a petiole.

subpolaire, subpolar.

subramifié, divided into very short branches.

subrégulariflore, with almost regular flowers.

subrégulariforme, almost regular.

subséquemment, subsequently.

subséquent, subsequent.

subsessile, subsessile, almost devoid of a stalk.

subside, *m.* subsidy.

subsidiaire, subsidiary, auxiliary.

subsidiairement, subsidiarily, in addition.

subsidier, to subsidize.

subsistance, *f.* subsistence, maintenance; *pl.* provisions.

subsister, to subsist, continue to exist, remain, live.

subspathacé, almost spathaceous.

substance, *f.* substance, matter; **— coriaire,** tannin material; **— isolante,** insulating substance, insulator; **— mère,** mother substance; **— rebelle,** refractory arc.

substantiel, -elle, substantial.

substantif, *m.* substantive, noun.

substituant, *m.* substitute.

substituant, substituting, substituent.

substituer, to substitute; **se — à,** to supersede.

substitut, *m.* surrogate.

substitution, *f.* substitution.

substrat, substratum, *m.* substratum, substrate.

subterrané, subterranean.

subtil, subtle, fine (dust); subtile, thin, keen.

subtilement, subtly.

subtilifolié, with linear leaves and leaflets.

subtilisation, *f.* volatilization, subtilization.

subtiliser, to volatilize, subtilize.

subtilité, *f.* subtlety, fineness, subtleness, subtility.

subtrilobé, almost divided into three lobes.

subtuberculé, with slightly pronounced tubercles.

subulé, subulate.

subulifolié, with subulate leaves.

subvelouté, slightly velvety.

subvenir, to provide (for), help, assist.

subventionner, to subsidize, subventionize.

suc, m. juice, sap, substance, pith, latex; — concret, dry juice; — de citron, lemon juice; — de levure, yeast juice; — de réglisse, licorice extract; — épaissi, viscous juice — gastrique, gastric juice; — laiteux, latex; — pancréatique, pancreatic juice; — végétal, vegetable juice, plant juice, sap.

succédané, m. substitute, succedaneum.

succédané, substitute, succedaneous.

succéder, to succeed, prosper, follow; se —, to succeed one another, follow each other.

succenturié, accessory.

succès, m. success, issue, prosperity, event, result.

successeur, m. successor.

successif, -ive, successive, successional.

succession, f. succession; — des coupes, felling series, succession of fellings.

successivement, successively, in succession.

succin, m. succin, yellow amber (resin).

succinate, m. succinate; — de potasse, potassium succinate.

succinct, succinct, brief.

succinctement, succinctly, briefly, concisely.

succiné, resembling amber.

succinique, succinic.

succion, f. suction, sucking.

succomber, to succumb, fail, fall, die.

succotrin, m. see SOCOTRIN.

succulence, f. succulence, juiciness.

succulent, succulent, juicy, nutritious, fleshy.

succursale, f. branch, branch establishment, suboffice.

sucement, m. sucking, suction.

sucer, to suck, suck in, absorb.

sucette, f. apparatus to drain off sirup, suction box.

suceur, m. sucker, nozzle.

suceur, -euse, sucking.

sucoir, m. sucker, cupping glass.

sucotrin, m. see SOCOTRIN.

sucrage, m. sugaring.

sucrase, f. sucrase, invertase, invertin.

sucratage, m. extraction of sugar.

sucrate, m. sucrate, saccharate.

sucratier, m. saccharate maker.

sucre, m. sugar; — blanc, white sugar; — brut, raw (coarse) sugar, unrefined sugar; — candi, sugar candy, rock candy; — de betterave, beet sugar, beetroot sugar; — de canne, cane sugar; — de (la) fécule, starch sugar; — de fruit, fruit sugar, glucose, levulose; — de gélatine, glycocol(l), gelatine sugar; — de lait, milk sugar, lactose, lactine.

sucre, — de raisin, grape sugar; — de réglisse, licorice sugar; — d'orge, barley sugar; — en bâtons, stick candy; — en pain(s), loaf sugar; — interverti, invert(ed) sugar; — pilée, crushed sugar, powdered sugar; — raffiné, refined sugar; — râpé, pounded sugar.

sucré, sugared, containing sugar, sugary, sweet.

sucrer, to sweeten, sugar.

sucrerie, f. sugar refinery, sugar mill; pl. sweets, confectionery, sweet things.

sucreur, m. sweetener (of wine).

sucrier, m. sugar maker, sugar bowl.

sucrier, -ère, sugar, of sugar.

sucro-carbonate, m. sucrocarbonate.

sud, m. south.

sud, south, southern.

sud-américain, South American.

sudation, f. sudation.

sudatoire, diaphoretic.

sud-est, m. southeast.

sud-est, southeast.

sudoral, -aux, sudoral, of sweat.

sudorifique, m. sudorific.

sudorifique, sudorific.

sudoripare, sudoriferous, sudoriparous.

sud-ouest, m. southwest.

sud-ouest, southwest.

Suède, f. Sweden.

Suédois, m. Swede.

suédois, Swedish.

suée, f. sweat, sweating.

suer, to sweat, ooze, perspire.

suerie, f. sweating, drying barn (for tobacco).

suette, f. — miliaire, rat-bite disease, miliary fever.

sueur, *f.* sweat, perspiration.

suffire, to suffice, be sufficient, be enough; **se —,** to be self-supporting; **il suffit à (sa tache),** he is equal to (his task); **il suffit de,** it requires only.

suffisamment, sufficiently, enough.

suffisance, *f.* sufficiency, adequacy, self-conceit; **à —, en —,** plenty, sufficiently, enough.

suffisant, sufficient, sufficing, enough, conceited.

suffocant, suffocating.

suffocation, *f.* suffocation, choking.

suffoquer, to suffocate.

suffrutescent, shrubby, woody at base.

suggérer, to suggest.

suggestif, -ive, suggestive.

suggestion, *f.* suggestion.

suie, *f.* soot; **— cristallisée,** shining soot; **— de bois,** wood soot.

suif, *m.* tallow, suet, candle grease, solid grease; **— de cerf,** deer tallow; **— d'os,** bone fat; **— fondu,** rendered tallow; **— minéral,** mineral tallow, hatchettite, mountain tallow; **mettre en —,** to tallow, grease.

suiffer, suifer, to tallow.

suiffeux, -euse, tallowy, greasy.

suiffier, *m.* tallow maker.

suint, *m.* suint, grease (of wool); **— de laine,** wool grease, suint.

suintement, *m.* oozing, seeping.

suinter, to ooze (out), seep, exude, sweat, run, leak.

suis (être, suivre), (I) am, (I) am following.

Suisse, *f.* Switzerland.

suisse, Swiss; **petit —,** small cream cheese.

suit (suivre), (he) is following.

suite, *f.* succession, series, continuation, sequence, sequel, consequence, result, issue, order, collection, suite, what follows; **à la — de,** in consequence of, following; **dans la —, par la —,** eventually, later on, afterward; **de —,** in succession, in line, at once; **et ainsi de —,** and so forth, et cetera; **par —,** consequently, therefore; **par — de,** as a result of, because of; **par la —,** later on, by what follows; **tout de —,** immediately, at once.

suivant, *m.* follower, attendant.

suivant, according to, following, followed, connected, along. next; **— que,** according as, as.

suivi, followed, continued, connected, coherent, persistent, popular.

suivre, to follow, go on, pursue, keep up with, result from, frequent, attend, hear, watch, succeed, come, happen; **se —,** to follow one another, become connected, be continuous; **faire —,** to cause to follow, please forward (on mail), run on.

sujet, *m.* subject, object, topic, reason, cause, individual matter, person, fellow; **— à la dîme,** tithable; **— d'avenir,** thriving tree; **à ce —,** in this connection; **au — de,** on the subject of, about.

sujet, subject, liable.

sujétion, *f.* subjection, obligation, constraint, inconvenience.

sulfacide, *m.* sulphacid, thio acid.

sulfanilique, sulphanilic.

sulfarsénique, thioarsenic, sulpharsenic.

sulfatage, *m.,* **sulfatation,** *f.* sulphating, treating with copper sulphate.

sulfate, *m.* sulphate, sodium sulphate; **— d'alumine,** aluminum sulphate; **— de baryte,** barium sulphate; **— de chaux,** calcium sulphate; **— de magnésie,** magnesium sulphate, bitter salt, Epsom salt; **— de potasse,** potassium sulphate; **— de soude,** sodium sulphate, Glauber's salt; **— mercurique,** mercuric sulphate.

sulfaté, sulphated.

sulfater, to sulphate, treat or dress (vines).

sulfatiser, to sulphatize, treat with copper sulphate.

sulférer, to sulphur.

sulfhydrate, *m.* sulfhydrate, sulphydrate, hydrosulphide.

sulfhydrique, sulphhydric, hydrosulphuric, sulphydric; **acide —,** hydrogen sulphide, sulphureted hydrogen.

sulfine, *f.* sulphine.

sulfiné, sulphinated.

sulfinique, sulphinic.

sulfitage, *m.* treatment of wines with a sulphite.

sulfite, *m.* sulphite; **— de potasse,** potassium sulphite.

sulfiter, to treat (vines) with a sulphite.

sulfitique, sulphite.

sulfo, sulpho.

sulfoantimoniate, *m.* thioantimonate, sulphantimonate.

sulfoarsénite, *m.* thioarsenite, sulpharsenite.

sulfo-bactéries, *f.pl.* sulphur bacteria.

sulfobase, *f.* sulphur base.

sulfobenzide, *m.* sulphobenzide.

sulfocarbimide, *f.* sulphocarbimide, isothiocyanic acid.

sulfocarbonate, *m.* thiocarbonate, sulphocarbonate.

sulfocarboné, containing thiocarbonic acid or thiocarbonates.

sulfocarbure, *m.* carbonic bisulphide or disulphide, carbosulphide, sulphocarbide.

sulfoconjugué, *m.* sulpho derivative.

sulfocyanate, *m.* thiocyanate, sulphocyanate.

sulfocyanique, thiocyanic, sulphocyanic.

sulfocyanure, *m.* sulphocyanide, thiocyanate, sulphocyanate.

sulfoléique, sulpholeic.

sulfonal, *m.* sulphonal.

sulfoné, sulphonated, sulphonic, sulpho.

sulfosel, *m.* sulpho salt.

sulfostéatite, *m.* sulphosteatite (a fungicide).

sulfotellurure, *m.* sulphotelluride.

sulfo-urée, *f.* thiourea, sulphourea.

sulfurabilité, *f.* capability of being sulphurized.

sulfurage, *m.* treatment with carbon disulphide.

sulfuration, *f.* sulphur(iz)ation.

sulfure, *m.* sulphide, sulphuret; — **d'ammonium,** ammonium sulphide; — **de carbone,** carbon disulphide, sulphocarbide; — **de sodium,** sodium sulphide; — **méthylique,** methylsulphide —**s alcalins,** sulphur alkalies, alkaline sulphides.

sulfuré, combined with sulphur, sulphur containing, sulphurized, sulphurated, sulphureted.

sulfurer, to combine or treat with sulphur, sulphurize, sulphurate.

sulfureur, sulphurizing.

sulfureux, -euse, sulphurous.

sulfurifère, containing sulphur.

sulfurique, sulphuric.

sulfurisation, *f.* sulphur(iz)ation.

sulfuriser, to sulphurize, treat with sulphuric acid.

sulfuryle, *m.* sulphuryl.

sumac, *m.* sumac; — **à perruque,** — **fustet,** Venetian sumac, wig tree, (*Cotinus. coggygria*); — **des corroyeurs,** tanner's sumac (*Rhus coriaria*); — **vénéneux,** poison ivy (*Rhus toxicodendron*); — **du Japon,** Japanese varnish tree (*Rhus verniciflus*).

sûmes (savoir), (we) knew.

super, to get stopped up, suck up, be plugged up.

superaxillaire, superaxillary, growing above an axil.

superbe, *f.* pride, haughtiness.

superbe, superb, magnificent, splendid, lofty, proud.

superbement, superbly, magnificently, proudly.

supère, superior, growing above.

superficiaire, superficial.

superficie, *f.* superficies, surface, area; — **de section,** sectional surface, cut surface.

superficiel, -elle, superficial, on the surface, shallow.

superfin, *m.* superfine quality.

superfin, superfine.

superflu, *m.* superfluity, excess.

superflu, superfluous, redundant.

superfluité, *f.* superfluity.

supérieur, *m.* superior.

supérieur, upper, higher, superior, greater (than), high.

supérieurement, in a superior manner.

supériorité, *f.* superiority.

superlatif, *m.* superlative.

superlatif, -ive, superlative.

superovarié, with superior ovary.

superpalite, *f.* trichloromethyl chloroformate.

superphosphate, *m.* superphosphate; — **de chaux,** superphosphate of lime, calcium superphosphate; — **d'os,** bone superphosphate.

superposable, superposable, coincident.

superposé, superposed, superincumbent.

superposer, to superpose; **se —,** to overlap.

superpositif, -ive, superposed, vertically over some other part.

superposition, *f.* superposition, superposing.

supersaturer, to supersaturate.

superstitieux, -euse, superstitious.

supervoluté, supervolute, rolled over.

supplanter, to supplant.

suppléance, *f.* substitution.

suppléant, *m.* substitute, assistant.

suppléant, substitute.

suppléer, to supply, make good, make up, take the place of, substitute; — **à,** to make up for.

supplément, m. supplement, addition.

supplémentaire, supplementary, extra, further.

supplice, m. punishment, torture, torment.

supplier, to beseech, supplicate, entreat.

support, m. support, stand, rest, prop, stay, base, block, holder, bracket, bearer, help, host (of a parasite); — à **burettes,** stand for burettes; — à **entonnoir(s),** funnel holder; — à **ratelier,** test-tube rack; — de **secteur,** segment base; — en **fil,** wire support stand; — en **fil de fer,** wire cage, wire stand; — **isolant,** insulator; — **porte-tubes,** tube rack; — **pour éprouvettes,** test-tube stand; — **pour tubes à essais,** test-tube rack.

supportant l'ombrage, shade-enduring.

supporter, to support, stand, bear, endure, hold, sustain.

supposé, supposed, alleged, assumed, false, fictitious.

supposer, to suppose, imply, presuppose, conjecture; **se —,** to suppose, be supposed, fancy oneself.

supposition, f. supposition, hypothesis, forgery.

suppositoire, m. suppository.

suppression, f. suppression, cutting out, removal, abolition, doing away with.

supprimer, to suppress, cancel, remove, discontinue, relieve (pressure), do away with, take away, abolish.

suppurant, suppurating, running.

suppuratif, m. suppurative.

suppuratif, -ive, suppurative.

suppuration, f. suppuration.

suppurer, to suppurate, run.

supputation, f. computation, calculation.

supputer, to compute, calculate, reckon, suppute.

supra-axillaire, supra-axillary.

suprarénine, f. suprarenine, adrenaline.

suprématie, f. supremacy.

suprême, supreme, highest, last; **au — degré,** in the highest degree, eminently.

sur, on, upon, over, to, by, above, out of, among, according to, about, concerning, towards.

sur, sour, tart.

sûr, sure, certain, safe, secure; **peu —,** insecure; **pour —,** certainly, surely; **pour le plus —,** to be on the safe side.

surabondamment, superabundantly.

surabondance, f. superabundance.

surabondant, superabundant, luxuriant, rank.

surabonder, to superabound, be very abundant.

suractivité, f. abnormal activity.

suraddition, f. superaddition.

suraffinage, m. overrefining.

suraffiner, to overrefine.

surajouté, secondary (infection).

surajoutement, m. superaddition.

surajouter, to superadd, add also.

suralcooliser, to add excess of alcohol to.

suralimentation, f. overfeeding, stuffing.

sur-andouiller, m. bay or bez antler.

suranné, expired, antiquated, old-fashioned, superannuated, lapsed.

surate, m. Indian cotton.

surbaissé, depressed, surbased, low, flattened, elliptical.

surboucher, to cap (bottle).

surcalciner, to overcalcine, overburn.

surcapitalisation, f. overcapitalization.

surcharge, f. overload, overcharge, additional load.

surcharger, to overload, overburden, overcharge.

surchauffage, m., **surchauffe,** f. overheating, superheating.

surchauffé, superheated.

surchauffer, to superheat, overheat; **se —,** to be overheated.

surchauffeur, m. superheater.

surchauffure, f. overheating, burnt spot.

surchoix, m. first choice, finest quality.

surcoloration, f. staining with intense stain.

surcrénelé, doubly crenate or creneled.

surcroissance, f. overgrowth.

surcroît, m. addition, increase, surplus; **par —,** in addition, besides.

surcuire, to overburn, overboil.

surcuit, overburned, overdone.

surculeux, -euse, surculose, producing suckers.

surculigère, surculigerous, bearing suckers.

surdité, f. deafness.

surdorer, to double gild.

sureau, m. elder (bush) tree (Sambucus); — **noir,** black-fruited elder (*Sambucus* *nigra);* — **rouge,** — **à grappes,** red-berried elder, scarlet elder (*Sambucus racemosa*).

surégaliser, to screen coarsely.

surégalisoir, *m.* coarse sieve.

surélévation, *f.* raising, rise, excessive increase.

surélevé, elevated.

surélèvement, *m.* raising, increase.

surélever, to raise high, elevate, force up, increase, be raised higher, built up

surelle, *f.* wood sorrel.

sûrement, confidently, surely, certainly, assuredly, forsooth.

surenchérir, to outbid.

surenchérissement, *m.* further rise in price.

surent (savoir), (they) knew.

surépaisseur, *m.* increased thickness.

surestimation, *f.* overestimate, overvaluation.

surestimer, to overestimate, overvalue.

suret, -ette, sourish, somewhat sour.

sûreté, *f.* safety, security, surety, sureness, soundness, certainty, reliability, guarantee, safekeeping; **de** —, safety.

surette, *f.* wood sorrel.

surexcitable, easily excited.

surexciter, to overexcite, overstimulate.

surface, *f.* surface, area, superficies; — **boisée,** wooded area; — **d'aile,** wing surface; — **d'appui,** bearing surface; — **de cassure,** surface of fracture; — **de chauffe,** heating surface; — **de fente,** cleavage surface or face; — **dénudée,** denuded area, cleared area.

surface, — **de section,** sectional area, cut surface; — **d'exploitation,** felling area; — **inégale,** uneven surface; — **phréatique,** water table, ground-water surface; — **refroidissante,** cooling surface; — **terrière,** basal area; — **unie,** even (smooth) surface.

surfaire, to overcharge, overvalue, overrate, surfeit.

surfin, superfine.

surfleurir, to flower again.

surfondre, to superfuse.

surfondu, superfused, supercooled, undercooled.

surforce, *f.* excess of alcohol.

surfusibilité, *f.* superfusibility, extreme fusibility.

surfusible, superfusible, extremely fusible.

surfusion, *f.* superfusion, supercooling, undercooling.

surge, *f.* raw wool.

surgeon, *m.* sucker offshot, scion.

surgir, to rise, arise, appear, spring up, crop up.

surglacé, supercalendered.

surgreffage, *m.* topgrafting, double grafting.

surhaussement, *m.* raising, elevation forcing up.

surhausser, to heighten, raise, elevate.

surhumain, superhuman.

surimposer, to increase the tax on, overtax.

surin, *m.* young apple-tree stock.

surintendance, *f.* superintendence, superintendent's office.

surintendant, *m.* superintendent, overseer.

surir, sûrir, to sour (wine), turn sour, turn.

surjacent, overlying.

sur-le-champ, immediately, at once.

surlendemain, *m.* day after tomorrow, second day after.

surmenage, *m.* great activity, overwork, overworked state.

surmener, to overwork.

surmesure, *f.* excess height, top cut off bole, excess measure.

surmeule, *f.* runner stone, runner.

surmoi, *m.* superego.

surmontable, surmountable.

surmonter, to surmount, top, rise above, surpass, overcome.

surmoulage, *m.* casting or molding from plate, retreading (of tires).

surmoule, *f.* mold made from a casting.

surmouler, to retread (tire), mold from existing plate.

surmoût, *m.* new must, unfermented grape juice.

surmulot, *m.* brown rat.

surnageant, supernatant, floating on the surface.

surnager, to float on (swim on) the surface, survive.

surnaturel, -elle, supernatural.

surnaturellement, supernaturally.

surnom, *m.* surname.

surnuméraire, *m.* supernumerary.

surnuméraire, supernumerary.

suroxydation, *f.* peroxidation.

suroxyde, *m.* peroxide.

suroxydé, peroxidized, peroxide of.

suroxyder, to peroxidize, overoxidize.

suroxygénation, *f.* peroxidation.

suroxygéner, to superoxygenate, peroxidize, superoxidize.

surpasser, to surpass, exceed, excel, pass beyond, be higher than.

surpayer, to overpay.

surplomber, to overhang, be out of plumb.

surplus, *m.* surplus, excess, rest, remainder; **au —,** besides, however; **de —,** over and above; **en —,** in excess.

surpoids, *m.* overweight, excess weight.

surprenant, surprising, astonishing, amazing, stupendous.

surprendre, to surprise, astonish, catch, overhear, detect, abuse.

surpression, *f.* overpressure, excess pressure.

surpris, surprised.

surprise, *f.* surprise, amazement.

surproduction, *f.* overproduction.

surrection, *f.* surrection, rising up.

surrénal, -aux, suprarenal.

sursaturation, *f.* supersaturation.

sursaturer, to supersaturate.

sursaut (en), with a start.

sursel, *m.* supersalt, acid salt.

sursélection, *f.* overselection.

sursemer, to oversow.

surseoir, to suspend, delay, stay, defer.

sursis, *m.* delay, stay, suspension, reprieve.

sursis, suspended.

surtaxe, *f.* surtax, excessive tax.

surtaxer, to surtax, overtax.

surtension, *f.* excess pressure, overvoltage.

surtonte, *f.* clipping (wool).

surtout, *m.* outer garment, wrap.

surtout, particularly, especially, principally, above all, mainly.

survécu (survivre), survived.

surveillance, *f.* supervision, watching, inspection, oversight, superintendence, attention.

surveillant, *m.* supervisor, superintendent, inspector, overseer, keeper, watchman.

surveillant, supervising.

surveiller, to supervise, superintend, inspect, survey, oversee, watch.

survenir, to happen, supervene, come unexpectedly.

survider, to reduce the contents of, pour off some of contents.

survie, *f.* survival, outliving.

survivant, *m.* survivor.

survivant, surviving.

survivre, to survive, outlive.

survolter, to boost, step up, raise the voltage of.

sus, above, come on, now then; **— à,** on, upon; **en —,** besides, extra, more; **en — de,** over and above, in addition to.

susceptibilité, *f.* susceptibility.

susceptible, susceptible, capable of, liable, fit.

susciter, to raise (up), stir up, create, give rise to, instigate.

suscription, *f.* address (on letters).

susdénommé, susdit, above-mentioned, aforesaid.

suspect, suspicious, doubtful, suspected, questionable.

suspecter, to suspect.

suspendre, to suspend, hang, stop; **se —,** to be suspended.

suspendu, suspension, suspended, on springs.

suspens, *m.* suspense, uncertainty, doubt; **en, —,** in suspense.

suspenseur, *m.* suspensor.

suspension, *f.* suspension, hanging support, interruption, suspensor; **— bifilaire,** bifilar suspension.

suspirieux, -euse, sighing.

susrelaté, above-mentioned, above-related.

sussent (savoir), (they) might know, knew.

sustentation, *f.* sustenance, sustentation, support, supporting.

sustenter, to sustain, support.

sut (savoir), (he) knew.

suturaire, presenting a suture.

suture, *f.* suture.

svelte, slender, slim, graceful, elegant.

sveltesse, *f.* slenderness.

sycomore, *m.* sycamore (fig) (*Ficus sycomorus*), sycamore (maple) (*Acer pseudoplatanus*).

syconium, *m.* syconium, a multiple hollow fruit.

syénite, *f.* syenite.

syénitique, syenitic.
syllabe, *f.* syllable.
sylvain, *m.* sylvan.
sylvestre, forest, wood, woody, sylvestrian.
sylvestrène, *m.* sylvestrene.
sylviculteur, *m.* silviculturist.
sylvicultural, -aux, sylvicultural.
sylviculture, *f.* silviculture.
sylvine, *f.* sylvite.
symbionte, *f.* symbiote, symbiont.
symbiose, *f.* symbiosis.
symbiote, symbiotic, relating to symbiosis.
symbiotique, symbiotic.
symbiotrophe, symbiotrophic, deriving nourishment by symbiotic relationship.
symbole, *m.* symbol, emblem, sign.
symbolique, symbolic(al), typical.
symbolisation, *f.* symbolization, symbolizing.
symboliser, to symbolize.
symbolisme, *m.* symbolism.
symétrie, *f.* symmetry.
symétrique, symmetric(al).
symétriquement, symmetrically.
sympathie, *f.* sympathy, congeniality.
sympathique, sympathetic(al).
sympathiquement, sympathetically.
sympérianthé, (plant) with corolla and calyx united into one staminiferous tube.
symphoricarpe, symphoricarpous.
symphyostémone, symphyostemonous, having the stamens united.
symphytanthéré, synantherous.
symphytothèle, symphytothelous, symphytogynous, (calyx and pistil) more or less adherent.
symplaste, *m.* symplast, assemblage of energids.
sympode, *m.* sympodium.
sympodial, -aux, indefinite.
symptôme, *m.* symptom.
synandrodien, *m.* synandrium, cohesion of anthers of each male flower in certain Aroideae.
synanthéré, synantherous, syngenesious.
synanthrose, *f.* levulin.
synapsis, *f.* synapsis.
synaptase, *f.* emulsin.
synaptospore, *f.* stylospore, spore borne on a filament.

synarche, *m.* synarch, fusion of two sexual cells.
synarmophyte, *m.* synarmophytus, plant with gynandrous flowers.
synbacteries, *f.pl.* myxobacteria.
syncarpe, *m.* syncarp (a fleshy multiple fruit).
syncarpé, syncarpous, composed of two or more united carpels.
syncarpie, *f.* syncarpy, accidental adhesion of several fruits.
synchrone, synchronique, synchronous, synchronic(al).
synchronisme, *m.* synchronism, isochronism.
synchroniser, to synchronize.
syncotylédoné, syncotyledonous, with coalesced cotyledons.
syncotylie, *f.* syncotyly, state of cohesion of cotyledons by one margin only.
syncyte, *m.* syncyte, structure derived from absorption of cell walls.
syndactile, web-fingered.
syndic, *m.* syndic, receiver, magistrate, assignee.
syndical, *m.* member of a syndicate.
syndical, -aux, syndical.
syndicat, *m.* syndicate, trusteeship, society.
syndiquer, to syndicate.
synécologie, *f.* synecology, study of plant communities.
synématique, forming the synema.
synème, *f.* synema.
synergide, *f.* embryonic sac.
syngamète, *f.* zygote.
syngamose, *f.* gapes (in poultry).
syngénèse, synantherous.
syngénésie, *f.* syngenesia, syngenesis.
synkaryon, *f.* syncaryon, nucleus formed by fusion of two nuclei.
synkaryophyte, *f.* gametophyte.
synochorium, *m.* carcerule, dry, indehiscent, many-celled, superior fruit.
synonyme, *m.* synonym.
synonyme, synonymous.
synopsis, *f.* synopsis.
synorrhize, synorhizous, having a radicle whose point is united to the albumen.
synovial, -aux, synovial (gland).
synovie, *f.* synovia.

synsépale, gamosepalous, (sepals) coalescent.

synspermie, *f.* synspermy, union of several seeds.

synstigmatique, synstigmaticous, (pollen mass) furnished with a retinaculum.

synstylé, with styles fused.

syntagma, *f.pl.* syntagma, bodies built up of tagmata.

synthèse, *f.* synthesis, building up.

synthétique, synthetic(al).

synthétiquement, synthetically.

synthétiser, to synthesize.

syntonine, *f.* syntonin.

synzygie, *f.* synzygia, point of contact of opposite cotyledons.

syphiline, extract of syphilitic liver.

syphilis, *f.* syphilis.

syphilitique, syphilitic.

syphon, *m.* see SIPHON.

Syrien, *m.* Syrian.

syrien, -enne, Syrian.

syringa, *m.* lilac (Syringa).

systellophyte, *m.* systellophytum, persistent calyx forming part of the fruit.

systématique, systematic(al).

systématiquement, systematically.

systématisation, *f.* systematizing.

systématiser, to systematize.

système, *m.* system, project, plan, method, formation; — crétacé, Cretaceous system; — des racines, root system; — inducteur, field; — nerveux, nervous system.

systole, *f.* systole.

systolique, systolic.

systrophe, *f.* systrophe, strong light causing chlorophyll grains to congregate into a few masses.

systylé, with styles fused in a kind of colony.

T

t., *abbr.* (tonne) ton; (tome) volume.

t'., see TE, TOI.

-t-, (used euphonically), a-t-il? has he? va-t-il? is he going?

ta, *f.* thy, your.

tabac, *m.* tobacco (plant), tobacco color; — à chiquer, chewing tobacco; — à fumer, smoking tobacco; — à mâcher, chewing tobacco; — écôté, unribbed tobacco; — en andouilles, roll tobacco; — en carotte, twist.

tabacal, -aux, tobacco, of tobacco.

tabanidés, *m.pl.* biting flies of the genus Tabanus.

tabanifère, (flower) resembling a gadfly.

tabaschir, tabashir, *m.* tabasheer, bamboo salt.

tabellaire, in the form of tables.

tabide hectic, consumptive, wasted.

tabiser, to tabby, water (silk).

table, *f.* table, plate, slab, tablet, face, register, index, bed, section; — à écrire, writing table, writing desk; — à égruger, mixing table, mealing table; — à ouvrage, worktable, workbench; — à scier, saw table; — de cubage, volumetric table; — de doreur, gilding bench; — de foyer, hearth plate.

table, — de laboratoire, laboratory table; — de manipulation, switchboard, operating table; — de manoeuvre, control table; — de production, yield table, experimental table; — des matières, table of contents; — de tiroir, slide face; — de travail, work-table; — rasc, clean sweep; — salante, basin or table in a sea-salt work.

tableau, *m.* board, table, list, catalogue, chart, portrait, painting, scenery, tableland, plateau, switchboard, section; — à boutons, push switchboard; — d'avis, notice board; — de distribution, switchboard; — de service, duty list, timetable; — des marchandises, list of goods; — en ardoise, slate board; — étincelant, sparking board, sparking table; — graphique, diagram; — noir, blackboard; — pour écoles, blackboard; — réclame, show card, poster, placard.

tabletier, *m.* maker or dealer in fancy articles (of ebony, etc.).

tablette, *f.* shelf, tablet, lozenge, troche, pill, pastil, plate, slab, cake; *pl.* writing tablet, sill, bench, board, breast molding, bracket, table board; — à clefs, keyboard; — de cachou, troche of gambir; — de ratanhia, rhatany lozenge; — de santonine, worm tablet; — de tannin, troche of tannic acid, tannin lozenge.

tabletterie, *f.* fancy goods industry.

tablier, *m.* apron, floor, platform, stand (for hives), flooring, roadway, superstructure.

tabouret, *m.* stool, footstool, shepherd's purse.

tabulaire, tabular.

tacamaque, *m.* tacamahac.

tache, *f.* spot, stain, blemish, speck; — **d'étain,** tin spot; — **du soleil,** — **solaire,** sun spot; — **médullaire,** medullary spot, pith fleck; —**s crustacées de l'érable,** maple tar spot; —**s noires,** black spot, leaf blotch; —**s pailletées,** aucuba mosaic.

tâche, *f.* task, job, piece wages.

tachéométrie, *f.* tachymetric survey.

tacher, to stain, spot, tarnish, blemish.

tâcher, to try, endeavor, attempt, strive.

tacheté, spotted, speckled, brindle, brinded.

tacheter, to spot, speckle, blotch.

tacheture, *f.* speck, speckle, spot.

tachinaire, *f.* tachinid fly.

tachymètre, tachomètre, *m.* tachometer, tachymeter, speed gauge.

taciturne, taciturn, silent.

tact, *m.* touch, feeling, sense of touch, tact.

tacticien, *m.* tactician.

tactique, *f.* tactics.

tadorne, *m.* sheldrake.

taffetas, *m.* taffeta.

tafia, *m.* tafia (rum).

tagma, *f.* aggregation of molecules.

taie, *f.* film, leucoma.

taillade, *f.* cut, gash, slash.

taillanderie, *f.* edge-tool industry, edge tools.

taillant, *m.* (cutting) edge.

taillant, cutting.

taille, *f.* cutting, pruning, trimming, dressing, cut, height, figure, shape, size, tally, coppice, copse, statue, waist, stature.

taillé, cut, carved.

taille-douce, *f.* copperplate.

taille-légumes, *m.pl.* vegetable slicer.

tailler, to cut, trim, prune, form, hew, shape, lop, dress, carve, engrave.

taille-racines, *m.pl.* vegetable cutter.

taillerie, *f.* gem cutting, lapidary work.

tailleur, *m.* cutter, tailor.

tailleuse, *f.* cutting machine, tailoress.

taillis, *m.* coppice, copse, thicket; — **à écorce,** oak-bark coppice; — **d'osier,** willow coppice; — **menu,** short-rotation coppice, bushwood; — **sarté,** oak coppice with field crops; — **simple,** coppice.

tain, *m.* silvering, tin amalgam (for mirrors), tin foil, tin bath.

taire, to be silent on, say nothing, not tell, suppress, silence; **se** —, to be silent, become silent, be quiet, be passed over, not be mentioned.

talc, *m.* talc, talcum.

talcaire, talcose, talcous.

talcique, talc, of talc.

talent, *m.* talent, capacity, ability.

tallage, *m.* tillering, throwing out suckers.

talle, *f.* sucker, teller.

tallipot, *m.* talipot.

taloche, *f.* plasterer's float, thump, thwack.

talon, *m.* heel, stub, clod, flange, collar, nose, talon (of birds).

talose, *f.* talose.

talquer, to talc.

talqueux, -euse, talcose, talcous.

talure, *f.* contusion, bruise.

talus, *m.* slope, declivity, bank, embankment, talus.

talutage, *m.* sloping.

taluter, to slope.

tamarin, *m.* tamarind (tree or fruit) (*Tamarindus indica*), tamarisk (Tamarix).

tamarinier, *m.* tamarind tree (*Tamarindus indica*).

tamaris, tamarisc, tamarix, tamarisque, *m.* tamarisk (Tamarix).

tambour, *m.* drum, tambour, barrel, roller, cylinder; — **classeur,** revolving screen.

tamis, *m.* sieve, sifter, screen, bolter, tray, riddle, wire gauze; — **antipoussière,** dust cap; — **de crin,** hair sieve; — **de soie,** silk sieve; — **de toile métallique,** wire-gauze sieve; — **oscillant,** swing screen, shaking sieve; **passer au** —, to sift.

tamisage, *m.* sifting, screening, bolting.

tamise, *f.* tammy cloth.

Tamise, *f.* Thames.

tamiser, to sift, pass through a sieve, screen, strain, bolt, filter, percolate.

tamiseur, *m.* sifter, strainer, bolter.

tamiseuse, *f.* sifter, straining apparatus.

tampon, *m.* plug, stopper, bung, wad, pad, tampon, buffer, bumper, cushion, tampion, tompion, dowel, peg, dabber, stamp; — **à collé,** glue stamp; — **à ressort,** spring buffer; — **de ouate,** cotton plug; — **de sûreté,** safety plug; — **réfractaire,** fire-clay plug.

tamponnement, *m.* plugging, dabbing.

tamponner, to plug, stop, pack down, tamp, pack, drub, dab, collide with, dust (with a powder pad).

tamtam, *m.* tomtom, Indian drum.

tan, *m.* tanbark, tan, tanning bark, oak bark.

tanaisie, *f.* tansy (Tanacetum).

tanche, *f.* tench.

tandis que, while, as long as, whilst, whereas.

tangence, *f.* tangency.

tangent, tangent, tangential.

tangentiel, -elle, tangential.

tangentiellement, tangentially, tangent.

tanghin, tanghen, *m.* tanghin (tree poison), ordeal bark.

tangibilité, *f.* tangibility.

tangible, tangible.

tangrum, *m.* fertilizer from fish guano.

tanière, *f.* den.

tanification, *f.* conversion into tannin.

tanin, *m.* tannin, tanning material.

tanique, tannic.

taniser, to treat with tannin.

tannage, *m.* tanning, tannage, dressing; — à l'huile, oil tanning; — au chrome, chrome tanning; — aux écorces, — en fosse, bark tanning.

tannase, *f.* tannase.

tannate, *m.* tannate.

tanne, *f.* spot (on leather), blackhead (on face).

tanné, tan colored, tanned.

tannée, *f.* spent tanbark, spent tan.

tanner, to tan.

tannerie, *f.* tannery, tan yard, tan house.

tanneur, *m.* tanner.

tannifère, containing tannin.

tannin, *m.* tannin.

tannique, tannic.

tannoïde, tannoid.

tant, so much, so, so long, so far, so many, to such a degree, presently, soon; — bien que mal, somehow or other, as long as, as far as, rather poorly; — et plus, more than enough; — mieux, so much the better; — pis, so much the worse; — plus, so much the more; — plus que moins, more or less.

tant, — que, as long as, while, so much as; — ... que, both .. and, as well ... as; — s'en faut que, far from it, so far from; — soit peu, ever so little, somewhat; — (il) y a que, the fact remains that, however it may be; en — que, in so far as, as far as; faire — que de, to do as much as; si — est que, if at all, supposing.

tantalate, *m.* tantalate.

tantale, *m.* tantalum.

tantalifère, containing tantalum.

tantalique, tantalic.

tantaliser, to tantalize.

tantalite, *f.* tantalite.

tantième, *m.* percentage, share, given part.

tantième, given, such a.

tantôt, soon, presently, just now, a short time ago, sometimes, now, then; à —, good-by for the present; sur le —, towards evening; tantôt ... tantôt, now ... now, at one time ... at another time, sometimes ... sometimes.

taon, *m.* horsefly, oxfly, cleg, breeze fly, gadfly, whame.

tapage, *m.* noise, racket, tapping.

tape, *f.* tap, rap, plug, stopper, bung, tompion, tampion.

tapé, stopped, dried (fruit).

tape-marteau, *m.* skipjack, spring beetle.

taper, to plug, stop up, tap, hit, slap, rap, knock, dab on (paint), stamp, prime (frame), lay on, spread (varnish).

tapette, *f.* tap, light blow, stopper.

tapioca, *m.* tapioca.

tapir (se), to squat, crouch.

tapirer, to redden.

tapis, *m.* cover, cloth, carpet, rug, belt, tapetum, tapis, blanket, carpeting; — végétal, surface cover, vegetable covering.

tapissé, veiled.

tapisser, to hang with tapestry, drape, cover, coat, carpet, paper.

tapisserie, *f.* tapestry, hangings, wallpaper.

tapissier, *m.* upholsterer.

tapoter, to pat, drive.

tapure, *f.* shrinkage, crack, fissure.

taque, *f.* cast-iron plate.

taqueret, *m.* fore plate of a forge hearth.

taquerie, *f.* fire door (of reverberatory furnace).

taquet, *m.* cleat, step, tappet, catch, stop, block, flange.

tarage, *m.* taring, allowance for tare, calibration (of spring).

taraison, *m.* tile disk for closing a glass furnace.

tarare, *m.* winnower, fanning machine, fanning mill.

taraud, *m.* (screw) tap.

taraudage, *m.* screw cutting, tapping, threading.

tarauder, to tap, cut, screw.

taraxacine, *f.* taraxacin.

taraxacum, *m.* dandelion (*Taraxacum officinale*).

tard, *m.* late hour.

tard, late; **au plus —,** at the latest, at farthest; **sur le —,** late in the evening, late in life; **tôt ou —,** sooner or later.

tarder, to delay, tarry, dally, be long in, be slow; **— à faire quelque chose,** to put off, delay something; **ne — pas de,** not delay to, soon, immediately; **sans plus —,** without further delay; **il me tarde de,** I long to.

tardif, -ive, slow, tardy, behindhand, late, backward.

tardiflore, late flowering.

tardivement, slowly, tardily.

tardiveté, *f.* lateness, backwardness.

tare, *f.* tare, loss, deficiency, depreciation, defect.

taré, tared, damaged, injured, spoiled.

tarer, to tare, damage, spoil, calibrate; **se —,** to deteriorate.

taret naval, *m.* teredo, shipworm, timber beetle.

targette, *f.* slide bolt, fastening.

tari, *m.* palm wine.

tari, dried up, exhausted.

tarier, *m.* whinchat.

tarière, *f.* auger, borer, terebra (form of ovipositor).

tarif, *m.* price list, tariff, rate; **— de cubage,** volumetric table.

tarifer, to fix prices, tariff.

tarin, *m.* siskin.

tarir, to dry up, drain, exhaust, cease, stop, be exhausted.

tarissable, exhaustible.

tarissant, drying up, running dry, near exhaustion.

tarissement, *m.* drying up, exhaustion.

tarmacadam, *m.* tar macadam.

tarmacadamiser, to pave with tar macadam.

taro, *m.* taro, elephant's-ear (*Colocasia esculenta*).

tarse, *m.* tarsus, instep.

tarsien, -enne, tarsal.

tartareux, -euse, tartareous, tartarous.

Tartarie, *f.* Tartary; — **rouge,** tall red rattle, marsh lousewort (*Pedicularis palustris*).

tartarique, tartaric.

tartariser, to tartarize.

tartrage, *m.* treatment (of wine) with tartaric acid or calcium tartrate.

tartrate, *m.* tartrate; — **antimonio-potassique,** antimonyl potassium tartrate, tartar emetic; — **d'antimoine et de potasse,** antimonyl potassium tartrate, tartar emetic; — **de chaux,** tartrate of lime; — **de potasse,** potassium tartrate.

tartre, *m.* tartar, scale; — **brut,** crude tartar, argol; — **de marc,** tartrate of lime; — **stibié,** tartar emetic.

tartreux, -euse, tartarous, tartareous.

tartrier, *m.* tartar maker.

tartrifuge, *m.* boiler compound.

tartrique, tartaric.

tartronique, tartronic.

tartrovinique, tartrovinic, ethyltartaric.

tas, *m.* heap, pile, mass, (hand) anvil, building under construction; — **de bois,** pile of wood, stack; — **de foin,** haycock.

tasse, *f.* cup; — **à café,** coffee cup; — **de café,** cup of coffee.

tasseau, *m.* bracket, cleat, strip, lug, clamp, block, hand anvil, tappet, brick foundation.

tassée, *f.* cupful.

tassement, *m.* packing, cramming, settling, sinking, subsidence of the soil; — **du sol,** subsidence or submergence of the ground.

tasser, to compress, pack, tamp, ram, heap up, pile, grow thick; **se —,** to sink, set, settle, pack, heap up, subside, become compressed.

tâtage, *m.* feeling, touching.

tâte, *f.* sample.

tâter, to feel, touch, try, taste, test, sound; — **le terrain,** to explore the ground, feel one's way.

tâte-vin, *m.* winetaster, wine sampler.

tâtonnement, *m.* groping, trial, attempt.

tâtonner, to grope, feel one's way.

tâtons (à), groping, gropingly.

tatouage, *m.* tattooing.

tatouer, to tattoo.

taupe, *f.* mole, moleskin.

taupe-grillon, *m.* mole cricket.

taupière, *f.* mole trap.

taupin, *m.* click or snapping beetle.

taupinière, *f.* molehill.

taure, *f.* heifer.

taureau, *m.* bull.

taurillon, *m.* young bull, steer.

taurine, *f.* taurine.

taurocholate, *f.* taurocholate.

taurocholique, taurocholic.

taurocolle, *f.* taurocol, taurocolla (glue).

tautomère, *m.* tautomer.

tautomère, tautomeric.

tautomérie, *f.* tautomerism, tautomery.

tautomérique, tautomeric.

taux, *m.* rate, price, proportion, amount, strength; — **d'accroissement,** rate of growth, of increment; — **de placement,** rate of interest; — **de rendement,** yield per cent; — **d'intérêt,** rate of interest; — **pour cent parties,** parts per hundred, percentage (of).

taveler, to spot, speckle.

tavelure, *f.* spots, speckles.

taxation, *f.* taxation, fixing of prices, valuation.

taxe, *f.* tax, fixed price, rate, charge.

taxer, to tax, regulate the price of, charge for.

taxicole, growing on the trunk of yews.

taxidermie, *f.* taxidermy.

taxiforme, taxiform, arranged distichously.

taxinées, *f.pl.* Taxineae.

taxinomie, *f.* taxonomy, classification.

taxisme, *m.* taxism, tendency of unicellular organisms to arrange themselves according to lines of force or stimulation.

taxonomie, *f.* taxonomy, classification.

tchèque, Czechic, Czech.

te, thee, to thee, you, to you.

té, *m.* T, T-shaped, T tube, T iron.

technicien, *m.* technician, expert, engineer.

technicité, *f.* technicalness, technicality.

technique, *f.* technique, technic, art, technical skill.

technique, technical.

techniquement, technically.

technologie, *f.* technology; — **forestière,** forest utilization.

technologique, technological.

technologue, *m.* technologist.

teck, *m.* teak (tree or wood).

tectonique, *f.* structure, tectonics.

tectonique, tectonic.

tegmen, *m.* tegmen.

tegminé, tegminatous, (nucellus) invested by a covering.

tégule, *f.* tegula.

tégument, *m.* tegmen, tegument.

tégumentaire, tegumentary.

teignant, dyeing, staining.

teigne, *f.* ringworm, moth, tineid moth; — **de la pomme de terre,** potato-tuber moth; — **des bourgeons de chêne,** oak moth; — **des grains,** wolf moth; — **du carvi,** caraway webworm; — **du mélèze,** larch miner moth.

teignent (teindre), (they) are staining, dyeing.

teignit (teindre), (he) stained.

teillage, *m.* breaking, stripping, swingling (of hemp).

teille, *f.* lime bast, Russian bast.

teiller, to scutch.

teilleur, *m.* scutcher.

teindre, to dye, color, stain, tinge; **se** —, to be colored, dye one's hair.

teint, *m.* dye, color, complexion.

teinte, *f.* tint, shade, tinge, hue, color; — **dure,** ground tint, priming; — **vierge,** unmixed tint.

teinté, toned.

teinter, to tint, color slightly.

teinture, *f.* dyeing, dye, color, hue, solution, dye solution, dyeing shop, tincture; — **composée,** compound tincture; — **de bourre de soie,** waste silk dyeing; — **de cachou,** tincture of catechu; — **de lobélie enflée,** tincture of lobelia; — **de noix de galle,** tincture of nutgall; — **de paille,** straw dyeing; — **de racine d'aconit,** tincture of aconite; — **d'essence d'anis vert,** spirit of anise; — **d'essence de cajeput,** spirit of cajuput.

teinture, — **d'essence de cannelle,** spirit of cinnamon; — **d'essence de genièvre,** spirit of juniper; — **d'essence de muscade,** spirit of nutmeg; — **d'étoffes,** fabric dyeing; — **de tournesol,** tincture of litmus; — **de valériane ammoniacale,** ammoniated tincture of valerian; — **d'iode,** tincture of iodine; — **d'opium camphrée,** camphorated tincture of opium, paregoric; — **éthérée,** ethereal tincture; — **simple,** simple tincture.

teinturerie, *f.* dyeing, dye works.

teinturier, -ère, *m.&f.* dyer.

teinturier, -ère, dyeing, relating to dyeing.

ŧek, *m.* see TECK.

tel, telle, such, like, as, so, so much, many a, a given, such and such; — **que,** such as, just as, so great that; — **quel, telle quelle,** such as it is, just as it is, as it is, any given, such as it was, any whatsoever, helter-skelter; — **qu'il est,** exactly as he is; — **un** (éléphant), like an (elephant); **de (en) telle sorte que,** so that, in such a way that; **tel** . . . **tel,** like . . . like, as . . . so; **tel(le) ou tel(le),** this or that, such and such; **un(e) tel(le),** such as.

télégonie, *f.* telegony.

télégraphe, *m.* telegraph.

télégraphie, *f.* telegraphy; — **sans fil,** wireless telegraphy.

télégraphier, to telegraph.

télégraphique, telegraphic, telegraph.

teléianthe, teleianthous, hermaphrodite (flower).

télémécanique, *f.* telemechanics.

télémètre, *m.* telemeter, range finder.

téléphone, *m.* telephone.

téléphonie, *f.* telephony; — **sans fil,** wireless telephony, radio.

téléphonique, telephonic, by telephone.

télescope, *m.* telescope.

télescopique, telescopic(al).

teleutosore, *m.* teleutosorus, aggregation of teleutospores.

teleutospore, *f.* teleutospore.

telle, see TEL.

tellement, so, in such a manner, so much, so far, of such a kind, to such a degree.

tellurate, *m.* tellurate.

tellure, *m.* tellurium.

telluré, containing tellurium, tellurized, telureted.

tellureux, -euse, tellurous.

tellurien, -enne, tellurian (earth), of the earth.

tellurifère, telluriferous.

tellurique, telluric.

tellurite, *m.* tellurite.

tellurure, *m.* telluride.

telophase, *f.* telophase.

téméraire, rash, daring, foolhardy.

témérairement, rashly

témérité, *f.* temerity, rashness.

témoignage, *m.* testimony, evidence, token, proof, testifying.

témoigner, to testify, give evidence of, show, evince, prove, bear witness to.

témoin, *m.* witness, proof, evidence, control experiment, trial piece, test button, border tree left standing.

tempe, *f.* temple.

tempérament, *m.* temperament, moderation, shade endurance, light requirement, constitution, tolerance of shade; — **robuste,** light-requirement, intolerance of shade; **à** — **délicat,** shade-demanding.

tempérant, temperate, tempering, sedative, moderating, cooling.

température, *f.* temperature; — **basse, low** temperature; — **d'ébullition,** boiling point; — **de congélation,** freezing point; — **de l'ambiant,** air temperature; — **de lavage,** sparging temperature; — **de recuit,** annealing temperature; — **de trempe,** hardening temperature; — **du rouge,** red heat; — **haute,** high temperature; — **superficielle,** surface temperature.

tempéré, *m.* medium temperature.

tempéré, tempered, temperate, mild, moderate.

tempérer, to temper, moderate.

tempête, *f.* tempest, storm, gale, tumult.

tempêtueux, -euse, tempestuous.

templine, *f.* silver fir (*Abies alba*).

temporaire, temporary.

temporairement, temporarily.

temps, *m.* time, term, weather, pause, period, while, season, occasion, period of time, cycle; — **d'arrêt,** pause, delay, halt; — **de pose,** time of exposure; — **du rut,** pairing season, mating season; — **prohibé,** closed season; **à** —, in time; **à deux** —, two-cycle; **à quatre** —, four cycle; **au** — **de,** in the days of; **au** — **jadis,** in former times, formerly; **au** — **où,** when.

temps, dans le —, formerly; **dans le** — **que** (où), at the moment when, while; **de** — **à autre, de** — **en** —, from time to time; **de tout** —, always, at all time; **du** — **de,** in the time of; **en** — **et lieu,** at the proper time and place, in due course; **en même** —, at the same time; **entre** —, meanwhile, in the interval; **sur le** —, at once.

tenable, tenable.

tenace, tenacious, adhesive, tough, cohesive, obstinate, sticky, stiff (soil), persistent (colors), clinging (plant).

tenacement, tenaciously.

ténacité, *f.* tenaciousness, tenacity, toughness, stickiness.

tenaille, *f.* tenailles, *f.pl.* pincers, pincher, tongs, pliers, vise; — à vis, hand vise.

tenant, *m.* champion, supporter, defender, adjacent part, related thing.

tenant, holding, sitting.

tendance, *f.* tendency, inclination.

tendant, stretching, directed toward, tending, bent, inclined, prone.

tendeur, *m.* stretcher, tightener, wire stay, layer, hanger, extension piece.

tendeur, stretching.

tendineux, -euse, tendinous.

tendon, *m.* tendon, sinew, string; — fléchisseur, flexor.

tendre, to stretch out, tend, draw out, tighten, strain, reach out, lay, set, hang, spread (out), hold out, aim, lead to, throw out; se —, to be stretched, become strained; — des pieges, lay snares, set a trap; — de noir, darken.

tendre, *m.* soft place, tender spot.

tendre, soft, tender, delicate, brittle, mild, affectionate, loving, fresh, new (bread), open, impressionable.

tendrement, tenderly.

+endresse, *f.* softness, tenderness, affection, love.

tendreté, *f.* tenderness (of meat).

tendron, *m.* shoot, tendril.

tendu, stretched, strained, tight, taut, tense, bent, strung (of bows), intent.

tendue, *f.* snaring.

ténèbres, *f.pl.* darkness.

ténébreux, -euse, dark, gloomy, obscure, tenebrous, secret.

ténelliflore, with small flowers.

ténesme, *m.* tenesmus.

teneur, *f.* content, tenor, terms, purport, course, contents, percentage, volume; *m.* holder; — en cendres, ash content, percentage of ash; — en eau, moisture content; — pondérale, content by weight.

tenir, to hold, keep, have, contain, take, possess, speak, get, consider, oblige, owe, hold fast, stick, adhere, hold out, keep up, hold good, stick together, be attached (to), be anxious, be adjacent (to), result, depend (on), relate (to), be like, be akin (to), partake (of), resemble, sit, be held, stand, stay, remain, be contained, take after, proceed, be connected with, be

desirous to, subsist, resist, occupy, be desirous of, be bent on, insist on, take care, result from (a).

tenir, — à, to be due to, to be anxious to; — compte de, to pay attention to, take into account, bear in mind; — en équilibre, to balance; — lieu de, to take the place of; être tenu de, to be obliged to; il ne peut plus y —, he can stand it no longer; se —, be held, contained, hold on, be adjacent, hold together, stick, cling, adhere (to), take after, be standing, be, stand (by), stand, stay, remain, keep, refrain, keep oneself, consider oneself.

tenir, cela tient à, that is due to; il tient à ce que (je la fasse), he insists on (my doing it); il tient bon, he holds fast; il tient (peu) compte de, he takes (little) account; il tient de son père, he takes after his father; il (lui) tient tête, he opposes (him); il ne tient qu'à (vous), it depends only on (you); il se tient là, he stays there; il se le tient pour dit, he considers it settled; il s'en tient à son projet, he sticks to his plan.

tenon, *m.* tenon, stud, projection.

tension, *f.* tension, pressure, tenseness, tightness, stretching, voltage; — aux bornes, terminal voltage; — d'allumage, ignition tension; — de distribution, service voltage; — de la vapeur, steam pressure; — de marche à vide, open-circuit voltage; — de régime, normal tension; — de traction, tensile stress; — de vapeur, vapor tension, vapor pressure; — d'origine étrangère, external voltage; — du sang, blood pressure; — superficielle, surface tension; en —, in series; sous —, on charge.

tentant, tempting, alluring.

tentateur, -trice, *m.&f.* tempter.

tentatif, -ive, tentative.

tentation, *f.* temptation.

tentative, *f.* attempt, effort, trial.

tente, *f.* tent, awning, cylindrical plug, (linen) probe.

tenter, to try, attempt, prove, risk, tempt.

tenthrède, *f.* — du frêne, ash-tree sawfly.

tenthrédinidé, *f.* sawfly.

tenture, *f.* hangings, tapestry, curtain.

tenu, held, bound, compact.

ténu, tenuous, thin, slender, minute, attenuated.

tenue, *f.* holding, session, dress, steadiness, firmness, behavior, bearing, deportment, bookkeeping.

ténuité, *f.* tenuity, thinness.

tépale, *f.* tepal, division of the perianth, sepal, or petal.

téphracanthe, with spines or whitish needles.

téphrophylle, with grayish, ash-colored leaves.

téphrosanthe, with dull-colored flowers.

tephrosie, *f.* Tephrosia.

tératogénie, *f.* teratogeny, the production of monsters.

tératologie, *f.* teratology, study of malformations and monstrosities.

tératologique, teratologic.

terbine, *f.* terbia.

terbium, *m.* terbium.

tercine, *m.* nucellus.

térébate, *m.* terebate.

térébène, *n.* terebene.

térébènique, terpene.

térébenthène, *m.* terebenthene.

térébenthine, *f.* turpentine; **essence de —,** oil (essence) of turpentine.

térébenthiné, containing turpentine.

térébinthe, *m.* terebinth, turpentine tree (*Pistacia terebinthus*).

térébinthiné, terebinthine, terebinthinate.

térébique, terebic.

térébrant, boring, keen, terebrating.

térébrants, *m.pl.* Terebrantia, borers.

térébrator, *m.* terebrator, trichogyne in Gyrophora.

térébrer, to bore, terebrate.

téréphthalique, terephthalic.

térétiacuminé, with tapering, cylindrical point.

téréticaule, having stalk with blunt angles.

térétifolié, with slender leaves.

tergéminé, tergeminate.

tergite, *m.* tergite.

terme, *m.* term, limit, end, end point, expression, member, time, quarter, three months, bound, boundary; *pl.* wording, terms, conditions; **— de chasse,** sporting term; **— de métier,** technical term; **— technique,** technical term; **aux —s de,** by the terms of; **en d'autres —s,** in other words.

terminaison, *f.* termination, ending, suffix.

terminal, -aux, terminal, apical.

terminé, terminated, ended, finished; **— en bec,** rostrate, beaked.

terminer, to terminate, end, finish, conclude, limit, complete; **se —,** to terminate, end, finish, come to an end.

terminologie, *f.* terminology.

terminus, *m.* terminus.

terminus, terminal.

termite, *m.* termite, white ant.

ternaire, ternary.

terne, dull, lusterless, tarnished, wan, ternate.

terné, ternate.

terniflore, with flowers arranged in threes.

ternifolié, with leaves whorled in threes.

ternir, to tarnish, dull, deaden, soil, fade.

ternissement, *m.* tarnishing, clouding, fading.

ternissure, *f.* tarnishing, dullness, blemish, stain, tarnish.

terpène, *m.* terpene.

terpénoïdes, *f.pl.* terpenoids. group of flower scents produced by terpenes.

terpine, *f.* terpinol, terpin.

terpinène, *m.* terpinene.

terpinol, terpinéol, *m.* terpineol, terpinol.

terpinolène, *m.* terpinolene.

terrage, *m.* scouring, claying.

terraille, *f.* earthenware.

terrain, *m.* ground, piece of ground plot of land, site, soil, rock, field, terrain, terrane, formation, (rock) system; **— accidenté,** hilly ground; **— argileux,** clay soil, loamy soil; **— crétacé,** chalk formation; **— d'alluvion,** alluvial ground, alluvial soil; **— défriché,** clearing, reclamation.

terrain, — détritique, weathered soil, disintegrated soil; **— en pente,** sloping ground; **— fragmentaire,** disintegrated soil; **— gazonné,** grassland; **— marécageux,** marshland, moorland, marshy ground; **— schisteux,** slate soil; **— tourbeux,** peat layer, turf bed; **— vaseux,** mud, bog, swamp.

terramare, *f.* terramara (fertilizer).

terras, *m.* resin mixed with earth.

terrasse, *f.* terrace, foreground, earthwork, banking.

terrassement, *m.* earthwork.

terrasser, to bank up, embank, fill in, floor, lay low, confound.

terrassier, *m.* navvy, laborer.

terre, *f.* earth, clay, ground, soil, land, world, dirt, field, loam, shore, coast, territory; **— à —,** sordid, commonplace; **— à bois,** woodland, land suitable for tree growth; — **à briques,** brick clay, brick loam; — **à cassettes, sagger** clay; — **à foulon,** fuller's earth; — **alcaline,** alkaline earth; — **à poterie(s),** potter's earth, potter's clay; — **cuite,** terra cotta.

terre, — **d'alumine,** alumina, aluminum oxide; — **de bruyère,** heather soil; — **des cornues,** lute; — **d'infusoires,** infusorial earth, kieselguhr; — **d'ombre,** umber; — **émiettée,** garden soil, fine earth; — **ferme,** terra firma, dry land; — **franche,** vegetable mold, garden mold, garden earth; — **glaise,** flucan, partially decomposed rock adjoining a vein, clay, potter's clay; — **grasse,** clayey soil; — **graveleuse,** gravel ground, grit ground.

terre, — **inculte,** waste land, barren land; — **labourable,** arable land; — **lehmeuse,** loamy soil; — **limoueuse,** clay, argillaceous earth; — **marécageuse,** fen land, marshland; — **mauvaise,** uncultivable land; — **mérite,** turmeric; — **meuble,** loose earth; — **original,** original soil, virgin soil; — **pesante,** heavy earth, baryta; — **pourrie,** rottenstone.

terre, — **rare,** rare earth; — **réfractaire,** fire clay, refractory earth; — **silicée,** siliceous earth, flinty earth; — **tourbeuse,** peat soil, peat land; — **vague,** wasteland, barren land; — **végétale,** vegetable mold, superficial soil; —**s basses,** lowland; —**s en frich,** wasteland; —**s vagues et vaines,** wasteland, barren land; **à —,** on land; **en —,** earthen, in the ground; **mettre à la —,** to ground; **par —,** on the ground, on the floor.

terreau, *m.* mold, humus, compost, vegetable mold.

Terre-Neuve, *f.* Newfoundland.

terre-plein, *m.* platform.

terrer, to scour, warp (field), clay (sugar), earth up, cover with earth, full (cloth).

terrestre, terrestrial, of the earth, land.

terreur, *f.* terror.

terreux, -euse, earthy, dull (color), dull, grubby, dirty.

terrible, terrible, awful, dreadful.

terriblement, terribly.

terrier, *m.* hole, burrow.

terrier, burrowing.

terrifier, to frighten, terrify.

terrine, *f.* pan, dish; — **pour évaporations,** evaporating dish.

territoire, *m.* territory, region.

terroir, *m.* soil, ground, earth.

tertiaire, tertiary, Tertiary.

tes, *pl.* thy, your.

tessellé, tessellate.

tesson, *m.* potsherd, fragment of glass, tile, or pottery.

test, *m.* (see TÊT), test, testa, shell, crucible, skin (of seed).

testicule, *m.* testicle, testis.

testif, *m.* camel's hair.

test-objet, *m.* test object.

têt, *m.* test, fragment of glass, tile, or pottery, small fire-clay cup; — **à gaz,** a fire-clay dish, beehive shelf; — **à rotir,** roasting dish, scorifier; — **de coupellation,** cupel.

tétanie, *f.* tetany; — **des herbes,** grass disease, grass tetany; grass staggers.

tétanine, *f.* tetanine.

tétanique, tetanic.

tétanisant, tetanising.

tétaniser, to produce tetanus in.

tétanolysine, tetanolysin.

tétanos, *m.* tetanus, lockjaw.

tétanotoxine, *f.* tetanotoxin.

têtard, *m.* tadpole, polliwog, pollard.

tétartoèdre, *m.* tetartohedral.

tétartoédrie, *f.* tetartohedrism.

tétartoédrique, tertartohedral.

tête, *f.* head, headpiece, brains, mind, top, front, first part, horns, antlers, capitulum, cluster, cap, attire; *pl.* first runnings; — **à —,** confab, in private, just two; — **à serrage,** grip head; — **carrée,** square head; — **d'alouette,** crown, crown knot; — **d'attelage,** coupler head; — **d'aube,** tip of blade.

tête. de bobine, coil end; — **de jetée,** mole head, pier; — **de soupape,** valve head; — **de touche,** key head; — **perdue,** flush head, dead head, feeding head, lost head; — **refaite,** completed attire, perfect head (of a deer); **de —,** of the head, from memory, by imagination; **eaux de —,** headwaters; **en —,** at the head.

têter, to suck.

têtier, *m.* header.

tétine, tetine, *f.* nipple, teat, udder.

téton, teton, *m.* breast, nipple, teat, projection, stud, end.

tétrabasique, tetrabasic.

tétracanthe, bearing four spines.

tétracarbonique. tetracarboxylic.

tétracarpe, bearing four fruits.

tétrachloré, tetrachloro.

tétrachlorure, *m.* tetrachloride· — de carbone, carbon tetrachloride.

tétracoque, tetracoccous.

tétracuprique, tetracupric.

tétracyclique, tetracyclic, (flower) composed of four whorls of organs.

tétrade, tetrad.

tétradène, bearing four glands.

tétradymite, *f.* tetradymite, native bismuth telluride.

tétradyname, tetradynamous.

tétraédral, -aux, tetrahedral.

tétraèdre, *m.* tetrahedron.

tétraèdre, -aux, tetrahedral.

tétraédrique, tetrahedral.

tétraéthylé, tetraethyl.

tétraéthylique, tetraethyl.

tétrafolié, tetrafoliatous, tetrafolious, fourleaved.

tétrafluorure, *m.* tetrafluoride.

tétragène, *f. Micrococcus tetragenus.*

tétragonal, -aux, tetragonal.

tétragone, *m.* tetragon; — cornue, New Zealand spinach, prolific spinach (*Tetragonia expansa*).

tétragone, tetragonal.

tétragonidie, *m.* tetragonidium, tetraspore.

tétragonocarpe, with tetragonal fruit.

tétragonolobé, with quadrangular lobes.

tétragyne, tetragynous.

tétrakène, *m.* schyzocarpic fruit formed by four akenes.

tétramère, tetramerous.

tétraméthylène-diamine, *f.* tetramethylenediamine, putrescine.

tétraméthylénique, tetramethylene.

tétrandre, tetrandrous.

tétrandrie, *f.* tetrandria.

tétrane, *m.* tetrane, butane.

tétranol, *m.* tetranol, butanol.

tétrapétalu, tetrapetalous.

tétraptère, tetrapterous, four-winged, four produced angles.

tétraquètre, tetraqueter, with four sharp angles.

tétrarche, tetrarch.

tétras, *m.* grouse.

tétrasépale, tetrasepalous, having four sepals.

tétrasperme, tetraspermous, with four seeds.

tétraspermé, with tetraspermous fruit.

tétraspore, *m.* tetraspore.

tétrastachyé, with flowers in quaternate spikes.

tétrastémone, with four stamens.

tétrastigmaté, with four-cleft stigma.

tétrastique, tetrastichous, in four vertical ranks.

tétrastyle, furnished with four styles

tétratomique, tetratomic.

tétravalence, *f.* tetravalence, quadrivalence.

tétravalent, tetravalent, quadrivalent.

tétrazine, *f.* tetrazine.

tétrazoïque, tetrazo.

tétrazol, *m.* tetrazole.

tétréthylé, tetraethyl.

tétronal, *m.* tetronal.

tétrose, *f.* tetrose.

tétroxyde, *m.* tetroxide.

tétryl, *m.* tetryl.

texte, *m.* text.

textile, textile.

textuel, -elle, textual, literal.

textuellement, textually, literally, word for word.

textulaire, textural.

texture, *f.* texture, structure; — du bois, texture of wood.

thalame, *m.* thalamus.

thalamiflore, thalmifloral.

thalamion, *m.* thalamus.

thalamostémone, with stamens inserted on the receptacle.

thalle, *m.* thallus.

thallieux, thalleux, -euse, thallous, thallious.

thalline, *f.* thalline.

thallique, thallic.

thallium, *m.* thallium.

thallome, *m.* thallus.

thallophytes, *f.pl.* thallophytes

thalweg, *m.* thalweg, channel.

thapsie, *f.* deadly carrot (*Thapsia gargantica*).

thapsoïde, resembling a white bubble

thé, *m.* tea, tea tree (Thea).

théâtral, -aux, theatrical.

théâtre, *m.* theater, stage, scene.

thebaïde, *f.* desert, solitude.

thébaïne, *f.* thebaine, paramorphin.

thébaïque, pertaining to opium.

thécaphore, *m.* attenuation of a carpel.

thécigère, thecigerous, theca-bearing, theciferous.

théine, *f.* theine.

thème, *m.* theme, topic, subject.

théobrome, *m.* cacao (*Theobroma cacao*).

théobromine, *f.* theobromine.

théophylline, *f.* theophylline.

théorème, *m.* theorem.

théorétique, theoretic(al).

théoricien, *m.* theorist.

théorie, *f.* theory, doctrine, speculation; — atomique, atomic theory; — cinétique, kinetic theory.

théorique, theoretic(al).

théoriquement, theoretically.

théoriser, to theorize.

thèque, *f.* ascus, fungus.

thérapeutique, *f.* therapeutics, therapy.

thérapeutique, therapeutic(al).

thérapeutiquement, therapeutically.

thérapique, therapeutical.

thériaque, *f.* theriaca, theriac.

thermal, -aux, thermal, warm, hot.

thermauxisme, *m.* action of heat on growth.

thermique, thermic, of heat.

thermite, *f.* thermite.

thermochimie, *f.* thermochemistry.

thermochimique, thermochemical.

thermochroïque, thermochroic.

thermocleistogamie, *f.* thermocleistogamy, flowers not expanding because of insufficient warmth.

thermodynamique, *f.* thermodynamics.

thermodynamique, thermodynamic.

thermo-électrique, thermoelectric(al).

thermogène, thermogenic, thermogenous.

thermogénèse, *f.* thermogenesis.

thermolabile, thermolabile.

thermolampe, *m.* thermolamp, heat- and light-giving lamp.

thermomètre, *m.* thermometer; — à air, air thermometer; — à gaz, gas thermometer; — à mercure, mercury thermometer; — contrôlé, calibrated thermometer; — de refroidissement, radiator thermometer;

— enregistreur, recording thermometer; — médical, clinical thermometer.

thermométrie, *f.* thermometry.

thermométrique, thermometric(al).

thermophile, thermophilic.

thermorégulateur, *m.* thermoregulator, thermostat.

thermosis, *f.* thermosis, change due to warmth upon an organism.

thermotropisme, *m.* thermotropism, curvature dependent upon temperature.

thèse, *f.* thesis, discussion.

thialdéhyde, *m.* thioaldehyde.

thiazinique, thiazine, of thiazine.

thiazol, *m.* thiazole.

Thibet, *m.* Tibet.

thiénone, *f.* thienone.

thigmomorphose,*f.* thigmomorphosis,change in original structure due to contact.

thille, *f.* thyllus.

thioacide, *m.* thio acid.

thioaldéhyde, *m.* thioaldehyde.

thiocyanate, *m.* thiocyanate.

thioflavine, *f.* thioflavin.

thionine, *f.* thionine.

thionique, thionic.

thionyl, *m.* thionyl.

thiophène, *m.* thiophene.

thiophénique, thiophene, of thiophene, thiophenic.

thiophtène, *m.* thiophthene.

thiosinamine, *f.* thiosinamine.

thiosulfate, *m.* thiosulphate.

thiourée, *f.* thiourea.

thioxène, *m.* thioxene.

thlaspi, *m.* pennycress (Thlaspi).

thon, *m.* tunny (fish).

thorax, *m.* thorax, chest.

thorine, *f.* thoria.

thorique, thorium, of thorium, thoric.

thorite, *f.* thorite.

thorium, *m.* thorium.

thran, *m.* fish oil, whale oil.

thrips, *m.* thrips, thysanopter.

thrombine, *f.* thrombin.

thrombose, *f.* thrombosis.

thulium, *m.* thulium.

thuya, thuia, *m.* arborvitae (Thuja).

thylle, *f.* tylosis.

thym, *m.* thyme (Thymus); — **bâtard,** wild thyme (*Thymus serpyllum*); — **commun,** garden thyme (*Thymus vulgaris*).

thymélée, *f.* Thymelaeaceae.

thymine, *f.* thymine.

thymique, thymic, thymol.

thymol, *m.* thymol.

thymonucléique, thymonucleic, thymus nucleic.

thymus, *m.* thymus gland.

thyroïde, thyroid.

thyroïdien, -enne, thyroidal, thyroid.

thyroïdine, *f.* thyroidin.

thyrsanthé, thyrsoid.

thyrse, *m.* thyrsus, panicle.

thyrsiflore, thyrsoid.

thyrsoïde, thyrsoid.

thysanocarpe, with fringed fruits.

tibia, *m.* shinbone, tibia.

tiède, tepid, lukewarm.

tièdement, coldly, lukewarmly.

tiédeur, *f.* lukewarmness, tepidity, tepidness.

tiédir, to become tepid or lukewarm, cool off, make tepid or lukewarm, take the chill off.

tien, tienne, thine, yours.

tiendra (tenir), (he) will hold.

tiers, *m.* third, third part.

tiers, tierce, third.

tiers an, *m.* boar three-years old.

tiers-point, *m.* three-cornered file, saw file.

tige, *f.* stem, stalk, scape, shaft, arm, shank, trunk, bar, rod, pin, stock, origin, suspender, pricker, neck; — **de commande (du tiroir),** valve stem; — **de communication,** connecting rod; — **de réglage,** adjusting rod; — **d'expérience,** test tree.

tige, — **du piston,** piston rod; — **en verre,** glass rod; — **type,** sample tree; — **végétale,** plant stalk; — **de chanvre,** hemp plant; **à** — **courte,** short stemmed; **à longue,** long stemmed; **basse-tige,** nursery-grown tree up to one meter.

tigelle, *f.* tigella, caulicle.

tigellé, furnished with a tigella.

tigellulaire, like a tigellula.

tigre, *m.* tiger.

tigré, striped.

till, *m.* sesame.

tillage, *m.* stripping, scutching

tille, *f.* linden bast, hemp harl, lime bast, Russian bast, hammer with an edged peen.

tilleul, *m.* linden, basswood, lime tree (Tilia); — **à petites feuilles,** small-leaved linden (*Tilia cordata*); — **à grandes feuilles** broad-leaved linden (*Tilia platyphyllos*); — **intermédiaire,** common lime tree (*Tilia vulgaris*); — **noir,** basswood, American linden (*Tilia glabra*).

timbrage, *m.* stamping, testing.

timbre, *m.* stamp, stamp office, ring, tone, bell, timbre, timber, crest, test plate.

timbre-poste, *m.* postage stamp.

timbrer, to stamp.

timide, timid, shy.

timidement, timidly.

timon, *m.* pole, shaft.

timonier, *m.* steersman, helmsman.

timoré, timorous, fearful.

timothée bacillus, *m.* timothy-grass bacillus.

tincal, *m.* tincal, crude (native) borax.

tinctorial, -aux, tinctorial, dyeing.

tine, *f.* water cask, tub.

tinette, *f.* a keg, small cask, small tub.

tinkal, *m.* see TINCAL.

tinne, *f.* water cask, tub, (clay) pugmill.

tint (tenir), (he) held.

tinté, tinted.

tintement, *m.* tinkling, ringing, tolling.

tinter, to tinkle, ring, ting, tingle, toll, support, prop, block, chock up.

tintin, *m.* clink, tinkle.

tion, *m.* tool for cleaning crucibles.

tique, *f.* tick, cattle tick.

tiqueté, spotted, speckled.

tir, *m.* fire, firing, shooting, target practice, shooting grounds.

tirage, *m.* pulling, dragging, drawing, draft, traction, impression (of book), proof, pull, printing, extension, towing; — **forcé,** forced draft, artificial draft.

tiraille, *f.* connecting rod.

tiraillement, *m.* pulling, twitching, disagreement, discord, friction.

tirailler, to pull about, tug, twitch.

tirant, *m.* tie beam, tie rod, stay rod, boot strap, shoelace, draft (of ship), tendon (in meat).

tirant, pulling, drawing.

tirasse, *f.* dragnet, drawnet.

tire, *f.* drawing, draw, pull(ing); **tout d'une —**, at one stroke, without stopping.

tiré, drawn, pulled, worn out, haggard.

tire-bouchon, *m.* corkscrew, ringlet.

tire-pièce, *m.* skimmer.

tire-point, *m.* piercing tool, stabbing awl.

tirer, to draw, pull, stretch, extract, derive, deduce, get, obtain, contract, attract, absorb, shoot, fire, print, strain, extricate, design, sketch, make for, go to, verge (on); **se—**, to be drawn, be pulled, get out of, recover, shrink, contract; — **parti de,** to take advantage of; **il tire au sort,** he draws lots; **il tire parti de,** he profits by; **il se tire d'affaire,** he gets out of trouble.

tireur, *m.* drawer, puller, printer, shooter, marksman.

tiroir, *m.* drawer, slide valve, slide, gate, turnout, extension of track.

tisane, *f.* infusion, decoction, tea; — **d'aloes composée,** compound decoction of aloes; — **d'angosture,** infusion of angostura; — **de champagne,** light champagne; — **d'écorce de cerisier sauvage,** infusion of wild cherry; — **de girofle,** infusion of cloves; — **de serpentaire,** infusion of serpentary; — **d'uva-ursi,** infusion of bearberry.

tisard, *m.* fire (door) hole.

tiser, to stir, poke, stoke (fire).

tiseur, *m.* stoker, fireman.

tison, *m.* (fire) brand, live coal.

tisonner, to stir, poke a fire.

tisonnier, *m.* poker.

tissage, *m.* weaving, cloth mill, texture.

tisser, to weave.

tisserand, tisseur, *m.* weaver.

tissu, woven.

tissu, *m.* tissue, cloth, fabric, texture, textile fabric; — **adipeux,** adipose tissue, fat tissue; — **aquifère,** water tissue; — **à mailles,** knitting, knitted fabric; — **cellulaire,** cellular tissue; — **collagène,** collagenous tissue; — **conjonctif,** connective tissue; — **de consolidation,** strengthening tissue; — **de formation,** cambium; — **de paille,** straw tissue.

tissu, — de réserve, fostering tissue, reserve tissue; — **fibreux,** fibrous tissue, sclerenchyma; — **fibrovasculaire,** fibrovascular tissue; — **filtrant,** filter cloth; — **imperméable,** waterproof cloth; — **ligneux,** ligneous tissue, woody tissue; — **métallique(s),** wire gauze, wire netting; —

nutritif, fostering tissue; — **subéreux,** corky tissue, periderm.

tissure, *f.* texture, web.

tistre, to weave.

titanate, *m.* titanate.

titane, *m.* titanium.

titané, containing titanium.

titanesque, titanic.

titaneux, -euse, titanous.

titanifère, titaniferous.

titanique, titanic, gigantic

titanite, *f.* titanite.

titanium, *m.* titanium.

titillation, *f.* tickling.

titrage, *m.* titration, testing, assaying; — **en retour,** back titration.

titrant, titrating.

titre, *m.* titer, standard, title, right, claim, title page, heading, document, warrant, bond, stock certificate, normality factor, concentration; — **pondéral,** gravimetric titer; **à — de,** by virtue of, by right of, in virtue of; **à — principal,** as the main product; **au — de,** as, on the score of.

titré, titrated, standard, tested, titled, standardized; **liqueurs —s,** standardized solutions.

titrer, to titrate, standardize, test, title.

titulaire, *m.* titular, holder of a title or grade, incumbent.

titulaire, titular, titulary.

toc, *m.* stop, catch, tappet, faked stuff, imitation gold.

tocage, *m.* adding of fuel.

tocane, *f.* new champagne.

toddy, *m.* toddy.

toi, thou, thee, you.

toile, *f.* cloth, linen, (wire) gauze, canvas, (stage) curtain, linen cloth, tissue, fabric, cloth, painting, picture, sparadrap, sail; — **à calquer,** tracing cloth; — **amiantine,** asbestos cloth; — **cirée,** oilcloth, wax cloth; — **d'amiante,** asbestos cloth.

toile, — de coton, cotton cloth, calico, cotton; — **d'emballage,** packing cloth, burlap, wrapping cloth, wrapper; — **grasse,** waterproof canvas; — **métallique,** wire gauze, wire netting, wire cloth; — **métallique amiantée,** asbestos wire gauze; — **vernie,** painted canvas, varnished canvas.

toilette, *f.* fine cloth, costume, toilet, toilet table, dressing table, wrapper.

toilier, *m.* linen-cloth maker.

toilier, -ère, linen.

toise, *f.* fathom (6.39 ft.).

toiser, to measure, survey, scrutinize.

toison, *f.* fleece.

toit, *m.* roof, house top, tegment; — à porcs, pigsty, piggery.

toiture, *f.* roofing, roof.

tokai, tokay, *m.* Tokay (wine).

tolane, *m.* tolane, diphenylacetylene.

tôle, *f.* sheet metal, sheet iron, iron plate; — d'acier, sheet steel, steel plate; — émaillée, enameled sheet iron;— forte, — ciab, coarse plate, thick plate; — galvanisée, galvanized sheet iron;— mince, sheet metal, thin sheet; — noire, black sheet iron; — ondulée, corrugated (sheet) iron; — zinguée, zinc sheet, galvanized sheet iron.

tolérable, bearable.

tolérablement, tolerably, bearably.

tolérance, *f.* tolerance, toleration.

tolérer, to tolerate, allow.

tôlerie, *f.* sheet-iron making or trade, covering of sheet iron.

tôlier, *m.* sheet-iron worker.

tolite, *f.* tolite, trinitrotoluene.

tolu, *m.* tolu, balsam of Tolu.

toluène, *m.* toluene, methyl benzene.

toluidine, *f.* toluidine.

toluifère, yielding balsam of Tolu.

toluique, toluic.

toluiser, to coat (pills) with balsam of Tolu.

toluol, *m.* toluene, toluole.

toluylène, *m.* toluylene.

tolylénique, tolylene, of tolylene.

tolysal, *m.* tolysal.

tomate, *f.* tomato.

tombac, *m.* tombac.

tombant, falling.

tombe, *f.* tombstone, tomb, grave, bed of leaf mold.

tombeau, *m.* tomb, sepulcher, tombstone.

tombée, *f.* fall; — de la nuit, nightfall.

tomber, to fall, drop, abate, fail, decline, subside, tumble, sink in, fall away; il tombe bien, he comes at the right time.

tombereau, *m.* tumbrel, dumpcart.

tome, *m.* volume, tome, part.

tomenteux, -euse, tomentose, densely pubescent with matted wool.

tomentum, *m.* tomentum.

tomiange, *m.* sporiferous sac of the mosses.

tomie, *f.* tomie, asexual reproductive body.

tomillares, *f.pl.* tomillares, sclerophyllous vegetation.

tomiogone, *m.* tomiogone, organ that produces tomies.

tomipare, tomiparous, (plants) reproducing by fission.

tomotocie, *f.* caesarian section.

ton, *m.* tone, shade, color, key.

ton, thy, your.

tonalité, *f.* tonality.

tonca, *m.* see TONKA.

tondage, *m.* shearing, clipping.

tondaison, *f.* sheep-shearing time.

tondelle, *f.* shearings of cloth, flocks.

tondeuse, *f.* grasscutter; — à gazon, lawn mower, roller.

tondre, to shear, shave, clip, trim.

tondu, shorn.

tondure, *f.* shearing, shearings.

tonesie, *f.* tonesis, ability of an organism to exhibit a strain.

tonicité, *f.* tonicity, elasticity.

tonifier, to tone, brace, give tone to.

tonique, tonic.

tonka, *m.* tonka bean, tonka (*Dipteryx odorata*).

tonnage, *m.* tonnage, burden, carrying capacity.

tonnant, detonating, thundering.

tonne, *f.* ton (metric ton of 1,000 kg.), tun, large cask, kibble, bucket, skip barrel, drum; — de fermentation, loose hogshead, puncheon, stillion.

tonneau, *m.* cask, tub, barrel, drum, tun, vat, bin, (nautical) ton; — d'arrosage, watering cart, water cart; — mélangeur, mixing barrel, drum.

tonneler, to tunnel.

tonnelet, *m.* small cask, keg, barrel, drum.

tonnelier, *m.* cooper.

tonnelle, *f.* arbor, bower, barrel vault. opening into the furnace.

tonnellerie, *f.* cooperage.

tonner, to thunder, detonate.

tonnerre, *m.* thunder, thunderbolt, explosion chamber.

tonomètre, *m.* tonometer.

tonoplaste, *f.* tonoplast, vacuolar living membrane controlling the pressure of cell sap.

tonosis, *f.* tonosis, changes in turgescence due to intercellular osmotic force.

tonotropisme, *m.* tonotropism, response to osmotic stimulus.

tonsille, *f.* tonsil.

tonte, *f.* sheep shearing, clip, shearing time, clipping, mowing (lawn), pruning.

tontisse, from shearing; **bourre —,** cropping flock.

tonture, *f.* shearings, flocks, cut grass, hay, clippings (from hedge).

tonus, *m.* tone, tonicity.

topaze, *f.* topaz.

topette, *f.* slender flask, sample bottle, phial.

topinambour, *m.* Jerusalem artichoke (*Helianthus tuberosus*).

topique, *m.* topic, local remedy.

topique, topical, local.

topographe, *m.* topographer; **— de détail,** surveyor.

topographie, *f.* topography.

topographique, topographical.

toquerie, *f.* fireplace, hearth (of forge).

toqueux, *m.* sugar refiner's poker.

torche, *f.* torch, link, mat, pad.

torchepot, *m.* nuthatch.

torcher, to wipe.

torcher, *m.* worker in cob.

torchis, *m.* cob, daub, clay mixed with straw.

torchon, *m.* cloth (for dusting), rag, dishcloth, duster, towel.

torchonner, to wipe, clean, dust.

torcol, *m.* wryneck (the bird).

tordage, *m.* twisting, twist, torsion.

tordeuse, *f.* **— du pin,** resin gall moth; **— du mélèze,** larch moth.

tordoir, *m.* oil press, ore crusher, wringer, twisting machine.

tordre, to twist, wrest, wring, torture, contort, writhe, distort.

tordu, twisted.

tore, *m.* torus.

torfacé, torfaceous, growing in bogs.

torifère, bearing tubercles.

tormentille, *f.* tormentil (*Potentilla tormentilla*).

toron, *m.* strand (of rope).

torpédo, *m.* open touring car.

torpeur, *f.* torpor.

torpide, torpid.

torpille, *f.* torpedo.

torpilleur, *m.* torpedo man, torpedo boat.

torque, *f.* torque, coil, twist (of tobacco).

torquer, to twist (tobacco).

torquet, *m.* trap.

torquette, *f.* twist, roll (of tobacco).

torréfacteur, *m.* roaster.

torréfacteur, *m.* -trice, roasting, parching.

torréfaction, *f.* torrefaction, roasting.

torréfier, to torrefy, roast, scorch, parch, burn, ensiccate, desiccate, dry.

torrent, *m.* torrent, hill torrent, stream; **à —s,** in torrents.

torrentiel, -elle, falling in torrents.

torride, torrid.

tors, *m.* twisting, twist.

tors, twisted, contorted, crooked.

torsade, *f.* twisted fringe, coil, twist joint.

torse, *m.* trunk, torso, wringing.

torsil, contorted.

torsion, *f.* torsion, twisting, wrenching; **— des fibres,** torse fiber, spiral grain, twisted growth.

torsoir, *m.* wringing pole.

tort, *m.* wrong, error, injury, fault, harm, detriment, mischief; **à —,** wrongly, wrongfully; **à — et à travers,** at random; **avoir —,** to be wrong; **faire — à,** to wrong, injure, harm.

torte, see TORS.

tortelle, *f.* hedge mustard (*Sisymbrium officinale*).

tortillard, *m.* dwarf elm (*Ulmus pumila*), crooked elm.

tortillement, *m.* twisting.

tortiller, to twist, wriggle.

tortipède, with twisted pedicel.

tortisme, *m.* tropism, curvature resulting from response to stimulus.

tortu, tortuous, twisted, crooked.

tortue, *f.* tortoise, turtle.

tortueux, -euse, tortuous, winding, twisted, crooked.

toruleux, -euse, nodose, knotty, torous.

torus, *m.* torus, receptacle.

toscan, Tuscan.

tôt, soon, shortly, early, promptly; — **après,** soon after; — **ou tard,** sooner or later; **au plus —,** as soon as possible.

total, -aux, total, whole, all, entire, sum total; **au —,** on the whole, in all.

totalement, totally, entirely, wholly.

totalisateur, *m.* adding machine, totalizer.

totalisateur, -trice, adding, totaling.

totaliser, to total up, add up, form a total.

totalité, *f.* totality, whole, sum total; **en —,** entirely, as a whole.

touage, *m.* warping, towing.

touchant, touching, moving, affecting, tangent, concerning, with regard to, about.

touchau, touchaud, *m.* touch needle, contact piece.

touche, *f.* touch, stroke, manner, key (of instruments), contact piece.

toucher, to touch (upon), adjoin, affect, touch on, dwell on, reach, assay, test, draw, receive (salary), be related to, drive, ink, move, cauterize, concern, be in contact with, feel, border upon, draw near, meddle, interfere; **se —,** to touch each other, touch, adjoin, meet, be related, be touched; **il touche au port,** it draws near to port.

toucher, *m.* touch, sense of feeling, feeling.

toucheur, *m.* cattle driver, drover.

touer, to tow, warp.

touffe, *f.* tuft, cluster, clump, bunch, wisp (of hay), hill (of potatoes), fascicle (of leaves).

touffer, to arrange (trees) in tufts, in clusters.

touffu, bushy, thick, wooded, leafy, tufted, massy, full of detail, bundled.

touillage, *m.* mixing, decanting, stirring, beating.

touille, *f.* mixing shovel.

touiller, to stir, mix, beat.

toujours, always, ever, still, anyhow, nevertheless.

touloucouma, *n.* tulucuna, kundah.

touloupe, *f.* lambskin.

toupet, *m.* forelock, tuft of hair.

toupie, *f.* top, spinning top, molding machine.

toupillage, *m.* spinning, whirling.

toupiller, to spin, whirl, shape.

tour, *f.* tower, turret.

tour, *m.* turn, revolution, trip, excursion, round, circuit, orbit, circumference, circle,

tour, journey, outline, reel, trick, feat, lathe, turning lathe, (potter's) wheel, winch, whim, windlass, revolutions per minute; — **à —,** by turns, in turn, one after another; — **à main,** hand lathe; — **à noyauter,** core turning lathe; — **à plateau,** surface lathe; — **à potier,** potter's wheel.

tour, — **à support,** slide lathe; — **de force,** feat; — **de main,** trick; — **de potier,** potter's wheel, potter's lathe; **à — de role,** in turn; **à son —, à leur —,** in turn; **de —,** in girth; **en un — de main,** in a moment; **faire le — de,** to go around.

touraillage, *m.* kiln drying (of malt).

touraille, *f.* malt kiln, kiln, oast.

tourailler, to kiln-dry (malt).

touraillon, *m.* malt dust, malt combs.

tourbage, *m.* peat digging, peat formation.

tourbe, *f.* peat; — **comprimée,** compressed turf; — **draguée,** bagger turf.

tourber, to dig peat, extract peat from.

tourbeux, -euse, peaty, turfy.

tourbier, *m.* peat worker.

tourbier, -ère, peaty (soil).

tourbière, *f.* peat bog, turf pit, fen, moor, peat bed, peat moss, peat; — **à cypéracées,** meadow moor, meadow bog; — **à sphaignes,** high peat bog, high moor; — **basse,** lowland moor, black bog, valley moor; — **de pente,** slope moor; — **de plaine,** flat bog.

tourbillon, *m.* vortex, eddy, whirlwind, whirlpool, whirl, air eddy; — **de neige,** driving snow or blizzard.

tourbillonnement, *m.* whirling, eddying.

tourbillonner, to whirl, eddy, swirl.

tourdillé, speckled.

tourelle, *f.* turret.

touret, *m.* small wheel, reel, drill.

tourie, *f.* carboy.

tourillon, *m.* spindle, journal, pivot (pin), trunnion, swivel pin, gudgeon (pin); — **de la manivelle,** crankpin.

tourisme, *m.* touring, automobiling.

touriste, *m.* tourist.

tourmaline, *f.* tourmaline.

tourmente, *f.* storm, tempest.

tourmenter, to torment, molest, toss, agitate; **se —,** to warp.

tournage, *m.* turning.

tournant, *m.* turn, double furrow, turning, bend, corner, elbow, water wheel (of mill), shift.

tournant, turning, revolving, turned, rotating.

tournassage, tournasage, *m.* finishing on the wheel, throwing.

tournasser, tournaser, to throw, shape on the wheel, turn.

tournassin, tournasin, *m.* finishing (shaping) tool.

tournassure, tournasure, *f.* shavings, turnings.

tourné, turned, twisted, changed, spoiled, sour (milk); — **en dehors,** retrorse.

tourne-à-gauche, *m.* wrench, handle, lever, saw set.

tournée, *f.* round, tour, circuit.

tourne-foin, *m.* haymaker.

tourner, to turn, fashion, shape, examine, revolve, get around, evade, rotate, swivel, shift, wheel, become sour or rancid (of milk), become turbid and flat (of wine); **se** —, to be turned, turn around, turn; — **à la graisse,** to become ropy, get oily; — **à vide,** to run idle, run light or without a load; — **la meule,** to turn the millstone, work the treadmill.

tournesol, *m.* litmus, turnsole, girasol, fire opal, sunflower (Helianthus); — **des teinturiers,** turnsole (*Chrozophora tinctoria*); huile de —, sunflower oil.

tournesolie, *f.* turnsole (*Chrozophora tinctoria*).

tournette, *f.* reel.

tourneur, *m.* turner.

tourneur, turning.

tournevis, *m.* screwdriver.

tourniquet, *m.* turnstile, roller, pulley, revolving stand, swivel.

tournis, *m.* turnsick, staggers, gid.

tournoiement, *m* turning, whirling spinning.

tournoyer, to turn round and round, whirl.

tournure, *f.* turning, turnings, turn, direction, shape, figure, appearance.

tourte, *f.* raised pie, press cake (from fruits or seeds), cake, torta (of silver ore), cake (of linseed for manure).

tourteau, *m.* round loaf, press cake, oil cake, disk, block; — **coprah,** cocoanut cake; — **d'arachide,** peanut or groundnut cake; — **de colza,** colza cake, rape cake; — **de fourrage,** cattle cake, oilcake; — **de graines de coton,** cottonseed cake; — **de lin,** linseed cake; — **de palmiste,** palm-nut cake; — **de sésame,** sesame-oil cake; — **de soja,** soybean cake.

tourterelle, *f.* turtle dove.

tous, *m.pl.* all; — **deux,** both; — **les deux,** both; — **les jours,** everyday; — **les trois,** all three.

tous-les-mois, *m.* pot marigold (*Calendula officinalis*).

tousser, to cough.

tout, all, each, whole, every, any, everything, everyone, wholly, quite, completely, entirely, just, fully, very; — **à coup,** suddenly; — **à fait,** quite, wholly, entirely; — **à l'heure,** soon, in a moment, just now, in a little while, a little while ago; — **au long,** at length; — **au moins,** at least, at the very least; — **au plus,** at most, at the very most, at best; — **autre,** quite different, any other; — **beau,** gently.

tout, — **d'abord,** from the very first, right at first; — **de bon,** for good; — **de même,** all the same, just the same; — **de suite,** at once, immediately, instantly; — **droit,** straight ahead; — **du long,** all along, wholly, entirely; — **d'un coup,** all at once, at one time; — **en (jouant),** while (playing); — **entier,** entire, whole, complete.

tout, — **fait,** ready made; — **le long de,** all along, all through; — **le monde,** everybody; — **le temps,** constantly; — **un,** —**e une,** a whole; à — **à l'heure,** I'll see you shortly; à — **prendre,** on the whole, all things considered; à — **propos,** at every instant.

tout, après —, after all; du —, not at all; en —, in all; en — et pour —, all in all; en — sens, in every direction; le —, the whole thing; pas du —, not at all; point du —, not at all; rien du —, nothing at all; sur le —, above all.

toute-bonne, *f.* clary (*Salvia sclarea*).

toute-épice, *f.* allspice, pimento (*Pimenta officinalis*).

toutefois, yet, nevertheless, still, however.

toutenague, *f.* tutenag.

toute-puissance, *f.* omnipotence.

toutes, *f.pl.* all.

tout-puissant, toute-puissante, omnipotent, almighty.

tout-venant, *m.* unsorted material (coal) of any kind, run of mine.

toux, *f.* cough, coughing.

toxalbumine, *f.* toxalbumin.

toxémie, *f.* toxemia.

toxicité, *f.* toxicity, poisonousness.

toxicodendron, *m.* poison oak (*Rhus toxicodendron*), toxicodendron.

toxicogène, toxicogenic.

toxicologie, *f.* toxicology.

toxicologique, toxicological.

toxicologue, *m.* toxicologist.

toxicose, *f.* toxicosis.

toxine, *f.* toxin.

toxique, *m.* poison, toxic substance, virus.

toxique, toxic, poisonous.

toxogénine, *f.* toxogenin.

toxoïde, *m.* toxoid.

toxone, toxone.

toxophore, toxophore, toxophoric.

toxophylle, (leaves) sagittate.

trabécules, *f.pl.* trabecula.

traçant, running, ramifying, repent, prostrate and rooting, stoloniferous.

tracas, *m.* bustle, din, footstep, trail, footprint, spur.

trace, *f.* trace, track, sketch, mark, sign, footprint; — **foliaire,** leaf trace.

tracé, *m.* outline, sketch, direction, curve, plan, tracing, design, profile, copy, line, route.

tracé, traced, drawn.

tracement, *m.* tracing, laying out, drawing, marking.

tracer, to trace, draw, sketch, lay out, mark out, outline, cut, plot.

trachée, *f.* trachea, windpipe, vessel.

trachéide, *f.* tracheid.

trachéome, *f.* tracheome, hydrome, water system of a vascular bundle.

trachycarpe, trachycarpous, rough to the touch.

trachystachyé, with rough spikes (heads) bristling with hairs.

trachyte, *m.* trachyte.

trachytique, trachytic.

trachytophyte, *f.* plant with leaves rough to the touch.

traçoir, *m.* tracing awl, marking tool.

tracteur, *m.* tractor; — **à chaîne,** caterpillar tractor; — **à roues,** wheeled tractor.

traction, *f.* traction, pull, effort; **en double** —, double-headed.

traditionnel, -elle, traditional.

traditionnellement, traditionally.

traducteur, *m.* translator.

traduction, *f.* translation.

traduire, to translate, convey, show, interpret, express, represent; **se** —, to be expressed, be manifested, be translated, appear, be shown, manifest.

traduisible, translatable.

trafic, *m.* traffic, trade, trading, communication, intercourse; — **d'amélioration,** process of improvement.

trafiquant, *m.* trader.

trafiquer, to traffic, deal, trade.

tragacanthe, *f.* tragacanth (*Astragalus gummifer*), milk vetch (*Astragalus glycyphyllos*).

tragédie, *f.* tragedy.

tragiquement, tragically.

tragopogonoïde, resembling salsify.

trahir, to betray, disclose.

trahison, *f.* treachery, treason, betrayal.

train, *m.* train, movement, pace, rate, speed, course, way, mood, series, rolls, herd (of cattle), quarters (of animal); — **d'atterrissage,** landing carriage, landing gear, undercarriage; — **de bois,** raft; — **de grande vitesse,** fast train; — **de voyageurs,** passenger train; — **finisseur,** finishing rolls; — **rapide,** flyer, express train; **bon** —, quickly; **en** — **de,** in the act of; **être en** — **de,** to be engaged in; **mettre en** —, to set going, throw into gear.

trainant, *m.* runner, sucker.

trainant, dragging, slow, flagging, trailing, heavy, dull.

trainasse, *f.* bent grass, hog weed.

traine, *f.* dragging, training.

traineau, *m.* sledge, sled, sleigh, drag.

trainée, *f.* trail, belt (of ore), track, train (of powder), runner, layer.

trainement, *m.* dragging, trailing.

trainer, to drag; trail, draw, pull, haul, tow, skid, lay behind, carry along, linger, drag out, delay, lengthen; **se** —, to crawl, creep, creep along.

traire, to milk.

trait, *m.* stroke, dash, line, mark, streak, trace, cut, trait, shaft, dart, arrow, ray, stripe, draft, flash, touch, feature, lineament, connection, relation, reference; — **de crayon,** pencil line; — **de jauge,** calibration mark; — **de repère,** reference mark, guiding line; — **de scie,** saw cut; — **d'union,** hyphen, bond, dash; — **plein,** solid line.

trait, wire-drawn, milked.

traitable, tractable, manageable, treatable.

traite, *f.* stage, stretch, trade, traffic, transport, milking, draft.

traité, *m.* treatise, essay, treaty, agreement, compact.

traitement, *m.* treatment, salary; — des bois, silviculture.

traiter, to treat, manage, negotiate, deal, take up, execute, call, style; — à la vapeur, to steam.

traître, -esse, traîtreux, -euse, treacherous, traitorous, vicious (animal).

trajectoire, *f.* trajectory.

trajet, *m.* journey, trip, passage, distance, path, line, course, way.

tramail, *m.* trammel net, dragnet.

trame, *f.* woof, weft, tram (silk) plot, network, stroma.

tramer, to weave, hatch, plot.

tramway, *m.* tramway, streetcar, tramcar.

tranchage, *m.* sawing up, cleaving.

tranchant, *m.* edge, cutter, chisel.

tranchant, sharp, sharp-edged, cutting, trenchant, decisive, peremptory, loud, glaring (colors).

tranche, *f.* slice, round (of beef), section, ridge, furrow, slab, face, plate, edge (of coin), surface cross section, end face, chisel.

tranché, sharp, clear-cut, decided, definite, cut, distinct.

tranchée, *f.* trench, drain, ditch, cutting, ride, excavation; *pl.* griping pain, colic; — garde feu, fire line, fire trace.

tranchefil, *m.* chain wire.

tranche-gazon, *m.* sod knife, turf cutter.

trancher, to cut, slice, cut off, cut short, settle, differentiate, curtail, solve (a question), contrast, clash.

tranchet, *m.* cutter, chisel, paring knife.

tranquille, quiet, calm, tranquil.

tranquillement, quietly, tranquilly.

tranquilliser, to quiet, soothe, calm; se —, to grow easy, calm oneself.

tranquillité, *f.* tranquillity, calmness, quiet, calm.

transaction, *f.* compromise, transaction; *pl.* dealings.

transbordement, *m.* transshipment.

transborder, to transship, convey, ferry.

transcendance, *f.* transcendency.

transcendant, transcendent, transcendental.

transcrire, to transcribe, copy.

transférable, transferable.

transfèrement, *m.* transfer.

transférer, to transfer.

transfert, *m.* transfer, assignment.

transfigurer, to transfigure.

transformable, transformable.

transformateur, *m.* transformer.

transformation, *f.* transformation, conversion; — en marécage, becoming marshy.

transformer, to transform, change; se —, to be transformed.

transformisme, *m.* transformism.

transformiste, transformistic.

transfuser, to transfuse.

transgresser, to transgress, contravene.

transi, chilled.

transiger, to compromise, compound, come to terms.

transir, to chill, benumb, paralyze.

transition, *f.* transition.

transitionnel, -elle, transitional.

transitoire, transitory, transient.

transitoirement, transitorily, transiently.

translateur, *m.* translator.

translation, *f.* transfer, translation, transmission, removal, revolution, movement, conveyance.

translucide, translucent, translucid, semitransparent.

translucidité, *f.* translucency, translucidity.

transmetteur, *m.* transmitter, carrier.

transmetteur, transmitting, sending, carrying.

transmettre, to transmit, forward, transfer; se —, to be transmitted.

transmission, *f.* transmission, inheritance.

transmuable, transmutable.

transmuer, to transmute.

transmutable, transmutable.

transmutateur, *m.* transmuter.

transmutation, *f.* transmutation.

transmutatoire, transmutative.

transmuter, to transmute.

transocéanien, -enne, transoceanic.

transparaître, to show through.

transparence, *f.* transparency, transparence, translucency, pellucidity.

transparent, *m.* transparency, black lines (to place under paper).

transparent, transparent.

transpercer, to transfix, pierce through.

transpiration, *f.* transpiration, perspiration.

transpirer, to transpire, perspire.

transplantation, *f.* transplanting.

transplanter, to transplant.

transplantoir, *m.* garden trowel.

transport, *m.* transport, conveyance, carriage, carrying, transfer, transmission, passage, education, exhaustion, removal, fetching, revolution, fit of delirium, ecstasy, rapture, transportation.

transportable, that may be conveyed or transported.

transporter, to convey, carry, transfer, enrapture, transport, bring, transmit; **se —,** to be conveyed, transport oneself, be carried.

transporteur, -trice, conveying.

transposer, to transpose.

transposition, *f.* transposition, rearrangement.

transsudat, *m.* transudate.

transsudation, *f.* transudation.

transsuder, to transude, ooze through.

transvasement, *m.* transfer, transference, decantation.

transvaser, to transfer, decant, change from one receptacle to another.

transvaseur, *m.* decanter (person or apparatus).

transvection, *f.* transvection.

transversal, -aux, transverse, transversal, cross (section), athwart, lying crosswise.

transversalement, transversely, crosswise.

transverse, transverse, crosswise.

transvider, to decant, empty from one vessel into another.

trapan, *m.* bridge, top landing.

trapèze, *m.* trapezoid, trapeze.

trapéziforme, trapezoidal, trapezoid.

trapézoèdre, *m.* trapezohedron.

trapézoïdal, -aux, trapezoidal.

trapezoïde, *m.* trapezium.

trapézoïde, trapezoid(al).

trapp, *m.* trap (rock).

trappe, *f.* trap, trap door, sliding door, hatch, blower.

trapper, to trap animals (for fur).

trapu, squat, stocky, dumpy, thickset.

traque, *f.* beating, enclosing, battue, drive, circular drive, enclosing game.

traquer, to surround, hem in, enclose, beat.

traquet, *m.* trap.

traqueur, *m.* beater.

trass, *m.* trass.

traumatisme, *m.* traumatism.

travail, *m.* work, effort, toil, occupation, employment, workmanship, process, labor, industry, task, operation, warping, strain, fatigue, childbirth, brake, care, anxiety parturition; — **à domicile,** homework; — **à la flexion,** bending stress or strain; — **à la main,** hand labor; — **champêtre,** agricultural labor; — **de frottement,** work or energy of friction.

travail, — mécanique, machine work; — **moteur,** work developed by a motor; — **résistant,** work necessary to overcome a resistance; — **utile,** useful work; **travaux d'entretien et de repeuplement,** cultural operations; **travaux graphiques,** drafting, drawing, designing; **trauvaux forcés,** compulsory labor.

travaillé, worked, wrought, finished, worn, weathered (rock).

travailler, to work, torment, fashion, handle, cultivate, till, toil, ferment, strain, swell, alter, fade (color); **se —** to be worked, be wrought, labor, strain, endeavor; — **à la meule,** to rub down, make smooth; **faire —,** to cause to work, strain, run.

travailleur, *m.* worker, workman, laborer.

travailleur, -euse, industrious, laborious.

travaux, *m.pl.* see TRAVAIL.

travée, *f.* truss, span, bay.

travers, *m.* breadth, thickness, crookedness, eccentricity, whim, broadside (of ship), cross crack; **à —,** **au — de,** through, across, athwart; **de —,** crooked, obliquely, wrong, askew, awry; **en —,** crosswise, transversely, across, athwart.

traverse, *f.* traverse, crosspiece, crossbeam, crossbar, stay, brace, bar, bull, beam, sleeper, tie, beam support, crosshead, impediment, girder, short cut, crossroad, obstacle; — **de chemin de fer,** railway sleeper, tie; — **du piston,** piston pin; **à la —,** in the way.

traversé, decussate.

traversée, *f.* traversement, *m.* passage, voyage, crossing; — **de voie à niveau** crossing on a level, grade crossing.

traverser, to cross, traverse, go through, thwart, penetrate, go over.

traversier, -ère, cross, crossing, transverse.

traversin, *m.* crossbar, crosspiece, beam (of balance).

travertin, *m.* travertine, calcareous sinter.

travesti, disguised, travestied.

travestir, to disguise.

trayage, *m.* — à l'essai, cow testing

trayon, *m.* dug, teat, nipple.

trébuchant, *m.* slight overweight.

trébuchant, staggering, stumbling, tipping the beam, full weight.

trébucher, to tip the beam, test for weight, turn the scale, tip, incline, stumble, trip.

trébuchet, *m.* small precision balance, prescription balance, bird trap, snare.

tréfilage, *m.* wiredrawing.

tréfiler, to wiredraw.

tréfilerie, *f.* wiredrawing, wire mill.

tréfileur, *m.* wiredrawer.

trèfle, *m.* trefoil, clover (Trifolium); — **blanc,** Dutch clover, white clover (*Trifolium repens*); — **commun,** — **des prés,** common red clover, meadow clover, (*Trifolium pratense*); — **couche,** hop clover hop trefoil (*Trifolium agrarium*); — d'eau, buck bean, bog bean, marsh trefoil (*Menyanthes trifoliata*); — **flectueux,** zigzag clover or trefoil, marl grass (*Trifolium medium*); — **hybride,** alsike clover (*Trifolium hybridum*); — **jaunâtre,** pale yellow clover; — **retourné,** buffalo clover (*Trifolium reflexum*).

tréflière, *f.* clover field.

tréfonds, *m.* subsoil, minerals underneath the ground, depth.

tréhala, *m.* trehala.

tréhalose, *f.* trehalose.

treillage, *m.* trellis, latticework.

treillager, to trellis, lattice.

treille, *f.* vine arbor.

treillis, *m.* lattice, trellis, latticework, grating, canvas, sacking, sackcloth; — **en espace,** latticework in space.

treizaine, *f.* about thirteen.

treize, thirteen, thirteenth.

treizième, thirteenth.

trèjetage, *m.* lading, transferring.

trèjeter, to lade, transfer (molten glass).

tremblaie, *f.* aspen grove.

tremblant, trembling, shivering, flickering.

tremble, *m.* **peuplier —,** aspen, aspen tree, trembling poplar (*Populus tremula*).

tremblé, wavy, waved, trembling.

tremblement, *m.* trembling, tremor, shivering, flickering, quaking, shaking; — **de terre,** earthquake.

trembler, to tremble, flicker, quake, shake, shiver, fear.

tremblette, *f.* quaking grass, trembling (*Briza media*).

trembleur, *m.* trembler, vibrator, interrupter.

tremblotant, quivering, tremulous.

trembloter, to tremble (slightly), quiver, twinkle, flicker, sparkle.

trémellacées, *f.pl.* Tremellaceae.

trémelle, *f.* Tremella.

trémie, *f.* hopper, mill hopper, funnel, feeding box.

trémolite, *f.* tremolite.

trémoussement, *m.* fluttering, frisking.

trémousser, to flutter; **se —,** to flutter about, frisk, stir about, fidget, bestir oneself.

trempage, *m.* soaking, steeping, wetting.

trempant, soaking, capable of being tempered.

trempe, *f.* steeping, soaking, wetting, temper (of steel), malting water; — **de la surface,** — **en paquet,** half converting, casehardening.

trempé, soaked, tempered, wet (to the skin).

tremper, to soak, steep, dip, mix, dilute (wine) with water, immerse, drench, dampen, wet, temper, soften, harden, be implicated; — **à blanc,** to temper when white hot.

trempeur, *m.* soaker, temperer, hardener.

trempis, *m.* pulp vat, steeping room, soaking water for salt fish.

trempoir, *m.*, **trempoire,** *f.* steeping (tub) vat.

trentaine, *f.* (about) thirty.

trente, thirty, thirtieth.

trente-deux, thirty-two.

trentième, thirtieth.

trépan, *m.* trepan, trepanning tool, boring bit, drill, boring.

trépaner, to bore into, drill into, trepan.

trépasser, to die, pass away.

trépidation, *f.* vibration, oscillation, tremor, jarring, trepidation.

trépied, *m.* tripod, three-legged stand, trivet.

trépigner, to tread, trample down, stamp.

tréponème, *f.* Treponema.

très, very, most, very much; de — bonne heure, very soon.

trésaille, trésallé, crackled.

trésaillure, *f.* crack, minute crack.

trésor, *m.* treasure, treasury; *pl.* riches.

trésorerie, *f.* treasury.

trésorier, *m.* treasurer, paymaster.

tressage, *m.* plaiting, braiding.

tressaillé, crackled.

tressaillement, *m.* sudden movement, shuddering, starting.

tressaillir, to start, throb, thrill, leap (up), jump, quiver.

tressaillure, *f.* crack.

tresse, *f.* plait, tress, braid, plat, thick brown paper.

tresser, to plait, braid, plat, weave, interweave, twist, form into tresses.

tréteau, *m.* trestle, support, horse, stand.

treuil, *m.* winch, windlass.

tri, *m.* sorting.

triable, worth (capable of) being sorted or picked.

triacanthe, with spines arranged in threes.

triacétine, *f.* triacetin.

triacontaèdre, triacontahedral.

triacrorhize, triacrorhize, (roots) arising from three initial cells or groups at the apex.

triade, *f.* triad.

triadelphie, triadelphous.

triadique, triad, triadic.

triage, *m.* sorting, separating, assorting, grading, beat (of forest guard), wet picking.

triailé, with three wings.

triandre, triandrous.

triandrie, *f.* triandria.

triangle, *m.* triangle; — acutangle, acute-angled triangle, oxygon; — isocéle, isosceles triangle; — obtusangle, obtuse-angled triangle, amblygon.

triangulaire, triangular.

triangulation, *f.* survey, surveying.

triannuel, -elle, triennial.

trianthe, trianthous, three-flowered.

triarche, *f.* triarch, a fibrovascular cylinder having three xylem strands.

triaristé, provided with three aristae (awns).

triasique, Triassic.

triatomique, triatomic.

triazine, *f.* triazine.

triazinique, triazine, of triazine.

triazol, *m.* triazole.

triaxifère, (inflorescence) presenting three degrees of vegetation.

tribarytique, tribarium.

tribasicité, *f.* tribasic character.

tribasique, tribasic.

tribiné, thrice binate (leaf).

triboluminescence, *f.* triboluminescence.

triboluminescent, triboluminescent.

tribord, *m.* starboard.

tribracté, (flower) with three bracts.

tribractéolé, (pedicels) having three bracteoles.

tribu, *f.* tribe, clan.

tribuloïde, tribuloid.

tribunal, *m.* tribunal, court.

tribune, *f.* rostrum, gallery, platform.

tribut, *m.* tribute.

tributaire, tributary.

tributyrine, *f.* tributyrin.

tricalcique, tricalcium, tricalcic.

tricamare, tricamarous, (fruit) composed of three loculi.

tricapsulaire, tricapsular.

tricarballylique, tricarballylic.

tricarbonique, tricarboxylic.

tricarpe, tricarpous, of three carpels.

tricétone, *f.* triketone.

trichanthe, with capillary flowers.

trichélostyle, with three stigmas and triangular seed.

trichine, *f.* trichina, threadworm.

trichite, *f.* trichite.

trichiure, *m.* trichiura, trichiurus.

trichloré, trichloro.

trichlorhydrique, trihydrochloric.

trichlorure, *m.* trichloride.

trichocalicé, with velvety calyx.

trichocarpe, trichocarpous, (fruit) covered with hairlike pubescence.

trichocaule, hairy stemmed.

trichophytique, trichophytal.

trichophyton, *m.* trichophyton.

trichotome, trichotomous, three-forked.

trichotoxine, *f.* trichotoxin.

trichocéphale, trichocephalous, (flowerheads) with hairlike appendages.

trichoclade, with hairy branches.

trichode, trichodes, resembling hair.

trichogyne, *f.* trichogyne.

trichome, *f.* trichome, hairlike outgrowth of the epidermis.

trichopétale, with velvety petals.

trichophylle, trichophyllous, hairlike, finely cut (leaves).

trichosépale, with velvety sepals.

trichosporange, *m.* trichosporangium.

trichospore, *f.* zoospore, swarm spore.

trichostémone, with velvety stamens.

trichotomie, *f.* trichotomy, division into threes.

trichroïsme, *m.* trichroism.

trichrome, three-color, trichrome.

trichromie, *f.* three-color process, threecolor photography.

triclinique, triclinic.

tricoque, tricoccous (fruit).

tricorne, three-horned (animal).

tricosté, tricostate, having three ribs.

tricot, *m.* knitting, knitted fabric.

tricoter, to knit.

tricotylédoné, tricotyledonous, having three cotyledons.

tricuspide, tricuspid.

tricyanhydrine, *f.* tricyanhydrin.

tricyclène, *m.* tricyclene.

tridactyle, with leaves composed of three leaflets.

tridenté, tridentate, three-toothed, tridentpointed.

tridigité, tridigitate, thrice digitate, ternate.

tridyname, tridynamous, with three stamens out of six that are longer.

trie, *f.* sorting.

trièdre, *m.* trihedron.

trièdre, trihedral.

triennal, -aux, triennial.

triennat, *m.* three-year period.

triépineux, -euse, bearing three spines.

trier, to pick out, sort out, choose, select, separate, classify.

trieur, -euse, *m.&f.* sorter, separator, winnower, grader.

trifarie, trifarious, in three vertical ranks.

trifide, three-cleft.

triflore, triflorous.

trifolié, trifoliate.

trifoliolé, trifoliate, trifoliolate.

trifolium, *m.* clover (Trifolium).

trifurcation, *f.* trichotomy, division into three.

trifurquer, to divide into three branches.

trigame, trigamous.

trigéminé, tergeminal, tergeminate.

trigemme, bearing three buds.

triglandé, with fruit composed of three glands.

triglochidé, *f.* thorn terminating in three bent points.

triglochin, *m.* arrow grass (Triglochin).

triglume, with three glumes.

triglycéride, *n.* triglyceride.

trigone, trigonal, triangle, trigone.

trigonelle, fenugreek (*Trigonella foenumgraecum*).

trigonelline, *f.* trigonelline.

trigonocarpe, trigonocarpous, (fruit) having three evident angles.

trigonométrie, *f.* trigonometry.

trigonométrique, trigonometric(al).

trigyne, trigynous.

trihalogéné, trihalogenated.

trihilaté, trihilatous, having three apertures.

trihydroxylé, trihydroxy.

trijumeau, triplet, trigeminal, trifacial.

trilatéral, -aux, three-sided.

trilépide, furnished with three scales.

trilinéaire, trilinear.

trilobé, trilobate.

triloculaire, trilocular.

triloupe, *f.* three-lens magnifying glass.

trimellique, trimellitic, trimellic.

trimère, *m.* trimer, polymer; les —s, the trimera.

trimère, trimeric.

trimésique, trimesic.

trimestre, *m.* quarter (of a year), trimester.

trimestriel, -elle, quarterly, trimestrial.

triméthylamine. *f.* trimethylamine.

triméthylène, *m.* trimethylene.

triméthylénique, trimethylene.

triméthylméthane, *m.* trimethylmethane, isobutane.

trimorphe, trimorphous.

trinervé, trinervate, three-nerved.

tringle, *f.* rod, strip, tringle, wire; — de manoeuvre, switch bar.

tringlette, *f.* small rod.

Trinité, *f.* Trinidad.

trinitration, *f.* trinitration.

trinitré, *m.* trinitro compound.

trinitré, trinitrated, trinitro.

trinitrine, *f.* trinitrin, nitroglycerin.

trinitrophénol, *m.* trinitrophenol.

trinitrotoluène, *m.* trinitrotoluene.

trinôme, trinomial.

trio, *m.* trio, three-high roll.

triode, *f.* triode, three-electrode (lamp).

trioecie, *m.* trioecism, polygamodioecism.

trioléine, *f.* triolein.

triomphal, **-aux,** triumphal, triumphant.

triomphant, excelling, triumphant, decisive.

triomphe, *m.* triumph.

triompher, to triumph, excel, exult.

trional, *m.* trional.

triostée, **trioste,** *m.* feverroot, feverwort (*Triosteum perfoliatum*).

trioxyde, *m.* trioxide.

trioxyméthylène, *m.* trioxymethylene.

tripalmitine, *f.* tripalmitin.

tripartit, tripartite, divided into three parts.

tripe, *f.pl.* entrails, intestines, tripe.

tripennatifide, thrice pinnatifid, tripinnate.

tripennatisqué, thrice pinnatisect, tripinnate.

tripenné, tripinnate.

tripeptide, *n.* tripeptide.

tripétale, **tripetalé,** tripetalous.

triphane, *m.* triphane, spodumene.

triphasé, triphase, three-phase.

triphénylamine, *f.* triphenylamine.

triphénylique, triphenyl.

triphénylméthane, *m.* triphenylmethane.

triphylle, triphyllous.

triphylline, *f.* triphylite, triphyline.

triple, triple, treble, threefold; — liaison, triple bond.

triplement, *m.* tripling.

triplement, triply, trebly.

tripler, to triple, treble.

triplication, *f.* triplication.

triplinervé, triple-nerved.

tripoli, *m.* tripoli, Tripoli.

tripolisser, tripolir, to polish with tripoli or rottenstone.

tripotage, *m.* mess, medley, intrigue.

tripoter, to mess, dabble, manipulate.

tripsac, *m.* gama grass (*Tripsacum dactyloides*).

trique, *f.* stick, cudgel.

trique-madame, *f.* white stonecrop (*Sedum album*).

triquètre, triquetrous (stem).

triquiné, triquinate, divided into three, then into five.

triquinoyle, *m.* triquinoyl.

trisaccharide, *m.* trisaccharide.

trisannuel, **-elle,** triennial (plant).

triscape, with three stems.

trisépale, trisepalous.

triséqué, with three segments.

trisoc, *m.* three-furrow plow.

trisperme, three-seeded.

trissement, *m.* twitter (of swallows).

tristachyé, tristachyous, three-spiked.

tristaminifère, bearing three stamens.

triste, sorrowful, sad, dull, cheerless, gloomy, poor, sorry, wretched.

tristéarine, *f.* tristearin.

tristement, sadly, poorly, wretchedly.

tristesse, *f.* sadness, melancholy, gloom, dullness.

tristigmaté, with three stigmas.

tristique, tristichous.

trisubstitué, trisubstituted.

trisulfure, *m.* trisulphide; — d'arsenic, sulphide of arsenic.

triternatiséqué, thrice divided into three segments.

triterné, triternate, thrice-ternate.

triticé, like wheat.

tritocérébron, *m.* tritocerebrum.

tritoxyde, *m.* tritoxide.

triturable, triturable.

triturateur, *m.* triturator, grinder.

trituration, *f.* trituration, grinding, crushing

triturer, to triturate, grind, masticate, pulverize.

trivalence, *f.* trivalence.

trivalent, trivalent.

trivalérine, *f.* trivalerin.

trivalve, trivalvular, three-valved.

trivial, -aux, trivial, ordinary, common.

troc, *m.* truck, barter, exchange.

trochanter, *m.* trochanter.

trochantin, *m.* trochantin.

trochée, *f.* brushwood.

trochereau, *m.* long-leaved pine, Georgia pine (*Pinus palustris*).

trochet, *m.* cluster (of fruits or flowers).

trochiscation, *f.* formation into tablets or pastils.

trochisque, *m.* tablet, cake, troche, trochiscus, lozenge, pastil.

trochisquer, to form into tablets or pastils.

trochléaire, trochlear.

trochléariforme, trochleariform, pulley-shaped.

troène, *m.* privet (*Ligustrum vulgare*).

troglodyte, *m.* wren.

trognon, *m.* core (of fruit), stump (of cabbage), stalk.

trois, three, third.

troisième, *m.* third, third floor; *f.* third class.

troisième, third.

troisièmement, thirdly, in the third place.

trois-pieds, *m.* tripod.

trois-quarts, *m.* triangular file, triangular rasp.

trois-six, *m.* proof spirit.

trolle, *m.* globeflower (Trollius).

trombe, *f.* waterspout.

trommel, *m.* trommel, revolving screen.

trompe, *f.* steam whistle, pump, trompe, horn, trunk (elephant), proboscis, probe (of insect), tube, aspirator, water pump, oviduct; — **à eau,** water-jet pump, water suction pump; — **à mercure,** mercury pump; — **aspirante,** suction pump; — **de Fallope,** Fallopian tube; — **d'Eustache,** Eustachian tube.

tromper, to deceive, mislead, beguile, cheat, elude; **se —,** to be mistaken, be deceived, deceive oneself, make a mistake.

tromperie, *f.* deceit, cheating, deception, fraud.

trompette, *m.* trumpeter.

trompette, *f.* trumpet.

trompeur, -euse, *m.&f.* deceiver, cheater beguiler.

trompeur, -euse, deceptive, delusive, deceitful, false.

trompeusement, deceptively, deceitfully.

tronc, *m.* trunk (of tree), main body, stump, stem, stock, frustum, thorax; — **d'arbre,** stump, stub, stool; — **de cône,** frustum of a cone, truncated cone; — **commun,** main line.

troncature, *f.* truncation.

tronce, *f.* butt, log.

tronche, *f.* stem-pruned tree.

tronçon, *m.* piece, fragment, stump, butt, portion, section.

tronconique, truncated, of the shape of a truncated cone.

tronçonner, to cut into pieces, lengths, or sections.

trône, *m.* throne.

tronqué, truncate.

tronquer, to truncate, mutilate.

troostite, *f.* troostite.

trop, *m.* too much, too many, excess.

trop, too, over, too much, too many, too long, very many; — **mur,** overripe; **de —,** too much, too many; **par —,** over much, too much, far too.

tropacocaïne, *f.* tropacocaine.

tropéolé, like a nasturtium.

tropéolées, *f.pl.* Tropaeolaceae.

tropéoline, *f.* tropaeolin.

tropéolum, *m.* tropaeolum.

trophée, *m.* trophy.

trophie, *f.* trophy, an unequal lateral growth of tissue.

trophophylle, *f.* trophophyll, vegetative leaf.

trophoplasme, *f.* trophoplasm, alveolar plasma.

trophosperme, *m.* trophosperm, placenta.

tropical, -aux, tropical.

tropilidène, *m.* tropilidene.

tropique, *m.* tropic.

tropisme, *m.* tropism.

tropophile, tropophilous, adapted to seasonal changes.

tropophylle, *f.* tropophyll, leaf of shrub or tree.

trop-plein, *m.* overflow pipe or basin, catchall.

troquer, to exchange, barter.

trot, *m.* trot.

trottant menu, moving about with little steps.

trotter, to trot.

trotteur, *m.* trotter.

trottoir, *m.* sidewalk, footpath.

trou, *m.* hole, opening, orifice, gap, cavity, foramen; — de coulée, taphole, discharge opening; — d'écoulement, outlet; — de graissage, oil hole; — de laitier, slag hole, cinder hole; — de remplissage, filling hole; — de ver, worm hole; — de vidange, emptying hole; — d'homme, manhole.

trouble, *m.* turbidity, cloudiness, affection, trouble, agitation, disturbance, confusion, disorder.

trouble, turbid, cloudy, muddy, troubled, thick, overcast, dull, dim.

troubler, to cloud, trouble, disturb, render turbid, agitate, make cloudy, perplex, confuse; se —, to cloud, become turbid, be confused, be disturbed, become dim, become troubled.

troué, full of holes, bored, pierced, perforated.

trouée, *f.* gap, opening, gap cutting, blank, glade.

trouer, to make a hole or holes, bore, perforate, pierce; se —, to get full of holes, open up.

troupe, *f.* troop, band, company, throng, party, number, herd, drove, flock, flight, muster, swarm.

troupeau, *m.* herd, flock, drove, troop.

trousse, *f.* bundle, packet, truss (of hay), case (of instruments).

trousseau, *m.* bunch (of keys), small bundle, fasciculus, outfit, trousseau.

trousser, to bundle up, turn up, tuck up, tie up, pack up, truss (fowl).

trouvable, that can be found, discoverable.

trouvaille, *f.* find, finding, discovery.

trouvé, found, new, original, felicitous.

trouver, to find, discover, come upon, detect, invent, meet with, deem, think, consider; se —, to be found, be, be present, feel, happen, turn out, happen to be, occur, prove to be, think oneself, meet with, find oneself; — à dire à, to find fault with; ça se trouve comme ça, it happens like that; il se trouve bien, he is well; il le trouve bon, he approves of it; il se trouve que, it happens that.

truc, *m.* truck, flatcar, trick, knack, skill.

truck, *m.* truck.

truellage, *m.* troweling.

truelle, *f.* trowel.

truellée, *f.* trowelful.

trueller, to trowel.

truffe, *f.* truffle.

truffière, *f.* truffle ground, truffle bed.

truflier, *m.* privet (*Ligustrum vulgare*).

truie, *f.* sow.

truite, *f.* trout; — arc-en-ciel, rainbow trout; — commune, common trout, brown trout; — des lacs, lake trout.

truité, red-spotted, speckled, mottled (iron), flea-bitten, finely crackled.

truncicole, growing and living on trunks of trees.

trust, *m.* trust.

truxilline, *f.* truxilline.

truxillique, truxillic.

tryma, *f.* tryma, drupaceous nut with dehiscent exocarp.

trypanolytique, trypanolytic.

trypanose, *f.* trypanosomiasis, disease caused by Trypanosoma.

trypanosome, *m.* trypanosome.

trypsine, *f.* trypsin.

trypsinogène, *f.* trypsinogen.

tryptique, tryptic.

tryptophane, *m.* tryptophan.

tu (taire), kept secret, silent.

tu, thou, you.

tuant, killing, tiresome.

tub, *m.* bathtub, bath.

tubage, *m.* tubing, casing, lining with a tube, intubation.

tube, *m.* tube, pipe; *pl.* tubing, duct; — à azote, nitrogen tube; — abducteur, delivery tube, exit tube; — à boule, bulbed tube; — à brome, bromine funnel, dropping funnel; — à combustion, combustion tube; — à condensation, condensation tube; — à dégagement, delivery tube, exit tube; — à dessécher, drying tube; — à éntonnoir, funnel tube; — à essai, test tube, boiling tube.

tube, — alimentaire, feed pipe; — allonge, lengthening tube; — à robinet, tube with stopcock; — à vaccin, vaccine tube; — à vide, vacuum tube, valve; — à vis, bottle with a screw top; — capillaire, capillary (tube); — d'alimentation, feed pipe; — d'arrivée, inlet tube; — de cuivre, copper tube, copper pipe; — de dégagement, delivery tube; — de laiton, brass tube, brass insulating conduit.

tube, — d'épreuve, — d'essai, test tube; — de sûreté, safety tube, safety funnel; — divisé, graduated tube; — doseur, measuring tube; — en acier, steel tube, steel pipe; — en biscuit, porcelain tube; — en caoutchouc, (India) rubber tube, rubber tubing; — en carton, pasteboard (paper) tube; — en fer, iron pipe; — en fer bouché à vis, iron tube with screw cap; — en Iéna, tube of Jena glass; — en papier, paper (tube) insulating conduit; — en U, U tube.

tube, — en verre, glass tube; — en verre d'Iéna, tube of Jena glass; — étiré, drawn tube; — fermé, closed tube; — latéral, side tube, side neck; — laveur, gas-washing tube; — métallique, metal tube, metal pipe; — ovarique, egg-tube, ovariole; — pollinique, pollen tube, pollen sack; — sans soudure, seamless tube; — soudé, welded (tube) pipe; — témoin, indicator tube.

tuber, to tube, case, line with a tube, intubate.

tubéracé, tuberaceous.

tubercule, m. tuber, tubercle, nodule, eminence.

tuberculé, swollen.

tuberculeux, -euse, tubercular, tuberculous, tuberous.

tuberculigène, productive of tuberculosis.

tuberculination, f. performance of the tuberculin test.

tuberculine, f. tuberculin.

tuberculinque, tuberculin.

tuberculisé, infected with tuberculosis.

tuberculoïde, resembling the tubercle bacillus.

tuberculose, f. tuberculosis.

tubéreuse, f. tuberose.

tubéreux, -euse, tuberous.

tubériforme, like a tubercle.

tubérisation, f. tuberisation.

tubérosité, f. tuberosity.

tubiflore, tubiflorous, (florets) tubular.

tubispathe, with tubular spathe.

tubulaire, tubular.

tubule, m. small tube, tubule, tubulus.

tubulé, tubulated.

tubuleux, -euse, tubulous, tubulose, tubular.

tubuli-forme, tubiform.

tubulure, f. tubulure, tubulature, neck, tubulus, nozzle, tubule, small tube, pipe, side neck.

tudesque, Gothic, Germanic, Teutonic.

tue-cafards, m.pl. beetle exterminator.

tue-chien, m. dogbane (Apocynum androsaemifolium), meadow saffron (Colchicum autumnale).

tue-fourmis, m.pl. ant exterminator.

tue-mouche, m. poison fly paper, fly agaric (Amanita muscaria).

tuer, to kill, slay, bore.

tue-vent, m. wind screen.

tuf, m. tufa, tuff, bottom, bedrock.

tufacé, tufaceous, tuffaceous.

tuffeau, tufeau, m. calcareous tufa.

tufier, m. tufa quarry.

tufier, -ière, tufaceous.

tuile, f. tile, misfortune, unlucky event; — creuse, gutter tile, hollow tile; — faîtière, ridge tile, hip tile; — plate, flat (plain) tile.

tuileau, m. fragment of a tile.

tuilerie, f. tileworks, tilery, tilemaking; pl. Tuileries.

tuilette, f. small tile.

tuilier, m. tilemaker.

tuilier, pertaining to tilemaking.

tulipacé, like a tulip.

tulipe, f. tulip.

tulipé, like a tulip.

tulipier, m. tulip tree (Liriodendron tulipifera).

tulle, m. tulle, fine silk net (fabric), netting.

Tultèques, m.pl. Toltecs.

tuméfaction, f. tumefication, tumefaction; — de l'écorce, cankerous growth.

tuméfier, to tumefy, swell.

tumeur, f. tumor, tumour, swelling; pl. crown wart.

tumide, tumid, turgid.

tumulte, m. tumult, turmoil, agitation.

tumultueux, -euse, tumultuous, violent, noisy, turbulent.

tumulus, m. tumulus, sepulchral mound.

tungstate, m. tungstate; — de soude, sodium tungstate.

tungstène, m. tungsten.

tungstique, tungstic.

tuniciers, m.pl. Tunicata.

tunicine, f. tunicin.

tunique, f. tunic, coat, layer, membrane envelope, tegument.

tuniqué, tunicate.

tunisien, -enne, Tunisian.

turanose, *f.* turanose.

turbide, turbid.

turbidité, *f.* turbidity, cloudiness.

turbinage, *m.* centrifuging.

turbinaire, turbinate.

turbine, *f.* centrifugal machine, centrifuge, turbine; — **à action,** impulse turbine; — **à réaction,** reaction turbine; — **à vapeur,** steam turbine; — **d'impulsion,** impulse turbine; — **parallèle centripète,** mixed-flow turbine.

turbiné, turbinate.

turbiniflore, (flowers) arranged turbinately.

turbith, *m.* turpeth, subsulphate of mercury; — **minéral,** turpeth mineral; — **végétal,** vegetable turpeth, Indian jalap (*Ipomoea turpethum*).

turbo-alternateur, *m.* turboalternator.

turbulence, *f.* turbulence.

turbulent, *m.* a rotating box, drum.

turbulent, turbulent, violent.

turc, *m.* Turkish, Turk.

turgescence, *f.* turgescence.

turgide, turgid, swollen.

turion, *f.* turion, scaly sucker.

turnep, turneps, *m.* field turnip, kohlrabi.

turque, Turkish.

turquet, *m.* maize, Indian corn (*Zea mays*).

Turquie, *f.* Turkey.

turquin, dark, designating a kind; **bleu —,** slate blue, bluish-gray (marble).

turquoise, *f.* turquoise.

tussack, *m.* tussock (grass).

tussah, tussau, tussah (silk).

tussigène, cough-provoking.

tussilage, *m.* coltsfoot (*Tussilago farfara*).

tussor, *m.* tussah, tussore silk.

tute, *f.* assay crucible.

tutelle, *f.* tutelage, guardianship.

tutenay, *n.* tutenag, crude zinc, spelter.

tuteur, *m.* guardian, protector, stake, prop.

tuteur, -trice, protecting.

tuthie, tutie, *f.* tutty, crude zinc oxide.

tuyau, *m.* pipe, tube, nozzle, hose, gutter, conduit, flue, (hollow) stem, stalk; — **à gaz,** gas pipe, gas tube; — **alimentaire,** feed pipe; — **à vapeur,** steam pipe; — **courbé,** bent pipe; — **d'arrivée,** intake pipe; — **d'arrosage,** garden hose; — **d'aspiration,** suction pipe, draft tube.

tuyau, — **de conduite,** conduit, main; — **de décharge,** discharge pipe, outlet pipe, waste pipe; — **de dégagement,** delivery (escape) pipe; — **de refoulement,** delivery (exhaust) pipe, rising pipe; — **de trop-plein,** overflow pipe; — **de vidange,** waste pipe, drain pipe; — **en caoutchouc,** rubber (hose) tube; — flexible, hose, (rubber) tubing.

tuyautage, *m.* piping, tubing.

tuyauterie, *f.* pipes, piping, tubing, pipeworks, factory, or trade.

tuyère, *f.* tuyère, blast pipe.

tympanite, *f.* tympanitis.

tympe, *f.* tymp.

type, *m.* type, style, standard.

type, standard, typical; **solution —,** standard solution.

typer, to stamp, be of a certain standard or type.

typhique, of typhus, typhous, typhoidal.

typhoïde, typhoid.

typhoïdique, *m.* typhoid patient.

typhus, *m.* typhoid fever.

typique, typical, symbolical.

typographie, *f.* typography.

typographique, typographic, typographical.

tyrannie, *f.* tyranny.

tyranniser, to tyrannize over.

tyrien, -enne, Tyrian.

tyrolien, -enne, Tyrolese.

tyrosinase, *f.* tyrosinase.

tyrosine, *f.* tyrosine.

U

U, en forme d'—, U-shaped.

ubiquiste, *f.* ubiquist, plant occurring on any type of geological formation.

ulceration, *f.* ulceration.

ulcère, *m.* ulcer, suppuration, ulceration, festering.

ulcéré, ulcerated.

ulcérer, to ulcerate.

ulcéreux, -euse, ulcerous.

ulex, *m.* furze, gorse, whin (*Ulex europaeus*).

ulexine, *f.* ulexine.

uliginaire, uligineux, -euse, uliginous, uliginal (plant), uliginose.

ulmacé, ulmaceous (plant).

ulmaire, *f.* meadowsweet (*Filipendula ulmaria*).

ulmine, *f.* ulmin.

ulmique, ulmic.

ulophylle, with curly leaves.

ulosperme, (seed) with curly ribs.

ulotrique, with curly hair.

ultérieur, ulterior, further. later, subsequent.

ultérieurement, later, subsequently.

ultime, ultième, last, ultimate, final.

ultimo, lastly, finally.

ultrafiltration, *f.* ultrafiltration.

ultramarine, *f.* ultramarine, lapis lazuli.

ultramicroscope, *m.* ultramicroscope.

ultramicroscopie, *f.* ultramicroscopy.

ultramicroscopique, ultramicroscopic(al).

ultra-terrestre, ultraterrestrial.

ultra-violet, -ette, ultraviolet.

ulve, *f.* green laver, sea lettuce (Ulva).

umbraculiforme, umbrella-shaped.

un, une, one, a, an; *pl.* some, the ones; — à —, one by one; l'— et l'autre, both; l'— l'autre, each other, reciprocally; **les —s,** some; **les —s et les autres,** all.

unanime, unanimous.

unanimement, unanimously.

unanimité, *f.* unanimity.

unciné, hook-shaped.

une, one, a, an.

uneicosane, *m.* heneicosane.

unguis, *m.* unguis.

uni, united, level, smooth, even, harmonious, uniform, plain, simple, calm, usual.

unibractété, with solitary bracts.

unicapsulaire, unicapsular.

unicaréné, with a single carina.

unicaule, uniaxial.

unicellulaire, unicellular, one-celled.

unicolore, one-colored, unicolor, whole-colored, self-colored.

uniembryoné, uniembryonatous, having one embryo.

unième, trente et —, thirty-first.

unifascié, marked by a single band.

unifeuille, unifoliate.

unification, *f.* unification, consolidation.

unifier, to unify, unite, standardize, amalgamate.

uniflore, uniflorous.

uniflorigère, with a single flower.

unifolié, unifoliate.

uniforme, uniform, even, regular, evenaged, monochromatic.

uniformément, uniformly, evenly.

uniformiser, to make uniform, standardize.

uniformité, *f.* uniformity, evenness.

unigemme, single-budded.

unijugué, unijugate.

unilabié, unilabiate.

unilatéral, -aux, unilateral.

unilobé, unilobate.

uniloculaire, unilocular.

uniment, smoothly, evenly, simply, plainly.

unimoléculaire, unimolecular, monomolecular.

uninervé, uninerviate, one-veined, oneribbed.

uninucléé, uninucleate, having a single nucleus.

union, *f.* union, unity, concord, linkage.

uniovulé, uniovulate, with a solitary ovule.

unipare, uniparous.

unipétale, unipetalous.

unipolaire, unipolar.

unique, sole, only. single, unique, singular, standing alone, unequaled.

uniquement, only, uniquely, solely, alone.

unir, to unite, join together, combine, level, smooth; s'—, to unite, pair, mate, come together, be united.

unisérié, uniseriate, in one horizontal row or series.

unisexué, unisexual, monoecious.

unisexuel, -elle, unisexual, of one sex.

unissant, uniting.

unisson, *m.* unison.

unitaire, of a unit, per unit, unitary.

unité, *f.* unit, unity; — de chaleur, unit of heat, thermal unit; — de débit, unit of output; — de longueur, unit of length; — de surface, unit of area; — thermique, heat or thermal unit.

univers, *m.* universe, world.

universaliser, to universalize.

universalité, *f.* universality, entirety, whole, generality.

universel, *m.* universal.

universel, -elle, universal, general, residuary.

universellement, universally, generally.

universitaire, *m.* university professor.

universitaire, university.

université, *f.* university.

univoque, unequivocal, unambiguous.

upas, *m.* upas (tree or juice) (*Antiaris toxicaria*).

uracile, uracil, *m.* uracil.

ural, *m.* ural.

uraminé, uramido, carbamido.

uranate, *m.* uranate.

urane, *m.* uranium, oxide of uranium.

uraneux, -euse, uranous.

uranifère, uraniferous.

uranique, uranic.

uranite, *f.* uranite.

uranium, *m.* uranium.

uranophane, *m.* uranophane.

uranyle, *m.* uranyl.

urao, *m.* urao.

urate, *m.* urate (fertilizer).

urazol, *m.* urazole.

urbain, urban, city.

urcéiforme, like a vase or goblet.

urcéolé, urceolate, urn-shaped.

uréase, *f.* urease.

urédinée, *f.* rust fungus.

urédo, *m.* uredo.

urédosore, *m.* uredosorus, group of uredospores.

urédospore, *f.* uredospore.

urée, *f.* urea.

uréide, *m.* ureide.

uréique, urea, pertaining to urea.

urémie, *f.* uremia, uraemia.

urémique, uremic.

uréomètre, *m.* ureameter, ureometer.

uréométrie, *f.* ureametry, ureometry.

uretère, *m.* ureter.

uréthane, *m.* urethane, ethyl carbamate.

uréthral, -aux, urethral.

uréthrite, *f.* urethritis.

urètre, *m.* urethra.

urgemment, urgently.

urgence, *f.* urgency.

urgent, urgent.

urginée, *f.* squill.

urinaire, urinary.

urine, *f.* urine.

uriner, to urinate.

urineux, -euse, urinous, urinose.

urinomètre, *m.* urinometer.

urique, uric.

urne, *f.* urn.

urobiline, *f.* urobilin.

urobilinogène, *m.* urobilinogen.

urochrome, *m.* urochrome.

urochs, *m.* aurochs, bison.

urogénital, -aux, urogenital.

urolithe, *m.* urolith, urinary calculus.

uromère, *f.* abdominal segment, uromere.

uromètre, *m.* urometer, urinometer.

urophage, urea-decomposing.

urophylle, (leaf) extended into a kind of tail.

uroscopie, *f.* uroscopy.

urotoxique, urotoxic.

urotropine, *f.* urotropine.

urticacées, *f.pl.* Urticaceae.

urticaire, *f.* urticaria, nettle rash, purples.

urticifolies, (leaves) resembling those of the nettle.

usage, *m.* use, usage, custom, habit, practice; à l'— de, for the use of; d'—, customary; hors d'—, out of use, obsolete.

usagé, used, old, worn.

usager, *m.* rightholder, commoner.

usé, worn out, threadbare, stale, obsolete.

user, to use up, consume, wear out, make use of, rub down, polish, impair; s'—, to be used up, wear out, waste, decay, be spent up, wear away, rub, deteriorate.

user, *m.* wear, wearing.

useur, *m.* grinder, polisher.

usinage, *m.* machining, machine finishing.

usine, *f.* works, factory, manufactory, mill, plant, shop; — à gaz, gasworks; — centrale, power station; — d'engrais, fertilizer factory; — frigorifique, refrigerating plant, cold-storage plant; — génératrice, power station; — hydraulique, water-power plant, waterworks; — marémotrice, tide motor power plant.

usiner, to machine, tool (casings), exploit, sweat.

usinier, *m.* manufacturer.

usité, used, in use, current, usual.

usquebac, *m.* usquebaugh (whisky).

ustensile, *m.* utensil, implement, tool; — de cuisine, kitchen utensil; — de ménage, household utensil.

ustilaginacées, *f.pl.* Ustilaginaceae.

ustilaginée, *f.* smut fungus.

ustion, *f.* burning, cauterization.

usuel, -elle, usual, common, customary.

usuellement, usually.

usufruit, *m.* usufruct, fruition.

usufruitier, *m.* usufructuary.

usure, *f.* wear, wearing, wear and tear, usury.

usurpateur, *m.* usurper.

usurpateur, -trice, usurping.

usurper, to usurp, encroach on.

utile, useful, serviceable, advantageous, convenient, profitable.

utilement, usefully, advantageously.

utilisable, utilizable, usable.

utilisation, *f.* utilization, use, using, sale, exploitation; — de déchets, waste utilization.

utiliser, to utilize, make use of, use, turn to account, sell.

utilitaire, utilitarian.

utilité, *f.* utility, use, profit, usefulness, value; sans —, unnecessarily.

utopique, Utopian.

utricule, *m.* utricle.

utrigère, bearing utricles.

uva-ursi, *m.* bearberry, uva-ursi (*Arctostaphylos uva-ursi*).

uvette, *f.* shrubby horsetail (Ephedra).

uvifère, grape-bearing.

uviforme, grape-shaped.

uvulaire, *f.* bellwort (Uvularia).

uvulaire, uvular.

V

V, en —, V-shaped.

va (aller), (he) is going; un va-et-vient, a coming and going.

vacance, *f.* vacancy; *pl.* vacation.

vacant, vacant.

vaccin, *m.* vaccine, lymph.

vaccinateur, *m.* vaccinator.

vaccinateur, -trice, vaccinating.

vaccination, *f.* vaccination.

vaccine, *f.* vaccinia, cowpox.

vacciner, to vaccinate.

vacciniées, *f.pl.* Vacciniaceae.

vaccinine, *f.* arbutin.

vaccinique, vaccine.

vaccinogène, lymph-producing.

vaccinotherapique, vaccinotherapic.

vache, *f.* cow, cowhide, leather; — laitière, milk cow; — pleine, cow in calf.

vachelin, vacherin, *m.* a kind of Gruyère cheese (soft).

vacher, *m.* milker.

vacherie, *f.* cow shed, barn.

vachette, *f.* kid, kidskin, calfskin.

vaciet, *m.* grape hyacinth (*Muscari botryoides*).

vacillant, unsteady, wavering, inconstant, vacillating, versatile (author).

vacillation, *f.* unsteadiness, wavering, vacillation.

vaciller, to vacillate, waver, be undecided, flicker, falter, stagger, fluctuate.

vacillité, *f.* waveringness, unsteadiness.

vacuité, *f.* vacuity, emptiness.

vacuolaire, vacuolar, vesicular.

vacuole, *f.* vacuole, vesicle.

vacuomètre, *m.* vacuum gauge.

vacuum, *m.* vacuum.

va-et-vient, *m.* backward and forward or up-and-down motion, seesaw motion, reciprocating motion.

va-et-vient, back-and-forth, reciprocating.

vagabond, *m.* vagabond, tramp.

vagabond, vagabond, vagrant, wandering.

vagabondage, *m.* vagrancy.

vagabonder, to tramp about, wander, rove.

vagiforme, vagiform, having no certai҄ figure.

vagile, wandering.

vagin, *m.* vagina.

vaginé, sheathed.

vaginervé, vaginervis, (veins) arranged without apparent order.

vaginifère, vaginiferous, furnished with a sheath.

vaginite, *f.* vaginitis.

vaginule, *f.* vaginula.

vagirameux, -euse, with scattered branches.

vagissement, *m.* squalling, mewling, wailing).

vagon, *m.* wagon.

vague, *f.* wave, billow, wandering, oar (for stirring mash); *m.* vagueness, waste land, empty space.

vague, vague, empty, waste, uncertain, untilled, not clearly defined.

vaguement, vaguely.

vaguer, to wander, ramble, rove, stir or mix (mash) with rake or oar.

vaillamment, valiantly, stoutly.

vaillant, *m.* property, possessions.

vaillant, valiant, brave, valorous, worth.

vain, vain, ineffectual, (of land) waste, fruitless, unoccupied; **en —,** in vain; **—e pâture,** common pasture, herbage.

vaincre, to vanquish, conquer, overcome, defeat, subdue, master, surpass, outdo.

vaincu, conquered.

vainement, vainly, in vain.

vainqueur, *m.* victor, conqueror.

vainqueur, victorious.

vainquit (vaincre), (he) conquered.

vais (aller), (I) am going.

vaisseau, *m.* vessel, receptacle, edifice, ship, duct, canal, tube; **— annelé,** annular vessel; **— criblé,** sieve tube; **— dorsal,** heart, pumping organ, dorsal vessel; **— lacticifère,** lacticiferous duct or tube; **— ponctué,** pitted tube; **— réticulé,** reticulate vessel; **— sanguin,** blood vessel, artery, vein.

vaisselle, *f.* dishes, tableware, plates and dishes; **— de porcelaine,** table china; **— d'étain,** pewter tableware.

vaissellerie, *f.* tableware.

vake, *f.* wacke.

val, *m.* valley, vale.

valable, valid, good.

valablement, validly.

valait (valoir), (it) was worth

valaque, Wallachian.

valence, *f.* valence, valency, atomicity; *cap.* Valencia.

valent (valoir), (they) are worth.

valérianacees, *f.pl.* Valerianaceae.

valérianate, valérate, *m.* valerate, valerianate.

valériane, *f.* valerian (Valeriana); **— officinale,** allheal, common valerian (*Valeriana officinalis*).

valérianelle, *f.* lamb's-lettuce, corn salad (*Valerianella locusta*).

valérianique, valérique, valeric, valerianic.

valéroamylique, éther —, amyl valerate.

valérylène, *m.* valerylene, pentine.

valet, *m.* valet, servant, clamp, holdfast, dog, claw, door weight, support, rest; **— de laboratoire,** stand for laboratory use; **— en bois,** wooden ring; **— en paille,** straw ring.

valeur, *f.* value, worth, ability, amount, quality, valuation, bill; *pl.* shares, security, valor, currency; **— actuelle,** present value; **— approchée,** approximate value; **— assurable,** insurable value; **— calorifique,** heating power, fuel value, calorific value; **— d'attente,** prospective or expectation value; **— d'avenir,** future value; **— de l'utilisation,** value of utilization; **— d'usage,** intrinsic value.

valeur, — limite, limiting value, limit (value); **— marchande,** market value, value in money; **— moyenne,** mean value, mean; **— naturelle,** natural food value; **— nutritive,** nutritive value, food value, value of consumption; **— pratique,** value of use; **— vénale,** sale value, sale price, market price; **en — absolue,** in exact figures; **mettre en —,** to enhance in value, improve.

valide, valid, good, healthy, vigorous.

validement, validly.

valider, to validate, make valid.

validité, *f.* validity.

valine, *f.* valine.

vallécule, *f.* vallecula, channel.

vallée, *f.* valley; **— encaissée,** narrow valley surrounded by hills.

vallon, *m.* small valley, dell, dale.

vallonée, *f.* valonia.

vallonie, *f.* knopper gall, oak gall.

vallonier, *m.* valonia oak (*Quercus aegilops*).

valoir, to be worth, set off, show to advantage, be of value, have the value of, be equal to, deserve, merit, bring, give, avail, cause, gain, win, procure; **à — sur,** on account of; **faire —,** to make the most of, improve, turn to account, emphasize, commend; **cela vaut mieux (que rien),** that is better (than nothing); **vaille que vaille,**

for better or worse, come what may, at all costs; — **mieux**, to be better.

valoné, *m.*, **valonée**, *f.* valonia.

valu (valoir), valued.

value, *f.* value.

valut (valoir), (it) was worth.

valvaire, valvate.

valve, *f.* valve, vaginal retractor.

valvé, valved, valvate.

valvoline, *f.* cylinder oil.

valvule, *f.* valvule, small valve, valvelet, valve.

van, *m.* winnow, van.

vanadate, *m.* vanadate; — **de soude**, sodium vanadate.

vanadeux, -euse, vanadious, vanadous.

vanadinite, *f.* vanadinite.

vanadique, vanadic.

vanadite, *m.* vanadite.

vanadium, *m.* vanadium.

vanadyle, *m.* vanadyl.

vandoise, *f.* dace.

vanesse, *f.* Vanessa (butterfly).

vanille, *f.* vanilla.

vanillé, flavored with vanilla.

vanillier, *m.* vanilla plant (*Vanilla planifolia*).

vanilline, *f.* vanillin.

vanillique, vanillic.

vanillisme, *m.* vanillism.

vanité, *f.* vanity.

vannage, *m.* fanning, winnowing, sluicing, regulation (of flow in a turbine).

vanne, *f.* gate valve, sluice gate, water gate, sash gate, sliding lock; — **d'air chaud**, hot-blast valve; — **d'écluse**, sluice door, sash gate; **eaux —s**, waste water.

vanneau huppé, *m.* lapwing, green plover, pewit.

vannelle, *f.* small valve, small sluice gate.

vanner, to winnow, fan, tire out, exhaust, sluice, gate.

vannerie, *f.* wickerwork, wattlework.

vannes, *f.pl.* winnowed corn.

vanneur, *m.* winnower.

vanneuse, *f.* winnowing machine, winnower.

vannure, *f.* winnowed impurities, husks, chaff.

vanter, to praise, extol, boast of, vaunt; **se —**, to boast, brag.

vantrer (se), to wallow in the mire.

vapeur, *f.* vapor, steam, fume, steamer, steamship; — **d'eau**, water vapor, steam; — **d'eau surchauffée**, superheated steam; — **d'échappement**, exhaust steam, waste steam; — **de décharge**, exhaust steam; — **humide**, moist vapor, moist steam; — **sèche**, dry vapor, dry steam; **à toute —**, at full steam; **bateau à —**, steamboat; **machine à —** steam engine.

vaporeux, -euse, vaporous, vapory.

vaporifère, *m.* steam generator.

vaporisateur, *m.* vaporizer, atomizer.

vaporisation, *f.* vaporization, atomization, vaporizing.

vaporiser, to vaporize, atomize, spray; **se —**, to vaporize, become vaporized.

vaquer, to be vacant, not meet; — **à**, to attend to, be occupied with.

vaquois, *m.* screw pine (Pandanus).

varaigne, *f.* tide gate.

varaire, *f.* white hellebore (*Veratrum album*).

varech, varec, *m.* varec, kelp, wrack, seaweed; — **vésiculeux**, bladder wrack (*Fucus vesiculosus*).

variabilité, *f.* variability.

variable, *m.* change.

variable, variable, changeable.

variant, variable, variant.

variante, *f.* variant, form arising from a variation.

variation, *f.* variation, change.

varice, *f.* varix, varicose vein.

varicelle, *f.* varicella, chicken pox.

varié, varied, variegated.

varier, to vary, diversify, change, be at variance.

variété, *f.* variety, change, subspecies.

variifolié, variifolious, possessing leaves of different forms.

variole, *f.* variola, smallpox.

variolé, adspersed, speckled, pocky.

variolite, *f.* variolite.

variosperme, with seed of diverse sizes.

varlope, *f.* large plane, jointer.

varloper, to try up (plank), plane, dress with a plane.

varlopeuse, *f.* planing machine.

Varsovie, *f.* Warsaw.

vasculaire, vascular, vasculose.

vasculose, *f.* vasculose, component of the vegetable skeleton of the cellulose group.

vase, *f.* mud, slime, ooze, mire, silt.

vase, *m.* vessel, vase, urn, beaker, receptacle; — **clos,** closed vessel; — **de sûreté,** safety vessel; — **en grès,** stoneware vessel; — **en Iéna,** vessel o. Jena glass; — **gradué,** graduated vessel; — **jaugé,** calibrated (graduated) vessel; — **poreux,** porous vessel, porous cell.

vasé, muddy, covered with mud.

vaseline, *f.* vaseline.

vaseux, -euse, muddy, slimy, marshy, miry.

vasiducte, *m.* vasiductus, raphe.

vaso-dilatateur, *m.* vasodilator.

vasque, *f.* pan.

vaste, vast, spacious, immense, extensive, great.

vastement, vastly.

vatérie, *f.* artificial amber tree.

vaucour, *m.* potter's bench.

vaudrait (valoir), (it) would be worth.

vauqueline, *f.* strychnine.

vaut (valoir), (it) is worth; **cela —mieux (que rien),** that is better (than nothing).

vautour, *m.* vulture.

vautrait, *m.* boar-chase equipage.

vautrer, to roll, wallow, welter.

vaux, *m.pl.* valleys; see VAL.

veau, *m.* calf, calfskin, veal; — **marin,** seal.

vecteur, *m.* vector, carrier; **radius —,** radius vector.

vectoriel, -elle, vectorial.

vécu (vivre), lived, real, actual, that has happened.

vécut (vivre), (he) lived.

végétal, *m.* vegetable, plant, tree; — **autotrophique,** autophyte; — **ligneux,** ligneous or woody plant.

végétaline, *f.* vegetable butter.

végétalisme, *m.* vegetarianism.

végétant, vegetating, vegetative.

végétarien, *m.* vegetarian.

végétarisme, *m.* vegetarianism.

végétatif, -ive, vegetative.

végétation, *f.* vegetation, growth; — **arbustive,** brushwood, bushes; — **inférieure,** undergrowth, underwood; — **rabougrie,** scrub.

végétaux, *m.pl.* see VÉGÉTAL.

végéter, to vegetate, grow, exist.

véhément, vehement, impetuous.

véhicule, *m.* vehicle, medium, carriage.

veillant, watchful, watching.

veille, *f.* watch, lookout, watching, wakefulness, eve, day before, nyctitropism, waking, vigil, night work.

veillée, *f.* evening, evening party, night nursing, vigil, sitting up at night.

veiller, to watch, stand by, attend to, look after, be awake, keep awake, sit up nights, take care, be visible; — **à,** to see to; — **à ce que,** to watch that, see that.

veilleur, *m.* watcher, watchman.

veilleuse, *f.* low burner, pilot light, night light, meadow saffron (*Colchicum autumnale*).

veillotte, *f.* meadow saffron (*Colchicum autumnale*), haycock.

veinage, *m.* veining, graining.

veine, *f.* vein, jet, streak of ore, stream, luck, fortune; — **liquide,** stream, jet; — **porte,** portal vein.

veiné, veined, veiny.

veiner, to vein, grain.

veineux, -euse, venous, veined, veiny.

vêlage, *m.* calving.

vélaminaire, velaminaris, (anther) dehiscing by rolling up one side of a cell from base to apex.

vélanède, *f.* valonia.

vélani, *m.* valonia oak (*Quercus aegilops*).

vélar, vélaret, *m.* hedge mustard (Sisymbrium); — **officinale,** *Sisymbrium officinale.*

vèle, *f.* female calf.

vêler, to calf.

vélin, *m.* vellum.

velléité, *f.* velleity, longing.

véloce, swift, rapid.

vélocité, *f.* velocity, swiftness, speed.

velours, *m.* velvet, velvety surface, prince's-feather (*Amaranthus hybridus* var. *hypochondriacus* or *Polygonum orientale*).

velouté, *m.* velvetiness, softness, velvety surface, bloom, velouté.

velouté, velvet, velvety, mellow, smooth, villous, rich.

velouter, to give a velvety, soft appearance to.

veloutine, *f.* cosmetic of rice powder containing bismuth.

veltage, *m.* gauging (casks).

velte, *f.* gauging stick.

velter, to gauge (casks).

velu, *m.* hairiness, villosity, roughness.

velu, hairy, villous, shaggy, hirsute, uncut, rough, pilose.

velvote, *f.* bastard toadflax (Comandra).

venaison, *f.* venison.

venant, *m.* comer.

venant, thriving, coming.

vendable, salable, marketable.

vendange, *f.* vintage, marsh, vine harvest, grapes.

vendanger, to gather the grapes, ravage, devastate, feather one's nest.

vendangeur, *m.* vintager.

vendeur, *m.* seller, vender, dealer.

vendre, to sell, vend, realize, turn something into money; à —, for sale, to be sold.

vendredi, *m.* Friday.

vendu, sold.

vené, slightly tainted (meat).

vénéneux, -euse, poisonous, venomous, deleterious.

vénénosité, *f.* poisonousness, toxicity.

vénérablement, venerably, reverently.

vénérer, to venerate, reverence.

vénerie, *f.* venery, woodcraft.

vénérien, -enne, venereal.

veneur, *m.* huntsman.

vénézuélien, Venezuelan.

vengeance, *f.* vengeance, revenge.

venger, to avenge; se —, to revenge oneself.

venimeux, -euse, venomous, poisonous.

venin, *m.* venom, poison.

venir, to come, occur, attain, reach, be suited (to), agree, fit, happen, arise, spring from, hail from, come up, grow up, thrive (with de and an infinitive) to have just; — à, to happen to; — à bien, to succeed, prosper; — à bout, to succeed; — à faire, to happen to do, chance to; — à rien, to come to nothing; — au contact, to come in contact; — de, to have just; — de faire quelque chose, to have just done something.

venir, — de fonderie, to be cast in one piece; — de fonte, to be cast on; — en prise, to mesh, engage, be caught; — en tour d'exploitation, to come under the ax; il en vient à bout, he is accomplishing his end; il vient de, he has just (been here); bien —, to thrive; en —, to arrive, come; en — à, to come to, go so far as, be reduced to; faire —, to send for, call in, summon,

suggest, grow; s'en —, to come away, come along.

vénitien, -enne, Venetian.

vent, *m.* wind, gale, blast, air, windage, clearance, flatus; — alizé, trade wind; — chaud, hot blast; — contraire, headwind, contrary wind; — de bout, headwind, contrary wind; — froid, cold blast, cold wind; — nul, calm, still air; au —, windward side; en plein —, in the open air; mettre au —, to expose to the air, air, spread out, have out to air.

vente, *f.* sale, cutting, felled timber, felling auction; — à l'amiable, — de gré à gré, sale by private contract or agreement; — à l'enchère, sale to the highest bidder, auction; — à l'unite des produits, sale by unit of produce; — après façonnage, sale after conversion; — au détail, retail sale.

vente, — au rabais, sale with abatement, Dutch auction; — au tarif, sale by royalty, sale at fixed tariff prices; — de bois, sale of wood; — par soumission cachetée, sale by sealed tender; — sur pied, sale of standing trees; de bonne —, de —, salable; en —, on (for) sale.

venter, to blow, be windy.

venteux, -euse, windy, blistered (casting), flatulent.

ventilateur, *m.* ventilator, ventilating aperture or fan.

ventilation, *f.* ventilation, valuation; à — forcée, with forced ventilation.

ventiler, to ventilate, value.

ventouse, *f.* cupping glass, sucker, nozzle, ventilator, air hole, vent, haustorium, patella.

ventral, -aux, ventral, anterior or inner.

ventre, *m.* belly, stomach, abdomen, swell, sag, bulge, paunch, body; à plat —, flat on the ground.

ventrée, *f.* litter, fall (of lambs), bellyful.

ventricule, *m.* ventricle.

ventru, big-bellied, corpulent, ventricose.

venu, *m.* comer; nouveau —, primier —, newcomer, first comer.

venu, come, arrived, grown, done, happened.

venue, *f.* coming, arrival, approach, growth.

venule, *f.* vein, nervuol.

vêpres, *f.pl.* vespers.

ver, *m.* worm, grub, larva, maggot; — à soie, silkworm; — blanc, larva of the cockchafer; — du fromage, skipper, jumper; — fil, wireworm; — luisant, glowworm.

véracité, *f.* veracity.

véraison, *f.* ripening.

vératrate, *m.* veratrate.

vératre, *m.* white (false) hellebore (Veratrum); — **blanc,** European white hellebore (*Veratrum album*); — **vert,** American hellebore, green hellebore (*Veratrum viride*).

vératrine, *f.* veratrine.

vératrique, veratric.

verbal, *m.* official report.

verbal, -aux, verbal.

verbalement, verbally.

verbaliser, to draw up an official report, make a formal statement.

verbe, *m.* word, verb, speech, voice, tone.

verdage, *m.* manure crop.

verdâtre, *m.* greenish color.

verdâtre, greenish.

verdaud, somewhat green, not fully ripe.

verdelet, greenish, tart, slightly acid (wine).

verdet, *m.* verdigris; — **de Montpellier,** true verdigris, basic copper acetate.

verdeur, *f.* greenness, sap (in wood), tartness, acidity (of wine), vigor, vitality.

verdier, *m.* green finch.

verdillon, *m.* crowbar, pinch bar.

verdir, to turn green, grow green, be green, be verdant, become coated with verdigris.

verdissage, *m.* coloring green.

verdissant, turning green, verdant, fresh.

verdissement, *m.* turning green.

verdoyant, verdant, growing green.

verdoyer, to become green, be green or verdant.

verdure, *f.* greenness, verdure, pot herbs, greens, leaves; — **d'hiver,** wintergreen; — **pour bouquets,** ornamental grasses.

véreux, -euse, wormy, bad, worm-eaten, maggoty, grubby.

verge, *f.* rod, wand, stick, bar, shaft, perch, pole, shank, handle, beam, spindle, penis, verge; — **à fruit,** fruit rod; — **d'or,** goldenrod.

vergé, laid (paper), streaky, badly dyed (cloth), virgate.

vergeoise, *f.* sugar-loaf mold.

verger, *m.* orchard.

vergette, *f.* brush, hoop.

vergeure, *f.* wire (of the mold for laid paper), wire marks (on laid paper), streakiness (of textile).

verglas, *m.* glaze of ice, glazed frost, rime.

vergne, *m.* alder tree (Alnus).

vergue, *f.* yard, sail yard.

vericle, *f.* imitation gem; **diamants de —,** imitation diamonds.

vérifiable, verifiable.

vérificateur, *m.* verifier, inspector, examiner, auditor, tester, gauge.

vérification, *f.* verification, examination, investigation, testing, checking, check; — **de la résistance,** testing for resistance; — **du vernissage,** bruising test.

vérifier, to verify, test, check, inspect, audit, confirm, check up, prove; — **si,** to make sure that.

vérin, *m.* jack, screw jack.

vérissime, very true, most veracious.

véritable, true, real, genuine, actual, veritable.

véritablement, truly, really, actually, indeed, veritably.

vérité, *f.* truth, verity, fact, sincerity; **à la —,** in truth, indeed; **en —,** really, truly, actually, indeed.

verjus, *m.* verjuice, sour grapes, very sour wine.

verjuté, acid, sour, tart, made with verjuice.

vermeil, *m.* vermeil, gilded silver, gilded bronze.

vermeil, -eille, vermilion, red, ruddy, rosy.

vermicelier, *m.* vermicelli maker.

vermicelle, vermicel, *m.* vermicelli (soup).

vermicellerie, *f.* manufacture of vermicelli.

vermicide, *m.* vermicide.

vermicide, vermicidal.

vermiculaire, *f.* stonecrop (Sedum).

vermifuge, *m.* vermifuge.

vermifuge, vermifugal.

vermillon, *m.* vermilion (color), bright red, cinnabar.

vermillonner, to vermilion, paint bright red.

vermination, *f.* damage done by worms.

vermine, *f.* vermin.

vermisseau, *m.* small earthworm.

vermouler (se), to become worm-eaten.

vermoulu, worm-eaten, decayed, fragile, brittle.

vermoulure, *f.* worm dust, wormhole dust.

vermout, vermouth, *m.* vermouth (wine).

vernal, -aux, vernal.

vernation, *f.* vernation, leafing.

verni, varnished, glazed, japanned.

vernier, *m.* vernier.

vernir, to varnish, polish, glaze, japan.

vernis, *m.* varnish, polish, gloss, glaze, glazing, japan, varnish tree (*Rhus vernicifera*); — **à la copale,** copal varnish; — **à l'alcool,** spirit varnish; — **à l'essence,** turpentine varnish; — **à l'huile,** oil varnish; — **au succin,** amber varnish; — **de plomb,** lead glaze; — **de silice,** silicate varnish; — **du Japon,** Japanese varnish tree, lacquer tree (*Rhus vernicifera*); — **gras,** oil varnish; — **siccatif,** quick-drying varnish.

vernissage, *m.* varnishing, enameling, lacquering.

vernissé, glazed.

vernisser, to glaze, varnish.

vernisseur, *m.* varnisher, glazer.

vernissure, *f.* varnishing, glazing.

vérole, *f.* pox, syphilis; **petite —,** smallpox; **petite — volante,** chicken pox.

vérolette, *f.* varicella, chicken pox.

vérolique, pocky.

véron, *m.* minnow.

Vérone, *f.* Verona.

véronique, *f.* speedwell (Veronica); — **à feuilles,** ivy-leaved speedwell (*Veronica hederaefolia*); — **de Virginie,** Culver's root (*Veronica virginica*); — **mâle,** common speedwell (*Veronica officinalis*).

verpunte, *m.* sugar loaf of inferior quality.

verra (voir), (he) will see.

verraille, *f.* small glassware.

verrain, *m.* jack.

verrat, *m.* boar.

verre, *m.* glass, jar; — **à boire,** drinking glass; — **à bouteilles,** bottle glass; — **à conserve,** packing jar; — **à expérience,** glass cylinder; — **à pied,** glass with a foot; — **ardent,** burning glass; — **armé,** wire glass; — **à vitre(s),** window glass; — **bombé,** convex glass; — **coulé,** cast glass; — **de Bohême,** Bohemian glass; — **de champ,** field lens.

verre, — de couleur, colored (stained) glass; — **de modèle standard,** standard jar; — **de montre,** watch glass; — **de pendule,** glass for pendulum clocks; — **de plomb,** lead glass, crystal glass, flint glass; — **dépoli,** ground glass, focusing screen; — **d'Iéna,** Jena glass; — **d'oeil,** eyepiece, eye lens; — **doublé,** flashed glass; — **en manchons,** cylinder glass, sheet glass; — **en tables,** plate (sheet) glass.

verre, — filigrané, filigree glass; — **grossissant,** magnifying glass; — **imprimé,** figured (printed) glass; — **moulé,** pressed glass, molded glass; — **mousseline,** muslin glass, enameled sheet glass; — **objectif,** object glass, objective; — **pilé,** crushed glass, glass wool; — **pulvérisé,** powdered glass, glass powder, glass meal; — **sablé,** sanded glass, glass cast in sand; — **soluble,** water glass; — **soufflé,** blown glass.

verré, coated with powdered glass; **papier —,** sandpaper.

verrée, *f.* glassful.

verrequartz, *m.* quartz glass, fused silica glass.

verrerie, *f.* glassmaking, glassworks, glassware; — **artistique,** artistic glass(ware): — **de laboratoire,** laboratory glassware; — **de pharmacie,** medicine glassware; — **d'Iéna,** Jena glassware; — **graduée,** graduated glassware.

verrier, *m.* glassmaker, glass stand, tray for glasses.

verrière, *f.* glass casing, stained-glass window.

verrine, *f.* glass casing, barometer tube, glass bell, bell glass.

verront (voir), (they) will see.

verroterie, *f.* small glassware, glass trinkets, glass beads.

verrou, *m.* bolt, switch lock, lock.

verrouiller, to bolt, lock.

verrucaire, *f.* wartwort (Verrucaria).

verrucosité, *f.* verrucosity, warty swelling.

verrue, *f.* wart, verruca.

verruqueux, -euse, warty, verrucose, warted.

vers, *m.* verse, line.

vers, toward, to, about.

versage, *m.* emptying, first plowing.

versant, *m.* side, bank, slope, declivity, sloping ground.

versant, pouring.

versatile, versatile, turning freely on its support.

verse, *f.* pouring, charcoal basket, lodging, laying, beating down (of grain); **sinus —,** versine, versed sine.

versé, beaten down, laid, lodged, poured, versed, proficient, skilled.

versement, *m.* pouring, payment, deposit.

verser, to pour, shed, spill, empty, overturn, upset, beat down, lodge, lay (grain), pay in, deposit, invest, turn over, plow, give, lavish, issue; **se** —, to be poured, flow, pour, discharge.

verseur, *m.* pourer.

versicolore, versicolor(ed), with several colors, variegated.

version, *f.* version.

verso, *m.* back, reverse, left-hand page.

versoir, *m.* moldboard, breast.

vert, *m.* green, green color, green grass, fresh vegetation, tartness, acidity; **prendre sans** —, to catch napping, take by surprise.

vert, green, fresh, unripe, immature, tart, somewhat acid (wine), sharp, severe; — **à l'aldéhyde,** aldehyde green; — **bouteille,** bottle green; — **cantharide,** iridescent green; — **clair,** light green, gaudy green; — **d'eau,** sea green.

vert, — **de chrome,** chrome green; — **de cuivre,** verdigris, chrysocolla; — **de gris,** verdigris; — **d'émeraude,** emerald green; — **de montagne,** mountain green, malachite green, basic copper carbonate, chrysocolla; — **de sève,** sap green, bladder green; — **des feuilles,** leaf green, chlorophyll; — **de vessie,** sap green, bladder green.

vert, — **d'herbe,** grass green, grass-colored; — **d'olive,** olive green; — **émeraude,** emerald green; — **épinard,** spinach green; — **foncé,** deep green, dark green; — **gazon,** grass-green; — **Guignet,** Guignet's green, emerald green, chrome green; — **jaune,** greenish yellow; — **malachite,** malachite green, benzaldehyde green; — **méthyle,** methyl green; — **naissant,** faint green, pale green — **pomme,** apple green.

vert-de-gris, *m.* verdigris.

vert-de-grisé, coated with verdigris.

vertébral, -aux, vertebral, spinal.

vertèbre, *f.* vertebra; **sans** —**s,** invertebrate.

vertébré, vertebrate.

vertébrés, *m.pl.* the vertebrates.

vertelle, *f.* sluice gate.

vertement, vigorously, sharply, energetically.

vertical, *f.* vertical.

vertical, -aux, vertical, perpendicular.

verticalement, vertically.

verticalité, *f.* verticality, verticalness.

verticillacanthe, with spines or thorns arranged verticilately.

verticillaster, *m.* false whorl.

verticille, *m.* whorl, verticil.

verticillé, verticillate, whorled.

verticilliflore, verticilliflorus, (whorls) having a spicate arrangement.

verticilliose, *f.* verticillium, wilt.

vertige, *m.* megrim, dizziness.

vertigo, *m.* vertigo, dizziness, staggers.

vertu, *f.* virtue, quality, property, force; **en** — **de,** by virtue of, in pursuance of.

verveine, *f.* vervain (Verbena); — **de l'Inde,** lemon grass (Cymbopogon); — **officinale,** common European vervain (*Verbena officinalis*).

verveux, *m.* hoop net.

vesce, *f.* vetch, tare (Vicia); — **commune,** common vetch (*Vicia sativa);* — **velue,** villous vetch, hairy vetch (*Vicia villosa*).

vésicant, vésicatoire, vesicatory, vesicant, blistering, vesicating.

vésiculaire, vesicular.

vésicule, *f.* vesicle, bladder; — **embryonnaire,** embryonic vesicle; — **séminale,** vesicula seminalis.

vésiculeux, -euse, vesiculose, apparently composed of small bladders.

vesou, *m.* cane juice.

vespéral, -aux, evening.

vespidés, *m.pl.* Vespidae, wasp family.

vesse-de-loup, *f.* puffball.

vessie, *f.* bladder, blister, vesicle; — **natatoire,** natatory vessel, air bladder.

vessigon, *m.* bog spavin, windgall.

veste, *f.* jacket, vest.

vestige, *m.* vestige, trace, track, footprint.

vésuvine, *m.* vesuvin, Bismarck brown.

vêtement, *m.* garment, clothing, raiment, tunic.

vétérinaire, *m.* veterinarian.

vétérinaire, veterinary.

vétille, *f.* trifle, squib, a small serpent.

vétilleux, -euse, delicate, particular.

vêtir, to clothe, dress; **se** —, to dress oneself, clothe oneself.

vétiver, *m.* vetiver, khuskhus grass (*Vetiveria zizanioides*).

vêtu, dressed, clad.

vétuste, old, decayed, decrepit, worn out.

vétusté, *f.* decay, decrepitude, rustiness, antiquity, oldness.

vétyver, *m.* see VÉTIVER.

veuf, veuve, widowed, bereft, minus, deprived.

veuille (vouloir), (he) may wish.

veule, soft, weak; plante —, sickly plant; terre —, poor soil.

veulent (vouloir), (they) wish.

veuve, f. widow; fleur de —, sweet scabious (*Scabiosa atropurpurea*).

vexant, provoking.

vexatoire, vexatious.

vexer, to vex, provoke.

vexillaire, vexillar, pertaining to the vexillum.

viabilité, f. viability.

viable, capable of living, viable.

viaduc, m. viaduct.

viager, m. life interest.

viager, -ère, for life.

viande, f. meat, flesh; — congelée, frozen meat; — de conserve, canned meat; — — fraîche, fresh meat;—frigorifieé, frozen meat; — fumée, smoked meat, hung beef; — réfrigérée, cold-storage meat; — salée, salt meat.

viander, to feed (of deer), browse.

viandis, m. food of deer.

vibrant, vibrating, vibrant.

vibrateur, m. vibrator.

vibratile, vibratile.

vibration, f. vibration, oscillation.

vibratoire, vibratory.

vibrer, to vibrate, oscillate.

vibreur, n. vibrator, make and break, buzzer.

vibrion, m. vibrio.

vibrionien, -enne, vibrionic.

vice, m. defect, flaw, imperfection, fault, unsoundness, vice.

vice-roi, m. viceroy.

viciation, f. vitiation.

vicié, vitiated, resembling a vetch.

vicier, to vitiate; se —, to become tainted, be vitiated.

vicieusement, viciously.

vicieux, -euse, vicious, faulty, abnormal, defective.

vicinal, -aux, parish, parochial.

vicinisme, m. vicinism, variation due to growth of other plants close by.

vicioïde, like a vetch.

victime, f. victim.

victoire, f. victory.

victorieuse ment, victoriously.

victorieux, -euse, victorious.

vidage, m. emptying, gutting, cleaning, drawing (of fowl).

vidange, f. emptying, draining, discharge, blowing off, removal (of earth), ditch, clearing (of felled timber), removal of night soil; pl. night soil, sediment, sludge (in boiler); en —, partly empty, not full; vouteille en —, opened bottle.

vindanger, to empty, drain, discharge, blow off.

vidé, emptied.

vide, m. vacuum, empty space, void, emptiness, opening, interstice, hollow, blank, fail place, chasm; — d'air, exhausted; à —, empty, in vacuo, in the air, unused; faire le —, to evacuate, create a vacuum.

vide, empty, vacant, vacuous, void, devoid (of), free, open, barren.

vide-bouteille(s), m. siphon for bottles.

videment, m. emptying.

vider, to empty, drain, clear (forest), clean, hollow out, exhaust (mind), blow off (a boiler), blow (an egg), draw (fowls), stone (fruit), vacate, quit, dismiss, reduce, settle, adjust; se —, to be emptied, empty, empty itself, become empty.

vidure, f. material cleaned out, openwork.

vie, f. life, living, path, way (in a salt garden), animation; à —, for life, life; en —, living, alive; jamais de la —, never; pour la —, for life; sans —, lifeless.

vieil, vieille, old; see VIEUX.

vieillard, m. old man, the aged.

Vieille-Californie, f. Old or Lower California.

vieillesse, f. age, old age, old people, oldness.

vieilli, grown old, aged, outworn.

vieillir, to grow old, become obsolete, become antiquated, go out of use.

vieillissement, m. ageing, growing old, senescence, obsolescence.

vieillot, -otte, oldish, antiquated, old-fashioned.

viendra (venir), (he) will come.

Vienne, f. Vienna, Vienne (in France).

viennent (venir), (they) are coming.

viennois, Viennese, Vienna.

vient (venir), (he) is coming.

vierge, virgin, untouched by the ax, pure; vigne —, Virginia creeper (*Parthenocissus quinquefolia*).

vieux, *m.* old thing, old man; **les —,** the old.

vieux, vieil, vieille, old, obsolete; **— de deux ans,** two years old; **vielles écorces,** old trees.

vieux-oing, *m.* cart grease, axle grease.

vif, *m.* living flesh, quick, heart (of tree), core, solid part, living person.

vif, vive, live, living, lively, bright, animated, vivid, brilliant, quick, ardent, sharp, keen, intense, strong, great, raw, spirited, zealous, earnest, severe, harsh; **roc —,** live rock, solid rock; **de vive force,** by main force; **de vive voix,** viva voce, by word of mouth; **eau vive,** living or spring water.

vif-argent, *m.* quicksilver, mercury.

vigie, *f.* lookout, watch.

vigilamment, vigilantly, watchfully.

vigilant, vigilant, watchful.

vigne, *f.* vine, grape, grapevine, vineyard; **— vierge,** Virginia creeper (*Parthenocissus quinquefolia*).

vigneron, *m.* vinegrower, vintager.

vignette, *f.* vignette, revenue stamp, label, meadowsweet (*Spiraea ulmaria*), clematis (*Clematis viticella*), mercury (*Mercurialis annua*).

vignoble, *m.* vineyard.

vignoble, grape-growing.

vigogne, *f.* vicuña (wool).

vigorite, *f.* vigorite.

vigoureusement, vigorously.

vigoureux, -euse, vigorous, strong, stout, lusty.

vigueur, *f.* vigor, vigour, strength, force.

vil, vile, low, base, mean.

vilain, villainous, nasty bad, vile, ugly, unpleasant.

vilainement, uglily.

vilebrequin, *m.* brace and bit, wimble, crankshaft.

vileté, *f.* cheapness, worthlessness, insignificance, mean action.

villa, *f.* bungalow, country home.

village, *m.* village.

ville, *f.* town, city.

villeux, -euse, villose, hairy.

villifère, villous, villose.

villosité, *f.* villus; **— intestinale,** intestinal villus.

vimaire, *f.* storm damage.

vin, *m.* wine; **— blanc,** white wine; **— bourru,** unfermented wine; **— de bulbe de** **colchique,** colchicum wine; **— de cerises aigres,** sour-cherry wine; **— de Champagne,** Champagne wine, champagne; **— de choix,** choice wine, fine wine; **— de fruits,** fruit wine; **— de groseilles,** currant wine, gooseberry wine; **— de liqueur,** liqueur, cordial, dessert wine; **— de mûres,** blackberry wine; **— de myrtilles,** bilberry wine.

vin, — de raisins secs, raisin wine; **— de seigle ergoté,** wine of ergot; **— d'Espagne,** Spanish wine, sherry; **— doux,** sweet wine; **— émétique,** wine of antimony; **— filant,** ropy wine; **— fort,** strong wine; **— médicinal,** medicated wine; **— mousseux,** sparkling wine, champagne; **— rouge,** red wine; **— sec,** dry wine; **— sucré,** sweet wine.

vinage, *m.* fortifying (of wine), addition of alcohol to wine, production of wine, making of wine.

vinaigre, *m.* vinegar; **— anhydre,** glacial acetic acid; **— cantharidé,** vinegar of cantharides; **— d'eau-de-vie,** brandy vinegar; **— de bois,** wood vinegar, pyroligneous acid; **— de plomb,** lead vinegar; **— de toilette,** aromatic vinegar, toilet vinegar; **— de vin,** wine vinegar, cider vinegar; **— épicé,** spiced vinegar; **— glacial,** glacial acetic acid.

vinaigrerie, *f.* vinegar factory, vinegar making, vinegar trade.

vinaigrier, *m.* vinegar maker, vinegar cruet sumac tree, tanner's sumac (*Rhus coriaria*).

vinaire, relating to wine.

vinasse, *f.* vinasse, residuary liquor, poor (washy) wine.

vindicatif, -ive, vindictive, revengeful.

vinéal, -aux, vinealis, growing in vineyards.

vinée, *f.* vintage, wine crop.

viner, to add alcohol to (wine).

vinerie, *f.* wine making.

vinette, *m.* sorrel (*Rumex acetosa*), barberry (*Berberis vulgaris*).

vinettier, vinetier, *m.* barberry (*Berberis vulgaris*).

vineux, -euse, vinous, of wine, wine-flavored (drink), (wine) high in alcohol, strong, wine-colored, rich in vintage.

vingt, *m.* twenty.

vingt, twenty, twentieth.

vingtaine, *f.* (about) twenty, score.

vingt-cinq, twenty-five.

vingtième, *m.* twentieth.

vingtième, twentieth.

vinicole, winegrowing, vinicultural.

viniculture, *f.* viniculture.

vinifère, wine-producing, wine-bearing.

vinificateur, *m.* vinificator.

vinification, *f.* vinification, production of wine.

vinique, vinic (alcohol).

vinomètre, *m.* vinometer, oenometer.

vinosité, *f.* vinosity, flavor and strength (of wine).

vin-pierre, *m.* winestone, tartar.

vinrent (venir), (they) came.

vioforme, *m.* vioform.

viol, *m.* rape.

violacé, violet-colored, purple, violaceous.

violacées, *f.pl.* Violaceae.

violacer, to assume a violet (purple) tint.

violariacées, *f.pl.* Violaceae.

violarié, resembling the violet.

violation, *f.* violation.

violâtre, purplish tending to violet, violescent.

violemment, violently.

violence, *f.* violence, force.

violent, violent.

violenter, to do violence to.

violer, to violate.

violet, *m.* violet.

violet, violet (color), violet-colored; — **bleu,** bluish violet; — **foncé,** dark violet; — **formyle,** formyl violet; — **rouge,** reddish violet.

violette, *f.* violet (*Viola*), pansy (*Viola tricolor*); — **des blés,** corn violet (*Specularia speculum-veneris*); — **odorante,** sweet violet (*Viola odorata*).

violier, *m.* wallflower (*Cheiranthus cheiri*), stock, gillyflower (*Matthiola incana*).

violon, *m.* violin, violinist.

violoncelle, *m.* violoncello, violoncellist.

viorne flexible, *f.* wayfaring tree (*Viburnum lantana*).

vioutte, *f.* dogtooth (violet) (*Erythronium dens canis*).

vipère, *f.* viper, adder.

vipérine, *f.* viper's bugloss, blueweed, blue thistle (*Echium vulgare*).

virage, *m.* turning, turning of color, toning (of proofs), account by transfer.

vire, *f.* whitlow, inflammatory tumor.

virée, *f.* turning, turn.

virées (par), by strips.

virement, *m.* turning, shifting, transfer.

virent (voir, virer), (they) saw, (they) are turning.

virer, to turn, veer, shift, turn color, change color, undergo toning, bank transfer (a sum), clear (checks); **se** —, to be turned, turn (color), be toned, be transferred, be cleared; — **de bord,** to tack about, go about.

virescence, *f.* virescence.

vireux, -euse, poisonous, virose, virous, malodorous, fetid, noxious, unpleasant.

virginal, -aux, virginal, maidenly.

virginité, *f.* virginity, freshness.

virgule, *f.* comma.

virgulte, *m.* virgultum, vigorous twig or shoot.

viridiflore, bearing green flowers.

viridifolié, green-leaved.

viril, virile, manly.

virole, *f.* ferrule, collar, hoop, sleeve, ring, thimble joint, chanterelle (*Cantharellus cibarius*).

virtuel, -elle, virtual, potential.

virtuellement, virtually.

virulence, *f.* virulence, malignity.

virulent, virulent.

virus, *m.* virus.

vis (vivre, voir), (I) am living, (I) saw.

vis, *f.* screw, thread; — **à ailettes,** wing(ed) screw, thumbscrew; — **à gauche,** left-handed screw; — **calante,** leveling screw; — **d'arrêt,** setscrew, clamping screw, stop screw; — **de pression,** setscrew, press(ing) screw; — **de réglage,** adjusting screw; — **de serrage,** setscrew, binding (clamping) screw; — **sans fin,** endless screw, worm.

visage, *m.* face, aspect, countenance, look.

vis-à-vis, *m.* person opposite, opposition, vis-à-vis.

vis-à-vis, opposite, over against, facing, with respect to, towards; — **de,** opposite, facing, with respect to.

viscères, *m.pl.* viscera, entrails.

viscidité, *f.* viscidity.

viscine, *f.* viscin.

viscosimètre, *m.* viscosimeter.

viscosité, *f.* viscosity, viscidity, stickiness.

vis-écrou, *m.* binding post, fuse plug.

visée, *f.* aim, sighting; *pl.* designs.

viser, to aim at, aspire, take aim at, sight on, have in view, visa.

viseur, *m.* aimer, finder, sighting tube.

viseur, -euse, sighting.

visibilité, *f.* visibility.

visible, visible, apparent, obvious, evident, manifest.

visiblement, visibly, obviously, manifestly.

visière, *f.* visor, (eye)shade, sight.

vision, *f.* vision, sight, hallucination.

visionnaire, *m.* visionary.

visionnaire, visionary.

visite, *f.* visit, examination, inspection, search.

visiter, to visit, explore, inspect, search.

visiteur, *m.* visitor, searcher, inspector.

visqueux, -euse, viscous, slimy, clammy, ropy, viscid, glutinous.

vis-robinet, *m.* screw faucet, screw cock.

vissage, *m.* screwing.

visser, to screw, screw in or on; — à fond, to screw tight.

visuel, -elle, visual.

visuellement, visually.

vit (vivre, voir), (he) is living, (he) saw.

vital, -aux, vital, essential.

vitalement, vitally, essentially.

vitalisme, *m.* vitalism.

vitalité, *f.* vitality.

vitamine, *f.* vitamin.

vitaux, *pl.* see VITAL.

vite, quick, fast, rapid, swift, prompt, quickly, rapidly, speedily, hastily; au plus —, as quickly as possible.

vitellin, vitelline.

vitelline, *f.* vitellin.

vitellus, *m.* vitellus, yolk.

vitement, quickly.

vitesse, *f.* velocity, speed, rapidly, swiftness, celerity, quickness; — angulaire, angular velocity; — de réaction, rate of reaction; — de régime, normal speed; — finale, terminal velocity, final speed; — initiale, initial velocity; — moyenne, mean (average) speed; — uniforme, uniform velocity; grande —, high speed.

viticole, viticultural, vine-(grape-)growing.

viticuleux, -euse, viticulose, sarmentose, producing viticulae.

viticulteur, *m.* viticulturist, winegrower.

viticulture, *f.* viticulture, vine culture, grape growing.

vitrage, *m.* glazing, glass, glasswork, panes of glass, glass partition, small curtain.

vitrail, *m.* stained-glass window.

vitraux, *m.pl.* see VITRAIL.

vitre, *f.* pane of glass, glass, window glass.

vitré, glazed, vitreous, of glass, glazen.

vitrer, to glaze, fit with glass.

vitrerie, *f.* glazier's work, window-glass making or trade.

vitrescible, vitrifiable, vitrescible.

vitreux, -euse, vitreous, glassy, glazen, hyaline.

vitrier, *m.* maker (seller) of window glass, glazier.

vitrifiable, vitrifiable.

vitrificateur, *m.* vitrifier.

vitrificateur, vitrifying.

vitrification, *f.* vitrification, vitrifaction.

vitrifié, vitrified.

vitrifier, to vitrify.

vitrine, *f.* glass case, showcase.

vitriol, *m.* vitriol, sulphuric acid; — blanc, white vitriol, zinc sulphate; — bleu, blue vitriol, copper sulphate; — de cuivre, blue vitriol, copper sulphate; — de zinc, zinc vitriol, zinc sulphate; — rouge, red vitriol, colcothar.

vitriolage, *m.* vitriolation, vitriolization, souring.

vitrioler, to vitriolate, treat with vitriol, vitriol, sour, vitriolize.

vitriolerie, *f.* vitriol factory or manufacture.

vitriolique, vitriolic; acide —, vitriolic acid, sulphuric acid.

vitrosité, *f.* vitreousness, glassiness.

vittine, *f.* vittin, substance found in the more watery vittae of Umbelliferae.

vivace, long-lived, hardy, perennial.

vivacité, *f.* liveliness, sprightliness, vividness, brilliancy (of colors), vivacity, vivaciousness, brightness, acuteness, hastiness.

vivant, *m.* living being or person, living organism, liver, lifetime; *pl.* the living.

vivant, living, live, alive, quick, lively.

vive, alive; see VIF.

vivement, briskly, sharply, forcibly, lively, intensely, greatly, quickly, brightly, actively, keenly.

vivier, *m.* fishpond, store pond.

vivifiant, vivifying.

vivifié, strengthened.

vivifier, to vivify, enliven, animate.

vivipare, viviparous.

viviparie, *f.* vivipary.

viviparité, *f.* viviparity.

vivisection, *f.* vivisection.

vivre, *m.* life, living, food; *pl.* provisions, victuals; —**s de conserve,** canned (tinned) foods.

vivre, to live, exist, subsist.

vocabulaire, *m.* vocabulary.

vocal, -aux, vocal, of the voice.

vocation, *f.* calling, vocation, talent.

vœu, *m.* vow, wish, desire, will.

vogue, *f.* vogue; **de la —,** up to date.

voguer, to row, float or toss upon.

voici, here is, here are, this is, these are, behold, ago; **me —,** here I am; — **qu'il vient,** here he comes.

voie, *f.* way, method, process, road, highway, track, line, path, trail, deerpath, tract, gauge (width), load, cartload, passage, canal, duct; — **aérienne,** air passage, airway, course of flight; — **d'accès** approach, leading in line; — **de bois,** half a cord; — **de charbon,** a hectoliter of wood charcoal; — **charretière,** wagon road; — **de service,** feeding track; — **étroite,** narrow gauge; — **ferrée,** railway (track), — **fluviale,** waterway.

voie, — **humide,** wet way; — **libre,** right of way, clear; — **normale,** standard gauge; — **pleine,** open track; — **publique,** thoroughfare, public road; — **routière,** public road; — **sèche,** dry way, dry method; **à deux —s,** double track; **en — de,** in course of; **la Voie lactée,** the Milky Way; **les —s digestives,** the digestive tract; **par —s de,** by means of; **par la — interne** internally.

voilà, there is, there are, that is, those are, that's the way it is; — **que,** all at once, suddenly; — **tout,** that's all there is to it; **en — une raison,** that's a nice reason; **le —,** here he comes; **le — venu,** he has come.

voile, *f.* sail; *m.* cover, cloak, obscurity, clouding, pellicle, scum, veil, mark, velum.

voiler, to cloud, cover, overcast, obscure, dim, conceal, muffle, veil; **se —,** to be clouded, be warped.

voilier, flying, sailing; **oiseau —,** long-flight bird.

voilure, *f.* airfoil, flying surface, sails.

voir, to see, behold, discern, look at, examine, overlook, face; **se —,** to be seen, see each other, meet, find oneself; **faire —,** **laisser —,** to show, let see; **cela se voit,** that is obvious.

voire, in truth; — **même,** even, (and) indeed.

voirie, *f.* administration of public ways, public ways, system of roads, offal, dust heap, garbage, common sewer.

voisin, *m.* neighbor.

voisin, near, next, neighboring, adjacent, closely related, similar, vicinal; — **de,** near, near to, bordering on.

voisinage, *m.* neighborhood, vicinity.

voisiner, to be near.

voiturage, *m.* carriage, conveyance (of goods).

voiture, *f.* vehicle, van, wagon, coach, carriage, car, automobile, conveyance, cost, fare, load, freight; — **à marchandises,** freight car, lorry; — **automobile,** automobile; — **de chemin de fer,** railway car, railway carriage.

voiturer, to convey, transport, carry, cart, haul.

voiturette, *f.* small (light) automobile, light carriage, runabout.

voix, *f.* voice, vote; **à — basse,** in a low tone; **à demi- —,** in an undertone; **à haute —,** aloud.

vol, *m.* flight, theft, robbery, soaring, flying; — **de bois,** theft of wood; — **plané,** volplane, glide; **à — d'abeille,** in a bee line; **à — d'oiseau,** in a straight line, bird's-eye (view).

volage, unsteady, fickle, flighty, inconstant.

volaille, *f.* poultry, fowls, birds; — **comestible,** table poultry.

volant, *m.* flywheel, steering wheel, loose leaf, detachable part, guiding wheel, shuttlecock, reserve supply; — **denté,** cogged flywheel.

volant, flying, volant, movable, loose, portable, of short duration, detachab'

volatil, volatile, light, ethereal.

volatile, *m.* winged creature.

volatilisable, volatilizable.

volatilisation, *f.* volatilization.

volatiliser, to volatilize.

volatilité, *f.* volatility.

volcan, *m.* volcano.

volcanique, volcanic.

volcanisme, *m.* vulcanism, volcanic activity.

volée, *f.* flight, volley, discharge of guns, round, splinter bar; **à la —,** in the air, flying, promptly, quickly, at random, broadcast.

volemite, *f.* volemite.

voler, to fly, soar, steal, rob.

volet, *m.* shutter, blind, damper, trap door, pigeon house, sorting board (for seeds), paddle; **— des étangs,** water lily.

voleter, to flutter.

volettement, *m.* fluttering.

voleur, *m.* thief, robber, burglar, flier.

volière, *f.* bird cage.

volige, *f.* small board, batten.

volis, *m.* windbreak.

volontaire, *m.* volunteer.

volontaire, voluntary.

volontairement, voluntarily, willfully.

volonté, *f.* will, willingness, desire; **à —,** at will.

volontiers, willingly, readily, easily, naturally, gladly, usually; **faire —,** to like to do.

volt, *m.* volt.

voltage, *m.* voltage.

voltaïque, voltaic, galvanic.

voltamètre, *m.* voltmeter.

voltige, *f.* vaulting, tumbling.

voltiger, to flutter.

volubile, winding.

volubilis, *m.* bindweed (Convolvulus).

voluble, winding.

volucelle, *m.* syrphus fly.

volume, *m.* volume, mass, bulk; **— apparent,** stacked contents, stacked measure; **— plein, — réel,** cubic contents, solid contents; **— sur pied,** standing stock or crop.

volumètre, *m.* volumeter.

volumétrie, *f.* volumetric analysis.

volumétrique, volumetric(al).

volumineux, -euse, voluminous, large, bulky.

volupté, *f.* pleasure, delight, voluptuousness.

voluptueusement, voluptuously.

volute, *f.* volute, scroil.

volutine, *f.* volutin.

vomique, *f.* vomica; **noix —,** nux vomica, vomit nut.

vomiquier, *m.* nux vomica tree (*Strychos nux-vomica*).

vomir, to vomit, emit, throw forth.

vomissement, *m.* vomiting, vomit.

vomitif, *m.* vomitive.

vomitif, -ive, vomitive, emetic.

vont (aller), (they) are going.

vorace, voracious, ravenous.

voracement, ravenously.

vos, *pl.* your.

Vosges (les), *f.pl.* the Vosges (mountains).

vosgien, -enne, Vosges, of the Vosges.

votant, *m.* voter.

votant, voting.

votation, *f.* voting.

vote, *m.* vote.

voter, to vote.

votre, your.

vôtre, yours.

voudra (vouloir), (he) will wish, want to.

vouer, to vow, consecrate, destine, dedicate.

voulant, willing.

vouloir, to will, be willing, like, want, wish, desire, consent, require, demand, mean, intend, be pleased, attempt, admit, allow; **— dire,** to mean; **bien —,** to be kind enough; **en — à,** to have a grudge against, have designs on, look for; **s'en —,** to be angry, vexed; **comment voulez-vous que . . . ?** how do you expect that . . . ? **il veut bien,** he is willing; **que voulez-vous?** what do you expect?

vouloir, *m.* will; **mauvais —,** ill will.

voulu, required, requisite, needful, willed, desired, due.

voulut, (he) wished.

vous, you, to you.

vous-même, yourself.

voussoir, vousseau, *m.* voussoir, archstone.

voussure, *f.* arching, vaulting, bulging, curve.

voûte, *f.* vault, arch, vaulted roof, fornix, archway; **— foliacée,** leaf canopy.

voûtelette, *f.* small vault.

voûter, to vault, arch; **se —,** to be bent, arch, become bent, begin to stoop.

voyage, *m.* journey (by land), trip, excursion, run, voyage (by sea), traveling, sojourn.

voyager, to travel, migrate, (of goods) be carried, journey (on land), voyage (on sea)·

voyageur, -euse, *m.&f.* traveler, passenger, voyager.

voyageur, -euse, traveling, migratory.

voyant, *m.* seer, conspicuousness, showiness, mark, signal, eyepiece, slide vane.

voyant, seeing, showy, conspicuous.

voyelle, *f.* vowel.

voyer, to run off, pour on (lye).

voyer, *m.* road supervisor.

vrac, *m.* rubbish; **en —,** in bulk, loose.

vrai, *m.* truth, true.

vrai, true, real, genuine, sincere, proper, right, truly, really, indeed; **à — dire,** to tell the truth; **au —, dans le —,** in truth, truly; **pas —?** is it not true? is it not so?

vraiment, truly, really, indeed, verily.

vraisemblable, probable, likely.

vraisemblablement, probably, likely.

vraisemblance, *f.* probability, likelihood.

vrille, *f.* gimlet, tendril, cirrus; **— articulée,** jointed tendril.

vriller, to bore (with a gimlet), ascend in a spiral.

vrillette, *f.* deathwatch (beetle).

vu, *m.* sight, inspection, examination, preamble.

vu, seen, regarded, seeing, considering, in view of, according to; **— que,** seeing that, considering that, since, whereas.

vue, *f.* view, sight, inspection, prospect; outlook, survey, intention, eyesight, window, opening; **— de côté,** side view; **— en bout,** end view; **à —,** after sight, at sight, **à — d'oeil,** visibly; **en — de,** with a view to, with the object of; **dans cette —,** with this object in view.

vulcanisation, *f.* vulcanization, vulcanizing.

vulcanisé, vulcanized.

vulcaniser, to vulcanize.

vulgaire, *m.* common man, ignorant man, common folk; **le — de,** the common run of.

vulgaire, common, vulgar, ordinary.

vulgairement, commonly, vulgarly, ordinarily.

vulgarisation, *f.* popularization.

vulgariser, to popularize.

vulnérable, vulnerable.

vulnéraire, *m.* vulnerary; *f.* kidney vetch, woundwort (*Anthyllis vulneraria*).

vulnéraire, vulnerary.

vulpin, *m.* foxtail grass (Alopecurus).

vulvaire, *f.* stinking goosefoot (*Chenopodium vulvaria*).

vulve, *f.* vulva.

W

wacke, *m.* wacke.

wad, *m.* wad, bog manganese.

wagage, *m.* river mud (used as fertilizer).

wagon, *m.* car, cart, carriage, wagon, (railroad) coach, flue tile; **— à houille,** coal car, coal wagon.

wagon-citerne, *n.* tank car.

wagon-lit, *m.* sleeping car, sleeper.

wagonnet, *m.* little car, truck.

wagon-réservoir, *m.* tank car.

waringa, *m.* benjamin fig tree (*Ficus benjamina*).

warranter, to warrant.

watt, *m.* watt.

watt-heure, *m.* watt-hour.

wattman, *m.* motorman.

wattmètre, *m.* watt-hour meter.

wiesnérie, *f.* water plantain (*Alisma plantago*).

willémite, *f.* willemite.

wistarie, *f.* wisteria (Wisteria).

withérite, *f.* witherite, native barium carbonate.

witloof, *m.* witloof, large-rooted Brussels chicory.

wolfram, *m.* wolfram, tungsten.

wolframine, *f.* tungstite, native tungsten trioxide.

wolframocre, *m.* wolfram ocher, tungstite, tungstic ocher.

woorari, *m.* curare, woorari (*Strychnos toxifera*).

wootz, *m.* wootz, India steel.

X

xalapa, *f.* see JALAP.

xanthe, *m.* burweed.

xanthéine, *f.* xanthein.

xanthiacé, resembling xanthium or burweed.

xanthié, resembling xanthium or burweed.

xanthine, *f.* xanthin(e)

xanthinine, *f.* xanthinine.

xanthique, xanthic.

xanthocarpe, with yellow fruits.

xanthochromique, xanthochromic.

xanthogène, *m.* xanthogen.

xanthone, *f.* xanthone.

xanthophylle, *f.* xanthophyll.

xanthoprotéine, *f.* xanthic base.

xanthoprotéique, xanthoproteic.

xanthorhize, with yellow roots.

xanthosperme, with yellow seed.

xanthoxyle, *m.* prickly ash (Zanthoxylum).

xanthoxylé, with wood of yellow color.

xémorphose, *f.* aitiomorphosis, change in shape caused by external factors.

xénie, *f.* xenia, direct influence of foreign pollen.

xénocarpe, (fruit) resulting from xenogamy.

xénocarpie, *f.* xenocarpy, producing fruit as result of xenogamy.

xénogamie, *f.* xenogamy, cross-fertilization.

xénogénèse, *f.* xenogenesis.

xénomorphose, *f.* xenomorphosis, change in shape caused by external factors.

xénon, *m.* xenon.

xérocléistogamie, *f.* xerocleistogamy, flowers remaining closed because of insufficient moisture.

xérophile, *f.* xerophile, plant growing in arid places.

xérophile, xerophilous, growing in arid places.

xérophyte, *f.* xerophyte, plant subsisting on little moisture.

xérosis, *f.* xerosis.

xéroté, growing on dry ground.

xiphifolié, with sword-shaped (ensiform) leaves.

xiphoïd, xiphoid, like a sword, ensiform.

xiphophylle, xiphophyllous, with ensiform leaves.

xyphoïd, see xiphoïd.

xylane, *m.* xylane.

xylème, *m.* xylem.

xylène, *m.* xylene.

xylidine, *f.* xylidine.

xylin, related to wood.

xylique, xylic.

xylite, *f.* xylite.

xyloïdine, *f.* xyloidin.

xylol, *m.* xylene, xylol.

xylomètre, *m.* xylometer.

xylomyce, growing on wood.

xylophage, xylophagous.

xylopode, *m.* xylopodium, a fruit like a nucule but lacking a cupule.

xylose, *m.* xylose.

xylotile, *f.* xylotile.

Y

y, there, to him, to her, to that, to it, to them, in it, in them, therein, here, thither, by it, by them, for it, for them, at it, at them; **il y a,** there is, there are, ago; **il y a lieu,** there is reason to.

yahourt, *m.* see yoghourt.

yatagan, *m.* yataghan, Turkish dagger.

yèble, *m.* dwarf elder, Danewort (*Sambucus ebulus*).

yeuse, *f.* holm oak, holly oak, evergreen oak (*Quercus ilex*).

yeux, *m.pl.* see oeil.

yoghourth, yogourt, *m.* yoghurt.

youfte, *m.* yufts, Russia leather.

ymnadiphyte, (organs) surrounded by a spathe.

ypérite, *f.* yperite (mustard gas).

ypréau, *m.*wych-elm (*Ulmus glabra*).

ypsoophytes, (stamens) inserted at top of peduncle.

ytterbium, *m.* ytterbium.

yttria, *m.* yttria.

yttrialite, *f.* yttrialite.

yttrifère, yttriferous.

yttrique, yttric, yttrium, of yttrium.

yttrium, *m.* yttrium.

yucca, *m.* yucca.

Z

zèbre, *m.* zebra.

zébrer, to stripe (like a zebra).

zébrure, *f.* striping, stripes.

zédoaire, *f.* zedoary.

zéine, *f.* zein.

Zélande, *f.* Zealand.

zèle, *m.* zeal, ardor.

zélé, *m.* zealot.

zélé, zealous.

zellulose, *m.* cellulose

zénith, *m.* zenith.

zéolithe, zéolite, *f.* zeolite.

zéolithique, zéolitique, zeolitic.

zéro, *m.* zero, cipher, naught.

zérotage, *m.* calibration, determination of the zero point.

zeste, *m.* peel (of citrus fruits), partition membrane (in nuts).

zester, to peel off the outer skin.

zeugite, *f.* zeugite, fungus spore from fusion of two nuclei.

zigzag, *m.* zigzag.

zigzaguer, to zigzag.

zinc, *m.* zinc; — carbonaté, smithsonite, zinc spar, cadmia; — laminé, sheet zinc, zinc plate.

zincage, *m.* covering (roof) with zinc, galvanizing.

zincifère, zinciferous, zinc-bearing.

zincique, zincic, of zinc.

zincographie, *f.* zincography.

zincographier, to reproduce by zincography.

zingage, *m.* covering (roof) with zinc, zincing, galvanizing.

zingiber, *m.* ginger.

zingibéracées, *f.pl.* Zingiberaceae.

zinguer, to zinc, galvanize, cover (roof) with zinc.

zinguerie, *f.* zinc works, ware, or trade.

zingueur, *m.* zinc worker.

zinguer, to galvanize.

zinquier, *m.* zinc founder.

zinzolin, *m.* reddish violet.

zinzolin, reddish violet or purple.

zinzoliner, to color (dye) reddish violet or purple.

zircon, *m.* zircon.

zircone, *f.* zirconia.

zirconique, zirconic, zirconium, of zirconium.

zirconium, *m.* zirconium.

zodiaque, *m.* zodiac.

zoïdiogame, zoidiogamous, with ciliated antherozoids.

zoïdiogames, *f.pl.* zoidogamae, plants pollinated by animals.

zone, *f.* zone, belt.

zoné, zoned, zonate.

zonule, *f.* small zone.

zoobiologie, *f.* zoobiology.

zoocécidie, *f.* cecidium (gall) by animal parasite.

zoochimie, *f.* zoochemistry.

zoocyste, *f.* zoocyst, cyst giving rise to ciliated or amoeboid zoogonidia.

zoogames, *f.pl.* zoogamae, plants with motile reproductive elements.

zoogamète, *f.* zoogamete. planogamete, mobile ciliated gamete.

zooglée, *f.* zoogloea.

zoologie, *f.* zoology.

zoologique, zoological.

zoologiste, zoologue, *m.* zoologist.

zoophobe, zoophobous, protecting against animals.

zoophyte, *m.* zoophyte, plantlike animal.

zoosporange, *m.* zoosporagium.

zoospore, *f.* zoospore.

zoosporocyste, *f.* zoocyst.

zostère, *f.* seaweed, sea wrack (Zostera).

zumatique, zumatic.

zumique, zymic.

zygolytique, zygolytic, separation of allelomorphic pairs of unit-characters.

zygomatique, zygomatic.

zygomorphe, zygomorphous, bilaterally symmetrical.

zygospore, *f.* zygospore, body produced by the coalescence of two similar gametes.

zygote, *f.* zygote.

zygozoospore, *f.* zygozoospore, motile zygospore.

zymase, *f.* zymase.

zymine, *f.* zymin.

zymique, zymic.

zymogène, *m.* zymogen.

zymogène, zymogenic.

zymohydrolyse, *f.* zymohydrolysis, hydrolysis by fermentation.

zymologie, *f.* zymology.

zymosimètre, *m.* zymosimeter, zymometer.

zymotechnie, *f.* zymotechnics.

zymotique, zymotic.

REVISED SUPPLEMENT OF TERMS IN AERONAUTICS, ASTRONAUTICS, ATOMIC ENERGY, AUTOMOBILE TECHNOLOGY, ELECTRONIC DATA PROCESSING, ELECTRONICS, NUCLEAR SCIENCE AND TECHNOLOGY, RADAR, RADIO, TELEVISION[1]

A

*abaisser, to step down, suppress (a carrier).

*abaisseur de tension, voltage step-down device, step-down transformer.

*abandonner l'avion, to bail out.

*abaque, alignment or nomographic chart, abacus.

*abattage, working.

*abattre un avion ennemi, to shoot down an enemy aircraft.

abernathyite, f. abernathyite.

*aberration chromosomique, chromosome aberration.

abmho, abmho.

*abondance, — cosmique, cosmic abundance; — isotopique, isotopic abundance.

*aborder, to collide.

*abréger, to abstract, summarize.

*abri, — antigaz, gasproof shelter; — collectif, public air-raid shelter; — contre avions, air-raid shelter; — contre le bombardment, bombproof shelter; — de sondeur, doghouse; — public, public air-raid shelter.

*abrité, dripproof.

*absence illégal, absence without leave.

*absent, away.

*absorbant, absorber; — de neutrons, neutron absorber; — sélectif, selective absorber.

*absorber l'énergie, to absorb energy.

*absorbeur, — de neutrons, neutron absorber; — étalonné, calibrated absorber, standard absorber; — faible, weak absorber; — oscillant, oscillating absorber.

absorbeur-neutralisateur, m. scrubber (astronautics).

*absorption, capture, uprake; — calorique, heat absorption; — dans le sol, ground absorption; — de mésons, meson absorption; — d'énergie, absorption of energy; — de neutrons, neutron absorption; — de neutrons de résonance, resonance absorption.

absorption, — de neutrons épithermiques, epithermal absorption; — de neutrons sans fission, nonfission neutron absorption; — de neutrons thermiques, absorption of thermal neutrons; — de photons, photon absorption; — de rayons gamma, gamma-ray absorption; — exponentielle, exponential absorption; — interne, internal absorption.

absorption, — mésique, meson absorption; — paramagnétique, paramagnetic absorption; — par effet photoélectrique, photoelectric absorption; — photonique, photon absorption; — sans fission, nonfission absorption; — sélective, selective absorption; — totale des neutrons, total absorption of neutrons.

*abstraction, faire —, not to mention, to leave out of consideration.

abukumalite, f. abukumalite.

abvolt, m. abvolt.

*académie aéronautique, air-force academy.

accastillage, m. upper works (naut.).

*accélérateur, — à grand énergie, high-energy accelerator, high-energy machine; — à induction, induction accelerator; — à ondes progressives, traveling-wave linear accelerator; — à tension constante, constant-potential accelerator; — de corpuscules, particle accelerator; — de particules, atomic accelerator, particle accelerator; — de particules à haute énergie, high-energy-particle accel-

[1] The asterisk denotes additional translations of words given in the main text.

erator; — **de particules chargées,** charged-particle accelerator; — **d'ions,** ion accelerator; — **linéaire,** linear accelerator, linac.

*accélération, — absolue, absolute acceleration; — centrifuge, centrifugal acceleration; — centripete, centripetal acceleration; — de la pesanteur, acceleration of gravity; — des ions, ion acceleration; — par saccades, jerky acceleration.

*accélérer, to activate.

*accepteur d'ions, ion acceptor.

*accès, — en parallèle, simultaneous access; — libre (à), open shop; — par file d'attente, queued access method; — restreint (à), closed shop; — sélectif, random access, direct access; — séquentiel, sequential access, serial access; — sériel, serial access, sequential access.

*accessoire, component; — d'avions, aircraft accessory.

*accident, detail (in a sound track); — au décollage, accident during take-off; — dê aux dépôts de givre, accident due to ice formation.

*accord, tuning; — aigu, sharp tuning; — aplati, flat tuning, broad tuning; — approximatif, coarse tuning, rough tuning; — d'antenne, antenna tuning, aerial tuning; — du circuit de chauffage, tuning of the filament circuit; — pointu, sharp tuning; — précis, fine tuning; — syndical, cartel contract; — unique, single-dial tuning, single-control tuning, unicontrol.

*accorder, to tune, resonate, cause to assume resonance.

accostage, *m.* docking (astronautics).

accoster, to board (naut.).

*accoudoir, armrest.

*accouplement, fastening, connection; — à baïonnette, clutch bayonet; — à bride, flanged coupling; — à serrage automatique, automatic coupling; — de rotule, ball-and-socket joint; — global, coupling in the large; — mécanique, ganging; — serré, tight or close coupling; — total, coupling in the large.

accourcissement, *m.* shortening, shrinking (materials).

accrétion, *f.* accretion.

accrochage, *m.* heapstead (min.), landing; — d'oscillations, starting or setting in of oscillations; — d'une machine asynchrone, crawling (of alternators); — d'une machine synchrone, paralleling (of alternators).

accroché, on hook.

*accrocher, to (start to) oscillate.

*accroissement, — d'énergie, energy gain; — d'impulsion, incremental impulse.

*accroître, to increment.

accu, *m.* accumulator.

*accumulateur, — à décalage, shifting accumulator; — à liquide immobilisé, unspillable accumulator; — de chauffage, filament accumulator; — d'énergie atomique, atomic-energy storage battery; — des poussières, dust collector; — électrique, battery; — électrique de chaleur, electric heat accumulator; — fernickel, iron-nickel accumulator, Edison storage battery.

accumulation, *f.* build-up.

*acétate d'uranyle, uranyl acetate.

*acheminement, routing.

*acide, — carbonique, carbon dioxide; — désoxyribonucléique, deoxyribonucleic acid; — fluorhydrique, hydrofluoric acid; — nucléique, nucleic acid; — residuaire, residual acid; — ribonucléique, ribonucleic acid; — sulfophénique, phenolsulphonic acid.

acido-résistant, acid-resistant.

*acier, — à haute résistance, high-grade steel; — allié, alloy steel; — au carbone, carbon steel; — autotrempant, air-hardening steel; — embouti (à la presse), pressed steel; — estampé, stamped steel; — estampé à froid, swaged steel; — étiré, drawn steel; — extradoux, dead-soft steel; — forgé par chocs, drop-forged steel; — forgé par pression, press-forged steel; — fritté, sintered steel; — hypertrempé, martempered steel; — Martin, open-hearth steel; — matricé, die-forged steel; — mi-dur, medium steel; — rapide, high-speed steel; — recuit, annealed steel; — refoulé, upset steel; — revenu, tempered steel; — spécial, alloy or special steel; — sur sole, open-hearth steel; — Thomas, basic Bessemer steel; — trempé à coeur, through-hardened steel.

*acieration, acierage.

*aciérie, rolling mill.

*à-coup, sudden stop; — de courant, surge of current; — de surcharge, sudden overload; —s, gear backlash.

*acquisition des données, data gathering, data collection.

*acquit de douane, customs clearance.

acrobatie, *f.* — de haute école, advanced aerobatics; — en planeur, glider aerobatics.

*acte, — constatant la nationalité, certificate of registry; — de propriété, certificate of ownership.

actinides, *m.pl.* actinides.

actinon, *m.* actinon.

actino-uranium, *m.* actinouranium.

*__action__, — __coulombienne entre particules,__ particle Coulomb interaction; — __d'arrondir,__ rounding off; — __de choc à courte distance,__ short-range collisional interaction; — __de choc à faible portée,__ short-range collisional interaction; — __de faible portée,__ short-range interaction; — __d'érosion,__ jetting action; — __de rétablissement,__ reset action; — __de . . . sur . . .,__ interaction.

__action, — due au noyau,__ nuclear interaction; — __en retour,__ feedback; — __entre noyaux,__ nuclear interaction; — __faible,__ weak interaction; — __mutuelle,__ mutual interaction; — __mutuelle des ondes électromagnétiques,__ Luxemburg effect, interaction of radio waves; — __par dérivation,__ rate action; — __par les roues d'arrière,__ rear drive; — __par les roues d'avant,__ front-wheel drive; — __réciproque,__ mutual interaction; __à__ — __differée,__ time-limit release or relay; __à__ — __instantanée,__ instantaneous release or relay.

*__actionné, — par des commandes rigides,__ operated by means of tie rods; — __par une commande differentielle,__ differentially operated.

activation, *f.* activation.

*__activité, — à saturation,__ saturated activity; — __engendrée,__ daughter activity; — __spécifique,__ specific activity; — __spécifique d'un radioélément,__ isotope specific activity; — __spécifique par gramme d'élément,__ gram element specific activity; __en__ —, in gear.

*__actualité,__ outside broadcast.

*__acuité d'accord,__ sharpness of tuning.

adac (A.D.A.C.: avion à décollage et atterissage court), *m.* Stol (S.T.O.L.).

adacport, *f.* stolport.

adaptateur, *m.* converter, adapter; — __de lampe,__ valve adapter; — __phonographique,__ phonograph pickup, pickup; — __sur secteur,__ socket power unit, power pack.

*__adaptation,__ matching.

*__adapter,__ to match.

adapteur, *m.* adapter, data adapter unit, interface; — __de canal d'entrée analogique,__ analog input data channel adapter; — __de commande de contact,__ contact operate adapter; — __de données,__ data adapter unit; — __de ligne,__ line adapter; — __sériel,__ serial data adapter; — __synchrone de données,__ synchronous data adapter.

additif, *m.* attachment.

*__addition,__ extension; — __sans report,__ addition without carry; — __transversale,__ crossfoot.

additionneur, *m.* adder.

additionneur-soustracteur, *m.* adder-subtracter.

adion, *m.* adion.

*__administration gérante,__ managing administration.

admissible, permissible.

*__admission,__ input.

*__admittance, — d'entrée,__ input admittance; — __de sortie,__ output admittance; — __imaginaire,__ susceptance.

adressage, *m.* addressing; — __indirect,__ indirect addressing; — __optimal,__ optimum addressing; — __spécifique,__ boundary alignment.

*__adresse, — absolue,__ absolute address, effective address; — __d'arrêt,__ address stop; — __d'instruction,__ instruction address; — __de base,__ base address; — __de niveau N,__ N level address; — __de piste,__ home address (EDP); — __de retour,__ return address; — __déterminée,__ integral boundary (EDP); — __deux-plus-un,__ two-plus-one address; — __directe,__ direct address, one-level address; — __immédiate,__ immediate address, zero-level address; — __indirecte,__ indirect address, multilevel address; — __modifiée,__ modified address; — __quatre-plus-un,__ four-plus-one address; — __réele,__ actual address; — __relative,__ relative address, floating address; — __symbolique,__ symbolic address; — __transposable,__ relocatable address; — __trois-plus-un,__ three-plus-one address; — __un-plus-un,__ one-plus-one address.

adresse-machine, *f.* machine address (EDP).

adsorbant, *m.* adsorbent.

adsorbat, adsorbate.

advection, *f.* advection.

*__aérateur orientable,__ swiveling ventilator.

aérien, *m.* antenna, aerial (conductor or wire); — __d'émission,__ transmitting antenna or aerial; — __directif,__ directional antenna, directive aerial.

aérodrome, *m.* airport; — __d'entrainment,__ training airport; — __douanier,__ customs airport.

aéroglisseur, *m.* hovercraft.

aérologie, *f.* aerology.

*__aéromètre,__ hydrometer.

aéronaute, *m.* airman.

aéronomie, *f.* aeronomy.

aéroport, *m.* airport; — __d'aviation sportive,__ club airport; — __d'hydravions,__ marine airport.

aéroportuaire, having to do with airports.

aérosol, *m.* spray.

aérosurface, *f.* temporary site for planes.

affaibli, attenuated, weakened.

*affaiblissement, damping, attenuation, fading; — **caractéristique,** attenuation equivalent; — **des freins,** brake fading; — **du rayonnement gamma,** gamma attenuation; — **linéique,** attenuation constant; — **résultant,** over-all or net attenuation, over-all transmission loss; — **transductique,** transmission efficiency.

*affectation, allocation, assignment; — **de mémoire,** space allocation, storage allocation; — **univoque de tampons** (EDP), simple buffering.

*affecter, to relegate, set aside, allot to, allocate, assign.

*affections, — **cancéreuses,** — **malignes,** malignant diseases.

*affichage, display, visual display, posting (prices).

affilé, keen, sharp.

*affinage électrolytique, electrolytic refining.

*affranchissement d'unités périphériques, device independence.

affrètement, *m.* chartering; — **de longue durée,** time charter.

*affût de canon, gun carriage, gun mounting.

*âge, — **chimique,** chemical age; — **de Fermi,** Fermi age; — **de procréation,** parental age; — **des neutrons,** neutron age; — **jusqu'à absorption,** age to point of absorption.

*agencement, disposition; *pl.* fittings.

*agent, — **absorbant,** getter, absorbing means, absorbent; — **comptable,** accountant; — **de liaison,** linkage editor (EDP); — **d'épuration,** cleaning agent; — **de refroidissement,** coolant; — **de rétention,** hold-back agent; — **mutagène,** mutagenic agent; — **taxateur,** counter clerk, window clerk.

AGI, Année géophysique internationale, International Geophysical Year (IGY).

*agitation thermique, thermal motion.

agrandisseur, *m.* enlarger; — **de mise au point automatique,** autofocus enlarger.

agranulocytose, *f.* agranulocytosis.

*agrégat à forte granulométrie, coarse aggregate.

*agrément, certification.

agression, *f.* aggression; — **aérochimique,** gas attack from the air.

*aigrette, brush discharge.

*aiguille sauteuse, bouncing pin.

*aiguilleur, electromagnetic relay (translator).

aiguillot, *m.* pintle (of a rudder).

*aile, airfoil, fender, wing; — **à fente,** slotted wing; — **de cornière,** leg (of angle); — **démontable,** detachable wing; — **de profil et d'épaisseur amincis,** wing tapering in plan form and thickness; — **d'hélice,** propeller blade; — **encastrée,** continuous wing.

aile, — **entièrement cantilever,** pure cantilever wing; — **épaisse,** thick wing; — **haute incurvée,** gull wing; — **inférieure,** lower wing; — **latérale,** outboard wing; — **mince,** thin-section wing; — **multilongeron,** multispar wing; **à —s decalées,** staggered (aviation); **à —s surelevées,** high wing.

*aileron, — **à fente,** slotted flap; — **de courbure,** trailing edge flap; — **encastré,** aileron inset from wing tip.

*ailette, — **de refroidissement,** cooling fin or vane (tube); **à —s longitudinales,** longitudinal finning.

*aimantation, — **auxilière,** superposed magnetization; — **remanente,** residual magnetization.

*air, — **agité,** moving air; — **à grains,** bumpy air; — **de refroidissement,** air coolant; **d'— équivalent,** air equivalent.

*aire, area; — **de lancement,** launch pad; — **de migration,** migration area; — **de ralentissement,** slowing-down area; — **de service,** service area; — **élémentaire,** areal element; — **superficielle,** surface area.

*ais, shelf.

ajout, *m.* addition (to contract, proof, etc.).

ajoutage, *m.* addition.

*ajustement, fit; — **à chaud,** shrink fit; — **à la presse,** press fit; — **au maillet,** drive fit; — **avec jeu,** clearance fit; — **avec serrage,** interference fit; — **bloqué,** wringing fit, snug fit; — **glissant,** sliding fit; — **incertain,** transition fit; — **libre,** free fit; — **serré,** tight fit; — **tourant,** running fit.

*ajutage, discharge or efflux tube; — **d'automaticité,** air compensator jet.

*alambic, — **à marche discontinue,** batch still; — **cylindrique,** shell still.

albédo, *m.* albedo, reflection coefficient.

*alerte aux avions, air-raid warning.

*alésage, caliber; — **direct dans le bloc-cylindres,** parent bore.

*alesoir, drill press.

*algèbre, — **booléenne,** Boolean algebra; — **de Boole,** Boolean algebra.

algorithme, *m.* algorism, algorithm.

*alimentateur d'antenna, transmission line, feeder.

*alimentation, excitation or energization (of coils), power supply, boiler feeding, input; — accélérée, high-speed feed; — continue, successive feed; — continue en énergie, continuous input of energy; — électrique, power supply; — en cartes, card feed; — en combustible, fuel feed; — en combustible par gravité, fuel feed by gravity; — en courant continu, direct-current supply; — en documents, bill feed; — manuelle, hand feed; — mauvaise, misfeed; — en oxygène, oxygen feed; — en surface, slash feed; — par les neufs, nineedge first; — retardée, delayed feed; — secteur, — sur le secteur, main supply.

allanite, *f.* allanite.

allège, *f.* lighter, tender (naut.).

allèle, *f.* allele; — de type uniforme, standard type allele; — favorable, advantageous allele; — modèle, standard allele; — mutant, mutant allele; — type, standard allele.

allélisme, allelism.

*aller au fond, to sink.

allobare, allobar.

allongeur d'impulsions, *m.* pulse lengthener.

allotement, allotissement, *m.* allocation.

*allumage, — à avance automatique, automatically advanced ignition; — à étincelle, spark ignition; — blindé, screened ignition; — en retour, arcing back, backfiring.

allume-cigare, *m.* cigarette lighter.

*allure, action; — en fonction du temps, timedependent behavior.

*alluvion, drift; — glaciaire, glacial drift.

*alphabet à cinq éléments, five-unit code.

alphanumérique, alphanumeric, alphameric.

alphatopique, alphatopic.

*altération, — de l'aptitude génétique, deterioration of genetic fitness; — superficielle, weathering.

*altéré, non —, unweathered.

*alternance, — de multiplicités, alternation of multiplicities; — génétique, genetic alternation.

alternat, *m.* two-way communication apparatus, walkie-talkie, air-air or air-ground communication (aviation).

*alternateur, alternating-current generator; — à fer tournant, inductor alternator with moving iron.

alternatif brut, *m.* raw, unsmoothed alternating current containing ripples.

*alternation, half-period, half-cycle.

altiport, *f.* mountain airport.

altisurface, *f.* temporary mountain site for planes.

*altitude, — au-dessus du point de départ, altitude above the starting point; — d'utilisation, operational height; à l' — de pleine admission, at maximum boost altitude; à l' — de sécurité, at a safe height.

*alumine hydratée, aluminum hydroxide.

*alvéole du train rentrant, hole for retractable undercarriage.

*amarrage, docking (astronautics).

*amas, cluster; — ionique, ion cluster.

*amateur de radio-diffusion, radio fan, radio listener.

ambulance de première ligne, *f.* first-aid station.

*âme (cartouche de combustible), core; — de longeron, web of spar.

améliorant, *m.* dope.

*aménagement, setup, installation, housekeeping (EDP); — de locaux, site preparation.

*amenée, leadin.

*amener les neutrons à l'état thermique, to thermalize neutrons.

américium, *m.* americium.

amerissage, *m.* landing or descent on water; — forcé, forced landing.

amers, *m.pl.* distinctly visible landmark.

amirauté, *f.* admiralty.

ammètre, *m.* ammeter; — à fil chaud, hot-wire ammeter; — d'arsonval, moving-coil ammeter.

amniographie, *f.* amniography.

*amodiataire, sublessee (of mining concession).

amodiation, *f.* farming out, subleasing.

*amont, monter en —, connected above.

*amorçage, start-up, excitation, energization, striking (of an arc); — de l'arc, arc ignition; — d'oscillations, release, starting, or setting in of oscillations.

*amorce, bootstrap, leader (film), initial instruction (EDP); — de bande, tape leader.

*amorcer, to excite, energize; — un trou, to countersink.

*amorçoir, center punch.

amorti, damped, attenuated, diminished.

*amortissement, decay, attenuation; — critique, exponential damping; — du train, springing of the undercarriage; — exponentiel, exponential damping.

*amortisseur, shock absorber; — à double effet, double-acting shock absorber; — de vibrations, vibration damper; — pneumatique, pneumatic shock absorber, compressed-air shock absorber, compressed-air landing leg; — télescopique, telescopic shock absorber.

ampangabéite, f. ampangabeite.

ampèreheuremètre, m. ampere-hour meter, quantity meter.

*ampèremètre, — à bobine mobile, — à cadre mobile, moving-coil ammeter; — à fer, moving-iron ammeter, soft-iron ammeter; — à fil chaud, hot-wire ammeter.

*amphibie, — à coque, amphibian flying boat; — à flotteurs, amphibian float plane.

*amplificateur, — (à) basse-fréquence, low or audio-frequency amplifier; — à correction de dérive, drift-corrected amplifier; — à courant continu, direct-current amplifier; — à haute fréquence, high-frequency amplifier; — à impédance, choke-coupled amplifier; — à inversion, inverting amplifier; — à lampes, tube amplifier; — à liaison directe, direct-coupled amplifier; — à plusieurs sensibilités, multirange amplifier; — à résistance, resistance-coupled amplifier; — à résonance, tuned amplifier; — à selecteur mécanique, chopper amplifier.

amplificateur, — à transformateur(s), transformer-coupled amplifier; — à un étage, one-stage amplifier; — d'arret, shutdown amplifier; — de courant photoélectrique, photocell amplifier; — de puissance, power amplifier; — de sécurité, shutdown amplifier; — de sortie analogique, analog driver amplifier; — de tension, voltage amplifier; — differentiel, — équilibré, push-pull amplifier.

amplificateur, — d'impulsions, pulse amplifier; — haute fréquence accordé, tuned radio-frequency amplifier; — intermédiaire, booster; — linéaire, linear amplifier; — logarithmique, logarithmic amplifier; — proportionnel, proportional amplifier; — symétrique, push-pull amplifier.

*amplification, — à réaction, regenerative amplification; — en courant, current amplification; — en milieu gazeux, gas amplification; — poussée, high-gain amplification; — produite par l'ionisation, gas amplification.

*amplitude, — de diffusion, scattering amplitude; — d'une grandeur alternative symétrique, amplitude of a symmetrical alternating quantity; — totale d'une grandeur oscillante, total amplitude of an oscillating quantity.

*ampoule, — du redresseur, rectifier bulb; — électro-ionique, gas tube, gas valve.

*analyse, scan, scanning, exploring, sweeping (out) (telev.), systems analysis, sensing; — chimique continue, continuous chemical analysis; — cristalline, crystal analysis; — de fer, iron testing (chem.); — de l'image, scanning, exploring, or analyzing the picture; — de marché, market analysis; — des cryptogrammes, cryptanalysis; — des moyens de protection, protection survey; — des ventes, sales analysis; — d'impulsions en fonction du temps, pulse-time analysis; — d'un faisceau ionique, ion-beam scanning; — granulométrique, screen analysis.

analyse, — isotopique, statistical separation of isotopes; — manuelle par instruction, instruction cycle process; — mathématique, mathematical analysis; — microchimique, microchemical analysis; — numérique, numerical analysis; — par activation, activation analysis; — par lignes entrelacées, staggered or interlaced scanning; — par radio-activation, radioactivation analysis; — radiochimique, radiochemical analysis; — radiométrique, radiometric analysis; — statistique, statiscal analysis; — thermique, thermal analysis; — unitaire, single-step process.

*analyser, to sense.

*analyseur, scanner; — à spot mobile, flying spot scanner; — à un seul canal, single-channel analyzer; — d'amplitude des impulsions, kick-sorter, pulse-height analyzer; — de minutage d'impulsions, time analyzer; — de réseau, network analyzer; — différentiel mécanique, mechanical differential analyzer; — en temps, time analyzer; — optique, optical code scanner, optical scanner.

*ancien, senior.

*ancrage, anchorage.

*ancre, guy, tie plate (R.R.); — flottante, drag anchor, sea anchor.

andersonite, andersonite.

androgénèse, f. androgenesis.

*anémie à hématies falciformes, sickle-cell anemia.

*angle, elbow, knee; — d'avance, lead angle; — de Bragg, glancing angle; — de braquage, lock angle (vehicle turning); — de carrossage, camber; — de chasse, castor angle; — de décalage, phase-displacement angle, dephasing angle, phase shift; — de décart,

deflection or deviation angle; — **de déphasage**, phase angle; — **de dépouille**, clearance
angle; — **de descente**, gliding angle; — **de
déviation**, angle of deflection; — **de diffusion**,
scattering angle; — **de dwell**, dwell angle;
— **de fermeture de came**, dwell angle (contact
breaker cam closing angle); — **de 45 grades**,
gliding angle; — **de la pointe**, angle of center;
— **de montée**, angle of climb, climbing angle;
— **d'entaillage de la surface**, cutting angle; —
de planement, gliding angle; — **de retard**,
lag angle; — **des balais**, angle of brush load;
— **limite**, critical angle; — **mort**, blind angle;
— **solide**, solid angle; **faible** — **de planement**,
flat gliding angle.

angloir, *m.* bevel square.

*****animateur**, disk-jockey.

*****anneau**, spider, ear; — **de cerclage**, bonding
ring; — **de fixation**, retaining ring; — **de
garde**, guard ring; — **de refroidissement**,
coolant annulus; — **de silence**, blind spot;
— **pare-huile**, oil slinger (on shafts); — **tourbillonnaire**, vortex ring.

*****année financière**, fiscal year.

année-personne, *f.* man-year.

anneroédite, *f.* annerodite.

*****annexe**, auxiliary.

annihilation, *f.* annihilation, pair conversion.

*****annonce**, warning.

*****annuaire des abonnés au téléphone**, tele-
phone directory.

*****annulation**, deletion.

annulé, void.

*****annuler**, to delete, erase.

*****anode**, — **d'allumage**, — **d'amorçage**, ignition
or exciting anode; — **divisée**, — **fendue**, split
anode; — **perforée de canon à électrons**, lens
disk, lens scanning disk.

*****anomalie**, malfunction; — **génétique pro-
voquée par irradiation**, radiation-induced
genetic defect.

*****antenne**, — **à accord multiple**, multiple-
tuned antenna; — **à cage**, cage antenna; — **à
double épingle à cheveux**, double-hairpin-
type antenna; — **à faisceau**, beam antenna;
— **à nappe**, sheet antenna; — **à ondes mo-
biles**, antenna for progressive waves; —
basse, ground or earth antenna; — **Beverage**,
Beverage or wave antenna; — **bifilaire**, two-
wire antenna; — **compensée**, balanced aerial.

antenne, — **d'appartement**, inside antenna;
— **de fortune**, auxiliary antenna; — **demi-
onde**, half-wave antenna; — **d'émission**,
transmitting antenna; — **de secours**, emergency antenna; — **dirigée**, directional an-

tenna; — **émettrice**, transmitting antenna;
— **en cadre**, frame antenna; — **en dents de
scie**, zigzag antenna; — **en grecque**, Bruce
(or Grecian) type antenna.

antenne, — **en losange**, diamond-shaped or
rhombic antenna; — **en nappe**, flat-top or
plane antenna; — **en parapluie**, umbrella
antenna; — **en sapin**, fishbone antenna; —
fantôme, — **fictive**, artificial antenna; — **fermée**, frame antenna; — **fixe**, fixed antenna;
— **flottante**, trailing (-wire) antenna; — **intérieure**, indoor antenna; — **multifilaire**,
multiple-wire antenna.

antenne, — **non-accordée**, untuned antenna;
— **non-directive**, omnidirectional or non-
directional antenna; — **ouverte**, open antenna; — **pendante**, trailing (-wire) antenna;
— **periodique**, tuned antenna; — **projecteur**,
beam antenna; — **quart d'onde**, quarter-
wave antenna; — **radiogoniométrique**, direc-
tion-finder antenna; — **réceptrice**, receiving
antenna; — **réseau**, mains or light line
antenna; — **traînante**, trailing antenna.

antiamortissement, *m.* antidamping.

antibalançant, counterbalance.

antibourrage, *m.* antiblocking.

antichevrotement, *m.* antiflutter suppressor.

anticoincidence, *f.* anticoincidence.

antidéflagrant, explosion-proof, fireproof,
gasproof.

antidétonant, antiknock.

antifading différé, *m.* delayed automatic
volume control.

antigel, *m.* antifreeze.

antigivrant, *m.* antifreeze mixture.

antigivreur, *m.* anti-icing device, deicer.

antiméson, *m.* antimeson.

*****antimoniate deplomb**, lead antimoniate.

antineutrino, *m.* antineutrino.

antineutron, *m.* antineutron.

antinoeud, *m.* antinode, loop; — **de courant**,
current antinode or loop.

antinucléon, *m.* antinucleon.

antiparallèle, antiparallel.

antiparasite, *m.* pesticide, suppressor; —**s**,
radio interference suppressor; **dispositif** —,
suppressor; **équilibre** —, noise balance.

antiparticule, *f.* antiparticle.

antiproton, *m.* antiproton.

antiréactivité des barres de réglage, *f.*
strength of control rods.

antirésonance, *f.* parallel resonance, anti-
resonance.

antiretour de flammes, *m.* flame trap.
antivol, *m.* ignition lock, locking device.
*__apériodique,__ deadbeat.
aplané, even.
*__aplatissement,__ oblate distortion; — **coulombien,** Coulomb oblateness; — **du flux,** flattening of the flux.
apoastre, *m.* apastron.
apogée, *m.* apogee (astr.), apogon (ich.).
apparaux, *m.pl.* nautical gear.
*__appareil,__ aircraft, camera; — **à aiguille,** indexed meter, needle instrument, pointer instrument; — **à ailes battantes,** flapping-wing machine; — **à aimant mobile,** moving-needle galvanometer; — **à cadre mobile,** moving-coil instrument; — **à champ tournant,** rotating-field instrument; — **à diffraction,** diffraction instrument; — **à dilatation,** expansion instrument; — **à fer mobile,** moving-iron instrument; — **à fil chaud,** hot-wire meter; — **à gouverner,** steering gear (naut.); — **à gouverner électrique,** telecontrol of steering gear.
appareil, — à lecture directe, direct-reading instrument, pointer instrument; — **à miroir,** mirror instrument; — **arythmique,** start-stop apparatus; — **automatique d'alarme,** auto-alarm; — **avertisseur,** monitron; — **cinématographique,** motion-picture camera; — **combiné,** handset (teleph.); — **d'analyse,** testing apparatus; — **de chauffage à arc,** arc-heating apparatus; — **de chauffage à induction,** induction-heating apparatus; — **de chauffage par résistance,** resistance-heating apparatus.
appareil, — de conduite de tir, predictor (artil.); — **d'écoute,** hydrophone (naut.); — **de détent,** expansion gear; — **de manipulation à distance,** master-slave manipulator; — **de mesure,** rate meter; — **de mesure amorti,** deadbeat or highly damped instrument; — **de mesure de la réactivité,** reactivity meter; — **de mesure du taux de comptage,** counting rate meter; — **de mise au point,** adjusting apparatus.
appareil, — de photographie aérienne nocturne, aerial night camera; — **de prise de courant,** current collector; — **de prise de vues,** camera; — **de prise de vues en séries,** aerial survey camera; — **de recombinaison des gaz,** gas recombiner; — **de recombinaison par catalyse,** catalytic recombiner; — **de repérage par le son,** acoustic detecting apparatus; — **de restitution,** rectifier (phot.); — **de séchage,** drier; — **de secours,** emergency apparatus.
appareil, — de séparation électromagnétique

des isotopes, electromagnetic isotope-separation unit; — **d'essaie à contrepoids,** deadweight tester; — **de surveillance,** monitor; — **de surveillance de la radioactivité de l'eau,** water monitor; — **de télégraphie sans fil,** wireless set, radio; — **d'ondes courtes,** shortwave radio; — **enregistreur,** teletype, recording apparatus; — **étalon,** reference or standard instrument; — **imprimeur,** printing telegraph, type printer; — **linéaire de mesure,** ratemeter, linear ratemeter.
appareil, — photographique grand champ, camera with wide-angle lens; — **plongeur,** diving gear; — **pour désintégrer les atomes,** atom smasher; — **pour lever de doute,** sense finder; — **pour mesurer l'expansion,** extensometer; — **pour remplacer la batterie de plaque,** B-battery eliminator; — **rectangulaire,** box camera; — **respiratoire,** breathing apparatus, respirator; — **sonore,** sound head (moving pictures); — **téléphonique,** telephone set; — **vaporifère,** steam generator.
*__appareillage,__ apparatus, switchgear; — **classique,** auxiliary equipment; — **général,** general apparatus and equipment.
*__appartenances,__ fittings.
*__appauvrissement,__ depletion; — **du combustible,** fuel depletion.
*__appel, — a réception auditive,__ sound-reading apparatus, acoustic call device; — **au superviseur,** supervisor call (EDP); — **de détresse,** distress signal; — **sélectif,** polling.
*__appeler,__ to fetch, poll.
*__appliquer,__ to impress upon (voltage).
*__appoint,__ trimming capacitor (for alignment).
*__appontement,__ wharf.
*__apprenti,__ junior.
*__apprentissage automatique,__ machine learning.
*__apprêt,__ gloss, editing (EDP).
*__apprêter,__ to edit (EDP).
appui-tête, *m.* headrest.
*__apte à toutes acrobaties,__ fully aerobatic.
*__aptitude à la polarisation,__ polarizability.
arbalétrier, *m.* sloping beam.
*__arbre, — à cames en tête,__ overhead camshaft; — **cannelé,** splined shaft; — **de commande de débrayage,** clutch pedal shaft; — **couchant,** horizontal shaft; — **de direction,** steering shaft; — **de marche arrière,** reverse-speed shaft; — **de pompe,** pump shaft; — **de renvoi,** countershaft, idling shaft, lay shaft; — **de roue,** hub shaft; — **d'essieu,** axle shaft; — **de transmission,** driving axle; — **d'hélice,** propeller shaft; — **fixe,** headstock (lathe); —

intermédiaire arrière, jackshaft; — **primaire,** input shaft (gearbox); — **principal,** line shaft; — **récepteur,** output shaft; — **secondaire,** output shaft, intermediate shaft.

*****arc, — de pointage en hauteur,** elevating arc (artil.); — **producteur d'ions,** ion-producing arc; — **servant de source d'ions,** arc source of ions.

*****arc-boutant,** abutment, arch, stretcher.

arc-boutements, *m.pl.* jamming.

*****arceau,** bow (auto body); — **de sécurité,** rollover bar; — **de verrière,** canopy arch; — **support de réacteur,** engine support arch.

*****arête,** intersection (of planes); — **de réfraction,** refracting edge (phys.); — **du pylône,** side of a mast.

*****argent battu,** beaten silver.

*****argenture électrique,** electroplating.

*****arithmétique, — binaire,** binary arithmetic; — **en longueurs multiples,** multiple-length arithmetic.

*****armature,** keeper (elec.); — **à disque,** disk armature; — **de condensateur,** condenser or capacitor coat or plate, member of capacitor (either rotor or stator); — **d'un aimant,** keeper (elec.); — **d'un condensateur,** plate of a condenser; — **d'un électro-aimant,** armature of an electromagnet.

*****arme, — aérienne,** air force, air arm; — **atomique,** atomic weapon.

*****armée, — de l'air,** air force; — **permanente,** regular army.

*****armement,** fittings.

*****armer,** to make ready.

*****armoire,** cabinet, enclosure (EDP).

armurier, *m.* gunsmith.

arpon, *m.* ripsaw.

*****arrache-clou,** claw bar.

arrache-coulisse, *f.* jar pocket.

arrache-cuvelage, *m.* casing dog, casing spear.

arrache-tube, *m.* bell socket; — **indécrochable,** bulldog spear.

arrache-tuyau, *m.* pipe dog.

*****arrangement, — en tableaux,** tabulation; — **stable de protons et neutrons,** stable arrangement of protons and neutrons.

*****arrêt,** locking (screw or nut), stay, halt; — **automatique,** self-stopping; — **automatique pour surpuissance,** overpower scram; — **brusque,** scram; — **de fonctionnement,** interruption of work; — **(de la pile),** shutdown (of the reactor); — **de poussée,** thrust cut off; — **d'urgence,** emergency shutdown; — **facultatif,** optional stop; — **instantané,** scram; — **par épuisement,** burn out; — **programmé,** controlled stop; — **sur adresse,** address stop (EDP).

*****arrêter brusquement,** to scram.

*****arrêtoir,** safety washer; **rondelle —,** safety washer, tab washer.

*****arrière, à l'—,** abaft; **en — à toute vitesse,** full speed astern.

arrière-garde, *f.* rear guard.

arrière-vue, *f.* back view.

*****arrimer,** to trim (cargo).

arrimeur, *m.* stevedore.

*****arrivée de courant,** current leadin, current input.

*****arrosage,** spray.

arsenal, *m.* dockyard.

*****art de relier,** bookbinding.

*****artère,** feeder, feeder line, main line; — **de retour,** return feeder; — **d'interconnexion,** interconnecting feeder, interconnector.

arthrographie, *f.* arthrography.

articulation, knuckle (mach.).

ascendance, *f.* — **de pente,** hillside upcurrent; — **thermique,** thermal upcurrent.

*****ascendant,** parent (nucleus).

*****ascenseur,** crane (mach.); — **monte-charge,** hoist.

*****asphalte naturel,** natural rock.

asphaltisation, *f.* inspissation.

aspirant de marine, *m.* midshipman.

*****aspirateur,** suction fan, vacuum cleaner, ventilator.

*****aspirer,** to attract (magnetism).

*****assemblage, — à tenon et mortaise,** mortiseand-tenon joint; — **de cartouches élémentaires,** subassembly.

*****assembleur,** *m.* assembler.

*****asservir,** to lock.

asservissement, *m.* operating mechanism; — **àl a tension de modulation de l'amorçage des oscillations,** subject starting of oscillations to control action of modulation voltage.

*****assiette,** trim, attitude (vehicle); — **à cabrer,** nose-up attitude; — **à piquer,** nose-down attitude.

Association américaine d'essai matériaux, *f.* American Society for Testing Materials.

*****assortiment de programmes,** software package.

*****assortir,** to adapt.

*****assurance aux tiers,** third-party insurance;

— **tous-risques**, insurance on hull and equipment.

assureur, *m*. underwriter.

astatine, *m*. astatine.

astronaute, *m*. astronaut.

astronautique, *f*. astronautics, space travel.

astronef, *m*. spacecraft.

*****asymétrie azimutale**, azimuthal asymmetry.

*****atelier**, — **de montage**, assembly plant; — **de réparation**, repair shop.

*****atmosphère**, — **grisouteuse**, explosive atmosphere; **à — gazeuse**, gas-filled.

*****atome**, — **à radioactivité naturelle**, naturally radioactive atom; — **bombardé**, struck atom, bombarded atom; — **de carbone**, carbon atom; — **dépouillé d'électrons**, stripped atom; — **d'hélium**, helium atom; — **d'hélium ionisé de grande vitesse**, high-speed ionized helium atom; — **excité**, excited atom; — **fissile**, fissionable atom.

atome, — **fortement excité**, hot atom; — **hydrogénoïde**, hydrogen-like atom; — **interstitiel**, interstitial atom; — **ionisé**, ionized atom; — **léger**, light atom; — **lourd**, heavy atom; — **marqué**, tracer atom; — **mésique**, mesonic atom; — **percuté**, bombarded atom.

*****attache**, — **gaufrée**, crimp lock; —**s de l'aile**, wing junctions.

*****attaque**, leaching, excitation, drive, energizing; — **aérienne du sol**, ground strafing; — **à la trace**, attack from the rear, tail attack; — **alcaline**, alkaline leaching; — **au carbonate**, carbonate leaching; — **corrosive**, etching; — **de front**, frontal attack.

attaque, — **en montée**, attack on the climb; — **en pique**, diving attack; — **en plein jour**, daylight raid, day raid; — **en vol rasant**, low-flying attack; — **par en dessous**, attack from below; — **par le haut**, attack from above.

*****attaquer**, to impress (upon), engage, work on, apply to, feed, act on, excite, energize.

*****attelage**, yoke, mooring, docking.

atténuateur, *m*. attenuator, fader.

*****atténuation**, weakening; — **atmospherique**, atmospheric attenuation; — **de rayons gamma**, gamma attenuation; — **neutronique**, neutron attenuation, neutron-beam attenuation.

atterrir, to land; — **dans le vent**, to land into the wind; — **moteur calé**, to land with engine stopped.

*****atterrissage**, — **à l'aveugle**, blind landing; — **brutal**, heavy landing; — **de fortune**, forced landing; — **de piste**, wheel landing; — **enfoncé**, pancake landing; — **forcé**, forced landing; — **hélice calée**, landing with the propeller stopped; — **par mauvais temps**, bad-weather landing; — **par temps de brume**, fog landing; — **sur les roues**, wheel landing; — **sur les trois points**, three-point landing; — **vent arrière**, downwind landing; — **vent de côté**, crosswind landing.

atterrisseur, *m*. undercarriage; — **à jambe unique**, single-strut undercarriage; — **à large voie**, wide-track undercarriage; — **à roues**, wheel undercarriage; — **à skis**, ski undercarriage; — **avec essieu d'une seule pièce**, cross-axle undercarriage; — **en position basse**, extended undercarriage; — **escamotable**, retractable landing gear; — **escamotable en partie**, partly retractable landing gear; — **sans essieu**, split undercarriage; — **trois-roues**, three-wheeled undercarriage, tricycle landing gear.

*****attraction**, — **des courants parallèles**, attraction of parallel currents; — **mutuelle**, mutual attraction.

attribution, *f*. allocation; — **de fréquences**, allocation of frequencies.

*****aubes de turbine**, impeller blades.

*****audion**, detector valve or tube; — **detecteur**, detector tube (radio).

*****auditeur**, radio fan.

*****audition**, quality of reception.

auditorium, *m*. studio.

auerlite, *f*. auerlite.

*****augmentation**, build-up; — **d'énergie**, energy gain; — **linéaire de réactivité**, linear increase in reactivity; — **nette d'énergie**, net energy gain.

aumonier, *m*. chaplain.

*****auréole pléochroïque**, pleochroic halo.

autoabsorption, *f*. self-absorption.

auto-adaptable, self-adapting.

autoallumage, *m*. self-ignition, automatic or compression ignition.

autoamorçage, *m*. self-excitation, self-oscillation.

*****auto blindée**, armored car.

autocar, *m*. interurban bus.

autocatalytique, autocatalytic.

autodiffusion, *f*. self-diffusion, self-scattering.

autoéchauffement, *m*. self-heating.

autoentretenu, self-maintaining, self-propagating, self-supporting, self-sustaining.

*****autoexcitation**, self-oscillation.

autoexcitatrice, self-excited (machine).

autogénérateur, *m*. oscillatory generator, self-excited generator or oscillator.

autogire, *m.* autogiro; — **à décollage et atterrissage vertical**, direct take-off and landing autogiro.

autointerférence, *f.* self-interference.

autoionisation, *f.* autoionization.

automatiquement protégé, fail-safe.

automatisation, *f.* automation; — **de fabrication** process control.

*automobile à vapeur**, steam car.

*automoteur**, automotive.

*automotrice**, motor coach.

*autonome**, off-line (EDP).

*autonomie**, range; — **en croisière**, range at cruising speed.

autophasage, autophasing.

autopolarisation, *f.* self-bias.

autopropagé, self-maintaining.

autoprotection, *f.* self-screening; **à —**, fail-safe.

autoradiation, *f.* self-radiation.

autoradiographie, *f.* autoradiograph, radioautograph, contact radiography.

autoréglage, *m.* inherent control, internal control, self-regulating.

autorégulation, *f.* inherent control, internal control, self-regulation.

autorétablissement, *m.* self-reset.

autostéréogramme, *m.* autostereogram.

autostriction, *f.* self-constriction.

autostructurant, self-organizing.

autosynchrone, selsyn.

auto-treuil, *m.* motorcar for winch launching.

autunite, *f.* autunite.

auvent, *m.* open shed, tent canopy, helmet visor, cowl of auto body, louver, shroud of instrument panel; — **d'éclairage**, glare shield of instrument panel.

*auxiliaire d'essai**, test translator (EDP).

*avalanche électronique**, avalanche.

*avance**, — **à dépression**, vacuum spark advance; — **à l'allumage**, advanced ignition; — **à main**, hand feed; — **centrifuge**, centrifugal spark advance; — **de phase**, lead, leading, leading of phase.

avancée, *f.* advance.

*avancement**, headway (naut.).

*avant, en** — **à toute vitesse**, full speed ahead; **en** — **doucement**, slow speed ahead.

*avantage**, asset.

avant-trou, *m.* guide hole, lead hole, pilot hole.

*avarie**, imperfect.

Avercovir, *m.* combined horn and light switch (Renault).

*avertissement météorologique**, weather message.

*avertisseur**, — **à deux sons**, two-tone horn; — **d'incendie automatique**, automatic fire alarm; — **lumineux**, headlight flasher; — **sonore**, buzzer, horn.

aviaire, *f.* aviation.

aviateur, *m.* aviator, airman; — **fameux**, ace airman.

*aviation**, — **de bombardement**, bomber units; — **de défense métropolitaine**, home-defense air force; — **de reconnaissance**, general reconnaissance units; — **embarquée**, flying from an aircraft carrier; — **pour tous**, popular aviation; — **prémilitaire**, premilitary flying instruction.

aviatrice, *f.* aviatrix, woman pilot.

*avion**, — **à ailes hautes**, high-wing monoplane; — **à cabine**, cabin plane; — **à catapulte**, catapult plane; — **accidenté**, damaged airplane; — **à cinq places**, five-seater plane; — **à conduite intérieure**, enclosed-cabin airplane; — **à deux flotteurs**, twin-float plane; — **aile**, all-wing airplane; — **à moteur**, power plane.

avion, — **amphibie**, amphibian plane; — **à puissance musculaire**, muscle-power plane; — **à skis**, aircraft on skis; — **à surface variable**, airplane with variable lifting surface; — **atterrissant vite**, fast-landing airplane; — **automobile**, road-going airplane; — **à vapeur**, steam-driven aircraft; — **à voilures tournantes**, rotating-wing aircraft.

avion, — **d'acrobatie**, airplane for aerobatics; — **d'après-guerre**, postwar airplane; — **d'avant-guerre**, prewar airplane; — **de bombardement en vol pique**, dive bomber; — **de bombardement lointain**, long-range bomber; — **de bord**, ship plane, ship-borne aircraft; — **de chasse**, fighter plane; — **d'école**, school plane; — **d'école de début**, beginner's training plane.

avion, — **de commande à la chasse**, fighter-command plane; — **de course**, racing plane; — **de défense du territoire**, home-defense aircraft; — **de grande reconnaissance**, long-distance reconnaissance plane; — **de gros bombardement**, heavy bomber; — **de lignes nourricières**, feeder-service airplane; — **de luxe**, luxury airplane; — **de moyen bombardement**, medium bomber; — **d'entrainement au pilotage sans visibilité**, blind-flying training plane.

avion, — **de série**, production airplane; — **de sport à postes découverts**, open sports plane;

— **des services-éclair,** flash-services plane; — **de tourisme,** touring plane; — **de transition,** intermediate training aircraft; — **de transport de quotidiens,** newspaper-service plane; — **de transport rapide,** express air liner; — **d'études,** experimental plane.

avion, — **emetteur de fumées insecticides,** airplane for spraying insecticides; — **en vraie grandeur,** full-sized airplane; — **grosporteur,** freight-carrier plane; — **remorqueur,** towing aircraft; — **sanitaire,** ambulance plane; — **sesquiplan,** sesquiplane; — **torpilleur,** torpedo plane; — **voilier,** soaring plane, sailplane; **par** —, by air mail.

avion-auto, *m.* road-going airplane.

avion-canard, *m.* canard (plane).

avion-canon de chasses, *m.* cannon fighter.

avion-cible, *m.* target airplane.

avion-couchette, *m.* air sleeper.

avion-estafette, *m.* aircraft for communication work.

avion-fusée, *m.* rocket airplane.

*****avis,** advertisement; — **aux navigateurs,** notices to mariners; — **d'appel téléphonique,** notice of telephone call, messenger toll service; — **de service,** service instructions; — **de tempête,** gale warning.

avitaillement, *m.* supplying.

avitailleur, *m.* refueling tanker, fueling vehicle, bowser.

*****avoir,** assets; — **de la bande,** to heel over (naut.).

*****axe,** — **à épaulement,** clevis pin; — **à manivelle,** crankshaft; — **avant,** front axle; — **central,** centerline; — **de bougie,** ignition pin; — **de culbuteurs,** rocker arm shaft; — **de décharge,** discharge, axis; — **de fusée,** king pin, fulcrum pin, steering knuckle pin; — **de la broche,** axis of the spindle; — **de papillon,** throttle spindle; — **de piston,** piston pin, wrist pin; — **de regulateur,** governor spindle; — **de révolution,** — **de rotation,** axis of rotation; — **des pointes,** axis of centers; — **du condensateur,** condenser spindle; — **libre,** floating gudgeon pin; — **postiche,** dummy pin; — **pour l'hélice,** propeller shaft; — **principal de programme,** general routine (EDP).

axé sur le faisceau, coaxial to beam.

azéotrope, *m.* azeotrope.

*****azimut observé par radio,** observed radio bearing.

B

*****bâbord, à** —, to port (turning).

*****bac,** — **à compartiments,** multiple container; — **aérien,** air ferry; — **d'accumulateur,** accumulator box or container; — **de batterie,** battery case; — **de classement,** tub file; — **d'élément,** cell box; — **de recette séparateur,** knockout box; — **de vidange,** drain tank; — **d'un interrupteur,** switch tank; — **interrupteur,** multipolar switch; — **unique,** single enclosure.

*****badigeonnage,** spreading, coating (of a cathode).

*****bague,** bushing; — **à cames,** camshaft; — **collectrice,** slip ring (generator); — **d'arrêt,** adjusting collar; — **d'arrêt d'huile,** oil slinger (on shafts); — **de butée de débrayage,** clutch release bearing ring; — **de contact,** slip ring; — **de friction,** brake band; — **de frottement,** slip ring; — **de palier,** bushing; — **de pied de bielle,** connecting rod small end bushing; — **de retenue,** balk ring (automatic transmission); — **de synchronisation,** synchronizing ring; — **d'étanchéité,** seal ring, seal washer; — **élastique,** flexible bushing; — **extérieure,** out race, cup (rolling bearings); — **intérieure,** inner race (cone rolling bearings).

*****baguette,** — **de sourcier,** divining rod; — **en verre,** stirring rod, glass rod.

*****baie,** bight.

baille, *f.* bucket, tub.

*****bain,** — **de cémentation,** casehardening bath; — **de fixage,** fixing bath (phot.); — **galvanoplastique,** electroplating bath.

*****baladeuse,** portable lamp.

*****balai,** — **de lecture,** read brush; — **d'essuieglace,** windshield wiper blade; — **de vérification,** control brush; — **électromécanique,** electromechanical brush; — **pneumatique,** airbrush; — **supérieur,** upper brush (EDP).

*****balance aérodynamique,** wind-tunnel balance.

*****balancer,** to compensate.

*****balayage,** sweep (EDP), hunting (elec.), scanning (of a picture), scavenging; — **intercalé,** interlaced scanning.

*****balayement,** scanning, scansion, exploring, analyzing.

balisage, *m.* beaconing; — **de routes aériennes,** airway lighting.

*****balise,** boundary mark, ranger (radar); —

d'atterrissage, landing beacon; — **répondeuse,** transponder beacon.

baliser, to mark out, beacon.

baliseur, *m.* beacon boat.

*__balle,__ — **du regulateur,** governor balls; — **fumigène,** smoke bomb; — **traceuse,** luminous projectile, tracer bullet.

*__ballon cerf-volant,__ kite balloon.

*__ballonet latéral,__ outboard float.

ballon-sonde, *m.* registering balloon, sounding balloon.

balourd, *m.* unbalance.

*__banc,__ — **à fusion,** welding bench; — **de glace,** ice field; — **de nuages,** cloud bank; — **de sable,** sand bank; — **d'essai,** benchmark session; — **d'essai de moteur,** engine test stand; — **de tour,** lathe bed.

bandage, *m.* bandage, bandaging, solid tire; — **à talons,** beaded tire; — **à tringles,** wired tire; — **sans boudin,** blank tire.

*__bande,__ tape, range, iron rail, list (naut.); — **à carte,** tape-to-card; — **à cartouches,** ammunition belt; — **à grandes performances,** hypertape; — **de cadmium,** cadmium strip; — **de déformation,** deformation band; — **de frottement,** friction band; — **de mise à jour,** change tape; — **de modification,** change tape; — **d'énergie des neutrons,** neutron energy range; — **de rotation,** rotational band; — **d'exploitation,** systems tape; — **écrêteuse,** peaker strip; — **éprouvée,** faultless tape; — **latérale,** side band; — **latérale unique,** single side band; — **magnétique,** magnetic tape; — **passante,** bandwidth, pass band; — **semiperforée,** chadless tape; — **sensible,** proportional band, throttling range.

bande-pilote, *f.* carriage tape, control tape.

bande-programmothèque, *f.* library tape (EDP).

bande-vidéo, *f.* video tape.

bandotrope, tape oriented.

*__bandoulière,__ bandoleer.

banquise, *f.* ice floe.

baptême de l'air, *m.* first flight (in a plane).

*__barbotage,__ sparging.

*__barillet,__ rotating axle, barrel (of a lens).

barn, *m.* barn.

*__barque,__ barge.

*__barre,__ rod, helm; — **à caractères,** print bar; — **absorbante,** absorbing rod; — **assujettie,** fix bar; — **conductrice,** guide bar; — **d'accouplement,** steering tierod; — **de cabestan,** capstan bar; — **de commande,** control rod; — **de commande excentrique,** eccentric control rod; — **de commande par la puissance,** power control rod; — **de compensation,** shim rod; — **de compensation et de sécurité,** shim-safety rod; — **de direction,** steering drag link; — **de distribution,** distributing bus bar; — **de dopage,** booster rod; — **de pilotage,** regulating rod; — **de poussée,** push rod.

barre, — **de réaction,** torque arm or rod; — **de réglage,** control rod; — **de réglage à grande sensibilité,** fine control rod; — **de réglage fin,** fine (regulating) rod; — **de réglage grossier,** coarse rod; — **de réglage noire,** black control rod; — **de réglage totalement absorbante,** black control rod; — **de saut,** skip bar; — **de sécurité,** safety rod, scram rod, shutoff rod; — **de sûreté,** guard bar; — **de torsion,** torsion bar (suspension); — **d'impression,** type bar; — **d'impression alphanumérique,** alphamerical type bar; — **d'impression numérique,** numerical type bar; — **directrice,** conducting rod; — **du gouvernail,** tiller; — **omnibus,** bus bar, collecting bar; — **stabilisatrice antiroulis,** antiroll stabilizer bar.

*__barreau,__ rod; — **actif,** fuel rod; — **à griffe,** crowfoot bar; — **de graphite amovible,** removable carbon rod; — **d'uranium,** uranium rod; — **refroidi par l'intérieur,** internally cooled slug; **barreax d'échelle,** ladder rungs.

*__barrer,__ to impound.

*__barrette métallique,__ metal bridge.

barretter, *m.* ballast resistor.

*__barrière,__ — **coulombienne,** Coulomb barrier; — **de Coulomb,** Coulomb barrier; — **de diffusion,** diffusion barrier; — **de fission,** fission barrier; — **d'énergie,** energy barrier; — **de potentiel,** potential barrier; — **de potentiel de Coulomb,** Coulomb potential barrier; — **poreuse,** porous barrier; — **thermique,** thermal barrier.

*__baryum radioactif,__ radioactive barium.

*__bas,__ — **niveau,** low level; — **peu profond,** shoal.

basculant, flip-flop, toggle.

*__bascule,__ trigger pair, tail slide.

*__basculeur,__ time-base means (telev.), trigger pair.

*__base,__ radix; — **de lancement,** range, launching base; — **de parcours,** flight path; — **de vol,** flight path; — **flottante,** floating air base; — **linéaire,** base (of a polygon); — **superficielle,** base (of a polyhedron); —**s multiples,** mixed radix.

*__basilique,__ basilica.

bassetite, bassetite.

*__bassin,__ — **à flot,** floating dock; — **de carénage,** graving dock; — **d'échouage,** tidal dock; —

de radoub, dry dock; — des carènes, hydro-dynamic tank (aviation); — de sédimentation, settling basin.

*batardeau, dam.

*bateau, tender (naut.); — à rames, rowboat; — de sauvetage, lifeboat; — grande vitesse, speedboat; — magasin, supply ship; — non ponté, open boat.

bateau-catapulte, m. catapult ship.

*bateau-glisseur, speedboat.

*bâti, rack, cradle, casing; — des commandes du moteur, throttle box; — d'une machine, bedplate; — en tubes d'acier, steel-tube mounting.

*bâtiment école, training ship.

*bâton de foc, jib (crane).

*battant, gate (EDP).

*batterie, — d'accumulateur, storage battery; — de chauffage, filament, heater, or A battery; — de grille, grid-bias or C battery; — de lampe de poche, flashlight battery; — de plaque, plate, anode, or B battery; — équilibrée, floating battery; — volante, buffer or boosting battery.

bavardage de singe, m. monkey-chatter interference.

bayleite, f. bayleyite.

beaupré, m. bowsprit.

*bec du profile, wing nose, leading edge.

becquerélite, becquerelite.

bélinographie, f. work with Belin apparatus.

bémolisée, in B-flat (mus.).

*benne, cart, truck; — à griffes, grab bucket (of crane).

*béquile-ressort, m. spring tail skid.

*béquille de capot, hood prop (auto).

ber de lancement, m. launching cradle.

berkélium, m. berkelium.

berline, f. cart, sedan (auto), truck.

berne, en —, halfmast.

berthon, m. collapsible boat.

béryllate, beryllate.

*béryllium, glucinium.

*besoins en énergie, power requirements.

bêtafite, f. betafite.

bêtatopique, adj. betatopic.

bêtatron, m. betatron.

*béton, — lourd, heavy concrete; — précontraint, prestressed concrete.

*beurre de brèche, cheese grease, whey butter.

bidirectionnel, full duplex.

bidonville, m. hooverville, shantytown.

*bielle, crankshaft; — de poussée, torque arm; — pendante, steering lever, Pitman arm.

*biellette, crank, rudder lever, aileron lever; — de changement de vitesses, gearshift rod; — de direction, steering tierod; — secondaire, secondary connecting rod.

biergol, m. bipropellant.

bigrille, f. double-grid tube, space-charge tetrode.

bilame, f. bimetallic strip or switch.

*bilan, — brut, trial balance; — des neutrons, neutron economy; — énergétique, energy balance, power balance; — énergétique net, net power balance; — matière, material balance; — neutronique, neutron balance.

*bille, — de verrouillage, lock ball (synchro); coussinet à —s, roulement à —s, ball bearing.

*billet, — d'aller et retour, round-trip ticket; — ouvert, open ticket.

billette, f. billet.

billetterie, f. sale and purchase of tickets, ticket office.

billiétite, f. billietite.

bimensuel, bimonthly.

bimoteur, twin-engined, bimotored.

*binaire, bit, binary digit; — par colonne, column binary; — par ligne, row binary.

bineutron, m. bineutron.

binome de Newton, m. binomial theorem.

bionique, f. bionics.

bipale, two-bladed.

*bipartition, f. fragmentation of nucleus.

bipasse, m. bypass.

biplace, f. two-seater.

biplan, m. biplane; — à ailes décalées, staggered biplane; — à ailes décalées avec mât unique, single-bay staggered biplane; — à fuselage, biplane fuselage model.

biquinaire, biquinary.

biréfringence, f. birefringence.

bisauté, beveled.

*biseau de déviation, whipstock.

bistable, bistable.

bit (unité binaire théorique d'information), bit (binary information theoretical unit); — de contrôle de parité, parity check bit; — de données, information bit; — de parité, parity bit; — de ponctuation, punctuation bit; — de protection de mémoire, storage protect bit;

— **de signe,** sign bit; — **de vérification,** check bit; — **de zone,** zone bit; — **modificateur,** modifier bit.

***blanc,** tinned.

***blanchiment,** deblooming.

***blindage,** shield, screen, masking (motion picture); — **de l'allumage,** ignition screening.

blindé, *m.* armored car.

blindé, shielded, screened, canned, armored.

***bloc,** traffic jam, shielding, can; — **cylindres,** cylinder block (auto); — **d'amortisseur,** buffer block (mach.); — **d'entrée,** input block; — **de décision,** decision box; — **de données,** physical record (EDP); — **de frottement,** brake shoe, brake; — **de pilotage,** auto pilot; — **de poudre,** grain (astronautics); — **de puissance,** power unit, power module; — **de schéma,** box (EDP); — **de sortie,** output block; — **d'essai,** metering stud; — **de tête,** header record; — **d'oscillation,** tank circuit; — **de plaques,** element; — **inaccessible,** unreachable block; — **moteur,** unit power plant (engine-gearbox assembly); — **optique,** light unit (sealed beam); — **tronqué,** short block.

***blocage,** rubble, traffic jam, locking (of nut or screw), cutoff (of a tube).

bloc-notes, *m.* scratch pad, data event control block (EDP); — **de tâches,** task control block (EDP).

bloc-pile, *m.* reactor block.

blomstrandite, *f.* blomstrandite.

bloqué, jammed.

***bloquer,** to hang up (EDP).

***bobinage,** — **de champ magnétique,** magnetic-field coil; — **de l'induit,** armature winding; — **en piles,** pile-wound coil.

***bobine,** — **à air,** air-core coil; — **à fiches,** plug-in coil; — **à noyau en poudre de fer,** iron-dust-core coil; — **à noyau plongeur,** sucking coil; — **à plots,** — **à prises,** tapped coil or inductance; — **d'allumage,** ignition coil; — **d'arrêt,** choke coil; — **de basses fréquences,** choke coil; — **de charge,** loading coil.

bobine, — **débitrice,** — **de déroulement,** — **dérouleuse,** feed, pay-out, delivery, or pull-down spool; — **de compensation,** compensating coil; — **de concentration,** focusing coil; — **d'écoulement,** drainage coil; — **d'enroulement,** take-up spool (film); — **d'excitation,** magnetic-field coil; — **en dérivation,** bridging coil; — **exploratrice,** search coil, pickup loop; — **inductrice,** magnetic-field coil; — **réceptrice,** take-up spool (film); —**s de déplacement de l'orbite,** orbit-shift coils.

bodenbendérite, *f.* bodenbenderite.

***bois,** stock (gun), xylem; — **d'aune,** alder wood; — **de fusil,** gunstock; — **de sapin,** deal wood; — **traité,** improved wood.

***boisage,** logging.

***boisseau,** cock.

***boîte,** — **à aiguilles,** pin-box; — **à eau,** header tank; — **à fusibles,** fuse box; — **à gants,** glove box; — **à graisse,** axle box; — **à mécanismes,** stunt box (EDP); — **à mouches,** blind-flying hood; — **à pont,** post-office box; — **d'accord,** tuning box; — **d'alimentation,** power pack, power unit; — **de circulation,** channel; — **de connexions,** connection box; — **de dérivation,** dividing box; — **de prises,** receptacle (elec.); — **d'embranchement,** junction box.

boite, — **de retour,** header box; — **de retour à bouchon cônique,** tape-plug header; — **de retour à bouchon fileté,** screw-plug header; — **de retour à étriers,** compound-plug header; — **de retour type,** mule-type header; — **d'essieu,** journal box; — **de transfert,** transfer gearbox; — **de vitesses à présélection,** pre-selector gearbox (semi-automatic transmission); — **de vitesses auxiliaire,** auxiliary gearbox; — **d'extrémité,** sealing end; — **médicale de secours,** first-aid kit; — **métallique,** tin can.

bôitier, *m.* — **additionneur,** adding unit; — **déphaseur,** phase-shifting unit; — **d'un récepteur,** receiver cabinet; — **intégrateur,** integrating unit.

bombance, *f.* blowout.

***bombardement,** — **atomique,** atomic bombardment; — **de l'uranium par des neutrons,** neutron bombardment of uranium; — **en vol pique,** dive bombing; — **neutronique,** neutron bombardment; — **neutronique de l'uranium,** neutron bombardment of uranium; — **nucléaire,** nuclear bombardment; — **par des neutrons,** neutron bombardment; — **par des particules alpha,** alpha-particle bombardment; — **par deutérons,** deuteron bombardment.

bombardier, *m.* bomb aimer, bombardier.

bombardier-mitrailleur, *m.* gunner-bomber.

***bombe,** — **à éclats,** fragmentation bomb; — **à hydrogène,** hydrogen bomb, H-bomb; — **atomique,** atomic bomb; — **d'exercice,** practice bomb; — **lumineuse,** flash bomb.

***bon de prélèvement,** bin card.

***bonhomme de verrouillage,** synchromesh locking plunger.

boral, *m.* boral.

***bord,** — **arrière,** card trailing edge; — **avant,** card leading edge; — **d'attaque,** leading edge, document reference edge; — **de fuite,** trailing edge; — **de marque, de segment,** stroke edge; — **des instruments,** instrument panel; — **du toit,** eaves.

bordée, *f.* broadside (naut.).

*****bordereau, — de paie,** payroll register.

*****bordure de trottoir,** curbstone.

*****borne,** pliers, connector, term; — **à vis,** binding post; — **de raccordement,** connecting terminal; — **inférieure d'énergie,** low energy limit; —**s d'entrée,** input terminals; —**s de sortie,** output terminals.

boson, *m.* boson.

*****bosselé,** having bumps or humps, irregular.

bossoir, *m.* davit.

*****bouche d'incendie,** fire hydrant.

*****bouchon,** rejector (circuit); — **de radiateur,** radiator cap; — **de remplissage,** filler cap; — **de vidange,** drain plug; — **d'inflammation,** firing key (naut.); — **femelle,** threaded cap.

*****boucle,** staple (hook), cycle of instruction, ear or lobe (of space pattern); — **à rétroaction,** feedback loop; — **chaude,** hot loop; — **d'arrêt,** shutdown loop; — **de circulation,** circulation loop; — **de couplage,** single-turn search coil; — **de fonctionnement,** operating loop; — **de retour,** feedback; — **fermée,** closed loop; — **ouverte,** open loop, feedforward.

*****boucler,** to connect, attach.

*****bouclier,** cowl or scuttle (auto body); — **à neutrons thermiques,** thermal shield; — **anti-explosion,** blast shield; — **biologique,** biological shield; — **de perçage,** shield (tunneling); — **(de protection) (pile),** shield; — **de protection biologique,** biological shield; — **en agrégat lourd,** heavy aggregate shield; — **en béton,** concrete shield; — **thermique,** thermal shield; — **transformable,** flexible shield.

*****boue,** slurry.

*****bouée culotte,** breeches buoy.

*****bouffée,** burst; — **de neutrons,** neutron burst.

*****bougie, — de préchauffage,** glow plug (Diesel); — **incandescente,** glow plug (Diesel).

*****bouilleur,** boiler; — **atomique,** atomic boiler; — **de chaudière,** heating tube.

*****bouillonnement,** bubbling; — **au repos,** gassing.

bouldozeur, *m.* bulldozer.

*****boule, — à collet carré,** square-neck bolt; — **de changement de vitesses,** gearshift lever knob; — **isolante,** insulator.

bouleversé, upside down.

*****boulon, — à ergot,** snug bolt; — **à queue de carpe,** Lewis bolt; — **à six pans,** hexagonal

head bolt; — **à tête creuse,** socket head bolt; — **à tête encastrée,** countersunk bolt; — **à tête plane,** flat-headed bolt; — **à tête plate,** flathead bolt; — **à tige conique,** taper shank bolt; — **d'assemblage,** holding bolt, tack bolt; — **d'attelage,** coupling pin; — **de carrosserie,** carriage bolt; — **de chape,** clevis bolt; — **de fixation,** holding-down bolt; — **de fondation,** anchor bolt; — **de scellement,** expansion bolt; — **encastré,** countersunk bolt; — **étrier,** U bolt; — **fileté,** screw bolt; — **rectifié,** machine bolt.

boulonnage, *m.* bolting.

boulonné, bolted.

*****bourgeon de pin,** pine needle.

*****bourrage,** jamming or buckling of film; — **de cartes,** card jam, card wreck.

*****bourrelet,** bead (tire).

*****boursoflure,** blister.

*****boussole, — de déclinaison,** declinometer; — **des sinus,** sine galvanometer.

boussole-mère, *f.* master compass.

*****bout,** shred; — **d'aile,** wing tip; — **de vergue,** yardarm (naut.); — **d'une solive,** beam end.

*****boutefeu,** shooter.

bouteillon, *m.* faucet.

bouterolle, *f.* backup plate (riveting), snap, ward (lock or key).

bouteroller, to snap (rivet), rivet over.

bouteur, *m.* bulldozer.

bout-fors, *m.* boom (naut.).

*****boutique franche,** duty-free shop.

boutisse, *f.* header, bond (constr.).

*****bouton, — d'arrêt instantané,** scram switch; — **de contact,** contact stud, push button; — **de déclenchement,** release knob; — **moleté,** milled, knurled, or thumb knob.

bouton-poussoir, *m.* push-button.

*****bouts morts de la bobine,** dead-end turns.

*****boyau à saucisse,** sausage casing; — **de sable,** shoestring sands.

bragite, bragite.

*****brancard, — à bretelles,** sling stretcher; — **de bât,** mule stretcher, hand barrow; — **de pavillon,** cantrail (roof); —**s de caisse,** lowest members of body framework (auto).

*****branche exponentielle,** exponential tail.

*****branchement,** service line, supply line; — **par fission,** fission branching.

*****brancher,** to switch on; — **sur,** to tap (a wire).

brannerite, brannerite.

braquage, *m.* aiming, deflection, steering; — **de tuyère,** nozzle swiveling.

*bras, — d'accès, access arm; — de direction, steering arm; — de retenue, steady brace; — d'essuie-glace, windshield wiper arm; — de suspension, suspension arm, wishbone; — du vilebrequin, crank web.

*brasage, — fort, brazing; — tendre, soldering.

break, m. station wagon.

*brevet de vol à voile, soaring certificate.

brévium, m. brevium.

*bride, bow (mach.), hoop; — d'espacement, filler piece; — pleine, blank flange.

*brin, strand (of filament or cathode).

*briquette, compact.

brise-caille, m. mill or curd breaker.

brise-glace, m. icebreaker (naut.).

brise-lames, m. breakwater.

briseur d'atomes, m. atom smasher.

*broche de culot, pin or prong of tube base.

broeggérite, bröggerite.

*bronze dur, gunmetal.

brouillage, m. jamming, interference, garbling (radio); —s atmosphériques, atmospherics; —s industriels, man-made static or interference; —s naturels, atmospherics.

*brouiller, to garble (speech or signals), jam.

*bruit, — d'agitation thermique, thermal noise; — de fond, additional or background noise, background count; — de fond ambiant, environmental background; — de fond de l'amplificateur, amplifier noise; — de fond dû au rayonnement, radiation background; —

de fond d'un appareil, set noise; — de fond d'une lampe, valve noise.

bruit, — de fond en général, background noise; — de fond gamma, gamma background; — de fond permanent, steady background; — de salle, room noise; — de secteur, mains or power-line hum; — parasite, disturbing or interfering noise.

*brûler, — dans une pile, to burn in a pile; — un combustible nucléaire, to burn nuclear fuel.

*brûlure par irradiation, radiation burn.

*brunir, to smooth.

*brut de fonderie, as cast condition.

*bulletin de versement, pay-in slip.

*bureau, — central de départ et arrivée, exchange of origin and destination; — de renseignement, intelligence division; — des traces, drafting office; — d'études, engineering department; — extrême, terminal office; — postal d'émission, issuing office.

*buse, — aspiratrice, suction nozzle; — d'air, choke tube, Venturi tube; — diffusante, jet diffuser; — d'injecteur, injector spray tip.

*butée, thrust; — à billes, ball thrust bearing; — à rouleaux, roller thrust bearing; — d'alignement, joggle plate; — de choc, shock absorber; — de débrayage, clutch release bearing; — de direction, steering stop; — de frein, brake stop; — de tassement, joggle plate; — de vilebrequin, crankshaft thrust bearing; vis de —, adjusting screw.

*but-en-blanc, point-blank.

*butoir de pare-chocs, bumper guard, overrider.

C

*cabane, center section, cabane (aviation).

cabillot, m. belaying pin (naut.).

*cabine, — avancée, forward control cab (trucks); — avec couchettes, sleeping cabin; — de commandement, control gondola; — de conduite, driver's cabin; — de transformateurs, transformer box; — fermée, enclosed cabin.

*cabinet d'aisance, privy.

*câblage, wiring; — de tableau, panel or board wiring.

*câble, — co-axial, concentric cable; — de cabestan, catline; — de commande de débrayage, clutch cable; — de connexion, connecting cable; — de lancement, launching rope; — de remorquage, towing cable; — d'extraction, hoisting rope; — krarupisé, contin-

uously loaded cable; — porteur, messenger or bearer cable; — pour lignes interurbaines, toll cable (teleph.); — pupinisé, coil-loaded cable; — sans fin, dead line.

*câbler, to wire.

câbliste, m. cableman (TV).

caboche, f. hobnail.

*cabochon, dimmer, color cap of lamp.

cabriolet, m. convertible (auto).

cache, f. mask, cover, guard, escutcheon (auto body), screen, stop, obturator, shadowing means.

cache-borne, m. terminal block cover.

cache-bouton, m. switch guard, knob cover.

cache-culbuteurs, m. cylinderhead cover, rocker cover.

cache-poussière, *m.* dust cap.

*****cadence,** rhythm, schedule, rate, frequency.

cadmiage, *m.* cadmium plating.

cadrage, *m.* framing; — **à droite,** right alignment.

*****cadran,** — **à vernier,** slow-motion dial; — **d'accord,** tuning scale; — **d'appel,** dial; — **démultiplicateur,** — **demultiplié,** slow-motion dial; — **de repère,** tuning scale.

*****cadre,** — **à droite,** right justified; — **actif,** action spot; — **croisé,** cross-coil antenna; — **d'amenée d'antenne,** antenna feeder or downlead coil; — **de fuselage,** transverse frame of the fuselage; — **de ventilateur,** fan cowl or shroud; — **équilibré,** balanced loop; — **escamotable,** loop of folding type; — **explorateur,** pickup loop; — **mobile pour radiogoniometrie,** rotating-loop or rotating-frame antenna; — **radiogoniométrique,** directional loop (radio); — **tournant,** rotating loop.

*****cadrer,** to frame (film), scale, ensure framing.

cadreur, *m.* cameraman (TV and movies).

caillebotis, *m.* grating, duckboards, walkway.

*****caisse de sûreté,** safe.

*****caisson,** tank; — **de pression,** pressurized casing.

*****calage,** setting shifter (of phase), sticking, jamming, stalling (engine), freezing; — **de distribution,** timing setting; — **des balais,** adjustment of brushes, brush lead (elec.).

*****calamine,** carbon deposit.

*****calandre,** radiator grill (auto).

calangue, *f.* bight.

calciosamarskite, *f.* calciosamarskite.

calciothorite, *f.* calciothorite.

*****calcul,** computing; — **approché au premier ordre,** first-order theory; — **des réseaux,** lattice calculation; — **infinitesimal,** calculus (math.); — **limité au premier degré,** linearized treatment; —**s en théorie à plusieurs groupes,** multigroup calculations.

*****calculateur,** — **analogique,** analogue computer; — **de cap et de distance,** course-and-distance calculator.

*****calculatrice,** — **à grande vitesse,** high-speed computer; — **arithmétique électronique,** electronic computer; — **numérique,** digital computer.

*****cale,** — **de blocage,** block, chock; — **de construction,** shipbuilding dock; — **de lancement,** shipway; — **d'epaisseur,** shim; — **de serrage,** square washer; —**s de raue,** chocks, shims; —**s minces pour paliers,** shims.

*****calendrier,** schedule.

*****calibre,** — **d'épaisseur,** feeler gage; — **de forme,** template; — **étalonné,** template; — mâchoires, jaw type or go no-go gage.

californium, *m.* californium.

caloporteur, fluide —, coolant.

*****calorifuge,** thermal insulation.

calorifugeage, *m.* thermal insulation.

*****calorimètre à fission,** fission calorimeter.

calorstat, *m.* thermostat.

*****calotte,** — **à échancrures,** bubble cap; — **de barbotage,** bubble cap.

calutron, *m.* calutron.

*****came,** — **de rupteur,** contact breaker cam; — **étoilée,** star wheel.

*****camion,** — **à benne basculante,** dump truck; — **à plateau,** platform truck; — **inclinable,** dump truck.

camion-citerne, *m.* tank truck.

camionnette, *f.* light truck, delivery truck.

camp de vol à voile, *m.* glider camp.

*****canal,** data channel, track (EDP); — **à essais,** test hole; — **à va-et-vient pneumatique,** shuttle hole; — **analogique,** analog channel; — **commun,** bus circuit (EDP); — **de chargement,** fuel channel; — **de détection des fuites,** leak test channel; — **de fission,** fission channel; — **de la pile,** reactor tube; — **de lecture-écriture,** read-write channel; — **de mémoire,** storage capacity, machine capacity, memory capacity; — **d'entrée analogique,** analog input channel; — **de mesure de la puissance,** power-level channel; — **de refroidissement,** coolant channel; — **multiple,** multiplexor channel.

canal, — **de remplacement des cartouches,** fuel-exchange tube; — **de temps,** time channel; — **des dispositifs de sécurité,** shutdown channel; — **d'irradiation pour animaux de laboratoire,** animal tunnel; — **logarithmique,** logarithmic channel; — **passant dans le coeur,** core channel; — **sélecteur,** selector channel.

*****canalisation d'essence,** gasoline pipeline.

*****canne de niveau,** depletion sensor, level sensor.

*****canon,** — **à électrons,** electron gun; — **à obus explosifs,** shell-firing gun; — **porteamarre,** line-throwing gun (naut.); — **raye,** rifled gun.

canonnière, *f.* gunboat.

canot-cible, *m.* target boat.

*****canot pneumatique,** inflatable boat.

*****canton,** block.

*****cantonnement,** block system.

*cantonner, to quarter (mil.).

cantonnier, *m.* road laborer.

*caoutchouc au plomb, lead rubber.

*cap, mettre le — sur, to steer or head for.

*capacité, — de coupure, circuit-breaking capacity, absorbing capacitor; — d'équilibre, trimming capacitor; — d'un circuit oscillant, tank capacity; — entre deux électrodes, interelectrode capacity; — maximale, maximum load; — parasite, stray capacity; — pour le poids, bearing capacity; — repartie du cablage, stray capacity of wiring.

*capitaine de corvette, lieutenant commander.

capitonnage, *m.* upholstery.

*capot, — annulaire à sorties d'air réglables, controllable flapped cowl, gilled cowling; — à réglage commandé, controllable cowling.

capote, *f.* overcoat, greatcoat, hood (chimney); — anglaise, rubber contraceptive; — pliante, collapsible top (auto).

capotage, *m.* — annulaire, cowling ring; — de moteur, cowling, motor hood.

capoter, to turn over.

capteur, *m.* sensor, captor; — d'impulsions, pulse chamber; — piézo électrique, piezo electric pressure gauge.

*capture, lock-on; — dans le ralentisseur, moderator capture; — d'électrons, electron capture; — de neutrons, neutron capture; — de neutrons de résonance, resonance capture; — de neutrons lents, slow-neutron capture; — dissociative, dissociative capture.

capture, — d'un électron K, K-electron capture; — d'un neutron de fission, fission capture; — électronique, electron capture; — parasite des neutrons, parasitic neutron capture; — radiative, radiative capture; — radiative de neutrons, radiative capture of neutrons; — sans fission, nonfission capture, nonproductive capture, parasitic capture.

*caractère, — codé binaire, binary coded character; — de commande, control character; — de contrôle, check character; — de masque, pattern character; — de remplissage, fill character; — de sélection de chiffre, digit-select character; — dominant, dominant character; — dominant autosomique, autosomal dominant; — récessif, recessive character; — récessif autosomique, autosomal recessive; — spécial, special character (EDP).

*caractéristique, — amplitude-fréquence, frequency-response curve; — courante, standard feature; — courant-lumière, light characteristic; — de fonctionnement, perfor-

mance characteristic, operational feature; — de fréquence, frequency-response characteristic; — de palier, plateau characteristic; — de transformation, transformation property; — nominale, rated characteristic; — particulière, special feature; —s techniques, technical data.

carburane, carburan.

*carburant, — à nombre d'octanes élevé, high-octane fuel; — anti-détonnant, antiknock fuel; — au plomb tétraéthyle, leaded fuel.

*carburateur, — à double corps, dual barrel carburetor; — avec réchauffage, heated carburetor; — horizontal, crossdraft carburetor, sidedraft carburetor; — inversé, downdraft carburetor; — pour le vol sur le dos, carburetor for inverted flying; — vertical, updraft carburetor, vertical carburetor.

carburéacteur, *m.* jet fuel.

*carcasse d'une bobine, coil former.

carcinogénèse, *f.* carcinogenesis.

carénage, *m.* fairing; — de l'atterrisseur, undercarriage fairing; — de roue, wheel fairing, wheel spats; — de roue de queue, tail-wheel fairing; — du fuselage, fuselage covering; — du groupe-motopropulseur, engine fairing.

*carène, bottom (of a ship).

caret, *m.* caret (EDP), reel (rope making); fil de —, rope yarn.

*cargaison, lading (naut.); — de bombes, bomb load.

carne, *f.* edge.

*carnet, — de billets, bulk travel voucher; — de vol, flight logbook.

carnotite, *f.* carnotite.

*carottage neutronique, neutronic logging.

*carré, mess (naut.); — de l'équipage, crew's compartment.

carrier, *m.* quarryman.

carrossage, *m.* rake of the axle pin; angle de —, camber angle.

*carrosserie, — aérodynamique, streamlined body; — autoporteuse, integral body construction.

*carrure, watch-case center.

*carte, — à bande, card-to-tape; — à bords perforés, verge-punched card; — active, current card; — à encoches, edge-notched card, marginally-punched card; — à graphiter, mark sensing card; — à perforations marginales, edge-punched card; — à talon, stub card; — à volet, stub card; — binaire, binary card; — de contrôle, control card; — de présence, time card; — des chromosomes, chromosome map; — du flux des neutrons de résonance, resonance contour; — en-tête, header card; — graphitée, mark sensed card; — grise, auto license; — maîtresse, master card; — perforée, punched card; — postfichier, trailer card; — préperforée, prepunched card; — récapitulative, summary card; — réduite, Mercator's chart; — routiére, road map; — vierge, blank card.

carte-document, f. dual-purpose card.

*carter, — cylindres, crankcase; — de direction, steering box; — de distribution, timing gear case; — d'embrayage fixe, clutch housing, bell housing (fixed); — d'essieu, axle housing; — inférieur, oil pan; — moteur, crankcase.

carte-solde, f. balance card.

carte-titre, f. header card.

*carton de jute, jute board.

cartotrope, card oriented.

*cartouche, cartridge (EDP), round (of ammunition); — active, fuel element; — active hermetiquement close, sealed-in fuel unit; — active métallique, metallic fuel element; — à matière fissile, seed unit; — de combustible, fuel element, fuel rod, fuel slug.

cartouche, — de combustible irradié, irradiated fuel element; — de fusée éclairante, signal cartridge; — de référence, count area (EDP); — en matière fertile, blanket subassembly; — factice, dummy fuel; — fertile, fertile element; — filtrante, filtering apparatus.

*cascade, — d'appareils de séparation, cascade of separating units; — de cellules de même débit, square cascade; — d'épuisement, — d'extraction, stripping cascade.

*case, stage (of radio), frame or chase (type founding); — de réception, stacker, card stacker; — de tri, pigeonhole.

*casier, card cabinet; — à carte, card rack.

*casque, headphone, earphone.

*casse, crack-up, crash (aviation).

*casserole, nose spinner.

*cassure, twisting off; — chromosomique, chromosome break; — du deutéron en vol, deuteron stripping.

catadioptre, m. reflector, cat's eye.

*catalyseur, catalyst.

cataphote, m. reflector, cat's eye.

catapultage, m. catapult launching.

catapulter, to catapult.

*caténaire, triatic.

*cathode, — asservie, — suiveuse, cathode follower; — froide, cold cathode.

cathodyne, f. cathode follower.

catimaran, m. catamaran.

*cavalier, — de jonction, staple; — de tabulatrice, tabular insert.

*cavité, — à micro-ondes, microwave cavity; — de retrait, shrinkage cavity; — noire, black box; — résonnante, cavity resonator.

*ceinture, — blindée, armored belt; — de sauvetage, life belt; — de sécurité, safety belt; — de tourelle, safety strap.

*cellule, air frame, unit, mesh, section (of filter or network); — à contact rectifiant, rectifier photocell; — à effet arrière, backeffect cell; — anaerobique, anoxic cell; — à oxyde de cuivre, copper oxide cell; — binaire, binary cell; — de diffusion à plusieurs étages, multistage diffusion unit; — de diffusion à simple passe, single-stage diffusion unit; — de filtre, filter section; — de haute activité, hot cell; — de manipulation, manipulation cell; — de mémoire, storage cell; — de pile, battery cell; — de réseau, lattice cell; — de séparation, separating unit, separative element.

cellule, — électrolytique, electrolytic cell; — élémentaire du réseau, lattice cell; — en T, midseries termination; — germinale, germ cell; — magnétique, magnetic cell; — photoélectrique à grille de commande, threeelectrode cell; — somatique, somatic cell; — sustentatrice, gasbag; —s épithéliales, epithelial cells.

*centilitres, quatorze —, gill (measure).

*centrage, trim compensation.

*centrale, — atomique, atomic power station; — de pointe, peak-load plant; — d'orientation, attitude control unit; — électrique, generating station; — hydraulique, hydroelectric generating station; — industrielle, industrial power station; — inertielle, inertial unit; — nucléaire, atomic power plant, atomic energy plant, nuclear power plant; — nucléaire avec pile à ralentisseur de graphite, graphitemoderated nuclear power plant; — nucléaire de base, base-load-type nuclear power plant; — téléphonique, telephone exchange; — thermique, heat generating station.

*centre, — atomique, atomic center; — commutation automatique, automatic switching center; — de calcul, computing center; — de direction, control center; — de gravité, center of gravity; — de guidage, guiding center; — des hautes altitudes, high-flying school;

— **d'essais,** test center; — **d'essais en vol**
(C.E.V.), flight test center; — **d'inertie,**
center of mass; — **instantané de rotation,**
instantaneous center of rotation; — **noir,**
bull's-eye.

*****centrifugeuse,** — **à contrecourant,** counter-
current centrifuge; — **à écoulement continu,**
concurrent centrifuge, flow-through centrifuge; — **à gaz,** gas(eous) centrifuge; — **à
godets,** cup-type centrifuge; — **à grande
vitesse,** high-speed centrifuge; — **évapora-
tive,** evaporative centrifuge; — **ionique,** ionic
centrifuge.

*****cercle,** — **de volant,** horn ring; — **primitif,**
pitch circle.

*****cerf-volant,** kite; — **cellulaire,** box kite.

cermet, *m.* ceramet, metal-ceramic.

*****chaînage,** chaining (EDP); — **de données,**
data chaining.

*****chaine,** — **de comptage d'impulsions,** pulse-
counting channel; — **de décroissance,** radioactive chain, decay chain; — **de désintégration,** decay chain; — **de distribution,** timing
chain; — **de montage,** assembly line; — **de
pilotage,** attitude control system; — **de réactions des protons sur les protons,** proton-
proton chain; —**s antidérapantes,** anti-skid
chains.

*****chaise,** bearing pillow (mach.).

chalcolite, *f.* chalcolite.

*****chaleur,** — **de combustion,** heat of combus-
tion; — **de désintégration radioactive,** heat
of radioactivity; — **de fission,** fission heat;
— **de formation,** heat of formation; — **de
fusion,** (latent) heat of fusion; — **de réaction,**
reaction heat.

chaleur, — **de vaporisation,** heat of vaporization; — **engendrée par la radioactivité,** radioactive heat; — **radiogénique,** radiogenic
heat; — **rémanente,** afterheat; — **résiduelle,**
afterheat; — **spécifique,** specific heat.

*****chaloupe canonnière,** gunboat.

chalutier, *m.* trawler.

chambrage, *m.* counterbore, recessing.

*****chambre,** bladder, blister (métal.); — **à
aimant,** magnet space; — **à bore,** boron
chamber; — **à bulles,** bubble chamber; —
acélératrice, accelerating chamber; — **à détente,** cloud chamber, Wilson chamber; — **à
extrapolation,** extrapolation chamber; — **à
turbulence,** turbulent mixing chamber; —
à vide, vacuum chamber; — **d'alimentation,**
plenum chamber; — **d'eau,** waterspace; — **de
catalyse,** catalyst chamber.

chambre, — **de combustion,** combustion
chamber; — **de compensation,** clearing

(EDP); — **de détente,** expansion chamber; —
— **de détente à basse pression,** low-pressure
cloud chamber; — **de précombustion,** pre-
combustion chamber (Diesel); — **de turbulence,** swirl chamber (Diesel); — **de Wilson,**
cloud chamber, Wilson chamber; — **de Wilson
à forte pression,** high-pressure cloud chamber; — **d'ionisation,** ionization chamber, ion
chamber; — **d'ionisation à air libre,** free-air
ionization chamber, open-air ionization
chamber; — **d'ionisation à cavité,** cavity ionization chamber; — **d'ionisation à extrapolation,** extrapolation ionization chamber; —
d'ionisation à impulsions, pulse ionization
chamber.

chambre, — **d'ionisation à intégration,** integrating ionization chamber; — **d'ionisation
compensée,** compensated ion chamber; —
d'ionisation de poche, pocket ionization
chamber; — **d'ionisation de surveillance,**
monitor ionization chamber; — **d'ionisation
différentielle,** differential ionization chamber; — **d'ionisation équivalente à l'air,** air-
wall ionization chamber; — **d'ionisation
équivalente à un tissu,** tissue-equivalent
ionization chamber; — **d'ionisation placée
dos à dos,** back-to-back ionization chamber;
— **proportionnelle d'ionisation,** proportional
ionization chamber; — **tubulaire,** gas anchor;
—**s d'équipage,** crew's quarters.

*****champ,** — **à potentiel scalaire,** scalar poten-
tial field; — **à potentiel vecteur,** vector-poten-
tial field; — **atomique,** atomic field; — **auto-
congruent,** self-congruent field; — **axial,**
axial field, cylindrical field; — **brouilleur,**
interference field; — **cinétique,** motional
field; — **coulombien,** Coulomb field; — **créé
par l'électron et le neutrino,** electron-neu-
trino field.

champ, — **créé par l'électron et le positron,**
electron-positron field; — **créé par un proton
et un neutrino,** proton-neutrino field; — **critique,** sparking potential; — **cylindrique,**
cylindrical field; — **d'accumulation,** amount
field; — **de carte,** card field; — **de chauffage,** heating field; — **de confinement,** confining field; — **de contrôle,** control field; — **de
Coulomb,** Coulomb field; — **de délimitation
(magnétique),** confining field; — **de désioni-
sation,** clearing field; — **de dipôle,** dipole
field.

champ, — **de forces centrales,** central field;
— **de freinage,** reflecting or retarding field;
— **de gravitation,** gravitational field; — **de
jonction interne,** internal-junction field; —
de lecture, scanning area; — **de mouvement
interne,** motional field; — **de pesanteur,** gravity field; — **de potentiel,** potential field; — **de

rayonnement, radiation field; — de sortie, exit portal.

champ, — de spineurs, spinor field; — de variation, range; — de vérification, control field; — d'image, frame (of film or picture), picture field; — dû aux charges spatiales, space-charge field; — dû aux nucléons, nucleonic field; — du noyau, nuclear field; — électrique, electric field; — électrique aléatoire, random electrostatic field; — électrique cinétique, motional electric field; — électrique de dipôle, dipole electric field; — électrique de mouvement, motional electric field.

champ, — électrique transitoire, transient electric field; — électrique transverse de focalisation, transverse focusing electric field; — électrique variable, varying electric field; — électromagnétique, electromagnetic field; — électromagnétique alternatif, alternating electromagnetic field; — électronique, electronic field; — électrostatique, electrostatic field; — extérieur, externally generated field; — hélicoïdal, corkscrew field; — libre, free field.

champ, — local, local field; — magnétique axial, axial magnetic field; — magnétique de dipôle, dipole magnetic field; — magnétique de striction, pinch magnetic field; — magnétique dipolaire, dipole magnetic field; — magnétique guide, guide field; — magnétique longitudinal, longitudinal magnetic field; — magnétique pulsant, pulsating magnetic field; — magnétique stationnaire, static magnetic field; — magnétique uniforme, uniform magnetic field.

champ, — magnétostatique, magnetostatic field; — mésique, meson field; — non-uniforme, inhomogeneous field; — plan, plane field; — pseudoscalaire, pseudoscalar field; — scalaire, scalar field; — sphérique, central field, spherical field; — tensoriel, tensor field; — transverse, transverse field; — uniforme, homogeneous field, uniform field; — variant en fonction du temps, time-varying field; — vectoriel, vector field.

*changement, — de base, radix conversion; — d'échelle, scaling; — de marche, reversal (mach.);— de pas, pitch control (aviation); — de phase, phase transition; — d'état, status switching; — de vitesse, variable gear; — du film, threading (up) of film.

*chape, staplelike support; — de trolley, trolley harp.

*chapeau, — de bielle, connecting rod cap; — de moyeu, hub cap; — de palier, bearing cap; en —, bell-shaped.

*char amphibie, amphibian tank.

*charbon, — bitumineux, Boghead or cannel coal; — homogène, solid carbon; — mineralisé, flame carbon; — mobile, movable carbon rod; — uranifère, uranium-bearing coal.

*charge, — atomique, atomic charge; — atomique efficace, effective atomic charge; — de flambage, buckling load; — de l'électron, electron charge; — de mutations, mutational load; — déplacée, displaced charge; — de pointe, peak load; — de rupture, ultimate tensile strength; — d'espace, space charge; — du noyau, nuclear charge.

charge, — effective, effective charge; — électrique élémentaire, elementary electric charge; — électronique, electron charge, electronic charge; — en oeuvre, quantity of material being processed; — ionique, ionic charge; — massique linéique, mass loading per unit length; — massique par unité de longueur, mass loading per unit length; — mobile, moving charge; — négative, negative charge; — par essieu, axle load; — positive, positive charge; — réelle, effective charge; — transitoire, transient load; — uniformément répartie, uniformly distributed load; — utile, payload; à pleine —, full load; avec... kilogrammes de —, with . . . kilograms of useful load.

*chargé, live (charged).

*chargement, lading (naut.); — actif, fuel charge; — de bande, tape load; — dispersé, scatter load; — en contiguïté, block load; — initial de programme, initial program load; — permettant la divergence, critical loading.

*charger, to instruct.

*chargeur, gun clip, magazine, system loader; — à disque, disk pack; — d'exploitation, systems pack.

chargeuse-pelleteuse, f. backhoe loader.

*chariot, — automatique, automatic carriage; — d'atterrissage, landing gear; — de lancement, catapulting cradle; — de manutention, dolly; — de tour, lathe carriage; — orientable, rotating bridge; —s à commandes indépendantes, dual feed carriage.

*charnière, link (EDP); — du volet, flap hinge.

*chasse, blowdown; — à percer, drift, punch; — de la direction, castor effect; angle de —, castor angle.

chasse-clef, m. key driver.

chasse-neige, m. snow plow.

*chasseur, fighter.

*châssis, cabinet, casing, housing; — à poutre centrale, central tube frame; — à rétrécissement AV & AR, joggled frame (trucks); — caisson, box frame (auto); — sans caisse, running gear; — **surbaissé,** dropped frame, underslung frame (auto).

*__château de plomb,__ lead cask.

*__chaudière,__ — à lessive, lye vat; — d'impregnation, impregnating pan; — en tôle ondulée, corrugated boiler.

*__chaudronnerie,__ boiler plant.

*__chaudronnier,__ boilermaker.

*__chauffage,__ — **atomique,** atomic heating; — **industriel,** industrial heating; — **par effet Joule,** Joule heating; — **par rayonnement,** radiant heating; — **par résistance,** ohmic heating.

*__chaussette,__ thimble (nuclear reactor).

chausson d'aviateur, m. flying boot.

*__chef,__ — de bloc, — d'îlot, air-raid warden; — pompier, fire inspector.

chéloïde, f. keloid.

*__chemin,__ — à côté de la chaussée, bypass; — de fer aérien, cable railway; — de fer à voie étroite, narrow-gauge railroad; — de fission, fission channel; — de guidage, course bearing, head (of a ship).

*__cheminement,__ flow, stream, creep.

chemisage, m. canning, cladding, sheathing; — des cartouches de ralentisseur, moderator can.

*__chemise,__ can; — amovible, replaceable cylinder liner; — en boral, boral liner; — humide, wet-type cylinder liner; — sèche, dry-type cylinder liner.

cheralite, f. cheralite.

*__chercheur,__ search coil; — à ressort, cat whisker (radio); — d'appel, call or line finder.

*__chevalement,__ derrick.

*__cheville,__ wedge.

*__chevron,__ joist.

chevrotement, m. trembling, quivering, (mechanical) flutter of film.

*__chicane annulaire,__ doughnut ring.

*__chiffre,__ dial; — **algébrique,** sign digit; — **binaire,** binary digit; — d'autocontrôle, self-checking digit; — de départ, source digit; — de vérification, check digit.

*__chiffrer,__ to digitalize, code.

*__chimie,__ — de la pile, reactor chemistry; — des éléments lourds, heavy-element chemistry; — des radio-éléments indicateurs, tracer chemistry; — des solutions de produits de fission, solution chemistry of gross fission products; — nucléaire, nuclear chemistry; — sous irradiation, radiation chemistry.

chinkolobwite, chinkolobwite.

*__choc,__ — à courte distance, close collision; — aléatoire, random encounter; — à l'ouverture, opening shock; — de basse fréquence, motorboating; — d'échange, exchange collision; — élastique, elastic collision; — électronique, electron impact; — inélastique, inelastic collision; — lointain, distant collision; — mutuel, mutual collision; — thermique, thermal collision, thermal shock.

chromatide, f. chromatid.

*__chromatine sexuée,__ sexual chromatin.

*__chromosome,__ — **sexuel,** sex chromosome; — s à segment inverse, inversion chromosomes; — s homologues, homologous chromosomes.

*__chronomètre,__ clock, interval timer, timer.

chronoscope, m. timer.

chronotron, m. chronotron.

*__chute,__ pitch (incline), crash; — **cathodique,** cathode fall of potential; — de tension, power failure; — linéaire de réactivité, linear decrease in reactivity; — mortelle, fatal crash; — relative de tension, relative voltage drop, inherent regulation.

*__cible à deutérium,__ deuterated target.

*__ciment pour joints hermétiques,__ hermetic sealing medium.

*__cimentation étagée,__ multiple-stage cementing.

*__cinétique de la pile,__ reactor kinetics.

*__circuit,__ circuit race, lap, circuit diagram; — à bascule, see-saw circuit, Eccles-Jordan circuit; — **accordé,** tuned circuit; — à coïncidences, gate circuit, coincidence circuit; — à déclencheur, trigger circuit; — à échelle de comptage, scaling circuit; — à lampe, tube circuit; — à moins que, inhibitor, except gate; — à pont, bridge circuit; — à réaction, feedback circuit; — à rétroaction, feedback circuit; — à un coup, one-shot circuit; — basculant, flip-flop circuit; — bouchon, rejector circuit.

circuit, — collecteur, pickup circuit; — combinant, side circuit; — combiné, phantom circuit; — d'accord, tuning circuit, resonance circuit; — d'alimentation, fuel system; — d'alimentation électrique, power circuit; — d'allumage, ignition circuit; — d'amorçage, start-up loop; — d'amplification, amplifier circuit; — d'annonce, warning circuit; — de commutation, switching circuit; — de compensateur, annulating or suppressing network.

circuit, — de compensation, absorber circuit; — d'écrêtage, clipper circuit; — de débit, output circuit; — de déblocage, line-free circuit;

— de **démarrage,** start-up loop; — de **fonctionnement,** operating loop; — de **l'échangeur de chaleur,** heat-exchange circuit; — de **refroidissement,** cooling system; — de **régénération de la vapeur par réchauffage,** reheat regenerative steam cycle; — de **retour,** back circuit; — d'**étouffement,** quenching circuit; — de **voie,** track circuit.

circuit, — de **voie à courant alternatif,** alternating-current track circuit; — d'**excitation,** field circuit; — d'**extinction,** quenching circuit; — **différentiateur,** differentiating circuit; — **directeur,** trunk line (teleph.); — **et,** *and* circuit, intersector; — **équivalent,** equivalent circuit; — **expérimental thermique,** hot loop; — **explorateur,** pickup loop; — **extrafonctionnel,** outside loop; — **fermé,** closed circuit, circulation loop; — **flip-flop,** flip-flop circuit.

circuit, — **imprimé,** printed circuit; — **interfonctionnel,** local loop; — **intégrateur,** scaling circuit; — **intermédiaire,** intermediate circuit, intermediate loop; — **limiteur de flux,** overflux circuit; — **logique,** decision element; — **mélangeur,** *or* circuit; — **multipoint,** mutipoint circuit; — **non,** inverter; — **oscillation témoin,** standard or reference oscillation circuit; — **ou,** *or* circuit; — **ouvert,** open circuit, open cycle; — **pilote,** monitor circuit; — **plaque accordé,** tuned-anode circuit; — **radiophonique,** ratiotelephone circuit; — **réel,** side circuit; — **transistorisé,** solid-state circuit; — **utilisation,** load circuit, consumer circuit, useful or signal circuit; —**s parallèles,** parallel circuits.

**circulaire, cyclic.

**circulaire, pamphlet.

circulant, circulating.

circulation, — du **gaz, gas circulation; — en **circuit fermé,** recirculation.

cire, — à **déformer, cobbler's wax; — à **parquets,** floor polish; — **minérale,** native paraffin.

**cisaillement du champ magnétique, shear of magnetic field.

cission, *f.* shear stress.

**citerne de décantation, settling basin.

**clair de terre, earthshine.

clairance, *f.* clearance (transfer of solutes).

clairon, *m.* bugle.

clapet, — à **bille, ball valve; — à **décharge,** dump or relief valve; — de **dérivation,** bypass valve; — de **laminage,** restrictor, throttle valve; — de **non-retour,** check valve; — de **retenue,** check valve; — **surpression**

d'**huile,** oil pump pressure regulating valve; — **taré,** calibrated valve.

claquage, *m.* breakdown; — de **l'isolant,** insulation breakdown; — **électrique,** electric breakdown, puncture, or discharge.

claquement, rattling (brakes); — de **manipulation, key click; — de **piston,** piston slap.

clarkéite, *f.* clarkeite.

**clarté du point, spot brightness (television).

**classer, to rank, file, sift.

**classeur à tamis, screen classifier.

classificateur, *m.* classifier.

**clavette, gib (mach.), wedge.

clavier, — **alphanumérique, alphameric keyboard; — de **fonctions programmées,** programmed function keyboard; — de **sélection,** classification keyboard.

clef, clé, — à **fourche, forging wrench; — **Allen,** Allen Key; — à **molette,** adjustable wrench; **— **anglaise,** universal wrench; — à **tubes,** pipe wrench; — d'**annulation,** cancel key; — de **chargement,** load key; — de **contact,** ignition switch key; — de **correction,** correct key; — de **garde,** protection key; — de **mémoire,** storage key; — de **voute,** keystone; — **dynamométrique,** torque wrench; — en **double,** duplicate key.

clévéite, *f.* cleveite.

clignoter, to flicker, blink.

climatisation, *f.* air conditioning, heating and ventilating.

climatisé, air-conditioned.

climatiseur, *m.* heater (auto), air conditioner.

clinomètre longitudinal, *m.* fore-and-aft level indicator.

cloc, *m.* click, clicking noise.

cloche, *f.* bubble cap; — à **plongeur, diving bell; — à **pression,** pressure vessel; — d'**appel,** call bell.

cloison, bulkhead; — **étanche, bulkhead; — **ondulée,** baffle plate; — **pare-feu,** fireproof bulkhead.

**clou à grosse tête, hobnail.

**coche, kink.

codage, *m.* coding; — **accéléré,** speed coding; — **algébrique,** algebraic coding; — **articulé,** skeletal coding; — **automatique,** automatic coding; — d'**aiguillage,** branch coding; — en **absolu,** absolute coding, actual coding; — en **chaîne,** chain code; — en **langage-machine,** absolute coding; — **fondamental,** basic coding; — **ordinolingue,** absolute coding; — **relatif,** relative coding; — **sans boucle,** straight line coding; — **sommaire**, skeletal coding; — **symbolique**, symbolic coding.

***code**, information; — **alphanumérique**, alphanumeric code; — **binaire**, binary code; — **binaire réflechi**, reflected binary code, Gray code; — **biquinaire**, biquinary code; — **correcteur d'erreur**, error correcting code, self-checking code; — **d'appel**, calling code; — **d'autocontrôle**, error detecting code; — **de carte**, card code; — **de distance minimale**, minimum distance code; — **de Gray**, Gray code, reflected binary code; — **de perforation**, card code; — **de retour**, return code; — **de ruban perforé**, punched tape code; — **détecteur d'erreur**, error detecting code; — **deux-de-cinq**, two-out-of-five code; — **d'impulsion**, pulse code; — **d'opération**, operation code, instruction code; — **d'ordinateur**, computer code; — **mnémonique élargi**, extended mnemonic code; — **non pondéré**, nonweighted code; — **plus trois**, excess-three code; — **pondéré**, weighted code; — **progressif continu**, cyclic permuted code, unit distance code.

code-machine, *m.* machine code.

code-opération, *m.* operation code, operation part.

coder, to code, encode, encipher.

code-retour, *m.* return code.

codeur, *m.* coder.

***coefficient**, — **cavitaire**, void coefficient; — **d'absorption**, absorption coefficient; — **d'absorption atomique**, atomic absorption coefficient; — **d'absorption de l'énergie**, energy absorption coefficient; — **d'absorption molaire**, molar extinction coefficient; — **d'affaiblissement**, attenuation constant; — **d'amortissement**, damping coefficient; — **de canal chaud**, hot channel factor; — **de capacité**, read-around ratio; — **de commutation**, switching coefficient; — **de compressibilité**, compressibility factor.

coefficient, — **de consummation**, burnup factor; — **de conversion interne**, internal conversion coefficient; — **de danger**, danger coefficient; — **de détection**, detection coefficient; — **de diffusion**, diffusion coefficient; — **de dilatation cubique**, coefficient of expansion; — **de distribution**, distribution ratio; — **de friction**, friction coefficient; — **de groupage**, blocking factor; — **d'émanation**, emanating power; — **de multiplication**, reproduction factor.

coefficient, — **de partage**, distribution coefficient, partition coefficient; — **de pénétration**, reciprocal of amplification factor; — **de plane**, gliding angle; — **de pression de la réactivité**, barometric coefficient of the reactivity; — **de puissance**, power coefficient; — **de puissance d'antiréactivité**, power coefficient of negative reactivity; — **de qualité**, figure of merit; — **de recombinaison**, recombination coefficient; — **de réflexion**, coefficient of reflection; — **de rendement d'un plateau**, plate efficiency factor.

coefficient, — **de rétention**, retention coefficient; — **de rétrécissement**, coefficient of contraction; — **de température**, temperature coefficient; — **de température de la densité**, density temperature coefficient; — **de température de la réactivité**, nuclear temperature coefficient, reactivity temperature coefficient; — **de transfert de masse**, mass transfer coefficient; — **de visibilité d'une radiation**, luminosity factor of a monochromatic radiation; — **d'extinction molaire**, molar extinction coefficient; — **d'extraction**, extraction coefficient — **d'ionisation**, ionization coefficient.

coefficient, — **du coin**, wedge constant (motion picture); — **global de transmission de chaleur**, over-all coefficient of heat transfer; — **laminaire de transmission**, film coefficient; — **linéaire d'absorption**, linear absorption coefficient; — **massique d'absorption**, mass absorption coefficient; — **stoechiométrique**, stoichiometric coefficient.

***coeur**, active core, core (reactor); — **à neutrons intermédiaires**, intermediate neutron assembly; — **de la pile**, fuel assembly, reactor core; — **du réacteur à circulation double**, two-pass core; — **fissile**, fissile core; — **sans réflecteur**, bare core; — **type**, reference core.

coffinite, *f.* coffinite.

***coffre**, housing (of amplifier); — **aux pavillons**, flag locker (naut.).

coffré, boarded.

cognement, *m.* knocking.

***coiffe**, fairing, shroud, nose cone.

***coin coupé**, corner cut.

coincé, jammed, wedged.

***coincement**, sticking.

***coïncidence**, — **accidentelle**, accidental coincidence, random coincidence; — **de deux neutrons**, neutron-neutron coincidence; — **d'impulsions**, pulse coincidence; — **fortuite**, accidental coincidence; — **neutron-neutron**, neutron-neutron coincidence; — **retardée**, delayed coincidence; —**s étant fonctions de l'angle**, angular dependence of coincidences.

coittes, *f.pl.* ways (naut.).

***col**, saddle point, throat (of horn).

colimaçon, *m.* helix.

***collage de l'armature**, armature adherence.

collant, sticky.

*****colle à froid**, wood cement.

collecte, *f.* — **de données**, data collection, data acquisition; — **des électrons**, electron collection; — **des poussières**, dust collection.

collecter, to focus (cathode-ray tube).

*****collecteur**, contact ring or maker, slip ring (generator), shunt, distributor, header, rotary switch; — **d'admission**, inlet manifold; — **d'échappement**, exhaust collector; — **de vapeur**, steam drum.

*****coller**, to attract, keep attracted.

*****collet de l'essieu**, axle washer.

*****collier**, — **d'arbre**, shaft collar; — **d'attache**, securing ring; — **de butée**, thrust collar; — **de fixation**, pipe clamp, clip, bracket; — **de frein**, brake band; — **de lunette**, telescope clamp; — **de masse**, bonding clip.

*****collimateur**, — **à focalisation**, focusing collimator; — **à neutrons**, neutron collimator; — **convergent**, focusing collimator.

*****collision**, — **à courte distance**, close collision; — **aléatoire**, random encounter; — **due à l'énergie thermique**, thermal collision; — **élastique**, elastic collision; — **en théorie à deux corps**, two-body collision; — **frontale**, head-on collision; — **inélastique**, inelastic collision; — **ionisante**, ionizing collision; — **lointaine**, distant collision; — **mutuelle**, mutual collision; — **sans capture**, noncapture collision.

collure, *f.* splice (of film); **bruit de** —, splice bump, blooping.

*****colmatage**, clogging; **bassin de** —, settling tank.

*****colonne**, — **à bande**, tape column; — **à garnissage**, packed column, packed tower; — **à plateaux**, plate column, plate tower; — **à vide**, vacuum column, tape column; — **conductrice**, conducting column; — **de carte**, card column; — **de contrôle**, control column; — **de direction**, steering column (auto); — **de distillation**, distillation column; — **de plasma**, plasma column; — **d'épuisement**, stripping column; — **d'épuration**, scrub column; — **de séparation**, separation column.

colonne, — **d'extraction**, extraction column; — **diffusante**, thermal column; — **d'ions**, ion column; — **échangeuse d'ions**, ion exchanger column; — **ionique**, ion column; — **lumineuse**, luminous column; — **plasmatique**, plasma column; — **thermique**, thermal column.

*****colorant à l'indanthrène**, indanthrene dye.

columbite, *f.* columbite.

*****combinaison**, overall, combination suit; — **à chauffage électrique**, electrically heated suit; — **de vol en cuir**, leather flying suit; — **du spin et de la parité**, spin-parity combination; — **imperméable**, protective suit.

*****combinateur**, multistep or multipoint switch.

*****combiné**, handset; — **à hélice propulsive**, propeller compound helicopter; — **de bord**, instrument cluster; — **de tête**, boom set; — **téléphonique**, telephone handset; **impression** —, composite shot (cinema).

*****combustible**, — **appauvri**, impoverished fuel; — **au deutérium**, deuterium fuel; — **chimique**, chemical combustion; — **de grande pureté nucléaire**, high-grade nuclear fuel; — **dénaturé**, denatured fuel; — **fissile**, fissile fuel; — **immobilisé**, holdup of fuel; — **ionisé**, fuel ion; — **nucléaire**, reactor fuel, nuclear fuel.

combustible, — **nucléaire concentré**, concentrated nuclear fuel; — **nucléaire dénaturé**, denatured nuclear fuel; — **nucléaire de qualité supérieure**, high-grade nuclear fuel; — **pour fusion**, fusion fuel; — **réagissant**, reacting fuel; — **recyclé**, recycled fuel; — **régénéré**, regenerated fuel; — **thermonucléaire**, thermonuclear fuel; — **usé**, spent fuel.

combustion, *f.* burn-up (nuclear); — **massique**, specific burn-up.

Comité d'étude et d'essai des carburants, *m.* Cooperative Fuel Research Committee.

*****commande**); — **à chaine**, chain drive (automobiles); — **à distance**, remote control; — **à l'aide de réflecteur**, reflector control; — **automatique**, automatic control; — **de canal**, channel control; — **de chariot**, carriage control; — **de contact**, contact operate; — **de contact maintenu**, electronic contact operate; — **de la barre de réglage**, control-rod drive; — **de sécurité**, safety control; — **de sélection**, selection control; — **directe**, direct control; — **direct par ordinateur**, direct digital control; — **du compresseur**, supercharger drive; — **d'un réacteur**, reactor control; — **en souffrance**, back order, backlog; — **multiple**, multiple control; — **numérique**, numerical control; — **par bande**, tape control; — **par directive à l'opérateur**, operator guide control; — **par intégrale**, integral control, proportional speed floating control; — **par proportion**, proportional control, ration control; — **par proportion et intégrale**, proportional plus reset control; — **proportion-intégrale-différentielle**, proportional plus reset plus rate action; — **proportionnelle**, floating control, proportional control; — **tout-ou-rien**, on-off control; — **unique**, single control, unicontrol; —**s en cascade**, cascade control.

commençant, *m.* learner.

*****commis aux vivres**, steward (naut.).

*commun, bus circuit.

*communication, — à grande distance, long-distance call; — bilatérale entre les avions, two-way communication between aircraft; — collective, conference call; — de mouvement, intermediate gear.

*commutateur, cutout, reverser, rectifier; — à bascule, toggle switch; — à manivelle, lever switch; — à plots, step switch; — code, dimmer (auto); — commandant la chute des barres, scram switch; — de cadence, rate switch; — d'éclairage et avertisseur, combined horn and light switch; — de gammes d'ondes, range switch; — de mise à la terre, ground switch; — de pile, battery commutator; — de sélection de mémoire, storage select switch; — de veille automatique, attent-unattent switch; — pas à pas, step-by-step switch; — tournant, rotating or rotary switch.

commutation, f. switching, circuit switching; — de bande, tape switching; — de messages, message switching, store-and-forward switching; — partielle, partial switching.

commuter, to switch, reverse (polarity).

*compact, close-packed.

compactage, m. compaction.

compandor, m. compandor.

*comparateur, dial indicator; — de ruban, tape comparator; — traducteur, comparator-translator.

*compas, — à charnière, firm-joint calipers; — à ressort, spring calipers; — étalon, standard compass; — intérieur, inside caliper (tool); — liquide, fluid or floating compass; — maître-à danser, inside calipers; — quart de cercle, quadrant compass.

*compatibilité, — ascendante, upward compatibility; — d'équipment, equipment compatibility; — descendante, downward compatibility; — mécanoïde, equipment compatibility.

*compensateur, pressure equalizer, trimming tab; — de carburateur, compensator jet; — de phase, phase equalizer; — de voltage, line drop.

compensé, compensated, balanced, equalized.

*compenser l'affaiblissment, to reduce damping.

*complément, extension; — de base, radix complement.

complexant, m. complexing agent.

*comportement, — aberrant, erratic behavior; — cinétique, kinetic behavior; — moyen, average behavior; — transitoire, transient behavior.

*composante, — de fréquence radio, radio-frequency component; — de grande énergie, high-energy component; — de la vitesse, velocity component; — de radio-fréquence, radio-frequency component; — déwatté, reactive, idle, or wattless component; — dure, hard component; — molle, soft component; — pénétrante, penetrating component.

*composé, — chélaté, chelate compound; — de chélation, chelate compound; — d'uranium, uranium compound; — électrovalent, electrovalent compound; — gazeux, gaseous compound; — marqué, labeled compound.

*composé-entraineur, carrier compound.

*composer, to dial, type.

*composition, alloy, format, configuration, layout; — isotopique, isotopic composition.

composteur, m. type setting stick; — de données, data cartridge.

*compresseur, supercharger.

*compression adiabatique, adiabatic compression, reversible compression.

*comprimé d'oxyde, pressed oxide.

*comprimer, to pack (EDP).

*comptage, — parasite, background count, background; — par coïncidences, coincidence counting.

*compte, — à rebours, countdown; — à recevoir, account receivable.

compte-secondes, m. timer.

*compteur, — à bore, boron counter tube; — à chambre d'ionisation, counting ionization chamber; — à coïncidences, coincidence counter; — à courant gazeux, gas-flow counter; — à cristal, crystal counter; — à dépassement, excess-power or current meter; — à dépassement totalisateur, excess-energy meter; — à détecteur, foil counter; — à électrons secondaires, secondary-electron counter.

compteur, — à étincelles, spark counter; — à fenêtre en bout, end-window counter; — à gaz, gas counter; — à halogène, halogen counter; — alpha, alpha counter; — à paroi-écran, screen-wall counter; — à paroi mince, thin-walled counter tube; — à plaques parallèles, parallel-plate counter; — à plaques planes, flat-plate counter; — à scintillations, photomultiplier counter, scintillation counter.

compteur, — autocoupeur, self-quenching counter; — chromographe, time-period counter; — cosinus à dépassement, over-compensated induction meter; — d'adresse, sequence control register; — de lecture-écriture, read-write counter; — de neutrons, neutron counter; — **de photons à sodium,** sodium photon counter; — **de poche,** pocket monitor; — **de radioactivité,** radiation counter, radiation counter tube, survey meter; — **de surveillance,** monitor counter; — **détecteur,** monitor counter.

compteur, — détecteur à fission, fission monitor counter; — **d'impulsions,** scaled radiation detector, pulse counter, pulse chamber, impulse counter, scaler; — **d'ions,** ion counter; — **double de fission à impulsion,** back-to-back fission pulse counter; — **électrodynamique à collecteur (Thomson),** Thomson meter; — **Geiger,** Geiger counter; — **gradué en roentgens,** roentgen meter; — **ionique,** ion counter.

compteur, — journalier, trip mileage counter; — **pendulaire,** clock meter; — **proportionnel,** proportional counter, proportional counter tube; — **spiral à fission,** spiral fission counter; — **télescope,** telescope counter; — **totalisateur,** odometer.

compteur-cloche, end-window counter.

compteur-ordonnateur, *m.* sequence control register.

compteur-solde, *m.* balance counter.

*****concassage,** milling.

concatétner, to concatenate.

*****concentration,** beaming, focusing, concentration process; — **d'équilibre,** equilibrium concentration; — **du minerai,** ore concentration; — **gravimètrique,** specific-gravity concentration; — **maximum admissible,** maximum permissible concentration (MPC); — **par gravité,** specific-gravity concentration; — **relative,** relative concentration.

*****concentrés finement broyés,** finely ground concentrates.

*****concentrique,** coaxial.

*****conception,** design; — **de carte,** card design; — **de la pile,** reactor design.

*****concession,** lease.

*****concordance,** conformity; — **des phases,** phase coincidence.

condamnation, *f.* door or window locking device.

*****condamner,** to batten down.

condensat, *m.* condensate.

*****condensateur,** capacitor; — **absorbant les ondulations,** smoothing condenser; — **à démultiplicateur,** vernier control, tuning condenser; — **adjustable,** trimming condenser; — **à lames semi-circulaires,** straight-line capacity condenser; — **à lames vibrantes,** vibrating condenser; — **à variation linéaire**

de longueur d'onde, square-law condenser; — **d'appoint,** trimming capacitor, padding condenser, aligning condenser.

condensateur, — d'arrêt, blocking condenser; — **de démarrage,** starting condenser; — **de grille,** grid capacitor, grid-blocking condenser; — **de raccourcissement,** short-wave condenser; — **d'extinction,** quench condenser; — **en dérivation,** bypass or shunting condenser; — **en ligne,** — **jumelés,** ganged condenser; — **neutrodyne,** neutralizing condenser; — **parabolique,** square-law condenser.

*****condenser,** to pack, abstract.

*****condenseur,** capacitor; — **à reflux,** reflux condenser; — **de buées à arrosage,** spray condenser; — **de couplage,** coupling condenser; — **de pot de recombinaison,** recombiner condenser.

*****condition, — composée,** compound condition; — **d'attente,** wait condition; — **de fréquence,** frequency condition; — **quantique,** quantum condition; — **résultante,** result condition; —**s aux limites,** boundary conditions; —**s critiques,** critical state; —**s de stabilité,** stability conditions; —**s initiales,** initial conditions; —**s normales de température et de pression,** normal temperature and pressure.

conditionneur, *m.* data gate.

*****conductance d'un tube,** anode alternating-current conductance.

*****conducteur, — d'amenée,** lead (elec.); — **de bouclage,** ring main; — **de chaleur,** heat conductor; — **d'électricité,** conductor of electricity; — **de protection,** guard wire; — **du courant,** lead (elec.); — **idéal,** perfect conductor; — **parfait,** perfect conductor; — **pilote,** pilot wire.

*****conductibilité électrique,** electrical conductivity.

*****conduction par les gaz,** gaseous conduction.

*****conductivité, — effective,** effective conductivity; — **électrique,** electrical conductivity; — **ionique,** ionic conductivity; — **par trous,** hole-type conductivity; — **réelle,** effective conductivity; — **thermique,** heat conductivity, thermal conductivity.

*****conduit, — collecteur,** shunt; — **de vidage,** drainpipe.

*****conduite, — d'amenée,** gathering line, supply pipe; — **d'eau,** water channel; — **de la guerre aérienne,** aerial warfare; — **de la pile,** reactor control; — **d'un réacteur,** reactor control; — **intérieure,** sedan car.

*****cone, — d'ablation,** ablating cone; — **érodable,** ablating cone.

confetti, *m.pl.* chip, chad (EDP).

*confermement, — d'équilibre magnétohydro-
dynamique, magnetohydrodynamic equilibrium configuration; — électronique, electronic configuration.

confinement, *m.* containment (of plasma); —
d'equilibre, equilibrium confinement; — longitudinal, longitudinal confinement; —
magnétique, magnetic confinement; — par
striction, pinch confinement; — pour un
réacteur, containment.

*confondre, to merge; se —, to be taken for.

*congé, clearance; — de raccordement, fillet.

*conicité, coning.

*conjugué, interlocked, ganged, synchronized.

*connexion, — de base, basic wiring; — élec-
trique des rails, rail bond; — en parallèle,
shunt connection.

*conseil de guerre, court-martial.

*conservation, — de la charge, conservation
of charge; — de la parité, conservation of
parity; — de l'énergie, conservation of
energy; — des termes du premier ordre,
linearization.

*conserver, to store.

*consommation, — à la vitesse de croisière,
cruising consumption; — d'énergie, wattage;
— en combustible, fuel consumption; —
horaire en vol, hourly consumption in flight.

constantan, *m.* constantan.

*constante, — d'adresse, address constant;
— d'affaiblissement, attenuation constant;
— de couplage, coupling constant; — de
déphasage, phase or wave-length constant;
— de désintégration, decay constant, disintegration constant; — de force, force constant;
— de la lampe électronique, parameter of a
thermionic valve; — d'élasticité, spring constant; — de recapture, recapture constant.

constante, — des gaz, gas constant; — de
temps, time constant; — diélectrique, dielectric constant, permittivity or specific inductive capacity; — diélectrique absolue, absolute permittivity; — nucléaire, nuclear constant; — numérique, numerical constant; —
radioactive, radioactive constant, transmutation constant; —s nucléaires, nuclear
data.

*constitution, — de dépôts permanents, per-
manent storage; — des circuits, circuit
design; — du noyau, structure of nucleus;
— électronique (de l'atome), electronic
structure (of atom).

constitutionellement, constitutionally, in-
herently, intrinsically.

constriction, *f.* constriction.

*construction de l'atome, atomic structure.

consultation, *f.* consultation; — de fichier,
file search; — de table, table lookup.

consummation, *f.* burnup.

*contact, — à fiche, plug; — binaire, contact
closure; — curseur, sliding contact; — de
repos et de travail, make-and-break contact;
— tout-ou-rien, contact closure; —s auto-
nettoyants, self-cleaning contact points
(ignition); —s de rupteur, contact points,
distributor points.

*contacteur, *m.* electrical contact switch; —
d'allumage, ignition switch; — de portière,
door-operated light switch; — des feux stop,
brake-operated stop-light switch; — vibrant,
vibrating contactor.

contaminant, *m.* contaminant.

*contamination, — du milieu, environmental
contamination; — radioactive, radioactive
contamination.

conteneur, *m.* container.

conteneurisable, containerizable.

contingentement, *m.* contingent or make-shift
condition.

*continuité, — de piste, track overflow (EDP);
— de ralentissement, continuous slowing
down (neutrons).

*contour, — de caractère, character outline;
— imprécis, diffuse boundary.

*contraction, — du plasma, contraction of
plasma; — électrodynamique, electrodynamic contraction.

*contrainte, unit stress; — admissible, design
or working stress; — de cisaillement, shear
stress; — mécanique, mechanical stress; —
thermique, thermal stress.

contre-amiral, *m.* rear admiral.

contre-bride, *f.* companion flange, adapter.

contrechâssis, *m.* blind window.

contre-clavette, *f.* cotter pin.

*contre-coup acoustique, acoustic feedback.

*contre-courant, *m.* reverse direction flow;
condensateur à —, reflux condenser.

*contre-fiche, socket.

*contre-plaque, plywood plate.

contre-pointe, *f.* back center.

contre-quille, *f.* keelson.

contre-rail, *m.* guardrail.

*contre-réaction, *f.* reverse feedback.

*contresens, à —, across the grain.

contre-tension, *f.* countervoltage.

contre-torsion, *f.* back twist.

contre-tours, *m.pl.* opposing winding.

*****contre-vapeur,** back steam.

*****contrôle,** survey; — **à distance,** remote control; — **automatique,** automatic check, built-in check; — **automatique de papier,** tape-controlled carriage; — **de division,** divide check; — **d'entrée-sortie,** input-output control; — **de parité,** cyclic check; — **de poste télégraphique,** telegraph terminal control; — **de protection de mémoire,** storage protection check; — **de séquence,** sequence checking; — **de signal,** signal indicator; — **de tolérance,** longitudinal redundancy check; — **de tolérance vertical,** vertical redundancy check; — **de totaux,** parallel balance; — **de validité,** forbidden combination check; — **d'imparité,** oddness control; — **du rayonnement,** radiation survey; — **interne,** built-in check; — **magnétoscopic,** magnetic-particle inspection; — **marginal,** marginal check, high-low bias check; — **modulé,** residue check; — **par écho,** echo check; — **par ressuage de liquide coloré,** dye-penetrant inspection; — **par ressuage de liquide fluorescent,** fluorescent penetrant inspection; — **programmé,** programmed check; — **radiologique,** radiological monitoring; — **résiduel,** residue check; — **sur marbre,** tear-down inspection.

*****contrôleur,** timekeeper; — **de position d'aiguille,** switch position indicator.

*****convection thermique,** thermal convection.

*****convergence,** focusing.

*****conversion, — de la chaleur en énergie,** conversion of heat into power; — **directe en électricité,** direct electrical conversion; — **directe en énergie électrique,** direct conversion into electrical energy; — **d'une paire,** pair conversion; — **interne,** inner conversion, internal conversion; — **para-ortho,** para-ortho conversion.

*****convertisseur,** data set, modem; — **analogique-numérique,** analog-to-digital converter; — **de couple,** torque converter; — **de données,** data adapter unit; — **en cascade,** motor converter; — **espace-temps,** dynamicizer; — **numérique-analogique,** digital-to-analog converter; — **sériel,** serial data adapter; — **série-parallèle,** staticizer; **groupe —,** motor converter.

coordonnateur, *m.* coordinator.

*****copie,** film blank; — **en clair,** hard copy.

coprécipitation, *f.* coprecipitation.

*****coque,** fuselage, body shell (auto); — **autoportante,** integral body construction (auto).

coquerie, *f.* galley (of ship).

*****coquille,** footboard (automobile).

coracite, *f.* coracite.

*****cordage, — à moufles,** block and tackle; —**s,** tackle.

*****corde, — de chanvre,** hemp rope; — **de déclenchement,** rip cord, release cord; — **de haubannage,** span rope; — **de soutien,** guy rope.

*****cordon,** lanyard, flexible cord; — **de soudure,** weld bead.

*****cordonner,** to knurl.

*****corne,** gaff (naut.); — **d'entrée,** leading-pole tip; — **de sortie,** trailing-pole tip.

*****cornière,** elbow.

*****corps,** unit, hull (naut.); — **cristallisé,** crystalline material; — **de chauffe,** heater; — **de garde,** coastguard; — **de métier,** guild; — **de piston,** piston body, piston skirt; — **de pompe,** pump barrel; — **de propulseur,** engine body, motor body, jet body; — **fissile,** fissionable material; — **formant l'ossature,** structural materials; — **isotrope,** isotropic body; — **ralentisseur,** moderator material.

*****corpuscule,** particle; — **de grande énergie,** high-energy particle; — **d'une gerbe en cascade,** cascade particle; — **élémentaire,** fundamental particle; — **expulsé,** ejected particle; — **formé,** product particle; — **incident,** incident particle; — **nucléaire,** nuclear particle.

*****correcteur, — automatique,** device error recovery; — **d'avance,** spark advance corrector; — **de pression,** pressure equalizer; — **d'evanouissement,** antifading device.

*****correction,** patch (EDP); — **d'avance à l'allumage,** ignition timing; — **de données,** data purification; —**s de réglage (d'un champ magnétique),** shimming (adjustment of magnetic field).

*****correlation, — angulaire,** angular correlation; — **en direction,** directional correlation.

*****correspondance par translation,** translation symmetry.

*****correspondant,** sending clerk (teleg.).

*****corriger,** to patch, debug.

*****corrosion,** etching; — **de surface,** crevice; — **par contact,** contact corrosion.

corsaire, *m.* raider (naut.).

*****cosinus phi,** power factor (elec.).

cosmotron, *m.* cosmotron.

*****cosse,** thimble, socket, lug; — **d'extrémité,** soldering lug.

*****cote, — de disparité,** signal distance, — **de disponibilité,** serviceability; — **de trusquinage,** edge distance.

*côte, pitch (incline), gradient, ascent, elevation; — de la mer, seacoast; — d'un angle, angle side.

*côté, — aspiration, suction end of pump; — refoulement, delivery end of pump.

*couche, base, foot; — complète, closed shell; — conductrice, conducting layer; — conjuguée, conjugate layer; — cylindrique du plasma, plasma cylinder; — cylindrique plasmatique, plasma cylinder; — d'arrêt, blocking or stopping layer, barrier layer; — de demi-absorption, half-value layer; — de demi-atténuation, half-value layer; — de produit raffiné, raffinate layer.

couche, — de récupération, blanket, blanket assembly, breeder blanket; — de récupération de la pile, blanket, envelope of the reactor; — de roulement, tire tread; — de transition, boundary layer; — d'extraction, extract layer; — du plasma, plasma layer; — électronique, electronic shell; — électronique saturée, closed shell; — en ébullition, boiling bed; — fertile, blanket, blanket assembly.

couche, — limite, boundary layer; — limite laminaire, laminar boundary layer; — luminescente, scintillation layer; — nucléaire, nuclear shell; — où se produit l'effet pelliculaire, skin layer; — plasmatique, plasma layer; — protectrice, protective coating; — régénératrice, breeding blanket; à —s multiple, à plusieurs —s, multilayer.

*couché, embedded.

*coude, kink; — à large rayon, swept bend; — de tête d'injection, gooseneck; — de vilebrequin, throw of crankshaft.

*coulage, break out (from mold), sinking (ship); — d'une palier, burnout of bearing; — en carapace, shell casting; — en coquille sous pression, die casting.

*couleur éclairante, luminous paint.

*coulis, fire clay.

*coulisse à détente variable, expansion link.

*couloir de franchissement, air corridor.

*coup, round; — de canon de semonce, warning shot; — parasite instrumental, spurious count; — portant, hit; — unique, single hit; —s par minute, counts per minute; à deux —s, double hit (dose).

*coupe, log, report, explosion; — du segment de piston, piston ring gap.

coupée, f. gangway.

coupe-flamme, m. flame arrester.

*coupelle, cup-shaped foot of radio tube in which leadins are sealed; — à bulles, bubble cap.

*couper l'allumage, to switch off the ignition.

coupe-tout, m. master switch.

*couplage, — automatique, automatic interlocking; — en étoile, star connection, Y connection; — en retour, feedback, regeneration; — entre lampes, intervalve coupling; — faible, weak coupling; — fort, tight, strong, or close coupling; — inductif, inductive coupling, flux linkage; — lâche, weak coupling; — pseudoscalaire, pseudoscalar coupling; — retroactif, regenerative or reaction coupling; — rhythm opposé, push-pull circuit scheme; — spin-orbite, spin-orbit coupling; — triangle, delta or mesh connection.

*couple, — actif, deflecting couple; — conique, bevel drive gears; — de forces, torque; — de rappel, restoring torque; — de renversement, propeller torque; — de serrage, tightening torque; — de torsion, torque; — gyroscopique, gyroscopic force of the propeller; — hypoïde, hypoid drive gears (final drive); — moteur, deflecting couple; — par capacité, capacity coupled.

*couplet, frame joint.

coupleur, m. make-and-break; — centrifuge, centrifugal clutch; — d'automatique d'antenne, automatic antenna tuner; — de faisceau longitudinal, longitudinal beam coupler; — de vol stationnaire, hover flight coupler; — hydraulique, hydraulic torque converter; — hydrocinétique, hydrokinetic torque converter; — moteur, engine torque.

*coupole du poste de pilotage, cockpit enclosure.

*coupure, gap break, cutoff; — à étincelle, spark gap.

*cour militaire, court-martial.

*courant, race; — absorbé, current input; — alternatif, alternate current; — à ondulations réduites, smoothed current; — d'autostriction, self-constricting current; — de chauffage, filament of heater current; — de crête, peak current; — décroissant, decaying current; — de décharge, discharge current; — demi-alternance, half-wave current; — de déplacement, dielectric displacement current; — de diffusion, diffusion current; — de flamme, chimney draft; — de Foucault, eddy current, Foucault current.

courant, — de Foucault induit, induced eddy current; — de fuite, leakage current; — de grille, grid current; — d'émission, electronic or emission current; — de pincement, pinch current; — de plaque, plate or anode current; — de plasma, plasma current; — de probabilité, probability current; — de repos, spacing or spacer current or waves, quiescent or no-signal current, anode feed current.

courant, — de retour, downflow; — de rupture, intermittent current; — de saturation, saturation current; — descendant, downflow; — de striction, pinch current; — de travail, marking or working current (signaling, teleg.); — déwatté, wattless current; — déwatté arrière, lagging wattless current; — déwatté avant, leading wattless current.

courant, — d'induit, armature current; — d'ionisation, ionization current; — d'ions, ion current; — d'oscillation d'image, framesweep current; — effectif, net current; — électrique, electric current; — électrique interne, internal electric current; — en avance, leading current; — en circuit ouvert, no-load current; — fortement chargé, heavyload current; — gazeux, gas current.

courant, — inférieur, undercurrent; — infinitésimal, infinitesimal current; — ionique, ion current; — latéral, side stream; — littoral, feeder current, alongshore current; — maximum, peak current; — net, net current; — nominal, rated current; — parasite, parasitic, leakage, dark, or stray current; — passant par l'arc, arc current; — perte à la terre, earth resistance, loss current to earth.

courant, — plasmatique, plasma current; — porteur, carrier current; — secondaire, induced current; — sinusoidal, sine-wave current; — tellurique, ground current; — unidirectionnel, unidirectional current; — vibré, interrupted current.

*courbe, cam; — à cloche, probability curve; — à un réseau paramétrique, contour; — caractéristique, working curve; — d'absorption, absorption curve; — d'activation, activation curve; — d'activité, activity curve; — d'altitude, altitude curve; — de chien, drift curve (airplane); — de croissance, growth curve.

courbe, — de décrochage des oscillations, breakoff diagram; — de décroissance, decay curve; — de descente, landing beam; — de distribution de la déviation, deflection distribution curve; — de fonctionnement aux diverses fréquences, frequency-response curve; — de fréquence, frequency response; — d'élution, elution curve; — d'énergie en fonction du parcours, energy-range curve; — de potentiel, potential curve; — de production en fonction de la masse, yield-mass curve; — de puissance, performance curve.

courbe, — de récupération, recovery curve; — de sensibilité spectrale, current-wavelength characteristic; — de survie, survival curve; — de transition, easement curve,

transition curve; — d'excitation, excitation curve; — d'ionisation spécifique, specific ionization curve; — dose-effet, dose-effect curve; — énergie-parcours, energy-range curve; — énergie-portée, energy-range curve; — gauche, space curve.

courbe, — lissée, smooth curve; — parcoursvitesse, velocity-range curve; — photométrique, polar curve of light distribution; — répartition-masse, yield-mass curve; — représentant la fraction de cohésion, packing-fraction curve; — représentant l'effect en fonction de la dose, dose-effect curve; — représentant le parcours en fonction de la vitesse, velocity-range curve; — représentative du nombre des fissions, fission-yield curve; — sigmoïde, sigmoid curve; — théorique, calculated curve.

*couronne, — de lancement, flywheel ring gear; — de sondage, core bit, core drill; core cutterhead; — et pignon, wheel and pinion; — porte-balais, brush ring or collar.

*couronné de succès, successful.

*courroie, — de ventilateur, fan belt; — trapezoïdale, v-belt.

*course, — d'admission, suction stroke; — d'allumage, ignition stroke; — d'echappement, exhaust stroke; — de compression, compression stroke; — de détente, expansion stroke; — motrice, firing stroke (automobile).

*court-circuit, — acoustique, acoustic feedback; — franc, dead short circuit.

*coussinet, — de palier, bearing liner; — d'un tourillon, pillow block; — échauffe, hot bearing; — en deux pièces, sleeve bearing; — en une pièce, bushing; — pour tige d'entrainement, drive bushing.

*cout, — de l'unité d'énergie, per unit-energy cost; — d'exploitation, operating costs; — du kilowattheure, per-unit-energy cost.

*couteau, — à tourner, tool bit (lathe); — d'alimentation, feed knife; — de poche, jackknife.

*couvage, breeding.

*couvercle, — de boite de circulation, channel cover; — d'éléments, cell cover; — insonorisé, hush cover, acoustic cover.

*couverture, overburden.

couvre-culbuteur, m. cylinder head cover, rocker cover.

*couvrir de bombes, to bomb.

Covir, m. light switch (on Renault).

crabot, m. direct-drive dog clutch.

crabotage, m. clutching, dog clutch; — du differentiel, four-wheel drive system.

*crachement, sparking, rattle (of loud-speaker).

*craindre, pas à —, no risk.

*crampon, dog (mach.).

*cran, dividing line, cog, sprocket; fond à —, milled bottom.

crantage, m. serration.

crapouillot, m. trench mortar.

craquage, m. cracking; — catalytique, cat cracking; — isomérisant, isocracking; — thermique, thermal cracking; four de —, cracking furnace.

craqueur, m. cracking plant, person given to telling tall tales; — catalytique, cat cracker.

crassier, m. dump.

*création, — d'un défaut, discomposition effect; — d'une paire, pair formation; — sous l'effet des photons, photoproduction.

*crème pour cuir, leather dressing.

*créneau de lancement, firing window, launching window.

*crête, peak; — de modulation, peak modulation.

crevaison de pneumatique, f. puncture (tires).

*cric, — à double engrenage, double-purchase jack; — à pignon et crémaillière, ratchet jack; — à une griffe, single-clawed jack; — de bord, vehicle jack; — fixé à demeure, in-built jack; — mécanique, screw jack; — rouleur, garage jack, trolley jack.

crique, creek, bay, cove, pipe, crack, split (in steel); — dans le lingot, ingot piping.

crique-soie de cochon, f. bristle.

*cristal, — embryonnaire, nucleus; — ionique, ionic crystal.

*cristallisoir, chiller.

crit, m. crit.

criticité, f. criticality.

*critique pour les neutrons retardés, delayed-critical (reactor).

*croc, claw.

*crochet, kink; — d'arrêt, catch (mach.); — de manipulation des poteaux, cant hook (elec.); — de suspension, suspension lug; — d'hélice, front rubber hook; — fixé, rubber fixing hook.

crocodile, m. crocodile (alarm contact).

*croisement, — consanguin, inbreeding; — entre germains, sib crossing.

*croisillon, cross wires, cross hairs, arm, branch (plumbing), spider (mach.).

croisillonnement souple, m. wiring cross bracing.

*crosse, — d'appontage, arrestor hook (air-craft carrier); — de piston, piston crosshead (hydraulic cylinder).

cryogénie, cryogenics.

*cuiller, bailer; — de contact, trolley shoe.

*cuire au four, to bake.

*cuisine, galley (of ship).

*cuisson, calcination (cement).

*culasse, — de l'aimant, magnet yoke; — mobile, breechblock.

culbutage, m. tumbling (aer.).

culbuterie, f. rocker arm assembly, set of rocker arms.

*culot, — à cinq broches, five-pin base; — de bougie, spark plug base; — de lampe, valve or tube base; —s d'ergols, bottom, base (astronautics).

cumuler, to accumulate, summarize.

cumulus, m. cumulus cloud.

cunette, f. heading (min.).

cupro-autunite, cuproautunite.

cupro-sklodowskite, cuprosklodowskite.

cupro-uranite, f. cuprouranite.

*cupule de concentration, concentrating cup.

*curer, to bail.

curie, m. curie.

curite, f. curite.

curium, m. curium.

*curseur, slide contact, cursor; — d'impri-mante, format control.

custode, f. rear quarter (auto body).

*cuve, calandria; — à niveau constant, con-stant-level chamber; — à tan, tanning vat; — à teindre, dyeing vat; — contenant le coeur, core tank; — de flotteur, float chamber (carburetor); — de la pile, reactor vessel; — de refroidissement, cooling jacket; — guil-loire, fermentation vat; — intérieure, core tank; — sous pression, pressure vessel; — tampon, buffer tank.

*cuvette, outer race (rolling bearing), sump (of a furnace); — d'huile, sump.

cybernétique, f. cybernetics.

cyclage thermique, m. thermal cycling.

*cycle, — complet, total cycle; — de base, time base; — de chauffage, heating cycle; — de la photosynthèse, photosynthetic cycle; — d'élution, elution cycle; — de machine, ma-chine cycle; — d'empoisonnement, poison-ing cycle; — d'épuisement du combustible, burnout cycle; — de retard, cycle delay; — de surgénération, breeding cycle; — de tabu-lation, non-listing cycle; — de transfert de chaleur, heat-transfer cycle; — de transformation du combustible nucléaire, nuclear fuel cycle.

cycle, — d'hystérésis, hysteresis loop; — d'oxydoréduction, oxidation-reduction cycle; — élémentaire, minor cycle; — multiple, compound cycle; — neutronique, neutron cycle; — ouvert, open cycle, open circuit; — principal, major cycle; — solaire de Bethe, Bethe cycle, carbon-nitrogen cycle; — thermique, thermal cycle; — thermonucléaire, thermonuclear cycle.

cyclogyre, m. paddle-wheel airplane.

*cyclone de dégazage, degasser cyclone.

cyclotron, m. cyclotron.

*cylindre, — creux, hollow cylinder; — de frappe, hammer platen; — de frein, recoil cylinder (artil.); — partagé, split cylinder; — récepteur, wheel cylinder (brakes); — récepteur (embrayage), slave cylinder (clutch); — sans réflecteur, bare cylinder; — soufflant, blast cylinder.

cyrtolite, f. cyrtolite.

cytoplasme germinal, m. germ plasm.

D

D, m. dee (cyclotron).

*dalots, m.pl. scupper.

daltonien, m. color-blind person.

dandinement, m. wobble, shimmy (of wheels, etc.).

*danger dû à l'irradiation, radiation hazard.

*danse, jigging.

*dartre, lump (scale).

datage radioactif, m. dating.

datation, f. dating; — par le radiocarbone, radiocarbon dating; — par les isotopes, isotopic dating.

date-cible, f. target date.

date-limite, f. deadline date.

davidite, f. davidite.

*dé, dee (cyclotron).

*débarquer, to discharge.

débattement, m. struggling, displacement, travel, deflection, spring movement; — de la roue, wheel hop.

débimètre, m. flowmeter.

*débit, stream, flow, throughput; — critique de sortie (à), output-bound; — de gaz, gas flow; — d'entrée, inlet; — de vapeur, steam flow; — en binaire(s), bit rate.

*debiter à travers, to work across.

debiteur, m feed spool, pulldown spool (film).

déblocage, m. unlocking (of screw, nut), freeing the line (R.R.).

débloquer, to break (a joint).

débloqueur de trépan, m. bit breaker.

débobiner, to uncoil, wind off.

*débordement, overrun condition; — de piste, track overrun; — inférieur, underflow; — supérieur, overflow.

*débouché, mouth (of loudspeaker).

*debout, ahead.

débrancher, to switch off, switch out.

débrayable, disconnectable.

*débrayage, disengagement, throwout (of a gear or clutch); — et embrayage automatique, automatic gearshift.

*début d'exécution, start-up.

décaèdre, m. decahedron.

*décalage, shift, slipping (of film), lag (elec., phys.); — arithmétique, arithmetic shift; — brusque, rapid phase change; — circulaire, end-around shift; — dans le temps, time lag; — de phase, phase shift; — des ailes, stagger of the wings; — logique, logical shift (EDP); — simple, single shift (EDP).

décalaminage, m. carbon removal, decoking, descaling.

décalaminant, anti-carbon; carburant additionnel —, anti-carbon fuel.

décalé, out of step, offset; — en phase sur le courant, differing in phase from the current.

decalescence, f. decalescence.

*décantateur, settling tank.

*décantation, settling.

*décapage, pickling; — électrochimique, electrochemical pickling.

décapotable, m. convertible car.

décélération, f. deceleration.

*décharge, disposal, head, volley (mil.); — à filament de plasma, linear pinch discharge; — à forte intensité, high current discharge; — à forte intensité dans un gaz, high-current gas discharge; — à lueur, glow discharge; — dans les gaz, gaseous discharge; — dans un gaz sous striction, pinched gas discharge; — en aigrette, brush discharge.

décharge, — en arc, arc discharge; — induite toroïdale, toroidal induced discharge; —

luminescente, glow discharge; — **pulsée,** pulsed discharge; — **pulsée de grande puissance,** high-power pulsed discharge; — **sans striction,** pinch discharge; — **terrestre,** ground disposal.

*****décharger,** to dump (EDP).

*****déchets,** waste disposal, tailings; — **à faible activité,** low-level waste, cool waste; — **chimiques,** chemical waste; — **liquides,** liquid waste; — **radioactifs,** radioactive waste.

déchiffrer, to decode.

*****décimal,** — **codé binaire,** binary coded decimal; — **condensé,** packed decimal.

*****déclaration à la sortie,** clearance.

déclenché, opened, tripped, off (of switch).

*****déclenchement,** tripping, trigger action; **tube thermionique de** —, thyratron type of tube, triggering or tripping tube.

*****déclencher,** to trip.

déclencheur, *m.* trigger release switch; — **d'alimentation,** feed control unit; — **jumelé,** trigger pair.

décliqueter, to trip.

décoder, to decode.

décodeur, *m.* decoder.

décohéreur, *m.* decoherer.

*****décollage,** take-off; — **court,** short-start run; — **vent arrière,** take-off with the wind.

*****décollement des filets d'air,** breakaway of flow.

décolleur, *m.* stripper tool.

*****décoloration par contact,** contact filtration.

*****décomposition (radioactive) en chaine,** chain decay (series decay).

*****décompression,** pressure relief.

décomprimer, to unpack (EDP), decompress.

déconnecteur, *m.* disconnecting link or switch.

décontamination, *f.* decontamination; — **de plutonium,** decontamination of plutonium.

*****découpage,** — **à la presse,** blanking, press cutting; — **au chalumeau,** flame cutting; — **en dés,** dicing (EDP); **matrices pour** —**s,** cutting dies.

*****découpeur,** chopper.

découplage, *m.* decoupling.

découpoir, — **à main,** hand punch (mach.); — **à tôle,** plate-punching machine.

*****découvert,** bare.

*****décrément,** decrement (math.); — **d'énergie,** energy decrement; — **logarithmique d'éner-**

gie, logarithmic energy decrement; — **moyen d'énergie,** average energy decrement.

décrémètre, *m.* decremeter.

*****décrochage,** deenergization, breaking step.

décroché, off-hook.

*****décrocher,** to stop (oscillations), come out of step, fall out of step (as a motor).

*****décroissance,** — **radiative,** radiative decay; — **radioactive,** radioactive disintegration, radioactive decay.

*****décroissement,** damping, attenuation, decrement.

*****décroître,** to shrink.

*****dedans, en** — **du bord,** inboard (naut.).

dee, *m.* dee (cyclotron).

*****défaut,** — **de Frenkel,** vacancy-interstitial pair (Frenkel defect); — **de masse,** packing loss, mass correction, mass defect; — **d'isolement,** fault; — **du réseau,** lattice defect.

*****défectueux,** unsound.

défense d'entrer! keep out!

*****défilage,** releasing (of brake).

*****déflecteur,** flow baffle, ventilator window, baffle; — **d'air,** cooling baffle; — **de ventilateur,** fan cowl, fan shroud.

*****déflexion, sensibilité de** —, deflection sensitiveness (of beam).

défonceuse, *f.* ripper, heavy plow, trenching plow; — **portée,** ripper; — **tractée,** rooter.

*****déformation,** — **au col,** deformation at saddle point; — **critique,** critical deformation; — **d'origine thermique,** thermal strain; — **du col,** point deformation saddle; — **en ellipsoïde allongé,** prolate distortion; — **en ellipsoïde aplati,** oblate distortion; — **limite,** critical deformation; — **résiduelle,** spring-back.

défournement, *m.* unloading.

défreinage, *m.* unlocking (screw or nut).

*****dégagement,** — **de chaleur,** heat production; — **d'énergie,** release of energy; — **d'énergie par réaction,** energy release per reaction; — **d'énergie se propageant par lui-même,** self-propagating release of energy; — **gazeux,** gas evolution; — **instantané,** inrush, sudden outburst.

*****dégager,** reset, roll out (EDP); — **la vapeur,** to blow off (steam); — **le grenier,** to clear the attic.

*****dégâts résultant de l'irradiation,** radiation damage.

dégauchi, flush (even).

dégazage, *m.* gas removal.

dégazeur, *m.* deaerator.

dégivrer, to de-ice.

dégivreur, *m.* de-icer.

*__dégradation,__ — **d'énergie,** degradation of energy; — **par irradiation,** radiation damage.

*__degré,__ quality; — **de liberté,** degree of freedom; — **de liberté du gaz,** gas kinetic degrees of freedom; — **de pénétrance,** degree of penetrance; — **d'ionisation,** fractional ionization, degree of ionization; **du second** —, quadratic; **n'étant pas du premier** —, nonlinearity.

degréer, to dismantle.

*__dégrossisseur,__ coarse filter.

dégroupement, *m.* dispersal, unpack (EDP).

*__dehors, en__ — **du service,** off duty.

déjanter, to remove tire from rim.

déjeté, warped (wood).

*__dejeter,__ to distort.

délayé, diluted.

délétion chromosomique, *f.* chromosome deletion.

déliasseur, *m.* decollator.

*__délimitation,__ confinement; — **longitudinale,** longitudinal confinement; — **magnétique,** magnetic constriction, magnetic confinement; — **par pincement,** pinch confinement; — **par striction,** pinch confinement.

delorenzite, delorenzite.

*__démanteler,__ to clear.

*__démarrage,__ start, releasing; — **à chaud,** warm start; — **à froid,** cold start; — **automatique,** self-starting.

dématérialisation, *f.* dematerialization, pair conversion.

demi-additionneur, *m.* half-adder.

demi-clef, *f.* half-hitch; — **à capeler,** clove hitch.

demi-coussinet, *m.* half-bearing (sleeve bearing).

demi-essieu, *m.* half-shaft; — **oscillant,** swinging half-shaft.

demi-impulsion, *f.* half-pulse.

demi-largeur, *f.* half-width.

demi-looping, *m.* half loop.

demi-onde, *f.* half wave length.

demi-produits, *m.pl.* intermediate products.

demi-solde, *f.* half pay.

demi-tonneau, *m.* half roll.

démodulateur, *m.* demodulator.

démoduler, to redress.

démonté, knocked down, disassembled, disconnected.

démoraliser, to demoralize.

démouleur, *m.* ingot stripper (metal.).

*__démultiplicateur,__ vernier arrangement.

démultiplication, *f.* division.

démultiplié, geared down.

*__dénaturation du combustible nucléaire,__ denaturation of nuclear fuel.

dénébulateur, *m.* fog dispersal device.

dénébuler, to disperse the fog.

*__dénominateur,__ **même** —, common denominator.

*__densité,__ specific gravity; — **de charge,** charge density; — **de charge électronique,** electron charge density; — **de charge ionique,** ion charge density; — **de collision,** collision density; — **de courant,** current density; — **de deutérons,** deuteron density; — **de l'énergie de fusion,** fusion power density; — **d'énergie,** energy density; — **d'énergie magnétique,** magnetic energy density.

densité, — **d'enregistrement,** packing density; — **de particules du plasma,** plasma density; — **de probabilité,** probability density; — **de ralentissement,** slowing-down density; — **de rayonnement,** radiation density; — **de remplissage,** packing density; — **des électrons,** electron density; — **des neutrons,** neutron intensity, neutron density; — **des particules,** particle density; — **des particules parasites,** background particle density; — **d'ionisation,** ionization density.

densité, — **du cliché photographique,** photographic density; — **du flux neutronique,** neutron flux density; — **du gaz,** gas density; — **électronique,** electron density; — **en binaire(s),** bit density; — **gravimétrique,** packing density; — **ionique,** ion density; — **par unité de surface,** areal density; — **surfacique,** surface density, areal density.

densitomètre, *m.* densitometer.

*__dent,__ sprocket; — **d'induit,** armature tooth; **en** — **de scie,** saw-tooth shaped; **tension de balayage en** —**s de scie,** saw-tooth sweep potential.

*__dent-de-loup,__ hand crank jaw, starting crank.

*__denture,__ pitch (gears).

dépannage, *m.* repairing, trouble shooting.

dépanneur, *m.* repairman, troubleshooter.

*__déparer,__ to trip, release.

*__départ,__ start; — **du synchronisateur de régénération,** start regeneration timer.

dépassement, *m.* excess, overshooting, overflow, overflow ejection; — **de capacité,** overflow; — **du taux normal,** manifold increase; **tarif à —,** overload tariff.

***dépendance, — de l'énergie,** energy dependence; **— du temps,** time dependence.

***dépendant des variables spatiales,** space dependent.

***dépense,** exhaustion, use; **— continue d'énergie,** continuous expenditure of energy; **—s en combustible,** fuel costs.

déphasage, m. phase lag, phase shift; — **caracteristique,** phase constant; **— en arrière,** lag; **— en avant,** lead; **modulation par —,** outphasing modulation, phase modulation.

déphaser, to dephase.

déphaseur, m. phase shifter, dephaser.

dépistage, f. detection, seek (EDP); **— des pannes,** trouble shooting; **— immédiat,** direct seek.

***dépister,** to seek (EDP).

***déplacement,** drift motion; **— de charge,** charge displacement; **— de termes,** term displacement; **— infinitésimal,** infinitesimal displacement; **— latéral infinitésimal,** infinitesimal lateral displacement; **— virtuel,** virtual displacement.

dépliant, m. pamphlet, folder, brochure; **carte —,** folding map.

déport, m. offset, off-centering, mismatch, retransmission (radar).

dépose, f. removal, demounting (motor), helicopter landing.

déposé, registered.

***déposer,** to precipitate, jettison.

***dépôt,** accumulation or storing (of charges), registration; **— électrolytique,** electrodeposition; **— emplisseur en vrac,** filling station; **— par électrolyse,** electrodeposition; **— radioactif,** radioactive deposit.

***dépotage,** discharge.

***dépoussiérage,** air cleaning, dust collection.

dépoussiéreur, m. dust extractor.

dépressuriser, to off-load.

déprimomètre, m. draft gauge.

***dérangement,** failure, dislocation (of service), state of being out of order.

déraper, to slip.

***dérivation,** tap, shunt, branch, parallel (connection); **en —,** abreast.

***dérive,** fin, leeway, discrepancy, disparity, drift motion; **— de phase,** phase shift; **— de puissance,** power drift; **— de reactivité,** reactivity drift; **— du centre de guidage,** guiding center drift; **— d'une particule,** par-

ticle drift; **— du plasma,** plasma drift; **— fissile,** fissionable derivative.

dérive, — génétique, genetic drift; **— gravifique,** gravitational drift; **— induite d'une particule,** induced particle drift; **— linéaire,** linear drift; **— par effet Doppler,** Doppler shift, Doppler displacement; **— radioactif,** daughter product; **— transversale,** transverse drift; **en —,** adrift.

***dérivée,** variation (of frequency); **— par rapport au temps,** time derivative.

***dériver,** to branch (off), bridge.

dériveur, m. centerboard (naut.).

dérivomètre, m. drift indicator.

déroule-cartes, m. roller map case.

***déroulement,** run (EDP).

***dérouler,** to feed(out), pay out, unwind or pull down (film).

dérouleur, m. handler, magnetic tape deck, magnetic tape unit; **— de bande,** tape drive, magnetic tape unit.

***désaccord,** detuning.

désaccordage, m. detuning, mistuning.

désaccordé, untuned.

***désaccorder,** to untune, cause to be off resonance.

désactivation du combustible, f. fuel cooling.

désalignement de la poussée, m. thrust misalignment.

***désamorçage,** stopping of oscillations, deenergization; **— d'une tuyère,** unpriming (astronautics).

***désarmé,** laid up (naut.).

***désarmorcer,** to deenergize.

désaxage, m. misalignment, throwing off center.

***descendance,** progeny.

***descendant,** daughter element, daughter product.

***descente,** downlead (radio); **— d'antenna,** antenna downlead; **— des rapports,** downshift; **— par la pesanteur,** extension under own weight.

désembuage, m. demisting.

désembuer, to demist.

désemparé, disabled.

déséquerrage, m. out of squareness.

désexcitation, f. deenergization, deexcitation.

désexciter, to deenergize.

déshabillage, m. stripping.

déshuileur, m. oil trap, oil separator.

*désinfectant, deodorant.

*désintégration, — alpha, alpha-particle disintegration, disintegration with emission of alpha particles; — artificielle des noyaux, artificial nuclear disintegration; — atomique, atomic disintegration; — avec émission de neutrons, neutron decay; — avec émission de posit(r)ons, positron decay, disintegration with emission of positrons; — β⁺, positron decay, positron disintegration, disintegration with emission of positrons; — bêta, beta decay, beta transformation, beta-particle disintegration, disintegration with emission of beta particles.

désintégration, — du noyau, nuclear disintegration; — en chaine, chain decay (series decay); — exponentielle, exponential decay; — nucléaire exothermique, exothermic nuclear disintegration; — radiative, radiative decay; — radioactive, radioactive decay, radioactive disintegration; — radioactive artificielle, artificial radioactive disintegration; — spontanée, spontaneous decay, spontaneous disintegration; —s successives, series decay.

desintoxication, f. decontamination.

desmodromique, liaison —, connection by positively acting means, connection under constraint.

desmogène, m. linkage editor (EDP).

désorption, f. desorption.

*desserrer, to unscrew.

desservi, served, furnished with.

*desservir la mitrailleuse, to operate the machine gun.

*dessin, blueprint, format, layout; — coté, dimensioned drawing; — de carte, card design, card layout (EDP); — de fiche, record layout; — linéaire, mechanical drawing.

dessinateur, m. draftsman, designer.

*dessous, en —, undershot (hydr.).

destructeur, m. destroyer (naut.).

désuet, out of date, outdated.

*détacher, to detail (mil.).

*détail, specification.

*détartreur, tube cleaner, coke knocker.

détectable, detectable; — par machine, machine sensible.

*détecteur, foil, detector, sensor, probe, scanner; — à cristal, crystal detector; — à feuille, foil detector; — antidéflagrant, flameproof monitor; — à seuil, threshold detector; — de fuites, leak detector, leakage detector; — de la radioactivité de l'air, air monitor; — de neutrons, neutron detector; — de radioactivité, monitor.

détecteur, — des particules de l'air, air particle detector; — donnant l'intensité du flux des neutrons, neutron level detector; — local, area monitor; — métallique, metal foil; — mince, thin foil; — ponctuel, point detector.

*détection, — de la radioactivité, detection of radioactivity; — de particules nucléaires, detection of nuclear particles; — des neutrons, detection of neutrons; — des particules radioactives en suspension dans l'air, airborne particle detection; — du rayonnement, detection of radiation.

détectrice à grille écran, f. screen-grid detector tube.

*détente, flash, pressure relief.

detenteur, m. winner.

*détérioration de l'aptitude génétique, deterioration of genetic fitness.

*détermination absolue, absolute assay.

détonance, f. knocking, detonation.

*détonateur, blasting cap.

*détonation, engine knocking; — prématurée, predetonation.

détruit, wrecked, destroyed, ruined.

deutérium, m. deuterium; — gazeux, deuterium gas.

deutéron, m. deuteron; — cible, target deuteron; — de grande énergie, high-energy deuteron; — jouant le rôle de cible, target deuteron.

deutérure, f. deuteride; — de lithium, lithium deuteride.

développement en série, m. series expansion.

déverminage, m. burn-in (astronautics).

*déversoir mobile, movable sprayer.

*déviateur, divertor; — de jet, flow baffle.

*déviation, alternate routing, drift, refraction; — de faisceau, beam deflection; — de l'image, frame sweep; — des ions, ion deflection; — du zéro, zero error; — linéaire, line sweep; — totale au comparateur, total indicator reading.

dévideur, m. feed spool, pull-down spool.

dévidoir, m. feed spool, aerial winch; — sur roues, hose reel on wheels.

*devis, valuation.

*dévisser, to break (a joint).

dévolteur, m. negative booster.

déwatté, wattless.

dewindtite, dewindtite.

*dextrine, artificial gum.

diabolo, m. dolly.

diagnographe, *m.* diagnotor.

*diagramme, — de charge, load curve; — de connexions, plugging chart; — de diffraction, diffraction pattern; — de directivité, directional pattern, space pattern, directional diagram or characteristic; — de l'indicateur, indicator graph; — de rayonnement, radiation pattern; — des phases, phase diagram; — des tensions et des allongements, stress-strain diagram; — vecteur, vector diagram.

*diamètre, — de collision, minimum distance of approach; — du noyau, nuclear diameter.

diapason, *m.* pitch (sound), tuning fork.

diapasonnage, *m.* modulation.

diaphonie, *f.* cross talk.

*diaphragme (de microscope), baffle.

diderichite, *f.* diderichite.

diergol, *m.* bipropellant.

dièse, *m.* sharp (mus.).

*différé (en), off-line.

*différence, — de marche, — de parcours, — entre les chemins des ondes, path-length difference; — de phase, phase lag; — de potentiel, potential difference; — de pression, pressure difference; — moyenne des températures, mean temperature difference (MTD); — moyenne logarithmique des température, logarithmic mean temperature difference.

différentiel, *m.* incremental, differential; — différentiel des essieux de pont AR, inter-axle differential (trucks).

*differer, to distinguish.

*diffraction des neutrons, neutron diffraction.

*diffuser, to scatter.

*diffuseur, acoustic radiator, diffusion nozzle, diffuser, scatterer, cone loud-speaker; — factice, dummy diffuser; — multicellulaire, multicellular horn; — ponctuel, point scatterer.

*diffusion, scattering; — aléatoire, random scattering; — ambipolaire, ambipolar diffusion; — classique, classical scattering; — Compton, Compton scattering; — de la chaleur, diffusion of heat; — de neutrons, scattering of neutrons; — de particules, scattering of particles.

diffusion, — des électrons, electron scattering; — des mésons pi par les nucléons, pion nucleon scattering; — des neutrons, diffusion of neutrons; — des neutrons de résonance, resonance scattering; — des neutrons thermique, thermal diffusion column; — désor-

donnée, random scattering, incoherent scattering; — des particules, particle diffusion; — élastique, elastic scattering.

diffusion, — élastique composée, compound elastic scattering; — élastique due à la forme, shape elastic scattering; — en avant, forward scattering; — étant fonction de l'angle d'incidence, angular dependence of scattering; — gazeuse, gaseous diffusion; — incohérente, incoherent scattering; — inélastique, inelastic scattering; — isotrope, isotropic scattering.

diffusion, — moléculaire, free molecule diffusion, molecular diffusion; — moyenne, average scattering; — multiple, plural scattering, multiple scattering; — paramagnétique, paramagnetic scattering; — par choc, collisional diffusion; — par collision, collisional diffusion; — quasi-élastique, quasi-elastic scattering; — radiale, radial diffusion; — simple, single scattering; — sous grand angle, large angle scattering; — thermique, thermal diffusion.

*digue, sea wall, breakwater.

*diguer, to impound (hydr.).

delacération, *f.* tearing.

*dilatation, thermal expansion; — du plasma, expansion of the plasma.

diluant, *m.* diluent.

*dilution isotopique, isotopic dilution.

*dimension, — linéaire, linear dimension; — radiale, radial dimention; à une —, one-dimensional; —s critiques, critical size; —s du réseau, lattice dimensions; —s hors-tout, overall dimensions; —s physiques, physical dimensions.

*diminution, — de l'intensité neutronique, neutron attenuation, neutron beam attenuation; — de pression, pressure relief; — linéaire de réactivité, linear decrease in reactivity; — relative de l'énergie, energy decrement.

dineutron, *m.* dineutron.

diode-tétrode, *f.* binode.

diode-triode, *f.* binode.

dipole, *m.* — électrique, electric dipole; — magnétique, magnetic dipole.

*direct (en), on-line.

*direction, flow direction; — à crémaillière, rack and pinion steering; — à recirculation de billes, recirculatory ball steering; — assistée, power steering; — à vis et galet, worm and sector steering; — du champ, field direction; — régionale, district headquarters.

*directive, declarative instruction (EDP); — d'opérateur, operator command.

*discontinuité, — d'absorption, absorption discontinuity, absorption edge, absorption limit; — énergétique, energy gap.

*discordance, unconformity.

discorde, *f.* word circuit.

discothèque, *f.* disk library, record library.

discotrope, disk oriented.

discriminateur de l'amplitude des impulsions, *m.* pulse-height discriminator.

*disjoncteur de l'alternateur, generator breaker.

*dislocation, — coin, edge dislocation; — partielle, partial dislocation.

dispensateur, *m.* distributor, dispenser; — de cartes, card hopper.

*dispersion, — aléatoire, statistical straggling, straggling; — aléatoire du parcours, range straggle; — aléatoire en direction, angle straggling; — classique, classical scattering; — diffuse, diffuse scattering; — frontale, fore scattering; — par diffraction, diffraction scattering; — potentielle, potential scattering.

*disponible, preadjusted.

*dispositif, gear, equipment, operating feature, contrivance; — à commutation, switchable device; — ajouté, attachment; — antibruit commandé par l'onde porteuse, codan; — antiparasite, interference-suppression device; — automatique de surveillance, monitoring device; — compensateur de l'évanouissement, automatic volume-control circuit or system; — d'alimentation, feeder device; — d'amplification pour conférences publiques, public-address system; — d'antigivrage, ice guard (aviation); — de branchement sur le réseau, power pack or unit, supply-line connecting means.

dispositif, — de calage, phasing or phase-shifter device; — de commande, steering mechanism, control device; — de commande finale, final control element; — de commutation, switch gear; — de compensation, compensation apparatus; — de comptage des impulsions, pulse counting system; — de conditionnement, signal conditioning element; — de dégagement, clearing device; — de dérivation de la vapeur, steam bypass; — de flottabilité, flotation gear; — de gonflement automatique, automatic inflation device; — de manoeuvre du changement de pas, pitch-control mechanism.

dispositif, — de mesure, measuring equipment; — de mise en réservation, hold facility (EDP); — de montage, circuit organization or arrangement; — de multiplication par une constante, constant multiplier coefficient unit; — d'encaissement, coin box; — d'enregistrement des données, data storage; — d'entraînement, feed device; — de préhension, gripper; — de rattrapage d'usure, wear takeup device (brakes); — de réglage, set point station; — de remplissage, refueling device; — de repliage, retracting mechanism; — de saut de colonne, column shift unit; — de sauvetage, lifesaving equipment; — de secours pour l'arrêt, emergency shutdown system; — de sécurité, safety device.

dispositif, — (de stabilisation) commandé par les courants vocaux, voice-operated device (antisinging); — de suralimentation, supercharger (aviation); — de synchronisation, timing system, timer; — de télémanipulation, remote handling equipment; — d'homme mort, dead-man's handle (elec.); — d'immobilisation, locking device; — d'intégration, integrating amplifier; — hypersustentateur, lift-increasing device; — pare-étincelles, jump-spark system; — pneumatique d'irradiation, "rabbit" (pneumatically operated sample tube).

dispositif-indicateur de colonne, *m.* column indication device.

*disposition, — à symétrie sphérique, spherical symmetry, spherical geometry; — de fiche, record layout; — du réseau, lattice design, lattice arrangement; — type, representative array.

disqualifier, to disqualify.

*disque, planchet; — à fente, slotted disk; — à lentilles, lens disk, lens-scanning disk; — analyseur, scanning disk; — d'appel, dial; — de balayage, scanning disk; — d'embrayage, clutch driven plate; — d'exploitation, systems pack; — entraîné, clutch driven plate; — explorateur à spirale multiple, multispiral scanning disk; — interrupteur perforé, chopper disk; — sustentateur, actuator disk, lifting disk (helicopter).

disque-programmothèque, *m.* library disk (EDP).

dissident, *m.* outsider, dissenter.

*dissipation, — de l'énergie, dissipation of energy; — volumique de la chaleur engendrée par effet Joule, Joule heat volume dissipation.

*dissociation, — des gaz, gaseous dissociation; — gazeuse, gaseous dissociation.

*dissolution, dissolving.

*distance, — à vol d'oiseau, distance as the crow flies; — d'atténuation, attenuation distance; — de Debye, Debye shielding distance; — de proximité, minimum distance of approach; — **entre deux points,** distance between two points; — **explosive,** sparking distance, spark length; — **restante,** residual range; à —, remote (control).

*__distillat de tête,__ overhead product.

*__distillation,__ — **atmosphérique,** topping (petroleum refining); — **azéotropique,** azeotropic distillation; — **discontinue,** batch distillation; — **en discontinu,** batch distillation; — **intermittente,** batch distillation; — **moléculaire,** molecular distillation.

*__distributeur rotatif,__ rotating feeder.

*__distribution,__ cast, casting (TV, movies, etc.); — **à la came,** cam gear; — **à main,** hand shift (automobile); — **angulaire,** angular distribution; — **asymptotique du flux,** asymptotic flux distribution; — **continue des vitesses,** continuous velocity distribution; — **d'admission,** inlet governor; — **de données,** data organization; — **de l'émission,** source distribution; — **de l'irradiation,** dose distribution; — **de Maxwell,** Maxwell distribution; — **des électrons froids,** cold electron distribution.

distribution, — **des électrons lents,** cold electron distribution; — **des niveaux,** level distribution; — **de vitesses,** velocity distribution; — **de vitesses de Maxwell,** Maxwell velocity distribution; — **du courant,** current distribution; — **du flux,** flux distribution; — **du flux à l'intérieur d'une cellule (du réseau actif),** intercell flux distribution; — **électrique de l'heure,** electrical distribution of time; — **en énergie,** energy distribution; — **énergétique,** energy distribution.

distribution, — **énergétique des neutrons,** neutron energy distribution; — **en étoile,** radial system; — **en fonction du temps,** time distribution; — **isotrope des vitesses,** isotropic velocity distribution; — **maxwellienne,** Maxwell distribution; — **maxwellienne des vitesses,** Maxwell velocity distribution; — **spatiale,** spatial distribution; — **stationnaire,** steady-state distribution; — **stationnaire des neutrons,** steady-state neutron distribution.

*__divergence,__ criticality, angular divergence, focusing, toe-out (wheels), deviation; — **d'un champ,** divergence of field; **entrer en** —, to start a chain reaction (nuclear reactor).

*__divergent sous l'effet des neutrons instantanés,__ prompt critical.

*__diviseur,__ — **de tension,** potentiometer; — **de tension à remplissage gazeux,** glow-gap divider.

*__division,__ splitting; — **de la chromatide,** chromatid break; — **réductrice,** reduction division.

djalmaïte, djalmaite.

dock, *m.* dock.

*__doigt,__ — **d'allumeur,** distributor arm; — **d'avance,** feed finger; — **de débrayage,** clutch release lever; — **de rétraction,** shortening hook; — **de verrouillage,** locking pin.

dollar, dollar (unit of reactivity).

*__domaine,__ area, range, array; — **des énergies thermiques,** thermal energy region; — **des instructions,** instruction area; — **de tir,** firing envelope; — **de vol,** flight envelope; — **limité,** closed array (EDP).

*__dôme,__ — **d'échelle,** companionway (naut.); — **de prise de vapeur,** plenum chamber.

*__dommages dus au bombardement,__ bombardment damage (by particles).

*__donnée,__ —**s brutes,** raw data; —**s de contrôle,** control data; —**s significatives,** control data.

*__donner,__ to turn on; — **du manche à droite,** to apply right bank; — **du manche à gauche,** to apply left bank.

*__dose,__ dosage; — **absorbée,** absorbed dose; — **admissible,** tolerance dose, permissible exposure, permissible dose; — **affectée du coefficient d'efficacité biologique relative (EBR),** — **(en rems),** RBE dose (relative biological effectiveness of radiation); — **à la peau,** skin dose; — **cumulative,** cumulative dose; — **cutanée,** skin dose; — **dans l'air,** air dose; — **de doublement,** doubling dose; — **de seuil,** threshold dose.

dose, — **de sortie,** exit dose; — **de tolérance,** tolerance dose; — **d'exposition,** exposure dose; — **d'irradiation,** exposure dose, radiation dosage; — **EBR,** RBE dose; — **érythématique,** erythema dose; — **globale,** integral dose, volume dose; — **globale absorbée,** integral absorbed dose; — **gonades,** gonad dose.

dose, — **hebdomadaire admissible,** permissible weekly dose; — **individuelle,** personal dose; — **intégrale,** integral dose; — **intégrale absorbée,** integral absorbed dose; — **létale,** lethal dose (LD); — **létale médiane,** median lethal dose, minimal lethal dose (MLD); — **maximum admissible,** maximum permissible exposure; — **profonde,** depth dose; — **tissulaire,** tissue dose.

*__doser,__ to adjust (as volume), graduate.

*__doseur,__ proportioner.

dosimètre, *m.* dosimeter; — **à film,** film dosimeter; — **à taux de comptage,** counting-rate dosimeter; — **de poche,** pocket meter; — **intégrateur,** integrating dosimeter.

*__dosimétrie physique,__ physical dosimetry.

*dossier, record, backrest; — inclinable, adjustable seatback.

*double, — allumage, dual ignition; — chape, twin clevis; — commande, dual control; — commande débrayable, disconnectable dual control; — désintégration bêta, double beta decay; — diabolo, four-wheel bogie; — empennage vertical, dual fins and rudders; — flux, by-pass section; — monoplan, tandem monoplane; — pas, double thread; — perforation, double punch; — plancher, false floor; — sensibilité, autopilot pitch sensitivity system.

*doublet, diplet, dipole (as an antenna); — isotopique, isotopic doublet; — quart d'onde, quarter-wave dipole; —s conjugués, matched doublets.

*douille, lampholder, eye, turnbuckle, box (mach.), jacket; — à emboîtement, bell socket; — de jack, jack bush; — taraudée, threaded insert; — tendeur, turnbuckle barrel.

*drainer, to deplete.

*dresser, — en tables, to tabulate; — selon la table alphabétique, to index, alphabetize.

drisse, f. halyard, cable, cord; pavillion à mi—, ensign hoisted partway.

*droit, true, exact.

*droit, — sur la vente, royalty (payment); au — de, opposite, in front of.

*droite, — de balisage, beacon course; à —, clockwise.

drôme des embarcations, f. boathouse.

drosophile, f. Drosophila.

dual, dual; — de contrôle de parité, parity check bit; — de données, information bit; — de modification, modifier bit; — de parité, parity bit; — de ponctuation, punctuation bit; — de protection de mémoire, storage protection bit; — de sign, sign bit; — de vérification, check bit; — de zone, zone bit.

dumontite, f. dumontite.

dunette, f. deckhouse (naut.).

duplex, à — effet, double-acting.

duplicateur, m. duplicating device.

*durcissement, — du spectre neutronique, neutron hardening; — par déformation, strain hardening; — par revenu, temper hardening; — par trempe, quench hardening; — par vieillisement, age hardening; — structural, age hardening; — superficiel, case hardening.

*durée, endurance; — comparative, comparative lifetime; — de battement, period of heat; — de croissance, growth time; — de la retombée, releasing time; — de l'état excité, lifetime of excited state; — de l'impulsion, pulse duration.

durée, — de l'irradiation, length of exposure, duration of irradiation; — de parcours des électrons, electron transit time; — de réaction, reaction time; — de retour, reset time; — d'escale, turn-around time; — de stockage, shelf life; — de suppression, decay time; — d'établissement, building-up time;— de vie des neutrons, neutron lifetime; — moyenne de la réaction, mean reaction time; — sclérométrique, abrasive hardness; — totale de vol, total flying time.

durite, f. hose, flexible connection.

dyade (operateur tensoriel), f. dyad.

*dynamique des plasmas, plasma dynamics.

dynamiseur, m. dynamicizer.

*dynamo, — à balai auxiliaire, third-brush generator; — composée, compound-wound generator; — de charge, battery-charging generator.

dynamo-démarreur, f. dynamotor, generator-starter.

dynamoteur, m. dynamotor.

E

*eau, — d'appoint, make-up water; — de gisement, oil water; — de glace, ice water; — de javelle, Javelle water, sodium hypochlorite; — de refroidissement, cooling water, water coolant; — d'hydratation, water of hydration; — forte, aqua fortis; — fossile, connate water; — liée, bound water.

eau, — lourde, mud, heavy water; — ordinaire, light water; — radioactive, radioactive water; — radioactivée, activated water; — résiduaires, liquid waste; — sous-jacente, bottom water; — sous pression, pressurized water; — sus-jacente, top water; — vannes, liquid waste.

*ébarbures, pl. erosion.

*ébauche, blank; — coulée, blank or rough casting; — forgée, blank or rough forging; — matricée, drop forging.

ébavurage, m. fettling, deburring, trimming; — au tonneau, tumbling.

*éblouissement, glare, blending (optics, acoustics); — acoustique, aural dazzling.

*ébullition superficielle, surface boiling.

*écaillement, spalling.

*écart, scarf; — de réglage, offset (electronics); — des vitesses de guérison, differential recovery rate;— de vitesse, speed range; — moyen quadratique, standard deviation, root mean square deviation; — permanent, offset (electronics); — tolérable, tolerance; — type, standard deviation; — transversal, guide margin.

*écartement, — de la voie, track gauge; — des contacts, contact clearance; — des contacts de rupteur, contact breaker points gap; — des électrodes, spark gap; — des niveaux, level spacing; — des plaques, plate spacing; — des plateaux, plate spacing; — des roues, wheel base.

*échafaud, trestle, frame.

*échange, — catalytique, catalytic exchange; — chimique, chemical exchange; — chromosomique, chromosome exchange; — de charge, charge exchange (in plasma); — d'énergie, energy exchange; — entre chromosomes, chromosome exchange; — ionique, ion exchange; — isotopique, isotope exchange.

échangeur, m. — de chaleur, heat exchanger; — de chaleur primaire, primary heat exchanger; — d'ions, ion exchanger; — secondaire de chaleur, secondary heat exchanger; — thermique, heat exchanger.

*échantillon, form; — de référence, representative sample; — moyen, average sample, all-level sample; — non-représentatif, nonrepresentative sample; — représentatif, representative sample; — témoin, check sample.

*échappement de vapeur, puff (of steam).

*échauffement sulfurique, acid heat test.

*échéance, deadline, due date, target date.

*échelle, — annulaire, ring scaler; — automatique, autoscaler; — binaire, binary scaler, scale of two; — circulaire, ring scaler; — de charge, dead-weight scale; — décimale, decade scaler; — de comptage, scaled radiation detector, scaler; — de corde, rope ladder; — de coupée, accommodation ladder; — de deux, scale of two; — de dix, decade scaler.

échelle, — de léthargie, lethargy scale; — de marée, tide gauge; — d'embarquement, entrance ladder; — de réduction, reduced scale; — des temps, time scale; — des temps contractée, fast-time scale; — des temps dilatée, extended time scale; — divisée en pouces, inch scale; — double, step-ladder; à l'—, according to scale.

*échoué, aground, failed.

éclaboussure, f. splash, dab.

*éclairci, illustrated.

*éclaireur, gas detector.

*éclatement, splitting; — du noyau, nuclear burst.

*éclater, to break (down), disrupt.

*éclateur, — à étincelles interrompues, quenched spark gap; — déchargeur, spark gap; — fixe, simple, plain spark discharge.

*écluse, air lock, gate (of electron microscope).

*économie, — due au réflecteur, reflector saving; — neutronique, neutron economy.

économiseur, m. economizer.

éconostat, m. economizer (carburetor).

*écorce, — de Panama, quillai bark; — de quinquina, cinchona bark; — électronique, outermost electronic shell.

écorché, m. sectional view, cut view.

*écoulement, — à contre-courant, countercurrent flow; — compressible, compressible flow; — continu, streamline flow; — continu à contre-courant, continuous countercurrent flow; — d'un fluide incompressible, incompressible flow.

écoulement, — gazeux, gas flow; — laminaire, laminar flow, streamline flow; — parallèle, parallel flow; — préférentiel, preferential flow; — tourbillonnaire, vortex-type flow; — turbulent, turbulent flow.

écouloir, m. discharge outlet.

écoute, f. sheet (of sail), listening post, sentinal, reception; — locale, sidetone; — sonar, sonar search; cabine d'—, monitor room (movies), listening booth (records).

écoutille, f. hatch (naut.).

écouvillon, m. flue brush, tube brush, sponge (ordn.).

*écran, shield, shutter (optics), filter, screening box, back; — acoustique, baffle board; — à gaz, gas-fire screen; — calorifuge, thermal barrier; — cathodique, cathode screen; — capacitif, electrostatic shield; — de cimentation, cimenting basket, fishing tool for catching loose pieces of iron; — de plomb, lead screening; — fluorescent, fluorescent screen; — multicellulaire, lamp screen; — protecteur, protective screen; — renforçateur, intensifying screen; — thermique, (thermal screen, heat shield.

*écrasement, collapsing.

ecrêter, to chop, clip (electricity).

écrêteur, m. amplitude limiter, chopper, clipper; — à ligne de retard, delay line clipper.

*écrire à la machine, to typewrite.

*écriture, — avec rassemblement, gather write; — en clair, plain writing; — rassemblée, gather write.

*écrou, — à créneaux, castle nut; — autobloquant, self-locking nut; — borgne, capnut, acorn nut; — crénelé, castellated nut; — de blocage, locknut; — de fixation, adjusting nut; — de réglage, adjusting nut; — de retenue, retainer nut; — de serrage, retainer nut; — de vis mère, clasp nut; — en cage, floating anchor nut; — haut, heavy or thick nut; — HKZ, castellated nut; — HKZ bis, thin castellated nut; — HZ bis, hexagonal nut to specification AN-364; — mobile, plunger nut; — moleté, knurled nut; — papillon, wing nut; — raccord, union nut; — taraudeur, die nut.

écrou-raccord, m. union nut.

écroui, cold worked, work-hardened.

écuanteur, m. dish (wheels).

écubier, m. hawsehole, hawsepipe, grommet; — d'amarrage, mooring ring; — d'embossage, cathole; — de remorque, towing chock.

*édifice nucléaire, nuclear structure.

*effacer, to clear, delete, erase, overwrite; — la mémoire, to clear storage, clear memory (EDP).

effectifs, m.pl. complement (naut.).

*effectuer la réaction, to couple back.

*effet, — biologique de l'irradiation, biological effect of radiation; — centrifuge, centrifugal effect; — chimique, chemical effect; — collectif, collective effect; — couplé, coupled effect; — d'altitude, altitude effect; — de blocage, blocking effect; — de bord, board effect, ship field error (in direction finding), quadrantal error (due to structural parts, metal); — de bruit Schottky, shot effect; — de canalisation, channeling effect.

effet, — de cohésion, packing effect; — de corona, — de couronne, corona discharge, corona effect; — de couplage, coupled effect; — d'écran, screening, screening effect, shielding effect; — d'écran électronique, electron screening; — de déviation, deflecting effect; — de dilution, dilution effect; — de fission par les neutrons rapides, fast fission effect; — de grêle, shot effect.

effet, — de grenaille, shot effect, shot noise; — de Kelvin, skin effect, Kelvin effect; — de l'irradiation, radiation effect; — de masque, masking; — de Schottky, Schottky effect, shot effect, shot noise; — de scintillation, flicker effect; — de surface, surface effect; — de talon, heel effect; — de tassement, packing effect; — de tension superficielle, surface tension effect.

effet, — de trainage, coupling hysteresis effect; — de transfert, transfer effect; — de transition, transition effect; — de volume, volume effect; — d'induction, induction effect; — d'inertie, inertial effect; — d'ionisation, ionization effect; — d'ombre, self-screening, shielding effect; — d'ombre relativement à l'énergie, energy self-shielding; — Doppler, Doppler effect, Doppler displacement, Doppler shift.

effet, — dû à la masse, mass effect; — dû à la parité, parity effect; — dû au spin, spin effect; — dû aux charges d'espace, space-charge effect; — dû aux charges spatiales, space-charge effect; — électrique transitoire localisé, localized electrical transient effect; — électrostatique de grande portée, long-range electrostatic effect; — hydrodynamique de choc, shock-hydrodynamic effect; — imputable aux forces de liaison, binding effect; — imputable aux liaisons chimiques, chemical binding effect.

effet, — inhibiteur, suppressor effect; — isotopique, isotope effect; — Kelvin, skin effect; — pelliculaire, skin effect; — photoélectrique, photoelectric effect; — photoélectrique dans les couches de barrage, barrier-layer photoelectric effect; — photonucléaire, photonuclear effect; — radiobiologique, radiobiological action, radiobiological effect; — soupape, valve effect; — synergétique, synergistic effect.

effet, — thermo-électrique, thermoelectric effect; — toxique, poisonous effect; — transitoire, transient effect; — tunnel, tunnel effect; — utile de compteur, counter efficiency; — Wigner, discomposition effect, Wigner effect.

effets, — d'un empoisonnement transitoire, transient poison effects; — électrocinétiques, electrokinetic effects; — génétiques, genetic effects; — génétiques de l'irradiation, — génétiques des rayonnements, radiation genetics, genetic effects of radiation; — physiologiques des neutrons, physiological effects of neutrons.

*efficace, active.

*efficacité, — biologique relative, relative biological effectiveness; — chimique, chemical efficiency; — du malaxage, mixing efficiency; — totale, over-all efficiency.

effilotage, m. unraveling (of threads).

effluence, f. exhaust.

*effluent, — à haute activité, high-level waste; — radioactif, radioactive effluent.

*effluve, glow discharge, corona, silent or dark discharge; — électrique, brush discharge.

*effort, — à la jante, tractive effort (autos); — à l'ouverture, shock load on opening (parachute); — de tirage, pull; — de traction du crocket, drawbar pull; — limite de cisaillement, critical shear stress; — longitudinal au manche cyclique, longitudinal cyclic stock load (helicopter); — mécanique, mechanical stress; — tranchant, vertical shear (beams); — tranchant mobile, running shear load; —s de commande de pas, pitch control loads (helicopter).

*égalisation, smoothing (a current).

égalisatrice à courant continu, f. direct-current balancer.

*égaliseur de pression, pressure equalizer.

*égalité de charge, charge equality.

*éjecter, to knock out.

*éjecteur, jet pump.

*éjection, extrusion; — de cartes, card release.

*élargir, to underream.

*élargissement, broadening (of spectral lines); — Doppler, Doppler broadening.

élargisseur, m. underreamer; — creux, hollow reamer.

*électricité, electric power; — d'origine atomique, atomic power; — provenant des centrales nucléaires, nuclear power.

électrisé, charged.

électrobate, m. electric automobile.

électrocardiogramme, m. electrocardiograph.

électrocinétique, m. electrokinetics.

*électrode, — collectrice, collecting electrode; — d'accélération, accelerating electrode; — de freinage, retarding (field) or reflecting electrode (in Barkhausen tube).

*électrolyse en circuit fermé, cyclic electrolysis.

*électromètre à lame vibrante, vibrating-reed electrometer.

*électron, negatron; — à augmentation continue d'énergie, runaway electron; — célibataire, lone electron; — Compton, Compton electron; — de conversion, conversion electron; — de conversion interne, internal conversion electron; — de désintégration bêta, beta-decay electron; — de désintégration nucléaire, disintegration electron; — de grand vitesse, high-speed electron; — de passage, transit electron.

électron, — de recul, recoil electron; — de valence, valence electron, bonding electron; — dextrogyre, right-polarized electron; — du plasma, plasma electron; — emballé, runaway electron; — extra-nucléaire, extranu-

clear electron; — lévogyre, left-polarized electron; — libre, free electron.

électron, — parasite, background electron; — périphérique, outer electron; — planétaire, planetary electron, orbital electron; — satellite, orbital electron, planetary electron; — secondaire, secondary electron; — tournant, spinning electron; — -volt, electron-volt; —s non apparies, unpaired electrons; —s non couples, unpaired electrons.

électronégativité, f. electronegativity.

électronique, f. electronics; — aérospatiale, avionics.

électrophorèse, f. electrophoresis.

électropositif, -ive, electropositive.

*électroscope protégé contre le rayonnement, shielded electroscope.

électrovanne, f. solenoid valve, electromagnetic sluice gate.

*élément, — à anticoïncidence, exclusive or element, non-equivalence element; — à coincidence, equivalence element; — actif, seed unit, "meat"; — à façon, custom element; — à l'état de traces, trace element; — à nombre magique de nucléons, magic number nuclide; — ascendant, parent element; — assemblé, package (electronics); — combustible métallique, metallic fuel element; — d'accumulateur, battery cell; — d'aire, areal element; — d'alliage, alloying element; — de batterie, battery cell; — de bloc, blockette; — de combustible, fuel unit; — de formation directe, shielded nuclide; — de groupe, group item; — de liaison, connector element; — de matrice, matrix element.

élément, — de pile, cell; — de programme, module; — de réduction, regulating cell; — de substitution, vicarious element; — détecteur, sensing element; — de volume, element of volume; — du dispositif de commande et de réglage, control element; — en dérivation d'un filtre, shunt element of a filter; — enfichable, plug-in unit; — engendré, daughter element; — fertile, fertile element.

élément, — instable, unstable element; — léger, light element; — linéaire, line element; — logique, logical element; — logique de combinaison, combinational logic element; — mécanoïde, module; — naturel, natural element; — Non, negater; — non rebutable, salvageable item; — précurseur, parent element; — protégé, shielded nuclide; — radioactif, radioactive element, radioactive nuclide; — rendu artificiellement radioactif, artificially radioactive element; — séparateur, separative element, separative unit; — simple, elementary item; — transuranien, transuranic element.

élément-filtre, *m.* filter.

élément-seuil, *m.* threshold element.

*__élévateur__, hydraulic lift; — à **fourche**, forklift truck; — de **tension**, voltage multiplier; — **mixte**, adaptable lift.

*__élévation instantanée de température__, instantaneous temperature rise.

élève-modelliste, *m.* pupil of a model-aircraft school.

*__élever au carré__, to square (math.).

élève-vélivoleur, *m.* sailplaning pupil.

eliasite, *f.* eliasite.

*__élimination__, disposal; — **définitive**, permanent disposal; — des **déchets**, waste disposal; — des **produits de fission**, fission product disposal.

*__éliminer__, to delete, filter out.

élingue, *f.* sling (naut.).

ellsworthite, ellsworthite.

*__élongation__, deflection (of needle).

éluant, eluent.

éluat, eluate.

élutriation, *f.* elutriation.

*__emballement__, racing (motor), runaway; — de la **pile**, nuclear runaway, reactor runaway.

embardé, involving error in magnetic compass (northern turning error).

embarder, to yaw (naut.).

*__emboîtement__, nesting; — à **force**, driving fit; — **libre**, sliding fit.

*__embouchure__, inlet, throat (or a horn).

*__embranchement__, multiple decay, multiple disintegration.

embrasure de chauffe, *f.* stokehole.

*__embrayage__, — a **béquille**, sprag clutch; — à **crabots**, claw clutch; — à **disque**, disk clutch; — à **endentures**, jaw clutch; — à **roue libre**, free wheel clutch; — à **roue libre à galets**, overrunning roller clutch; — **centrifuge**, centrifugal clutch; — du **P.A.**, autopilot engagement; — **magnétique**, magnetic drive; — **monodisque à sec**, single-disk dry clutch; — **par cônes de friction**, cone clutch; — **par engrenage**, gear drive.

*__embrèvement__, dimpling, dimple, groove-and-tongue joint; — à **chaud**, hot dimpling process; — à **froid**, cold dimpling process.

*__embrocher__, to connect (elec.).

*__émeraude__, emerald.

*__émetteur__, antenna, emitter; — à **étincelles étouffées**, quenched-spark or gap transmitter; — à **exploration par rayon lumineux**, lightbeam or spotlight transmitter; — à **grande puissance**, high-power transmitter; — **alpha**, alpha radiator, alpha emitter; — à **ondes entretenues**, continuous-wave transmitter; — à **prise de vue directe**, daylight television transmitter; — à **relever**, beacon from which bearings are to be taken.

émetteur, — **brouilleur**, jamming station; — de **fumée**, smoke generator; — **d'électrons**, electron emitter; — **demi-point**, half-time emitter; — de **neutrons retardés**, delayed-neutron emitter; — de **positons**, positron emitter; — de **rayonnement**, radiator; — de **télécinema**, film transmitter.

émetteur, — de **télévision à film intermédiaire**, intermediate film transmitter (telev.). — **dirigé**, directional transmitter; — **intérieur**, internal emitter; — **interne**, internal emitter; — **plan**, plane source; — **radiophonique**, radiotelephone transmitter; — **stable**, stable emitter.

émetteur-récepteur de données, *m.* data transceiver.

*__émettre__, to radiate, radio.

*__émiettement du noyau__, pulverization of the nucleus.

émissaire, *m.* discharge outlet.

*__émission__, — **alpha**, alpha-particle emission; — **anisotrope**, anisotropic emission; — **corpusculair associée**, associated corpuscular emission; — **d'électrons**, electron emission; — **d'électrons secondaires**, secondary electron emission; — de **mésons**, meson emission; — **d'énergie**, emission of energy; — de **neutrons**, emission of neutrons, production of neutrons; — de **positons**, positron emission.

émission, — de **rayons bêta**, beta-ray emission; — **dirigée**, beam emission; — **du champ**, field emission; — **d'une paire**, pair emission; — **mésique**, meson emission; — **photonique**, photon emission; — **quantique**, quantum emission; — **successive de particules** β, successive emission of beta particles; — **thermoélectronique**, thermionic emission; — **s non-essentielles**, spurious radiation.

*__emmagasinement__, stockpiling.

*__emmanché à chaud__, hard driven.

*__emmanchement__, fit; — **dur**, interference or tight fit; — **sapin**, fir-tree root (motor blade).

*__emmener un passager__, to take up a passenger.

empan, *m.* span (of bridge).

*__empattement__, wheelbase.

*__empennage__, tail unit.

empilement, *m.* stack.

*emplacement, — **de perforation,** code position; — **de pièce,** gun emplacement; — **de signe,** sign position; —**s du réseau dépourvus de barreaux,** vacant lattice positions; —**s protégés,** protected locations; —**s vacants du réseau,** vacant lattice positions.

*emplâtre, tire patch.

*empoisonnement, poisonous effect; — **de la pile,** pile poisoning, poisoning of reactor; — **de l'uranium,** uranium poisoning; — **dû aux neutrons de résonance,** resonance poisoning; — **dû aux neutrons instantanés,** prompt poison; — **par les produits de fission,** fission-product poisoning.

*émulsion, — **au gelatino-bromure d'argent,** silver bromide emulsion; — **nucléaire,** nuclear emulsion; — **photographique,** photographic emulsion; — **sous l'effet des photons,** photoemulsion.

émulsionneuse, *f.* emulsifier.

énalite, *f.* enalite.

*encadrer, to span.

encastré, countersunk.

*encastrement, encasing.

*encastrer, to secure, hold, anchor (in a matrix material).

*enceinte, — **de confinement,** containment (nuclear reactor); — **de protection de télé-thérapie,** teletherapy shield; — **étanche,** pressure vessel; — **étanche aux gaz,** gastight housing.

*enchaîner, to concatenate, catenate.

enchevêtrement des signaux, *m.* interlocking of signals.

enclenché, engaged, closed, on (of switch).

*encliquetage, latch, locking device; — **à dents,** ratchet.

*encollage pour apprêts, finishing.

*encoller, to solder.

*encombrement, over-all dimension, floorspace, space requirement; — **hors-tout,** over-all dimensions.

*encre, — **corrosive,** etching ink; — **de chine,** India ink; — **magnétique,** magnetic ink.

endogame, inbred.

*endogamie, inbreeding.

endommagement, *m.* damage.

endostéum, endosteum.

*enduit, — **lait de ciment,** grout; — **tendeur,** stiffening dope (aviation).

*énergie, — **accumulée,** stored energy; — **atomique,** atomic energy; — **au repos,** rest energy; — **au repos d'une particule,** particle

rest energy; — **calorifique,** heat energy, thermal energy; — **chimique,** chemical energy; — **cinétique angulaire,** angular kinetic energy; — **cinétique d'agitation thermique moyenne,** average thermal kinetic energy; — **cinétique moyenne,** average kinetic energy, mean kinetic energy.

énergie, — **cinétique relative,** relative kinetic energy; — **cinétique relative initiale,** initial relative kinetic energy; — **cinétique totale,** total kinetic energy; — **correspondant à la température cinétique,** kinetic temperature energy; — **critique,** critical energy; — **d'absorption des neutrons de résonance,** energy of resonance absorption; — **d'activation,** activation energy; — **d'attraction,** attractive energy; — **de bombardement,** bombarding energy; — **de chaleur,** thermal energy.

énergie, — **d'échange,** exchange energy; — **de déformation limite,** critical deformation energy; — **de déplacement,** translational energy; — **de désintégration à l'état fondamental,** ground-state disintegration energy; **de désintégration nucléaire,** disintegration energy; — **de dissociation,** dissociation energy; — **de fission,** fission energy, fission power; — **de fusion,** fusion energy, fusion power; — **de la réaction,** reaction energy, reaction power; — **de liaison,** binding energy (of nucleus).

énergie, — **de liaison de neutron,** neutron binding energy; — **de liaison nucléaire,** nuclear binding energy; — **de liaison par nucléon,** packing fraction; — **de liaison par particule,** binding energy per particle; — **de particularisation,** specialization energy; — **de rayonnement,** radiation energy, radiated power; — **de répulsion,** repulsive energy; — **de rotation,** rotational energy; — **des ions,** ion energy.

énergie, — **des neutrons,** neutron energy; — **des neutrons de résonance,** resonance energy; — **des neutrons épithermiques,** epithermal energy; — **de sortie,** output energy; — **de translation,** translational energy; — **de vibration,** vibrational energy; — **d'excitation,** excitation energy; — **d'excitation du plasmon,** plasmon energy; — **d'injection,** injection energy; — **d'interaction,** interaction energy.

énergie, — **d'ionisation,** ionization energy, ionizing energy; — **du col,** saddle-point energy; — **d'une particule,** particle energy; — **électrique,** electric power; — **électromagnétique,** electromagnetic energy; — **émise,** radiated power; — **emmagasinée,** stored energy; — **globale des particules,**

total particle energy; — **gravifique**, gravitational energy; — **initiale**, initial energy.

énergie, — **interne**, latent energy, internal energy; — **intrinsèque**, intrinsic energy; — **ionique**, ion energy; — **latente**, latent energy; — **libérée**, transmitted energy; — **libre**, free energy; — **moyenne**, average energy, mean energy; — **moyenne de la particule**, mean particle energy; — **moyenne de la réaction**, mean reaction energy; — **moyenne de l'électron**, mean electron energy; — **nucléaire**, nuclear energy, nuclear power.

énergie, — **perdue**, energy loss; — **photonique**, photon energy; — **potentielle**, potential energy; — **potentielle coulombienne**, Coulomb potential energy; — **potentielle mutuelle de Coulomb**, mutual Coulomb potential energy; — **propre**, characteristic energy, proper energy, self-energy; — **radiante**, radiant energy; — **rayonnante**, radiant energy; — **relative**, relative energy; — **relative des particules**, relative particle energy; — **solaire**, solar energy.

énergie, — **spécifique**, specific energy; — **spécifique de fonctionnement**, working power density; — **spécifique de réaction**, reaction power density; — **spécifique totale de la réaction**, total reaction power density; — **stellaire**, stellar energy; — **superficielle**, surface energy; — **totale**, total energy; — **totale de liaison**, total binding energy; — **transformée**, converted energy; — **volumique**, volume energy.

enficher, to plug in.

*é**nfoncé**, embedded.

***enfoncement**, pressing down (of push button).

***enfouissement**, (ground) disposal.

***engagé**, waterlogged.

***engager**, to roll in (EDP).

***engendrer**, to send, emit.

***engin**, — **spatial**, spacecraft; — **tête chercheuse**, homing missile.

***engrenage**, — **à cremaillère**, rack and pinion, rack gear; — **pour étagement de vitesse**, reduction gear.

enjambement, *m.* crossing over.

enjoliveur, *m.* hubcap; — **de feu**, light rim; — **de moyeu**, hubcap.

enjolivure, *f.* small embellishment.

***enlever**, to hew; — **la pièce**, to clip; — **les dépôts d'une chaudière**, to scale a boiler.

***énoncé**, — **de problème**, problem description.

***enregistrement**, — **à densité constante**, variable-width recording; — **à densité variable**, variable-density recording; — **absent**, no record found; — **ajouté**, addition record; — **d'annulation**, deletion record; — **d'appoint**, overflow record; — **de largeur d'impulsion**, pulse-width recording; — **de longueur**, variable field; — **direct**, direct recording; — **du son**, sound recording; — **fractionné**, rerecording, multiplay-back, multiplay; — **magnétique**, magnetic recording; — **sonore**, sound recording; — **sur bande**, tape-recording; — **sur disques**, sound-on-disk recording; — **transversal**, variable-width track (sound film); — **-s de longueurs variables**, variable-length records; **passer un —**, to play a recording.

***enregistrer**, to store.

***enregistreur mécanique**, mechanical register.

***enregistreur-régulateur**, recorder-controller.

enrôler, to enlist.

***enroulement**, — **à pas entier**, drum winding with diametral pitch; — **à pas fractionnaire**, drum winding with fractional pitch; — **à pas raccourci**, drum winding with shortened pitch; — **d'inducteur**, magnetic field coil; — **imbrique**, lap winding; **à double —**, double-wound (elec.).

enrouleur, *m.* take-up spool (film), jockey pulley, winding wheel, reel; — **à inertie**, inertia reel, shoulder harness reel.

***enseignement**, — **automatisé**, computer assisted instruction; — **micrograduё**, programmed instruction; — **séquentiel**, programmed instruction.

ensellement, *m.* saddle (low part of anticline).

***ensemble**, assembly, set, series, array, system, equipment; — **actif**, core assembly; — **à maille rectangulaire**, rectangular array; — **critique**, critical assembly; — **de coeur**, core assembly; — **de données**, data set; — **de données concaténé**, concatenated data set; — **de grilles**, grid assembly; — **de la cible**, target assembly; — **de recombinaison**, recombiner assembly; — **des barreaux**, rod assembly; — **électrocomptable**, tabulating equipment; — **électronique**, electronic data processing machine; — **électronique industriel**, control system; — **local**, in-house system; — **muni d'un réflecteur**, reflected assembly; — **spécialisé**, dedicated system.

***entail**, aperture, slit (film).

***entassement**, embankment.

***en-tête**, header, heading; — **de section**, section header.

enthalpie, *f.* heat content.

***entonnoir à décantation**, separatory funnel.

*entraînement, drive, pulling into tune, entrainment, pull-in, practice, training; — **à cliquet,** pawl and ratchet drive; — **au vol,** flight training; — **de l'hélice,** propeller drive; — **des barres,** rod drive; — **du film,** film feed; — **par ergots,** pin-feed platen; — **par vis sans fin,** worm drive (automobile).

*entraîneur, carrier; — **de rétention,** holdback carrier; — **isotopique,** isotopic carrier; — **non-isotopique,** nonisotopic carrier; **sans** —, carrier-free.

entrait, *m.* tie beam.

entrecycle, *m.* intercycle.

*entrée, input; — **à labyrinthe,** radiation maze; — **analogique,** analog input; — **analogique dissymétrique,** single ended analog input; — **de filet,** thread lead-in; — **différée,** defferred entry; — **différentielle,** differential input; — **dissymétrique,** single ended input; — **étanchée,** hermetic leadin, seal; — **manuelle,** manual input; — **numérique,** digital input; — **prioritaire,** process interruption.

entrée-sortie, *f.* input-output; — **directe,** direct control; — **industrielle,** process input-output, front-end input-output.

*entrefer, pole gap.

entrefiche, *f.* inter-block gap, inter-record gap, record gap.

entrepont, *m.* between decks (naut.).

*entrer, — **en collision,** to collide; — **en jeu,** to become operative.

*entretien, program maintenance (EDP); — **correctif,** corrective maintenance; — **de fichier,** file maintenance; — **de programme,** program maintenance; — **d'oscillations non-désirées,** singing (elec.); — **préventif,** preventive maintenance; — **sous l'eau,** underwater maintenance; — **systématique,** preventive maintenance, scheduled maintenance.

*entretoise, girth, girt, spacer; — **articulée,** flexible stay bolt; — **de grille,** grid spacer.

*enveloppe, can; — **conductrice,** conducting layer; — **étanche aux gaz,** gastight casing; — **régénératrice,** breeder blanket.

envideur, *m.* **envidoir,** *m.* take-up spool or reel (film).

*envol par vent de côté, cross-wind take-off (aviation).

*épaisseur, — **d'arrêt équivalente,** stopping equivalent; — **d'eau,** water wall; — **d'eau constituant l'obturateur,** water shutter; — **d'eau formant bouclier,** water shield; — **de plomb équivalente,** lead equivalent.

épaisseur-moitié, *f.* half-thickness, half-value thickness.

épandeuse, *f.* spreader.

*épanouissement polaire, pole piece.

éphantinite, *f.* ephantinite.

épicadmique, apicadmium.

épilamen, *m.* boundary film.

*épilation, epilation.

épiphyse, *f.* epiphysis.

épissoir, *m.* marlinespike.

épissure, *f.* splice.

épithélioma cutané, *m.* cutaneous epithelioma.

épontille, *f.* stanchion (naut.).

*épreuve, image (motion pictures), negative (of film), copy; — **de rendement,** benchmark problem; — **saute-mouton,** leap-frog test; —**s de tournage,** rushes (TV and movies).

*éprouver, to check out, taste; — **avec le frein,** to check.

*épuisement, stripping, fatigue.

épurage, *m.* debugging.

*épuration, editing; — **de l'air,** air cleaning; — **des gaz,** gas cleaning, gas scrubbing.

*équation, — **asymptotique de la pile,** asymptotic reactor equation; — **aux dérivées partielles,** partial differential equation; — **aux différences,** difference equation; — **canonique,** canonical equation; — **cinétique,** kinetic equation; — **critique,** critical equation; — **de continuité,** equation of continuity; — **de cyclotron,** cyclotron equation; — **de diffusion,** diffusion equation; — **de diffusion en fonction de l'âge,** age-diffusion equation.

équation, — **de gain,** gain equation; — **de l'âge,** age equation; — **de l'hydrodynamique,** hydrodynamical equation; — **de masse relativiste,** relativistic mass equation; — **de mouvement,** equation of motion; — **de réaction,** nuclear-reaction equation, reaction formula; — **d'état,** equation of state; — **d'induction électromatique,** transformer equation; — **d'onde,** wave equation; — **du deuxième ordre,** second-order equation.

équation, — **du premier ordre,** first-order equation; — **du transformateur,** transformer equation; — **en théorie à plusieurs groupes,** multigroup equation; — **linéaire,** linear equation; — **macroscopique,** macroscopic equation; — **mise sous forme linéaire,** linearized equation; — **non-linéaire,** nonlinear equation; — **par rapport au temps,** secular equation; — **ramenée au premier degré,** linearized equation; — **relative à la pile,** reactor equation; — **séculaire,** secular equation.

*équerre, — à épaulement, back square; — en fer, iron bracket.

*equilbrage, equalization (generator-starter), balancing forces; — des roues, wheel balancing.

*équilibre, — céto-énolique, keto-enol equilibrium; — de pression, pressure equilibrium; — de pression cinétique, kinetic pressure balance; — hydrostatique, hydrostatic equilibrium; — isotopique, isotopic equilibrium; — métastable, metastable equilibrium; — mutation-sélection, mutation-selection balance; — passager, transient equilibrium.

équilibre, — radiatif, radiation equilibrium; — radioactif, radioactive equilibrium; — séculaire, secular equilibrium; — stable, stable equilibrium; — thermique, temperature equilibrium, thermal equilibrium; — thermodynamique, thermodynamic equilibrium; — thermodynamique cinétique, kinetic thermodynamic equilibrium; — transitoire, transient equilibrium.

équilibreur, m. equalizer, balance.

*équipage, moving element; — astatique, astatic pair; — mobile, moving-coil system.

équipartition de l'énergie, f. equipartition of energy.

*équipe d'extinction, fire-fighting party.

*équipement, hookup; — à arbre à cames, camshaft gear; — à combinateur, control gear; — auxiliaire, auxiliary equipment; — classique, conventional equipment; — courant, standard feature; — de contrôle, checkout system; — de survie, survival kit; — de vie, life support equipment; — normal, standard feature; — périphérique, peripheral equipment.

equipement-satellite, m. peripheral equipment.

*équivalence, — de l'énergie et de la masse, mass-energy equivalence; — masse-énergie, mass-energy equivalence.

*équivalent, bi-conditional; — d'affaiblissement, attenuation equivalent; — du roentgen, roentgen equivalent; — énergétique, energy equivalent; — en masse, mass equivalent; — en plomb, lead equivalent.

erf (fonction d'erreur), erf (error function).

ergol, m. propellant.

ergolier, m. fuel man.

*ergot, pin, catch, dowel, spigot; — conique, taper spigot; — d'entrainement, drive pin; — de positionnement, locating pin.

*erreur, — d'arrondi, rounding error, truncation error; — d'échantillonage, aliasing error; — de dérive au faisceau, stand-off beam error; — de parité, bus-out check; — de tri, missort; — équilibrée, balanced error; — nocturne, night error; — reportée, inherited error; — statistique, statistical error.

erreur-machine, f. machine bug.

*éruption, blowout, gusher.

escadre, f. wing, squadron; — de bombardement, bomber wing; — de chasse, fighter wing; — de ligne, battle fleet.

escadrille, f. squadron; — de protection, home-defense squadron.

escale, f. intermediate landing (aviation); — obligatoire, compulsory landing place.

*escalier, — en colimaçon, — en hélice, winding stairs; — mobile, escalator.

escamotable, retractable.

escamotage, m. intermittent or shuttle mechanism (in motion pictures); — par plaques tombantes, drop system of changing plates.

éschynite, f. eschynite.

*espace, — à aimant, magnet space; — de groupement, drift space, phase-focusing space; — de phase, phase space; — des charges, charge space; — des paramètres, parameter space; — d'ionisation, ionized space; — du spin isobarique, isobaric space, isotopic space, isotopic spin space; — extra-atmosphèrique, outer space; — interstitiel, pore space; — libre, clearance; — lointain, deep space; — mort, clearance; — neutre, dead volume; — réduit, adjacency.

*espacé, spaced.

*espacement arrière, backspace.

espace-temps, m. space-time.

*espèce, — atomique, atomic species; — de fission, fission species; — nucléaire, nuclide, nuclear species.

*espérance, — de fission itérée, iterated fission expectation; — de vie, life expectancy.

esquichage, m. squeeze job.

esquiche, f. squeeze, forced injection.

esquicher, to squeeze, inject under force.

*essai, — au banc, test-bench running; — au chalumeau, blowpipe analysis; — aux tiges, drill-stem test; — chimique, chemical assay; — climatique, environmental test; — d'adhérence, bonding test; — d'arrachement, peeling test; — de consommation, consumption test; — de durée, endurance test; — de dureté à l'indentation, indention hardness test; — de dureté par rebondissement, rebound hardness test; — de maquette, model test; — **de masse,** bulk test (of shield); — **de réception,** acceptance test; — **des matériaux,** materials testing; — **d'étanchéité,** leak testing; — **magnétique à la limaille,** magnetic-particle inspection.

essai, — **de volume,** bulk test (of shield); — **d'homologation,** acceptance test; — **en pile,** in-pile test; — **en pique,** diving test (aviation); — **en soufflerie,** wind-tunnel test; — **en vraie grandeur,** full-scale test; — **par lot,** sampling; — **physique,** physical assay; **en cours d'** —s, undergoing tests; — **saute-mouton modifié,** crippled leap-frog test.

essai-type, *m.* test case.

*****essence,** — **lourde,** naphtha, (highly) volatile hydrocarbon; — **très volatile,** wild gasoline.

*****essieu,** — **AR,** rear axle; —**AV,** front axle; — **coudé,** dropped axle; — **de poulie,** block pin; — **fixe,** dead axle; — **flottant,** floating axle; — **moteur,** drive axle; — **tournant,** live axle.

essuie-glace, *m.* windshield wiper.

*****estampe,** die.

estomper, to stump.

*****établissement,** mounting.

*****étage,** stage (of a cascade); — **pilote,** master stage; — **séparateur,** buffer stage, isolator stage; — **symétrique,** push-pull stage; — **théorique,** theoretical stage; — **unique de diffusion,** single diffusion stage; **à plusieurs** —s, multistage, multistep; **à un seul** —, single-stage.

*****étain marcassite,** soldering lead.

*****étalement,** debranching (of beam); — **des raies,** Doppler broadening.

*****étalon,** — **à bouts,** end standard; — **à traits,** marking line standard; — **de fréquence,** frequency generator; — **de radioactivité,** radioactive standard.

*****étalonnage,** — **de la bande,** band spread; — **d'erreur,** severity code; — **en réactivité,** reactivity calibration; — **relatif de sources de neutrons,** intercalibration of neutron sources.

*****étalonner,** to adjust.

étambot, *m.* post, stern (naut.).

*****étampe,** boss (mach.), form.

*****étanche,** leakproof; — **à l'eau,** impermeable; — **aux neutrons,** neutron-tight.

*****étançon,** stanchion.

*****étape,** step, program step; — **de travail,** job step.

*****état,** — **critique,** criticality; — **d'arrêt,** stopped state; — **d'attente,** wait state; — **de déséquilibre,** nonequilibrium state; — **de fonctionnement,** operating state, running state; — **de la question,** progress report; — **de la situation,** progress report; — **de marche,** operating state; — **de nonéquilibre,** nonequilibrium state; — **d'équilibre,** equilibrium state; — **de rotation du noyau,** nuclear rotational state; — **d'excitation,** excited state; — **d'interruption,** interrupt mode; — **d'oxydation,** oxidation state; — **du plasma,** plasma state; — **dynamique,** dynamic state.

état, — **fondamental,** ground state; — **interruptible,** interruptable state; — **masqué,** masked state; — **métastable,** metastable state; — **opérationnel,** running state; — **où il se produit des réactions en chaine,** chain reacting state; — **plasmatique,** plasma state; — **quantique,** quantum state; — **stationnaire,** stationary state, steady state; — **transitoire,** transient state; **en bon** —, in order; **en** — **de vol,** ready to fly; — **un,** *one* state; — **zéro,** *zero* state, reset state; —**s en attente,** stacked status.

état-problème, *m.* problem state.

état-programme, *m.* program state.

état-superviseur, *m.* supervisor state.

étayage, *m.* shoring.

*****étendre l'antenne,** to lower the antenna, to wind out the antenna.

*****étendue de mesure,** effective part of scale, instrument range.

ethylglycol, *m.* ethylene glycol.

*****étincelle,** — **d'allumage,** ignition spark; — **de trainage,** creeping or sneaking spark; — **disruptive,** flashover; — **éclatante,** jump spark.

*****étiquette,** tag, sticker.

*****étiré à dur,** hard-drawn.

*****étoile nucléaire,** nuclear star.

étoquiau, *m.* spring bolt or pin, stop pin.

*****étouffoir,** spark blowout.

*****étoupage,** lining (mach.); — **du piston,** leather gasket.

*****étrangeté,** strangeness number.

*****étranglement,** throat (Venturi tube); — **du courant,** current constriction.

*****étrier,** clevis, yoke; — **d'attache,** U-bolt; — **de fixation,** U-bolt; — **de levage,** link; — **de ressort,** spring U-bolt, spring clamp.

*****étude,** survey; — **technique de la pile,** reactor design; —**s à l'aide de traceurs,** tracer studies; —**s à l'aide d'indicateurs,** tracer studies; —**s et recherches sur les procédés de traitement,** process development.

*****étuve sechoire,** drying kiln.

euxénite, *m.* euxenite.

*évacuation, disposal; — définitive, permanent disposal; — des déchets, waste disposal.

*évaluation de la longévité, life expectancy.

*évanouissement des signaux, fading.

*évaporation nucléaire, nuclear evaporation.

*évasement, flare.

*événement ionisant, ionizing event.

éventail, m. fan, fanlight; en —, fan-shaped.

*examens, — globulaires, globules examinations; — hématologiques, hematological examinations.

*excédent, — d'énergie, excess energy; — de neutrons, isotopic number, neutron excess; — de réactivité, excess reactivity; — de réactivité dû aux neutrons instantanés, excess prompt reactivity; — de réactivité dû aux neutrons retardés, excess delayed reactivity; — de section efficace, excess cross section.

*excitation, excitation; — discrète, discrete excitation; — en dérivation, shunt excitation; — par choc, impulse or shock excitation; — thermique, thermal excitation.

excitatrice, f. exciter.

exclusivité, f. scoop (journalism).

*exécution, — autonome, autonomous work; — séquentielle, sequential control.

*execution du procédé, carrying method into effect or practice.

*exemple, instance, illustration; — chiffre, numerical example.

*exercice, drill (mil.); — de defense aérienne, air-raid drill; — de tir, firing practice (mil.); — du tir, gunnery.

*exogamie, outbreeding.

*expéditeur, dispatcher.

*expérience, — critique, critical experiment; — de détection des fuites, leak testing; — exponentielle de pile, exponential pile experiment; — exponentielle sur les neutrons rapides, fast exponential experiment; — homéostatique, machine learning.

*exploitation, — d'une pile, pile operation; — hydraulique, hydraulic mining; — minière, mining.

*explorateur, analyzer, scanner.

*exploration, scanning (telev.); — avant, forward scan; — de mémoire, storage scan; — géophysique, geophysical prospecting; — inverse, reverse scan; — par diffusion, reflected-light principle of scanning; — par lignes contigues, progressive scanning.

*explorer, to scan, look up, read.

*explosion, — atomique, atomic explosion; — de fiches, record scatter; — nucléaire, nuclear explosion.

exposition, — à un rayonnement, aux rayonnements, exposure to radiation; — professionnelle, occupational exposure; — prolongée, chronic exposure.

*expulser, to knock out.

*expulsion, emission, eduction; — d'un atome, discomposition effect.

*extensible, open-ended.

*extension en phase, phase space.

extensomètre, m. extensometer, strain gage.

extensométrie, f. strain measurement.

*extincteur à mousse, foam extinguisher.

*extinction, burn out, black out, quenching; — auditive, aural nulling; — complète, blackout, complete darkening; — de chaux, lime slaking; — par excès d'air, lean die-out; — par excès de carburant, rich blow-out.

*extra-courant, — de rupture, break impulse, doubling effect.

*extracteur de gaz, gas separator.

*extraction, recovery, stripping; — automatique, auto-abstract; — de cartes, card pulling; — de chaleur, heat removal; — de l'uranium, recovery of uranium; — du minerai, ore extraction; — en discontinu, batch extraction; — liquide-liquide, liquid-liquid extraction; — par absorption, absorption extraction; — par distillation, extractive distillation; — par l'éther, ether extraction process; — par partage, liquid-liquid extraction; — par solvant, solvent extraction.

extrapoler, to extrapolate.

*extrémité, — de faisceau, bundle termination (electricity); — de serrage, wrench end.

F

*fabrication, — d'éprouvettes, sampling action; — en série, mass production.

*face, — de chargement, charging face, load face; — de défournement, unloading face; — d'enfournement, inlet face, load face; — latérale, lateral face, facet, side face (geom.).

*facilité, installation.

façonné par marteau à chute, drop-forged.

*facteur, operand; — antitrappe, resonance escape probability; — d'affaiblissement, attenuation factor; — d'amortissement, damping factor; — d'échelle, scale factor; —

de cliquets, Klirr factor, nonlinear harmonic distortion factor, blur or rattle factor; — de cocarcinogénèse, cocarcinogenesis factor; — de comptage, scaling factor; — de contraction, shrinkage factor; * — de conversion, conversion factor; — de décontamination, decontamination factor; — de désavantage, disadvantage factor; — de diffusion atomique, atomic scattering factor.

facteur, — de diminution de flux, disadvantage factor; — de fission par les neutrons rapides, fast multiplication factor, fast fission effect factor, fast multiplication effect; — de forme, geometry factor; — de forme d'une grandeur alternative symétrique, form factor of a symmetrical alternating quality; — de multiplication, multiplication factor; — de multiplication des neutrons, neutron multiplication factor; — de multiplication en milieu infini, infinite multiplication constant; — de multiplication excessif, excess multiplication constant; — d'enrichissement, enrichment factor, separation factor; — d'enrichissement d'un procédé à simple passe, simple process factor.

facteur, — d'enrichissement total par etage, over-all enrichment per stage; — de pénétration, penetration coefficient; — de pointe, amplitude factor; — de proportionnalité, scale factor; — de régénération du combustible, fuel regeneration factor; — de répétition, duplication factor; — de reproduction, reproduction factor; — de rétrécissement, shrinkage factor; — de similitude, scale factor; — d'inhomogénéité, channeling effect factor; — d'ombre, self-shielding factor.

facteur, — dominant autosomique, autosomal dominant; — d'utilisation des neutrons thermiques, thermal utilization factor; — effectif de multiplication, effective multiplication factor; — génétique, genetic entity; — intrinsèque de séparation, intrinsic separation factor; — létal, lethal factor; — récessif autosomique, autosomal recessive; — théorique d'enrichissement dans un traitement à une seule passe, ideal simple process factor; — théorique de séparation, ideal separation factor.

*factice, dummy.

facturation, f. billing, invoicing; service de —, billing department.

*facultatif, selective.

*faculté de porter, bearing capacity (mach.).

*faible temps de montée, short period of rise.

*faiblement radioactif, low-level radioactive, slightly radioactive, weakly radioactive.

*faire, — branle-bas de combat, to clear for action; — cesser la vrille, to extricate the machine from the spin; — déjecter, to warp; — demarrer, to kick off; — des sondages, to take soundings; — eau, to leak; — escale, to make an intermediate landing; — la garde, to watch; — le plan de, to plot (geom.).

faire, — le projet, to design; — le tracé de, to lay out; — marcher, to propel; — saillie, to jut; — sauter, to blast, explode, blow up; — une prise, to tap; — une voie d'eau, to spring a leak; — un palier près du sol, to flatten out, hold off; — un point fixe, to run up the engine; — un tour de piste, to do a circuit.

*faisceau, beam, pencil, brush (telev.), bank; — à collimation rigoureuse (de neutrons), highly collimated beam (of neutrons); — cathodique, cathode beam; — cathodique explorateur, scanning electron beam; — de balayage, scanning sweep; — de cyclotron, cyclotron beam; — de détection incendie, fire detecting wire; — de deutérons, beam of deuterons; — d'électrons, electron beam; — d'éloignement, outbound beam; — de neutrons, beam source of neutrons, neutron beam.

faisceau, — de neutrons bien délimité, collimated beam of neutrons; — de neutrons diaphragmé, collimated beam of neutrons; — de neutrons parfaitement délimité, highly collimated beam of neutrons; — de plaques, group; — de radiateur, radiator core; — de rayonnement, radiation beam; — diffusé, scattered beam; — d'ions, ion beam; — d'ions positifs, positive ion beam; — émis, radiation beam.

faisceau, — explorateur, scanning beam, pencil of light; — fonction, unblanked beam; — hors fonction, blanked beam; — incident, incident beam, impinging beam; — ionique moléculaire, molecular ion beam; — moléculaire, molecular beam; — parallèle de particules incidentes, parallel beam of incident particles; — uniforme de neutrons, uniform beam of neutrons; — utile, useful beam.

*famille, — de l'actinium, actinium series; — radioactive, decay chain, radioactive chain, radioactive family, radioactive series, series decay.

*fanal, stop light.

fantôme, m. phantom; — d'eau, m. water phantom.

faradiser, to screen, shield.

*fardeau, — d'allèles mutants, load of mutant alleles; — de mutations, mutational load.

*fardier, truck.

fargue, f. gunwale.

fausse, — **attaque,** *f.* feint; — **équerre,** *f.* bevel square; — **nervure,** *f.* stiffening rib.

fausse-carène, *f.* false keel.

fausse-ligne, *f.* dummy line.

*__faux, fausse,__ — **alignement,** misalignment; **fausse attaque,** *f.* feint; — **axe,** dummy pin; — **châssis,** underframe; — **contact,** bad electrical contact; — **équerrage,** out of squareness; **fausse équerre,** *f.* bevel square; **fausse nervure,** *f.* stiffening rib; — **rond,** out of round.

faux-longeron, *m.* false spar.

feeder de descente, *m.* downlead, lead-in feeder.

fêlé, unsound, cracked.

femelle, *f.* bearing (mach.), spindle bearing.

fendement global, *m.* over-all efficiency.

*__fenêtre,__ — **de lancement,** firing window; — **d'irradiation,** beam hole.

*__fente,__ air gap; — **centrale,** central gap; — **d'aération,** vent hole; — **d'aile,** wing slot; — (**d'une pile**), gap (of a magnet) (of a reactor).

*__fer,__ — **à cornières,** angle iron; — **à répasser,** flatiron; — **à répasser électrique,** electric iron; — **d'angle,** corner bracket, angle bar; — **de ferraille,** soldering iron; — **de lingot,** ingot iron; — **d'induit,** armature iron; — **en T-double,** double-T gland; — **en U,** channel iron; — **façonné,** angle bar; — **feuillard,** band iron; — **fritté,** sintered iron; — **homogène,** soft steel; — **sulfuré,** iron pyrites.

ferghanite, *f.* ferganite.

fergusonite, *f.* fergusonite.

fer-hydrogène, iron-hydrogen (resistance).

*__fermé,__ enclosed, secure.

*__ferme en arbalète,__ truss frame.

*__fermer avec une latte,__ to batten down.

*__fermeture éclair,__ zip fastener, zipper.

fermion, *m.* fermion.

fermium, *m.* fermium.

*__ferraille,__ soldering iron, junk.

ferraillement, *m.* clanking, rattling, energetic squabble.

ferrailler, to rattle (metal), reinforce concrete, duel, squabble.

ferrailleur, *m.* junk dealer.

ferromagnétique, ferromagnetic.

ferrouter, to carry piggy back.

*__ferrure,__ metal fitting; —**s de catapultage,** catapult fittings.

*__feu, feux,__ gunfire; — **à eclipses,** flashing light; — **blanc,** white light; — **d'artifice,** fireworks; — **de forge,** forge; — **de recul,** backup light; — **de route postérieur,** taillight, rear light; **feux croisés,** crossfire; **feux de croisement,** dipped lights; **feux de direction,** turn signal lights; **feux de position AR,** tail lights; **feux de position AV,** pilot lights; **feux de stationnement,** parking lights; **feux rouges,** tail lights.

feuil, *m.* film.

*__feuille,__ ear or lobe (of space pattern); — **de barographe,** barograph chart; — **de laiton,** brass plate; — **de présence,** time sheet; — **de route,** waybill; — **de tôle,** sheet iron, plate; — **métallique,** metal foil; — **mince,** thin foil.

fiabilité, *f.* reliability; — **pour les départs,** dispatch reliability.

*__fibre de bois,__ excelsior.

*__fiche,__ record, key plug, hinge, prong (of radio tube); — **absente,** no record found; — **additionnelle,** addition record; — **à jack,** jack plug; — **banane,** banana-type plug; — **complémentaire,** trailer record; — **d'accompagnement,** transmittal sheet or slip; — **d'appoint,** overflow record; — **de base,** home record, master record; — **de consignes,** operation sheet; — **de contact,** hub, jack, contact plug; — **de culot,** pin or prong of tube base; — **de directives,** operation sheet; — **de mise à jour,** change record; — **de modification,** change record; — **de présence,** attendance card; — **descriptive,** descriptor record, track descriptor record, leader; — **de tournage,** dope sheet (TV and movies); — **indirecte,** non-home record; — **intermédiaire,** adapter, adapter plug; — **logique,** logical record; — **magnétique,** magnetic record; —**s de longueurs,** variable-length records.

fichier, *m.* filing cabinet, file (of computer); — **actif,** active file; — **central,** master file; — **courant,** active file, transaction file; — **de cartes perforées,** punched card file; — **principal,** master file; — **répertorié,** card index system; — **sur bande,** tape file; — **sur cartes,** card file.

*__fidèle,__ orthophonic (of sound reproduction).

*__fidélité,__ reproducibility (tests, etc.).

*__figure,__ mantissa, fixed point part.

*__fil,__ — **à contact mobile,** slide wire; — **à ligature,** binding wire; — **de connexion,** jumper (wire); — **de contact,** contact-wire antenna; — **de fer barbelé,** barbed wire; — **de lin,** thread; — **d'entrée,** leadin (radio); — **de pont,** slide wire; — **de sûreté,** wire fuse; — **divisé,** litz wire, stranded wire; — **double,** split wire; — **du bois,** grain (of wood); — **émaillé,** enameled wire; **en** — **indien,** single file (marching); — **magnétique,** magnetic wire.

*filage, — à la presse, extrusion; — sous pression, extrusion.

*filament, — chauffant, heater; — de plasma, plasma filament; — plasmatique, plasma filament.

*filature de jute, jute mill.

*file, — d'attente, queue; — d'attente de tâches, task queue; — de travail d'entrée, input work queue.

*filet, — d'air, streamline; — renverse, left-handed thread.

*filetage, (male) screw thread; — cylindrique, straight thread; — femelle, box thread.

fil-frein, m. locking wire.

filiale, f. subsidiary.

*filiation radioactive, radioactive relationship.

*filière, procedure, wiredrawing die, diestock; — simple, diestock.

filiforage, m. slim hole (boring).

film, m. film; — copié, blank film; — détecteur, film badge; — original, master film; — rationnel, film bearing sound track.

*filtration, — des aérosols, aerosol filtration; — inhérente, inherent filtration.

*filtre, — à air, air cleaner; — à air à bain d'huile, oil-bath air cleaner; — à entraînement, entrainment filter; — coupe-bande, band-rejection filter; — d'arrivée, input filter; — de bande, band-pass filter; — de départ, output filter; — de la soufflerie, honeycomb filter.

filtre, — d'entrée, inlet filter; — de sortie, output filter, outlet filter; — éliminateur de bruit, hum eliminator; — passe-bande, band-pass filter; — passe-bas, low-pass filter; — passe-haut, high-pass filter; — séparateur, band filter.

*filtrer, to mask.

*fin, — de bande, runout; — de course, end of travel; — de fichier, end of file; — de filet, thread runout; — de film, trailer; — de message, sign off; — de non-recevoir, negative acknowledgment; — de rayon, radius runout.

fines, f.pl. filler.

*fini, machined.

finition, f. finishing, top coat.

*fiole à décoction, flask (chem.).

*fissile, fissionable.

fission, f. — de l'uranium, uranium fission; — due aux neutrons rapides, fast neutron fission; — du noyau atomique, fission of atomic nucleus; — du plutonium, plutonium fission;

— du thorium, thorium fission; — émettrice, source fission; — explosive, explosive fission; — initiale, original fission; — nucléaire, fission, nuclear fission; — par les photons, photofission.

fission, — par neutrons lents, slow-neutron fission; — par neutrons rapides, fast-neutron fission, high-energy fission; — par neutrons thermiques, thermal-neutron fission, thermal fission; — primaire, original fission; — provoquée, induced fission; — provoquée par les neutrons rapides, fast-neutron fission; — provoquée par un neutron, fission induced by neutron; — spontanée, spontaneous fission; — ternaire, ternary fission.

*fissure, break.

*fixation, dispositif de—, holder device, socket (of lamp).

*flambage, buckling.

*flamber, to buckle (bend).

*flanc, sidewall (tire); — arrière d'une impulsion, pulse trailing edge; — avant d'une impulsion, pulse leading edge; — menant, driving side.

*fleau, hinge, planchet.

*flèche, — apparente, sag; — longitudinale, sweep back (aviation); — normale, dip.

*fléchissement, — de courant, partial power failure.

flexibloc, m. flexible bushing.

*flocon de neige, snowflake.

*floculant, flocculant.

*flot, à —, afloat.

flottabilité, f. buoyancy.

flottation, f. flotation.

*flottement, flutter, shimmy (wheels).

*flotteur, carburetor, float gauge, pontoon.

flou, flat, without contrast; fuzzy, soft, blurred.

fluage, f. creeping, creep; — tertiaire, tertiary creep.

*fluctuation statistique, statistical fluctuation.

fluidbloc, m. flexible bushing.

*fluide, — corrosif, corrosive fluid; — de l'échangeur thermique, heat-exchanger fluid; — de refroidissement, coolant; — évacuateur de chaleur, heat-transfer fluid.

*fluidifiant, fluxing.

fluidifier, to fluidize.

fluorégène, fluorescigenic.

*fluorescence par collision, impact fluorescence.

fluorimètre, *m.* fluorometer, fluorophotometer.

fluorocarbone, *m.* fluorocarbon.

fluorographie, *f.* fluorography.

fluoroscope, *m.* fluoroscope.

fluoroscopie, *f.* fluoroscopy.

fluoroscopique, fluoroscopic.

fluoruration, *f.* hydrofluorination.

fluotournage, *m.* rotary extrusion, hydrospinning, power spinning, roll extrusion, shear forming, spin forging.

***flûte,** — **marine,** streamer (geophysics).

***flux,** — **axial maximum,** axial peak flux; — **calorifique,** heat flux; — **de chaleur,** heat flux; — **d'énergie,** energy flux; — **de neutrons,** neutron flux; — **de neutrons épithermiques,** epithermal flux; — **de neutrons lents,** slow flux; — **de neutrons thermiques,** thermal flux.

flux, — **de rayonnement,** radiation flux; — **de rayonnement intense,** high radiation flux; — **des neutrons de résonance,** resonance flux, resonance neutron flux; — **directionnel,** directional flux; — **énergétique,** energy flux; — **intense de neutrons rapides,** high fast flux; — **isotrope,** isotropic flux; — **laminaire,** laminar flow.

flux, — **magnétique,** magnetic flux; — **maximum dans la direction perpendiculaire à l'axe,** radial peak flux; — **neutronique excédentaire,** excessive neutron flux; — **neutronique intégré,** integrated neutron flux; — **neutronique relatif,** relative neutron flux; — **neutronique vierge,** virgin neutron flux; — **vierge,** virgin flux.

foc, *m.* jib, crane.

focale, *f.* focal length or distance.

focalisation, *f.* focusing.

***fonction,** — **adjointe,** adjoint function; — **à gradins,** step function; — **de distribution,** distribution function; — **de distribution de retard,** delay distribution function; — **de la hauteur,** height dependence; — **d'énergie à variation lente et régulière,** smooth function of energy; — **de pondération,** weighting function; — **d'erreur,** error function; — **de transfert,** transfer function; — **d'excitation,** excitation function.

fonction, — **d'importance,** importance function; — **d'onde,** wave function; — **Et,** *and* function; — **intégrée,** integrand; — **linéaire,** linear dependence; — **non-linéaire,** nonlinear dependence; — **propre,** eigenfunction; **en — de la fréquence,** frequency dependence; **en — de l'énergie,** energy dependence.

fonctionnant au point critique, operating at critical.

***fonctionnement,** — **à l'allure de régime,** steady-state operation; — **d'une pile,** pile operation; — **en parallèle,** parallel operation; — **local,** home loop operation; — **sans surveillance,** unattended time; — **simultané,** simultaneous operation; — **spécialisé,** line loop operation.

fonction-saut, step function.

***fond,** — **à cran,** milled bottom (of watch); — **bombé,** dished head; — **continu,** background; — **de cale,** bilge; — **de piston,** piston head; — **de rayonnement gamma,** background gamma radiation; — **de réservoir,** tank bottom, unusable fuel; — **parasite,** background.

fondu, *m.* fade, fading, cast, shunt; — **au noir,** fade out; — **enchaîné,** cross fading; — **sonore,** sound fading; **fermeture en —,** fade out; **ouverture en —,** fade in.

***fonte durcie,** casehardened casting.

***forage,** — **à injection,** hydraulic circulating system; — **à la grenaille,** shot drill; — **double,** dual system; — **d'exploration, de reconnaissance,** wildcatting; — **sauvage,** wildcatting.

***force,** — **compensatrice,** balancing force; — **contrebalançante,** balancing force; — **d'attraction à courte distance,** short-range attractive force; — **d'échange,** exchange force; — **de cohésion,** cohesive force; — **de Coulomb,** Coulomb force; — **de désintégration,** decay strength; — **de faible portée,** short-range force; — **de grande portée,** long-range force; — **d'entraînement,** drag loading.

force, — **d'équilibre électrostatique,** electrostatic restoring force; — **de rappel,** restoring force; — **de rappel électrostatique,** electrostatic restoring force; — **de répulsion,** repulsive force; — **d'inertie,** inertial force; — **disruptive,** fission-producing force, disruptive force; — **électrique,** electric force; — **électrodynamique,** electrodynamic force.

force, — **électromagnétique,** electromagnetic force; — **équilibrante,** balancing force; — **faible,** weak interaction; — **interatomique,** interatomic force; — **intermoléculaire,** intermolecular force; — **intra-atomique,** intraatomic force; — **intranucleaire,** intranuclear force; — **ionique,** ionic strength; — **magnétique,** magnetic force; — **mécanique,** mechanical force.

force, — **motrice,** motive power; — **moyenne,** mean; — **non-centrale,** noncentral force; — **nucléaire,** nuclear force; — **nucléaire attractive,** nuclear attraction; — **pseudogravitique,**

pseudogravitational force; — **radiale**, radial force; — **répulsive de Coulomb**, Coulomb repulsive force; — **tensorielle**, tensor force; — **vive**, kinetic energy.

forerie verticale, *f.* vertical drill.

*****foret**, — **à centrer**, bit, countersinking; — **à cuiller**, center bit; — **américan**, twist drill, auger bit; — **hélicoïdal**, auger bit.

*****forger**, to hammer.

formage, *m.* shaping, molding, forming; — **au four**, oven forming; — **par étirage**, forming by drawing.

formanite, formanite.

*****formation**, — **de bulles**, bubbling; — **de noyaux**, nucleation; — **sous l'effet des photons**, photoproduction.

*****forme**, — **de bonne pénétration**, streamline shape; **en bonne** —, in order.

*****formule**, form (EDP); — **aux différences finies**, finite difference formula; — **d'aile**, wing shape; — **de la masse**, mass formula; — **d'équivalence de la masse et de l'énergie**, mass-energy conversion formula; — **de transformation**, transformation formula; — **leucocytaire**, differential white (blood) count; — **semi-empirique relative à la masse**, semi-empirical mass formula.

*****fosse**, — **d'effondrement**, rift valley; — **de la pile**, reactor pit.

*****fouille**, — **binaire**, binary search, dichotomizing search.

*****four**, — **à cémenter**, casehardening furnace; — **à fumée**, smudge fire; — **à fusion**, melting furnace; — **à induction**, induction furnace; — **atomique**, atomic furnace; — **basculant**, tilting furnace; — **de recuisson**, annealing furnace.

*****fourche**, cradle; — **de roue**, wheel fork; — **élastique**, spring fork.

*****fourchette à désembrayage**, gab.

*****fourgon**, rake (coal).

fourgonner, to stoke.

fourmariérite, *f.* fourmarierite.

*****fourneau**, — **à réverbère**, air furnace; — **de chaudière**, boiler furnace; — **de fusion**, melting furnace.

*****foyer**, — **à soufflage**, forced-draft furnace; — **d'incendies**, furnace; — **intérieur**, internal flue.

*****fraction**, cut; — **d'échange**, exchange fraction; — **de cohésion**, packing fraction; — **de conversion**, conversion fraction; — **molaire**, molar fraction, mole fraction.

fractionne, *m.* rerecording, multiplayback, multiplay, duoplay.

*****fractionnement de la dose**, dose fractionation.

*****fractionner**, to segment.

*****fragilité**, embrittlement.

*****fragment**, — **de fission**, fission fragment; — **de spallation**, spallation fragment; — **du noyau**, nuclear fragment; — **nucléaire**, nuclear fragment; — **rapide**, fast fragment.

*****fragmentation**, splitting; — **du noyau**, fragmentation of nucleus.

*****frais**, — **de fabrication**, production costs; — **de garage**, garage fees; — **d'entretien**, maintenance charge; — **d'établissement**, establishment charge; — **d'utilisation**, operating costs; — **généraux**, overhead cost.

*****fraiser**, to cut.

*****fraises pour perceuses**, countersinker.

francium, *m.* francium.

franc-jeu, *m.* fairplay.

franco quai, free alongside ship.

*****frange**, lobe or ear (of directional pattern).

*****frein**, lock washer; — **à disques**, disk brake unit (landing gear); — **à ergot**, tab washer; — **à main**, hand brake; — **à patte**, tab washer; — **à pédale**, foot brake; — **à ruban**, band brake; — **à tambour**, drum brake; — **d'écrou**, lock washer; — **de secours**, emergency brake; — **de stationnement**, parking brake; — **plat**, lockplate; — **supplémentaire**, emergency brake.

freinage, *m.* locking (screw or nut), safetying; — **brutal**, sharp braking; — **électrique**, electric braking; — **électrique par récupération**, regenerative braking; — **électrique rhéostatique**, rheostatic braking; — **par accumulation**, energy-storage braking; — **par récupération**, regenerative braking; **électrode de** —, retarding or reflecting electrode.

*****fréquence**, — **angulaire**, angular frequency; — **angulaire de cyclotron**, cyclotron angular frequency; — **audible**, audio frequency; — **audio**, audio frequency; — **d'apparition**, rate of occurrence; — **de base**, clock frequency; — **de battements**, beat frequency; — **de collisions entre particules**, interparticle collision frequency; — **de coupure**, cutoff frequency; — **de cyclotron**, cyclotron frequency; — **de denture**, slot-ripple frequency.

fréquence, — **de relaxation**, relaxation frequency; — **de rotation**, frequency of revolution, gyrofrequency; — **des chocs**, collision rate; — **des chocs par particule**, collision rate per particle (in plasma); — **des collisions**, collision rate; — **des grains d'image**, video frequency (telev.); — **des mutations**, mutation rate; — **d'image,** frame frequency; — **d'impulsions,** pulse repetition rate.

fréquence, — d'ondulation, ripple frequency; **— du plasma,** plasma frequency; **— intermédiaire,** intermediate frequency; **— linéaire,** line frequency; **— nominale,** assigned frequency; **— porteuse,** carrier frequency; **— propre,** natural frequency; **— ultra-acoustique,** supersonic frequency; **— unitaire de collision,** collision rate per particle (in plasma).

*****frère et sœur,** sib.

*****fret sur le vide,** dead freight.

frettage, *m.* shrinking-on, hooping, concrete reinforcement.

*****frette,** ironwork.

freyalite, *f.* freyalite.

frison, *m.* chip.

*****frittage,** sintering.

friture, *f.* grinders (radio); **— du microphone,** mike stew, frying noise.

fritzchéite, *f.* fritzscheite.

*****front, — de chargement,** charging face; **— d'onde,** wave front.

*****frontière, — diffuse,** diffuse boundary; **— diffuse d'allure exponentielle, — limite floue d'allure exponentielle,** exponentially diffuse boundary.

*****frottement, sliding; — superficiel,** skin friction.

*****frotteur,** collector shoe gear.

*****fuite, — de neutrons, — des neutrons,** escape of neutrons, neutron leakage.

*****fumées,** flue gas.

*****furet,** rabbit (for radioactive sample).

*****fuseau horaire,** standard time zone.

fuseau-moteur, *m.* engine nacelle.

*****fusée,** steering knuckle, rocket motor; **— de pale,** blade cuff or spindle; **— de roue,** wheel axle; **— de signalisation,** flare; **— instantanée,** percussion fuse; **— porteuse,** booster rocket, launch vehicle; **— sonde,** probe rocket.

fusée-sonde, *f.* probe, sounding rocket.

*****fuselé,** streamlined.

fusible, *m.* fuse.

*****fusil, — à vent,** air gun; **— se chargeant par la culasse,** breech-loading gun.

fusillade, *f.* gunfire.

*****fusion,** match and merge, fluxing; **— au four,** furnace fluxing; **— chimique,** chemical fusion; **— contrôlée,** controlled fusion; **— (d'un métal),** smelting; **— equilibrée,** balanced merge; **— ménagée,** controlled fusion; **— nucléaire,** nuclear fusion; **— thermonucléaire,** thermonuclear fusion.

*****fusionner,** to merge.

fût, cylinder barrel; **— perdu,** ontrip barrel, lost barrel.

G

gabare, *f.* tender (naut.), lighter.

*****gabarit,** coil form, jig, template, overall dimensions; **— à fraiser,** milling jig; **— à percer,** drill jig; **— à raboter,** planing jig; **— de chargement,** loading gauge; **— de voie,** track gauge; **— pour aléser,** boring jig.

gaillard, *m.* **— d'arrière,** *m.* quarter-deck; **— d'avant,** forecastle.

*****gain,** amplification; **— anodique,** surface glow on the anode; **— de conversion,** conversion gain; **— de surrégénération,** breeding gain; **— de travail,** mechanical advantage; **— effectif d'énergie,** net energy gain.

gainage, *m.* cladding, canning: **— en acier inoxydable,** stainless-steel cladding.

*****gaine,** casing, can, braid, duct, curve; **— d'air,** air gap; **— de thermomètre,** thermometer well; **— souple,** conduit, sleeve, tubing; **— tressée,** braiding.

gainé, canned.

galée, *f.* galley (print.).

*****galerie, — d'écoulement,** adit; **— de fond,** deep level (of mine); **— de toit,** roof luggage rack for auto.

*****galet,** sprocket; **— de came,** cam follower, tappet roller; **— de réservoir,** feed roll; **— suiveur,** cam follower.

galopin, *m.* hand truck.

*****galvanomètre, — à aiguille,** needle galvanometer; **— à bobine,** coil galvanometer; **— à boucle,** loop galvanometer; **— à corde,** string galvanometer; **— à équipage mobile,** moving-coil galvanometer; **— à miroir,** reflecting galvanometer; **— apériodique,** deadbeat galvanometer; **— à vibration,** string galvanometer; **— cuirassé,** shielded galvanometer; **— de résonance,** vibration galvanometer.

galvanostegie, *f.* electroplating.

gammagraphie, *f.* gamma radiography, gammagraphy.

*****gamme,** range, series.

*****ganglion lymphatique,** lymph node.

garage, *m.* garage, nesting store, shunting, dock; — **à sec,** dry basin.

garde, *f.* blank page; — **à la pédale,** pedal free travel; — **au sol,** ground clearance; **récipient de** — , storage or stand-by vat.

garde-boue, *f.* mudguard.

garde-frein, *m.* brakeman.

garde-poumons, *m.* respirator.

gardien, *m.* guard.

***gare d'embranchement,** junction (R.R.).

gargouille, *f.* gargoyle.

***garnir,** — **de graisse,** to pack with grease; — **un frein,** to line a brake.

garnison, *f.* garrison.

***garnissage,** hard facing.

***garniture,** gasket; — **antifriction,** bearing lining; — **de chanvre,** hemp packing (naut.); — **de col,** throat liner, nozzle insert; — **de coton,** cotton packing; — **de frein,** brake lining; — **d'embrayage,** clutch lining; — **de presse-étoupe,** stuffing-box packing; — **en cuivre rouge,** copper gasket; — **intérieure,** interior trim (auto body); — **intermédiaire,** bushing.

gasoil, *m.* gas oil.

gastunite, *f.* gastunite.

***gateau de cire,** wax disk (in sound recording).

***gauchissement des plaques actives,** fuel-plate warpage.

gaufré, ribbed, corrugated.

gauge bêta, beta gauge.

gauss, *m.* gauss.

***gaz,** — **à deux dimensions,** two-dimensional gas; — **à trois dimensions,** three-dimensional gas; — **à un seul paramètre,** one-dimensional gas; — **calorifère,** heating gas; — **complètement ionisé,** completely ionized gas; — **conducteur,** conducting gas; — **de combat,** war gas; — **de deutérium,** deuterium gas; — **de gazogène,** producer gas; — **dégénéré,** degenerate gas.

gaz, — **d'électrons,** electron gas; — **de pétrole,** casing-head gas; — **de queue,** residual or tail gases; — **des fumées,** stack gas, flue gas; — **diffusé,** dialyzate, diffusate; — **du paradis,** — **hilarant,** laughing gas; — **électronégatif,** electronegative gas; — **électronique,** electron gas; — **inerte,** inert gas, noble gas.

gaz, — **ionique,** ion gas; — **ionisé,** ionized gas; — **lacrymogène,** tear gas; — **neutre,** neutral gas; — **nondégénéré,** nondegenerate gas; — **organique d'extinction,** organic quenching gas; * — **pauvre,** water gas; — **radioactif,** radioactive gas.

gaz, — **rare,** inert gas, noble gas, rare gas; —

raréfié, rarefied gas; — **résiduel,** residual gas; — **riche,** wet gas, casing-head gas; — **sous striction,** pinched gas; — **tonnant,** oxyhydrogen gas; — **toxique,** poison gas, poisonous gas; — **traité,** process gas; — **utilisé,** working gas, process gas; — **vesicants,** vesicant gases, blister gases.

gazé, *m.* gassed person.

gazoduc, *m.* pipeline.

gazole, *m.* gas oil.

***gazomètre,** gas tank (city supply).

***gel de silice,** silica gel.

***gélatine,** — **chromifère,** chrome gelatin; — **en feuilles,** sheet gelatin; — **explosive,** blasting gelatin; — **pour la photographie,** photographic gelatin.

gélifié, jellied.

gène, *m.* gene; — **autonome unique capable de mutation,** single autonomous mutable gene; — **fonctionnel complexe,** complex functional gene; — **indicateur,** indicator gene; — **marqueur,** gene marker; — **mutant,** mutant gene.

***générateur,** generating routine (EDP); — **à étincelles,** spark transmitter; — **à ondes carrées,** square-wave oscillator; — **à tension constante,** constant-voltage generator; — **à tubes thermioniques,** thermionic valve transmitter; — **automatique de programmes** (GAP), report program generator (RPG); — **de caractères,** character generator; — **de mesure,** test oscillator; — **de neutrons,** neutron generator; — **de rapports,** report generator; — **d'harmoniques,** harmonic generator; — **d'impulsions,** impulse generator; — **électrostatique,** electrostatic generator; — **en cascade,** cascade generator; — **Van de Graaff,** Van de Graaff machine.

***generation,** — **automatique de symboles,** auto-fill character generation; — **de données,** data origination; — **transitoire,** transient generation; — **de noyaux,** generations of nuclei; — **s successives de noyaux,** successive generations of nuclei.

***génératrice,** — **atomique,** power reactor; — **du courant de chauffage,** filament generator; — **du courant de plaque,** anode generator; — **électronucléaire à solution aqueuse,** aqueous homogeneous power reactor; — **excitée en dérivation,** shunt dynamo; — **nucléaire,** nuclear power reactor.

génératrice, — **nucléaire à fission,** fission power reactor; — **nucléaire à neutrons rapides,** fast power reactor; — **nucléaire transportable,** package power reactor; — **polymorphique,** multiple-current generator; —

shunt, shunt-wound generator; — unipolaire, homopolar generator.

*génétique, — cellulaire, cellular genetics; — générale, formal genetics.

*génie, — atomique, nuclear engineering; — chimique, chemical engineering; — civil, civil engineering; — nucléaire, reactor engineering.

génome, *f.* genome.

*genouillère, knuckle joint, swing joint.

*géométrie, — à deux dimensions, plane geometry; — cylindrique, cylindrical geometry.

gerbage, *m.* racking, stacking, piling, binding, sheaving.

*gerbe, shower (of cosmic rays); — de noyaux et d'electrons, nucleus-electron shower; electron-nuclear shower; — en cascade, cascade shower; — étendue, extensive shower; — étroite, narrow shower; — mésique, meson shower; — pénétrante, penetrating shower.

germain, *m.* sib.

*germe, nucleus.

*gestion, — des données, data management; — des tâches, task management; — des travaux, job management.

getter, *m.* getter (electronics).

*gicler, to jet.

*gicleur, restrictor, (carburetor) jet, jet pump; — calibré, metering jet; — d'air de freinage, air compensator jet; — de ralenti, idling jet; — principal, main jet.

gigannée, *f.* billion years.

gilbert, *m.* gilbert.

gilpinite, gilpinite.

*girouette à godets, Robinson's anemometer, cup anemometer.

*gisement, — alluvial, alluvial deposit; — de minerai, ore deposit; — d'origine secondaire, deposit of secondary origin.

*gîte, — filonien uranifère, uranium-bearing vein deposit; — minéral, ore deposit.

givrage du carburateur, *m.* freezing up of the carburetor.

*glissade, sideslip (aviation).

*glissement, slip; espace de — , drift space, phase focusing space (in beam tube).

*glissière, coal shoot; — de forage, drilling jar; — de glace, window guide rail (auto); — de siège, seat rail (auto).

*global, integral.

*globule, — blanc du sang, amoebocyte; — rouge, erythrocyte.

gloglotement, *m.* rumbling of brakes.

*godet, — de mercure, mercury switch; — graisseur, grease box.

*golfe, bay.

*gomme indigène, artificial gum.

*gomme-laque, shellac.

gondolement, *m.* buckling.

*gonie, gonial cell.

*goniomètre, position finder.

*gorge, pass (of rolls); — coulissante, keyway, sliding slot; — de l'essieu, bearing neck.

*gouffre, heat sink.

*goujon, (stud) bolt; — à ressort, spring stud; — de centrage, dowel; — de charnière, hinge pin; — épaulé, shouldered stud; — prisonnier, stud bolt.

*goulotte de chargement, charge chute.

*goupille, cotter pin; — de sûreté, safety pin (tech.); — fendue, cotter pin.

*gouttière, drip molding, drip rail body.

*gouvernail, — de direction, rudder; — de fortune, jury rudder; — de gauchissement, lateral controls; — de profondeur, elevator; — marin, water rudder.

*gradient, gradient; — de champ, field gradient; — de flux, flux gradient; — de la densité, density gradient; — de potentiel, potential gradient; — de pression, pressure gradient; — du champ positif, positive field gradient; — négatif, negative gradient.

*graduation, scale.

*gradué, calibrated.

*graduer, to divide.

*grain, grain; — de plomb, shot (pellet); — d'un cliché photographique, grain of a photographic negative.

*graissage à carter sec, dry-sump lubrication.

*graisse, — d'adhésion, adhesive grease; — d'impregnation des cables, cable compound; — pour les essieux, axle grease; — pour voitures, wagon grease.

*graisseur, grease cup.

grand, — axe, major axis; —e capacité, large scale; — chemin, highway; — ligne, trunk line (R.R.); — pas, coarse pitch; —e route, highway; —e tenue, full dress; —e vitesse, high speed.

grand-angulaire, wide-angled.

*grandeur, — naturelle, full scale; — vectorielle, vector quantity.

*granule, — basophile, basophil granule; — neutrophile, neutrophil granule.

granulocyte, granulocyte.

granulométrie, *f.* screen analysis, particulate size distribution.

granulopénie, *f.* granulopenia.

*__graphique d'isodoses,__ isodose chart.

graphitage, *m.* graphitizing, graphitization, mark sensing.

graphitation, *f.* graphitization.

graphiter, to mark-sense (EDP).

*__grappin,__ grapnel, anchor, tie (R.R.).

*__graveur en creux,__ stamp die.

gravicélération, *f.* swing by.

gravidéviation, *f.* swing by.

gravillonneuse, *f.* gravel spreader.

gravisphère, *f.* sphere of activity.

graviton, *m.* graviton.

*__gravure,__ illustration.

gréé à traits carrés, square-rigged (naut.).

grelin, *m.* hawser.

*__grenade sous-marine,__ depth charge.

grenaillage, *m.* shot peening, blasting.

grès cérame, *m.* stoneware.

*__grésil,__ pea coal.

*__grésillement,__ small shot noise (due to Schottky effect, in radio tubes).

*__griffe à feu,__ fire grate.

*__grillage,__ gridding.

*__grille,__ — à barreaux, wreckage guard, trash rack; — à doubles bouts, double-ended grid; — blindée, screened grid; — d'arrêt, suppressor grid; — de charge d'espace, space-charge grid; — de chaudière, boiler grade; — de commande, control grid, signal grid; — de rejet, — de retenue, suppressor grid; — écran, screen grid; — rassembleuse, phase-focusing grid, buncher (to group electrons into bunches).

grimpement, *m.* creeping (of an electrolyte).

gros porteur, *m.* jumbo jet.

groupage, *m.* blocking (EDP); — d'électrons, bunching of electrons; — d'ions, bunching of ions.

*__groupe,__ pool (reporters, stenographers, etc.); —combinable, phantom group; — convertisseur, motor generator; — d'aviation d'artillerie, artillery spotting unit; — d'aviation de bombardement, bomber group; — de fiches, record block; — de haute pression, high-pressure unit; — d'essai, test job; — d'excitation, excitation set; — d'isotopes, isotope group; — d'ondes, wave packet; — motopropulseur, drive train (auto), power plant; — thermique, heat engine set.

*__groupement,__ packing; — cinématique des electrons, cinematic focusing of electrons; — de pistes, band (EDP); — en phase, phase bunching.

*__grouper,__ to pack (EDP).

groupiste, *m.* groupman, man in charge of generating set.

*__grue mobile,__ jenny (mach.).

*__guérison,__ annealing.

guetteur, *m.* watchman.

*__guidage,__ guidance; — azimut, azimuth guidance; — par itération, iterative guidance; — vertical, elevation guidance.

*__guide,__ ledge, border, guide, manual; — de bande, caster; — d'ondes, wave guide; route de —, beacon course, beam.

guidon, *m.* sight (instrument).

guignol, *m.* traveling block, crank, rudder lever.

guillaume, *m.* rabbet plane (carp.).

*__guipage,__ braiding.

gummite, *f.* gummite.

gutter, to gum, glue.

gyro-compas, *m.* gyroscopic compass.

gyromagnétique, gyromagnetic.

gyroscope de direction, *m.* directional gyroscope.

H

*__habillage,__ upholstering, fitting-up; — extérieur, exterior trim (auto body).

habitabilité, *f.* roominess (auto, etc.).

habitacle, *m.* car interior, cockpit, binnacle; — du pilote, pilot's compartment.

*__hache à poing,__ hatchet.

hafnium, *m.* hafnium.

halage, *m.* haulage.

halde, *m.* heap.

halètement, *m.* chuffing.

*__hall,__ — d'assemblage, assembly building, preparation building; — de montage, assembly shop.

halogénure alcalin, *m.* alkali halide.

hamac, *m.* hammock.

hantise, *f.* frequentation.

*harmonique sphérique, spherical harmonic.

harnachement, *m.* harness; **— de parachute,** parachute harness.

*harnais, elevator (min.).

harpon, *m.* fishing hook.

hatchettolite, *f.* hatchettolite.

*hauban, — en corde, span rope; **— en fil,** stay or guy wire.

haubaner, to stay.

haubanné, — par fil, wire-braced; **— rigidement,** strut-braced.

hausse optique, *f.* telescopic sight.

haussière, *f.* hawser.

*haut, de — en bas, up and down; **en — tout le monde,** all hands on deck.

*haut-parleur, — à basses et à hautes fréquences, duplex loud-speaker uniting treble and woofer in one unit; **— à cornet,** horn loud-speaker; **— à diaphragme conique,** cone loud-speaker; **— à pavillon,** horn loud-speaker; **— dynamique,** moving-coil speaker.

*hauteur, hill, pitch; **— apparente,** virtual or effective height; **— critique,** critical height; ***— d'eau,** head of water; **— de la barrière,** barrier height; **— d'élévation,** lift; **— de l'unité de transfert,** height of transfer unit (HTU).

hauteur, — des nuages, cloud height; **— du baromètre,** barometer reading; **— du métacentre,** metacentric height; **— d'une dent,** cog or tooth depth; **— du pas d'une vis,** screw-thread pitch; **— limite,** critical height; **— queue au sol,** tail-down height.

haveuse, *f.* cutting machine.

hayon, *m.* tailgate (station wagon, truck, etc.).

*hélice, — à deux valeurs de pas, two-position controllable propeller; **— à toutes positions,** fully controlled propeller; **— démultipliée,** geared-down propeller; **— placée en drapeau,** feathered propeller; **— propulsive,** pusher-type propeller; **— rigide,** fixed-pitch propeller; **— sustentatrice,** lifting propeller; **— tractive,** tractor-type propeller, puller propeller.

hélicoptère, *m.* helicopter.

héligare, *f.* helicopter station.

héliographe, *f.* heliograph.

héliostat, *m.* heliograph.

héliport, *m.* heliport.

hélistation, *f.* helistop.

hélisurface, *f.* temporary helicopter site.

helvine, *f.* helvite.

helvite, *f.* helvite.

heptode changeuse de fréquence, *f.* pentagrid converter.

*herbe de chanvre indien, Indian hemp.

*hérédité — cytoplasmique, cytoplasmic inheritance; **— liée au sexe,** sex linkage.

*hermétique, explosion-proof, leakproof.

hertz, *m.* hertz, cycle per second.

*hétérodyne, *m.* heterodyne receiver; **— de service,** test oscillator.

hétérosis, *f.* heterosis.

hétérotopique, heterotopic.

*hétérozygote, heterozygous; **— d'inversion,** inversion heterozygote.

hétérozygotisme, *m.* heterozygosity.

*heure, — de fonctionnement, running hour; **—s supplémentaires,** overtime.

heuristique, heuristic.

heurtement, *m.* jar, shock.

*heurtoir, cog.

*hexafluorure d'uranium, uranium hexafluoride.

hexamoteur, six-engined.

hexode, — à pente variable, variable-mu hexode; **— changeuse de fréquence,** mixing hexode; **— oscillatrice-modulatrice,** hexode mixing value.

hielmite, *f.* hjelmite.

*hisser à bord, to hoist aboard.

hodoscope, hodoscope.

holonome holonomic.

homéostatique, homeostatic.

homologation, *f.* homologation, certification, approval, probate (will).

homozygotisme, homozygosity.

*horaire, schedule; **— d'été,** summer timetable; **— d'hiver,** winter timetable.

*horizon, — naturel, real horizon; **— repère,** key horizon.

*horizontal, level.

*horloge, — à temps réel, real time clock; **— interne,** interval timer; **— mère,** master clock.

horloger, *m.* watchmaker.

hors de portée, out of range.

hors-ligne, *m.* off-line, patch of land outside boundary of projected highway.

hors-texte, *m.* zone (EDP), inset plate.

*hotte, — à fumée, fume hood; **— à vapeur,** cap (mach.); **— de cheminée,** chimney breeching.

*houille grasse, bituminous coal.

*houle de fond, ground swell (naut.).

*housse, slipcover, protective covers of various types.

*hublot, — à double paroi transparente, dual pane window; **— largable,** jettisonable window.

*huile, — à broches, spindle oil; **— antirouille,** flushing oil; **— à tarauder,** screwcutting oil; **— de cade,** cade oil; **— de genièvre,** juniper oil; **— de goudron de houille,** coal-tar oil; **— de pierre,** rock oil, coal oil; **— de rinçage,** flushing oil; **— de suif,** tempering oil; **— de trempe,** quench oil, tallow oil.

*huit, — horizontal, figure-of-eight turn; **— vertical,** vertical figure of eight.

*humidité d'équilibre, equilibrium water.

hurlement, *m.* howling.

huttonite, *f.* huttonite.

*hybridation, cross.

*hydraulique, hydraulic lift.

hydravion, *m.* seaplane; **— à coque,** flying boat; **— à deux coques,** twin-hulled flying boat; **— à flotteurs,** floatplane; **— d'école,** school or training seaplane; **— de course,** racing seaplane; **— de haute mer,** seagoing aircraft.

hydrobase, *f.* seaplane base.

hydrocraquage, *m.* hydrocracking.

hydrocraqueur, *m.* hydrocracker.

*hydrogène, — gazeux, gaseous hydrogen; **— lourd,** heavy hydrogen.

hydromagnétique, hydromagnetics.

hydroplaneur, *m.* hydro-sailplane or glider.

hydroscale, *f.* seaplane airport.

*hydrure lourd, deuteride.

hyperfréquence, *f.* ultra high frequency; **—s,** microwaves; **réceptuer d'—s,** microwave receiver.

hypergol, *m.* hypergol.

hypermicroscope, *m.* supermicroscope, ultra microscope, electron microscope.

hypéron, *m.* hyperon; **— de cascade,** cascade hyperon.

hypersustentation, *f.* increase of lift.

hypertrempé, re-annealed.

hypsomètre, *m.* transmission-level meter.

HZ, HZZ, hexagonal nut.

I

ianthinite, *f.* ianthinite.

*identification, ascertainment; **— chimique,** chemical assay; **— de caractères,** character recognition; **— de motif,** pattern recognition; **feu d'—,** identification light.

*identité, identical equation.

idiostatique, idiostatic.

illogisme, *m.* inconsistency, illogicality.

*image, — de carte, card image; **— inexposée,** unexposed negative (phot.); **— latente,** latent image; **— magnétique,** magnetic spectrum; **— photographique,** photographic image, photographic trace; **— secondaire,** echo image.

imagé, ornate.

imbrication, *f.* overlapping, imbrication, nest.

*imbriqué, interlocked, interlaced.

immoler, to sacrifice.

imparité, *f.* imparity, odd parity (EDP).

*impédance, — cinétique, kinetic impedance; **— de bouclage,** feedback impedance; **— de charge,** load resistance; **— d'entrée,** input impedance; **— de rétroaction,** feedback impedance; **— en dérivation,** shunt impedance.

*imperméable, leakproof.

imperméabliser, to waterproof (textiles).

implanter, to implement.

implication, *f.* implication, contradiction; **— conditionnelle,** if-then operation, conditional operation; **— logique,** if-then operation.

*impôt sur le revenu, income tax.

*impracticable, impassable.

*impression, printout, plain writing; **— au vol,** on-the-fly printing; **— de contenu de mémoire,** storage print-out, core dump; **— de mémoire,** core dump; **— détaillée,** detail printing; **— de totaux,** total printing; **— par ligne,** line printing.

imprimante, *f.* printer (of computer); **— à chaîne,** chain printer; **— à clavier,** keyboard printer; **— à grande vitesse,** high-speed printer; **— à la volée,** hit-on-the-fly printer; **— à ligne,** parallel printer; **— à pointes,** stylus printer, wire printer, matrix printer; **— à roues,** wheel printer; **— à stylets,** stylus printer; **— de pupitre,** console printer, console typewriter; **— en série,** serial printer.

*imprimé, form; **— en continu,** continuous form.

*imprimer, — en liste, to list.

*imprimeur, printing telegraph; — de circulaires, job printer; — télégraphique, teletype.

impulseur, *m.* impeller.

*impulsion, flash (of light), pulsation; — angulaire, angular impulse, impulsive moment; — complète, full drive pulse; — de coïncidence, coincidence pulse; — de courant, pulse of current; — de décharge, discharge pulse; — d'égalité, equal pulse; — d'horloge, clock pulse; — d'inégalité, unequal pulse; — d'inscription, full write pulse; — de lecture, full read pulse; — de position, P pulse; — de positionnement, set pulse; — de pression, pressure pulse; — de remise à zéro, reset pulse; — de sortie, pulse output; — partielle d'inscription, partial write pulse; — partielle de commande, partial drive pulse; — partielle de lecture, partial read pulse.

impulsion, — de recul, impulse of recoil; — de tension, potential pulse; — d'ionisation, ionization pulse, pulse of ionization; — neutronique, neutron pulse; — orbitale, orbital moment; — parasite instrumentale, spurious pulse.

*impureté, contaminant; — correspondant aux atomes, impurity atom; — correspondant aux éléments, impurity element.

*incidence rasante, glancing angle.

*inclinaison, bank, pitch (screw threads), fall; — de la ligne, line skew (EDP); — d'un caractère, character skew (EDP); — du pivot de fusée, king pin inclination; — latérale, bank (navigation); — longitudinale, pitch; — transversale, bank.

inclinomètre, *m.* inclinometer.

*incohérence, inconsistency.

*incompatibilité, inconsistency.

incorporation, *f.* uptake.

incorporé, built-in.

increvable, puncture proof.

incrusté, embedded, encrusted, inlayed.

indanthrène, *m.* indanthrene.

*indépendance par rapport à la charge, charge independence.

*indépendant, off-line; — de la charge, charge independent; — des variables spatiales, space independent.

*index, directory; — de support, volume index.

indexage, *m.* indexing; — automatique, auto-indexing.

*indicateur, flag, timetable, tracer, radioactive indicator; — aérien, air-traffic guide; — à lames vibrantes, tuned-reed indicator; — au repos, indicator off; — chimique, chemical tracer; — d'accord, tuning indicator; — d'acheminement, routing indicator; — d'eau, glass gauge; — de battements, beat indicator; — de changement, escape character (EDP); — de crête, peak indicator; — de débit, orifice meter.

indicateur, — de dérive et de vitesse par rapport au sol, drift and ground-speed indicator; — de dernière carte, last card indicator; — de fin d'opération, completion code; — de fin de fichier, file gap; — de limite, reflective spot (tape); — de niveau, level gauge; — de niveau tout ou rien, cutoff level gauge; — de pente transversale, bank indicator, cross level; — de position, position indicator; — de rayonnement, radiation indicator; — de report, carry indicator; — de suppression, S trigger (EDP); — de virage, turn indicator; — de vitesse relative, kymograph (aviation); — de zone significative, significant start character.

indicateur, — du niveau de coupure, cutoff level gauge; — excité, indicator on; — intégreur chimique, chemical integrating indicator; — isotopique, isotopic tracer; — lumineux d'accord, visual-tuning indicator; — moléculaire, molecular tracer; — physique, physical tracer; — radioactif, isotopic tracer, radioactive tracer.

indicatif, *m.* code letter, key, label, internal label; — d'appel, call signal; — d'ensemble de données, data set label, data set control block; — de priorité, limit priority; — de transfert, branch code; — du navire, call letter or signal; — d'un office, code name or indication; — normalisé, standard label; — régional, area code.

*indication, — d'appel du navire, call letter, code signal; — visuelle directe, direct-reading indication.

*indice, — de décontamination, decontamination index; — de réfraction, index of refraction; — de structure, structural ratio; — inférieur, subscript; — supérieur, superscript.

*inductance, — d'arrêt, choke coil; — de protection, protective reactance coil; — de syntonisation, tuning coil; — propre, self-inductance.

*induction, — cinétique, motional induction; — homopolaire, homopolar induction; — nucléaire, nuclear induction; — parasite, parasitic induction; — statique, condensing (elec.); — unipolaire, homopolar induction, unipolar induction.

*industrie, — à l'ombre, shadow industry; — de base, key industry; — des cellules, air-frame industry.

*inertie, — de l'oeil, persistence of vision; — électromagnétique, electromagnetic inertia; — mécanique, mechanical inertia; — thermique, thermal inertia.

inexposé, unexposed (phot.).

infanterie, f. infantry; — aérienne, air infantry.

infecteur, m. injector.

*infériorité, disadvantage.

infirmerie, f. sick bay, hospital.

*inflammabilité nucléaire, nuclear inflammability.

*infléchir, to hand (a curve).

*inflexion, shoulder (of a curve).

informaticien, m. computer expert, systems analyst, data processor.

informaticien-conseil, m. systems engineer, consultant.

*information, data; — de contrôle, check information; — massive, mass data; —s, data; —s analogiques, analog data.

informatique, f. data processing; — absolue, integrated data processing; — commerciale, business data processing; — industrielle, industrial data processing.

inframicroscopique, submicroscopic.

*infrastructure, internal structure, ground equipment (in direction finding).

ingéniérie, m. engineering, study and development of projects, project-study department.

*ingestion opaque, opaque meal.

ingouvernable, unmanageable.

*inhalateur d'oxygène, oxygen apparatus.

*inhibiteur, liner (astronautics), inhibitor, except gate (EDP); — de corrosion, corrosion inhibitor.

*inhibition de le réaction, inhibition of reaction.

"inhour," "inhour" (inverse hour reactivity).

*injecteur, burner, atomizer; — de combustible, fuel injector; — de pompe, pump injection nozzle (carburetor).

innavigable, unseaworthy.

inox, m. stainless steel.

inscripteur, m. recorder.

*inscription, register (telegr.), recording, track (of sound); — machine, idiomatic feature.

*insérer une fiche, to plug in (teleph.).

*insigne, badge.

insonorisation, f. soundproofing, silencing.

insonorisé, soundproof.

*inspecteur, customer engineer, field engineer.

*inspection radiographique, radiographic inspection.

*instabilité, hunting; — à coude, kink instability; — de la striction, pinch instability; — hydromagnétique, hydromagnetic instability.

*installation, — de charge, battery charger; — d'énergie, power plant; — d'énergie électrique, generating plant; — d'enrichissement, enrichment plant; — d'épuration de l'air, air-conditioning plant; — de séparation par diffusion thermique, thermal diffusion plant; — d'essai semi-industrielle, pilot plant; — électrogène, generator plant; — frigorifique, ice plant; — industrielle, full-scale system; — pilote, pilot plant; — pour la masse des coefficients globaux de protection, bulk shielding facility.

*instruction, command, order; — à N adresses, N address instruction; — à répétition, repetition instruction; — à une, deux, . . ., n adresses, one, two, . . ., n address instruction; — canonique, declarative instruction; — d'appel, cue; — de commande de travail, job control statement; — de contrôle de compilation, compiler directing statement; — de logique, logic instruction; — d'en-tête, header statement; — de référence, data definition statement; — de saut, jump instruction; — de sélection, extract instruction; — d'ordinateur, computer instruction; — d'origine, source statement; — en langage-machine, machine instruction, actual instruction; — factice, dummy instruction, no-operation instruction, instructional constant; — imbriquée, nested statement; — logique, logical instruction, decision instruction; — ordinolingue, actual instruction; — provisoire, presumptive instruction; — sans adresse, zero address instruction; — terminale, trailer statement; — théorique, ground instruction; —s d'enchaînement, encascade, en chaîne, enchaînées, chained instructions, command chaining.

instruction-machine, f. instruction code, computer instruction, machine instruction.

instruction-patron, f. model statement.

instruction-prototype, f. prototype statement.

*instrument, — de contrôle, survey instrument; — de mesure, instrumentation; — étalon, reference instrument, calibration instrument.

instrumentation, f. instrument equipment.

*intégrale, integral quantity; — de phase,

phase integral; — **de résonance,** resonance integral; — **effective de résonance,** effective resonance integral; — **par rapport au temps,** time integral.

intégrateur, *m.* scaled radiation detector, scaler; — **de doses,** integrating dose meter; — **logarithmique,** log counting rate meter; — **par la méthode des rectangles,** rectangular integration.

*__intégration des impulsions,__ pulse integration.

*__intendance des bâtiments,__ bureau of construction.

intensimètre, *m.* counting rate meter, rate meter, intensitometer; — **linéaire,** linear rate meter; — **logarithmique,** log counting rate meter.

*__intensité,__ — **acoustique subjective,** loudness; — **d'activation,** intensity of activation; — **de champ électrique,** electric field strength; — **de dose,** dose rate; — **de la pesanteur,** acceleration of gravity; — **de la radioactivité,** intensity of radioactivity; — **de la source,** source strength; — **de la source de neutrons,** neutron source strength; — **des forces agissantes,** interaction strength; — **d'impression,** print contrast ratio.

intensité, — **d'irradiation,** dose rate; — **du champ,** field strength; — **du champ magnétique,** magnetic field strength; — **du courant de tension,** fusing current; — **du faisceau cathodique,** intensity of the electron beam; — **d'une source,** strength; — **du rayonnement,** radiation level, radiation rate, radiation intensity; — **effective du rayonnement,** effective radiation rate; — **dynamique de caractères,** dynamic character intensity; — **dynamique de vecteurs,** dynamic vector intensity; — **relative du rayonnement,** relative intensity of radiation.

interaction, *f.* interaction; — **coulombienne,** Coulomb interaction; — **due aux forces de Coulomb,** Coulomb interaction; — **électromagnétique,** electromagnetic interaction; — **nucléaire,** nuclear interaction; — **tensorielle,** tensor interaction.

intercepteur, *m.* interceptor fighter.

*__interchangeabilité de tampons,__ exchange buffering.

interclasser, to collate, merge (EDP).

*__interdiction,__ restriction; — **de partir,** flying restriction.

*__interdire,__ to inhibit.

interface, interface; — **d'entrée-sortie,** input-output interface.

*__interférence,__ jamming (radio).

interféromètre à micro-ondes, *m.* microwave interferometer.

interfiche, *f.* record gap.

*__interligne,__ pitch, row pitch, line space.

*__intermédiaire,__ jobber (com.); **par — de,** as a function of (mach.).

intermodulation, *f.* cross modulation.

interrangée, *f.* row pitch.

*__interrogation,__ inquiry, polling (EDP); — **automatique,** autopolling.

*__interroger,__ to poll.

*__interrompre,__ to intermit.

*__interrupteur,__ — **à basculeur,** tumbler switch; — **à couteaux,** knife switch (elec.); — **à marteau,** hammer break (elec.); — **à rupture brusque,** quick-break switch; — **de choc,** impact switch; — **de délestage,** load shedding switch; — **de fin de course,** limit switch; — **de section,** section circuit-breaker; — **horaire,** time switch; — **rapide,** ticker.

*__interruption,__ inrush, sudden outburst, interrupt, suspension, trap; — **d'entrée-sortie,** input-output interrupt; — **de programme,** program interrupt; — **de traitement,** process interruption; — **en attente,** pending interruption; — **en suspens,** pending interruption; — **externe,** external interruption; — **irrégulière,** abend, abnormal end.

intersecteur, *m.* intersector, data gate.

*__intersection,__ junction, conjunction.

*__interstice,__ gap.

*__intervalle,__ gap, range; — **de décharge,** discharge gap; — **de livraison,** turnaround time; — **de marche normale,** power range; — **de masse,** mass range; — **de mesure,** instrument range, period range; — **d'énergie,** energy range; — **de proportionnalité,** proportional band; — **de proportionnalité limitée,** region of limited proportionality; — **de puissance,** power range.

intervalle, — **de recouvrement,** overlap region; — **d'erreur,** range of error; — **des grandes énergies,** high-energy region; — **des neutrons rapides,** fast neutron region, fast region; — **des vitesses,** velocity range; — **des vitesses des neutrons,** range of neutron velocity; — **de temps moyen,** mean time; — **énergétique des neutrons,** neutron energy range; — **énergétique des neutrons intermediaires,** intermediate neutron energy region; — **moyen de temps entre collisions,** mean collision time.

*__intervention dans la lutte terrestre,__ ground strafing, air-to-ground fighting.

intraduire, to lead in, admit.

intra-fonctionnement, *m.* home loop operation.

invariance, *f.* invariance; — **de charge,** charge invariance.

*invariant, re-enterable routine.

*inventaire, schedule; — **de matériel,** device table (EDP), material inventory; — **des rechanges,** list of spares.

*inventeur, designer.

*inverseur, commutator, information gate, change over switch, double throw; — **code,** dimmer; — **de poussée,** thrust reverser; — **de réenclenchement,** reset switch; — **de sélection démarrage,** starter selector switch; — **de sens de marche,** reverser; — **deux positions,** toggle switch; — **manuel,** sense switch; — **marche-arrêt,** on-off switch; — **programmé,** program switch.

*inversion, — **chromosomique,** chromosome inversion; — **des raies,** reversion of lines.

*ion, — **à charge unique,** singly charged ion; — **amphotérique,** amphoteric ion; — **de grande vitesse,** high-speed ion; — **du combustible,** fuel ion; — **gazeux,** gaseous ion; — **primaire,** primary ion; — **secondaire,** secondary ion; —**s formés,** ion yield.

*ionisation, — **cumulative,** avalanche; — **des gaz,** gas ionization; — **en chaîne,** avalanche; — **en colonne,** columnar ionization; — **globale,** total ionization; — **minimum,** minimum ionization.

ionisation, — **par choc,** — **par chocs,** collisional ionization, impact ionization; — **par collision,** collisional ionization; — **par les photons,** atomic photoelectric effect, photoionization; — **primaire,** primary ionization; — **spécifique,** specific ionization; — **spécifique primaire,** primary specific ionization.

ioniser les gaz, to ionize gases.

ionosphère, *f.* ionosphere.

iontophorèse, iontophoresis.

irinite, *f.* irinite.

*irradiation, radiation exposure; — **admissible,** permissible exposure; — **aiguë,** acute exposure; —**artificielle,** artificial irradiation; — **chronique,** chronic exposure; — **des gonades,** gonad irradiation; — **maximum admissible,** maximum permissible exposure; — **professionnelle,** occupational exposure, occupational irradiation; — **solaire,** insolation (meteor.); — **unique,** single exposure.

*irrégularité, abnormality.

irrigateur, *m.* water wagon.

irriguer, to irrigate.

ishikawaïte, ishikawaite.

isobare, isobar.

isodiaphère, isodiaphere.

*isolateur, — **d'arrêt,** strain insulator; — **d'entrée,** lead-in insulator; — **de traversée,** bushing insulator; — **en caoutchouc durci,** rubber insulator; — **ovoïde,** egg insulator; — **rigide,** pin insulator.

*isolement thermique, thermal insulation.

*isomorphe, isostructural.

isoploïde, isoploid.

isostérique, isosteric.

isotone, isotone.

*isotope, — **235 de l'uranium,** U-235; — **instable,** unstable isotope; — **lourd,** heavy isotope; — **plus lourd,** heavier isotope; — **radioactif,** radioisotope, radioactive isotope; — **sans entraîneur,** carrier-free isotope; — **stable,** stable isotope.

itérer, to repeat.

J

jack, *m.* jack (elec.), socket; — **à rupture,** break jack; — **d'écoute,** listening-in jack; — **général,** multiple jack; — **local,** calling jack.

*jaillissement, gusher; — **en source,** flowing by heads.

*jalon, turning point; — **piquetage,** surveying rod; —**s de vérification,** audit trail.

*jambe, — **de force,** strut, radius arm; — **du train,** undercarriage leg; — **élastique,** compression leg, spring leg; — **oléo-pneumatique,** oleo-pneumatic shock-absorber strut; — **oléo-ressort,** oleo strut.

*jante, — **à base creuse,** drop-center rim, well-base rim; — **à rebord,** beaded rim; — **à talon,** clincher rim; — **de pneu,** tire bead; — **de roue,** wheel rim.

*jauge, — **à huile,** oil gage, dipstick; — **de niveau d'essence,** fuel gauge; — **d'épaisseur,** feeler gauge, thickness gauge; — **pour fils métalliques,** wire gauge.

jerricane, *m.* jerrycan.

*jet, run; — **d'eau,** drip molding (auto body); — **de la mer,** jetsam.

jetée, *f.* pier, layer of road grit, finger (aer.).

*****jeter de côté,** to junk.

*****jeu,** deck, set; — **algo,** source deck; — **algorithmique,** source deck; — **axial,** end play; — **de bobines,** coil assembly; — **de caractères,** character set; — **de caractères universel,** universal character set; — **de cartes ordinolinques,** condensed deck; — **de cartes résultant condensé,** condensed deck; — **de denture,** backlash (mechanical); — **de marche,** operational play or clearance; — **d'entreprise,** business; — **dents,** backlash; — **de palier,** bearing clearance; — **d'essai,** test deck; — **des soupapes,** value clearance; — **diamétral,** radial clearance; — **d'instructions,** instruction set; — **d'instructions fondamental,** standard instruction set; — **d'instructions universel,** universal instruction set; — **en bout,** end play; — **latéral,** side play or clearance; — **longitudinal,** end play; — **perdu,** end play (screw); — **résultant,** object deck; — **transversal,** side play or clearance; — **x et tolérances,** fits and clearances.

johannite, johannite.

*****joint,** link, fastening, hinge, gasket, coupling, connection; — **à bride,** flange joint; — **à brides à emboitement,** recessed flange joint; — **à clin,** lap joint; — **à dés,** trunnion coupling; — **à goujons,** dowel joint; — **à rotule,** knuckle joint; — **compensateur,** expansion joint; — **de cardan,** universal joint; — **de culasse,** cylinder-head gasket; — **de fibre,** fibre packing; — **de grains,** grain boundary.

joint, — **de rail,** rail joint; — **d'étanchéité,** gasket seal; —**élastique,** flexible coupling; — **en queue d'aronde,** dovetail joint; — **étanche,** water-tight joint; — **glissant,** expansion joint; — **homocinétique,** constant-velocity universal joint; — **ignifuge,** fireseal; — **metalloplastique,** asbestos or filled-metal jacketed gasket; — **métal sur métal,** ground joint; — **pour manches de pompe,** hose coupling; — **sphérique,** ball-and-socket joint; — **thermique,** thermal bond; — **torique,** O-ring; — **tournant,** rotary seal.

*****jointure croisée,** cross joint.

jole, *m.* dinghy (naut.).

*****jonc,** nailing strip, locking or retaining ring; — **à ergot,** snap ring; — **d'arrêt,** retaining ring; — **de retenue,** retaining, snap, or spring ring.

*****jonction,** linkage, interface; — **de base,** basic linkage; — **de revetement,** skin joint; — **des demi-voilures,** wing center attachment; — **différentielle,** hybrid junction (wave guides); — **soudée,** soldered joint; — **unique,** unifunction.

*****jouer,** to blow out (elec.).

*****joueur de disques,** record player, turntable.

*****jour,** — **à plomb,** skylight; — **d'un arc,** span (arch.).

*****journal,** — **d'exploitation,** operation log; — **de machine,** machine log; — **de marche,** operation log.

jumelage, coupling; — **de mitrailleuse,** twin guns.

jumelle, *f.* bed (mach.), yoke; — **d'arrêt,** stop yoke, retaining ring; — **de liaison,** connecting twin yoke; — **de ressort,** spring shackle.

jury, *m.* prize jury.

K

kahlérite, *f.* kahlerite.

karman, *m.* fillet; — **de mât réacteur,** stub fillet; — **d'empennage,** stabilizer fillet; — **de raccordement,** wing fillet; — **de voilure,** wing fillet.

kasolite, *f.* kasolite.

kenotron, *m.* kenotron.

khlopinite, *f.* hlopinite, khlopinite.

kilocurie, *f.* kilocurie.

*****kilomètres parcourus,** kilometers covered.

kilopériode, *f.* kilocycle.

kilovarheure, *f.* kilovar-hour.

kilovolt, *m.* kilovolt.

kinescope, *m.* kinescope (telev.).

*****kiosque,** — **des cartes,** chart room; — **d'un sousmarin,** conning tower.

klakson, klaxon, *m.* horn.

klauber, to buck, cob, sort.

klecksographie, *f.* ink-blot test.

klystron, *m.* klystron.

krarupisation, *f.* Krarup loading, continuous loading.

kryptoscope, *m.* kryptoscope.

L

*labile, unstable (labile).

*laboratoire, — d'analyses sous haute activité, hot analytical laboratory; **— de haute activité,** hot laboratory; **— de recherche,** research laboratory; **— non-radioactif,** cold laboratory; **— obscur,** darkroom (phot.).

*lâcher vol seul, to send up solo.

lacque, *m.* japan (paint.).

*lacune, vacancy; **— d'un ion negatif,** negative-ion vacancy.

*laine, — de fer, iron wool; **— de verre,** glass wool.

*laisser, — chauffer le moteur, to warm up the engine; **— passer,** to pass, transmit.

lamage, *m.* spot facing.

lambertite, *f.* lambertite.

lambrissage, *m.* wainscoting.

*lame, reed (of tuning fork); **— de collecteur,** collector or commutator segment; **— de rabot,** iron (plane); **— de raboteuse,** plane knife (mach.); **— de tôle,** iron sheet; **— d'un ressort,** leaf of a spring.

*lamelle de raccord, lug.

*laminage de tôles empillés, pack rolling.

laminographie, *f.* laminography.

*laminoir, — à tôle, plate mill; **— de serrage,** blooming mill (metal.).

*lampant ordinaire, stove distillate.

*lampe, valve; **— à arc,** arc light; **— à atmosphère gazeuse,** gas-filled lamp; **— à braser,** blowtorch; **— à champ retardé,** brake-field valve; **— acorn,** acorn valve; **— à decharge lumineuse,** glow-discharge tube, glow lamp; **— à filament dans le vide,** vacuum lamp; **— à filament thorie,** thoriated-filament valve; **— à flamme,** flame lamp.

lampe, — à grille mixte, Wunderlich valve; **— à pente variable,** variable-mu tube or valve; **— à source de lumiére ponctuelle,** point lamp; **— au pentane,** pentane lamp; **— à vide pousse,** high-vacuum tube or valve; **— bigrille,** double-grid valve or tube tetrode; **— changeuse de fréquence,** mixing valve, frequency-changer valve.

lampe, — de conversion, mixer tube, converter tube; **— de mine,** safety lamp (min.); **— mineur,** safety lamp (min.); **— d'émission,** transmitting tube or valve; **— démontable,** demountable valve or tube, assembled-parts tube; **— de néon,** neon light; **— dépolie,** frosted lamp; **— de puissance,** power valve or tube; **— dure,** hard valve; **— électro-**

mètre, electrometer tube; **— en vase clos,** enclosed lamp.

lampe, — excitatrice, exciter lamp; **— lumière du jour,** daylight lamp; **— luminescente,** fluorescent tube; **— secteur alternatif,** alternating-current valve; **— tare,** comparison lamp; **— témoin,** telltale, pilot, or control lamp; **— tempête,** hurricane lamp; **— thermoélectronique,** thermionic tube; **— tous courants,** alternating current–direct current valve; **— trigrille,** pentode.

lampe-témoin, *f.* test lamp.

*lance à feu, stoker (tool).

*lance-bombes, — èlectrique, electrical bomb-release gear; **— sous les ailes,** bomb racks under the wings.

*lancement, — au sandow, catapult launching, launching by means of a rubber rope; **— de bombes,** release of bombs, bomb dropping; **— remorqué,** towed start.

*lancer sur le marché, to put on the market.

lance-torpilles, *m.pl.* torpedo crutch.

lanceur, *m.* starting drive mechanism, booster rocket, launch vehicle; **— d'affaires,** business promoter.

lancière, *f.* leather lace (mach.).

*langage, — algo, — algorithmique, algorithmic language, source language; **— à machine,** machine-oriented language; **— à problémes,** problem-oriented language; **— à procédés,** procedure-oriented language; **— artificiel,** artificial language; **— chiffré, — convenue,** cipher or coded language; **— commun,** common language; **— d'arrivée,** object language; **— de départ,** source language; **— de programmation,** programming language; **— d'exécution,** object, effective, or target language; **— motivé,** high-level language; **— naturel,** natural language; **— ordinationnel,** machine language; **— ordinotrope,** computer-oriented, machine-oriented language; **— symbolique,** symbolic language; **— synthétique,** synthetic language.

langage-machine, *m.* machine language.

langage-source, *m.* source language.

*languette, feather.

*lanterne, projector (optics); **— de laboratoire,** safelight (phot.); **—s AR,** tail light clusters; **—s AV,** front light clusters.

lanthanide, *m.* lanthanide.

laplacien, *m.* buckling factor, buckling, Laplacian operator; **— critique,** critical buckling;

— du réseau, Laplacian of lattice; — géométrique, geometric buckling; — matière, material buckling.

*largeur, — de bande, bandwidth; — de la zone de silence, skipped or skip distance; — de marque, stroke width; — de segment, stroke width, width; — des raies spectrales, line width, spectral line width; — due à l'effet Doppler, Doppler width; — du niveau d'énergie, level width; — radiative, radiation width; — sur angles, width across corners; — sur pans, plats, width across flats.

largeur-moitié, m. half-width.

larguage, m. release (mach.).

*larmier, baffle, eaves.

*lavage des gaz, gas scrubbing.

lave-glace, m. windshield washer.

lave-vitre, m. windshield washer.

lèche-vitre, m. weatherstrip (between door and window of auto).

*lecteur, pickup, scanner, read-out device; — de bande magnétique, magnetic tape reader; — de cartes, card reader; — de phonographe, phonograph tone arm; — de ruban, paper tape reader, tape reader; — de ruban perforé, punched tape reader; — magnétique, magnetic character reader; — optique, optical character reader, visual scanner.

lecteur-interpréteur, m. reader-interpreter.

lecteur-perforateur de cartes, m. card reader-punch unit.

lecteur-scripteur, m. read-write head.

*lecture, — à rebours, read backward; — arrière, read backward; — au son, sound or aural reception; — avancée, read release; — avec éclatement, scatter read; — avec effacement, destructive read; — de caractères magnétiques, magnetic character reading; — de cartes graphitées, mark sense reading; — magnétique, magnetic reading; — non substitutive, non-destructive read; — optique, mark scanning optical character recognition; — par contact, contact sense; — régénératrice, regenerative reading; — sans effacement, non-destructive read; — substitutive, destructive read; — totale, full indicator reading (FIR).

lecture-perforation, f. read punch.

lente, f. nit egg, louse.

*lentille, — de sable pétrolifère, oil lens; — électrostatique, electrostatic lens; — magnétique, magnetic lens.

lepton, m. lepton.

*lésion, — de chromosomes, chromosome aberration; — due à l'irradiation, radiation injury; — histologique latente, latent tissue injury; — somatique, somatic injury; — tissulaire latente, latent tissue injury.

*lessivage, flushing, leaching; — alcalin, alkaline leaching.

*léthargie, neutron lethargy.

*lettre officielle, office action (of patent office).

*leucocyte, leucocyte; — polynucléaire, polymorphonuclear leucocyte.

*levé, — aérophotogrammétrique, aerial survey, air survey; — terrestre, ground survey.

*levée, knob (mach.), cog; — de came, cam lift; — de soupape, valve lift; — d'une prime, exercise of an option.

*lever, — du doute, sense finding; — le doute, correction or compensation of error or deviation, sense finding.

*levier, — brisé, angle lever, bent lever, elbow lever; — d'amorçage, priming lever of pump; — de changement de vitesses, gearshift lever; — de commande, operating lever; — de commande de débrayage, clutch operating lever; — de direction, Pitman arm; — d'embrayage et de désembrayage, gear lever; — intermédiaire de direction, steering relay lever; — oscillant, rocking lever; — sélecteur, selector on automatic transmission; — séparateur de frappe, hammer split.

*liaison, communication, hookup (radio); — à pivot, pin-jointed linkage; — bilatérale, two-way communication; — chimique, chemical bond; — conductrice de la chaleur, heat-conducting bond; — covalente, covalent bond, homopolar bond; — d'électrovalence, valence bond; — desmodromique, connection or coupling by positively acting means (under constraint), interlock.

liaison, — entrechaînée, cross linking; — entre les gènes, genetic linkage; — ionique, heteropolar bond, ionic bond; — métallique, metallic bond; — thermique, heat-conducting bond; — transversale, cross linking.

*liant, binding material, tie, union, bond.

*libelle, level (mech.).

*libellé, name, identifier, tag, trade description; — de macro, macro definition.

libérable, releasing.

*libération, — contrôlée d'énergie, — menagée d'énergie, controlled release of energy; — de canal, channel end; — de chaleur, heat production; — de lecteur, reader runout; — d'énergie, energy release, release of energy; — des contraintes, stress relief.

*libérer, to eject.

libre parcours moyen, *m.* mean (free) path; — **d'absorption,** absorption mean free path; — **de la réaction,** reaction mean free path; — **de transport,** transport mean free path.

libre temps moyen de réaction, *m.* reaction mean free time.

liebigite, *f.* liebigite.

*****lien,** link, linkage, connector.

*****ligature de fil,** wire connection, binding joint.

*****ligne,** row (EDP); — **adiabatique,** adiabatic line; — **aérienne,** air line, overhead line; — **à points,** dotted line; — **à retard,** delay line; — **bidirectionnelle,** duplex line; — **catenaire,** catenary construction; — **continue,** full line; — **d'action du vecteur,** vector line; — **d'alimentation,** feeder; — **d'atterrissage,** glide path, landing line; — **de conversion,** conversion line; — **de distribution,** distributor main; — **de drisse,** halyard.

ligne, — **de foi,** guide line; — **de force,** field line, line of force; — **de force circulaire,** circular line of force; — **de force électrique,** electric line of force; — **de force magnétique,** magnetic field line, magnetic line of force; — **d'embranchement,** junction line (R.R.); — **d'entrée-sortie,** input-output trunk; — **de poste aérienne,** air-mail route; — **des fibres invariables,** neutral axis (of rails); — **des temps,** time vector; — **de traction,** railway line; — **de transport,** transmission line.

ligne, — **d'exploration,** scanning line; — **d'images,** scanning line; — **d'inclinaison,** line of dip; — **interurbaine,** trunk line; — **médiane d'un segment,** stroke center line; — **privée,** tie-line; — **réservée,** leased line; — **secondaire,** feeder air line; — **supplémentaire,** extension; —**s d'impression,** sound track; —**s entrelacées,** interlaced lines (telev.); —**s noires du spectre solaire,** absorption lines.

*****lignée consanguine,** inbred line.

*****lime,** — **forte,** arm file; — **mordante,** grater; — **obtuse,** blunt file; *****— **plate,** hand file; — **rude,** coarse file.

liminaire, limit(ed).

*****limitation lambda,** lambda limiting process.

*****limite,** constraint, deadline, contour; — **apparante d'élasticité,** yield point; — **conventionelle d'élasticité,** yield strength; — **d'endurance,** endurance limit, fatigue limit; — **de résistance à la fatigue,** endurance limit; — **du réseau,** lattice boundary; — **élastique,** yield strength; — **extérieure de la zone de silence,** skip distance; — **extrapolée,** extrapolated boundary; — **floue d'allure exponentielle,** exponentially diffuse boundary; — **spécifique,** boundary (EDP); — **supérieure,** upper limit.

*****limité,** — **par le temps d'entrée,** input-bound; — **par le temps de traitement,** process-bound, computer-limited.

*****limiteur,** signal limiter; — **de course,** stop; — **d'effort,** force-limiting device; *****— **de tension,** voltage-limiting device, circuit breaker.

limiteur-répartiteur de freinage, *m.* brake pressure limiting valve.

limnimètre, *m.* limnimeter, level meter.

*****limon,** loam.

linéarisation, *f.* linearization.

linéariser, to linearize.

*****lingot,** billet, bar, lump; — **d'uranium,** uranium slug.

*****linguet,** drive tab (clutch).

linkage génétique, *m.* genetic linkage.

*****liqueur-mère,** feed, pregnant solution.

*****liquidation,** settlement.

*****liquide,** — **extincteur,** fire-extinguishing fluid; — **minéralisateur,** ore-forming fluid.

*****lire,** to sense.

liseur, *m.* pickup, scanner (of phonograph).

*****liseuse,** map light.

*****lisse,** stringer (chassis).

*****liste,** schedule; — **de contrôle,** checklist; — **de vérification,** checklist; — **directe,** push-up list (EDP); — **refoulée,** push-down list (EDP).

lithergol, *m.* lithergol.

littéral, *m.* code letter.

*****livre anglaise du commerce,** avoirdupois weight.

*****livre de bord,** logbook.

livre-pied, *m.* foot-pound.

livreur, *m.* deliveryman.

*****lixivation,** — **alcaline,** alkaline leaching; — **au carbonate,** carbonate leaching; — **par filtration,** percolation leaching.

*****localisation,** location, tracking (astronautics).

loci, *m.* locus, loci.

*****locomotive,** — **à tambour dévidoir,** locomotive with cable drum; — **à treuil,** locomotive with hauling drum; — **du fond,** mining locomotive; — **pour front de taille,** working face locomotive.

*****locus unique,** single locus.

*****logement,** housing (mach.), cable bedding; — **de béquille,** socket; — **de rotule,** socket pad for jack, swivel joint housing; — **de train,** landing gear well; — **fileté,** threaded recess.

logomètre, *m.* electric log (measuring vessel speed), logometer.

*loi, — adiabatique, adiabatic law; — d'action de masse, law of mass action; — de déplacement, displacement law; — de la défense passive, air-precaution legislation; — de Mariotte, Boyle's law; — des sinus, sine law.

*longeron, spar; — à âme pleine, solid spar; — arrière, rear spar; — avant, front spar; — caisson, box spar; à un seul —, monospar.

*longue portée, long-range.

*longeur, — de Debye, screening length; — de diffusion, diffusion length; — de la rainure, key bed; — de migration, migration length; — de mot, word length; — de radiation, radiation length; — de ralentissement, slowing-down length; — de registre, register length; — de relaxation, relaxation length; — de trajectoire, path length; — d'extrapolation, extrapolation distance, augmentation distance.

longeur, — d'extrapolation linéaire, linear extrapolation distance; — d'instruction, instruction length; — d'onde, wave length; — d'onde du neutron, neutron wave length; — d'onde effective, effective wave length; — d'onde propre (d'une bobine), natural wave length; — d'onde quantique, quantum-mechanical wave length; — du rayon d'action de l'effet d'écran, screening length; — montée, laid length; — utile, effective length (brake lining, etc.).

looping, m. loop; — vers l'avant, outside loop.

*loquet, hasp; — à ressort, jack latch.

*lot, batch; — à bon marché, job lot; — d'envoi, consignment; — en réception, testing batch; gagner le gros —, to be lucky.

*loupe, bloom (metal.).

*lourd, en —, dead weight.

louvoyage, m. tacking (naut.).

lové, coiled, wound.

*lubrifiant, pour forte pression, high-pressure lubricant.

lucite, f. lucite.

*lueur, flicker (in film); — négative, negative glow.

lumen heure, f. lumen-hour.

*lumière, — cendrée, earthshine; — ombragée, indirect lighting.

luminance, f. luminance, brightness; — énergétique spectrale, spectral radiance or steradiancy; — énergétique totale, radiance, steradiancy.

*luminosité, speed (of a lens), brightness, glow.

*lunette, bezel (bezel seal rim or watch lid), watch-crystal rim; — AR, back light (auto); — d'approche, telescope; — de glace mobile, glass rim with removable crystal; — de visée, sight tube; — équipée, sight assembly; — méridienne, transit (surv.); —s d'aviation, flying goggles; —s de custode, rear quarter windows of auto; —s de sûreté, eye protectors; —s protectrices, goggles.

*lustre, gloss.

*luter, to cement.

lux, m. lux.

lymphopénie, f. lymphopenia.

lyndochite, f. lyndochite.

lyre, f. lyre, adjusting plate; — commandes de vol, lyre-shaped bellcrank; — d'accrochage, finger-grip clip; — de dilation, loop (circuit, etc.); — de raccordement, connecting pipe.

M

*machine, — à balancier, beam engine; — à calculer à grande vitesse, high-speed computer; — à calculer analogique, analogue computer; — à calculer arithmétique, digital computer; — à calculer arithmétique électronique, electronic digital computer; — à calculer électronique, electronic computer; — à cintrer les tôles, brake (sheetmetal shop).

machine, — à épuisement, pumping machine; — à fabriquer le mortier, pugging mill; — à fileter, screw cutter; — à froid, freezer; — à perforer, punching machine; — à pilon, vertical engine; — à plier les tôles, bending machine; — à rainer, grooving machine.

machine, — à repasser électrique, ironing machine; — à soutirer, bottling machine; — bâbord, port engine; — calorique, hot-air engine; — comptable, accounting machine; — comptable alphanumérique, alphabetic accounting machine; — comptable électrique, electrical accounting machine; — de boulangerie, baking machine; — de renfort, auxiliary engine.

machine, — électronucléaire, electronuclear machine; — électrostatique à grande différence de potentiel, high-tension static machine; — frigorifique, ice machine; — locomotive tender, tender (R.R.); — thermique, heat engine; — tribord, starboard engine.

*mâchoire, — de filière, die; — de frein, brake shoe; —s àtordre, splicing clamps.

mackintoshite, *f.* mackintoshite.

maclage, *m.* twinning.

macroinstruction, *f.* macroinstruction; — imbriquée, inner macroinstruction.

macromolécule, macromolecule.

macroparticle, *f.* particulate; — en suspension dans l'air, airborne particulate.

*magasin, — à pellicule, film magazine; — à plaques, plateholder; — d'alimentation, feed hopper, card hopper, input magazine.

*magnétique, shear (of magnetic field).

*magneto, — blindée, screened magneto; — d'allumage, ignition magneto; — de départ, starting magneto.

magnétohydrodynamique, magnetohydrodynamic.

magnéton, *m.* magneton; — de Bohr, Bohr magneton; — électronique, electronic magneton; — nucléaire, nuclear magneton.

magnétostriction, *f.* magnetostriction.

magnétron, *m.* magnetron; — à temps de parcours électronique, transit-time magnetron.

*maille spatiale, spatial mesh.

main-courante, *f.* railing.

*maintien, — du vide, vacuum tightness; — en position, station keeping (astronautics).

*maison, — de construction, manufacturing firm; — de la radio, broadcasting house.

maitlandite, maitlandite.

maître-à-danse, maître de danse, *m.* caliper.

maître-oscillateur, *m.* master oscillator.

maître-ouvrier, *m.* foreman.

maîtresse-tige, *f.* drill stem.

*mal, — aéré, unventilated; — bâti, jerrybuilt; — de l'air, air sickness; — de mer, seasickness; — des rayons, radiation sickness.

*malaxeur, kneading machine.

*mâle de repêchage, male fishing trap.

*mamelon, — allongé, pipe nipple; — double, double union; —restreint, swaded nipple.

*manche, — à balai, control stick, control column; — à droite, control column over to right; — à gauche, control column over to left; — à incendie, fire hose; — au milieu, control column central; — au ventre, control column right back; — à vent, wind stocking; — de remplissage, filler neck; — en arrière, control column back; — remorquée, towed target.

*manchon, — à friction, expanding chuck; — à glissière, sleeve; — à soupape, casing float collar; — biconique, coupling with plain ends; — biconique à encoches, coupling with recessed ends; — de collecteur, sleeve of commutator; — de débrayage, clutch release sleeve; — incandescent, incandescent mantle; — protecteur, pipethread protector.

*mandrin, coil form, bobbin, collet, core; — à l'anneau, elastic chuck; — à mâchoires, jaw chuck; — brisé, expanding or elastic chuck; — de serrage, chuck; — de tour, lathe chuck; — de tour en l'air, jaw chuck; — pour tubes, tube expander.

*mandriner, to roll.

*manège, capstan.

*manette, multipoint switch; — d'accord, tuning knob; — d'arrêt, emergency-pull switch; — de blocage des commandes, gust lock control handle; — de changement de marche, reversing handle; — de combinateur, controller handle; — de commande, control handle; — de secours, emergency-pull switch; — des gaz, throttle lever.

manganine, *f.* manganin.

*manifestation de modèles réduits, model flying meeting.

*manipulateur, — asservi, master-slave manipulator; — télécommandé, master-slave manipulator; — universel, general-purpose manipulator.

*manipulation, keying (telegr.); — à distance, remote manipulation.

*manipuler, to key (telegr.).

*manivelle, — de lève-glace, window regulator handle; — de mise en marche, hand crank.

manocontact, *m.* pressure switch; — alarme de pression, pressure warning switch; — baisse de pression, pressure drop warning switch; — de pression, pressure indication switch; — de surpression, overpressure warning switch.

*manoeuvre, drill (mil.), seamanship.

*manoeuvrer le cadran d'appel, to dial.

*manomètre, vacuum gauge; — enregistreur, manograph.

*manque, — de development, underdevelopment (phot.); — de transparence, blackness.

mantisse, *f.* mantissa.

manuel, *m.* manual, handbook.

*manutention sous l'eau, underwater handling.

*maquette, mock-up, scale model, dummy; — de bois, solid scale model (made of solid wood); — **d'études aérodynamiques,** wind-tunnel model; — **volante,** flying-scale model.

*marche, run, operation, action; — **à deux positions,** two-position action; — **au ralenti,** idling of the engine, ticking over; — **du rayon,** ray path; — **en arrière,** backstroke, back motion; — **oscillante,** hunting; — **productive,** on stream; — **sans recyclage,** once-through operation.

*marchepied, running board.

marégraphe, *m.* tide gauge.

*marge, range; — **d'erreur,** error range; — **de sécurité,** lead time.

*marin, offshore.

*marquage multiple, multiple labeling.

*marque, flag, mark, stroke; — **de bande,** tape mark; — **de fiche,** record mark, item mark; — **de fin de bobine,** end-of-reel mark; — **de fin de fichier,** end-of-file mark or indicator; — **de group,** group mark; — **de mot,** word mark; — **d'enregistrement,** record mark, item mark; — **de segment,** segment mark; — **terminale,** end mark; —**s à chaud,** branding (on wood); —**s de matrissage,** die marks.

*marron, kernel core.

*marteau, — **à dents,** claw hammer; — **à deux mains,** sledge hammer; — **à panne bombée,** ball-peen hammer; — **à pilon,** steam hammer; — **de forgeron,** sledge hammer; — **d'impression,** striking hammer.

marteau-perforateur, *m.* drill hammer.

martellement, *m.* hammering, knocking (automobile).

martellerie, *f.* hammer mill.

*masque, visor, dummy section; — **de programme,** program mask.

*masse, — **atomique,** atomic mass, nuclidic mass; — **atomique chimique,** chemical atomic weight; — **au repos,** rest mass; — **critique,** critical mass; — **de l'électron au repos,** electron rest mass; — **de minerai,** ore body; — **d'un corpuscule,** particle mass; — **d'une particule,** particle mass; — **du noyau,** nuclear mass; — **électromagnétique,** electromagnetic mass.

masse, — **inerte,** inertial mass; — **isotopique,** isotopic mass, isotopic weight; — **moléculaire,** molecular mass; — **non suspendue,** unsprung weight; — **nucléaire vraie,** true nuclear mass; — **nulle au repos,** zero rest mass; — **polaire,** pole shoe; — **réduite,** reduced mass; — **sous-critique,** subcritical mass; — **spécifique apparente,** packing density; — **suspendue,** sprung weight; **tiré de la —,** cast as a single block; — **totale,** mass sum, total mass; —**s à pans,** sledge hammers.

*masselotte, flyweight, bob weight; — **d'embrayage,** shoe of centrifugal clutch.

*massicot, edge cutter.

*mastic, lute; — **réfractaire,** fireproof cement.

masuyite, *f.* masuyite.

*mât, — **de charge,** derrick (naut.); — **de fortune,** jury mast; — **de pavillon,** flagstaff; — **en treillis,** lattice mast; — **haubané,** stayed mast; — **ombilical,** umbilical mast (astronautics).

mâtage, *m.* calking.

*matelot de pont, common seaman.

mâter, to step (a mast).

mâtereau, *m.* spar (naut.).

matérialisation, *f.* materialization.

*matériaux du coeur, core material.

*matériel, hardware; — **classique,** unit record equipment.

*matière, — **active,** active material; — **à l'état cristallisé,** crystalline material; — **appauvrie,** depleted material, impoverished material; — **brute,** source material; — **constituant le coeur,** core material; — **cristalline,** crystalline material; — **epuisée,** depleted material; — **équivalente à un tissu,** tissue-equivalent ionization chamber, tissue-equivalent material; — **explosive,** explosive material; — **fertile,** breeder material, fertile material.

matière, — **filtrante,** filter medium; — **fissile,** active material, core material, fissionable material, seed; — **fusile,** fusionable material; — **fusionnable,** fusionable material; — **ou substance fortement radioactive,** highly radiative material; — **première,** feed, unprocessed material, unrefined material; — **radioactive contaminante,** contaminating radioactive material; — **réfractaire,** refractory material; — **soumise à une irradiation par les neutrons,** neutron-irradiated material; — **usante,** abrasive; —**s radioactives en suspension dans l'air,** airborne radioactivity.

*matrice, die bed (mach.), stamp; — **de choix,** adaptation matrix; — **de commutation,** matrix switch; — **de perforation,** punch die assembly; — **de réserve,** mother matrix; — **de transfert,** transfer matrix; — **de transition,** transition matrix; — **inférieure,** die block.

matricule, *f.* roll, register, employee number.

mâture en V, *f.* V struts (aviation).

*mauvaise géométrie, bad geometry.

*maximum, peak; — **de diffraction,** diffraction peak.

maxwell, *m.* maxwell.

*mazout, mazut.

mcurie, *f.* millicurie, mc.

*mécanicien, engine driver, fitter (mach.); — inspecteur, chief engineer (naut.).

*mécanique, — corpusculaire, elementary-particle mechanics; — des particules élémentaires, elementary-particle mechanics; — des systèmes non conservatifs, dissipative mechanics, nonconservative mechanics; — newtonienne, Newtonian mechanics; — ondulatoire, wave mechanics; — quantique, quantum mechanics; — statistique quantique, quantum statistical mechanics.

*mécanisme, apparatus, gear equipment; — d'arrêt, shutdown mechanism; — de déclenchement, trigger mechanism; — de déroulement, tape transport mechanism; — de distribution, valve mechanism; — d'entrainement, control drive mechanism.

mécano, *m.* (aircraft) mechanic.

mécanographie, *f.* punched card system, data processing, computer section, use of office machines; — à cartes perforées, punched card methods.

mécanoïde, *m.* hardware (EDP).

*mèche, — à centrer, countersinking bit; — de sûreté, safety fuse, electrical fuse.

medjitite, medjitite.

mégawattannée, *f.* megawatt-year.

megohm, *m.* megohm.

megohmmètre, *m.* megger, megohmmeter.

méiose, *f.* meiosis, reduction division.

*mélange, — alcoolisé, alcohol-blended fuel; — de sels fondus, fused-salt mixture; — détonant, explosive mixture; — d'isotopes, mixture of isotopes; — gazeux, gaseous mixture, gas mixture; — hétérogène, heterogeneous mixture; — réagissant, reactive mixture.

*membrane, diaphragm.

*membre de l'équipage, member of the crew.

*mémoire, data storage, storage device; — à accès rapide, rapid access storage; — à accès sélectif, direct access storage, random access storage; — à adresses, address memory; — à attente nulle, zero access storage; — à bande magnétique, magnetic tape storage; — acoustique, acoustic memory; — active, active, storage; — à disque magnetique, magnetic film storage; — adressable, addressable storage; — adressée, addressed storage; — à ferrites, core storage; — à feuillets, data cell; — à film magnétique, magnetic film storage; — à ligne à retard, delay line storage; — à matrice, matrix storage; — à mercure, mercury storage; — à mots, word organized storage; — annexe, bump storage;

— associative, associative storage, parallel search storage; — à tambour, drum storage; — auxiliaire, secondary storage; — chimique, chemical storage; — circulante, circulating memory; — commune, shared memory; — cyclique, dynamic storage; — d'appoint, auxiliary storage; — de contrôle, control memory; — de grande capacité, mass storage; — de masse, mass storage, mass memory; — de réserve, backing storage; — de transit, local storage; — de travail, work storage; — dormante, inactive storage; — dynamique, dynamic storage; — effaçable, erasable storage; — électrostatique, electrostatic storage; — en parallèle, parallel storage; — éphémère, temporary storage, volatile storage; — externe, external storage; — fixe, dead storage; — inaltérable, fixed storage, read-only storage; — intermédiaire, buffer storage; — interne, internal storage; — magnétique, magnetic storage; — matricielle, coordinate storage; — permanente, non-erasable storage; — physico-chimique, physical-chemical storage; — primaire, main storage; — primaire de grande capacité, bulk core storage; — principale, main storage, main memory; — rapide, quick-access storage; — régénérative, regenerative storage; — stable, non-volatile storage; — statique, static storage; — universelle, general storage.

mémoire-tampon, *f.* buffer storage.

mémorisation, *f.* memorizing, storage (EDP); — d'appel, clamp-on; tore de —, storage core.

mémoriser, to store.

mémostatique, resident.

mendéléefite, mendeleefite, mendelyeevite.

mendélévium, mendelevium.

mené, drawn (as of lines), led, conducted, guided.

*mention, instruction (telegr.).

menuisier, *m.* carpenter.

*mer, — agité, choppy sea; — debout, head sea; — de travers, beam sea; en —, offshore; — houleuse, heavy sea; — libre, open sea.

méson, *m.* meson; — chargé, charged meson; — léger, light meson; — lourd, heavy meson; — neutre, neutral meson; — pi, pion; — pseudoscalaire, pseudoscalar meson; — scalaire, scalar meson.

*message, — collectif, collective (or multiple-address) report; — d'avertissement, warning notice; — de prévision, weather forecast; — d'observation au sol, surface-observation report.

*mesure, pitch, gauge; — d'après la méthode

des couches sphériques, spherical-shell measurement; *— de capacité, cubic measure; — de laplacien, buckling measurements; — de l'irradiation, radiation measurement; — d'un yard, yardstick; — globale, integral measurement; — physique, physical measurement.

*métal, — à coussinets, babbitt metal, bearing metal; — alcalin, alkali metal; — employé, expanded metal; — laminé à chaud, hot-rolled metal; — liquide, liquid metal; — spongieux, biscuit.

*métallurgie des matériaux de la pile, reactor metallurgy.

métamictisation, f. metamictization.

métastase, f. metastasis.

*méthode, — à grand pouvoir séparateur, high-resolution method; — d'accès fondamentale, basic access method; — d'approche, approach (EDP); — de battements, beat method; — de battements zéro, zero-beat method; — de calcul, treatment; — de calcul à un seul groupe, one-velocity method; — de centrifugation, centrifuge method; — de différence des photons, photon difference method; — de radio-résonance, radioresonance method; — des éléments marqués, tracer technique.

méthode, — de séparation des isotopes par la spectrographie de masse, mass spectrographic method of isotope separation; — des longueurs, length method; — des particules associées, associated-particle technique; — des sphères d'action, d'influence, matched conics technique (astronautics); — des traceurs, tracer technique; — d'itération, iteration method; — du temps de vol, time-of-flight method; — exponentielle, exponential method; — réticulaire, mesh method; — souterraine, mining method.

*métre, — droit, straightedge; — pliant, folding ruler.

*mettre, — à la mer, to launch; — à la terre, to earth; — à l'échelle, to scale; — à l'épreuve, to tax, strain; *— au point, to focus; — des zéros, to zeroize; — en circuit, to switch in, switch on, connect in circuit; — en évidence, to bring out, ascertain, make evident; — en fonction, to enable; — en perce, to tap off (liquid); — en place, to implement.

mettre, — en régime, to put on stream; — en réserve, to pre-store; — en veilleuse, to bank up fires; — hors circuit, to cut out (elec.);— hors fonction, to disable; — l'altimètre au zéro, to set the altimeter; — les gaz, to open the throttle; — plein gaz, to open the throttle fully.

micanite, f. micanite.

microampère, m. microampere.

microcurie, f. microcurie.

microdureté, f. microhardness.

microfarad, m. microfarad.

microfissure, f. hair crack.

microhenry, m. microhenry.

micro-interrupteur, m. microswitch.

microlite, f. microlite.

micromanipulateur, m. micromanipulator.

micromètre-jauge, m. micrometer.

micro-onde, f. microwave.

*microphone électrodynamique, moving-coil microphone.

microphonicité des lampes, microphone effect.

microphotomètre, m. microphotometer.

microprogrammation, f. microprogramming.

microradiographie, f. microradiography.

microseconde, f. microsecond.

micro-siemens, m. micromho.

*migration, — des ions, — ionique, ion migration.

*mileu, peristasis; — absorbant, absorbing medium; — actif, active core, active section, core, seed; — actif à circulation double, two-pass core; — conducteur, conducting medium; — fertile, blanket, blanket assembly, breeder blanket, breeding blanket, envelope (of the reactor); — fissile, fissile core; — gazeux, gaseous medium; — hétérogène, inhomogeneous medium.

milieu, — multiplicateur, active section, core, core assembly, reactor core; — multiplicateur à l'uranium, uranium assembly; — multiplicateur sans reflecteur, unreflected assembly; — multiplicateur sous-critique, subcritical assembly; — multiplicateur type, reference core; — nonabsorbant, nonabsorbing medium; — nonhomogène, inhomogeneous medium; — sans source(s), source-free medium.

milliampèreseconde, f. milliampere-second.

millicurie, f. millicurie, mc.

*millième de l'unité de masse, millimass unit.

millirads, mrad units.

milliroentgen, mr, m. milliroentgen, mr unit.

mi-marée, f. half tide.

mi-mat, à —, half-mast.

*minerai, — à faible teneur, low-grade ore; — brut, crude ore; — concentré, mineral concentrate, ore concentrate; — de thorium, thorium ore; — de vanadium, vanadium ore;

— **d'uranium,** uranium ore; — **d'uranium à haute teneur,** high-grade uranium ore; — **pulvérulent,** powder ore; — **secondaire,** secondary mineral, secondary ore.

*****minimum de relèvement,** position of minimum signal.

minuterie, *f.* counting train, minute works (in a watch), timer.

minuteur, *m.* timer.

*****miroir magnétique,** magnetic mirror.

mis à la terre, earthed or grounded.

*****mise,** — **à jour superposée,** update-in-place; — **à l'air libre,** venting; — **à la masse,** grounding; — **au niveau,** leveling (float); — **au point,** tuning, aligning, focusing; — **au point des instruments de mesure,** instrumentation; — **en page,** framing (telev.), make-up, layout (print.); — **en phase,** phase setting or adjustment; — **en place,** set-up, fitting; — **en pression,** pressure buildup; — **en réserve,** storage; — **en route,** starting; — **en séquence,** sequencing; — **en service,** power on; — **hors circuit,** power off; — **sous tension,** power on.

mitotique, mitotic.

mitraillage, *m.* machine gunning.

*****mitraille,** small-shot effect, Schottky effect (in radio tubes).

mitrailleur en avion, *m.* air gunner.

mitrailleuse photographique, *f.* camera gun.

mitre onglet, *m.* 45-degree angle.

*****mixture,** gatch.

mnémotechnie, *f.* mnemonics, mnemotechny.

*****mobilité ionique,** ionic mobility.

*****mode,** — **arythmique,** start-stop mode; — **continue,** burst mode; — **d'accès,** access mode; — **de passage à l'état thermique,** thermalization process; — **d'exécution de réalisation,** exemplified embodiment of an invention; — **discontinue,** byte mode; — **opératoire,** procedure; — **par multiplet,** byte mode; — **substitutif,** substitute mode.

*****modèle,** template; — **à deux groupes,** two-group theory; — **à plusieurs groupes,** multi-group model; — **à un seul groupe,** one-group model; — **collectif du noyau,** collective nuclear model; — **d'appartement,** indoor flier, indoor model; — **d'après la théorie de l'âge,** age-theory model; — **de fichier,** file layout; — **de la goutte liquide,** liquid drop model; — **de la théorie à un groupe,** one-velocity model; — **de record,** record-holding model.

modèle, — **du noyau à frontière nette,** sharp-boundary model; — **du noyau à l'aide des particules alpha,** alpha-particle model of nucleus; — **du noyau à particules indépendantes,** independent-particle model; — **en couches,** shell model; — **expérimental,** experimental model; — **mathématique,** mathematical model; — **monocinétique,** one-velocity model; — **optique du noyau,** optical model of the nucleus; — **réduit,** mock-up; — **réduit d'une pile,** model reactor; — **unifié du noyau,** unified nuclear model.

*****modification,** — **d'adresse,** address modification; — **de commande,** change of control; — **des gènes,** gene change; — **des structures,** structural change; **grande** — **de structure,** gross structural change.

modulateur-démodulateur, *m.* modem, modulator-demodulator.

*****modulation,** — **de vitesse,** velocity modulation; — **du temps,** constant frequency; — **par déphasage,** outphasing modulation, phase modulation; — **par variation de la tension de plaque,** choke modulation; — **par variation de la tension-grille,** grid-bias modulation.

*****module,** — **à segments superposables,** overlay module; — **de départ,** source module; — **de glissement,** modulus of rigidity; — **d'élasticité,** modulus of elasticity; — **d'élasticité au cisaillement,** shearing modulus of elasticity; — **de rigidité,** modulus of rigidity; — **de rupture,** modulus of rupture; — **d'exploitation,** systems pack; — **diamétral,** diametral pitch; — **d'inertie,** section modulus; — **d'origine,** source module; — **ordinolingue actualisé,** load module; — **résultant,** object module; — **superposable,** overlay module.

*****modulé à fréquence musicale,** tone-modulated.

module-objet, *m.* object module.

*****moelle osseuse,** bone marrow.

molalité, molality.

molarité, molarity.

*****môle,** jetty.

*****molécule,** — **gazeuse ionisée,** ionized gas molecule; — **ionisée,** ionized molecule; — **marquée,** labeled molecule, tagged atom; — **neutre,** neutral molecule.

molette, *f.* head wheel, pulley.

mollissement, *m.* softening.

*****moment,** — **cinétique,** angular momentum; — **cinétique de rotation,** spin; — **cinétique nucléaire,** nuclear angular momentum; — **cinétique orbital,** orbital angular momentum; — **cinétique relatif au mouvement orbital,** orbital angular momentum; — **de dipôle,** dipole moment; — **de flexion, fléchissant,** bending moment; — **de multipôle,** multipole moment; — **de rotation,** torque; — **de torsion,** torque; — **d'inertie,** moment of inertia.

moment, — dipolaire, dipole moment; — **magnétique,** magnetic moment; — **magnétique anomal,** anomalous magnetic moment; — **magnétique à spin supérieur,** extraspin magnetic moment; — **magnétique de dipôle,** magnetic dipole moment; — **magnétique de spin,** spin magnetic moment; — **magnétique dipolaire,** magnetic dipole moment; — **nucléaire,** nuclear moment; — **quadripolaire,** quadrupole moment; — relatif au mouvement orbital, orbital moment.

monergol, *m.* monopropellant.

*****moniteur, — de file d'attente,** queue control block; — **de superposition,** overlay supervisor; — **de vol à voile,** gliding instructor; — **en chef,** chief flying instructor.

monitrice de perforation, *f.* chief keypunch operator.

monochromateur à cristal, *m.* crystal monochromator.

*****monochrome,** monochrome pencil, bands, or rays (in color phot.).

monoénergétique, monoenergetic.

monolithique, monolithic.

monomoteur, *m.* single-engined.

monoplace, *m.* single-seater; — **blindé,** armored single-seater; — **de bombardement en pique,** single-seater dive bomber; — **de combat,** single-seater fighter; — **de poursuite,** single-seater pursuit plane.

*****montage,** set-up, mount, jig, fixture, circuit; — **à neutrons intermédiaires,** intermediate assembly; — **à réaction,** regenerative circuit; — **critique,** critical experiment; — **de bande,** tape loading; — **de glace,** iceberg; — **de l'avion,** assembly of the airplane; — **d'ensemble des cartouches de combustible,** fuel assembly; — **des armes,** installation of armament; — **du barreau,** rod assembly.

montage, — encastré, flush mounting; — **en contre-temps,** push-pull arrangement (of tubes); — **en étoile,** star circuit, star connection; — **en surface,** parallel connection; — **en tampon,** buffer circuit; — **équilibré,** quiescent push-pull circuit; — **flottant,** shock mounting; — **instantané,** instantaneous assembly; — **symétrique,** push-pull circuit.

*****montant,** bed (mach.), stay (of grid); — **algébrique,** signed field; — **de renforcement,** buckstay (for condenser box); — **du train,** undercarriage strut.

*****monte-charge,** crane (mach.).

*****montée, — en accélération,** acceleration buildup (astronautics); — **en caisson,** chamber ascent; — **en chandelle,** zooming up; — **linéaire de réactivité,** linear increase in reactivity.

*****monter,** to reeve (a wire line), climb, fit; — **sur charnières,** to hinge.

montmorillonite, *f.* montmorillonite.

*****montre à arrêt,** stop watch.

*****monture,** stock (gun).

mordage, *m.* clamp.

*****mort due à l'irradiation,** radiation death.

mortaiseuse, *f.* slotting machine.

*****mortier, — asphaltique,** sand carpeting; — **peu épais,** slurry.

*****mort-terrain,** cover.

*****mot,** cell (EDP); — **complet,** full word; — **d'adresse de canal,** channel address word; — **de commande,** control word; — **de commande de canal,** channel control word; — **de commande d'organe,** unit control word; — **de commande d'unité,** unit control word; — **de longueur variable,** variable word length; — **de passe,** password code; — **d'état de canal,** channel status word; — **d'état de programme,** program status word; — **d'ordinateur,** computer word; — **double,** double word; — **modificateur,** index word; — **significatif,** keyword.

mot-clé, *m.* keyword, docuterm.

mot-instruction, *m.* instruction word.

mot-machine, *m.* machine word.

*****moteur, — à ailettes,** air-cooled engine; — **à arbre creux,** quill-drive motor; — **à bagues de frottement,** slip-ring motor; — **à cage (d'écureuil),** squirrel-cage motor; — **à carburateur,** spark ignition engine; — **à compresseur,** blower-fed engine; — **à enroulements auxiliaires,** split field series motor; — **aérobie,** air-breathing engine; — **à excitation,** compound-wound motor; — **à induit bobiné,** wound-rotor motor; — **à mouvement alternatif,** reciprocating engine; — **anaérobie,** non-airbreathing motor.

moteur, — à réaction, jet-propulsion motor; — **à réducteur,** geared engine; — **à régime lent,** slow-running engine; — **à soupapes latérales,** L-head engine; — **à soupapes latérales opposées,** T-head engine; — **asynchrone synchronisé,** synchronous induction motor; — **autosynchrone,** selsyn motor; — **à vapeur,** steam motor; — **avec compresseur pour basse altitude,** ground supercharged engine; — **de croisière,** sustainer engine; — **demi-suralimenté,** medium supercharged engine; — **des volets de courbure,** wing flap motor.

moteur, — **en étoile,** radical engine; — **en étoile sans reducteur,** ungeared radial engine; — **nu,** basic engine; — **phonique,** phonic drum motor; — **refroidi par liquide chaud,** hot-cooled engine; — **sans compresseur,** naturally aspirated engine, unsupercharged motor; — **supercarré,** oversquare engine; — **suralimenté,** supercharged engine; — **surcomprimé,** supercompression engine; — **suspendu élastiquement,** flexibly mounted engine; — **thermique,** heat engine.

moteur-fusée, *m.* rocket engine.

***motif,** design, pattern.

moto-ballon, *m.* motor balloon, motorized balloon.

moto-planeur, *m.* motor glider.

moto-ventilateur, *m.* motor-driven fan.

mouflage, reeving.

***moufle fixe,** crown block.

***moufler,** to reeve (a wire line).

***moule,** template; — **à creux,** heavy-cored casting.

***moulin,** curd breaker.

***moulure,** — **de ceinture,** waist molding (auto body); **rabot à** —, molding plane.

***mouvement,** transaction; — **aléatoire,** random motion; — **alternatif,** reciprocating motion; — **axial,** axial motion; — **d'agitation thermique,** thermal motion; — **de dérive,** drift motion; — **de lacet,** lurching movement; — **de rotation,** rotational motion; — **désordonné,** disordered motion, random motion; — **de soufflet,** bellows movement; — **des terres,** ground leveling.

mouvement, — **de translation,** translational motion; — **hélicoïdal,** helical motion; — **hydrodynamique,** hydrodynamic motion; — **irrégulier,** disordered motion; — **macroscopique,** macroscopic motion; — **ordonné,** ordered motion; — **pendulaire,** hunting; — **propre,** background; — **sinusoïdal,** harmonic motion; — **vol à voile,** gliding movement.

***moyen d'extinction,** fire-extinguishing medium.

***moyenne,** — **arithmétique,** arithmetic mean; — **de la puissance spécifique totale de la réaction,** mean total reaction power density; — **par rapport à la variable spatiale,** spatial average; — **pondérée,** weighted average.

***moyeu,** boss (propeller); — **à charnière,** door hinge hub; — **de came,** camb hub; — **de carter d'entrée d'air,** inlet case hub; — **fretté,** shrunk-on boss.

multi-adresse, multi-address; **instruction** —, multiple address code.

***multicolore,** polychromate.

multicontrôle, multiple control.

multiperforatrice, *f.* keypunch.

multiperforer, to lace.

multiplet, *m.* multiplet, byte; — **d'analyse,** sense byte; — **d'état,** status byte; — **(de spin) isobarique,** isotopic multiplet.

multiplex, *m.* multiplex, multiplexing; — **à répartition dans le temps,** time-division or time-shared multiplex; — **à répartition de fréquences,** frequency division multiplex.

***multiplicateur d'électrons,** secondary-emission multiplier.

***multiplication,** — **des neutrons,** neutron multiplication; — **du levier,** lever advantage; — **effective,** effective multiplication.

multiplicatrice numérique, *f.* digital multiplicator.

multiprogrammation, *f.* multiprogramming.

multitraitement, *m.* multiprocessing, concurrent processing.

multitransformeur, *m.* multiprocessor.

mumétal, *m.* Mumetal.

muon, muon.

***muqueuse,** mucosa.

***mur,** footwall; — **d'eau,** water wall; — **de chaleur,** thermal barrier; — **de son,** sound barrier; — **de soutenement,** retaining wall; — **massif,** fireproof wall; — **thermique,** thermal barrier.

***murage en briques,** brickwork, brick wall.

mutant, *m.* mutant.

mutateur, *m.* mutator.

***mutation,** — **dans la lignée germinale,** germinal mutation; — **dominante,** dominant mutation; — **en retour,** back mutation; — **génique,** gene mutation; — **létale,** lethal mutation; — **liée au sexe,** sex-linked mutation; — **naturelle,** natural mutation; — **nouvelle,** fresh mutation.

mutation, — **ponctuelle,** point mutation; — **provoquée par les rayonnements,** radiation-induced mutation; — **radio induite,** radiation-induced mutation; — **récessive,** recessive mutation; — **récessive nuisible par irradiation,** radiation-induced deleterious recessive mutation; — **réverse,** back mutation; — **spontanée,** spontaneous mutation.

mycalex, *m.* Mycalex.

myélographie, *f.* myelography.

N

naëgite, *f.* naegite.

***nageoire,** sponson, seawing.

nanofarad, *m.* nanofarad.

***nappe,** array (of antennas), sheet antenna; — **phréatique,** ground water table; — **triangulaire,** triangular flat-top system.

***narcose,** electric sleep.

***navigation,** — **aérienne,** aeronautics; — à **voile,** sailplaning, sailflying, aerial sailing; — **cosmique,** astronautics.

***navire,** — **abandonné,** derelict; — **piège,** decoy (naut.).

navire-citerne, *m.* tanker.

***nébulosité,** cloudiness.

***nécessaire de réparation,** repair kit.

nef, *f.* nave (arch.).

***neige carbonique,** dry ice.

néoplasme, *m.* neoplasm, neoplastic cell.

neper, *m.* neper.

neptunium, *m.* neptunium.

***nerf,** rib.

***nervé,** corrugated.

***nervure,** vane; — à **âme pleine,** solid rib; — **d'écartement,** spacer rib; — **de rive,** end rib; — **de séparation,** spacer rib; — **en arc,** crescent-shaped rib; — **en contreplaqué,** plywood rib; — **en treillis,** lattice rib; — **étanche,** sealed rib; — **forte,** bulkhead rib, main rib; — **glissière,** flap track rib; — **guignol,** bellcrank rib; — **naca,** naca profile; — **triangulée,** braced rib.

***netteté,** sharpness, definition.

***nettoyage à jet de sable,** sandblasting.

***nettoyeuse à vide,** vacuum cleaner.

neunérite, neunerite.

***neutralité électrique,** electrical neutrality.

neutretto, *m.* neutretto.

neutrino, *m.* neutrino.

neutrodynage, *m.* neutrodyning.

neutron, *m.* neutron; — **absorbé,** absorbed neutron; — **appartenant au noyau,** nuclear nuetron; — à **spin dirigé vers le bas,** down-spin neutron; — à **spin dirigé vers le haut,** up-spin neutron; — **de faible énergie,** low-energy neutron; — **de fission,** fission neutron; **de grande énergie,** high-energy neutron; — **de grande vitesse,** high-speed neutron; — **d'énergie thermique,** thermal-energy neutron; — **de résonance,** resonance neutron.

neutron, — **diffusé,** scattered neutron; — **épithermique,** slow neutron; — **froid,** cold neutron; — **incident,** incident neutron; — **instantané,** prompt neutron; — **instantané de fission,** instantaneous neutron of fission; — **intermédiaire,** intermediate neutron; — **lent,** slow neutron; — **libre,** free neutron; — **naturel,** natural neutron; — **n'ayant pas subi de chocs,** uncollided neutron.

neutron, — **non-diffusé,** uncollided neutron; — **non-ralenti,** unmoderated neutron; — **parasite,** background neutron; — **rapide,** fast neutron; — **rapide de fission,** high-speed fission neutron; — **retardé,** delayed-fission neutron, delayed neutron; — **secondaire,** secondary neutron; — **subthermique,** cold neutron; — **thermique,** thermal neutron; — **vagabond,** stray neutron,

neutronique, *f.* neutron physics.

neutronographie, *f.* x-ray made by means of neutrons.

neutronothérapie, *f.* neutron therapy.

nicolayite, *f.* thorogummite.

***nid de pie,** crow's-nest (naut.).

***nitrate,** — **de thorium,** thorium nitrate; — **d'uranyle,** uranyl nitrate.

***niveau,** — **de contrôle inférieur,** minor control; — **de contrôle supérieur,** major control (EDP); — **d'énergie,** energy level, power level, quantum level, quantum state; — **d'énergie nulle,** zero-energy level; — **de noir,** black level (TV); — **de programme,** program level; — **de tension,** voltage level; — **d'interruption,** interrupt level; — **d'irradiation,** amount of radiation; — **énergétique de résonance,** resonance level; **énergétique de rotation,** rotational energy, rotational level; — **énergétique du noyau,** nuclear energy level.

niveau, — **énergétique quasi-stationnaire,** quasi-stationary level; — **excité,** excited level (of nucleus); — **métastable,** metastable level; — **non-lié,** unbound level; — **virtuel,** virtual level; **au-dessous du** —, sublevel.

niveleuse, *f.* grader, motorgrader; — **automotrice,** motorgrader.

nivellomètre, *m.* liquid content gage.

nivénite, *f.* nivenite.

nivomètre, *m.* snow gauge.

nobélium, *m.* nobelium.

***noeud,** — **du réseau,** lattice node; — **radial,** radial node.

nohlite, *f.* nohlite.

*noir comme du jais, jet-black.

noircie, blackened, darkened, made opaque.

*noircissement de la plaque photographique, blackening of photographic plate.

*noix, — d'articulation, hinge yoke; — de cardan, yoke (helicopter); — de démarreur, starter jaw; — de lancement, starter (auto).

noliser, to charter.

*nom, — circonstancié, qualified name (EDP); — de mécanoïde, hardware name; — d'unité, hardware name.

*nombre, — absolu des fissions produites, absolute fission yield; — aléatoire, random number; — binaire, binary number; — de charge, atomic charge, atomic number; — de charge effectif, effective atomic number; de charge total, total charge number; — de fissions, fission rate; — de masse, atomic mass, mass number, nucleon number, nuclidic mass; — de masse total, total mass number; — de neutrons, neutron number; — de neutrons emis par fission, yield of neutrons per fission.

nombre, — de noyaux de la cible par cm², number of target nuclei per cm²; — de particules par unité de volume, number density; — de photons émis, yield of radiations; — de plateaux théoriques, number of theoretical plates (NTP); — des fissions produites, fission yield; — des neutrons émis par unité de temps, neutron rate; — des quanta émis, quantum efficiency, quantum yield; — des rayons gamma produits, gamma yield; — d'étalonnage, calibration number.

nombre, — d'étrangeté, strangeness number; — de transport des ions, transport number of ions; — d'occupation, occupation number; — d'onde, wave number; — d'unités de transfert, number of transfer units (NTU); — entier, integral number, integral quantity; — isotopique, isotopic number; — magique, magic number; — ordinal, ordinal number; — propre, characteristic number.

nombre, — quantique, quantum number; — quantique azimutal, azimuthal quantum number; — quantique de spin, spin quantum number; — quantique de spin isobarique, isobaric spin quantum number; — quantique magnétique, magnetic quantum number; — quantique orbital, orbital quantum number; — quantique principal, principal quantum number; — quantique radial, radial quantum number; — quantique secondaire, orbital quantum number, secondary quantum number; à — d'octanes élevé, high-octane.

*nomenclature, directory, arts list, schedule;

— de catalogue, catalog index; — d'instructions, instruction repertory.

nomination, f. appointment (to a position).

nomogramme, m. nomogram, alignment chart; — à points alignés, alignment-type nomogram.

non-condensé, unpacked (EDP).

non-conjonction, f. not-both operations (EDP).

non-conservation de la parité, f. nonconservation of parity.

non-consommable, trapped (fuel).

non-étanche, nonsealed.

non-fonctionnement, m. failure to operate.

non-hydrophile, nonabsorbent.

non-identification, f. failure of ascertainment.

non-linéaire, nonlinear.

non-pelucheux, lintless.

non-perturbé, undisturbed.

non-recensement, m. failure of ascertainment.

non-suspendu, unsprung.

non-travaillant, unstressed.

normaliser, to standardize, normalize.

*norme, — de tolérance, tolerance standard; — d'identification, standard of recognition.

*notation, — à base mixte, mixed-base notation, mixed-radix notation; — binaire, binary notation.

*note, — de battements, beat note; — harmonique, overtone (acoustics).

*notice, — constructeur, manufacturer's handbook; — de fonctionnement, operational manual; — d'entretien, maintenance manual; — technique, technical handbook.

*nourrice d'essence, gravity tank.

*nouveau traitement, reprocessing.

*nouvel arrangement du noyau, nuclear rearrangement.

noyage, m. flooding.

*noyau, hollow key, core or packet (of sheets or laminations); — à charge positive, positively charged nucleus; — à radioactivité artificielle, artificially radioactive nucleus; — à radioactivité naturelle, naturally radioactive nucleus; — atomique, atomic nucleus; — atomique stable, stable atomic nucleus; — bombardé, struck nucleus; — central, core; — cible, bombarded nucleus, target nucleus.

noyau, — composé, compound nucleus, intermediate nucleus; — condensé, condensed nucleus; — de commutation, switching core; — de l'intégrale de diffusion, scattering kernel; — **de l'intégrale de ralentissement,** slowing-down kernel; — **de l'intégrale de transport,** transport kernel; — **de recul,** recoil nucleus; — **dérivé,** resultant nucleus; — **d'hélium,** helium nucleus; — **d'hydrogène,** hydrogen nucleus; — **du combustible, fuel** nucleus.

noyau, — **d'uranium,** uranium nucleus; — **émetteur de particules alpha,** alpha-radioactive nucleus; — **engendré,** daughter nucleus; — **excité,** excited nucleus; — **fissile,** fissionable nucleus; — **formé,** product nucleus; — **générateur,** generative nucleus; — **impair-impair,** odd-odd nucleus.

noyau, — **impair-pair,** odd-even nucleus — **instable,** unstable nucleus; — **intégral de diffusion,** diffusion kernel; — **isobare,** nuclear isobar; — **isomère,** nuclear isomer; — **léger,** light nucleus; — **lourd,** heavy nucleus; — **miroir,** mirror nuclide.

noyau, — **non-magique,** nonmagic nucleus; — **nu,** bare nucleus; — **pair-impair,** even-odd nucleus; — **pair-pair,** even-even nucleus; — **particule de recul émise par radioactivité,** radioactive recoil; — **particule radioactive de recul,** radioactive recoil; — **précurseur,** parent nucleus; — **radioactif,** radioactive nucleus.

noyau, — **résiduel,** daughter nucleus; — **résultant,** product nucleus, resultant nucleus; — **résultant d'une fusion,** fusion nucleus; — **sans electrons,** bare nucleus; — **stable,** stable nucleus; — **très lourd,** super-heavy nucleus.

***nuage,** — **d'électrons,** electron cloud; — **de mésons,** meson cloud; — **d'ions,** ion cloud; — **mésique,** meson cloud.

***nuance,** grade (steel, etc.).

nucléation, f. nucleation.

nucléogenèse, nucleogenesis.

nucléon, m. nucleon; —**s non-appariés,** unpaired nucleons; —**s non-couples,** unpaired nucleons.

nucléoprotéine, nucleoprotein.

***numération,** — **des globules blancs,** white blood count; — **des globules rouges,** red blood count; — **des plaquettes,** platelet count; — **globulaire,** blood count; — **leucocytaire totale,** total white count; — **positionnelle,** positional notation; — **thrombocytaire,** platelet count.

***numéro,** — **atomique,** atomic number; — **atomique effectif,** effective atomic number; — **de montage,** build number; — **d'ensemble supérieur,** next assembly drawing; — **de plan,** drawing number; — **de sortie,** issue number; — **de stock,** stock number; — **de vol,** flight number; — **d'immatriculation,** registration number.

***numérotage,** dialing (teleph.).

numérotation, f. numbering, classification allocation; — **d'un livre,** paging of a book.

nu-pied, barefooted.

O

***objectif,** — **à diamètre à gradin,** objective with echeloned diameter, Fresnel or echelon lens; — **aérien,** aerial target; — **grand champ,** wide-angle lens.

***objet,** — **spatial,** object in space; —**s trouvés,** lost property.

obligé d'atterrir, forced to land.

***obtenir du plutonium par régénération,** to regenerate plutonium.

***obturateur,** mask, sealing means, sealer, baffle plate; — **au cadmium,** cadmium shutter; — **opaque aux neutrons,** neutron shutter.

obturé, blanked off.

***obturer,** to eclipse, mask, shutter.

***obus,** — **à mitraille,** shrapnel; — **de rupture,** armor-piercing shell.

***occupants carbonisés,** occupants burnt to death.

océlite, m. ocelit.

octet, m. octet, eight-bit byte.

***oeil,** ear, hook eye; — **normal,** average eye (the standard eye for photometry).

***oeillet,** — **de cable,** thimble; — **de sondage,** soldering tab; — **fileté,** eyebolt.

oersted, m. oersted.

***officer,** — **d'état-major,** staff officer; — **du génie,** military engineer; — **en second,** executive officer; — **général,** flag officer (naut.).

***oisif,** idle.

oléoduc, m. pipeline.

oléoréseau, m. hydrant system (aer.).

oléoserveur, m. servicer (aer.).

oligoélément, m. trace element.

***ombre dans l'espace,** spatial self-shielding.

***omnibus automobile,** motorbus.

***onde,** — **associée,** associated wave; — **carrée,** square wave; — **d'appel,** calling wave; — **de base,** fundamental wave; — **de choc,** shock wave; — **de choc cylindrique,** cylindrical shock wave; — **de choc oscillante,** oscillating shock wave; — **de compression,** compression wave; — **de pression,** pressure wave; — **de radiodiffusion,** broadcast wave.

onde, — **de repos,** spacing wave; — **d'espace,** sky wave; — **de tension,** voltage surge; — **de travail,** operating, marking, or working wave; — **diffusée,** scattered wave; — **électromagnétique,** electromagnetic wave; — **électrostatique,** electrostatic wave; — **entretenu,** continuous or sustained wave; — **fondamentale,** ordinary wave; — **hydromagnétique,** magnetohydrodynamic wave, hydromagnetic wave.

onde, — **incidente,** incident wave; — **limitrophe,** medium high-frequency wave; — **longitudinale,** longitudinal wave; — **mobile,** progressing or moving wave; — **moyenne,** medium wave; — **ordinaire,** ordinary wave; — **partielle,** partial wave; — **plane,** plane wave; — **porteuse,** carrier wave; — **progressive,** traveling wave, progressive wave.

onde, — **radio-électrique,** radio wave; — **rectangulaire,** rectangular wave; — **saccadée,** steep-fronted wave, surge; — **simple,** ordinary wave; — **sinusoidale,** sine wave; — **stationnaire,** standing wave; — **torsionnelle,** torsional wave; — **transversale,** transverse wave; — **ultracourte,** microwave, ultrashort wave; — **vibratoire transversale,** transverse vibrational wave; —**s dirigées,** beam transmission.

ondemètre, *m.* ondometer, wavemeter.

ondulateur, *m.* undulator.

*****onduler,** to pulsate.

onduleur, *m.* inverter.

*****opacité,** blackness.

*****opalescence laiteuse,** milky opalescence.

opaque, — **aux neutrons,** neutron-tight; — **aux radiations,** radiopaque.

opérande, *m.* operand, augend, augmend; — **B,** B-register, B-line; — **muet,** blank operand.

*****opérateur,** operator; — **binaire,** binary operator; — **booléen,** Boolean connective, logical operator; — **commutatif,** commutative operator; — **de division,** divider; — **de relation,** relational operator; — **laplacien,** Laplacian operator; — **logique,** logical operator, Boolean connective; — **unaire,** unary operator.

*****óperation,** — **à cycle fixe,** fixed-cycle operation; — **à tâches multiples,** multitask operation; — **à travaux multiples,** multijob operation; — **arithmétique,** arithmetical operation; — **binaire,** binary operation; — **comptable,**

bookkeeping operation; — **connexe,** auxiliary operation; — **d'aménagement,** red-tape operation; — **d'échantillonnage,** sampling action; — **d'éclusage,** air-locking, grating (in electron microscope); — **en parallèle,** parallel operation; — **éphémère,** one-shot operation; — **Et,** *and* operation; — **fondamentale,** basic operation; — **latérale,** auxiliary operation; — **logique,** logical operation; — **monadique,** monadic operation, unary operation; — **multipiste,** multiple track operation; — **Ni,** *nor* operation, zero-match; — **Ni L'un Ni L'autre,** *neither-nor* operation; — **Non,** *not* operation; — **Ou,** *or* operation; — **Ou Exclusif,** *exclusive-or* operation, nonequivalence operation; — **préférentielle,** privileged operation; — **privilégiée,** privileged operation; — **répétitive,** repetitive operation; — **séquentielle,** sequential operation; — **sérielle,** serial operation; — **simultanée,** simultaneous operation; — **Sinon,** *not-if-then* operation; — **sur deux opérandes,** dyadic operation.

*****opercule,** *m.* disk.

*****opposition rhythmée,** phase opposition, pushpull.

optimisation, *f.* optimization.

optimiser, to optimize.

*****optique neutronique,** neutron optics.

*****orange de balayage,** scanner.

orangite, *f.* orangite.

*****orbite,** — **à ensoleillement constant,** orbit giving constant sunlight ratio; — **circulaire,** circular orbit; — **d'attente,** parking orbit (astronautics); — **de l'électron,** electron orbit; — **de transfert,** transfer orbit; — **directe,** direct orbit; — **électronique,** electron orbit; — **elliptique,** elliptic orbit; — **equatoriale,** equatorial orbit; — **excentrique,** eccentric orbit; — **externe,** outer orbit; — **helicoïdale,** helical orbit; — **ionique,** ion orbit; — **képlérienne,** Keplerian orbit; — **non perturbée,** Keplerian orbit; — **perturbée,** disturbed orbit; — **polaire,** polar orbit; — **quasi parabolique,** nearly parabolic orbit; — **retrograde,** retrograde orbit; — **stable,** stable orbit; — **stationnaire,** stable orbit.

ordinateur, *m.* computer, data processing machine, stored program computer; — **à cycle variable,** variable cycle computer; — **à grande vitesse,** high-speed computer; — **à incréments,** incremental computer; — **à mots de longeurs variables,** variable word length computer; — **à mots isométriques,** fixed word length computer; — **asynchrone,** asynchronous computer; — **d'exécution,** object computer; — **spécialisé,** special purpose computer; — **synchrone,** synchronous computer; — **universel,** general purpose computer.

ordinateur-moniteur, *m.* processor controller.

ordination, *f.* computing.

ordinogramme, *m.* flowchart, block diagram, process chart.

ordonnancement, *m.* scheduling, sequencing, timing.

ordonnateur, *m.* scheduler (EDP).

*ordre, — du jour, agenda; — inférieur, low order; — supérieur, high order; —s en cascade, command chaining; —s en chaîne, command chaining.

*oreille de jonction, splicing ear.

*orfèvre, silversmith.

*organe, — analyseur, scanning device, scanner; — critique, critical organ; — de calcul, arithmetic unit; — de commande, control unit; — d'entrée, input unit; — de sortie, output unit; — de traitement, central processor, processing unit; — essentiel, critical organ; — synthétiseur, re-creator device (telev.), picture reproducer; — transitorisé, solid state component.

*organisation, — de mémoire, storage allocation (EDP).

oriel, *m.* oriel window.

*orientation, — des spins nucléaires, nuclear spin alignment; — dominante, preferred orientation; — préférentielle, preferred orientation.

original, *m.* master copy, master card; — de travail, working master.

orthodiagraphie, *f.* orthodiagraphy.

orthohydrogène, orthohydrogen.

oscillateur, *m.* oscillator; — à battment de fréquence, beat frequency oscillator; — à contréaction, feedback oscillator; — à dents de scie, saw-tooth oscillator; — à inductance saturable, reactance-controlled oscillator; — à inductance variable, inductance-controlled oscillator; — à ondes carrées, square-wave oscillator; — à quartz, crystal oscillator; — à réactance variable, reactance-controlled oscillator; — à résistance variable, variable resistance oscillator; — à tension variable, voltage-controlled oscillator; — commandé par quartz, quartz-crystal-controlled oscillator; — d'appoint, booster oscillator; — de balayage, sweep oscillator; — de fréquence auxiliaire, quenching oscillator; — de haute, fréquence, high-frequency oscillator; — de pile, pile oscillator; — pilote, master-drive or pilot oscillator, exciter tube; — symétrique, push-pull oscillator.

*oscillation, cycling; — amortie, stable oscillation; — amplifiée, unstable oscillation; — auto-entretenue, self-sustained oscillation; basculante, relaxation or time-base oscillation, saw-tooth or ratchet oscillation; — de dérapage, skid oscillation; — de lacet, snaking; — de résonance, resonant oscillation; — des électrons dans le plasma, electronic plasma oscillation; — des ions positifs, positive-ion oscillation; — des lignes, line sweep; — d'image, frame sweep; — d'orientation, orientational oscillation.

oscillation, — du plasma, plasma oscillation; — électrique, electrical transient; — électronique, pendulum motion of electron; — électrostatique du plasma, electrostatic plasma oscillation; — entretenue, sustained oscillation; — hydromagnétique, hydromagnetic oscillation; — ionique, ionic oscillation; — longitudinale, longitudinal oscillation.

oscillation, — pendulaire, hunting; — périodique, periodic oscillation; — propre, natural oscillation; — radiale, radial oscillation; — sinusoïdale, sine oscillation; — transversale, transverse oscillation; — turbulente du plasma, turbulent plasma oscillation.

*oscillographe, — à boucle, loop oscillograph; — à fer doux, soft-iron oscillograph.

oscilloscope, *m.* oscilloscope; — à rayons cathodiques, cathode-ray oscilloscope.

ostéogénique, osteogenic.

ostéosarcome, *m.* osteosarcoma.

ostéotrope, *m.* bone seeker.

*ou, — exclusif, exclusive *or*; — inclusif, inclusive *or*.

*ouie, louver, inlet, ear (ventilator).

*outil, — à charioter, lathe tool; — de programmation, software package; — de tour, lathe tool.

*outillage, hand tool.

ouvert à la circulation aérienne publique, open to public use.

*ouverture, — d'observation, scanning hole; — d'un canal expérimental, beam hole; — relative, relative aperture.

*ouvrage en terre, earthwork.

*ouvreaux, refractory blocks, burner blocks.

*ouvrier endurci, roughneck (slang).

*ouvrir, to turn on.

ovalisation, *f.* out of round, elongation.

oxycoupage, *m.* flame cutting, oxyacetylene cutting out.

*oxyde, — **brun d'uranium,** brown uranium dioxide; — **de béryllium,** beryllia (beryllium oxide); — **de deutérium,** deuterium oxide; — **de référence,** standard oxide; — **d'uranium** uranium oxide; — **noir d'uranium,** black uranium dioxide; — **pressé,** pressed oxide.

*oxygène lourd, heavy oxygen.

oza, gelatine reproductible.

P

pacte aérien, *m.* air pact.

paie, *f.* wage, payroll; **jour de** —, payday.

*paillette, keeper, armature, blade, reed.

*pair, de — **avec,** together with.

*paire, — **d'ions,** ion pair; — **formée par l'électron et le positon,** electron-positron pair; — **lacune-interstitiel,** vacancy-interstitial pair; —**s d'ions formées,** ion-pair yield.

*palan à chaine, chain hoist.

*palette, shovel; — **de commutateur,** switch blade; — **d'un relais,** tongue of a relay.

palettisable, palletizable.

palettiser, to palletize.

*palier, plateau, bracket, spindle bearing, cushion (mach.); — **à anneau graisseur,** oil ring; — **à chapeau,** pillow block; — **à collet** collar; — **à graissage automatique par bande mobile,** ring-lubrication bearing; — **d'essieu,** axle bearing; — **du vilebrequin,** crankshaft bearing; — **lisse,** plain bearing; — **suspendu,** hangar.

palonnier, *m.* rudder bar, spreader, sling bar, swingle bar, whipple tree; — **du frein,** compensator.

palpeur, *m.* sensor, probe, testing spike, cam follower; — **à signal lumineux,** rheostatic heat switch; pecker (EDP).

palplanches, *f.pl.* timber lining.

panmixie, *f.* panmixia.

*panne, breakdown; — **aléatoire,** random failure; — **cataleptique,** catastrophic failure; — **de courant,** power failure; — **de post-combustion,** failed after burner; — **de réacteur,** engine failure; **en** —, hove to; — **franche,** breakdown, straight failure; — **prévisible,** wear-out failure; **tomber en** — **d'essence,** to run out of gasoline.

*panneau, rack; — **acoustique,** baffle board; — **de commande,** control panel; — **de contrôle,** control panel; — **de sécurité,** hatch (relief and gauging).

*papier, forms; — **à cartes,** card stock; — **à dessin,** bristol board; — **au bromure,** bromide paper; — **cache collant,** masking tape; — **diagrammé,** coordinate, cross-section, or ruled paper; — **en accordéon,** continuous form, folding paper; — **fort,** cartridge paper; —

indicateur, litmus paper; — **millimetré,** graph paper; — **potée,** emery paper; — **sulfurisé,** greaseproof paper.

papillotage, *m.* **effet de** —, flicker effect.

papillotement, *m.* flicker.

*paquet, deck of cards, burst.

par tout ou rien, "on-off."

parabiose, *f.* parabiosis.

*parachèvement, trimming (mach.).

parachute, *m.* parachute; — **dossier,** backtype parachute; — **extracteur,** auxiliary parachute.

parachute-siège, *m.* seat-type parachute, seatpack parachute.

parachutisme, *m.* parachute jumping.

parachutiste, *m.* parachutist.

parahydrogène, *m.* parahydrogen.

*parallèle, en — **par bit,** parallel by bit; **en** — **par dual,** parallel by bit.

*parallélépipède rectangle, rectangular parallelepiped.

*paramagnètisme nucléaire, nuclear paramagnetism.

*paramètre, — **cinétique de la réaction,** reaction-rate parameter; — **de dispersion aléatoire,** straggling parameter; — **de la collision,** collision parameter; — **de la divergence,** criticality factor; — **de puits,** well parameter; — **de réglage,** control variable.

paramètre, — **dont on peut fixer la valeur,** adjustable parameter; — **du choc,** impact parameter; — **du réseau,** lattice parameter; — **numérique de fissilité,** dimensionless fissionability parameter; — **prédéterminé,** preset parameter; — **statique,** static parameter.

*parapet, hand railing.

parapher, to initial (a document).

parapositonium, *m.* parapositronium.

*parasite, idling gear, interference, stray; *pl.* static (radio), parasitic disturbance; —**s industriels,** man-made static.

*parc, — **à bétail,** stockyard; — **de stockage,** tank farm.

*parcelle assimilable à un point, point slug.

*parcours, access path, routing, stream, flowline, path length, range; — **d'amplificateur**, amplifier channel; — **de carte**, card track; — **d'une particule**, range of particle; — **extrapolé**, extrapolated range; — **inverse de diffusion**, inverse diffusion length; — **maximum**, maximum range; — **moyen**, mean range; — **moyen d'arrêt**, mean stopping path; — **rectiligne**, linear range; **temps de —**, transit time.

pare-brise, *m.* windshield.

pare-chocs, *m.* bumper.

*pare-étincelles, arcing contact.

pare-flammes, *m.* flame arrester, flame grease, flame trap.

parhelium, *m.* parhelium.

*parité intrinsèque, intrinsic parity.

*paroi en colombage, framework.

parsonite, *f.* parsonite.

*partage, bridging; — **de l'energie**, partition of energy; — **de temps**, time-sharing.

*partant, en angle de — , from transit angle.

*particule, — **accélérée**, accelerated particle; — **à faible parcours**, short-range particle; — **alpha**, alpha particle; — **alpha de grande énergie**, high-energy alpha particle; — **alpha résultante**, resultant alpha particle; — **arrêtée**, stopped particle; — **bêta**, beta particle; — **β-**, disintegration electron; — **bombardée**, bombarded particle; — **capturée**, trapped particle.

particule, — **chargée**, charged particle; — **chargée de grande vitesse**, high-speed charged particle; — **cible**, target particle; — **constituante**, constituent particle; — **coordonnée**, coordinated particle; — **de désintégration**, disintegration particle; — **de faible énergie**, low-energy particle; — **de grande énergie**, high-energy particle; — **de grande vitesse**, high-speed particle; — **de la gerbe**, shower particle.

particule, — **d'épreuve**, test particle; — **de recul de fission**, fission recoil particle; — **de recul émise par radioactivité**, radioactive recoil; — **d'essai**, test particle; — **diffusée**, scattered particle; — **du champ**, field particle; — **électrisée**, charged particle; — **élémentaire**, fundamental particle, subatomic particle; — **émise**, ejected particle, emerging particle; — **émise par le noyau**, nuclear recoil.

particule, — **étrange**, strange particle; — **expulsée**, ejected particle; — **fondamentale**, fundamental particle; — **formant**, structural particle; — **formée**, product particle; — **incidente**, bombarding particle, colliding particle, incident particle; — **initiale**, initial par-

ticle, original particle; — **injectée**, injected particle; — **ionisante**, ionizing particle.

particule, — **neutre**, neutral particle, uncharged particle; — **non-ionisante**, nonionizing particle; — **nucléaire**, nuclear particle; — **ponctuelle**, point particle; — **radioactive de recul**, radioactive recoil, — **radioactive en suspension dans l'air**, radioactive airborne particle; — **relativiste**, relativistic particle; — **retardée**, delayed particle; — **ultrarapide**, ultra-high-speed particle; — **utilisée**, consumed particle, spent particle.

*partie, — **active**, "meat" (of sandwich-type flue element); — **centrale (plaque)**, core; — **centrale de la bombe**, bomb core; — **fixe**, status portion; — **imaginaire**, imaginary part; — **réelle**, real part; — **supplémentaire**, attachment.

partie-adresse, *f.* address part.

partir, — **de l'eau**, to rise from the water; — **en vrille**, to fall into a spin.

*pas, program step, step length; — **à droite**, RH thread; — **à fixe**, fixed pitch (aviation); — **à gauche**, LH thread; — **à variable**, variable pitch (aviation); — **au collecteur**, commutator pitch; — **aux encoches**, slot pitch (of a drum winding); — **collectif**, attitude control system; — **de case**, track pitch; — **d'engrenage**, pitch (gears); — **dentaire**, tooth pitch; — **des spires**, pitch of turns; — **de tir**, launching pad; — **de Whitworth**, Whitworth thread (mach.); — **d'image**, frame gauge (moving-picture film); — **du réseau**, lattice pitch, lattice spacing.

*passage, run, pass (EDP), transparency or penetration (factor of radio tubes), leadin (conductor), ferry; — **à direct**, streamline; — **à niveau**, level crossing; — **de données**, data flow; — **défectueux**, misfeed; — **de roue**, wheel well; — **des neutrons à l'état thermique**, neutron thermalization; — **des neutrons à travers la matière**, passage of neutrons through matter; — **d'un électron (d'une couche à une autre)**, electron transfer; — **inférieur**, underground crossing; — **par un foyer**, focusing; — **réversible**, bi-directional flow.

*passe, coat or application (paint); — **d'essai**, test run; — **de travail**, manufacturing stage or phase.

passe-balle, *m.* caliber gauge.

passe-fil, *m.* grommet.

passepoil, *m.* edging, braid, piping (clothes).

*passer, — **à gué**, to ford; — **à l'encre**, to ink; — **au crible**, to screen; — **une commande**, to place an order.

*passerelle, jetway; — des instruments de mesure, instrument bridge.

*pâte, — à fourneaux, stove polish; — à roder, grinding compound.

*patin, skid; — de frein, brake pad.

patinage, m. skating, slipping, skid, skidding; — de courroie, belt slip; — de l'embrayage, clutch slippage; — des roues, wheel spin.

*patrimoine génétique, gene pool.

*patron, skipper.

patrouille, f. flight.

*patte, — de fixation, bracket; — de moteur, engine bearer; —s d'araignée, oil groove; —s d'une ancre, flukes of an anchor.

*pauvre, lean.

*pavillon, roof (auto); — d'écouteur, earpiece (telephone); — d'entrée, air intake duct; — microphone, mouthpiece (telephone); — national, ensign.

pavois, m. bulwark (naut.).

*pechblende, nasturan.

*pédale, — code, light dipping pedal; — d'accélérateur, accelerator pedal; — d'alternat, press-to-talk pedal; — de frein, brake pedal.

*pédoncule, squash, press (of a radio tube).

*pegmatite granitique, granite pegmatite.

*peigne, die, spray point.

*peinture, — au pistolet, spray painting; — de camouflage, shadow painting.

*pelle, blade.

*pellicule, — d'huile, oil film; — magnétique, magnetic thin film.

palvimétrie, f. pelvimetry.

pendage, m. hade (geol.).

pendante latérale, f. queen post.

*pendule, m. dropper (of a catenary line), pendulum.

pénétrance, f. penetrance.

*pénétrant, acute.

*pénétration, — de la barrière, barrier penetration; — de potentiel, barrier penetration; — du champ magnétique, magnetic-field penetration; — quantique, quantum leakage.

pénétromètre, m. penetrometer.

pentamoteur, five-engined.

pentaploïdie, f. pentaploidy.

*pente, mutual conductance, essentially linear portion of grid-voltage plate-current characteristic, grid-plate transconductance; —de conversion, conversion conductance; —

de la veine, dip slope (min.); — relative du palier, relative plateau slope.

pentode finale, f. output pentode.

*penture, — d'arrêt, strap hinge, check strap; — et gond, hook and hinge; —s du gouvernail, rudder bands.

peptisation, f. peptization.

*percement, cutting across, breakdown.

perce-meule, m. stone drill.

*percer, to disrupt.

*perche, boom.

perchiste, m. pole vaulter, boom man, perchman.

*perçoir à rochet, ratchet drill.

*perdre connaissance, to faint.

perfocalculateur, m. calculating punch, multiplying punch, multiplier.

*perforateur, casing gun; — à main, hand punch; — de bande, paper tape punch.

*perforation, — codée, code punching; — de carte graphitée, mark sense punching; — de vérification, control hole; — de zone, overpunch; — Douze, twelve punch, Y punch; — en grille, lace punching; — en série, gang punch; — erronée, error punch; — hors texte, overpunch, zone punch; — intercalée, interstage punch; — multiple, multiple punch; — numérique, numeric punch; — Onze, eleven punch, X punch; — par blocs, block punch; — significative, code hole; — Y, twelve punch.

*perforatrice, perforator, keypunch; — à clavier, keypunch; — manuelle, hand punch.

perforécepteur, reperforator.

perforeur, m. punch; — accouplé, summary punch; — de ruban, paper tape punch; — de sortie, output punch; — électrique, electric punch; — numérique, numeric punch.

perforeuse, f. keypunch operator; — de ruban, paper tape punch.

perforeuse-imprimante, f. printing punch.

performance, f. performance; — globale, overall performance; —s non-indiquées, performance figures not given.

pergelisol, m. permafrost.

périastre, m. periastron.

*périmer, to supersede.

*période, half-life, half-value period; — biologique, biological half-life; — comparative, comparative lifetime; — d'arrêt, down time; — d'échange, half time of exchange; — de démarrage, start-up time; — de désintégration, period of decay; — de doublement, doubling time.

periode, — de fission spontanée, spontaneous fission half-life; **— de fonctionnement,** operation period, sensitive time; **— de la pile,** pile period, reactor period; **— de latence,** latent period; **— de vibration collective,** period of collective oscillation; **— d'extinction,** decay time; **— d'impulsion,** pulsation period.

période, — du neutron, neutron period; **— effective,** effective half-life; **— glaciare,** ice age; **— radioactive,** radioactive half-life, half-value period, half-life; **— transitoire,** transient period; **— transitoire finale,** dying-down time (time of decay); **— transitoire initiale,** building-up time; **de courte —,** short-lived; **de longue —,** long-lived.

périodemètre, *m.* period meter.

*périodicité nucléaire, nuclear periodicity.

péristase, *f.* peristasis.

péritectique, peritectic.

perluette, *f.* ampersand.

permagel, *m.* permafrost.

Permalloy, *m.* Permalloy.

*perméabilité, *f.* (magnetic) permeability.

permeamètre, *m.* permeameter.

*permis, certificate; **— de conduire,** driver's license.

permittivité, permittivity.

*permutation, transportation; **— circulaire,** circular shift, cycle shift.

*persistance retinienne, visual persistence.

*personnel, — de pont, deck hands; **— d'exploitation,** operating staff; **— non-navigant,** ground personnel.

*perspective d'oiseau, bird's-eye view.

*perte, drop, leakage (elec.), casualty; **— de chaleur,** thermal loss; **— de courants de Foucault,** eddy-current loss; **— de l'accord,** detuning; **— d'énergie,** energy loss; **— d'énergie par ionisation,** ionization loss; **— d'énergie par rayonnement de freinage,** bremsstrahlung loss; **— d'énergie superficielle,** surface-energy loss.

perte, — de neutrons rapides, fast leakage; **— des neutrons par fuite,** loss of neutrons by escape; **— de vitesse,** stall; **— directe d'énergie,** direct energy loss; **— due à un nouveau traitement,** reprocessing loss; **— en neutrons thermiques,** thermal leakage; **— indirect d'énergie,** indirect energy loss.

perte, — irréversible d'énergie, irreversible energy loss; **— Joule en courant continu,** copper (Joule) losses with direct current; **— linéaire d'énergie,** linear energy transfer; **— par courants parasites,** eddy-current losses; **— par effet Joule,** resistance loss,

Joule loss; **— par grillage,** loss on ignition; **— par rayonnement,** radiation loss; **— supplémentaire,** stray loss.

pertinax, *m.* pertinax.

*perturbation, interference, static; **— à coude,** kinking perturbation; **— d'instabilité,** unstable perturbation.

*petit, — axe, minor axis; **— bras de mer,** inlet (geog.); **— dirigeable,** blimp (aviation); **— pas,** fine pitch; **—e boucle,** pellet; **—e poulie,** snatch block.

pétrolier, *m.* tanker.

*phare, — à éclats, flashing beacon; **— d'atterrissage,** landing light; **— de roulement,** taxi light; **— encastré,** flush headlight; **— flottant,** lightship; **—s code et route,** combined driving and dipped lights.

*phase, — aqueuse, aqueous phase; **— de même,** cophasal; **— d'exploration,** scan period; **— d'obturation,** dark interval (in projection); **— gazeuse,** gaseous phase; **— solide,** solid phase.

phasemètre, *m.* phasemeter, power factor.

Phasotron, Phasotron.

phénocopie, *f.* phenocopie.

*phénomène, — à coup unique, single-hit event; **— à deux coups,** double-hit event; **— chimique,** chemical process; **— continu,** continuous process; **— de battements,** beating; **— de collision,** collision process; **— de concentration,** concentration process; **— de dégradation de l'énergie,** energy degradation process; **— de dispersion aléatoire du parcours,** range straggling.

phénomène, — de duplication, replicative process; **— de fission,** fission process; **— de soudre,** rejoining process; **— de vieillissement,** age-hardening; **— électrique transitoire,** electrical transient; **— élémentaire,** unit process; **— provoquée par des neutrons,** neutron-induced process; **— réel de diffusion,** effective scattering event; **— transitoire,** transient process.

phlébographie, *f.* phlebography.

phonique, phonic.

phonogénique, suitable for sound recording.

phonon, *m.* phonon.

*phosphate d'uranyle, uranyl phosphate.

phosphuranylite, phosphuranylite.

phot, *m.* phot.

photocathode, *m.* photocathode.

photodésintégration, *f.* photodisintegration; **— du noyau,** nuclear photoelectric effect, photodissociation.

photodeutéron, *m.* photodeuteron.

photoélectron, *m.* photoelectron.

photoémission, *f.* photoproduction.

photoémulsion, *f.* photoemulsion.

photofission, *f.* photofission.

photoionisation, *f.* photo-ionization.

photolecture, *f.* mark scanning.

photoméson, *m.* photomeson.

*photomètre, — à papillottement,** flicker photometer; — à tâche d'huile, grease-spot photometer.

photomultiplicateur, *m.* multiplier phototube, photomultiplier tube.

photon, *m.* photon, light quantum; — d'annihilation, annihilation gamma quantum; — de dématérialisation, annihilation photon; — gamma, gamma-ray photon, gamma quantum.

photoneutron, *m.* photoneutron.

photoproton, *m.* photoproton.

photosensible, photosensitive.

phototélégramme, *m.* phototelegram.

phototélégraphie radioélectrique, *f.* picture transmission (radio).

*physique, — de l'état solide,** solid-state physics; — de neutron, neutron physics; — de plasma, plasma physics; — de santé, health physics; — nucléaire, nuclear physics; — nucléaire expérimentale, experimental nuclear physics.

*pic, — de diffraction,** diffraction peak; — de résonance, resonance peak; — photoélectrique, photo peak.

picofarad, *m.* picofarad, micromicrofarad.

picot, *m.* sprocket (of film feed).

*pièce, — active,** fuel slug; — brute, blank; — corrective, patch; — de campagne, field gun; — de fusion, seal (fused in glass); — d'électron coulée, electron casting; — de siège, siege gun; — détachée, spare part; — en fonte matricée, die casting; — sample; — forgée, forging; — justificative, voucher; —s embouties, pressings.

*pied, — à profondeur micrométrique,** micrometer depth gauge; — de centrage, locating dowel.

pied-de-biche, *m.* hammer claw, jimmy.

pied-de-boeuf, *m.* lie key.

*pied-de-chèvre, jimmy.

*piédestal, basis, base.

*piège, — à condensation,** cold trap; — froid, cold trap.

*pierre, — à affûter,** oilstone; — angulaire, cornerstone; — à plâtre, plaster of paris; — d'azur, azure stone.

*pieux, *m.pl.* piling.

pige, *f.* feeler gauge, measuring rod; — de niveau d'huile, dipstick; — micrométrique, stick micrometer; travail à la —, piece work.

*pignon, — à chevrons,** herringbone gear; — à ergot, pick-up gear; — à queue, shaft gear; — cloche, bell gear; — conique, bevel gear; — conique d'attaque, driving input bevel gear; — d'angle, bevel gear; — d'attaque, driving gear, input bevel pinion; — de chaîne, sprocket wheel, chain sprocket; — denté (câble), sprocket; — différentiel, differential gear; — droit, spur gear; — fou, idler gear; — planétaire, sun wheel; — principal d'engrenage epicycloïdal, sun gear; — réducteur, reducing gear; — satellite, planet gear; — silencieux, quiet gear.

pilbarite, pilbarite.

*pile, reactor; — à chaîne convergente de fissions,** convergent reactor; — à combustible circulant, circulating reactor; — à combustible dilué, diluted reactor; — à combustible enrichi, enriched pile; — à combustible en suspension, slurry reactor, suspension reactor; — à combustible fluidifié, fluidized reactor; — à combustible métallique liquide, liquid-metal fuel reactor; — à combustible solide fluidifié, fluidized solid-fuel reactor; — à deux fins, dual-purpose reactor; — à eau bouillante, boiling-water reactor, water-solution reactor.

pile, — à eau lourde, heavy-water pile; — à eau sous pression, pressurized-water reactor; — à ébullition, boiling reactor; — à échelle réduite, small-scale reactor; — à fission, fission reactor; — à flux intense, high-flux reactor; — à graphite, graphite reactor; — à graphite refroidie à l'air, air-cooled graphite reactor; — à l'état critique, critical reactor; — à liquide immobilisé, unspillable cell.

pile, — à neutrons de resonance, resonance reactor; — à neutrons rapides, fast-neutron reactor, fast reactor; — à neutrons thermiques, thermal-neutron reactor, thermal reactor, slow reactor; — APPR, package power reactor; — à plutonium, plutonium reactor; — à ralentisseur en graphite, graphite-moderated pile, graphite-moderated reactor; — à ralentisseur hydrogéné, hydrogen-moderated reactor; — à réactions en chaîne, chain-reacting pile; — à réseau, lattice reactor; — à suspension aqueuse, aqueous-slurry reactor.

pile, — **à thorium,** thorium reactor; — **atomique,** pile, atomic reactor, nuclear reactor; — **atomique à fissions,** chain reactor; — **à trés haute puissance,** superpower reactor; — **à uranium,** uranium furnace, uranium pile; — **à uranium et à graphite,** uranium-graphite pile; — **autorégulatrice,** self-stabilizing reactor; — **chaude (fission),** high-temperature reactor; — **chaudière,** boiling reactor, water-boiler reactor; — **chaudière expérimentale,** experimental boiling reactor; — **convertisseuse,** converter.

pile, — **couveuse,** breeder reactor, breeder; — **couveuse à neutrons rapides,** fast breeder, fast-neutron breeder reactor; — **cylindrique de hauteur finie,** finite cylindrical reactor; — **dangereuse,** unsafe pile, dangerous pile; — **de faible puissance,** low-power pile; — **de grande puissance,** high-power pile, high-energy level pile; — **d'épaisseur constante dépourvue de réflecteur,** bare slab reactor; — **dépourvue de réflecteur,** bare reactor (unreflected reactor); — **de puissance nulle,** zero-power reactor; — **de recherche,** research reactor.

pile, — **de recherche à caractéristiques poussées,** high-performance research reactor; — **d'essai,** test reactor; — **destinée à des études de détail,** development reactor; — **en état stationnaire,** stationary reactor; — **expérimentale,** experimental reactor; — **expérimentale à sodium,** sodium experimental reactor; — **fonctionnant dans la région de résonance,** resonance reactor; — **génératrice d'énergie,** secondary pile, power pile; — **hétérogène,** heterogeneous reactor; — **homogène,** homogeneous reactor.

pile, — **homogène à ébullition,** boiling homogeneous reactor; — **homogène à solution aqueuse,** aqueous homogeneous reactor; — **industrielle,** engineering reactor, industrial reactor; — **infinie à épaisseur constante,** slab reactor, infinite slab reactor; — **mobile,** mobile reactor; — **munie d'un réflecteur,** reflected reactor; — **non-empoisonnée,** clean reactor; — **perturbée,** distorted pile; — **plutonigène,** plutonium-producing reactor, production reactor; — **pourvue d'un ralentisseur,** moderated reactor.

pile, — **presque surrégénératrice,** near-breeder; — **primaire,** primary pile, primary reactor; — **productrice de chaleur industrielle,** process heat reactor; — **productrice de matière fissile,** production pile; — **prototype,** ptototype reactor; — **réacteur à ralentisseur organique,** organic-moderated reactor; — **refroidie à l'air,** air-cooled reactor; — **régénératrice,** regenerative reactor; — **sans**

ébullition, nonboiling reactor; — **secondaire,** secondary reactor.

pile, — **sigma,** sigma pile; — **surrégénératrice,** breeder, breeder reactor; — **surrégénératrice à neutrons de résonance,** resonance breeder; — **surgénératrice à neutrons rapides,** fast breeder, fast-neutron breeder reactor; — **surgénératrice à neutrons thermiques,** thermal breeder; — **surrégénératrice productrice d'énergie,** power breeder, power breeder reactor; — **thermoélectrique à neutrons,** neutron thermalization, neutron thermopile; — **transportable,** mobile reactor; — **utilisant l'eau comme ralentisseur,** water-moderated reactor; — **utilisant l'eau lourde comme ralentisseur,** heavy-water-moderated reactor.

pile, — **utilisant le béryllium comme ralentisseur,** beryllium-moderated reactor; — **utilisant le sodium comme réfrigérant,** sodium-cooled reactor; — **virtuelle,** image reactor, virtual reactor.

pile-piscine, *f.* swimming-pool-type reactor, open pool reactor.

pile-plaque, *f.* slab reactor; — **sans reflecteur,** bare slab reactor.

*****pilotage,** steering, attitude control, handling; — **automatique,** automatic control.

*****pilote,** master, drive, or pilot oscillator, exciter tube; — **de ligne,** route pilot; — **de vol à voile,** glider pilot; — **moniteur,** flying instructor.

*****piloter,** to stabilize (aeronautics).

*****pilotis,** framing.

*****pilules allumantes,** fuse pellets.

*****pince,** gripper; — **à commande flexible,** flexible shaft grapple; — **à fusible,** fuse tongs.

*****pinceau,** pencil, beam (of electrons); — **d'électrons,** jet (of cathode-ray oscillograph).

*****pincement,** pinch.

pinte, *f.* pint.

pion, *m.* pion.

pipelinier, *m.* pipeliner.

*****piqué à la verticale,** vertical dive.

pisékite, *f.* pisekite.

*****piste,** loop, track, runway; — **cimentée,** tarmac; — **courante,** primary track; — **d'atterrissage,** landing strip; — **de circulation,** taxiway; — **de décollage,** runway; — **dégagement,** secondary runway; — **d'envol,** runway; — **de référence,** library track; — **de secours,** alternate track; — **principale,** primary track; — **sonore,** sound track; — **supplémentaire,** alternate track; **en —,** on the line.

piste, true (exact).

*pistolet, — **de dessinateur,** curve (draw.); — **lance-fusée,** signal pistol, Very pistol; — **pneumatique,** spray gun.

*piston, swab; — **aspirant,** valve piston; — à **tête plate,** flat-topped piston; — **foré,** hollow piston.

*piton, staple.

pittinite, *f.* pittinite.

*pivot, spindle, swivel, gudgeon, axle end, journal (mach.).

*place, — **en dehors,** offset; — **vacante,** vacancy.

plafonnier, *m.* dome light (auto).

*plage, band space, field of vision.

*plan, — **coté,** dimensioned drawing; — **d'archives,** stock shot (TV, movies, etc.); — **de cablage,** wiring scheme, wiring plan; — **de clivage,** joint; — **de composition,** configurator; — **de glissement,** shore plane; — **de joint,** mating face, parting line, parting plane, gasket face; — **de l'orbite,** orbital plane; — **de mémoire,** storage map; — **de montage,** assembly drawing; — **de pose,** layout; — **de rejet,** slip plan; — **de réseau actif,** core lattice design; — **de rotation de l'hélice,** propeller disk.

plan, — **de stratification,** bedding plane; — **des supports,** media planning; — **d'exécution,** working drawing, "as fitted" drawing; — **en vraie grandeur,** full-size drawing; — **fixe,** horizontal fin (aviation); — **fixe horizontal,** fixed-tail plane; — **fixe réglable,** adjustable-tail plane; — **fonctionnel,** functional design; **gros** —, close-up; — **horizontal,** ground plan; — **serré,** close-up.

*planche, — **d'ambiance,** general layout drawing; — **de bord,** dashboard, instrument panel; — **de débarquement,** gangplank.

planchéiage, *m.* planking, boarding.

*plancher, — **surélevé,** false floor; **element de** —, floor panel (aircraft, etc.).

planéité, *f.* flatness, parallelism.

planette, *f.* drawknife.

*planeur, — **Chanute,** hanging glider; — **d'école,** elementary training glider; — **de début,** primary trainer glider; — **d'exercice,** intermediate glider; — **marin,** hydrosailplane; — **remorqué,** towed glider.

planification, *f.* planning; — **chaînée,** chained scheduling; — **des naissances,** family planning; — **en chaîne,** chained scheduling.

*planimétrie, plane geometry.

*plaque, badge, plate, sign, notice; — à **caissons,** box plate; — à **rosettes,** rosette plate;

— **collectrice,** pick-up plate; — **contenant de la matière fissile,** seed fuel plate; — **d'ancrage,** backstay; — **de centrage,** spacer plate; — **de déviation,** deflecting plate; — **de jointement,** seal plate; — **d'embase (de jointement),** bottom plate.

plaque, — **de séparation,** spacer plate; — **de tour,** lathe faceplate; — **en bois,** veneer; — **enchassée,** wall plate; — **frontale du déversoir,** weir dam plate; — **infinie,** infinite slab; — **sandwich,** sandwich plate; — **signalétique,** rating plate; — **tournante,** turntable.

plaque-chicane, *f.* baffle plate.

*plaquette, platelet; — **de frein,** brake pad.

*plasma, — à **autoréaction,** self-reacting plasma; — à **décharge gazeuse,** gas-discharge plasma; — à **deutérium,** deuterium plasma; — à **haute température,** high-temperature plasma; — **asymétrique,** unsymmetric plasma; — **chaud,** hot plasma; — **combustible,** fuel plasma; — **complètement ionisé,** fully ionized plasma.

plasma, — **confiné,** confined plasma; — **confiné par un champ magnétique,** magnetically confined plasma; — **délimité,** confined plasma; — **électronique,** electronic plasma; — **en état de non-équilibre,** nonequilibrium plasma distribution; — **faiblement ionisé,** weakly ionized plasma; — **hydrogénique,** hydrogenic plasma; — **ionique,** ionic plasma.

plasma, — **isolé,** isolated plasma; — **isotherme,** isothermal plasma; — **magnétiquement délimité,** magnetically confined plasma; — **non-turbulent,** nonturbulent plasma; — **réagissant,** reacting plasma; — **sous autostriction,** self-pinched plasma; — **surchauffé,** hot plasma.

plasmagène, *f.* plasmagene.

plasmatique, plasma physics.

plasmoïde, plasmoid.

plasmon, plasmon.

plat-bord, *m.* gunwale.

*plateau, faceplate, distributor; — à **calottes de barbotage,** bubble plate; — à **chicane,** baffle plate; — **annulaire,** doughnut disk plate; — **d'embrayage, d'entraînement,** drive plate (clutch), pressure plate; — **oscillant,** swash plate; — **perforé,** sieve plate.

*platine, — **de commande,** control panel (of radio equipment).

*plâtre cuit, burnt gypsum.

plâtrier, *m.* plasterer.

pleurographie, *f.* pleurography.

pliodynatron, *m.* pliodynatron.

ploïdie, ploidy.

*plomb, fuse; — **en saumons,** pig lead; — **filé,** lead wire; — **tetraéthylique,** tetraethyl lead.

*plomber, to calk, plug.

*plongement, pitch.

*plongeon, dive.

*plot, contact stud, hub; — **d'entrée,** entry hub, input hub, normal entry; — **de mise au repos normal,** normal dropout hub; — **de sortie,** exit hub, output hub, normal exit, outlet; — **mort,** dead contact; — **positif,** *plus* hub.

plumboniobite, *f.* plumbonicbiote.

*plume à dessiner, drawing pen.

plutonium, *m.* plutonium.

*pneu, cover; — **à bavette,** water deflecting tire; — **à ceinture, ceinturé,** radial-ply tire; — **clouté,** spiked tire; — **de rechange,** spare tire; — **diagonal,** bias tire.

*poche, lock (vapor); — **d'air,** air lock; — **de dégazage,** venting pocket (carburetor).

pochoir, *m.* template, stencil.

*poids, — **à sec,** dry weight; — **atomique brut,** rough atomic weight; — **au cheval vapeur,** weight per horsepower; — **à vide,** weight empty; — **de construction,** structural weight; — **du train routier,** gross train weight; — **lourd,** heavy-duty truck; — **total autorisé en charge,** gross vehicle weight; — **utile,** effective weight.

*poignée, — **de déclanchement,** release handle; — **de largage,** jettison handle; — **dynamométrique,** torque handle; — **escamotable,** retractable handle.

poilier, *m.* spindle.

*poinçon, spot punch, stamper (in record making), king post; — **de réserve,** master negative (in record making); — **emboutisseur,** punch die.

*poinçonner, to gauge.

*poinçonneuse, keypunch.

*point, — **à bombarder,** bombing target; — **cathodique,** cathode point; — **critique,** critical point; — **d'alimentation,** feed point; — **d'amorçage,** working point; — **d'application d'une force,** working point; — **d'arrêt programme,** breaking point; — **d'atterrissage,** landing spot; — **de bonification,** good point.

point, — **de branchement,** tapping point; — **d'ébullition,** boiling point; — **de charge,** spray point; — **de contrôle,** checkpoint; — **de coupure,** cutoff point; — **de croisement,** crossover point; — **de déclanchement,** trip point; — **de fonctionnement,** operating point; — **de fusion,** melting point.

point, — **de départ,** load point (EDP); — **de jonction,** matching point; — **d'engorgement,** flooding point; — **de passage des vitesses,** speed shift point; — **de pénalisation,** penalty; — **de pliage,** plait point; — **de prise,** tapping point; — **de régime,** normal operating point; — **de réglage,** set point; — **de repère,** control point; — **de reprise,** restart, rerun point; — **d'étanchéité,** sealing ring.

point, — **d'exploration,** scanning spot; — **dur,** friction point; — **estimé,** dead reckoning (naut.); — **focal,** focal spot; — **fractionnaire,** radix point; — **milieu,** midpoint, center tap (of a winding); — **mort,** backlash, home position, dead center; — **mort haut,** top dead center; — **mort inférieur,** bottom dead center; —s **brillants,** high lights; —s **et traits,** dots and dashes.

*pointage, classification by points.

*pointe, center punch, brad; — **de chaleur,** thermal spike; — **de fusée-sonde,** nose cone; — **percutante,** firing pin; — **thermique,** thermal spike.

*pointeau de flotteur, carburetor needle.

*poire, swedge.

*poison, contaminant, — **de fission,** fission poison; — **de fission gazeux,** poison gas; — **nucléaire,** reactor poison; — **radioactif,** radioactive poison.

*poix, — **d'asphalte,** bituminous pitch; — **de goudron de hêtre,** beech-tar pitch; — **pour brasseurs,** brewer's pitch.

polarisabilité, *f.* polarizability.

*polarisation, bias, biasing.

*pôle auxiliaire, interpole, commutating pole.

*polissage électrolytique, electrolytic polishing.

polycrase, *f.* polycrase.

polycristallin, polycrystalline.

polycythémie, *f.* polycythemia.

*polyèdre tronqué, frustum (geom.).

polyénergétique, polyenergetic.

polyglobulie essentielle, *f.* polycythemia.

polygonale, *f.* mesh (a series of branches forming a complete loop in a network).

polyploïdie, polyploidy.

polyréseau, *m.* composite lattice.

pompage, *m.* cyclic variation, hunting, exhauston; — **combine,** back-crank pumping; — **magnétique,** magnetic pumping.

*pompe, — **à balancier,** beam pump; — **accélératrice,** jet pump; — **à diffusion,** diffusion pump, vacuum diffusion pump; — **à feu,** fire engine; — **à induction,** induction pump; —

alimentaire, boiler feeder, donkey pump; — à membrane, diaphragm or surge pump; — à vide, vacuum pump.

pompe, — d'alimentation, supply pump, fuel pump; — de diffusion à un seul étage, single-stage diffusion pump; — de drain, bilge pump; — d'épreuve, boiler tester; — d'épuisement, sump pump; — de reprise, accelerating pump; — rotative, rotating pump; — rotative à enciente étanche, canned rotor pump; — rotative blindée, canned rotor pump.

*pomper, to exhaust.

pomperie, f. pump station, water works.

*pont, bracket, drive axle; — à poutres, girder bridge; — AR, rear axle; — AR à deux essieux, tandem drive axle of truck; — AV, front axle; — avant, foredeck; — de bateaux, pontoon bridge; — de chevalets, trestle bridge; — de mesure à contact glissant, slide-wire bridge; — d'envol, flying deck; — en encorbellement, cantilever bridge; — inférieur, lower deck; transformateur à deux —s, two-leg transformer.

pontage, m. cross-linking.

pontée, f. deck cargo.

pontet, m. trigger guard, cell connector (battery).

pont-levis, m. drawbridge.

pont-promenoir, m. promenade deck.

*populations mendéliennes, Mendelian populations.

*port, bearing capacity; — d'attache, port of registry; — de guerre, naval port; — de lettre, postage; — d'escale, port of call; — du couloir, film gate.

portance, m. lift.

*porte, — de visite, inspection door; — largable, push-out door.

porte-à-faux, m. overhang, cantilever.

porte-agrafe, m. ear, hook.

porte-avions, m. aircraft carrier.

porte-bombes, m. bomb rack, bomb carrier.

porte-cartes, m. map carrier.

*portée, firing range, shoulder, offset, transmission range; — d'affaiblissement, attenuation distance; — de fusil, gun range (distance); — de l'arbre, journal (mechanical engineering); — de soupape, valve collar; — d'une particule, range of particle; — efficace, effective range; — extrapolée, extrapolated range; — maximum, maximum range; — moyenne, mean range; — optique, visual or optical range; — résiduelle, residual range; à grande —, at long range.

porte-fusée, m. steering knuckle.

porte-fusibles, m. fuse holder.

porte-lames, m. saw frame.

porte-moteur, m. motor bearing.

*porte-objet, specimen holder.

*porte-outil, boring block.

porte-scie, f. blade holder, saw blade.

*porteur de charge, charged carrier.

*portique, crossbar.

*pose, — continue, time exposure; — de bit, de dual, bit settling (EDP); — de marque de mot, set word mark; — instantanée, instantaneous exposure, snapshot; — volante, laying of temporary track; train de —, track-laying train.

*poser, to land, let down (airplane).

pose-tubes, f. side-boom.

*position, — de marquage, mark position, response position; — de mémoire, storage location (EDP); — de perforation, punch position; — de zéro, home position; — d'impression, print position; — émission, transmit position; — excitée de sélecteur, X side of a selector; — retard, retard position (of pile); — sur un rayon, radial position; — travail, rest contacts, on position, operating configuration.

positionnement, m. registration, positioning; bras de —, positioning arm (EDP).

positionner, to index, set, position, calculate account balance.

positon, m. positron.

positonium, m. positronium.

*poste, m. equipment; — à distance, remote terminal, data terminal; — alternatif-continu, all-mains receiver; — d'affichage, display station (EDP); — de cantonnement, block post; — de commande, flight controller; — de lancement, blockhouse; — de lecture, read station; — de lecture à balais, brush station; — de pilotage, cockpit; — de pompage, pumping station; — de réception, receiving set (radio); — de secours, first-aid station, emergency set.

poste, — de tir, gun position, firing post; — de vigie, crow's-nest (naut.); — d'interrogation, inquiry station (EDP); — éloigné, remote terminal, data terminal; — émetteur, radio station; — émetteur-récepteur, transmitting and receiving set, two-way radio; — garde-côte, coast-guard station; — intermédiaire, way station; — prioritaire, hot-line terminal; — spécialisé, job-oriented terminal.

*pot, — circulaire d'échappement, exhaust collector ring; — d'échappement, muffler; — silencieux, exhaust muffler.

*poteau, en treillage, lattice mast or pole.

potelet, *m.* rigid support, small post, strut, prop.

*potentiel, voltage; — **de claquage,** breakdown potential, breakdown voltage; — **de forces centrales,** central potential; — **de pénétration,** penetration potential; — **d'excitation,** excitation potential; — **d'interaction,** interaction potential; — **d'ionisation,** ionization potential; — **explosif,** disruptive voltage.

potentiel, — **gravifique,** gravitational potential; — **mésique,** meson potential; — **nucléaire,** nuclear potential; — **scalarie,** scalar potential; — **vecteur,** vector potential.

*potentiomètre, — **à plots,** step potentiometer; — **de puissance,** volume control.

*poterie d'isolation, insulating tiles.

Pou du Ciel, *m.* sky louse (aviation).

*poudre, solid propellant (astronautics); — **à braser,** soldering powder; — **à canon sans fumée,** smokeless powder; — **à écurer,** scouring powder; — **de guerre,** gunpowder; — **de mine,** blasting powder; — **d'or,** gold dust; — **pour cuir,** baking powder.

*poudrette, bone black.

*poulie, spool, drum (mach.); — **à bord,** flange pulley; — **à courroie,** band wheel; — **à deux gorges,** double-sheave pulley; — **à émerillon,** swivel block; — **à gradins,** stepped pulley.

poulie, — **conductrice,** guide pulley; — **courante,** running (hoisting) block; — **de commande,** driving or operating pulley; — **de curage,** sand sheave; — **de forage,** crown pulley; — **de renvoi,** guide roller or pulley; — **de ventilateur,** fan pulley; — **étagée,** stepped pulley; — **mouflée,** pulley block; — **portante,** bearing pulley.

poupe, *f.* stern (naut.).

*poupée, — **coulissante,** lathe tailstock; — **motrice,** lathe headstock.

*pourcentage, — **de la dose profonde,** depth-dose percentage; — **des fissions conduisant directement à un produit donné,** independent fission yield, primary fission yield; — **des pertes,** percentage loss; — **de temps utile,** operating ratio.

*poursuite, tracking (astronautics).

*poussée, high gain, overrun (as a tube).

*pousser, to overrun, overvolt.

pousseur, *m.* booster (astronautics).

*poussier de coke, coke dust.

*poussière, — **de carneau,** flue dust; — **d'houille,** coal dust; — **radioactive,** radioactive dust.

*poussoir, push-pull rod (jet engine), follower mach.); — **à impulsion,** beep-switch; — **à tige, tappet;** — **à vis,** screw plunger; — **bipolaire,** two-pole push button; — **de blocage gyro,** gyro push to cage button; — **de contact,** contact button; — **de jaugeur,** capacitor unit test button; — **d'enclenchement,** cut-in push button; — **de sélection,** selector push button; — **de vide-vite,** dump valve button; — **et jaugeur,** capacitor and test button; — **manette,** auto manual push button.

*poutre, — **à âme pleine,** plate gland; — **transversale,** crossbeam, sleeper.

*pouvoir, — **absorbant,** absorptive power, absorptivity; — **antidétonant,** antiknock rating; — **calorifique,** calorific power, calorific value; — **d'arrêt,** stopping power; — **d'arrêt linéaire,** linear stopping power; — **d'arrêt massique,** mass stopping power; — **de pénétration,** penetrating power; — **de résolution,** power of resolution, resolving power; — **d'ionisation,** ionizing power.

pouvoir, — **émanateur,** emanating power; — **émissif,** emissivity; — **grossissant,** magnifying power; — **magnétique rotatoire,** rotatory magnetic power; — **ralentisseur,** slowing-down power; — **relargant,** slating strength; — **séparateur,** resolving power, power of resolution, separative power; **grand — de résolution,** high resolving power; **grand — séparateur,** high resolving power.

*pratique courante, usual standard.

*préalable, biased, biasing.

préamplificateur, *m.* preamplifier.

préavis, *m.* notification.

précédés, *m.pl.* ways and means, method of procedure.

préchargé, preloaded.

*précipitation avec entraîneur, precipitation with a carrier.

*précipité, — **entraîneur,** — **préformé,** preformed precipitate.

*précipité, abrupt.

*précis, — **de fichier,** data control block; **tir —,** accurate artillery fire.

précontrainte, *f.* prestress.

*précurseur, parent isotope, parent nuclide; — **radioactif,** radioactive percursor.

préenquête, *f.* pre-testing.

préfichier, *m.* header label.

*préfixe, code.

*prélèvement, collection, gathering (by an electrode in klystron), sampling (action).

premier, — secours, first aid; —s soins médicaux, first aid.

*première pierre, foundation stone.

*prendre, — de la vitesse, to gain speed; — des relèvements, to take bearings (naut.); — le terme moyen, to average; — un relèvement, to bear.

*préparation, — decolorante, decolorizer (chem.); — de mémoire intermédiaire, buffer scheduling (EDP); — mécanique des minerals, ore dressing.

prépayer, to prepay.

*préposé, (service-station) salesman; — à la réception, receiving clerk; — au port pétrolier, wharfinger.

*prepositionnement, m. preset.

préréacteur, m. experimental reactor.

*prescriptions techniques, specifications.

présérie, f. pre-production run, prototype production.

*pressage, — à chaud, hot-pressing; — à froid, cold-pressing.

*presse à paçage, veneer press.

presse-cartes, m. card weight (EDP).

presse-étoupe, m. stuff joint, stuffing box, gland, (bootleg) packer.

*pression, pressure head; — à la vanne d'admission, throttle pressure; — cinétique, kinetic pressure; — cinétique de gaz, particle pressure; — cinétique du gaz, gas kinetic pressure; — d'écoulement, open pressure; — de fonctionnement, working pressure; — de gisement, closed pressure, rock pressure.

pression, — de plasma, plasma pressure; — de souffle, blast pressure; — du deutérium, deuterium pressure; — du gaz, gas pressure; — magnétique, magnetic pressure; — moyenne du gaz, mean gas pressure; — radiative, radiation pressure.

presspan, m. pressboard.

*preuve, — d'addition, summation check; — par solde nul, zero balancing.

*prévision du temps, weather forecast.

*principle, — de conception, design principle; — d'exclusion, exclusion principle; — d'incertitude, uncertainty principle; — d'indétermination, uncertainty principle; — du tout ou rien, all-or-none basis; —s de fonctionnement, principles of operation.

*priorité, — effective, dispatching priority.

*prise, outlet, socket, plug; — à broches, plug (fitted with prongs or pins); — d'air, air intake; — de baladeuse, utility outlet, portable lamp plug; — de bobine, coil tap; — de courant à fiches, plug socket; — de mousse, foam connection; — de mouvement, power take-off, main drive pinion; — de repos, dummy connector.

prise, — d'essence, gasoline pump; — de tachymètre, speedometer drive; — de terrain, approach; — de terre, ground terminal, ground plate; *— de vues, taking or shooting (pictures), photographic camera; — directe, direct drive; — médiane, center tap (ping); — téléphonique de coque, telebriefing receptacle.

*prisme tronqué, frustum (of a prism), truncated prism.

*prix, — affiché, posted price; — d'honneur, prize cup; — en espèces, cash prize.

*probabilité, — d'adhérence, "sticking" probability; — d'échapper à la perte par fuite, nonleakage probability; — de désintégration, probability of disintegration; — de fission, fission probability; — de pénétration, penetration probability, transmission coefficient; — de réaction, probability of reaction; — de transition, transition probability; — d'ionisation, ionization probability.

*procédé, — à fournées, batch process; — à simple passe, single-stage process; — chimique, chemical process; — d'échantillonnage, sampling system; — de choc, collision process; — de collision, collision process; — de concentration, concentration process; — de duplication, replicative process; — d'enregistrement à longueur fixe, fixed-length record system; — de radiochimie, radiochemical process; — de régénération, regenerative process.

procédé, — de traitement discontinu, batch process; — d'itération, iteration method; — en continu, continuous process; — en discontinu, batch process; — intermittent, batch process; — réversible, reversible process; — trichrome, three-color process; — unitaire, unit process; —s concurrents, competing processes; —s rivaux, competing processes.

processeur, m. processor.

*production, — constante, constant yield; — continue d'énergie atomique, steady production of atomic power; — d'énergie interne, internal energy generation; — de neutrons par la fission, fission production of neutrons; — de paires, pair production; — de sons différentiels, intermodulation; — éruptive, flush production.

*produire, to generate.

*produit, — de condensation, condensate; **— de désintégration,** decay product; **— de désintégration radioactive,** radioactive decay product; **— de filiation,** daughter element, daughter product; **— de fission,** fission product; **— de fission de courte période,** short-lived fission product; **— de fission de longue période,** long-lived fission product; **— de fission gazeux,** gaseous fission product.

produit, — de fission radioactif, radioactive fission product; **— de fusion,** fusion product; **— de la réaction,** reaction product; **— d'empoisonnement,** contaminant; **— électrisé de la réaction,** charged reaction product (in plasma); **— final,** final product; **— intermédiaire,** intermediate product; **— résidu,** by-product; **—s de queue,** bottoms.

*profile, streamline wire.

programmateur, m. programmer, scheduler; **— de travaux,** job scheduler; **— principal,** master scheduler.

programmation, f. programming; **— automatique,** automatic programming; **— de canaux,** channel scheduling; **— en absolu,** absolute programming; **— en parallèle,** multiprogramming; **— linéaire,** linear programming; **— optimisée,** optimum programming; **— par bloc,** modular programming; **— symbolique,** symbolic coding.

*programme, schedule; **— à chargement efficace,** load-to-go program; **— algorithmique,** source program; **— assembleur,** assembly program; **— coulisse,** background program; **— d'arrière-plan,** background program; **— d'assemblage,** assembler, assembly program; **— de chargement,** load program; **— de commande,** control program; **— de commande de travail,** job control program; **— de contrôle,** check program; **— de détection,** tracing program; **— d'entretien,** service routine; **— de premier plan,** foreground program; **— de programmothéque,** library program; **— de service,** utility program; **— desmogène,** linkage editor; **— de sortie,** output routine; **— d'essai,** test program; **— de traitement,** processing program; **— de transfert de support,** dump-restore program; **— de vérification,** check program; **— directeur,** master control program; **— enregistré,** stored program; **— éphémère,** one-shot program; **— généralisé,** generalized routine; **— interprétatif,** interpretive program; **— manuel,** single cycle process; **— mémorisé,** stored program; **— optimisé,** optimum program; **— ordinolingue,** object program, target program; **— ordinolingue actualisé,** load module; **— principal,** master program; **— résultant,** object program, target program; **— sur bande,** program tape; **— traducteur,** interpreting program; **— utilitaire,** general program.

programmer, to program; **— sans boucles,** to unwind (EDP).

programmerie, f. software.

programmeur, m. programmer.

programmothécaire, m. librarian (EDP).

programmothéque, f. program library; **— à bande,** program tape; **— d'appoint,** private library; **— de transposables,** relocatable library; **— d'usager,** private library; **— figée,** core image library.

projecteur, m. projector, discharger; **— d'atterrissage,** landing light; **— de piste,** runway floodlights; **— de virage,** wide-beam headlight; **— orientable,** spotlight.

*projectile, — atomique, atomic projectile; **— nucléaire,** nuclear projectile.

*projet de pile, reactor design.

prométhéum, m. promethium.

*propagation, — des ondes électromagnétiques, electromagnetic wave propagation; **— d'ondes,** wave propagation; **— sans atténuation,** unattenuated propagation.

propergol, m. propellant.

*proportion, — des isotopes du carbone, carbon-isotope ratio; **— des sexes,** sex ratio.

*proportionnalité limitée, limited proportionality.

*propriété, — chimique, chemical property; **— de ralentissement,** moderating property; **— du noyau,** nuclear property; **— nucléaire,** nuclear property; **— physique,** physical property.

propulsé par fusées, rocket-driven.

*propulseur, engine, motor, jet, thrust; **— auxiliaire,** booster; **— électrothermique,** resistojet.

*propulsion, drive; **— à quatre roues motrices,** four-wheel drive; **— arrière,** rear-wheel drive.

prospection, f. surveying, prospecting; **— aérienne,** aerial survey, air survey; **— au sol,** ground survey; **— géologique au sol,** geological ground survey.

protactinides, proactinides.

protactinium, m. protoactinium.

*protection, — contre la radioactivité, radiological protection; **— contre les rayonnements,** health physics, shielding against radiation; **— contre l'irradiation,** protection against radiation.

protégé, protected, secure.

protium, m. protium.

proton, *m.* proton; — **appartenant au noyau**, nuclear proton; — **de recul**, recoil proton; — **incident**, incident proton; — **primaire**, primary proton.

protoxyde, *m.* protoxide, suboxide.

*****provision de carburant**, fuel supply, tankage.

provoqué, induced.

pseudo-image, *f.* pseudoimage.

pseudo-vecteur, *m.* pseudovector.

publipostage, *m.* mailing.

*****puisage au gaz**, gas lift.

*****puisard**, pit, drain pit, catch basin.

*****puiser**, to bail.

*****puissamment armé**, heavily armed.

*****puissance**, potency; — **absorbée**, input; — **apparent**, normal power; — **calorifique nominale en MW**, nominal rated output (megawatts: thermal); — **commandé**, output power; — **débitée**, power output; — **de commande**, driving power; — **de coupure**, cutoff input; — **de décollage**, take-off power; — **de fonctionnement**, operating power.

puissance, — **de pointe**, peak output power; — **de rayonnement**, radiated power; — **de régime**, operating power; — **d'obscuration**, cutoff input; — **du levier**, leverage; — **électrique aux barres**, net electric capacity; — **électrique nette**, net electric capacity; — **équivalente sur arbre**, equivalent shaft horsepower; — **fournie**, output.

puissance, — **installée**, power capacity; — **instantanée de la réaction**, instantaneous reaction power; — **massique**, power-to-weight ratio; — **nominale**, power rating, rated capacity, rated power; — **nulle**, zero power; — **spécifique**, power density, specific power; — **spécifique du rayonnement**, radiation power density; — **théorique prévue**, design power; — **utile**, useful power; — **vocal**, acoustical speech power.

*****puits**, — **à bras**, dug well; — **à calories**, thermal pit; — **à chaleur**, heat sink; — **à potentiel sphérique**, spherical well; — **carré**, square well; — **carré de potentiel central**, square spherical well; — **carré de potentiel complexe**, complex square well; — **d'aérage**, air shaft.

puits, — **de bougie**, spark-plug recess; — **de cendrier**, ashpit; — **de potentiel**, potential box, potential well; — **de potentiel arrondi**, smooth potential well; — **de potentiel carré de dimensions finies**, finite square well; — **d'extraction**, hoisting shaft; — **exponentiel**, exponential well; — **jaillissant**, gusher, spouter; — **rectangulaire**, rectangular well; — **thermique**, thermal pit, thermal bubble.

*****pulpe d'attaque acide**, acid leach slurry.

*****pulsation**, angular velocity or frequency.

*****pulvérisation**, spray; — **anodique**, anode sputtering; — **cathodique**, cathode sputtering.

pupinisation, *f.* pupinization, coil loading.

pupiniser, to pupinize.

pûpitre, *m.* console, desk; — **de commande**, control panel.

*****purge**, blowdown, bleeding (fluids), dump (EDP).

purger, to clear, dump (EDP).

*****purgeur de vapeur**, steam trap.

purpura, *m.* purpura.

pyélographie, *f.* pyelography.

*****pylône**, *m.* pole, mast, pylon; — **antenne**, antenna of tower-aerial type.

*****pyramide tronquée**, frustum (of a pyramid).

pyrochlore, *m.* pyrochlore.

pyrométallurgie, *f.* pyrometallurgy.

*****pyrotechnie**, powder factory.

pyrotron, mirror machine, pyrotron.

Q

*****quadrilatère**, — **articulé**, Ackerman quadrilateral system; — **déformable**, variable quadrangle (auto suspension); — **gauche**, skew quadrilateral.

*****quadrillage**, screen.

quadrimoteur, four-engined.

quadrinôme, *m.* quadrinomial.

quadripole, *m.* quadripole.

quadruplace, *f.* four-seater.

quaiche, *f.* ketch.

qualificateur, *m.* qualifier.

qualimètre, *m.* penetrometer.

quantification, *f.* quantization; — **d'un signal**, quantization, quantizing; — **spatiale**, space quantization.

quantifier, to quantize.

quantimètre, quantitomètre, *m.* quantimeter.

*****quantité**, — **d'électricité produite**, power output; — **de mouvement**, momentum; — **de mouvement relative au mouvement orbital**,

orbital moment; — **d'énergie directement perdue**, direct energy loss; — **d'énergie libérée**, energy release; — **d'énergie libérée par réaction**, energy release per reaction; — **d'énergie produite**, energy yield; — **de neutrons émis**, neutron yield.

quantité, — **discrète**, intergral quantity; — **dissipée**, dissipation rate; — **économique**, economical order quantity; — **fournie au système**, input; — **fournie débitée**, output; — **irrationnelle**, surd (math.); — **minimale par emballage**, standard package quantity; — **perdue par diffusion**, diffusion loss rate; — **scalaire**, scalar (quantity); — **vectorielle**, vector quantity.

*****quantum**, — **d'action**, quantum of action; — **de grande énergie**, high-energy quantum; — **de lumière**, light quantum; — **d'énergie**, quantum of energy; — **du champ**, field quantum; — **gamma d'annihilation**, annihilation gamma quantum; — **virtuel**, virtual quantum.

quarte, *f.* quad.

*****quartier général**, headquarters (mil.).

quartier-maître, *m.* quartermaster (naut.).

*****quarts du compas**, points of the compass.

quasi-statique, quas-static.

quatorze centilitres, gill (measure).

quétone, *m.* ketone.

*****queue**, plate lug; — **d'aronde**, dove tail; — **de carpe**, trapezoid cup; — **de l'avion**, aircraft tail unit; — **de lion**, tilting lever; — **de pignon conique d'attaque**, input bevel pinion shaft; — **de poussée**, thrust decay; — **de rat**, rat tail; — **de rivet**, rivet shank end; — **de rotule**, jacking pad end (jack); — **de roue conique**, bevel ring flanged shaft; — **de soupape**, valve system.

*****queue-d'aronde**, dovetail.

queues, *f.pl.* tailings.

queuzot, *m.* pumping lead (of vacuum pump), exhaust tube, exhaust vent.

*****quille**, hull, quill, ventral fin; — **d'angle**, chine; — **d'echouage**, keel; — **de poussée**, thrust member; — **de voilure**, wing bolster beam; — **horizontale**, even keel.

quinconce, en —, staggered.

quinteux, -euse, peevish, fretful.

R

rabattable, hinged, folding.

*****rabattage des bords**, flaring mouth, bell-mouth.

*****rabot debout**, jack plane.

rabotage, *m.* planing (carp.); — **incliné**, angle planing.

raboteur, *m.* clamper.

raboutage, *m.* stubbing.

*****raccord**, hose coupling, adapter, patch attachment; — **à démontage rapide**, quick disconnect; — **à planter**, plug-in union; — **articulé**, swivel fitting; — **à serre-câble**, adapter; — **auto-obturant**, self-sealing coupling; — **auto-obturateur**, presslock coupling; — **banjo**, banjo union; — **coaxial**, coaxial connector; — **coudé**, elbow union; — **croisé**, crossing; — **d'aération**, breather connection; — **de décharge**, overboard fitting; — **de déclenchment**, teat of pneumatic release; — **de graissage**, grease fitting; — **de marques**, stroke centerline; — **d'enveloppe**, case fitting; — **de réduction**, pipe reducer; — **de segments**, stroke centerline; — **de tuyauterie**, fitting; — **fileté**, screw-threaded union, film splice; — **orientable**, swivel connection; — **oscillant**, banjo connection; — **presse-étoupe**, gland (mach.); — **réducteur**,

restricting union; —**torique**, circular union, cylindrical union; — **trois pieces**, union.

*****raccorder**, to patch, make flush, splice, blend out; — **à la masse, à la terre**, to ground; — **transversalement**, to cross-connect.

raccorderie, *f.* fittings.

raccroché, on-hook.

*****raccrocher**, to hang up (teleph.).

*****racine**, radix (math.).

*****raclette**, squeegee.

*****racloir de pile**, battery knife.

rad, rad.

radar, *m.* radar; — **à balayage latéral**, sideways-looking radar; — **à impulsions**, pulse radar; — **d'alerte lointaine**, early-warning radar; — **d'altitude**, beavertail; — **de bord**, airborne radar; — **de conduite de tir**, fire control radar; — **de designation d'objectif**, acquisition radar; — **de recherche**, search radar; — **de surveillance**, airport surveillance radar (ASR); — **de tenue de poste**, station keeping radar; — **d'obstacle**, terrain avoidance radar; — **météo**, weather radar; **autoguidage par** —, radar homing; **balise** —, radar beacon.

radariser, to equip with radar.

radariste, _m. & f._ radar operator.

radian, _m._ radian.

***radiateur,** heater; — à ailettes, finned or ribbed radiator; — à nid d'abeilles, cellular radiator; — d'huile, oil cooler.

***radiation,** — d'annihilation, annihilation radiation; — de courte période, short-lived radiation; — électromagnétique, electromagnetic radiation; — ionisante, atomic radiation, ionizing radiation; — pénétrante, penetrating radiation; — primaire, primary radiation; — radioactive, radioactive radiation; — secondaire, secondary emission, secondary radiation.

***radical libre,** free radical.

radioactinium, _m._ radioactinium.

***radioactivité,** — ambiante, environmental background radioactivity; — artificielle, artificial radioactivity; — des matières en suspension dans l'air, airborne radioactivity; — naturelle, natural radioactivity; — naturelle ambiante, natural radiation background; — provoquée, induced radioactivity.

radio-alignement, _m._ directional signal.

radioantécédents, _m.pl._ radiation history.

radiobalise, _f._ radio beacon, marker beacon.

radiobaliser, to equip (a route) with a radionavigation system.

radiobiologie, _f._ radiobiology.

radiocanal, _m._ radio channel.

radiocarbone, radioactive carbon.

radiochimie, _f._ radioactive chemistry, radiochemistry.

radiocolloïde, radiocolloid.

radiocommandé, radio-controlled.

radiocompas, _m._ homing device (radio), direction finder.

radioconducteur, _m._ coherer.

radiodermite, _f._ radiodermatitis.

***radiodiffusion,** radio broadcast.

radio-électricité, _f._ radio engineering.

radio-élement, _m._ radionuclide, radioisotope, radioelement, nuclear species, nuclide; — d'activation, activating isotope; — indicateur, indicator element; — naturel, naturally occurring radionuclide.

radiogénique, radiogenic.

radiogoniomètre, _m._ radiogonometer; — de bord, ship direction-finding station.

radiogramme, _m._ radiogram, radiograph, skiagram.

***radiographie,** roentgenography; — neutronique, neutron radiography.

radioguidage pour navigation, _m._ radio aids to navigation.

radioiode, radioactive iodine.

radiolésion, _f._ radiation injury.

radiologique, radiological.

radiologiste, _m._ radiologist.

radioluminescence, _f._ radioluminescence.

radiolyse, radiolysis.

radionuclide naturel, _m._ naturally occurring radionuclide.

radiophare, _m._ radio beacon, radiophare.

radiophone guidage, _m._ homing device, guidance device.

radiophonographe, _m._ radio-phonograph.

radiophysique médicale et sanitaire, _f._ health physics.

radiorésistance, _f._ radiation resistance, radioresistance; — acquise, acquired radioresistance.

***radioscopie,** fluoroscopy.

radiosensibilité, _f._ radiosensitivity.

radiosensible, radiation-sensitive.

radiosonde, _f._ radio meteorograph.

radiostérilisation, _f._ radiation sterilization.

radiotechnicien, _m._ radio engineer.

radiotechnique, _f._ radio engineering.

radiotélégramme, _m._ radiogram.

radiotélégraphie par ondes amorties, _f._ spark telegraphy.

radiotélégraphier, to radiotelegraph.

***radiothérapie,** radiation therapy; — superficielle, contact radiation therapy.

radiotoxémie, _f._ radiation sickness.

radon, _m._ niton, radon, radium emanation.

***raffinage du minerai,** ore refining.

ragréage, _m._ refitting, trimming, polishing; — d'une façade, polishing or cleaning a facade.

ragréer, to trim, repair, polish, clean; — les filetage, to chase threads.

***raideur,** toughness.

raidisseur, _m._ stiffener, wire stretcher, stringer; hauban —, straining tie.

***raie,** — d'étincelle, spark radiation; — monochromatique, monochromatic spectrum; — spectrale, spectral line.

***rail,** — conducteur, third rail; — mobile, switch (R.R.).

***rainure,** — de calage, keyway; — de clavette, key seat; — de dudgeonnage, groove hole (in tube sheet); — de l'induit, armature slot; —

de repêchage, catching groove; — de segment, ring groove.

*ralenti, au —, idling (mach.).

*ralentir, to lessen.

*ralentissement, slowing down; — aux vitesses d'agitation thermique, reduction to thermal velocities; — des neutrons, slowing down of neutrons.

*ralentisseur, additive, inhibitor, engine brake, moderator; — en équilibre, stationary moderator; — hydrogéné, hydrogeneous moderator.

*rame, string (boring); la dernière — du métro, last subway train for the night.

*ramener au premier degré, to linearize.

ramicelles, f.pl. streamers (in sparking).

*ramification, multiple decay, multiple disintegration.

*rampe, ascent; — d'allumage, ignition harness; — de chargement, loading ramp, loader; — de culbuteurs, rocker-arm system; — de graissage, lubricating rack; — de lancement, launching ramp; — de réception, card stacker (EDP); — de soufflage, hot air gallery (air conditioning); — de verrouillage, locking ramp; — électrique, strip light; — lumineuse d'atterissage, landing lights (airfield); — montante, upgrade; — trainante, trailing apron; —s d'épandage, spraying bars; —s de saupoudrage, dust spreaders.

randite, f. randite.

*rang, grade, quality.

*rapidité de guérison, recovery rate.

*rappel, cross reference.

*rapport, — cadmique, cadmium ratio; — cadmium, cadmium ratio; — cavitaire, void fraction; — d'absorption différentiel, differential absorption rate; — d'allongement, aspect ratio (aviation); — d'atténuation, ratio of attenuation; — de chargement, cargo-intake certificate.

rapport, — de conversion, conversion ratio; — de décontamination, decontamination factor; — de démultiplication, reduction ratio; — de grille, grid ratio; — de la tension à la déformation, stress/strain ratio; — de leviers, leverage; — d'embranchement, branching ratio; — de multiplication, gear ratio; — d'endurance, endurance ratio; — d'enroulements, ratio of the windings; — de ralentissement, moderating ratio; — de reflux, reflux ratio.

rapport, — de résistance à la fatigue, endurance ratio; — des nombres d'atomes, atomic ratio; — de surgénération, breeding ratio; — d'expansion, expansion ratio; — du volume à

la superficie, volume/surface ratio; — gyromagnétique, gyromagnetic ratio; — intérimaire d'activité, progress report; — isotopique, abundance ratio, isotopic ratio; — journalier de forage, log, boring journal; — molaire, molar ratio; — pondéral, mass ratio; — réducteur, gear ratio at end of travel, step-down ratio; — signal/bruit, signal/noise ratio; — volumétrique, compression ratio.

*rapt, pick-up (nuclear reaction).

*rarefier, to evacuate.

rassembleur, m. collector space, buncher, rhumbatron, klystron.

ratissette, f. drag hook.

*rattrapage de jeu des soupapes, valve adjustment.

rauvite, rauvite.

ravitaillement, m. refueling; — par avion, supply dropping by aircraft.

*rayon, radial dimension; — alpha, alpha ray; — bêta, beta ray; — bêta de haute énergie, strong beta ray; — caractéristique, fluorescent radiation; — cathodique, cathode ray; — cosmique, cosmic ray; — cosmique primaire, primary cosmic ray; — critique, critical radius; — critique réel du coeur, actual critical loading radius.

rayon, — d'action, radius of action, range of action; — d'action des forces nucléaires, range of nuclear forces; — d'action efficace, effective range; — d'action efficace infini, infinite effective range; — de Bohr, Bohr radius; — de braquage, turning radius; — de Bucky, grenz ray; — de collision de neutron, neutron collision radius; — de courbure, radius of curvature; — de cyclotron, cyclotron radius.

rayon, — de Debye, Debye shielding distance; — de giration, radius of gyration; — de la barrière de Coulomb, Coulomb barrier radius; — de l'effet d'écran, cutoff distance; — d'inertie, radius of gyration; — doux, soft radiation; — du noyau, nuclear radius; — dur, hard radiation; — électrostatique, electrostatic radius.

rayon, — fluorescent, fluorescent radiation; — gamma de capture, capture gamma radiation; — gamma de courte période, short-lived gamma ray; — gamma de grande énergie, high-energy gamma radiation; — gamma dur, strong gamma ray; — gamma instantané, prompt gamma ray; — indirect, space ray; — mou, soft radiation; — positif, canal ray, positive ray.

*rayonnement, — ambiant, radiation background, background radiation; — atomique, atomic radiation; — calorifique, thermal radiation; — **caractéristique**, characteristic radiation; — **corpusculaire**, corpuscular radiation; — **cosmique**, cosmic radiation; — **dangereux**, dangerous radiation; — **de collision**, impact radiation; — **de courte période**, short-lived radiation.

rayonnement, — **de freinage**, bremsstrahlung; — **de grande énergie**, high level radiation; — **de longue periode**, long-lived radiation; — **de multipôle**, multipole radiation; — **de neutrons**, neutron radiation; — **diffusé**, scattered radiation; — **direct**, direct radiation; — **dû aux fuites**, leakage radiation; — **du corps noir**, black-body radiation; — **du plasma**, plasma radiation; — **dur**, hard radiation.

rayonnement, — **électromagnétique**, electromagnetic radiation; — **gamma ambiant**, background gamma radiation; — **gamma naturel**, gamma background; — **gamma parasite**, background gamma radiation; — **hétérogène**, heterogeneous radiation; — **homogène**, homogeneous radiation; — **ionisant**, atomic radiation, ionizing radiation; — **monochromatique**, monoenergetic radiation, monochromatic radiation; — **monoénergétique**, monoenergetic radiation; — **neutronique du plasma**, plasma neutron radiation.

rayonnement, — **neutronique polyénergétique**, polyenergetic neutron radiation; — **noir**, black-body radiation; — **nucléaire**, nuclear radiation; — **parasite**, spurious radiation, background radiation; — **pénétrant**, penetrating radiation; — **primaire**, primary radiation; — **radioactif**, radioactive radiation; — **rétrodiffusé**, back-scattered radiation; — **secondaire**, secondary radiation; — **vagabond**, stray radiation; —**s de radiofréquence**, radiofrequency radiations.

réa, *m.* pulley wheel; — **à rais**, spokesheave; — **plein**, disc sheave.

reactance, *f.* reactance.

réacteur, *m.* reactor, reaction member, stator, jet engine; — **à chaine de fissions divergente**, divergent reactor; — **à double flux**, dual flow jet engine; — **à eau bouillante**, boiling-water reactor; — **à faible flux de neutrons**, low-flux reactor; — **à fission**, fission reactor; — **à fusion**, fusion reactor; — **à fusion contrôlée**, controlled-fusion reactor; — **à fusion ménagée**, controlled-fusion reactor; — **à haute température** (**fusion**), high-temperature reactor; — **à neutrons intermédiaires**, intermediate reactor.

reacteur, — **à neutrons rapides**, fast reactor, fast-neutron reactor; — **à réfrigérant gazeux**, gas-cooled reactor; — **à simple flux**, single flow jet engine; — **atomique**, atomic reactor,

pile; — **à uranium naturel**, natural-uranium reactor; — **commandé**, controlled reactor; — **convertisseur**, converter; — **cylindrique**, cylindrical reactor; — **de fusion d'énergie nulle**, zero-energy thermal apparatus; — **de radiobiologie**, biomedical reactor; — **d'essai des matériaux**, materials testing reactor (MTR).

réacteur, — **en bout de pale**, pressure jet, tip jet; — **expérimental**, experimental reactor; — **hétérogène**, heterogeneous reactor; — **homogène**, homogeneous reactor; — **industriel**, industrial reactor, engineering reactor; — **nucléaire**, nuclear reactor; — **pilote**, master engine; — **plutonigène**, production reactor; — **poreux**, porous reactor; — **préfabriqué**, package reactor; — **prototype**, prototype reactor; — **réduit**, engine at idle; — **surgénérateur**, breeder reactor; — **thermonucléaire**, thermonuclear reactor.

*****réaction**, feedback, regenerative coupling (of amplifiers), retraction; — **à la chaleur**, thermal response; — **auto-entretenue quant à l'énergie**, energetically self-sustaining reaction; — **binaire**, binary reaction (two-body collision); — **capacitive**, electrostatic feedback; — **chimique**, chemical process, chemical reaction; — **de capture**, capture reaction; — **d'échange**, exchange reaction; — **d'échange catalytique**, catalytic exchange reaction.

réaction, — **d'échange entre l'hydrogène et l'eau**, hydrogen-water exchange reaction; — **de collision**, collision process; — **de fission**, fission reaction; — **de fission explosive**, explosive fission reaction; — **de fusion**, fusion reaction; — **de fusion autoentretenue**, self-sustaining fusion reaction; — **de fusion contrôlée**, controlled fusion reaction, controlled thermonuclear reaction; — **de fusion ménagée**, controlled fusion reaction, controlled thermonuclear reaction; — **de type solaire**, solar-type reaction; — **différentielle**, rate action, derivative action; — **due aux neutrons rapides**, fast-neutron reaction.

réaction, — **en chaîne automodérée**, self-limiting chain reaction; — **en chaîne contrôlable**, controllable chain reaction; — **en chaîne de fissions**, fission chain reaction; — **en chaîne génératrice d'énergie**, power chain reaction; — **en chaîne nucléaire autoentretenue**, self-maintaining nuclear chain reaction; — **en chaîne par neutrons lents**, slow-neutron chain reaction; — **en chaîne par neutrons thermiques**, thermal-neutron chain reaction; — **en chaîne provoquée par les neutrons**, neutron chain reaction; — **endothermique**, endothermic reaction; — **entre . . .**, interaction.

réaction, — **entretenue,** sustained reaction; — **exothermique,** exothermic reaction; — **explosive,** explosive reaction; — **générale,** systemic reaction; — **intégrante,** reset action; — **interne,** internal reaction; — **nucléaire,** nuclear process, nuclear reaction; — **nucléaire contrôlée,** controlled nuclear reaction; — **nucléaire en chaîne,** nuclear chain reaction; — **nucléaire exothermique,** exothermic nuclear disintegration; — **nucléaire ménagée,** controlled nuclear reaction.

réaction, — **nucléaire provoquée,** induced nuclear reaction; — **photonucléaire,** photonuclear reaction; — **positive,** positive regeneration; — **provoquée par des (les) neutrons,** neutron reaction, neutron-induced process; — **secondaire de fusion,** secondary fusion reaction; — **thermonucléaire,** thermonuclear reaction; — **thermonucléaire ménagée,** controlled fusion reaction, controlled thermonuclear reaction; — **thermonucléaire stationnaire,** stationary thermonuclear reaction; — **thermonucléaire transitoire,** transitory thermonuclear reaction, momentary thermonuclear reaction.

réactivité, *f.* reactivity; — **due aux neutrons instantanés,** prompt reactivity; — **en fonction de la température,** temperature dependence of reactivity; — **intrinsèque,** built-in reactivity; — **résiduelle,** residual reactivity.

réaléser, to bore out, rebore, counterbore.

*****réalisation,** embodiment (of an invention and its underlying idea).

réarmer, to recommission, rearm.

réarrangement, *m.* rearrangement; — **chromosomique,** chromosome rearrangement; — **des électrons périphériques des atomes,** rearrangement of the outer electronic structures of the atoms.

rebobiner, to rewind.

*****rebroussement, point de** —, return, turning, or reversal point.

*****rebuts,** *pl.* tailings.

récapitulatif, summarized.

récapitulation, *f.* summary.

*****recensement,** ascertainment.

*****réceptacle,** output hopper, pocket, receiver; — **d'attache,** fastener receptacle; — **de cartes à décalage,** offset stacker.

*****recepteur,** — **à amplification directe,** straight receiver; — **à changement de fréquence,** heterodyne receiver set; — **alimenté par le réseau,** — **alimenté par le secteur,** mains-operated receiver; — **à secteur,** all-electric set, supply-line or mains-connected receiver (including power pack or unit); — **à valve d'oscillation,** tube detector.

récepteur, — **balise,** marker receiver; — **de cartes,** card stacker; — **de garde,** standby frequency receiver; — **de jaugeur,** fluid level indicator; — **de poche,** pocket radio; — **double,** dual indicator; — **pneumatique,** pneumatic fitting (crop spraying); — **populaire,** people's receiver; — **reflex,** dual receiver; — **sur secteur,** mains receiver; — **sur secteur alternatif,** alternating-current operated receiver; — **sur secteur continu,** direct-current operated receiver; — **tous-courants,** all-mains receiver.

*****réception,** — **au casque,** headphone reception; — **au son,** aural reception; — **dirigée,** directive reception; — **imprimée,** printing reception; — **lointaine,** distant reception; — **sur antennes espaces,** diversity reception; **à la** —, at the receiving end.

rechaper, to retread (tires).

*****réchauffeur instantané,** flashed heater.

*****recherche,** — **appliquée,** industrial research; — **binaire,** binary search, dichotomizing search; — **de filons,** prospecting; — **des pannes,** troubleshooting; — **documentaire,** information retrieval; — **et mise au point,** research and development; — **opérationnelle,** operations search; — **par dichotmie,** binary search; — **pétrolifère,** oil prospecting; — **pure,** fundamental research; —**s nucléaires,** nuclear research.

*****récipient,** — **à chemisage double,** double-clad vessel; — **en acier au carbone,** carbon-steel vessel; — **fermé,** sealed bulb, container, or vessel.

recirculation, *f.* recirculation; — **de l'eau,** water recirculation.

recollement, *m.* rejoining process.

*****recoller,** to pull up (armature).

recombinaison radiative, *f.* radiative recombination.

récomposition d'une image, *f.* picture recreation or reproduction.

reconditionnement, *m.* reconditioning, reworking, work over (shaft).

reconstituant, *m.* tonic.

*****record homologue,** recognized record.

recoupement, *m.* cross-check, stepping (of wall, etc.); **méthode du** —, resection (surveying); **moyens de** —, cross-checks.

*****recouvrement,** scarf, overlay.

*****recouvrer,** to retrieve.

*****recouvrir,** to overlap (of adjoining lines, telev.).

recristallisation, *f.* recrystallization.

*****rectifier,** to strip.

rectifieuse pour cylindres, *f.* cylinder grinder.

recto-au-dessus, face up.

recto-en-dessous, face down.

*****recueillir,** to absorb, take off (potential).

*****recul,** recoil, falling back, retreat, slip (aviation), kick (gun); — **d'hélice,** propeller slip; — **moleculaire,** aggregate recoil.

*****récupération,** — **de l'uranium,** recovery of uranium; — **d'énergie,** power recovery; — **des déchets,** waste recovery; — **des déchets d'uranium,** recovery of waste uranium; — **par voie chimique,** chemical recovery.

*****récurrance,** recursion.

recyclage, *m.* recycling, recirculation; — **à simple passe,** single-stage recycle; — **continu,** continuous recycling of dissolved fuel.

recycler, to recycle.

*****redan,** step; **à un —,** single-step.

redevance, *f.* royalty, rent, dues.

redondance, *f.* redundancy; **contrôle par —,** redundancy check (EDP).

redondant, redundant; **code —,** redundant code (EDP).

*****redressement de puissance,** power rectification.

*****redresser,** to pull out.

*****redresseur,** rectifier; — **à arc,** mercury-arc rectifier; — **à cathode incandescente,** hot-cathode rectifier; — **à couche d'arrêt,** blocking-layer or barrier-film rectifier (specifically, copper oxide rectifier); — **à deux alternances,** full wave rectifier; — **à effluves,** glow-discharge rectifier; — **à lame vibrante,** tuned-reed rectifier; — **à lamelle,** vibrating-reed or blade rectifier; — **à une alternance,** half-wave rectifier; — **de charge continue,** trickle charger; — **de courant,** current rectifier.

*****réducteur,** reduction gear, potential divider; — **double,** double regulating switch.

*****réduction,** — **au premier degré,** linearization; — **chromatique,** chromatic reduction; — **mâle et femelle,** bushing; **en —,** scaleddown.

*****réduire le moteur à fond,** to close the throttle.

réembrayage, *m.* reengagement (autopilot).

réenclenchement, *m.* reswitching, reconnection, or reclosure (of a circuit).

réenrichessement, *m.* reenrichment.

réexpédier, to retransmit (telegr.).

*****refaire,** to improve.

*****référence,** tag, name, identifier, reference, docuterm; — **d'appel,** call number; — **de données,** data definition name; — **externe,** external reference (EDP); — **fabricant,** manufacturer's part number; — **fuselage,** fuselage datum line; **point de —,** datum or reference point.

référentiel, *m.* frame of reference; — **du laboratoire,** laboratory frame of reference; — **relatif,** moving coordinate system.

*****réflecteur,** floodlight, reflector, tamper; — **non-absorbant,** noncapturing reflector; **sans — latéral,** radially bare.

*****reflet,** bloom or cast.

*****réflexion,** — **d'un faisceau de neutrons,** reflection of neutron beam; — **spatiale,** space reflection.

reforer, to drill out or up to.

reformage, *m.* reforming (petroleum).

reformeur, *m.* reformer (petroleum refining).

refouloir, *m.* rammer (artil.).

*****réfractaire,** fireproof, heatproof.

*****réfraction d'un faisceau de neutrons,** refraction of neutron beam.

*****réfrigérant,** coolant; — **de l'eau,** water coolant; — **du ralentisseur,** moderator coolant; — **métallique liquide,** liquid-metal coolant.

*****réfrigération,** — **à passage unique,** once-through cooling, direct-flow cooling, uniflow cooling; — **par liquide,** liquid cooling.

refroidi par vapeur d'eau, steam-cooled.

*****refroidir par dilatation,** to cool by expansion.

*****refroidissement,** — **au gaz,** gas cooling; — **de l'eau,** water cooling; — **par écoulement direct,** once-through cooling, direct-flow cooling, uniflow cooling; — **par l'air,** air cooling; — **par l'eau,** water cooling; — **par le bismuth fondu,** molten-bismuth cooling; — **par l'hélium,** helium cooling; — **par liquide,** liquid cooling; — **par métal liquide,** liquid-metal cooling.

*****refroidisseur,** coolant.

regâcher, to mix, or temper (cement).

régale, *m.* babbitt metal.

*****régénération,** breeding; — **des combustibles nucléaires,** regeneration of nuclear fuels; — **des neutrons,** regeneration of neutrons; — **d'organes,** reduplication of organs; — **du combustible,** fuel regeneration; — **par neutrons thermiques,** thermal breeding.

*****régime,** — **critique,** critical speed; — **de décollage,** take-off rating; — **de compteur,** counter range; — **de descent du trépan,** feed; — **définitif,** steady state; — **de fonc-**

tionnement, operating conditions; — de marche aux instruments, instrument range; — de marche sous l'effet de la source de neutrons, source range (of reactor operation); — d'équilibre, steady state; — du moteur, speed of the engine.

régime, — économique, economical cruising conditions; — élevé, high speed; — lent, slow speed; — non-stationnaire, nonstationary operation; — permanent, steady state; plein —, full throttle; — porteur, carrier-wave operation; — pulsé, pulsed operation; — stationnaire, steady state; — transitoire, transient operation; au grand —, at maximum speed.

*région, — conductrice, conducting region; — contenant la matière fissile, seeded area; — d'attribution synchronique, dynamic area; — de faible radioactivité, cold area; — de Mach, Mach region; — d'énergie du continu, continuum energy region; — de proportionnalité, proportional region; — de réaction, reaction region.

région, — de résonance, resonance region; — des instructions, instruction area; — des neutrons thermiques, thermal region; — de turbulence, turbulent region; — du continu, continuum region; — euchromatique, euchromatic region; — interfaciale, interface region; — périphérique, peripheral region; — radioactive, active region.

*registre, soundtrack (film); — à décalage, shift register, circulating register; — à ligne à retard, delay line register; — à point flottant, floating point register; — B, B register, B line; — de base, base register, B line; — de commande, control instruction register; — de modification, modifier register; — de retour, return code register; — d'index, index-register, B box, index word; — d'instruction, instruction register, program register, location counter; — d'opération, operation register; — général, general purpose register; — magnétique à décalage, magnetic shift register; — mémoire, storage register; — multiplicateur-quotient, multiplier-quotient register.

registre-adresse, m. address register.

registre-tampon, m. B register, B line.

*réglage, — approximatif, coarse adjustment; — automatique, automatic control; — d'artillerie, artillery observation; — de la configuration, configuration control; — de la pile, reactor control; — des fusées, fuse setting; — en temps, phasing, timing; — interne, internal control, inherent control; — nucléaire, nuclear control; — par absorption des

neutrons, absorption control; — précis, fine control; — silencieux, squelch.

règle, — de l'effet cumulatif de l'exposition aux radiations, additivity of irradiation effect; — de sélection, selection rule.

*réglementer, to prorate.

*régler, to adapt.

*régulateur, — altimétrique, barometric controller; — au cadmium, cadmium regulator; — automatique, automatic controller; — de base, master clock; — de fréquence, frequency stabilizer or regulator; — de roulis, lateral trim; — de tangage, longitudinal trim; — de tension, adjustable voltage control; — isodrome, speed governor; — par tout ou rien, hit-or-miss regulator.

*regulation, control; — d'orientation, attitude control; — thermique, thermal control.

*rejet, — vertical, throw (of a fault); grille de —, suppressor grid.

rejeu, m. playback (geophysics).

*relais, transducer, switch, shift; — à armature bobinée, motor-type relay; — à cadre mobile, moving-coil relay; — à mercure, mercury-wetted relay; — anémométrique, IAS control unit; — avertisseur de givrage, ice detecting relay; — batteur, impulse relay; — d'arrêt, stop relay; — de cantonnement, block-system relay; — de manipulation, key relay (telegr.).

relais, — de prise, contact relay, holding relay; — des émissions radiophoniques, rebroadcast (radio); — de surintensité, overcurrent relay; — de surtension, over-voltage relay; — différé, delayed or time-lag relay; — gardé, restraining or guard relay; — inverseur bipolaire, double-pole double-throw relay; — ionique, gas-filled relay; — manométrique, anti-surge valve; — modulateur, chopper, vibrating contactor, vibrator; — pneumatique, differential pressure relay; — retardeur, time-delay relay.

*relation, — avec la fréquence, frequency dependence; — avec la hauteur, height dependence; — entre l'énergie et la portée, energy-range relation; — entre le parcours et l'énergie, range-energy relation; — exponentielle, exponential dependence; — fonctionelle, functional dependence; — linéaire, linear dependence; — non-linéaire, nonlinear dependence.

*relevage du train, retraction of the undercarriage.

relève, f. relief, shift, direction finding.

*relèvement radiogoniométrique, radiogoniometer work, position finding.

*relief acoustique, binaural hearing, reproduction, stereophonic condition (with auditory perception).

rem, rem (roentgen equivalent, man).

rémanence, *f.* persistence of vision, remanence; — **sur l'écran radar,** image retention, radar scope afterglow.

remblayeuse, *f.* backfiller.

rembobiner, to rewind.

*remède, help, aid.

*remettre, — **à l'état initial,** to reset; — **à zéro,** to clear, reset; — **en circulation,** to recycle; — **en état,** to rebuild.

remisage, *m.* dead storage.

remorquage, *m.* towing, hauling.

*remorque, trailer coach; — **à cabine,** driving trailer.

remorqueur, *m.* tug, tender (naut.).

rémouleur, *m.* knife grinder.

remous, *m.* eddy; — **de la mer,** backwash.

*remplissage, padding (EDP); **instruction de** —, dummy instruction; **robinet de** —, feedwater cock; **zone de** —, filler.

renardite, renardite.

*rendement, — **absolu de fission,** absolute fission yield; — **constant,** constant yield; — **de fission,** fission yield; — **de malaxage,** mixing efficiency; — **d'énergie,** power output; — **du compteur,** counter efficiency; — **énergétique,** energy yield; — **en mutations ponctuelles relevées,** yield of recovered point mutations.

rendement, — **en neutrons,** neutron yield; — **en rayons gamma,** gamma yield; — **fractionnaire,** fractional yield; — **global,** over-all efficiency, total efficiency, total output; — **matière,** material efficiency; — **moyen,** average rating; — **quantique,** photoelectric efficiency, quantum efficiency, quantum yield; — **thermique,** heat efficiency, thermal efficiency; — **thermodynamique,** thermodynamic yield, thermodynamic efficiency; **grand** — **d'énergie,** high energy yield.

*rendre étanche, to calk.

renfermé, enclosed.

*renflement, belly, upset.

*renforçateur à opposition rhythmée, push-pull amplifier.

*renforcé, heavy duty.

*renforcement, intensification (phot.).

*renouvellement, turnover.

*rénovation, — **des combustibles nucléaires,** regeneration of nuclear fuels; — **du combustible,** fuel reprocessing.

*rentrer un pavillon, to strike (a flag).

*renversé, left-handed (mach.).

renvideur, take-up spool (film).

*renvoi, cross-reference; — **de mouvement par des arbres,** shafting; **galet de** —, feed roll.

rep, rep (roentgen equivalent physical).

*réparation des effets somatiques, reversibility of the somatic effects.

*répartiteur, header, dispatcher, distribution frame; — **de canaux,** channel scheduler (EDP); — **de charge,** load dispatcher; — **de compas gyrostatique,** indicator for gyrostatic compass; — **d'entrée,** main distribution frame; — **de ressource,** resource manager; — **de tâches,** task dispatcher (EDP); — **de travaux,** job scheduler; — **principal,** master scheduler.

*répartition, — **de données,** data organization; — **de Maxwell,** Maxwell distribution; — **des sources,** source distribution; — **de vitesse de Maxwell,** Maxwell velocity distribution; — **en fonction du temps,** time distribution.

*repasser, to rerun (EDP).

repêchage, *m.* fishing, instrument equipment.

*repérage, position finding, logging (of stations), search (disk); — **d'erreur programmé,** program sensitive fault; — **d'erreur structuré,** pattern sensitive fault; — **par des couleurs,** color code (wires, etc.).

*repère, reference letter or numeral, symbol (in a drawing), flag, spot, sentinel, checkpoint, bloop, basis of reference; — **mobile,** moving coordinate system; **organe de** —, setting adjusting, or tuning means, position finding.

*repérer, to read (an instrument), calibrate.

*répertoire, directory; — **de fichiers,** volume table on contents (disk); — **de transposition,** relocatable library dictionary, relocation dictionary (EDP); — **d'instructions,** instruction repertory.

répertorier, to list.

répétiteur, *m.* — **de compas,** auxiliary compass; — **selsyn,** selsyn.

réplétion, *m.* mascon.

*réplique, cue.

report, *m.* amount carried forward, carry; — **accéléré,** high-speed carry; — **à ligne fixe,** fixed line posting; — **aux neuf,** standing-on-nines carry; — **circulaire,** end-around carry; — **complet,** complete carry; — **des points,** plot; — **en cascade,** cascaded carry; — **partiel,** partial carry; — **simultané,** simultaneous carry; **signal de fin de** —, carry-complete signal; **table de report,** plotting board.

****reporter,** to transfer (of film).

reporteur d'image, *m.* reporter-cameraman.

reporteuse, *f.* posting machine.

repose-pieds, *m.* footrest.

***repousser,** to enlarge, flare; — **les protons,** to repel protons.

***reprendre,** to continue, rerun.

***représentation,** notation; — **binaire équivalente,** equivalent binary digits; — **décimale codée,** binary coded decimal representation; — **graphique,** graph, plot; — **positionnelle,** positional representation, positional notation; — **symbolique,** symbolic representation.

***reprise,** rework, rerun, restart; — **à blanchir,** clean-up (machining); — **du moteur,** engine pickup; — **d'une voiture,** trade-in allowance on a car; — **en main par le pilote,** reversion to manual piloting.

repriser, to mend.

***reproduire,** to copy.

***répulsion,** — **coulombienne,** Coulomb repulsion; — **mutuelle,** mutual repulsion; — **nucléaire,** nuclear repulsion.

reradier, to reradiate.

rescrit, *m.* patent-office action (answer of examiner).

***réseau,** — **à barreau unique,** single-rod lattice; — **à barreaux creux,** hollow-slug lattice; — **à barreaux groupés,** cluster lattice; — **à barreaux pleins,** solid-rod lattice; — **actif,** active lattice; — **à excédent de ralentisseur,** overmoderated lattice; — **à faible pas,** closely spaced lattice; — **à ralentisseur en graphite,** graphite-moderated lattice; — **à refroidissement par l'eau,** water-cooled lattice; — **à uranium et à graphite,** uranium-graphite lattice.

réseau, — **à uranium métallique,** uranium metal lattice; — **automatique,** dial exchange; — **avertisseur,** spotting system (air defense), warning system; — **composé,** composite lattice; — **cristallin,** crystal lattice; — **cubique,** cubic lattice; — **cubique à faces centrées,** face-centered cubic lattice; — **cubique centré,** body-centered cubic lattice; — **d'antennes,** antenna array.

réseau, — **de bord,** aircraft mains, aircraft power system; — **de commutation,** switching network; — **de diffraction,** diffraction grating; — **d'effacement,** wash-out network; — **de réflecteur,** reflector lattice; — **des barreaux,** rod lattice; — **des caractéristiques,** family of characteristics; — **d'ordinateurs,**

computer network; — **double,** dual lattice; — **d'uranium à ralentisseur de graphite,** graphite uranium lattice; — **élémentaire,** basic lattice; — **en losanges,** rhombic lattice; — **hétérogène,** heterogeneous lattice; — **immergé,** wet lattice.

réseau, — **infini,** infinite lattice; — **lâche,** widely spaced lattice; — **local,** in-plant system; — **mixte,** dual lattice; — **peu serré,** widely spaced lattice; — **ponctuel,** point lattice; — **ralentisseur,** moderator lattice; — **réciproque,** reciprocal lattice; — **rhombique,** rhombic lattice; — **sans ralentisseur,** unmoderated lattice; — **sec,** dry lattice; — **très compact,** closely packed lattice; —**x conjugués,** paired lattices.

***réserve,** stock, blank, pool; — **de constantes,** literal pool; — **de tampons,** buffer pool; — **de vocables,** literal pool; — **d'usinage,** machining allowance.

***réservoir,** — **à vase,** catch pit; — **d'admission,** plenum chamber; — **de convoyage,** ferry fuel tank; — **de décharge,** dump tank; — **de détente,** blow-off tank; — **de secours,** reserve tank; — **de soutirage,** sump tank; — **des toilettes,** aseptic tank.

réservoir, — **de stockage,** storage tank; — **de trop-plein,** sump tank; — **de vidange,** dump tank; — **en charge,** gravity tank (automobile); — **hermétique,** seal tank; — **largable,** drop tank, pylon tank; — **protégé,** self-sealing tank; — **souterrain d'aéroport,** fuel hydrant; — **supplémentaire,** auxiliary tank.

réservoir-tampon, *m.* ballast tank, surge tank.

résident, *m.* resident, system skeleton.

résidus, *m. pl.* waste, scrap, tailings; — **à haute activité,** high-level waste; — **chimiques,** chemical waste; — **radioactifs,** radioactive waste; — **très acides,** high-acid waste.

résilience, *f.* impact strength.

résine échangeuse d'ions, *f.* ion-exchange resin.

***résistance,** resistor, endurance; — **à curseur,** slide rheostat; — **à la corrosion,** corrosion resistance; — **à la rupture,** ultimate strength; — **à la traction,** tensile strength, ultimate tensile strength; — **à l'écrasement,** compressive strength; — **à l'irradiation,** radiation resistance; — **à l'obscurcissement,** dark resistance; — **au choc,** impact strength; — **aux rayonnements,** radioresistance.

résistance, — **de charge,** load resistance, resistor; — **de compensation,** balancing resistance; — **de frottement,** frictional force; — **de fuite,** leakage resistance, grid leak.

résistance, — **de haute fréquence,** radiofrequency resistance; — **de la bobine,** coil resistance; — **de pertes,** wasteful resistance; — **d'équilibrage,** balancing resistor; — **d'extinction,** quenching resistor; — **du circuit de grille,** input resistance of a thermionic valve; — **d'utilisation,** useful or signal resistance; — **électrique,** electric resistance; — **en série d'un voltmètre,** range multiplier; — **ferhydrogène** (tube), iron-filament ballast lamp.

résistance, — **hydrodynamique,** water resistance; — **inductive,** inductive resistance; — **interne,** plate resistance, anode resistance; — **magnétique spécifique,** reluctivity; — **ohmique,** ohmic resistance; — **parallèle,** shunt resistance; — **parasite,** parasite drag; — **ramenée,** concentrated resistance; — **superficielle,** surface resistance; **à haute** —, highgrade (metal.).

résistivité, *f.* resistivity.

*__résolution,__ — **angulaire,** angular resolution; — **en temps,** time resolution.

*__résonance,__ — **aiguë,** sharp resonance; — **amortie,** flat resonance; — **magnétique de spin,** spin magnetic resonance; — **magnétique nucléaire,** nuclear magnetic resonance.

résonateur, *m.* resonator; — **à coupure,** Hertz spark-gap resonator.

*__ressac,__ breakers (surf).

*__resserrement,__ pinch; — **du courant,** current constriction; — **électrodynamique,** electrodynamic contraction; — **électromagnétique,** electro-magnetic constriction; — **magnétique,** magnetic constriction; — **stabilisé,** stabilized pinch.

*__ressort,__ *— **à boudin,** helical or coil spring; — **à lames,** leaf spring; — **annulaire,** annular spring; — **à paillette,** leaf spring; — **à spirale,** coil spring; — **bandé,** coil spring; — **bilame,** bimetallic spring, twin-leaf spring (with differential expansion); — **d'allumage,** ignition spring; — **de buttoir,** buffer spring; — **de soupape,** valve spring.

ressortissant, *m.* national, citizen.

ressoudage, *m.* rewelding.

*__reste,__ residuum, balance.

*__résultat,__ answer; — **obtenus,** performance; **—s,** data.

*__résumé,__ abstract.

*__rétablir,__ to initialize, reset; — **à N,** to reset to N.

*__retard,__ time lag; — **de phase,** phase lag; — **externe,** external delay; **à** — **indépendant,** inverse time-limit relay.

retardant freinant, lagging.

*__retarder la discussion,__ to defer prosecution (of a case).

*__retassure,__ pinhole.

*__rétention initiale par le corps,__ initial body retention (of radioactive material).

réticulation, *f.* cross linking.

retirant, retractable.

*__rétombée,__ fallout; — **radioactive,** radioactive fallout.

*__retoucher,__ to alter, rework.

*__retour,__ — **à la piste,** turn-back, return to ramp; — **à la terre,** ground connection, earthing; — **bache,** reservoir return line; — **de chariot,** carriage return; — **de flamme,** backfire; — **de manivelle,** kickback (starting handle); — **des rechanges attribués,** return of assigned spares; — **d'information,** information feedback; — **en arrière,** flashback; — **inverse dans le temps,** time reflection.

retournement, *m.* reversal.

retraire, se —, to shrink.

retraitement, *m.* reprocessing; — **du combustible,** fuel reprocessing.

retransmission, *f.* regenerative repeater.

*__retravailler,__ to improve.

*__rétrécissement,__ constriction, necking down; — **thermique,** thermal constriction.

rétroaction, *f.* feedback; — **négative,** negative feedback.

rétrochargeuse, *f.* back loader.

rétrodiffusion, *f.* backscattering, backdiffusion.

*__rétrogradation,__ reflux (in a condenser).

rétroviseur, *m.* rearview mirror.

*__réunion instantanée,__ instantaneous assembly.

réutiliser, to recycle.

revendeur, *m.* dealer, jobber.

*__revêtement,__ sheath; — **en boral,** boral liner; — **métallique,** cladding; **à** — **travaillant,** with stressed skin.

*__revision du moteur,__ engine overhauling.

rhénium, *m.* rhenium.

*__rhéostat,__ rheostat; — **de chauffage,** filament rheostat; — **d'excitation,** field rheostat.

rhéostriction, *f.* pinch effect, rheostriction.

rhéotron, *m.* rheotron.

*__rhythme opposé,__ push-pull, phase opposition.

rhythmeur, *m.* interrupter.

ribord, *m.* waist (of a ship).

richetite, *f.* richetite.

*__ridé,__ ribbed.

*rideau, — ***à rayonnement longitudinal,** end-on directional array; — **de fumée,** smoke screen.

*rigidité, — **diélectrique,** dielectric or disruptive strength; — **magnétique,** magnetic rigidity.

*rigole, sluice.

*risque, — **biologique,** health hazard; — **radiologique,** radiological hazard; — **sanitaire,** health hazard.

ristourne, f. rebate.

*river à froid, to edge.

*rivet, — **à tête fraisée,** — **à tête perdue,** countersunk rivet.

riveteur, m. riveter.

rivoir, m. riveting hammer.

*robinet, — **à eau,** water faucet; — **d'épreuve,** gauge cock; — **de prise d'eau à la mer,** sea cock; — **de refoulement,** feed valve; — **indicateur,** gauge cock; — **purgeur,** petcock.

robot manipulateur, m. master-slave manipulator.

*robustesse, rigidity.

*roche, — **clastique,** clastic rock; — **encaissante,** host rock; — **mère,** native rock; — **native,** native rock.

*rodage, breaking-in (engine, auto, etc.).

*roder, to reseat.

roentgenologie, f. radiology.

roentgenthérapie, f. roentgen therapy.

rogersite, f. rogersite.

*rôle, — **refoulé,** pushdown list (EDP); — **réintégrant,** warparound list.

*rond, collar (mach.).

*rondelle, planchet; — **biseautée,** bevelled washer; — **chambrée,** recessed washer; — **collerette,** flanged washer; — **contre-rivure,** rivet washer; — **d'appui,** thrust washer; — **d'arrêt,** lock washer; — **de Belleville,** cup washer; — **d'écoulement d'huile,** oil ring; — **de raccordement RK4,** dished washer; — **de sertissage,** crimping washer; — **éventail,** fan-lock washer; — **Grower,** spring washer, split washer.

rondelle-arrêtoir, f. safety washer.

ronéoter, to roneo, mimeograph.

ronéotyper, to roneo, mimeograph.

*ronflement, hum, motorboating, alternating current pickup (noise in radio).

ronfleur, m. buzzer.

ronronnement, m. humming noise.

*rosace, escutcheon (auto body).

*rose des vents, f. compass card, wind rose.

*roseau, cane (bot.).

*rosette, collar (mach.), indicator disk.

*rotation, rotational motion; — **de l'axe,** circular motion.

rotationnel, curl.

*rotor, armature; — **anti-couple,** anti-torque rotor; — **d'hélicoplane,** paddle-wheel rotor; —**s engrenants,** meshing rotors.

rotothérapie, f. rotation therapy.

*roue, — **à aubes,** impeller; — **à augets,** bucket wheel; — **à chaine,** chain pulley, sprocket wheel; — **à couronne,** face wheel, cogwheel; — **à déclic,** ratchet wheel; — **à libre,** free wheeling; — **à vis sans fin,** worm wheel; — **à voile ajouré,** perforated disk wheel; — **à voile plein,** solid disk wheel; — **carénée,** spatted wheel; — **conductrice,** driving wheel; — **conique,** bevel wheel.

roue, — **d'angle,** angular wheel, bevel wheel; — **de secours,** spare wheel; — **de transmission,** idler (gear); — **d'impression,** print wheel, type wheel; — **directrice,** guide wheel; — **élastique à amortisseur incorporé,** internally sprung wheel; — **endentée,** cogwheel; — **et pignon,** gear and pinion; — **fil,** wire-spoke wheel; — **planétaire,** — **satellite,** epicyclic gear; — **sur l'arbre,** wheel and axle.

*rouet d'antenne, antenna winch or reel.

rouf, m. deckhouse (naut.).

*roulage, conveyance, haulage.

*rouleau, spool (film); — **debiteur,** feed spool (film).

rouleau-écraseur, m. steam roller.

*roulement, — **à aiguilles,** needle bearing; — **à galet,** roller bearing; — **à l'atterrissage,** landing run; — **à rotule sur billes,** self-aligning ball bearing; — **à rotule sur rouleaux,** spherical race bearing; — **à rouleaux,** roller bearing; — **de guidage,** clutch pilot bearing.

*rouler, to go (mach.), ride over.

roulet d'antenne, m. aerial winch.

roulotte, f. trailer (automobile).

*route, beam (in direction finding).

*ruban, tape; — **de cadmium,** cadmium strip; — **de combustible,** fuel tape; — **de fil adhésif,** adhesive tape; — **de frein,** brake lining, brake band; — **encré,** typewriter ribbon; — **gomme,** adhesive tape; — **goudronné,** tarred tape; — **guide,** pad roller; — **para,** rubber tape; — **perforé,** punched tape.

*rubrique, heading, breakout (documentation); — **de données,** data division; — **de**

milieux, environment division (EDP); — **de références,** identification division; — **d'opérations,** procedure division.

rupteur, *m.* circuit breaker; — **de charge,** load cut-out; — **de ligne,** line breaker; — **thermique,** thermal switch.

****rupture,** breakdown, disruption; — **chromosomique,** chromosome break; — **d'aile,** wing failure; — **de courant,** power failure; — **de gaine d'une cartouche,** slug burst; — **de**

l'isolant, insulation breakdown; — **du deutéron en vol,** deuteron stripping; — **en vol du deutéron,** stripping; — **fragile,** brittle fracture; **charge de —,** breakdown load.

rustine, *f.* inner-tube patch.

rutherford, *m.* rutherford.

rutherfordine, *f.* rutherfordine.

rutherfordite, rutherfordite.

rythmeur, *m.* timer.

S

sablage, *m.* sand blasting, spreading sand; — **humide, liquide,** vapor blasting.

****sable, — boulant,** quicksand; — **de monazite,** monazite sand; — **fin,** — **peu perméable,** close sand; **—* **mouvant,** quicksand.

****sabord d'entrée,** gangway (naut.).

saborder, to scuttle.

****sabot, — à soupape,** casing-float shoe; — **de frein,** brake shoe.

sabugalite, *f.* sabugalite.

****sac, — à terre,** sandbag; — **du parachute,** parachute pack; — **postal,** mailbag.

****saccadé,** discontinuous.

sacoche, *f.* kit (tool bag).

saillie d'une roue, *f.* wheel flange.

****saillir,** to jet.

****sain,** clean.

****saisie, — des données,** data acquisition; — **d'une hypothèque,** mortgage foreclosure.

saisissante, *f.* arrester.

saléite, saleite.

****salle, — de régie,** instrument room; — **bleine,** full house.

saluer du pavillon, dip (a flag).

salve, *f.* salvo, volley, burst (of neutrons).

samarium, *m.* samarium.

sandow, *m.* launching elastic, bungee cord.

sangle, *f.* strap, band, belt, girth.

****sans, — adresse,** zero address (EDP); — **collage,** tack-free; — **contact à court-circuit,** break before make; — **objet,** not applicable.

****sas,** airlock, gate (in electron microscope); — **à air,** air lock.

satellite, *m.* satellite, planet wheel (planetary gears), bevel gear, peripheral equipment; — **géostationnaire,** geostationary satellite; — **géosynchrone,** earth synchronous satellite; — **sous synchrone,** subsynchronous satel-

lite; **engrenage à —,** planet gear, sun-and-planet gear.

satineuse, *f.* glazing rollers.

****saumon,** detachable wing tip; — **de dérive,** fin tip, vertical stabilizer tip; — **de pale,** blade tip cap, blade tip fairing; — **de plan fixe,** horizontal stabilizer tip.

****saut,** step; — **quantique,** quantum jump, quantum transition.

saut, skip, branch; — **conditionnel,** conditional branch; — **de capacité,** overflow ejection; — **de papier,** paper skip; — **de puissance,** power excursion; — **de puissance de la pile,** power excursion of reactor; — **de réactivité,** step change in reactivity.

****sauter,** to blow (a fuse).

****sauterelle,** snap fastener, bevel square.

****sauteur,** hopper.

sautillement, *m.* lilt, slow change of signal frequency.

****sauvegarde,** life line.

****savon, — doux,** neutral soap; — **métallique,** antifriction grease.

scalaire, scalar.

scaphandre, *m.* diving suit; — **aérien,** high-altitude flying suit.

****schéma, — de branchement,** connection diagram; — **de câblage,** wiring diagram; — **de carte,** card design; — **de circulation,** flow chart, flow sheet; — **de composition,** bill of material; — **de graissage,** lubrication chart; — **de logique,** logic diagram; — **de mémoire,** storage layout (EDP); — **de montage,** wiring scheme or plan; — **de principe,** skeleton diagram, flow sheet; — **général,** block diagram; — **synoptique,** block diagram.

****schiste aluneux,** alum shale.

schoepite, schoepite.

schoopage, *m.* metal spraying, Schoop process.

schroeckingerite, schroeckingerite.

*scie, — à arc, jigsaw; — à chantourner, bow saw; — à chassis, jigsaw; — à lame sans fin, endless saw; *— à métaux, hack saw; — à plaçage, veneer saw, frame saw; — contournée, fret saw; — de travers, — passe-partout, crosscut saw; — rotative, disk saw.

scié en quatre parties, quarter-sawed.

*scier, — à culer, — aux avirons, to backwater.

scintillateur, m. — liquide, liquid scintillator; — liquide à deux canaux, dual-channel liquid scintillator.

scintillation, f. scintillation, flickering.

scintillement, m. flickering, scintillation, sparkling; impression de —, flicker effect, flickering.

*scintiller, to flicker.

scintillomètre pour faible intensité, m. low-level scintillation counter.

*scorie, scum (metal.).

*sculpture, tread (tire).

*séance, — d'abonnement, subscription conversation; — d'évaluation, benchmark session.

*séchage, de — dur, hard-drying.

*second, mate (naut.).

*secondaire, auxiliary; voie —, side track, secondary route (EDP).

*secours, de —, back-up, standby; convoi de —, breakdown train; porte de —, sally port; roue de —, spare wheel; terrain de —, emergency landing field.

*secteur, supply line, network.

*section, — de configuration, configuration section; — de fichiers, file section; — d'entrée-sortie, input-output section; — de rectification, enriching section; — d'extraction, stripper; — droite du canal, channel area.

section efficace, cross section; — d'absorption, absorption cross section; — d'absorption des neutrons, neutron-absorption cross section; — d'activation, activation cross section; — d'arrêt, atomic stopping power, stopping cross section; — d'arrêt équivalente, equivalent-stopping cross section; — de capture, capture cross section; — de capture des neutrons, neutron-capture cross section; — de capture radiative, radiative-capture cross section.

section efficace, — de collision, collision cross section; — de diffusion, scattering cross section; — de diffusion de l'atome lié, bound-atom scattering cross section; — de diffusion élastique, elastic-scattering cross section; —

de diffusion élastique due à la forme, shape elastic cross section; — de diffusion inélastique, inelastic-scattering cross section; — de fission, fission cross section; — de fission par neutrons de grande vitesse, high-speed neutron fission cross section; — de fission par neutrons rapides, fast-neutron fission cross section; — de fission par neutrons thermiques, thermal-fission cross section, thermal-neutron fission cross section.

section efficace, — de production, yield cross section; — de transport, transport cross section; — d'interaction, interaction cross section; — du noyau, nuclear cross section; — effective d'absorption pour les neutrons de résonance, effective resonance absorption cross section; — effective de capture, effective capture cross section; — élémentaire, differential cross section; — équivalente, equivalent cross section.

section efficace, — macroscopique, macroscopic cross section; — macroscopique de capture, macroscopic capture cross section; — macroscopic effective, effective macroscopic cross section; — macroscopique totale, bulk cross section; — microscopique, microscopic cross section; — parasite, parasitic cross section; — pondérée suivant la valeur du flux, flux-weighted cross section; — pour la fission de l'uranium, cross section for uranium fission; — pour la fission nucléaire, cross section for nuclear fission.

section efficace, — pour la réaction, reaction cross section; — pour les neutrons, neutron cross section; — pour les neutrons de résonance, resonance cross section; — pour les neutrons thermiques, thermal cross section; — pour les photons, gamma cross section; — pour les rayons gamma, gamma cross section; — pour une réaction nucléaire, cross section of a nuclear reaction; — totale, total cross section.

sectionneur, m. disconnecting switch, isolator.

*sécurité, — des piles, reactor safety; — du fonctionnement, reliability; de —, back-up, stand-by.

*sédimentation, settling.

*segment, piston ring, stroke (EDP); — de base, root segment; — d'embrayage, clutch shoe; — d'étanchéité, compressing ring; — racleur, scraper ring; — refouleur, top ring (piston); —s compatibles, inclusive segments (EDP); —s incompatibles, exclusive segments.

segmentation, f. quantization, partitioning, segmentation; — spirale, spiral cleavage.

*segregation, — de solidification, freezing (metallurgy); — par gravité, gravity segregation.

séisme, *m.* earthquake.

*sel, — anglais, Epsom salts; — d'uranium, uranium salt; — fondu, fused salt.

sélecter, to choose; — par égalité, to select (EDP); — par non égalité, to edit (EDP).

*sélecteur, selector; — à grande vitesse, fast chopper; — à large faisceau, large-beam chopper; — à petit faisceau, small-beam chopper; — à trois voies, three-way selector; — auxiliaire, co-selector; — de cap, heading selector; — d'écoute, audio switch; — de fonction, flight reference selector; — de frein, selector valve; — de l'amplitude des impulsions, pulse-height selector; — de neutrons, neutron chopper.

sélecteur, — de vitesse, velocity selector; — de vitesse des neutrons, neutron velocity selector; — d'impulsions, pulse chopper; — manuel, dial (EDP); — mécanique (de neutrons), chopper; — mécanique pour faibles vitesses, slow chopper; — prise statique, static air selector; — stroboscopique, chopper.

sélecteur-pilote, *m.* pilot selector (EDP).

*sélection, area search (EDP).

self, *f.* induction coil; — à basse fréquence, low-frequency choke; — à fer, iron-cored choke coil; — à saturation, saturable inductance coil; — d'amortissement, quenching choke; — de circuit oscillatoire, oscillation-circuit inductance; — de filtrage, smoothing choke; — de filtre du redresseur, ripple filter choke.

*sellette, bosun's chair.

sémantique, *f.* semantics.

sémaphore, *m.* stop signal.

*semelle, — de longeron, spar flange; — de palier, ground plate.

semi-additionner, to half-add.

semi-additionneur binaire, *m.* binary half-adder.

semiconducteur, *m.* semiconductor.

semi-fermé, semienclosed.

semi-létal, semilethal.

semi-protégé, semiprotected.

semi-soustracteur, *m.* one-digit subtracter.

sengiérite, *f.* sengierite.

*sens, normal direction flow; — anti-horaire, counterclockwise; — anti-trigonométrique, clockwise; — de l'exploration, direction of scanning; — direct, counterclockwise; — d'un

vecteur, sense (of a vector); — horaire, clockwise; — opposé, opposite direction; — rétrograde, negative direction; — trigonométrique, counterclockwise.

*sensibilité, response; — de déviation, deflectability; — spectrale, color sensitivity (of a photocell).

*séparateur, delimiter (EDP), resolver; — à un seul étage, single-stage separator; — de colonne, column split (EDP); — de mots, word separator (EDP); — électromagnétique des isotopes, electromagnetic mass separator; — isotron, isotron separator; — magnétique d'isotopes, magnetic separator; — pour matières entraînées, entrainment separator.

*séparation, — chimique, chemical separation; — des charges, charge separation (in plasma); — des isotopes, isotope separation; — des masses, mass separation; — d'isomères, isomer separation; — électromagnétique, electromagnetic separation; — électrostatique, electrostatic separation; — isotopique, isotope separation; — par gravité, -gravity separation.

séquence, *f.* sequence; — à pas de pèlerin, backstep sequence (welding); — d'appel, calling sequence; — de comparaison, collating sequence; — de nombres pseudo-aléatoire, pseudo-random number sequence; — de tri, collating sequence; — pseudo-aléatoire, pseudo-random sequence.

séquenceur, *m.* sequencer (astronautics).

séquentiel, serial, sequential; — indexé, indexed sequential.

*série, string (EDP), production run.

*sériel, serial; — par bit, serial by bit; — par dual, serial by bit.

*serpentin, helix; — bouilleur, reboiler coil; — de refroidissement, cooling coil; — plat, pancake coil; — réfrigérant, condensing coil.

serre-câbles, *m.* cable clips, drilling clamps.

*serre-fils, *m.* pliers.

serre-flan, *m.* blank holder, draw ring.

serre-glace, *f.* window pane retainer.

serre-lame, *f.* terminal (elec.).

*serrement, dam.

*serrer le frein, to brake.

serre-tête, *m.* earphone.

*sertir, — par gaufrage, to swage; — par replis, to crimp.

sertissage, *m.* crimping, setting (jewels, glass in lead), swaging, spot stamping; — de rotule, spot stamping of ball-joint.

servants d'une pièce, *m.pl.* gun crew.

*service, — de formation professionnelle, education center; — des archives, filing department; — des méthodes, systems department; — d'exploitation, operation department; — d'informatique, data processing department; — d'inspection, customer engineering; — mécanographique, punched card department; — navette, shuttle service; —s après ventes, product support division; — service, in service, on stream.

servo-actionneur de trim, *m.* trim actuator.

servo-amplificateur de trim, *m.* trim servo amplifier.

servocommande, *f.* servo-control; excitation d'une —, servo input; résponse d'une —, servo output.

servo-direction, *f.* power steering.

servodistributeur, *m.* servo valve.

servo-frein, *m.* servo brake.

servograisseur, *m.* pressure oiler.

servointégrateur, *m.* decision integrator.

servomécanisme, *m.* servo-control mechanism, servo; — fonctionnant par tout ou rien, on-off type of servo.

servomultiplieur, *m.* potentiometer multiplier.

*seuil, — d'énergie cinétique, threshold kinetic energy; — d'énergie pour la fission, threshold energy of fission; — de fission, fission threshold; — de fission par neutrons, neutron-fission threshold; — de fréquence, frequency threshold; — de photodissociation, photodissociation threshold; — de photofission, photofission threshold; — de réaction, reaction threshold; — de tension, voltage threshold; — photoélectrique, photoelectric threshold.

*sève, pith.

sharpéite, *f.* sharpite.

shuntage, *m.* bridging.

sialographie, *f.* sialography.

siège-baquet, *m.* bucket seat.

siège-couchette, *m.* reclining seat.

*sièges décalés, staggered seats.

siemens, *m.* mho.

sifflant, sibilant, fricative.

*sifflement, singing.

*signal, — binaire, bit stream; — d'action en retour, feedback signal; — d'attaque, input signal; — de cantonnement, block signal; — de circumpropagation, round-the-world signal; — de commande, control signal; — de compensation, compensation signal; — de

contrôle, control signal; — de ligne occupée, audible busy signal; — de manoeuvre, ready tone, dial tone, audible alarm; — d'entrée, input signal; — de numérotage, dial hum; — de repos, interval or spacing signal; — de retour perturbé, disturbed response signal (EDP); — de rétroaction, feedback signal; — de sécurité, guard signal; — de sortie, output signal; — de travail, marking signal; — d'horloge, clock pulse; — d'images, video signal (telev.); — d'interdiction, inhibiting signal; — d'occupation, busy signal; — parasite, interfering signal; signaux enclenchés, interlocked signals; — sonore, ready tone, dial tone, audible alarm.

*signe, pulse, impulse.

silencieux, *m.* muffler (auto).

silentbloc, *m.* flexible bushing.

sillage, *m.* wake (of a boat).

simbleau, *m.* centering bridge (tool).

simulateur, *m.* simulator; — de pile, model reactor; — de pilotage, link trainer; — d'etude de réseaux, network analyzer; — de vol, flight simulator; — mathématique de calculateur, Turing machine.

*simultané, real time, concurrent, full duplex.

*simultanéité, overlap (EDP); — d'entrée-sortie, input-output overlap; erreur de —, coincidence error.

*singularité ponctuelle, point singularity.

singulet, singlet.

*siphon, — de fermeture hydraulique, liquid seal; — thermique, thermal siphon.

sirène, *f.* siren.

skiagramme, *m.* radiograph.

skiographe, *m.* skiograph.

sklodowskite, sklodowskite.

*socle rond, floor flange.

soddyite, *f.* soddite.

*sodium de refroidissement, cooling sodium.

soldat, simple —, private (soldier).

*solde, — net, net balance; — nul, zero balance; pour — de tout compte, in full payment.

*solidaire, locked with, fixedly connected with, integral with.

*solide, body.

*solidité, durability.

*sollicitation, application (for a position), stress, stressing, straining.

*solubilité, resolving power.

*solution, — active, fuel solution; — asymptotique, asymptotic solution; — de combustible, fuel solution; — **électrolytique,** electrolytic solution; — **liquide,** liquid solution.

solution, — mère, pregnant solution; — **minéralisante,** ore-bearing solution; — **relative à l'état stationnaire,** steady-state solution; — **solide,** solid solution; — **stationnaire,** steady-state solution; — **toute prête,** prepared solution (phot.).

*solvant d'extraction, extractant.

*somme des masses et des énergies, mass-energy total.

*son, — additionnel, summation tone; — d'un vibrateur, buzzing tone.

*sondage, — à balancier, beam well; — d'épreuve, test hole; — en contrebas, bottom-side sounding; — en contre-haut, topside sounding; — intérieur, borehole made from galleries; — ionosphèrique vertical, vertical sounding.

*sonde, — amplificatrice à autocoupage, quenching probe unit; — de mesure, probe for measurements; — spatiale, space probe; — thermométrique, temperature scanner.

*sonder, to dive.

*sondeur, sounding device.

sonomètre, *m.* noise meter.

*sonore, audio.

sorbonne, *f.* fume hood.

sorption, *f.* sorption.

*sortie, leadin (of a tube), output; — analogique, analog output; — d'air de dégrivrage, de-icing air outlet; — de registre, register output; — de secours, emergency exit; — différé, deferred exit; — d'impression, print selection common exit; — écart (VOR/LOC), deviation output; — en direct, on-line output; — en ligne, on-line output; — hors-ligne, off-line output; — numérique, digital output; — par impulsion, pulse output; — tout-ou-rien maintenue, latched contact output; — traversée fuselage, feed through; — Un, *one* output.

*sortir, to recover; — de caractéristiques essentielles, depart from scope or spirit (of an invention).

*soubassement, underbody (vehicle).

*souche, stack (smoke), strain.

*soudage, — à clin, lap welding; — à molette, seam welding; — à l'arc sous flux, submerged arc welding; — aluminothermique, thermit welding; — autogène, fusion welding; — des goujons, stud welding; — par bossages, projection welding; — par étincelage, flash welding; — par points, spot welding; —

par points continus, stitch welding; — par pression, pressure welding.

*soudé, sealed, welded, shut, fused.

soudo-brasage, *f.* braze welding, bronze welding.

*soudure, — arasée, flush weld; — chromosomique, chromosome rejoining; — d'épinglage, tack weld; — par points, spot welding; — par rapprochement, butt welding; — par recouvrement, joint welding.

*soufflante, turboblower.

*souffle de l'hélice, propeller slipstream.

*soufflement, blasting.

*soufflerie, wind tunnel; — à veine libre, free jet tunnel; — hypersonique, supersonic wind tunnel.

*souffleur, glass blower.

*souffrance, en —, backlog; travail en —, work awaiting attention.

*soulager, to discharge.

*soulever, to heave.

*soumissionnaire, contractor.

*soupape, bailer; — à gorge, butterfly valve; — aiguille, needle valve; — à tiroir, slide valve; — de détente, relief valve; — de sûreté, relief valve; — en champignon, poppet valve; —s en tête, overhead valves.

*souplesse de transformation, flexibility.

*source, — (ionique) à fente, slit source (of ions); — de chaleur, source of heat; — d'émission, emitting source; — d'énergie, power source, source of energy; — de neutrons, neutron source; — de neutrons au radium béryllium, radium-beryllium source of neutrons; — d'ions, ion source; — d'ions non-ponctuelle, extended ion source.

source, — émettrice, emitting source; — en couche cylindrique, cylindrical shell source (of neutrons); — énergétique de faible puissance, low-power source of energy; — en forme de disque, disk source; — étalonnée, standard source; — linéaire, linear source, line source; — négative de chaleur, heat sink.

source, — non-ponctuelle, distributed source; — plane, plane source; — plane infinie, infinite plane source; — ponctuelle, point source; — pulsée, pulsed source; — radioactive de grande intensité, high-level radiation source; — superficielle sphérique, spherical surface source; — virtuelle, image source.

*sources, headwaters.

souricière, *f.* mousetrap, trap, ambush, combination socket, fishing socket.

sous-canal, *m.* subchannel.

sous-critique, subcritical.

sous-ensemble, *m.* sub-assembly, sub-system, sub-unit, subset (EDP).

sous-espace, *m.* subspace.

sous-gamme, *f.* subband.

sous-groupe d'essai, *m.* test set (EDP).

sous-harmonique, *m.* subharmonic.

*sous le vent, leeward.

sous-oeuvre, *m.* groundwork, foundation.

sous-poutre, *f.* tie, sleeper (R.R.), girder.

*sous-produit radioactif, radioactive by-product.

sous-programme, *m.* subroutine (EDP); — **à niveau unique,** one-level subroutine; — **à utilisations coincidantes,** reenterable routine; — **à utilisations successives,** serially reusable routine; — **comptable,** accounting routine; — **d'assemblage,** assembly routine; — **de contrôle,** checking routine; — **de détection,** tracing routine; — **de diagnostic,** diagnostic routine; — **d'entrée,** input routine; — **de point de controle,** check-point routine; — **de premier ordre,** first order routine; — **d'épuration,** error detection routine; — **de service,** utility routine; — **de traitement d'interruptions,** interrupt handling routine; — **de vérification,** checking routine; — **directeur,** executive routine; — **fermé,** closed routine; — **interprétatif,** interpretive routine; — **invariant,** reenterable routine; — **ouvert,** open routine; — **traducteur,** translating routine.

sous-virage, *m.* understeering.

*soutien, *m.* stay (antenna).

spallation, *f.* spallation.

*spatialiser, to adapt to space conditions.

spationaute, *m.* astronaut.

spationautique, *f.* astronautics, space travel.

spationef, *f.* spacecraft.

*spatule, spud, bumper.

*spécification, rating; — **de tête,** header label.

*spectre, — **bêta,** beta spectrum; — **continu,** continuous spectrum; — **de bandes,** band spectrum; — **de coïncidence,** coincidence spectrum; — **de fission,** fission spectrum; — **de masse,** mass spectrum; — **de micro-ondes,** microwave absorption spectrum, microwave spectrum.

spectre, — **de molécule,** molecular spectrum; — **de radiofréquences,** radiofrequency spectrum; — **de raies,** line spectrum; — **de rayonnement de freinage,** bremsstrahlung spectrum; — **des chemins,** channel spectrum;

— **des neutrons,** neutron spectrum; — **des neutrons de pile,** reactor spectrum.

spectre, — **des réactions possibles,** channel spectrum; — **des voies,** channel spectrum; — **du rayonnement gamma,** gamma-ray spectrum; — **en amplitude des impulsions,** pulse-height spectrum; — **énergétique,** energy spectrum; — **énergétique de fission,** fission energy spectrum; — **moléculaire,** molecular spectrum; — **total des électrons dégradés,** total degraded-electron spectrum.

*spectrographe, — **à quartz,** quartz spectrograph; — **de masse,** mass spectrograph.

*spectromètre, — **à cristal,** crystal spectrometer; — **alpha,** alpha-ray spectrometer; — **à neutrons,** neutron spectrometer; — **à scintillations,** scintillation spectrometer; — **bêta,** beta-ray spectrometer.

spectromètre, — **de masse,** electromagnetic mass separator, mass analyzer, mass spectrometer; — **de masse à trajectoire cycloïdale,** trochoidal mass analyzer; — **gamma,** gamma-ray spectrometer; — **magnétique,** magnetic spectrometer; — **magnétique à paires,** magnetic pair spectrometer; — **optique,** optical spectrum.

*spectroscopie gamma, gamma spectroscopy.

spermatogénèse, *f.* spermatogenesis.

*sphère, — **aplatie aux pôles,** oblate spheroid; — **d'action,** sphere of activity; — **de Debye,** screening sphere; — **de l'effet d'écran,** screening sphere.

sphérique, *m.* spherical balloon.

spider, *m.* roadster (vehicle), rumble seat.

spin, *m.* spin quantum number, spin; — **de la voie,** channel spin; — **de l'électron,** electron spin; — **demi-entier,** half-integral spin; — **du photon,** photon spin; — **isobarique,** isobaric spin, isotopic spin; — **nucléaire,** nuclear spin.

spineur, *m.* spinor.

spinthariscope, *m.* spinthariscope.

*spire, spring coil; — **de court-circuit,** shading ring; — **morte,** idle coil turn; — **utile,** effective coil; —**s jointive,** solid length (spring).

spot mobile, *m.* flying spot.

*stabilisation, attitude control; — **par champ magnétique axial,** axial magnetic field stabilization.

*stabilité, — **du confinement,** containment stability; —**dynamique,** dynamic stability; — **nucléaire,** nuclear stability; — **propre,** inherent stability; — **statique,** static stability; — **thermique,** thermal stability.

*stagiaire, trainee, junior.

starter, *m.* choke, signal giver (race).

stasite, stassite.

*__station,__ terminal; — **centrale,** electric power plant; — **de caissier,** teller terminal; — **de pompage,** pumping station; — **de terre,** terrestrial station; — **en envergure sur la pale,** blade spanwise station; — **libre-service,** self-service station; — **mobile terrestre,** mobile land station; — **orbitale,** orbital station; — **radio,** ground radio station; — **radiométéorologique,** weather station (radio); — **spatiale,** space station; — **terrienne,** earth station.

station-aval, *f.* downrange station.

*__statistique quantique,__ quantum statistics.

statolimnimètre, *m.* static level meter.

statoréacteur, *m.* ram-jet engine.

statut, *m.* statute, ordinance, rule, regulation, status, state.

stellarateur, *m.* stellarator.

stéréographique, stereographic.

*__stéréométrie,__ solid geometry.

stéréoradioscopie, *f.* stereofluoroscopy.

*__stérilisation par irradiation,__ radiation sterilization.

stilb, *m.* stilb.

stock, *m.* stock, inventory; — **de sécurité,** safety inventory of spares.

*__stockage,__ stockpiling, storage; — **permanent,** permanent storage; — **sous l'eau,** underwater storage; — **souterrain,** underground storage.

stocker, to stock (goods), store (data).

strapontin, *m.* jump seat, folding seat.

*__stratification entrecroisée,__ false bedding.

stratifié, *f.* layer structure.

stratus, *m.* stratus cloud.

striction, *f.* reduction of area, pinch; — **du courant,** current constriction; — **électrodynamique,** electrodynamic contraction; — **électromagnétique,** electromagnetic constriction; — **magnétique,** magnetic constriction; — **quasi-stationnaire,** quasi-stationary pinch effect, simple pinch effect; — **stabilisée,** stabilized pinch; — **thermique,** thermal constriction; — **transitoire,** transient pinch effect.

stripage, *m.* stripping (nuclear).

stroboscope, *m.* stroboscope.

*__structure,__ — **atomique,** atomic structure; — **de logique,** logic design; — **d'instruction,** instruction format; — **du noyau,** structure of nucleus; — **du noyau en couches,** nuclear shell structure; — **du réseau,** lattice struc-ture; — **électronique,** electronic structure; — **en couche,** layer structure; — **en couches,** shell structure; — **en nids d'abeilles,** honeycomb structure; — **en spirale double,** double helical structure.

structure, — **extranucléaire,** extranuclear structure; — **fine,** fine structure; — **génique,** gene structure; — **grossière,** gross structure; — **hexagonale compacte,** hexagonal close-packed structure; — **hyperfine,** hyperfine structure; — **nucléaire,** nuclear structure; — **réticulaire,** lattice structure, reticular structure; — **soudée,** welded structure.

studio, *m.* studio, study, radio studio.

studtite, *f.* studtite.

*__style,__ stylus.

stylo-dosimètre, *m.* pocket meter.

*__suage,__ bleeding.

*__sublimation,__ sputtering.

*__subordonné,__ — **à,** limited; — **au temps de calcul,** compute-limited; — **au temps d'entrée,** input-bound; — **au temps de sortie,** output-limited; — **au temps de traitement,** process-bound, computer-limited.

*__substance,__ — **absorbante,** absorbing material; — **adsorbée,** adsorbate; — **condensée,** condensate; — **de base,** key substance; — **de ralentissement,** moderator material; — **fissile,** fissionable material; — **ne permettant pas la surgénération,** nonbreeding material; — **radioactive de courte période,** short-lived radioactive substance; — **réflectrice-retarda-trice,** tamper material; — **réfractaire,** refractory material.

substruction, *m.* substructure.

subvention, *f.* subsidy.

*__succession,__ repetition.

*__suite,__ string (EDP); — **d'instructions,** control sequence.

suiveur, *m.* tracker (astronautics).

*__sujétion,__ — **d'unités périphériques,** device dependence.

supercarburant, *m.* premium grade gasoline, supergasoline.

*__supérieur,__ *m.* senior.

*__supériorité,__ advantage.

supermodulation, *f.* overmodulation.

super-polissage, *m.* mirror finish.

*__superposer,__ to overlay, overwrite.

*__superposition,__ overlay; — **auto-consistante,** self-consistent superposition.

super-réaction, *f.* superregeneration.

superviseur, supervisor (EDP); — **de segments de recouvrement,** overlay supervisor.

*****support,** girder, volume, medium, cradle, carrier; — **à berceau,** saddle pipe; — **amortisseur,** cushioned mounting; — **cornier,** angle bracket; — **cranté,** notched angle; — **d'appui,** bearing block; — **de collier,** clamp block; — **de lampe,** socket; — **de l'arbre,** bearing journal; — **de mémoire,** storage medium; — **de mine,** pit prop (min.); — **de tiges de forage,** beam hangar; — **de train avant,** nose gear saddle; — **de travail,** job program; — **de tubage à coins,** casing spider; — **de tube,** conduit support; — **d'exploitation principal,** system residence volume; — **d'hélice,** propeller nosepiece; — **d'information,** recording medium, recording support (EDP); — **gauchissement,** later cyclic control support (helicopters); — **pendant,** hangar; — **programmoïde,** package (EDP); — **réproductible,** master copy; — **sphérique,** socket pad (jack); — **vierge,** blank medium (EDP); —**s régulateurs,** adjustable bearings (mach.).

*****supposer,** to assume, underpin.

*****suppression,** — **de l'amortissement par frottement,** frictional antidamping; — **des parasites,** noise suppression.

supraconducteur, *m.* superconductor.

*****suralimentation,** supercharging (aviation).

suramorti, overdamped.

*****surcharge,** overstress, overrun (EDP).

surcontrastes, *m.pl.* bleeding whites.

surcritique, supercritical.

surdimensionné, oversize.

*****surépaisseur,** weld reinforcement; — **d'usinage,** machining allowance.

surestarie, *f.* demurrage (naut.).

surfaçage, *m.* leveling, surfacing, facing, spot facing, mechanical polishing.

*****surface,** — **à usure,** — **de fatigue,** bearing surface; — **de contact (poudres),** surface area; — **de niveau,** level surface; — **de séparation,** boundary, interface; — **de séparation du réseau,** lattice boundary; — **de séparation entre le coeur et la couche de récupération,** core-blanket interface; — **écrouie localement,** single shot surface; — **efficace de cible,** effective target area; — **élémentaire d'analyse,** cell area (EDP); — **émettrice,** emitting surface.

surface, — **équipotentielle,** equipotential surface, level surface; — **extérieure,** boundary; — **extérieure de coeur,** core boundary; — **extérieure de la cellule,** cell boundary; — **extérieure du réflecteur,** outer boundary of the reflector; — **magnétique,** magnetic surface; — **photométrique,** polar surface of light distribution; — **portante,** airfoil surface; — **radiante,** cooling surface; **monter en —,** parallel connection.

surgénération, *f.* breeding; — **en neutrons rapides,** fast-neutron breeding.

surimpulsion, *f.* burst.

surintensité, *f.* excess current.

surmodulation, *f.* overmodulation.

surmoduler, to overmodulate.

surmultiplicateur, *m.* overdrive.

suroscillation, *f.* overshoot (radio).

surraffiné, overrefined.

surrégénérateur, réacteur —, breeder reactor.

surréservation, *f.* overbooking.

*****surtension,** voltage rise, increment of voltage, voltage step-up, supertension, overvoltage, supervoltage; — **du compteur,** counter overvoltage.

*****surveillance,** monitoring, survey; — **des rayonnements nucléaires,** nuclear control; — **du personnel,** personnel monitoring; — **neutronique,** neutron monitoring.

*****surveillant,** timekeeper.

*****surveillante,** — **de perforation,** chief keypunch operator.

survieillissement, *m.* overaging (metallurgy).

survirage, *m.* oversteering.

survoler, to fly over.

*****survolter,** to overrun, overload, overvolt (a tube or lamp).

survolteur, *m.* booster (elec.).

survolteur-dévolteur, *m.* positive and negative booster.

*****suspens, en —,** pending.

*****suspension,** slurry; — **à bras tirés,** trailing arms suspension; — **aqueuse,** aqueous slurry, water slurry; — **AV à roues indépendantes,** independent front suspension; — **dans l'eau,** water slurry; — **dans un liquide,** wet suspension; — **sèche,** dry suspension.

suspentes, *f.pl.* cords of a parachute.

*****symbole,** character, report-form (EDP), indicator; — **de décision,** decision box (EDP); — **de logique,** logic symbol; — **d'organigramme,** flowchart symbol; — **éliminatoire,** ignore character; — **externe,** external symbol; — **littéral,** literal (EDP); — **logique,** logical symbol; — **supplémentaire,** additional character.

*****symétrie,** — **axiale,** axial symmetry; — **par rapport à un plan,** reflection of symmetry; — **par réflexion,** reflection symmetry; — **par**

retour inverse dans le temps, time-reflection symmetry; — **spatiale par rapport à une droite,** rotation symmetry; **ayant la — de révolution,** ring geometry.

synchro, *m.* synchro; — **de recalage,** slaving synchro; — **d'érection,** gyro leveling synchro; — **détecteur de roulis,** roll detector synchro; — **détecteur de tangage,** pitch detector synchro; — **répétiteur,** synchro transmitter; — **transmetteur,** resolver.

synchrocyclotron, *m.* synchrocyclotron.

synchronisation, *f.* phasing, timing, synchronization; — **de la pile,** reactor synchronizing; — **dépendante,** step-by-step synchronizing; — **indépendante,** independent time control; — **par réseau,** mains synchronization.

synchroniseur, *m.* synchronizer, synchromesh device.

synchroniseuse, *f.* film synchronizer.

synchrophasotron, *m.* synchrophasotron.

synchrotransmetteur, *m.* selsyn motor.

synchrotron, *m.* synchrotron; — **à protons,** proton synchrotron.

*****syndicat corporatif,** trade union.

syndrome, *m.* syndrome; — **aigu du mal des rayons,** acute radiation syndrome.

synergie, *f.* synergy.

*****synthèse de l'image,** reproduction of the image.

synthétiseur, *m.* scanning device at the receiving end; — **de voix,** voice code translator; **organe —,** recreator device, reproducing device (telev.).

syntonisation, *f.* tuning (radio).

syntoniser, to tune, resonate, cause to be in resonance.

*****système, — à action différée,** time-limit protection; — **à bande latérale unique,** single-band system; — **à base donnée,** radix notation, radix scale; — **à cartes perforées,**

punched card system; — **à combustible circulant,** circulating fuel system; — **à courants porteurs,** carrier system (EDP); — **à décharges pulsées,** pulsed system; — **à deux phases,** two-phase system; — **à ondes porteuse,** carrier system (EDP); — **à retour de puissance,** reverse-power protection; — **autogénérateur,** bootstrap operation; — **avec neutre à la terre,** earthed neutral system.

système, — d'antennes régulièrement placées, array of antenna; — **d'autoguidage,** homing system; — **d'avertissement,** warning device; — **d'axes,** frame of reference; — **d'axes du laboratoire,** laboratory system; — **de commande,** control system; — **de coordonnées,** coordinate system; — **de démarrage réacteur,** engine starting system; — **de flottabilité de secours,** emergency flotation gear; — **de numération binaire,** binary number system; — **de particules,** system of particles.

système, — de programmation, programming system; — **de refroidissement,** cooling system; — **de relevé d'exceptions,** exception principle system; — **de répartition prioritaire,** priority scheduling system; — **de répartition séquentielle,** sequential scheduling system; — **de sécurité de dépression,** vacuum safety system; — **de téléphone automatique,** dial telephone; — **de transmission secrète,** privacy system; — **d'exploitation,** operating system, monitor system; — **d'inertie,** inertial system; — **d'informatique,** data processing system; — **d'interruption,** system alert (EDP); — **flotteur,** float undercarriage; — **hydromagnétique,** hydromagnetic system; — **indépendant,** basic programming system (EDP); — **intégré de gestion,** management information system; — **mobile d'axes de référence,** moving coordinate system; — **numérique,** number representation, numeral system; — **polycyclique,** multifrequency system; — **pulsé d'énergie nulle,** pulsed zero energy system; — **quantifié,** quantized system; — **retardateur,** lagging system.

T

*****table, — à dessiner,** drawing board; — **à secousses,** jig table; — **de charnières,** link library; — **de concentration,** concentration table; — **de contrôle,** control dictionary; — **de lancement,** launcher; — **des matières,** volume table of contents (EDP); — **de symboles externes,** external symbol dictionary; — **de vérité,** truth table; — **d'harmonie,** sounding board; — **périodique,** periodic table; — **traçante,** plotting board (EDP).

*****tableau, — de bord,** dashboard; — **de charges,** planning board; — **de commande,** control panel, plugboard; — **de connexions,** control panel, wiring board, jack panel; — **de correspondance,** cross-reference table; — **d'épreuves,** test panel, testing schedule; — **des coefficients de pose,** exposure table; — **de signalisation,** indicating panel; — **des vents,** wind scale; — **de voies,** track diagram; — **reproducteur de sons,** growler board.

tabler sur, to reckon with.

*****tablier,** bulkhead (auto), dashboard (loosely used), scooter footrest, hearth of forge; — **sans fin,** apron feeder.

tabulatrice, *f.* tabulator, tabulating machine; — **alphanumérique,** alphabetic accounting machine; — **numérique,** digital tabulator.

tabuler, to tabulate.

*****tache,** hot spot (cinematography); — **d'encre au recto (TAD),** extraneous ink front; — **d'encre au verso (TAD),** extraneous ink back; — **lumineuse,** light spot; — **magnétique,** magnetic spot.

*****taffetas gomme,** adhesive plaster.

*****taille critique,** critical size.

*****tailler en pointe,** to taper.

*****tailleur de limes,** file cutter.

*****talc,** French chalk.

*****talon,** tire bead; — **de souche,** check stub.

*****tambour,** — **à inertie,** inertia reel; — **à picot,** sprocket drum; — **de frein,** brake drum; — **de grue,** hoisting drum; — **d'embrayage,** clutch drum; — **de tension,** tensator drum; — **de treuil,** winch drum; — **magnétique,** magnetic drum.

*****tamis à secousses,** vibrating screen.

*****tamisage,** straining.

*****tampon,** — **amortisseur,** shock absorbing pad; — **à rondelles de caoutchouc,** rubber washer (mach.); — **atmosphérique,** air buffer; — **de butée,** thrust pad (vehicle suspension); — **de vapeur,** vapor lock; — **élastique,** snubber (auto); — **étalon,** master ring gage; — **imbibé,** polishing swab; **monté en** —, forming a buffer.

tangage, *m.* pitch (vehicle or ship).

tantale, *m.* tantalum; **lampe à filament de** —, **lampe au** —, tantalum lamp.

*****taper,** to drive (a nail).

*****tapisserie,** upholstery.

taraudeuse, *f.* screw cutter, tapping machine, thread cutter.

*****tarer,** to normalize, standardize.

*****tarif,** — **à dépassement,** overload tariff; — **binôme,** two-part tariff; — **de douane,** customs tariff; — **dégressif,** diminishing tariff; **plein** —, day rate (teleph.).

*****tas de ferraille,** junk pile.

tasselotte, *f.* head (casting).

tâte-ferraille, *m.* junk feeler.

tâtement, *m.* scanning (telev.).

tâteur, *m.* scanner, scanning device, test prod,

tip jack, weft-replenishing control of automatic loom.

*****tâtonnement,** trial and error, by tentation, fishing.

taud, *m.,* **taude,** *f.* awning (naut.).

*****taux,** — **d'absorption différentiel,** differential absorption rate; — **d'accroissement d'énergie,** rate of energy gain; — **d'activité,** activity ratio; — **d'amortissement,** damping ratio; — **d'amplitude,** peak-to-valley ratio; — **de branchement,** branching ratio; — **de combustion,** burnup; — **de compression,** compression ratio; — **de comptage,** counting rate; — **de comptage du bruit de fond,** background counting rate; — **de comptage parasite dû à l'instrument,** spurious counting rate.

taux, — **de comptage spontané,** background; — **de consommation,** burnup; — **de consommation du combustible,** fuel burnup; — **de conversion,** conversion ratio; — **de croissance,** rate of growth; — **de décroissance,** disintegration rate; — **de diffusion,** diffusion rate; — **de fission,** fission rate.

taux, — **de modulation,** modulation ratio; — **de multiplication des neutrons instantanés de fission,** prompt neutron multiplication rate; — **de mutation,** mutation rate; — **de pannes,** failure rate; — **de production d'énergie,** rate of power generation; — **de transfert d'énergie,** energy transfer rate; — **de variation,** rate of change; — **d'extraction de chaleur,** cooling rate; — **d'irradiation,** radiation rate; — **d'ondes stationnaires (T.O.S.),** standing wave voltage ratio; — **d'ondulations résiduelles,** ripple ratio.

té, — **double,** crossover tee; — **oblique,** lateral tee.

technétium, *m.* technetium.

*****technique,** procedure, technology; — **de génération de nombres aléatoires,** randomizing technique; — **de la construction des piles,** reactor engineering; — **de la haute tension,** high-voltage technique.

*****teinte,** *f.* shading, gradation (pictures).

télécinema, *m.* film television.

télécinématographie, *f.* film television.

télécommande, *f.* telemetering, remote control.

télécommunications, *f.pl.* communication.

télé-compas, *m.* remote compass, telecompass.

télédétection, *f.* remote sensing.

télédiffusion, *f.* wire broadcasting.

télé-entrée, *f.* remote job entry.

télégramme-mandat, *m.* telegraphic (cable) money order.

télégraphe-imprimeur, *m.* type-printing telegraph.

*__télégraphie harmonique,__ voice-frequency telegraphy.

télégraphiste, *m.* telegrapher.

téléimprimeur, *m.* telewriter, teleprinter.

télémaintenance, *f.* housekeeping (astronautics).

télémanipulation, *f.* remote manipulation.

télémesure, *f.* telemetering, telemetry.

télémétrie, *f.* range finding.

télémoin, standard, reference.

téléobjectif, *m.* telephoto lens.

*__téléphone,__ — à cadran, rotary dial telephone; — de bord, interphone, intercommunication system; — de piste, telebriefing installation.

téléphote, *m.* televiser, television apparatus.

téléphotie, *f.* television.

télépointage, *m.* telecontrol of guns.

téléscripteur, *m.* teletype, teleprinter.

télétraitement, *m.* teleprocessing.

télétransmission, *f.* data transmission, data communication.

téléviseur, *m.* — à tambour à lentilles, lens-drum television scanner.

télévision, *f.* radiovision, television.

*__témoin,__ indicator, mark, non-machined area, signal light; — d'allumage, ignition warning light; — de charge, charging indicating light; — de starter, choke warning light; — des clignotants, direction indicator warning lights; — d'opérations, event control block; **appartement** —, model apartment; **denrée** —, basic commodity; **échantillon** —, check sample.

*__température,__ — absolue, absolute temperature; — cinétique, kinetic temperature; — critique, critical temperature; — d'allumage, ignition temperature; — de Debye, Debye temperature; — de fonctionnement, operating temperature; — de fusion, fusion temperature; — de régime, operating temperature; — des réactions de fusion, thermonuclear temperature; — de transformation, transformation temperature.

température, — d'ignition, ignition temperature; — du corps noir, black-body temperature; — électronique, electron temperature; — et pression normales (TPN), normal temperature and pressure (NTP); — ionique, ion temperature; — limite, critical temperature; — minimum d'allumage, minimum igni-

tion temperature; — minimum d'ignition, minimum ignition temperature; — ultra-haute, ultra-high temperature.

temporiser, to temporize, delay, retard.

*__temps,__ stages; — d'accélération, acceleration time; — d'accès, access time; — d'addition, add time (EDP); — d'amorçage, start-up time; — d'attente, wait time, latency, rotational delay; — d'augmentation, growth time; — d'autocollision, self-collision time; — de battement, float time; — d'échange de l'énergie, energy exchange time; — de chute des barres, scram time; — de confinement, confinement time; — de correction, correction time, settling time; — d'écrêtage, clipping time; — de croissance, growth time, time of growth, time of response, rise time.

temps, — de décroissance, decay time; — de déflection, deflection time; — de dépannage, repair time; — de désactivation, cooling time; — de diffusion, diffusion time; — de fonctionnement, running time; — de génération, generation time; — de génération des neutrons instantanés, prompt generation time; — d'élaboration de programme, program development time; — de montée de l'instabilité, instability growth time; — de multiplication des neutrons, neutron multiplication time; — d'entretien, maintenance time; — de passage d'entrefiches, cross-gap time; — d'équilibre, equilibrium time; — d'équipartition, time of equipartition.

temps, — de ralentissement, slowing-down time; — de réaction, reaction time, response time; — de récupération, recovery time; — de relaxation, relaxation time; — de relaxation de choc, collisional relaxation time (of plasma); — de relaxation de collision, collisional relaxation time (of plasma); — de réponse, response time; — de résolution, resolving time; — de rétention, retention time; — de retour, flyback time.

temps, — de séjour, holdup time, retention time; — de sensibilité, sensitive time; — de tenue, clipping time; — de transit, transit time; — de vol, flying time, total time (TT); — d'inspection, maintenance time, engineering time; — létal moyen, median lethal time; — logistique, AOG time, logistic time; — minimum de réponse, minimum response time; — mort, idle time; — moyen de retard, mean delay time; — perdu, time lag; — réel, real time (EDP); — utile, productive time, effective time; en fonction du —, time dependence.

tenancier, *m.* tenant.

*__tenant le vide,__ vacuum-tight.

*__tendeur,__ eccentric clamp, turnbuckle, bridle irons, wire stretcher; — à vis, turnbuckle;

— **de chaîne,** chain tensioner; — **de courroie,** idler.

***teneur, — en uranium,** uranium content; — **isotopique,** isotopic abundance; — **relative,** abundance ratio.

teneurmètre, *m.* content meter; — **de minerai,** ore content meter; — **en uranium,** uranium content meter.

tenseur, *m.* tensor; — **d'albédo,** albedo tensor; — **de déformation,** strain tensor; — **de masse effective,** effective mass tensor; — **de sollicitation,** stress tensor; — **de tension,** stress tensor.

***tension,** elasticity; — **accélératrice,** accelerating voltage; — **constante équivalente,** equivalent constant potential; *— **d'allumage,** striking potential, striking voltage; — **d'amorçage,** priming potential, starting voltage; — **de blocage,** cutoff potential, grid cutoff voltage; — **de chauffage,** filament voltage; — **de cisaillement,** shear stress; — **de coupure,** cutoff voltage; — **de crête,** crest or peak voltage.

tension, — de fonctionnement, working voltage; — **de percement,** breakdown voltage, breakdown potential; — **de pointe,** crest or peak potential; — **de polarisation,** bias; — **de référence de précision,** reference voltage; — **de rupture,** breakdown potential, breakdown voltage; — **de saturation,** saturation voltage; — **de sortie,** output voltage; — **de verrouillage,** cutoff voltage, locking voltage; — **du circuit coupé,** recovery voltage.

tension, — du secteur, mains voltage; — **d'utilisation,** working voltage; — **nominale,** rated voltage; — **opposée,** inverse voltage; — **plaque retardatrice,** retarding plate voltage; — **porteuse,** carrier voltage; — **potentiel de claquage,** breakdown voltage, breakdown potential; — **préalable,** — **préliminaire,** biasing or bias potential; — **quantique,** quantum voltage; **appareil à haute —,** high-voltage apparatus; **haute —,** high tension, high voltage.

***tenue,** maintenance; — **à jour de fichier,** file maintenance; — **d'assiette,** attitude hold; — **de cap,** heading hold; — **de route,** road stability (vehicle); — **de route en côte,** gradeability; — **de stationnaire,** hover control; — **de stock,** inventory control; — **mécanique,** mechanical aspect.

***terme,** operand, minuend; — **d'addition,** addend; — **isotopique,** composition term; — **mixte,** cross term; — **relatif à l'émission extérieure des neutrons,** source term; — **relatif aux neutrons émis par les sources,** source term.

***terrain collant,** sticky formation.

***terrassier,** excavator.

***terre, — activée,** activated fuller's earth; — **à porcelaine,** porcelain clay; — **décolorante,** clay; **à —,** aground.

***territoire national,** homeland.

***test d'allélisme,** test of allelism.

***tête, — d'avertisseur de givrage,** ice probe; — **d'écriture,** write head; — **de cuvelage,** casing head; — **d'effacement,** erase head; — **de lecture,** read head; — **de pont,** bridgehead; — **d'éruption,** flow head; — **de série,** pre-production run; — **de sonde,** casing head; — **d'essuie-glace,** wiper head; — **de visée,** gunsight head; — **double de percussion,** dual discharge head; — **du manche pilote,** control column boss; — **HF,** HF module; — **magnétique,** magnetic head.

tétrode, *f.* tetrode; — **à pente variable,** variable tetrode.

teugue, *f.* deckhouse (naut.).

thalassémie, *f.* thalassemia.

***théorie, — à deux groupes,** two-group theory; — **approchée au premier ordre,** first-order theory; — **approchée du transport,** transport approximation; — **à un seul groupe,** one-group model; — **de la cible,** target theory, hit theory; — **de la diffusion à un (ou plusieurs) groupe(s),** group diffusion theory (one- or multi-).

théorie, — de l'équilibre, balance theory; — **de l'information,** information theory; — **des files d'attente,** queueing theory; — **des perturbations,** perturbation theory; — **du réseau,** lattice theory; — **du transport,** transport theory; — **monocinétique,** one-velocity method.

thermistance à tige, *f.* rod-type thermistor.

thermocontact, *m.* thermal switch, water temperature switch.

thermoduricissable, thermosetting.

thermoélectron, *m.* thermion, thermoelectron.

thermofission, *f.* thermofission.

thermofusion, *f.* thermofusion.

***thermomètre, — à cadran,** heat gauge; — **d'eau,** water temperature gauge; — **mouillé,** wet-bulb thermometer; — **sec,** dry-bulb thermometer.

thermomètre-fronde, *m.* sling thermometer.

thermophone, *m.* thermophone.

thermoplongeur, *m.* immersion heater.

thorianite, *f.* thorianite.

thoride, thoride.

thorié, thoriated (as a filament).

*thorine, thorium dioxide.

*thorium, au —, thoriated (chem.).

thorogummite, f. thorogummite, nicolayite.

thoron, m. thorium emanation, thoron.

thorotungstite, f. thorotungstite.

thrombocyte, m. platelet.

thucholite, thucholite.

thyratron, m. thyratron.

*tige, — à joints lisses, flush-joint drill pipe; — coulissante, slide rod; — de culbuteurs, pushrod; — de direction, steering drag rod; — de repérage, centering pin; — de soupape, valve system; — graduée, dividing rod; — témoin, warning rod (fuel tank).

tige-poussoir, f. pushrod, tappet-rod.

tikker, m. ticker, interrupter.

tillac, m. deck (naut.).

*timbre, maximum allowable working pressure; — de la poste, postmark.

timonerie, f. steering (ship); — de direction, steering linkage (auto); — de frein, brake linkage.

*tir, m. gunfire, gunnery range.

*tirant, wall clamp; — d'eau, draft; — de faible d'eau, shallow draft; — de grand d'eau, deep draft.

tire-clou, m. nail puller.

tire-fond, m. lag screw.

tire-ligne, m. ruling or drawing pen.

*tirer, to tap off (liquids); — d'une carrière, to quarry.

tiret, m. dash, hyphen.

tirette, f. pullrod, flue damper.

tirez! courez! lachez! walk! run! release!

*tiroir, valve.

*tissu, — chauffant, heating fabric; — germinal, germinal tissue.

titromite, f. titromite.

*toc, cog; — de tour, lathe dog.

toddite, f. toddite.

*toile, — à voiles, sailcloth; — d'émeri, emery cloth; — huilée, varnished cambric.

*toit, — amovible, hardtop (auto); — découvrable, ouvrant, sliding roof, sun roof.

*tôle, — déflectrice, baffle plate; — en acier, sheet steel; — feuilletée, laminated plates or sheets, magnetic lamination.

tôle-plancher, f. floor tray (auto).

*tolérance, — admise, allowance; avec une — de, subject to a tolerance of.

*tomber, — à la mer, to fall overboard; — en flammes, to go down in flames.

*tombereau, dumper.

*tonalité, shading value (telev.), timbre, ready tone, dial tone.

*tonnage en lourd, dead weight.

*tonneau, roll (aviation); — à droite, roll to the right; — arraché, flick roll; — lent, slow roll.

top, m. brief (synchronizing) pulse or impulse, crest or peak, brief signal, time tap.

torbernite, f. torbernite.

*tore, — activé, magnetized core; — au repos, unmagnetized core; — magnétique, magnetic core.

tornade, f. tornado.

toroïde, torus.

toronné, stranded.

torpillage, m. shooting, torpedoing.

*torpilleur, destroyer (naut.).

*total, — de contrôle, de vérification, check total, hash total, gibberish total, proof total; — élémentaire, minor total; — mélé, hash total; — par liasse, batch total; — par lot, batch total; — principal, major total; totaux cumulatifs, rolling totals.

total-bidon, m. hash total, gibberish total.

*totaliser, to accumulate.

*totaliseur, accumulator, odometer.

*touche, segment, bar (of commutator); — d'appel, request key; — de restauration de système, system reset key; — d'intervention, request key.

*toucher, to hit (mil.); — sur, to impinge.

*toucheur, (sound) pickup.

*tour, — à canons, cannon-boring lathe; — à fileter, thread-cutting lathe, engine lathe; — à gabarit, copying lathe; — à pédale, foot lathe; — à pointes, center lathe; — à tablier, apron lathe.

tour, — de condensation, condensation tower; — de fourage, derrick; — de fractionnement, fractionating tower; — de lancement, launching rail; — de lavage, scrubber; — de montage, servicing tower; — d'épuration, scrubber; — de refroidissement, cooling tower; — de remplissage, umbilical tower (rocket fueling); — de sondage, derrick; — d'établi, bench lathe; — en l'air, chuck lathe; —s par minute, revolutions per minute.

tour-alesoir, m. boring lathe.

*tourbe, turf.

tourbillonnaire, m. vortex.

*tourillon, bearing, wrist pin, axle end; — à **boulet,** ball-and-socket joint.

*tourné, provided with screw thread.

tourne-disque, m. record player, disk drive, disk storage drive.

*tourner, to gyrate; — **un filin,** to belay.

tournisse,f. baluster.

*tournoyer, to gyrate, to resolve.

*tourteau, filter cake, cornice.

traçage, m. tracing, opening up a coal seam; laying out of roads.

*trace, layout; — **des villes,** town planning; — **ionique,** ionization path.

*tracé, graph, layout, diagram; — **de points,** point plotting; — **des villes,** town planning; — **de vecteurs absolus,** absolute vector graphics.

*tracer, to delineate, recreate, or reproduce a picture (telev.).

traceur, m. tracer; — **de courbes,** plotter; — **d'organigrammes,** autochart; — **moléculaire,** molecular tracer; — **physique,** physical tracer; — **radioactif,** radioactive tracer.

*traction, propeller thrust, drive; — **avant,** front-wheel drive.

*traducteur, assembly program, end instrument, transducer; — **de tension,** voltage-controlled oscillator; — **tension-fréquence,** voltage-controlled oscillator.

*train, — **à pantalon,** trousered undercarriage; — **AR (arrière),** rear axle; — **automoteur,** self-propelling landing gear; — **AV (avant),** front axle; — **blindé,** armored train; — **caréné,** faired undercarriage; — **d'accès,** access group; — **de bits,** bit stream; — **de sonde,** drill-pipe string; — **de travaux,** job stack; — **de travaux d'entrée,** input job stream; — **d'impulsions,** pulse train; — **épicycloïdal,** epicyclic gear train; — **verrouillé rentré, sorti,** gear up and locked, down and locked.

trainage, m. coupling hysteresis effect.

*trainée, drag (airplane).

*trait, stand; — **d'une scie,** kerf.

*traitement, processing; — **à haute température,** high temperature processing; — **anodique,** anodizing; — **à plusieurs phases,** multistage process; — **automatique des listes de matériel,** bill of material processor; — **chimique,** chemical process, chemical processing; — **chimique de récuperation,** chemical reprocessing; — **de détente,** stress relief treatment; — **de l'information, des données, des informations,** data processing; — **de mise en solution,** solutionizing; — **en différé,** off-

line operation; — **en direct,** on-line operation; **en file d'attente,** stacked job processing; — **en ligne,** on-line operation; — **en parallèle,** parallel processing; — **hors-ligne,** off-line operation; — **immédiat,** inline processing; — **intégré de l'information,** integrated data processing; — **mathématique,** mathematical process (EDP); — **multi-taches, multi-travaux,** multi-job operation (EDP); — **par les neutrons,** neutron therapy; — **par lots,** batch processing, stacked job processing; — **par voie sèche,** dry process; — **simultané,** processing overlap; — **thermique,** heat-treatment.

*traiter, to process, handle, compute.

trajectographie, f. tracking, trajectography.

*trajectoire, — **d'une particule,** particle trajectory; — **du vol,** flight path; — **ionique,** ionization path, ion path; — **semicirculaire,** semicircular path; — **spirale,** spiral path.

*trajet, track; — **aléatoire,** random walk; — **cycloidal,** cycloidal "walk" (of particles in plasma).

*trame, scan, scanning field, ruled screen grating.

traminot, m. tramway agent.

*trancher, to clip.

*tranchet, hardy.

trancheuse,f. ditcher.

transbordeur, m. car ferry; **bac —,** transporter ferry; **chariot —,** traverser; **pont —,** transporter bridge.

transcoder, to translate (EDP).

transcripteur, m. output writer (EDP).

transcription,f. transcription, copy, registration, posting, transliteration.

transducteur, m. transductor, transducer.

transduction bactérienne,f. bacterial transduction.

*transférer, to switch over, assign (goods, etc.).

*transfert, branch, re-recording, jump, move; — **avec marque de not,** load mode (EDP); — **de chaleur,** heat transfer; — **d'électrons,** electron transfer; — **de magnétisation,** magnetic printing; — **de masse,** mass transfer; — **d'énergie,** power transfer; — **d'énergie par choc,** collisional energy transfer; — **de quantité de mouvement,** momentum transfer; — **des efforts,** load carrying; — **de température,** heat transfer; — **direct,** direct control (EDP); — **électronique,** electron transfer; — **moyen d'énergie,** mean energy transfer.

*transformateur, — abaisseur, step-down transformer; — accordé, tuned transformer; — à circuit magnétique fermé, closed-core transformer; — à colonnes, core-type transformer; — à deux ponts, two-leg transformer; — à noyau, core-type transformer; — basse-fréquence, audio-frequency transformer; — cuirassé, shell-type transformer; — d'adaptation, matching transformer; — de chauffage, filament transformer; — de coordonnées, resolver; — de fréquence, frequency transformer.

transformateur, — de liaison, intervalve transformer; — de mesure, instrument transformer; — de mesure compensé, phase-compensating transformer; — d'énergie, transducer; — d'entrée, input transformer; — de réseau, mains transformer; — de sortie, output transformer; — différentiel, hybrid coil, hybrid transformer; — d'intensité, current transformer; — élévateur, step-up transformer; — redresseur, transformer rectifier; — réducteur, step-down transformer; — suceur, negative boosting transformer; — toroïdal, ring transformer.

*transformation, — allotropique, phase transformation; — canonique, canonical transformation; — cyclique du combustible, fuel cycle; — cyclique du combustible avec surgénération, fuel breeding cycle; — de coordonnées, transformation of coördinates; — de la chaleur en énergie, conversion of heat into power; — de mésons, meson transformation; — faisant apparaître la symétrie, symmetrization.

transformation, — irréversible, irreversible process, nonadiabatic process; — mésique, meson transformation; — non-adiabatique, irreversible process, nonadiabatic process; — nucléaire, nuclear transformation; — nucléaire artificielle, artificial nuclear transformation; — nucléaire artificiellement provoquée, artificially induced nuclear transformation; — nucléaire en cascade, nuclear cascade process; — nucléaire non-contrôlée, uncontrolled nuclear transformation.

transformation, — nucléaire spontanée, spontaneous nuclear transformation; — par similitude, similarity transformation; — provoquée, induced transformation; — réversible, reversible process; — rotationnelle, rotational transform; — structurale, structural change; — tautomère, tautomeric transformation; — thermonucléaire, thermonuclear transformation.

transformée de Laplace, f. Laplace transform.

*transformer, to process.

transformeur, m. processor.

transistor, m. transistor; — à effet photoélectrique, photistor; — à gradient de champ, drift transistor; — à jonction, junction transistor; — à pointes, point contact transistor; — au germanium, germanium transistor; — planaire, planar transistor; à —, transistorized.

*transition, — de phase, phase transition; — interdite, forbidden transition; — isomérique, isomeric transition; — nucléaire, nuclear transition; — permise, allowed transition; — radiative, radiative transition.

translocation chromosomique, f. chromosome translocation.

*transmetteur, transducer; — d'asservissement, servo transmitter; — de débit, flow transmitter; — de direction, rudder follow-up; — de données, data link; — de flux magnétique, flux gate transmitter; — de freinage, hydrastic brake transmitter; — de gauchissement, aileron follow-up; — de jaugeur, fuel level transmitter; — de pas, pitch transmitter; — de position de volets, wing flap follow-up; — de poussée, thrust transmitter; — de pression, pressure transmitter; — de profondeur, elevator follow-up; — de synchro détection, synchro transmitter; — récepteur HF, HF transceiver.

*transmission, gear (mach.); — à crémaillère et pignon, rack and gear mechanism; — à distance, remote control, remote reading; — arrière, tail rotor drive; — arythmique, start-stop transmission; — aux roues, final drive; — de chaleur, heat transfer; — de données, data communication; — d'énergie, energy transfer, power transfer; — en parallèle, parallel transmission; — en série, serial transmission; — oblique, inclined drive shaft; — rotor arrière, tail rotor drive shaft.

*transmutation, — artificielle des éléments, artificial transmutation of elements; — d'éléments, transmutation of elements; — des espèces nucléaires, transmutation of nuclei.

*transparence de grille, inverse amplification factor.

*transport, — d'énergie, energy transfer; — d'énergie électrique, power transfer; — effectif, net transport.

transposabilité, relocatability.

*trappe, — à calories, heat sink; — à faisceau, beam catcher.

*traumatisme, trauma.

*travail, — à froid, cold work; — d'extraction, work function; — en parallèle, time-sharing; — en temps réel, real time operation (EDP);

— simultané, time-sharing; **— utile,** capacity (phys.).

travail-machine, *m.* machine load (EDP).

traversier, *m.* triatic (tower), bearer cable or wire (electric traction), spreader.

***trembleur,** buzzer (elec.), ticker, vibrating contactor.

***trémie,** feeder; **— de chargement,** bucket.

***trempé,** hardened (metal.).

tremper, to pickle (metal.), sprinkle.

***trépan aléseur,** reaming bit.

trépan-benne, *m.* hammer-grab.

***treuil,** cathead, hoist; **— auxilaire,** pulling machine.

***tri,** sort; **— magnétique,** magnetic sort; **— par grands groupes, par lots,** block sort; **— sur band,** tape sort.

***triage,** dressing (of mineral ores).

***triangle de vecteurs,** vector triangle.

***tribord la barre,** starboard the helm (naut.).

triergol, *m.* tripropellant.

***tringle,** bead core (tire).

tringlerie, *f.* linkage, system of rods; **— de direction,** steering linkage; **— de frein,** brake linkage; **— d'injecteur,** injector control linkage.

tripartition spontanée, *f.* spontaneous triple fission.

triplace, three-seater.

triplage, *m.* triplication.

triplan, *m.* triplane.

triplet, *m.* triplet; **— (de spin) isobarique,** isotopic triplet.

tripleur, *m.* tripler.

tritium, *m.* tritium.

triton, *m.* triton; **— de recul,** recoil triton.

***trituration,** attrition.

tritureuse, *f.* pulvimixer.

trochotron, trochotron.

troegerite, *f.* troegerite.

trolite, *m.* trolit.

***trompe,** ejector; **— de brume,** foghorn.

***trompette,** flared axle tube; **— de pont,** rear axle flared tube.

tropadyne, *m.* tropadyne.

troposphère, *m.* troposphere.

***trou,** drop-out (radio); **— à l'avancement,** advance borehole, protection hole, pilot borehole; **— borgne,** blind hole; **— de bonde,** bunghole; **— débouchant,** through hole; **— de**

clavette, pilot hole; **— de fixation,** attachment hole; **— d'énergie,** energy gap; **— d'entraînement,** feed hole, sprocket hole; **— de raidissement,** flanged hole; **— de serrure,** keyhole; **— détrompeur,** locating hole; **— d'homme,** manhole; **— d'usinage,** tooling pick-up; **— embouti,** punched hole, stamped hole; **— noyauté,** cored hole; **— ovalisé,** elongated hole, out of round hole; **— traversant,** through hole.

***trouble, — dû à l'irradiation,** radiation damage; **— phytohormonal,** phytohormone damage.

***trousse,** set or assembly (of lenses).

***tube,** hose, barrel; **— à barbotage,** bubble tube; **— à décharge,** discharge tube; **— à décharge dans les gaz,** gas discharge tube; **— à pilotage,** drive pipe; **— à résistance ballast,** ballast tube; **— cannelé,** grooved tube; **— cathodique,** cathode ray tube, display tube; **— convergent,** focusing tube.

tube, — d'accélération, accelerating tube; **— d'allumage,** incandescent tube; **— de caoutchouc,** bicycle-valve tubing; **— de chauffe,** heating tube; **— de cuvelage,** casing pipe; **— de décomposition d'image,** image or picture dissector, scanning tube; **— de direction,** steering column tube; **— de flux,** flux tube, tube of flux; **— de focalisation,** focusing tube.

tube, — de maturation, soaker tube; **— de niveau,** gauge column; **— de paroi,** wall tube; **— de repêchage,** horn socket; **— de repêchage à frottement,** friction socket; **— de repêchage fondu,** cherry picker; **— de spectrographe,** spectrograph tube; **— de transmission,** torque tube; **— d'extraction,** pull tube.

tube, — d'objective, objective mount or barrel (electron microscope); **— électronique,** valve; **— évasé,** flared tube; **— luminescent,** glow tube, fluorescent tube; **— miniature,** bantam tube; **— multiplicateur d'électrons,** electron multiplier tube; **— profile,** tube of streamlined section; **— thermoélectronique,** thermionic tube; **— toroïdal,** "doughnut" (toroidal tube), flux converter.

tube-compteur, *m.* counter tube.

tube-guide, *m.* conductor string.

***tubulure,** (tubular) nipple piece, tubulation (of exhaust fumes); **— d'admission,** induction manifold; **— d'échappement,** exhaust manifold; **— de jaugeage,** gauging hatches, haulage; **— sans bride,** pipe riser.

***tuf volcanique,** volcanic tuff.

tulipage, *m.* gyro gauging.

***tumeur, — bénigne,** benign tumor; **— maligne,** malignant tumor.

*tungstène, wolfram.

tunnel aérodynamique, *m.* wind tunnel; — à retour, wind tunnel of the closed-circuit type; — vertical de vrille, spinning tunnel, vertical wind tunnel for spinning tests; — vraie grandeur, full-scale wind tunnel.

turbine-pompe, *f.* torque converter pump.

*turbo-alternateur, turbogenerator.

turbo combustible, *m.* turbine fuel.

turbocompresseur, *m.* turbocharger, turbocompressor.

turbo-hélice, *f.* turbo prop.

turbo-propulseur, *m.* turbo-fan.

turboréacteur, *m.* reaction jet propeller.

turbo-réfrigérateur, *m.* turbine cooler unit.

turbo-ventilateur, *m.* turbo-fan.

*tuyau flambeur, flue.

*tuyauterie, — d'arrivée, inlet tube; — de

dérivation, by-pass line; — de purge, flushing line; — de sortie, delivery tube; — liquide avant, fluid tube (cock-to-boom); — liquide départ, fluid tube (tank-to-cock); — rigide, rigid pipe; — souple, flexible pipe, hose.

tuyauteur, *m.* pipe fitter.

*tuyère, nozzle, jet pipe, tail pipe; — à corps central, plug nozzle; — à noyau, plug nozzle; — d'échappement, exhaust pipe assembly; — d'éjection des gaz, jet pipe propelling nozzle; — de PC, reheat tail pipe nozzle; — orientable, swiveling nozzle.

*type, — de croisements, breeding habits; — de fission, fission channel; — d'unions (homme), breeding habits; — périmé, obsolete type; de — sauvage, wild type.

typhon, *m.* typhoon.

tyrite, tyrite.

tysonite, *f.* tysonite.

tyuyamunite, tyuyamunite.

U

udomètre, *m.* udometer.

ulrichite, *f.* ulrichite.

ultracentrifugeuse, *f.* ultracentrifuge.

ultramicrochimie, *f.* ultramicrochemistry.

ultrasensible, ultra-sensitive.

ultra-son, *m.* ultrasonics; essai à , ultrasonic testing.

ultrasonique, ultrasonic, supersonic.

ultravide, *m.* ultra-high vacuum.

umohoite, umohoite.

*un, — conditionnel, elusive one (EDP).

uni-adresse, *f.* single address (EDP).

unicité, *f.* uniqueness.

unidimensionnel, one-dimensional.

unidirectionel, simplex (telecommunication).

unifilaire, unifilar.

union instantanée, *f.* instantaneous assembly.

unions au hasard, *f.pl.* panmixia.

uniphasé, single-phase, monophase.

*unipolaire, homopolar, single-pole (switch); — à deux directions, single-pole double throw; — à une direction, single-pole single throw.

*unité, — centrale de traitement, central processing unit (EDP); — d'affichage, visual display unit (EDP); — de chaleur centésimale, centigrade heat unit; — de masse atomique, atomic mass unit, physical mass unit; — de mémoire, storage unit (EDP); —

de poids atomique, atomic weight unit; — de réponse verbale, audio response unit; — de ruban, paper tape unit; — de télécommande, transmission control unit; — de trame, raster unit; — de transfert, transfer unit; — de volume, unit volume; — enfichable, plug-in unit; — fondamentale de masse, fundamental unit of mass; — homogène, consistent unit; — Mache, Mache unit; — motrice, power plant; — motrice complète, propulsion unit; — physique, physical unit; — terminale, device terminal; — X, X unit.

*universel, general purpose.

uraconite, uraconite.

*uranate de sodium, sodium uranate.

*urane oxydule, pitchblende.

uranide, uranide.

*uranium, — alpha, alpha uranium; — gamma, gamma uranium; — enrichi, enriched uranium; — métallique, uranium metal; — non-enrichi, unenriched uranium; — -235, actino uranium, U-235.

uranochalcite, uranochalcite.

uranocircite, *f.* uranocircite.

uranomolybdate, uranomolybdate.

uranopilite, *f.* uranopilite.

uranospathite, uranospathite.

uranosphérite, *f.* uranosphaerite.

uranospinite, uranospinite.

uranothallite, *f.* uranothallite, flutherite.

uranothorite, uranothorite.

uranotile, *f.* uranotile.

uranotite, uranotite.

usager, (of things) in daily use; **effets —s,** personal effects.

*usine, — chimique, chemical plant; **— de décontamination,** decontamination plant; **— de production,** production plant; **— de raffinage,** refinery; **— de séparation isotopique,** separation plant; **— d'extraction,** extraction plant; **— métallurgique,** metallurgical plant.

usiné, machined.

*usure, attrition; **— de frottement,** galling, frictional wear; **— par corrosion,** fretting.

utilisateur, *m.* user.

*utilisation, — à grande échelle, large-scale utilization; **— des neutrons thermiques,** thermal utilization; **— d'une pile,** pile operation; **— industrielle,** large-scale utilization; **— médicale des rayonnements,** medical applications of radiation.

uvanite, uvanite.

V

*vaisseau, — amiral, flagship; **— de ligne,** liner (naut.); **— spatial,** spaceship, spacecraft.

*vaisselle de terre, earthenware dishes.

*valeur, — absolue, absolute value, magnitude; **— au seuil,** threshold value; **— caractéristique de l'énergie,** energy eigen-value; **— de l'énergie,** quantum level, energy level, power level; **— de pas indiquée,** indicated pitch angle; **— de pointe,** peak value; **— effective,** effective value; **— erratique,** freak value; **— expérimentale du laplacien,** experimental buckling; **— la plus probable,** most probable value; **— limite de l'énergie de fission,** critical fission energy; **— moyenne du carré de la distance,** mean square distance; **— moyenne du carré de la longueur de ralentissement,** mean square length of moderation.

valeur, — moyenne du carré de la variation de la quantité de mouvement, mean square momentum change; **— moyenne du carré de la vitesse,** mean square velocity; **— moyenne quadratique de la vitesse,** root mean square velocity; **— probable,** probable value; **— propre,** eigenvalue; **— propre de l'énergie,** energy eigenvalue; **— réelle,** effective value; **—s approchées aux différences finies,** finite difference approximations; **—s expérimentales,** data.

*validité, applicability, effectivity; **— avion,** aircraft effectivity; **— ouverte,** open-ended effectivity.

vandenbrandeite, vandenbrandite.

*vanne, valve, damper; **— à passage direct,** gate valve (straight-way valve); **— à pointeau,** needle valve; **— d'admission,** throttling valve; **— d'arrêt,** trip valve; **— de décharge,** dump valve; **— de réglage,** control valve; **— de retenue à flotteur,** float check valve; **— d'éruption,** blowout preventer; **— de sec-**tionnement, block valve; **— d'étranglement,** throttling valve.

vanoxite, vanoxite.

*vapeur, — charbonnier, collier (naut.); ***— d'échappement,** dry steam; **— indirecte,** closed steam; **— radioactive,** radioactive steam; **— saturée,** saturated steam; **— sursaturée,** supersaturated vapor.

vaprocraquage, *m.* steam cracking.

vaprocraqueur, *m.* steam cracker.

var, *m.* var.

vareuse, *f.* jumper.

*variable, — arbitraire, controlled variable; **— spatiale,** spatial variable; **à une seule —,** one-dimensional; **—s canoniques conjuguées,** canonically conjugate variables.

*variante, modification, modified form.

*variation, gradient; **— à long terme de la réactivité,** long-term reactivity change; **— apériodique,** aperiodic variation; **— de densité,** density gradient; **— de la densité,** gradient of the density; **— d'énergie par atome,** energy change per atom; **— dépendante,** dependent variable; **— de pression,** pressure gradient; **— de réactivité,** change of reactivity; **— de relèvement,** variation in apparent bearing; **— indépendante,** independent variable; **— modifiable,** controllable variable; **— non modifiable,** uncontrollable variable.

variation, — du flux, flux gradient; **— en fonction du volume,** volume effect; **— génique,** genovariation; **— globale de la quantité de mouvement,** total momentum change; **— relative,** fractional change; **— spatiale du flux,** spatial variation of flux; **— théorique de réactivité,** calculated reactivity change; **— unitaire d'énergie,** energy change per atom.

vario-alternateur, *m.* constant frequency alternator.

variocoupleur, *m.* variocoupler.

variomètre, *m.* variometer.

***vase,** — **d'expansion,** expansion chamber, surge tank; — **poreux,** porous cell.

vasistas, *m.* transom.

***vecteur,** radiant; — **à quatre coordonnées,** four-dimensional vector; — **axial,** axial vector (pseudovector); — **de spin isobarique,** isobaric spin vector; — **laplacien,** buckling vector; — **polaire,** polar vector; — **quantité de mouvement,** momentum transfer, momentum vector; — **spatial,** space vector; — **spin,** spin **·vector;** — **unitaire,** unit vector; — **vitesse,** velocity vector.

vecteur-colonne, column vector.

vecteur-courant de neutrons, neutron current density.

vecteur-force, force vector.

vectographe enregistreur, *m.* vectorial recorder.

vedette, *f.* patrol boat.

***vehicle,** — **à chenilles,** tracked vehicle; — **spatial,** space vehicle; — **tous terrains,** cross-country vehicle.

***vent frais,** gale.

***ventilateur,** — **débrayable,** declutching fan; — **de climatisation,** heating fan; — **refroidissement,** cooling fan.

***ventre,** antinode, loop.

venturi, *m.* venturi tube, choke tube.

***venu,** — **de fonderie,** as-cast condition; **plante bien —e,** healthy plant.

***verificateur de gain,** gain checker.

***verification,** — **arithmétique,** arithmetic check; — **automatique,** autobalance; — **avant le vol,** preflight checkout; — **de codes d'opération,** code check; — **de la sensibilité,** accuracy test; — **de perforation,** hole-count check; — **double,** twin check; — **interne,** internal auditing; — **modulo N,** modulo N check; — **par double opération,** duplication check; — **par redondance,** redundancy check; — **sans dépose,** *in situ* inspection.

vérificatrice, *f.* verifier (machine); — **à clavier,** key verifier.

***verifier,** to examine, ascertain.

vérifieur, vérifieuse, *m. & f.* verifier operator.

***vérin,** ram; — **à bille des volets,** flap jack; — **à billes,** ball screw jack; — **à galet,** roller jack; — **à simple effet,** single action cylinder; — **contrefiche,** truss actuating jack; — **correcteur d'effort,** pitch corrector unit (flight controls); — **de calage,** stabilizing jack; — **de chasse,** starting ram (launching ways); — **de**

compensation, compensating cylinder; — **de décrochage escalier,** stairway unlocking cylinder; — **de décrochage train,** landing gear unlocking cylinder; — **de déverrouillage,** unlocking cylinder; — **de direction du train avant,** nose wheel steering cylinder; — **de levage,** hydraulic jack; — **de poussée,** starting ram; — **de rappel dans l'axe,** compensating cylinder.

vérin, — **de relevage de train,** landing gear actuating cylinder; — **de rentrée,** retraction actuator; — **de repliage de pales,** blade folding cylinder; — **des aubes de guidage,** intake guide van ram; — **d'escamotage,** retraction actuator; — **de sécurité verrouillage train bas,** landing gear down lock safety cylinder; — **de sensation musculaire,** artificial feel cylinder; — **de serrage,** strut jack; — **de servo-commande (rotor),** servo-jack (rotor); — **de volet,** flap control; — **d'orientation train avant,** steering cylinder (landing gear); — **électrique,** electric actuator; — **hydraulique,** hydraulic actuating cylinder; — **mobile,** mobile jack; — **servodyne trappes,** main landing-gear door-actuating cylinder; — **zéro direction,** rudder trim jack.

***vernis,** — **à la nitrocellulose,** nitrocellulose dope; — **d'asphalte,** black japan.

***vernissage,** finish.

***verre,** windowpane; — **à file de fer,** wire glass; — **anti-éblouissant,** tinted glass, anti-glare glass; — **coloré,** stained glass; — **de lampe,** lamp chimney; — **de sécurité,** safety glass; — **gradué,** graduate; — **ne brunissant pas,** nonbrowning glass; — **sans éclats,** splinterproof glass.

***verrier,** glass blower.

***verrou,** — **à platine,** slip bolt; — **à ressort,** spring lock; — **de dzus,** dzus fastener; — **de frappe,** hammerlock; — **de pale,** blade lock; — **de repliage,** folding hinge lock; — **de tuyère réacteur,** keep plate; — **d'interdiction de relevage,** L/G safety lock, retraction lock; — **électromagnétique,** scrubber; — **magnétique,** magnetic clutch.

verrouillage, *m.* locking or clamping device, interlock; — **de clavier,** keyboard lockout; — **de la commande de repliage de pales,** blade folding control lock; — **de la source,** source interlock; — **de lecture,** read interlock; — **en position train rentré, sorti,** unlocking, down locking (landing gear); — **hydraulique,** hydraulic locking; — **mécanique,** mechanical locking.

***vers,** against.

***versant,** watershed.

***verso,** outer face; — **de carte,** back of card.

vestibule, *m.* hallway.

*vêtement parachute, parasuit.

*vibrateur, buzzer.

*vibration, jar, shock.

*vibrer, to interrupt (a current).

*vibreur, chopper (radio), stickshaker, vibrating contactor; — de départ, magneto booster coil; — de signalisation, buzzer; — pour vérification des induits, armature tester; crayon —, engraving tool.

*vidage, dump (EDP), non-process runout.

*vidange rapide du réservoir, fuel jettisoning.

*vide, — poussé, high vacuum; — spatial, space vacuum; à —, idle, no load.

· vide-mémoire, *m.* core dump program, storage print; — autopsie, postmortem dump; — dynamique, dynamic dump; — sélectif, selective dump.

vide-poches, *m.* glove compartment, cubby hole.

*vider, to jettison, clear, dump (EDP).

vide-vite, *m.* fuel jettison, fuel dump valve.

vie moyenne, *f.* average life, mean lifetime.

*vierge, blank; zone —, clear area.

vietinghofite, *f.* vietinghofite.

vif, au rouge —, red-hot.

*vigie, crow's-nest (naut.).

*vilebrequin, — à contrepoids, counterweighted crankshaft; — à plusieurs coudes, multi-throw crankshaft.

*virage, turn, curve, banking (airplane).

*virement, transfer, clearing; — de bord, tacking (naut.).

virer vent devant, to tack.

*virgule, — binaire, binary point; — fixe, fixed point; — flottante, floating point; — fractionnaire, radix point.

*vis, — à crépine, filter screw; — à crochet, hook screw; — à métaux, machine screw; — à tête cylindrique, cheesehead screw; — à tête cylindrique bombée, fillister head screw; — à tête fraisée, countersunk head screw; *— calante, wedging screw, clamp screw; — de butée, adjusting screw; — de conduite, guide screw; — de purge, bleed screw; — de rallonge, temper screw; — globique, hourglass wormscrew; — imperdable, captive screw; — pierre, jewel support; — platinées, contact points; — tangente, worm screw; — taraud, self-tapping screw.

viscoréducteur, *m.* visbreaker.

viscoréduction, *f.* viscosity breaking, visbreaking.

*viser, to cover, disclose (as a patent).

*viseur, view finder (phot.), gunsight.

*visiblité, outlook, view; — nocturne, noctovision; — nulle, zero visibility.

*visiter, to examine.

visotéléphonie, *f.* television, telephony, video telephony.

*visser, to fasten.

*vitesse, — absolue, absolute velocity; — aréolaire, areal velocity; — constante, uniform velocity; — cosmique, cosmic velocity; — critique, critical velocity; — d'absorption, uptake rate; — d'agitation thermique, thermal velocity; — d'approche, rate of approach; — d'augmentation, rate of increase; — d'échange isotopique, isotope exchange rate; — de changement de phase, phase velocity; — de comptage, counting rate.

vitesse, — de contraction, velocity of contraction; — de croissance, rate of increase, rate of growth; — de décantation, settling velocity; — de décharge, rate of discharge; — de décharge d'un électroscope, rate of discharge of electroscope; — de décroissance, rate of decrease; — de décroissance radioactive, decay rate; — de dérive, drift velocity; — de désintégration, rate of disintegration; — de désintégration radioactive, rate of radioactive decay.

vitesse, — de diffusion, diffusion velocity, rate of diffusion; — de diffusion par choc, collisional diffusion velocity; — de diminution, rate of decrease; — de dissipation, dissipation rate; — de formation, rate of formation.

vitesse, — de la particule, particle velocity; — de la réaction totale, total reaction rate; — de liberation, escape velocity; — de mutation, mutation rate; — de pleine marche, full speed.

vitesse, — de propagation, velocity of propagation; — de pulsation, angular frequency; — de réaction chimique, rate of chemical reaction; — de réglage de la pile, reactor control rate; — de sédimentation, settling velocity; — des neutrons, neutron speed; — de tabulation, accumulating speed; — de transfert de masse, mass velocity; — de transfert d'énergie, energy transfer rate; — de transport, transport velocity; — d'extraction, rate of withdrawal.

vitesse, — d'impression, list speed; — d'incorporation, uptake rate; — d'onde, wave velocity; — horaire, speed per hour; *— initiale, muzzle velocity; — intermédiaire, intermediate speed; — la plus probable, most probable velocity; — locale de dérive, local drift velocity; — **macroscopique**, macroscopic velocity.

vitesse, — **moyenne de l'électron**, mean electron velocity; — **non-uniforme**, variable velocity; — **orbitale**, orbital velocity; — **periphérique**, tip speed; — **quadratique**, four-velocity; — **radiale**, radial velocity; — **relative**, relative velocity; — **relativiste**, relativistic velocity; — **sur base**, speed over a straight-line course; — **variable**, variable velocity.

vitrose, *f.* cut film.

vocable, *m.* language element, literal symbol.

voglianite, *f.* voglianite.

voglite, *f.* voglite.

*****voie**, — **alternante**, half-duplex channel; — **bidirectionnelle**, duplex channel, full duplex channel; — **de désintégration**, decay channel, transformation constant; — **de fission**, fission channel; — **dérivée**, shunt circuit; ***— de service**, siding (R.R.); — **omnibus**, common bus, bus; — **simplex**, simplex channel; — **téléphonique**, voice grade channel; — **unidirectionnelle**, simplex channel.

voilage, *m.* buckling (wheel, etc.), lateral runout, out of true.

*****voile**, *m.* fog (film), spider (of a wheel).

*****voiler**, to buckle, bend.

voilerie, *f.* sail loft.

voilier, *m.* sailmaker.

*****voilure**, surface of a parachute, canopy, wing unit.

*****voiture**, — **à rémorquer**, trailer (automobile); — **d'arrosage**, sprinkler (street); — **décapotable**, convertible (auto); — **de course**, racing car; — **d'occasion**, used car; — **hors série**, custom-built car.

voix, — **chiffrée**, digitally coded voice; — **dans le champ**, voice in; — **hors champ**, voice off.

*****vol**, — **aller**, outward flight, outgoing flight; — **pique**, dive (aviation); — **sur le dos**, invert

flying (aviation); **dispositif de — au but**, homing device.

*****volant**, — **à main**, **de manoeuvre**, handwheel, — **d'aileron**, aileron control wheel; — **de commande**, derrick wheel; — **de direction déformable**, collapsible steering wheel; — **de réglage**, handwheel adjuster; — **de serrage**, friction wheel; — **d'inertie**, flywheel; — **moteur**, flywheel.

*****volatil**, unstable.

*****volatilité relative**, relative volatility.

*****voler seul à bord**, to go solo.

*****volet**, — **d'aération**, ventilator flap; — **d'appel**, call indicator; — **d'atterrissage**, landing flap; — **de centrage**, trim flap; — **de départ**, choke flap; — **de fente avant**, slat; — **de regard**, peephole door; — **d'intrados**, split flap; **—s de radiateur**, radiator shutter (auto); **à —s**, undershot.

*****voltage accélérateur**, accelerating voltage.

voltmètre, *m.* — **de crête**, peak voltmeter, vacuum-tube or slide-back voltmeter; — **électrostatique à corde**, string electrometer.

*****volume**, size, flow; — **de la réaction**, reacting volume; — **élémentaire**, element of volume; — **infini d'épaisseur constante**, infinite slab; — **réagissant**, reacting volume; — **sensible**, sensitive volume.

vorticule, *m.* vortex.

*****voyant**, light, signal light; — **d'appel**, call indicator.

vraquier, *m.* bulk carrier.

vrillage, *m.* twisting action, thread snarl.

*****vrille**, spin (aviation).

vrombissement, *m.* humming, throbbing, purring (engines, etc.).

*****vue**, — **éclatée**, exploded view; — **en crevé**, cut-away view; — **en élévation**, elevation (view); — **fantôme**, cut-away view; — **plan**, plan view.

vumètre, *m.* volume meter.

W

*****wagon-citerne**, tank car 10,000 kilograms; — **couvert**, — **fermé**, boxcar; — **plat**, gondola car.

walpurgite, walpurgite.

watt-heure, *f.* watt-hour.

wattheuremètre, *m.* watt-hour meter, energy meter.

wiikite, wiikite.

X

xénotime, *f.* xenotime.

Y

yole, *f.* yawl (naut.).

youyou, *m.* dinghy (naut.).

*ytterbium, aldebaranium.

yttrocrasite, yttrocrasite.

yttroersite, yttroersite.

yttrogummite, *f.* yttrogummite.

yttrotantalite, *f.* yttrotantalite.

Z

*zéro, — absolu, absolute zero; **— non significatif,** leading zero; **— reader,** zero reader (ZR).

zinquer, to scratch (of loud-speaker).

zippéite, *f.* zippeite.

zirkelite, *f.* zirkelite.

zonage, *m.* zoning.

*zone, field, region, zone band, band, area; **— active,** active region; **— à matière active,** seed area; **— axiale,** axial region; **— de balayage,** scanning area; **— de brouillage,** interference area, nuisance area (radio); **— de contact,** contact face; **— de contrôle,** control block, check field; **— de contrôle de marquage,** check field; **— de mémoire,** storage block (EDP); **— d'entrée,** input area; **— d'entrée-sortie,** input-output area; **— de perforation,** punch field; **— de réaction,** reaction region; **— de sortie,** output area; **— de tolérance,** neutral zone, dead zone; **— de transaltion,** relocatable library (EDP); **— de transit,** transient area, variable area; **— libre,** clear area.

zone, — d'indifférence, neutral zone; **— infectée,** gassed area, contaminated area; **— interdite,** prohibited zone, exclusion area; **— morte,** blind spot; **— plastique,** plastic range; **—s de l'arrière,** back areas, hinterland.

ABBREVIATIONS

Abbr.	French	English
a.	ampère(s)	ampere(s)
A.	association	association, fellowship
a.	accélération	acceleration
a.	are(s)	are (119.60 square yards)
aa.	Ana	of each the same quantity
ab.	abandonné	relinquished (right, etc.)
a.c.	argent comptant	ready money
ac.	acompte	payment on account
A.C.L.	assuré contre l'incendie	insured against fire
act.	action	share
add.	addatur	add
adm., admin.	administration	administration
ad., adr.	adresse	address
A.d.S.	Académie des sciences	Academy of Science
A. & M.	arts et métiers	arts and crafts
aff.	affluent	tributary
Ag	argent	silver
agr.	agriculture	agriculture
Ah.	ampère(s)-heure	ampere-hour(s)
ah., aH.	ampère(s)-heure	ampere-hour(s)
Al	aluminium	aluminum
a.m.	ante meridiem	before noon
amp.	ampère(s)	ampere(s)
anal.	analyse, analytique	analysis, analytic, analytically
anc.	ancien	old
angl.	anglais	English
anme.	anonyme	limited liability (company)
appt.	approvisionnement	provisioning
ap(r). J.-C.	après Jésus-Christ	anno Domini, A.D , in the year of Our Lord
a.r.	(ascensio recta) ascension droite	right ascension
AR., A.R.	arrière	head end, crank end, back
Ar	argon	argon
arr.	arrondissement	district
ar.-g., arr.-g.	arrière-garde	rear guard
art.	artillerie	artillery
As	arsenic	arsenic
a/s	au soins de	care of
asse.	assurance	insurance
a.-t.	ampère-tours	ampere turns
atm.	atmosphère	atmosphere
à.t.p.	à tout prix	at any cost
Au	or	aurum, gold.
auj.	aujourd'hui	today
av.	avenue	avenue
avdp.	avoirdupoids	avoirdupois
av.J.-C.	avant Jésus-Christ	before Christ, B.C.
B	bore	boron
B.	boréal	northern

b.	bougie(s)	candle(s), candle power
Ba	baryum	barium
bacc. en dr.	baccalauréat en droit	Bachelor of Laws (degree)
bacc. ès l.	baccalauréat ès lettres	Bachelor of Letters (degree)
bacc. ès sc.	baccalauréat ès sciences	Bachelor of Science (degree)
balce.	balance	balance
Banq.	banque	bank
b.à p.	billet à payer	bill payable
b.à r.	billet à recevoir	bill receivable
barr.	barrique	barrel
bat.	bata llon	battalion
batt.	batterie	battery
bce.	balance	balance
bd.	boulevard	boulevard
bde.	brigade	brigade
Be	beryllium	beryllium
beau.	bordereaux	memorandum, invoice
B. en Dr.	Bachelier en Droit	Bachelor of Laws
B. ès A.,	Bachelier ès Arts	Bachelor of Arts
B. ès L.	Bachelier ès Lettres	Bachelor of Letters
B. ès S.	Bachelier ès Sciences	Bachelor of Science
Bi	bismuth	bismuth
Bib.	bibliothèque	library
bie.	batterie	battery
biv.	bivouac	bivouac
bl.	baril	barrel
bld.	boulevard	boulevard
blle.	bouteille	bottle
Bo	bore	boron
bon.	bataillon	battalion
bot.	ballot	bale
boul.	boulevard	boulevard
B.P., BP., b.p.	basse pression	low pressure, low tension
B.P.F.	bon pour francs	value in francs
bque.	barrique, barque	barrel, bark
br.	broché	sewn
Br	brome	bromine
brig.	brigade	brigade
bt.	brut, boucaut, billet	gross, barrel, bill
bté.	breveté	patent
btn.	bataillon	battalion
C	carbone	carbon
C., c.	capacité	capacity, ability, efficiency
C.	coulomb(s)	coulomb(s)
C.	grande(s) calorie(s)	kilocalorie(.)
C.	centigrade	centigrade
c.	petite(s) calorie(s)	small calorie(s)
c.	centimètre(s)	centimeters
c.	coupon	coupon
c.	cube, cubique, cubiform	cube, cubic, cubical, cube-shaped
Ca	calcium	calcium
c.a.	courant alternatif	alternating current
ca.	centiare(s)	centiare, centare (10.7639 square feet)
c.-à-d.	c'est-à-dire	that is to say
cage.	courtage	brokerage
caire.	commissionnaire	agent
cap.	capital, capitale	capital
Cal.	grande calorie	(large, great) calorie

cal.	calorie(s)	calorie(s)
cal.	calibre	caliber
Cb	columbium	columbium
c.c.	courant continu	direct current
c/c.	compte courant	current account
cc.	centimètre(s)	centimeter(s)
Cd	cadmium	cadmium
c. de f.	chemin de fer	railway
c. de g.	centre de gravité	center of gravity
Ce.	compagnie	company
Ce	cérium	cerium
cent.	centime, centième, centiaire	centime, hundredth, centiare
centig., centigr.	centigramme, centigrade	centigram, centigrade
centil.	centilitre(s)	centiliter(s)
centim.	centimètre(s)	centimeter(s)
cg.	centigramme(s)	centigram(s)
c.g.s.	centimètre-gramme-seconde	centimeter-gram-second
ch. (chx, pl.)	cheval (chevaux, pl.)	horse, horsepower
ch., chap.	chapitre	chapter
ch. de f.	chemin de fer	railway, railroad
ch. eff.	chevaux effectifs	effective horsepower
ch. vap. (chx. vap)	cheval-vapeur (chevaux-vapeur)	horsepower
Cie	compagnie	company
Cl	chlore	chlorine
cl.	centilitre(s)	centiliter(s)
cm.	centimètre(s)	centimeter(s)
C.M.	carat(s) métrique(s)	metric carat(s)
c/m., ^{c}/m	centimètre(s)	centimeter(s)
cm²., cm^q.	centimètre(s) carré(s)	square centimeter(s)
cm³., cm^c.	centimètre(s) cube(s)	cubic centimeter(s)
cm.:s.	centimètre(s) par seconde	centimeter(s) per second
cm.:s².	centimètre(s) par seconde par seconde	centimeter(s) per second per second
Co	cobalt	cobalt
Co.	compagnie	company
cochl.	cochleare	spoonful
cochleat.	cochleatim	by spoonfuls
colat	colature	colature
com.	commission	commission
Comptes Rend.	Comptes rendus de l'académie des sciences	Proceedings of the Academy of Science
conf.	conferatur	compare
confl.	confluent	tributary
conv.	converti	converted
coq.	coquatur	cook
cor.	corollaire	corollary
cort.	cortex	bark
cos.	cosinus	cosine
coséc.	cosécante	cosecant
cot.	cotangente	cotangent
coul.	coulomb(s)	coulomb(s)
coup.	coupon, coupure	coupon, denomination
cour.	courant	instant
court.	courtage	brokerage
C.Q.F.D.	ce qu'il fallait démonstrer	Q.E.D.
cr.	crédit	credit
Cr	chrome	chromium
cs.	cours	quotation
Cs	césium	cesium
csse.	caisse	case

ct.	courant, centilitre	instant, centiliter
ctg.	courtage, cotangente	brokerage, cotangent
cts.	centimes	centimes
Cu	cuivre	copper
cub.	cube, cubique	cube, cubic, cubical, cubiform.
cuill.	cuillerée	spoonful
cum.	cumulatif	cumulative
cyat.	cyathus	cupful, glassful
d.	densité, dureté	density, specific gravity, hardness
d.	déclinaison	declination
d.	droit	right
d.	demande	demand
Dag., dag.	décagramme	decagram
Dal., dal.	décalitre	decaliter
Dam., dam.	décamètre	decameter
d.d.	donné dans	given in
déb.	débit	debit
dec.	decoctio	decoction
déc.	décédé, décembre, décime	deceased, died, December, ten centimes
décagr., dag.	décagramme(s)	decagram(s)
décal., dal.	décalitre(s)	decaliter(s)
décam., dam.	décamètre(s)	decameter(s)
décbre.	décembre	December
décig., décigr.	décigramme	decigram
déclin.	déclinaison	declination
dem.	demain	tomorrow
der., derr.	dernier	ultimo
dg., décig.	décigramme(s)	decigram(s)
diam.	diamètre	diameter
dig.	faites digérer	digest
dil.	faites diluer	dilute
dist.	distillez	distill
dl., décil.	décilitre(s)	deciliter(s)
dm., décim.	décimètre(s)	decimeter(s)
dm.	dimanche	Sunday
D.M.	Docteur Médicin	Doctor of Medicine
dm²., dmq.	décimètre(s) carré(s)	square decimeter(s)
dm³., dmᶜ.	décimètre(s) cube(s)	cubic decimeter(s)
do.	dito	ditto
dol., doll.	dollar	dollar
douz.	douzaine	dozen
d.q.	dernier quartier	last quarter
Dr.	docteur	doctor
dr.	débiteur	debtor
dynam.	dynamique	dynamic, dynamical
dyne: cm².	dyne par centimètre carré	dyne per square centimeter
dz.	douzaine	dozen
e.	force électro-motrice	electromotive force
E.	est	east, eastern, easterly
E.	équivalent mécanique de la chaleur	mechanical equivalent of heat
éd(it)	édition	edition
E.-N.-E.	est-nord-est	east-northeast
E.-S.-E.	est-sud-est	east-southeast
élec., électr.	électricité, électrique	electricity, electric, electrical
ens.	ensemble	together
env.	environ	about
eq.	équivalent	equivalent
Er	erbium	erbium

erg: s.	erg par seconde	erg per second
esc., escte.	escompte	discount
étabt.	établissement	establishment
etc.	et caetera	et cetera
E.-U.	États-Unis	United States
E.V.	en ville	local
ex.	exemple	example, exercise, year's trading
ex. att.	exercice attaché	cumulative dividend
F	fluor	fluorine
F.	farad(s)	farad(s)
F.	force	force, power
F.	Fahrenheit	Fahrenheit
f.	franc(s)	franc(s)
f.	fréquence	frequency
f.	féminin	feminine
fab.	fabrication	make
fasc.	fasciculus	fascicule
fco.	franco	free of charge, carriage paid
Fe	fer (ferrum)	iron
f.é.-m.	force électro-motrice	electromotive force
fév.	février	February
fig.	figure	illustration, figure
filt.	filtrez	filter
fl.	florin, fleurs	florin, flowers
Fl	fluor	fluorine
fl.	fleuve	river
fo., fol.	folio	folio
fol.	folia	leaves
follic.	follicule(s)	follicule(s)
fque.	fabrique	make
fr., franç.	français	French
fr(s).	franc(s)	franc(s)
Fr.	France	France
fruct.	fructus	fruit
F.S.	faire suivre	please forward
f.s.a.	fiat secundum artem	prepare according to formula
G.	conductance	conductance
g.	gauche	left
g.	gramme(s)	gram
g.	gravité	gravity
Ga	gallium	gallium
Ge	germanium	germanium
géol.	géologie	geology
géom.	géométrie	geometry
gr.	gramme(s)	gram(s)
gr.	graines	seeds
gt., gtte., gutt.	goutte	drop
H	hydrogène	hydrogen
h.	hypothèque	mortgage
h.	heure(s)	hour(s)
h.	hier	yesterday
ha., hect.	hectare(s)	hectare(s)
haut.	hauteur	height, elevation
He	helium	helium
hect.	hectare	hectare
hectogr.	hectogramme	hectogram
hectol.	hectolitre	hectoliter
hectom.	hectomètre	hectometer

Hf	hafnium	hafnium
Hg	mercure (hydrargyrum)	mercury
hg., hectogr.	hectogramme(s)	hectogram(s)
hl., hectol.	hectolitre(s)	hectoliter(s)
hm., hectom.	hectomètre(s)	hectometer(s)
hon.	honorée	letter
h.p.	haute pression	high pressure
hydraul.	hydraulique	hydraulic
hydros.	hydrostatique	hydrostatic(s), hydrostatical
hyp(oth).	hypothèque	mortgage
I	iode	iodine
I.	intensité	intensity
I., i.	intensité de courant, courant	intensity of current, current
I.	moment d'inertie	moment of inertia
I. (Is. pl.)	île	island
ibid.	ibidem	ibidem
id.	idem	idem
In	indium	indium
ind.	industrie	industry
inéd.	inédit	unpublished
in-f(o)., in-fol.,	in-folio	folio
ing., ingén.	ingénieur	engineer
inj.	injection	injection
in-pl.	in-plano	broadsheet
int., intér.	intérêt	interest
inv.,	invenit	(he) designed (it)
Io	iode	iodine
Ir	iridium	iridium
J.	jour	zero day
J.	joule(s)	joule(s)
J.	intensité d'aimantation	intensity of magnetization
j.	janvier	January
janv.	janvier	January
jce.	jouissance	cumulative dividend
j/d.	jours de date	days after date
Je.	jeune	junior
jer.	janvier	January
jet.	juillet	July
jr.	jour	day
K.	potassium	potassium
K.	moment d'inertie	moment of inertia
K.	(échelle de) Kelvin	absolute temperature scale
k.	susceptibilité,	susceptibility
kg., kilogr., kil.	kilogramme(s)	kilogram(s)
kgr.	kilogramme	kilogram
kg.: cm².	kilogramme(s) par centimètre carré	kilograms per centimeter squared
kgm.	kilogrammètre(s)	kilogrammeter(s)
kgm.: s.	kilogrammètre(s) par seconde	kilogrammeter per second
kil., kilo.	kilogramme	kilogram
kilog., kilogr.	kilogramme	kilogram
kilom.	kilomètre	kilometer
kl.	kilolitre(s)	kiloliter(s)
km., kil.	kilomètre(s)	kilometer(s)
km.: h.	kilomètres par heure	kilometers per hour
km².	kilomètre(s) carré(s)	square kilometer(s)
Kr	krypton	krypton
kVa.	kilovolt-ampère	kilowatt

kW.	kilowatt	kilowatt
kWh.	kilowatt(s)-heure	kilowatt-hour(s)
L., l.	inductance	inductance
L., l.	longueur	(wave)length
l.	livre	pound
l.	litre(s)	liter(s)
l.	lieue	league
l.a.	lege artis	as directed
La	lanthane	lanthanum, lanthanium
larg.	largeur	width
lat.	latitude	latitude
lb.	livre	pound
l.c.	loco citato, lieue carrée	at the place cited, square league
lég.	légation	legation
L. en D.	Licencié en Droit	Bachelor of Laws
L. ès Sc	Licencié ès Sciences	Master of Science (approx.)
Li	lithium	lithium
lin.	liniment	liniment
liq.	liqueur	solution
lit.	litre(s)	liter(s)
liv(r).	livraison	delivery
liv. st.	livre sterling	pound sterling
l.l.	loco laudato	at the place indicated
loc. cit.	loco citato	at the place cited
log.	logarithme	logarithm
long.	longitude	longitude
long.	longueur	length
Lu	lutécium	lutecium, lutetium
M.	monsieur, majesté, mars	Mr., Majesty, March
M.	moment magnétique	magnetic moment
M.	masse	mass
M.	midi	south
M.	intensité de pôle,	pole strength
m.	mort, mon, masculin	died, my, masculine
m.	misce	mix
m.	mètre(s)	meter(s)
m.	minute(s)	minute(s)
m²., m^q ,	mètre(s) carré(s)	square meter(s)
m³., m^c.	mètre(s) cube(s)	cubic meter(s)
mA.	milliampère(s)	milliampere(s)
m.à m.	mot à mot	word for word
man.	manipulus	handful
mat. méd.	matière médicale	materia medica
mb.	millibar	millibar
Mcin.	médecin	doctor
m.d.	main droite	right hand
méc., mécan.	mécanique	mechanical
menuis., men.	menuiserie	joinery
mét.	métier	trade, craft, profession
métall.	métallurgie	metallurgy
météor.	météorologie	meteorology
Mg	magnésium	magnesium
mg., milligr.	milligramme(s)	milligram(s)
m.g.	main gauche	left hand
mill.	millième	millieme
min.,	mines	mines
min.,	minute(s)	minute(s)
minér., minéral.	minéralogie	mineralogy
ml.	millilitre(s)	milliliter(s)
mle.	modèle	pattern

Mlle	Mademoiselle	Miss
Mlles	Mesdemoiselles	the Misses
m/m., m/m	millimètre(s)	millimeter(s)
mm.	mégamètre	megameter
mm., millim	millimètre(s)	millimeter(s)
mm²., mmq.	millimètre(s) carré(s)	square millimeter(s)
mm³., mmᶜ.	millimètre(s) cube(s)	cubic millimeter(s)
MM.	Messieurs	Messrs.
Mme	Madame	Mrs.
Mmes	Mesdames	Mesdames
Mn	manganèse	manganese
Mo	molybdène	molybdenum
mol.	molécule	molecule
M.-P.	mandat-poste	post-office order
Mrs.	Messieurs	Messrs.
MS.	manuscript	manuscript
ms	moins	less
m.: s.	mètre(s) par seconde	meter(s) per second
m.: s².	mètre(s) par seconde par seconde	meter(s) per second per second
msin.	magasin	store
MSS.	manuscripts	manuscripts
myg.	myriagramme	myriagram
myriam.	myriamètre	myriameter
N	nitrogène	azote, nitrogen
N.	nom	name
N.	nord	north, northern
n.	notre, neutre	our, neuter
Na	sodium	sodium
Nb	niobium, columbium	columbium, niobium
Nd	néodyme	neodymium
Ni	nickel	nickel
No.	numéro	number
novbre.	novembre	November
n.pl.	notre place	our town
O	oxygène	oxygen
O.	officier	officer
O.	ouest	west
ol.	oleum	oil
O/O, 0/00	pour cent	per cent
O/OO, 0/00	pour mille	per mille, per mill, per thousand
op. cit.	opere citato	in the work quoted
opt.	optique	optical, optics
ord.	ordinaire	ordinary
ortho.	orthochromatique	orthochromatic
Os	osmium	osmium
ov.	ovum	egg
P.	protesté, protêt	protested, protest
P	phosphore	phosphorus
P.	puissance électrique	electric power
P.	puissance	power, strength
p.,	page, par	page, per
P. an.	par an	per annum
p. 0/00	pour cent, pied, pouce	per cent, foot (measure), inch
p.	pression	pressure, tension, voltage
p.	pouls	pulse
p.	pour	per
p.	puissance	power
p.	poids	weight

p.	prenez	take
p.æ.	partes æquales	equal parts
pass.	passim	in various places
p. at.	poids atomique	atomic weight
Pb	plomb	lead
p. 100	pour cent	per cent
p.c.	pour cent	per cent
p/c.	pour compte	on account
P.C.C.	pour copie conforme	true copy
Pd	palladium	palladium
p.d.	port dû	carriage forward
pd.	pied	foot
P.é.	parts égales	equal parts
per.	premier	first
persp.	perspective	perspective
p. et ch.	ponts et chaussées	bridges and highways
p. et m.	poids et mesures	weights and measures
pétrol.	pétrologie	petrology
p. ex.	par exemple	for example
p.g.	(on telegram), pour garder	to be called for, poste restante
p.g.c.d.	plus grand commun diviseur	greatest common divisor
Ph	phosphore	phosphorus
photogr.	photographie	photography
photom.	photométrie	photometry
phys.	physique	physics
pil.	pilule	pill
P.L.	pleine lune	full moon
pl.	planche	full-page illustration
pl. ou m.	plus ou moins	more or less
p.m.	post meridiem	afternoon
p.mol.	poids moléculaire	molecular weight
po.	pouce	inch
p.p.,	port payé	carriage paid
P.P.C.	pour prendre congé	to take leave
p.p.c.m.	plus petit commun multiple	least common multiple
ppo.	pouces	inches
ppté.	précipité	precipitate
P.Q.	premier quartier	first quarter
pr.	pour, prochain, précédent	per, for, next (month), prox(imo),
		preceding
Pr	praséodyme	praseodymium
préc.	précédent	preceding
préf.	préfecture, préférence	prefecture, preference
priv., privil.	privilégié	preferential (share)
prov.	province	province
p.s.	poids spécifique, pur sang	specific gravity, thoroughbred
P.S.	post scriptum	postscript
P.ß	poids spécifique	specific gravity
p. suiv.	page suivante	following page
Pt	platine	platinum
pte.	perte	loss
P.T.T.	télégraphes et téléphones	the General Post Office
pts.	parts	parts
pug.	pugillus	pinch
pulv.	pulvis	powder
P.V.	procès-verbal, petit vitesse	(official) report, per goods train
Q., q.	quantité d'électricité	quantity of electricity
Q.	éclairage	light
q.	carré	square
q	quintal métrique	metric quintal

q.p.	quantum placet	as desired
qq.	quelques, quelqu'un	some, someone
qqf.	quelquefois	sometimes
q.s.	quantité suffisante	sufficient quantity
qt.	quintal	quintal
q.v.	quantum vis	as much as desired
qx.	quintaux	quintals
R., r.	résistance	resistance
R.	retarder	slow
R.	Réaumur	Reaumur
R.	rivière	river
r.	rue, respiration, recipe, recommandé	road, breathing, take, registered
Ra	radium	radium
rac.	racine	root
Rb	rubidium	rubidium
r.d.	rive droite	right bank
rd.-vs.	rendez-vous	meeting place
rel.	relié	bound
résid.	résident	resident
R.F.	République Française	(the) French Republic
r.g.	rive gauche	left bank
Rh	rhodium	rhodium
ro.	recto	recto
r.p.	réponse payée	reply prepaid
rse,	remise	discount
R.S.V.P.	réponse s'il vous plaît	the favor of an answer is requested
R.-T.	radio-téléphonie	radiotelephony
Ru	ruthénium	ruthenium
S	soufre	sulphur
S., s.	surface	surface, area, space
S.	sud	south
s.	son, signé, seconde, soir, surface	his, signed, second, evening, surface
s.	seconde(s)	second(s)
s.	stère(s)	stere(s)
Sa	samarium	samarium
Sb	antimoine	antimony
Sc	scandium	scandium
s.d.	sans date	no date
Se	sélénium	selenium
sec.	seconde(s)	second(s)
séc.	sécante	secant
sec., sect.	section	section
sem.	semence	seed
sept.	septentrional	northern
septbre.	Septembre	September
serrur.	serrurerie	locksmithing
S.F.	sans frais	no expenses
S.G.D.G.	sans garantie du gouvernement	(patent) without government guarantee (of quality)
sh.	schelling	shilling
Si	silicium	silicon
sin.	sinus	sine
sir.	sirop	syrup
S.L.F.	selon la formule	according to formula
s.l.n.d.	sans lieu ni date	of no address and no date
Sn	étain	tin
solv.	dissolvez	dissolve
sq.	sequentia	following

sqq.	sequentiaque	and the following
Sr	strontium	strontium
sr.	sieur, successeur	Mr., successor
ss.	sequenti.	following
ssq.	sequentia	following
st.	stère	cubic meter
Sté.	société	company
suiv.	suivant	following
sum.,	summitates	tops (of plants)
surv.	surveillant	overseer, superintendent
S.V.P.	s'il vous plaît	if you please
svt.	suivant	following
syr.	sirop	syrup
T.	période	period
T., t.	temps	time
T.	température absolue	absolute temperature
t.	titre	security, stock
t.	tonneau	barrel
t.	tome	volume
t.	tonne(s)	metric ton(s)
t.	toque	cap
t.	tour(s), révolution(s)	revolution(s)
t.	température	temperature
t.	temps	time
Ta	tantale	tantalum
tang.	tangente	tangent
Tb	terbium	terbium
t.c.	télégramme collationné, toutes coupures	repetition paid (telegram), **all** denominations (of bank notes)
Te	tellure	tellurium
te	tonne	metric ton
tech.	technique	technical, technics
tél.	télégraphique, téléphone	telegraphic; telephone
télégr.	télégraphie	telegraphy
téléph.	téléphonie	telephony
t.f.	travaux forcés	hard labor
tg.	tangente	tangent
Th	thorium	thorium
thal.	thaler	thaler
th.	théorème	theorem
Ti	titane	titanium
tinct.	tinctura	tincture
tit.	titre	security, stock
Tl	thallium	thallium
t.m.	télégramme multiple	multiple-address **telegram**
t.:m., t.p.m.	tour(s) par minute,	revolutions per minute
Tm	thulium	thulium
tom.	tome	volume
topogr.	topographie	topography
t.-p.	timbre-poste	stamp
t.p.	tout payé	all expenses paid
t.p.m.	tours par minute	revolutions per minute
T.P.S.F.	téléphonie sans fil	wireless telephony
Tr	terbium	terbium
tr.	traite	draft
tra.	tinctura	tincture
trav. publ.	travaux publics	public works
trs.	traites	drafts
T.S.F.	télégraphie sans fils	wireless telegraphy
T.S.V.P.	tournez s'il vous plaît	please turn over, **P.T.O.**
.t.	transfert télégraphique	transfer by telegraph

Tu	tungstène	tungsten
Tu.	thulium	thulium
tx.	tonneaux	barrels
U	uranium	uranium
U., u.	différence de potentiel	potential difference
un.	unité	unit
ung.	unguentum	ointment
U.P.	union postale	Postal Union
V	vanadium	vanadium
V.	volt(s)	volt(s)
V.	volume	volume, mass, bulk
v.	vitesse	speed, velocity, swiftness
v.	voyez, votre, volt, volume, vendez, voir	see, your, volt, volume, sell, see
VA.	volt(s)-ampère(s)	volt-ampere(s), watt(s)
val.	valeur	security, stock
val. déc.	valeur déclarée	declared value
vap.	vapeur	vapor, fume, steam, steamer
v/c.	votre compte	your account
Vd	vanadium	vanadium
Ve.	veuve	widow
vg(e).	village	village
v.h.	votre honorée	your letter
vo.	verso	back of the page
vol.	volume	volume, mass, bulk
vte.	vente	sale
v/v	votre ville	your town
W	tungstène	tungsten
W.	watt(s)	watt(s), volt-ampere(s)
W.	énergie	travail, energy, power
W.C., w.c.	water-closet	water closet
Wh.	watt(s)-heure	watt-hour(s)
W.L.	wagons-lits	sleeping cars
W.R.	wagons-restaurants	dining cars
wtt.	watt	watt
X.	réactance	reactance
X.	nom	name
x.bon.	ex-bonification	ex-bonus
Xbre.	décembre	December
x.c., x.coup.	ex-coupon	ex-coupon
x.d.	ex-dividende	ex-dividend
x.dr.	ex-droits	ex-rights
Xe	xénon	xenon
X.P.	exprès payé	express paid
Y	yttrium	yttrium
Yb	ytterbium, néo-ytterbium	ytterbium, neo-ytterbium, alde baranium
Z.	impédance	impedance, virtual resistance
Zn	zinc	zinc
Zr	zirconium	zirconium
7bre	septembre	September
8bre	octobre	October
9bre	novembre	November
o/o	pour cent	per cent
o/oo	pour mille	per thousand

GRAMMATICAL GUIDE FOR TRANSLATORS

THIRD-PERSON CONJUGATION OF COMMON VERBS

Present	Imperfect	Past	Pluperfect	Future	Conditional
ALLER—go					
il va	allait	alla	est allé	ira	irait
ils vont	allaient	allèrent	sont allés	iront	iraient
AVOIR—have					
il a	avait	eut	a eu	aura	aurait
ils ont	avaient	eurent	ont eu	auront	auraient
CONDUIRE—conduct					
il conduit	conduisait	conduisit	a conduit	conduira	conduirait
ils conduisent	conduisaient	conduisirent	ont conduit	conduiront	conduiraient
CONNAITRE—know					
il connâit	connaissait	connut	a connu	connaîtra	connaîtrait
ils connaissent	connaissaient	connurent	ont connu	connaîtront	connaîtraient
COURIR—run					
il court	courait	courut	a couru	courra	courrait
ils courent	couraient	coururent	ont couru	courront	courraient
CRAINDRE—fear					
il craint	craignait	craignit	a craint	craindra	craindrait
ils craignent	craignaient	craignirent	ont craint	craindront	craindraient
CROIRE—believe					
il croit	croyait	crut	a cru	croira	croirait
ils croient	croyaient	crurent	ont cru	croiront	croiraient
CROITRE—grow					
il croît	croissait	crût	a crû	croîtra	croîtrait
ils croissent	croissaient	crûrent	ont crû	croîtront	croîtraient
DEVOIR—owe, must, have to, be obliged					
il doit	devait	dut	a dû	devra	devrait
il doivent	devaient	durent	ont dû	devront	devraient
DIRE—say					
il dit	disait	dit	a dit	dira	dirait
ils disent	disaient	dirent	ont dit	diront	diraient
ECRIRE—write					
il écrit	écrivait	écrivit	a écrit	écrira	écrirait
ils écrivent	écrivaient	écrivirent	ont écrit	écriront	écriraient
ENVOYER—send					
il envoie	envoyait	envoya	a envoyé	enverra	enverrait
ils envoient	envoyaient	envoyèrent	ont envoyé	enverront	enverraient
ETRE—be					
il est	était	fut	a été	sera	serait
ils sont	étaient	furent	ont été	seront	seraient
FAIRE—make, do					
il fait	faisait	fit	a fait	fera	ferait
ils font	faisaient	firent	ont fait	feront	feraient
FALLOIR—be necessary					
il faut	fallait	fallut	a fallu	faudra	faudrait

Present	Imperfect	Past	Pluperfect	Future	Conditional
FINIR—finish					
il finit	finissait	finit	a fini	finira	finirait
ils finissent	finissaient	finirent	ont fini	finiront	finiraient
METTRE—place, put					
il met	mettait	mit	a mis	mettra	mettrait
ils mettent	mettaient	mirent	ont mis	mettront	mettraient
OUVRIR—open					
il ouvre	ouvrait	ouvrit	a ouvert	ouvrira	ouvrirait
ils ouvrent	ouvraient	ouvrirent	ont ouvert	ouvriront	ouvriraient
PARAITRE—appear					
il paraît	paraissait	parut	a paru	paraîtra	paraîtrait
ils paraissent	paraissaient	parurent	ont paru	paraîtront	paraîtraient
PARTIR—leave, depart					
il part	partait	partit	est parti	partira	partirait
ils partent	partaient	partirent	sont partis	partiront	partiraient
POUVOIR—be able					
il peut	pouvait	put	a pu	pourra	pourrait
ils peuvent	pouvaient	purent	ont pu	pourront	pourraient
PRENDRE—take					
il prend	prenait	prit	a pris	prendra	prendrait
ils prennent	prenaient	prirent	ont pris	prendront	prendraient
RECEVOIR—receive					
il reçoit	recevait	reçut	a reçu	recevra	recevrait
ils reçoivent	recevaient	reçurent	ont reçu	recevront	recevraient
RENDRE—give back					
il rend	rendait	rendit	a rendu	rendra	rendrait
ils rendent	rendaient	rendirent	ont rendu	rendront	rendraient
SAVOIR—know					
il sait	savait	sut	a su	saura	saurait
ils savent	savaient	surent	ont su	sauront	sauraient
SERVIR—serve					
il sert	servait	servit	a servi	servira	servirait
ils servent	servaient	servirent	ont servi	serviront	serviraient
SUIVRE—follow					
il suit	suivait	suivit	a suivi	suivra	suivrait
ils suivent	suivaient	suivirent	ont suivi	suivront	suivraient
TENIR—hold					
il tient	tenait	tint	a tenu	tiendra	tiendrait
ils tiennent	tenaient	tinrent	ont tenu	tiendront	tiendraient
TROUVER—find					
il trouve	trouvait	trouva	a trouvé	trouvera	trouverait
ils trouvent	trouvaient	trouvèrent	ont trouvé	trouveront	trouveraient
VENIR—come					
il vient	venait	vint	est venu	viendra	viendrait
ils viennent	venaient	vinrent	sont venus	viendront	viendraient
VIVRE—live					
il vit	vivait	vécut	a vécu	vivra	vivrait
ils vivent	vivaient	vécurent	ont vécu	vivront	vivraient
VOIR—see					
il voit	voyait	vit	a vu	verra	verrait
ils voient	voyaient	virent	ont vu	verront	verraient
VOULOIR—wish, want					
il veut	voulait	voulut	a voulu	voudra	voudrait
ils veulent	voulaient	voulurent	ont voulu	voudront	voudraient

IDIOMATIC USES OF VERBS

Devoir. The verb *devoir*, in addition to its literal meaning of *to owe*, has a number of idiomatic uses that should be carefully distinguished.

a. The present indicative expresses what is to be or is logically necessary.

il *doit* me le donner demain	he is to give it to me tomorrow
il a travaillé toute la journée; il *doit* être fatigué	he has been working all day; he must be tired

b. The imperfect indicative expresses what was to be and also a variety of past obligations.

il *devait* partir à midi	he was to leave at noon
il ne savait pas ce qu'il *devait* faire	he did not know what he should (it was his duty to) do

c. The past indefinite expresses an absolute or a logical necessity in the past.

il *a dû* lui rendre l'argent plus tôt qu'il ne pensait	he had to return the money to him sooner than he expected
il *a dû* pleuvoir cette nuit	it must have rained last night

d. The pluperfect indicative expresses a past necessity or obligation accruing before a time in the past.

il *avait dû* résoudre le problème tout seul	he had had to solve the problem all alone

e. The future expresses a future necessity.

il *devra* me le payer	he will have to pay me for it

f. The conditional expresses a present moral obligation.

il *devrait* payer cette dette	he ought to pay that debt

g. The conditional anterior expresses a past moral obligation.

il *aurait dû* le lui dire tout de suite	he ought to have told him so at once

Falloir

il lui faut partir	he must (it is necessary for him to) go
il me faudra travailler	I shall have to (it will be necessary for me to) work
il ne faut pas parler ici	we (one) must not talk here
il faut que vous restiez ici	you must remain here
il est nécessaire que vous restiez ici	it is necessary for you to remain here

il faut que vous parliez	you must speak
il faudra que vous parliez	you will have to speak

Other Verbs

il *vient de recevoir* une lettre	he *has just received* a letter
il *va recevoir* une lettre	he *is going to receive* a letter
nous supposons	we are supposing
supposons	supposing
il a donné	he has given, he gave, he did give
il a fini	he has finished, he finished, he did finish
il a vendu	he has sold, he sold, he did sell
il a eu	he has had, he had, he did have
il a été	he has been, he was
il est venu	he came, has come
ils sont allés	they went, have gone
ils étaient arrivés	they have arrived
elle serait partie	she would have gone
il se lève	he is getting up
il s'est levé	he got up
nous nous sommes levés	we got up
elles se sont levées	they got up
il se sert de ce livre	he is using this book
il s'est servi de ce livre	he (has) used this book
nous nous en sommes servi	we (have) used it
ils se sont mis à	they started out
il se met à	he sets out
nous nous sommes mis à	we started out

NEGATIVES

ne . . . guère que	almost only, scarcely, hardly except
ne . . . jamais que	never but
ne . . . pas encore	not yet
ne . . . pas même	not even
ne . . . pas non plus	neither, not either
ne . . . plus rien	nothing more, no longer anything
ne . . . plus que	only, nothing but
ne . . . pas plus que	not more than
ne . . . encore que	still only, yet only

il *ne* me reste *plus que* vingt francs	I have *only* twenty francs left (not more than)
il *n'*y avait *plus qu'*à mourir	there was *nothing* left *but* to die
je *ne* bois *jamais que* de l'eau	I *never* drink anything *but* water
ce môt *n'*est *guère* usité *qu'*à Lyon	the word is *hardly* used *except* at Lyon

PARTITIVES

$$\left.\begin{array}{l} \text{de } + \text{ le } = \text{du} \\ \text{de } + \text{ la } = \text{de la} \\ \text{de } + \text{ le or la } = \text{de l'} \\ \text{de } + \text{ les } = \text{des} \end{array}\right\}$$
mean *some* or *any* or may not be translated before a noun in English

il mange du pain	he is eating some bread
avez-vous de l'argent?	have you any money?
il a de la craie	he has (some) chalk
il désire de l'eau	he wants water
ils achètent des livres français	they are buying French books

If an adjective stands before a noun or if the verb is negative *de* alone means *some* or *any*.

ils achètent de grand livres	they are buying large books
il ne mange pas de viande	he does not eat any meat

The pronoun for *some* or *any* with a noun is expressed by *en; en* may also be translated with *of it, of them.*

avez-vous de l'argent?	oui, j'*en* ai	yes, I have (*some*)
avez-vous de la craie?	j'*en* ai beaucoup	I have a great deal *of it*
achètent-ils des livres?	ils *en* achètent quatre	they are buying four *of them*

DEPUIS

depuis quand êtes-vous ici?	how long have you been here?
je suis ici *depuis* longtemps	I have been here for a long time
je vous attends *depuis* dix minutes	I have been waiting for you for ten minutes

COMPARISON OF ADJECTIVES

grand	great	bien	well	bon	good
plus grand	greater	mieux	better	meilleur	better
le plus grand	greatest	le mieux	the best	le meilleur	the best

ADJECTIVES USUALLY FOLLOW THEIR NOUNS

le tableau noir	the blackboard
le livre français	the French book
une histoire intéressante	an interesting story
une fenêtre fermée	a closed window

PRONOUN OBJECTS USUALLY PRECEDE THE VERB

il me le donne	he gives it to me
je ne vous la donne pas	I do not give it to you
je la lui donne	I give it to him
il leur en donne	he gives them some
j'y vais	I am going there
il y en a	there are some

RELATIVE PRONOUNS

voici le professeur *qui* a écrit cette lettre	here is the professor *who* wrote that letter
voilà les livres *qui* vous seront utiles	there are the books *which* will be useful to you

la personne *que* je vois	the person *whom* I see
la tante *chez qui* je demeure	the aunt *with whom* I live
les amis *dont* vous parliez	the friends *of whom* you were speaking

la maison *dont* je vois *les fenêtres*	the house *whose windows* I see (of which I see the windows)

CE QUI, CE QUE, CE DONT

je sais *ce qui* vous interesse	I know *what* interests you
je vois *ce que* vous faites	I see *what* you are doing
ce que vous cherchez est ici	*that which* you are looking for is here
ce dont j'ai besoin	*what* I need

ON

on dit qu'elle est très jeune	*they* say that she is very young
où va-t-*on* maintenant?	where are *we* going now?
on ne doit pas faire cela	*you* must not do that
a-t-*on* jamais vu une telle chose?	has *one* ever seen such a thing?
ici on parle français	French is spoken here

IDIOMATIC EXPRESSIONS

quelle que soit la température	whatever the temperature may be
les mêmes enfants	the same children
les enfants mêmes	even the children
il nous a même insultés	he even insulted us
au moins	at least
de même	likewise
tout à fait	entirely, completely
afin de	in order to
bien des	many
de sorte que	so that
ne . . . guère	scarcely
ne . . . que	only
il n'a écrit que'une lettre	he wrote only one letter
il s'agit de	**it** concerns, it is a question of